FEATURES AND BENEFITS OF *GEOMETRY FOR DECISION MAKING*

1. Comprehensive real-life coverage of traditional basic geometry concepts that encourages students to explore, experiment, and discover. See table of contents on pages III-IV.

2. Computer and calculator usage is integrated throughout the text so that students will understand how to use these tools to solve problems. See page 138 for an example.

3. Each chapter begins with stated learning objectives to aid the student in understanding what will be learned in the chapter. See page 42 for an example.

4. Class participation problems (class activities) provide the link between narrative learning materials and conceptual understanding of each topic. See page 509 for an example.

5. Each chapter contains a discovery activity assignment which enables students to explore in cooperative groups possible solutions to real-life problems. See page 214 for an example.

6. Each chapter contains a project that enables students to use a variety of strategies and problem-solving techniques to solve non-routine problems, or a report to further explore related topics. See page 302 for an example.

7. A wide variety of help is available at the end of the text. Included are tables (squares, square roots, trigonometry ratios), a list of theorems and postulates with references, an English-to-Metric conversion table, answers to selected problems, a glossary, and an index.

8. The extensive ***MicroExam*** testing package enables the instructor to develop tests quickly and efficiently.

9. Teacher's Annotated Edition contains a wealth of activities, help, and suggestions for the teacher. This text contains strategies and suggestions for teaching each section of the text, section objectives, materials needed list, prerequisite skills, additional questions and examples, references to ancillaries, quizzes, exercise comments, extra practice problems, consideration for individual differences, calculator activities, enrichment activities, and overprinted achievement tests with answers.

TEACHER'S ANNOTATED EDITION

SOUTH-WESTERN

GEOMETRY FOR DECISION MAKING

JAMES E. ELANDER
North Central College
Naperville, Illinois

MG01AW
PUBLISHED BY
SOUTH-WESTERN PUBLISHING CO.
CINCINNATI, OH DALLAS, TX LIVERMORE, CA

Program Consultants:

Richard Sgroi
Assistant Professor
Mathematics Department
State University of New York
New Paltz, New York

Edgal Bradley
Lloyd High School
Erlanger, Kentucky

Editorial Advisers:

Mike Coit
Lisle High School
Lisle, Illinois

Tamilea Daniel
Educational Consultant
Mandeville, Louisiana

Tommy Eads
North Lamar High School
Paris, Texas

Tony Martinez
Leander High School
Leander, Texas

Cynthia Nahrgang
The Blake School
Hopkins, Minnesota

Emily Perlman
Educational Consultant
Houston, Texas

ISBN: 0-538-60296-1

2 3 4 5 6 7 8 9 0 RN 0 9 8 7 6 5 4 3 2 1

Printed in the United States of America

CONTENTS

Teacher's Edition

The complete Contents for the Student Edition begins on page T17.

The complete Contents for the Student Edition begins on page T17.

Teach Geometry from a Unique Angle

Introduce your math students to a practical understanding of geometry and how it relates to real life with *Geometry For Decision Making.* This unique text is ideal for informal, basic, and investigative geometry courses. Or for any geometry teacher interested in a practical, innovative approach.

Geometry For Decision Making emphasizes intuitive development rather than formal two-column proofs. Students become enthusiastic about geometry as they explore, experiment, and discover geometric concepts through a variety of construction and measurement activities and other practical exercises.

The text implements the National Council of Teachers of Mathematics (NCTM) *Curriculum and Evaluation Standards*, teaching your students traditional geometry concepts and more.

Integrated calculator and computer activities teach students how to use technology for problem-solving and discovery. Special reteaching and enrichment exercises and additional manipulatives assist you in developing your students' inductive reasoning and practical geometry skills.

Find everything you need to get your students interested in and excited about geometry with the complete *Geometry For Decision Making* package.

Draw Your Students into Geometry...

Choose the Geometry Text that Measures Up

Objectives begin each section to keep students focused on the learning goals.

The **simple-to-complex** presentation provides students with a strong foundation and confidence before tackling the more challenging points.

Practical examples and **real-life situation problems** emphasize the value of geometry to solve everyday problems.

Discovery Activities let the class participate in important conceptual conclusions.

Home Activities provide independent practice that students need with a variety of exercises, including **critical thinking** problems.

Appendices provide helpful lists of theorems and postulates with references; metric, square root, and other tables; answers to selected problems; and a glossary of terms.

Unparalleled Student Edition

Equip your students with the problem-solving skills and practical geometry principles that will benefit them throughout their lives. *Geometry For Decision Making* measures up to the NCTM *Standards*. And more.

Your students become excited about geometry as they see the applications to their everyday lives. And hands-on exercises throughout the text and a variety of thinking exercises keep students reading, talking, and applying geometry.

412 ▼ Chapter 11 ▼ Figures in Space

HOME ACTIVITY

In this activity, use your calculator as needed. Find the volume of each prism. Answer to the nearest whole unit.

1. 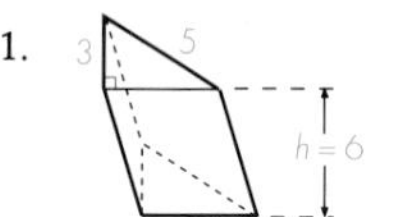

2. 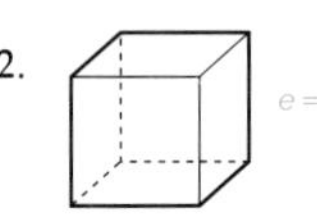

3. 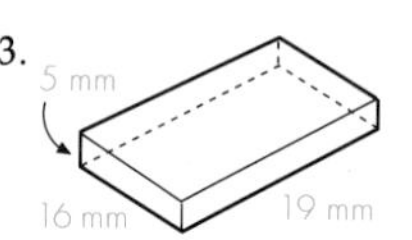

4. 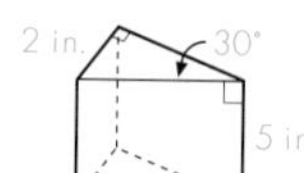

Find the volume of each cylinder.

5. 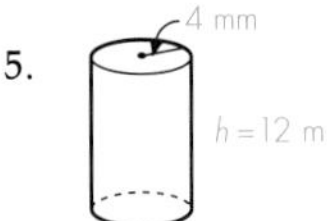

6. 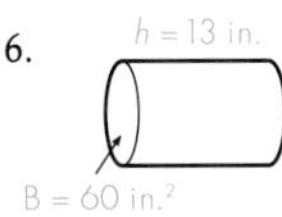

7. 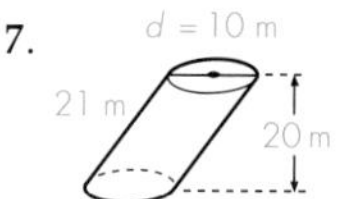

8. 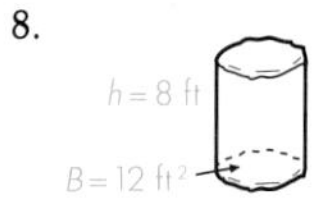

9. Find the volume of the gasoline storage tank at the

10. Find the volume of the barn.

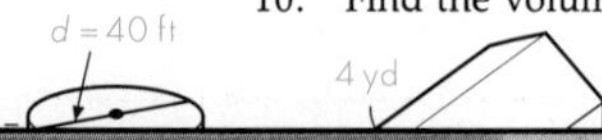

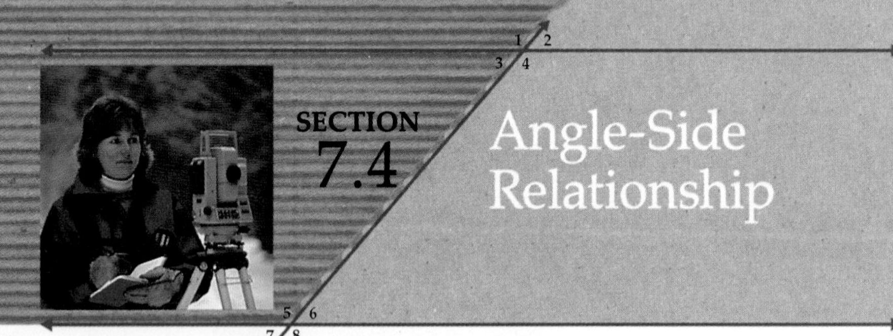

After completing this section, you should understand

- ▲ that the side opposite the greatest angle of a triangle is the longest side.
- ▲ that the angle opposite the longest side of a triangle is the greatest angle.
- ▲ applications of triangle inequalities.

Surveyors use triangles and indirect measurement to help them establish the boundaries of large pieces of land. Very often the sides of the a triangle are not of equal length nor the angles of equal measure. In such cases, how are the sides related to the angles? In this section, you will learn one way in which they are related.

UNEQUAL ANGLES IN A TRIANGLE

In Section 3.2, you learned that the sum of two sides of a triangle is greater than the third side. In this section you will learn about another inequality involving triangles.

Volumes of Practice in the Student Supplement

Give your students the additional practice they need with their own *GEOMETRY FOR DECISION MAKING* student workbooks. An additional reteaching, enrichment, and computer/calculator activity for each section, as well as tessellations are included.

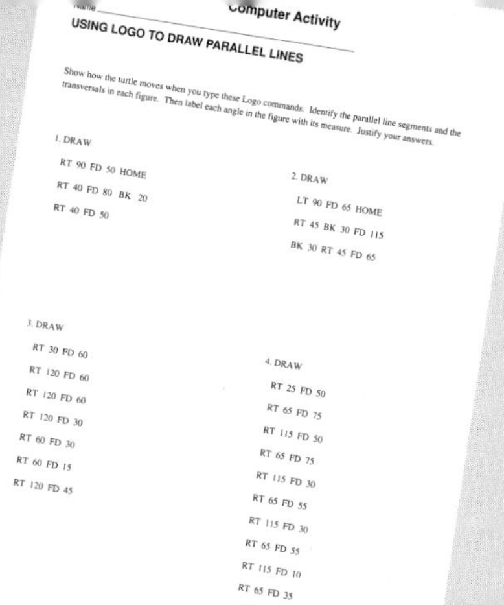

Name ________ **Computer Activity**

USING LOGO TO DRAW PARALLEL LINES

Show how the turtle moves when you type these Logo commands. Identify the parallel line segments and the transversals in each figure. Then label each angle in the figure with its measure. Justify your answers.

1. DRAW
   RT 90 FD 50 HOME
   RT 40 FD 80 BK 20
   RT 40 FD 50

2. DRAW
   LT 90 FD 65 HOME
   RT 45 BK 30 FD 115
   BK 30 RT 45 FD 65

3. DRAW
   RT 30 FD 60
   RT 120 FD 60
   RT 120 FD 60
   RT 120 FD 30
   RT 60 FD 30
   RT 60 FD 15
   RT 120 FD 45

4. DRAW
   RT 25 FD 50
   RT 65 FD 75
   RT 115 FD 50
   RT 65 FD 75
   RT 115 FD 30
   RT 65 FD 55
   RT 115 FD 30
   RT 65 FD 55
   RT 115 FD 10
   RT 65 FD 35
   RT 115 FD 10

ALGEBRA SKILLS

Example: Solve for x.

$$
\begin{aligned}
6(8x - 3) &= 5x + 3(5 - x) - 10 \\
48x - 18 &= 5x + 15 - 3x - 10 \\
48x - 18 &= 2x + 5 \\
48x - 18 - 2x &= 2x + 5 - 2x \\
46x - 18 &= 5 \\
46x - 18 + 18 &= 5 + 18 \\
46x &= 23 \\
\frac{46x}{46} &= \frac{23}{46} \\
x &= \frac{1}{2}
\end{aligned}
$$

Example: Solve for x.

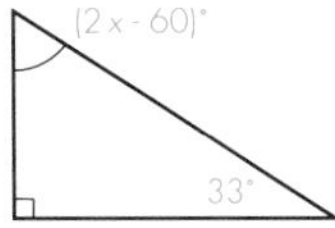

$$
\begin{aligned}
(2x - 60) + 33 + 90 &= 180 \\
2x + 63 &= 180 \\
2x &= 117 \\
x &= 58\tfrac{1}{2}
\end{aligned}
$$

1. Solve for x.

a. $x - 14 = 27$ b. $x + 33 = 81$ c. $4x = 34$ d. $-5x = 35$

e. $3x - 1 = 50$ f. $53 = 14x + 7$ g. $\frac{x}{5} = \frac{17}{6}$ h. $\frac{x-1}{8} = \frac{1+x}{5}$

i. $0.6x + 1.2 = 4.8$ j. $5x - 17 = 28 - 3x$

Using the same setting, place the point on A, and mark an arc in the center. Do the same from B. Label the vertex C and the intersecting arcs D. $\overrightarrow{CD}$ is the angle bisector.

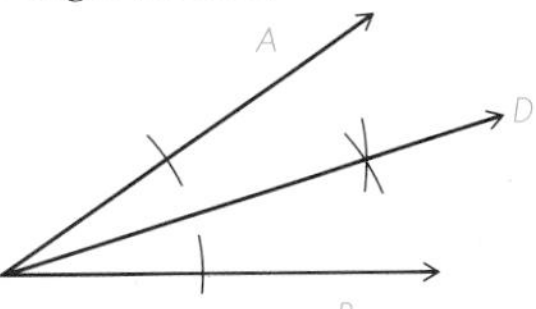

REPORT 6-1-1 The compass we use today can hold a setting until it is changed. Euclid used a collapsible compass. He lost the setting every time he picked the compass up to move it! Research how Euclid could bisect

1. a line segment. 2. an angle.

Suggested reference:
Cajori, F. *A History of Elementary Mathematics*. New York: Macmillan Publishing Co., 1905.

CLASS ACTIVITY

1. Show that $\overline{CD}$ is really the bisector of $\angle ACB$. (Hint: Show that $\triangle CAD \sim \triangle CBD$.)

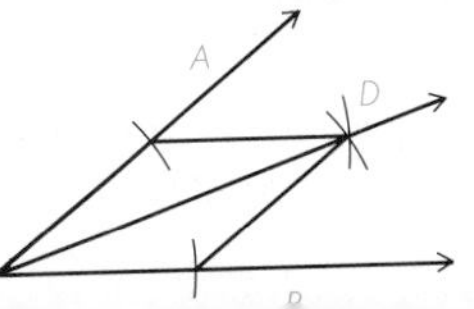

Integrated **computer activities,** using Logo software, **and calculator activities**, using regular and graphing calculators, train students to use today's technology for measurement and important mathematical discoveries.

At the end of each chapter, **Chapter Summaries** consolidate concepts from the chapter. **Chapter Reviews** provide important maintenance of skills and allow you to assess mastery.

Preview of algebraic principles ensure that students thoroughly understand the algebra principles they need for success in the next section.

Reports and Projects in each chapter connect geometry with writing and research, and encourage students to solve non-routine problems.

Class Activities promote cooperative learning and allow guided practice that is essential for thorough comprehension.

Construct a Teaching System Built for Success...

Shape Up Your Geometry Class

Solid Teacher's Annotated Edition

Geometry For Decision Making provides a wealth of teaching assistance with an easy-to-use, Teacher's Annotated Edition. Reduced student pages with answers in place and valuable margin notes provide all the teaching aids you need for great lessons.

3

CHAPTER 3 OVERVIEW

Triangle Basics

Suggested Time: 12–14 days

The quote at the beginning of the chapter was selected because it not only states what occurs in the chapter—discovery activities or experiments leading to a conclusion, then justification of the conclusion—but also the objectives of geometry. The first theorem in the chapter tells us that the sum of the measures of the angles of a triangle is 180°. This theorem is derived from the famous Euclidean 5th postulate, which is concerned with the number of lines parallel to another line through a point not on the line. Another key theorem in this chapter states that the sum of the lengths of any two sides of a triangle is greater than the length of the third side. You may wish to read Chapter 2 of Morris Kline's *Mathematics in the Physical World*, a very informative chapter on proof.

My design . . . is not to explain the properties by [assuming them], but to prove them by reason and experiments.

Isaac Newton

CHAPTER 3 OBJECTIVE MATRIX

	End of Chapter Items by Activity				Student Supplement Items by Activity		
Objectives by Number	Review	Test	Computer	Algebra Skills	Reteaching	Enrichment	Computer
3.1.1	✓	✓	✓		✓		✓
3.1.2	✓	✓		✓	✓	✓	
3.2.1	✓					✓	
3.2.2	✓	✓		✓	✓		
3.2.3	✓	✓		✓	✓		

Objectives matrices in each chapter prologue let you ensure your students' progress throughout the text. And **teaching strategies and suggestions** for each section assist in lesson planning.

A valuable **timeline** suggests how much time you should plan to spend on each chapter in order to keep your class progressing efficiently.

Reduced student text pages with teaching notes in the margin put teaching assistance right where you need it. **Teaching suggestions, objectives**, list of **materials** you'll need for each lesson, **vocabulary**, and **prerequisite skills** for your students help you better prepare for each lesson.

SECTION
3.1

Resources
Reteaching 3.1
Enrichment 3.1

Objectives
After completing this section, the student should understand
▲ the definition of a triangle.
▲ the sum of the measures of the angles of a triangle is 180°.

Materials
protractor
calculator
library resources

Vocabulary
collinear points
noncollinear points
triangle
parallel lines
theorem
postulate

GETTING STARTED

Quiz: Prerequisite Skills
1. What term is used to describe three points on the same line? [collinear]
2. Draw two parallel lines intersected by a nonperpendicular transversal. Mark the acute alternate interior angles *a* and *b*. [Student drawings should look like the following.]

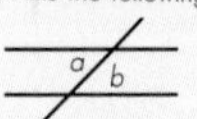

Warm-Up Activity
This paper-folding activity may be used to create physical models of triangles. Students mark the four corners of a sheet of paper with the letters *A, B, C,* and *D*. They fold the paper twice, along each of the two diagonals. Have them mark the intersection point of

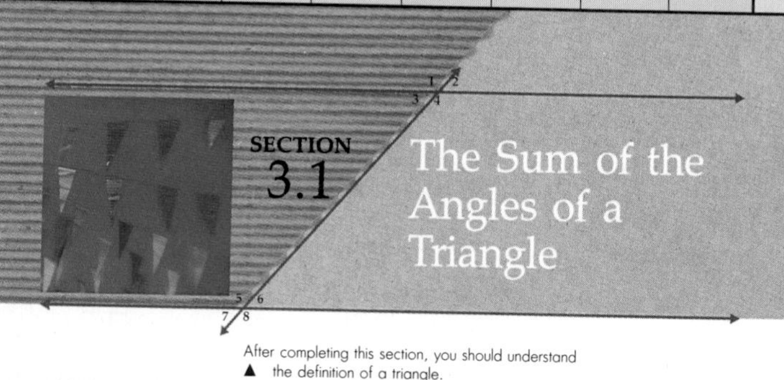

After completing this section, you should understand
▲ the definition of a triangle.
▲ the sum of the measures of the angles of a triangle is 180°.

Recall the definitions of collinear points and noncollinear points.

Lines, line segments, and rays are determined by two points. According to Postulate 1-1-1, two points determine exactly one line. Given three points, the points can be collinear or noncollinear. If noncollinear, then three lines are determined instead of one line.

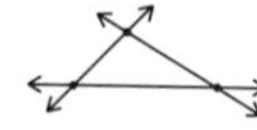

A closed figure is determined by the three lines.

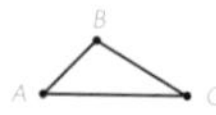

Observe the following about figure *ABC*.

Vertex	Side	Angle
A	$\overline{AB}$	∠*A*, or ∠*BAC*, or ∠*CAB*
B	$\overline{BC}$	∠*B*, or ∠*ABC*, or ∠*CBA*
C	$\overline{AC}$	∠*C*, or ∠*ACB*, or ∠*BCA*

82

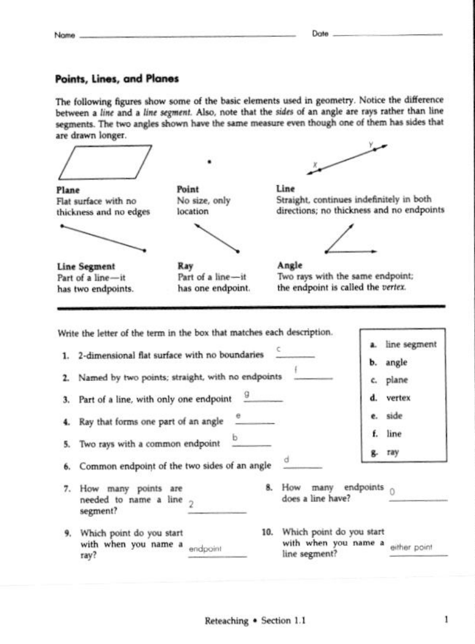

Name ______________________ Date ______________

Points, Lines, and Planes

The following figures show some of the basic elements used in geometry. Notice the difference between a *line* and a *line segment.* Also, note that the *sides* of an angle are rays rather than line segments. The two angles shown have the same measure even though one of them has sides that are drawn longer.

Plane Flat surface with no thickness and no edges

Point No size, only location

Line Straight, continues indefinitely in both directions; no thickness and no endpoints

Line Segment Part of a line—it has two endpoints.

Ray Part of a line—it has one endpoint.

Angle Two rays with the same endpoint; the endpoint is called the *vertex.*

Write the letter of the term in the box that matches each description.

1. 2-dimensional flat surface with no boundaries c
2. Named by two points; straight, with no endpoints f
3. Part of a line, with only one endpoint g
4. Ray that forms one part of an angle e
5. Two rays with a common endpoint b
6. Common endpoint of the two sides of an angle d
7. How many points are needed to name a line segment? 2
8. How many endpoints does a line have? 0
9. Which point do you start with when you name a ray? endpoint
10. Which point do you start with when you name a line segment? either point

a. line segment
b. angle
c. plane
d. vertex
e. side
f. line
g. ray

Reteaching • Section 1.1 1

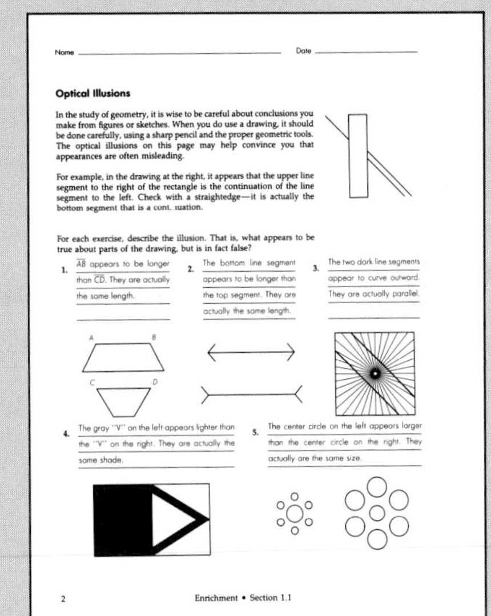

Name ______________________ Date ______________

Optical Illusions

In the study of geometry, it is wise to be careful about conclusions you make from figures or sketches. When you do use a drawing, it should be done carefully, using a sharp pencil and the proper geometric tools. The optical illusions on this page may help convince you that appearances are often misleading.

For example, in the drawing at the right, it appears that the upper line segment to the right of the rectangle is the continuation of the line segment to the left. Check with a straightedge—it is actually the bottom segment that is a cont. ıuation.

For each exercise, describe the illusion. That is, what appears to be true about parts of the drawing, but is in fact false?

1. $\overline{AB}$ appears to be longer than $\overline{CD}$. They are actually the same length.
2. The bottom line segment appears to be longer than the top segment. They are actually the same length.
3. The two dark line segments appear to curve outward. They are actually parallel.
4. The gray "V" on the left appears lighter than the "V" on the right. They are actually the same shade.
5. The center circle on the left appears larger than the center circle on the right. They actually are the same size.

2 Enrichment • Section 1.1

The figure has three vertices, three sides, and three angles. The prefix "tri" means "three." Thus, the figures below are all triangles.

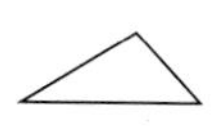

NOTEBOOK DEFINITION 3-1-1 **A TRIANGLE is a figure consisting of three noncollinear points and their connecting line segments.**

A triangle is named by its vertices.

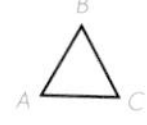

$\triangle ABC$

CLASS ACTIVITY

Tell whether each figure is a triangle. If the figure is not a triangle, explain why not.

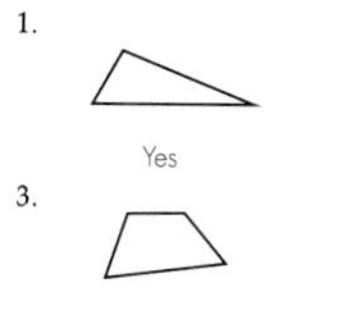

1. Yes
2. No; the sides are not connecting.
3. No; there are 4 sides.
4. No; the 3 points are collinear.

THE SUM OF THE MEASURES OF ANGLES

Mathematicians have studied the properties of triangles for many years. About 300 B.C. a man named Euclid studied a property of triangles, the sum of the measures of the angles of the triangle.

Chalkboard Example
Draw the following triangle on the chalkboard.

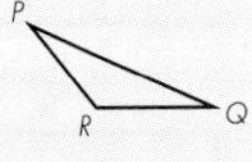

Have students name each vertex, each side (two ways), and each angle (three ways).

TEACHING COMMENTS

Using the Page
During the discussion of Definition 3-1-1, have students give their own definitions for a triangle. Discuss which ones are valid and which ones are not. For the definitions that are not valid, have students work in groups to come up with counterexamples.

Extra Practice
Tell whether each figure is a triangle. If the figure is not a triangle, explain why not.
1. [Yes]

2. [No; the sides are not connecting.]

Courtesy United Technology Corporation

Within the postlogue, you'll find all the **answers** to the **Reteaching** Activities, **Enrichment** Activities and **Calculator/Computer** Activities as well as **answers** to the **test bank** on reduced pages with answers overlayed.

Additional questions and **examples, warm-up activities, extra practice,** and **extension** and **enrichment** exercises let you provide as much practice as students need.

Quizzes allow you to assess mastery, while **references to the supplementals** show you how the entire program fits together to accomplish the lesson objectives.

Order the Geometry Package that Encompasses All Your Needs . . .

Take Geometry to a New Dimension

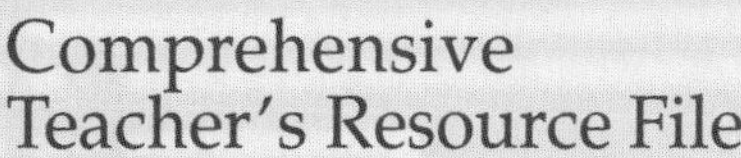

Comprehensive Teacher's Resource File

All the supplemental support materials you need for a dynamic geometry class are available in this sturdy crate that keeps the essential materials at your fingertips.

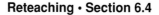

Reteaching • Section 6.4

Name ____________________

Constructions Involving Triangles

Altitude
Connects a vertex with opposite side. Must be perpendicular to the side.

Median
Connects a vertex with opposite midpoint.

Angle Bisector
Divides an angle of the triangle into two equal parts.

B
A C

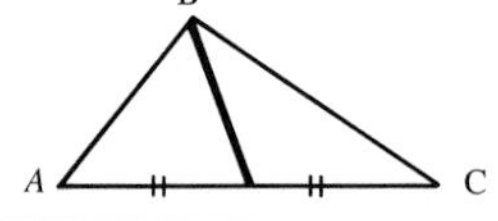

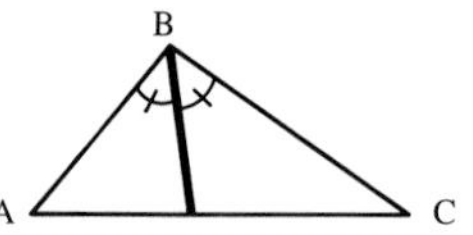

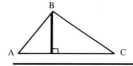

For each triangle, write whether the thick line is an **altitude**, **median**, or **angle bisector**. For some triangles, there is more than one answer.

1. ________ 2. ________ 3. ________ 4. ________

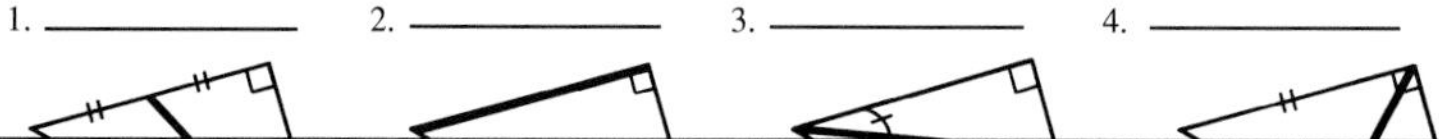

Enrichment • Section 6.2

Name ____________________

Constructing Regular N-Gons

There are many ways to construct regular polygons using just a compass and straightedge. The drawings on this page involve only circles and perpendicular lines.

In each figure, the heavy line segments show the beginning of a regular n-gon. Construct each design without using a protractor. Complete the n-gon and write its name. (Hint: For Problem 4, the name of the figure starts with the letters *do*.)

1. ____________________

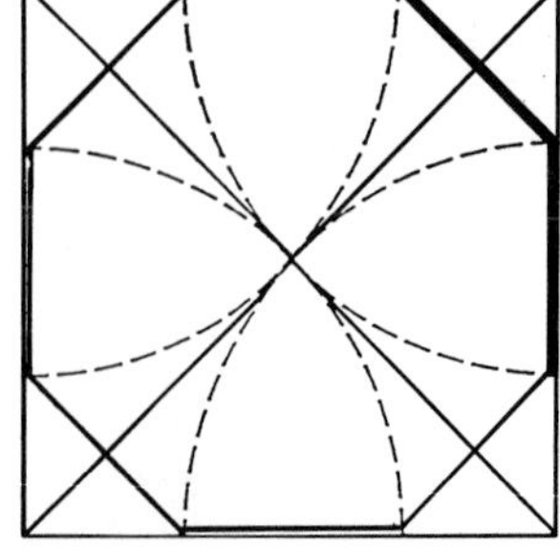

2. ____________________

3. ____________________

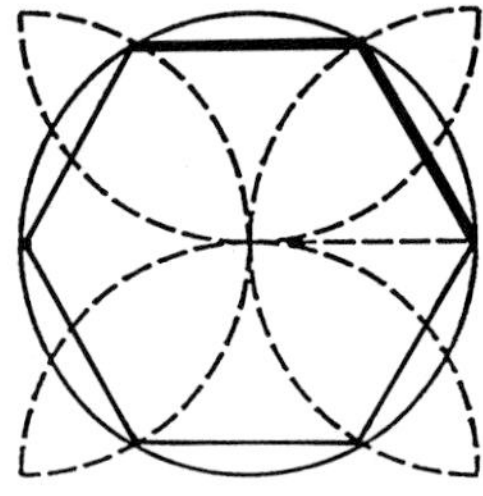

4. ____________________

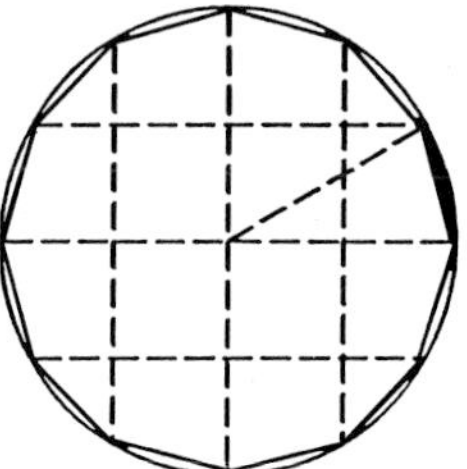

5. ____________________

Reteaching Activities let you provide additional reinforcement and on-going cumulative review for students who need it.

■

Enrichment Activities let you offer your students the opportunity to explore alternative approaches to non-routine problems.

■

Chapter- and topic-organizer files keep your materials organized for quick reference.

■

Overhead Transparency Masters for each lesson provide additional visual support to reinforce geometric principles.

■

Effective Testing Program

Assess your students' mastery with *Geometry For Decision Making's* testing package, which contains one 2-page test for each chapter and two 2-page final exams.

Or use the innovative ***MicroExam*** software — available for Apple, IBM, or Macintosh — to customize your own tests with original or randomly selected questions.

Get the Text That's Right in Line.

Author James Elander has provided a unique, practical approach to geometry that keeps students enthusiastic about learning. The text focuses on inductive reasoning to develop problem-solving and critical-thinking skills, as well as your students' geometry abilities.

Contact South-Western Publishing Co. today to order this innovative approach to Geometry.

1-800-543-7972

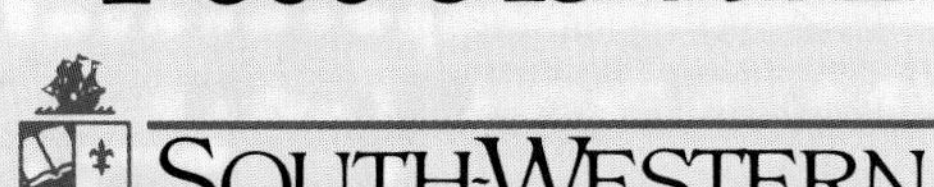

Your Mathematics Publisher for the '90s

* *An 80-column card is required to run* ***MicroExam*** *on the Apple IIe.*

Order the complete *Geometry For Decision Making* package today!

- *Geometry For Decision Making* Student Text
- Teacher's Annotated Edition
- Student Supplement
- Teacher's Resource File
  - *Reteaching Activities
  - * Enrichment Activities
- Printed Achievement Tests
- Overhead Transparencies
- ***MicroExam*** Diskettes for Apple IIe* and Apple IIc; ***MicroExam*** II Diskettes for IBM PC and Macintosh.

Let *Geometry For Decision Making* Set Your Reference Points...

Professional References and Resources

Courtesy Greater Lexington Convention and Visitors Bureau

Suggested Readings for Supplementary Instruction

Abbott, E. A. ***Flatland**: **A Romance in Many Dimensions.*** 5th rev. ed., New York: Barnes and Noble Books, 1963

Bell, E. T. ***Men of Mathematics***. New York, NY: Simon and Schuster, 1965

Bitter G. et al. ***Activities Handbook for Teaching the Metric System***. Boston, MA: Allyn and Bacon, 1976

Brandes, Louis G. ***Geometry Can Be Fun***. Portland, ME: J. Weston Walch, 1958

Bronowski, J. ***The Ascent of Man***. Boston, MA: Little, Brown and Co., 1973

Cajorie, F. ***History of Elementary Mathematics***. New York, NY: The MacMillan Co., 1905

Dantzig, T. ***The Bequest of the Greeks***. New York, NY: Charles Scribner's Sons, 1955

Davis, P. and Hersh, R. ***The Mathematical Experience***. Boston, MA: Houghton Mifflin Co., 1981

Davis, P. and Hersh, R. ***Descartes' Dream***. Boston, MA: Harcourt Brace Jovanovich, 1986

Dunham, W. ***Journey Through Genius—The Great Theorems of Mathematics***. New York, NY: John Wiley & Sons, 1990

Eves, H. ***Great Moments in Mathematics Before 1650***. Washington, D.C.: The Mathematical Association of America, 1983

Florman, S. C. ***Engineering and the Liberal Arts***. New York, NY: McGraw-Hill Book Co., 1968

Gardner, M. ***Mathematical Carnival***. New York, NY: Alfred A. Knopf, 1975

Hogben, L. ***The Wonderful World of Mathematics***. Garden City, NY: Garden City Books, 1955

Johnson, D. and Rising, G. ***Guidelines for Teaching Mathematics***. Belmont, CA: Wadsworth Publishing Co., 1967

Kline, M. ***Mathematics and the Physical World***. New York, NY: Thomas Y. Crowell, 1959

Kasner, E. and Newman J. ***Mathematics and the Imagination***. New York, NY: Simon and Schuster, 1940

Kramer, E. ***The Nature and Growth of Modern Mathematics***. New York, NY: Hawthorn Books, 1970

Lasserre, F. ***The Birth of Mathematics in the Age of Plato***. Larchmont, NY: American Research Council, 1964

Lieber, L. ***The Education of T. C. Mits***. New York, NY: W. W. Norton and Co., 1954

Lieber, L. ***Mits, Wits, and Logic***. New York, NY: Institute Press, 1954

Northrup, E. ***Riddles in Mathematics***. Princeton, NJ: D. Van Nostrand Co., 1944

O'Daffer, P. and Clemens, S. ***Geometry: An Investigative Approach***. Menlo Park, CA: Addison-Wesley Publishing Co., 1977

Peterson, I. ***The Mathematical Tourist***. New York, NY: W. H. Freeman and Co., 1988

Polya, G. ***Mathematical Discovery Volume 2***. New York, NY: John Wiley & Sons, 1965

Reid, C. ***A Long Way from Euclid***. New York, NY: Thomas Y. Crowell, 1963

Rucker, R. ***The 4th Dimension—Toward a Geometry of a Higher Reality***. Boston, MA: Houghton Mifflin Co., 1984

Steinhaus, H. ***Mathematical Snapshots***. New York, NY: Oxford University Press, 1950

Weber, R. L. ***A Random Walk in Science***. New York, NY: Crane, Russak & Co., 1973

Wilder, R. ***Evolution of Mathematical Concepts: An Elementary Study***. New York, NY: John Wiley and Sons, 1968

Yearbooks: National Council of Teachers of Mathematics, 1906 Association Drive, Reston, VA

13th—***Nature of Proof***
27th—***Enrichment Mathematics for the Grades***
28th—***Enrichment Mathematics for High School***
1987—***Learning and Teaching Geometry, K-12***

Professional Journals

Mathematics Teacher. National Council of Teachers of Mathematics, 1906 Association Drive, Reston, VA

Arithmetic Teacher. National Council of Teachers of Mathematics, 1906 Association Drive, Reston, VA

School Science and Mathematics. School Science and Mathematics Association, 126 Life Science Building, Bowling Green State University, Bowling Green, OH

The College Mathematics Journal. Mathematical Association of America, 1529 Eighteenth Street. N.W., Washington, D.C.

PREFACE

Using the Teacher's Edition

The Prologue for each chapter contains three distinct parts: an Overview, an Objective Matrix, and the Perspectives. The Overview outlines the mathematical content of the chapter and frequently mentions the overall approach taken to the material. The Objective Matrix lists objectives for the chapter by number and indicates which of the objectives are stressed in specific end-of-chapter material and student supplements. The matrix also indicates which objectives make use of algebra skills taught in the chapter or in earlier Algebra Skills lessons. The Perspectives section of the Prologue discusses the content of each section. Often, more general teaching suggestions are made, although more specific suggestions are usually reserved for side copy material found with the reduced student page.

The side copy for each section begins with a listing of **Resources** (Reteaching, Enrichment, and/or Transparency Masters), **Objectives** (for that section), **Materials,** and **Vocabulary.** This is followed by a GETTING STARTED section which contains a **Quiz of Prerequisite Skills,** a **Warm-Up Activity,** and/or a **Chalk Board Example.** All are designed to introduce the students to the material of the section in the most appropriate way possible. The GETTING STARTED section is followed by TEACHING COMMENTS. These include suggestions for **Using the Page, Additional Answers,** and **Extra Practice.** The side copy for each section ends with a FOLLOW-UP section that includes an **Assessment** and individualization suggestions for **Reteaching** and **Enrichment.** The Postlogue for each chapter contains reduced copies with annotations of the Reteaching, Enrichment, Calculator or Computer masters and Achievement Test for the chapter.

Student Notebook

In the letter to the student it is suggested that as a space traveler it would be wise to keep a log of the important facts students observe and any discoveries they might make. This allusion to a space traveler's log refers to a technique the author has found to be extremely helpful to his students in their study of geometry. This technique is the maintenance of a notebook in which the student records definitions, postulates, and theorems along with outlines of proofs or informal justifications for the acceptance of geometric ideas. Informal geometry stresses the use of geometric concepts. Frequent review of their notebooks helps students internalize the concepts and enables them to use the material more effectively in problem-solving situations.

Objectives

The following charts list each objective by number. These charts may be used with the Objective Matrix section of each Prologue to identify specific objectives.

After completing each chapter, the student should understand:

Chapter 1

1.1.1	the basic elements of geometry–points, lines, line segments, rays, and planes.
1.2.1	how numbers are matched with points on a number line.
1.2.2	how to measure line segments and angles.
1.2.3	how to classify angles.
1.3.1	how to find the midpoint of a line segment.
1.3.2	how to find the bisector of an angle.
1.3.3	the meaning of betweenness.
1.4.1	the meaning of supplementary angles.
1.4.2	the meaning of complementary angles.
1.5.1	the meaning of inductive reasoning.
1.5.2	the pitfalls involved in inductive reasoning.
1.6.1	how to use some LOGO commands to draw geometric figures.
1.6.2	how to use a computer and geometric drawing tools to explore geometric concepts.

▲ **Chapter 2**

2.1.1	collinear, noncollinear, coplanar, and noncoplanar points.
2.1.2	intersecting, parallel, and skew lines.
2.2.1	parallel lines and transversals.
2.2.2	corresponding angles.
2.3.1	how deductive reasoning is used.
2.3.2	vertical angles.
2.3.3	alternate interior angles.
2.4.1	the concept of error in measurement.
2.4.2	how to convert from one unit of measurement to another.
2.4.3	how to add and subtract measurements.
2.5.1	alternate exterior angles.
2.5.2	perpendicular lines.
2.6.1	how a three-dimensional figure appears when drawn on a two-dimensional surface.
2.6.2	the relationship between a three-dimensional figure and its layout.

▲ **Chapter 3**

3.1.1	the definition of a triangle.
3.1.2	the sum of the measures of the angles of a triangle is 180°.
3.2.1	indirect reasoning.
3.2.2	the sum of the lengths of two sides of a triangle is greater than the length of the third side.
3.2.3	the length of any side of a triangle is greater than the difference of the lengths of the other two sides.
3.3.1	how triangles are classified by the length of their sides.
3.4.1	how to classify triangles by the measure of their angles.
3.5.1	the triangle exterior angle theorem.
3.6.1	the x-y (Cartesian) coordinate plane.
3.6.2	how the Cartesian coordinate plane is used in geometry.

▲ **Chapter 4**

4.1.1	how to map figures.
4.1.2	how to write correspondences.
4.2.1	ratios and proportions.
4.2.2	how to solve problems involving ratios and proportions.
4.3.1	what similar triangles are.
4.3.2	how to solve problems involving similar triangles.
4.4.1	scale drawings.
4.4.2	applications of scale factors.
4.5.1	how to justify conclusions involving similar triangles.
4.5.2	how to solve for missing lengths, using similar triangles.
4.6.1	how to solve real-world problems involving similar triangles.

▲ **Chapter 5**

5.1.1	what a set is.
5.1.2	what Venn diagrams are.
5.1.3	what set operations are.
5.1.4	how to solve problems using sets.
5.2.1	the meaning of congruent triangles.
5.2.2	how to prove triangles congruent.
5.3.1	corresponding parts of congruent figures.
5.3.2	how to determine lengths of segments and measurements of angles.
5.4.1	the meaning of converse.
5.4.2	how a statement and its converse are related.
5.5.1	applications of congruent triangles.
5.5.2	applications of the converse.

▲ **Chapter 6**

6.1.1	how to copy a line segment using the tools of geometry.
6.1.2	how to copy an angle using the tools of geometry.
6.1.3	how to bisect a line segment.
6.1.4	how to bisect an angle.
6.2.1	how to construct a perpendicular to a line from a point on the line.
6.2.2	how to construct a perpendicular to a line from a point not on the line.
6.3.1	how to construct a line parallel to another line.
6.4.1	the medians of a triangle.

6.4.2	the altitudes of a triangle.
6.4.3	the angle bisectors of a triangle.
6.4.4	the perpendicular bisectors of the sides of a triangle.
6.5.1	how to divide a line segment into n equal parts.
6.6.1	how constructions lead to conclusions.

Chapter 7

7.1.1	the Pythagorean Theorem.
7.1.2	squares and square roots of numbers.
7.2.1	the properties of isosceles right triangles.
7.2.2	how to solve problems using isosceles right triangles.
7.3.1	theorems about 30°–60° right triangles.
7.3.2	how to solve problems using 30°–60° right triangles.
7.4.1	that the side opposite the greatest angle of a triangle is the longest side.
7.4.2	that the angle opposite the longest side of a triangle is the greatest angle.
7.4.3	applications of triangle inequalities.
7.5.1	how the Pythagorean Theorem is applied to the coordinate plane.
7.5.2	the Distance Formula.

Chapter 8

8.1.1	the properties that determine a quadrilateral.
8.1.2	the difference between convex and concave quadrilaterals.
8.2.1	the properties of trapezoids and isosceles trapezoids.
8.2.2	how to apply the theorems about isosceles trapezoids and medians of trapezoids.
8.3.1	the properties of a parallelogram.
8.3.2	how to apply the theorems about parallelograms.
8.4.1	the properties of a rectangle.
8.5.1	the properties of a rhombus.
8.5.2	the properties of a square.

Chapter 9

9.1.1	the definitions of perimeter and area.
9.1.2	how to find the perimeter and area of a rectangle.
9.2.1	how to find the perimeter and area of a triangle.
9.2.2	how to solve problems involving the area of a triangle.
9.3.1	how to find the perimeter and area of a parallelogram.
9.4.1	how to find the perimeter and area of a trapezoid.
9.5.1	how to find the area of a quadrilateral with no parallel sides.
9.5.2	the relationship between area and perimeter of similar figures.

Chapter 10

10.1.1	the definition of a regular polygon.
10.1.2	how to find the measure of the interior angles of a polygon.
10.1.3	how to find the perimeter of a polygon.
10.2.1	how to find the area of a regular polygon.
10.3.1	the definition of a circle and its parts.
10.3.2	how to find the circumference of a circle.
10.3.3	how to find the area of a circle.
10.4.1	how to identify central and inscribed angles and their intersected arcs.
10.4.2	how to find the measures of angles intersecting circles.
10.4.3	how to identify a tangent to a circle.
10.5.1	the definitions of a chord and a secant.
10.5.2	how to use the theorems about chords and secants to solve problems.
10.6.1	how to solve problems using theorems about regular polygons and circles.

Chapter 11

11.1.1	what a prism and a cylinder are.
11.1.2	how to find the area of prisms and cylinders.
11.2.1	what volume is.
11.2.2	how to find the volume of prisms and cylinders.
11.3.1	what a pyramid is.
11.3.2	what a cone is.
11.3.3	how to calculate the surface area of a pyramid.

11.3.4	how to calculate the surface area of a cone.
11.4.1	how to find the volume of a pyramid.
11.4.2	how to find the volume of a cone.
11.5.1	how prisms and cylinders are used in everyday life.
11.5.2	how pyramids and cones are used in everyday life.
11.6.1	how to find the volume of a sphere.
11.6.2	how to find the area of a sphere.

Chapter 12

12.1.1	the definition of geometric locus.
12.1.2	how to determine a locus that satisfies given conditions.
12.2.1	how to find the translation image of a figure.
12.2.2	how to find the reflection of a figure and locate the line of reflection.
12.2.3	how to find the rotation image of a figure.
12.3.1	how to determine whether a figure has line symmetry.
12.3.2	how to find lines of symmetry.
12.3.3	how to determine whether a figure has point or rotation symmetry.
12.4.1	how to solve problems by using definitions and theorems about loci and transformations.
12.5.1	how a conditional statement is related to its converse, inverse, and contrapositive.
12.5.2	how to determine the truth value of a statement.

Chapter 13

13.1.1	how the term *point* is defined in coordinate geometry.
13.1.2	how equations are used to define *lines*.
13.2.1	the slope of a line.
13.2.2	how to find the slope of a line given two points on the line.
13.2.3	the y-intercept of a line.
13.3.1	how the slopes of parallel lines are related.
13.3.2	how the slopes of perpendicular lines are related.
13.4.1	how to write an equation for a circle whose center is the origin.
13.4.2	how to graph equations of the form $x^2 + y^2 = r^2$.
13.5.1	how to find and graph an equation for a parabola.
13.6.1	how to prove theorems using coordinate geometry.

Chapter 14

14.1.1	how to find sine ratios.
14.1.2	how to find cosine ratios.
14.2.1	how to use the tangent ratio.
14.2.2	how to use trigonometry to solve triangle problems.
14.3.1	what an angle of rotation is.
14.3.2	how the sine and cosine fuctions can be extended.
14.3.3	how to graph the sine function.
14.4.1	how to use the law of sines.
14.4.2	how to use the law of cosines.
14.5.1	how to use trigonometric functions to solve practical problems.
14.6.1	some basic identities for the sine, cosines, and tangent.
14.6.2	how to recognize a trigonometric identity.

Cumulative Reviews

Cumulative Reviews are found after Chapter 5, Chapter 10, and Chapter 14 in the text. The following chart indicates the number of items in each review and provides a conversion factor to be used to convert the raw score to a percentage.

	Items	Conversion Factor
Cumulative Review, Chapters 1–5	71	1.41
Cumulative Review, Chapters 1–10	61	1.64
Cumulative Review, Chapters 1–14	53	1.89

If you decide to eliminate certain items from the reviews, you may convert the raw score to a percentage by dividing the number of correct responses by the number of questions assigned and multiplying by 100.

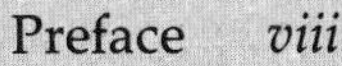

CONTENTS

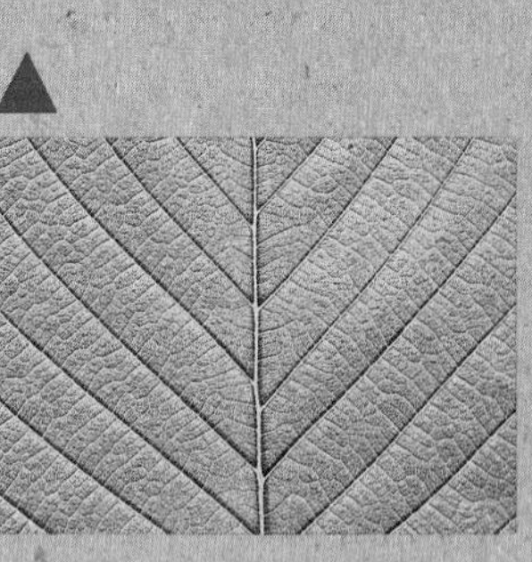

4 SIMILAR TRIANGLES 123

5 CONGRUENT TRIANGLES 165

6 CONSTRUCTIONS 205

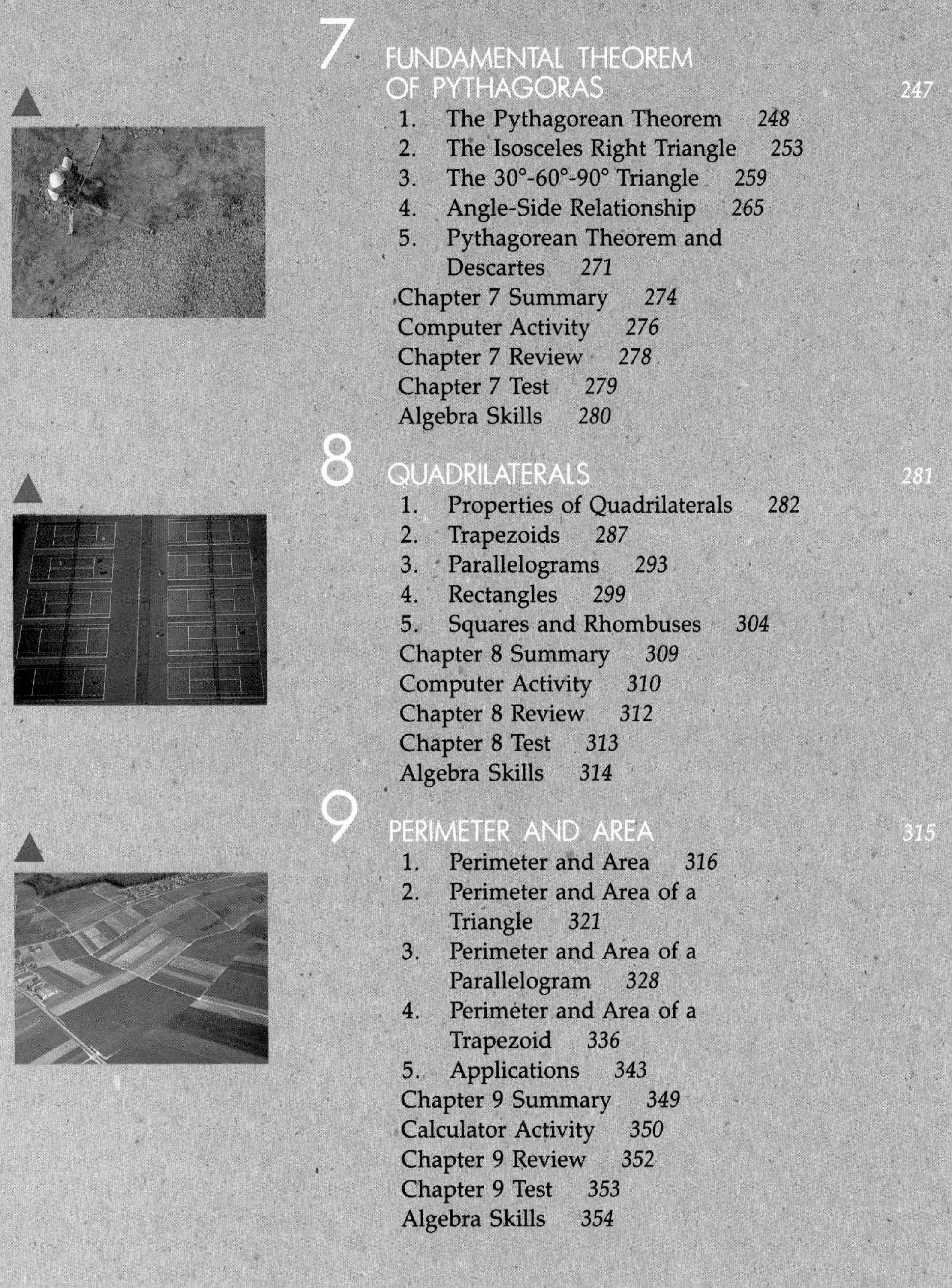

7 FUNDAMENTAL THEOREM OF PYTHAGORAS 247

8 QUADRILATERALS 281

9 PERIMETER AND AREA 315

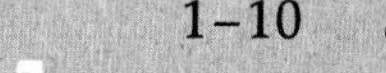

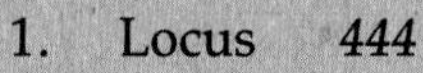

PREFACE

Geometry is the mathematical study of space. But your interpretation of space depends on your focus. Look at an object through different kinds of lenses and the image varies. Investigate space with different sets of postulates or Premises and different worlds result. Euclidean geometry is just one approach. Others exist, but the Euclidean view seems more realistic in terms of everyday experience. Nevertheless, even ancient navigators had to bend a few of Euclid's rules when traveling long distances on the surface of a sphere they called Earth.

GEOMETRY

Whether you live in a large city or a remote, rural or underdeveloped area, you are surrounded by a vast number of examples or models of geometric concepts. Some are the results of human toil, others occur naturally. You can find geometry in the construction of complex highway systems and in the markings of a fragile butterfly. You need only look.

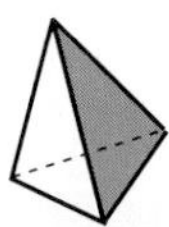

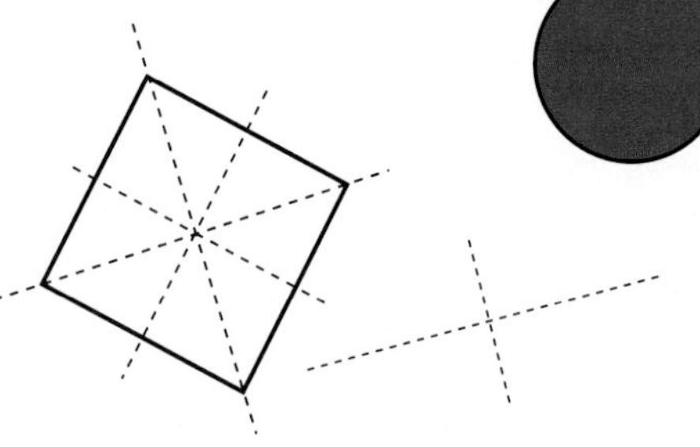

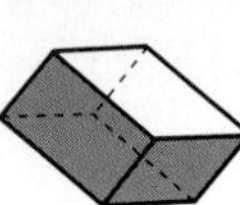

Geometry is all around you.

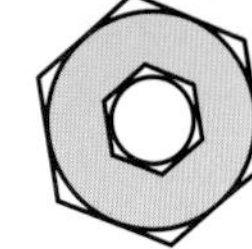

. . . In Theater

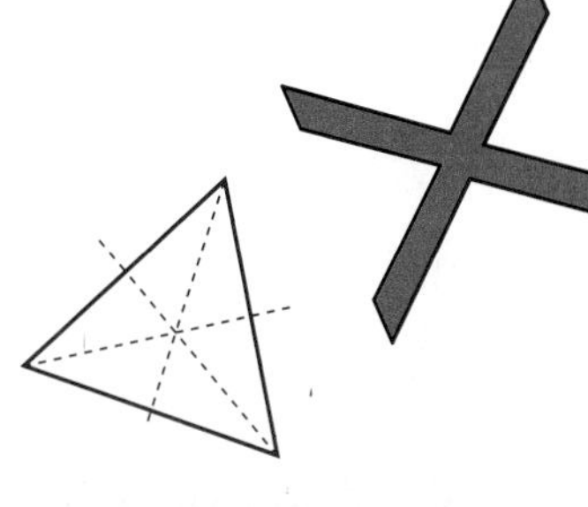

The Lyric Opera of Chicago premier of Gluck's timeless masterpiece, *Alceste*, starring Jessye Norman in the title role and Chris Merritt as Admète opened Lyric's 1990-91 season. Robert Wilson created this stunning new production.

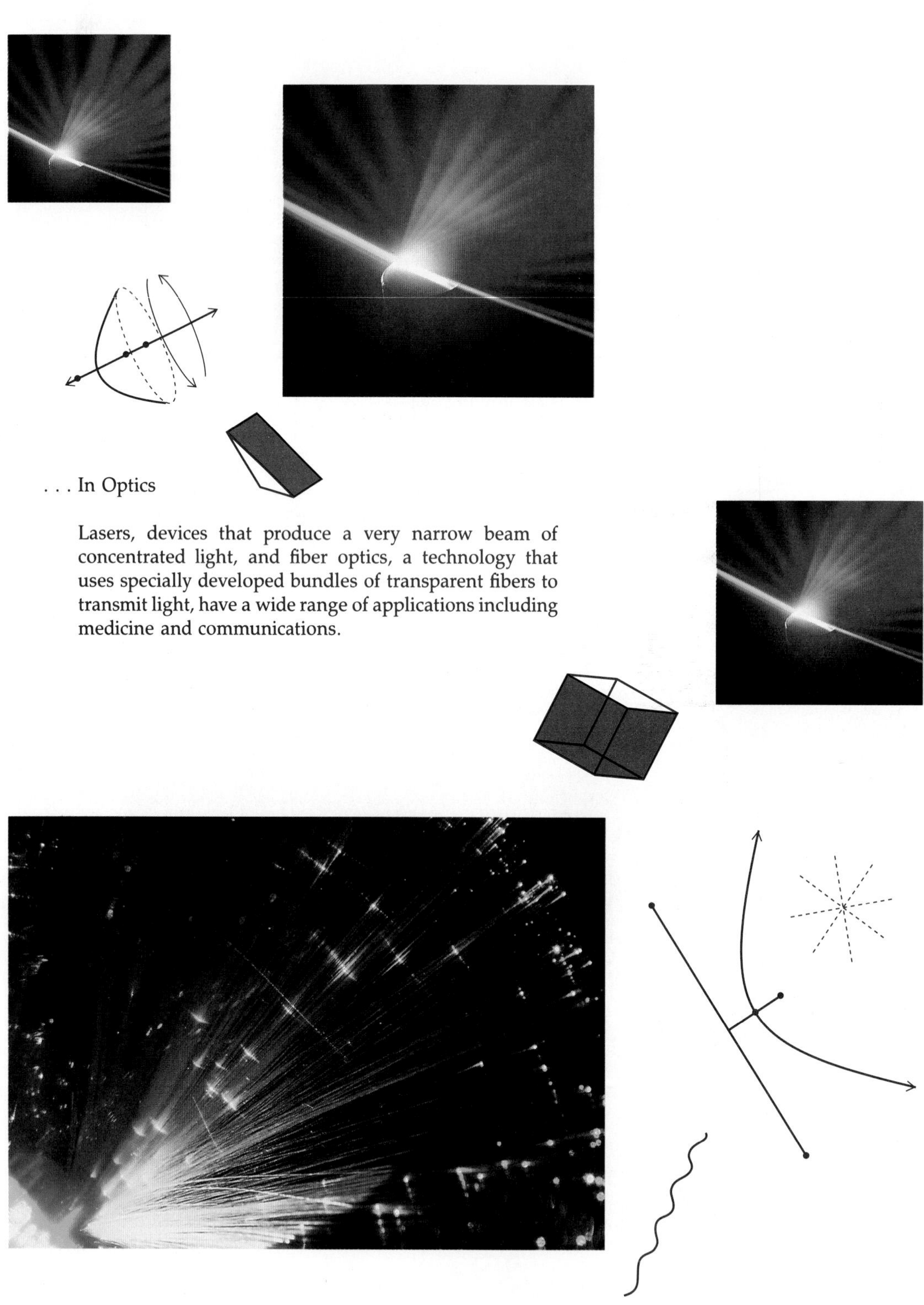

... In Optics

Lasers, devices that produce a very narrow beam of concentrated light, and fiber optics, a technology that uses specially developed bundles of transparent fibers to transmit light, have a wide range of applications including medicine and communications.

. . . In Chemistry

Mineralogists often use the geometric shape of crystals as one means of identification. Jewelers then cut some crystals geometrically to enhance their beauty. The abstraction below was created naturally. It depicts frost forming on a window pane.

. . . In Ecology

The experimental solar energy collector in New Mexico and wind turbines shown at sunset display a myriad of geometric shapes. Ecologists throughout the world are experimenting with a variety of techniques to produce energy in a way that avoids contamination of the environment. These endeavors and an educated public will preserve our world for future generations.

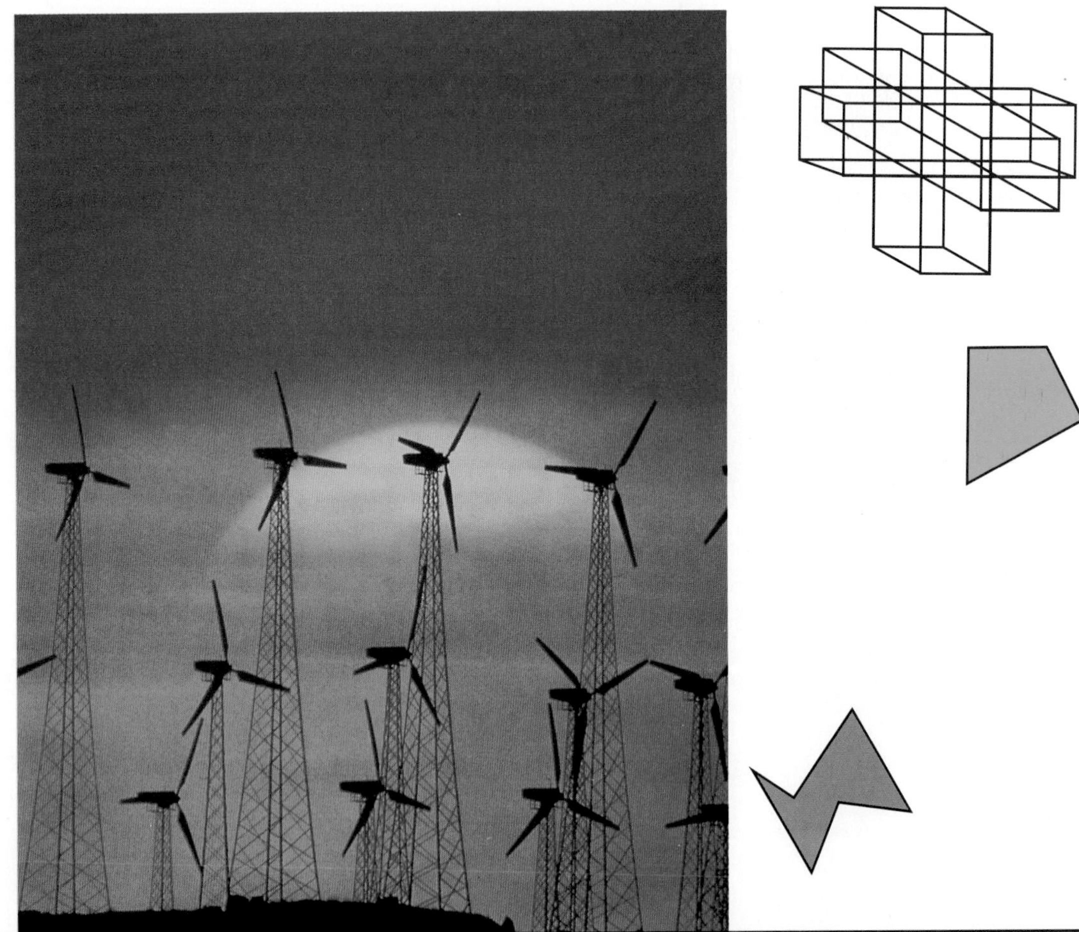

. . . In Architecture

This view of the East Wing of the National Gallery in Washington, D.C., shows how architects make use of geometric shapes to enhance the beauty and utility of their structures.

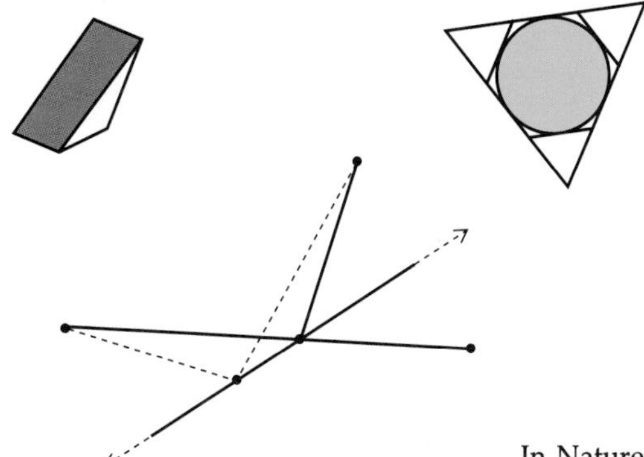

. . . In Nature

The mystery of the honeycomb still defies us. Bees produce these hexagonally shaped cells to store honey and reproduce their species.

. . . In Agriculture

Terracing was used by the early Babylonians to prevent soil erosion. The method is still practiced today in many areas of the world.

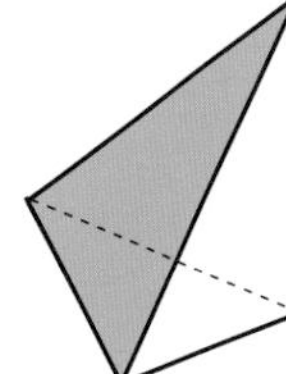

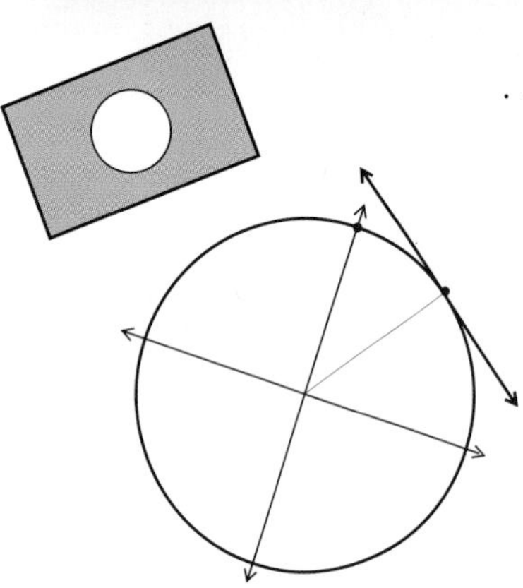

. . . In Sports

This photograph of the opening ceremony of the 1988 Summer Olympics in Seoul, Korea, illustrates how photographers position geometric shapes in a manner that creates a dramatic effect. But in sports many facets of geometry emerge. Measurement of distance and time is fundamental to Olympic events. The study of forces and vectors enable contestants to improve their techniques. Geometry is everywhere for all to see. You need only look.

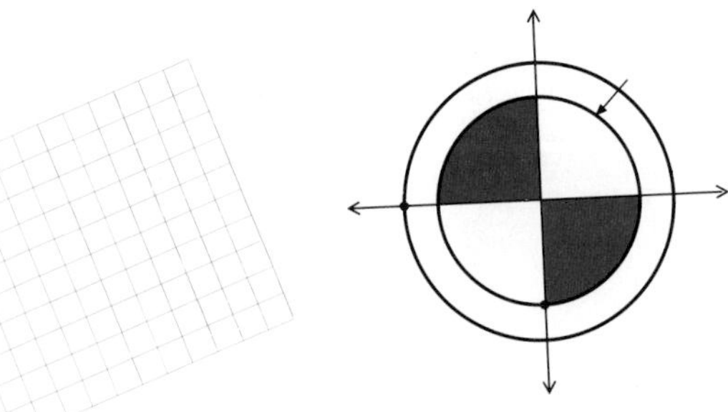

Letter to the Student

You are about to embark on a voyage through space. In your travels you will encounter objects so small that they have only one characteristic, position, and distances so large that to become meaningful they must be translated into time references, light years.

As a space traveler it would be wise to keep a log of the important facts you observe and any discoveries you might make. Each discovery can and should lay the groundwork for further discovery. Frequent review of your log will strengthen your hold on the material and make using it much easier.

When you encounter new objects or concepts in your travels, try to relate the new to the old. Relating new facts or relationships to familiar ideas will help you internalize the new and will make things easier to remember.

By the end of your journey you will have collected a vast amount of information, but it will be organized in a way that will stress relationships. A good understanding of these relationships will enable you to appreciate more fully the space around you, to discover additional relationships that exist in your world, and to use those relationships to solve real problems.

We hope that you will enjoy this trip through space and that you might find it sufficiently interesting to consider taking other excursions. Perhaps in the future you might consider trips to different kinds of space, where the few who dare visit are rewarded handsomely. They return home from alien lands with an appreciation for these new worlds, often with a greater understanding of their own world, but more importantly with a greater appreciation of the world around them.

James E. Elander
North Central College
Naperville, Illinois

James Elander, assistant professor of mathematics at North Central College, Naperville, Illinois, is the former mathematics department chair of Oak Park and River Forest High School, Oak Park, Illinois. He was named winner of the 1988 Burlington Northern Foundation Faculty Achievement Award for meritorious achievement in professional scholarship.

Mr. Elander served as chairman of the Illinois Section Mathematical Association of America Geometry Committee and served that organization's High School Lecture Program. He has helped organize programs for the National Council of Teachers of Mathematics, Illinois Council of Teachers of Mathematics, and the School Science and Mathematics Association.

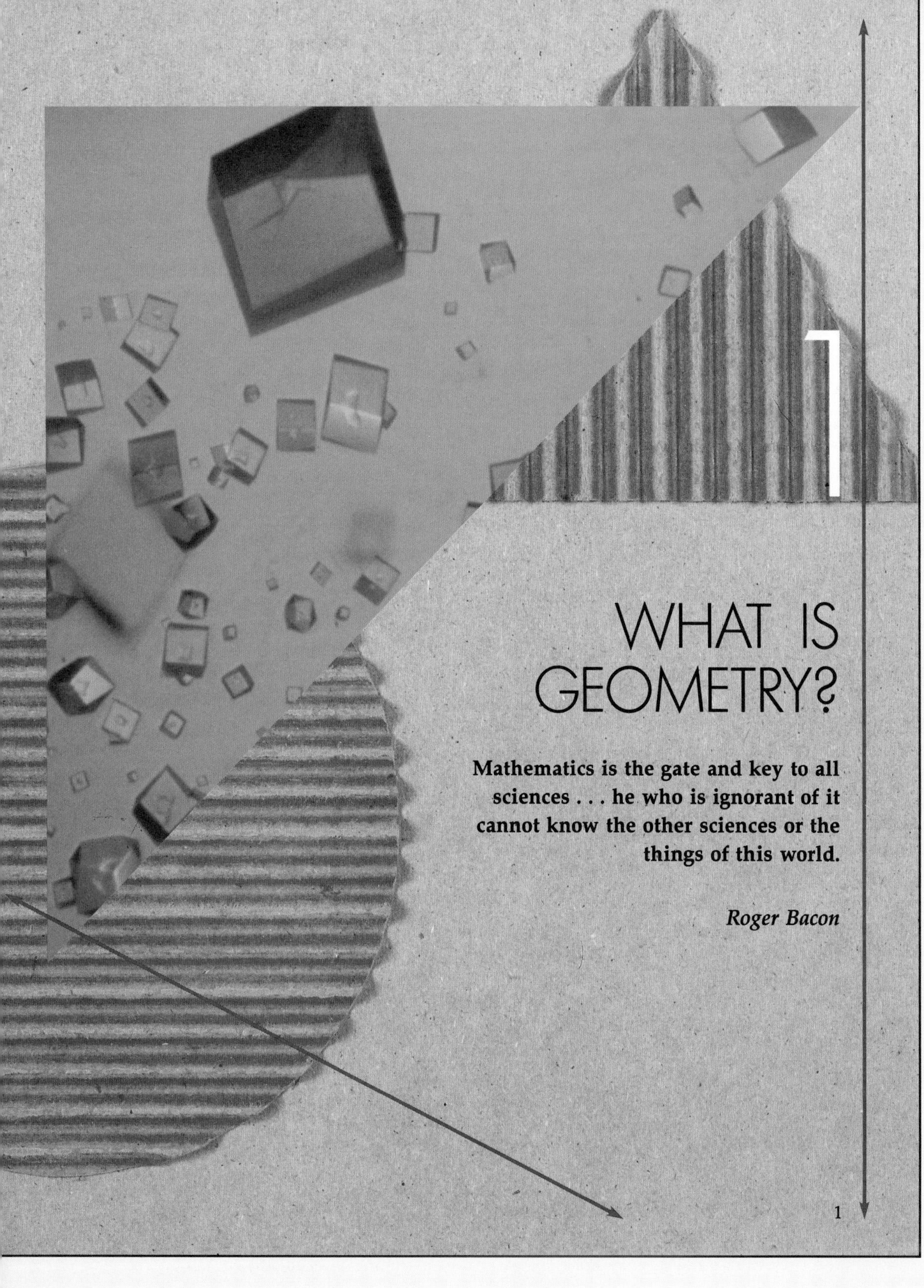

1 WHAT IS GEOMETRY?

Mathematics is the gate and key to all sciences . . . he who is ignorant of it cannot know the other sciences or the things of this world.

Roger Bacon

1

1

CHAPTER 1 OVERVIEW

What is Geometry?

Suggested Time: 12–14 days

Geometry is a term that comes from the ancient Greeks. Geometry encompasses a deductive system of mathematical ideas that has fascinated and inspired thinkers in other fields throughout history. Astronomy, architecture, physics, and biochemistry are among the many fields in which geometry has had a wide range of applications. For example, astronomers determined the spherical shape of the earth, charted the orbits of planets, and discovered the positions and motions of stars and galaxies by applying geometric ideas.

Mathematics is the gate and key to all sciences . . . he who is ignorant of it cannot know the other sciences or the things of this world.

Roger Bacon

CHAPTER 1 OBJECTIVE MATRIX

	End of Chapter Items by Activity			Student Supplement Items by Activity		
Objectives by Number	**Review**	**Test**	**Algebra Skills***	**Reteaching**	**Enrichment**	**Computer**
1.1.1	✓	✓		✓	✓	✓
1.2.1	✓	✓	✓			✓
1.2.2	✓	✓	✓	✓		✓
1.2.3	✓	✓		✓		
1.3.1	✓	✓	✓	✓		
1.3.2	✓	✓	✓	✓		
1.3.3	✓	✓	✓			
1.4.1	✓	✓	✓	✓		✓
1.4.2	✓	✓	✓	✓		
1.5.1	✓	✓	✓	✓		
1.5.2					✓	
1.6.1	✓	✓		✓		
1.6.2						

*A ✓ beside a Chapter Objective under Algebra Skills indicates that algebra skills taught within the chapter or in previous Algebra Skills lessons are used.

CHAPTER 1 PERSPECTIVES

▲ Section 1.1

Points, Lines, and Planes

Students are introduced to the basic elements of geometry: points, lines, planes, line segments, and rays. They learn the postulates and definitions that describe the properties of each element and distinguish one from another. Through a Discovery Activity, students find that two points determine a line. In a Project, they interview a person working in a field in which geometry is applied.

▲ Section 1.2

Measurement of Segments and Angles

Students are introduced to the one-to-one matching of real numbers and the points on a number line. Next, they learn how to find the degree measure of angles using a protractor. They classify angles as acute, right, obtuse, or straight. Although most students should have mastered the use of rulers and protractors by this point in their careers, some students may need additional help in aligning the zero point and reading the scale. This is especially true of the protractor. Help students to determine which scale to read for various orientations of the protractor.

▲ Section 1.3

Midpoints of Segments and Angle Bisectors

Students are introduced to the concept of a midpoint through the real world example of the center of a rope in a tug-of-war contest. In the first Class Activity, students measure segments to determine the midpoints and solve problems involving midpoints of segments. Students are asked to copy, measure, and classify given angles, and then use protractors to draw the bisectors of each. Then a road map is used to introduce the concept of betweenness. The topics of midpoints and angle bisectors lend themselves to a number of paper folding activities that can be used with students having difficulties.

▲ Section 1.4

Supplements and Complements

Through Discovery Activities, students investigate pairs of angles that are supplementary and pairs that are complementary. They learn to solve algebraic equations related to pairs of angles that are supplementary or complementary.

Because students so frequently use supplementary and complementary angles in situations where the angles share a common side and vertex some students have difficulty identifying supplementary and complementary angles when they are *not* adjacent. Take every opportunity to point out these situations. An excellent example of complementary angles are the acute angles of a right triangle. Students can cut out right triangles. Clip off the acute angles and place them side by side to form a right angle.

▲ Section 1.5

Inductive Reasoning

Students are introduced to the process of inductive reasoning as they discover certain geometric patterns. However, through doing the Discovery Activity, and from examining some examples from real life, students are guided to realize that there may be pitfalls in inductive reasoning. They conclude that some patterns that can be detected in a limited number of cases may not necessarily hold for all cases.

▲ Section 1.6

LOGO and Geometric Drawing Tools

Students are introduced to the computer language of LOGO and to programming using its turtle graphics. In a Class Activity they use a geometric drawing tool to draw and label an angle, and to explore the concept of bisecting an angle.

Information about LOGO programs can be obtained from:

Terrapin Software; 400 Riverside St.; Portland, ME 04103.

As of this writing, the two most commonly used geometric drawing tools are the Geometric Supposer series and the GeoDraw program from the Geometry Series. The Geometric Supposer can be ordered directly from:

Sunburst Communications; 39 Washington Avenue; Pleasantville, NY 10570-2898

Information about ordering the Geometry Series can be obtained from:

IBM Direct; PC Software Department; One Culver Road; Dayton, New Jersey 07645

SECTION 1.1

Resources
Reteaching 1.1
Enrichment 1.1

Objectives
After completing this section, the student should understand
▲ the basic elements of geometry—points, line, line segments, rays, and planes.

Materials
ruler

Vocabulary
plane
point
line
postulate
line segment
ray
endpoint
angle
side
vertex

GETTING STARTED

Warm-up Activity
Write the words *point*, *line*, and *plane* on the chalkboard. Ask students to attempt to write a definition of these terms. Also have students draw a representation of each term and to name objects that suggest each, if possible, from within the classroom.

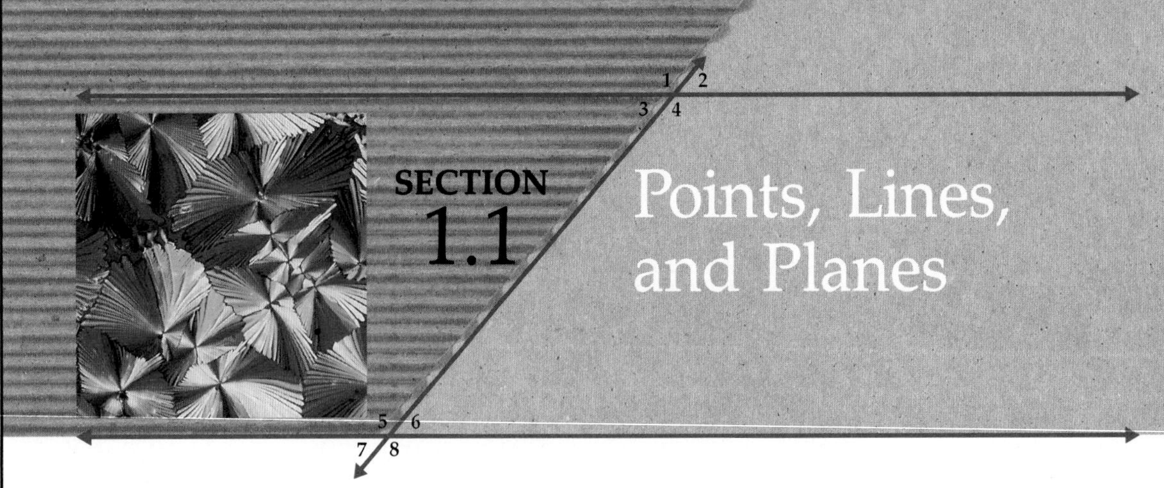

SECTION 1.1 Points, Lines, and Planes

After completing this section, you should understand
▲ the basic elements of geometry—points, lines, line segments, rays, and planes.

When you first begin to play a board game, you open the box, check the contents, and read the instructions. Then you discuss the rules, identify the players, and begin to play.

As a result of reading the instructions and beginning to play, you soon acquire the game's vocabulary and are able to communicate with all the other players. Communication is also very important in geometry. You need to learn its rules, definitions, and terminology.

To study geometry, you need to understand three basic terms—point, line, and plane. They are important because they are used to define other geometric terms.

PLANES, POINTS, AND LINES

A flat surface, such as a table top, suggests the idea of a **plane.** A table top is limited in size, but a geometric plane extends endlessly in all directions.

The flat surfaces pictured above suggest geometric planes. Unlike the flat surfaces in the photos, a plane has no boundaries.

The following drawings suggest how part of a plane, flat surface might look from different points of view.

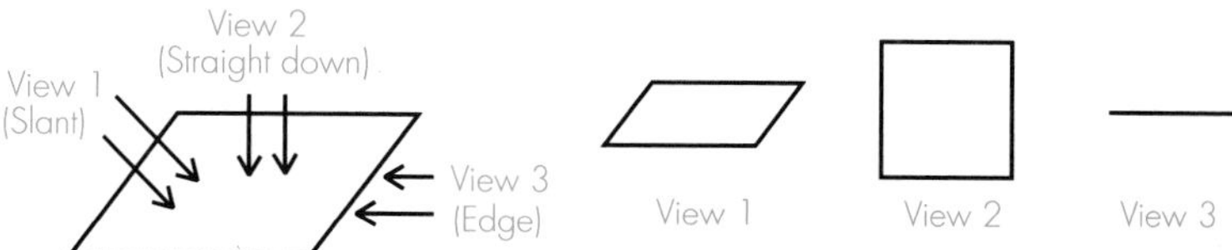

Planes are made up of points. The picture on a television screen consists of many small dots packed closely. You can think of each dot as representing a point. An individual point is only a location, but you can represent it by a dot.

In the diagram, A and B are two points in plane P. Passing through A and B is a line called **line** $\boldsymbol{AB}$. The line is straight and continues endlessly.

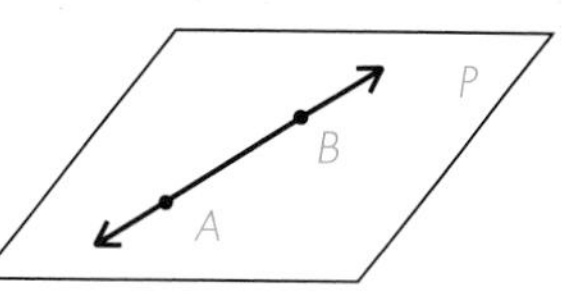

Like planes, lines are made up entirely of points. Lines and planes contain infinitely many points. Both of the figures below contain a infinite number of points. However, the figure to the right is not a line, because it is not straight.

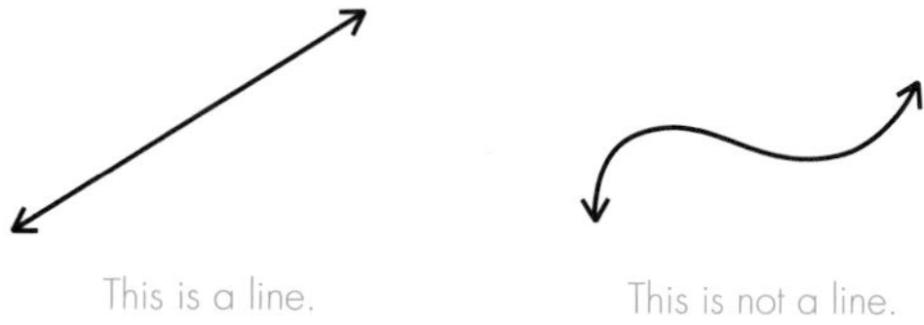

This is a line.

This is not a line.

DISCOVERY ACTIVITY

On a sheet of paper, draw a plane. On the plane, mark and label two points S and T, as shown.

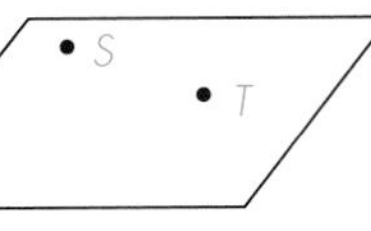

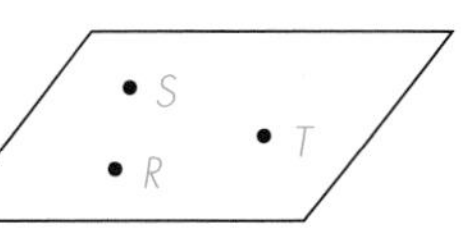

1. How many lines can you draw through point S? through point T? Infinite number
2. How many lines can you draw that contain *both* points? One
3. On the plane, mark and label a third point R, as shown. How many lines can you draw that contain all three points? None

TEACHING COMMENTS

Using the Page

As students look at the representations and read the descriptions of planes, points, and lines, be sure they understand that each of these terms is undefined and is given meaning by the relationships that are assumed to exist among them.

Be sure students understand that a point has no length, width, or thickness, and serves only to indicate a definite location.

Using the Page
After students have read Postulate 1-1-1, ask them what the arrows in the drawing suggest. [The line goes on indefinitely in those directions.]

After you have introduced Definition 1-1-1, ask students to explain why line segment *AB* is the same as line segment *BA*. [*AB* and *BA* have the same endpoints, and both name the identical part of the line, including all points between *A* and *B*.]

Using the Page
After students have learned the definition of a ray, ask them to draw a representation of $\overrightarrow{RQ}$.

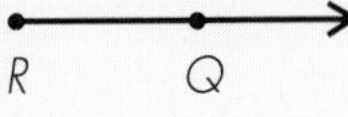

You probably found that you were able to draw one and only one line through points *S* and *T*. Also, you probably discovered that no line can be drawn that passes through all three points.

In geometry, a statement that is assumed to be true in all cases is called a **postulate.**

NOTEBOOK **POSTULATE 1-1-1:** **Two points determine exactly one line.**

According to Postulate 1-1-1, if *A* and *B* are any two points, there is one and only one line that goes through both. The written name for this line is $\overleftrightarrow{AB}$ or $\overleftrightarrow{BA}$, read "line *AB*" and "line *BA*", respectively.

LINE SEGMENTS, RAYS, AND ANGLES

Line segments and **rays** are parts of a line. This diagram shows the line segment connecting points *A* and *B*.

NOTEBOOK **DEFINITION 1-1-1:** *A* and *B* are the **endpoints** of line segment *AB*. You can name the line segment in symbols by writing $\overline{AB}$ or $\overline{BA}$.

A LINE SEGMENT is a subset of a line, consisting of two points *A* **and *B* on the line and all points between *A* and *B*.**
How would you name this line segment? $\overline{CD}$ or $\overline{DC}$

This is ray *AB*. Point *A* is called the **endpoint** of ray *AB*.

Ray *AB* starts at point *A* and continues indefinitely in the direction of point *B*. The symbol for ray *AB* is $\overrightarrow{AB}$. When naming a ray, always name the endpoint first.

NOTEBOOK **DEFINITION 1-1-2:** **A RAY is a subset of a line consisting of a point *A* on the line and all points of the line that lie to one side of point *A*.**

How would you name this ray? $\overrightarrow{QR}$

Two rays that have the same endpoint form a special figure called an **angle.**

NOTEBOOK

DEFINITION 1-1-3: **An ANGLE is formed by two rays that have a common endpoint.**

The two rays are called the **sides** of the angle. Their common endpoint is called the **vertex** of the angle. The diagram shows the angle formed from $\overrightarrow{BA}$ and $\overrightarrow{BC}$. The angle is called angle *ABC* or angle *CBA*—in symbols, $\angle ABC$ or $\angle CBA$. Notice that when you name an angle, you name the vertex in the middle. The other two letters name points on the two sides of the angle. If there is only one angle with a given vertex, you can simply write the angle symbol and the letter for the vertex.

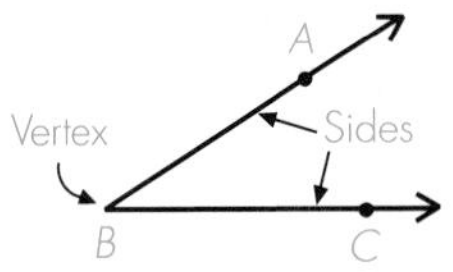

Example: The angle in the diagram can be called $\angle LMN$, or $\angle NML$, or $\angle M$. The sides are *ML* and *MN*. The vertex is point *M.*

How would you name this angle?
$\angle WTS$, or $\angle STW$, or $\angle T$

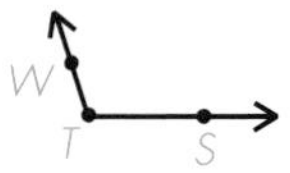

Sometimes a number is used to name an angle.

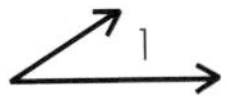

This is angle 1.

These are angles 2 through 5.

CLASS ACTIVITY

Name each of the following. Use symbols. See margin.

1. 

2. 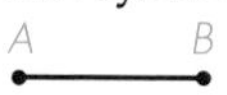

3. 

4. 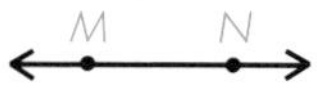

5. 

6. 

7. 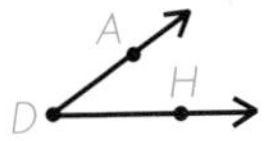

8. 

9. 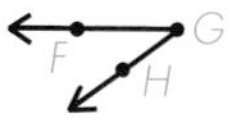

Draw and label each of the following. See students' drawings.

10. line PQ
11. ray WR
12. angle BMT
13. line segment FG
14. angle TRG
15. ray ZX

Draw, label, and name each of the following. See students' drawings.

16. $\overleftrightarrow{AB}$ containing a third point C
17. $\overrightarrow{XY}$ and $\overrightarrow{YX}$
18. $\overrightarrow{GF}$, point *K* not on $\overrightarrow{GF}$, and $\overrightarrow{KG}$
19. $\angle HIJ$ and $\overrightarrow{HJ}$

Using the Page

As you introduce angles, point out that angles and line segments are used to name and classify other geometric figures such as triangles and rectangles.

Additional Answers

1. Point P
2. $\overline{AB}$ or $\overline{BA}$
3. $\overrightarrow{GE}$
4. $\overleftrightarrow{MN}$ or $\overleftrightarrow{NM}$
5. $\overrightarrow{KL}$
6. $\overline{PO}$ or $\overline{OP}$
7. $\angle ADH$, or $\angle HDA$, or $\angle D$
8. $\angle TSR$, or $\angle RST$, or $\angle S$
9. $\angle FGH$, or $\angle HGE$, or $\angle G$

Extra Practice

Draw the angle below on the chalkboard. Ask students to name it in four different ways.

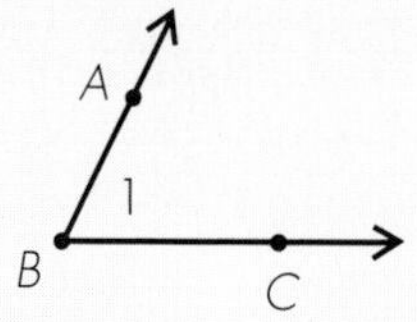

[$\angle 1$, $\angle B$, $\angle ABC$, $\angle CBA$]

Additional Answers

16. 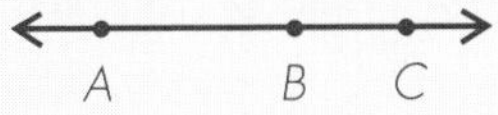

17. 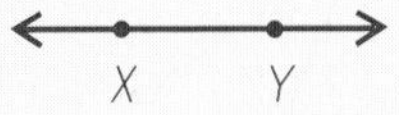

18. 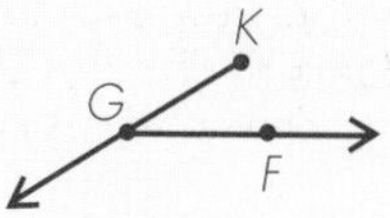

or

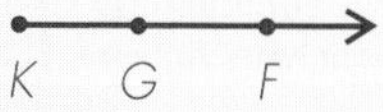

19. 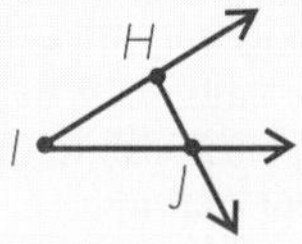

FOLLOW-UP

Assessment

1. Draw and label line segment *GH*.

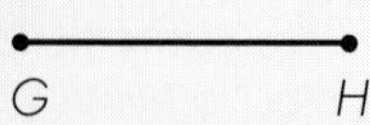

2. Draw, label and name $\angle ABC$ and $\overrightarrow{CA}$.

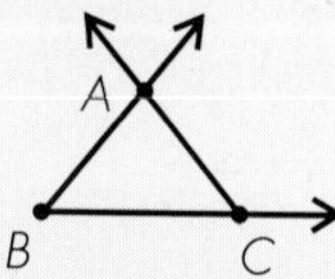

3. How many different angles are formed by the four rays from point *R*? Name them.

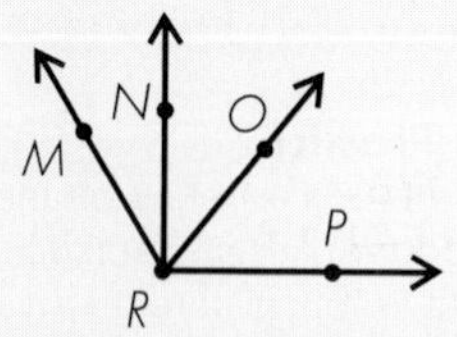

[6 angles: *MRP, MRO, MRN, NRO, NRP, ORP*]

14.

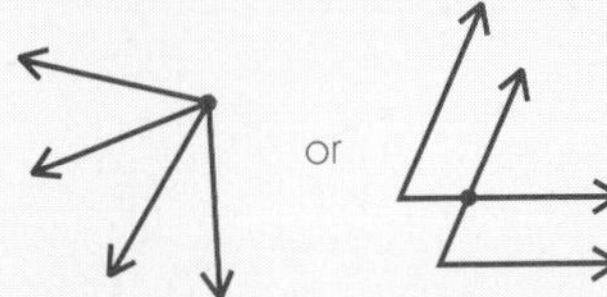

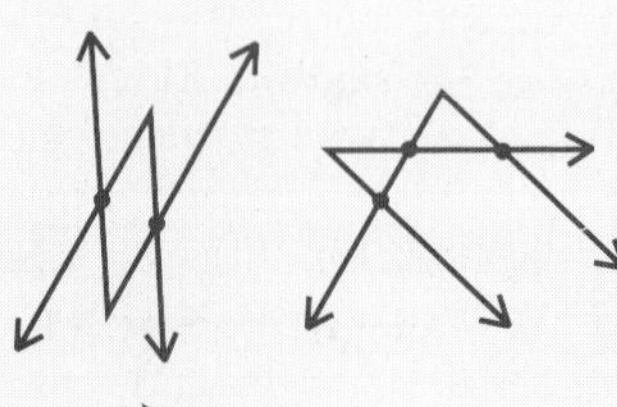

Reteaching
Students who have had difficulty with this section may benefit from Reteaching Activity 1.1.

Enrichment
For students who have mastered the material in this section, you may wish to assign Enrichment Activity 1.1.

PROJECT 1-1-1

People in many careers use geometry daily in their work. Highway engineers use geometry when surveying. Choreographers use it to plan dance movements on a stage. Graphic artists, architects, package designers, and many others have frequent opportunities to put geometry to use on the job.

Choose a profession or career that interests you. Find someone who is working in that field and talk to them. Find out things such as the following:

- what a typical work day is like.
- what education and training are required.
- how geometry or other kinds of mathematics are used in the work.

Take notes during your discussion. Share your findings with the class.

HOME ACTIVITY

Name each of the following. Use symbols.

1. 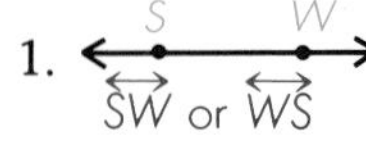

$\overleftrightarrow{SW}$ or $\overleftrightarrow{WS}$

2. 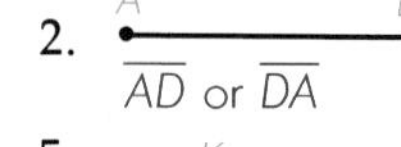

$\overline{AD}$ or $\overline{DA}$

3. 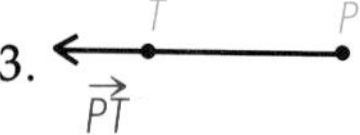

$\overrightarrow{PT}$

4. 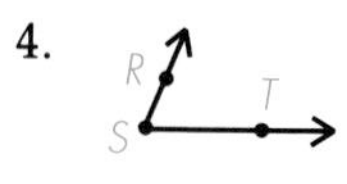

$\angle RST$ or $\angle TSR$

5. 

$\angle JKL$ or $\angle LKJ$

6. 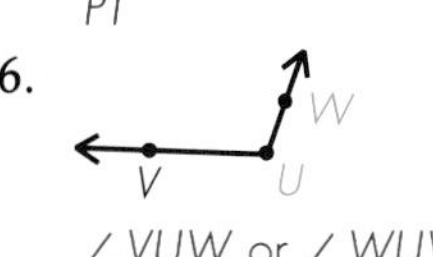

$\angle VUW$ or $\angle WUV$

Which of the geometric figures in this section do the following represent?

7. the surface of a still lake plane
8. a pencil line segment
9. a beam of sunlight ray
10. clock hands showing 7:45 angle

Use the diagram. Name the following. Answers will vary; samples given.

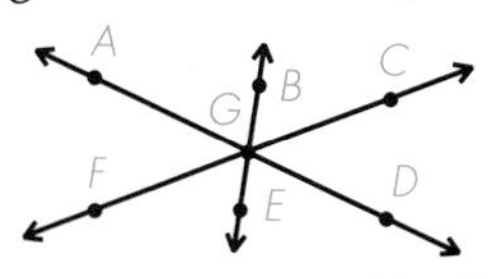

11. 2 rays $\overrightarrow{GA}$, $\overrightarrow{GC}$
12. 2 lines $\overleftrightarrow{AD}$, $\overleftrightarrow{BE}$
13. 2 segments $\overline{GC}$, $\overline{AD}$

CRITICAL THINKING

14. Draw diagrams to show that two angles can have exactly one, two, three, or four points in common. Is it possible for two angles to have exactly five points in common? See margin.

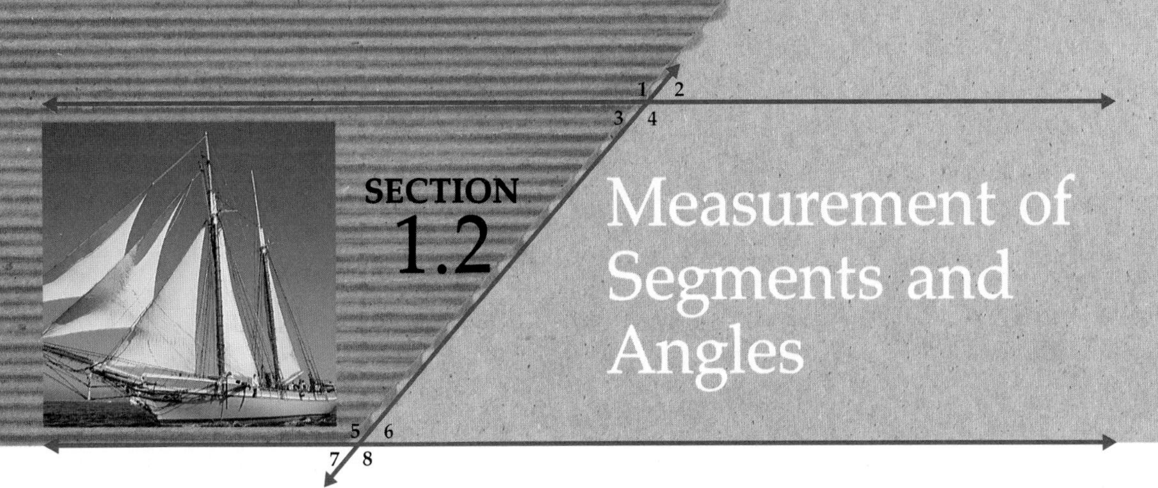

SECTION 1.2 Measurement of Segments and Angles

After completing this section, you should understand
- ▲ how numbers are matched with points on a number line.
- ▲ how to measure line segments and angles.
- ▲ how to classify angles.

The number markings on a football field determine the exact placement of movements on the field between the two end zones. Football fields are marked in 1-yard, 5-yard, and 10-yard segments. The officials measure to see whether one team has advanced the ball far enough to gain a first down.

Some of the ideas that the football officials are using are also used in geometry.

THE NUMBER LINE

You may recall from algebra that the set of all positive and negative numbers, together with zero, make up the set of **real numbers.** You may also recall that the set of real numbers can be matched one-to-one with the points on a line to obtain a **number line.** The number matched with a point is called the **coordinate** of the point.

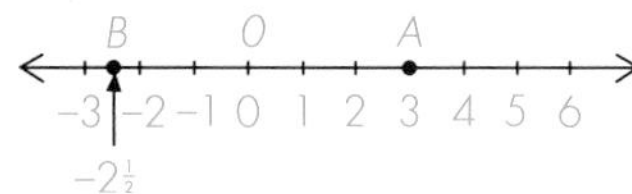

For the number line shown above, the coordinate of point A is 3 and the coordinate of point B is $-2\frac{1}{2}$. The point O whose coordinate is 0 is called the **origin.**

SECTION 1.2

Resources
Reteaching 1.2
Enrichment 1.2

Objectives
After completing this section, the student should understand
- ▲ how numbers are matched with points on a number line.
- ▲ how to measure line segments and angles.
- ▲ how to classify angles.

Materials
centimeter and inch ruler
protractor

Vocabulary
real numbers
number line
coordinate
origin
degree measure
protractor
acute angle
right angle
obtuse angle
straight angle

GETTING STARTED

Quiz: Prerequisite Skills
Draw the figure below on the chalkboard.

1. Name 3 lines. [$\overleftrightarrow{DE}$, $\overleftrightarrow{EF}$, $\overleftrightarrow{FD}$]

2. Name 2 rays from point E. [$\overrightarrow{ED}$, $\overrightarrow{EF}$]

3. On what two lines is point F? [$\overleftrightarrow{DF}$ and $\overleftrightarrow{EF}$]

4. Name a point that is on $\overleftrightarrow{DE}$ and $\overleftrightarrow{DF}$. [D]

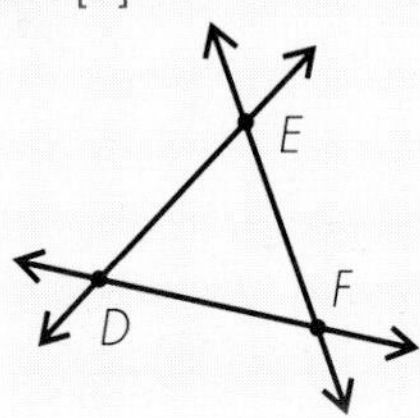

TEACHING COMMENTS

Using the Page
As you introduce the number line, emphasize that each point can be paired with a real number *and* each real number can be paired with a point.

Using the Page
Postulate 1-2-1 is known as the *Ruler Postulate.*

Be sure students understand that a point has no length, width, or thickness, and serves only to indicate a definite location.

Euclid defined a point as "that which as no part".

POSTULATE 1-2-1: There is a one-to-one matching between the points on a line and the real numbers. The real number assigned to each point is its coordinate. The distance between two points is the positive difference of their coordinates. A and B are two points with coordinates a and b such that $a > b$, then distance $AB = a - b$.

Notice that $\overline{AB}$ (with a bar above the letters) means the line segment joining A and B. The symbol AB (no bar) means the distance between A and B, that is, the *length* of $\overline{AB}$.

Example: On the number line, -3 is the coordinate of point A, 4 is the coordinate of point B, and 8 is the coordinate of point C. The length of $\overline{AB}$ is $AB = 4 - (-3)$, or 7.

A B C
−5 −4 −3 −2 −1 0 1 2 3 4 5 6 7 8

What is BC? What is AC? 4; 11

MEASURING LINE SEGMENTS

When you use a ruler to measure a line segment, you are using the ruler as part of a number line. The number you assign will depend on which measurement you are using. The number you get for the length depends on whether the unit you are using is feet, inches, centimeters, meters, and so on.

Example: What is the length of $\overline{AB}$ in inches?

A B

Place your inch ruler so that the mark that corresponds to 0 is at point A. Read the number at the other end of the segment.

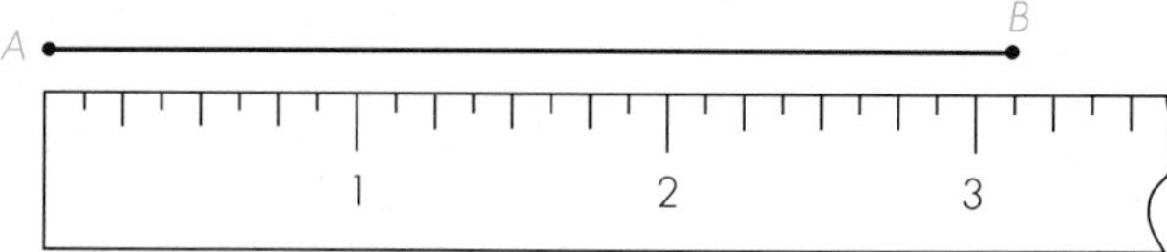

The segment measures $3\frac{1}{8}$ inches.
What is its measure in centimeters? about 8 cm

CLASS ACTIVITY

Use the number line to help you answer the questions.

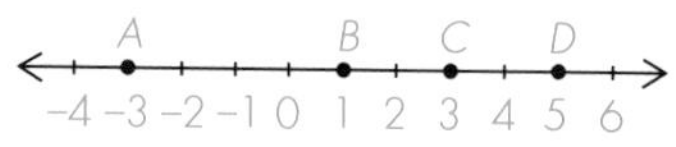

1. What are the coordinates of points A, B, C, and D? −3; 1; 3; 5
2. What is the length of $\overline{AB}$? of $\overline{BD}$? AB = 4; BD = 4

Use the number line below to help you answer the questions.

3. What is the coordinate of R? −2
4. What is the coordinate of T? 4
5. How many units longer than $\overline{QT}$ is $\overline{WR}$? 3 units

Measure each of the following segments in inches.

6. A ———— B 3 in.
7. C ———— B 4 in.

Measure these segments in centimeters.

8. E ———— F 5 cm
9. G ———— H 8 cm

MEASURING ANGLES

Postulate 1-2-1 assures you that each line segment has one particular number as its length. The next postulate accomplishes something similar for angles.

POSTULATE 1-2-2: **Let O be a point on $\overleftrightarrow{XY}$ such that X is on one side of O and Y is on the other side of O. Real numbers from 0 through 180 can be matched with $\overrightarrow{OX}$, $\overrightarrow{OY}$, and all the rays that lie on one side of $\overleftrightarrow{XY}$ so that each of the following is true:**

(1) 0 is the number matched with $\overrightarrow{OX}$.

(2) 180 is the number matched with $\overrightarrow{OY}$.

(3) If $\overrightarrow{OA}$ is matched with a and $\overrightarrow{OB}$ is matched with b and $a > b$, then the number matched with $\angle AOB$ is $a - b$.

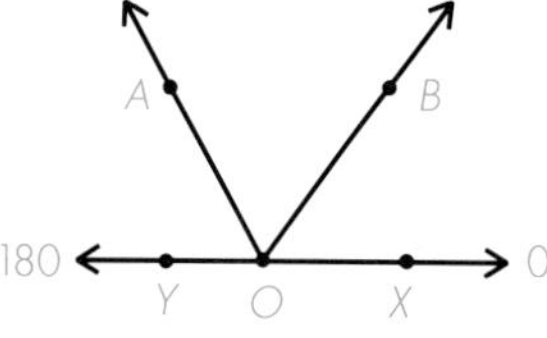

The number matched with $\angle AOB$ is called the **measure** or the **degree measure** of $\angle AOB$. In symbols, you can write $m\angle AOB$. For the angle shown in the diagram for the postulate, $m\angle AOB = a - b$.

Using the Page
Postulate 1-2-2 is known as the *Protractor Postulate.*

Use of the number 360 to measure angles can be traced to the Babylonians, more than 4,000 years ago. The *degree* is the unit based on this number.

Using the Page
As students are following the steps to measure an angle with a protractor, point out that they may need to use the straightedge of the protractor to extend a side of the angle they are measuring so that it crosses the protractor's scale.

Using the Page
You may wish to point out that some protractors are circular, and that those are convenient for measuring angles greater than 180°.

To measure angles, you use a protractor. The diagram shows a typical protractor. There is an outside scale and an inside scale, but both use numbers from 0 to 180. The unit of measure is shown with a small, raised °. $m\angle AOB = 115° - 50° = 65°$.

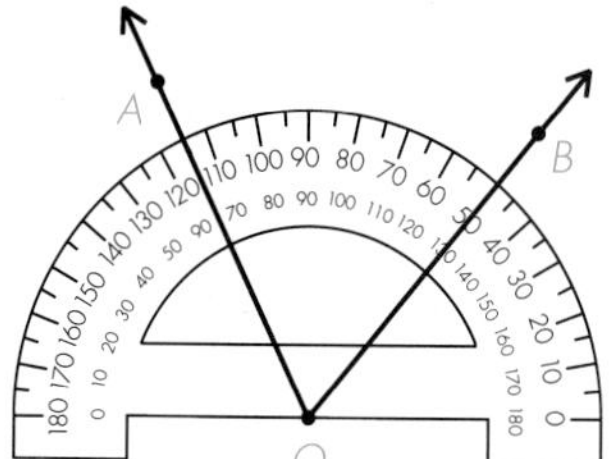

Follow these steps to measure an angle with a protractor:

1. Place the protractor over the angle so that the dot at the bottom center of the protractor is exactly over the vertex.
2. Make sure one side of the angle crosses the inside or outside scale at the 0-mark.
3. Follow around along that scale until you get to the place where the other side crosses the scale. The number for that place gives the degree measure of the angle.

Example: 1. Measure $\angle ABC$.

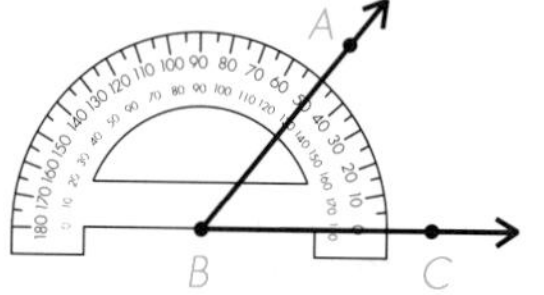

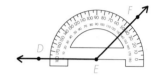

$\overrightarrow{BC}$ is at the 0-mark on the *outside* scale. So find the number where $\overrightarrow{BA}$ crosses the outside scale: $m\angle ABC = 50°$.

2. Measure $\angle DEF$.

$\overrightarrow{ED}$ is at the 0-mark on the *inside* scale. So find the number where $\overrightarrow{EF}$ crosses the inside scale: $m\angle DEF = 135°$.

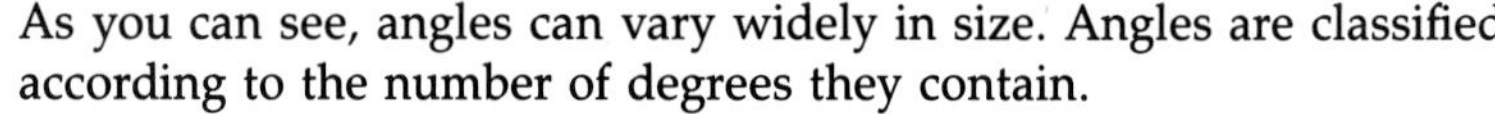

CLASSIFYING ANGLES

As you can see, angles can vary widely in size. Angles are classified according to the number of degrees they contain.

DEFINITION 1-2-1: **If the number assigned to an angle is between 0° and 90°, then the angle is called an ACUTE angle.**

Example: You can trace $\angle ABC$ on a sheet of paper and check that $m\angle ABC = 35°$. Since 35° is between 0° and 90°, $\angle ABC$ is acute.

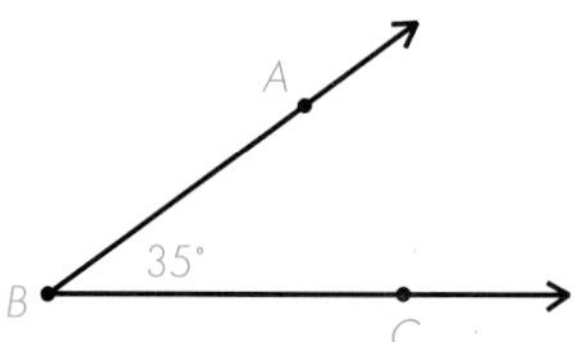

DEFINITION 1-2-2: **If the number assigned to an angle is exactly 90°, then the angle is called a RIGHT angle.**

Example: You can trace $\angle ABC$ and check that $m\angle ABC = 90°$. This means that $\angle ABC$ is a right angle.

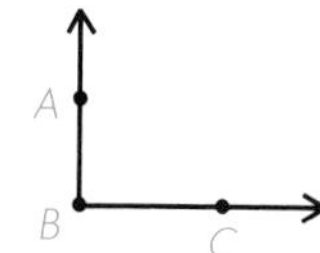

NOTEBOOK

DEFINITION 1-2-3: **If the number assigned to an angle is between 90° and 180°, then the angle is called an OBTUSE angle.**

Example: You can trace $\angle TUV$ and check that $m\angle TUV = 130°$. This means that $\angle TUV$ is obtuse.

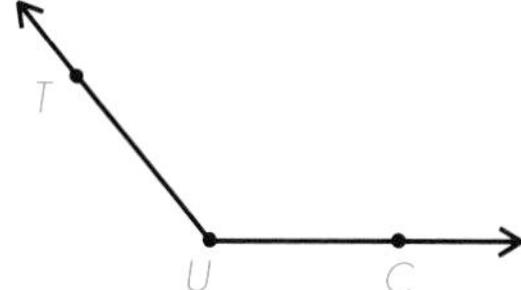

NOTEBOOK

DEFINITION 1-2-4: **If the number assigned to an angle is exactly 180°, then the angle is called an STRAIGHT angle.**

Example: $m\angle ABC = 180°$, so $\angle ABC$ is a straight angle.

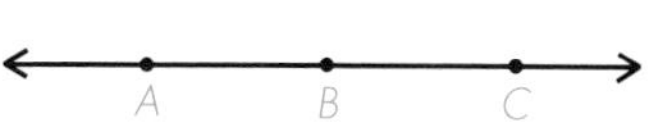

As you can see, if an angle is a straight angle, its sides form a line. Look around the classroom. Find examples of acute, right, obtuse, and straight angles.

CLASS ACTIVITY

Classify each angle as acute, right, obtuse, or straight.

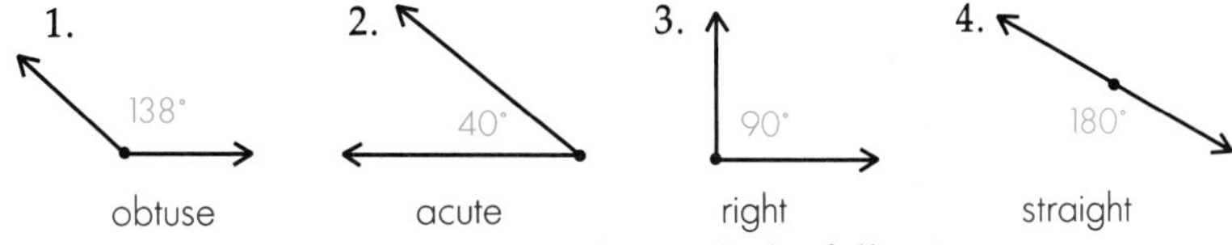

obtuse acute right straight

Use your protractor to draw angles with the following measures.

5. 60° 6. 100° 7. 20° 8. 145°

See students' drawings.

Measure and classify each angle.

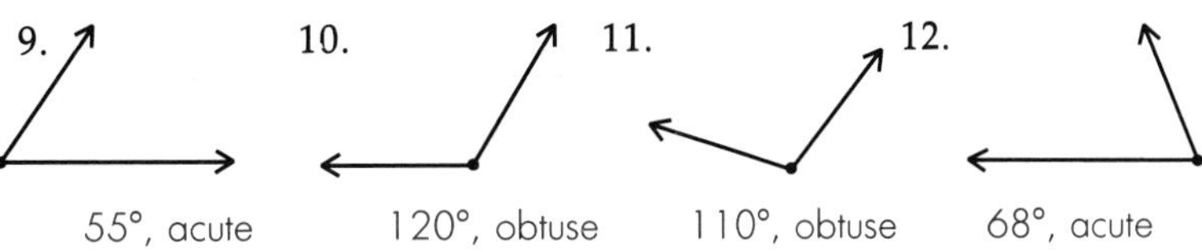

55°, acute 120°, obtuse 110°, obtuse 68°, acute

Using the Page

Draw several angles of different size on the chalkboard. Have students estimate the number of degrees each contains.

Students who can easily distinguish between acute and obtuse angles can use that ability to choose the sensible scale measure on the protractor when measuring an angle. Encourage students to estimate the degree measure of an angle before finding the actual measure.

For students who have difficulty in estimating the measure of an angle, suggest that they put the corner of an index card at the vertex of the angle. They can then tell quickly whether the measure of the angle is between 0° and 90°, about 90°, or between 90° and 180°.

Using the Page

Point out that a small square in the vertex of an angle will be used to indicate that it is a right angle.

39

Extra Practice

1. Measure $\overline{XY}$ in inches and in centimeters.

[$1\frac{3}{4}$ in.; 4.5 cm]

2. Find m∠PQR − m∠OQR.

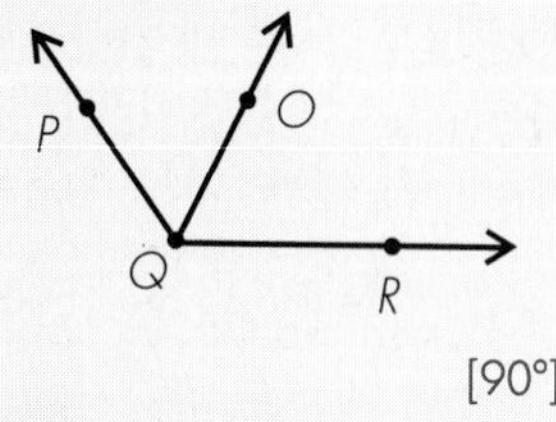

[90°]

FOLLOW-UP

Assessment

1. What is the length of $\overline{CD}$ in the number line below? [CD = 6]

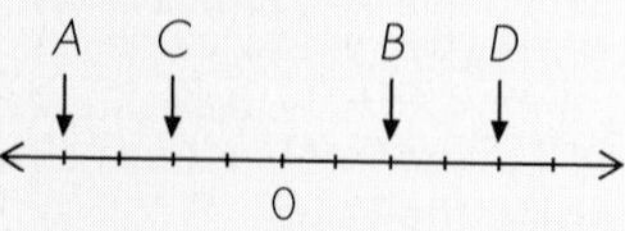

2. Find the measure of each angle in the figure below.

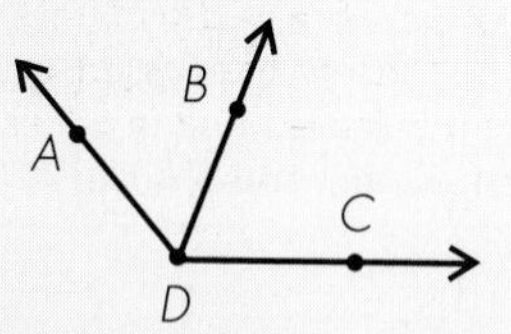

[m∠ADC = 130°; m∠ADB = 60°; m∠BDC = 70°]

Reteaching

Students who have had difficulty with this section may benefit from Reteaching Activity 1.2.

Enrichment

For students who have mastered the material in this section, you may wish to assign Enrichment Activity 1.2.

HOME ACTIVITY

Use the number line to help you answer the questions.

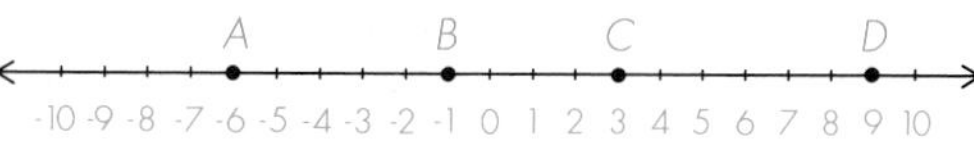

1. What are the coordinates of points *A*, *B*, *C*, and *D*? −6, −1, 3, and 9
2. What is the length of $\overline{AB}$? of $\overline{BD}$? AB = 5; BD = 10
3. What is the distance of each of the labeled points from the origin, zero? For A, B, C, and D, the distances are 6, 1, 3, and 9, respectively.

Use the number line to find each of the following.

R V P
−5 0 5

4. the coordinate of *V*. −1
5. the coordinate of *P* 4
6. the length of $\overline{VR}$ 3
7. the length of $\overline{RP}$ 8

Classify each angle as acute, right, obtuse, or straight.

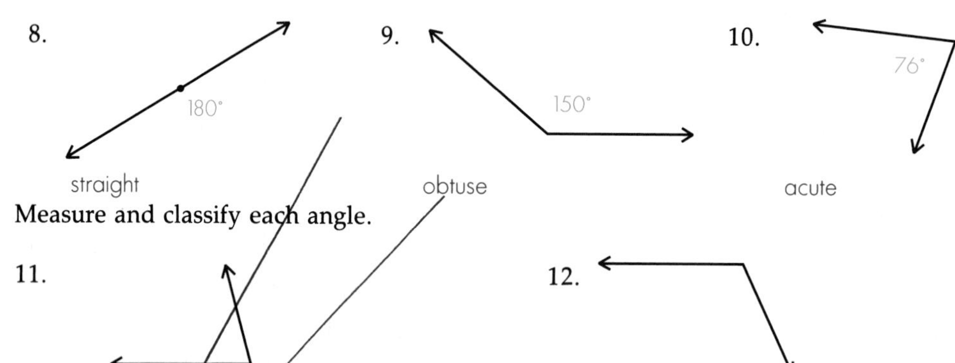

Measure and classify each angle.

11. 75°, acute

12. 116°, obtuse

List all the angles in each figure. Then measure and classify them.

13. ∠AOB, 15°, acute
∠AOC, 45°, acute
∠BOC, 30°, acute

14. ∠PMQ, 90°, right
∠QMR, 90°, right
∠PMR, 180°, straight

CRITICAL THINKING

Draw a figure consisting of three rays, $\overrightarrow{OA}$, $\overrightarrow{OB}$, and $\overrightarrow{OC}$, such that any pair of rays forms an angle that measures 120°.

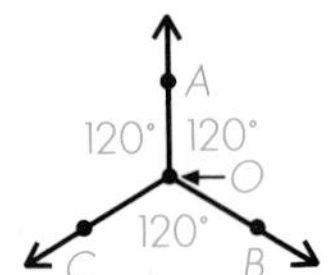

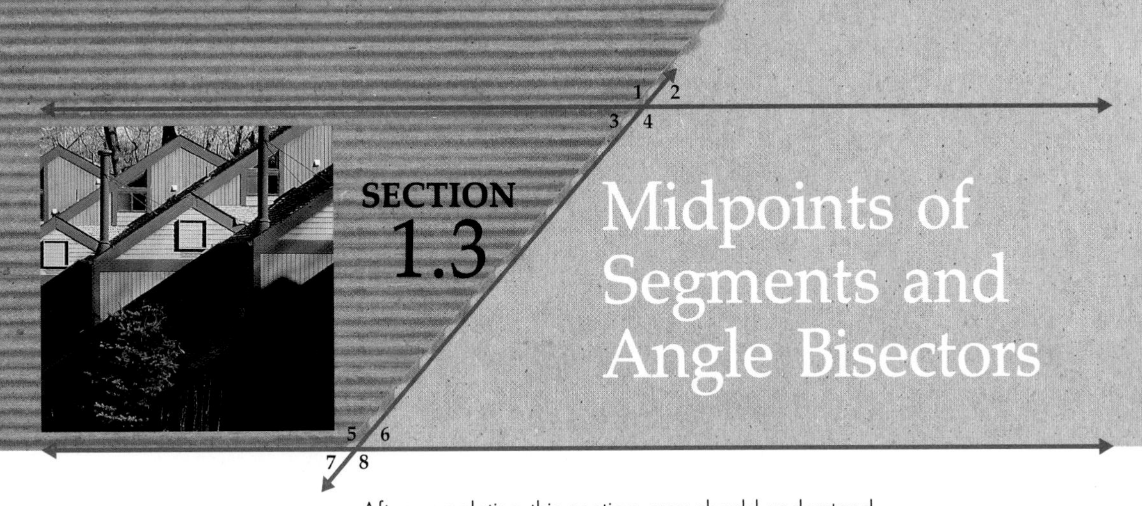

Midpoints of Segments and Angle Bisectors

After completing this section, you should understand
▲ how to find the midpoint of a line segment.
▲ how to find the bisector of an angle.
▲ the meaning of betweenness.

In a tug-of-war contest, a length of rope is marked in the middle. Players position themselves on opposite sides of this middle point and at equal distances from it.
The teams stand on opposite sides of a line on the ground. They pull the rope tight and situate the middle point straight above the line on the ground. Then, at the signal; the tugging begins.

MIDPOINT OF A LINE SEGMENT

In the tug of war contest, the rope is like a line segment. The middle point of the rope corresponds to the **midpoint** of the line segment.

DEFINITION 1-3-1: The MIDPOINT of a line segment is the point which divides the segment into two equal segments. C is the midpoint of $\overline{AB}$ if $AC = BC = \frac{1}{2}AB$.

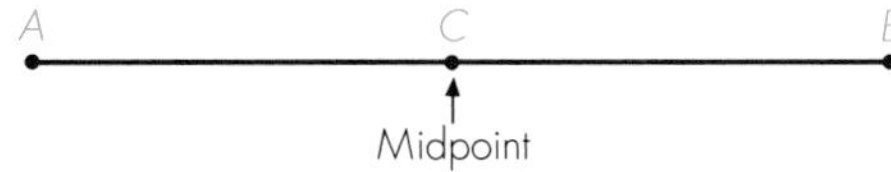

Since there is only one point that divides a given line segment equally, each line segment has one and only one midpoint.

SECTION
1.3

Resources
Reteaching 1.3
Enrichment 1.3

Objectives
After completing this section, the student should understand
▲ how to find the midpoint of a line segment.
▲ how to find the bisector of an angle.
▲ the meaning of betweenness.

Materials
centimeter and inch ruler
protractor

Vocabulary
midpoint
angle bisector
betweenness

GETTING STARTED

Quiz: Prerequisite Skills

1. Measure the following segment in inches and in centimeters. [2 in., about 5 cm]

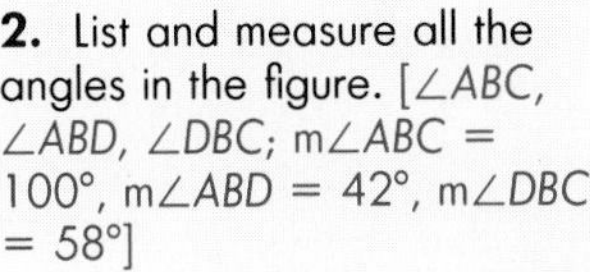

2. List and measure all the angles in the figure. [$\angle ABC$, $\angle ABD$, $\angle DBC$; $m\angle ABC = 100°$, $m\angle ABD = 42°$, $m\angle DBC = 58°$]

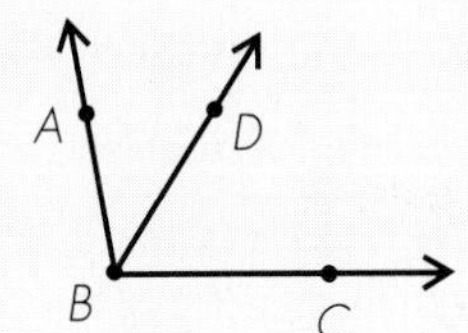

TEACHING COMMENTS

Using the Page
After students have looked at the examples, draw the following diagram on the chalkboard.

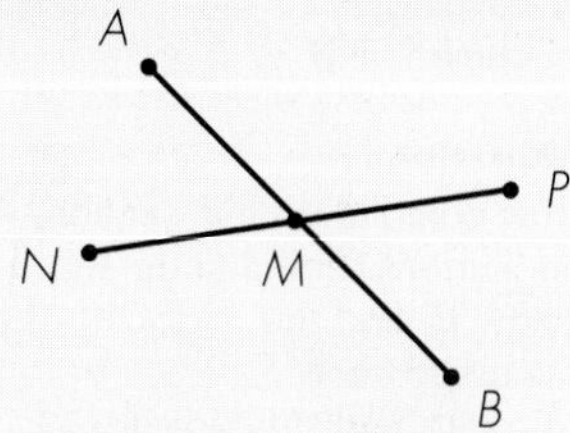

Tell students that M is the midpoint of $\overline{AB}$. Ask them to identify two segments that have the same length. [$\overline{AM}$ and $\overline{MB}$]

Examples:

1. Suppose the length of $\overline{AB}$ is 4 in. and that C is the midpoint of $\overline{AB}$. Then $AC = BC = \frac{1}{2} \times 4 = 2$ in.

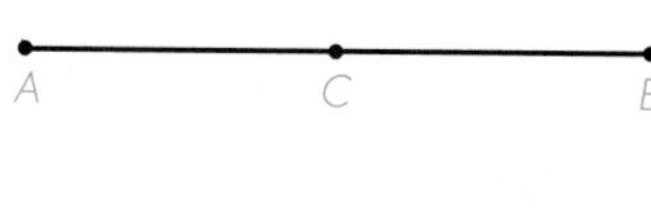

2. Suppose that $PQ = 13$ cm, $PM = 6.5$ cm, and $QM = 6.5$ cm. You can conclude that M is the midpoint of $\overline{PQ}$, since both PM and QM are equal to $\frac{1}{2}PQ$.

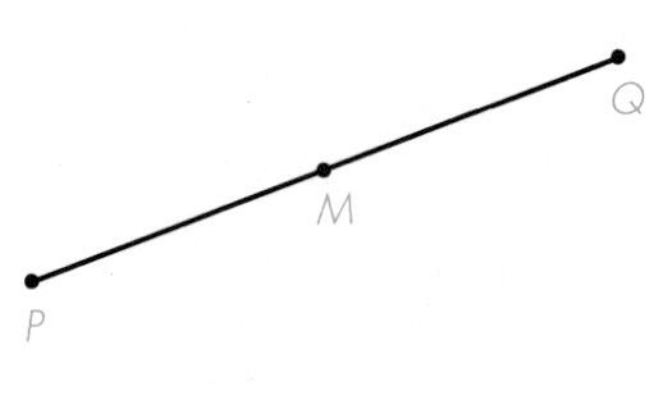

If point C is the midpoint of $\overline{AB}$ and $AB = 8.5$ cm, what is the length of $\overline{AC}$? of $\overline{CB}$? 4.25 cm; 4.25 cm

CLASS ACTIVITY

Use your ruler to find the length of each segment in inches. Tell how far the midpoint of each segment is from the endpoints.

1. M ———— N 3 in; 1.5 in.

2. R ———— S 3.5 in; 1.75 in.

Measure each segment in centimeters. Tell how far from the left-hand endpoint the midpoint of the segment will be.

3. J ———— K 8 cm; 4 cm

4. O ———— P 7 cm; 3.5 cm

5. If point R is the midpoint of $\overline{QS}$, and $RS = 4$ in., how long is $\overline{QS}$? How long is $\overline{QR}$? 8 in.; 4 in.

In the figure, $AB = 2$ cm, $BC = 1$ cm, and $AD = 6$ cm.

6. How long is $\overline{CD}$? 3 cm
7. Which point is the midpoint of $\overline{AD}$? C
8. Draw a line segment $\overline{XY}$ that is $3\frac{1}{2}$ in. long. Then mark the midpoint M of the segment you drew. See students diagrams.

ANGLE BISECTORS

Using the Page
As you introduce angle bisectors, you may wish to point out that a line, line segment, or ray that passes through the midpoint of a line segment is called the bisector of the segment.

The midpoint of a line segment divides the line segment into two equal segments. An **angle bisector** divides an angle into two equal angles.

DEFINITION 1-3-2: An ANGLE BISECTOR of a given angle is a ray with the same vertex that separates the given angle into two angles of equal measure. $\overrightarrow{BD}$ is the angle bisector of $\angle ABC$ if m$\angle ABD$ = m$\angle DBC$ = $\frac{1}{2}m\angle ABC$.

Example: Suppose that $\angle BAC = 50°$ and that $\overrightarrow{AD}$ bisects $\angle BAC$. You can conclude that $m\angle BAD = m\angle DAC = \frac{1}{2} \times 50° = 25°$.

In the figure, $\angle RST$ has a measure of 110° and $\overrightarrow{SP}$ bisects $\angle RST$. What is the measure of $\angle PST$? 55°

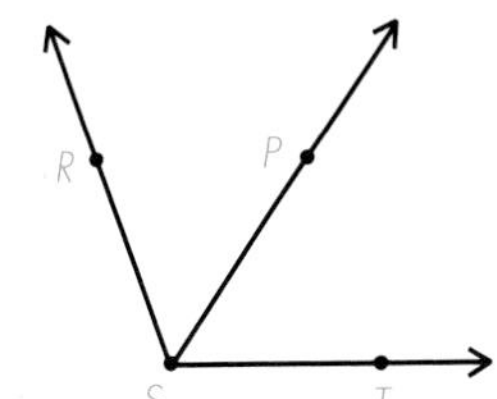

CLASS ACTIVITY

In each figure, $\overrightarrow{RS}$ is an angle bisector.

1. 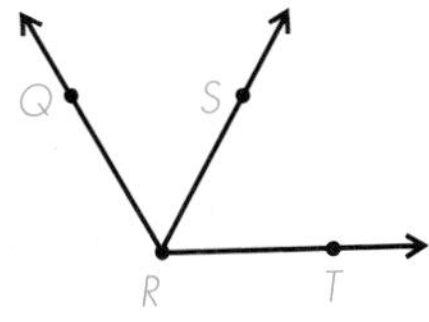

$m\angle QRS = 60°$
$m\angle QRT = $ 120°

2. 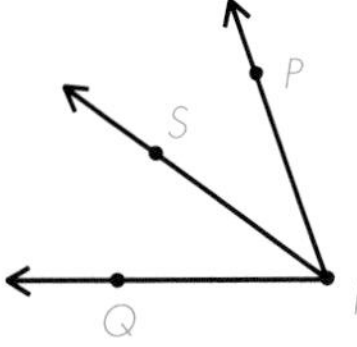

$m\angle PRS = 35°$
$m\angle SRQ = $ 35°

Use your protractor. Measure each angle and draw an angle of the same size. Classify each as acute, right, obtuse, or straight. With the aid of your protractor, draw the bisector of each angle. Give the measures of the two new angles you create. See students' diagrams.

3. 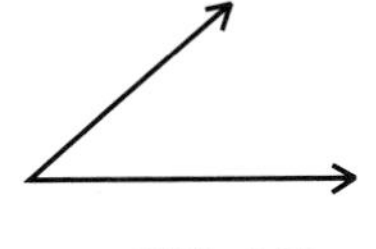

acute; 20°, 20°

4. 

acute; 40°, 40°

5. 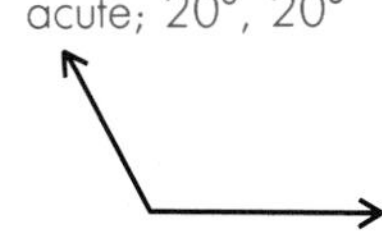

obtuse; 60°, 60°

6. 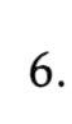 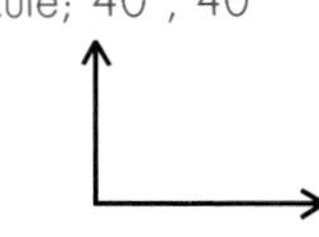

right; 45°, 45°

7. 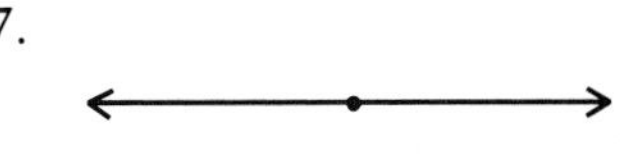

straight; 90°, 90°

8. 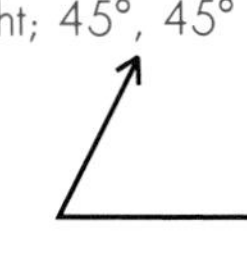

acute; 32.5°, 32.5°

Using the Page
After students have read the definition of angle bisector, ask them how many lines of symmetry an angle has. [1] Then ask how many bisectors an angle can have. [1] Next, ask them to describe the relationship between the bisector of an angle and the angle's line of symmetry. [The angle bisector determines the line of symmetry.]

Have students find midpoints of segments and bisectors of angles by paper-folding. They should observe that they find the midpoint of a segment by folding so that the endpoints coincide, and they find the bisector of an angle by folding so that the sides coincide.

BETWEENNESS

Route 36 crosses the entire state of Kansas. Baileyville is between Fairview and Marysville along that route.

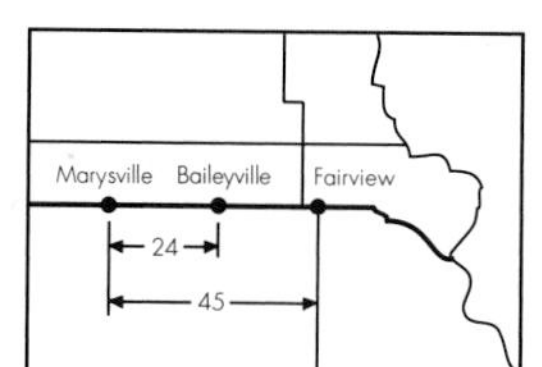

Route 36 goes in an east-west direction, so Baileyville is to the west of Fairview *and* to the east of Marysville. The distance from Fairview to Marysville is 45 miles. It is 24 miles between Baileyville and Marysville. How far is it from Baileyville to Fairview?

On the number line, if you know the coordinates of three points, it is easy to tell which point is between the other two.

DEFINITION 1-3-3: **On a number line, point C is BETWEEN points A and B if the coordinates of A, B, and C (a, b, and c) meet the condition that $a < c < b$ or $a > c > b$.**

A C B
a c b

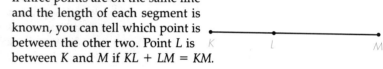

If three points are on the same line and the length of each segment is known, you can tell which point is between the other two. Point L is between K and M if $KL + LM = KM$.

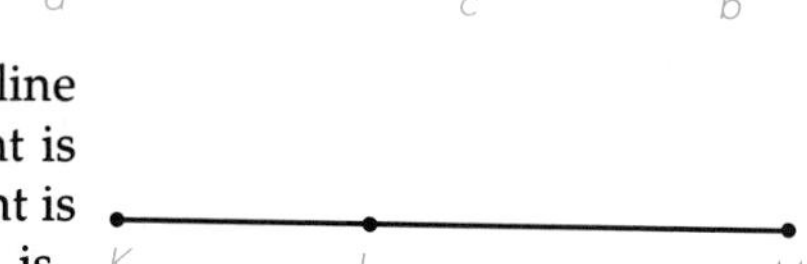

Using the Page
You may wish to point out to students that the statement, "Point L is between K and M if $KL + LM = KM$," is known as the *Betweenness of Points Theorem*.

Examples: In the figure, the coordinates of P, Q, and R are $2\frac{1}{2}$, 5, and 6 respectively. Since $2\frac{1}{2} < 5 < 6$, point Q is between P and R.

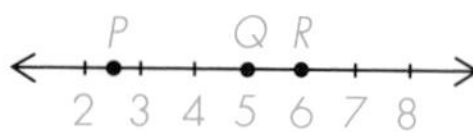

Extra Practice
Use the figure below to complete the equations.

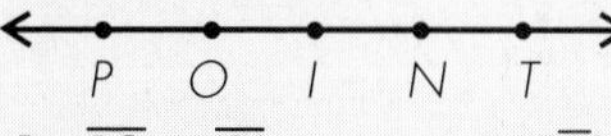

1. $\overline{PO} + \overline{OI} =$ ________ [$\overline{PI}$]
2. $\overline{PI} + \overline{IN} =$ ________ [$\overline{PN}$]
3. $\overline{PT} = \overline{PO} +$ ________ [$\overline{OT}$]

CLASS ACTIVITY

Points P, J, and K are on a number line. Their coordinates are 2, 8, and 3, respectively.

P K J
2 3 4 5 6 7 8

1. Draw a figure to illustrate this.
2. Which point is between the other two? K is between P and J.
3. Complete the following inequality relating the coordinates of the three points: $2 <$ __3__ $<$ __8__

Find each of the following lengths.

4. $PK =$ __1__
5. $KJ =$ __5__
6. $PJ =$ __6__
7. The sum of the lengths of which two line segments equals the length of the third segment?
The sum of the lengths of $\overline{PK}$ and $\overline{KJ}$ equals the length of $\overline{PJ}$.

DISCOVERY ACTIVITY

You have learned under what conditions one point on a line is between two others. Under what conditions is a ray between two other rays?

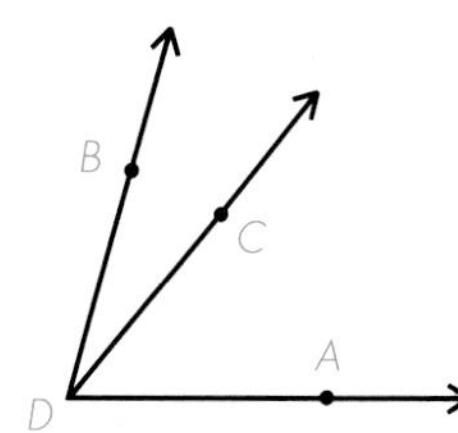

The diagram shows $\overrightarrow{DA}$, $\overrightarrow{DB}$ and $\overrightarrow{DC}$.
Measure $\angle BDA$ with your protractor. $m\angle BDA = 75°$
Then measure $\angle BDC$ and $\angle CDA$.
$m\angle BDC = 25°$, $m\angle CDA = 50°$
Is $m\angle BDC < m\angle BDA$? Is $m\angle CDA < m\angle BDA$? Yes; Yes
What is the sum of the measures of $\angle BDC$ and $\angle CDA$? 75°

Would you say that $\overrightarrow{DC}$ is between rays $\overrightarrow{DB}$ and $\overrightarrow{DA}$? Explain why.

Answers may vary. Possible answer: Yes; the rays have a common endpoint, two of the angles have measures less than the measure of the third, and the sum of the measures of the two smaller angles equals the measure of the third.

In the Discovery Activity, $\overrightarrow{DC}$ can be thought of as between $\overrightarrow{DA}$ and $\overrightarrow{DB}$. The rays have a common vertex, and the sum of the measures of two of the angles they form is equal to the measure of the third angle.

CLASS ACTIVITY

1. Use your protractor to help you make an accurate copy of $\angle RST$. Then draw a ray $\overrightarrow{SQ}$ between $\overrightarrow{SR}$ and $\overrightarrow{ST}$ such that $m\angle QST = 50°$. What is $m\angle QSR$? 40°

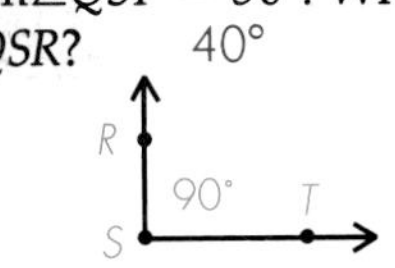

2. Use your protractor to help you make an accurate copy of $\angle DEF$. Then draw a ray $\overrightarrow{EG}$ between $\overrightarrow{EF}$ and $\overrightarrow{ED}$ such that $m\angle GED = 70°$. What is $m\angle GEF$? 15°

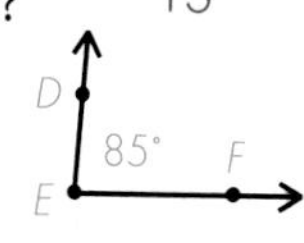

Extra Practice
In addition to the problems in the Class Activity, have students use a protractor to draw 70° angle ABC, and $\overrightarrow{BD}$ so that $\angle ABD = 45°$.

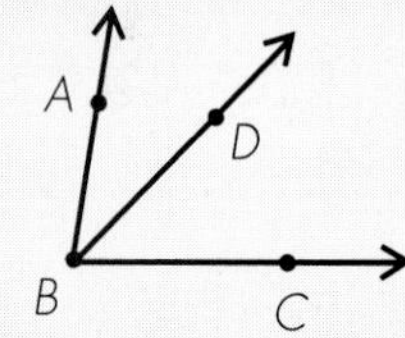

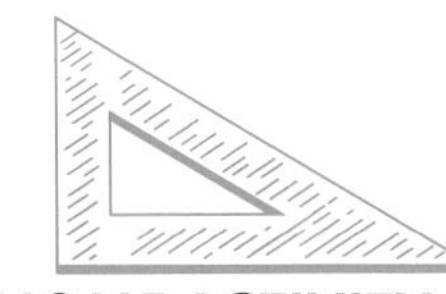

HOME ACTIVITY

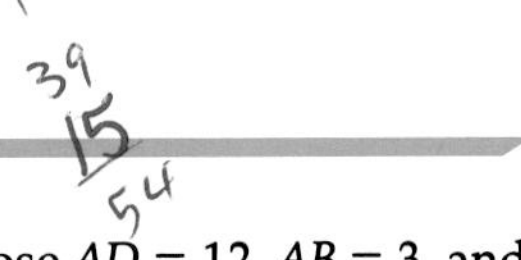

In the figure, B is between A and D, and C is between B and D. Suppose $AD = 12$, $AB = 3$, and $BC = 5$.

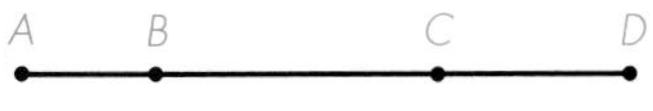

1. $CD =$ 4

2. Is C the midpoint of $\overline{AD}$? No

FOLLOW-UP

Assessment

1. M is the midpoint of $\overline{CD}$. $\overline{CD}$ = 11 mm. What is the length of $\overline{CM}$? [$\overline{CM}$ = 5.5 mm]

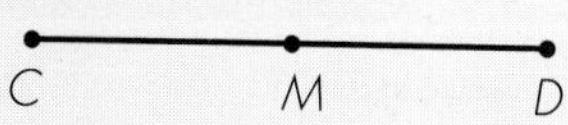

2. $\overrightarrow{BE}$ bisects ∠ABC. Find the measure of each angle in the figure below. [m∠ABC = 80°; m∠ABD = 12°; m∠ABE = 40°; m∠DBE = 28°; m∠DBC = 68°; m∠EBC = 40°]

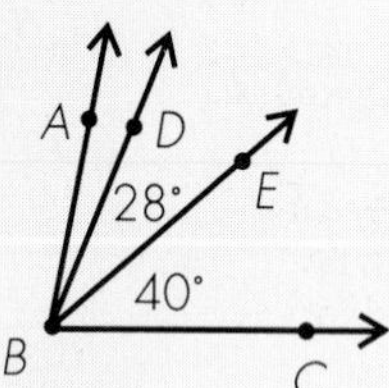

Reteaching

Students who have had difficulty with this section may benefit from Reteaching Activity 1.3.

Enrichment

For students who have mastered the material in this section, you may wish to assign Enrichment Activity 1.3.

In the figure, S is the midpoint of $\overline{QT}$, R is the midpoint of $\overline{QS}$, and $QS = 9$.

3. QT = 18

4. QR = 4.5

Measure in inches and make an accurate copy of each segment. Mark and label the midpoint M and tell how far it is from the endpoints. See students' diagrams.

Use your protractor. Measure each angle and draw an angle of the same size. Classify each as acute, right, obtuse, or straight. With the aid of your protractor, draw the bisector of each angle. Give the measures of the two new angles you create.

7.

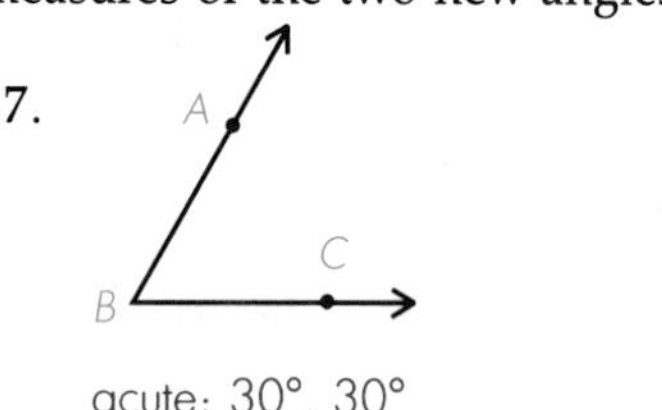

acute; 30°, 30°

8.

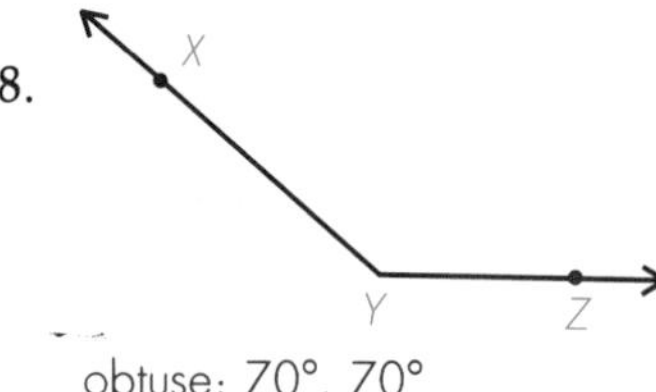

obtuse; 70°, 70°

In the figure, BD bisects ∠ABC.

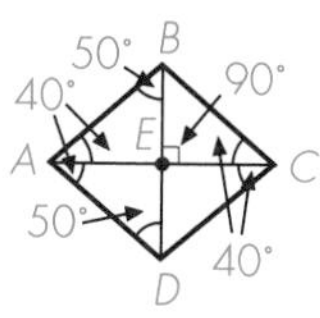

9. Name a right angle. ∠BEC
10. Name the bisector of ∠BCD. $\overrightarrow{CA}$
11. What is the measure of ∠CBD? 50°

Use your protractor to make an accurate copy of the given angle. Then draw a ray *between* the sides of the given angle to make an angle as indicated below the figure. Give the measure of the other new angle that is formed.

12.

Draw $\overrightarrow{GK}$ to make $m\angle KGH = 20°$. 25°
See students' diagrams.

13.

Draw $\overrightarrow{PK}$ to make ∠CPK = 100°. 20°

CRITICAL THINKING

Milt walks the same route to and from school each day. He always passes the fruit market, the post office, the park, the video store, and the cleaners, but not in that order. Going to school he passes the park before the video store, but after the post office. He passes the cleaners first. Going home, he passes the fruit market second. List the five places Milt passes as he walks to school. List them in the order in which he passes them.
cleaners, post office, park, fruit market, video store

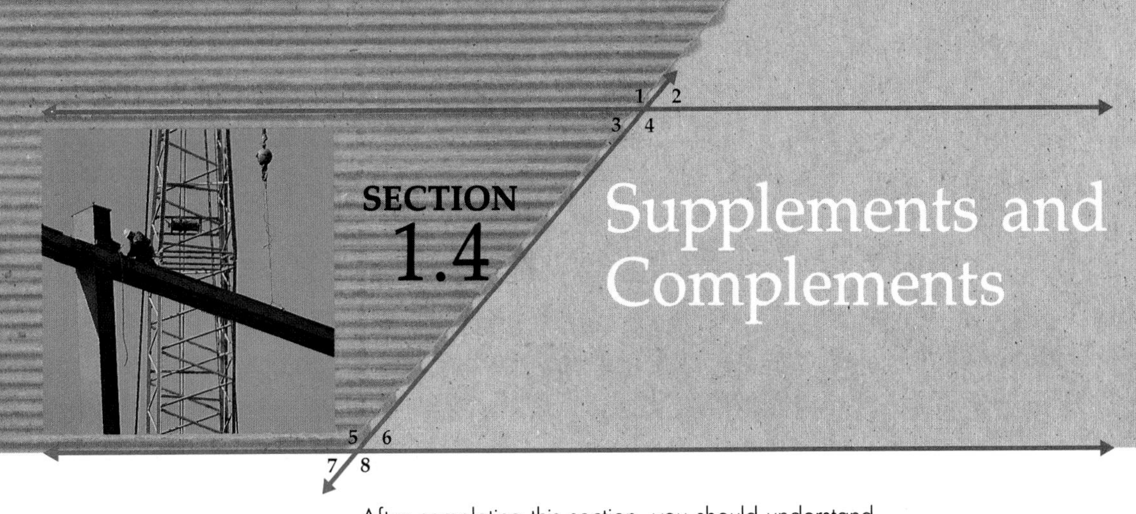

Supplements and Complements

After completing this section, you should understand
▲ the meaning of supplementary angles.
▲ the meaning of complementary angles.

Since angle measures are numbers, they can be added and subtracted. In this section you will learn about pairs of angles whose measures have sums of 90° or 180°.

DISCOVERY ACTIVITY

Draw $\overleftrightarrow{AB}$ on a sheet of paper.
Choose a point along the segment between *A* and *B* and label it C. See students's diagrams.
Next, draw and label $\overrightarrow{CF}$ where *F* is not on line *AB*.
Using your protractor, measure ∠*ACF* and ∠*BCF*. Answers may vary.
Add the measures of the two angles. What do you notice? The sum is 180°.
Did the result surprise you? Explain. Explanations may vary.

You probably concluded that the two angles in the Discovery Activity have a sum of 180°, since ∠*ACB* is a straight angle.

DEFINITION 1-4-1: **Two angles are SUPPLEMENTARY if their measures add to 180°.**

If two angles are supplementary, then each is called a **supplement** of the other.

Two angles do not have to have a common side in order to be supplementary.

SECTION 1.4

Resources
Reteaching 1.4
Enrichment 1.4
Transparency Master 1-1

Objectives
After completing this section, the student should understand
▲ the meaning of supplementary angles.
▲ the meaning of complementary angles.

Materials
protractor
straightedge

Vocabulary
supplementary angles
complementary angles

GETTING STARTED

Warm-Up Activity
Have students use a protractor and a straightedge to draw a right angle and a straight angle. Then ask them to divide each angle into two angles by drawing a ray from the vertex. Ask students what they notice about the two pairs of angles they have drawn. [Possible answer: one is a pair of acute angles, and the other is either a pair of right angles or an acute angle and an obtuse angle.]

TEACHING COMMENTS

Using the Page
Ask students what they can say about two angles that are supplements of the same angle. [The two angles are equal.]

Using the Page
Before have students work through the Example, you may wish to quickly review the steps in solving an equation in which the variable appears in more than one term.

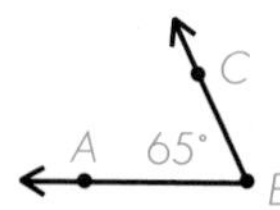

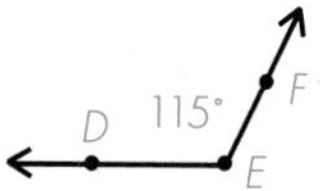

$\angle ABC$ and $\angle DEF$ are supplementary because $65° + 115° = 180°$. Knowing that two angles are supplementary is often helpful in solving problems.

Example: In the diagram, $\angle PQR$ and $\angle RQS$ are supplementary. Find the number of degrees in each.

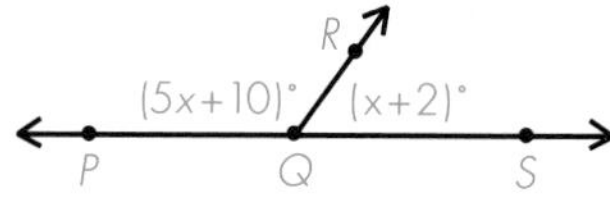

You can write an equation to express the fact that the angles are supplementary. Solve for x and then find the angle measures.

Use the given information. Remove parentheses. Combine the like terms. Subtract 12 from both sides. Divide both sides by 6.

$$(5x + 10) + (x + 2) = 180$$
$$5x + 10 + x + 2 = 180$$
$$6x + 12 = 180$$
$$6x = 168$$
$$x = 28$$

Substitute the value 28 for x in the expressions for the angle measures.
$(5x + 10)° = 150°$; $(x + 2)° = 30°$
$5x + 10 = 5 \times 28 + 10 = 140 + 10 = 150$ and $x + 2 = 28 + 2 = 30$
The measures of the angles are 150° and 30°.

CLASS ACTIVITY

1. Identify three examples of supplementary angles in your classroom. Answers will vary.
2. If $m\angle A = 55°$, what is the supplement of $\angle A$? 125°
3. What is the measure of a supplement of an angle of 145°? of an angle of 95°? 35°; 85°
4. Complete the table to find the measure of a supplement of an angle having the given measure.

Angle Measure	Equation	Supplement
30°	$30 + s = 180$	150°
45°	$45 + s = 180$	135°
80°	$80 + s = 180$	100°
90°	$90 + s = 180$	90°
135°	$135 + s = 180$	45°
149°	$149 + s = 180$	31°
$x°$	$x + s = 180$	$(180 - x)°$
$(x - 40)°$	$x - 40 + s = 180$	$(220 - x)°$
$(2x + 20)°$	$2x + 20 + s = 180$	$(160 - 2x)°$

5. In the diagram, $\overleftrightarrow{CG}$ and $\overleftrightarrow{DF}$ intersect at E and $m\angle CED = 48°$. Find the measures of every other angle.

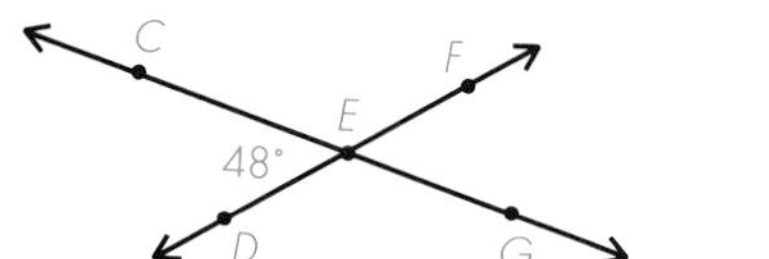

$m\angle CEF = 132°$
$m\angle FEG = 48°$
$m\angle CED = 132°$

In Exercises 6 and 7, $\angle PIO$ and $\angle VEG$ are straight angles. Find the measures of the supplementary angles.

6.

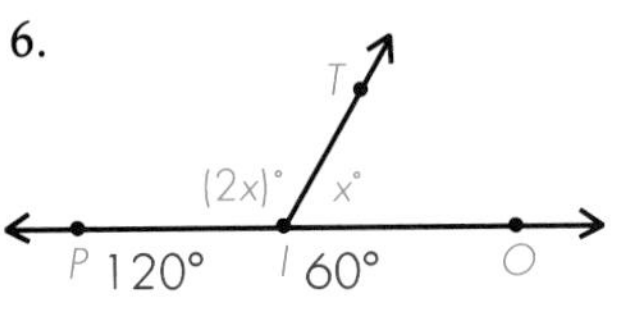

7.

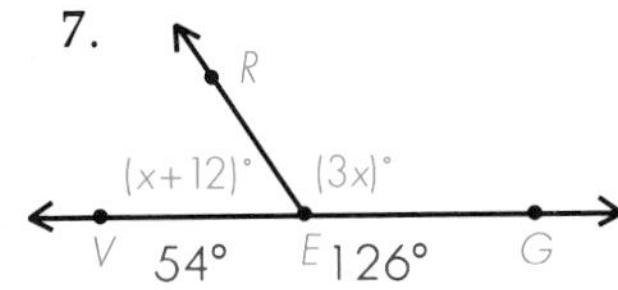

DISCOVERY ACTIVITY

Draw a right angle on a sheet of paper. Label it $\angle ACB$. From the vertex, C, draw a ray $\overrightarrow{CD}$ between $\overrightarrow{CA}$ and $\overrightarrow{CB}$. Use your protractor to measure $\angle ACD$ and $\angle DCB$. Add the two measures together. What do you notice? The sum is 90°.

In the Discovery Activity, you probably found that the sum of the measures of the two new angles is 90°.

DEFINITION 1-4-2: **Two angles are COMPLEMENTARY if their measures add to 90°.** If two angles are complementary, then each is called a **complement** of the other. Are these two angles complementary?

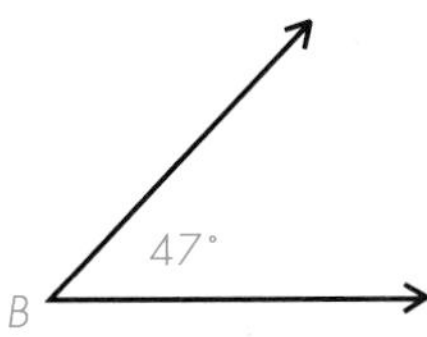

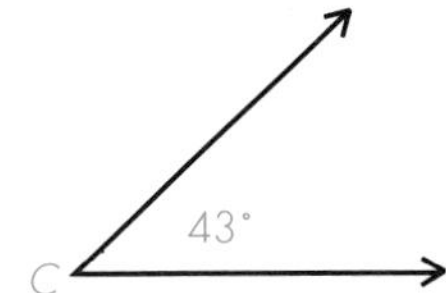

$\angle B$ and $\angle C$ are complementary, because $47° + 43° = 90°$.

Example: In the diagram, $\angle APB$ and $\angle BPC$ are complementary. Find the number of degrees in each angle.

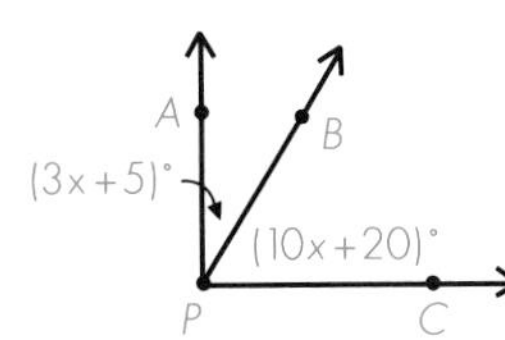

Using the Page
After students complete the Class Activity, ask them to give the supplement of an angle of n degrees. [$(180 - n)°$]

Using the Page
After students have been introduced to complementary angles, ask them what they can say about two angles that are complementary to the same angle. [They are equal.]

Using the Page
Some students have difficulty visualizing geometric relationships. To help such students improve their abilities, use exercises like the following.

1. What are two ways to draw the supplement of $\angle ABC$?

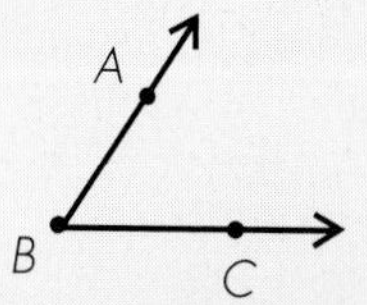

[Draw the opposite ray to $\overrightarrow{BA}$, or draw the opposite ray to $\overrightarrow{BC}$.]

2. What are two ways to draw the complement of $\angle DEF$?

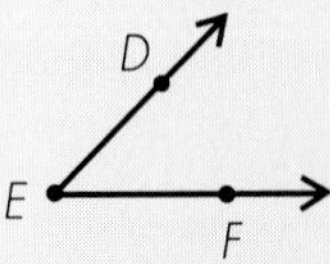

[Draw a ray with endpoint E that is perpendicular to $\overrightarrow{EF}$, with $\overrightarrow{ED}$ between $\overrightarrow{EF}$ and the new ray; or draw a ray with endpoint E that is perpendicular to $\overrightarrow{ED}$, with $\overrightarrow{EF}$ between $\overrightarrow{ED}$ and the new ray.]

Using the Page
Transparency 1-1 can be used to review adjacent and nonadjacent supplementary and complementary angles.

Extra Practice
An angle measures $(x + 25)°$. What is the measure of the complement (c) of that angle? [$c = (65 - x)°$]

Additional Answers
1. 55°

You can write an equation expressing the fact that the angles are complementary. Solve the equation and then find the angle measures.

Use the information. Remove parentheses. Combine like terms. Subtract 25 from both sides. Divide both sides by 13.

$$(3x + 5)° + (10x + 20)° = 90°$$
$$3x + 5 + 10x + 20 = 90$$
$$13x + 25 = 90$$
$$13x = 65$$
$$x = 5$$

Substitute the value 5 for x in the original expressions.

$$3x + 5 = 3 \times 5 + 5 = 15 + 5 = 20$$
$$10x + 20 = 10 \times 5 + 20 = 50 + 20 = 70$$

The measures of the angles are 20° and 70°.

CLASS ACTIVITY

1. If $m\angle A = 35°$, what is the measure of a complement of $\angle A$? 55°
2. What is the measure of a complement of an angle of 25°? of an angle of 75°? 65°, 15°
3. Complete the table to find the measure of a complement of an angle having the given measure.

Angle Measure	Equation	Complement
30°	$30 + c = 90$	60°
45°	$45 + c = 90$	45°
10°	$10 + c = 90$	80°
80°	$80 + c = 90$	10°
85°	$85 + c = 90$	5°
14°	$14 + c = 90$	76°
29°	$29 + c = 90$	61°
$x°$	$x + c = 90$	$(90 - x)°$
$(x - 40)°$	$x - 40 + c = 90$	$(130 - x)°$
$(2x + 20)°$	$2x + 20 + c = 90$	$(70 - 2x)°$

In Exercises 4 and 5, $\angle BAZ$ and $\angle SRV$ are right angles. Find the measures of the complementary angles.

4. 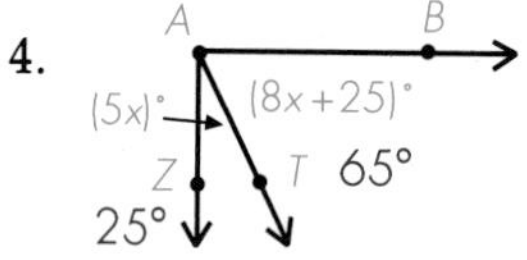

5. 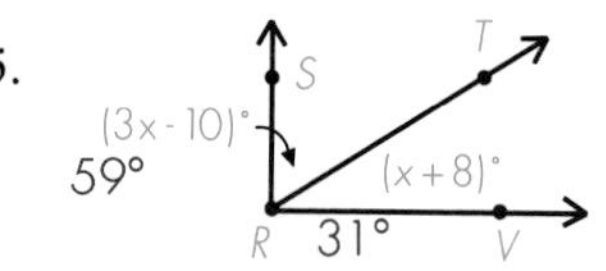

In each of the following, assume that the two angles are complementary. Find the measures of both angles using the given information.

6. $m\angle T$ is twice $m\angle R$ 60°, 30°
7. $m\angle R$ is 10° more than $m\angle T$ 50°, 40°
8. $m\angle R$ is 40 less than $m\angle T$ 25°, 65°
9. $m\angle T$ is 4 times $m\angle R$ 72°, 18°

21
36

HOME ACTIVITY

In the figure at the right, $m\angle GOA = m\angle GOE = 90°$ and $\angle FOB$ is a straight angle. $\overrightarrow{EO}$ bisects $\angle FOC$. Name the angles described. See margin.

1. a pair of complementary angles
2. three acute angles
3. two pairs of supplementary angles
4. two obtuse angles
5. a straight angle other than ∠FOB
6. a right angle

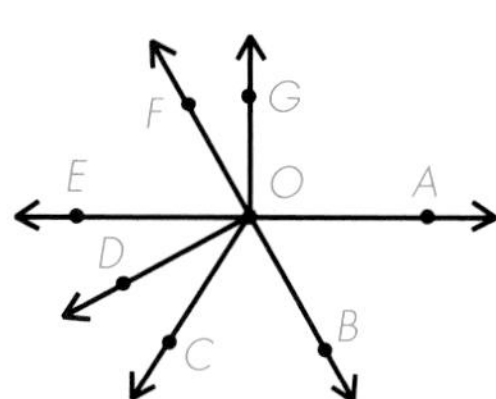

Find the measure of an angle supplementary to an angle having the given measure.

7. 30° 150°
8. 87° 93°
9. 161.4° 18.6°
10. $x°$ $(180 - x)°$

Find the measure of an angle complementary to an angle having the given measure.

11. 40° 50°
12. 77° 13°
13. 6.05° 83.95°
14. $x°$ $(90 - x)°$

Tell which words you should use in the blanks—*acute, obtuse, right,* or *straight*—to make true statements.

15. A complement of an acute angle is ________. acute
16. A supplement of a right angle is ________. right
17. A supplement of an acute angle is ________. obtuse
18. ________ angles do not have supplements. straight
19. ________ angles have acute supplements. obtuse
20. Any pair of ________ angles are supplementary. right

In each of the following figures, the indicated pair of angles are either complementary or supplementary. Find the degree measure of each angle.

21. 95° 85°

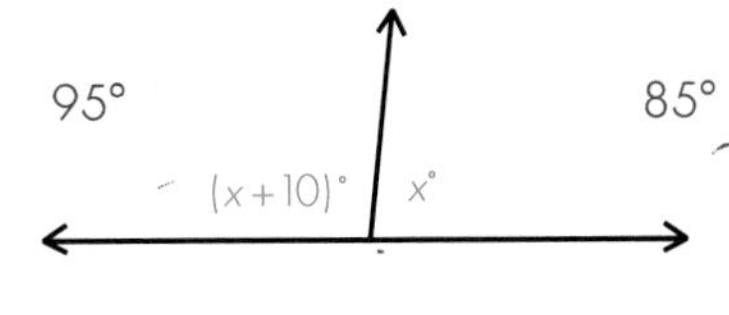

22. 72° 18°

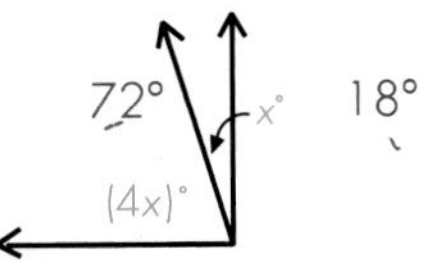

23. 30° 60°

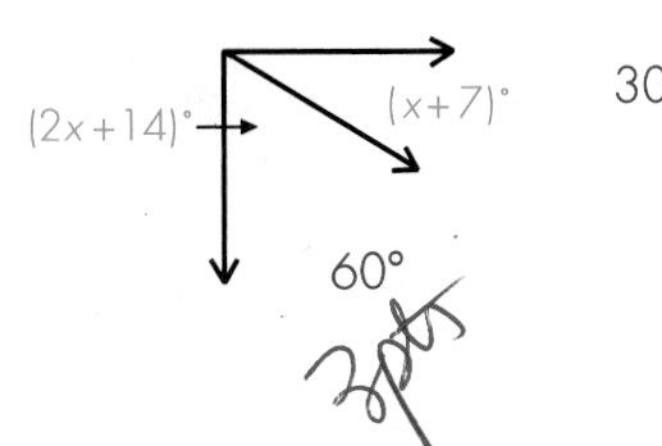

24. 120° 60°

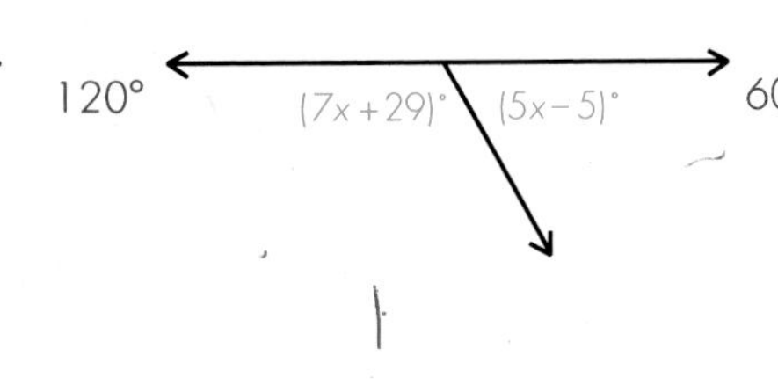

3pts

Additional Answers

1. ∠EOF and ∠FOG
2. Possible answer: ∠FOG, ∠EOF, ∠EOD; other answers are possible.
3. Possible answers: ∠AOB and ∠AOF, ∠EOC and ∠FOA; other answers are possible.
4. Possible answer: ∠AOF and ∠DOG; other answers are possible.
5. ∠EOA
6. ∠GOA or ∠GOE

FOLLOW-UP

Assessment

1. What is the supplement of a 68° angle? What is the complement of a 68° angle? [112°; 22°]

2. $\angle ABC$ is a straight angle. Find the measure of $\angle 1$. [$m\angle 1 = 25°$]

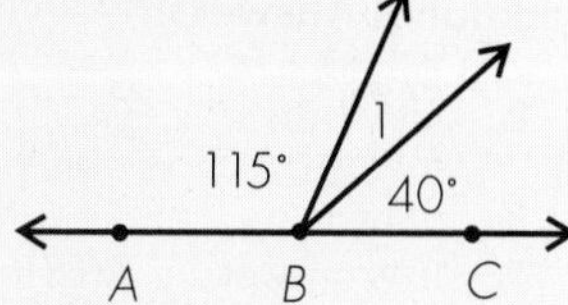

Find the value of x in each of the following. Assume that the lines that look like straight lines are straight lines.

25.

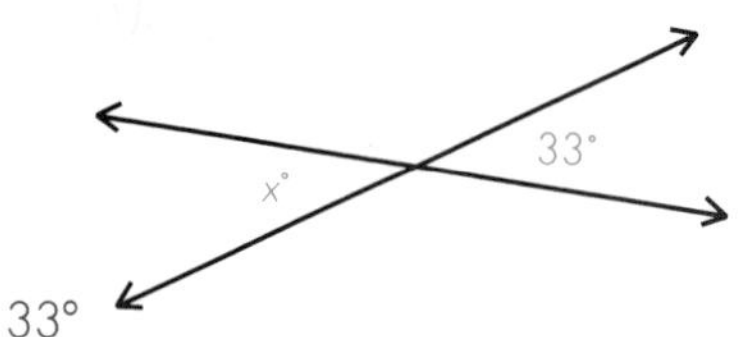

33°

26.

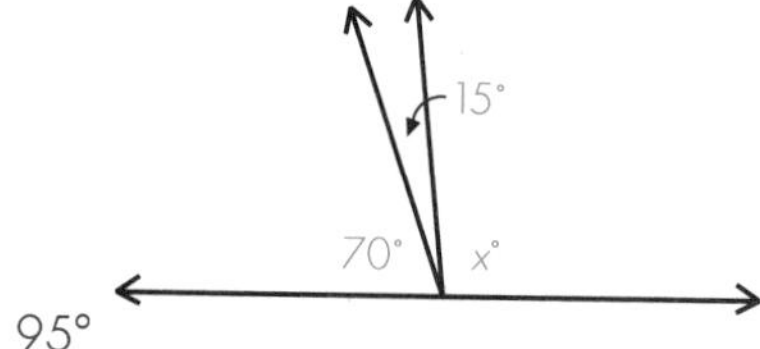

95°

27.

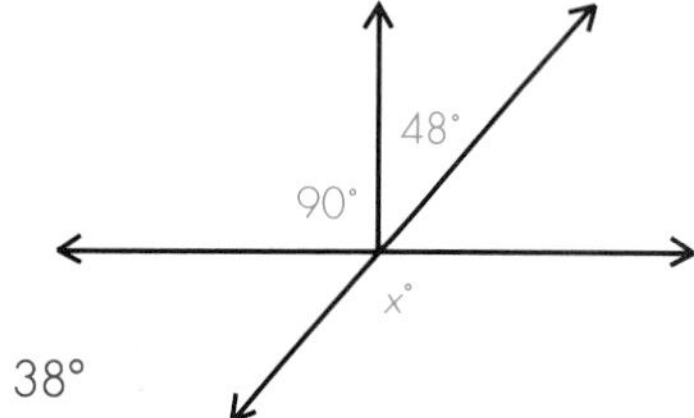

138°

28.

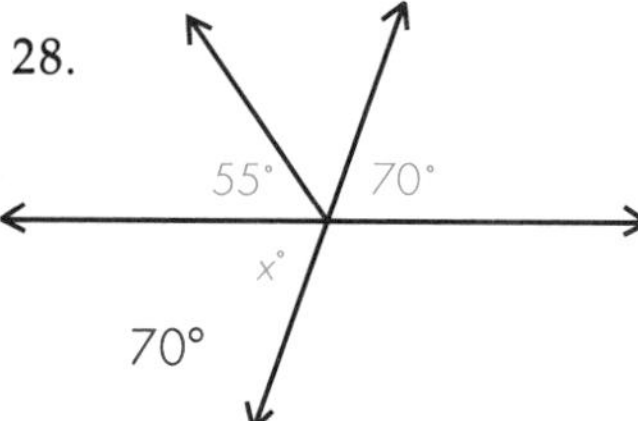

70°

Draw and label a diagram to show each of the following.

29. $\overrightarrow{CR}$ is between $\overrightarrow{CB}$ and $\overrightarrow{CA}$ in right angle ACB, and $m\angle ACR = 5\ m\angle BCR$. Drawing shows complementary angles of 75° and 15°.

30. $\overrightarrow{BD}$ bisects $\angle ABC$ and $m\angle ABC = 90°$. Drawing shows two 45° angles.

CRITICAL THINKING

The team equipment manager was counting the number of baseballs the team had. She noticed that if she counted the balls two at a time, there was 1 left over. If she counted them three at a time or four at a time there was also 1 left over. If she counted them five at a time, there were none left over. How many baseballs does the team have if they have between 75 and 100 baseballs? 85 baseballs

Reteaching

Students who have had difficulty with this section may benefit from Reteaching Activity 1.4.

Enrichment

For students who have mastered the material in this section, you may wish to assign Enrichment Activity 1.4.

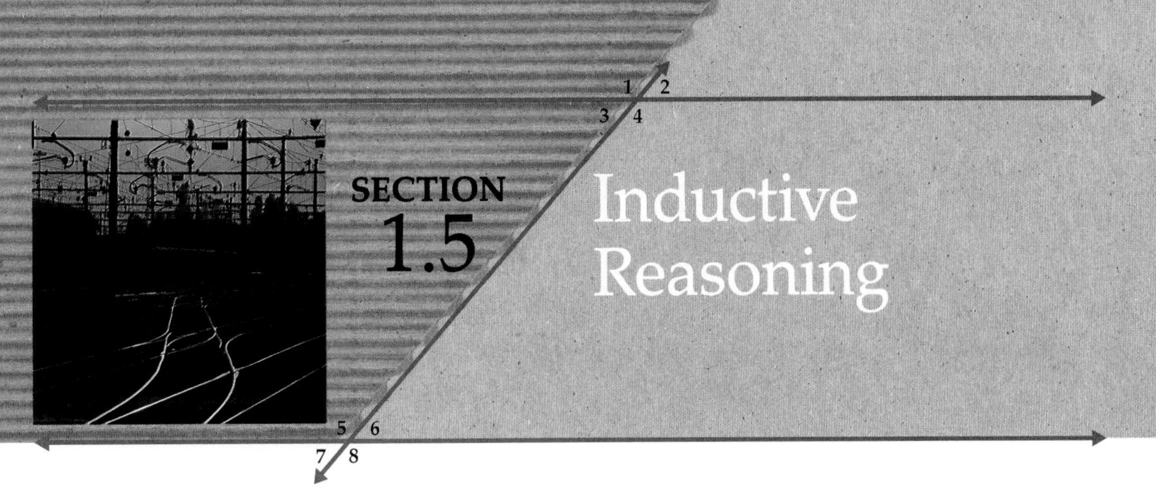

SECTION 1.5

After completing this section, you should understand
- ▲ the meaning of inductive reasoning
- ▲ the pitfalls involved in inductive reasoning

Suppose it has rained everyday for the past week. On the basis of that information, is it reasonable to predict that it will rain tomorrow? Answers will vary.

In your study of geometry, you will be discovering things that are true for all geometric figures of a certain kind. Also you will look for ways to justify your belief that the things you discover are *always* true.

In your attempts to discover things about geometric figures, you will often use what is known as **inductive reasoning.**

WHAT IS INDUCTIVE REASONING?

Inductive reasoning is the process of using a limited number of cases to arrive at a conclusion that will hopefully be true in *all* cases, including the cases you have not yet examined.

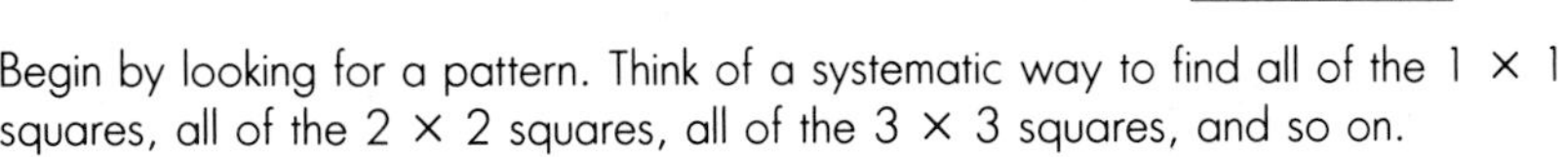

DISCOVERY ACTIVITY

How many squares of all sizes are there in the 5 × 5 square shown at the right? 55

Begin by looking for a pattern. Think of a systematic way to find all of the 1 × 1 squares, all of the 2 × 2 squares, all of the 3 × 3 squares, and so on.

What pattern did you notice? You can add the squares of the whole numbers from 1 to 5 to get the number of squares of all sizes.

SECTION 1.5

Resources
Reteaching 1.5
Enrichment 1.5

Objectives
After completing this section, the student should understand
- ▲ the meaning of inductive reasoning.
- ▲ the pitfalls involved in inductive reasoning.

Materials
compass
straightedge

Vocabulary
inductive reasoning

GETTING STARTED

Warm-up Activity
Tell students that Ed has stacked cans of soup in a display at a market. He has stacked 100 cans, with 1 can in the top row, 3 in the next row, 5 in the next row, and so on. Ask: How many rows of cans are in the display? How many cans are in the bottom row? [10 rows, 19 cans in bottom row]

TEACHING COMMENTS

Using the Page
The pattern observed in the Discovery Activity would work for an 8 × 8 square. Ask students how many squares of all sizes would be in an 8 × 8 square. [64 + 49 + 36 + 55 = 204 squares]

You may have observed that a pattern involving sums of square numbers could be used to answer the question in the Discovery Activity. In a 5 × 5 square, there are twenty-five 1 × 1 squares, sixteen 2 × 2 squares, nine 3 × 3 squares, four 4 × 4 squares, and one 5 × 5 square, for a total of 55 squares of all sizes. Do you think this pattern would work if you had started with an 8 × 8 square? Here is another example where you can use inductive reasoning.

Example: If you mark one point, C, on between A and B on $\overline{AB}$, 3 line segments are formed: $\overline{AC}$, $\overline{CB}$, and $\overline{AB}$.

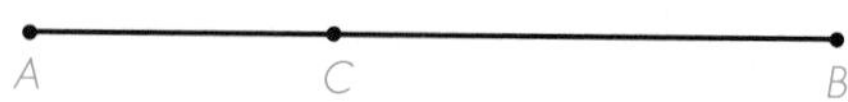

If you mark two points, C and D, between A and B, you get 6 line segments: $\overline{AC}$, $\overline{AD}$, $\overline{AB}$, $\overline{CD}$, $\overline{CB}$, $\overline{DB}$.

Marking three points between A and B gives you 10 segments. Continue the pattern. How many segments do you predict will be created when you mark four points between A and B? five points?
15 segments; 21 segments

Can you describe a pattern that would let you find the number of line segments formed for whatever number of points you wish to mark between A and B? See margin.

Using the Page
Some students may solve the Class Activity problem by adding 1 per side for every multiple of 3, starting with 5 per side for 12, 6 per side for 15, and so on.

CLASS ACTIVITY

1. Triangular dot numbers are so named because that number of dots can be used to form a triangle with an equal number of dots on each side. Examine the following triangular dot numbers.

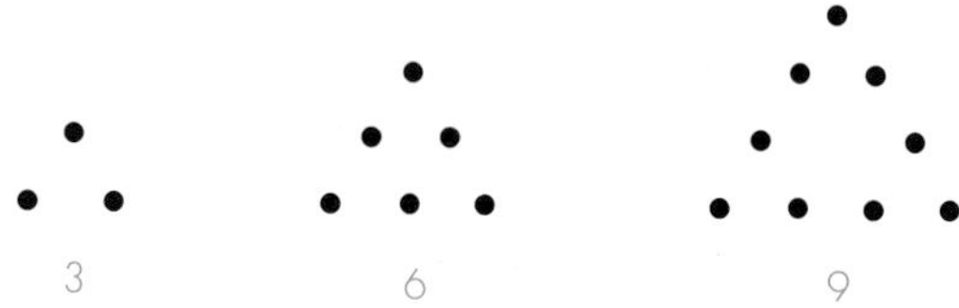

Predict what triangular dot number will have 10 dots on a side. Explain how you arrived at your prediction.
27; Possible answer: 1 less than number of dots per side times 3—continue the pattern: 2 per side (2 − 1) × 3 = 3; 3 per side (3 − 1) × 3 = 6; 4 per side (4 − 1) × 3 = 9, and so on.

DISCOVERY ACTIVITY

Draw a circle and mark two points on it, as shown. Connect the points with a line segment and count the regions within the circle. 2 regions

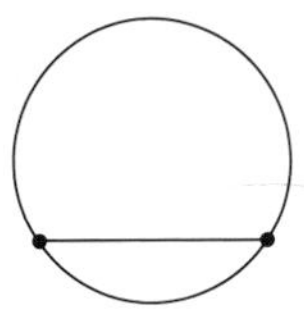

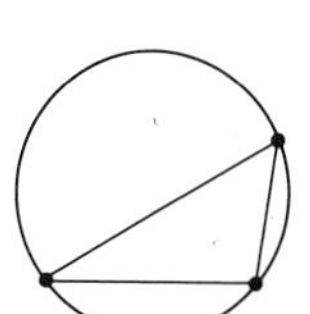

Draw another circle and mark three points, as shown. Connect the points and count the regions. 4 regions

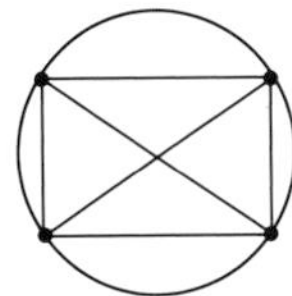

Draw another circle and mark four points. Again, connect the points and count the regions. 8 regions

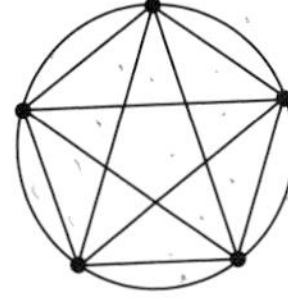

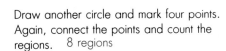

Draw a circle and mark five points as shown. Connect the points. First predict the number of regions. Then count the regions to check your prediction. 16 regions

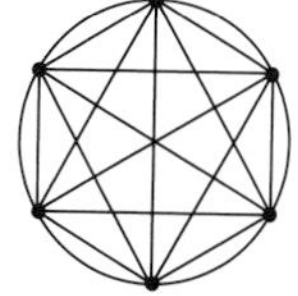

Now draw a circle and mark six points. Connect the points and predict the number of regions formed. Count the regions to check your prediction. What did you notice?
There are 31 regions. (Students may have predicted 32.)

This Discovery Activity points out a weakness in inductive reasoning. Did you predict that when you connected six points on the circle you would have 32 regions? In fact, there are only 31. You have discovered that a pattern detected in a limited number of cases does not necessarily hold up in all cases.

4+3+2
5+4+3
10+9+8

Using the Page
In the Discovery Activity, make sure students use a compass and straightedge to connect each point to all other points.

Using the Page
It is important for students to understand that a generalization arrived at through inductive reasoning is invalidated by a single counterexample. Ask students what kind of numbers result when n in the expression $2n^2 - 1$ is replaced by integers greater than 1. For $n = 2, 3, 4$, they might conclude that the result will always be prime. Be sure they see that when they find that $n = 5$ results in a composite number, the earlier generalization is immediately invalidated.

Using the Page
Ask students to give other examples from everyday life of the shortcomings of inductive reasoning.

Inductive reasoning may not always lead you to the right conclusion. Often this is because important factors have been overlooked. Here is an example from real life.

Example: Karla knows that the Tigers beat the Panthers in football and that the Panthers beat the Lions. The Tigers are playing the Lions in football today and he predicts the Tigers will win.

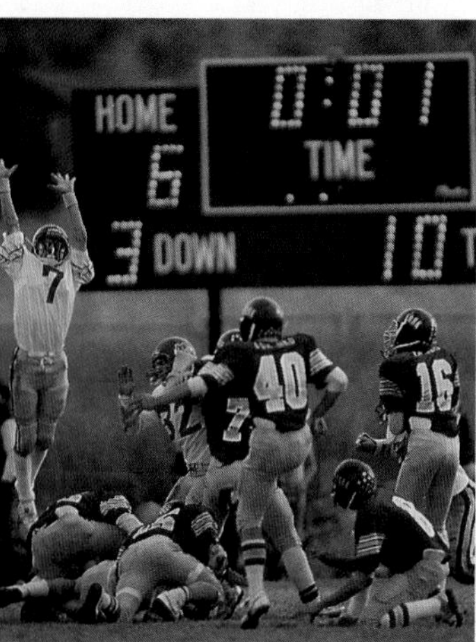

Although the Tigers may win this game, the reasoning is based on too few cases. Also, Karla is not thinking about some key things such as the line-up of players, the condition of the field, and so on.

Extra Practice
Fill the blanks. Describe the pattern.

1. 2, 3, 5, 9, ____, ____, ____ [17, 33, 65; pattern: $2n - 1$]
2. 1, 4, 10, 22, 46, ____, ____, ____ [94, 190, 382; pattern: $2n + 2$]

CLASS ACTIVITY

Fill the blanks. Describe the pattern you found.

1. 1, 3, 7, 13, 21, 31, 43
   The numbers are increasing by even numbers: 2, 4, 6, 8, and so on.
2. 1, 3, 6, 10, 15, 21, 28, 36 The numbers are increasing by consecutive whole numbers: 2, 3, 4, 5, 6, and so on.
3. 1, 3, 7, 15, 31, 63, 127, 255
   To get from one term to the next, multiply by 2 and add 1.

Use a calculator for Exercises 4 and 5.

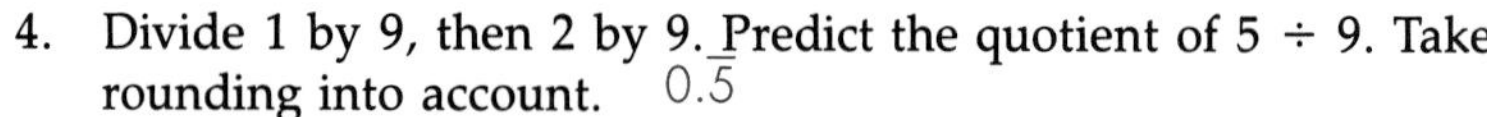

4. Divide 1 by 9, then 2 by 9. Predict the quotient of 5 ÷ 9. Take rounding into account. $0.\overline{5}$
5. Divide 1 by 11, then 2 by 11. Predict the quotient of 4 ÷ 11 and 9 ÷ 11. Take rounding into account. $0.\overline{36}$; $0.\overline{81}$
6. These are examples of square dot numbers.

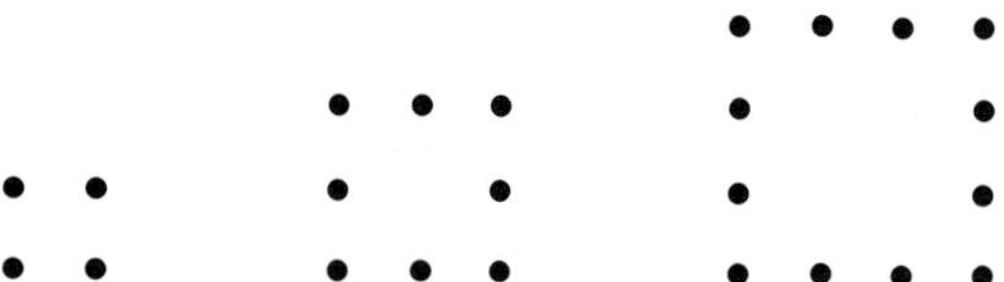

How many dots are on each side of the square dot number 28? 8

HOME ACTIVITY

Fill the blanks. Describe the pattern you found.

1. 2, 5, 8, 11, 14, 17, 20, 23
2. 1, 2, 6, 24, 120, 720, 5040, 40,320
3. 2, 3, 5, 9, 17, 33, 65, 129
4. $1, \frac{1}{2}, \frac{1}{4}, \frac{1}{8}, \frac{1}{16}, \frac{1}{32}, \frac{1}{64}$

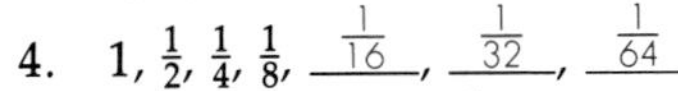

5. What fraction of the square region in the figure at the right is shaded? *Hint:* Discover the pattern of the shading.
$\frac{1}{4} + \frac{1}{16} + \frac{1}{32} + \frac{1}{64} = \frac{23}{64}$ shaded

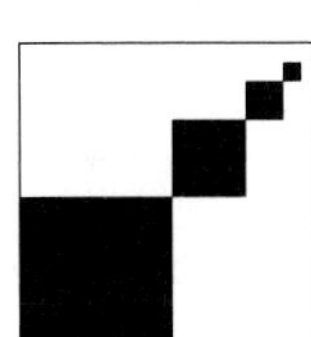

Complete a table like the one below to show how many line segments are determined by n points that are evenly spaced around a circle.

Number of Points	Diagram	Number of Segments	Number of Points	Diagram	Number of Segments
2		1	3		3
4		6	5		6. 10
6		7. 15	7		8. 21
		$n \to$ 9. $\frac{1}{2}n(n-1)$			

CRITICAL THINKING

10. Hannah's bank has been charging her a monthly service charge of \$3 plus \$.10 for each check she writes. Now the bank is changing its policy to charge a \$4 monthly fee and \$.06 a check. The bank officer tells Hannah these changes will save her money. How many checks must Hannah write each month for this to be true? at least 26 checks

FOLLOW-UP

Assessment

1. What is the sum of the first 2 odd numbers? the first 3 odd numbers? the first 4? the first 10? the first 100? What is the sum of the first *n* odd numbers? [4, 9, 16, 100, 10,000, n^2]

2. Inez piled 165 cans of pet food so that 1 can was on top, 3 cans were in the next layer, 6 cans were in the next layer, and so on. How many layers were in the pile? [10 layers]

Reteaching

Students who have had difficulty with this section may benefit from Reteaching Activity 1.5.

Enrichment

For students who have mastered the material in this section, you may wish to assign Enrichment Activity 1.5.

SECTION 1.6

Resources
Reteaching 1.6
Enrichment 1.6
Transparency Master 1-2

Objectives
After completing this section, the student should understand
- ▲ how to use some LOGO commands to draw geometric figures.
- ▲ how to use a computer and geometric drawing tools to explore geometric concepts.

Materials
computer
geometric drawing tool

Vocabulary
LOGO
turtle graphics
LOGO commands
LOGO procedures

GETTING STARTED

Warm-Up Activity
Have students write specific instructions to each other describing how to draw a particular geometric figure on graph paper, such as a rectangle with its length equal to twice its width. Encourage students to be as careful as possible when formulating their instructions.

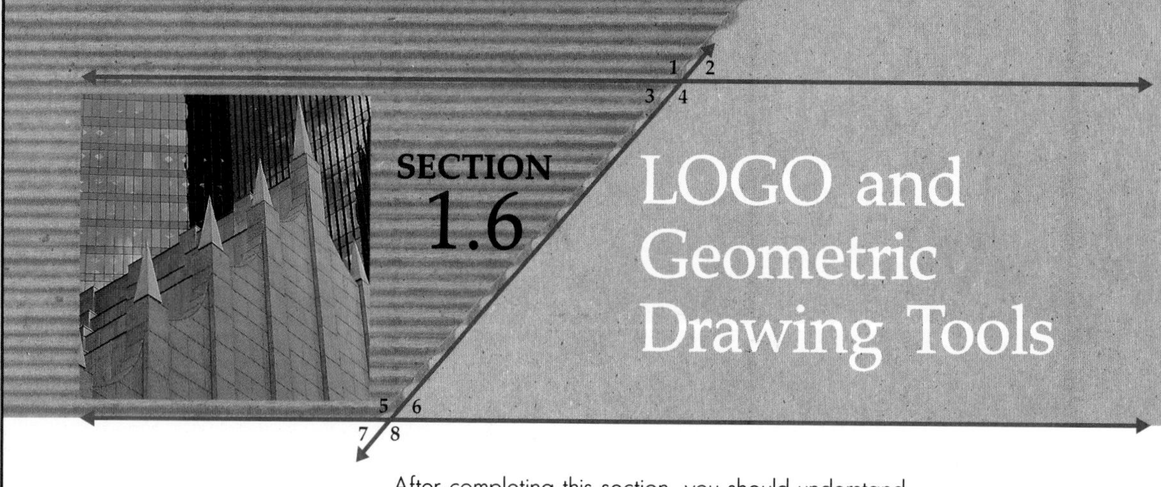

SECTION 1.6

LOGO and Geometric Drawing Tools

After completing this section, you should understand
- ▲ how to use some Logo commands to draw geometric figures.
- ▲ how to use a computer and geometric drawing tools to explore geometric concepts.

The computer language Logo was designed for use in education and artificial intelligence research. Today, Logo is used primarily in schools to introduce programming to students using turtle graphics. You can use turtle graphics to draw geometric figures by using relatively simple commands.

MOVING THE TURTLE

The commands used in turtle graphics, sometimes called primitives, manipulate a small triangular figure called a turtle. When you type the Logo command DRAW, the screen is cleared and the turtle is placed in its home position in the center of the screen, pointing up. The turtle shown in the first drawing on the next page is in home position.

To draw a picture using Logo, you use special commands that move the turtle around the screen. The turtle will leave a trail on the screen as it moves from one position to the next. After the turtle has been moved, it can be returned to its home position by typing the command HOME.

The command FORWARD or FD moves the turtle forward the number of spaces you specify. For example, FORWARD 6Ø or FD 6Ø tells the turtle to move 6Ø spaces forward. The command BACK or BK moves the turtle back the number of spaces you specify. For example, BACK 55 or BK 55 tells the turtle to move 55 spaces back. The command RIGHT or RT turns the turtle to the right the number of degrees you specify. For example, RIGHT 8Ø or RT 8Ø tells the turtle to turn to the right 80°. The command LEFT or LT turns the turtle to the left the number of degrees you specify. For example, LEFT 115 or LT 115 tells the turtle to turn to the left 115°.

Example: Show how the turtle moves when you type these Logo commands.

DRAW — The screen is cleared and the turtle is placed in its home position.

RT 9Ø — The turtle turns 90° to the right.

FD 12Ø — The turtle moves forward 120 spaces.

LT 45 — The turtle turns 45° to the left.

BK 8Ø — The turtle moves back 80 spaces.

TEACHING COMMENTS

Using the Page
If students will be using a color monitor, you may want to introduce the commands BG and PC. BG changes the background color and PC changes the pen color. Colors are coded with the numbers 0 through 5 as follows: black is 0, white is 1, green is 2, violet is 3, orange is 4, and blue is 5. BG1 sets the background to white and PC3 sets the pen color to violet.

Using the Page
Students can complete the Class Activity without a computer in a manner similar to that shown in the example on page 31.
Use Transparency Master 1-2 to show the commands and the sketches for Exercises 1-3 and for Exercises 1 and 2 on page 34.

Additional Answers

1.

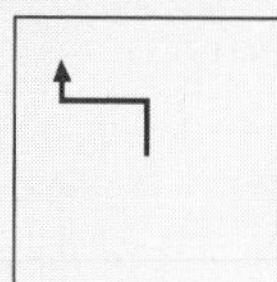

2.

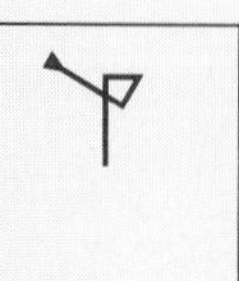

3.

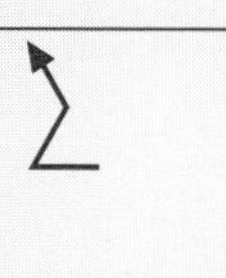

CLASS ACTIVITY

Type these Logo commands on a computer. Make a sketch of the figure drawn by the turtle. See margin.

1. DRAW
   FD 5Ø
   LT 9Ø
   FD 8Ø
   LT 9Ø
   BK 3Ø

2. DRAW
   FD 8Ø
   RT 9Ø
   FD 3Ø
   RT 12Ø
   FD 3Ø
   LT 9Ø
   BK 8Ø

3. DRAW
   RT 9Ø
   BK 25
   LT 6Ø
   FD 25
   LT 6Ø
   FD 25

Write Logo commands that tell the turtle how to draw each figure. Use a computer to check your answers. Answers may vary.

4. 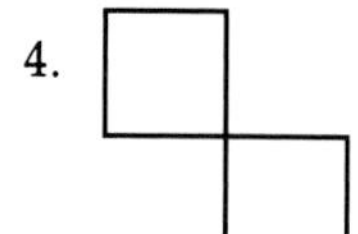

5. 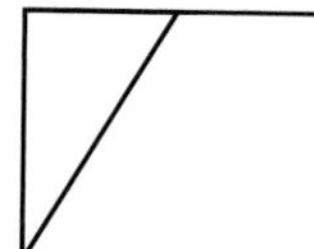

6. a 75° angle

DISCOVERY ACTIVITY

1. Type the commands from the example on page 31 on one line. What happens?
   The figure is the same as before.
2. Type the commands from the example on page 31 as shown below. What happens?
   The figure is the same as before.
   DRAW
   RT 9Ø FD 12Ø
   LT 45 BK 8Ø
3. What do you think would happen if you typed the commands from the example as shown below? Use a computer to check your answer.
   The figure would not change.
   DRAW
   RT 9Ø
   FD 12Ø LT 45 BK 8Ø
4. Write a generalization that seems to be true.
   More than one command can be typed on one line.

You should have discovered that more than one command can be typed on one line.

MORE COMMANDS

Sometimes you may want to move the turtle without leaving a trail. The command PENUP or PU will accomplish this task. The command PENDOWN or PD tells the turtle to start leaving a trail again.

Extra Practice
Write LOGO commands that tell the turtle to draw two diagonal segments that intersect at right angles. Check the answer with a computer. [Answers will vary.]

Example: Show how the turtle moves when you type these Logo commands.

```
DRAW
PU
LT 9Ø
FD 5Ø RT 9Ø
FD 5Ø RT 9Ø
PD
FD 1ØØ RT 9Ø
FD 1ØØ RT 9Ø
FD 1ØØ RT 9Ø
FD 1ØØ RT 9Ø
```

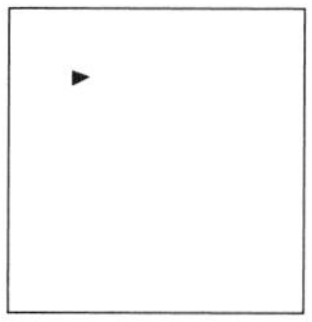

The commands in the first six lines move the turtle to this position without leaving a trail.

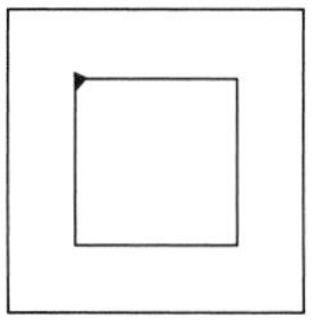

The result for the remaining commands is a square that is centered on the screen.

In order to draw the square in this last example, several groups of the same commands were typed over and over again. The REPEAT command can be used to avoid this.

```
DRAW
PU
LT 9Ø
REPEAT 2[FD 5Ø RT 9Ø]
PD
REPEAT 4[FD 1ØØ RT 9Ø]
```

In a REPEAT command, the number in front of the first bracket tells the turtle how many times to execute the command in the brackets. In the example above, the turtle will repeat the commands in the brackets in the fourth line 2 times and the commands in the brackets in the last line 4 times.

CLASS ACTIVITY

Write Logo commands that tell the turtle how to draw the figure described or shown. Use a computer to check your answers. Answers may vary.

1. a horizontal segment and a vertical segment that intersects the horizontal segment at its midpoint

2. a horizontal segment and a nonvertical segment that intersects the horizontal segment at its midpoint

3.

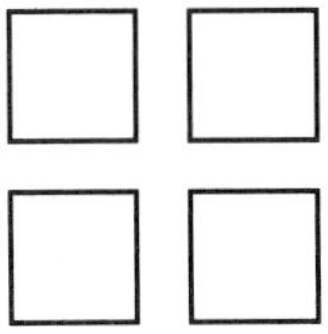

4. 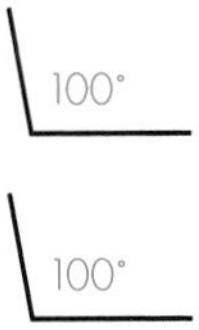 100° 100°

Additional Answers
The RECTANGLE procedure would be executed. Then the turtle would be returned to its home position. Then the TRIANGLE procedure would be executed.

Additional Answers

1.
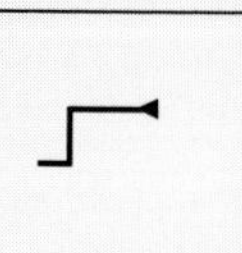

2.
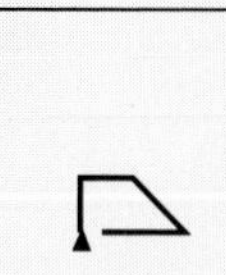

3.
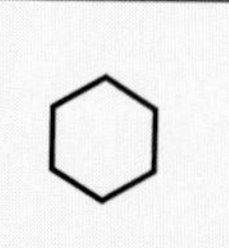

4.

Extra Practice
Write LOGO commands that tell the turtle how to draw the figure below. [Answers may vary.]

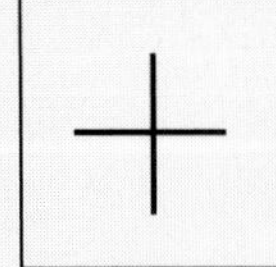

PROCEDURES

In Logo, you can write procedures, or programs, that contain commands needed to complete a certain task. A name is assigned to the commands in the procedure. To execute the procedure, you simply type its name.

The commands in the example on page 33 can be assigned the name SQUARE as shown below.

```
TO SQUARE
PU
LT 9Ø
REPEAT 2[FD 5Ø RT 9Ø]
PD
REPEAT 4[FD 1ØØ RT 9Ø]
END
```

DISCOVERY ACTIVITY

1. Write a procedure named RECTANGLE to draw a rectangle that is centered on the screen. Use a computer to check your answer. Answers may vary.
2. Write a procedure named TRIANGLE to draw a triangle that is centered on the screen. Use a computer to check your answer. Answers may vary.
3. What do you think would happen if you typed the following commands? Use a computer to check your answer. See margin
   RECTANGLE HOME TRIANGLE

CLASS ACTIVITY

Show how the turtle moves when you type these Logo commands. See margin.

1. DRAW
   PU RT 9Ø BK 9Ø PD
   FD 3Ø LT 9Ø
   FD 5Ø LT 9Ø
   BK 8Ø
2. DRAW
   PU BK 8Ø RT 9Ø PD
   FD 1ØØ LT 135
   FD 7Ø LT 45
   FD 5Ø LT 9Ø
   FD 5Ø

Write Logo commands that tell the turtle how to draw each figure. Answers may vary.

3.

4.
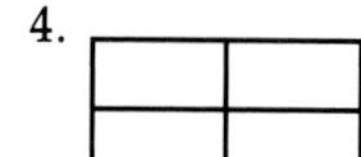

5.

6.
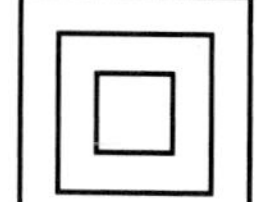

GEOMETRIC DRAWING TOOLS

There are several geometric drawing tools that have been developed for the computer. These tools enable you to use the computer to explore some of the concepts you will study in this course.

To use one of these tools, simply follow the instructions that will be given to you on the screen. Be sure to record all important measurements so that you can quickly refer to them at a later time.

Example: Use a geometric drawing tool and follow the steps below to explore the concept of betweenness of points as presented in Definition 1-3-3.

(1) Draw a segment of any length. Label the endpoints *A* and *B*.

Suppose you tell the computer to draw a segment that has a length of 8. The segment will appear on the screen as shown below.

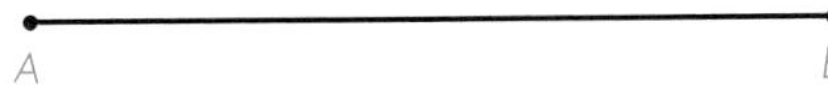

(2) Draw any point between points *A* and *B*. Label the point *C*.

The segment will now look something like this.

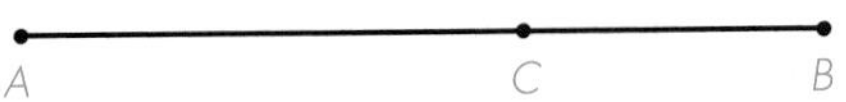

(3) Find the lengths of $\overline{AC}$, $\overline{CB}$, and $\overline{AB}$.

Tell the computer to measure $\overline{AC}$, $\overline{CB}$, and $\overline{AB}$. These measures will appear on the screen. Record the measures.

(4) Find the sum of *AC* and *CB*. Compare the sum with the length of $\overline{AB}$.

Suppose point *C* was drawn so that $AC = 5$ and $CB = 3$. Then the sum of *AC* and *CB* would be 8. In comparing this sum with the length of $\overline{AB}$, we see that they are the same. Record this result.

(5) Repeat steps 1–4 several times. Study the results and write a generalization that seems to be true.

After completing this step, you will see that when point *C* is between points *A* and *B*, the sum of the lengths of $\overline{AC}$ and $\overline{CB}$ is the same as the length of $\overline{AB}$.

FOLLOW-UP

Assessment

1. Describe what happens when you type the DRAW command. [The screen is cleared and the turtle is placed in the home position.]

2. Describe the difference between a command and a procedure in LOGO. [Commands move the turtle about the screen; procedures are programs containing commands.]

3. Write LOGO commands that tell the turtle to draw the figure shown below. [Answers will vary.]

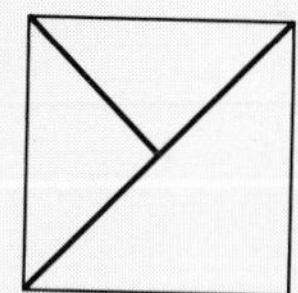

Reteaching

Students who have had difficulty with this section may benefit from Reteaching Activity 1.6.

Enrichment

For students who have mastered the material in this section, you may wish to assign Enrichment Activity 1.6.

CLASS ACTIVITY

Use a geometric drawing tool and follow the steps below.
Answers may vary.

1. Draw an acute angle.
2. Label the angle *ABC*, placing points *A* and *C* so that $BA = BC$.
3. Draw $\overline{AC}$.
4. Draw point *X* on $\overline{AC}$ so that *X* is the midpoint of $\overline{AC}$.
5. Draw $\overline{BX}$.
6. Measure $\angle ABX$ and $\angle XBC$.
7. Compare the measures of $\angle ABX$ and $\angle XBC$. What is $\overline{BX}$?
   $m\angle ABX = m\angle XBC$; $\overline{BX}$ is the bisector of $\angle ABC$.

HOME ACTIVITY

Show how the turtle moves when you type these Logo commands.

1. DRAW
   REPEAT 6 [FD 2Ø RT 6Ø]
2. DRAW
   REPEAT 3[FD 5Ø RT 12Ø]
   RT 6Ø
   REPEAT 2[FD 5Ø RT 12Ø]
3. DRAW
   REPEAT 3[FD 50 RT 12Ø]
   LT 90
   REPEAT 3[FD 50 RT 90]

Write Logo commands that tell the turtle how to draw each figure.

4. 5. 6. 7.

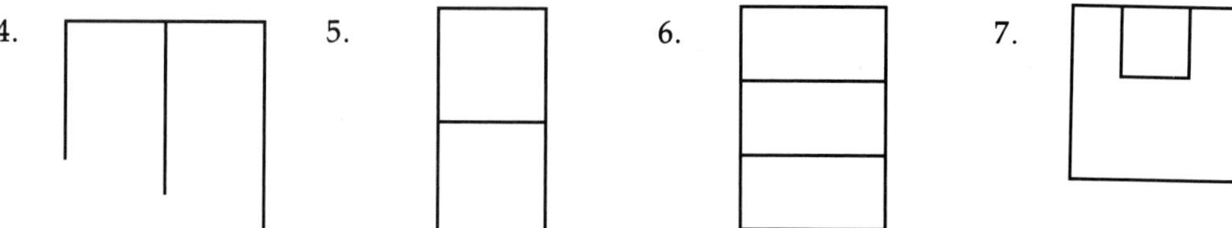

Write procedures that tell the turtle how to draw each figure. Give each procedure the indicated name. Answers may vary.

8. an acute angle; ACUTE
9. an obtuse angle; OBTUSE
10. a 130° angle and its bisector; ANGLEBISECTOR

CRITICAL THINKING

Write a paragraph explaining why it is important to give precise, detailed instructions to the computer when want to use it to draw geometric figures. Would it be necessary to give similar instructions to a live human being who is going to draw a figure that you describe? Answers will vary.

CHAPTER SUMMARY

1.1 A **plane** is a flat surface that extends endlessly in all directions. Two points determine exactly one **line.** A **line segment** is part of a line and consists of two points A and B and all points between them. A **ray** consists of a point and all points to one side of that point. Two rays with a common endpoint form an **angle.**
In the diagram, plane P contains line *CD* ($\overleftrightarrow{CD}$), line segment *AB* ($\overline{AB}$), rays *ED, EF,* and *EC* ($\overrightarrow{ED}$, $\overrightarrow{EF}$, and $\overrightarrow{EC}$), and angles *DEF, FEC,* and *DEC* ($\angle DEF$, $\angle FEC$, and $\angle DEC$).

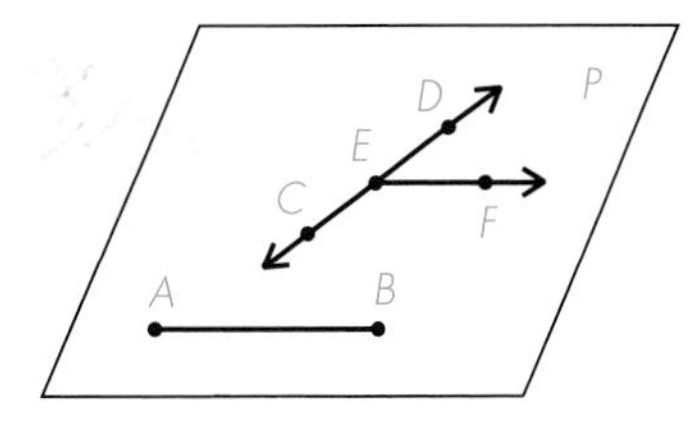

1.2 There is a one-to-one correspondence between points on a line and the real numbers. The distance between two points on a number line is the positive difference of their coordinates.
Acute angles have measures less than 90°, **right** angles measure exactly 90°, **obtuse** angles measure between 90° and 180°, and **straight** angles measure exactly 180°.

The diagram shows acute angles *BCD* and *DCE,* right angles *BCE* and *BCA,* obtuse angle *ACD,* and straight angle *ACE.*

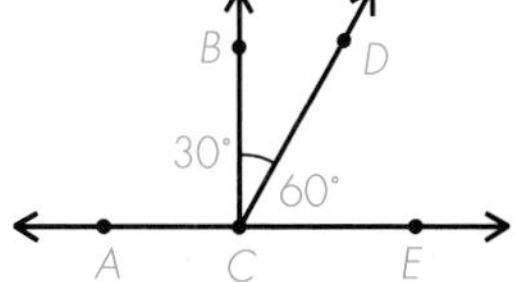

1.3 The **midpoint** of a line segment is the point which divides the segment into two equal segments. The **bisector** of an angle is a ray with the same vertex that separates the angle into two angles, each of which has a measure equal to half of the original angle.

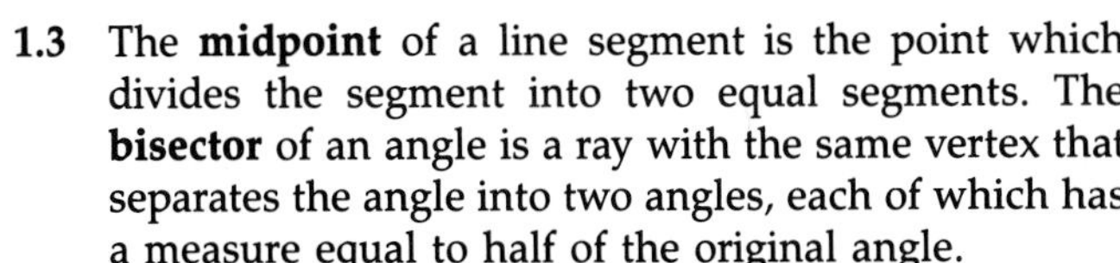

In the diagrams, C is the midpoint of $\overline{AB}$. $\overrightarrow{EF}$ bisects $\angle DEG$. $\overrightarrow{EF}$ is **between** $\overrightarrow{ED}$ **and** $\overrightarrow{EG}$**.**

A C B
$3\frac{1}{2}$ $3\frac{1}{2}$

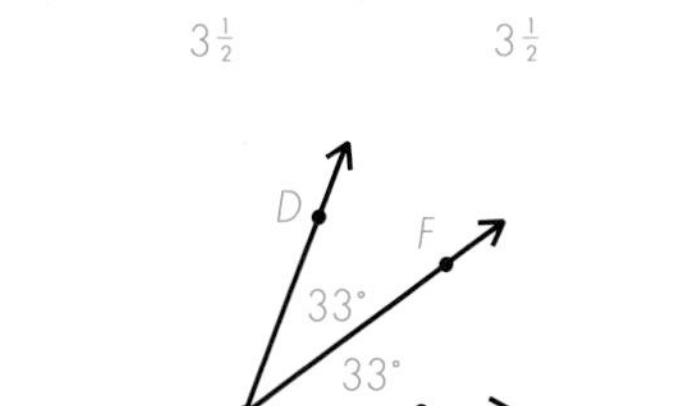

1.4 Two angles are **supplementary** if their measures add to 180°. Two angles are **complementary** if their measures add to 90°.

In the diagrams, $m\angle ABD = 180°$ and $m\angle EFG = 90°$, $\angle ABC$ is supplementary to $\angle CBD$, and $\angle EFH$ and $\angle HFG$ are complementary.

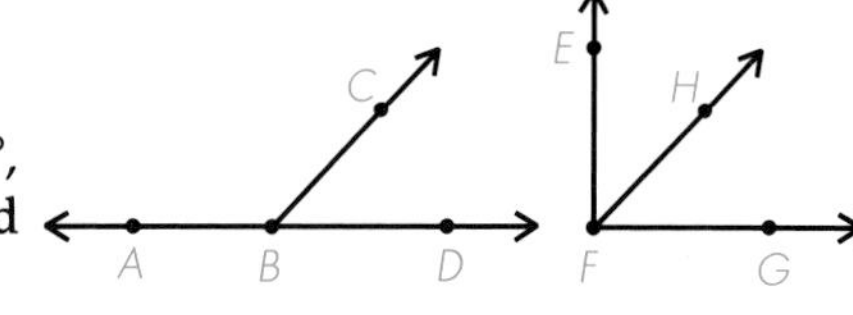

1.5 **Inductive reasoning** is the process of using a limited number of cases to arrive at a conclusion that you expect to be true in all cases.

1.6 **LOGO** is a computer language that can be used to investigate geometric properties.

Chapter Vocabulary
acute angle
angle
angle bisector
betweenness
complementary angles
coordinate
degree measure
endpoint
inductive reasoning
line
line segment
LOGO
LOGO commands
LOGO procedures
midpoint
number line
obtuse angle
origin
plane
point
postulate
protractor
ray
real numbers
right angle
side
straight angle
supplementary angle
turtle graphics
vertex

37

Using the Page
Students can work individually or in groups to study the Chapter 1 Review. If you have students work in groups, they should work together to make sure all members of the group know the material in the chapter. Informal assessment such as interviews, classroom observations, or a review of student portfolios can be used instead of paper-and-pencil tests.

Name symbols to name each of the following.

1. **2.** **3.**

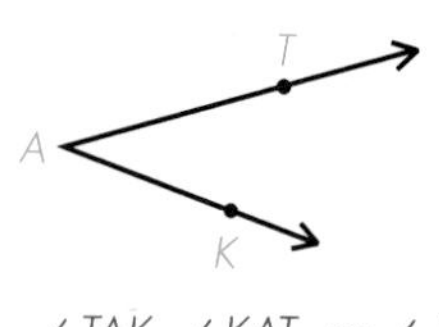

$\overline{AB}$ or $\overline{BA}$ $\quad$ $\overrightarrow{BA}$ $\quad$ $\angle TAK$, $\angle KAT$, or $\angle A$

4. Measure the segment to the nearest inch and then to the nearest centimeter. $3\frac{1}{2}$ in; 9 cm

A ———————————— B

5. Suppose $\overline{EG}$ has a length of 15.8 cm. If M is the midpoint of EG, what is the distance from M to G? 7.9 cm

Measure each angle and classify it as acute, obtuse, right, or straight.

6.

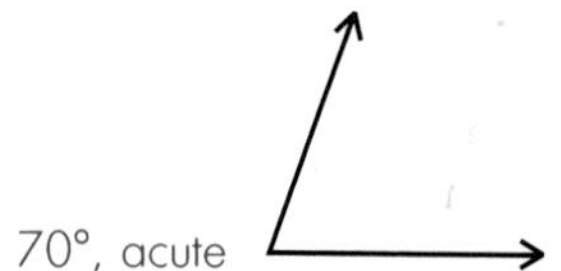

70°, acute

7.

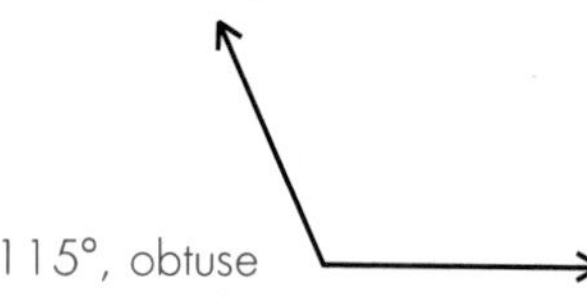

115°, obtuse

8. Use your protractor to draw an angle whose measure is 76°. Then draw its bisector.
See students' drawings. The smaller angle should measure 38°.

Refer to the number line at the right.

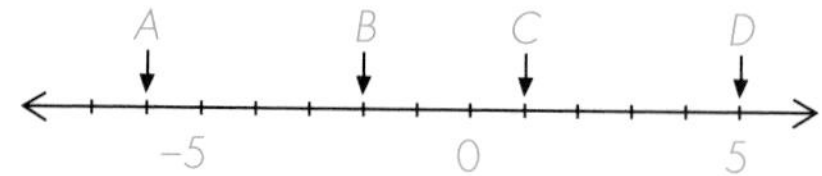

9. What are the coordinates of points A, B, C, and D? -6; -2; 1; 5

10. What are the lengths of $\overline{AB}$ and $\overline{BD}$? 4; 7

11. Points J, K, and L are on the same number line. Their coordinates are 7, 3, and 4. Draw a diagram to show this and tell which point is between the other two.

Each of the following pairs of angles is either complementary or supplementary. Find the measure of each angle.

12.

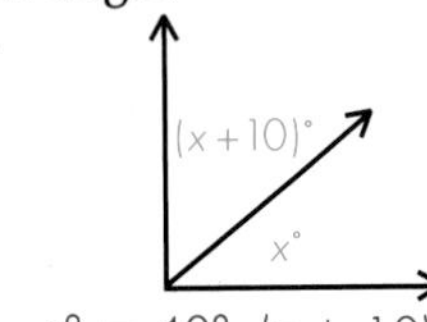

$x° = 40°$, $(x + 10)° = 50°$

13.

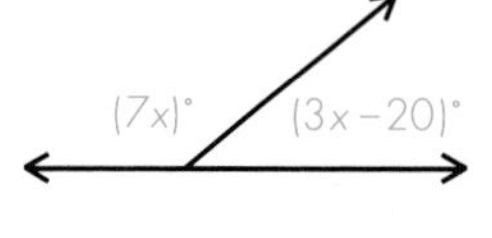

$(7x)° = 140°$, $(3x - 20)° = 40°$

14.

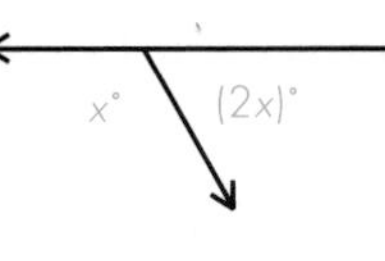

$x° = 60°$, $(2x)° = 120°$

15. Look for a pattern. Then fill in the blanks.
a. 1, 3, 6, 10, 15, __21__, __28__, __36__
b. 2, 6, 18, 54, __162__, __486__, __1458__

16. Write LOGO commands for drawing a vertical line segment intersected at its midpoint by a horizontal line segment. Answers will vary.

38

CHAPTER TEST

1. Complete the following table.

Angle	Complement	Supplement
41°	49°	139°
86°	4°	94°
5°	85°	175°
$(4x)°$	$(90 - 4x)°$	$(180 - 4x)°$
$(2x - 20)°$	$(110 - 2x)°$	$(200 - 2x)°$

Measure each angle and classify it as acute, obtuse, right, or straight.

2. 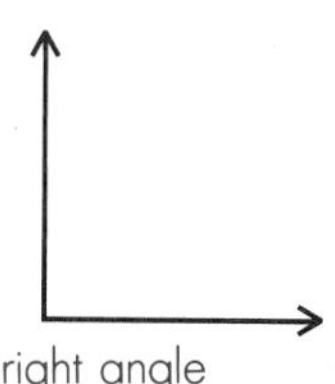

90°; right angle

3. 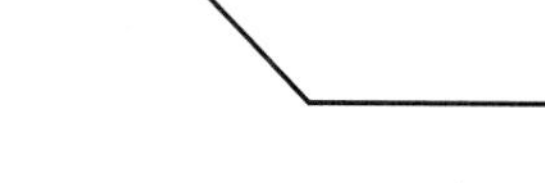

135°; obtuse angle

4. What is the measure of each angle formed by the bisector of a 90° angle? 45°

Tell what kind of figure each of the following is. Use symbols to name it.

5. A B ray; $\overrightarrow{AB}$

6. 

line; $\overleftrightarrow{AB}$ or $\overleftrightarrow{BA}$

7. A B line segment; $\overline{AB}$ or $\overline{BA}$

8. 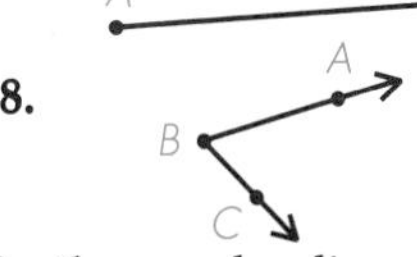

angle; $\angle ABC$, $\angle CBA$, or $\angle B$

On the number line at the right, O is the origin. P and Q have coordinates −4 and 6, respectively, and $PS = 2$.

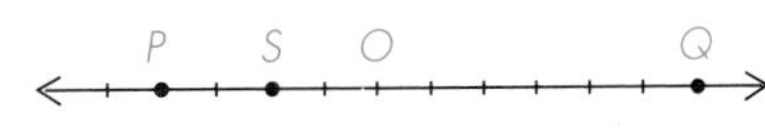

9. What is the coordinate of S? −2
10. What is the length of $\overline{PQ}$? of $\overline{SQ}$? 10; 8

Suppose B is between A and P. Point K is the midpoint of $\overline{BP}$. $AP = 8$ and $KP = 1\frac{1}{2}$.

11. Draw a diagram to show how A, B, K, and P are situated.
12. What is the measure of $\overline{BK}$? of $\overline{AB}$? $1\frac{1}{2}$; 5

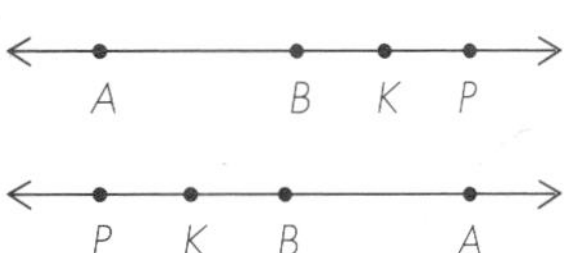

Look for a pattern, then fill in the blanks.

13. 1, 3, 7, 15, 31, 63, 127, 255
14. −8, −5, −2, 1, 4, 7
15. Write LOGO commands for drawing a 120° angle and its bisector. Answers will vary.

Assessment Resources
Achievement Tests pp. 1-2

Test Objectives
After studying this chapter, students should understand
- the basic elements of geometry—points, lines, line segments, rays, and planes.
- how numbers are matched with points on a number line.
- how to measure line segments and angles.
- how to classify angles.
- how to find the midpoint of a line segment.
- how to find the bisector of an angle.
- the meaning of betweenness.
- the meaning of supplementary angles.
- the meaning of complementary angles.
- the meaning of inductive reasoning.
- how to use LOGO commands to draw geometric figures.

Algebra Review Objectives
Topics reviewed from algebra include
- evaluating an algebraic expression.
- solving an equation in one variable.
- writing equations to solve problems about geometry.

ALGEBRA SKILLS

An algebraic expression consists of numbers and variables combined with operation signs (+, −, ×, ÷, and so on). To evaluate an algebraic expression means to find its value given specific values for its variables.

To evaluate an expression, do the following things in this order.

1. Replace all variables by their values.
2. Simplify any expressions within parentheses.
3. Simplify any powers or roots.
4. Multiply or divide from left to right.
5. Add or subtract from left to right.

Example:
Evaluate $2x + y$ for $x = 10$ and $y = -2$.
$2(10) + (-2) = 20 - 2 = 18$
$2x + y = 18$

To solve an equation means to find the number or numbers that can replace the variable to make the equation true. Such a number is called a *solution* of the equation. To solve an equation, you isolate the variable in one member of the equation by using inverse operations. Your goal, in other words, is to get the variable alone on one side of the equal sign to find its value.

Example:
Solve for x.
$2x + 6 = 34$
$2x + 6 - 6 = 34 - 6$
$2x = 28$
$2x \div 2 = 28 \div 2$
$x = 14$
Check: $2(14) + 6 = 34$
$28 + 6 = 34$
$34 = 34$

Evaluate the following expressions for $x = 12$ and $y = -20$.

1. $x + y$ −8
2. $x - y$ 32
3. xy −240
4. $\frac{y}{5}$ −4
5. $y - x$ −32
6. y^2 400
7. $\frac{x}{y} - 1$ $-1\frac{3}{5}$
8. $x - (1 - y)$ −9

Evaluate these expressions for $x = \frac{1}{2}$ and $y = \frac{2}{3}$.

9. $x + y$ $1\frac{1}{6}$
10. $x - y$ $-\frac{1}{6}$
11. $y - x$ $\frac{1}{6}$
12. $\frac{x}{y}$ $\frac{3}{4}$
13. xy $\frac{1}{3}$
14. x^2 $\frac{1}{4}$
15. y^3 $\frac{8}{27}$
16. $2x + y$ $1\frac{2}{3}$

Solve for *x*.

17. $4x - 7 = 53$ $x = 15$
18. $2.5x + 8 = 23$ $x = 6$
19. $\frac{x}{2} + 15 = 40$ $x = 50$
20. $\frac{3}{4}x - 5\frac{1}{2} = 3\frac{1}{2}$ $x = 12$
21. $12 - (2x - 5) = 9$ $x = 4$
22. $3x = 8 - 5x$ $x = 1$
23. Two angles are complementary. One is 5 times as great as the other. What is the measure of each angle? 75°, 15°
24. Two angles are supplementary. The measure of one is 15° more than twice the measure of the other. What is the measure of the smaller angle? 55°

Name ______________________ Date ______________

Points, Lines, and Planes

The following figures show some of the basic elements used in geometry. Notice the difference between a *line* and a *line segment.* Also, note that the *sides* of an angle are rays rather than line segments. The two angles shown have the same measure even though one of them has sides that are drawn longer.

Plane
Flat surface with no thickness and no edges

Point
No size, only location

Line
Straight, continues indefinitely in both directions; no thickness and no endpoints

Line Segment
Part of a line—it has two endpoints.

Ray
Part of a line—it has one endpoint.

Angle
Two rays with the same endpoint; the endpoint is called the *vertex.*

Write the letter of the term in the box that matches each description.

a. line segment
b. angle
c. plane
d. vertex
e. side
f. line
g. ray

1. 2-dimensional flat surface with no boundaries c
2. Named by two points; straight, with no endpoints f
3. Part of a line, with only one endpoint g
4. Ray that forms one part of an angle e
5. Two rays with a common endpoint b
6. Common endpoint of the two sides of an angle d
7. How many points are needed to name a line segment? 2
8. How many endpoints does a line have? 0
9. Which point do you start with when you name a ray? endpoint
10. Which point do you start with when you name a line segment? either point

Name ______________________ Date ______________

Optical Illusions

In the study of geometry, it is wise to be careful about conclusions you make from figures or sketches. When you do use a drawing, it should be done carefully, using a sharp pencil and the proper geometric tools. The optical illusions on this page may help convince you that appearances are often misleading.

For example, in the drawing at the right, it appears that the upper line segment to the right of the rectangle is the continuation of the line segment to the left. Check with a straightedge—it is actually the bottom segment that is a continuation.

For each exercise, describe the illusion. That is, what appears to be true about parts of the drawing, but is in fact false?

1. $\overline{AB}$ appears to be longer than $\overline{CD}$. They are actually the same length.

2. The bottom line segment appears to be longer than the top segment. They are actually the same length.

3. The two dark line segments appear to curve outward. They are actually parallel.

A B C D

4. The gray "V" on the left appears lighter than the "V" on the right. They are actually the same shade.

5. The center circle on the left appears larger than the center circle on the right. They actually are the same size.

Name ______________________ Date ______________

Measurement of Segments and Angles

Right Angle
An angle that measures 90°—the corner of a sheet of paper approximates a right angle.

Acute Angle
An angle that measures between 0° and 90°—if you fold the corner of a sheet of paper in half, you get an acute angle.

Obtuse Angle
An angle that measures between 90° and 180°—an obtuse angle is greater than a right angle.

Straight Angle
An angle that measures 180°—it looks like a line; however, the vertex and two other points are often named.

For each figure, write whether the angle shown in dark lines is *right, acute, obtuse,* or *straight.*

1. acute **2.** obtuse **3.** right **4.** straight

5. Write a number next to each statement so that the steps for measuring an angle with a protractor are in the correct order.

1 If necessary, extend the sides of the angle so that they reach the scale on the protractor.

3 Use the correct scale and read the number of degrees.

2 Line up your protractor so that the vertex of the angle is at the proper mark and one side of the angle is at the 0° mark.

6. Use the figure below. Name two angles of each type. Other answers are possible.

a. right $\angle APQ$ and $\angle PQB$

b. acute $\angle APB$ and $\angle BPQ$

c. obtuse $\angle PBR$ and $\angle QBC$

Name ______________________ Date ______________

Moebius Strips

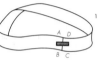

Figure 1 at the right is a Moebius strip. It has many surprising properties.

1. To make a Moebius strip, take a strip of paper (adding machine tape works well) and give the strip a half twist. Match point *A* with point *D*, and point *B* with point *C* as shown in Figure 1.

2. Make another Moebius strip. But this time, start by coloring a checkerboard pattern on one side of the paper as shown in Figure 2. Leave the other side plain. The original strip has two sides, one checked and one plain. How many sides does the Moebius strip have?
just one side

3. Draw a line down the center of the first Moebius strip you made, as shown in Figure 3. Then cut along this line. Describe the result. Is it what you expected?
one large band with four half-twists rather than two separate bands

4. As interesting result is obtained when you trisect a Moebius strip. To do so, start cutting one-third of the way in from the edge as shown in Figure 4. Before you begin, predict what you will get. Then describe the results.

Prediction: Answers will vary.

Results: Two linked bands—one is a narrow version of the original Moebius strip and the other is the same as the long strip produced in Exercise 3.

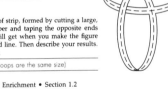

5. Figure 5 shows another type of strip, formed by cutting a large, thin "X" from a sheet of paper and taping the opposite ends together. Predict what you will get when you make the figure and then cut along the dashed line. Then describe your results.

Prediction: Answers will vary.

Results: a square (if the two loops are the same size)

Name ______________________ Date ______________

Midpoints of Segments and Angle Bisectors

The *midpoint* of a line segment divides it into two parts of equal length. The *bisector* of an angle divides it into two new angles of equal measure.

Use the figure at the right. Write *true* or *false* for each statement.

1. Point *C* is the midpoint of segment *AB*. true
2. Point *B* is the midpoint of segment *DE*. false
3. Point *C* is the midpoint of segment *DG*. true
4. Segment *BC* bisects angle *DCE*. false
5. Segment *DE* bisects angle *FDC*. true
6. Segment *CE* bisects angle *GCB*. false

Use a protractor and a centimeter ruler for Exercises 7 and 8.

7 a. Measure one side of the square below. What is the length in centimeters? 3 cm
b. Point *M* is the midpoint of side *QR*. What is the length of *QM*? 1.5
c. What is the measure of angle *PSQ*? 45°
d. What is the measure of angle *RSQ*? 45°
e. Does $\overline{SQ}$ bisect angle *PSR*? yes

8. a. Measure the width of the rectangle below. What is the width in centimeters? 3 cm
b. Point *M* is the midpoint of side *QR*. What is the length of *QM*? 1.5 cm
c. What is the measure of angle *PSQ*? about 62°
d. What is the measure of angle *RSQ*? about 28°
e. Does $\overline{SQ}$ bisect angle *PSR*? no

Name ______________________ Date ______________

Origami—Constructions with Paper Folding

Origami is the Japanese art of paper folding. To make the construction on this page, use thin crisp paper that will take a sharp crease. Wrapping paper that is white on the back works well.

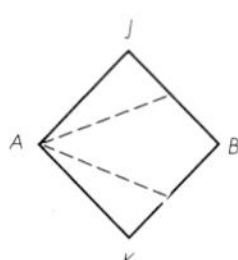

Start with a square at least 8 inches on a side. With the white side of the paper facing you, fold edges $\overline{AJ}$ and $\overline{AK}$ to the center line (AB) of the paper as shown at the right. Then follow the steps below.

1. Mountain-fold edges *CB* and *DB* underneath to the center line *AB* in back. This type of line shows a mountain fold: —·—·—

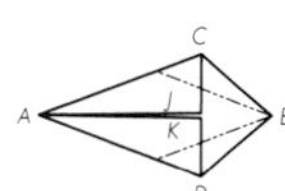

2. Valley-fold the tiny flaps up and outward from the center line. This type of line shows a valley fold: - - - - - - - -

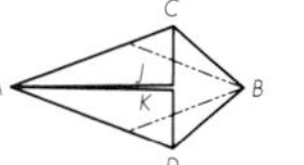

3. Valley-fold the model in half, bringing the top half forward and down over the bottom half.

4. Cut two inches along the top fold as indicated by the dashed line at the left below. Open out the pocket on the side and adjust it downward to the position shown in Step 5. Repeat for the pocket on the other side.

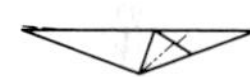

5. Fold the front half of the tail upward and bring the back half downward. Fold flap *x* upward in front, and repeat on the other side.

6. The fish is reversible. If you turn it inside out the colors will be different.

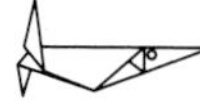

Name ______________________ Date ____________

Supplements and Complements

If the sum of the measures of two angles is 90°, they are called *complementary* angles.

If the sum of the measures of two angles is 180°, they are called *supplementary* angles.

You might think of the angles as if they were corners torn from a paper triangle.

Start with a sheet of paper that is $8\frac{1}{2}$ by 11 inches. Make the two folds.

1. Assume the corners of the paper are right angles. Write four pairs of complementary angles located at the corners.
   ∠XWO and ∠ZWO, ∠WZO and ∠YZO, ∠ZYO and ∠XYO, ∠YXO and ∠WXO

2. Write four pairs of supplementary angles created by the diagonals.
   ∠XOW and ∠XOY, ∠WOZ and ∠YOZ, ∠XOW and ∠ZOW, ∠XOY and ∠ZOY

3. Write two pairs of complementary angles found in the figure at the right.
   ∠M and ∠Q, ∠ONP and ∠OPN

4. Write two pairs of supplementary angles.
   ∠QPN and ∠NPO, ∠MNP and ∠PNO

5. Write two pairs of supplementary angles found in the figure at the right.
   ∠AEC and ∠DEC, ∠BCE and ∠DCE

6. Show that there are no pairs of complementary angles.
   There are no right angles, so neither triangle has a pair of complementary angles. Student may also measure all the angles.

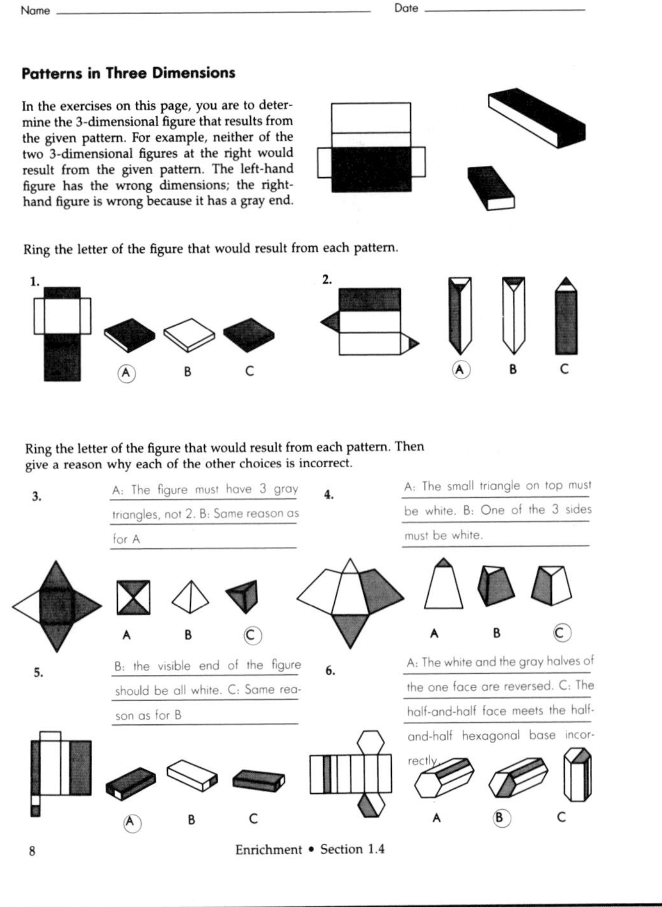

Name ______________________ Date ____________

Patterns in Three Dimensions

In the exercises on this page, you are to determine the 3-dimensional figure that results from the given pattern. For example, neither of the two 3-dimensional figures at the right would result from the given pattern. The left-hand figure has the wrong dimensions; the right-hand figure is wrong because it has a gray end.

Ring the letter of the figure that would result from each pattern.

1. (A) B C

2. (A) B C

Ring the letter of the figure that would result from each pattern. Then give a reason why each of the other choices is incorrect.

3. A: The figure must have 3 gray triangles, not 2. B: Same reason as for A
   A B (C)

4. A: The small triangle on top must be white. B: One of the 3 sides must be white.
   A B (C)

5. B: the visible end of the figure should be all white. C: Same reason as for B
   (A) B C

6. A: The white and the gray halves of the one face are reversed. C: The half-and-half face meets the half-and-half hexagonal base incorrectly.
   A (B) C

Name ______________________ Date ____________

Inductive Reasoning

When you find a pattern in a series of numbers, a set of figures, or a sequence of events, the pattern can be used to predict what will happen next. The prediction is an "educated guess." You cannot be absolutely certain that your prediction will come true.

For each series of numbers, give two different numbers that might logically come next. Explain what patterns you are using.

Answers will vary. Sample answers are given.

1. **a.** 2, 4, 6, 8, 10, 12 increasing even numbers
   b. 2, 4, 6, 8, 10, 8 even numbers increase to 10; then decrease
2. **a.** 1, 2, 4, 8 each number is twice the previous number
   b. 1, 2, 4, 7 add 1, add 2, add 3
3. **a.** 10, 100, 1000 increasing powers of ten
   b. 10, 100, 190 add 90

For each series of figures, give two different figures that might logically come next. Explain what patterns you are using.

Answers will vary. Sample answers are given.

4. A B
   A: alternates top to bottom
   B: rotates 180°
5. A B
   A: one less square each time
   B: shaded figures alternate
6. A B
   A: alternate 4 shaded squares and 3 shaded squares
   B: one less square each time

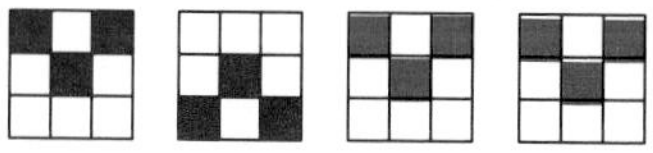

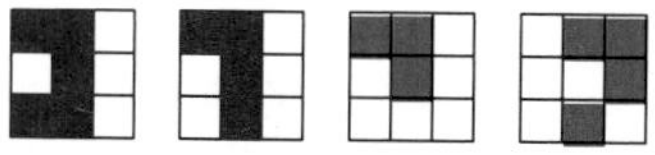

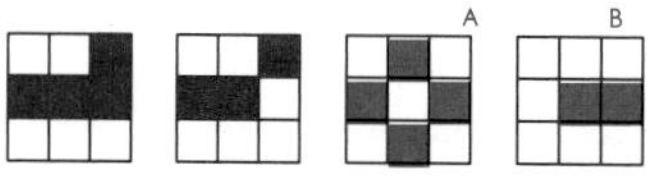

Name ______________________ Date ____________

Impossible Figures

Among the most interesting drawings are those of impossible figures. These are figures that could never be built in 3-dimensional space. In the figure at the right, edge *AB* would have to be *behind* edge *CD* for the cube to actually exist.

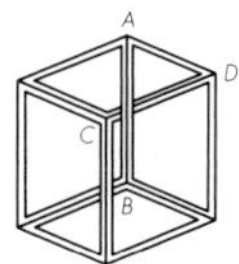
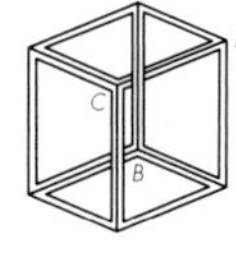

For each drawing, describe what parts would need to be changed for the figure to actually exist. Then sketch the new figure on a separate sheet of paper.

Answers will vary. Sample answers are given.

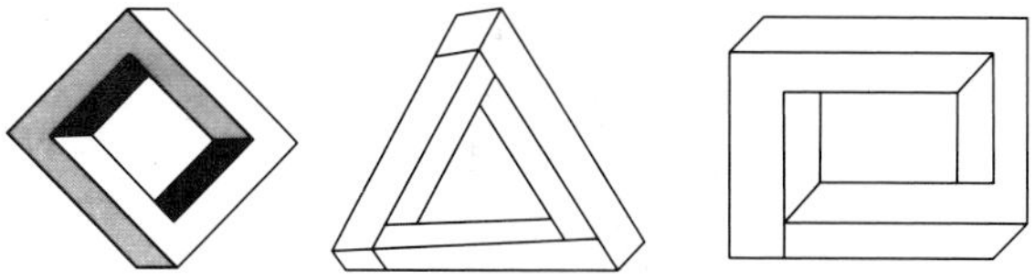

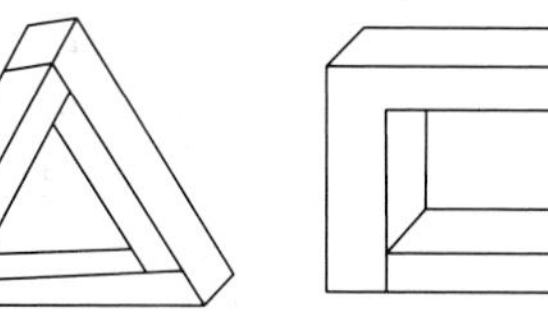

1. Redraw the right side of the figure, putting a shaded diamond on top.
2. Start with the center triangle. Draw a left and bottom strip to show depth.
3. Redraw either the left or right side, putting a white frame-shaped face on top.

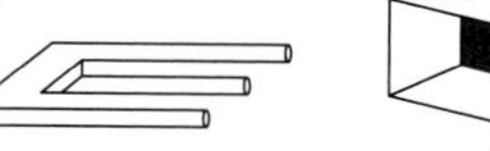
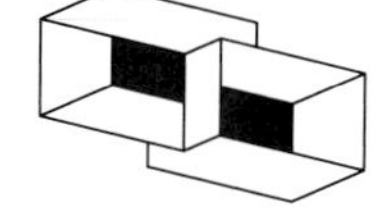
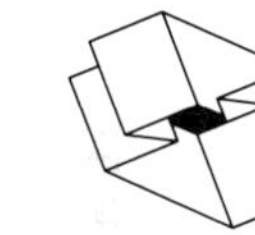

4. Redraw the right side so there are only two prongs instead of three.
5. Either box alone is okay. You could draw two separate boxes.
6. Redraw either the left or right side so the folds are reversed.

Name ______________________ Date ______________

LOGO and Geometric Drawing Tools

There are many different computer programs that you can use to draw geometric, as well as other, figures. One computer *language* that was specifically designed for students is called *LOGO*. After you have used LOGO for a while, the questions on this page will help you determine how well you understand the language.

1. Fill in each blank with one of the words in the box.

turtle	home	left	right	drawing
back	forward	geometric	degrees	penup

LOGO is a drawing program designed to help you learn about geometric figures. To draw a figure, you move a triangle that is called a turtle around the screen of the computer. The command forward will make the turtle move in the direction that it is pointing. The command back will make it go the opposite way. The direction that the turtle is facing can be changed by typing either left or right, and then typing the number of degrees you want the turtle to turn. To make the turtle move without drawing, you must type penup. The command home will always make the turtle go back to the center of the screen.

Will the two programs give the same picture? Write *yes* or *no*.

2. yes

```
DRAW     DRAW
RT 90    LT 90
FD 100   BK 100
```

3. yes

```
DRAW     DRAW
FD 50    FD 50
RT 90    LT 270
FD 50    FD 50
```

4. no

```
DRAW     DRAW
PU       FD 60
FD 60    PU
PD       FD 60
FD 60
```

5. no

```
DRAW     DRAW
FD 90    FD 90
RT 30    RT 30
BK 90    BK 90
DRAW     HOME
```


Name ______________________ Date ______________

Patterns Involving Triangles

If the three sections of an equilateral triangle like these are each colored with one of three colors, you get a set of 11 different triangles. Rotations are not considered different.

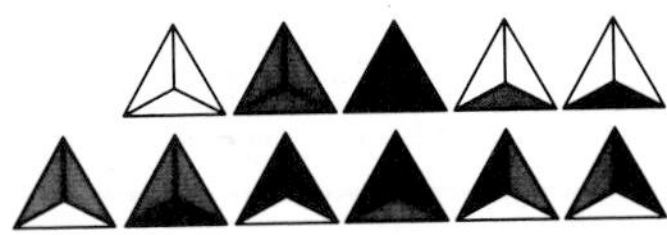

Refer to the 11 triangles shown. How many of the triangles are exactly

1. one-third white 4
2. three-thirds dark gray 1
3. two-thirds white 2
4. three-thirds light gray 1
5. three-thirds white 1
6. two-thirds dark gray and one-third light gray 1
7. two-thirds dark gray 2
8. two-thirds light gray and one-third dark gray 1
9. Two of the triangles that are one-third white are shown at the right. Describe how they are different.
The light gray and dark gray shadings are reversed. The triangles are mirror images on each other.

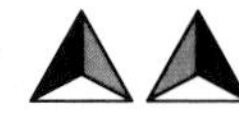

10. If the three sections of an equilateral triangle are each colored with one of *two* colors, you get a set of four different triangles. Show the four triangles by shading the triangles at the right.

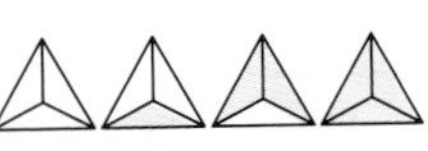

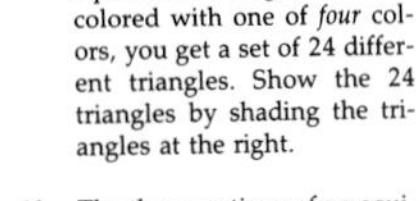

11. If the three sections of an equilateral triangle are each colored with one of *four* colors, you get a set of 24 different triangles. Show the 24 triangles by shading the triangles at the right.

12. The three sections of an equilateral triangle are each colored with one of *five* colors. How many different triangles do you get? 45

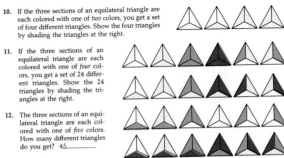

Name ______________________ Date ______________

Writing Procedures

Pentominoes are figures formed by putting five squares of the same size together so that each square has a common edge with at least one other square. Three pentominoes are shown below.

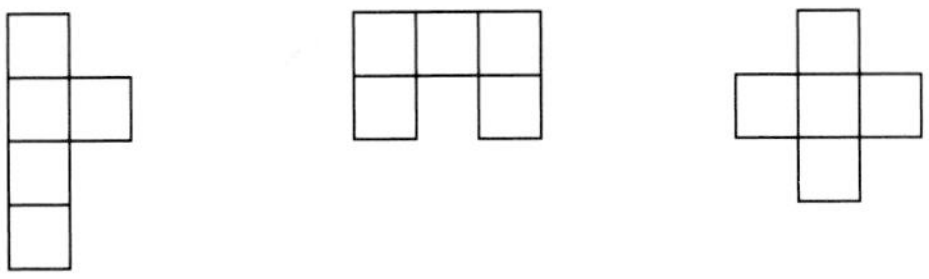

1. Write a LOGO procedure to tell the turtle to draw each pentomino shown above. Answers may vary.

2. Draw three other pentominoes of your own. Write a LOGO procedure to tell the turtle to draw each pentomino. Answers may vary.

3. How are the procedures for drawing each pentomino the same? How are the procedures different?
Same: Each procedure involves drawing five squares. Different: The squares in each procedure are in different positions.

Name ______________________ Date ______________

Straight Angles

Use a computer and a geometric drawing tool to draw the figure described in each exercise. Make a sketch of each figure in the space provided and record all important measurements.

1. Draw any segment AB. Draw $\overline{XY}$ so that X is on $\overline{AB}$ and Y is not on $\overline{AB}$. Measure $\angle AXY$ and $\angle BXY$. Answers may vary.

2. What is the sum of the measures of $\angle AXY$ and $\angle BXY$? 180°

3. Repeat Exercise 1 several times, changing the position of point Y each time. What is the sum of the measures of $\angle AXY$ and $\angle BXY$ each time? 180°

4. What kind of angles were the pairs of angles drawn in Exercises 1 and 3? supplementary

5. What kind of angle is formed by the two non-common sides of each pair of angles in Exercises 1 and 3? straight angle

6. Study the results of Exercises 1–5 and write a generalization that seems to be true.
The non-common sides of two adjacent supplementary angles form a straight angle.

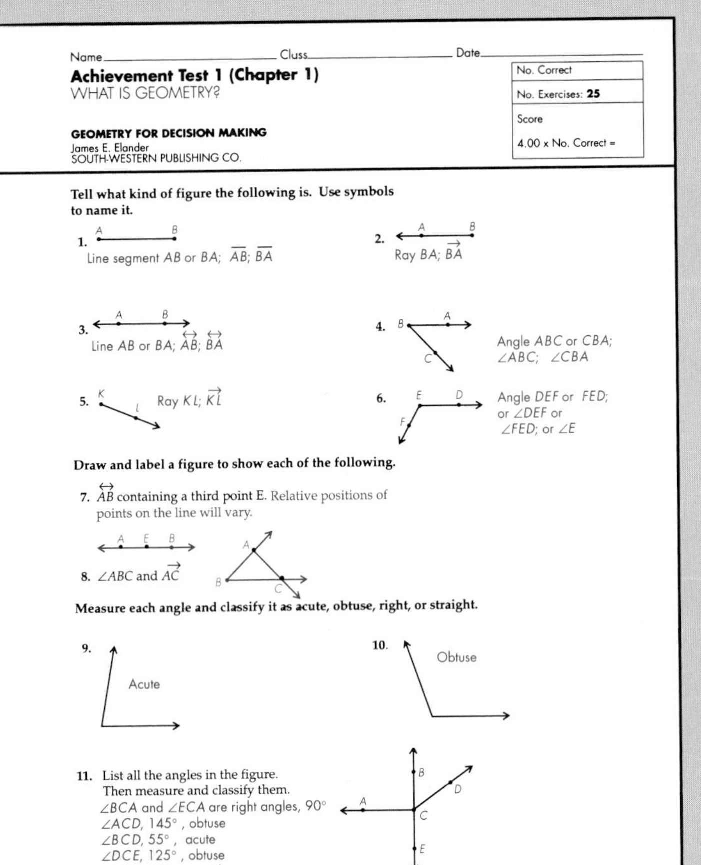

Name________________ Class________________ Date________________

Achievement Test 1 (Chapter 1)
WHAT IS GEOMETRY?

No. Correct
No. Exercises: **25**
Score
4.00 x No. Correct =

GEOMETRY FOR DECISION MAKING
James E. Elander
SOUTH-WESTERN PUBLISHING CO.

Tell what kind of figure the following is. Use symbols to name it.

1. Line segment AB or BA; $\overline{AB}$; $\overline{BA}$
2. Ray BA; $\overrightarrow{BA}$
3. Line AB or BA; $\overleftrightarrow{AB}$; $\overleftrightarrow{BA}$
4. Angle ABC or CBA; $\angle ABC$; $\angle CBA$
5. Ray KL; $\overrightarrow{KL}$
6. Angle DEF or FED; or $\angle DEF$ or $\angle FED$; or $\angle E$

Draw and label a figure to show each of the following.

7. $\overleftrightarrow{AB}$ containing a third point E. Relative positions of points on the line will vary.
8. $\angle ABC$ and $\overrightarrow{AC}$

Measure each angle and classify it as acute, obtuse, right, or straight.

9. Acute
10. Obtuse
11. List all the angles in the figure. Then measure and classify them.
$\angle BCA$ and $\angle ECA$ are right angles, 90°
$\angle ACD$, 145°, obtuse
$\angle BCD$, 55°, acute
$\angle DCE$, 125°, obtuse
$\angle BCE$, 180°, straight
12. What is the measure of each angle formed by the bisector of a 60° angle? 30°

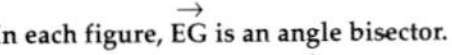

In each figure, $\overrightarrow{EG}$ is an angle bisector.

13. $m\angle DEF$ = 50°
$m\angle GEF$ = 25°

14. 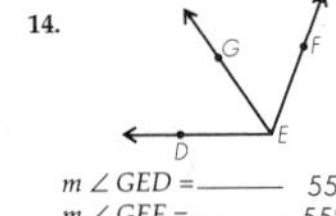
$m\angle GED$ = 55°
$m\angle GEF$ = 55°

On the number line at right, O is the origin. P and S have coordinates −5 and 4, respectively, and $PT = 3$.

15. What is the coordinate of T? −2
16. What is the length of $\overline{TS}$? of $\overline{PS}$? $TS = 6$; $PS = 9$

Suppose C is between A and E. Point M is the mid point of $\overline{CE}$. $AE = 12$ and $ME = 2\frac{1}{2}$.

17. Draw a diagram to show how A, C, M, and E are situated.
18. What is the length of $\overline{CE}$? of $\overline{AC}$? $CE = 5$; $AC = 7$

19. Complete the following table.

Angle	Complement	Supplement
32°	58°	148°
88°	2°	92°
15°	75°	165°
$(3x)°$	$(90 - 3x)°$	$(180 - 3x)°$
$(2x - 10)°$	$(100 - 2x)°$	$(190 - 2x)°$

Find the value of x in each of the following. Assume that the lines that look like straight lines are straight lines.

20. 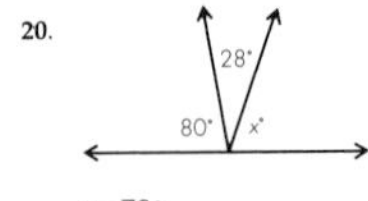

$x = 72°$

21. 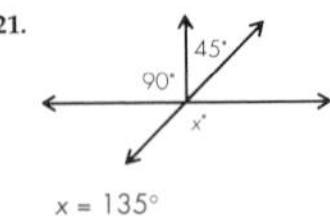

$x = 135°$

Look for a pattern. Then fill in the blanks.

22. 1, 3, 6, 10, 15, ___, ___, ___ 21, 28, − 36
23. −6, −1, 6, 15, 26, ___, ___, ___ 39, 54, 71
24. 2, 5, 11, 23, 47, ___, ___, ___ 95, 191, 383
25. Write LOGO commands for drawing a 120° angle and its bisector.

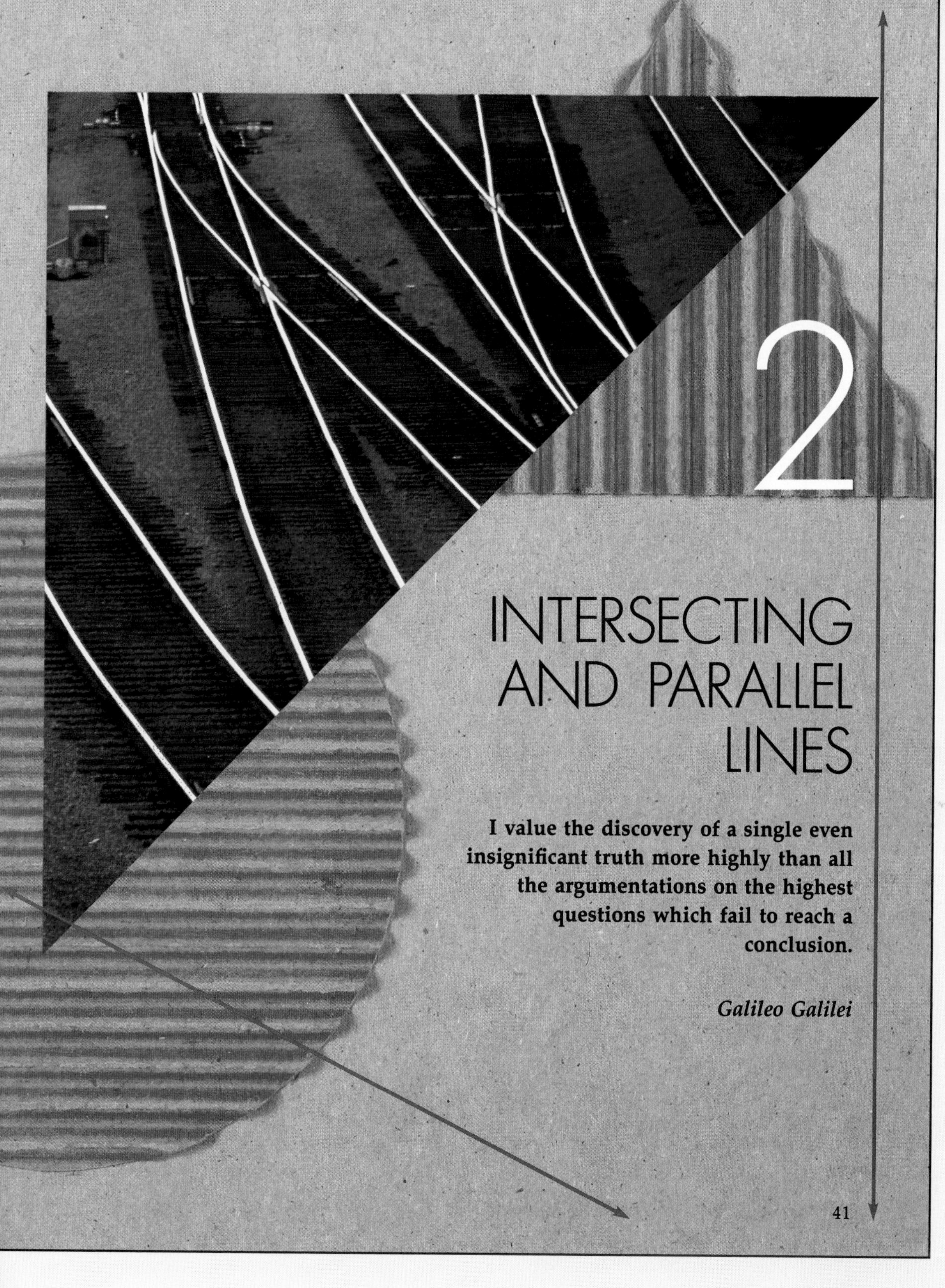

2

INTERSECTING AND PARALLEL LINES

I value the discovery of a single even insignificant truth more highly than all the argumentations on the highest questions which fail to reach a conclusion.

Galileo Galilei

41

CHAPTER 2 OVERVIEW

Intersecting and Parallel Lines

Suggested Time: 12-14 days

About 2,300 years ago a mathematician named Euclid wrote the *Elements*, which contained much of what was known about mathematics at that time. A unique feature of this book was that it organized it information by first presenting a few basic principles, and then deriving all else from them. The quote at the beginning of the chapter was selected because it supports the use of deductive reasoning. In this chapter students start with ideas they already have learned and employ deductive reasoning and discovery as a basic strategy for drawing conclusions about other geometric relationships.

I value the discovery of a single even insignificant truth more highly than all the argumentations on the highest questions which fail to reach a conclusion.

Galileo Galilei

CHAPTER 2 OBJECTIVE MATRIX

	End of Chapter Items by Activity				Student Supplement Items by Activity		
Objectives by Number	**Review**	**Test**	**Computer**	**Algebra Skills**	**Reteaching**	**Enrichment**	**Computer**
2.1.1	✓	✓	✓		✓		
2.1.2	✓	✓	✓		✓	✓	✓
2.2.1	✓	✓	✓		✓	✓	✓
2.2.2	✓	✓	✓	✓	✓		✓
2.3.1		✓				✓	
2.3.2	✓	✓		✓	✓		
2.3.3	✓	✓	✓	✓	✓		✓
2.4.1	✓	✓		✓	✓	✓	
2.4.2				✓	✓		
2.4.3				✓	✓		
2.5.1	✓	✓		✓	✓		
2.5.2	✓			✓	✓		
2.6.1	✓	✓			✓	✓	
2.6.2		✓			✓		

*A ✓ beside a Chapter Objective under Algebra Skills indicates that algebra skills taught within the chapter or in previous Algebra Skills lessons are used.

CHAPTER 2 PERSPECTIVES

▲ Section 2.1

Points and Lines in Space

Students are introduced to collinear and coplanar points, and are asked to determine whether given points are collinear, noncollinear, coplanar, or noncoplanar. Through a Discovery Activity, they conclude that three noncollinear points determine a plane. Then students distinguish among intersecting lines, parallel lines, and skew lines. Students use LOGO procedures to draw different kinds of lines.

2

▲ Section 2.2

Corresponding Angles

Students examine what happens when two or more lines are intersected by another line. They are introduced to the term *transversal*, and are asked to identify lines that are or are not transversals. After corresponding angles are introduced and defined, students use a Discovery Activity to draw conclusions about the degree measures of corresponding angles formed when parallel lines are intersected by a transversal. Students are asked to identify and then find the degree measures of corresponding angles. In so doing, they call upon their knowledge of algebra and their understanding of supplementary angles to solve the problems.

The Home Activity includes an exercise in which students write a LOGO procedure to draw two parallel lines and a transversal, given algebraic expressions for the measures of the supplementary angles. Then students find the degree measures of all angles formed.

▲ Section 2.3

Vertical and Alternate Interior Angles

Students are introduced to the process of deductive reasoning and the form and content of conditional statements. In the first Class Activity, students draw conclusions by using deductive reasoning. Then vertical angles and the vertical angle theorem are introduced. But before they are actually presented with the vertical angles theorem, students use deductive reasoning to show that vertical angles are equal. Next, students are introduced to alternate interior angles and discover that these angles are equal when a transversal intersects two parallel lines.

The Home Activity includes exercises in which students identify hypotheses and conclusions and exercises in which they are asked to draw conclusions given conditional statements.

▲ Section 2.4

Measurement and Estimation

Students are introduced to the idea that measurements are never absolutely exact. Then they learn, partially by discovery, about how error in measurement is affected by the unit of measure used. Students learn that error in measurement is half the unit used to measure an object or angle. Next, students learn how to convert form one measure to another *within* the metric and customary systems and to convert *between* systems. The common metric prefixes *milli-*, *centi-*, and *kilo-* are introduced. The Home Activity includes exercises calling for measurement to the nearest fractions of inches and to the nearest degree. It also includes exercises in which students add or subtract units of measure.

▲ Section 2.5

Alternate Interior Angles

Students are introduced to alternate exterior angles. Through measuring in the Discovery Activity, they conclude that when two parallel lines are intersected by a transversal, the alternate interior angles are equal. Then they use deductive reasoning to arrive at the same conclusion, which is presented as Theorem 2-5-1. Next, the concept of perpendicular lines is presented. After the definition is given, students practice identifying and labeling lines that are perpendicular. The problems in the Class Activity and the Home Activity require students to combine all of what they have learned thus far about the relationship between parallel lines and the angles formed when they are intersected by a transversal.

▲ Section 2.6

Visualizing Three-Dimensional Figures

Students are introduced to the important real-world technical skill of representing three-dimensional figures on a two-dimensional surface. They see how the illusion of depth is created in two-dimensional drawings. Then they practice visualizing and drawing objects from different points of vision. Next, students describe three-dimensional figures when given flat layouts for those figures.

One problem in the Home Activity involves an optical illusion. It requires students to examine a figure that appears to be composed of stacked cubes and to describe it in more than one way.

SECTION 2.1

Resources
Reteaching 2.1
Enrichment 2.1

Objectives
After completing this section, the student should understand
- ▲ collinear, noncollinear, coplanar, and noncoplanar points.
- ▲ intersecting, parallel, and skew lines.

Materials
cardboard
pencils

Vocabulary
collinear
coplanar
noncollinear
noncoplanar
intersecting lines
parallel lines
skew lines

GETTING STARTED

Quiz: Prerequisite Skills

1. How many points determine a line? [2]

2. Draw a plane with points *M* and *N* on it. How many lines can be drawn through either of these points? [an infinite number] How many lines can you draw through both points? [one]

3. Place point O on the plane so that it is not on the same line as points *M* and *N*. How many lines can you draw that contain all three points? [none]

Warm-Up Activity
In the figure below, which line is the continuation of line *r*? [line *t*]

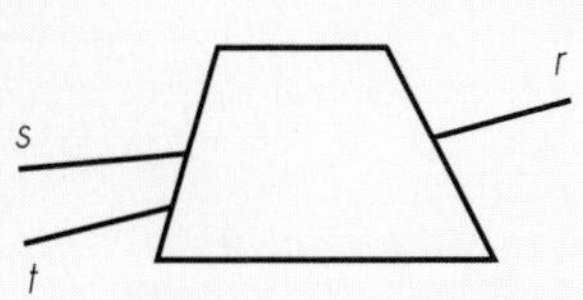

SECTION 2.1

Points and Lines in Space

After completing this section, you should understand
- ▲ collinear, noncollinear, coplanar, and noncoplanar points.
- ▲ intersecting, parallel, and skew lines.

In the diagrams below you see two groups of points. The group on the left is special because its points all lie on the same line. The group on the right is special because all the points are on the same plane (flat surface).

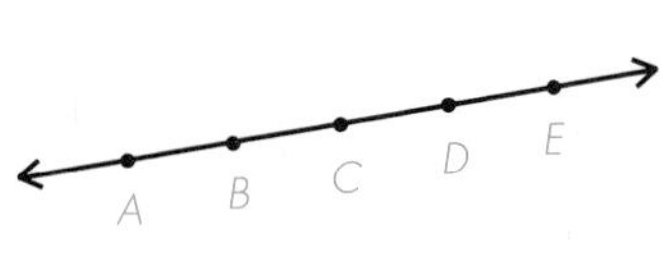

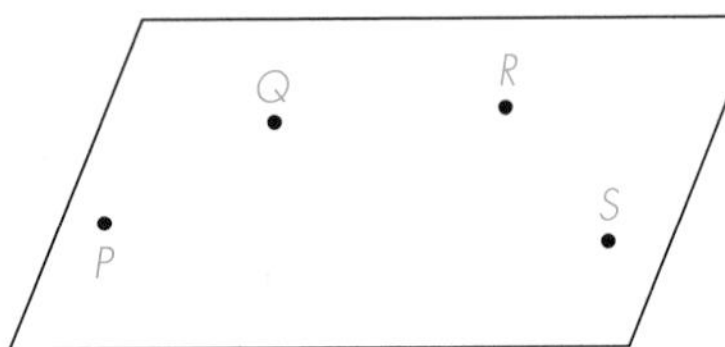

COLLINEAR AND COPLANAR POINTS

In geometry there are special terms for groups of points like those in the diagrams above.

NOTEBOOK **DEFINITION 2-1-1:** **COLLINEAR POINTS are points on the same line.**

NOTEBOOK **DEFINITION 2-1-2:** **COPLANAR POINTS are points on the same plane.**

Points that are not on the same line are *noncollinear.*
Points that are not on the same plane are called *noncoplanar.*

Examples: Refer to the diagram.

1. *A, B,* and *D* are collinear.
2. *C, D,* and *E* are coplanar.
3. *C, D,* and *E* are noncollinear.
4. *A, C, D,* and *E* are noncoplanar.

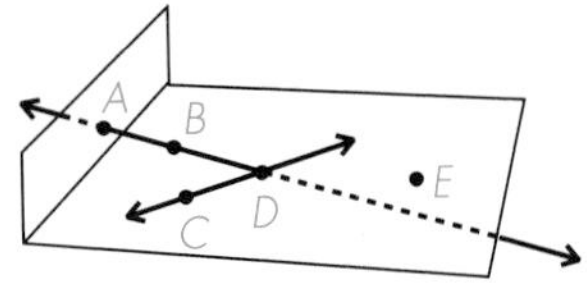

DISCOVERY ACTIVITY

You have learned that two points determine a line. You can experiment to find out how many points determine a plane.

Get a flat surface, such as a piece of cardboard, and some sharpened pencils. Using various numbers of pencils, try to balance the flat cardboard on the pencil points.

1. What was the least number of pencil points needed to balance the cardboard? 3

2. Were the pencil points collinear or noncollinear? noncollinear

NOTEBOOK

POSTULATE 2-1-1: **Three noncollinear points determine a plane.**

CLASS ACTIVITY

Match the terms on the right with the descriptions on the left.

1. a flat surface b	a. noncoplanar points
2. points on the same line c	b. plane
3. points on the same plane e	c. collinear points
4. points not on the same plane a	d. noncollinear points
5. points not on the same line d	e. coplanar points

Tell whether each of the following statements is true or false.

6. Any three points are coplanar. True
7. Any two points are collinear. True
8. Two points can determine a plane. False

Identify the following points as *collinear, coplanar,* or *noncoplanar.*

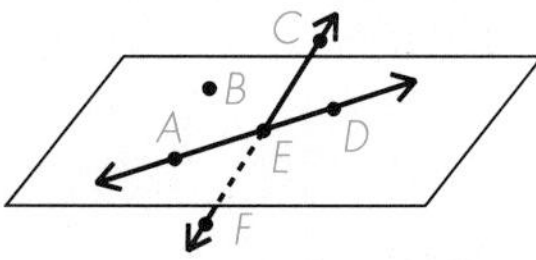

9. *B, E,* and *C* coplanar	10. *B, E,* and *D* coplanar
11. *A, E,* and *D* collinear and coplanar	12. *C, D,* and *F* coplanar
13. *A, B, C,* and *D* noncoplanar	14. *A, B, C,* and *F* noncoplanar

TEACHING COMMENTS

Using the Page
After doing the Discovery Activity, ask students whether the three pencil points that balance the cardboard are coplanar. [yes] Then ask them whether any set of three points determine a plane. [Only three noncollinear points determine a plane.]

Extra Practice
In addition to the True-False questions given in the Class Activity, add: Any two points are coplanar. [True]

INTERSECTING LINES AND PARALLEL LINES

There are lines everywhere you look. Is the road shown at the left representative of a line or a line segment? Explain. Answers may vary.

On a sheet of paper, draw several pairs of lines. Will every pair you can draw intersect?

For two lines in a plane there are only two possibilities: They will intersect or they will not.

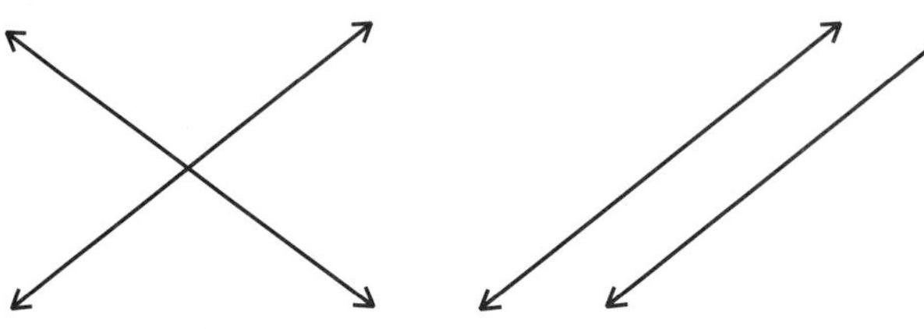

In Figure 1 the two lines intersect at one point. In Figure 2 the two lines will never intersect.

NOTEBOOK **POSTULATE 2-1-2:** **If two lines intersect, then they intersect at exactly one point.**

NOTEBOOK **DEFINITION 2-1-3:** **Two lines are PARALLEL if they are coplanar and do not intersect.**
The symbol ∥ means "is parallel to."

Look around your classroom. List three pairs of intersecting lines and three pairs of parallel lines. Examples will vary.

Are all the lines represented in the photograph in the same plane? No

Using the Page
After students have discussed Postulate 2-1-2, draw a sketch of intersecting planes, as shown below, on the chalkboard.

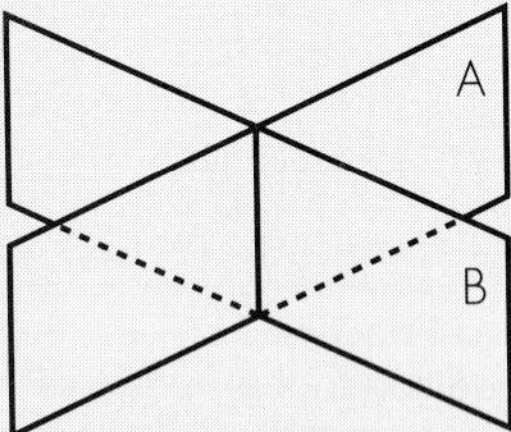

Have students compare the intersection of planes with the intersection of lines. Challenge them to complete the following statement: If two planes intersect, their intersection is always a(an) __________. [line]

DISCOVERY ACTIVITY

Place two pencils on your desk to illustrate a pair of intersecting lines. Then lift the top pencil straight up into the air. Do the two lines intersect? Are they parallel? No, no

SKEW LINES

You have learned that in a plane, two lines will either intersect or be parallel. However, in 3-dimensional space it is possible to have two lines that are not parallel and do not intersect. The figure shows how this is possible. Lines of this type are called **skew lines.**

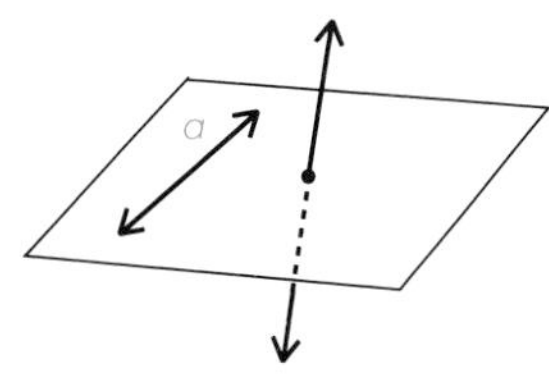

DEFINITION 2-1-4: **SKEW LINES are lines in three dimensions which do not intersect and are not parallel.**

CLASS ACTIVITY

Classify each pair of lines as parallel or intersecting. Assume the lines are in the same plane.

1. a b intersecting

2. 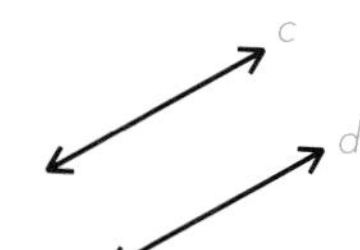

parallel

3. 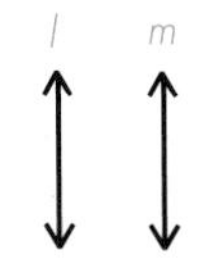

parallel

4. 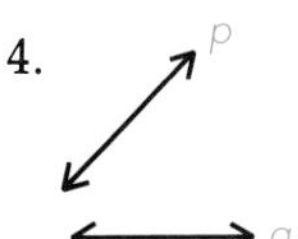

intersecting

Use the figure at the right. Identify a pair of lines of each type.

5. intersecting lines $\overleftrightarrow{BF}$, $\overleftrightarrow{AC}$
6. parallel lines $\overleftrightarrow{AC}$, $\overleftrightarrow{DE}$
7. skew lines $\overleftrightarrow{BF}$, $\overleftrightarrow{DE}$

Write true or false.

8. Intersecting lines are coplanar. True
9. Skew lines are coplanar. False
10. Parallel lines are noncoplanar. False

B A C D E F

Using the Page
Students who have difficulty with the exercises in the Class Activity should be encouraged to work in pairs and be given the opportunity to discuss their answers. Ask students why lines *AC* and *DE* are not skew lines. [They are in the same plane.]

HOME ACTIVITY

Tell whether each set of points is collinear.

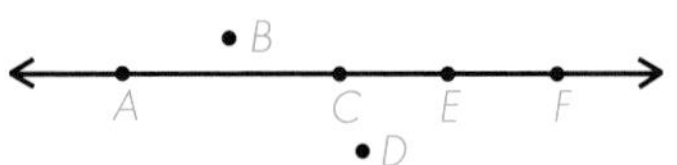

1. *A, C, E* Yes
2. *B, C, F* No
3. *B, C, D* No
4. *B, C* Yes

Use the drawing at the right. State whether each set of points is coplanar.

5. *B, E, G* Yes
6. *C, H, D* Yes
7. *G, C, F* Yes
8. *B, C, E, D* No

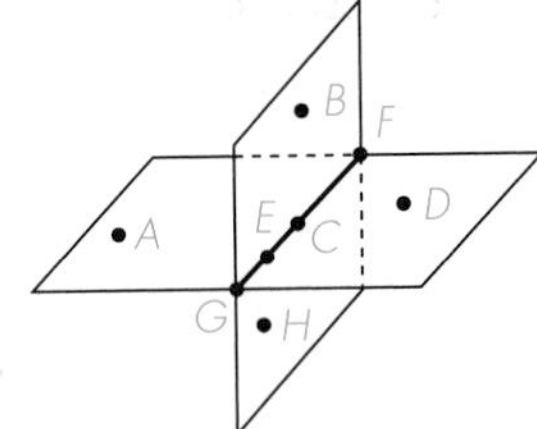

Use the drawing at the right. Identify the following.

9. parallel lines $\overleftrightarrow{AB}$, $\overleftrightarrow{EF}$
10. skew lines $\overleftrightarrow{CD}$, $\overleftrightarrow{AB}$
11. intersecting lines $\overleftrightarrow{CD}$, $\overleftrightarrow{EF}$

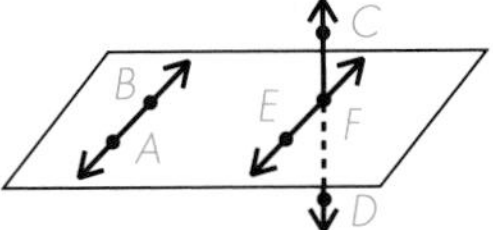

Write Logo procedures to tell the turtle to draw a figure representing a plane. Then tell the turtle to draw each pair of lines described below. Answers may vary.

12. intersecting lines on the plane
13. parallel lines on the plane
14. a line on the plane and a line skew to that line

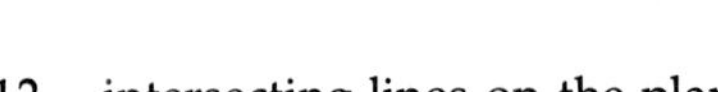

CRITICAL THINKING

15. Draw a figure that shows six points such that no three of the points are collinear. Draw all the lines possible through pairs of these points. How many of these lines are there?

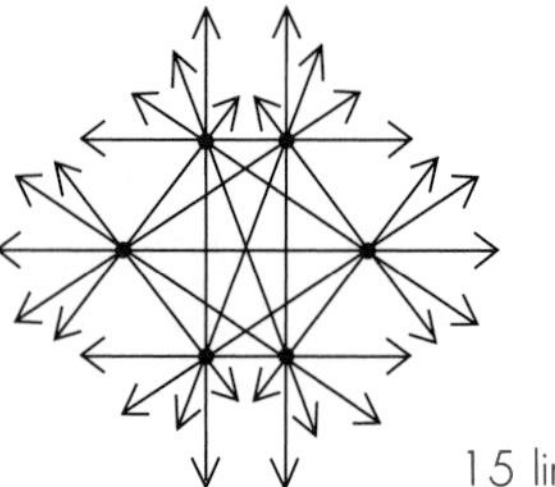

15 lines

FOLLOW-UP

Assessment
Use the drawing. Identify the following:

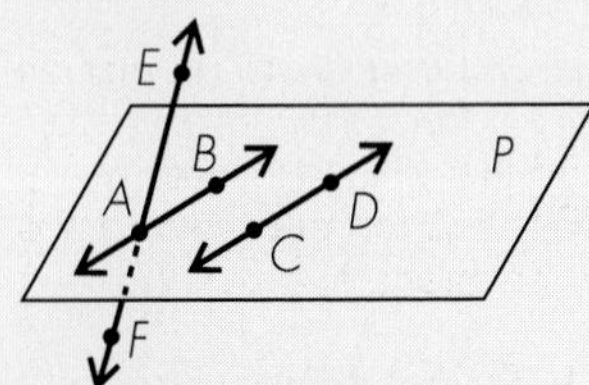

1. a pair of parallel lines [$\overleftrightarrow{AB}$ and $\overleftrightarrow{CD}$]

2. a pair of intersecting lines [$\overleftrightarrow{AB}$ and $\overleftrightarrow{EF}$]

3. two noncoplanar points [Possible answers: *E* and *B*, *C*, or *D*; *F* and *B*, *C*, or *D*.]

Reteaching
Students who have had difficulty with this section may benefit from Reteaching Activity 2.1.

Enrichment
For students who have mastered the material in this section, you may wish to assign Enrichment Activity 2.1.

SECTION 2.2 Corresponding Angles

After completing this section, you should understand
- ▲ parallel lines and transversals.
- ▲ corresponding angles.

You have learned that two lines are parallel if they are coplanar and do not intersect. Points in evenly spaced rows and columns form a **lattice.**

How many lines can be drawn through two or more points of this lattice that are parallel to the given line? 16

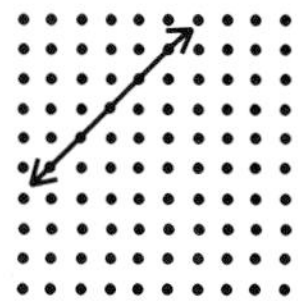

PARALLEL LINES AND TRANSVERSALS

Suppose $\overleftrightarrow{AB}$ and $\overleftrightarrow{CD}$ are parallel. If you draw a third line $\overleftrightarrow{EF}$ in the same plane, either it will be parallel to the other two, or it will intersect them.

Look at the examples below.

$\overleftrightarrow{EF}$, $\overleftrightarrow{AB}$, and $\overleftrightarrow{CD}$ are parallel.

$\overleftrightarrow{EF}$ intersects $\overleftrightarrow{AB}$ and $\overleftrightarrow{CD}$.

In the figure on the right, $\overleftrightarrow{EF}$ is called a **transversal** of $\overleftrightarrow{AB}$ and $\overleftrightarrow{CD}$.

SECTION 2.2

Resources
Reteaching 2.2
Enrichment 2.2

Objectives
After completing this section, the student should understand
- ▲ parallel lines and transversals.
- ▲ corresponding angles.

Materials
protractor
compass
straightedge

Vocabulary
lattice
transversal
corresponding angles

GETTING STARTED

Quiz: Prerequisite Skills

1. Use a protractor to draw a 45° angle and a 110° angle.

2. Two angles are supplementary. One measures 85°. What is the degree measure of the other angle? [95°]

3. Two angles are supplementary. One measures $(x + 50)°$ and the other measures $(x - 30)°$. Find the degree measure of each angle. [130°, 50°]

Chalkboard Example
Ask students to tell which lines are parallel in a chalkboard diagram like the one below. [$\overleftrightarrow{RS}$ and $\overleftrightarrow{TV}$]

TEACHING COMMENTS

Using the Page
A transversal that intersects two lines forms eight angles. Showing that certain pairs of these angles are or are not equal will enable students to prove whether or not the lines transversed are parallel.

NOTEBOOK

DEFINITION 2-2-1: **A TRANSVERSAL is a line which intersects two or more coplanar lines at different points.**

Examples:

1. Line t is a transversal of lines a and b, because t intersects a and b in two different points.

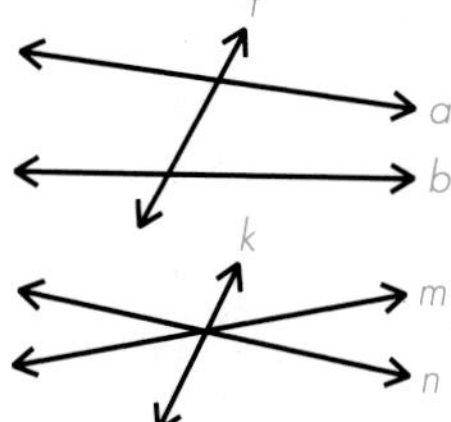

2. Line k is not a transversal of lines m and n, because it intersects them in only one point.

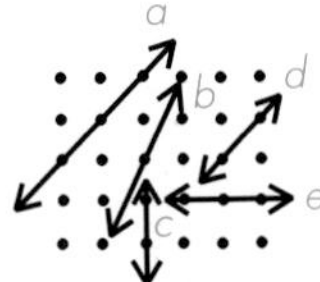

CLASS ACTIVITY

1. Which lines are parallel?
   a and d

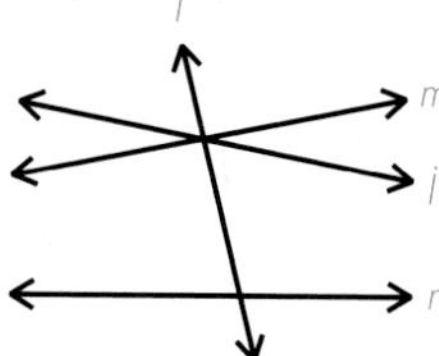

2. Name a pair of lines for which line l is a transversal.
   m and n; or j and n

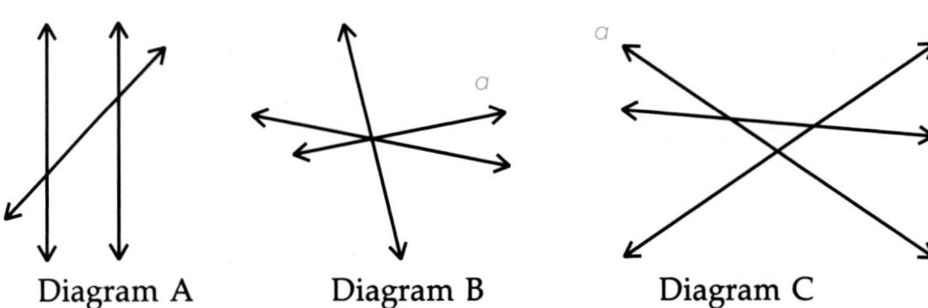

3. In which diagram is line a *not* a transversal of the other two lines? Figure B

4. At the points where transversal a intersects lines b and c, how many angles are formed? 8 angles

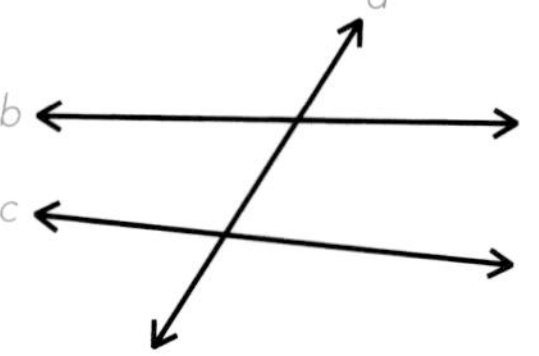

Extra Practice
Transversal t intersects lines a, b, and c. How many angles are formed? [12 angles]

CORRESPONDING ANGLES

Take a sheet of paper and trace the diagram at the right, in which line t is a transversal. Notice that lines a and b are not parallel lines. Number the angles formed 1 through 8, as shown.

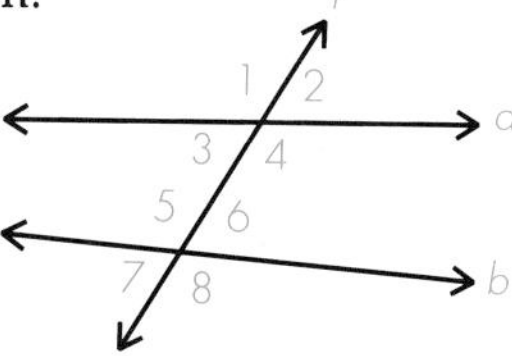

Notice that in your figure ∠1 and ∠5 are both on the left of the transversal and above the two lines cut by the transversal. You can say that ∠1 **corresponds** to ∠5.

Notice that ∠4 and ∠8 are both on the right of the transversal and below the two lines cut by the transversal. So ∠4 **corresponds** to ∠8.

What angle corresponds to ∠3 in the figure? What angle corresponds to ∠2? ∠7; ∠6

DEFINITION 2-2-2: **CORRESPONDING ANGLES are angles which are in the same relative position with respect to the two lines cut by a transversal and the transversal.**

Use a protractor to find the degree measures of each of the eight angles in the diagram. Do any two corresponding angles have the same measure? No

DISCOVERY ACTIVITY

Using both edges of a ruler, draw two parallel lines. Label the lines r and s.
Next, use your ruler to draw transversal t.
Then label all the angles formed by the transversal.
List all pairs of corresponding angles.
Finally, use a protractor to find the degree measure of each angle in the pair.
What do you notice? Compare your results with those of a classmate.

Diagrams and measurements will vary.

If you drew the lines correctly and measured carefully with your protractor, you probably discovered that the degree measures of the corresponding angles were the same.

POSTULATE 2-2-1: **If two parallel lines are intersected by a transversal, then the corresponding angles are equal in measure.**

Using the Page
After students have discussed Definition 2-2-2, have them describe the features of corresponding angles in their own words.

Using the Page
After students have completed the Discovery Activity, ask them what they can surmise about two lines that form equal corresponding angles with a transversal. [The lines are parallel.]

Using the Page
Ask students how they know that angles 1 and 2 are supplementary. [The angle formed by the intersection of the transversal and $\overleftrightarrow{AB}$ is a straight angle.]

Ask students to name the other pairs of supplementary angles in the diagram. [∠2, ∠3; ∠3, ∠4; ∠4, ∠1; ∠6, ∠5; ∠5, ∠7; ∠7, ∠8; and ∠8, ∠6]

You can use what you know about corresponding angles to solve problems.

In the diagram at the right, $\overleftrightarrow{AB} \parallel \overleftrightarrow{CD}$. Line t is a transversal. What are the measures of angles 3 to 8 if $m\angle 1 = (x + 10)°$ and $m\angle 2 = (3x + 50)°$?

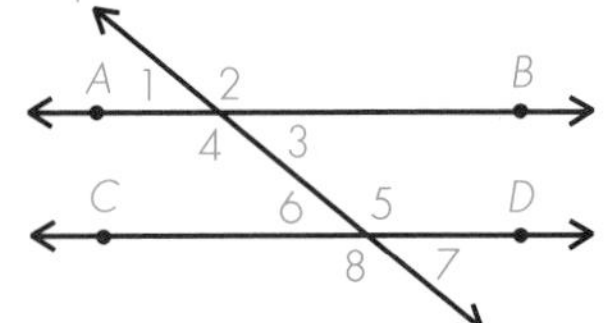

Since ∠1 and ∠2 are supplementary, their measures have a sum of 180°.

First find x.

$$(x + 10) + (3x + 50) = 180$$
$$4x + 60 = 180$$
$$4x = 120$$
$$x = 30$$

Then substitute.

$(x + 10)° = (30 + 10)°$ or 40°, so $m\angle 1 = 40°$.
$(3x + 50)° = (90 + 50)°$ or 140°, so $m\angle 2 = 140°$.

Using Postulate 2-2-1 and the meaning of supplementary angles:

$$m\angle 1 = m\angle 6 = m\angle 3 = m\angle 7 = 40°$$
$$m\angle 2 = m\angle 5 = m\angle 4 = m\angle 8 = 140°$$

CLASS ACTIVITY

In the diagram at the right, lines b and c are parallel. Identify each of the following.

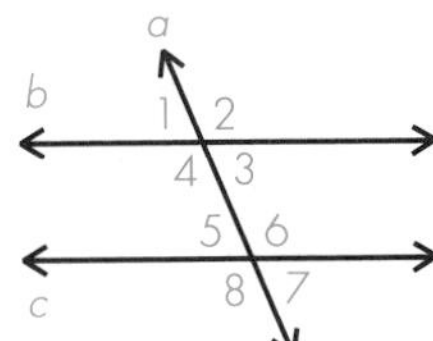

1. a transversal   a
2. two parallel lines   b and c
3. four pairs of corresponding angles
   ∠1, ∠5; ∠4, ∠8; ∠2, ∠6; ∠3, ∠7

Refer to the diagram and complete the following.

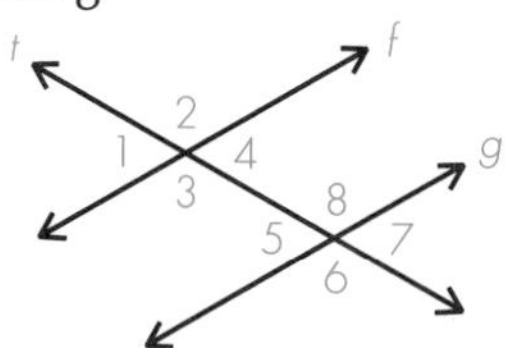

4. ∠1 corresponds to ∠?   5
5. ∠2 corresponds to ∠?   8
6. ∠3 corresponds to ∠?   6
7. ∠4 corresponds to ∠?   7

In the diagram, $p \parallel q$. Complete the following.

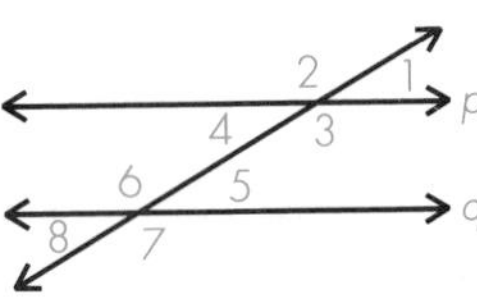

8. If $m\angle 1 = 30°$, then $m\angle 5 = ?$ 30°
9. If $m\angle 2 = 150°$, then $m\angle 6 = ?$ 150°
10. If $m\angle 3 = 150°$, then $m\angle 7 = ?$ 150°
11. If $m\angle 4 = 30°$, then $m\angle 8 = ?$ 30°

12. In the diagram, $j \parallel k$ and $m\angle 1 = 37°$. Find the measures of the other seven angles.

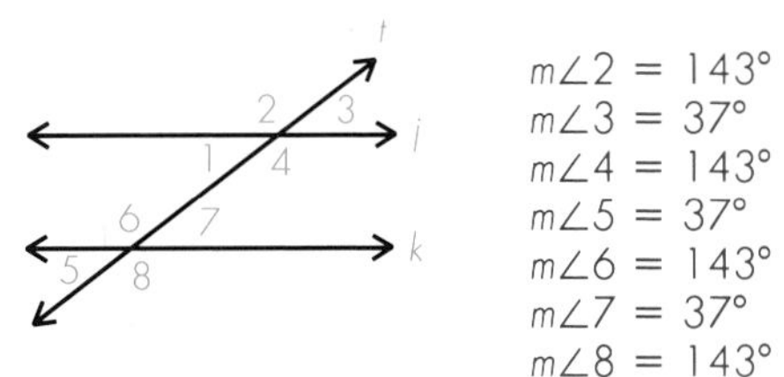

$m\angle 2 = 143°$
$m\angle 3 = 37°$
$m\angle 4 = 143°$
$m\angle 5 = 37°$
$m\angle 6 = 143°$
$m\angle 7 = 37°$
$m\angle 8 = 143°$

13. In the diagram, $r \parallel s$, $m\angle 1 = (2x - 10)°$, and $m\angle 2 = (x + 40)°$. Find the measures of all eight angles.
All angles have a measure of 90°.

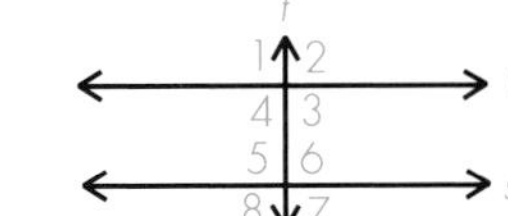

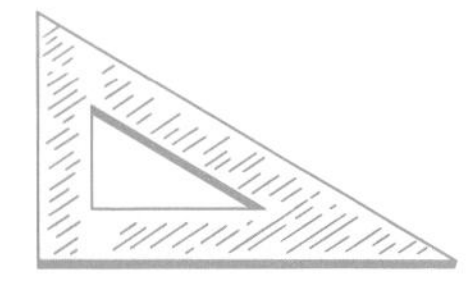

HOME ACTIVITY

In the diagram at the right, $a \parallel b$. Identify each of the following.

1. Two lines for which a is a transversal l and k
2. Two lines for which l is a transversal a and b
3. Two lines for which k is a transversal a and b
4. Eight pairs of corresponding angles
   $\angle 1, \angle 9; \angle 4, \angle 12; \angle 2, \angle 10; \angle 3, \angle 11; \angle 5, \angle 13; \angle 8, \angle 16; \angle 6, \angle 14; \angle 7, \angle 15$

In the diagram, $r \parallel s$ and $m\angle 1 = 67°$. Complete the following.

5. $\angle 1$ corresponds to $\angle$?, which measures ?. $\angle 5$; 67°
6. $\angle 2$ corresponds to $\angle$?, which measures ?. $\angle 6$; 113°
7. $\angle 3$ corresponds to $\angle$?, which measures ?. $\angle 7$; 67°
8. $\angle 4$ corresponds to $\angle$?, which measures ?. $\angle 8$; 113°

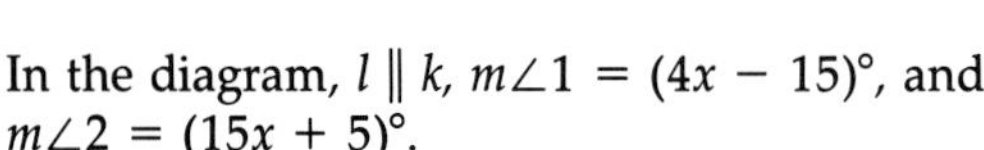

In the diagram, $l \parallel k$, $m\angle 1 = (4x - 15)°$, and $m\angle 2 = (15x + 5)°$.

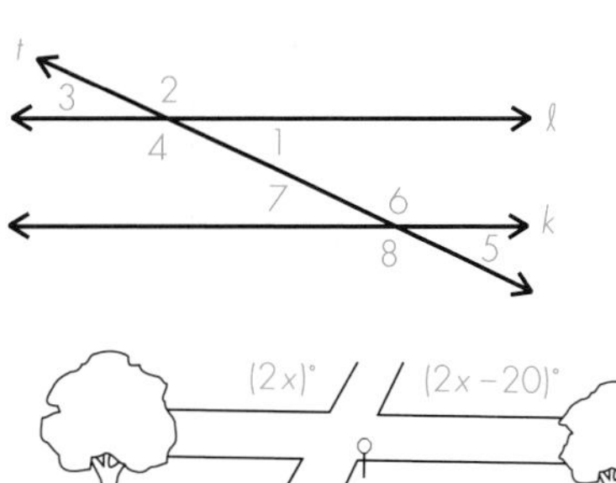

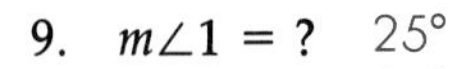

9. $m\angle 1 = ?$ 25°
10. $m\angle 2 = ?$ 155°
11. $m\angle 3 = ?$ 25°
12. $m\angle 4 = ?$ 155°
13. $m\angle 5 = ?$ 25°
14. $m\angle 6 = ?$ 155°
15. $m\angle 7 = ?$ 25°
16. $m\angle 8 = ?$ 155°

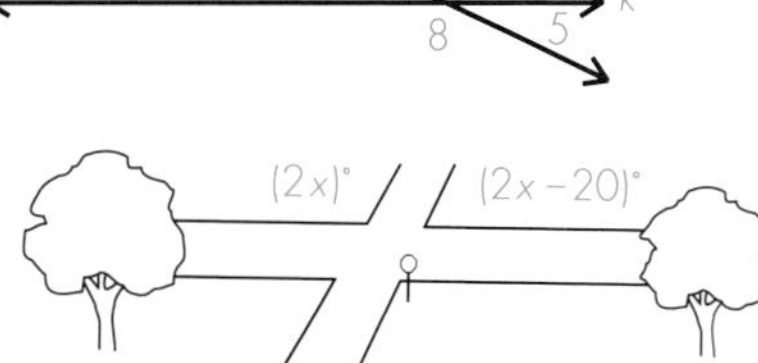

17. The diagram represents a street intersection. Calculate the measures of the indicated angles.
    100°, 80°

Extra Practice
Have students name the sets of corresponding angles in the diagram.

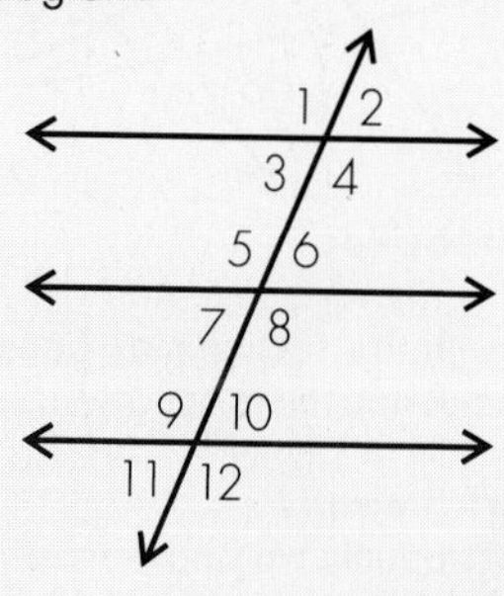

[$\angle 1, \angle 5, \angle 9; \angle 2, \angle 6, \angle 10; \angle 3, \angle 7, \angle 11; \angle 4, \angle 8, \angle 12$]

Extra Practice
After students have completed the Critical Thinking activity, have them use their protractors to draw a 70° angle. Then, using a compass and a straightedge, they should copy the angle by following the procedure shown.

FOLLOW-UP

Assessment

1. In the diagram, $h \parallel j$. What is the measure of $\angle 6$? [60°]

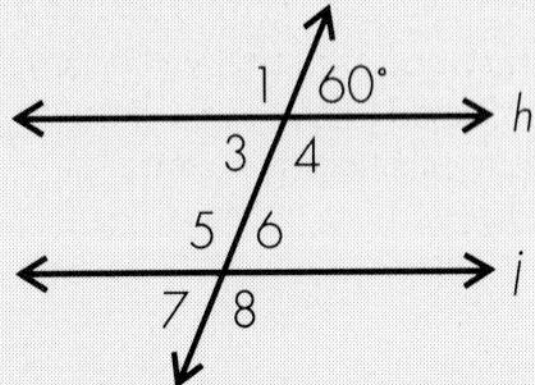

2. In the diagram, $k \parallel l$. Find the measure of $\angle 1$. [$m\angle 1 = 115°$]

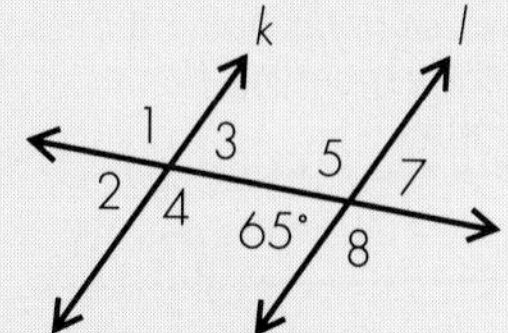

Reteaching
Students who have had difficulty with this section may benefit from Reteaching Activity 2.2.

Enrichment
For students who have mastered the material in this section, you may wish to assign Enrichment Activity 2.2.

18. Write a Logo procedure to tell the turtle to draw two left-to-right parallel lines and a transversal so that the two supplementary angles formed by the top line and the transversal have measures of $(3x + 10)°$ and $(2x + 20)°$. Show how the turtle will move when you type in your procedure. Then label the angles and find the degree measure of all eight angles formed.
Procedures may vary. $(3x + 10)° = 100°$, $(2x + 20)° = 80°$

CRITICAL THINKING

You can use a compass and a straightedge to copy an angle. Here's how.

Given angle **Construction**

1. Draw a ray with endpoint P.
2. Construct an arc with its center at A. Then construct an arc with its center at P. Use the same compass opening.
3. Place your compass point at X. Adjust the opening so that you can draw an arc passing through Y. Then, using the same compass opening, construct an arc with its center at R.
4. Draw $\overrightarrow{PQ}$ with your straightedge. $\angle YAX = \angle QPR$.

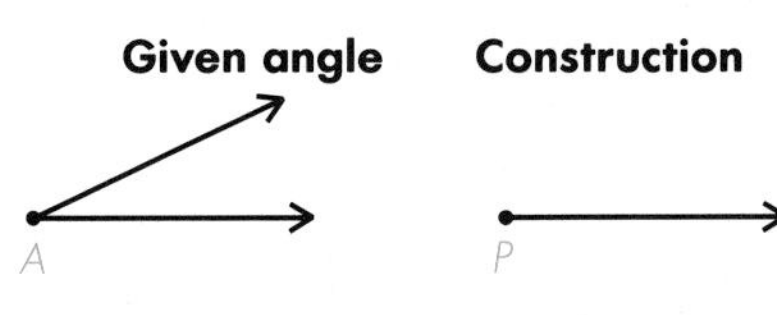

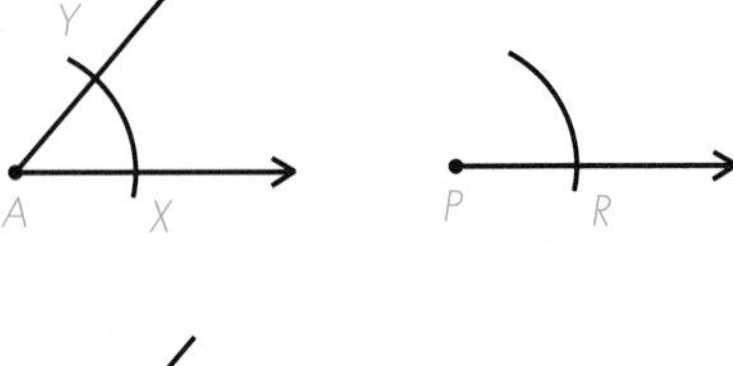

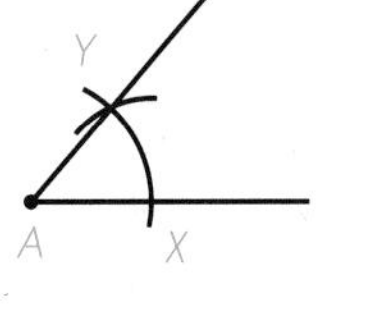

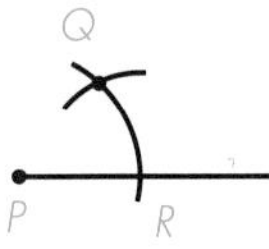

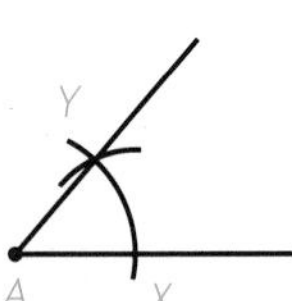

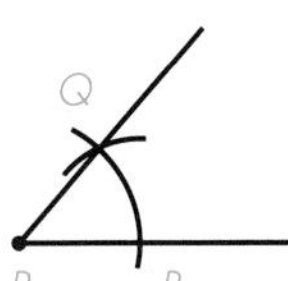

Suppose you have a line r and a point P not on r. Study this procedure for drawing a line s through P and parallel to r.

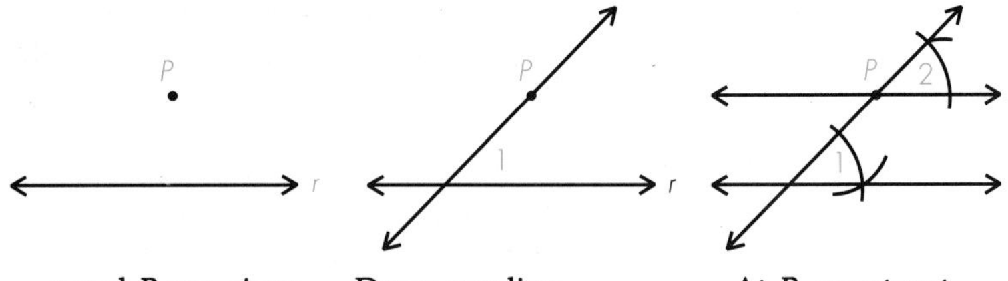

r and P are given. Draw any line through P that intersects r. Label $\angle 1$. At P, construct $\angle 2$ equal to $\angle 1$.

19. Explain why it is reasonable to think that $s \parallel r$.
$\angle 1$ and $\angle 2$ have equal measures and are corresponding angles.

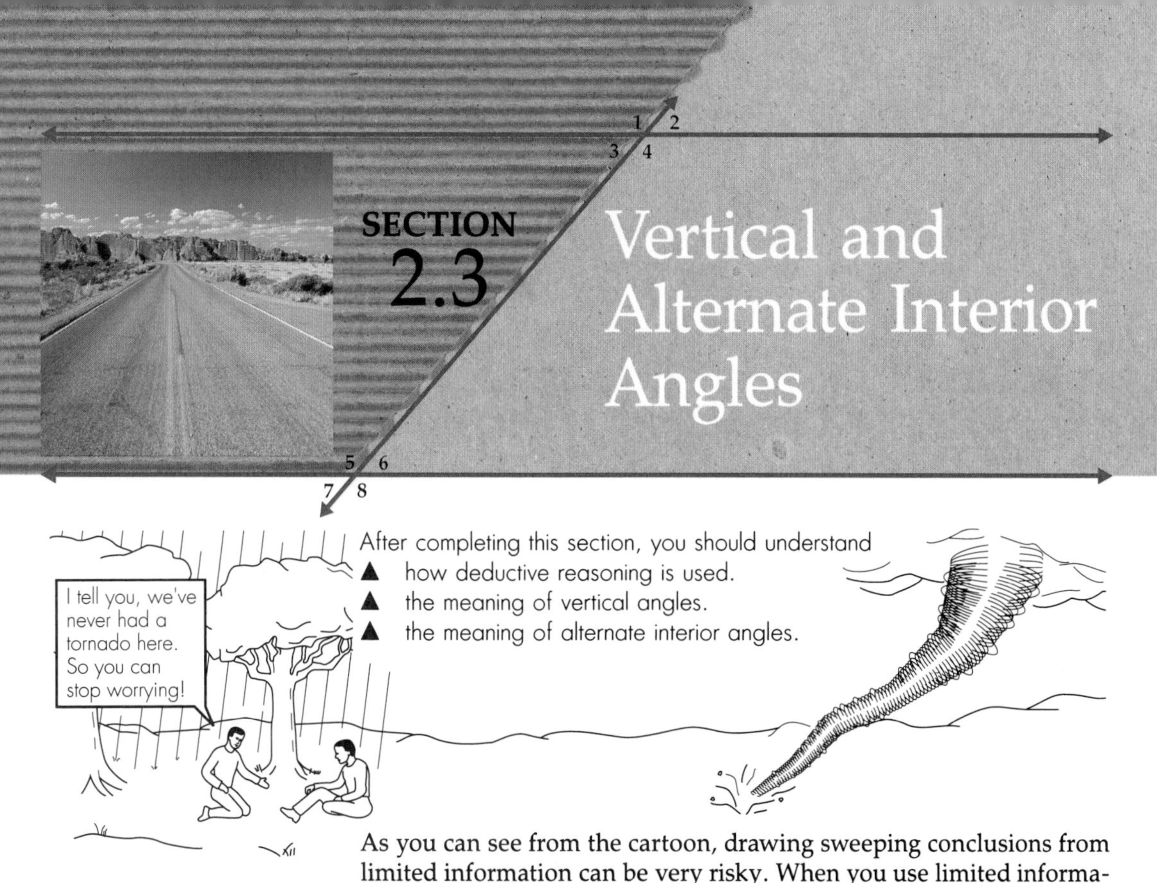

SECTION
2.3
Vertical and Alternate Interior Angles
I tell you, we've never had a tornado here. So you can stop worrying!

SECTION 2.3 Vertical and Alternate Interior Angles

After completing this section, you should understand
- ▲ how deductive reasoning is used.
- ▲ the meaning of vertical angles.
- ▲ the meaning of alternate interior angles.

As you can see from the cartoon, drawing sweeping conclusions from limited information can be very risky. When you use limited information to arrive at a general conclusion, you are using **inductive** thinking. Everyone uses inductive thinking, but you have to be careful with it. In geometry you can use inductive thinking to arrive at good hunches about what is always true, but to **prove** that your hunches are correct, you need **deductive reasoning.**

DEDUCTIVE REASONING

To use deductive reasoning, you start with some given information. Then you present a series of statements, backed up by reasons, that lead logically, step-by-step to a final conclusion. The idea is to show that anyone who accepts the given information will have to accept the final conclusion.

Example: In the figure $\overrightarrow{XB}$ bisects $\angle AXC$ and $\overrightarrow{XC}$ bisects $\angle BXD$. Suppose you want to show that $m\angle 1 = m\angle 3$. $m\angle 1 = m\angle 2$, because of what *bisects* means. $m\angle 2 = m\angle 3$, for the same reason. So $m\angle 1 = m\angle 3$, by properties of algebra.

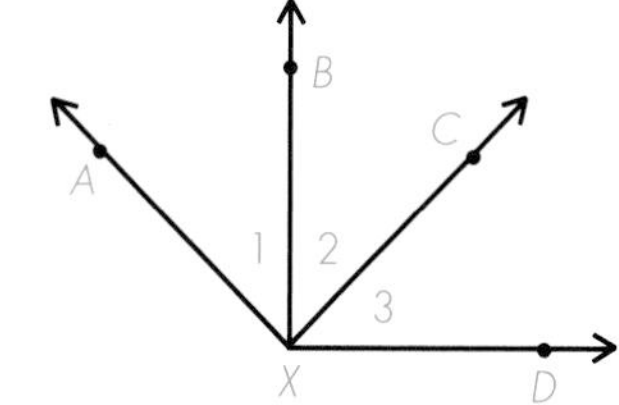

SECTION 2.3

Resources
Reteaching 2.3
Enrichment 2.3

Objectives
After completing this section, the student should understand
- ▲ how deductive reasoning is used.
- ▲ vertical angles.
- ▲ alternate interior angles.

Materials
protractor

Vocabulary
inductive thinking
deductive reasoning
conditional statement
hypothesis
conclusion
vertical angles
theorem
alternate interior angles

GETTING STARTED

Warm-Up Activity
Have students give additional examples of inductive thinking, particularly some in which incorrect conclusions may be drawn.

TEACHING COMMENTS

Using the Page
After you have discussed conditional statements with students, provide a few statements in the "If . . . , then" form for which students must provide the conclusion. For example: If it is raining, then I will wear my raincoat. It is raining.
Conclusion: [I will wear my raincoat.]
If you are on the team, then you have at least a "C" average. Carlos is on the team.
Conclusion: [Carlos has at least a "C" average.]

Extra Practice
Have students express the following statement as a conditional statement: All residents of Tucson live in Arizona. [If a person is a resident of Tucson, then that person lives in Arizona.]

Additional Answers
7. Points L, M, and N are noncollinear. Points L, M and R determine a plane.

In geometry you will often encounter statements of the form "If . . . , then . . .", where certain statements are in place of the dots. Statements of this kind are called **conditional** statements. Within a conditional statement, the statement after "if" is called the **hypothesis.** The statement after "then" is called the **conclusion.**

conditional statement

If $m\angle A$ is between 0° and 90°, then $\angle A$ is acute.

hypothesis (If $m\angle A$ is between 0° and 90°) — conclusion ($\angle A$ is acute.)

Suppose you know that a conditional statement and its hypothesis are true. You can conclude the conclusion of the conditional statement must also be true.

True → If (statement A), then (statement B).
True → Statement A
You conclude → Statement B is true.

Example: Consider the figure shown here. Suppose a is parallel to b. This statement is true because it is a given. If two parallel lines are intersected by a transversal, then the corresponding angles are equal in measure.

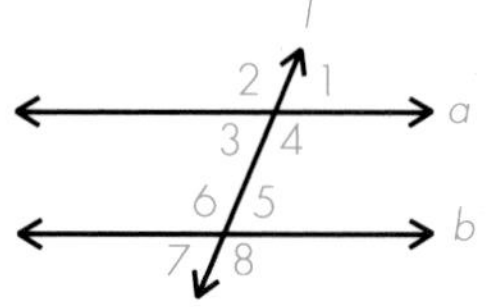

You know that $a \parallel b$ and l is a transversal. You can conclude that the corresponding angles are equal in measure.

CLASS ACTIVITY

Copy each statement. Underline the hypothesis and ring the conclusion.

1. If you live in Utah, then you live in the United States.
2. If you are in London, then you are in England.
3. If your pet is a terrier, then it is a dog.
4. If $x + 5 = 7$, then $x = 2$.

Suppose each pair of statements is true. What can you conclude?

5. If an angle is obtuse, then its measure is greater than 90°. $\angle PQR$ is obtuse. $m\angle PQR > 90°$
6. If two angles are supplementary, then the sum of their measures is 180°. $\angle X$ and $\angle Y$ are supplementary. $m\angle X + m\angle Y = 180°$
7. If three points are noncollinear, then they determine a plane. See margin.
8. If four points are collinear, then any three of them are collinear. Points A, B, C, and D are collinear. Any three of A, B, C, and D are collinear.

VERTICAL ANGLES

Recall that two lines in a plane will either be parallel or intersect. When two lines intersect, they form four angles.

In the diagram at the right, the angles formed by the intersection of $\overleftrightarrow{PY}$ and $\overleftrightarrow{XZ}$ are labeled 1, 2, 3, and 4.

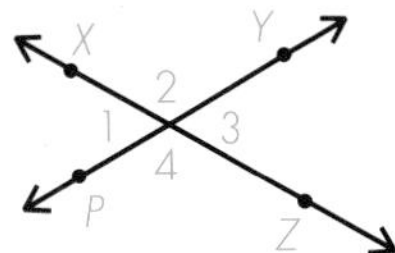

Notice that angles 1 and 3 are opposite or across from each other in the diagram. They are opposite or vertical angles. Angles 2 and 4 are another pair of vertical angles.

DEFINITION 2-3-1: VERTICAL ANGLES are the angles across from each other when two lines intersect.

From the diagram, you can see that $\angle 1$ and $\angle 3$ look about equal in size. So do $\angle 2$ and $\angle 4$. By using deductive reasoning, you can show that they really are equal. First, notice that $\angle 1$ and $\angle 2$ are supplementary by the definition of supplementary angles (Definition 1-4-1)

$$m\angle 1 + m\angle 2 = 180.$$

But $\angle 3$ and $\angle 2$ are also supplementary, so that

$$m\angle 3 + m\angle 2 = 180.$$

Using the algebraic properties of equality, you can see that

$$m\angle 1 + m\angle 2 = m\angle 3 + m\angle 2$$

Now subtract $m\angle 2$ from both sides.

$$\begin{array}{rcl} m\angle 1 + m\angle 2 & = & m\angle 3 + m\angle 2 \\ - m\angle 2 & = & - m\angle 2 \\ \hline m\angle 1 & = & m\angle 3 \end{array}$$

You have come to the conclusion that vertical angles 1 and 3 are equal. You can use the very same kind of reasoning to show that angles 2 and 4 are equal.

The same reasoning would work for *any* intersecting lines and *any* vertical angles. When deductive reasoning has been used to show that a statement is true in all cases, the statement is called a **theorem.**

THEOREM 2-3-1: If two lines intersect, then the vertical angles are equal.

This theorem is called the *vertical angle theorem.*

Using the Page
After the students have used deductive reasoning to conclude that vertical angles are equal, have them measure those angles with their protractors as a real-world check of the conclusion.

Using the Page
Point out to students that theorems are statements that have been proved using previously accepted defined and undefined terms, assumptions, and other theorems. For students having difficulty with this concept, guide them to see that they relied upon definitions of straight and supplementary angles to conclude that $m\angle 1 = m\angle 3$.

Once you have proved a theorem, you can use it in any situation where it applies.

Example: $\overleftrightarrow{AB}$ and $\overleftrightarrow{CD}$ intersect to form $\angle APC$, having a measure of 75°. What is the measure of $\angle DPB$?

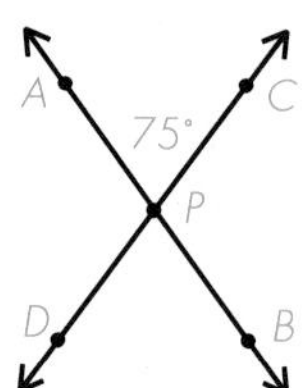

The vertical angle theorem tells you that $\angle APC$ and $\angle DPB$ have the same measure. Since $m\angle APC = 75°$, you can conclude that $m\angle DPB = 75°$.

Using the Page
Have students look about the classroom for examples of vertical angles.

DISCOVERY ACTIVITY

Draw two parallel lines *m* and *n* intersected by transversal *t*. Label the angles formed using the numbers 1 through 8, as shown.

1. List all pairs of vertical angles.

   $\angle 1$ and $\angle 3$ $\qquad$ $\angle 2$ and $\angle$? 4
   $\angle 5$ and $\angle$? 8 $\qquad$ $\angle 6$ and $\angle$? 7

2. Recall that corresponding angles are equal. List all of these pairs of equal angles.

   $\angle 1$ and $\angle 5$ $\qquad$ $\angle 2$ and $\angle$? 6
   $\angle 3$ and $\angle$? 8 $\qquad$ $\angle 4$ and $\angle$? 7

3. Use your results from Exercises 1 and 2. What can you conclude about $\angle 3$ and $\angle 5$? about $\angle 4$ and $\angle 6$? Both pairs of angles are equal.

Using the Page
After students have completed the Discovery Activity, ask them to give another explanation for why m∠4 = m∠7, other than because they are corresponding angles. [m∠2 = m∠6 because they are corresponding angles; ∠4 and ∠2 are vertical angles, as are ∠7 and ∠6. By deductive reasoning, equals can be substituted for equals. Therefore, m∠4 = m∠7.]

CLASS ACTIVITY

Use a computer and a geometric drawing tool to complete Exercises 1-5. Answers for Exercises 1 through 4 may vary.

1. Draw and label two intersecting lines.

2. Measure each angle formed.
3. Which angles have equal measures?
4. Repeat Exercises 1 through 3 several times.
5. Do your findings from Exercises 1 through 4 support Theorem 2-3-1? yes

Notice that ∠3 and ∠5 are on opposite or **alternate** sides of the transversal. They are also between the parallel lines. Such pairs of angles are called **alternate interior angles.** Which other pair of angles in the your diagram are alternate interior angles?

NOTEBOOK **DEFINITION 2-3-3:** **ALTERNATE INTERIOR ANGLES are angles on opposite sides of the transversal and between the parallel lines.**

NOTEBOOK **THEOREM 2-3-2:** **If two parallel lines are intersected by a transversal, then the alternate interior angles are equal.**

You can use this theorem to help you solve problems.

Example: In the figure, t is a transversal of parallel lines a and b. $m\angle 1 = 110°$ and $m\angle 2 = 70°$. Find $m\angle 3$ and $m\angle 4$.

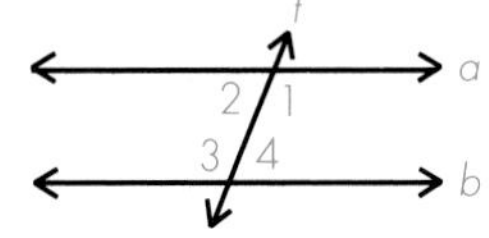

Use Theorem 2-3-2. You can conclude that $m\angle 1 = m\angle 3 = 110°$ and $m\angle 2 = m\angle 4 = 70°$.

CLASS ACTIVITY

In the diagram, $a \parallel b$ and t is a transversal.

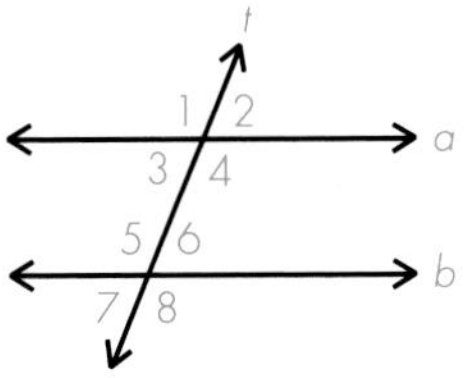

1. Identify a transversal
2. Identify a pair of parallel lines. a, b
3. Identify four pairs of vertical angles. See margin.
4. Identify the corresponding angles. See margin.
5. Identify two pairs of alternate interior angles
∠3 and ∠6, ∠4 and ∠5

Each diagram shows two parallel lines and a transversal. Find the measure of each numbered angle.

6.

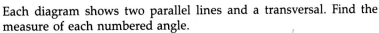

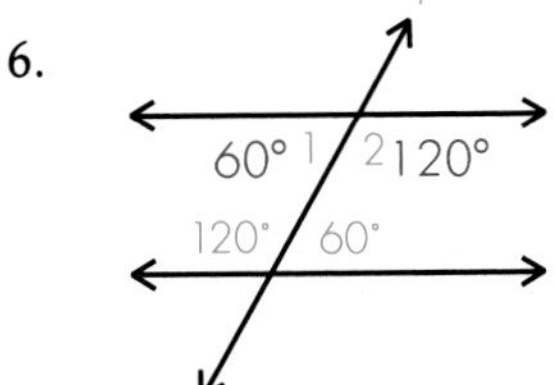

7.

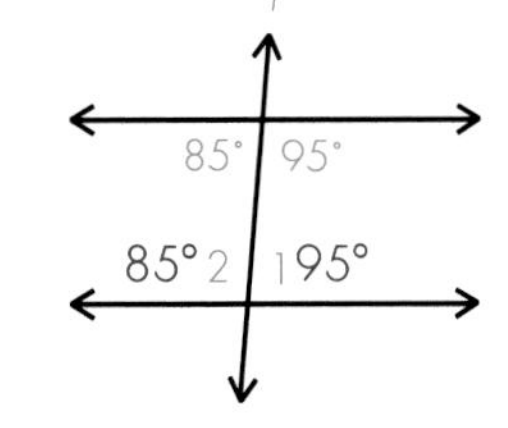

Using the Page
You may wish to point out to students that they can easily spot the alternate interior angles in a diagram by looking for the "Z-shaped" figure, or its mirror image, as shown below.

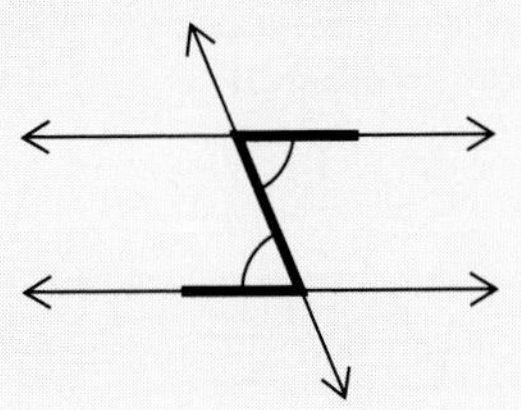

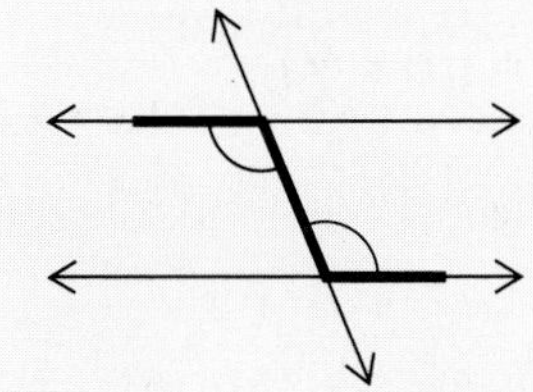

Additional Answers
3. ∠1 and ∠4, ∠2 and ∠3, ∠5 and ∠8, ∠6 and ∠7
4. ∠1 and ∠5, ∠2 and ∠6, ∠3 and ∠7, ∠4 and ∠8

Extra Practice
Write the converse of the conditional statement that is not true.

1. If an angle is a right angle, then it has 90°.

2. If a number is divisible by 3, then it is divisible by 6. [If a number is divisible by 6, then it is divisible by 3.]

FOLLOW-UP

Assessment

1. Identify the hypothesis and the conclusion for the statement: If n is a prime number greater than 2, then $n + 1$ is a composite number. [hypothesis; conclusion]

2. In the diagram, $m \parallel n$, t is a transversal, and $m\angle 2 = 48°$. Find the measures of the remaining angles. [Numbers in brackets are annos.]

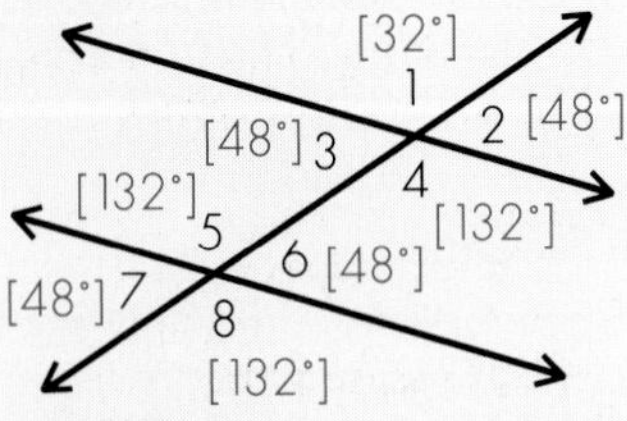

Reteaching
Students who have had difficulty with this section may benefit from Reteaching Activity 2.3.

Enrichment
For students who have mastered the material in this section, you may wish to assign Enrichment Activity 2.3.

Use a computer and a geometric drawing tool to complete Exercises 8–12. Answers for Exercises 8–11 may vary.

8. Draw and label two parallel lines and a transversal.
9. Measure each interior angle formed.
10. Which angles have equal measures?
11. Repeat Exercises 8–10 several times.
12. Do your findings from Exercises 8–11 support Theorem 2-3-2? yes

HOME ACTIVITY

Identify the hypothesis and conclusion for each statement.

1. If two angles are vertical angles, then they are equal.
2. If $n + 1$ is odd, then n is even.

Suppose each pair of statements is true. What can you conclude?

3. If a person is a taxi driver, then that person knows the city. Pete is a taxi driver. Pete knows the city.
4. If an angle is straight angle, its sides form a straight line. $\angle ABC$ is a straight angle. The sides of $\angle ABC$ form a straight line.
5. If a number is greater than 2, then its square is greater than 4. The number 2.001 is greater than 2. The square of 2.001 is greater than 4.

In each figure t is a transversal of two parallel lines. Find the measure of each numbered angle.

6. 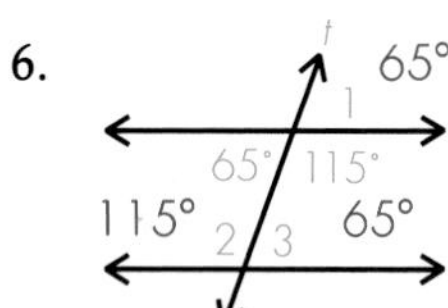

7. 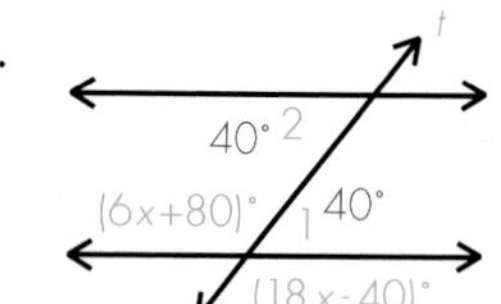

8. 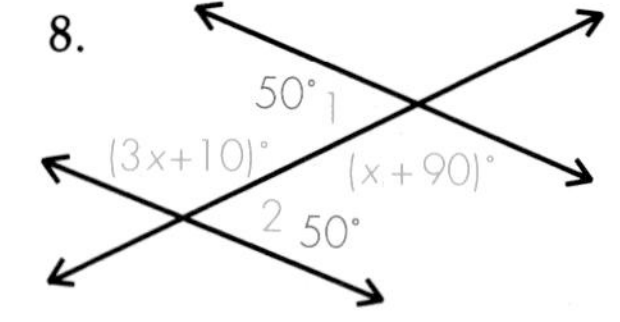

9. In the diagram, $\overleftrightarrow{AB} \parallel \overleftrightarrow{CD}$, s and t are transversals, $m\angle 1 = 37°$, $m\angle 2 = 87°$, and $m\angle 12 = 124°$. Find the measures of the other numbered angles.

$m\angle 3 = 93°$	$m\angle 6 = 93°$	$m\angle 9 = 143°$
$m\angle 4 = 56°$	$m\angle 7 = 37°$	$m\angle 10 = 56°$
$m\angle 5 = 87°$	$m\angle 8 = 143°$	$m\angle 11 = 124°$

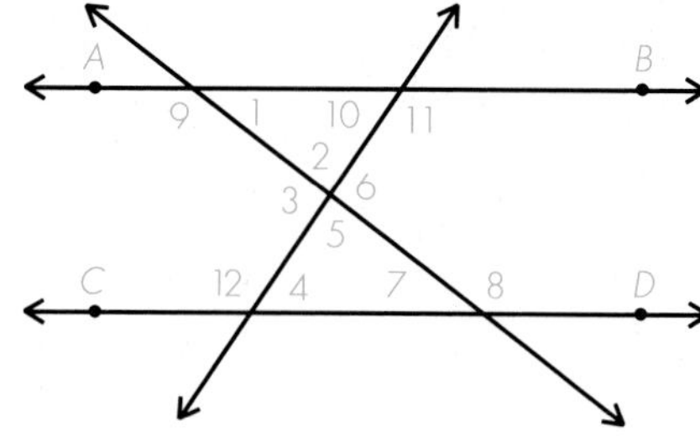

CRITICAL THINKING

10. It is true that if two angles are vertical angles, then their measures are equal. If $\angle A$ and $\angle B$ are supplementary angles and vertical angles, what are their measures? $m\angle A = m\angle B = 90°$

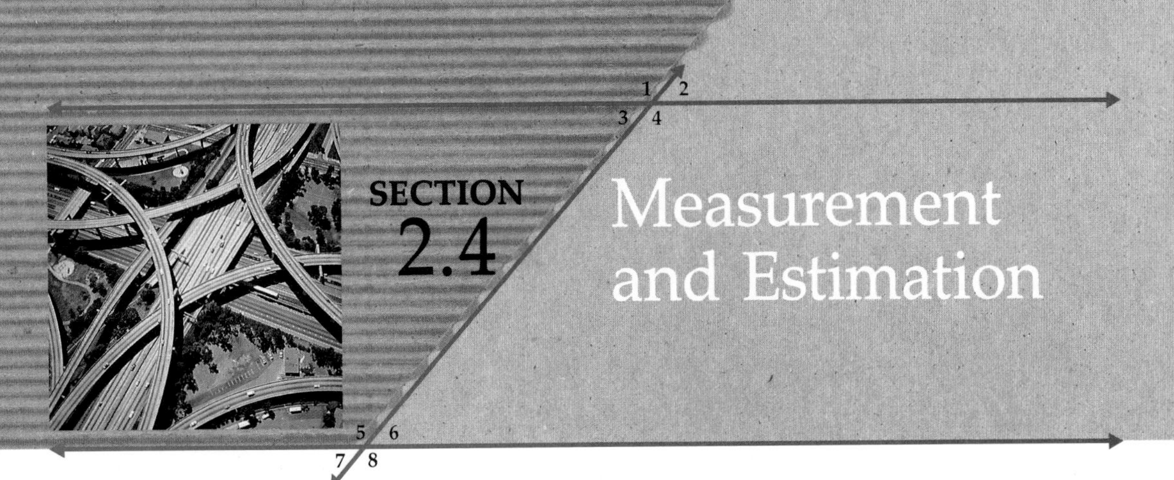

SECTION 2.4 Measurement and Estimation

After completing this section, you should understand

- ▲ the concept of error in measurement
- ▲ how to convert from one unit of measurement to another
- ▲ how to add and subtract measurements

Measurement is important in everyday life, and in fields such as mathematics, science, and economics. So far in this book, you have encountered two kinds of measures—length and angle measure.

ERROR IN MEASUREMENT

In everyday life you routinely put up with many inaccuracies and small errors. If you go to a cafeteria with a friend, you probably wouldn't ask the person serving rice to count the number of grains to be sure you and your friend get the same number. On the other hand, if you were cooking stew and used two teaspoons of salt instead of one, your might ruin the stew. If you are cutting wood for a book shelf, a hundredth of an inch more or less probably won't make much difference. But an automobile mechanic who ground a hundredth of an inch extra from the wall of a motor cylinder would probably be told to look for another job.

Absolutely exact measurements are never possible. The instruments in a high school science lab may permit more accurate measures of length than a tape measure in your house, but they wouldn't be good enough for a NASA engineer.

SECTION 2.4

Resources
Reteaching 2.4
Enrichment 2.4

Objectives
After completing this section, the student should understand
- ▲ the concept of error in measurement.
- ▲ how to convert from one unit of measurement to another.
- ▲ how to add and subtract measurements.

Materials
inch and centimeter ruler
protractor
calculator

Vocabulary
measurement error
subunit
minute
second
convert
milli-
centi-
kilo-

GETTING STARTED

Quiz: Prerequisite Skills
Choose the sensible measure.

1. The height of your classroom might be about ________.
[4 m]
4 cm 4m 4 km

2. The distance from Phoenix to Tucson might be about ________.
[175 km]
175 m 175 dam 175 km

Warm-Up Activity
Have students give examples of when exact measurements are needed and other examples of when estimated measurements make the most sense.
[Answers will vary]

Using the Page
Have students examine their rulers before beginning the Discovery Activity. Ask them to identify the relationship between the successive ruler markings given in steps 1-5. [Each marking designates a unit half as great as the preceding unit.] Then ask students to explain the relationship between the successive markings and the precision in measurement they afford. [Each marking allows for a measurement twice as precise as the previous one.]

The way measures are reported tells you how accurate the person making the measurement was trying to be.

DISCOVERY ACTIVITY

Take your ruler and measure the segment below.

A ———————————— B

1. First, measure $\overline{AB}$ to the nearest inch. 3 in.
2. Next, measure $\overline{AB}$ to the nearest $\frac{1}{2}$ inch. $2\frac{1}{2}$ in.
3. Next, measure $\overline{AB}$ to the nearest $\frac{1}{4}$ inch. $2\frac{3}{4}$ in.
4. Next, measure $\overline{AB}$ to the nearest $\frac{1}{8}$ inch. $2\frac{6}{8}$ in.
5. Finally, if your ruler has markings of $\frac{1}{16}$ inch, measure AB to the nearest $\frac{1}{16}$ inch.
6. When you measured to the nearest $\frac{1}{8}$ inch, what is the greatest error you could have made? $\frac{1}{16}$ in.
7. If you measured to the nearest $\frac{1}{16}$ inch, what is the greatest amount of error likely? $\frac{1}{32}$ in.

When you report a length to the nearest unit or subunit, the *error* in measurement is half that unit or *subunit.* It tells you how much uncertainty there is in the measurement.

Example: To the nearest $\frac{1}{4}$ inch, $\overline{CD}$ measures $1\frac{3}{4}$ inch. The measurement is to the nearest $\frac{1}{4}$ inch, so the error is $\frac{1}{8}$ inch.

C ———— D

Similar ideas apply when you measure angles. Your protractor is marked off in degrees. When you use it to measure angles, the error is $\frac{1}{2}°$.

For more accurate measure of angles, degrees are divided into *minutes* and *seconds.*

1 degree = 60 minutes. The symbol for minutes is $'$.
1 minute = 60 seconds. The symbol for seconds is $''$.

Example: For a measurement of an angle to the nearest degree, the error is $\frac{1}{2}°$ or $30'$. For a measurement to the nearest minute, the error is $\frac{1}{2}'$ or $30''$.

CLASS ACTIVITY

Measure each of these segments to the nearest inch and to the smallest subunit on your ruler. State the error for each measurement.
See margin.

1. A ——————————————— B

2. C ——————————————— D

Measure each of these segments to the nearest centimeter and to the nearest millimeter. State the error for each measurement.

3. E ——————————— F

4. G ——————————— H

Measure each angle to the nearest 5 degrees and to the nearest degree. State the error for each measurement. See margin.

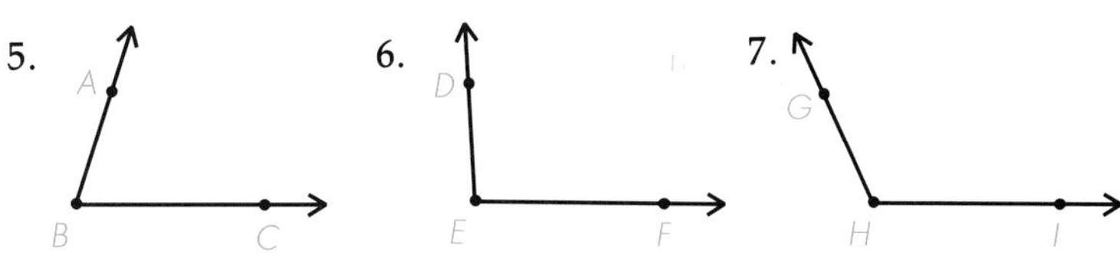

CONVERTING FROM ONE UNIT OF MEASUREMENT TO ANOTHER

You can convert (change) from one unit of length to another in the customary or metric system. To do this you need to know how the units are related to one another.

The following equations show the relationships between some of the most common units of length in the customary and metric systems.

Customary Measures	Metric Measures
1 foot (ft) = 12 inches (in.)	1 centimeter (cm) = 10 millimeters
1 yard (yd) = 3 feet	1 meter (m) = 100 centimeters
1 mile (mi) = 5280 feet	1 kilometer (km) = 1000 meters
1 mile = 1760 yards	

Customary and Metric
1 in. = 2.54 cm
1 yd = 0.91 m
1 mi = 1.6093 km

The prefixes *milli, centi,* and *kilo* mean one-thousandth, one-hundredth, and one thousand, respectively.

Extra Practice

Have each student use a ruler to make a segment that is between 3 and 5 inches long. Then students exchange segments. Each student then measures the segment he or she receives to the nearest inch, nearest $\frac{1}{4}$ inch and nearest $\frac{1}{8}$ inch.

Additional Answers

1. nearest inch: 4 in.; $\frac{1}{2}$ in.; nearest $\frac{1}{8}$ inch: $4\frac{1}{8}$ in., error: $\frac{1}{16}$ in.; nearest $\frac{1}{16}$ inch: $4\frac{1}{16}$ in. or $4\frac{2}{16}$ in.; error: $\frac{1}{32}$ in.

2. nearest inch: 4 in.; error: $\frac{1}{2}$ in.; nearest $\frac{1}{8}$ inch: $3\frac{7}{8}$ in.; error: $\frac{1}{16}$ in.; nearest $\frac{1}{16}$ inch: $3\frac{14}{16}$.; error: $\frac{1}{32}$ in.

3. nearest cm: 5 cm; error: $\frac{1}{2}$ cm or 5 mm; nearest mm: 52 mm; error: $\frac{1}{2}$ mm

4. nearest cm: 7 cm; error $\frac{1}{2}$ cm or 5 mm; nearest mm: 68 mm; error: $\frac{1}{2}$ mm

5. nearest 5 degrees: 70°; error: $2\frac{1}{2}$°; nearest degree: 72°; error: $\frac{1}{2}$°

6. nearest 5 degrees: 95°; error: $2\frac{1}{2}$°; nearest degree: 94°; error: $\frac{1}{2}$°

7. nearest 5 degrees: 115°; error: $2\frac{1}{2}$°; nearest degree: 116°; error: $\frac{1}{2}$°

REPORT 2-4-1: Prepare a report on the history of the metric system that includes the following information.

1. When and how the metric system was first established.
2. What kinds of quantities other than length the metric system permits people to measure.
3. How the new International System (SI) units have refined the way basic metric units were originally defined.

Source:
The New Encyclopaedia Britannica, 15th edition, Encyclopaedia Britannica, Inc., Chicago: "The Metric System of Weights and Measures" from the Macropaedia article, "Measurement and Observation".

Using the Page
When examining the examples on this page and when doing the activities on the next page, some students may benefit from referring to the equations that show relationships among commonly used units of the metric and customary systems given on page 61.

Using the Page
Remind students that 1 degree = 60 minutes. Then challenge students to find the number of minutes in 3.6 degrees. [216 minutes or 3 degrees 36 minutes]

To convert from one unit to another, you multiply or divide.

Examples:

1. A student's height is 70 in. What is her height in feet and inches? Divide to find the answer.

$$\begin{array}{r} 5 \leftarrow \text{feet} \\ \text{Inches in 1 foot} \rightarrow 12\overline{)70} \\ \underline{60} \\ 10 \leftarrow \text{inches} \end{array}$$

   The student's height is 5 feet 10 inches.

2. How many meters is 124 centimeters?
   100 cm = 1 m, and so 1 cm = $\frac{1}{100}$m or 0.01 m. Multiply both sides of the second equation by 124. 124 cm = 1.24 m.

3. How many inches are equal to 124 centimeters? 1 in = 2.54 cm, and so 1 cm = $\frac{1}{2.54}$.
   Multiply both sides of the second equation by 124.
   124 cm = 124 × $\frac{1}{2.54}$ in.
   Use your calculator and you should find that 124 cm = 48.8 in (to the nearest tenth of an inch).

4. What is the total length of two boards that measure 5 ft 7 in., and 7 ft 8 in.? Add.

$$\begin{array}{r} 5 \text{ ft } \ 7 \text{ in.} \\ + \ 7 \text{ ft } \ 8 \text{ in.} \\ \hline 12 \text{ ft } 15 \text{ in.} \end{array} = 12 \text{ ft} + 1 \text{ ft } 3 \text{ in.} = 13 \text{ ft } 3 \text{ in.}$$

CLASS ACTIVITY

Use a calculator to help you answer the questions.

See margin.

1. Tina is 58 inches tall. What is her height in feet and inches?
2. A car seat is 142 centimeters wide. What is its width in meters? In inches? 1.42 m; 55.9 in.
3. A basketball hoop is 10 feet above the floor. What is this height in inches? In meters? 120 in.; 3.05 m
4. A motel is 60 kilometers from Seattle. How far is it from Seattle in miles? 37.3 mi

HOME ACTIVITY

Measure each of these segments to the nearest $\frac{1}{4}$ inch, the nearest $\frac{1}{8}$ inch, and the nearest centimeter. State the error for each measurement. See margin.

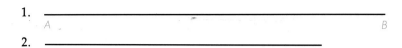

Measure each angle to the nearest degree. State the error for each measurement.

3. 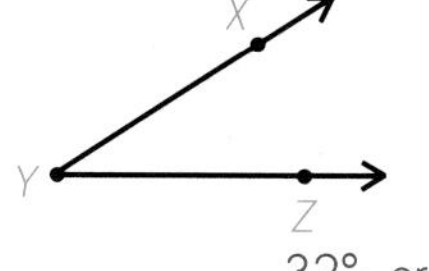

32°; error: $\frac{1}{2}$°

4. 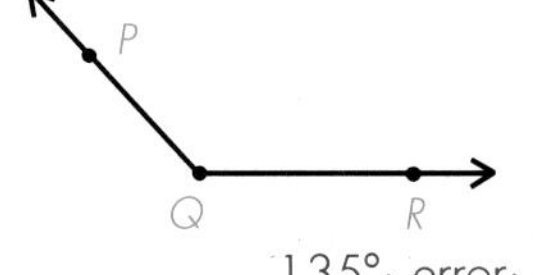

135°; error: $\frac{1}{2}$°

5. 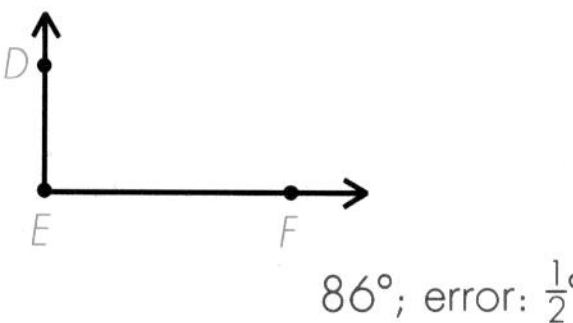

86°; error: $\frac{1}{2}$°

Measure the sides of these figures to the nearest $\frac{1}{8}$ in. Then add to find the total distance around each figure.

6. 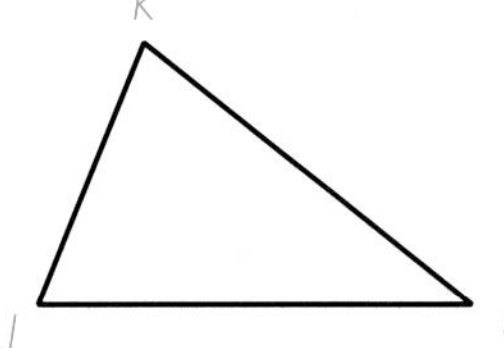

about $6\frac{3}{8}$ in.

7. 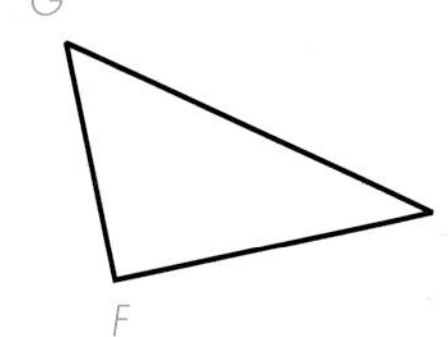

about $6\frac{5}{8}$ in.

8. 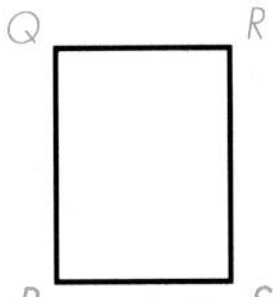

about $4\frac{7}{8}$ in.

9. 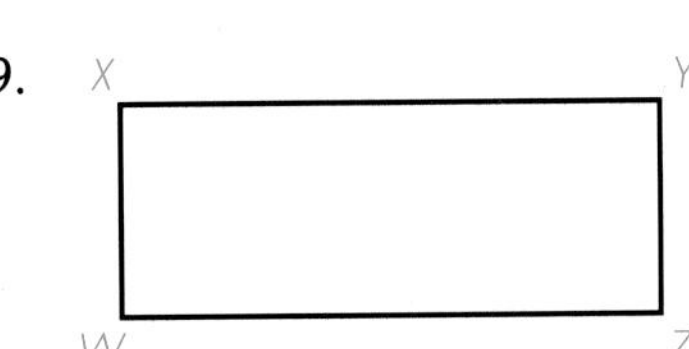

about $7\frac{2}{8}$ in.

Additional Answers

1. 4 ft 10 in.

Using the Page

Ask students to give Tina's height in centimeters. [147.32 cm]

Additional Answers

1. nearest $\frac{1}{4}$ in.: $4\frac{1}{4}$ in.; error: $\frac{1}{8}$ in. nearest $\frac{1}{8}$ in.: $4\frac{2}{8}$ in; error: $\frac{1}{16}$ in. nearest cm: 11 cm; error: 0.5 cm

2. nearest $\frac{1}{4}$ in.: $3\frac{2}{4}$ in.; error: $\frac{1}{8}$ in. nearest $\frac{1}{8}$ in.: $3\frac{3}{8}$ in.; error: $\frac{1}{16}$ in. nearest cm: 9 cm; error: 0.5 cm

Using the Page
Remind students to rename 1 degree as 60 minutes when subtracting angle measures that require renaming.

Find the sum of the angles of each figure.

10.

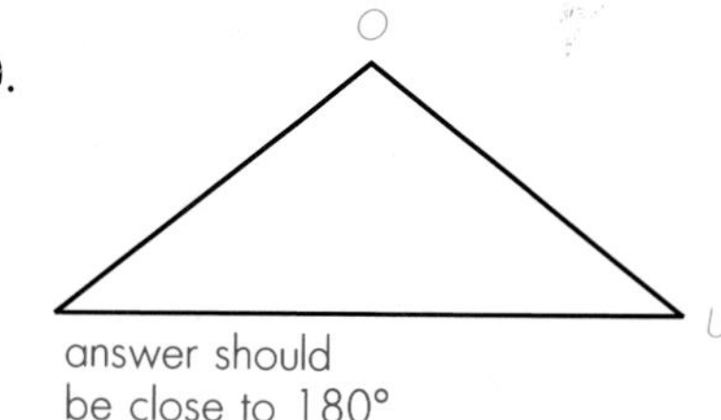

answer should be close to 180°

11.

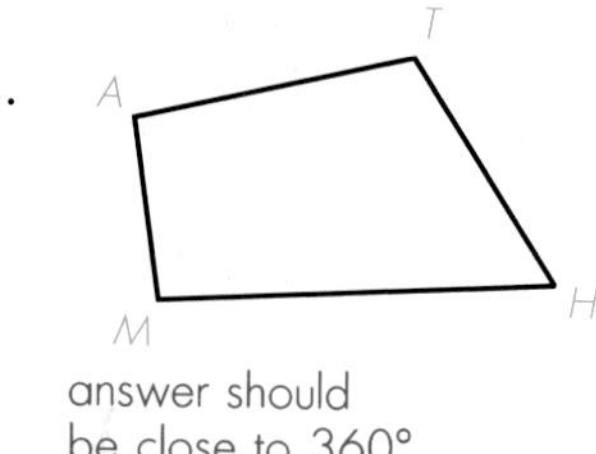

answer should be close to 360°

12. The distance between two towns is 57 mi. What is the distance in kilometers? 91.7 km

13. In millimeters, how long is a segment that measures 60.9 cm? 609 mm

14. A highway crew is going to pave a road that is 53,470 yd long. To the nearest tenth of a mile, how long is the road? 30.4 mi

Add or subtract the measures.

15. 12 yd 1 ft
+2 yd 2 ft
15 yd

16. 12 yd 1 ft
−7 yd 2 ft
4 yd 2 ft

17. 7 ft 10 in.
+5 ft 6 in.
13 ft 4 in.

Recall that $1° = 60'$. Thus $73\frac{1}{2}° = 73°\ 30'$. Write each of the following measures in degrees and minutes.

18. $76\frac{1}{3}°$ 76°20′

19. $158\frac{3}{4}°$ 158°45′

20. 89.1° 89°6′

21. $118\frac{2}{3}°$ 118°40′

Add or subtract the angle measures.

22. 17° 25′
+36° 30′
53° 55′

23. 114° 38′
+26° 50′
141° 28′

24. 56° 40′
−42° 50′
13° 50′

25. 147° 35′
−80° 40′
66° 55′

CRITICAL THINKING

26. What is the measure of the angle formed by the hands of a clock when the time is 2:00? When it is 6:00? 60°; 180°
Through how many degrees does the hour hand of a clock move in one hour? 30°

FOLLOW-UP

Assessment

1. A segment measures $16\frac{5}{8}$ in. when measured to the nearest $\frac{1}{8}$ inch. State the error for this measurement. $[\frac{1}{16}$ in.$]$

2. How many centimeters long is a box that is 2 ft 2 in. long? [66.04 cm]

3. Subtract the measures.

10 yd 1 ft
− 4 yd 2 ft
[5 yd 2 ft]

24° 30′
− 18° 45′
[5° 45′]

Reteaching
Students who have had difficulty with this section may benefit from Reteaching Activity 2.4.

Enrichment
For students who have mastered the material in this section, you may wish to assign Enrichment Activity 2.4.

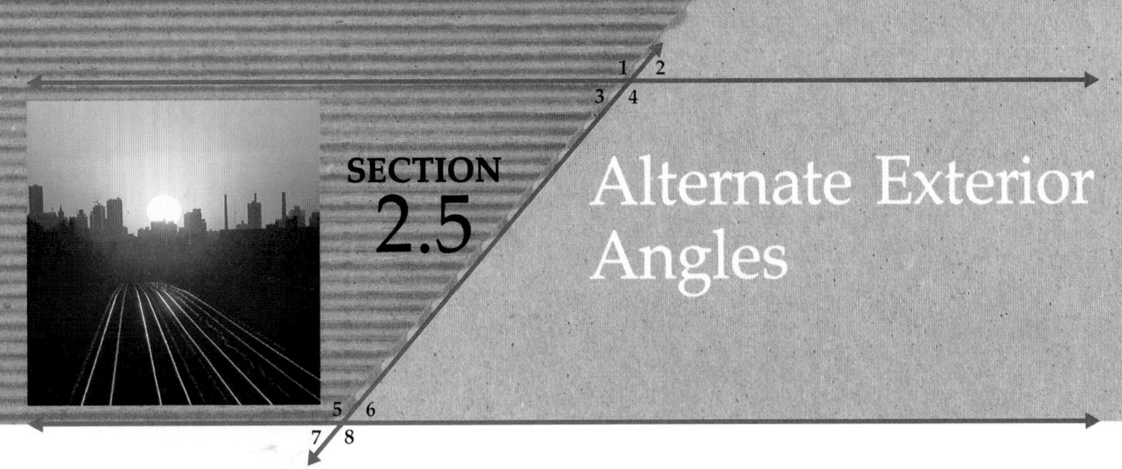

SECTION 2.5 Alternate Exterior Angles

After completing this section, you should understand
- ▲ alternate exterior angles.
- ▲ perpendicular lines.

Breakthrough designs using triangular braces have made it possible for architects to envision skyscrapers half a mile high. The Citicorp Building in Manhattan, for example, exerts less than half as much pressure on each square foot of ground below than the Empire State Building, which is only a few hundred feet taller.
In many modern buildings you can easily observe important relationships between the angles formed of parallel lines and a transversal.

ALTERNATE EXTERIOR ANGLES

In the diagram at the right, *a* and *b* are parallel lines intersected by transversal *t*. The angles formed are labeled 1 through 8.

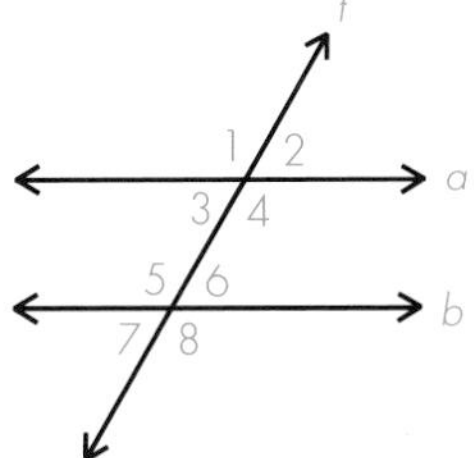

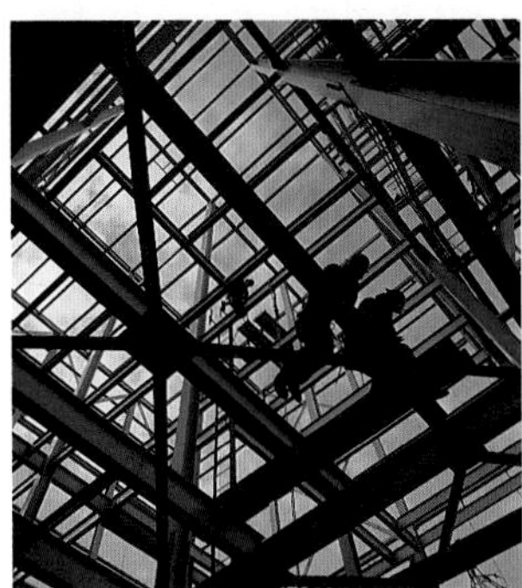

Look at angles 1 and 8. They are both *outside* the parallel lines. For this reason they are called **exterior** angles. Notice also that these two angles are on opposite or **alternate** sides of the transversal. Angles that are on opposite sides of the transversal and are outside the two parallel lines are called **alternate exterior angles.**

SECTION 2.5

Resources
Reteaching 2.5
Enrichment 2.5
Transparency Master 2-1

Objectives
After completing this section, the student should understand
- ▲ alternate exterior angles.
- ▲ perpendicular lines.

Materials
protractor

Vocabulary
alternate exterior angles
perpendicular lines

GETTING STARTED

Quiz: Prerequisite Skills

1. Find the measure of ∠1. [m∠1 = 65°]

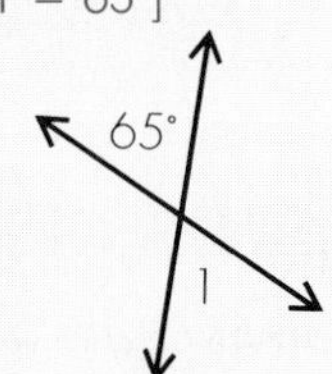

2. The diagram shows two parallel lines and a transversal. Find the measure of each numbered angle. [m∠1 = 105°; m∠2 = 75°; m∠3 = 75°; m∠4 = 105°]

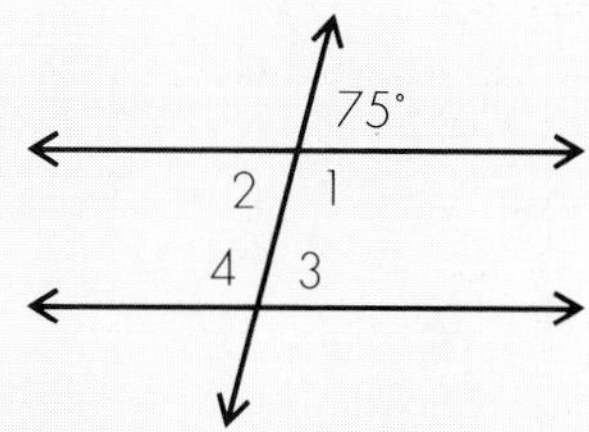

TEACHING COMMENTS

Using the Page
After students have completed the Discovery Activity, ask them to give examples of how they have previously used deductive reasoning to draw conclusions about relationships among angles.

DEFINITION 2-5-1: ALTERNATE EXTERIOR ANGLES are angles on opposite sides of the transversal and outside the parallel lines.

Name another pair of angles in the diagram that are alternate exterior angles. ∠2 and ∠7

DISCOVERY ACTIVITY

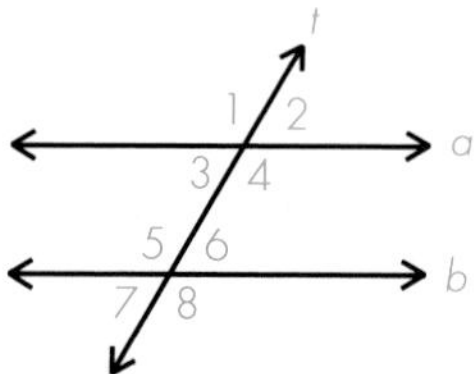

Draw a diagram of two parallel lines and a transversal. Label the angles as in the diagram above.

1. Measure angles 1, 2, 7, and 8 with your protractor. Measurements will vary.
2. What relationship do you notice between the measures of alternate exterior angles 1 and 8? of alternate exterior angles 2 and 7? The measures should be equal.
3. Write your conclusions. Students may conclude that alternate exterior angles formed by two parallel lines and a transversal are equal.

You may have concluded that alternate exterior angles formed by two parallel lines and a transversal are equal. By using known facts and deductive reasoning, you can prove that this is so.

Here, again, is the diagram.

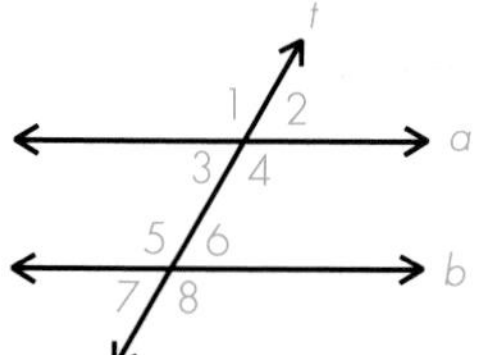

Step 1. $m\angle 1 = m\angle 4$ Reason: Vertical angles are equal.
Step 2. $m\angle 4 = m\angle 5$ Reason: Alternate interior angles are equal.
Step 3. $m\angle 5 = m\angle 8$ Reason: Vertical angles are equal.
Step 4. Since $m\angle 1 = m\angle 4$, $m\angle 4 = m\angle 5$, and $m\angle 5 = m\angle 8$, you can conclude that $m\angle 1 = m\angle 8$.

The *transitive property of equality,* which you learned when you studied algebra, states that if $a = b$ and $b = c$, then $a = c$. It is this property that allows the conclusion in Step 4. Similar reasoning will allow you to conclude that the other pair of alternate exterior angles are equal. How do the results you have obtained through deductive reasoning compare with the results you obtained through measurement in the DISCOVERY ACTIVITY? Answers may vary, but the measurements should confirm the results of the reasoning.

THEOREM 2-5-1: **If two parallel lines are intersected by a transversal, then the alternate exterior angles are equal.**

PERPENDICULAR LINES

If two lines intersect so that the angles formed are right angles, the two lines are said to be **perpendicular.**

In the diagram, line t is perpendicular to line a.

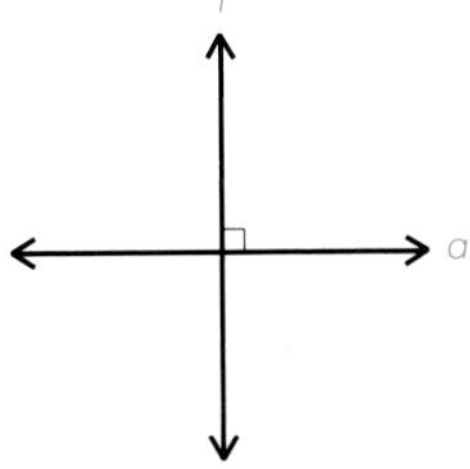

DEFINITION 2-5-2: **PERPENDICULAR LINES are lines that intersect to form right angles.**

The symbol ⊥ means, "is perpendicular to." For the diagram above, you can write $a \perp t$ or $t \perp a$. The small ⊾ in the diagram, at the point where a and t intersect, is used to show that the angle indicated is a right angle.

Examples:

1. The lines in which the floor and walls of the room meet are perpendicular.
2. The horizontal and vertical sides of the picture frame shown here are perpendicular.

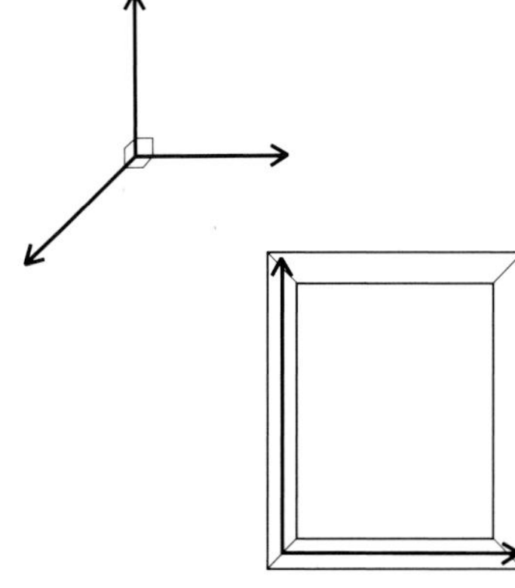

Using the Page

After perpendicular lines have been defined, use Transparency Master 2-1. Point out that the corners of the room shown in the transparency are right angles, even though they do not appear to be so. Refer students to the corners of the classroom to verify that this is true. Also point out the right angles formed in the other two figures shown on the transparency.

Then ask students if it is possible for a line to be perpendicular to more than one other line. Have students use a straightedge to draw examples of one line that is perpendicular to two others. Be sure students include the symbols to indicate the right angles in their drawings. If students have difficulty with this concept, have them draw a line perpendicular to one line on lined paper. Ask them if the line they drew is perpendicular to the other lines on the sheet.

3. At 3 o'clock, the hands of this clock are perpendicular.

Look around you and see how many pairs of perpendicular lines you can see right where you are.

CLASS ACTIVITY

1. For an ordinary mailing envelope, how many pairs of parallel edges are there? How many perpendicular edges? 2; 4

In the diagram, $a \parallel b$ and t is a transversal. For each of the following angles, name three angles equal to it. See margin.

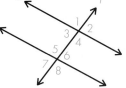

2. $\angle 1$
3. $\angle 6$
4. $\angle 5$
5. $\angle 4$
6. $\angle 2$
7. $\angle 3$
8. $\angle 7$
9. $\angle 8$

In the diagram, lines a and b are parallel and t is a transversal.

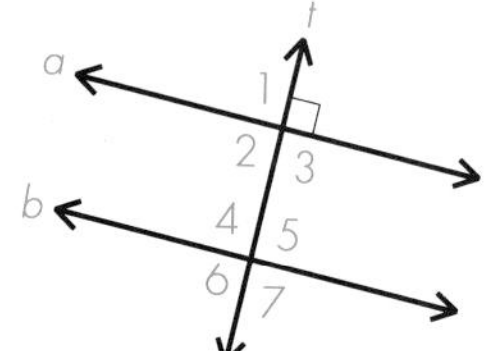

10. What is the measure of $\angle 1$? of $\angle 2$? and of $\angle 3$?
    90°, 90°, 90°
11. What is the measure of each angle where t and b intersect?
    90°
12. Is t perpendicular to b? Explain. They meet at right angles.

In the diagram, $j \parallel k$ and l is a transversal. Find the measures of angles 1 through 6.

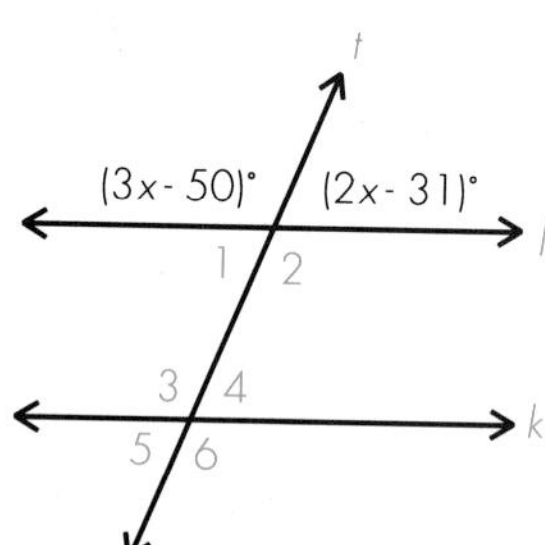

13. $m\angle 1 =$ 73.4°
14. $m\angle 2 =$ 106.6°
15. $m\angle 3 =$ 106.6°
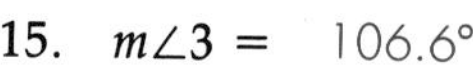
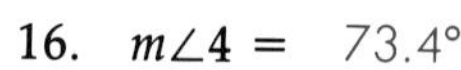
16. $m\angle 4 =$ 73.4°
17. $m\angle 5 =$ 73.4°
18. $m\angle 6 =$ 106.6°

Additional Answers

1. $\angle 4$, $\angle 5$, $\angle 8$
2. $\angle 1$, $\angle 4$, $\angle 8$
3. $\angle 3$, $\angle 6$, $\angle 7$
4. $\angle 2$, $\angle 3$, $\angle 6$
5. $\angle 2$, $\angle 3$, $\angle 7$
6. $\angle 1$, $\angle 5$, $\angle 8$
7. $\angle 2$, $\angle 6$, $\angle 7$
8. $\angle 1$, $\angle 4$, $\angle 5$

12. Yes. By using the Theorems about vertical angles, alternate interior angles, and alternate exterior angles, you know that all the angles where b and t intersect are right angles. So, $b \perp t$.

In the diagram, $a \parallel b$ and $x \parallel y$. Name all the pairs of angles that fit the description. See margin.

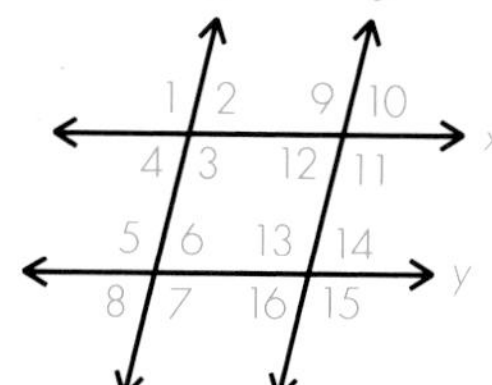

19. Alternate exterior angles with respect to a and b and transversal x.

20. Alternate exterior angles with respect to x and y and transversal b.

In the diagram, $p \parallel q$ and $f \parallel g$. Name all the angles that are equal to the given angle. See margin.

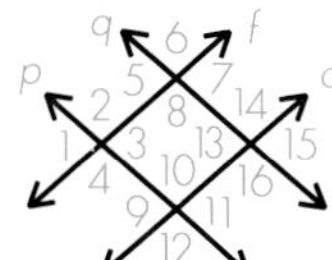

21. $\angle 1$

22. $\angle 16$

In the diagram, $k \parallel l$. Find the measures of the numbered angles.

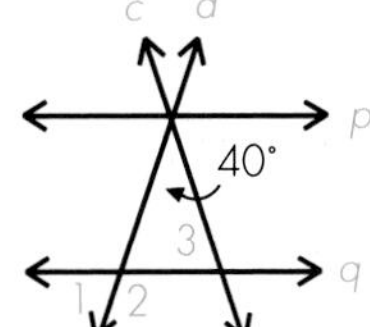

23. $\angle 1$ 70°
24. $\angle 2$ 110°
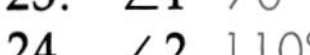
25. $\angle 3$ 70°
26. $\angle 4$ 70°

HOME ACTIVITY

In the diagram, t is a transversal of lines a and b. Name the following.

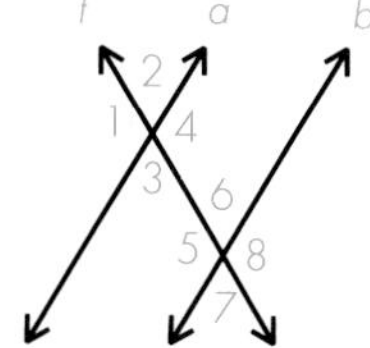

1. two pairs of alternate exterior angles.
   $\angle 1$ and $\angle 8$; $\angle 2$ and $\angle 7$
2. four pairs of vertical angles
   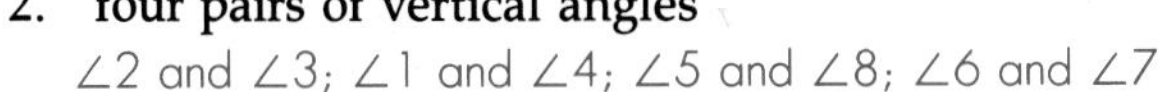
   $\angle 2$ and $\angle 3$; $\angle 1$ and $\angle 4$; $\angle 5$ and $\angle 8$; $\angle 6$ and $\angle 7$
3. two pairs of alternate interior angles
   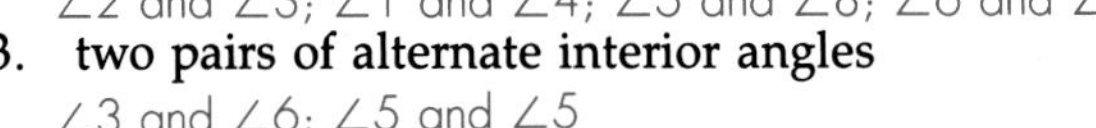
   $\angle 3$ and $\angle 6$; $\angle 5$ and $\angle 5$
4. four pairs of corresponding angles
   $\angle 1$ and $\angle 5$; $\angle 2$ and $\angle 6$; $\angle 3$ and $\angle 7$; $\angle 4$ and $\angle 8$

In the diagram, $e \parallel f$. Find the degree measure of each numbered angle.

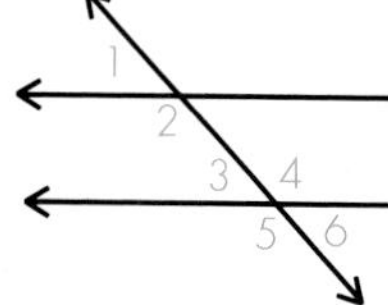

5. $\angle 1$ 45°
6. $\angle 5$ 135°
7. $\angle 4$ 135°
8. $\angle 2$ 135°
9. $\angle 3$ 45°
10. $\angle 6$ 45°

Additional Answers

19. $\angle 1$ and $\angle 11$; $\angle 4$ and $\angle 10$
20. $\angle 9$ and $\angle 15$; $\angle 10$ and $\angle 16$
21. $\angle 3$, $\angle 5$, $\angle 7$, $\angle 9$, $\angle 11$, $\angle 13$, $\angle 15$
22. $\angle 2$, $\angle 4$, $\angle 6$, $\angle 8$, $\angle 10$, $\angle 12$, $\angle 14$

Extra Practice

1. $a \parallel b$. Find the value of x.

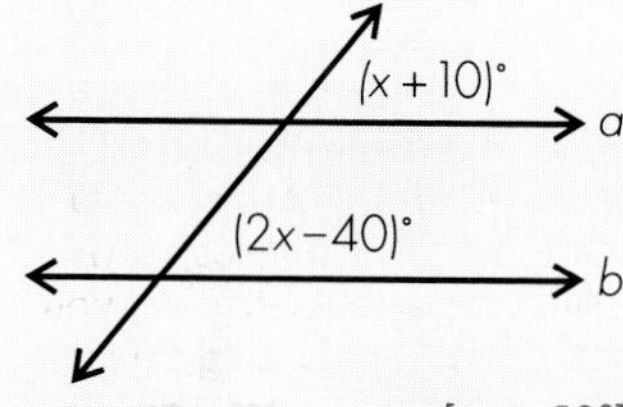

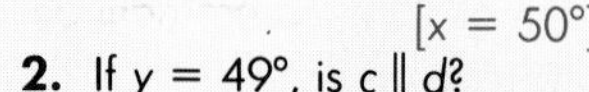
[x = 50°]

2. If $y = 49°$, is $c \parallel d$?

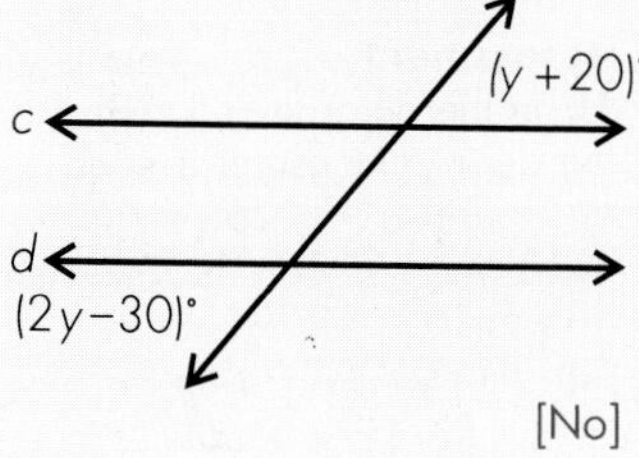

[No]

In the diagram, $c \parallel d$. Find the measures of the numbered angles.

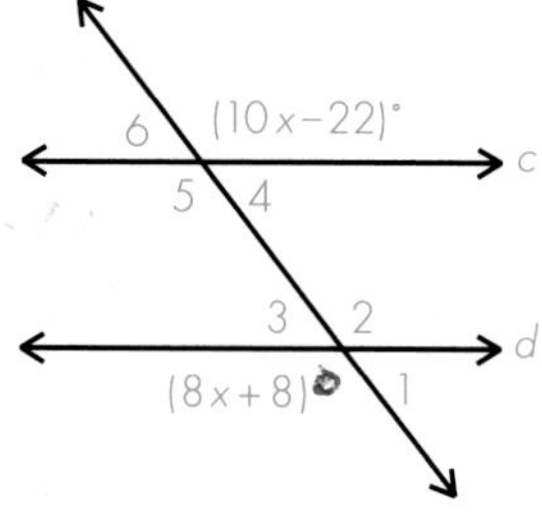

11. ∠1 52°
12. ∠2 128°
13. ∠3 52°
14. ∠4 52°
15. ∠5 128°
16. ∠6 52°

In the diagram, $p \parallel q$. Find the degree measure of each angle.

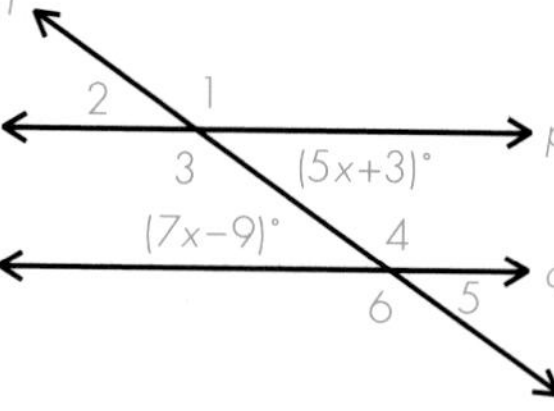

17. ∠1 147°
18. ∠5 33°
19. ∠4 147°
20. ∠2 33°
21. ∠3 147°
22. ∠6 147°

23. In the diagram, $c \parallel d$. Find the measure of ∠1. 95°

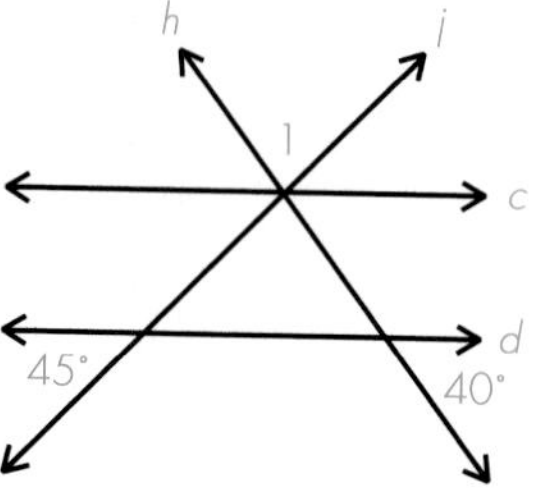

In the diagram, $s \parallel t$ and $v \parallel w$. Find the measures of the numbered angles.

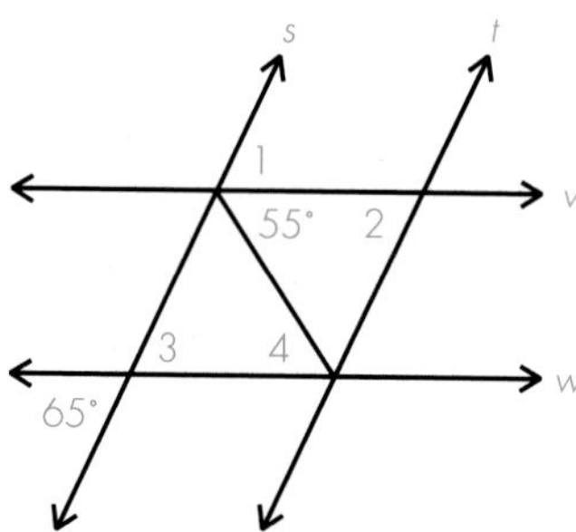

24. ∠1 65°
25. ∠2 65°
26. ∠3 65°
27. ∠4 55°

CRITICAL THINKING

28. In a diagram showing two parallel lines *a* and *b*, Karen numbered the angles from 1 through 8. As labeled, ∠1 and ∠2 were vertical angles, ∠1 and ∠6 were alternate exterior angles, and ∠4 and ∠6 were supplementary angles. Name also another pair of angles that you can be sure are supplementary. ∠2 and ∠4

FOLLOW-UP

Assessment

1. In the diagram, $r \parallel s$. Find the degree measure of each angle. [m∠1 = 125°; m∠3 = 125°; m∠4 = 55°; m∠5 = 55°; m∠6 = 125°]

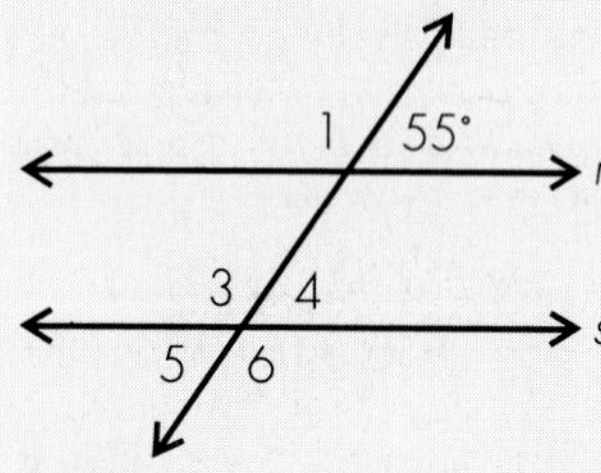

2. What is the degree measure of ∠1? [90°]

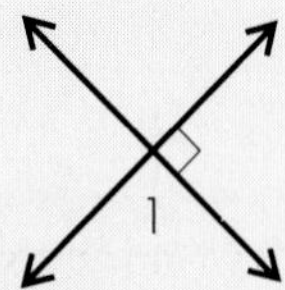

Reteaching

Students who have had difficulty with this section may benefit from Reteaching Activity 2.5.

Enrichment

For students who have mastered the material in this section, you may wish to assign Enrichment Activity 2.5.

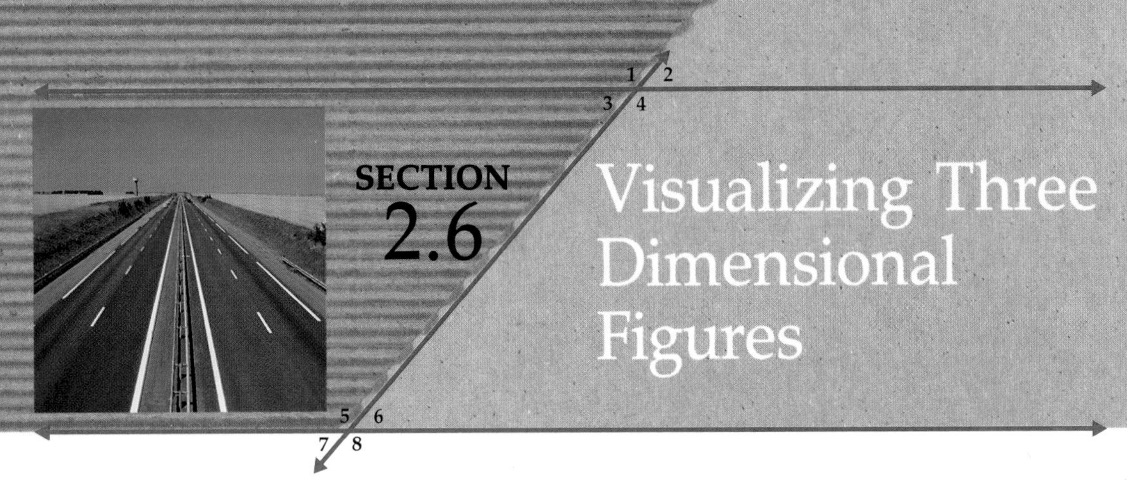

SECTION 2.6 Visualizing Three Dimensional Figures

After completing this section, you should understand
- ▲ how a three-dimensional figure appears when drawn on a two-dimensional surface.
- ▲ the relationship between a three-dimensional figure and its layout.

Highway systems, bridges and tunnels, buildings in apartment complexes and universities, and parks are planned by means of carefully conceived two-dimensional drawings. The ability to visualize a three-dimensional object and to represent it on a two-dimensional surface is an important technical skill for planners and designers of thousands of things around you.

DRAWINGS OF THREE-DIMENSIONAL FIGURES

All geometric figures (lines, rays, angles, and so on) consist of points.
A line is one-dimensional, because a line has only length.
A plane is two-dimensional, because figures on a plane can have length and width.
Space is three-dimensional, because figures in space can have length, width, and depth.

Examples:

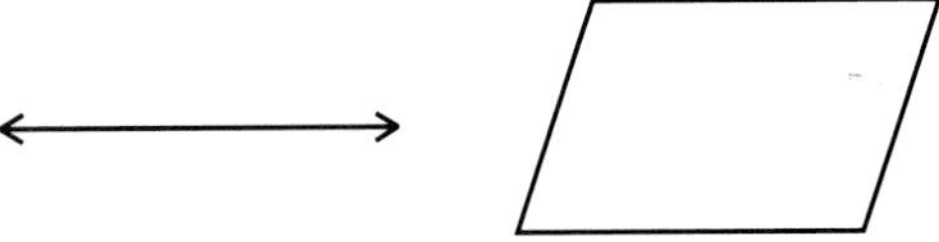

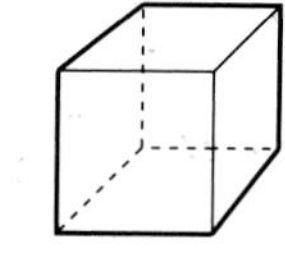

one-dimensional or 1-D | **two-dimensional or 2-D** | **three-dimensional or 3-D**

Notice how the illusion of depth is created in the drawing of the three-dimensional figure. The sides of the figure that are in full view have edges drawn with solid lines. To indicate the part not in view, you make use of dotted lines.

SECTION 2.6

Resources
Reteaching 2.6
Enrichment 2.6
Transparency Masters 2-2 and 2-3

Objectives
After completing this section, the student should understand
- ▲ how a three-dimensional figure appears when drawn on a two-dimensional surface.
- ▲ the relationship between a three-dimensional figure and its layout.

Materials
ruler

Vocabulary
two-dimensional
three-dimensional
layout

GETTING STARTED

Warm-Up Activity
Ask students to draw the following objects as accurately and realistically as they can: a shoebox, a can of dog food, and any classroom object with three dimensions.

Chalkboard Example
Copy the layout below on the chalkboard. Ask students to guess what geometric figure is formed if this layout is folded along the dotted lines. [a cube]

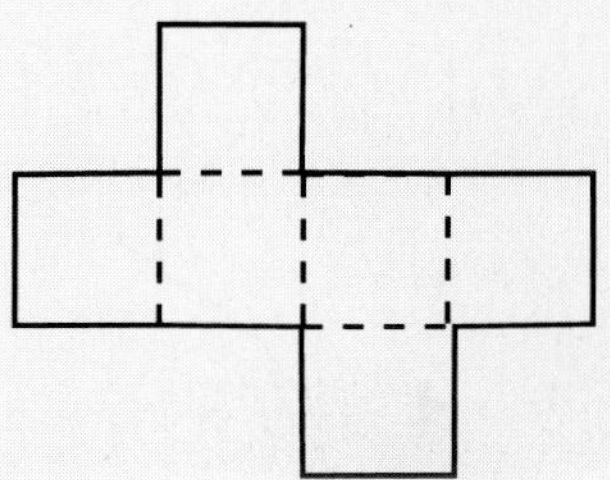

The following drawing shows a plane, a line on the plane, and another line, not on the plane, that intersects the first line at point *A*.

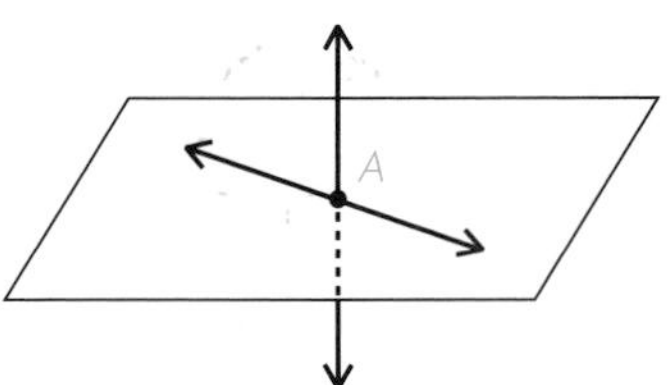

Again notice how a dotted line is used to create a three-dimensional effect.

The next drawing can be used to show three intersecting planes. The use of dotted lines is a little more complicated. Still, if you use your imagination, you can probably see how this drawing pictures something similar to three sheets of glass that intersect in the line *AB*.

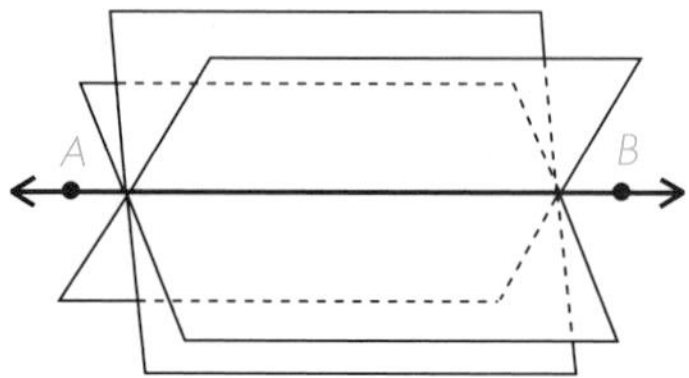

You can become good at drawing figures such as these. Practice picturing them and all their parts in your mind's eye.

DISCOVERY ACTIVITY

The drawing shows a three-dimensional figure called a pyramid. Imagine it as a big pyramid. It has three faces and a bottom. There are three people standing around it, looking at it from different points of view. If person B drew the figure from his point of view it would look much as it does in the drawing above.

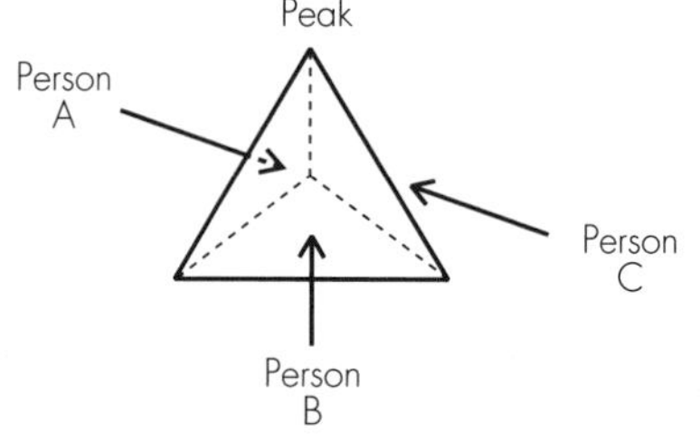

1. Use solid lines and dotted lines to show the pyramid from person A's point of view.
2. Draw the pyramid from person C's point of view.
3. Imagine someone passing directly over the pyramid in a helicopter. Draw the pyramid from that point of view (looking down at the peak).

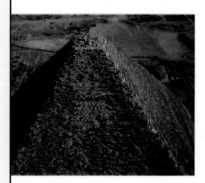

TEACHING COMMENTS

Using the Page
Use Transparency Master 2-2 to show the drawing of the three intersecting planes on this page.

Using the Page
Use Transparency Master 2-2 to show the pyramid in the Discovery Activity. Ask students what figure they would see if they were to view the pyramid from high in the air directly above it. [a triangle]

Then ask them what they would see from above if they were looking at the second pyramid shown on the transparency. [a square]

LAYOUTS OF THREE-DIMENSIONAL FIGURES

If you have ever eaten at a cafeteria you have probably had your food served in a cardboard pop-up container. Examine the flat **layout** and the container that it forms.

Layout

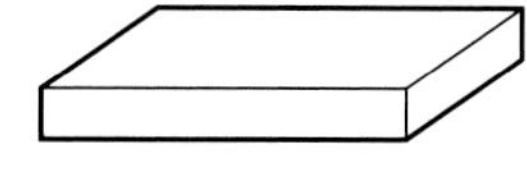

Container

How would you describe the closed figure that will be formed by this layout? In fact, it will be a triangular prism. Accept any reasonable descriptions the students come up with.

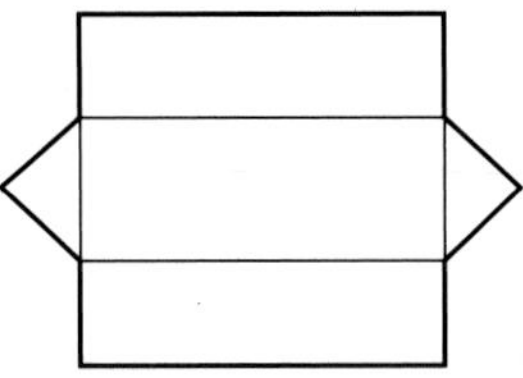

CLASS ACTIVITY

1. Draw the three-dimensional figure this layout would form.

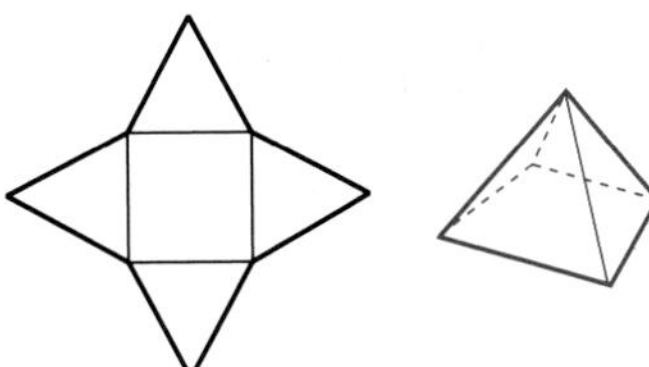

2. Draw a layout for this round, can-shaped figure.
   Other layouts are possible.

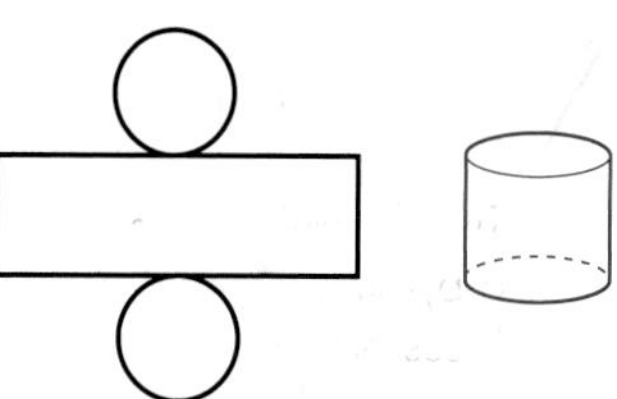

3. Draw the following views of a quarter placed on a tabletop.

   a. Looking directly down on it from above

   b. Looking at it from eye level with the table

   c. Looking at it from a standing position 3 feet from the table

Using the Page
Ask students how they would change the layout for the pop-up container so that it would be able to hold more food. [possible answers: increase the length and/or width of the bottom, or increase the height of the sides and ends]

Using the Page
Use Transparency Master 2-3 for showing the layouts of the triangular prism and the square pyramid presented on this page.

Extra Practice
Use Transparency Master 2-3 to show the figure below. Ask students what they see when they look at it.

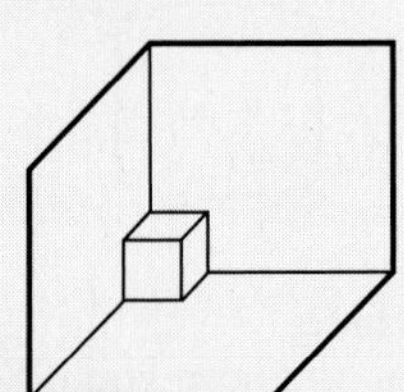

[possible answers: small cube in corner of a room; small cube stuck on outside of larger solid block; large block with small cube removed from corner]

FOLLOW-UP

Assessment
1. Draw the top view, end view, and side view for this figure.

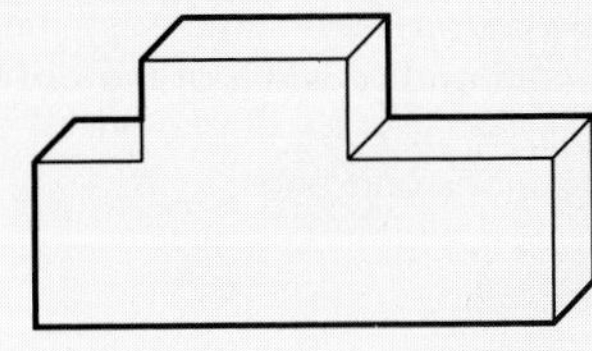

top view:

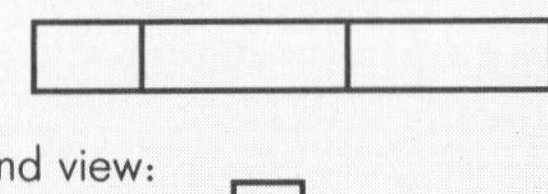

end view:

side view:

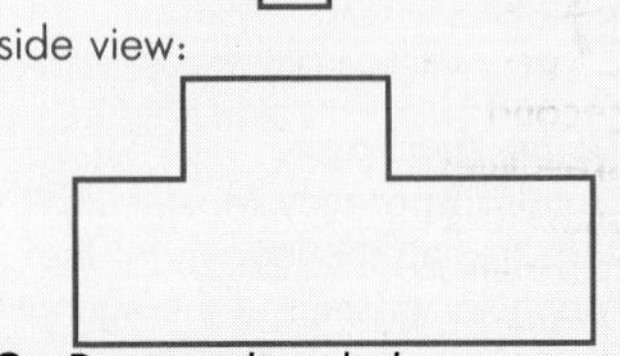

2. Draw a closed, three-dimensional figure for this layout.

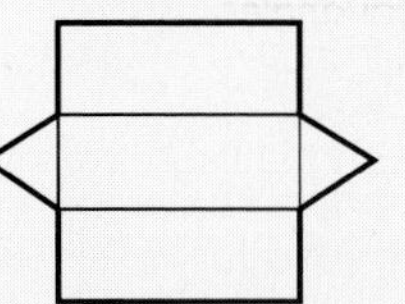

[Students' drawing should be a triangular prism.]

Additional Answers
5. Possible answers: cubes piled in all 6 corners; 4 cubes (one unseen) piled in corner; star within a hexagon
8.

Reteaching
Students who have had difficulty with this section may benefit from Reteaching Activity 2.6.

Enrichment
For students who have mastered the material in this section, you may wish to assign Enrichment Activity 2.6.

HOME ACTIVITY

1. Draw plane P with line $\overleftrightarrow{BC}$ on it. Draw two perpendicular lines intersecting at B, one line on the plane, and the other not in the plane. See student's drawings.

2. Fold a piece of paper in half. Open it. You now have two intersecting planes. What conclusion can you draw about the intersection of two planes? It is a line.

3. Write a Logo procedure to tell the turtle to draw a figure similar to the one shown at the right. Answers may vary.

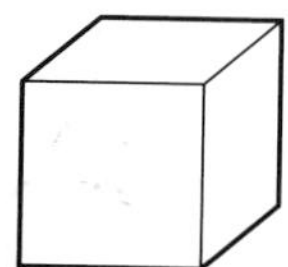

4. Draw the layout for this three-dimensional figure. Other layouts are possible.

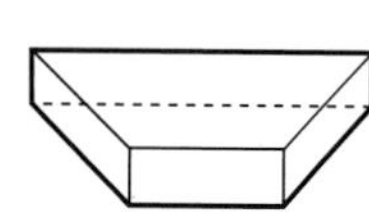

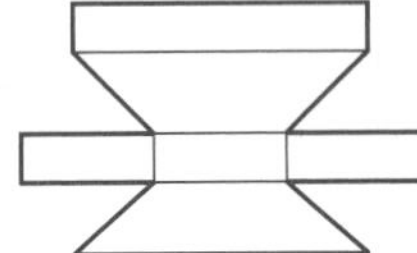

5. This drawing can be thought of as showing different figures. Describe as many as you can.

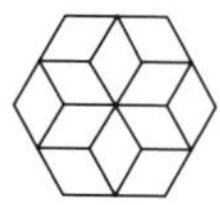

Draw closed, three-dimensional figures for each of these layouts.

6.

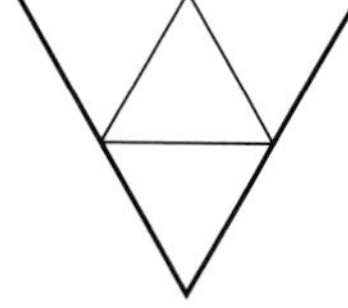

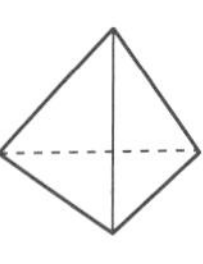

7.

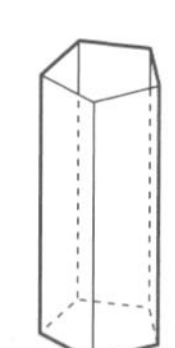

8. Draw the three-dimensional view of the object with the following views:

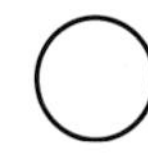

Top view   Side view   Front view

9. Draw the top view, end view, and front-side view of this building.

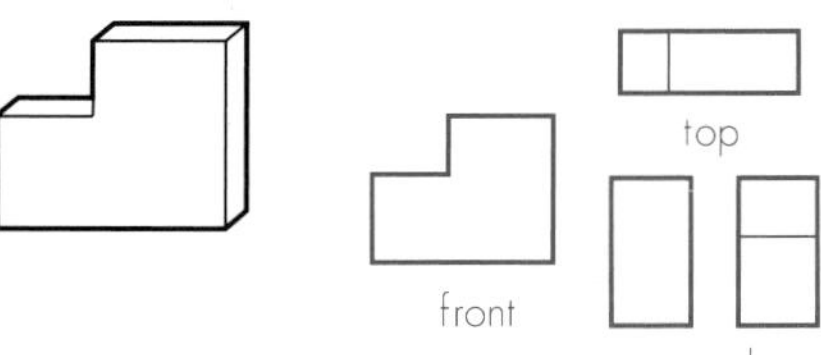

CRITICAL THINKING

10. If you were to fold each of these two layouts, which one would form a cube?
Both will form a cube.

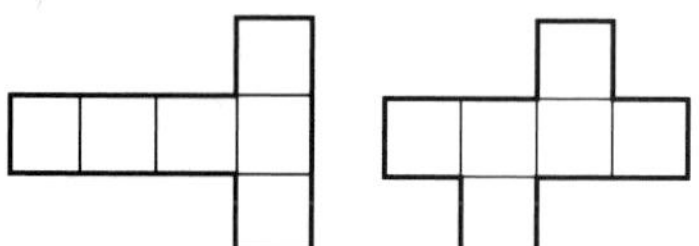

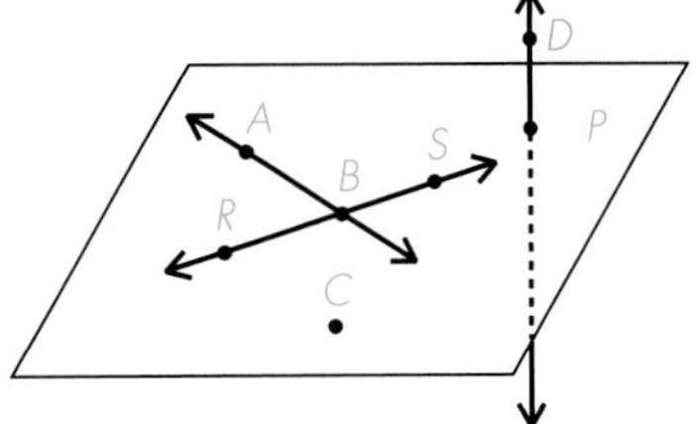

2.1 Collinear points are points on the same line. Noncollinear points are points not on the same line. Coplanar points are points on the same plane. Noncoplanar points are points that are not on the same plane. If two lines intersect, they intersect at exactly one point. In the diagram, *R*, *B*, and *S* are collinear. Points *R*, *S*, and *C* are noncollinear. Points *A*, *B*, and *C* are coplanar. Points *A*, *B*, *C*, and *D* are noncoplanar. Lines $\overleftrightarrow{AB}$ and $\overleftrightarrow{RS}$ intersect at point *B*.

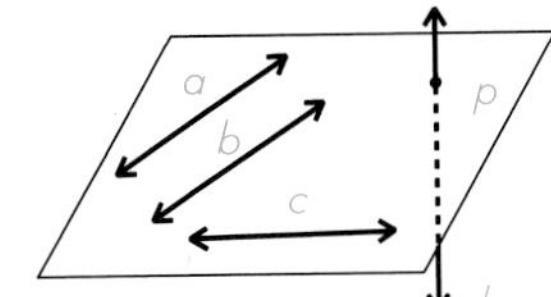

Two lines are parallel if they are coplanar and do not intersect. Skew lines are lines that do not intersect and are not parallel. In the diagram, lines *a* and *b* are parallel. Lines *c* and *d* are skew lines.

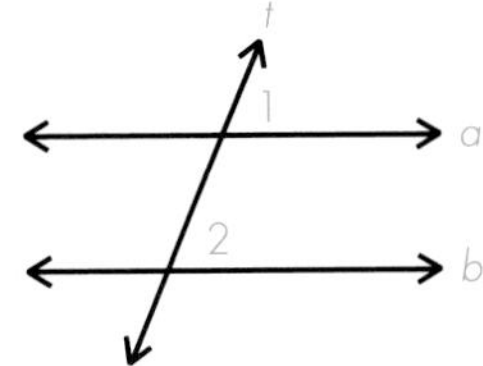

2.2 A transversal is a line that intersects two or more coplanar lines in different points. For two lines and a transversal, two angles in the same relative position are corresponding angles. If the two lines are parallel, the corresponding angles are equal. Line *t* is a transversal of parallel lines *a* and *b*. Since $\angle 1$ and $\angle 2$ are corresponding angles, they are equal.

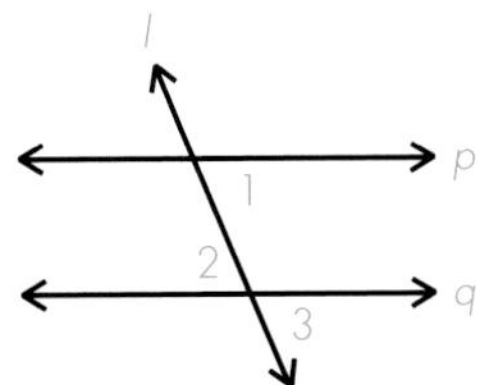

2.3 For two intersecting lines, opposite angles are vertical angles and are equal. For two parallel lines and a transversal, angles on opposite sides of the transversal and between the parallel lines are alternate interior angles. Alternate interior angles are equal. In the diagram, $p \parallel q$. $\angle 2$ and $\angle 3$ are vertical angles and $\angle 1$ and $\angle 2$ are alternate interior angles. $m\angle 1 = m\angle 2$ and $m\angle 2 = m\angle 3$.

2.4 The error measurement is one-half the size of the unit or subunit to which you are measuring. A measurement to the nearest $\frac{1}{8}$ inch has an error of $\frac{1}{16}$ inch.

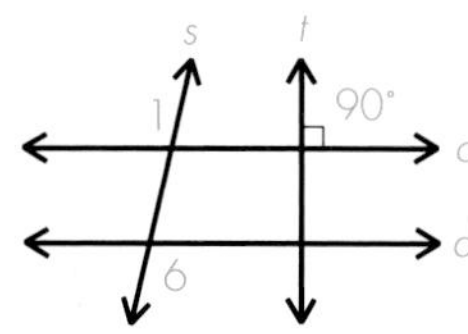

2.5 In the figure, $c \parallel d$. Lines *s* and *t* are transversals. $\angle 1$ and $\angle 6$ are on opposite sides of *s* and neither is between the parallel lines. Such angles are alternate exterior angles and are equal. Since lines *t* and *c* form 90° angles, they are perpendicular.

Vocabulary
alternate exterior angles
alternate interior angles
centi-
collinear
conclusion
conditional statement
convert
coplanar
corresponding angles
deductive reasoning
hypothesis
intersecting lines
kilo-
lattice
layout
measurement error
milli-
minute
noncollinear
noncoplanar
parallel lines
perpendicular lines
second
skew lines
subunit
theorem
three-dimensional
transversal
two-dimensional
vertical angles

Using the Page
In this Computer Activity the student first learns to write a simple LOGO program to draw a line segment called LINESEG. This program is then incorporated into a program called PARALLEL which enables the student to draw parallel segments.

COMPUTER ACTIVITY

Once a Logo procedure has been defined, it can be used in other procedures. It is very convenient to be able to use a procedure within a procedure when the same figure needs to be drawn many times. In the example below, the procedure LINESEG is defined and then used in the procedure PARALLEL to draw parallel lines.

Example: Write a procedure named PARALLEL to tell the turtle to draw two parallel line segments.

First, write a procedure called LINESEG to draw one line segment.

```
TO LINESEG
RT 90
FD 60
END
```

Then use LINESEG in a procedure called PARALLEL to tell the turtle to draw two parallel line segments.

TO PARALLEL	
LINESEG	The turtle draws a horizontal line segment 60 spaces long.
PU HOME FD 50 PD	The turtle moves back to its home position and then forward 50 spaces.
LINESEG	The turtle draws another horizontal line segment that is 60 spaces long and is parallel to the first line segment drawn.
END	

The procedure PARALLEL can now be used in another procedure to draw two parallel line segments and a transversal.

Example: Write a procedure named TRANS to tell the turtle to draw two parallel line segments and a transversal.

TO TRANS

PARALLEL — The turtle draws two parallel line segments.

PU LT 90 FD 30 LT 135 PD — The turtle moves into position to draw the transversal.

FD 100 — The turtle draws the transversal.

END

Modify the procedure LINESEG and the procedure PARALLEL to tell the turtle to draw each figure. Answers may vary.

1. three parallel horizontal line segments
2. two parallel vertical line segments
3. two parallel line segments that are neither horizontal not vertical

Use the modified procedures from Exercises 1 through 3 and modify the procedure TRANS to tell the turtle to draw each figure. Answers may vary.

4. three parallel horizontal line segments and a transversal
5. two parallel vertical line segments and a transversal
6. two parallel line segments that are neither horizontal nor vertical and a transversal

Modify any of the procedures from the examples and Exercises 1 through 6 and use them in procedures to tell the turtle to draw each of the following figures. The line segments that appear parallel are parallel. Answers may vary.

7.

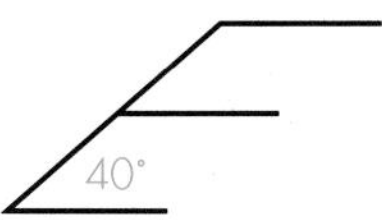

8.

9.

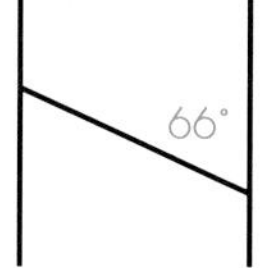

10.

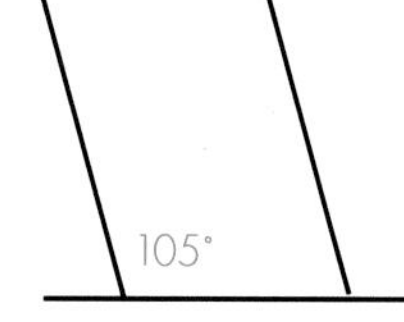

Using the Page

Here the student incorporates the PARALLEL program into a program called TRANS. This enables the student to draw parallel lines crossed by a transversal. In the exercises the student modifies the LINESEG, PARALLEL, and TRANS programs to draw parallel line segments and transversals in different positions on the screen. Students then use these basic programs to create more elaborate images.

In the diagram, $\overleftrightarrow{AB}$ and $\overleftrightarrow{DC}$ do not intersect. Also, $m\angle FEA = 90°$. Identify the following.

1. intersecting lines See margin.
2. a transversal $\overleftrightarrow{DH}$, $\overleftrightarrow{BC}$
3. perpendicular lines $\overleftrightarrow{FG}$ and $\overleftrightarrow{AB}$
4. parallel lines $\overleftrightarrow{AB}$ and $\overleftrightarrow{DC}$
5. skew lines $\overleftrightarrow{FG}$ and $\overleftrightarrow{DC}$, $\overleftrightarrow{DH}$, $\overleftrightarrow{BC}$
6. coplanar lines See margin.
7. three collinear points See margin.
8. three noncollinear points
A, E, C (other answers are possible)

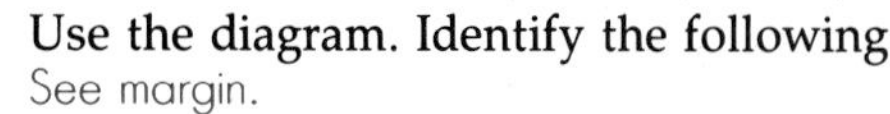

Use the diagram. Identify the following.
See margin.

9. corresponding angles
10. alternate interior angles
11. alternate exterior angles
12. vertical angles

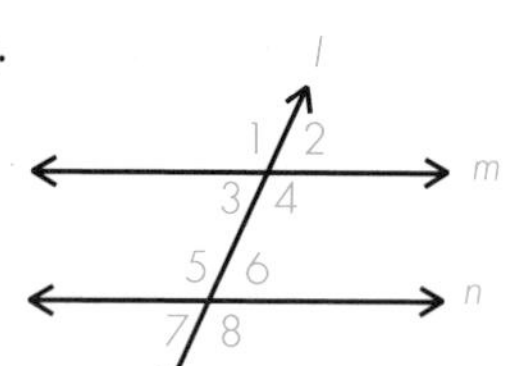

In the diagram, $x \parallel y$.

13. $\angle 1$ 50°
14. $\angle 2$ 130°
15. $\angle 3$ 130°
16. $\angle 4$ 50°
17. $\angle 5$ 50°
18. $\angle 6$ 130°

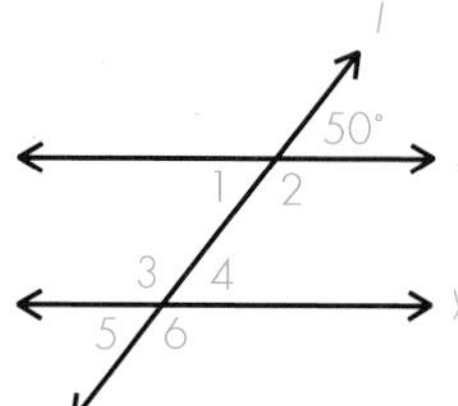

Measure the segment as indicated. State the error in each measurement.

19. to the nearest $\frac{1}{2}$ inch 2 in.; error: $\frac{1}{4}$ in.

20. to the nearest $\frac{1}{8}$ inch $2\frac{3}{8}$ in.; error: $\frac{1}{16}$ in.

Measure each angle to the nearest degree. State the error in each measurement.

21.

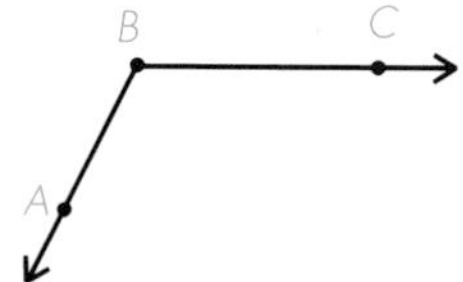

118°; error: $\frac{1}{2}°$

22.

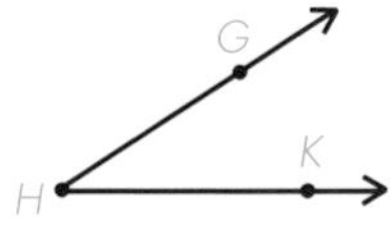

33°; error: $= \frac{1}{2}°$

23. Draw a three-dimensional view of a shoebox with the top off. Check students' diagrams.

78

Additional Answers

1. $\overleftrightarrow{AB}$, $\overleftrightarrow{FG}$; $\overleftrightarrow{DH}$, $\overleftrightarrow{AB}$, $\overleftrightarrow{DH}$, $\overleftrightarrow{DC}$; $\overleftrightarrow{BC}$, $\overleftrightarrow{AB}$; $\overleftrightarrow{BC}$, $\overleftrightarrow{DC}$

6. $\overleftrightarrow{AB}$, $\overleftrightarrow{BC}$, $\overleftrightarrow{DC}$, and $\overleftrightarrow{DH}$; $\overleftrightarrow{FG}$ and $\overleftrightarrow{AB}$

7. F, E, G; A, E, B (other answers are possible)

9. $\angle 1$ and $\angle 5$, $\angle 2$ and $\angle 6$, $\angle 3$ and $\angle 7$, $\angle 4$ and $\angle 8$

10. $\angle 3$ and $\angle 6$, $\angle 4$ and $\angle 5$

11. $\angle 1$ and $\angle 8$, $\angle 2$ and $\angle 7$

12. $\angle 1$ and $\angle 4$, $\angle 2$ and $\angle 3$, $\angle 5$ and $\angle 8$, $\angle 6$ and $\angle 7$

Using the Page

Students can work individually or in groups to study the Chapter 2 Review. If you have students study in groups, they should work together to make sure all members of the group know the material in the chapter. Informal assessment such as interviews, classroom observations, or a review of student portfolios can be used instead of paper-and-pencil tests.

CHAPTER TEST

In the figure, $\overleftrightarrow{GH}$ and $\overleftrightarrow{EF}$ do not intersect. Identify the following: Answers may vary.

1. three collinear points *E, I, F*
2. three noncollinear points *E, I, J*
3. four coplanar points *E, F, G, H*
4. two parallel lines $\overleftrightarrow{EF}$, $\overleftrightarrow{GH}$
5. two skew lines $\overleftrightarrow{AB}$, $\overleftrightarrow{CD}$
6. three lines that intersect in one point $\overleftrightarrow{EF}$, $\overleftrightarrow{IJ}$, $\overleftrightarrow{AB}$

Use the diagram. Identify the following.
Answers may vary.

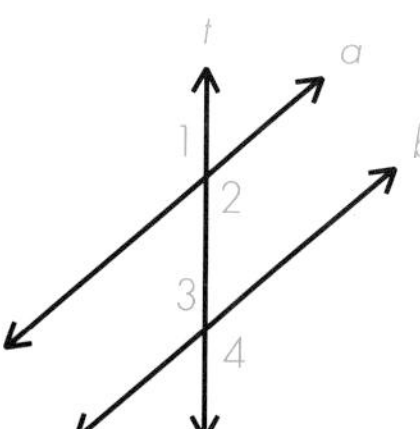

7. a pair of alternate interior angles ∠2 and ∠3
8. a pair of alternate exterior angles ∠1 and ∠4
9. a pair of corresponding angles ∠1 and ∠3
10. a transversal *t*
11. a pair of vertical angles ∠1 and ∠2

In the diagram, $p \parallel q$ and t is a transversal. Give the degree measures of each angle.

12. ∠1 140°
13. ∠2 140°
14. ∠3 140°
15. ∠4 140°
16. ∠5 40°
17. ∠6 40°

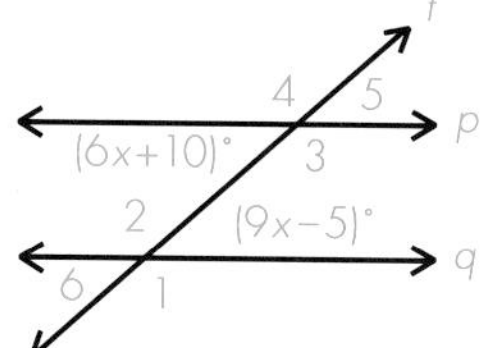

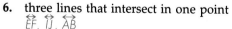

18. Draw the layout for this figure.
Layouts may vary.

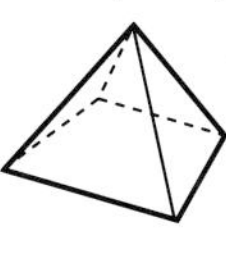

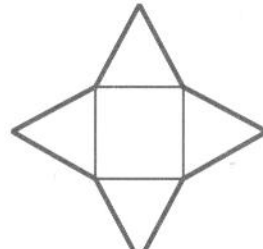

19. Draw the three-dimensional figure this layout would form.
Views of the figure may vary.

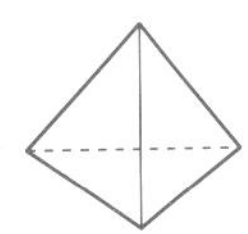

20. What is the error in a measurement made to the nearest $\frac{1}{32}$ inch? $\frac{1}{64}$ inch

21. If a person studied hard for the test, then the person's score was high. Sarah studied hard for the test. What can you conclude about Sarah's score? Sarah's score was high.

Additional Answers

7. ∠2 and ∠3
8. ∠1 and ∠4
9. ∠1 and ∠3; ∠2 and ∠4
10. *t*
11. ∠1 and ∠2; ∠3 and ∠4
18.

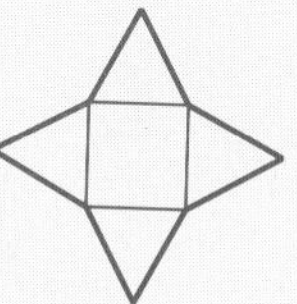

19.

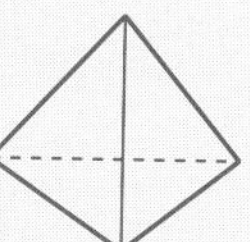

Assessment Resources
Achievement Tests pp. 3-4

Test Objectives
After studying this chapter, students should understand

- collinear, noncollinear, coplanar, and noncoplanar points.
- intersecting, parallel, and skew lines.
- parallel lines and transversals.
- corresponding angles.
- how deductive reasoning is used.
- vertical angles.
- alternate interior angles.
- the concept of error in measurement.
- alternate exterior angles.
- how a three-dimensional figure appears when drawn on a two-dimensional surface.
- the relationship between a three-dimensional figure and its layout.

Algebra Review Objectives
Topics reviewed from algebra include
- evaluating algebraic expressions.
- solving equations.
- using algebra to solve geometric problems.

ALGEBRA SKILLS

To evaluate an algebraic expression is to find its value given specified values for its variables.

To evaluate an expression:
1. Replace the variables by their values.
2. Simplify any expressions within parentheses.
3. Simplify any powers or roots.
4. Do all multiplications and divisions, from left to right.
5. Do all additions and subtractions, from left to right.

Example:
Evaluate $3x + y$ for $x = 12$ and $y = -6$.

$$3x + y$$
$$3(12) + (-6)$$
$$36 - 6$$
$$30$$

To solve an equation means to find the number or numbers that can replace the variable to make the equation true. Such numbers are called solutions of the equation. Recall that solving equations is the process of isolating the variable in the equation in one member of the equation by using inverse operations.

Example:
Solve for x.

$$7x = 63 - 2x$$
$$7x + 2x = 63 - 2x + 2x$$
$$9x = 63$$
$$x = 7$$

Evaluate the following expressions for x = 36 and y = −59.

1. $x + y$ −23 **2.** $x - y$ 95 **3.** xy −2124 **4.** $\frac{x}{12}$ 3

5. $y - x$ −95 **6.** x^2 1296 **7.** $\frac{y-1}{x}$ $-1\frac{2}{3}$ **8.** $2x - y$ 131

Evaluate these expressions for $x = \frac{1}{4}$ and $y = -\frac{2}{3}$

9. $x + y$ $-\frac{5}{12}$ **10.** $x - y$ $\frac{11}{12}$ **11.** xy $-\frac{1}{6}$ **12.** $\frac{x}{y}$ $-\frac{3}{8}$

13. $y - x$ $-\frac{11}{12}$ **14.** y^2 $\frac{4}{9}$ **15.** $\frac{y}{x}$ $-2\frac{2}{3}$ **16.** $\frac{x+y}{2}$ $-\frac{5}{24}$

Solve for x.

17. $5x - 35 = 55$ x = 18

18. $\frac{-3}{x-4} = -\frac{3}{5}$ x = 9

19. $\frac{4}{5} = \frac{2x}{20}$ x = 8

20. $\frac{8}{3} = \frac{21}{3x}$ $x = 2\frac{2}{3}$

21. $10 - 2(3x - 5) = 80$ x = −10

22. $12 - 5(2x - 20) = 3(2x - 8)$ x = 8.5 or $8\frac{1}{2}$

23. Two angles are supplementary. One is 3 times as great as the other. What is the measure of the larger angle? 135°

24. Angle 1 and angle 2 are corresponding angles. The measure of angle 1 is $(6x - 5)°$. The measure of angle 2 is $(x + 60)°$. What is the measure of each angle? 73°

25. Angle 3 and angle 4 are alternate interior angles. The measure of angle 3 is $(2x - 15)°$ and the measure of angle 4 is $(85 - 3x)°$. What is the measure of each angle? 25°

Name ______________________ Date ______________

Points and Lines in Space

Points on the same line are **collinear. Noncollinear** points are not on the same line.

Points in the same plane are **coplanar. Noncoplanar** points are not in the same plane.

Points *E, F,* and *G* are collinear. Points *H, E,* and *F* are noncollinear.

Points *A, B,* and *C* are coplanar. Points *A, B,* and *D* are noncoplanar.

When two lines are in the same plane and do not intersect, the lines are **parallel.**

When two lines are not in the same plane and do not intersect, the lines are **skew.**

Lines *a* and *b* are parallel.

Lines *c* and *d* are skew.

Refer to the figure at the right.

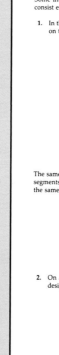

1. Are points *R, S,* and *T* collinear or noncollinear? Collinear
2. Are points *A, R,* and *X* collinear or noncollinear? Noncollinear
3. Are points *A, R,* and *T* coplanar or noncoplanar? Coplanar
4. Are points *X, R, T,* and *Y* coplanar or noncoplanar? Coplanar
5. Are points *X, R, T,* and *A* coplanar or noncoplanar? Noncoplanar

Refer to the drawing of the house at the right.

6. Name two parallel lines. a and b
7. Name two intersecting lines. a and c
8. Name two skew lines. b and c

Name ______________________ Date ______________

Line Designs

Some interesting line designs can be created by the following technique. Although the designs consist entirely of straight lines, they seem to contain curves.

1. In the figure below, draw line segments to connect point 1 on the top line segment to point 1 on the bottom line segment, point 2 to point 2, and so on.

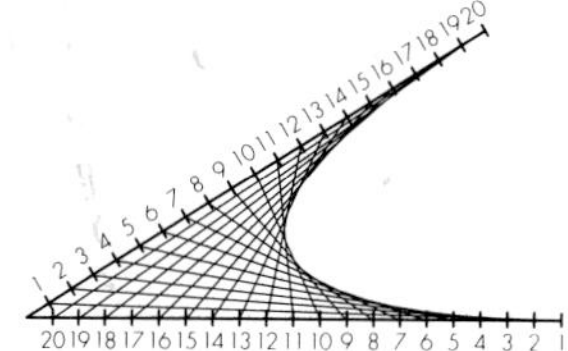

The same technique can be used to create more complicated designs by using more than two line segments. Note that you can use as many points as you want on each segment as long as you use the same number of points on each.

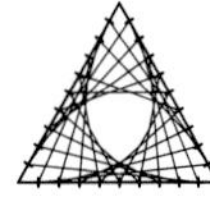

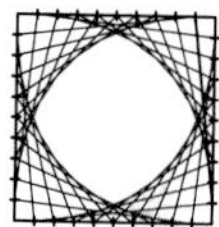

2. On another sheet of paper, make up a line design of your own. You may choose to make a design for your name. A sample is shown below.

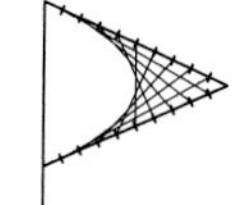

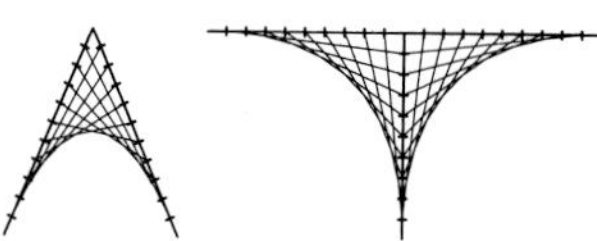

Name ______________________ Date ______________

Corresponding Angles

In Figure 1, line c is a transversal of lines a and b.

There are four pairs of corresponding angles:

$\angle 1$ and $\angle 5$, $\angle 2$ and $\angle 6$, $\angle 3$ and $\angle 7$, $\angle 4$ and $\angle 8$.

When two parallel lines are cut by a transversal, the corresponding angles are equal in measure.

In Figure 2, line d and line e are parallel. Therefore,

$m\angle 9 = m\angle 13$, $m\angle 10 = m\angle 14$, $m\angle 11 = m\angle 15$, $m\angle 12 = m\angle 16$.

Example: In Figure 2, if $m\angle 9 = 110°$, find $m\angle 14$.

You know $\angle 9$ and $\angle 10$ are supplementary.

$$110° + m\angle 10 = 180°$$
$$m\angle 10 = 70°$$

Since $m\angle 10 = m\angle 14$, you know $m\angle 14 = 70°$.

In Figure 3, line g is parallel to line h.

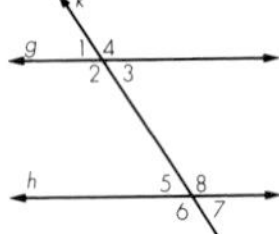

1. Which line is a transversal of lines g and h? Line k
2. In the figure, name four pairs of corresponding angles. $\angle 1$ and $\angle 5$, $\angle 2$ and $\angle 6$, $\angle 3$ and $\angle 7$, $\angle 4$ and $\angle 8$
3. Is line g a transversal of lines h and k? No

In Figure 3, $m\angle 1 = 41°$. Find the following angle measures.

4. $m\angle 2$ 139°
5. $m\angle 3$ 41°
6. $m\angle 4$ 139°
7. $m\angle 5$ 41°
8. $m\angle 6$ 139°
9. $m\angle 7$ 41°

In Figure 3, suppose $m\angle 1 = (3x + 13)°$ and $m\angle 2 = (4x - 1)°$. Find the following.

10. $m\angle 1$ 85°
11. $m\angle 8$ 95°

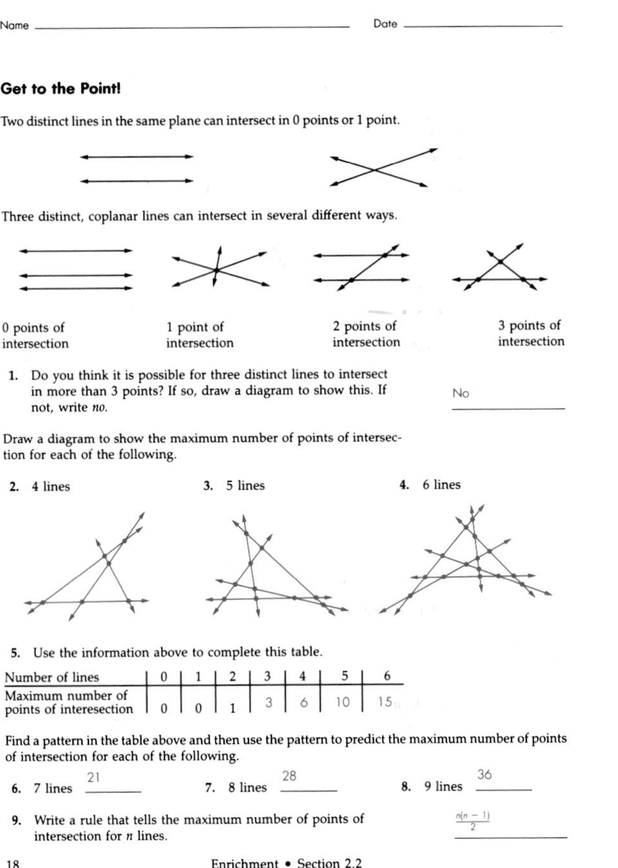
Name ______________________ Date ______________

Get to the Point!

Two distinct lines in the same plane can intersect in 0 points or 1 point.

Three distinct, coplanar lines can intersect in several different ways.

0 points of intersection | 1 point of intersection | 2 points of intersection | 3 points of intersection

1. Do you think it is possible for three distinct lines to intersect in more than 3 points? If so, draw a diagram to show this. If not, write *no*. No

Draw a diagram to show the maximum number of points of intersection for each of the following.

2. 4 lines
3. 5 lines
4. 6 lines

5. Use the information above to complete this table.

Number of lines	0	1	2	3	4	5	6
Maximum number of points of interesection	0	0	1	3	6	10	15

Find a pattern in the table above and then use the pattern to predict the maximum number of points of intersection for each of the following.

6. 7 lines 21
7. 8 lines 28
8. 9 lines 36
9. Write a rule that tells the maximum number of points of intersection for n lines. $\frac{n(n-1)}{2}$

Name ____________ Date ____________

Vertical and Alternate Interior Angles

When two lines intersect, there are two pairs of **vertical angles.** Vertical angles are equal in measure.

$\angle 1$ and $\angle 3$ are vertical angles.
$\angle 2$ and $\angle 4$ are vertical angles.

So $m\angle 1 = m\angle 3$ and $m\angle 2 = m\angle 4$.

When two parallel lines are cut by a transversal, the **alternate interior angles** are equal in measure. In the figure at the right, lines *c* and *d* are parallel.

$\angle 7$ and $\angle 10$ are alternate interior angles.
$\angle 8$ and $\angle 9$ are alternate interior angles.

So $m\angle 7 = m\angle 10$ and $m\angle 8 = m\angle 9$.

Refer to the figure at the right.

1. Name four pairs of vertical angles in the figure.
   $\angle 1$ and $\angle 3$, $\angle 2$ and $\angle 4$, $\angle 5$ and $\angle 7$, $\angle 6$ and $\angle 8$
2. Name two pairs of alternate interior angles in the figure.
   $\angle 2$ and $\angle 8$, $\angle 3$ and $\angle 5$

In the figure, lines *a* and *b* are parallel. Find the angle measures.

3. $m\angle 1$ 51°
4. $m\angle 2$ 129°
5. $m\angle 7$ 129°
6. $m\angle 8$ 51°
7. $m\angle 5$ 51°
8. $m\angle 6$ 129°

In the figure, lines *m* and *n* are parallel. Find the angle measures.

9. $m\angle 9$ 85°
10. $m\angle 10$ 95°
11. $m\angle 12$ 95°
12. $m\angle 11$ 85°

In the figure, lines *RS* and *AB* are parallel, $m\angle 1 = 20°$, $m\angle 2 = 110°$, and $m\angle 3 = 50°$. Find the angle measures.

13. $m\angle 4$ 70°
14. $m\angle 5$ 110°
15. $m\angle 6$ 70°
16. $m\angle 7$ 20°

Name ____________ Date ____________

Is it Logical?

Think about the following true statement.

If a person lives in France, then the person lives in Europe.
Now consider the following arguments.

AFFIRMING THE HYPOTHESIS

Pierre lives in France.
Therefore, Pierre lives in Europe.

AFFIRMING THE CONCLUSION

Pierre lives in Europe.
Therefore, Pierre lives in France.

The first argument is valid and is the basic type of argument used in deductive reasoning. The second type of argument is not valid. Clearly, Pierre might live in some other European country besides France. Be careful that you do not use this faulty type of reasoning.

Two other types of reasoning are sometimes used. The first is valid and the second is faulty.

DENYING THE CONCLUSION

Pierre does not live in Europe. Therefore, Pierre does not live in France.

DENYING THE HYPOTHESIS

Pierre does not live in France. Therefore, Pierre does not live in Europe.

Tell whether each argument is valid or faulty.

1. If you live in New York, you live in America.
   Al lives in America.
   Therefore, Al lives in New York.
   Faulty
2. If you live in Puerto Rico, you live on an island.
   Maria lives in Puerto Rico.
   Therefore, Maria lives on an island.
   Valid
3. If two angles are supplementary, the sum of their angle measures is 180°.
   The sum of the measures of $\angle 1$ and $\angle 2$ is not 180°.
   Therefore, $\angle 1$ and $\angle 2$ are not supplementary.
   Valid
4. If two angles are right angles, the angles are supplementary.
   $\angle 1$ and $\angle 2$ are not right angles.
   Therefore, $\angle 1$ and $\angle 2$ are not supplementary.
   Faulty

Name ______________________ Date ______________

Measurement and Estimation

Angles are measured in degrees. When a more precise measurement is needed, you can use *minutes* and *seconds.*

1 degree = 60 minutes
1 minute = 60 seconds

To measure length you can use many different units of measurement. In this country, we mainly use customary measurements. However, the metric system of measurement is widely used in other countries.

Customary measurements	*Metric measurements*
1 foot = 12 inches	1 centimeter (cm) = 10 millimeters
1 yard = 3 feet	1 meter (m) = 100 centimeters
1 mile = 5,280 feet	1 kilometer = 1,000 meters

In some situations, an estimated measurement is appropriate. In other situations, you need to use a measuring device.

Estimate the measure of each angle in degrees. Then measure with a protractor. Estimates will vary.

1.

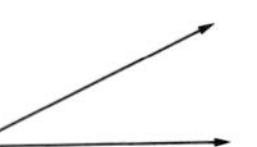

2.

Estimate: ______ Actual: 25° ______ Estimate: ______ Actual: 132° ______

Estimate the length of each segment to the nearest inch and to the nearest centimeter. Then measure to the nearest $\frac{1}{8}$ inch and to the nearest millimeter. Estimates will vary.

3. ______________________ 4. ______________________

Estimate: ___ in. Estimate: ___ cm Estimate: ___ in. Estimate: ___ cm

Actual: $1\frac{7}{8}$ in. Actual: 47 mm Actual: $3\frac{1}{8}$ in. Actual: 79 mm

Complete the following:

5. 3 miles = 15,840 feet
6. 60 yd = 180 ft
7. 8 km = 8,000 m
8. 750 cm = 7.5 m
9. 9.2 cm = 92 mm
10. 600 in. = 50 ft

Name ______________________ Date ______________

Greatest Possible Error

Any measurement is approximate. The **precision** of a measurement depends upon the **unit of measurement.** The smaller the unit of measurement, the more precise is the measurement. For example, a measurement of $12\frac{1}{16}$ inches is more precise than a measurement of $12\frac{1}{2}$ inches. For $12\frac{1}{16}$ inches, the unit of measurement is $\frac{1}{16}$ inch. For $12\frac{1}{2}$ inches, the unit of measurement is $\frac{1}{2}$ inch.

The **greatest possible error** of a measurement is half of the unit of measurement.

Example:

Measurement	Unit of measurement	Greatest possible error
$15\frac{1}{4}$ inches	$\frac{1}{4}$ inch	$\frac{1}{8}$ inch
3.1 km	0.1 km	0.05 km

Give the unit of measurement and the greatest possible error for each measurement.

	Unit of measurement	Greatest possible error
1. $2\frac{1}{2}$ inches	$\frac{1}{2}$ in.	$\frac{1}{4}$ in.
2. $5\frac{1}{4}$ feet	$\frac{1}{4}$ ft	$\frac{1}{8}$ ft
3. 27 cm	1 cm	0.5 cm
4. 58.1 cm	0.1 cm	0.05 cm
5. 8.46 km	0.01 km	0.005 km
6. $3\frac{1}{16}$ inches	$\frac{1}{16}$ in.	$\frac{1}{32}$ in.

Which measurement is more precise?

7. 8.1 km or 12 km 8.1 km
8. 8 ft or 85 in. 85 in.
9. 5 cm or 12 mm 12 mm
10. 9 m or 87 cm 87 cm
11. $16\frac{1}{2}$ in. or $3\frac{1}{2}$ ft $16\frac{1}{2}$ in.
12. 3 km or 3 cm 3 cm
13. 18 yd or 1.7 mi 18 yd
14. 42 mm or 1.8 km 42 mm

Name ______________________ Date ______________

Alternate Exterior Angles

When two parallel lines are cut by a transversal, the **alternate exterior angles** are equal in measure. In the figure at the right, lines *a* and *b* are parallel.

$\angle 1$ and $\angle 8$ are alternate exterior angles.
$\angle 2$ and $\angle 7$ are alternate exterior angles.
So $m\angle 1 = m\angle 8$ and $m\angle 2 = m\angle 7$.

Perpendicular lines are lines which form a 90° angle (a right angle). The symbol for "is perpendicular to" is $\perp$. In the figure at the right, line $c \perp$ line d.

1. In the figure at the right, lines *p* and *q* are parallel. Name two pairs of alternate exterior angles.
   $\angle 1$ and $\angle 8$, $\angle 3$ and $\angle 6$
2. Name two pairs of perpendicular lines in the figure.
   p and r, q and r
3. Name all the right angles in the figure.
   $\angle 1$, $\angle 2$, $\angle 3$, $\angle 4$, $\angle 5$, $\angle 6$, $\angle 7$, $\angle 8$

In the figure, lines *e* and *f* are parallel. Find the angle measures.

4. $m\angle 1$ 30°
5. $m\angle 2$ 150°
6. $m\angle 3$ 30°
7. $m\angle 4$ 30°
8. $m\angle 5$ 150°
9. $m\angle 6$ 30°

In the figure, lines *m* and *n* are parallel. Find the angle measures.

(5x + 10)° (4x − 10)°

10. $m\angle 7$ 70°
11. $m\angle 8$ 110°
12. $m\angle 9$ 110°
13. $m\angle 10$ 70°
14. $m\angle 11$ 70°
15. $m\angle 12$ 110°

Name ______________________ Date ______________

The Reflexive, Symmetric, and Transitive Properties

Three important properties of equality are the *reflexive, symmetric,* and *transitive* properties.

Reflexive property	$a = a$
Symmetric property	If $a = b$, then $b = a$.
Transitive property	If $a = b$ and $b = c$, then $a = c$.

If you replace = with < in each statement above, you have:

a < a	◀Not true
If a < b, then b < a.	◀Not true
If a < b and b < c, then a < c.	◀True

Only the transitive property is true for "is less than."

Example: Determine which of the three properties above are true for "is supplementary to."

$\angle A$ is supplementary to $\angle A$..	◀Not true
If $\angle A$ is supplementary to $\angle B$, then $\angle B$ is supplementary to $\angle A$.	◀True
If $\angle A$ is supplementary $\angle B$ and $\angle B$ is supplementary to $\angle C$, then $\angle A$ is supplementary to $\angle C$.	◀Not true

Only the symmetric property is true for "is supplementary to."

Determine which properties are true for each relation.

1. is greater than Transitive
2. is perpendicular to Symmetric
3. is taller than Transitive
4. is the brother of Transitive
5. is on the same team as Reflexive, Symmetric, Transitive
6. is the first cousin of Symmetric
7. is younger than Transitive

Name ______________________ Date ______________

Perspective Drawings

Artists sometimes use a technique called perspective drawing to give the illusion that a 2-dimensional figure is actually a 3-dimensional figure. Two common types of perspective drawing involve *1-point perspective* and *2-point perspective.*

1-point perspective

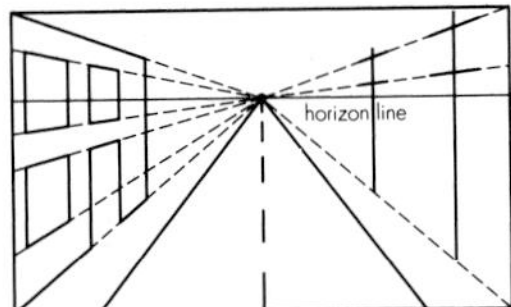

2-point perspective

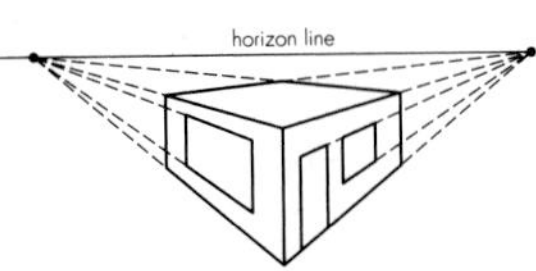

In the first drawing, there is only one **vanishing point** where all the lines seem to meet. In the second drawing, there are two vanishing points. Vanishing points are always located on a horizontal line called the **horizon line.** The placement of the horizon line depends on the eye level of the viewer. The higher the eye level of the viewer, the higher is the horizon line.

1. Trace the following drawings and locate the vanishing point or points.

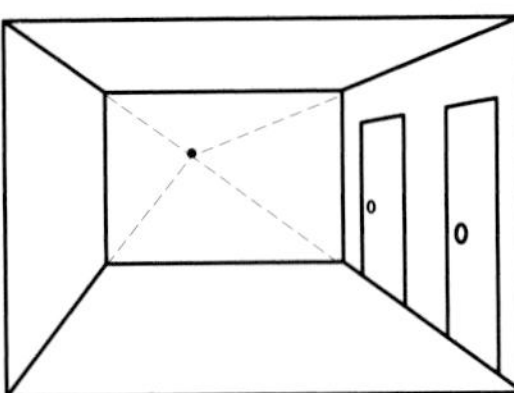

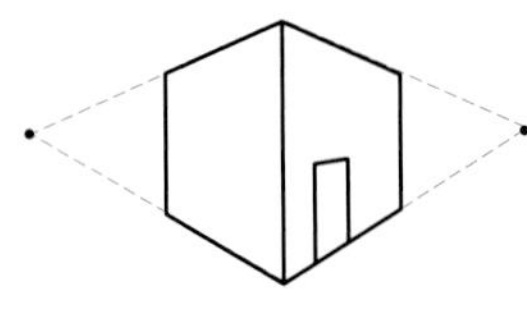

2. In the first drawing, add a window to make the drawing look like the interior of a room.
   Student drawings will vary.
3. In the second drawing, add some more details to make the drawing look more like a building. Student drawings will vary.

Name ______________________ Date ______________

Visualizing Three-Dimensional Figures

Certain tricks can be used to "fool the eye" into perceiving a 2-dimensional figure as a 3-dimensional figure. For example, right angles are sometimes drawn larger or smaller than a right angle, and dashed lines are used to imply that the lines are hidden from view.

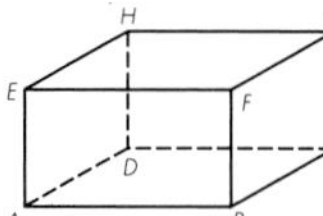

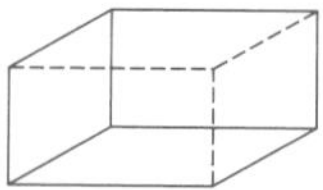

Refer to the drawing of a cardboard box shown above.

1. Name all angles that are drawn larger than a right angle.
   ∠ABC, ∠ADC, ∠EFG, ∠EHG, ∠BFG, ∠BCG, ∠ADH, ∠AEH
2. Name all angles that are drawn smaller than a right angle.
   ∠DAB, ∠DCB, ∠HEF, ∠HGF, ∠FBC, ∠FGC, ∠EAD, ∠EHD
3. Name all angles that are drawn as a right angle.
   ∠EAB, ∠ABF, ∠BFE, ∠FEA, ∠DCG, ∠CGH, ∠GHD, ∠HDC
4. In the actual box which the drawing represents, which angles are right angles?
   All of the angles
5. How many lines are hidden when you view the box from this viewpoint?
   3
6. The drawing on this page shows what the box looks like when you look down at the box from above and to the right. Draw a similar figure to show what the box looks like when you look up at it from below and to the left. HINT: The dashed lines will be located in different positions from those in the figure above. Draw your figure next to the one at the top of the page.

Name ______________________ Date ______________

Using LOGO to Draw Parallel Lines

Draw a diagram in the space below each exercise to show how the turtle moves when you type these LOGO commands. Identify the parallel line segments and the transversal(s) in each diagram. Then label each angle in the diagram with its measure. Justify your answers.

1. DRAW
RT 90 FD 50 HOME
RT 40 FD 80 BK 20
RT 40 FD 50

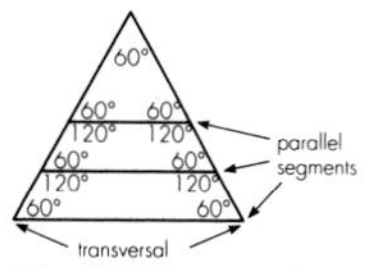

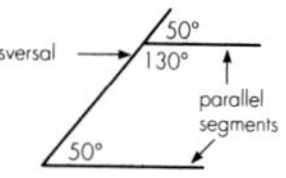

The 50° angles are corresponding.
The 50° angle and the 130° angle are supplementary.

2. DRAW
LT 90 FD 65 HOME
RT 45 BK 30 FD 115
BK 30 RT 45 FD 65

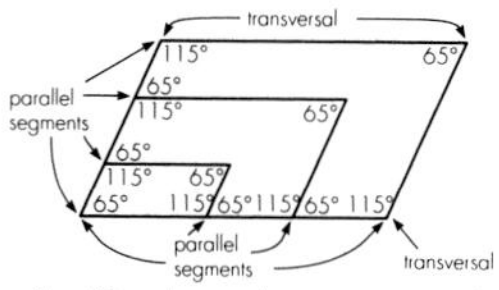

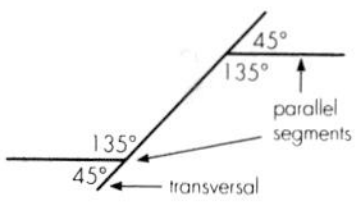

The 45° angles are alternate interior angles. The 135° angles are alternate interior angles.

3. DRAW
RT 30 FD 60
RT 120 FD 60
RT 120 FD 60
RT 120 FD 30
RT 60 FD 30
RT 60 FD 15
RT 120 FD 45

All 60° angles are corresponding angles. All 120° angles are corresponding angles.

4. DRAW
RT 25 FD 50
RT 65 FD 75
RT 115 FD 50
RT 65 FD 75
RT 115 FD 30
RT 65 FD 55
RT 115 FD 30
RT 65 FD 55
RT 115 FD 10
RT 65 FD 35
RT 115 FD 10

The 65° angles formed by both sets of parallel segments are corresponding angles. The 115° angles formed by both sets of parallel segments are corresponding angles.

Name________________ Class________________ Date________________

Achievement Test 2 (Chapter 2)
INTERSECTING AND PARALLEL LINES

No. Correct
No. Exercises: **39**
Score
2.56 x No. Correct =

GEOMETRY FOR DECISION MAKING
James E. Elander
SOUTH-WESTERN PUBLISHING CO.

Match the terms on the right with the descriptions on the left.

1. points on the same line b
2. points not on the same plane d
3. a flat surface a
4. points on the same plane c

a. plane
b. collinear points
c. coplanar points
d. noncoplanar points

Tell whether the statement is true or false.

5. Intersecting lines are not coplanar. False
6. Any two points are collinear. True
7. Parallel lines are coplanar. True

In the figure, $\overleftrightarrow{AB}$ and $\overleftrightarrow{CD}$ are coplanar lines that do not intersect. Identify the following.

8. three collinear points A, E, B, or C, F, D
9. three noncollinear points A, E, C ; other answers possible
10. four coplanar points A, B, C, D; other answers possible
11. two parallel lines $\overleftrightarrow{AB}$, $\overleftrightarrow{CD}$
12. two skew lines $\overleftrightarrow{GH}$, $\overleftrightarrow{AB}$
13. the point at which three lines intersect F

In the diagram, $r \parallel s$. Identify the following.

14. a transversal t
15. a pair of vertical angles $\angle 1$, $\angle 2$ or $\angle 3$, $\angle 4$
16. a pair of corresponding angles $\angle 1$, $\angle 3$ or $\angle 2$, $\angle 4$
17. a pair of alternate interior angles $\angle 2$, $\angle 3$
18. a pair of alternate exterior angles $\angle 1$, $\angle 4$

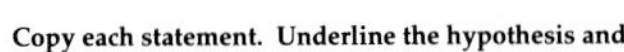

Copy each statement. Underline the hypothesis and ring the conclusion.

19. If you live in Rome, then you live in Italy.
20. If your pet is a collie, then it is a dog.
21. If $y + 3 = 8$, then $y = 5$.

MG01AG

Suppose each pair of statements is true. What can you conclude?

22. If an angle is acute, then its measure is less than 90°. $\angle ABC$ is acute. $m\angle ABC < 90°$
23. If two angles are complementary, then the sum of their measures is 90°. $\angle 1$ and $\angle 2$ are complementary. $m\angle 1 + m\angle 2 = 90°$

In the diagram, $p \parallel q$ and t is a transversal. Give the degree measure of each angle.

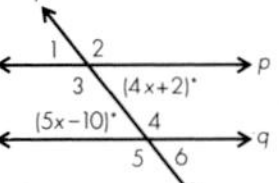

24. $\angle 1$ 50°
25. $\angle 2$ 130°
26. $\angle 3$ 130°
27. $\angle 4$ 130°
28. $\angle 5$ 130°
29. $\angle 6$ 50°

Use a calculator to help you answer the questions.

30. Maria is 64 in. tall. What is her height in feet and inches? 5 ft 4 in.
31. An industrial park is 38,623 m from Fort Worth. How far is it in kilometers? 38.623 km
32. To the nearest tenth of a mile, how long is a path that is 6,653 yd long? 3.8 mi

Measure each segment to the nearest inch and $\frac{1}{8}$ inch. State the error for each measurement.

33. A ______ B nearest inch: 2 in. error: $\frac{1}{2}$ in.; nearest $\frac{1}{8}$ in.: $2\frac{3}{8}$ in. error: $\frac{1}{16}$ in.
34. C ______ D nearest inch: 3 in. error: $\frac{1}{2}$ in.; nearest $\frac{1}{8}$ in.: $3\frac{3}{8}$ in. error: $\frac{1}{16}$ in.

Measure each angle to the nearest 5 degrees and to the nearest degree. State the error for each measurement.

35.
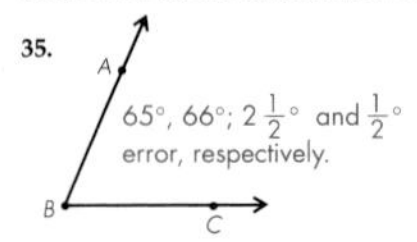

65°, 66°; $2\frac{1}{2}$° and $\frac{1}{2}$° error, respectively.

36.
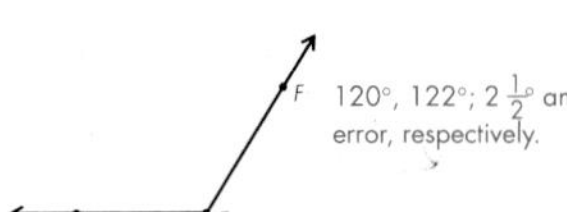

120°, 122°; $2\frac{1}{2}$° and $\frac{1}{2}$° error, respectively.

Draw a layout for each figure.

37.

38.
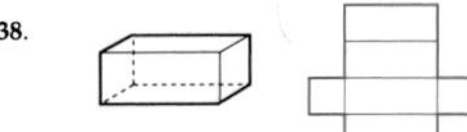

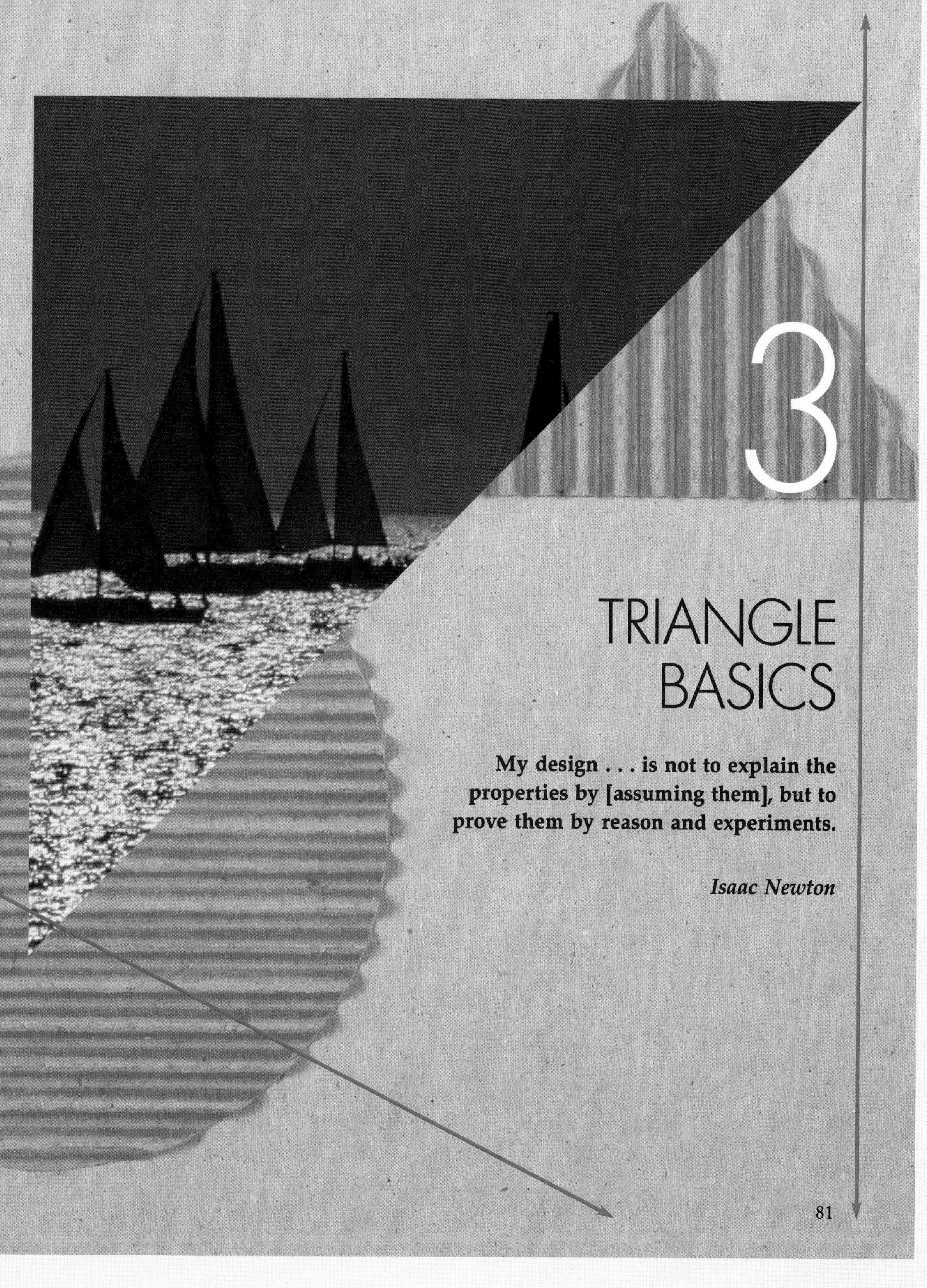

3

TRIANGLE BASICS

My design . . . is not to explain the properties by [assuming them], but to prove them by reason and experiments.

Isaac Newton

3

81

CHAPTER 3 OVERVIEW

Triangle Basics

Suggested Time: 12–14 days

The quote at the beginning of the chapter was selected because it not only states what occurs in the chapter—discovery activities or experiments leading to a conclusion, then justification of the conclusion—but also the objectives of geometry. The first theorem in the chapter tells us that the sum of the measures of the angles of a triangle is 180°. This theorem is derived from the famous Euclidean 5th postulate, which is concerned with the number of lines parallel to another line through a point not on the line. Another key theorem in this chapter states that the sum of the lengths of any two sides of a triangle is greater than the length of the third side. You may wish to read Chapter 2 of Morris Kline's *Mathematics in the Physical World*, a very informative chapter on proof.

My design . . . is not to explain the properties by [assuming them], but to prove them by reason and experiments.

Isaac Newton

CHAPTER 3 OBJECTIVE MATRIX

	End of Chapter Items by Activity				Student Supplement Items by Activity		
Objectives by Number	**Review**	**Test**	**Computer**	**Algebra Skills**	**Reteaching**	**Enrichment**	**Computer**
3.1.1	✔	✔	✔		✔		✔
3.1.2	✔	✔		✔	✔	✔	
3.2.1	✔					✔	
3.2.2	✔	✔		✔	✔		
3.2.3	✔	✔		✔	✔		
3.3.1	✔	✔	✔		✔		
3.4.1	✔	✔	✔		✔		
3.5.1	✔	✔		✔	✔	✔	
3.6.1	✔	✔	✔		✔	✔	✔
3.6.2	✔	✔	✔	✔	✔	✔	✔

*A ✔ beside a Chapter Objective under Algebra Skills indicates that algebra skills taught within the chapter or in previous Algebra Skills lessons are used.

CHAPTER 3 PERSPECTIVES

▲ Section 3.1

The Sum of the Angles of a Triangle

The term *triangle* is defined. Then, through the Discovery Activity, the class should conclude by inductive reasoning that the sum of the measures of the angles is 180°. Help the students to realize that the conclusion is not a certainty because of the inherent weakness of inductive reasoning. A report is suggested for groups of students to prepare as extra credit and present when the chapter has been completed. The author has found that students working together in groups promotes confidence and teamwork, and creates an interesting and informative spirit in the classroom.

The Home Activity will reinforce the important theorem on the sum of the angle measures of a triangle, as well as review theorems from previous chapters. Continue to encourage students to keep notebooks for recording definitions, postulates, theorems, and examples.

▲ Section 3.2

Two Triangle Inequalities

The two triangle inequalities that follow are presented in this section. A third inequality will be presented in another chapter.

- The sum of the lengths of two sides of a triangle is greater than the length of the third side.
- For any given triangle, the length of any side is greater than the absolute value of the difference of the lengths of the other two sides.

The proof of the first inequality involves indirect reasoning. The indirect method will be used throughout the textbook in both geometrical problems and those of a non-mathematical nature. The latter—problems that are 'real-world' applications—are often more interesting to students, as they can discuss them with their peers and families.

The Home Activity includes exercises showing students that the third side of a triangle is between two easily determined numbers.

▲ Section 3.3

Classification of Triangles

Classifying triangles by sides is approached through a Discovery Activity, and then definitions for the terms *equilateral*, *isosceles*, and *scalene* are introduced. Encourage students to rewrite the definitions in their own meaningful ways, as long as their definitions are correct and complete.

The Home Activity includes measuring the sides of triangles in both the metric and customary measurement systems. Exercises 7–21 may need to be discussed in class.

▲ Section 3.4

Classification of Triangles by Angles

In this section, the definitions for the terms *equiangular, right, acute,* and *obtuse* are presented. Point out that triangles can be described by both the sides and the angles: for example, a right isosceles or an obtuse scalene triangle.

Have the students contrast the four terms: For equiangular and acute triangles, all the angles are equal or acute, respectively. For right and obtuse triangles, just one angle is involved in the definition.

▲ Section 3.5

Exterior Angles of a Triangle

The terms reviewed at the beginning of this section should be entered in the students' notebooks. Then the Discovery Activity leads to a new theorem: The measure of an exterior angle of a triangle equals the sum of the measures of the two non-adjacent interior angles. Note that the inductive method is used here.

Two-column proof is introduced for the first time in this section. Students should realize that the quality of their reasoning is more important than the set-up of the argument.

The Home Activity provides practice with the new theorem. Exercises 13–15 might be used as a take-home activity to involve the family with the practical uses of geometry and indirect measurement.

▲ Section 3.6

Triangles on the *x*-*y* Plane

The purpose of this section is to refresh the students' memories about graphing concepts. Remind them that any point on the Cartesian plane can be shown with an ordered pair of numbers. The two numbers in the pair are called the coordinates of the point. Be sure that students are reminded of which axis is the *x*-axis and which is the *y*-axis. Chapter 12 will continue the discussion of the Cartesian plane.

As the students do the problems in the Home Activity, encourage neat drawings on graph paper.

SECTION
3.1

Resources
Reteaching 3.1
Enrichment 3.1

Objectives
After completing this section, the student should understand
- ▲ the definition of a triangle.
- ▲ the sum of the measures of the angles of a triangle is 180°.

Materials
protractor
calculator
library resources

Vocabulary
collinear points
noncollinear points
triangle
parallel lines
theorem
postulate

GETTING STARTED

Quiz: Prerequisite Skills

1. What term is used to describe three points on the same line? [collinear]

2. Draw two parallel lines intersected by a nonperpendicular transversal. Mark the acute alternate interior angles *a* and *b*. [Student drawings should look like the following.]

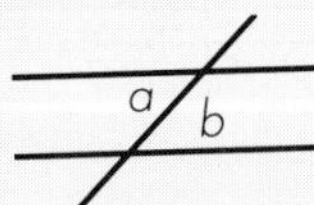

Warm-Up Activity
This paper-folding activity may be used to create physical models of triangles. Students mark the four corners of a sheet of paper with the letters *A*, *B*, *C*, and *D*. They fold the paper twice, along each of the two diagonals. Have them mark the intersection point of the folds as point O. Ask for informal descriptions of the four shapes formed by the folds.

SECTION
3.1

The Sum of the Angles of a Triangle

1 2 3 4 5 6 7 8

After completing this section, you should understand
- ▲ the definition of a triangle.
- ▲ the sum of the measures of the angles of a triangle is 180°.

Recall the definitions of collinear points and noncollinear points.

E • G • F • H

Lines, line segments, and rays are determined by two points. According to Postulate 1-1-1, two points determine exactly one line. Given three points, the points can be collinear or noncollinear. If noncollinear, then three lines are determined instead of one line.

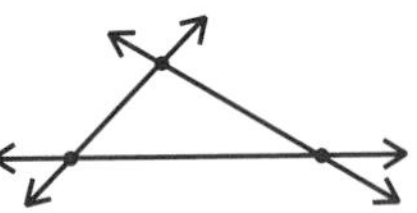

A closed figure is determined by the three lines.

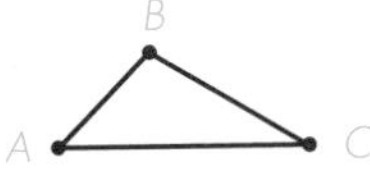

Observe the following about figure *ABC*.

Vertex	Side	Angle
A	$\overline{AB}$	$\angle A$, or $\angle BAC$, or $\angle CAB$
B	$\overline{BC}$	$\angle B$, or $\angle ABC$, or $\angle CBA$
C	$\overline{AC}$	$\angle C$, or $\angle ACB$, or $\angle BCA$

The figure has three vertices, three sides, and three angles. The prefix "tri" means "three." Thus, the figures below are all triangles.

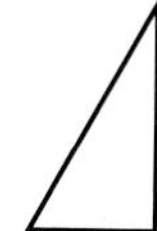

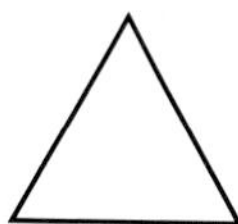

A TRIANGLE is a figure consisting of three noncollinear points and their connecting line segments.

A triangle is named by its vertices.

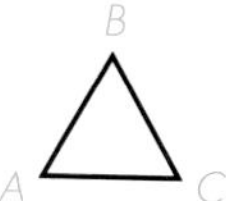

$\triangle ABC$

CLASS ACTIVITY

Tell whether each figure is a triangle. If the figure is not a triangle, explain why not.

1. 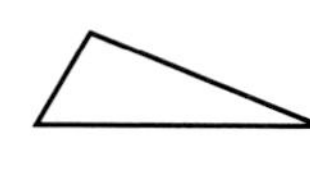
Yes

2. 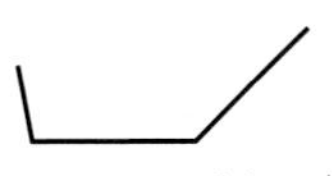
No; the sides are not connecting.

3. 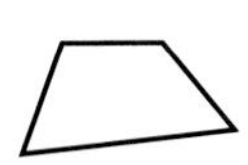
No; there are 4 sides.

4. 
No; the 3 points are collinear.

THE SUM OF THE MEASURES OF ANGLES

Mathematicians have studied the properties of triangles for many years. About 300 B.C. a man named Euclid studied a property of triangles, the sum of the measures of the angles of the triangle.

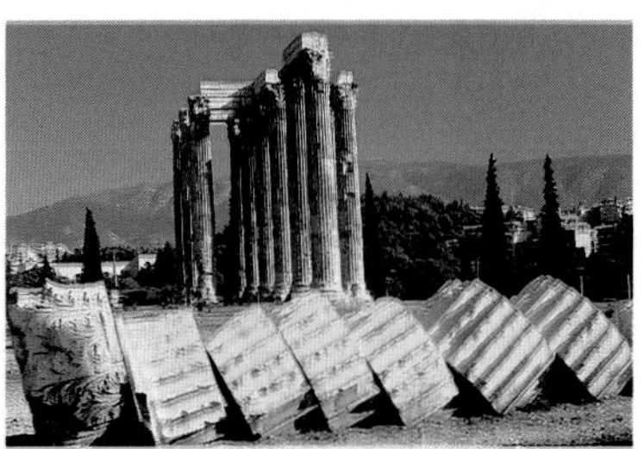

Chalkboard Example
Draw the following triangle on the chalkboard.

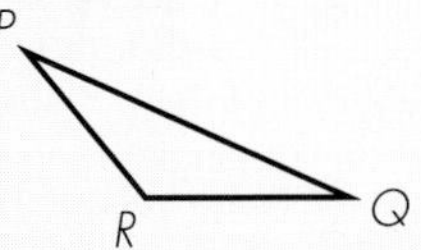

Have students name each vertex, each side (two ways), and each angle (three ways).

TEACHING COMMENTS

Using the Page
During the discussion of Definition 3-1-1, have students give their own definitions for a triangle. Discuss which ones are valid and which ones are not. For the definitions that are not valid, have students work in groups to come up with counterexamples.

Extra Practice
Tell whether each figure is a triangle. If the figure is not a triangle, explain why not.

1. [Yes]

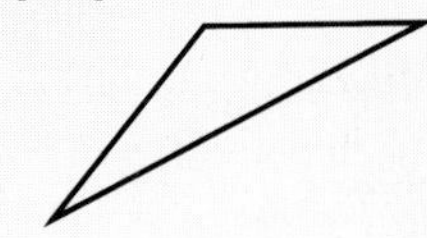

2. [No; the sides are not connecting.]

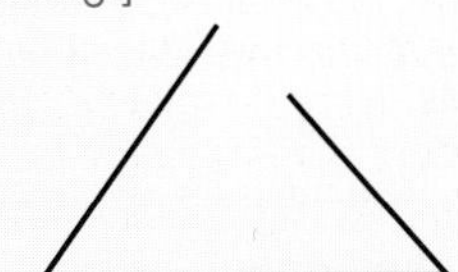

Using the Page
While using the Discovery Activity, you may wish to have students try to form a figure other than a triangle using three noncollinear points. They should conclude that a triangle is the only possibility.

Additional Answers
For the final step in the Discovery Activity, students may write any number of valid generalizations. Help them realize that the desired response is a generalization about the sum of the measures of the angles in any triangle.

Using the Page
Go through each step of the proof of Theorem 3-1-1 and make sure the students understand the reason for each step. This could also be set up as a statement=reason type proof.

Using the Page
Postulate 3-1-1 could be another activity for the students. Challenge a student to try a draw a second line through point A that is also parallel to the first.

DISCOVERY ACTIVITY

1. Mark three noncollinear points on a sheet of paper and label them *A*, *B*, and *C*. Connect the points to form a triangle.
2. Measure each angle, then find the sum of the angles.
3. Have one person record the sum for each student's triangle on the chalkboard. Using a calculator, find the average sum.
4. Draw a different triangle, and then find the sum of the angles for this triangle.
5. Write a generalization that seems to be true. The sum of the measures of the angles of a triangle is 180°.

You probably found that the sum of the angles is close to 180°.

Euclid gave the following as a proof that the sum of the measures of the angles of a triangle is 180°.

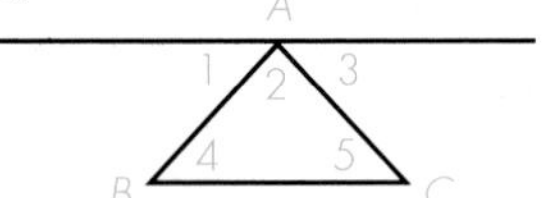

Draw a line through *A* parallel to $\overline{BC}$. $m\angle 1 = m\angle 4$ and $m\angle 3 = m\angle 5$. If parallel lines are intersected by a transversal, then the alternate interior angles are equal in measure. $m\angle 1 + m\angle 2 + m\angle 3 = 180°$, by definitions of straight angle and angle addition. Then $m\angle 4 + m\angle 2 + m\angle 5 = 180°$ by substitution of equal numbers.

THEOREM 3-1-1 **If the figure is a triangle, then the sum of the measure of the angles is 180°.**

Euclid's proof for Theorem 3-1-1 depended on the fact that only one parallel line can be drawn through point *A* parallel to $\overline{BC}$. Euclid presented this as a postulate.

POSTULATE 3-1-1 **Through a point not on a line there is only one line parallel to the given line.**

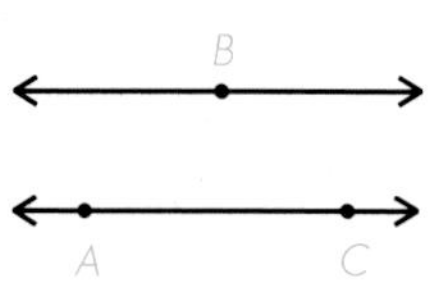

REPORT 3-1-1 The first geometry book was written by Euclid. His greatest contribution to geometry was the organization of theorems into a logical order. Use resources in the library to answer these questions about Euclid.

1. What is the name of his geometry book? *Elements*
2. How many sections, or "books," did his text contain? 13
3. What is Euclid's famous fifth postulate? Parallel postulate
4. Who was one of his teachers? Plato
5. In which book can you find the Pythagorean Theorem? 1

Suggested sources:
Boyer, Carl B. *A History of Mathematics.* Princeton, NJ: Princeton University Press, 1985.

SOLVING FOR ANGLES

Example: Find the measure of each angle.

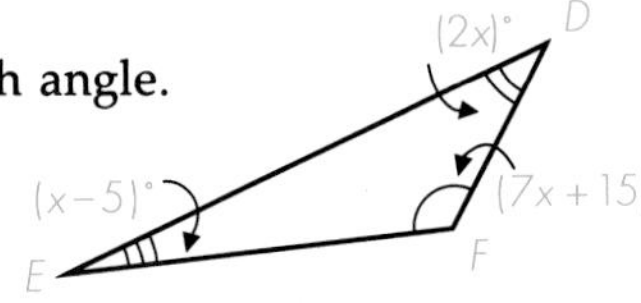

$$2x + (7x + 15) + (x - 5) = 180$$
$$10x + 10 = 180$$
$$10x = 170$$
$$x = 17$$

$\angle D$: $2(17) = 34°$
$\angle E$: $17 - 5 = 12°$
$\angle F$: $7(17) + 15 = 134°$

CLASS ACTIVITY

Find the measure of each unknown angle.

1. 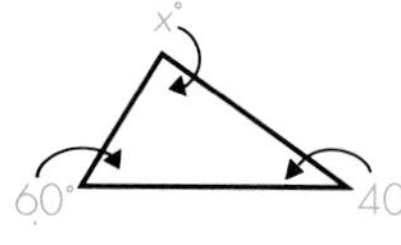

$x = 80°$

2. 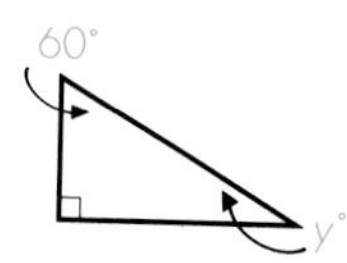

$y = 30°$

3. 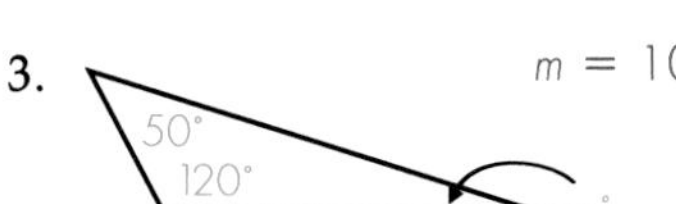

$m = 10°$

4. 60° B (3x)° A (5x+40)° C
$\angle A = 30°$
$\angle C = 90°$

Using the Page
Students who have difficulty with the problems in the Class Activity might first find the missing angles by measuring. Some additional measuring experiences may help to convince them that the sum of the measures of the angles is always 180°.

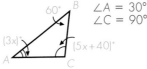

Extra Practice
Find the measure of each unknown angle.

1. [20°; 30°]

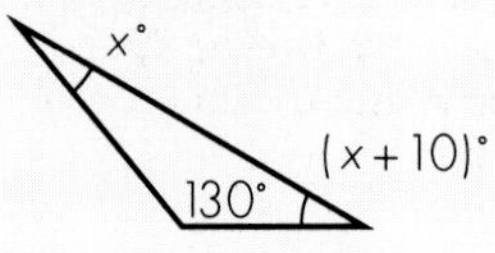

2. [20°; 50°; 110°]

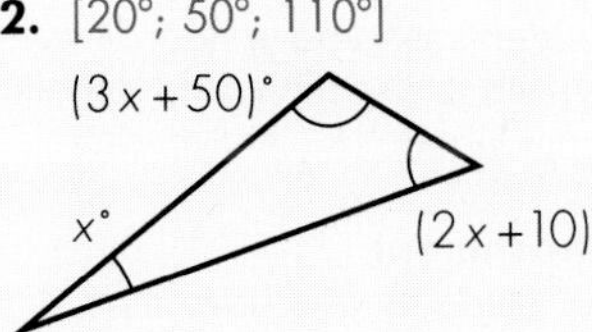

FOLLOW-UP

Assessment

1. Explain why each figure is *not* a triangle.

a. [The figure does not have three sides.]

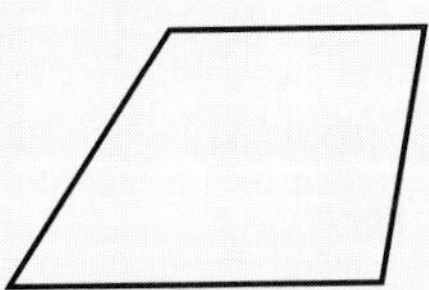

b. [The figure is not closed.]

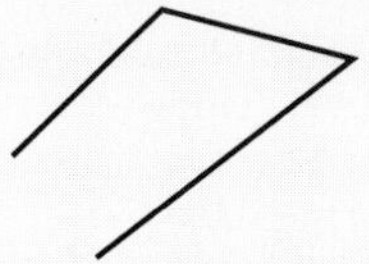

2. Find the measurement of the angle marked x in this triangle. [45°]

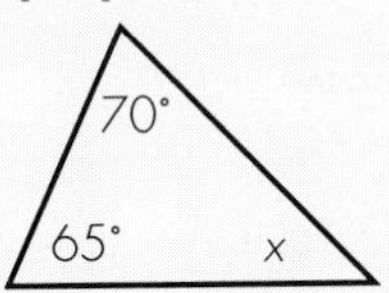

Additional Answers

1. **a.** $\triangle ABC$ **b.** $\overline{AB}$, $\overline{BC}$, $\overline{AC}$ **c.** A, B, C **d.** $\angle BAC$, $\angle ABC$, $\angle ACB$
2. **a.** $\triangle DEF$ **b.** $\overline{DE}$, $\overline{EF}$, $\overline{DF}$ **c.** D, E, F **d.** $\angle EDF$, $\angle DEF$, $\angle DFE$
7. $y = 60°$, $a = 40°$, $x = 80°$, $z = 120°$, $c = 140°$
8. $x = 40°$, $y = 60°$, $z = 80°$, $a = 100°$, $b = 60°$, $w = 100°$, $c = 40°$, $d = 20°$

12. Yes; $x + (x + 1) + (x + 2) = 3x + 3 = 3(x + 1)$; one factor is a multiple of 3, therefore the product is divisible by 3.

Reteaching

Students who have had difficulty with this section may benefit from Reteaching Activity 3.1.

Enrichment

For students who have mastered the material in this section, you may wish to assign Enrichment Activity 3.1.

HOME ACTIVITY

For each triangle, a. name the triangle, using three letters; b. name the sides; c. name the vertices; d. name the angles (three-letter method).

1. 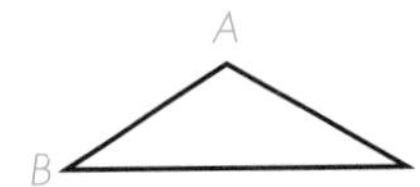

See margin.

2. 

See margin.

Use the information in each figure to find the measure of each unknown angle. Write the letter of the theorems or definition below that you used to solve the exercise.

A. If the figure is a triangle, then the sum of the measures of the angles is 180°.
B. If two parallel lines are intersected by a transversal, then the alternate interior angles are equal.
C. If two lines intersect, then the vertical angles are equal.
D. Two angles are supplementary if their measures add to 180°.

3. 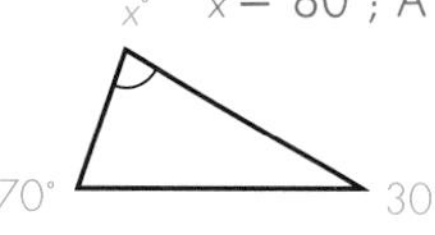

$x = 80°$; A

4. 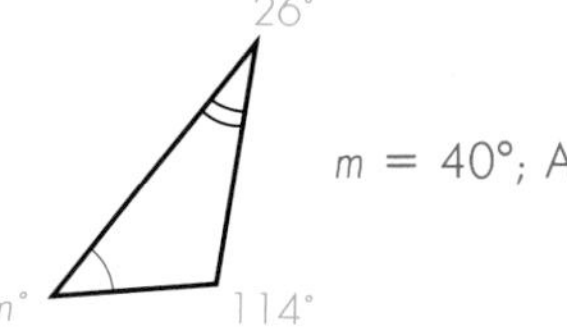

$m = 40°$; A

5. 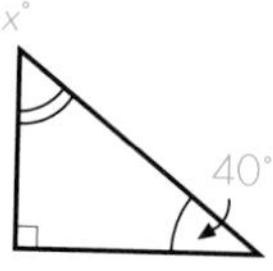

$x = 50°$; A

6. 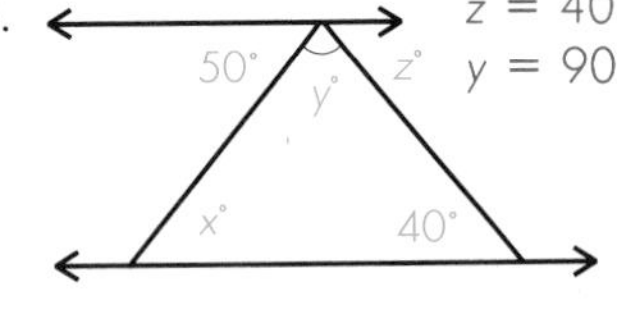

$x = 50°$; B
$z = 40°$; B
$y = 90°$; A

7. See margin.

8. 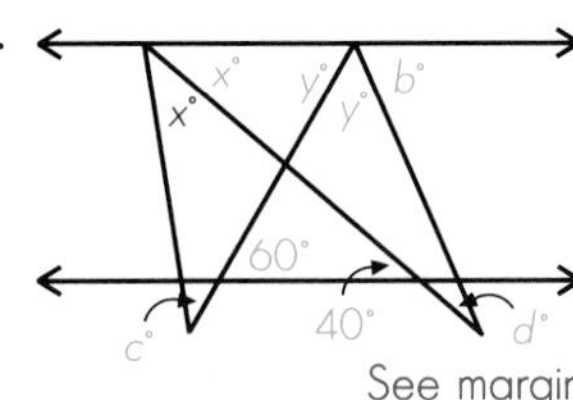

See margin.

9. Can a triangle have two obtuse angles? Explain your answer. No; the sum of angles would be greater than 180°.

Write Logo commands to draw each triangle described below. Use pencil and paper to find the measurement of the third angle. Answers may vary.

10. $\triangle XYZ$ with $m\angle X = 30°$ and $m\angle Y = 90°$
$m\angle Z = 60°$

11. $\triangle ABC$ with $m\angle A = 60°$ and $m\angle B = 60°$
$m\angle C = 60°$

CRITICAL THINKING

12. Randy and Jan predicted by inductive reasoning that the sum of any three consecutive counting numbers is divisible by three. Do you think they are right? Try a few cases and look for a pattern. See margin.

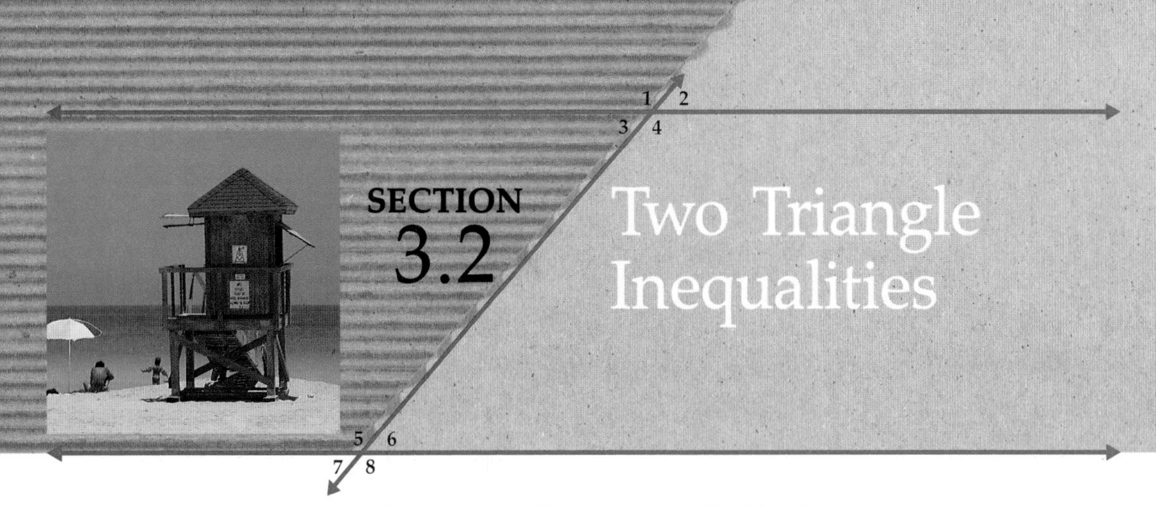

SECTION 3.2 Two Triangle Inequalities

After completing this section, you should understand

- ▲ indirect reasoning.
- ▲ the sum of the lengths of two sides of a triangle is greater than the length of the third side.
- ▲ the length of any side of a triangle is greater than the difference of the lengths of the other two sides.

In Section 3.1, the theorem about the sum of the measures of the angles of a triangle was presented. A proof given by Euclid was outlined to justify the theorem. Two more theorems about triangles appear in this section.

Jaime and his family live in a house with a corner lot. Jaime's dog is at *A*, and Jaime is at *B*. If Jaime calls his dog, do you think the dog will follow the sidewalk, or cut across the yard to Jaime? Why?

DISCOVERY ACTIVITY

On a sheet of paper, draw a large triangle. Measure the length of each side in inches or centimeters, and record the lengths.

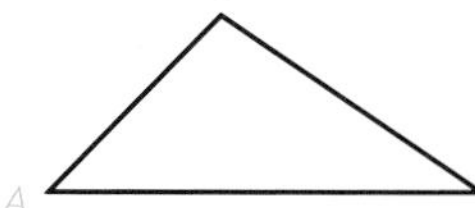

AB = __________ *BC* = __________ *AC* = __________

SECTION 3.2

Resources
Reteaching 3.2
Enrichment 3.2

Objectives
After completing this section, the student should understand

- ▲ indirect reasoning.
- ▲ the sum of the lengths of two sides of a triangle is greater than the length of the third side.
- ▲ the length of any side of a triangle is greater than the difference of the lengths of the other two sides.

Materials
ruler

Vocabulary
indirect reasoning

GETTING STARTED

Quiz: Prerequisite Skills
Solve each inequality.

1. $9 + 14 > x$ [$x < 23$]
2. $5 + x > 17$ [$x > 12$]
3. x 6 > 2 [$x > -4$]
4. Simplify $|26 - 18|$ and $|18 - 26|$. [8; 8]

Warm-up Activity
Have each student draw three line segments of different lengths on a sheet of paper. Students should measure each line segment to the nearest $\frac{1}{16}$ inch and to the nearest millimeter. Have students trade papers and measure the three line segments they now have.

Chalkboard Example
Draw a large obtuse triangle, $\triangle XYZ$, on the chalkboard. Measure the length of each side to the nearest inch. Use these measurements to illustrate Theorem 3-2-1 and Theorem 3-2-2.

TEACHING COMMENTS

Using the Page
In the Discovery Activity, students' measurements will vary, but the conclusion should be the same. The sum of the lengths of any two sides is greater than the length of the third side.

Using the Page
Students may need help listing all the possibilities for the proof by indirect reasoning. Restate the Trichotomy Law: for two numbers a and b, $a = b$, $a > b$, or $a < b$. One of the three *must* hold. Once the possibilities are listed, make sure that students know which case is to be proved to be true. All the other cases are proved to be false, until only the one case remains. This case *must* be true.

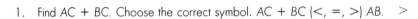

1. Find $AC + BC$. Choose the correct symbol. $AC + BC$ (<, =, >) AB. >
2. Find $AB + BC$. Choose the correct symbol. $AB + BC$ (<, =, >) AC. >
3. Find $AC + AB$. Choose the correct symbol. $AC + AB$ (<, =, >) BC. >
4. What is your conclusion in each case? The sum of the lengths of 2 sides of a triangle is greater than the length of the third side.

THE RELATIONSHIP BETWEEN THE SIDES OF A TRIANGLE

In the Discovery Activity, you may have found that the sum of the lengths of any two sides of a triangle is greater than the length of the third side.

This statement can be classified as a theorem if it can be justified. One method used to justify a theorem is indirect reasoning. Indirect reasoning involves listing all the possibilities for a situation, then showing that all but one of the possibilities are impossible.

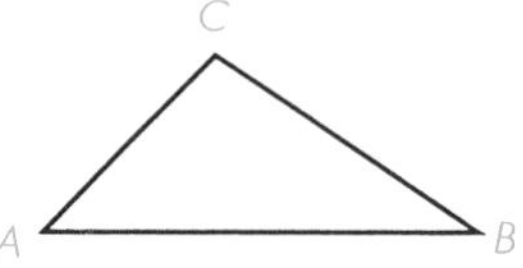

For the first part of the Discovery Activity, comparing $AC + BC$ to AB, the possibilities are
1. $AC + BC > AB$. 2. $AC + BC = AB$. 3. $AC + BC < AB$.

Case 2 says $AC + BC = AB$. If this were true, then by Definition 1-3-3, C is on $\overline{AB}$ between A and B. Therefore, all three points are collinear, and no triangle exists. This contradicts the given fact that figure ABC is a triangle. Therefore, Case 2 is impossible.

Case 3 says $AC + BC < AB$. This assumption implies that there is a shorter distance from A to B than the straight line path $\overline{AB}$. Therefore we must reject Case 3.

Case 1 which says $AC + BC > AB$ is the only possibility left.

This result can now be written as a theorem.

THEOREM 3-2-1 **The sum of the lengths of two sides of a triangle is greater than the length of the third side.**

Example: Use the figure to solve for all possible values for x.

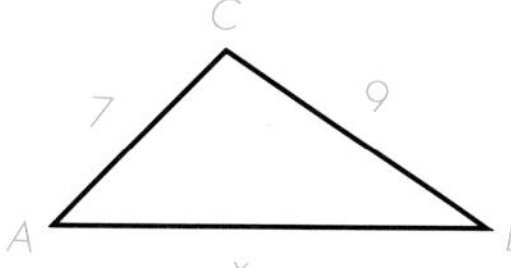

Use Theorem 3-2-1 to write three inequalities.

1. $AC + BC > AB$ 2. $AC + AB > BC$ 3. $AB + BC > AC$

Substitute the known values, and then solve the resulting inequalities for x.

1. $AC + BC > AB$	2. $AC + AB > BC$	3. $AB + BC > AC$
$7 + 9 > x$	$7 + x > 9$	$x + 9 > 7$
$16 > x$	$x > 2$	$x > -2$
$AB < 16$	$AB > 2$	$AB > -2$

Since length must be positive, $AB > 2$ and $AB < 16$ are possible. Thus, AB is between 2 and 16. The range of possible values is $2 < x < 16$.

A SECOND TRIANGLE INEQUALITY

Another relationship can be developed from Theorem 3-2-1 and the example.

DISCOVERY ACTIVITY

1. Draw a triangle ABC on a sheet of paper. Let $\overline{AB}$ be the longest side of the triangle.

2. Draw a line ℓ. Measure $\overline{AB}$ and mark the length of $\overline{AB}$ on ℓ.

A B l

Using the Page
$AB > -2$ is not considered, since line segments cannot have negative lengths. Students should be aware of this and eliminate this possibility each time it occurs.

3. Measure $\overline{BC}$. Mark the length of $\overline{BC}$ on ℓ so that you start at B, and C is between A and B.

4. What does AC on ℓ represent? $AB - BC$
5. Measure $\overline{AC}$ on the triangle and $\overline{AC}$ on line ℓ. Which length is longer? $\overline{AC}$ on the triangle
6. Mark the length of $\overline{BC}$ on a second line, m. Subtract AB from this. Compare AC and CA on the lines. They are the same.

The difference of the segments AB and BC is either $AB - BC$ or $BC - AB$. The value is the same, except for the sign. Since distance is always positive, the difference can be written as $|AB - BC|$.

THEOREM 3-2-2 **For any given triangle, the length of any side is greater than the absolute value of the difference of the other two sides.**

Example: Given $\triangle ABC$ as marked, BC is greater than what value?

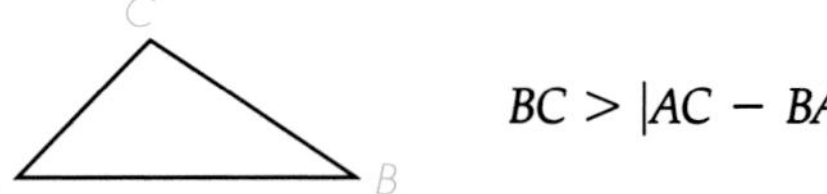

Using the Page

Using the example on this page, have students tell what value AC is greater than and what value AB is greater than. [$AC > |BC - AB|$; $AB > |AC - BC|$]

Extra Practice

Find the range of values for x from the information in each diagram.

1. [$2 < x < 12$]

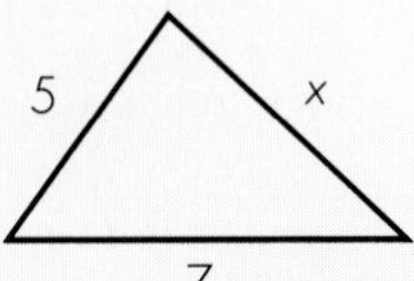

2. [$0 < x < 4$]

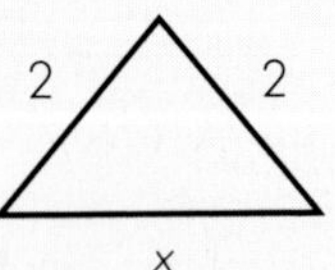

3. MN is greater than what value? [4]

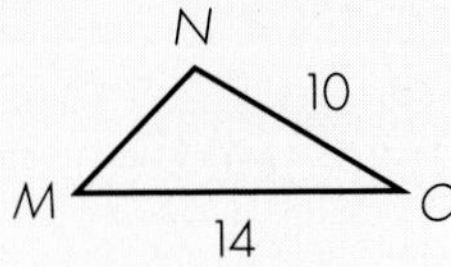

CLASS ACTIVITY

1. Find the range of values for x from the information in the diagram. $3 < x < 17$

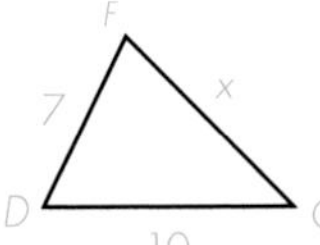

2. AC is greater than what value? $AC > 7$

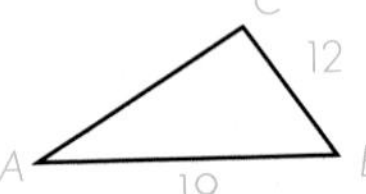

Use a computer and a geometric drawing tool to draw the two sides of each triangle described below. Then use the computer to find the range of values for the third side.

3. $\triangle ABC$ with $AB = 3$ and $BC = 5$ $2 < AC < 8$
4. $\triangle MNP$ with $MN = 2$ and $NP = 6$ $4 < MP < 8$
5. $\triangle RST$ with $ST = 15$ and $RT = 25$ $10 < RS < 40$

This result can now be written as a theorem.

THEOREM 3-2-1 **The sum of the lengths of two sides of a triangle is greater than the length of the third side.**

Example: Use the figure to solve for all possible values for x.

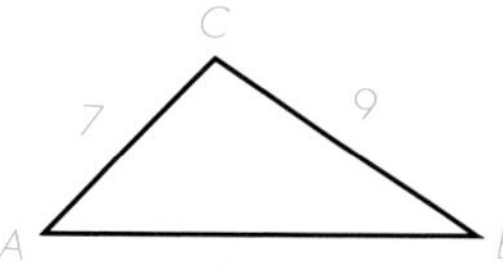

Use Theorem 3-2-1 to write three inequalities.

1. $AC + BC > AB$ 2. $AC + AB > BC$ 3. $AB + BC > AC$

Substitute the known values, and then solve the resulting inequalities for x.

1. $AC + BC > AB$	2. $AC + AB > BC$	3. $AB + BC > AC$
$7 + 9 > x$	$7 + x > 9$	$x + 9 > 7$
$16 > x$	$x > 2$	$x > -2$
$AB < 16$	$AB > 2$	$AB > -2$

Since length must be positive, $AB > 2$ and $AB < 16$ are possible. Thus, AB is between 2 and 16. The range of possible values is $2 < x < 16$.

Using the Page
$AB > -2$ is not considered, since line segments cannot have negative lengths. Students should be aware of this and eliminate this possibility each time it occurs.

A SECOND TRIANGLE INEQUALITY

Another relationship can be developed from Theorem 3-2-1 and the example.

DISCOVERY ACTIVITY

1. Draw a triangle ABC on a sheet of paper. Let $\overline{AB}$ be the longest side of the triangle.

2. Draw a line ℓ. Measure $\overline{AB}$ and mark the length of $\overline{AB}$ on ℓ.

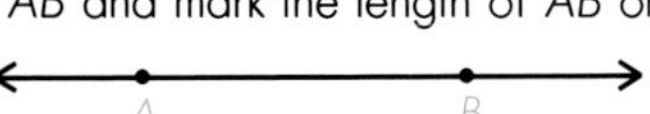

3. Measure $\overline{BC}$. Mark the length of $\overline{BC}$ on ℓ so that you start at B, and C is between A and B.
4. What does AC on ℓ represent? AB − BC
5. Measure $\overline{AC}$ on the triangle and $\overline{AC}$ on line ℓ. Which length is longer? $\overline{AC}$ on the triangle
6. Mark the length of $\overline{BC}$ on a second line, m. Subtract AB from this. Compare AC and CA on the lines. They are the same.

The difference of the segments AB and BC is either AB − BC or BC − AB. The value is the same, except for the sign. Since distance is always positive, the difference can be written as $|AB - BC|$.

NOTEBOOK THEOREM 3-2-2 **For any given triangle, the length of any side is greater than the absolute value of the difference of the other two sides.**

Example: Given △ABC as marked, BC is greater than what value?

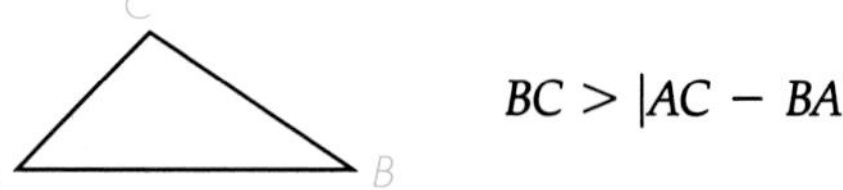

Using the Page
Using the example on this page, have students tell what value AC is greater than and what value AB is greater than. [$AC > |BC - AB|$; $AB > |AC - BC|$]

Extra Practice
Find the range of values for x from the information in each diagram.

1. [$2 < x < 12$]

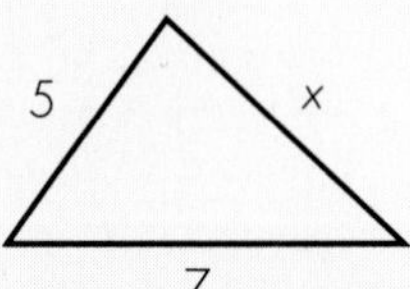

2. [$0 < x < 4$]

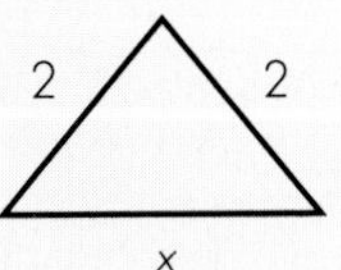

3. MN is greater than what value? [4]

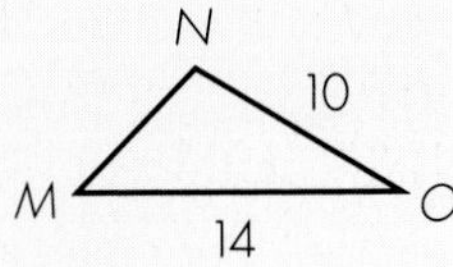

CLASS ACTIVITY

1. Find the range of values for x from the information in the diagram. $3 < x < 17$

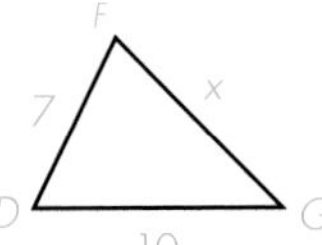

2. AC is greater than what value? $AC > 7$

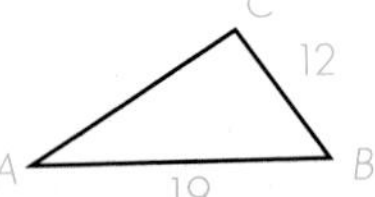

Use a computer and a geometric drawing tool to draw the two sides of each triangle described below. Then use the computer to find the range of values for the third side.

3. △ABC with AB = 3 and BC = 5 $2 < AC < 8$
4. △MNP with MN = 2 and NP = 6 $4 < MP < 8$
5. △RST with ST = 15 and RT = 25 $10 < RS < 40$

HOME ACTIVITY

Could the following be three sides of a triangle?

1. 3, 4, 5 Yes
2. 4 cm, 5 cm, 7 cm Yes
3. 10 ft, 20 ft, 30 ft No
4. 2 m, 4 m, 6 m No
5. 300, 500, 600 Yes
6. 1.5 ft, 27 in., 3.5 ft Yes

The lengths of the sides of a triangle are given. Find the range of values of x for each.

7. 3, 5, x $2 < x < 8$
8. 20 ft, 60 ft, x ft 40 ft $< x <$ 80 ft
9. 50 yd, 100 yd, x yd 50 yd $< x <$ 150 yd
10. 4.5 ft, 30 in., x ft 2 ft $< x <$ 7 ft
11. 60 km, 45 km, x km 15 km $< x <$ 105 km
12. Two sides of a triangular lot measure 70 feet and 90 feet. Between what two numbers is the measure of the third side? 20 ft and 160 ft
13. BC is between what two values? 16 and 24

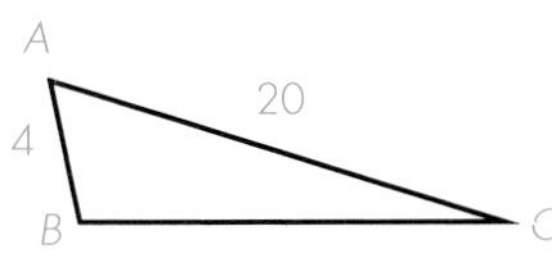

14. If Y, R, and E are three cities, what is the maximum number of miles you save by taking road $\overline{RE}$ instead of route $\overline{RY}$ and then $\overline{YE}$? 44 miles

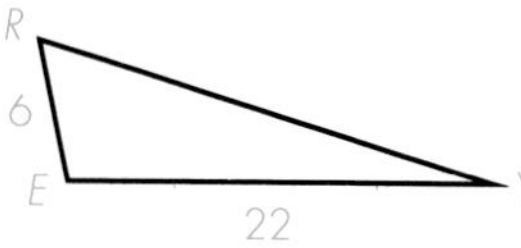

15. Find the range of values for HJ. $4 < HJ < 44$

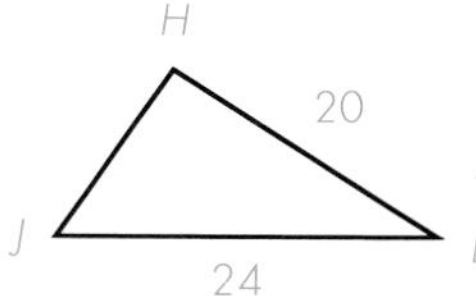

16. Use the diagram to complete the following: ______ $< SO <$ ______ 29; 93

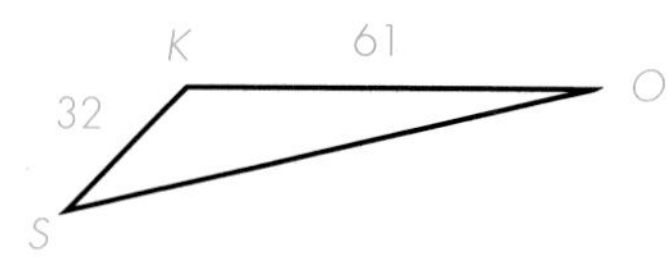

Using the Page
In Exercises 1-6, students should realize that no matter which two measurements are picked, the sum must be greater than the third measurement in order for a triangle to exist.

Using the Page
Exercises 20-22 involve writing inequalities for individual restrictions on a variable, then combining these restrictions into one set of restrictions for the variable that will fit all individual cases.

17. If the measurements are in kilometers, what is the range of values for VJ?
$9 \text{ km} < VJ < 29 \text{ km}$

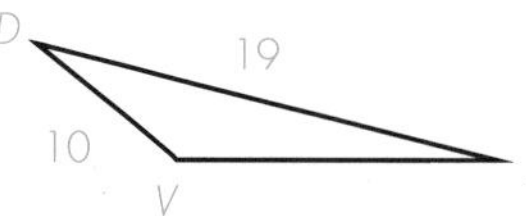

18. Is figure PRY a possible triangle? Justify your answer.
No; $10 + 6$ is not greater than 18.

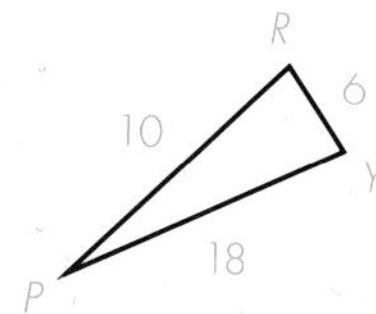

19. A city park designer was planning a new park that would be in the shape of a triangle. The designer informed the city council members that the park would measure 150′, 180′, and 340′. One of the council members directed the designer to go back and measure again. Why?
$150 + 180 \not> 340$; the triangle is not possible.

Additional Answers
22. The shortcut and sidewalks form a triangle. The shortcut is shorter than the sum of the lengths of the sidewalks.

20. Are the measurements on the following drawing possible? If they are, determine the possible values for AK from both triangles. Yes, for $6 < AK < 15$

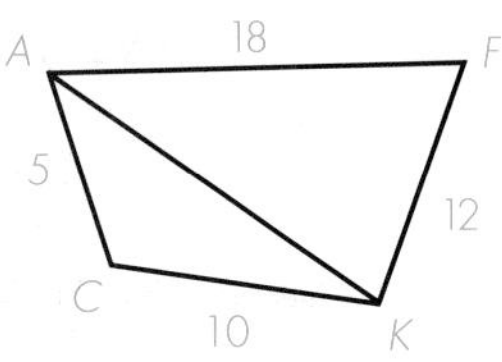

FOLLOW-UP

Assessment
1. In $\triangle KLM$, KL must be greater than what number? KL must be less than what numbers? [7; 25]

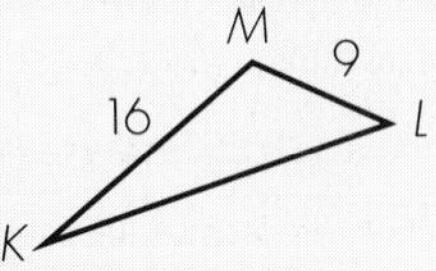

21. Find the range of values for x that will make figure MNO a triangle. $16 < x < 36$

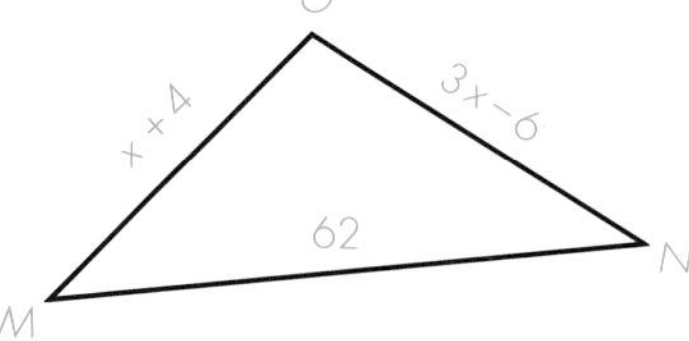

2. Find the range of values for x. [$5 < x < 19$]

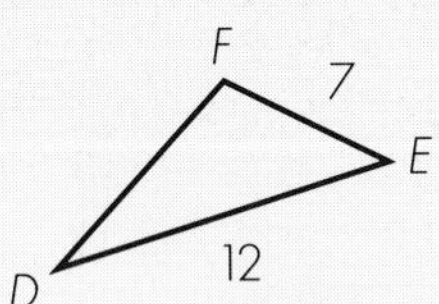

CRITICAL THINKING

22. Using the theorems in this section, explain why it is shorter for a dog to cut across the yard than to follow the sidewalk.
See margin.

Reteaching
Students who have had difficulty with this section may benefit from Reteaching Activity 3.2.

Enrichment
For students who have mastered the material in this section, you may wish to assign Enrichment Activity 3.2.

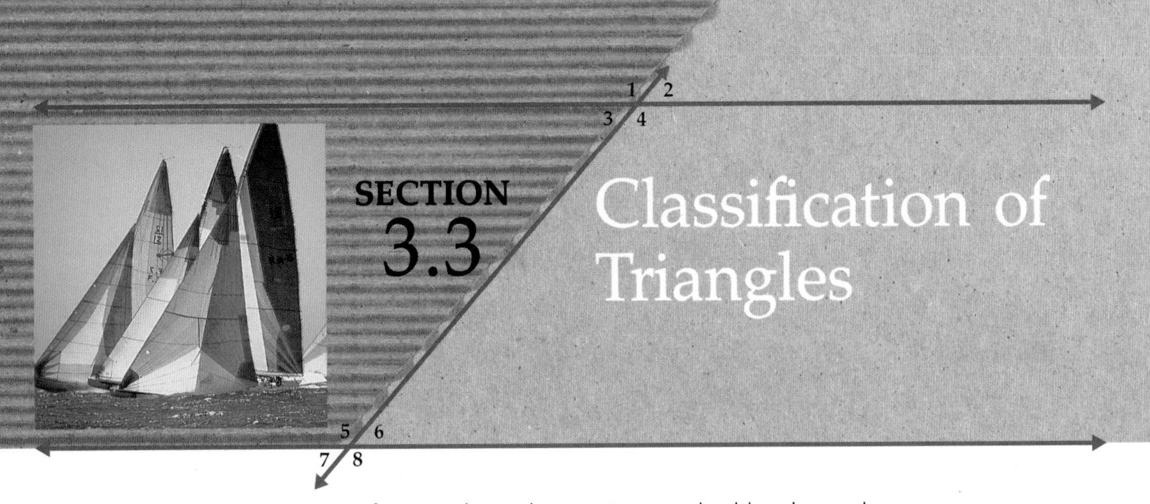

SECTION 3.3 Classification of Triangles

After completing this section, you should understand
▲ how triangles are classified by the length of their sides.

Thus far, three theorems that apply to all triangles have been presented. These theorems are:

The sum of the measures of the angles of a triangle equals 180°.

The sum of the lengths of any two sides of a triangle is greater than the length of the third side.

The length of any side of a triangle is greater than the absolute value of the difference of the lengths of the other two sides.

DISCOVERY ACTIVITY

Cut five strips of paper (no more than $\frac{1}{4}$-inch wide). Cut three of the strips to a length of 6 inches, one to the length of 5 inches, and one to the length of 4 inches. Using any three of the strips at a time, how many different shaped triangles can you form?

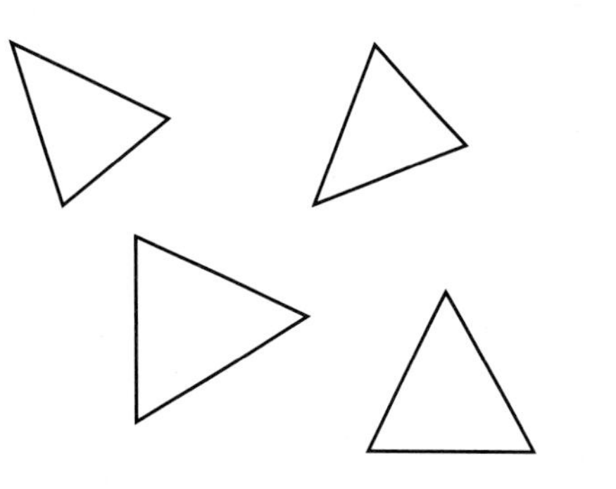

You should have found that you can make four different triangles. Certain triangles have special characteristics. One characteristic relates to the length of the sides of a triangle.

SECTION 3.3

Resources
Reteaching 3.3
Enrichment 3.3

Objectives
After completing this section, the student should understand
▲ how triangles are classified by the lengths of their sides.

Materials
ruler
scissors

Vocabulary
equilateral triangle
isosceles triangle
legs of triangle
base of triangle
scalene triangle

GETTING STARTED

Quiz: Prerequisite Skills

1. Draw two line segments that each measure 2 inches. Do the line segments have equal measure? [Yes]

2. Draw two line segments, one $1\frac{1}{2}$ inches long and one 3 inches long. Do the line segments have equal measure? [No]

Warm-Up Activity
Use the following triangle to review the second and third theorems listed on this page.

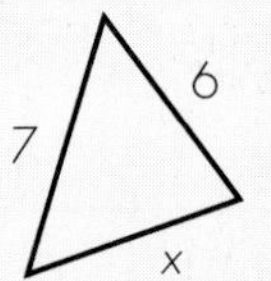

Ask students for the range of values for x. [$1 < x < 13$] Then ask students for a value of x that would make two of the lengths equal. [6 or 7]

Chalkboard Example

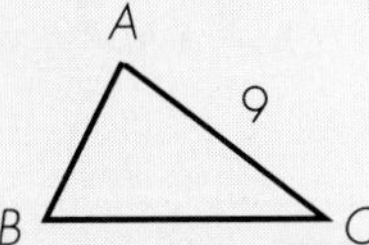

Have students give a number for the length of $\overline{AB}$ and of $\overline{BC}$ so that all sides would have the same length. [9; 9]

Have students give a number for the length of $\overline{AB}$ and of $\overline{BC}$ so that two of the sides have the same lengths, but the third side has a different length. [Answers may vary.]

Have students give a number for the length of $\overline{AB}$ and of $\overline{BC}$ so that all three sides have different lengths. [Answers may vary.]

TEACHING COMMENTS

Using the Page
Students found four different triangles in the Discovery Activity on Page 93. Have students sort the triangles according to how many sides have the same length, and then discuss the three categories on this page.

Using the Page
When discussing the slash marks, emphasize that one slash mark indicates one set of equal sides, two slash marks indicate another set of equal sides, and so on. Use this figure to have students name the sets of equal sides. [$\overline{AB}$ and $\overline{AF}$, $\overline{BC}$ and $\overline{EF}$, and $\overline{CE}$ and $\overline{CD}$]

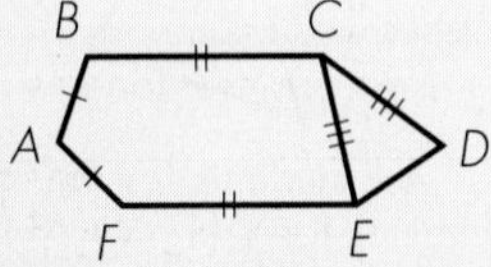

TYPES OF TRIANGLES

Triangles can be classified according to the lengths of their sides. Three possibilities exist for the lengths of the sides of a triangle.

1. All of the sides could be the same length.

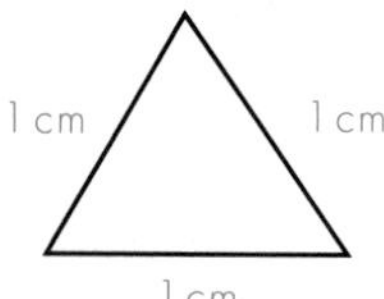

2. Two of the sides could be the same length.

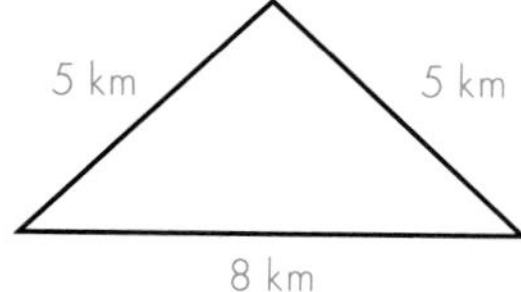

3. None of the sides are the same length.

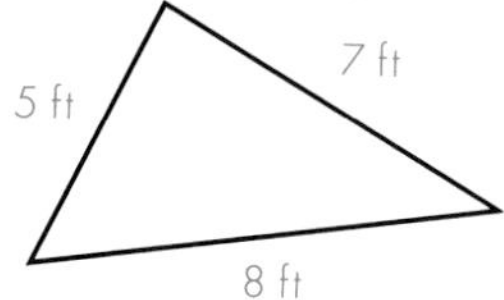

In the first type of triangle, all the sides are equal. This triangle is called equilateral.

DEFINITION 3-3-1 **An EQUILATERAL TRIANGLE is a triangle with all three sides having the same measure or length.**

In this drawing, a side note indicates that all three sides have equal measure. When notes are not included, you cannot assume from a drawing that lengths are equal. To indicate equal lengths, slash marks (/) will be included on the drawing.

Example:

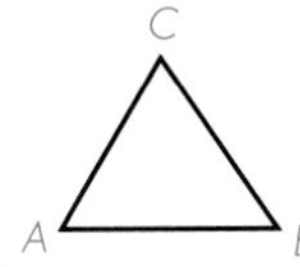
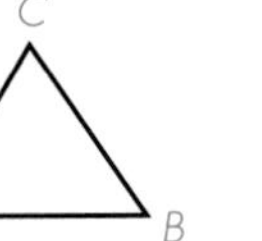

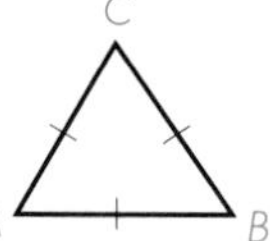

$AB = BC = AC$

In the second type of triangle, two sides are of equal length. This triangle is called isosceles. *Isosceles* is derived from a Greek word meaning two equal sides.

DEFINITION 3-3-2 **An ISOSCELES TRIANGLE is a triangle with two sides equal in measure or length.**

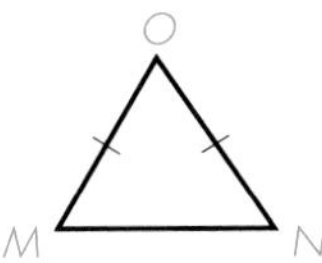

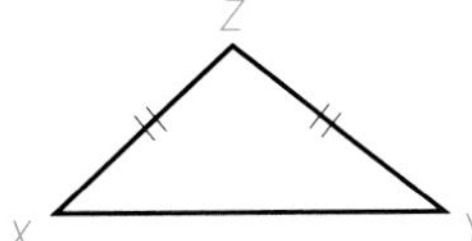

ΔMNO and ΔXYZ are isosceles triangles.

In an isosceles triangle, the two sides of equal length are called the LEGS of the triangle. The third side is called the BASE of the isosceles triangle.

The third type of triangle has sides that all have different lengths. The name given to this triangle is *scalene*—Greek for uneven.

DEFINITION 3-3-3 **A SCALENE TRIANGLE is a triangle with no equal sides.**

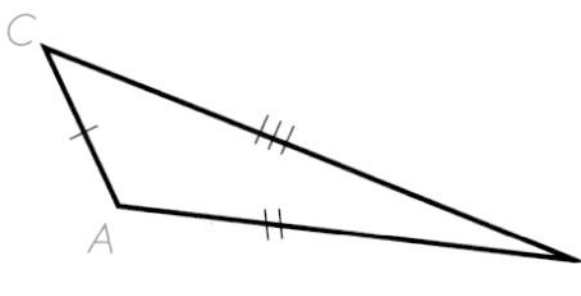

$AB \neq BC \neq AC$

Different numbers of slash marks on segments in a drawing indicate different lengths.

CLASS ACTIVITY

Classify each triangle.

1.

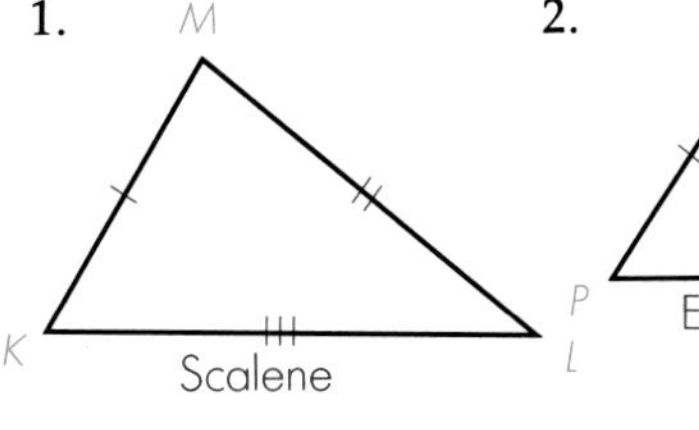

2.

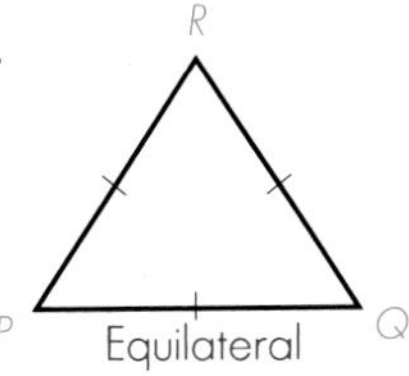

3.

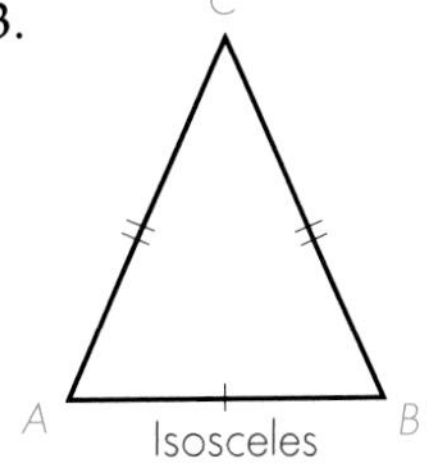

Using the Page
You may want to have some students research the history of the words *equilateral*, *isosceles*, and *scalene*. Have students report their findings to the class.

Extra Practice
Use figure *PQRS* to identify

1. a scalene triangle. [$\triangle PQT$, $\triangle QTR$, $\triangle RTS$, $\triangle PTS$, $\triangle PQR$, or $\triangle PRS$]
2. an isosceles triangle. [$\triangle PQS$ or $\triangle QRS$]
3. an equilateral triangle. [none]

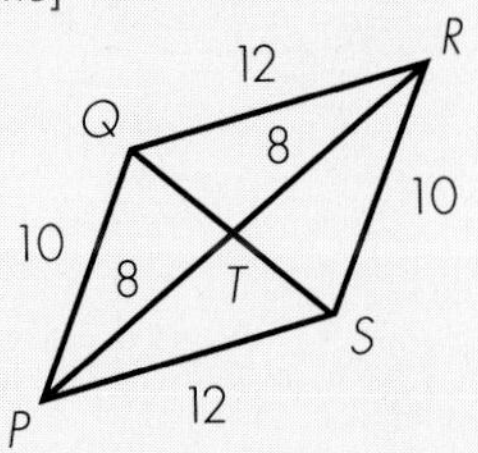

Using the Page
You may want to have students measure each triangle using both the English system and the metric system. This should reinforce the idea that the classification does not change just because the measuring system changes.

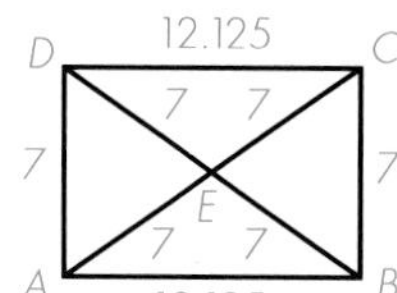

Use figure *ABCD* to identify

4. a scalene triangle. $\triangle ADC$, $\triangle ABC$, $\triangle DAB$, or $\triangle BCD$
5. an isosceles triangle. $\triangle CED$ or $\triangle ABE$
6. an equilateral triangle. $\triangle ADE$ or $\triangle CBE$
7. the total number of triangles shown. 8

Choose the lengths of the sides for the type of triangles described. Then use a computer and a geometric drawing tool to draw each triangle. Compare your triangles with your classmates' triangles.
Answers may vary.

8. Isosceles triangle
9. Scalene triangle
10. Equilateral triangle

HOME ACTIVITY

Classify the triangles according to the markings.

1. 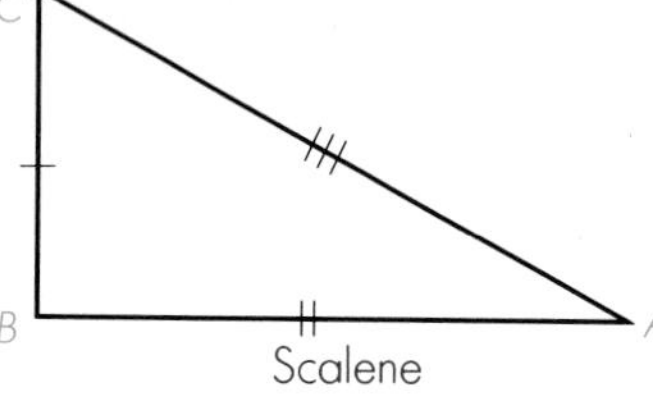

Scalene

2. 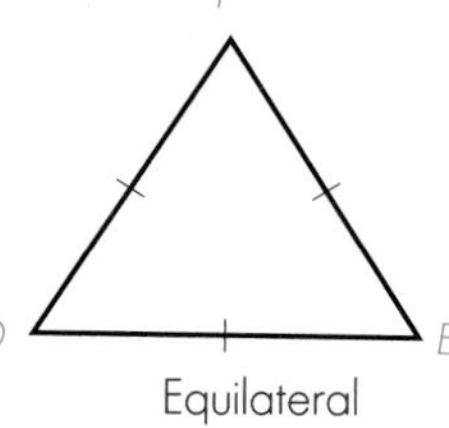

Equilateral

3. 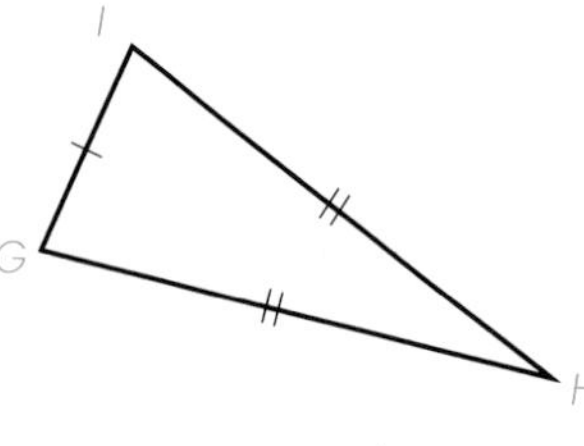

Isosceles

4. 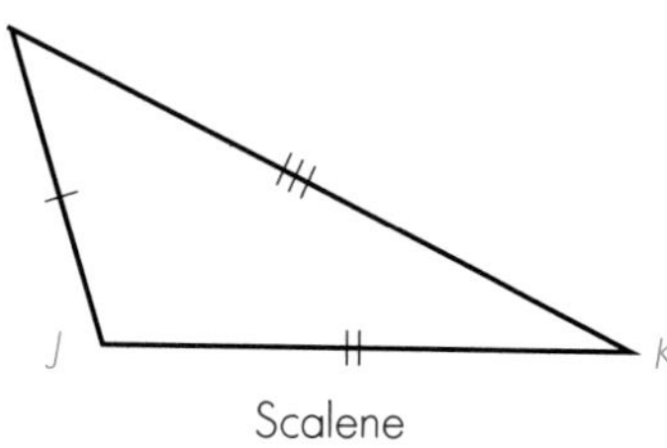

Scalene

5. 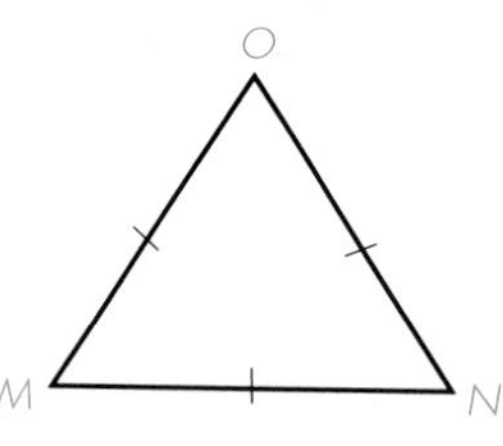

Equilateral

6. 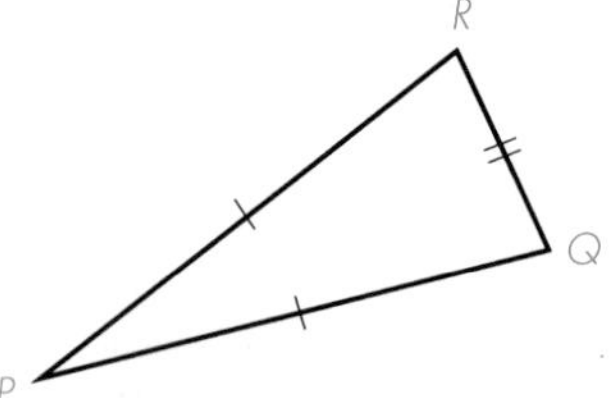

Isosceles

Classify each triangle based on figure *ABCD*.

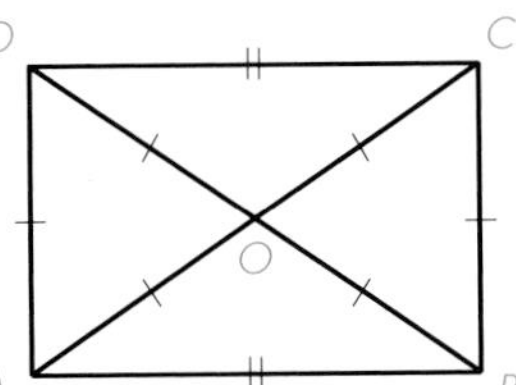

7. $\triangle ABC$ Scalene
8. $\triangle ADB$ Scalene
9. $\triangle ADO$ Equilateral
10. $\triangle AOB$ Isosceles

Classify each triangle based on figure *PQRT*.

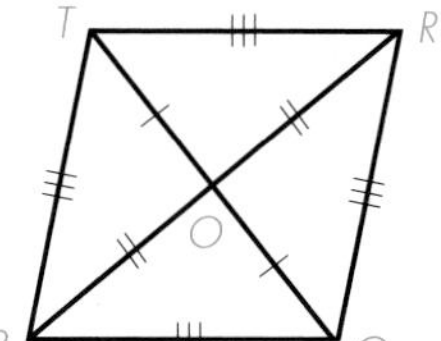

11. $\triangle PTR$ Isosceles
12. $\triangle ROT$ Scalene
13. $\triangle QOR$ Scalene
14. $\triangle PTQ$ Isosceles
15. How many isosceles triangles can be found?
8

Classify each triangle based on figure *ABCD*.

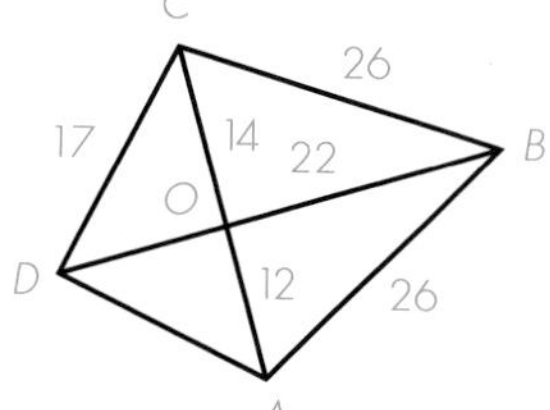

16. $\triangle ACB$ Equilateral
17. $\triangle COB$ Scalene
18. $\triangle ABO$ Scalene

Lines *AB* and *DE* are parallel. Lines *AE* and *DB* are perpendicular.

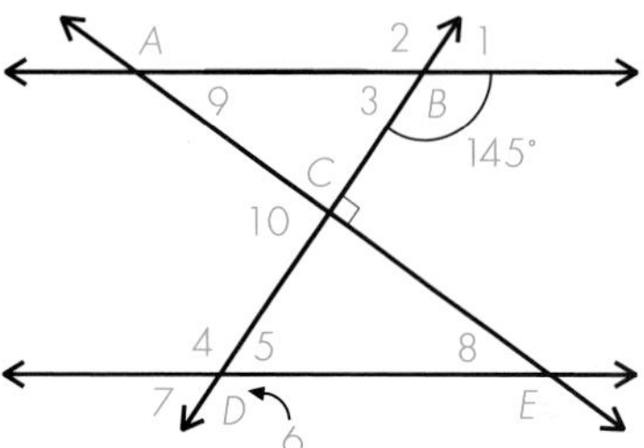

19. Find the measures of angles 1–10.
See margin.
20. Is $m\angle 9 + m\angle 3 = m\angle 10$? Yes
21. What do you know about $\angle ACB$ and $\angle DCE$?
They measure 90°.

Additional Answers

19. $m\angle 1 = 35°$; $m\angle 2 = 145°$; $m\angle 3 = 35°$; $m\angle 4 = 145°$; $m\angle 5 = 35°$; $m\angle 6 = 145°$; $m\angle 7 = 35°$; $m\angle 8 = 55°$; $m\angle 9 = 55°$; $m\angle 10 = 90°$

Using the Page
Exercises 23-24 contain important concepts. Make sure students understand that all equilateral triangles are isosceles, but not all isosceles triangles are equilateral.

FOLLOW-UP

Assessment

1. What kind of triangle has sides of all different lengths? [Scalene]

2. What kind of triangle has sides all the same length? [Equilateral]

3. What kind of triangle has two sides the same length? [Isosceles]

Reteaching
Students who have had difficulty with this section may benefit from Reteaching Activity 3.3.

Enrichment
For students who have mastered the material in this section, you may wish to assign Enrichment Activity 3.3.

22. What is the measure of $\angle A$? 29°

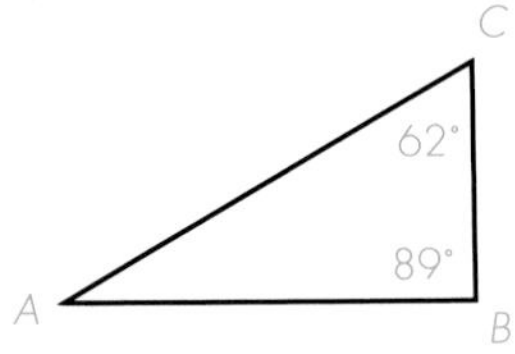

23. Is an equilateral triangle an isosceles triangle? Justify your answer.
Yes; any two sides chosen are of equal length.

24. Is an isosceles triangle an equilateral triangle? Justify your answer.
Not always; only two sides must be of equal length.

CRITICAL THINKING

25. The distances from the island of Sandwich to the island of Whitebread and the ports at Rye and Pumpernickel are correctly marked on the figure. The cargo boat *Robin* starts at the island of Whitebread and sails by the most direct route to Rye and Pumpernickel and then back to Whitebread. The skipper figures at the end of the trip that the *Robin* has traveled 25 kilometers altogether. Is this possible? Give a reason for your answer.
No; 25 km would exceed the total of the maximum possible lengths of the legs of the trip.

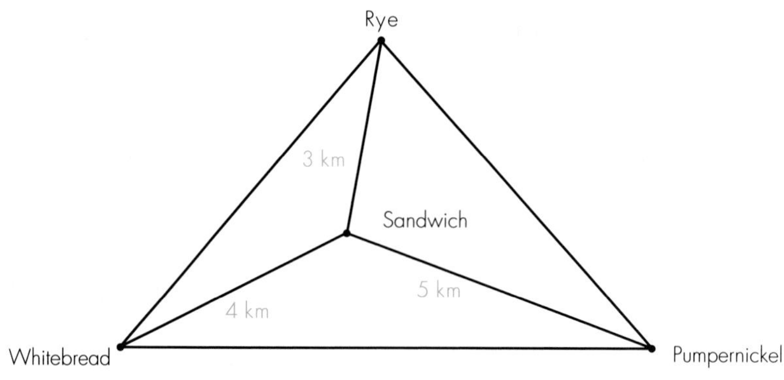

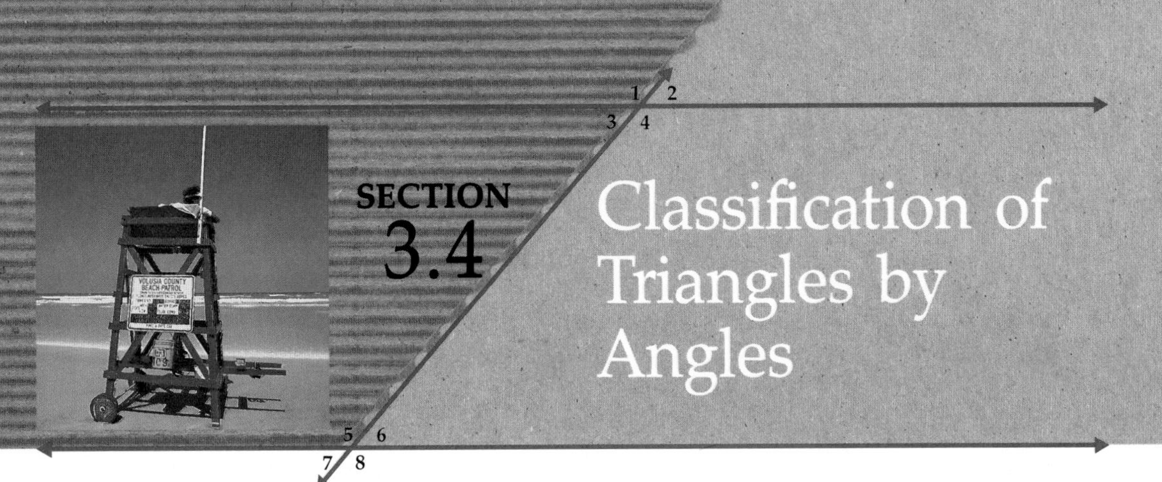

SECTION 3.4 Classification of Triangles by Angles

After completing this section, you should understand
▲ how to classify triangles by the measure of their angles.

Triangles were classified by the lengths of their sides in Section 3.3.

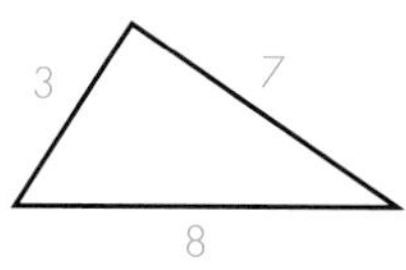

Scalene

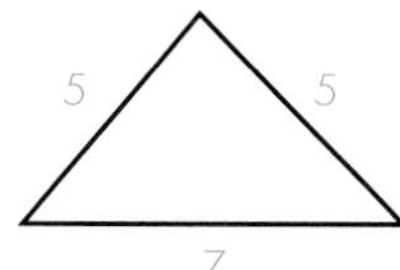

Isosceles

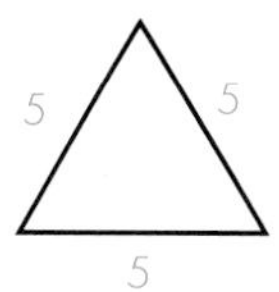

Equilateral

CLASSIFYING TRIANGLES BY ANGLES

Triangles can also be classified according to the measure of their angles. Triangles with similar angles share special properties independent of the lengths of their sides.

Case 1: All the angles are equal in measure.

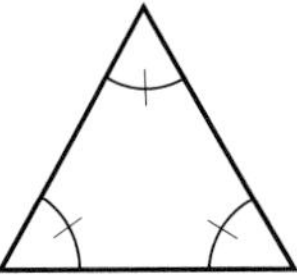

The arcs with slash marks indicate equal angles. The name given to this triangle is equiangular triangle. Since the sum of the measures of the angles of a triangle is 180°, each angle measures 60°.

99

SECTION 3.4

Resources
Reteaching 3.4
Enrichment 3.4

Objectives
After completing this section, the student should understand
▲ how to classify triangles by the measure of their angles.

Materials
protractor
ruler

Vocabulary
equiangular triangle
right triangle
acute triangle
obtuse triangle

GETTING STARTED

Quiz: Prerequisite Skills
1. What is the measure of the unknown angle? [72°]

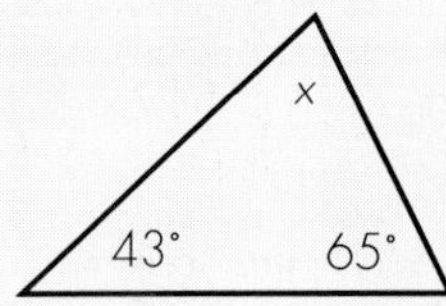

2. Solve for x.
$2x + 3x + (2x + 40) = 180$
[$x = 20$]

Warm-Up Activity
Have students measure the angles of the three triangles at the top of the page. List the measures for each angle of each triangle on the chalkboard. Ask students what they notice about the measures.

Chalkboard Example
You may wish to work another example similar to the second example on page 101:
A triangle has two equal angles. The third angle is 60 more than 4 times one of the equal angles. What is the measure of each angle? [20°; 20°; 140°]

TEACHING COMMENTS

Using the Page
You may want to draw several triangles on the chalkboard. Make some acute, some right, some obtuse, and some equiangular triangles. Challenge students to devise a method of sorting the triangles by using the angles.

Using the Page
Point out that equiangular triangles are also acute triangles.

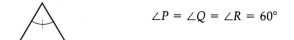

NOTEBOOK **DEFINITION 3-4-1** **An EQUIANGULAR TRIANGLE is a triangle with the measure of each angle equal to 60°.**

$\angle P = \angle Q = \angle R = 60°$

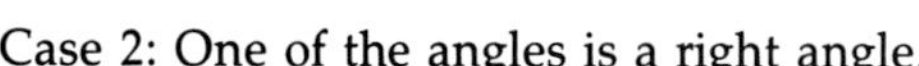

Case 2: One of the angles is a right angle.

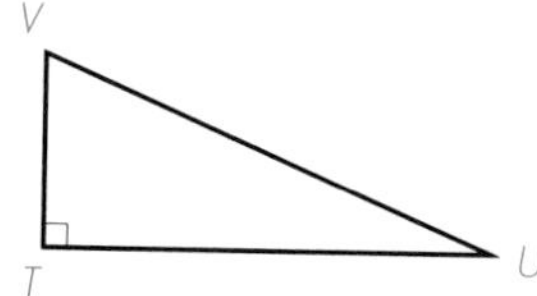

The symbol ⊾ is used to indicate a right angle. $\triangle TUV$ is a right triangle, where $\angle T$ is the right angle. Recall that right angles measure 90°.

NOTEBOOK **DEFINITION 3-4-2** **A RIGHT TRIANGLE is a triangle with a 90° angle.**

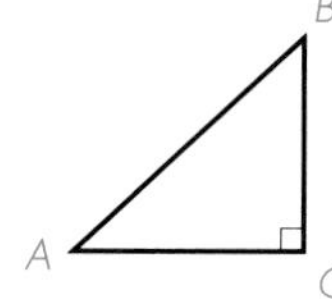

$\angle C = 90°$

Case 3: All of the angles measure less than 90°. Recall that an angle measuring less than 90° is an acute angle. In this case the triangle is called an acute triangle.

NOTEBOOK **DEFINITION 3-4-3** **An ACUTE TRIANGLE is a triangle with the measures of all three angles less than 90°.**

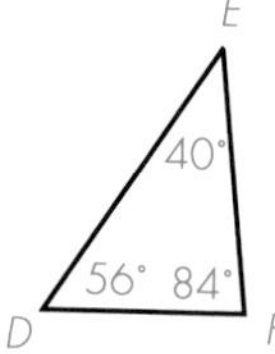

$\triangle DEF$ is an acute triangle.

Case 4: One angle measures greater than 90°. Recall that an angle measuring greater than 90° is an obtuse angle. In this case the triangle is called an obtuse triangle.

NOTEBOOK

DEFINITION 3-4-4 **An OBTUSE TRIANGLE is a triangle with one angle measuring greater than 90°.**

$\angle A$ is obtuse, therefore $\triangle ABC$ is an obtuse triangle.

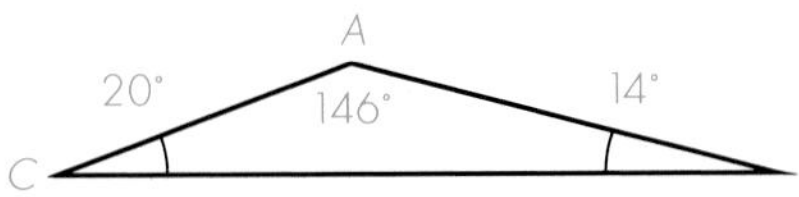

Example: Refer to figure *ABCFD*. Figure *ABCD* is a rectangle. Classify each triangle by its angles.

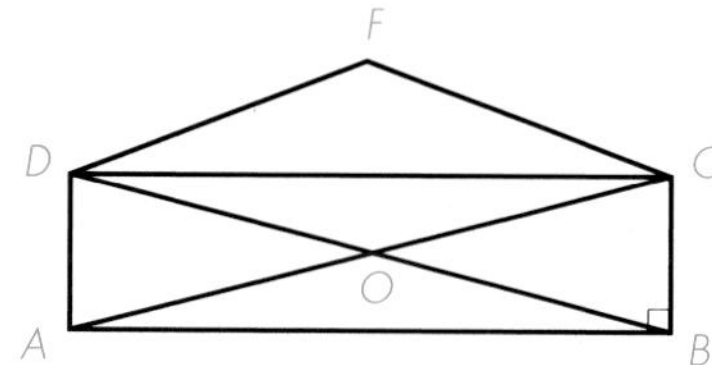

$\triangle ABC$ is right. $\triangle DAO$ is acute.
$\triangle DOC$ is obtuse. $\triangle DCF$ is obtuse.
There are nine triangles in all.

Example: A triangle has two equal angles. The third angle is seven times one of the equal angles. What is the measure of each angle?

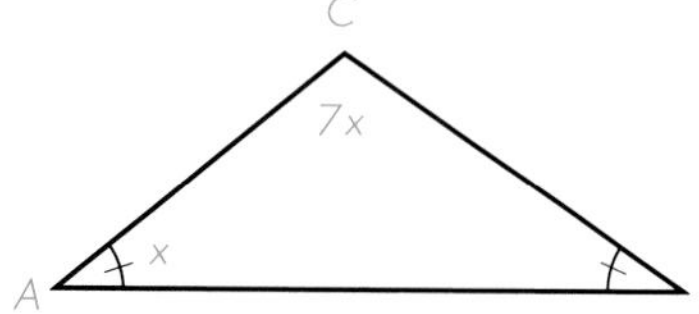

Solution:

Step 1: Draw a diagram.
Step 2: Label the angles.
x is the measure of one of the equal angles.
$7x$ is the measure of the largest angle.
Step 3: Write and solve an equation.

$$x + x + 7x = 180$$
$$9x = 180$$
$$x = 20$$

Step 4: Interpret the answer to the equation.
$x = 20°$ the measure of each equal angle
$7x = 140°$ the measure of the largest angle
Step 5: Check.
$20° + 20° + 140° = 180°$

Using the Page
In the first example, remind students that right angles should not be assumed. Either a right angle will be marked as such, or information will be given that will enable a student to determine that an angle is a right angle. Students should be able to estimate an obtuse or an acute angle.

Using the Page
Emphasize Step 5 of the second example. Checking the answer will help students check for correctness and reasonableness of answers.

CLASS ACTIVITY

Solve for each unknown angle. Then classify each triangle according to its angles.

1.

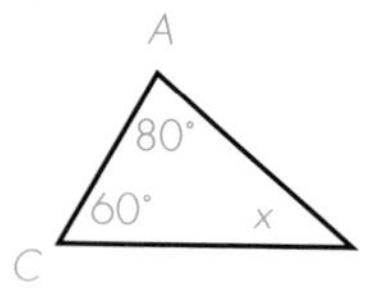

m∠x = 40°; acute

2.

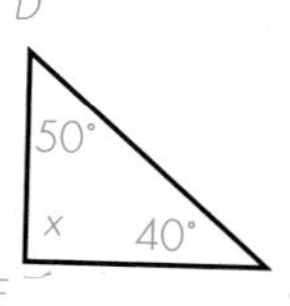

m∠x = 90°; right

3.

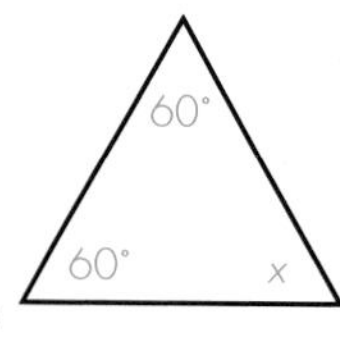

m∠x = 60°; equiangular

4.

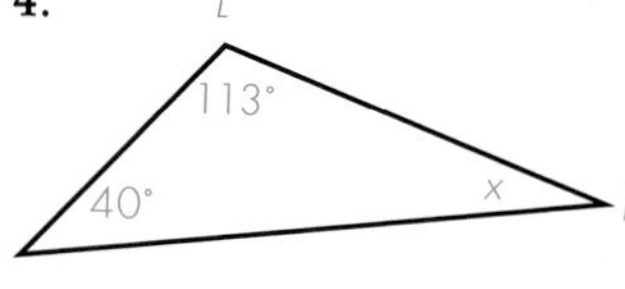

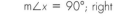

m∠x = 27°; obtuse

5. Solve for the measures of angles *A*, *B*, and *C*. Then classify the triangle according to its angles.

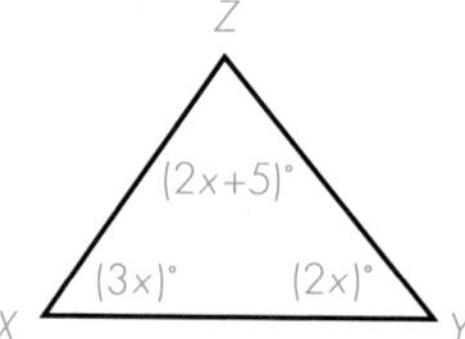

m∠X = 75°; m∠Y = 50°; m∠Z = 55°; acute

Use a computer and a geometric drawing tool to complete.

6. Draw an isosceles triangle. Measure each angle. Repeat this activity several times, using different lengths for the sides each time. How can an isosceles triangle be classified according to the measure of its angles? Acute, right, obtuse
7. Draw an equilateral triangle. Measure each angle. Repeat this activity several times, using different lengths for the sides each time. How can an equilateral triangle be classified according to the measure of its angles? Equiangular
8. Draw a scalene triangle. Measure each angle. Repeat this activity several times, using different lengths for the sides each time. How can a scalene triangle be classified according to the measure of its angles? Acute, right, obtuse

Extra Practice
Solve for the measures of angles *A*, *B*, and *C*. Then classify the triangle according to its angles.

1. [m∠A = 36°; m∠B = 117°; m∠C = 27°; obtuse]

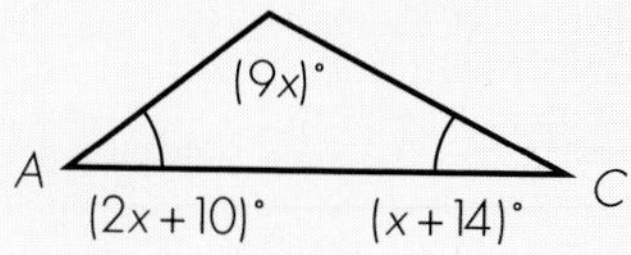

2. [m∠A = 35°; m∠B = 80°; m∠C = 65°; acute]

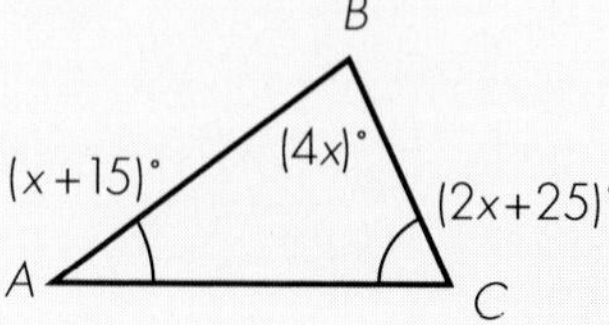

HOME ACTIVITY

1. Solve for the measures of the unknown angles in $\triangle ABC$, then classify the triangle according to its angles.
   m∠A = 60°; m∠C = 60°; equiangular

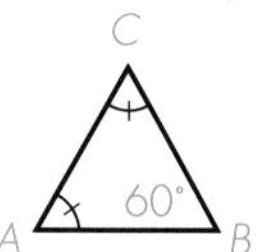

2. In the sketch of the ramp, what is the measure of the unknown angle? The ramp illustrates the case of what kind of triangle?
   90°; right

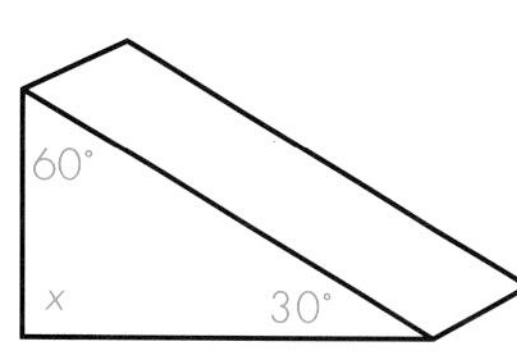

3. A city surveyor measured the angles of the triangular-shaped lot, but forgot to record the measure of ∠R. Find the measure of ∠R, and classify the triangle.
   m∠R = 74.2°; acute

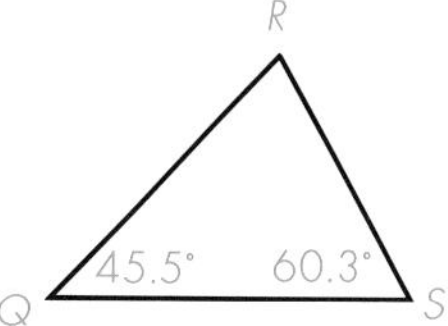

Refer to triangle *ABC*. Find the measures.

4. m∠*ABC* 80°

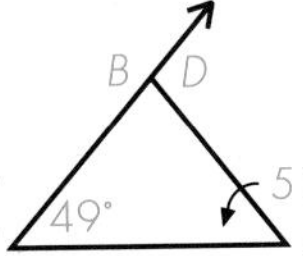

5. m∠*ABD* 100°

6. m∠*A* + m∠*C* 100°

7. Compare the answers to Exercises 5 and 6.
   Same

8. Refer to *TUV*. Solve for the measures of the angles of the triangle, then classify the triangle.

   m∠*T* = 24°;
   m∠*U* = 86°;
   m∠*V* = 70°; acute

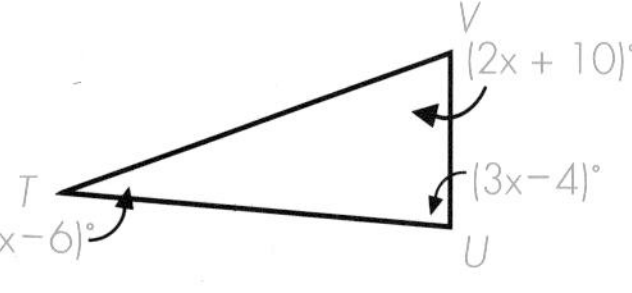

9. Refer to $\triangle JTO$. Solve for the measures of the angles, then classify the triangle.

   m∠*J* = 30°;
   m∠*O* = 60°;
   m∠*JTO* = 90°; right

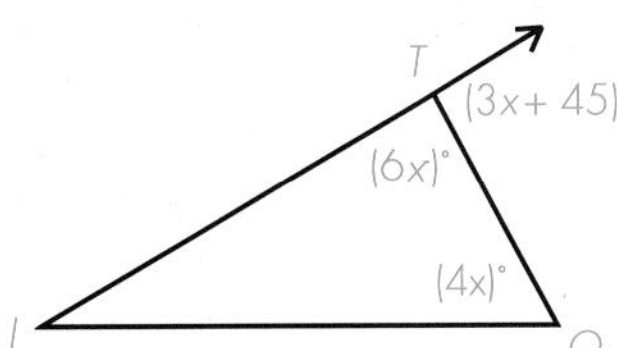

Using the Page
Exercises 4-7 lead in to the Exterior Angle Theorem of Section 3.5.

Using the Page
In Exercise 9, students should use the straight angle to solve for the value of x. Once x is known, the values of the measures of the angles of the triangle can be found.

Refer to $\triangle ABC$. $\overline{BE}$ and $\overline{CD}$ are angle bisectors.

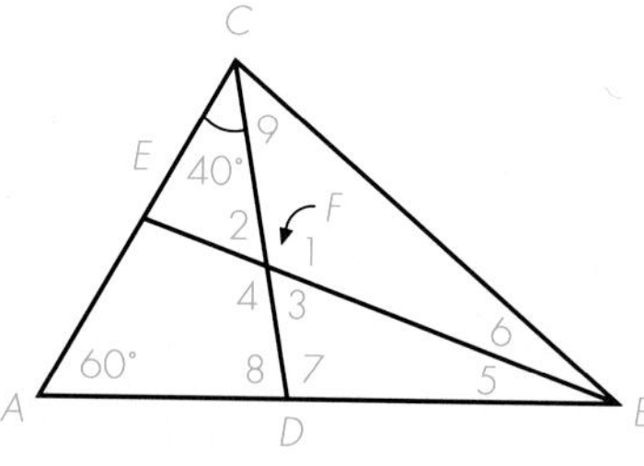

10. Find the measures of angles 1 to 9. See margin.

11. Use the three-letter method to name all the triangles in the figure. $\triangle ABC$, $\triangle CEF$, $\triangle ACD$, $\triangle AEB$, $\triangle BFD$, $\triangle BCD$, $\triangle BEC$, $\triangle BCF$

12. Name one obtuse triangle. $\triangle BCF$, $\triangle BDF$, or $\triangle CBD$

CRITICAL THINKING

Answer true or false for Exercises 13–25.

13. A triangle can have three acute angles. True
14. A triangle can have two acute angles. True
15. A triangle can have one acute angle. False
16. A triangle can have no acute angles. False
17. A triangle can have no obtuse angles. True
18. A triangle can have one obtuse angle. True
19. A triangle can have two obtuse angles. False
20. A triangle can have three obtuse angles. False
21. A triangle can have no right angles. True
22. A triangle can have one right angle. True
23. A triangle can have two right angles. False
24. A triangle can have an acute angle and a right angle. True
25. A triangle can have an obtuse angle and a right angle. False
26. Draw a right triangle that has sides or legs of length 10 cm and 3 cm. The length of the third side is between what two integers? What is the actual length of the third side? What is the name of the third side? $7 \text{ cm} < AC < 13 \text{ cm}$; $\sqrt{109}$ cm; hypotenuse

Additional Answers

10. $m\angle 1 = 120°$; $m\angle 2 = 60°$; $m\angle 3 = 60°$; $m\angle 4 = 120°$; $m\angle 5 = 20°$; $m\angle 6 = 20°$; $m\angle 7 = 100°$; $m\angle 8 = 80°$

Using the Page

Exercise 26 requires the use of the Pythagorean Theorem. Students may be familiar with this theorem from work in previous math courses.

FOLLOW-UP

Assessment

1. What kind of triangle has angles each measuring 60°? [equiangular]

2. What kind of triangle has an angle measuring 90°? [right]

3. What kind of triangle has an angle measuring greater than 90°? [obtuse]

4. What kind of triangle has angles all measuring less than 90°? [acute]

Reteaching

Students who have had difficulty with this section may benefit from Reteaching Activity 3.4.

Enrichment

For students who have mastered the material in this section, you may wish to assign Enrichment Activity 3.4.

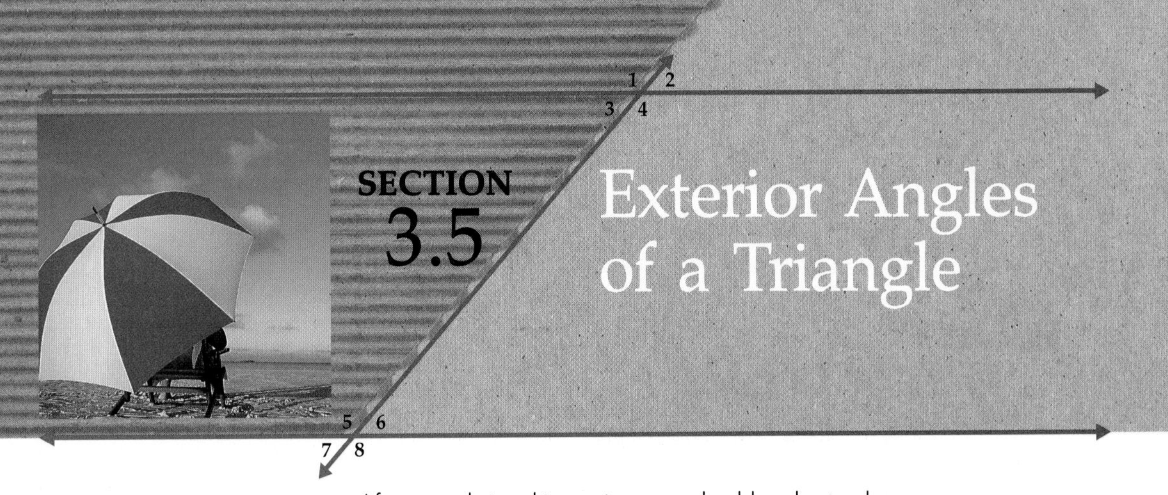

SECTION 3.5 Exterior Angles of a Triangle

After completing this section, you should understand
▲ the triangle exterior angle theorem.

The concepts in geometry have so far been divided into four different categories.

Category	Examples
1. Undefined term	point, line, plane
2. Defined term	segment, ray, angle, triangle
3. Postulate	If two parallel lines are intersected by a transversal, then the corresponding angles are equal in measure. Two points determine exactly one line.
4. Theorem	If the figure is a triangle, then the sum of the measures of the angles is 180°. If two lines intersect, then the vertical angles are equal in measure.

Recall that theorems must be justified.

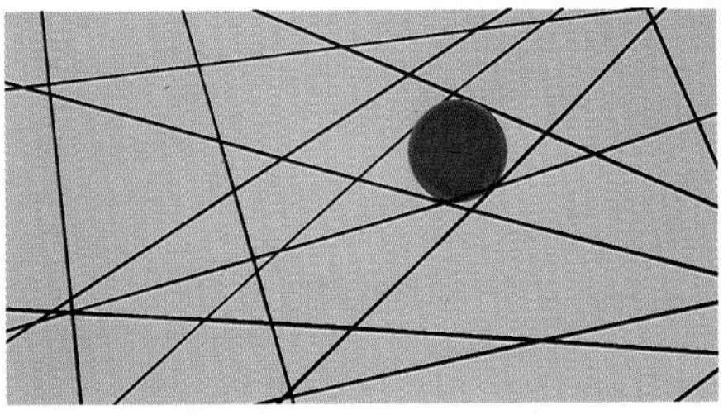

SECTION 3.5

Resources
Reteaching 3.5
Enrichment 3.5
Transparency Master 3-1

Objectives
After completing this section, the student should understand
▲ the Triangle Exterior Angle Theorem.

Materials
protractor
ruler

Vocabulary
exterior angle
adjacent interior angle
remote interior angle
conclusion
given
two-column proof

GETTING STARTED

Quiz: Prerequisite Skills
1. Find the sum. 56° + 89° [145°]
2. Give the number of degrees in a straight angle. [180]
3. What is the sum of the measures of two supplementary angles? [180°]

Warm-Up Activity
Have each student draw an angle, then extend one ray in the opposite direction so that a straight angle is formed. Have students find the measure of the smaller angle, then find the sum of the measures.

Chalkboard Example

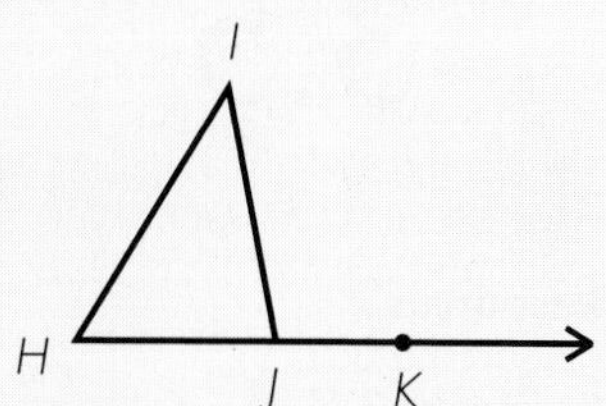

Find the measure of $\angle IJK$ if $m\angle H = 58°$ and $m\angle I = 43°$. [101°]

DISCOVERY ACTIVITY

1. On a sheet of paper, draw and label a triangle ABC. Extend segment $\overline{AB}$ past point B, and indicate point D on the extension.
2. Use a protractor to measure each of the following angles. Record the measurements.

   $m\angle A$ ________ $m\angle C$ ________ $m\angle A + m\angle C$ ________ $m\angle CBD$________

3. Compare $m\angle A + m\angle C$ to $m\angle CBD$. They are equal.
4. Compare your results in Part 3 with the results of your classmates. Were the results the same? They should be.
5. Draw a different triangle and repeat Steps 1–4. Were the results in Step 4 the same for everyone again? They should be.

TEACHING COMMENTS

Using the Page
After students complete the Discovery Activity, have them use $\triangle ABC$, the first triangle drawn. Students should trace the triangle, cut it out, and then tear off $\angle A$ and $\angle C$. Place $\angle A$ and $\angle C$ on the original drawing to show that $m\angle A + m\angle C = m\angle CBD$.

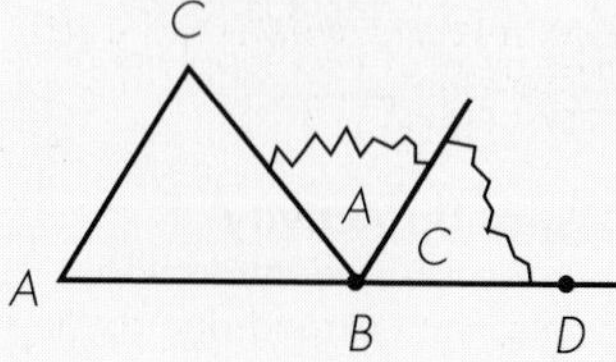

Using the Page
Using $\triangle ABC$, have students draw the other exterior angles and name the remote interior angles for each exterior angle.

The Discovery Activity shows another way to justify a theorem—using an inductive approach. With the inductive approach, several cases are examined, a pattern is established, and a conclusion is drawn from this pattern about all cases, not just those that were checked.

Some angles from the Discovery Activity need to be defined before the next theorem is introduced.

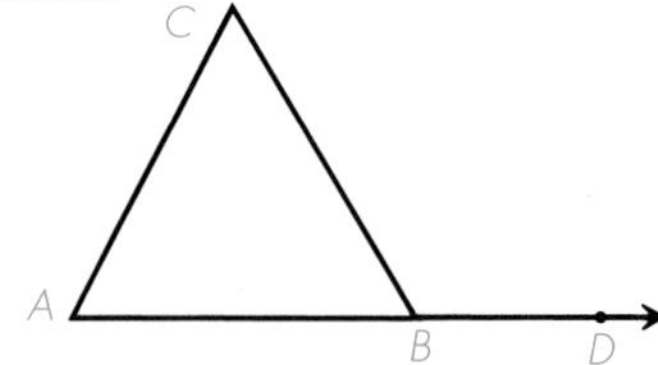

$\angle CBD$ is called an exterior angle of a triangle.

DEFINITION 3-5-1 **An EXTERIOR ANGLE of a triangle is the angle less than 180° in measure, formed by extending one side of the triangle.**

$\angle A$ and $\angle C$ are the remote, or non-adjacent, interior angles for the exterior angle, $\angle CBD$.

You should have observed from the Discovery Activity that the measure of the exterior angle of a triangle seems to always be equal to the sum of the measures of the two remote interior angles. This fact is the CONCLUSION of the observation. The drawing that you started with showed the GIVEN facts.

Here is an outline of the justification of the conclusion.

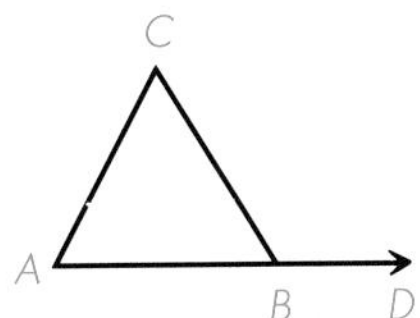

1. $m\angle A + m\angle B + m\angle C = 180°$
   Reason: Sum of the measures of the angles of a triangle is 180°.
2. $m\angle ABD = 180°$
   Reason: Definition of straight angle
3. $m\angle ABC + m\angle CBD = 180°$
   Reason: Definition of supplementary angles
4. $m\angle ABC + m\angle CBD = m\angle A + m\angle B + m\angle C$
   Reason: Sums equal to 180° are equal to each other.
5. $m\angle ABC = m\angle B$
   Reason: Identity
6. $m\angle CBD = m\angle A + m\angle C$
   Reason: Equal numbers (measurements) can be subtracted from both sides of an equality.

Step 6 shows the conclusion in the form of a statement. This outline serves as a proof of Theorem 3-5-1.

Using the Page
The proof at the top of this page is less formal than others. It may help students who have trouble with proofs in organizing their thoughts.

THE TWO-COLUMN PROOF

Proofs are often given with the statements shown in one column and the reasons shown in another column. This is called a two-column proof. Theorem 3-5-1 follows, with a two-column proof to justify it.

THEOREM 3-5-1 **The measure of an exterior angle of a triangle is equal to the sum of the measures of the two non-adjacent, or remote, interior angles.**

Given: $\triangle ABC$ with side $\overline{AB}$ extended to D.
Prove: $m\angle A + m\angle C = m\angle CBD$

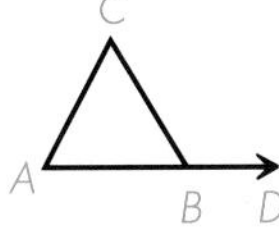

Statements	Reasons
1. $m\angle A + m\angle B + m\angle C = 180°$	1. Theorem 3-1-1. Sum of the measures of the angles of a triangle = 180°.
2. Extend $\overline{AB}$ to D.	2. Two points determine a line.
3. $\angle ABD$ is a straight angle.	3. $\overline{BD}$ drawn to extend $\overline{AB}$
4. $m\angle ABD = 180°$	4. Definition of straight angle
5. $m\angle ABC + m\angle CBD = 180°$	5. Definition of supplementary angles
6. $m\angle ABC + m\angle CBD = m\angle A + m\angle B + m\angle C$	6. Both sides equal to 180°.
7. $m\angle B = m\angle ABC$	7. Same angle
8. $m\angle B + m\angle CBD = m\angle A + m\angle B + m\angle C$	8. Substitution
9. $m\angle CBD = m\angle A + m\angle C$	9. Subtract $m\angle B$ from both sides.

Using the Page
The two-column proof is sometimes called a T=proof. Make sure students understand each step of the proof.

Extra Practice

1. Find the sum of m∠1 and m∠2. [130°]

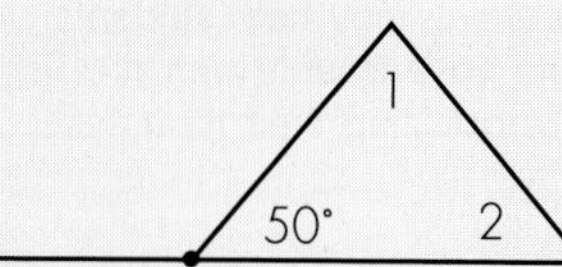

2. In Exercise 1 above, find m∠1 and m∠2 if m∠1 = $(x + 30)°$ and m∠2 = $(2x + 10)°$. [m∠1 = 60°; m∠2 = 70°]

Additional Answers

1.

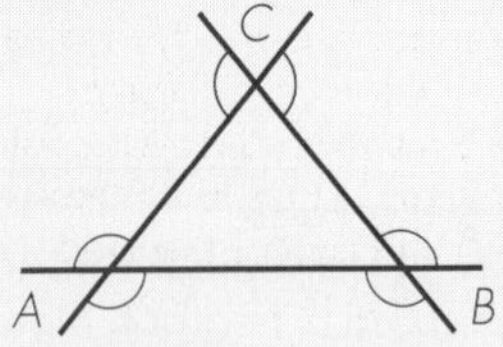

2. m∠BCA = m∠DCE [Vertical angles]
3. m∠A + m∠B = m∠E + m∠D [Subtraction of equals]
4. $\overline{BE} \perp \overline{AB}$, $\overline{DE} \perp \overline{AD}$ [Given]
5. ∠B, ∠D are right angles. [Definition of right angles]

CLASS ACTIVITY

What is the sum of m∠1 and m∠2 for each triangle?

1. 120°

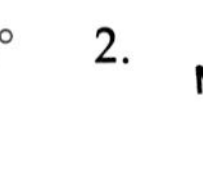

2. 88°

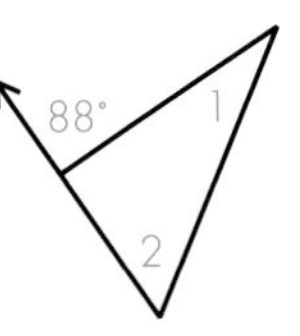

3. 110°

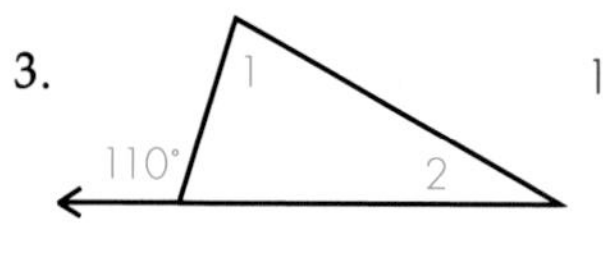

4. 60°

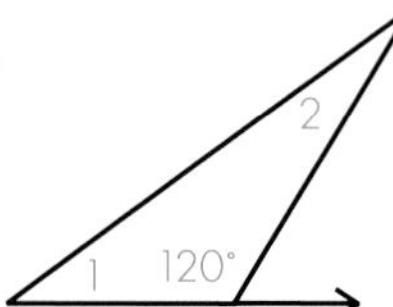

5. For Exercise 4, find m∠1 and m∠2 if m∠1 = $(x + 20)°$ and m∠2 = $(5x + 10)°$.
m∠1 = 25, m∠2 = 35

HOME ACTIVITY

1. Draw a triangle, △ABC. Extend the sides. Mark each exterior angle with an arc. How many exterior angles are there? 6; see margin for art.

2. Using Theorem 3-5-1, write three equations involving the exterior angles marked in the figure.

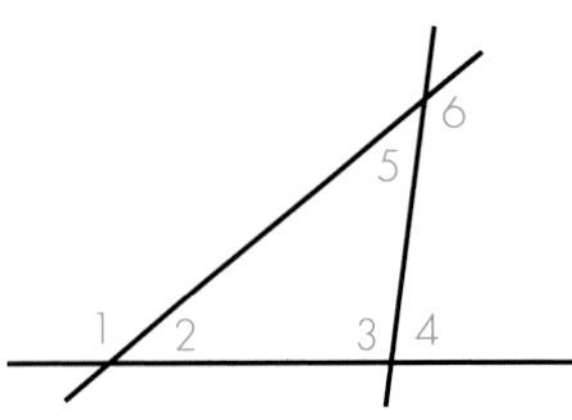

m∠6 = m∠2 + m∠3;
m∠1 = m∠3 + m∠5;
m∠4 = m∠2 + m∠5

Find the sum of m∠1 and m∠2.

3. 166°

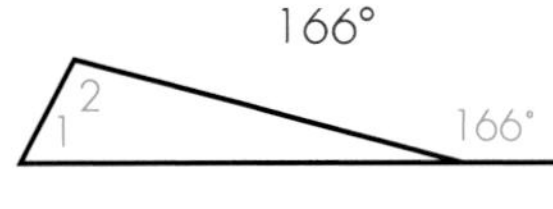

4. 96°

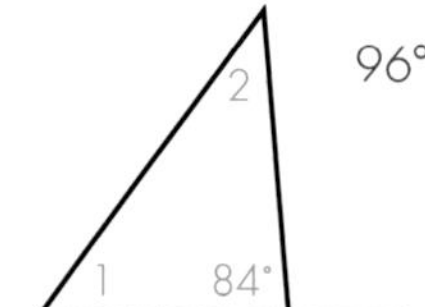

5. 130°

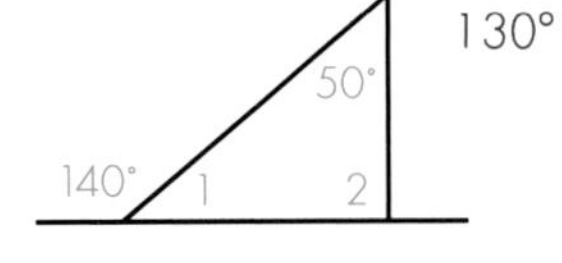

6. Find m∠1 and m∠2.

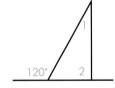

$m\angle 1 = (3x)^\circ$ 30°

$m\angle 2 = (6x + 30)^\circ$ 90°

7. Find each angle measurement if $l_1 \parallel l_2$.

m∠1 = 40°; m∠2 = 140°;
m∠3 = 80°; m∠4 = 60°;
m∠5 = 120°

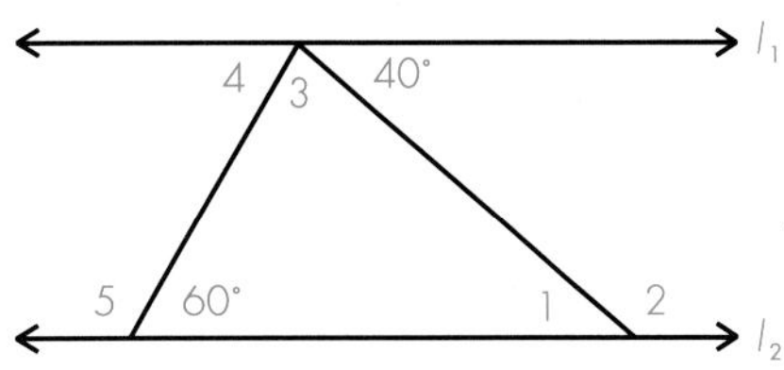

8. Show that m∠1 = m∠2. $\overline{BE} \perp \overline{AB}$, $\overline{DE} \perp \overline{AD}$ See margin.

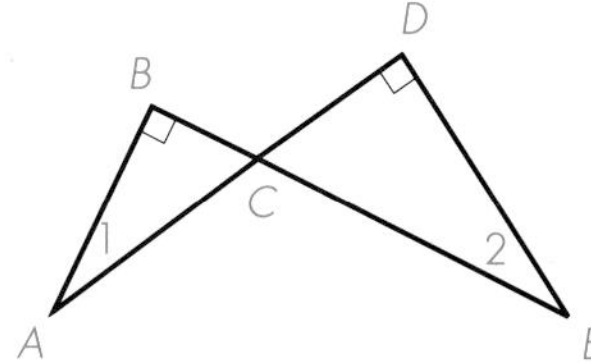

9. What is the sum of the measures of the six exterior angles of a triangle? 720°

10. What kind of angles are an exterior angle and its adjacent interior angle of a triangle? Supplementary

11. Can the measure of an exterior angle of a triangle ever equal the measure of its adjacent interior angle? Explain. Yes; both can be 90°.

12. Can the measure of an exterior angle of a triangle ever equal the measure of one of the remote interior angles? Explain. No; there would have to be two 90° angles in the triangle.

Refer to △*ABC*. $\overline{AC}$ represents a shortcut on a path.

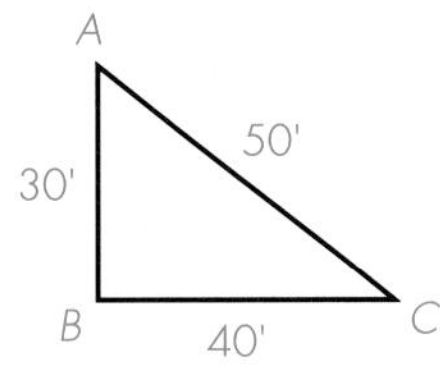

13. How many feet are saved by taking the shortcut? 20′

14. If a student takes the shortcut twice a day, how many feet will she save in 5 days? in 20 days? in a school year (180 days)? 200′; 800′; 7200′

15. Convert the answer for the school year in Exercise 14 to miles. 1.36 miles

Additional Answers

6. m∠*B* = m∠*D* [All right angles measure 90°.]
7. m∠*A* = m∠*E* [Subtraction of equals] Therefore, m∠1 = m∠2.
8. 1. m∠*A* + m∠*B* + m∠*BCA* = 180°
m∠*DCE* + m∠*D* + m∠*E* = 180°
[Sum of measures of angles of triangle is 180°.]

Using the Page

Use Transparency Master 3-1 to prepare students for Home Activity Exercise 9. They should measure each exterior angle and compare the measure to the sum of the measures of the two remote interior angles.

Using the Page

Students may need to use a table of conversion factors in Exercise 15. Encourage research to find this factor.

Additional Answers
To accomplish the task in Exercises 16-19, the turtle must be turned through the exterior angle.

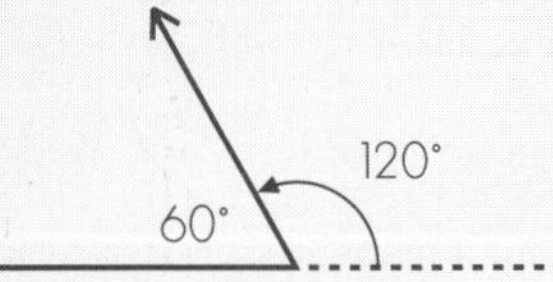

Using the Page
The exercises under CRITICAL THINKING use indirect reasoning. Students can list all possibilities, showing one person at a time telling the truth (or lying). Only one possibility will not lead to a contradiction in facts.

FOLLOW-UP

Assessment
1. Name the exterior angle shown, and then name the two remote interior angles for this exterior angle. [∠EFG; ∠D, ∠E]

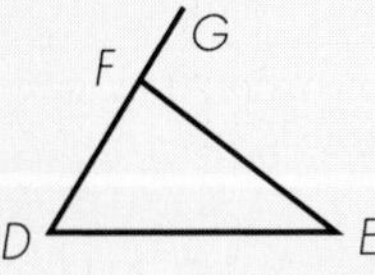

2. In the figure above, if m∠D = 50° and m∠E = 45°, what is m∠EFG? [85°]

Reteaching
Students who have had difficulty with this section may benefit from Reteaching Activity 3.5.

Enrichment
For students who have mastered the material in this section, you may wish to assign Enrichment Activity 3.5.

Write Logo procedures to draw each triangle described below by moving the turtle forward and to the right each time.
Answers may vary. See margin.

16. Equiangular $\triangle MNP$
17. $\triangle XYZ$ with $m\angle X = 40$, $m\angle Y = 30$, $m\angle Z = 110$
18. $\triangle TUV$ with $m\angle T = 125$, $m\angle U = 20$, $m\angle V = 35$
19. $\triangle CDE$ with $m\angle C = 62$, $m\angle D = 90$, $m\angle E = 28$

Complete the two-column proof.

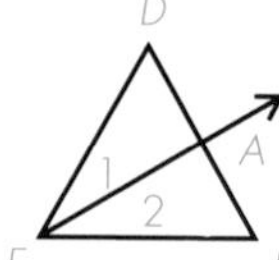

20. Given: $\overrightarrow{EA}$ bisects $\angle DEF$; $m\angle 1 + m\angle F = 90°$

Prove: $\angle 2$ and $\angle F$ are complementary.

Statements	Reasons
a. $\overrightarrow{EA}$ bisects $\angle DEF$.	a. Given
b. $\angle 1 = \angle 2$	b. If angle is bisected, it is divided into equal angles.
c. $m\angle 1 + m\angle F = 90°$	c. Given
d. $m\angle 2 + m\angle F = 90°$	d. Substitution
e. $\angle 2$ and $\angle F$ are complementary.	e. If sum of measures of 2 angles = 90°, the angles are complementary.

CRITICAL THINKING

21. Roger, Bob, and Ruth are arguing over the number of points their bowling team scored during the season. They each made one statement. Only one person is telling the truth. Who is telling the truth? Bob

Roger: The team scored at least 2800 points.
Bob: The total was not as great as Roger said.
Ruth: The team scored at least 1500 points.

22. A newspaper was discarded before everyone had a chance to read it. Four siblings, one of whom committed this act, made the following statements when questioned.

Alan: Doris did it.
Doris: Trina did it.
Gary: I didn't do it.
Trina: Doris lied when she said I did it.

a. If only one of these four statements is true, who was guilty? Gary
b. If only one of these four statements is false, who was guilty? Doris

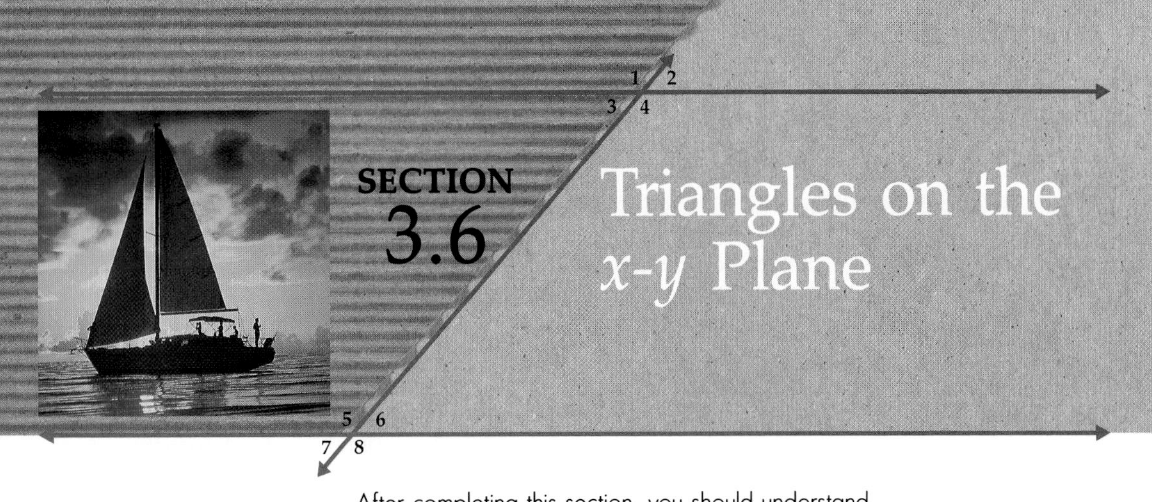

SECTION 3.6 Triangles on the x-y Plane

After completing this section, you should understand
▲ the x-y (Cartesian) coordinate plane.
▲ how the Cartesian coordinate plane is used in geometry.

In the early 1600's, a man named René Descartes (pronounced day cart) adapted plane geometry to a plane surface with a numbered grid. The numbered grid made locating points much easier for people.

DISCOVERY ACTIVITY

Oldtown

1. You are at the drugstore and need to make a delivery to Ms. Hsu's house. Can you describe how to get to her house? Why? No; you don't know from the drawing where you are or where Ms. Hsu lives.

Drug Store

Oldtown

Ms. Hsu

2. If you were shown this much information on a map, could you find Ms. Hsu's house? Yes

SECTION 3.6

Resources
Reteaching 3.6
Enrichment 3.6
Transparency Master 3-2

Objectives
After completing this section, the student should understand
▲ the x-y (Cartesian) coordinate plane.
▲ how the Cartesian coordinate plane is used in geometry.

Materials
protractor
ruler
grid paper

Vocabulary
coordinate plane
origin
x-coordinate
y-coordinate
ordered pair

GETTING STARTED

Quiz: Prerequisite Skills

1. Draw two lines that are perpendicular. [Typical drawing given]

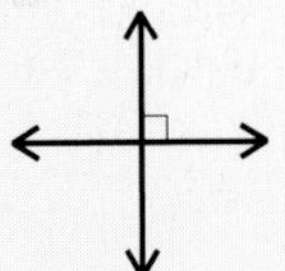

2. Complete the number line shown.

−5 ? ? ? ? 0 ? ? ? ? 5

[−4, −3, −2, −1, 1, 2, 3, 4]

Warm-Up Activity
Use a map of your neighborhood, city, or state. Have students locate various points, such as street corners, parks, buildings, or towns.

Chalkboard Example
Draw a grid on the board. Place a point at (−3, −5). Ask students for the coordinates of the point.

TEACHING COMMENTS

Using the Page
For Part 2 of the Discovery Activity on Page 111, ask students how they could describe to someone else how to get to Ms. Hsu's house. Students might respond that they could give left turn-right turn instructions, along with number of blocks until the next turn. Then they could ask neighbors which house Ms. Hsu lived in, or use a description of the house to find the right one.

Using the Page
Some students may know the x-axis as the horizontal axis and the y-axis as the vertical axis.

3. What would make it easier for someone who didn't know the town to go from the drugstore to Ms. Hsu's house? Street names and address numbers

You should have found from the Discovery Activity that having labels such as street names and address numbers help people find specific points or locations.

THE CARTESIAN PLANE

Descartes' Cartesian plane consists of two number lines that are perpendicular to each other. The horizontal number line is the x-AXIS and the vertical number line is the y-AXIS. The point of intersection is the ORIGIN.

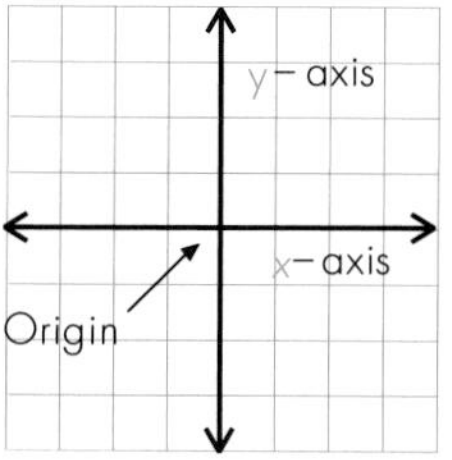

The positive numbers are to the right on the x-axis and up on the y-axis. Negative numbers are to the left on the x-axis and down on the y-axis. The arrows indicate that the axes extend indefinitely.

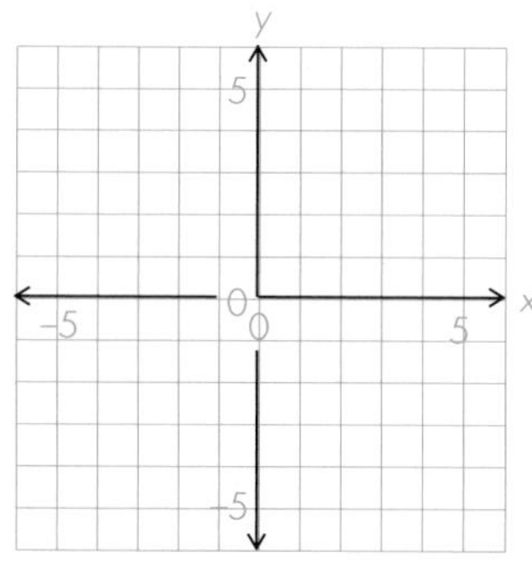

Two numbers are needed to locate a point. These numbers are called the coordinates of the point. The x-coordinate is given first, then the y-coordinate. The two coordinates form an ordered pair, a pair of numbers that follow a particular order.

DEFINITION 3-6-1 **A point on the x-y plane is an ORDERED PAIR of numbers, (x,y).**

Example: Locate the point (3,−4).

Start at the origin. Move 3 positive units (to the right) on the x-axis and 4 negative units (down) on the y-axis.

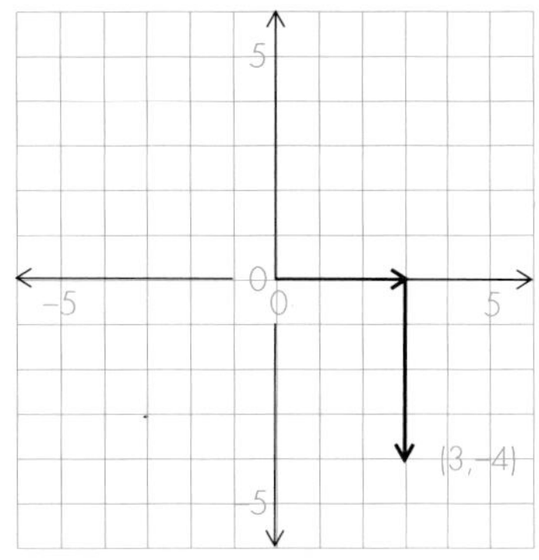

Example: On the x-y plane, locate the following points:

$A(2,4)$
$B(-5,1)$
$C(2,-5)$
$D(-1,-6)$
$E(0,0)$
$F(0,3)$
$G(-4,0)$
$H(0,-3)$
$I(7,0)$
$J(\frac{1}{2},\frac{1}{2})$

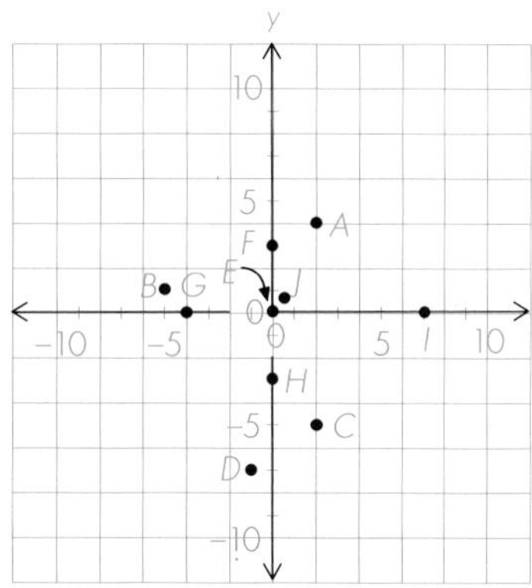

Using the Page
Using Transparency Master 3-2 of the Cartesian plane should make it easier for students to visualize locating points on the plane. You could also use a computer and software that will allow you to plot points if these are available.

Extra Practice
1. Plot the points (0,2), (2,0), and (2,2) on the x-y plane. Connect the three points, and describe the triangle formed. [right, isosceles]

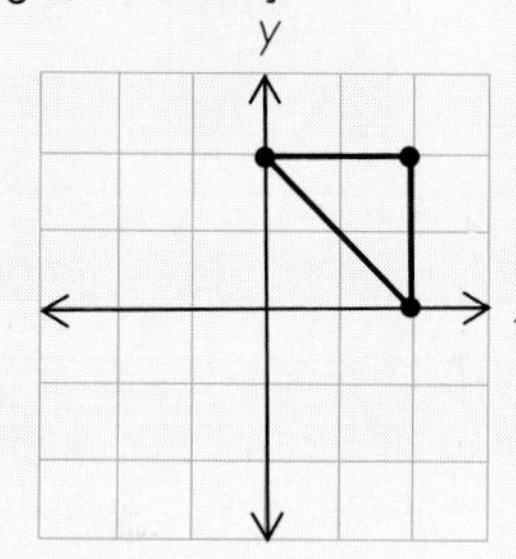

CLASS ACTIVITY

Draw the x-y plane and label the axes from −10 to 10. Locate the following points. See margin.

1. $A(2,6)$	2. $B(3,5)$	3. $C(4,0)$	4. $D(3,-2)$
5. $E(-5,3)$	6. $F(-6,-1)$	7. $G(7,-4)$	8. $H(-4,0)$
9. $I(-2,4)$	10. $J(0,9)$	11. $K(0,0)$	12. $L(8,-4)$
13. $M(-4,-7)$	14. $N(5,0)$	15. $O(0,-3.5)$	16. $P(-2.7,-7.3)$

The coordinates of the vertices of a triangle are given. Use a graphing calculator to graph each triangle. Then classify each triangle according to its sides and angles.

17. $A(0,5)$, $B(5,0)$, $C(0,0)$ Isosceles, right
18. $D(0,9)$, $E(7,0)$, $F(4,8)$ Scalene, obtuse
19. $G(-1,-1)$, $H(4,0)$, $I(-1,2)$ Scalene, acute
20. $J(5,-7)$, $K(1,-1)$, $L(8,-1)$ Scalene, acute

Additional Answers
1-16.

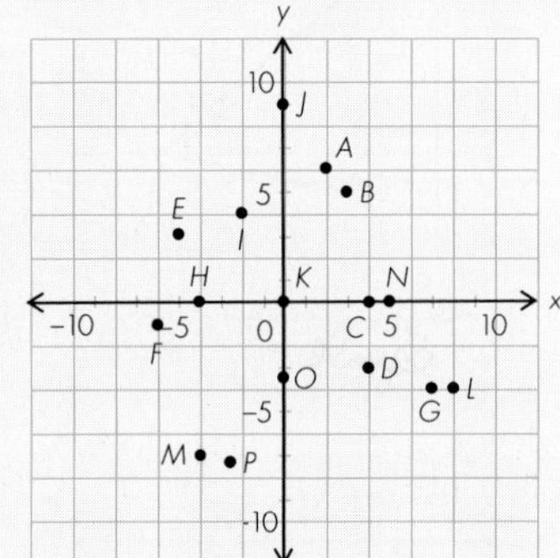

Additional Answers

22. Acute, isosceles

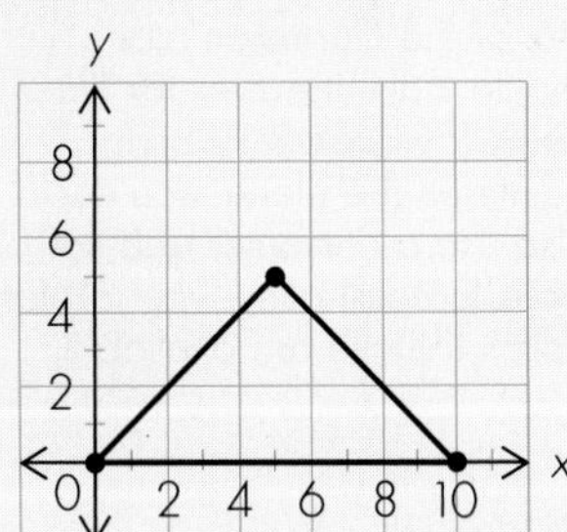

1. Isosceles, right

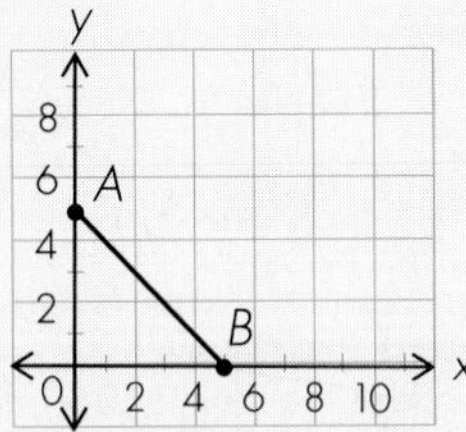

2. Scalene, obtuse

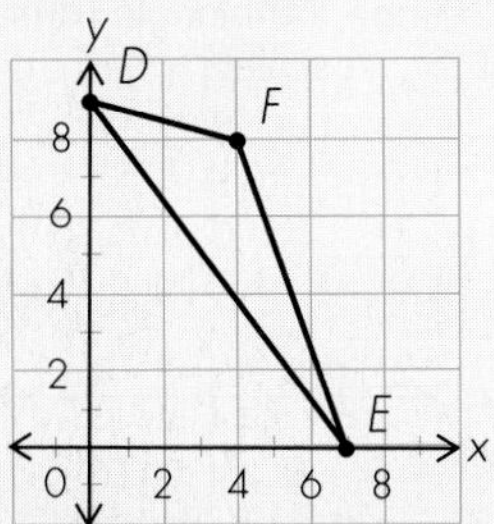

3. Scalene, acute

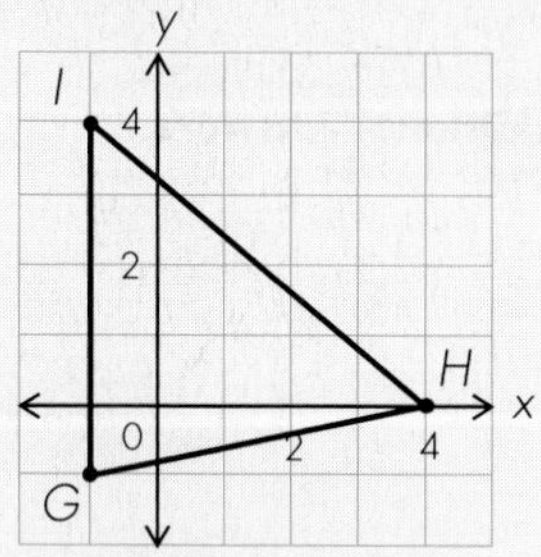

4. Scalene, acute

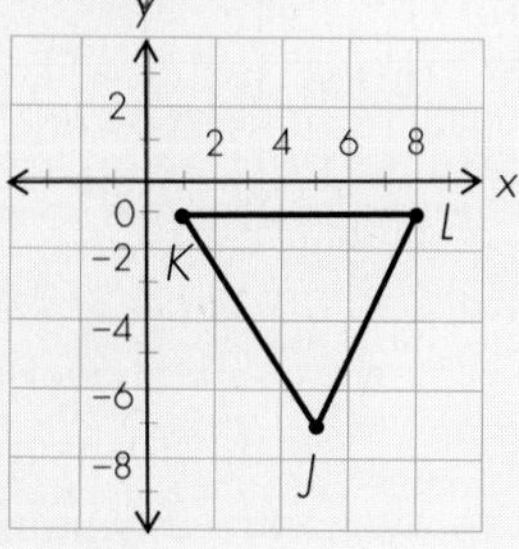

21. Give the coordinates of each point.

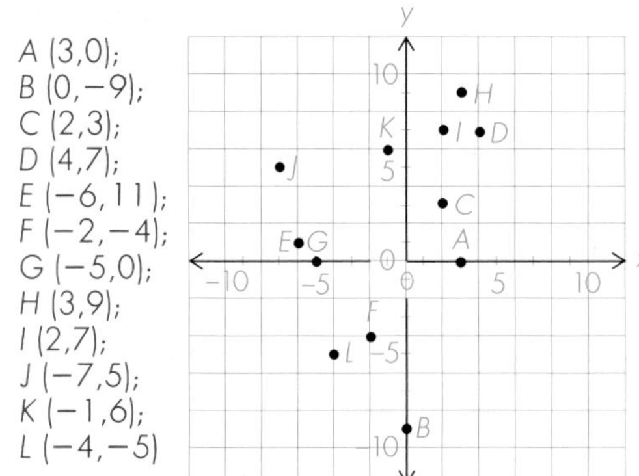

A (3,0);
B (0,−9);
C (2,3);
D (4,7);
E (−6,11);
F (−2,−4);
G (−5,0);
H (3,9);
I (2,7);
J (−7,5);
K (−1,6);
L (−4,−5)

22. Plot the points (0,0), (5,5), and (10,0) on the x-y plane. Connect the three points, and describe the triangle formed. See margin.

HOME ACTIVITY

Graph each set of ordered pairs on a separate coordinate plane.

1. $A(0,5)$, $B(5,0)$, $C(0,0)$ See margin.
2. $D(0,9)$, $E(7,0)$, $F(4,8)$ See margin.
3. $G(-1,-1)$, $H(4,0)$, $I(-1,4)$ See margin.
4. $J(5,-7)$, $K(1,-1)$, $L(8,-1)$ See margin.

Connect the three points with line segments. Classify each triangle according to its sides and its angles.

5. Give the coordinates of each point.

A (2,3); B (0,0); C (7,9); D (−10,0);
E (−3,2); F (−7,0); G (0,6); H (0,−4);
I (−1,7); J (2,−6); K (5,−9); L (10,0);
M (0,−9); N (7,−8); O (−10,4); P (8,−5)

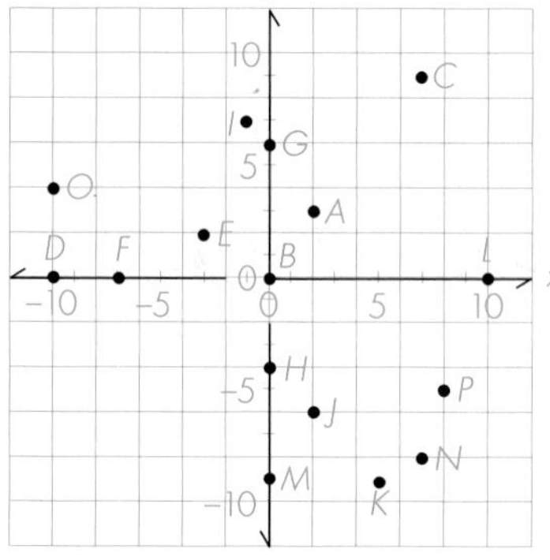

6. Draw the x-y plane and label the axes from −10 to 10. Locate the following points. See margin.

$A(7,0)$	$B(1,2)$	$C(0,0)$	$D(-8,6)$
$E(4,6)$	$F(1,-10)$	$G(-2,-3)$	$H(-6,-6)$
$I(4.5,-6.5)$	$J(-4,4)$	$K(6,4)$	$L(2,-3)$
$M(-\frac{2}{3},\frac{7}{3})$	$N(-8.5,-3)$	$O(0,6)$	$P(-7,0)$

Name the type of figure formed if each set of points is connected in order, left to right, as given. Use as many descriptive names as possible (i.e., use sides and angles to describe triangles).

7. (0,5), (0,0), (7,0), (0,5)
Right triangle, scalene triangle

8. (0,0), (6,6), (7,7)
Line segment

9. (−8,7), (0,0), (5,0), (−8,7)
Scalene triangle, obtuse triangle

10. (1,−5), (0,0), (−5,1), (1,−5)
Isosceles triangle, obtuse triangle

11. All points that have an x-coordinate of 0 and a positive y-coordinate represent what geometric figure? Ray

12. What is the minimum number of noncollinear points needed to determine a plane? Three

13. Is the point (3,4) the same as the point (4,3)? Explain. No; the order of the coordinates is different.

14. Plot these points and identify the type of triangle outlined: (−3,0), (3,0), (0,5.2)
Equilateral, acute, equiangular

Give the relationship between each pair of lines and the coordinates for any point of intersection.

15. The line determined by points (0,3) and (0,−1), and the line determined by the points (2,4) and (2,−3) Parallel

16. The lines determined by points (3,3) and (−4,−4), and the line determined by the points (−3,3) and (2,−2) Perpendicular; intersection point (0,0)

17. The line determined by the points (−4,3) and (−1,−1), and the line determined by the points (3,−1) and (4,3) Intersecting, at (2,−5)

18. The line determined by the points (0,1) and (2,2), and the line determined by the points (−6,−2) and (−2,0) Same line

19. The figures in the coordinate plane at right share a number of properties. Give the coordinates of the sides of each figure, and describe two properties the figures all share.

See margin.

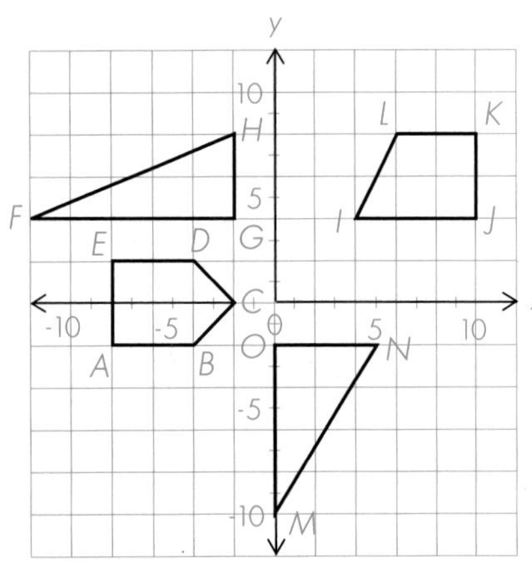

Additional Answers
(for Page 114)
6.

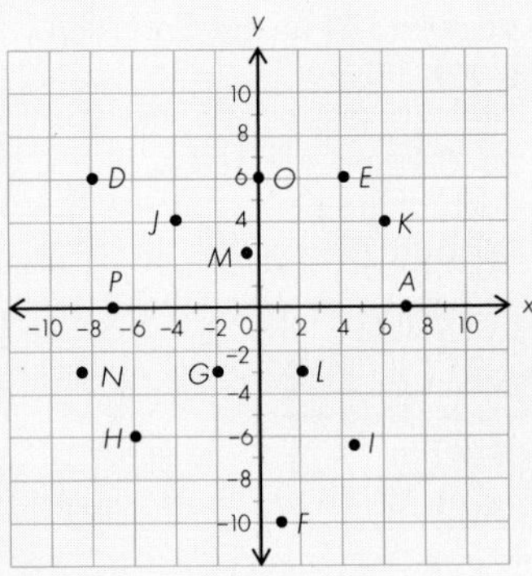

19. A(−8,−2), B(−4,−2), C(−2,0), D(−4,2), E(−8,2); F(−12,4), G(−2,4), H(−2,8); I(4,4), J(10,4), K(10,8), L(6,8); M(0,−10), N(5,−2), O(0,−2). Answers may vary; possible answers: all have at least one right angle, all have areas of 20, all have at least one side parallel to x-axis, all have at least one side parallel to y-axis.

FOLLOW-UP

Assessment

1. Explain how to locate the point (−3,2) on the Cartesian plane. [Move 3 units to the left of the origin, then 2 units up.]

2. Give the coordinates of each of the points. [A(−2,3), B(0,6), C(3,−5)]

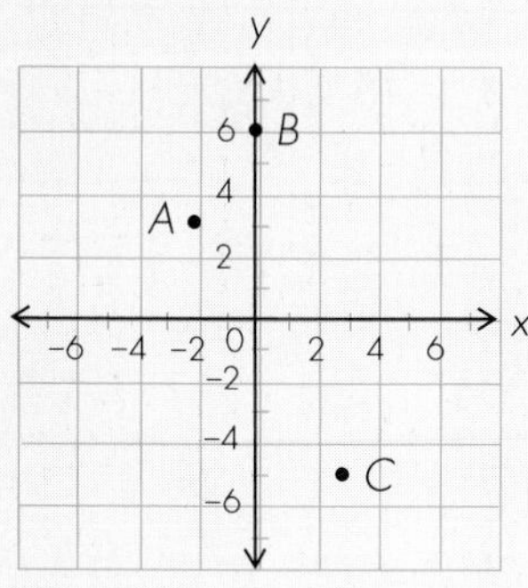

CRITICAL THINKING

20. Give 12 points such that when the coordinates of each point are reversed, the same point results. (1,1), (2,2), (3,3), . . .,(12,12)

21. Write an equation that relates the x-coordinate and y-coordinate of each point in Exercise 20. What geometric figure is formed when the points are connected? $y = x$; line

A triangle is a rigid form. If you fasten three strips of cardboard together at their ends, the triangle that results holds its shape, even when it is moved. This property of triangles is the secret of the strength of many structures.

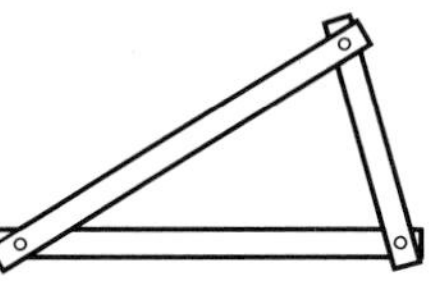

A quadrilateral, made with four strips fastened together, is flexible and can be changed into a variety of shapes.

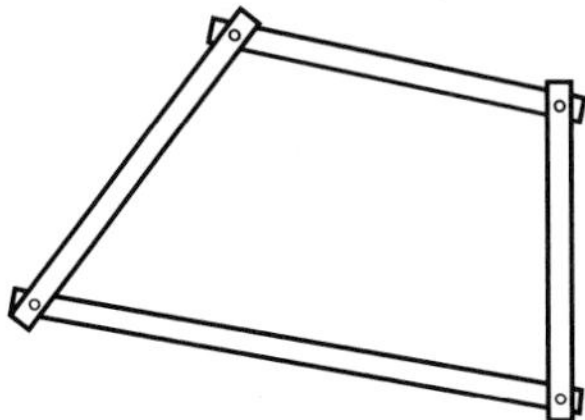

Decide whether the constructions shown below would be rigid or flexible.

22. Rigid

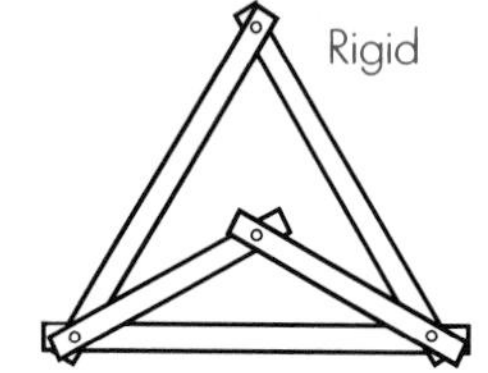

23. Flexible

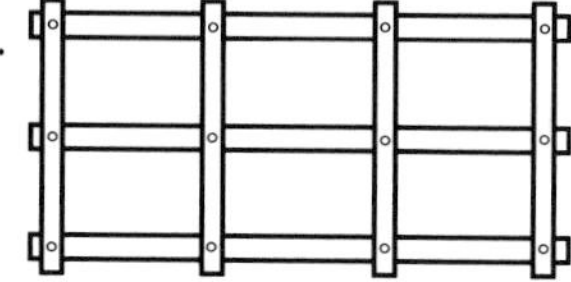

24. Rigid

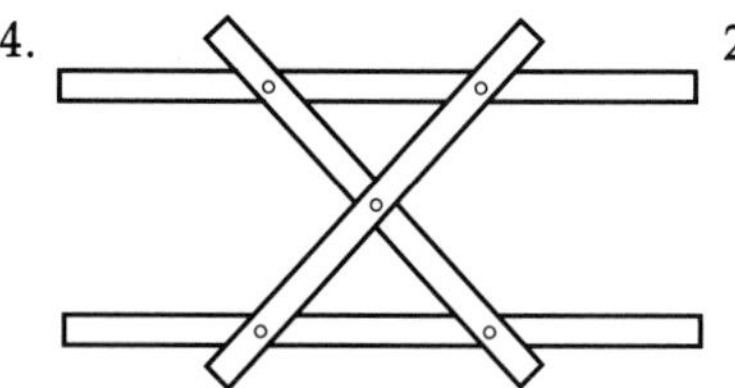

25. Rigid

Reteaching

Students who have had difficulty with this section may benefit from Reteaching Activity 3.6.

Enrichment

For students who have mastered the material in this section, you may wish to assign Enrichment Activity 3.6.

CHAPTER SUMMARY

3.1 Figure RST is a triangle.

$$m\angle R + m\angle S + m\angle T = 180°$$

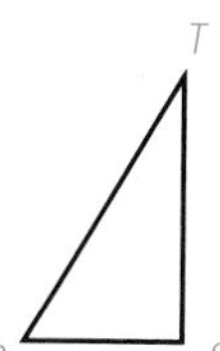

3.2 The lengths of the sides of any triangle have the following relationships:

$RS + ST > TR$	$\lvert RS - ST\rvert < TR$
$ST + TR > RS$	$\lvert ST - TR\rvert < RS$
$RS + TR > ST$	$\lvert RS - TR\rvert < ST$

3.3 $\triangle ABC$ is an equilateral triangle; all three sides are equal in length.

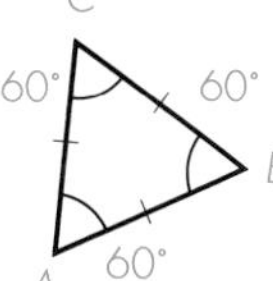

$\triangle DEF$ is an isosceles triangle; two sides are equal in length.

$\triangle GHI$ is a scalene triangle; all three sides are of different length.

3.4 $\triangle ABC$ is an acute triangle; all three angles measure less than 90°. It is also equiangular.

$\triangle DEF$ is a right triangle; it has one 90° angle.

$\triangle GHI$ is an obtuse triangle; one angle measures more than 90°.

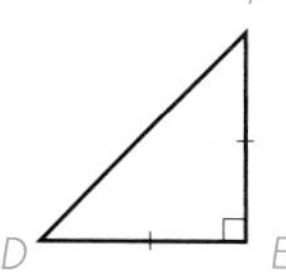

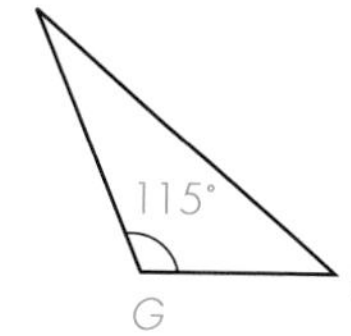

3.5 $\angle QST$ is an exterior angle of $\triangle QRS$. The measure of an exterior angle of a triangle is the sum of the remote interior angles.

$$m\angle R + m\angle Q = m\angle QST$$
$$70° + 30° = 100°$$

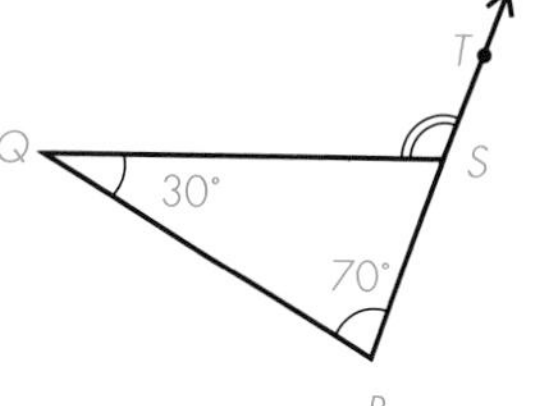

3.6 Point F on the x-y plane can be identified by an ordered pair of numbers, $(-3,4)$.

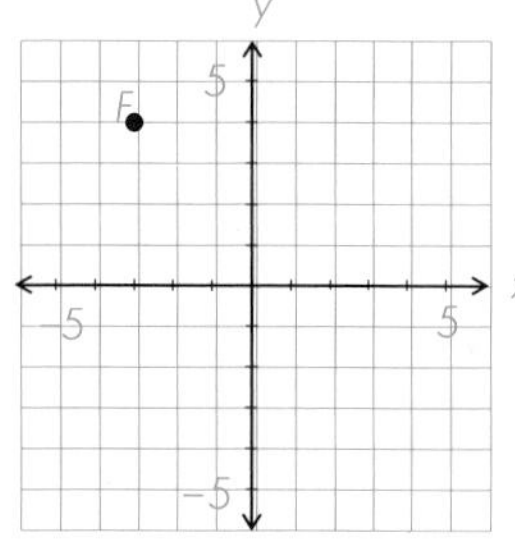

Chapter Vocabulary

acute triangle
adjacent interior angle
base of triangle
collinear points
conclusion
coordinate plane
equiangular triangle
equilateral triangle
exterior angle
given
indirect reasoning
isosceles triangle
legs of triangle
noncollinear points
obtuse triangle
ordered pair
origin
parallel lines
postulate
remote interior angle
right triangle
scalene triangle
slash mark
theorem
triangle
triangle inequality
two-column proof
x-coordinate
y-coordinate

117

Using the Page
This Computer Activity uses LOGO to create a portion of the x-y plane and to plot points thereon. If students have difficulty, work through the steps with them at the computer. Then have them perform the procedure on their own.

COMPUTER ACTIVITY

When using Logo, you can think of the computer screen as an x-y plane in which the turtle's home position is the origin (0,0). Every location on the screen can be represented by an ordered pair of numbers (x,y).

There are special Logo commands that can be used to move the turtle to any location on the screen's x-y plane. The command SETX moves the turtle horizontally to the x-coordinate that you specify. For example, SETX 10 tells the turtle to move horizontally to the x-coordinate 10. The command SETY moves the turtle vertically to the y-coordinate that you specify. For example, SETY 6 tells the turtle to move vertically to the y-coordinate 6. The command SETXY moves the turtle to the location with coordinates (x,y) that you specify. For example, SETXY 3 8 tells the turtle to move to the location with coordinates (3,8). If the y-coordinate in the SETXY command is negative, you must put parentheses around it.

Example: Write a procedure named XYPLANE that tells the turtle how to draw the x-axis and the y-axis with the origin at the turtle's home position.

```
TO XYPLANE

  SETX 139          These commands tell the
  HOME              turtle to draw the x-axis.
  SETX −139
  HOME

  SETY 119
  HOME              These commands tell the
  SETY −119         turtle to draw the y-axis.
  HOME

  END
```

The screen will appear as shown below.

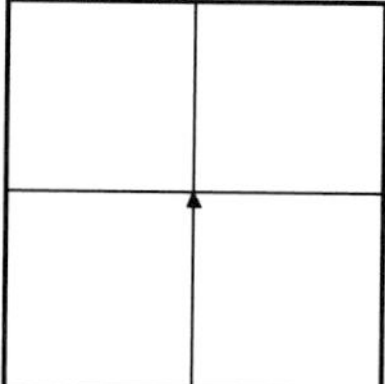

118

The procedure XYPLANE can now be used in other procedures to show the graph of ordered pairs on the x-y plane.

Show how the turtle moves when you type these Logo commands. Classify each triangle according to its sides and its angles. See margin.

1.	2.	3.	4.
XYPLANE	XYPLANE	XYPLANE	XYPLANE
PU	PU	PU	PU
SETXY 1 1	SETXY −1 2	SETXY −2 (−2)	SETXY 4 (−1)
PD	PD	PD	PD
SETXY 3 1	SETXY −2 4	SETXY 2 (−2)	SETXY 0 2
SETXY 2 4	SETXY −2 2	SETXY 2 2	SETXY −4 (−1)
SETXY 1 1	SETXY −1 2	SETXY −2 (−2)	SETXY 4 (−1)

Use the procedure XYPLANE in procedures that tell the turtle to draw each figure. Give each procedure the indicated name. Answers may vary.

5.

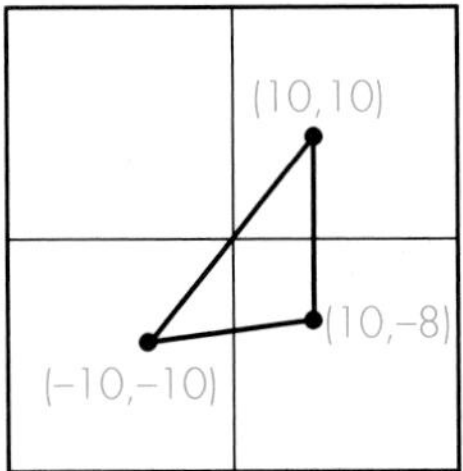

TRIANGLE1

6.

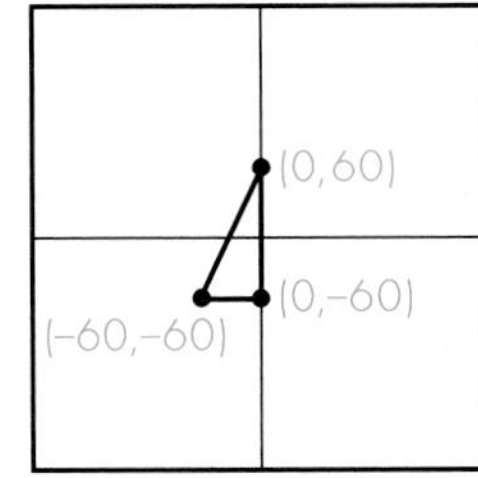

RTTRIANGLE1

7.

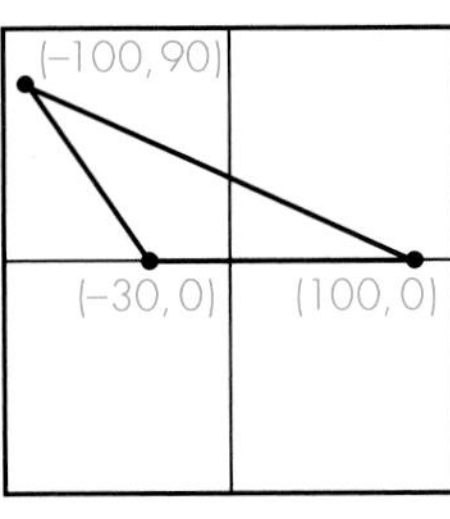

TRIANGLE2

8.

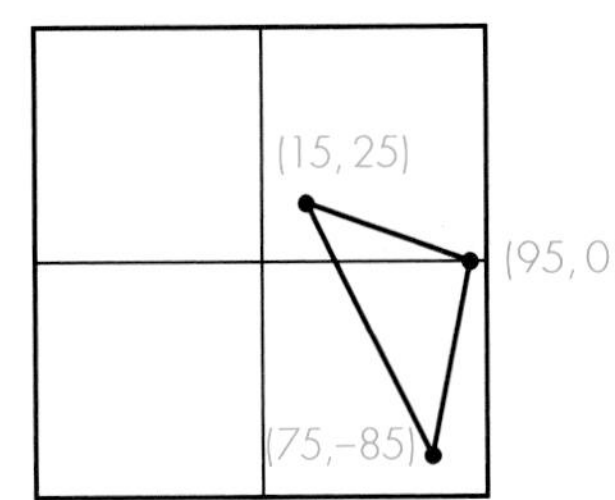

TRIANGLE3

9. A right triangle with the right angle at the origin; RTTRIANGLE2

10. An isosceles triangle with the midpoint of the base at the origin; ISOSCELES

Additional Answers

1. Isosceles, acute

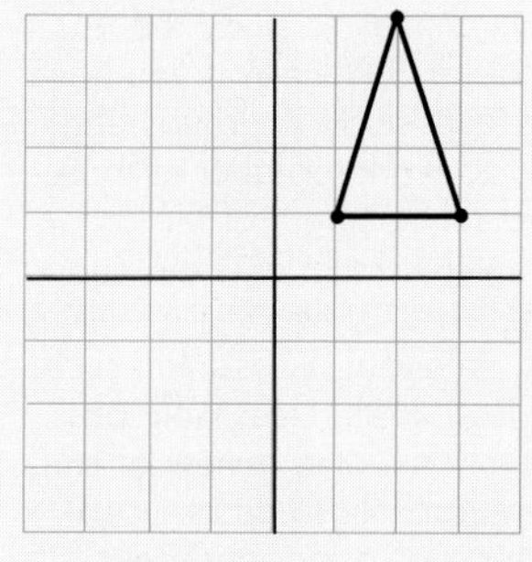

2. Scalene, right

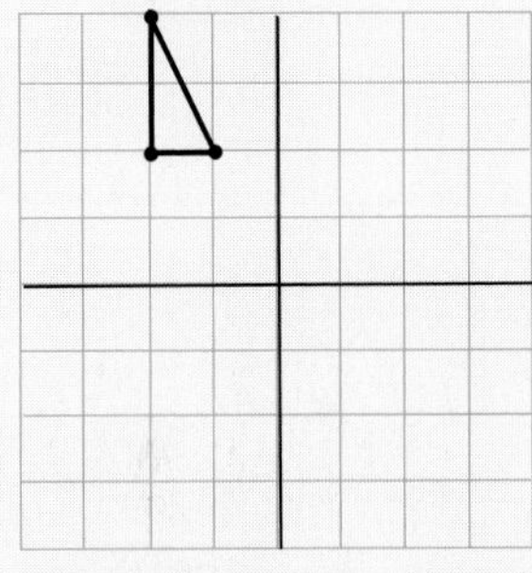

3. Isosceles, right

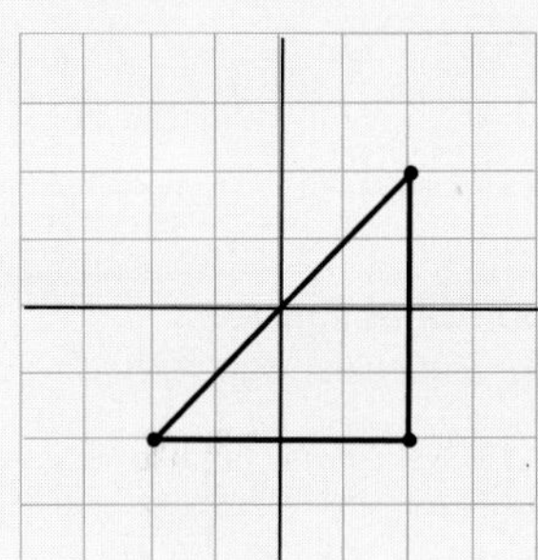

4. Isosceles, obtuse

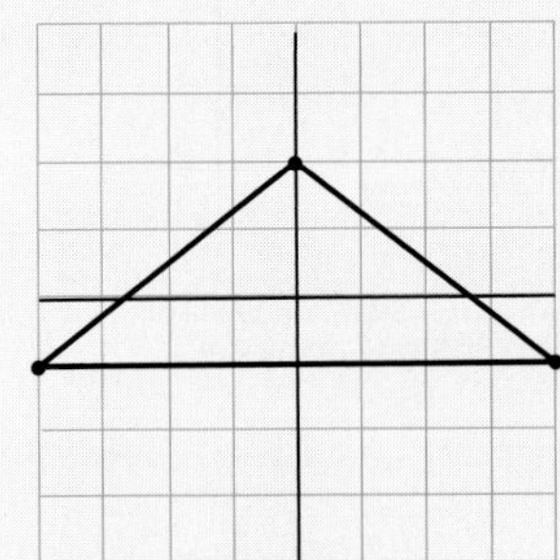

Using the Page
Students can work individually or in groups to study the Chapter 3 Review. If you have students study in groups, they should work together to make sure all members of the group know the material in the chapter.

Informal assessment such as interviews, classroom observations, or a review of student portfolios can be used instead of paper-and-pencil tests.

Match the types of triangles on the left with the figures on the right. Each term may describe more than one triangle. List as many as possible. Triangles may be used more than once.

1. Equilateral 5

2. Isosceles 2, 4, 5

3. Scalene 1, 3, 6, 7

4. Equiangular 5

5. Right 2, 3

6. Acute 1, 4, 5, 7

7. Obtuse 6

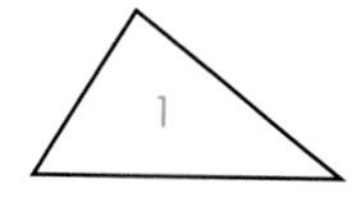

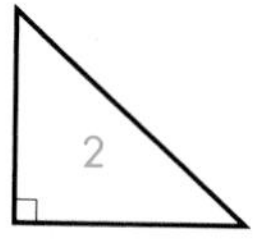

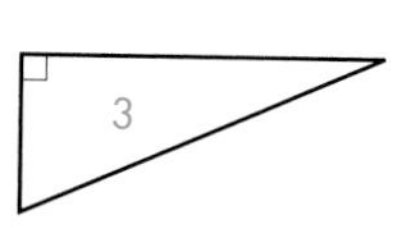

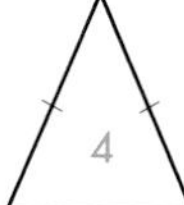

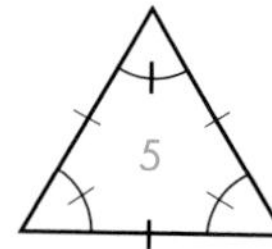

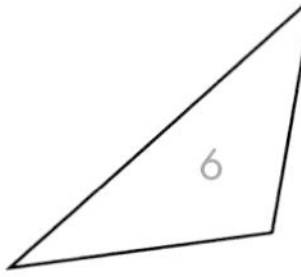

8. If a figure is a triangle, then the sum of the measures of the angles is ________. 180°

Refer to $\triangle ACD$. Find the measure of each angle.

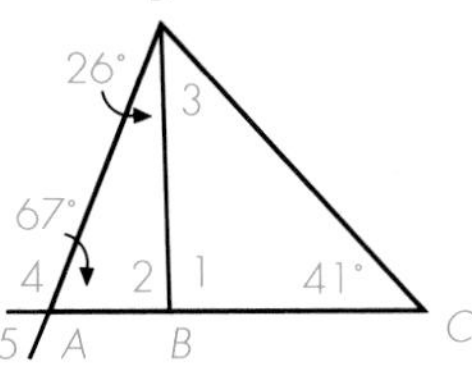

9. $\angle 1$ 93° **10.** $\angle 2$ 87°

11. $\angle 3$ 46° **12.** $\angle 1 + \angle 3$ 139°

13. $\angle 5$ 67° **14.** $\angle 4$ 113°

Refer to $\triangle RAT$. If $RT = 37$, $RB = 29$, and $BA = 20$, then find the limits for the line segment.

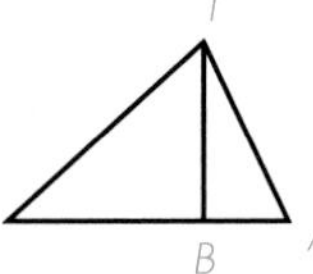

15. $\underline{8} < TB < \underline{66}$ **16.** $\underline{12} < TA < \underline{86}$

17. Use indirect reasoning to reach the proper conclusion. A guidance counselor questioned three students, trying to determine which one had damaged a car and which one had damaged a bike. Only one of the following statements is true. Who damaged the car, and who damaged the bike? Carl damaged the car, Bev the bike.

Alice: Carl did not damage the car. Carl: Bev did damage the bike. Bev: Carl is lying.

Plot the following points on the x-y plane and identify the type of triangle that is formed when the points are connected. See margin.

18. (0,−5), (−5,0), (0,0) **19.** (5,7), (1,1), (−8,1)

Additional Answers

18. Isosceles, right

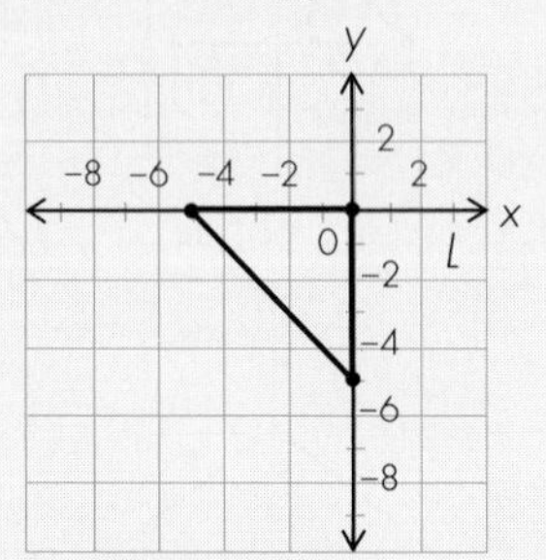

19. Scalene, obtuse

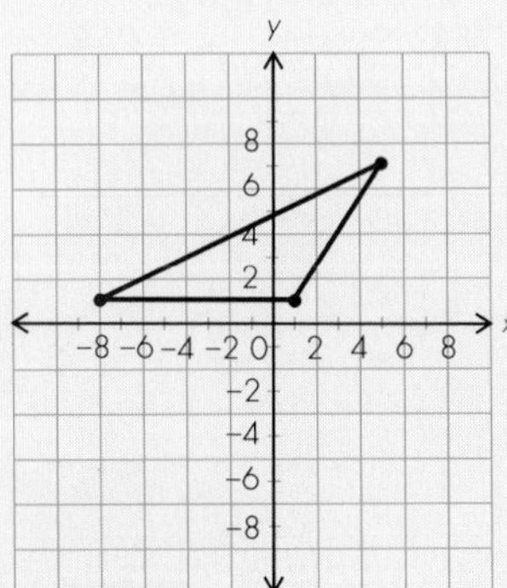

CHAPTER TEST

Write the letter of each triangle that fits the description. There may be more than one triangle that fits, and triangles may be used more than once.

1. Acute triangle   a, d, h
2. Scalene triangle   a, c, f, g
3. Equilateral triangle   d
4. Obtuse triangle   b, f
5. Right triangle   c, e, g
6. Isosceles triangle   b, d, e, h
7. Equiangular triangle   d

a. 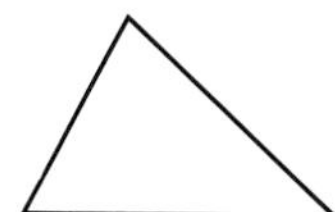

b. 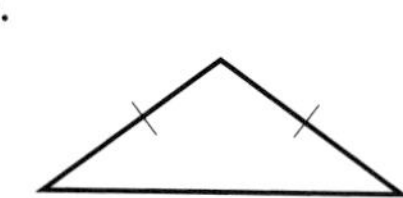

c. 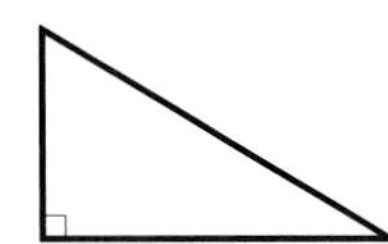

d. 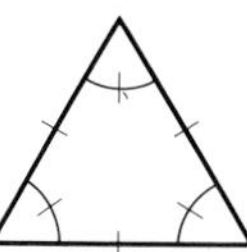

e. 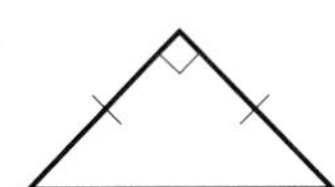

f. 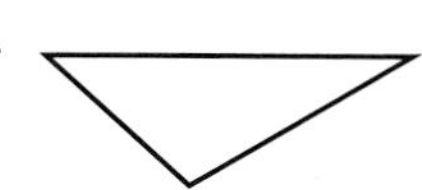

g. 

h. 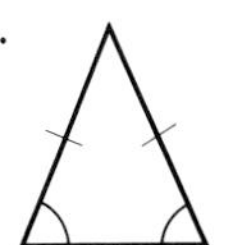

8. The length of the unknown side in the triangle is between what two numbers? $13 < x < 45$

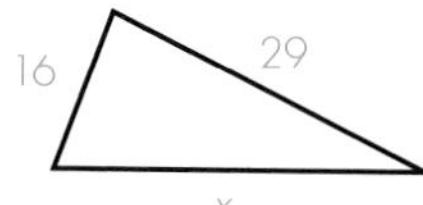

9. What is the measure of $\angle B$?   75°

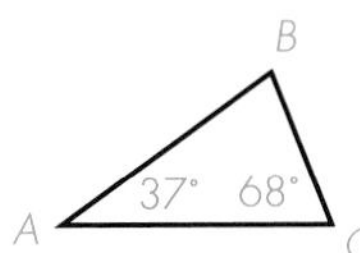

10. Given the following triangle and the information indicated, find the measure of angles 1–5.
$m\angle 1 = 110°$; $m\angle 2 = 70°$; $m\angle 3 = 150°$; $m\angle 4 = 140°$; $m\angle 5 = 40°$

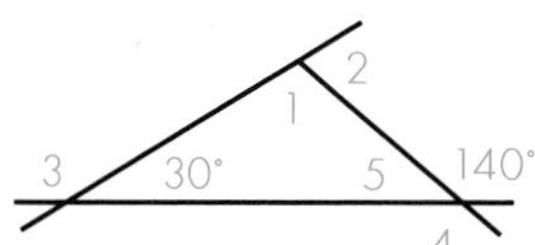

11. Plot and connect the following points. Classify the type of triangle formed. $A(0,5)$, $B(-5,5)$, $C(-5,0)$ See margin.

Assessment Resources
Achievement Tests pp. 5-6

Test Objectives
After studying this chapter, students should understand
- the sum of the measures of the angles of a triangle is 180°.
- the sum of the lengths of two sides of a triangle is greater than the length of the third side.
- the length of any side of a triangle is greater than the difference of the lengths of the other two sides.
- how triangles are classified by the length of their sides.
- how to classify triangles by the measure of their angles.
- the triangle exterior angle theorem.
- the x-y (Cartesian) coordinate plane.
- how the Cartesian coordinate plane is used in geometry.

Additional Answers
11. Right, isosceles

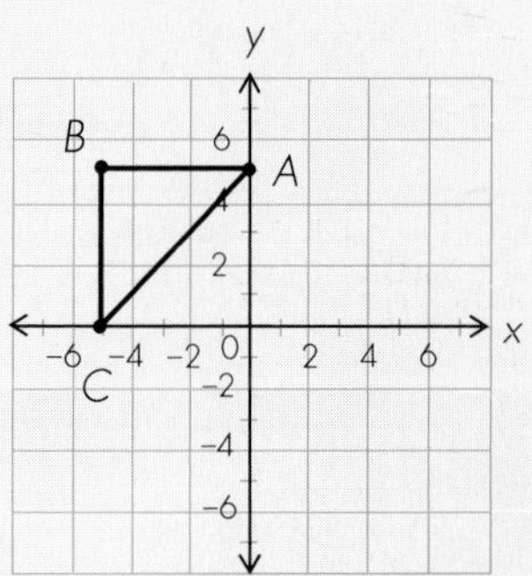

Algebra Review Objectives
Topics reviewed from algebra include

- using decimals to evaluate two-variable expressions.
- using fractions to evaluate two-variable expressions.
- solving proportions and equations.
- solving problems using ratio and proportion.

ALGEBRA SKILLS

Example:

Evaluate $x + 3y$ using $x = 1.6$ and

$y = 2.8$.

$1.6 + 3(2.8)$

$1.6 + 8.4$

10

Example:

Solve for x.

$3x - 5 = 7x + 9$

$3x - 14 = 7x$

$-14 = 4x$

$x = -3.5$

In Exercises 1–8, evaluate each expression using $x = 2.1$ and $y = -5.2$.

1. $x + y$ -3.1 **2.** $x - y$ 7.3 **3.** xy -10.92 **4.** $x/7$ 0.3

5. $y - x$ -7.3 **6.** x^2 4.41 **7.** $\frac{y-1}{2}$ -3.1 **8.** $|y|$ 5.2

In Exercises 9–16, evaluate each expression using $x = \frac{1}{5}$ and $y = -\frac{2}{3}$.

9. $x + y$ $-\frac{7}{15}$ **10.** $x - y$ $\frac{13}{15}$ **11.** xy $-\frac{2}{15}$ **12.** $\frac{x}{y}$ $-\frac{3}{10}$

13. $y - x$ $-\frac{13}{15}$ **14.** y^2 $\frac{4}{9}$ **15.** $\frac{y}{x}$ $-3\frac{1}{3}$ **16.** $|y|$ $\frac{2}{3}$

In Exercises 17–22, solve each equation for x.

17. $\frac{2}{x} = \frac{54}{27}$ $x = 1$ **18.** $\frac{-3}{x-4} = \frac{45}{27}$ $x = 2.2$ **19.** $\frac{5}{4} = \frac{2x}{20}$ $x = 12.5$ **20.** $\frac{3}{8} = \frac{21}{2x}$ $x = 28$

21. $20 - 2(3x - 5) = 90 - 2x$ $x = -15$ **22.** $12 - 5(2x - 20) = -3(2x + 8)$ $x = 34$

Solve each problem.

23. If the ratio of two angles is $\frac{2}{3}$ and the angles are complementary, then what is the measure of the larger angle? 54°

24. If $m\angle A$ in a given triangle is twice as great as $m\angle B$, and $m\angle B$ is three times as great as $m\angle C$, then what is the measure of each angle? $\angle A = 108°$, $\angle B = 54°$, $\angle C = 18°$

Name  Date

The Sum of the Angles of a Triangle

The sum of the angle measures of any triangle is 180°. You can use this fact to solve for the unknown angles in a triangle. First, write an equation to state that the sum of the anlges is 180°. Then solve for the unknown.

Example: Find the measure of $\angle C$.

Since the measure of $\angle C$ is unknown, you can call it x. Then write an equation.

$27 + 85 + x = 180$
$112 + x = 180$
$x = 68$ So $m\angle C = 68°$.

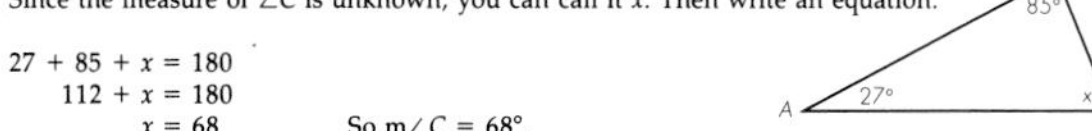

1. On a separate sheet of paper, draw a large triangle and label the angles D, E, and F. Then cut off the corners of the triangle and arrange them so that the three angles form a straight angle. Will the angles always form a straight angle? What is the degree measure of a straight angle? Yes; 180°

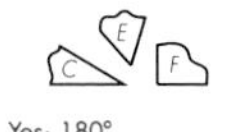

Find the measure of each unknown angle.

2. 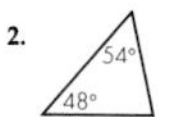 78° 
3. 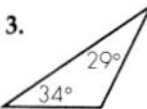 117° 
4. 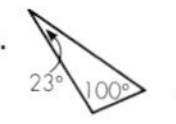 57°
5. 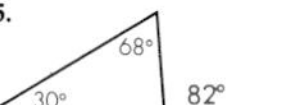 82°
6.  95° 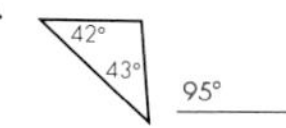
7.  66
8. 53°, 63°, 64°
9. 30°, 60°, 90° 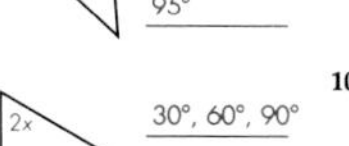
10. 50°, 53°, 77° 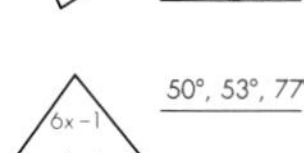

In each figure, $\overline{AB} \parallel \overline{CD}$. Find x and y.

11. 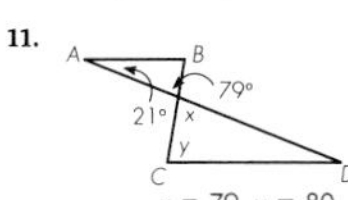 $x = 79, y = 80$
12. 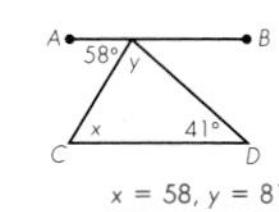 $x = 58, y = 81$
13. 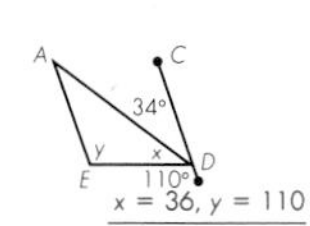 $x = 36, y = 110$

Name Date

More About Angle Sums

The sum of the measures of the angles of a triangle is 180°. Suppose, however, that a figure has more than three sides. Do you think the angle sum will still be 180°? Look at the 4-sided figure and 5-sided figure shown below. Suppose you draw the dashed lines and label the figures as shown.

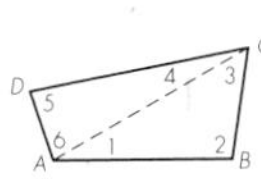

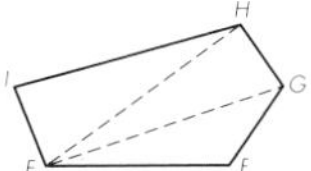

1. In the figure on the left, how many triangles are formed when you draw the dashed line? 2
2. What is $m\angle 1 + m\angle 2 + m\angle 3$? 180°
3. What is $m\angle 4 + m\angle 5 + m\angle 6$? 180°
4. What is $m\angle 1 + m\angle 2 + m\angle 3 + m\angle 4 + m\angle 5 + m\angle 6$? 360°
5. How does $m\angle A + m\angle B + m\angle C + m\angle D$ compare with the angle sum in Exercise 4? They are equal.
6. In figure $EFGHI$, how many triangles are formed by the dashed lines? 3
7. What is the sum of all the angle measures of the triangles in figure $EFGHI$? 3 × 180° or 540°
8. How does $m\angle E + m\angle F + m\angle G + m\angle H + m\angle I$ in figure $EFGHI$ compare with the answer to Exercise 7? They are equal.
9. Suppose a figure has 6 sides. How many triangles are formed by drawing all the possible lines from one vertex to each of the other vertices? 4
10. How does the number of triangles formed in each case above compare with the number of sides of the figure? The number of triangles is 2 less than the number of sides.
11. For a 6-sided figure, what is the sum of all the angle measures? 4 × 180° or 720°
12. Suppose a figure has n sides. Write a formula for finding the sum of all the angle measures. $s = (n - 2)\ 180°$

Name ______________________ Date ______________

Two Triangle Inequalities

In any triangle, the sum of the lengths of any two sides is always greater than the length of the third side. The difference of the lengths of any two sides is always less than the length of the third side. For example, in $\triangle ABC$,

$6 + 10 > 8$	$10 - 6 < 8$
$6 + 8 > 10$	$8 - 6 < 10$
$10 + 8 > 6$	$10 - 8 < 6$

Example: Find the range of values of x if the lengths of the three sides of a triangle are 14, 8, and x.

Since the sum of the lengths of any two sides is always greater than the length of the third, you know

$$14 + 8 > x \text{ or } x < 22$$

Since the difference of the lengths of any two sides is always less than the length of the third side, you know

$$14 - 8 < x \text{ or } x > 6$$

Decide if the given lengths could be lengths for the sides of a triangle. Write *yes* or *no*.

1. 12 m, 18 m, 10 m Yes
2. 16 mm, 5 mm, 22 mm No
3. 150 ft, 50 ft, 175 ft Yes
4. 18 cm, 9 cm, 9 cm No
5. 12 km, 1.5 km, 3 km No
6. 90 cm, 80 cm, 140 cm Yes

The lengths of the sides of a triangle are given. Find the range of values for x.

7. 6 in., 18 in., x in. 12 in. < x < 24 in.
8. 19.5 km, 25 km, x km 5.5 km < x < 44.5 km
9. 10 in., 20 in., x in. 10 in. < x < 30 in.
10. 70 yd, 30 yd, x yd 40 yd < x < 100 yd
11. 41 mm, 34 mm, x mm 7 mm < x < 75 mm
12. 1.9 m, 3.6 m, x m 1.7 m < x < 5.5 m
14. The length of $\overline{AB}$ is between what two values? 4 and 36

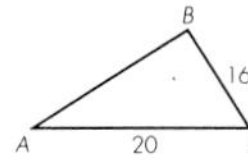

Name ______________________ Date ______________

Tessellations

A group of figures of the same shape and size is said to form a **tessellation** when the figures completely cover a flat surface with no overlapping. The figure that is used to cover the plane is said to *tessellate the plane*. For example, the square and the triangle shown below tessellate the plane, but the circle does not.

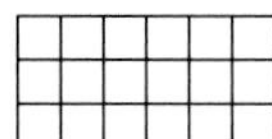

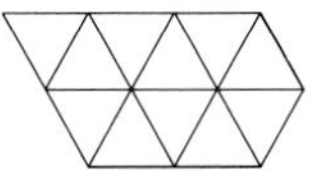

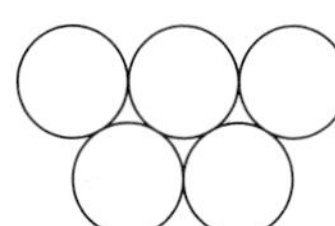

Make copies of each of the following triangles on dot paper or graph paper. Show that each triangle tessellates the plane by using at least 20 repetitions of each triangle. (NOTE: The triangles can be turned or flipped.) Drawings will vary.

1.
2.
3.

4. Do you think that any triangle can be used to tessellate the plane? Draw another triangle of any size on a separate sheet of paper and illustrate your answer. Yes Drawings will vary.

Make copies of each of the following figures on dot paper or graph paper. Show that each figure can be used to tessellate the plane. Drawings will vary.

5.
6.
7.
8.
9.
10.
11.
12.
13.

Name ______________________ Date ______________

Classification of Triangles

Triangles can be classified according to the lengths of their sides.

An **equilateral triangle** has three sides of equal length. $\triangle ABC$ is an equilateral triangle.

An **isosceles triangle** has two sides of equal length. $\triangle DEF$ is an isosceles triangle.

A **scalene triangle** has no sides of equal length. $\triangle GHI$ is a scalene triangle.

Classify each triangle as *equilateral, isosceles,* or *scalene.*

1. $\triangle GRE$ Isosceles
2. $\triangle XYZ$ Equilateral
3. $\triangle NMT$ Scalene
4. $\triangle KOG$ Scalene
5. $\triangle FGK$ Isosceles
6. $\triangle HKZ$ Equilateral
7. $\triangle POL$ Isosceles
8. $\triangle POR$ Scalene

Measure the sides of each triangle to the nearest millimeter. Then classify the triangle as *equilateral, isosceles,* or *scalene.*

9.

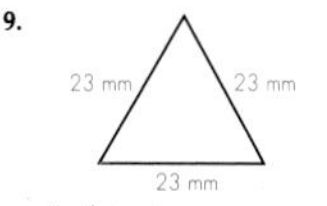

Equilateral

10.

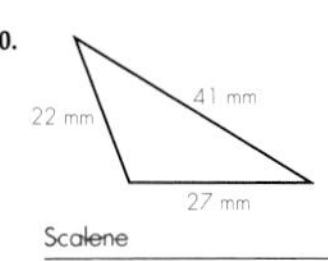

Scalene

11.

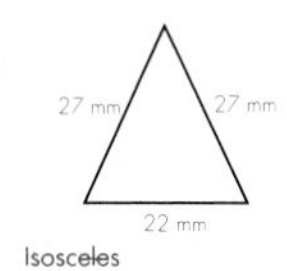

Isosceles

12. How many isosceles triangles are in the figure at the right?
8

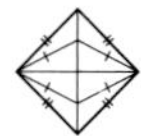

Name ______________________ Date ______________

Triangles and More Triangles

Many common puzzles and brainteasers involve counting the number of triangles hidden within a figure. Sometimes the figures contain many more triangles than you might think because the triangles overlap.

Count the number of triangles in each of the following figures.

1.

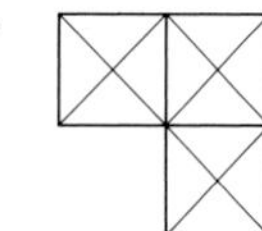

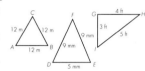

29 triangles

2.

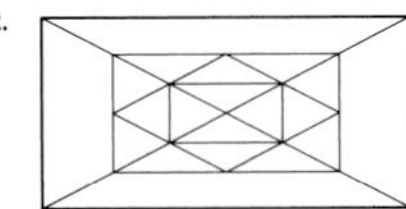

40 triangles

3.

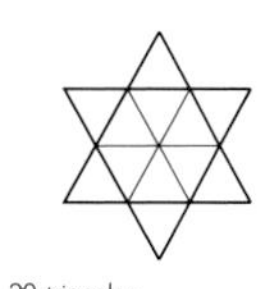

20 triangles

4.

41 triangles

5. How many triangles can you draw by connecting the points on the circle?

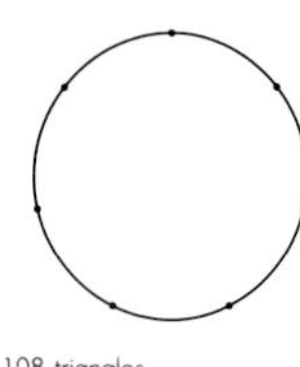

108 triangles

6. How many different isosceles triangles can you draw by connecting the dots?

136 triangles

Name ______________________ Date ______________

Classification of Triangles by Angles

Triangles can also be classified according to their angles.

An **equiangular triangle** has three equal angles, each measuring 60°. $\triangle ABC$ is an equiangular triangle.

A **right triangle** has one right angle. $\triangle DEF$ is a right triangle.

An **acute triangle** is a triangle with all three angles measuring less than 90°. $\triangle XYZ$ is an acute triangle.

An **obtuse triangle** is a triangle with one angle measuring more than 90°. $\triangle LMN$ is an obtuse triangle.

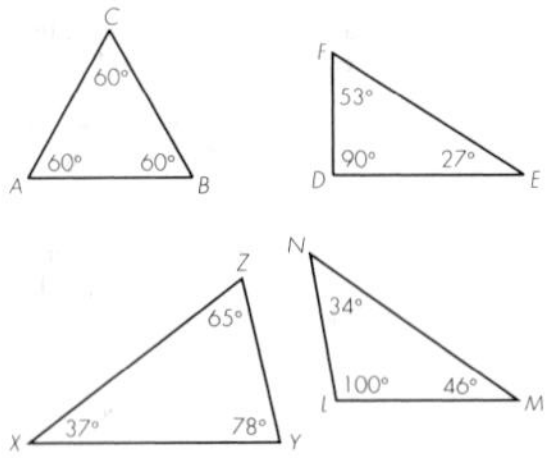

Classify each triangle as *equiangular, right, acute,* or *obtuse.*

1. 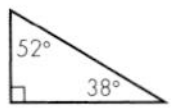

Right

2. 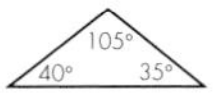

Obtuse

3. 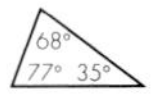

Acute

4. 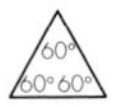

Acute or equiangular

Solve for each unknown angle measure. Then classify each triangle according to its angles.

5. 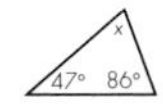

x = 47°; Isosceles

6. 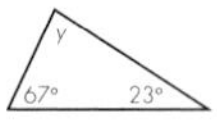

y = 90°; Right

7. 

y = 108°; Obtuse

8. 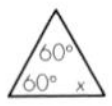

x = 60°; Acute or equiangular

Use the figure at the right for Exercises 9–12. Find each angle measure in degrees.

9. m∠*ABC* 96°

10. m∠*ACB* 32°

11. m∠*BAC* 52°

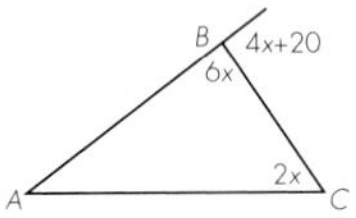

12. Classify the triangle according to the angle measures. ~~Acute~~ Obtuse

Name ______________________ Date ______________

The Shortest Distance

A plot of land is shaped like an equilateral triangle and is surrounded by three highways. The owner wants to build a cabin somewhere within the plot and then build three roads connecting the highways. The goal is to locate the cabin so that the total length of the three roads is as small as possible. This will allow the roads to be built at the least cost.

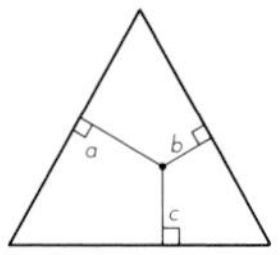

The road engineer said, "It doesn't matter where you locate the cabin as long as each road is perpendicular to the highway. The total $a + b + c$ will always be the same." Was the engineer correct?

Use the figure at the right. Measure each distance to the nearest millimeter.

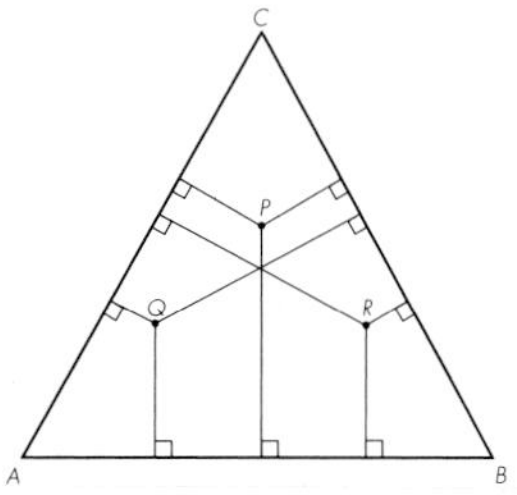

1. Distance from *P* to side *AB* 30 mm

2. Distance from *P* to side *BC* 13 mm

3. Distance from *P* to side *AC* 16 mm

4. What is the sum of the three distances in Exercises 1–3? 59 mm

5. What is the sum of the distances from point *Q* to each of the sides? 59 mm

6. What is the sum of the distances from point *R* to each of the sides? 59 mm

7. Pick another point within the triangle. Draw three lines from that point so that each line is perpendicular to one side of the triangle. Drawings will vary.

8. Measure the three segments you drew in Exercise 7 and find the sum of the lengths. 59 mm

9. Do you think that the road engineer was correct? Yes

Name ______________________ Date ______________

Exterior Angles of a Triangle

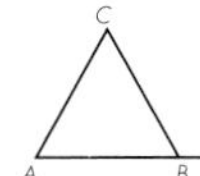

An exterior angle of a triangle is formed by extending one side of the triangle. In the figure at the right, $\angle CBD$ is an exterior angle of $\triangle ABC$, formed by extending $\overline{AB}$ through point B.

An exterior angle is equal to the sum of the measures of the two non-adjacent, or remote, interior angles. In the figure, $m\angle CBD = m\angle A + m\angle D$.

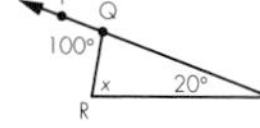

Example: Find the measure of $\angle R$.

Since $\angle PQR$ is an exterior angle, $m\angle PQR = m\angle R + m\angle S$.

So $100 = x + 20$ or $x = 80$. So $m\angle R = 80°$.

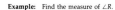

1. There are six exterior angles for any triangle. For each picture of $\triangle DEF$ below, draw a different exterior angle.

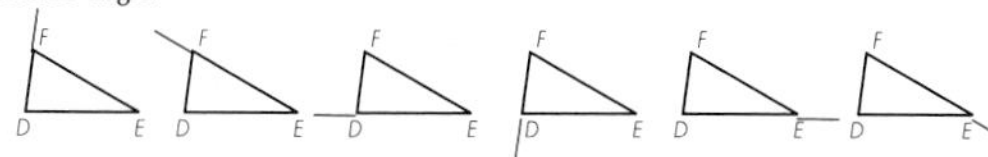

2. To show that the measure of an exterior angle is equal to the sum of the measures of the remote interior angles:

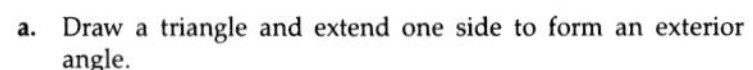
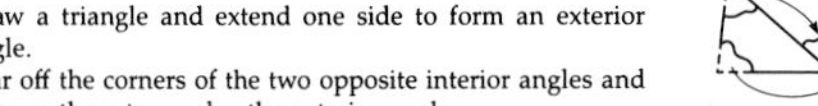
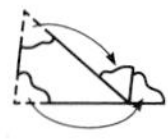

   a. Draw a triangle and extend one side to form an exterior angle.
   b. Tear off the corners of the two opposite interior angles and arrange them to overlay the exterior angle.

For each exercise, find $m\angle 1 + m\angle 2$.

3. 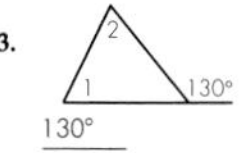

130°

4. 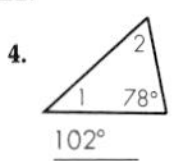

102°

5. 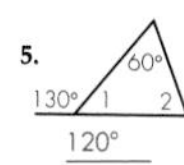

120°

Find the measure of each unknown angle.

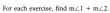

6. 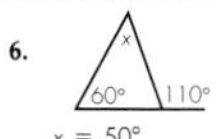

x = 50°

7. 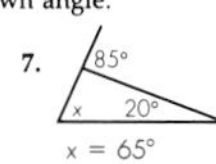

x = 65°

8. 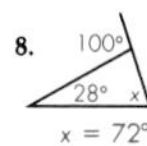

x = 72°

Name ______________________ Date ______________

More About Exterior Angles

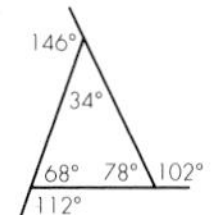

You know that the sum of the measures of the interior angles of a triangle is 180°. Suppose you form three exterior angles by extending the three sides as shown in the figure at the right. Do you think the sum of the three exterior angles will be 180°?

For the given triangle, the sum is 102° + 146° + 112°, or 360°. Do you think the sum will be 360° for any triangle?

1. Draw a triangle with angles whose measures are 50°, 70°, and 60°. Then extend the sides and find the measures of the three exterior angles. What is the sum? 360°

Suppose a figure has more than three sides, for example, 4 or 5 sides as shown in the following figures.

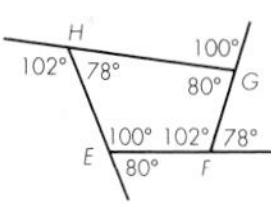

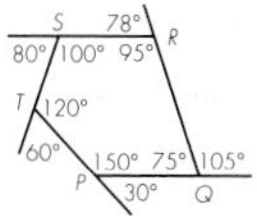

2. What is the sum of the measures of the exterior angles in figure *EFGH*? 360°

3. What is the sum of the measures of the exterior angles in figure *PQRST*? 360°

4. What do you notice about the sums of the measures of the exterior angles for a 3-sided figure, a 4-sided figure, and a 5-sided figure? Each sum is 360°.

5. What do you think the sum of the measures of the exterior angles for a 6-sided figure would be? 360°

6. On a separate sheet of paper, draw a large 6-sided figure. Label the angles as shown in the diagram at the right. Then cut out the six exterior angles and arrange them with the vertices at a common point to show that the sum of the measures of the angles is 360°.

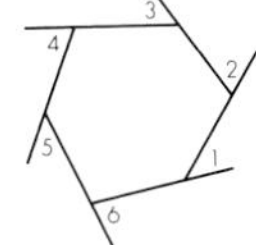

Name ____________________ Date ____________

Triangles on the x-y Plane

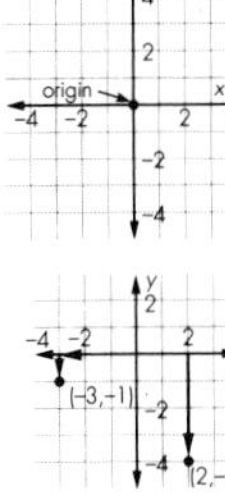

In the *x-y* plane, there are two perpendicular lines called the *x*-axis and the *y*-axis. Each axis is labeled with positive and negative numbers as shown at the right. The two axes intersect at the origin.

To locate a point, you use an ordered pair of the form (*x*, *y*). The first number is called the *x*-coordinate and the second number is called the *y*-coordinate.

Example: Locate the points (2, −4) and (−3, −1).

To locate (2, −4), start at the origin and go 2 units in the positive direction (to the right). Then go 4 units in the negative direction (down).

To locate (−3, −1), start at the origin and go 3 units to the left. Then go 1 unit down.

Locate the following points on the *x-y* plane at the right.

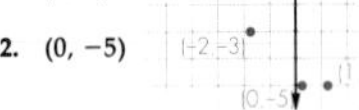

1. (2, 3)	**2.** (3, 2)	**3.** (1, 5)
4. (−2, 3)	**5.** (−2, −3)	**6.** (1, −5)
7. (5, −5)	**8.** (−5, 5)	**9.** (4, 5)
10. (−5, 0)	**11.** (0, 5)	**12.** (0, −5)

Give the coordinates of each point.

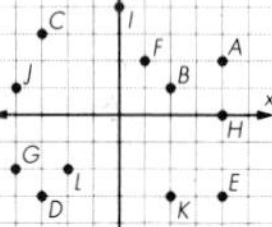

13. A (4, 2)	**14.** B (2, 1)	**15.** C (−3, 3)
16. D (−3, −3)	**17.** E (4, −3)	**18.** F (1, 2)
19. G (−4, −2)	**20.** H (4, 0)	**21.** I (0, 4)
22. J (−4, 1)	**23.** K (2, −3)	**24.** L (−2, −2)

Name ____________________ Date ____________

The Midpoint Formula

You know that the **midpoint** of a line segment is equally distant from both endpoints of the segment.

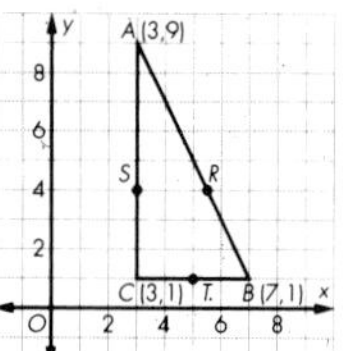

In △*ABC* at the right, the midpoints of the sides are *R*, *S*, and *T*.

When you know the coordinates of two points on the *x-y* plane, you can find the coordinates of the midpoint of the segment joining the points as follows:

Find the average of the *x*-coordinates.
Find the average of the *y*-coordinates.

For example, in △*ABC*, the coordinates of *S*, the midpoint of AC, are $\left(\frac{3+3}{2}, \frac{1+9}{2}\right)$ or (3, 5).

The method shown above suggests a formula for finding the midpoint of a line segment.

> MIDPOINT FORMULA
>
> Let *P*(*a*, *b*) and *Q*(*c*, *d*) be the endpoints of a line segment. The midpoint of *PQ* is $\left(\frac{a+c}{2}, \frac{b+d}{2}\right)$.

Use the Midpoint Formula to find the coordinates of each midpoint.

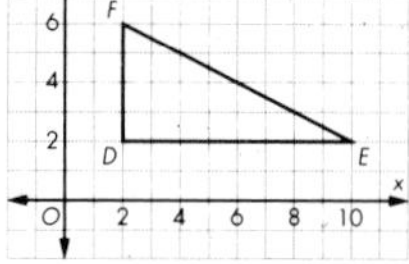

1. Midpoint *R* in △*ABC* above (5, 5)
2. Midpoint *T* in △*ABC* above (5, 1)
3. Midpoint of $\overline{DE}$ in △*DEF* (5, 2)
4. Midpoint of $\overline{DF}$ in △*DEF* (2, 4)
5. Midpoint of $\overline{EF}$ in △*DEF* (6, 4)

Use the Midpoint Formula to find the coordinates of the midpoint of the segment that joins the given points.

6. (2, 5) and (4, 7) (3, 6)	**7.** (0, 4) and (6, 6) (3, 5)
8. (−2, 1) and (8, −5) (3, −2)	**9.** (7, −4) and (−2, 1) $\left(\frac{5}{2}, -\frac{3}{2}\right)$
10. (8, 0) and (−4, 1) $\left(2, \frac{1}{2}\right)$	**11.** (16, 5) and (−8, 3) (4, 4)

Name ______________________ Date ______________

Using the SETXY Command

Refer to the set of commands below for Exercises 1-3. The procedure XYPLANE is explained on page 118 in the textbook.

```
XYPLANE
PU
SETXY 0 0
PD
SETXY 3 4
SETXY 2 5
SETXY 0 0
```

1. Which lines in the set of commands are not necessary? Explain.
   PU, SETXY 0 0, PD; These lines tell the turtle to go to the location with coordinates (0, 0) which is where the turtle always starts.

2. Which line in the set of commands moves the turtle to the same position as does the HOME command?
   SETXY 0 0

3. Rewrite the set of commands, eliminating the unnecessary commands and using the HOME command.
   XYPLANE
   SETXY 3 4
   SETXY 2 5
   HOME

In Exercises 4-6, modify the set of commands in the answer to Exercise 3 to tell the turtle to draw a triangle with the given vertices.

4. (0, 0), (4, 5), (8, 10)
   XYPLANE
   SETXY 4 5
   SETXY 8 10
   HOME

5. (0, 0), (12, 19), (15, 25)
   XYPLANE
   SETXY 12 19
   SETXY 15 25
   HOME

6. (5, 18), (52, 10), (0, 0)
   XYPLANE
   SETXY 5 18
   SETXY 52 10
   HOME

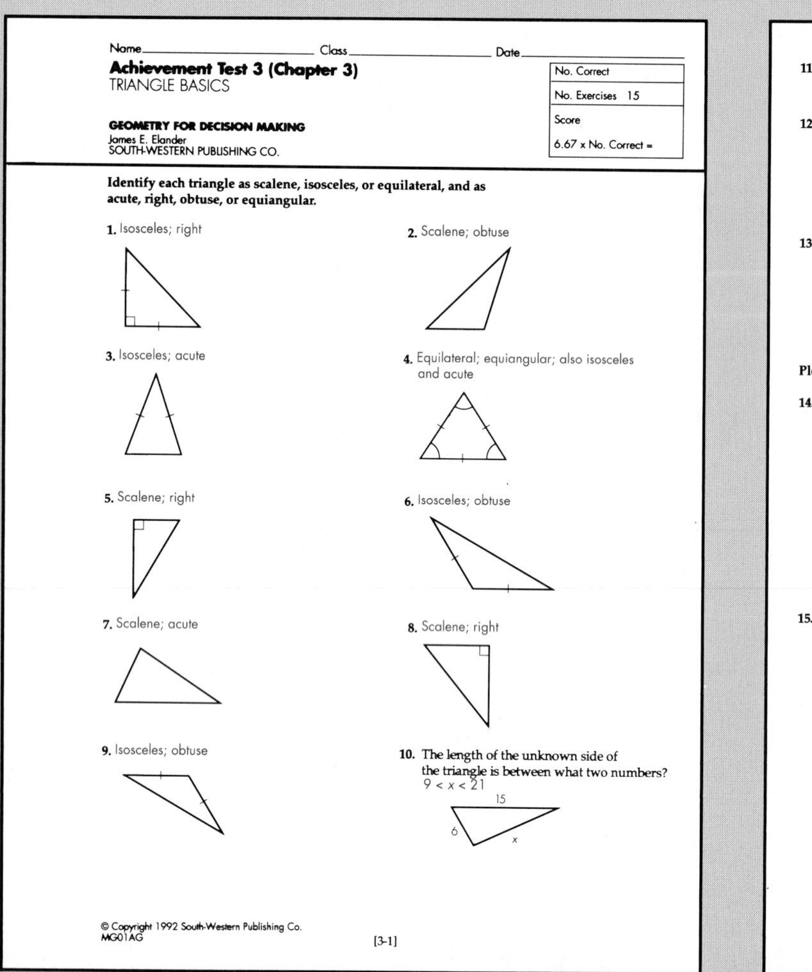

Name________________ Class________________ Date________________

Achievement Test 3 (Chapter 3)
TRIANGLE BASICS

No. Correct
No. Exercises 15
Score
6.67 x No. Correct =

GEOMETRY FOR DECISION MAKING
James E. Elander
SOUTH-WESTERN PUBLISHING CO.

Identify each triangle as scalene, isosceles, or equilateral, and as acute, right, obtuse, or equiangular.

1. Isosceles; right
2. Scalene; obtuse
3. Isosceles; acute
4. Equilateral; equiangular; also isosceles and acute
5. Scalene; right
6. Isosceles; obtuse
7. Scalene; acute
8. Scalene; right
9. Isosceles; obtuse
10. The length of the unknown side of the triangle is between what two numbers? $9 < x < 21$

MG01AG

[3-1]

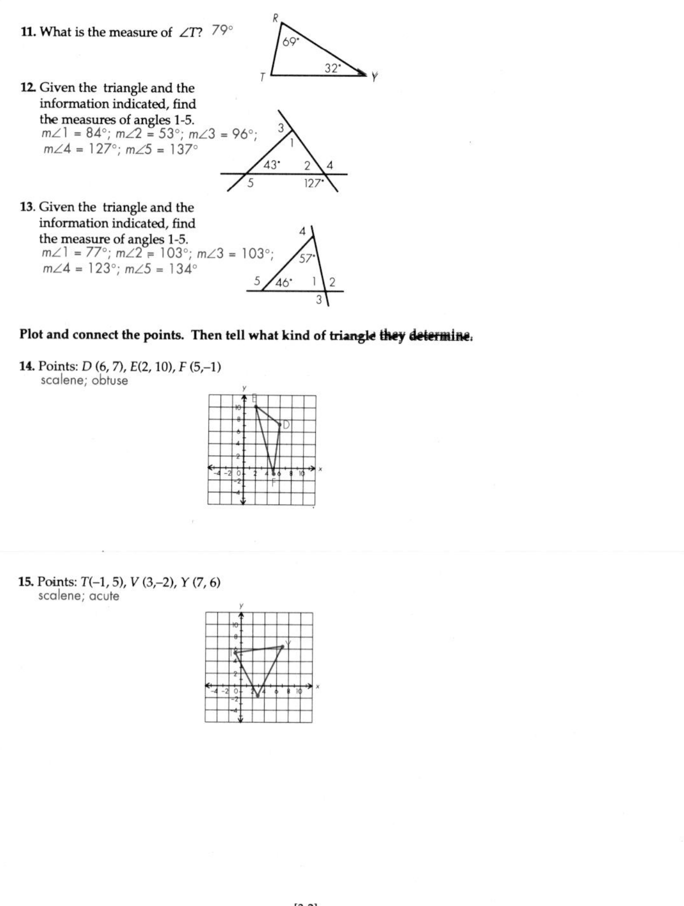

11. What is the measure of $\angle T$? 79°

12. Given the triangle and the information indicated, find the measures of angles 1-5.
$m\angle 1 = 84°$; $m\angle 2 = 53°$; $m\angle 3 = 96°$; $m\angle 4 = 127°$; $m\angle 5 = 137°$

13. Given the triangle and the information indicated, find the measure of angles 1-5.
$m\angle 1 = 77°$; $m\angle 2 = 103°$; $m\angle 3 = 103°$; $m\angle 4 = 123°$; $m\angle 5 = 134°$

Plot and connect the points. Then tell what kind of triangle they determine.

14. Points: $D(6, 7)$, $E(2, 10)$, $F(5, -1)$
scalene; obtuse

15. Points: $T(-1, 5)$, $V(3, -2)$, $Y(7, 6)$
scalene; acute

[3-2]

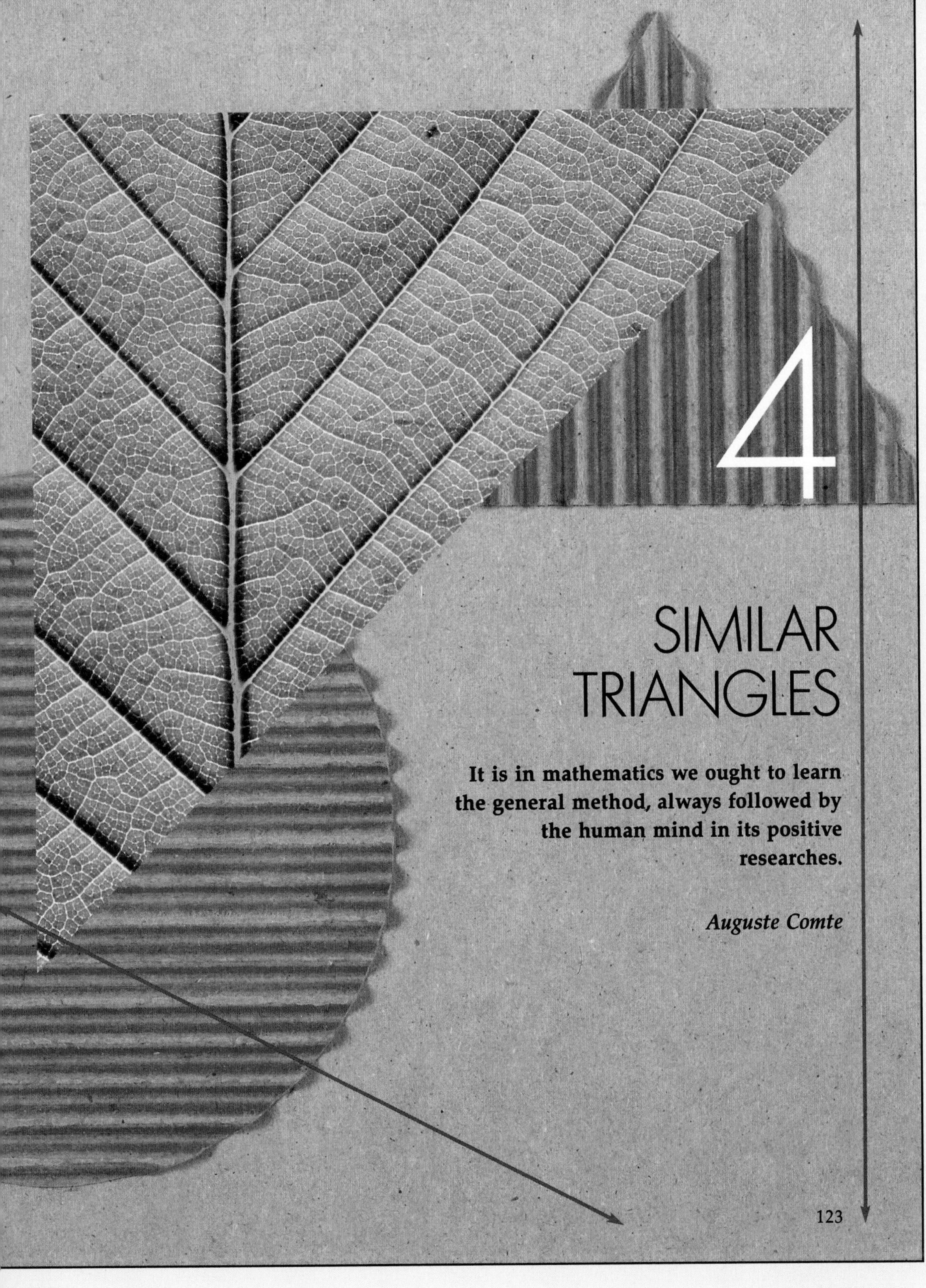

4

SIMILAR TRIANGLES

It is in mathematics we ought to learn the general method, always followed by the human mind in its positive researches.

Auguste Comte

123

CHAPTER 4 OVERVIEW

Similar Triangles

Suggested Time: 12–14 days

The quote at the beginning of the chapter applies to the learning process of this chapter. After learning how to tell if figures have the same shape, properties of similar triangles are studied. Note the similar figures that are evident in the photo that shows a magnification of an elm leaf. Ratios and proportions are included, since ratios of corresponding sides of similar triangles are equal. From there, the chapter looks at applications of these properties in problem solving.

It is in mathematics we ought to learn the general method, always followed by the human mind in its positive researches.

Auguste Comte

CHAPTER 4 OBJECTIVE MATRIX

	End of Chapter Items by Activity				Student Supplement Items by Activity		
Objectives by Number	**Review**	**Test**	**Computer**	**Algebra Skills**	**Reteaching**	**Enrichment**	**Computer**
4.1.1	✓	✓			✓		
4.1.2		✓			✓		✓
4.2.1	✓	✓		✓	✓	✓	✓
4.2.2	✓	✓		✓	✓	✓	
4.3.1	✓	✓	✓		✓		✓
4.3.2	✓	✓	✓	✓	✓		✓
4.4.1	✓	✓		✓	✓	✓	
4.4.2	✓	✓		✓	✓	✓	
4.5.1	✓	✓	✓	✓	✓	✓	✓
4.5.2	✓	✓	✓	✓	✓		
4.6.1	✓	✓		✓	✓	✓	

*A ✓ beside a Chapter Objective under Algebra Skills indicates that algebra skills taught within the chapter or in previous Algebra Skills lessons are used.

CHAPTER 4 PERSPECTIVES

▲ Section 4.1

Mappings and Correspondences

Mappings are introduced through the use of maps representing actual areas on the earth. Once the idea of maps representing something of a different size has been discussed, mapping of triangles is introduced. This is done so that students can easily identify corresponding vertices, angles, and sides. Students are shown that they may have to turn or flip one of the triangles before a mapping can be done or correspondences listed.

The Home Activity includes two exercises that list correspondences of vertices. The students are free to draw any mapping that would fit the given correspondences. A series of three exercises ask students to map and write correspondences of three triangles. Continue to encourage students to keep notebooks for recording definitions, postulates, theorems, and examples.

▲ Section 4.2

Ratios and Proportions

The terms *ratio* and *proportion* are defined. After ratio is defined, three different ratios not in lowest terms are written in lowest terms. For students having difficulty with this concept, display real objects. For example, display 3 green marbles and 6 red marbles. The ratio of green to red is 3 to 6. Then loop the 3 green marbles. Show the 6 red marbles looped in sets of 3. The ratio of the looped marbles is 1 to 2.

After proportions are defined, students are given examples of proportions. Some contain numbers only on each side of the equal sign, and some contain a variable. Proportions with a variable are then solved.

▲ Section 4.3

Similar Triangles

Through a Discovery Activity, students find that they can map two triangles when corresponding angles are equal in measure and the ratios of corresponding sides are equal in length. The definition of *similar triangles* is then introduced.

The Home Activity includes exercises in which students must analyze the given information to determine if two triangles are similar.

▲ Section 4.4

Similar Figures and Scale Drawings

Similar figures are defined, and both two-dimensional and three-dimensional examples are given. Scale drawings as a means of reducing or enlarging the size of an object are introduced. Through a Discovery Activity, students find the scale factor for a given situation and then use this scale factor to predict other lengths. For students having difficulty with this concept, use local road maps and have students find the actual distance between towns by measuring the distance on the map and using the conversion factor or scale.

The Home Activity provides practice in using scale factors to determine unknown lengths.

▲ Section 4.5

Similar Triangles and Conclusions

Conjectures are introduced, with ways that conjectures are made. After conjectures are proposed, a conclusion is made and then justified. During a Discovery Activity, students locate the midpoints of two sides of a triangle, measure the lengths of the segment joining these midpoints and the third side of the original triangle, and make a conjecture. This activity leads to a new theorem: If a segment connects the midpoints of two sides of a triangle, then the length of the segment is equal to $\frac{1}{2}$ the length of the third side.

A second Discovery Activity leads to a second theorem: If a line is parallel to one side of a triangle and intersects the other sides at any points except the vertex, then the line divides the sides proportionally. To help students internalize this concept, have them draw their own triangles and draw a line parallel to one of the sides. Have them measure the segments on the sides intersected by the parallel and show that the ratios are equal.

▲ Section 4.6

Solving Problems Involving Similar Triangles

This section introduces situations where similar triangles can be used to solve practical applications. The problems deal with situations where direct measurement is impracticable.

The Home Activity includes problems with projections, photography, and measuring across bodies of water.

SECTION

4.1

Resources
Reteaching 4.1
Enrichment 4.1
Transparency Master 4-1

Objectives
After completing this section, the student should understand
- ▲ how to map figures.
- ▲ how to write correspondences.

Materials
overhead projector
scissors
ruler

Vocabulary
mapping
correspondence
corresponding vertices
corresponding sides
corresponding angles

GETTING STARTED

Quiz: Prerequisite Skills

1. What could you look at to find how far it is from one city to another city? [map]

2. Which two angles have equal measure? [$\angle M$ and $\angle X$]

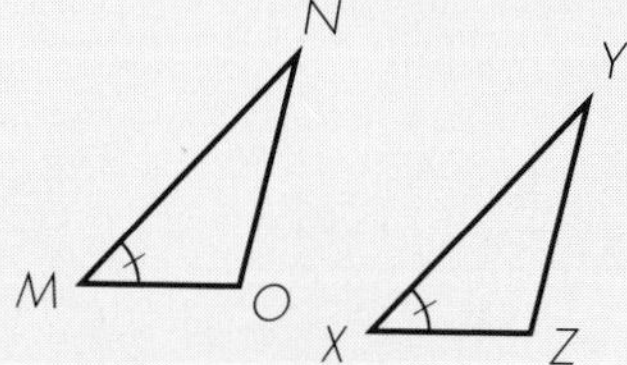

Warm-Up Activity
Put a drawing on the board or overhead projector that has rectangles representing the desks in your room. Ask students how they could show where each person sits. Ask them how they could show where windows, doors, or cabinets are located.

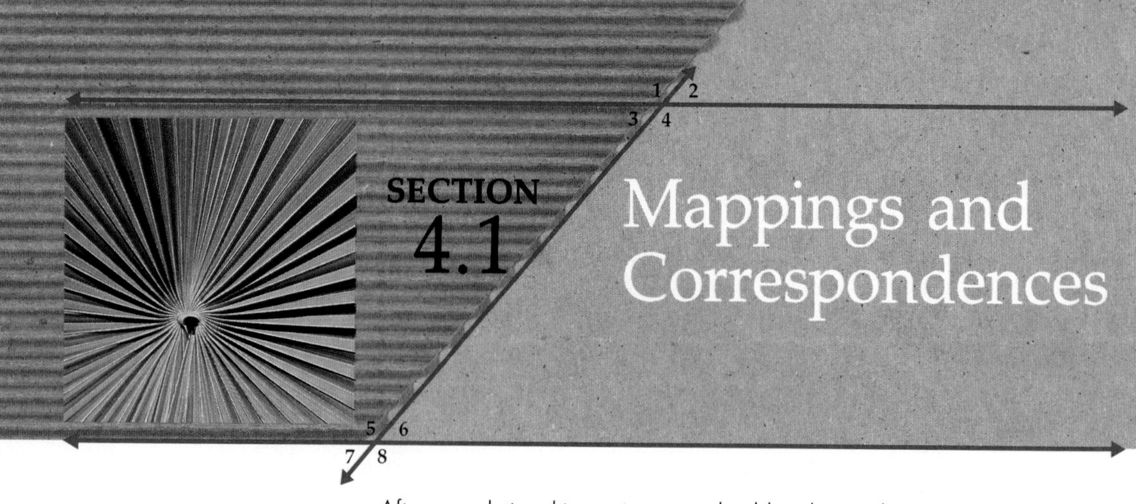

SECTION 4.1 Mappings and Correspondences

After completing this section, you should understand
- ▲ how to map figures.
- ▲ how to write correspondences.

In order to solve problems found in different professions, it is often useful to be able to map triangles (and other geometric figures). Mapping can help you find corresponding parts of two triangles.

MAPPINGS

Maps of cities or states are used to help people locate key points, such as streets or borders. Maps are drawn so that you can compare the map to the actual area where you are, and see how to get to the key point. Each feature of the area is keyed to one point on the map.

If a state were viewed from far above the earth, the cities would look like the dots representing them on the map. They would appear to be the same distances apart, even though one area encompasses square kilometers and the other square centimeters. The correspondence between the earth and the map would not depend on the actual size—they appear to have the same shape.

This idea of fitting two look-alike figures on top of each other and matching corresponding point to corresponding point can be applied to triangles or other geometric figures.

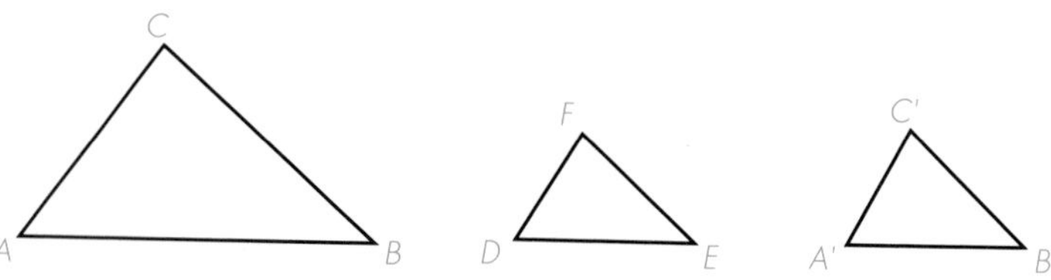

Two methods may be used to label triangles that are being mapped. The first method uses new letters to label the second triangle.

$\triangle ABC$ is mapped to $\triangle DEF$.

The second method uses the same letters as the first triangle, but with a mark following the letter. A', read A prime.

$\triangle ABC$ is mapped to $\triangle A'B'C'$.

WRITING CORRESPONDENCES

When mapping triangles, you must be given enough information to make sure that the shapes of the triangles will be the same.

Example: In the following triangles, $\angle A \cong \angle D$ and $\angle B \cong \angle E$.

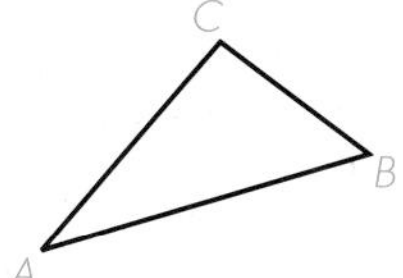

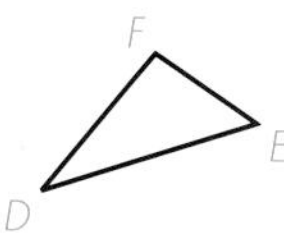

Nine correspondences can be written.

Vertices	Sides	Angles
A corresponds to D.	$\overline{AB}$ corresponds to $\overline{DE}$.	$\angle A$ corresponds to $\angle D$.
B corresponds to E.	$\overline{BC}$ corresponds to $\overline{EF}$.	$\angle B$ corresponds to $\angle E$.
C corresponds to F.	$\overline{CA}$ corresponds to $\overline{FD}$.	$\angle C$ corresponds to $\angle F$.

Chalkboard Example
Draw the following triangles on the board. Have students list the corresponding vertices, angles, and sides.

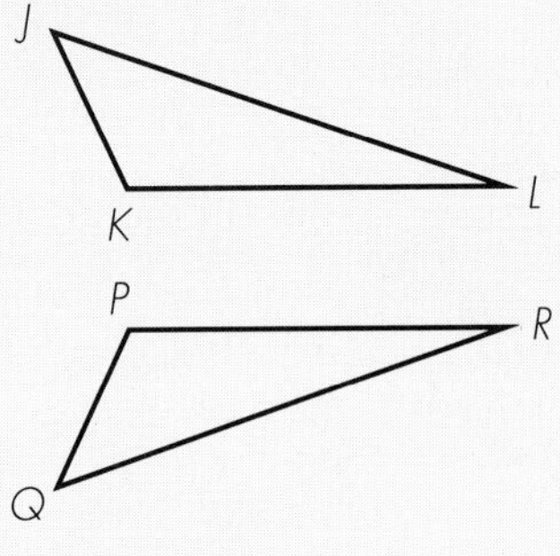

TEACHING COMMENTS

Using the Page
Point out that figures cannot be mapped unless the shapes are the same.

Using the Page
When one figure is mapped onto another figure, the order in which the vertices are listed is important. "$\triangle ABC$ is mapped to $\triangle DEF$" means that $\angle A \leftrightarrow \angle D$, $\angle B \leftrightarrow \angle E$, and $\angle C \leftrightarrow \angle F$. Use Transparency Master 4-1. Have students identify the corresponding vertices and sides in the similar figures.

To abbreviate the correspondences, a double arrow is used.

$A \leftrightarrow D$	Read: A corresponds to D.
$\overline{BC} \leftrightarrow \overline{EF}$	Read: $\overline{BC}$ corresponds to $\overline{EF}$.
$\angle C \leftrightarrow \angle F$	Read: $\angle C$ corresponds to $\angle F$.

DISCOVERY ACTIVITY

1. Draw a dotted line on a piece of paper as shown. Label the angles, then cut along the dotted line.

2. Map the two triangles. What did you have to do before one triangle could be put on top of the other one? Turn one of the triangles.

You may have discovered that sometimes you have to turn one of the triangles until a mapping can be done.

CLASS ACTIVITY

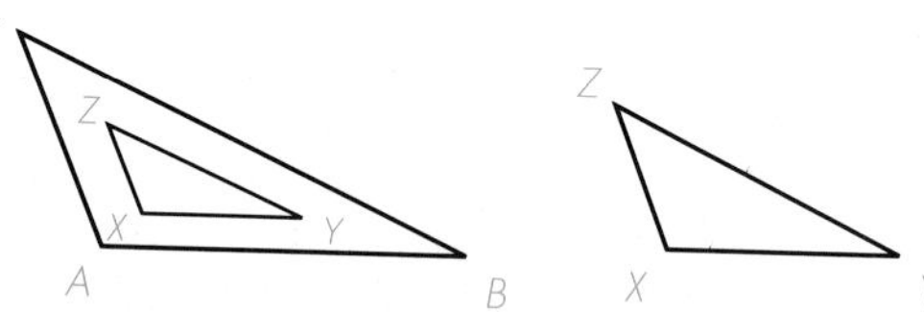

1. Trace the triangles on a sheet of paper so that one is fitted on top of the other (mapped).
2. Write the vertices that correspond to each other.
$A \leftrightarrow X$, $B \leftrightarrow Y$, $C \leftrightarrow Z$
3. Write the angles that correspond to each other.
$\angle A \leftrightarrow \angle X$, $\angle B \leftrightarrow \angle Y$, $\angle C \leftrightarrow \angle Z$
4. Write the sides that correspond to each other.
$\overline{AB} \leftrightarrow \overline{XY}$, $\overline{BC} \leftrightarrow \overline{YZ}$, $\overline{AC} \leftrightarrow \overline{XZ}$.
5. What condition is necessary in order to map two figures?
Same shape.

Using the Page
In Step 2 of the Discovery Activity, students could also answer that they flipped one triangle over and could then do the mapping.

Extra Practice

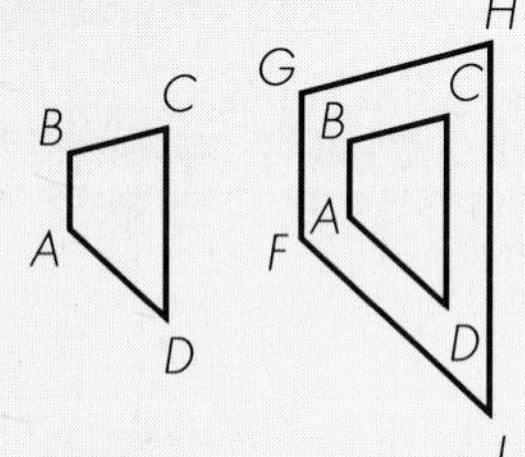

1. Map the figures.
2. Write the vertices that correspond to each other. [$A \leftrightarrow F$, $B \leftrightarrow G$, $C \leftrightarrow H$, $D \leftrightarrow I$]
3. Write the angles that correspond to each other. [$\angle A \leftrightarrow \angle F$, $\angle B \leftrightarrow \angle G$, $\angle C \leftrightarrow \angle H$, $\angle D \leftrightarrow \angle I$]
4. Write the sides that correspond to each other. [$\overline{AB} \leftrightarrow \overline{FG}$, $\overline{BC} \leftrightarrow \overline{GH}$, $\overline{CD} \leftrightarrow \overline{HI}$, $\overline{DA} \leftrightarrow \overline{IF}$]

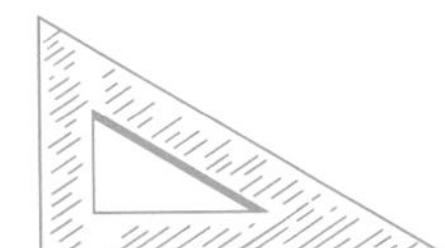

HOME ACTIVITY

Refer to triangles *PQR* and *ABC*.

1. Map the two triangles.
2. Write the vertices, sides, and angles that correspond to each other. See margin.

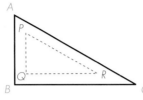

Refer to triangles *DEF* and *D′E′F′*.

3. Map the two triangles.
4. Write the vertices, sides, and angles that correspond to each other. See margin.

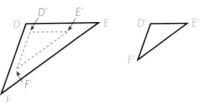

Refer to triangles *ABC* and *SVT*.

5. Map the two triangles.
6. Write the correspondences. See margin.

Use the XYPLANE procedure from page 118 and the command SETXY to write a procedure to draw the triangles whose coordinates are given. Then map the triangles and write correspondences.
Answers for procedures may vary. See margin for correspondences.

7. $\triangle ABC$ where $A(10,10)$, $B(50,10)$, $C(10,40)$
   $\triangle DEF$ where $D(0,-15)$, $E(20,-15)$, $F(0,0)$
8. $\triangle MNP$ where $M(0,0)$, $N(30,0)$, $P(15,20)$
   $\triangle XYZ$ where $X(-3,-4)$, $Y(3,-4)$, $Z(0,0)$

9. Given these correspondences, draw the two figures after they are mapped: $A \leftrightarrow B$, $C \leftrightarrow F$, $D \leftrightarrow X$. Answers will vary.

10. Given these correspondences, draw the two figures after they are mapped: $A \leftrightarrow C$, $B \leftrightarrow X$, $E \leftrightarrow Y$, $D \leftrightarrow Z$. Answers will vary.

Refer to triangles *ABC, DFG,* and *EFG*. You may want to trace all three and cut the two smaller ones apart.

11. Draw a picture of the three triangles mapped.
12. How many correspondences can be written? 27
13. Write the correspondences for the sides. See margin.

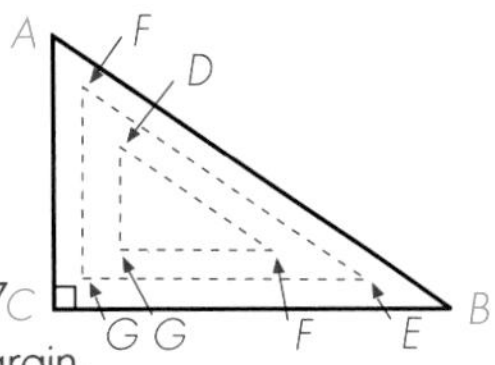

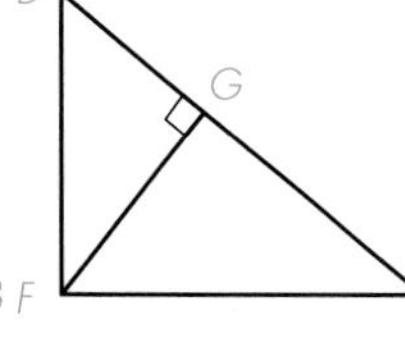

Additional Answers

2. $A \leftrightarrow P$, $B \leftrightarrow Q$, $C \leftrightarrow R$, $\overline{AB} \leftrightarrow \overline{PQ}$, $\overline{BC} \leftrightarrow \overline{QR}$, $\overline{AC} \leftrightarrow \overline{PR}$, $\angle A \leftrightarrow \angle P$, $\angle B \leftrightarrow \angle Q$, $\angle C \leftrightarrow \angle R$

4. $D \leftrightarrow D'$, $E \leftrightarrow E'$, $F \leftrightarrow F'$, $\overline{DE} \leftrightarrow \overline{D'E'}$, $\overline{EF} \leftrightarrow \overline{E'F'}$, $\overline{DF} \leftrightarrow \overline{D'F'}$, $\angle D \leftrightarrow \angle D'$, $\angle E \leftrightarrow \angle E'$, $\angle F \leftrightarrow \angle F'$

6. $A \leftrightarrow V$, $B \leftrightarrow S$, $C \leftrightarrow T$, $\overline{AB} \leftrightarrow \overline{VS}$, $\overline{AC} \leftrightarrow \overline{VT}$, $\overline{BC} \leftrightarrow \overline{ST}$, $\angle A \leftrightarrow \angle V$, $\angle B \leftrightarrow \angle S$, $\angle C \leftrightarrow \angle T$

7. $A \leftrightarrow D$, $B \leftrightarrow E$, $C \leftrightarrow F$, $\overline{AB} \leftrightarrow \overline{DE}$, $\overline{BC} \leftrightarrow \overline{EF}$, $\overline{AC} \leftrightarrow \overline{DF}$, $\angle A \leftrightarrow \angle D$, $\angle B \leftrightarrow \angle E$, $\angle C \leftrightarrow \angle F$

8. $M \leftrightarrow X$, $N \leftrightarrow Y$, $P \leftrightarrow Z$, $\overline{MN} \leftrightarrow \overline{XY}$, $\overline{NP} \leftrightarrow \overline{YZ}$, $\overline{MP} \leftrightarrow \overline{XZ}$, $\angle M \leftrightarrow \angle X$, $\angle N \leftrightarrow \angle Y$, $\angle P \leftrightarrow \angle Z$

13. $\overline{AC} \leftrightarrow \overline{FG}$, $\overline{AC} \leftrightarrow \overline{DG}$, $\overline{FG} \leftrightarrow \overline{DG}$, $\overline{AB} \leftrightarrow \overline{FE}$, $\overline{AB} \leftrightarrow \overline{DF}$, $\overline{FE} \leftrightarrow \overline{DF}$, $\overline{CB} \leftrightarrow \overline{GE}$, $\overline{CB} \leftrightarrow \overline{GF}$, $\overline{GE} \leftrightarrow \overline{GF}$,

Additional Answers

16.

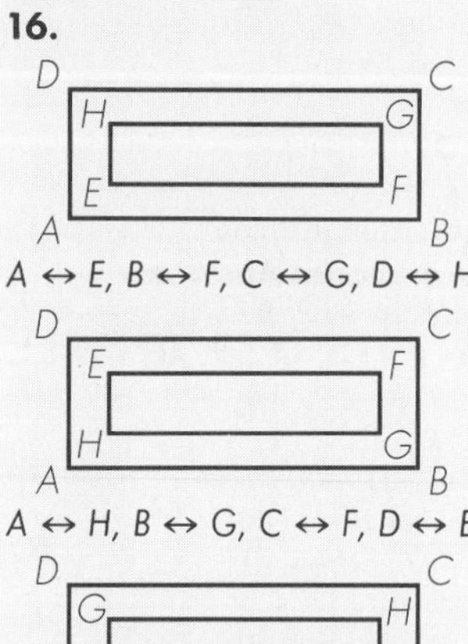

A ↔ E, B ↔ F, C ↔ G, D ↔ H

A ↔ H, B ↔ G, C ↔ F, D ↔ E

A ↔ F, B ↔ E, C ↔ H, D ↔ G

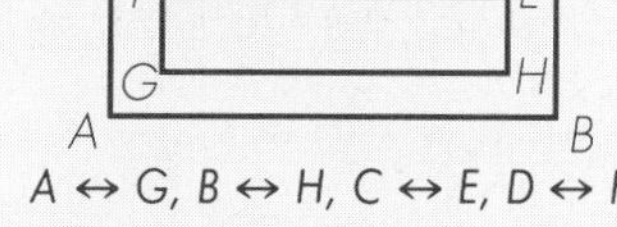

A ↔ G, B ↔ H, C ↔ E, D ↔ F

FOLLOW-UP

Assessment

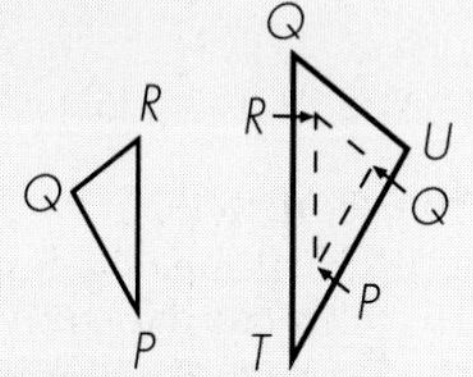

1. Map the two triangles.
2. Which vertex corresponds to *P*? [*T*]
3. Which angle corresponds to ∠*U*? [∠Q]
4. Which side corresponds to $\overline{PQ}$? [$\overline{TU}$]

Reteaching
Students who have had difficulty with this section may benefit from Reteaching Activity 4.1.

Enrichment
For students who have mastered the material in this section, you may wish to assign Enrichment Activity 4.1.

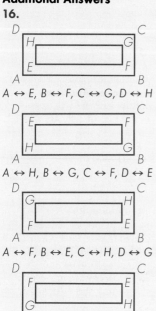

Two isosceles triangles are shown, with $\overline{AC} \cong \overline{BC}$, and $\overline{DF} \cong \overline{EF}$. $\angle C \cong \angle F$.

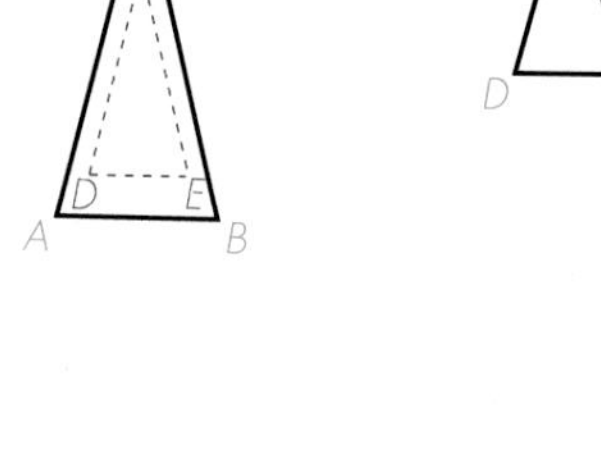

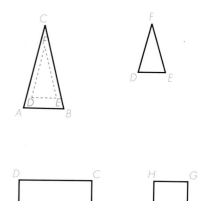

14. Map the two triangles and write the angle correspondences.
∠A ↔ ∠D, ∠B ↔ ∠E, ∠C ↔ ∠F
15. Maria said there was another way to map the two triangles. Explain. Write the angle correspondences.
Flip △FED over. ∠A ↔ ∠E, ∠B ↔ ∠D, ∠C ↔ ∠F
16. Wayne said that the rectangles on the right could be mapped four ways. Show the four ways, and write the vertex correspondences for each. See margin.

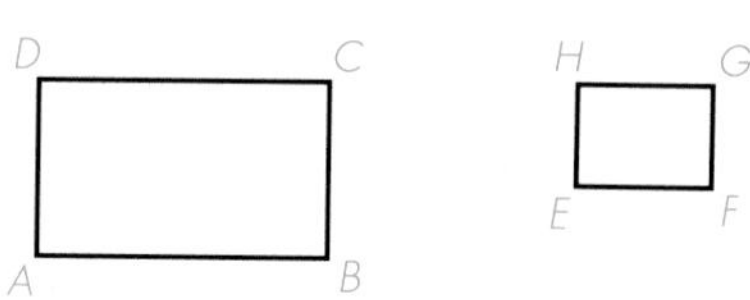

17. Name pairs of objects in your classroom that could be mapped, and explain their correspondences. Answers will vary.

18. Name one pair of objects in your home that could be mapped. Answers will vary.

19. Draw two triangles that can be mapped by sliding one on top of the other.
Answers will vary.

Refer to triangles *ABC* and *DEF*.

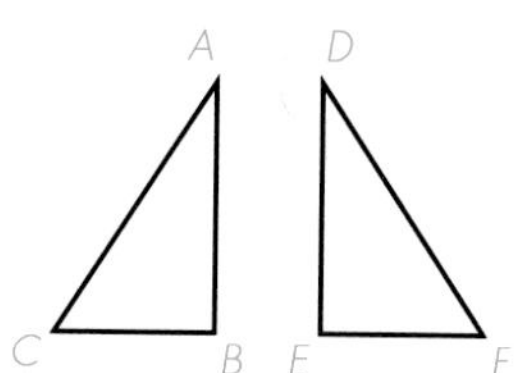

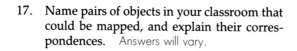

20. Try to map the two right triangles by sliding one on top of the other. Trace one triangle and cut it out. It cannot be done.

21. Map the two triangles if you are permitted to turn and flip one triangle.
Flip △ABC onto △DEF, or vice versa.

CRITICAL THINKING

22. On the physical education field, the coach noticed that a number of dogs had joined his P.E. class. He counted 70 legs and 21 heads. How many dogs and how many students were on the field? 14 dogs, 7 students.

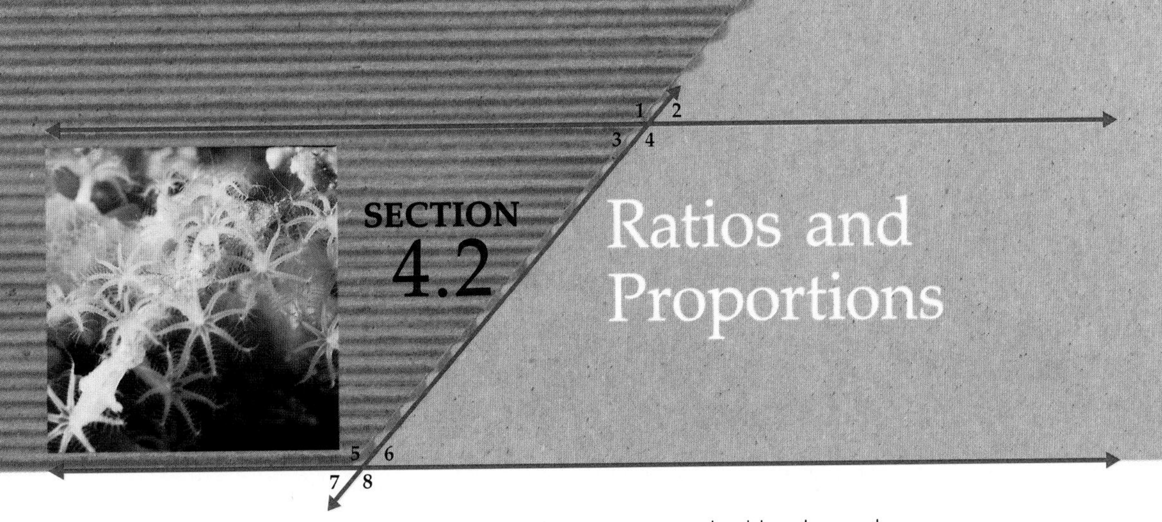

Ratios and Proportions

After completing this section, you should understand
- ▲ ratios and proportions.
- ▲ how to solve problems involving ratios and proportions.

You have probably encountered ratios and proportions many times each day, when you talk about sports statistics with your friends or look at a recipe in a cookbook.

RATIOS

In algebra, you learned about rational numbers, numbers that can be written in the form $\frac{\text{integer}}{\text{integer}}$, with the denominator not equal to 0. The word *ratio* is found in *rational.* A ratio is a comparison of two numbers by division.

DEFINITION 4-2-1 **A RATIO is one number divided by another number, or a fraction N/D where D is not zero.**

Example: Some ratios: $\frac{2}{3}$, 34/62, 3:4, and 5 to 7
Not a ratio: $\frac{5}{0}$

WRITING A RATIO IN LOWEST TERMS

In the example above, $\frac{2}{3}$, 3:4, and 5 to 7 are written in lowest terms. The ratio $\frac{34}{62}$ is not in lowest terms.

Example: Write $\frac{34}{62}$ in lowest terms.

$$\frac{34}{62} = \frac{\not{2}(17)}{\not{2}(31)} = \frac{17}{31}$$

Example: Write 0.5 in lowest terms.

$$0.5 = \frac{5}{10} = \frac{\not{5}(1)}{\not{5}(2)} = \frac{1}{2}$$

SECTION 4.2

Resources
Reteaching 4.2
Enrichment 4.2

Objectives
After completing this section, the student should understand
- ▲ ratios and proportions.
- ▲ how to solve problems involving ratios and proportions.

Vocabulary
rational number
ratio
lowest terms
proportion

GETTING STARTED

Quiz: Prerequisite Skills

1. Which of the following are rational numbers?

a. $\frac{2}{3}$ [Yes]
b. 5 [Yes]
c. $\frac{7}{0}$ [No]

2. Which of the following are equations?
a. $x + 7 = 9$ [Yes]
b. $x - 3 + 2x$ [No]
c. $x - 5 < 2x + 1$ [No]

Warm-Up Activity
Have students count the number of males and the number of females in the room. Are there more males or females? Repeat this with other items in the room, such as windows and doors, or chairs and people.

Chalkboard Example

Write $\frac{54}{72}$ in lowest terms.

$\frac{54}{72} = \frac{3 \cdot 18}{4 \cdot 18}$
$= \frac{3}{4}$

TEACHING COMMENTS

Using the Page
See the example on page 129. Point out that rational numbers like 5 or 1 should be written as $\frac{5}{1}$ or $\frac{1}{1}$. The definition says that a ratio is a comparison of *two* numbers.

Using the Page
The example $\frac{(7 - x)}{12} = \frac{20}{36}$ is a proportion, since one ratio appears on each side of the equal sign. The equation $\frac{7}{12} - \frac{x}{12} = \frac{20}{36}$ would have the same solution as the first equation, but would not be called a proportion. Even though the left-hand side contains ratios, the equation has two terms on that side of the equal sign.

Using the Page
Challenge students to write the definition of a proportion in their own words. Read some of the definitions to the class. Discuss whether each definition is valid. Have students work in small groups to find a counterexample for any definition that is not valid.

Example: Write $\frac{2}{3}/\frac{3}{4}$ in lowest terms.

$$\frac{\frac{2}{3}}{\frac{3}{4}} = \frac{2}{3} \div \frac{3}{4} = \frac{2}{3} \cdot \frac{4}{3} = \frac{8}{9}$$

PROPORTIONS

You know that an equation is a statement that says two expressions are equal.

$x = 560$ $\quad x + 350 = 468$ $\quad |x - 27| = 68$

A proportion is a special type of equation. Here are some examples:

$\frac{1}{2} = \frac{6}{12}$ $\quad \frac{3}{4} = \frac{x}{5}$ $\quad 1{:}2 = 7{:}14$ $\quad 15{:}x = 30{:}6$

$\frac{15}{25} = \frac{60}{y}$ $\quad \frac{x}{0.5} = \frac{11}{10}$ $\quad \frac{(7 - x)}{12} = \frac{20}{36}$

NOTEBOOK

DEFINITION 4-2-2 **A PROPORTION is an equation consisting of two equal ratios.**

Notice that proportions are written using a fraction bar or a colon (:). Each equation consists of two equal fractions. When one of the ratios contains a variable, you can solve for the value of the variable that makes the statement true.

Read $\frac{3}{4} = \frac{x}{5}$ as 3 is to 4 as x is to 5.

Example: Solve the proportion.

$\frac{4}{5} = \frac{x}{15}$ $\quad \frac{4}{5} \cdot 15 = \frac{x}{15} \cdot 15$
$12 = x$

CLASS ACTIVITY

1. Which of the following are ratios?

   a. $\frac{2}{3}$ b. $\frac{5}{7}$ c. 4:9 d. 0.8 e. $\frac{6}{0}$ f. $\frac{0}{6}$
   All except e.

2. Which of the following are proportions?

   a. $\frac{2}{3} = \frac{4}{6}$ b. $\frac{x}{3} = \frac{4}{6}$ c. $\frac{2}{3} = \frac{5}{6}$ d. $0.8 = \frac{x}{10}$
   All except c.

In Exercises 3–7, write each ratio in simplest form.

3. $\frac{16}{64}$ $\frac{1}{4}$
4. $\frac{128}{64}$ $\frac{2}{1}$
5. 0.25 $\frac{1}{4}$
6. $\frac{256}{1024}$ $\frac{1}{4}$
7. $\frac{\frac{5}{7}}{\frac{10}{14}}$ $\frac{1}{1}$
8. A volleyball team has 5 coaches and 75 players. What is the ratio of players to coaches? $\frac{15}{1}$

Solve each proportion using your calculator.

9. $\frac{2}{x} = \frac{10}{12}$ 2.4

10. $\frac{5x}{16} = \frac{72}{160}$ 1.44

11. $\frac{37}{78} = \frac{7x}{81}$ 5.489

SOLVING PROBLEMS WITH PROPORTIONS

You can use the rules you've already learned for solving equations to solve proportions. Remember that when you perform an operation on one side of the equal sign, you must perform the same operation on the other side of the equal sign. Recall that you cannot divide by zero.

Example: If 3 footballs cost $110, what will 5 footballs cost at the same price per football?

3 footballs are to 5 footballs as $110 is to $x.

$\frac{3}{5} = \frac{110}{x}$ (Multiply by $5x$.)

$5x\left(\frac{3}{5}\right) = 5x\left(\frac{110}{x}\right)$

$3x = 550$

$x = 183.3\overline{3}$

Rounded to the nearest penny, 5 footballs will cost $183.33.

Notice that the proportion used to solve the problem could also have been set up as $\frac{3}{110} = \frac{5}{x}$. The number of footballs appears in the numerators, with the appropriate cost in the denominators.

Extra Practice

1. Write each ratio in lowest terms.

a. $\frac{4}{16}$ $\left[\frac{1}{4}\right]$

b. 0.6 $\left[\frac{3}{5}\right]$

c. $2/7 \div \frac{5}{14}$ $\left[\frac{4}{5}\right]$

2. Solve each proportion.

a. $\frac{3}{4} = \frac{x}{24}$ $[x = 18]$

b. $\frac{5}{x} = \frac{35}{49}$ $[x = 7]$

c. $\frac{x}{21} = \frac{3}{7}$ $[x = 9]$

Using the Page

In the example, point out that there are times when $183.3\overline{3}$ would be an acceptable answer. Students should always reread the problem to see if they have answered the question or not. Since the solution for x is in terms of money, it must be rounded to the nearest cent.

Using the Page
Have students write an alternate proportion comparing inches to miles that they could use in the example.

Using the Page
The answer in the example was rounded because an answer should not be given with greater precision than the original measures.

Example: A map has a scale that shows 1 inch on the map equals 4.5 miles. Two cities are 2.3 inches apart on the map. How far apart are the cities in miles?

$$\frac{1 \text{ inch}}{2.3 \text{ inches}} = \frac{4.5 \text{ miles}}{x \text{ miles}}$$

$$\frac{1}{2.3} = \frac{4.5}{x} \quad \text{(Multiply by } 2.3x\text{)}$$

$$x = 2.3(4.5) = 10.35 \text{ miles}$$

Since the problem gave measurements to the nearest tenth, round 10.35 miles to 10.4 miles.

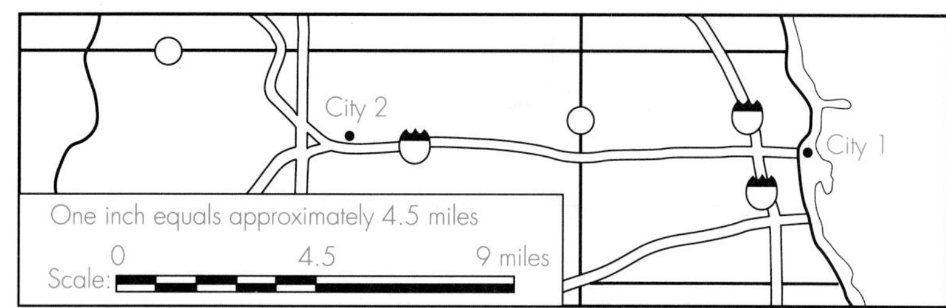

CLASS ACTIVITY

Write each ratio in the form N/D in lowest terms.

1. $\frac{16}{28}$ $\frac{4}{7}$
2. $\frac{5}{\frac{5}{6}}$ $\frac{6}{1}$
3. 34:48 $\frac{17}{24}$
4. $\frac{0.2}{\frac{1}{5}}$ $\frac{1}{1}$
5. If the ratio of boys to girls is 2 to 3 (2 boys for every 3 girls), then what would a ratio of 5 to 7 mean? 5 boys for every 7 girls.
6. The ratio of teachers to students is 1:16. How many teachers are needed for 560 students? 35
7. If a car gets 35 miles per gallon of gas (35 to 1), then how many gallons of gas will be consumed on a trip of 1435 miles? 41 gallons.

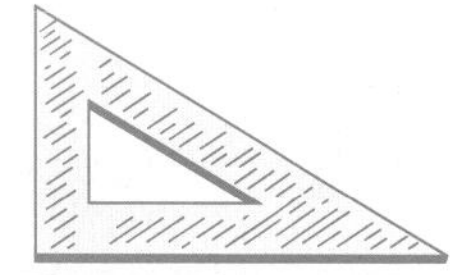

HOME ACTIVITY

Write each ratio in the form N/D in lowest terms.

1. $\frac{2}{4}$ $\frac{1}{2}$
2. $\frac{10}{16}$ $\frac{5}{8}$
3. $\frac{20}{32}$ $\frac{5}{8}$
4. $\frac{39}{65}$ $\frac{3}{5}$
5. $\frac{2730}{3094}$ $\frac{15}{17}$
6. $\frac{\frac{4}{7}}{\frac{20}{35}}$ $\frac{1}{1}$

Refer to the drawing below. Write each ratio in the form N/D, in lowest terms.

7. Triangles to squares $\frac{3}{2}$

8. Circles to triangles $\frac{1}{3}$

9. Circles to all figures $\frac{1}{6}$

10. Squares to triangles $\frac{2}{3}$

11. Triangles to circles $\frac{3}{1}$

12. Squares to all figures $\frac{1}{3}$

13. Circles to squares $\frac{1}{2}$

14. Squares to circles $\frac{2}{1}$

15. Triangles to all figures $\frac{1}{2}$

16. Circles and squares to triangles $\frac{1}{1}$

17. Squares and triangles to all figures $\frac{5}{6}$

18. Write a sentence to explain what Exercise 17 means.
Squares and triangles make up $\frac{5}{6}$ of the figures.

Solve each proportion for x using your calculator.

19. $\frac{3}{8} = \frac{x}{12}$ 4.5

20. $\frac{9}{4} = \frac{2x}{12}$ 13.5

21. $\frac{15}{x} = \frac{3}{4}$ 20

22. $\frac{(2x + 4)}{15} = \frac{40}{30}$ 8

23. $\frac{40}{(3x - 2)} = \frac{5}{21}$ $56\frac{2}{3}$

24. $\frac{(x + 3)}{6} = \frac{3}{4}$ $\frac{3}{2}$

25. $\frac{3x}{6} = \frac{(4x + 1)}{5}$ $-\frac{2}{3}$

26. $\frac{(2x - 3)}{8} = \frac{126}{(-6)}$ -82.5

CRITICAL THINKING

27. Which is a better buy, an 18-oz economy-size box of cereal for \$2.56, or a 13-oz box of the same cereal for \$1.82? 13-oz box.

28. What is the sum of the first 50 even counting numbers? Look at the cases below, and continue the table until you see a pattern.

Case	Numbers	Sum	
1	2	2	1(2)
2	2 + 4	6	2(3)
3	2 + 4 + 6	12	3(4)

50(51) = 2550

Using the Page
For Exercise 27, students should find the unit cost, or the cost per ounce, of each box of cereal. Challenge students to go to a grocery store and find another example where the larger container does not mean the better buy.

Using the Page
For case n, the sum of the numbers is $n(n + 1)$. You may want to show students the following reasoning:

$$\begin{array}{r} 2 + 4 + \ldots + 98 + 100 \\ 100 + 98 + \ldots + 4 + 2 \\ \hline 102 + 102 + \ldots + 102 + 102 \end{array}$$

There are 50 sums of 102. The sum is 50(102). Divide this by 2 so that the sum of the first 50 even counting numbers is found only once. 50(102)/2 = 50(51) = 2550

FOLLOW-UP

Assessment
1. Why is 0.9 a ratio? [It can be written as $\frac{9}{10}$.]
2. Explain why $\frac{4}{5} = \frac{7}{10}$ is not a proportion. [The ratios are not equal.]
3. Solve the proportion. $\frac{27}{x} = \frac{36}{8}$ [x = 6]

Reteaching
Students who have had difficulty with this section may benefit from Reteaching Activity 4.2.

Enrichment
For students who have mastered the material in this section, you may wish to assign Enrichment Activity 4.2.

SECTION 4.3

Resources
Reteaching 4.3
Enrichment 4.3

Objectives
After completing this section, the student should understand
- ▲ what similar triangles are.
- ▲ how to solve problems involving similar triangles.

Vocabulary
similar triangles

GETTING STARTED

Quiz: Prerequisite Skills

1. Are the ratios $\frac{35}{14}$ and $\frac{15}{6}$ equal? [Yes]

2. Write three pairs of angles with equal measures. [∠P and ∠B, ∠A and ∠T, ∠K and ∠W]

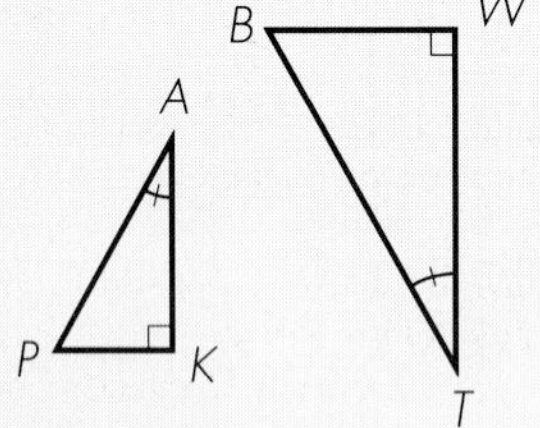

Warm-Up Activity
Have students write the ratios $\frac{8}{18}$, $\frac{16}{36}$, and $\frac{20}{45}$ in lowest terms. Ask students if they notice anything about the ratios. [All equal $\frac{4}{9}$.]

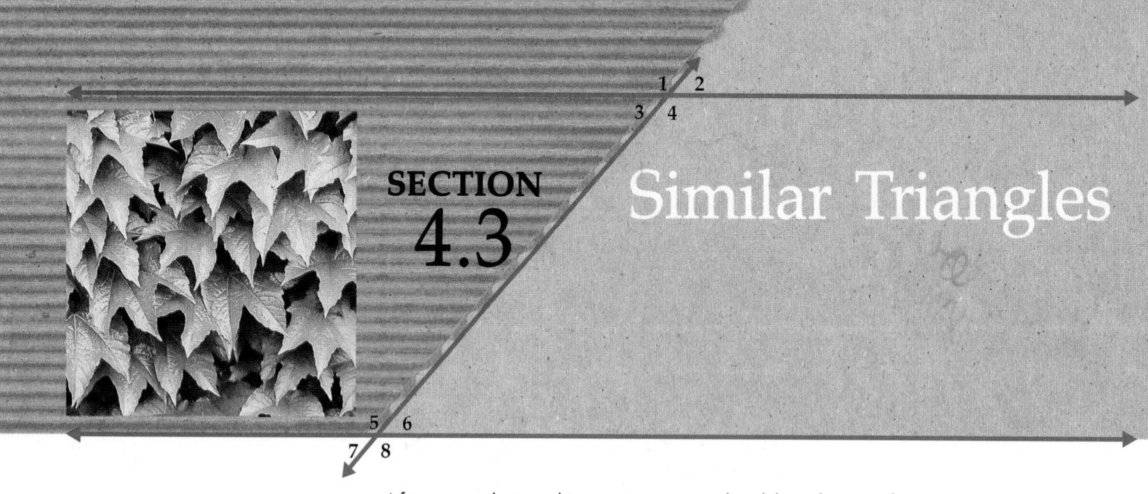

SECTION 4.3 Similar Triangles

After completing this section, you should understand
- ▲ what similar triangles are.
- ▲ how to solve problems involving similar triangles.

So far in this chapter, you have studied mappings, correspondences, ratios, and proportions. These ideas are now all tied together in work with triangles.

DISCOVERY ACTIVITY

m∠A = 40°	m∠X = 40°
m∠B = 55°	m∠Y = 55°
m∠C = 85°	m∠Z = 85°
AB = 12	XY = 6
BC = 8	YZ = 4
AC = 10	XZ = 5

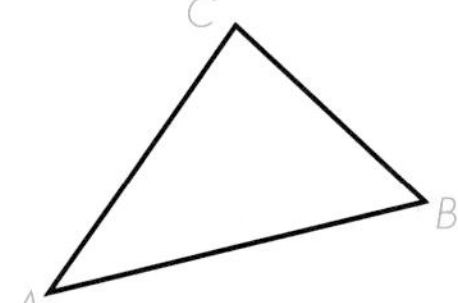

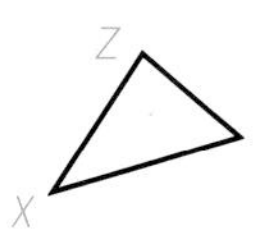

1. Compare m∠A to m∠X. Equal
2. Compare m∠B to m∠Y. Equal
3. Compare m∠C to m∠Z. Equal
4. Find the ratio AB/XY. $\frac{2}{1}$
5. Find the ratio BC/YZ. $\frac{2}{1}$
6. Find the ratio AC/XZ. $\frac{2}{1}$
7. Compare the ratios in questions 4–6. Equal
8. Can you map △ABC and △XYZ? Yes

SIMILAR TRIANGLES

In the Discovery Activity, you may have found that the corresponding angles were all equal in measure, and that the ratios of the corresponding sides were all equal. The two triangles are similar triangles.

DEFINITION 4-3-1 **Two triangles are SIMILAR if**

a. the corresponding sides have the same ratio;

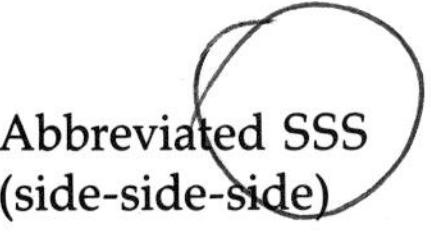

Abbreviated SSS
(side-side-side)

$\frac{6}{3} = \frac{10}{5} = \frac{8}{4} = \frac{2}{1}$

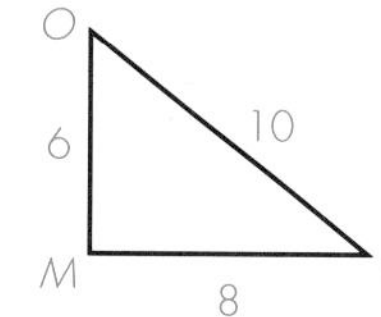

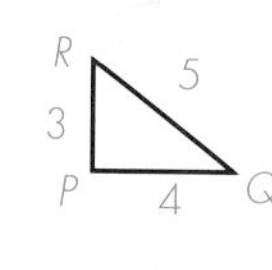

or

b. two angles of one triangle are equal in measure to two angles of another triangle;

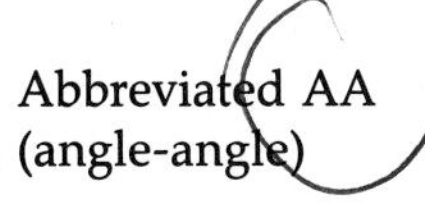

Abbreviated AA
(angle-angle)

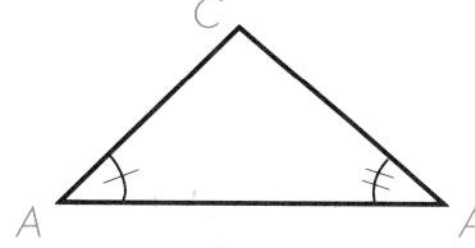

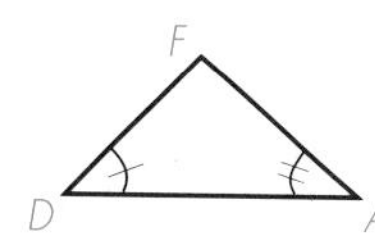

or

c. one angle of a triangle is equal in measure to an angle of the other triangle and the corresponding sides that include the equal angles are in the same ratio.

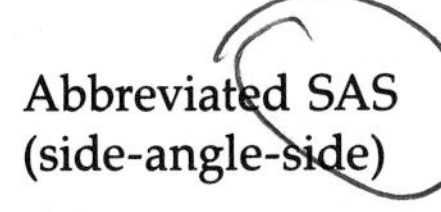

Abbreviated SAS
(side-angle-side)

$m\angle A = m\angle M$
$\frac{10}{5} = \frac{8}{4} = \frac{2}{1}$

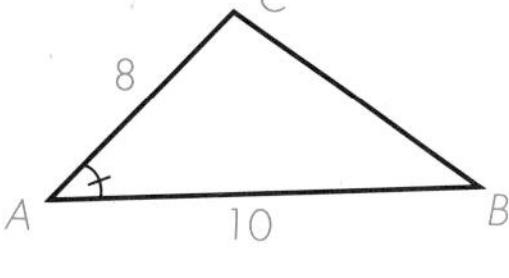

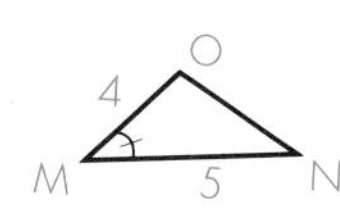

The symbol ~ is read *is similar to.* The drawing in part c of the definition could be labeled

$$\triangle ABC \sim \triangle MNO.$$

Chalkboard Example
Are the two triangles similar?
$\left[\text{No; } \frac{AB}{VU} = \frac{AC}{VW} \neq \frac{BC}{UW}\right]$

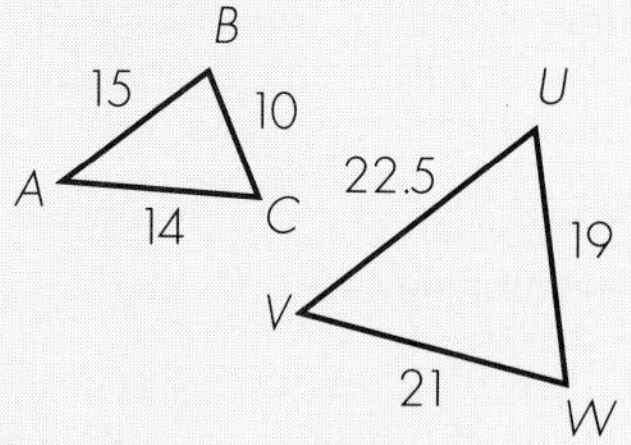

TEACHING COMMENTS

Using the Page
The last question of the Discovery Activity (on Page 134) asks if the triangles can be mapped. Just before Definition 4-3-1, you might mention that only similar figures can be mapped.

Using the Page
In part *a* of Definition 4-3-1, you might mention that properties of trigonometry can be used to show that
a. corresponding angles are equal in measure.
b. the ratios of corresponding sides are equal.

Using the Page
In part *b* of the Definition, ask students why $m\angle C = m\angle F$. [The sum of measures of the angles of a triangle = 180°; subtracting equal angle measures means the third pair of angle measures are also equal.]

Using the Page
Make sure students understand that only one part of the definition must be shown to be true in order for triangles to be similar. If enough information is given to show triangles congruent in more than one way, let students choose which method they use.

Using the Page
The reverse of Definition 4-3-1 will be used to solve for missing lengths of sides or missing measures of angles. (See the second example on page 137.)

It is important to list the vertices of each triangle in the correct order. Corresponding vertices must match when writing similarities.

Definition 4-3-1 has three parts separated by the word *or*. This means that only one of the three parts must be satisfied in order to have similar triangles. You do not need to show that all three parts are true, only one. Two, or even three, of the parts may be true, but only one is necessary.

Remember that a valid definition is true when reversed. From Definition 4-3-1, if two triangles are similar, then the corresponding angles are equal in measure and the corresponding sides are in equal ratio. Reversing the definition,

if the corresponding sides are in equal ratio (SSS),

or

if two angles of one triangle are equal in measure to two angles of another triangle (AA),

or

if one angle of a triangle is equal in measure to an angle of the other triangle and the corresponding sides that include the equal angles are in the same ratio (SAS),

then

the two triangles are similar.

Example:

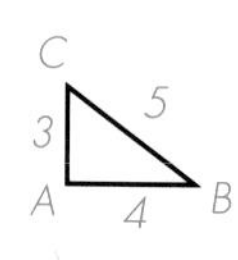

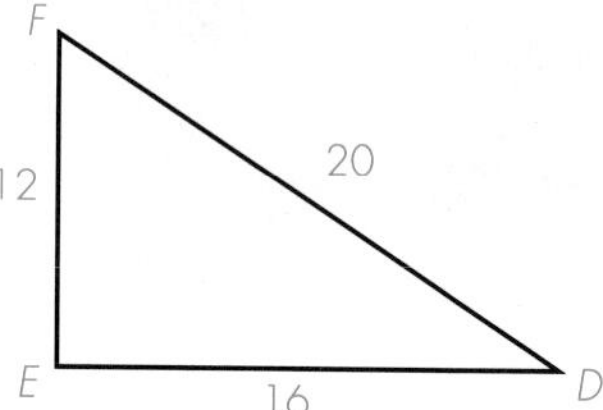

Find the ratio for each pair of corresponding sides.

$$\frac{4}{16} = \frac{1}{4}; \frac{3}{12} = \frac{1}{4}; \frac{5}{20} = \frac{1}{4}$$

Are the triangles similar? Yes, by SSS.

$\triangle ABC \sim \triangle EDF$

The side correspondences are $\overline{AB} \leftrightarrow \overline{ED}$, $\overline{BC} \leftrightarrow \overline{DF}$, and $\overline{AC} \leftrightarrow \overline{EF}$.

Example:

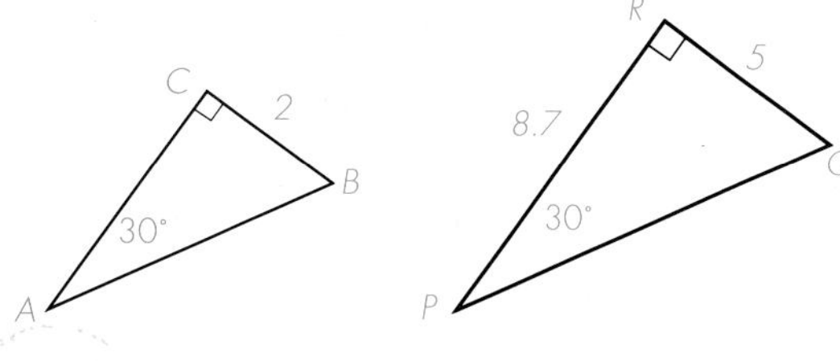

Two of the angles of $\triangle ABC$ are equal in measure to two of the angles of $\triangle PQR$. One pair of corresponding angles are 30°, and one pair are right angles and therefore of equal measure.

Therefore, $\triangle ABC \sim \triangle PQR$, by AA.

The angle correspondences are $\angle A \leftrightarrow \angle P$, $\angle B \leftrightarrow \angle Q$, and $\angle C \leftrightarrow \angle R$.

Find AC.

$RQ/CB = \frac{5}{2}$, therefore $PR/AC = \frac{5}{2}$.

$$\frac{8.7}{AC} = \frac{5}{2}$$

$$AC = 3.5$$

CLASS ACTIVITY

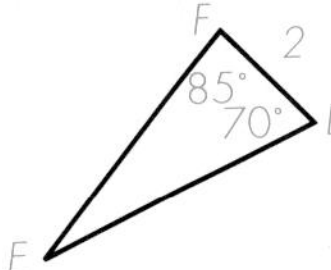

Refer to triangles DEF and ABC.

1. Why are the triangles similar?
   By AA
2. Map the triangles with the correct correspondences.
3. Write the angle correspondences.
   $\angle A \leftrightarrow \angle F$, $\angle B \leftrightarrow \angle D$, $\angle C \leftrightarrow \angle E$
4. Write the side correspondences.
   $\overline{AB} \leftrightarrow \overline{FD}$; $\overline{AC} \leftrightarrow \overline{FE}$; $\overline{BC} \leftrightarrow \overline{DE}$
5. Solve for EF and ED.
   $EF = 4.4$, $ED = 4.8$

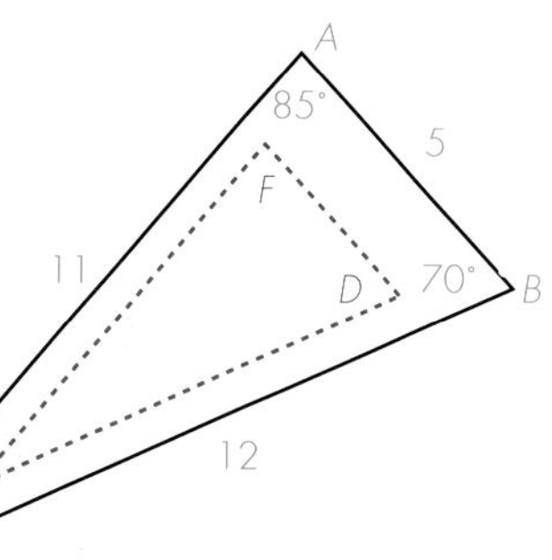

Using the Page
Point out that there are actually two ways to write ratios of corresponding sides. The ratios $\frac{16}{4} = \frac{4}{1}$, $\frac{12}{3} = \frac{4}{1}$, and $\frac{20}{5} = \frac{4}{1}$ could also be used to show that the triangles are similar. Caution students to use the same ordering in writing the ratios; the first terms of all three ratios must come from the same triangle, and the second terms must come from the other triangle.

Using the Page
See Exercise 5. To solve for EF, students may use the proportion $\frac{EF}{AC} = \frac{FD}{AB}$. To find ED, students should use the proportion $\frac{ED}{CB} = \frac{FD}{AB}$, not $\frac{ED}{CB} = \frac{EF}{AC}$. Emphasize that one ratio should always include two given measurements. Explain that this avoids the possibility of the value they found for EF being incorrect, causing the value they find for ED to also be incorrect.

Additional Answers

6. Yes; SSS; $\angle G \leftrightarrow \angle R$, $\angle B \leftrightarrow \angle K$, $\angle M \leftrightarrow \angle J$; $\overline{GB} \leftrightarrow \overline{RK}$, $\overline{BM} \leftrightarrow \overline{KJ}$, $\overline{MG} \leftrightarrow \overline{JR}$

Extra Practice

1. $\triangle FGH \sim \triangle IJK$. Find x and y. [$x = 16$; $y = 18$]

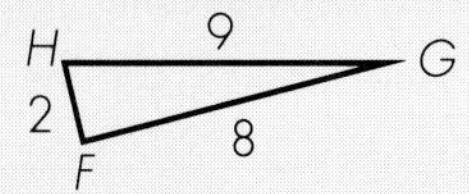

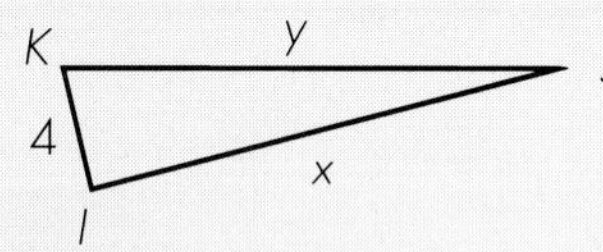

2. $\triangle TUV \sim \triangle OMP$. Find a and b. [$a = 5$; $b = 15$]

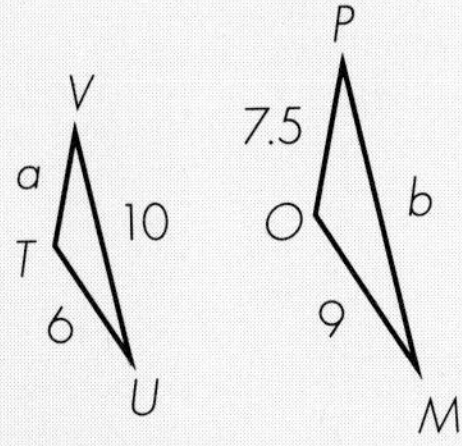

Additional Answers

1. $\angle A \leftrightarrow \angle D$, $\angle B \leftrightarrow \angle E$, $\angle C \leftrightarrow \angle F$; $\overline{AB} \leftrightarrow \overline{DE}$, $\overline{BC} \leftrightarrow \overline{EF}$, $\overline{AC} \leftrightarrow \overline{DF}$

2. $\angle M \leftrightarrow \angle Z$, $\angle O \leftrightarrow \angle Y$, $\angle N \leftrightarrow \angle X$; $\overline{MO} \leftrightarrow \overline{ZY}$, $\overline{ON} \leftrightarrow \overline{YX}$, $\overline{MN} \leftrightarrow \overline{ZX}$

Refer to triangles *GBM* and *RKJ*.

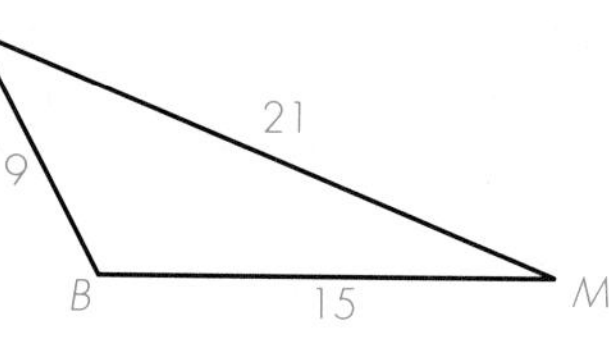

6. Are the triangles similar? Explain. If yes, identify the corresponding angles and sides. See margin.

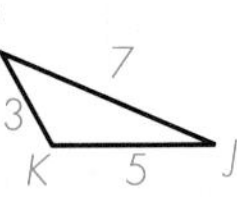

Use a computer and a geometric drawing tool to complete the exercises. Answers may vary.

7. Draw any $\triangle ABC$.
8. Draw $\overline{DE}$ such that D is on $\overline{AC}$ and E is on $\overline{BC}$ and $\overline{DE} \parallel \overline{AB}$.
9. Measure each side in each triangle.
10. Is $\triangle ABC \sim \triangle DEC$? If so, why? Yes; AA
11. If they are similar, map the triangles.
12. If they are similar, write the ratio of the corresponding sides.

HOME ACTIVITY

The two triangles are similar. Write the correspondences of the angles and of the sides.

1.

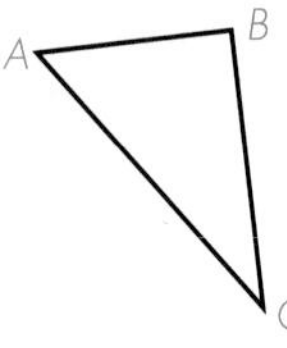

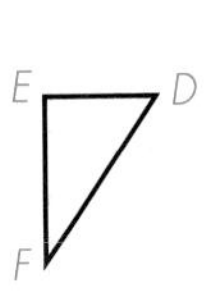

$\triangle ABC \sim \triangle DEF$

See margin.

2.

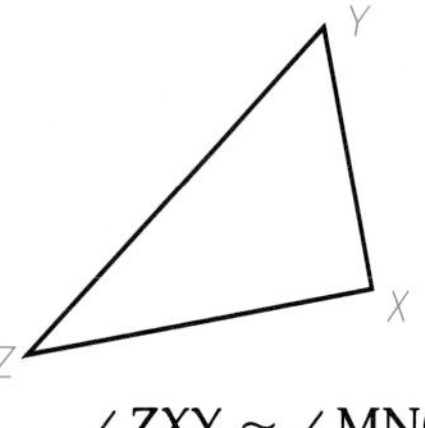

$\angle ZXY \sim \angle MNO$

See margin.

Refer to triangles *TAB* and *PRM*.

3. Are the triangles similar? Yes
4. Why or why not? By AA

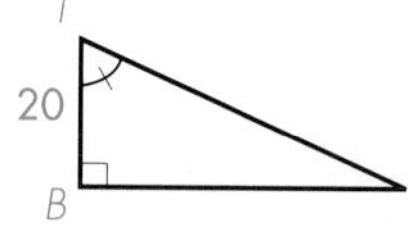

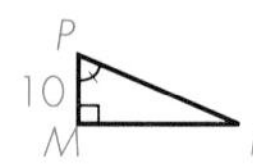

Refer to triangles *OHM* and *ALP*.

5. Why are these triangles similar? By SAS

6. Find the ratio of the sides. 2:1

Refer to triangles *MAP* and *ABC*.

7. Are the triangles similar?
Not enough information to tell

8. Why or why not?

Refer to triangles *ADC* and *PTQ*.

9. Are the triangles similar? Yes

10. Why or why not? SSS

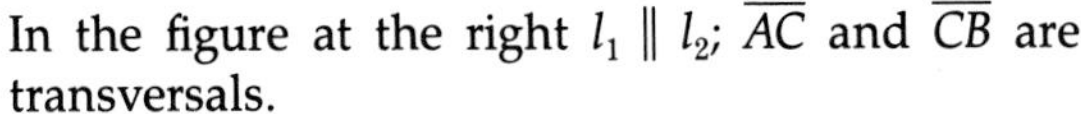

In the figure at the right $l_1 \parallel l_2$; $\overline{AC}$ and $\overline{CB}$ are transversals.

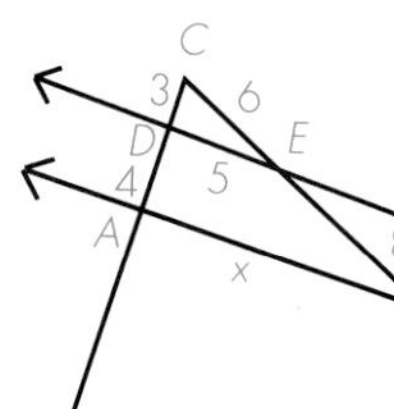

11. $\triangle ABC \sim \triangle DEC$. Why? By AA

12. Map $\triangle ABC$ and $\triangle DEC$. See margin.

13. Write the ratio of the corresponding sides of $\triangle ABC$ and $\triangle DEC$, and solve for *x*.
3:7; $x = 11\frac{2}{3}$

CRITICAL THINKING

14. *AG* is a flagpole that casts a shadow to *H*. I have only a tape measure, a pencil, and a piece of paper. Explain how I can figure out the height of the flagpole. (Hint: I know my own height.)
Measure the length of the pole's shadow (*HA*). I stand so the top of my shadow just reaches *H*, then measure the distance from my feet to *H* (call it *HB*). Solve the proportion.
$\frac{\text{length of } HA}{\text{length of } HB} = \frac{\text{length of } AG}{\text{my height}}$.

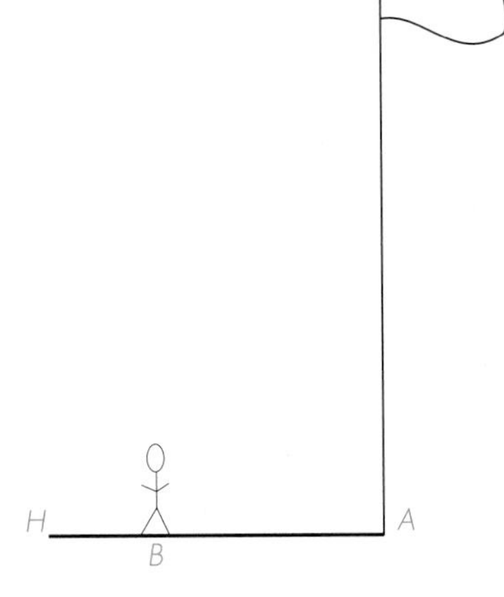

Additional Answers
12.

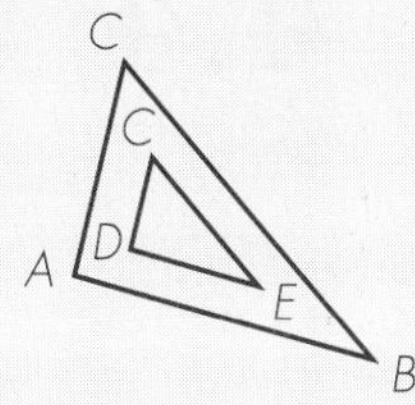

FOLLOW-UP

Assessment

1. Explain why each pair of triangles is similar.

a. [AA]

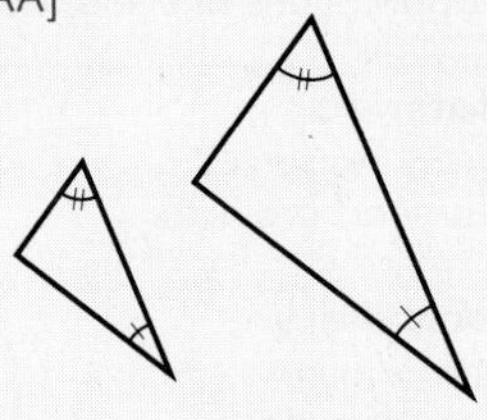

[SSS]

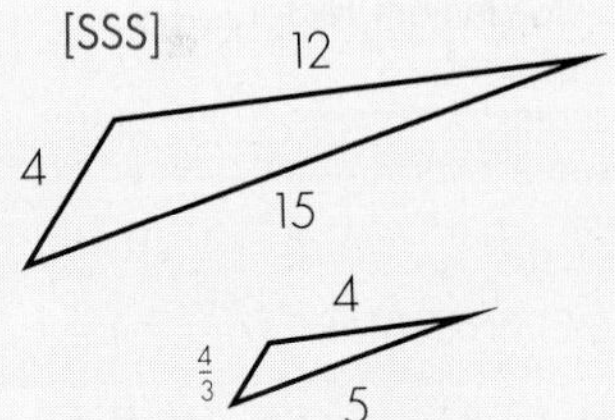

[SAS]

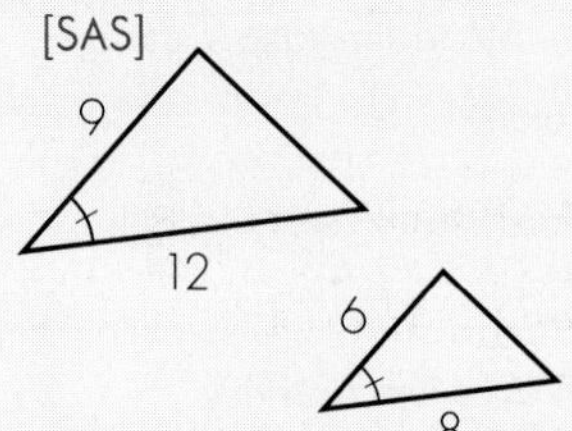

2. $\triangle BKJ \sim \triangle TWO$. Find *BK*. [7.5]

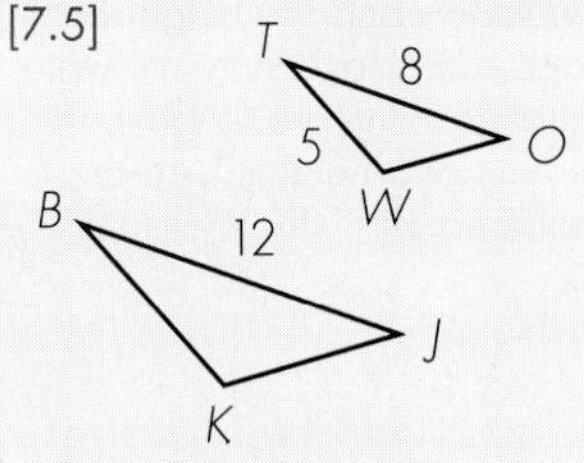

Reteaching
Students who have had difficulty with this section may benefit from Reteaching Activity 4.3.

Enrichment
For students who have mastered the material in this section, you may wish to assign Enrichment Activity 4.3.

SECTION 4.4

Resources
Reteaching 4.4
Enrichment 4.4
Transparency Master 4-2

Objectives
After completing this section, the student should understand
▲ scale drawings.
▲ applications of scale factors.

Materials
library resources
overhead projector

Vocabulary
similar figures
scale drawing
enlargement factor
multiplier
scale
scale factor

GETTING STARTED

Quiz: Prerequisite Skills

1. Write the ratio $\frac{4}{3}$ using a colon. [4:3]

2. Solve the proportion. $\frac{1}{2.5} = \frac{x}{7}$ [x = 2.8]

Warm-Up Activity
Bring in two shoe boxes. Have students measure the length, the width, and the height of each box. Have students write ratios to compare lengths, widths, and heights. Are the ratios equal?

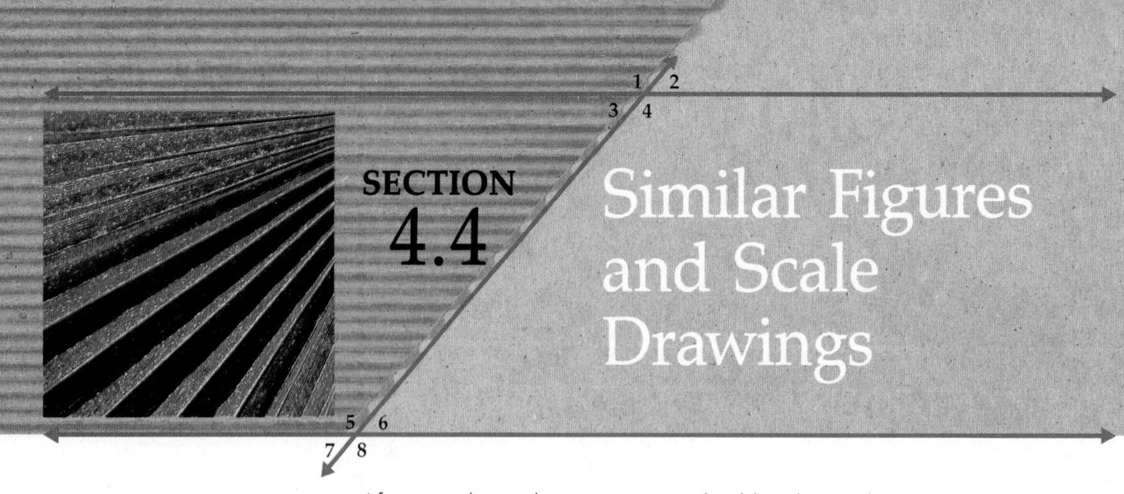

SECTION 4.4 Similar Figures and Scale Drawings

After completing this section, you should understand
▲ scale drawings.
▲ applications of scale factors.

In Section 4.3, you studied the conditions for similar triangles and wrote correspondences for angles and sides of the similar figures. The work you have done will now be expanded to applications in both two dimensions and three dimensions.

SIMILAR FIGURES

As with similar triangles, similar figures have equal corresponding angles and corresponding sides that are in equal ratio.

Examples: Two-dimensional

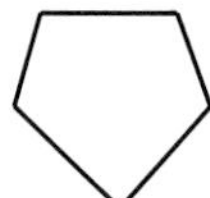
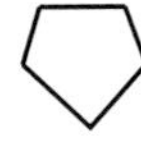
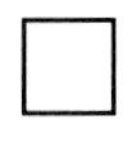
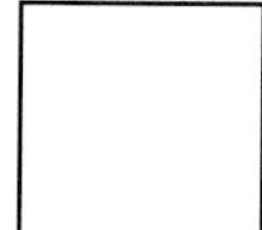

Three-dimensional

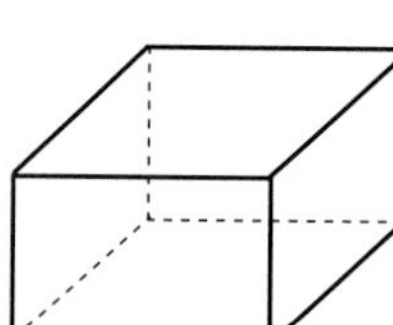
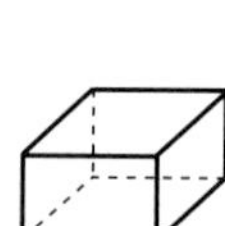
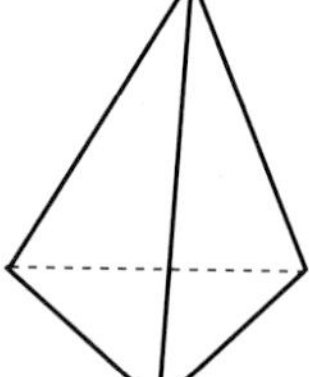
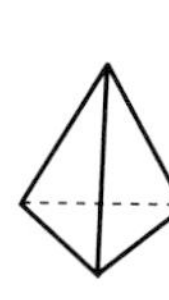

NOTEBOOK

DEFINITION 4-4-1 **If two figures are SIMILAR, then**

a. the measures of the corresponding angles are equal

and

b. the corresponding sides are in equal ratio.

Note the word *and* in the definition. Both conditions must exist for figures to be similar.

REPORT 4-4-1 Prepare a report explaining and illustrating the following subjects.

1. Lambert's projection
2. Mercator's projection
3. Topographical maps
4. Contour maps
5. The latitude and longitude readings for your community
6. A pantograph

Sources:

Bergamini, David, *Mathematics*. New York: Time-Life Books, 1963.
Gardner, Martin, *Time Travel and Other Mathematical Bewilderments*. New York: W. H. Freeman, 1987.
Newman, James R., *Encyclopedia of Science*. Edinburgh, Scotland: Nelson, 1963.

SCALE DRAWINGS

One use of similar figures is in scale drawings. Scale drawings can reduce the size of a large object in its representation, or increase the size of a small object. You see an example of scale drawings when a teacher uses the overhead projector or when you attend a movie. What you see on the screen is a projection of the picture on the transparency or the film.

Chalkboard Example
Write the proportion
$\frac{1 \text{ cm}}{25 \text{ km}} = \frac{x \text{ cm}}{540 \text{ km}}$
on the chalkboard. Have students solve for x. [$x = 21.6$]

TEACHING COMMENTS

Using the Page
Use the Warm-Up Activity to discuss whether the shoe boxes are similar figures.

Using the Page
Point out the word *and* in Definition 4-4-1. Ask students for an example of non-similar figures in which the measures of the corresponding angles are equal, but the corresponding sides are not in equal ratio. [One example: a square and a rectangle] Then ask students for an example of non-similar figures in which the corresponding sides are in equal ratio, but the measures of the corresponding angles are not equal. [One example: a square and a rhombus] Thus, both conditions must hold if the figures are similar. Display Transparency Master 4-2 to illustrate the necessity of both conditions of Definition 4-4-1 being true.
Assign small groups or individuals to prepare Report 4-4-1 and to report their findings to the class.

Using the Page
Scales are typically given in the form 1 inch = 20 miles or 1 inch:20 miles. Challenge students to find other ways of indicating scales on maps.

The picture on the screen is larger than the one on the film. This enlargement factor, or multiplier, is the same as the ratio number you worked with in Section 4.3. A ratio number also appears on maps and is identified as the scale or scale factor.

Scale: one inch = approximately 13.7 miles or 22.5 kilometers

The scale number is not always given, as in the case of the overhead projector or the movie. You can solve for these scale numbers by using ratios and proportions.

Using the Page
The Discovery Activity gives students a method of finding the scale factor when one is not given. Ask students why it was important that the overhead projector not be moved.

DISCOVERY ACTIVITY

Note: Do not move the overhead projector during this activity.

1. On a transparency, draw a line segment one centimeter in length.
2. Project the line segment on the screen.
3. Measure the segment on the screen.
4. What is the scale factor? Answers will vary.
5. Draw a scalene triangle on the transparency. Measure and record the lengths of the sides in centimeters.
6. Project the triangle on the screen.
7. Predict the length of the sides of the triangle on the screen, using the scale factor from question 4.
8. Measure the sides of the triangle on the screen and compare the lengths with the prediction.

You may have predicted the correct lengths of the sides of the triangle. Scale factors can be determined by careful measurement, then used to predict unknown lengths.

APPLICATION USING SCALE DRAWING

Many people use scale drawings in their work, including drafters, architects, and toy manufacturers.

The directions for a model car may state that the scale is 1:18. This means the model will be similar to the actual car, with 1 inch on the model car representing 18 inches on the actual car. Scales can also be written as 1 inch to x feet, such as 1 inch to 1.5 feet.

Example: A model car was built to the scale 1:18. If the model is 14.5 inches long, what is the length of the actual car?

$\frac{1}{18} = \frac{14.5}{x}$ (Multiply both sides by $18x$.)
$x = 18(14.5)$
$x = 261$ inches, or 21.75 feet
The length of the actual car is 21.75 feet.

Example: The actual car is 5.4 feet wide across the inside. What is the width of the model, if the scale is 1 inch to 1.5 feet?

$\frac{1}{1.5} = \frac{x}{5.4}$ (Multiply both sides by 1.5(5.4).)
$5.4 = 1.5x$ (Divide both sides by 1.5.)
$x = 3.6$ inches
The inside width of the model is 3.6 inches.

CLASS ACTIVITY

On a road map, the distance from Houston to San Francisco measured 11.5 inches. The scale used was 1 inch to 150 miles.

1. How far apart are the two cities? 1725 miles

2. A chart on the map listed the mileage between the two cities as 1953 miles. Explain why the two distances might be different. Hills, curves in the road.

3. The shadow of your hand on your desk is a projection. Name some other examples. Answers will vary.

Use a computer, a geometric drawing tool, and $\triangle WXY$ to complete the exercises.

4. Draw a similar triangle with sides twice as long (scale factor of 2).
Triangle with sides 13 cm, 5 cm, and 12 cm.

5. Draw a similar triangle with a scale factor of $\frac{1}{2}$.
Triangle with sides 3.25 cm, 1.25 cm, and 3 cm.

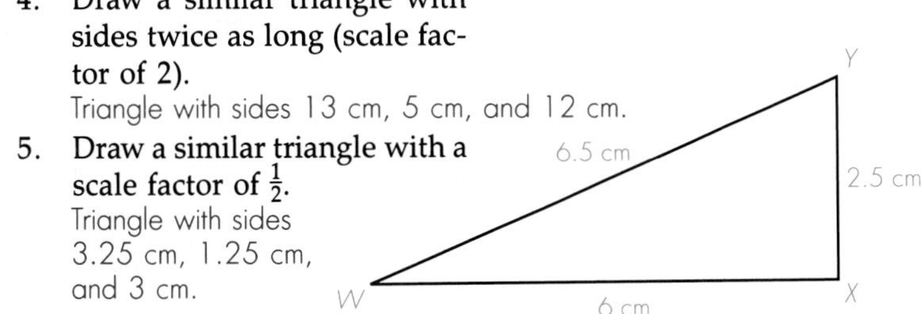

Using the Page

Many hobbies use scale factors. Ask students for other examples. One such example is a handicraft book that shows a pattern for a stuffed animal laid out on grids. The instructions say that each grid represents $\frac{1}{2}$ inch.

Extra Practice

1. A model of a building was built to the scale 2 cm = 3 m. If the model is 35 cm tall, how tall is the actual building? [52.5 m]
2. If the length of the building is 48 m, what is the length of the model? [32 cm]

HOME ACTIVITY

1. Find the scale for the projection illustrated. 1:19.2
2. A line segment on an overhead transparency is 2.5″ in length. The same segment on the screen is 9.5″. What is the scale? 1:3.8

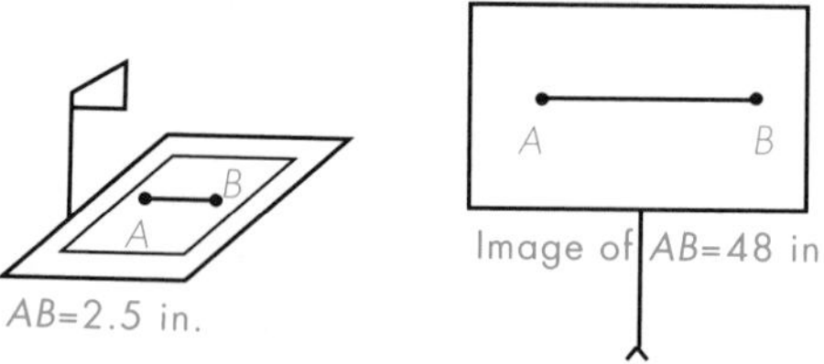

3. A model airplane kit indicates the scale is 1 inch to 4 feet. If the wingspan is 34 feet, what is the wingspan on the model? 8.5 in.

The scale for a map of Texas is 1 inch to 63 miles.

4. If the longest east-west segment on the map is 14.9 inches, what is the distance in miles? 938.7 mi
5. If the north-south distance is 710 miles, how many inches is this on the map? Round to the nearest tenth. 11.3 in.

A model of a steam locomotive was made to the scale of 1:120.

6. Explain what 1:120 means. 1 unit on model = 120 units on actual locomotive.
7. If the model is 8.4 centimeters long, what is the length of the actual steam locomotive in meters? 10.08 m
8. A chuck wagon used by the early settlers measured 10.5 feet in length. If the scale for the model is 1 to 16, then how many inches long is the model? 7.875 in.
9. A battleship is 860 feet long. What is the length of the model if the scale is 1 inch to 53.75 feet? 16 in.

Using the Page
See Exercises 6 and 7. The ratio 1:120 implies that the units of measurement are the same. Thus, students should first find the length of the steam locomotive in centimeters, then change this measurement to meters.

10. Copy and complete the following table. Use your calculator to solve the proportions.

Inches	1	2	3	4	5	7.5	10
Centimeters	2.54	5.08	7.62	10.16	12.70	19.05	25.40

11. What is the scale for converting inches to centimeters? 1:2.54

Determine the scale for the indicated conversions.

12. Feet to yards 3:1

13. Feet to miles 5280:1

14. Inches to feet 12:1

15. Yards to meters 1:0.9

16. Centimeters to meters 100:1

17. Yards to feet 1:3

18. Miles to feet 1:5280

19. Feet to inches 1:12

20. Meters to yards 1:1.1

21. Meters to centimeters 1:100

$\triangle ABC$ has sides of length 5 cm, 3 cm, and 4 cm.

22. Draw a similar triangle with sides twice as long (scale factor of 2).
Triangle with sides of length 10 cm, 6 cm, and 8 cm.

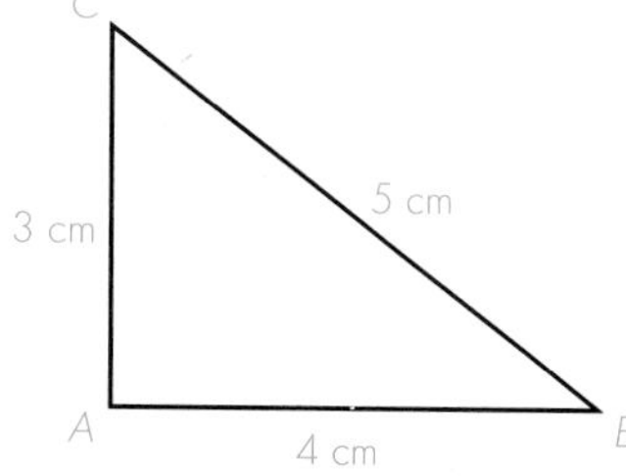

23. Draw a similar triangle with a scale factor of $\frac{1}{2}$.
Triangle with sides of length 2.5 cm, 1.5 cm, and 2 cm.

CRITICAL THINKING

24. Julie found a shortcut for reducing 16/64 to lowest terms. She crossed out the 6's.

$$\frac{1\not{6}}{\not{6}4} = \frac{1}{4}$$

Lucy didn't think that this would always work. Give one other example where Julie's method will work, and one where it won't work.
Answers will vary.
Possible answers: $\frac{7\not{0}}{1\not{0}} = 7$; $\frac{2\not{8}}{\not{8}} \neq 2$

Using the Page
Exercises 10-21 involve comparing units of measurement: metric to metric, standard to standard, and metric to standard or vice versa. Students may need to use library resources to complete these problems. For a project, you may want to have students write conversion factors for weights (masses) and then visit a grocery store. Many packages are labeled with both grams and ounces. Have students write pairs of numbers from packages and then verify their conversion factors.

FOLLOW-UP

Assessment
1. Explain why these figures are not similar. [Corresponding angles are not equal.]

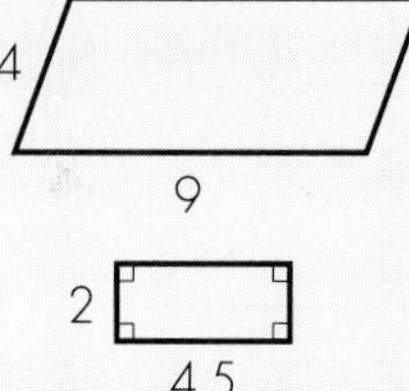

2. Explain how you can find a scale factor for two similar figures. [Write the ratio of the lengths of two corresponding sides.]
3. A model was made using the scale 1 cm = 2.5 m. If the length of the model is 6 cm, what is the length of the actual object? [15 m]

Reteaching
Students who have had difficulty with this section may benefit from Reteaching Activity 4.4.

Enrichment
For students who have mastered the material in this section, you may wish to assign Enrichment Activity 4.4.

SECTION 4.5

Resources
Reteaching 4.5
Enrichment 4.5

Objectives
After completing this section, the student should understand
- ▲ how to justify conclusions involving similar triangles.
- ▲ how to solve for missing lengths, using similar triangles.

Materials
ruler

Vocabulary
conjecture

GETTING STARTED

Quiz: Prerequisite Skills

1. If C is the midpoint of $\overline{AB}$, and $AB = 8$ inches, how long is $\overline{AC}$? [4 inches]

2. Draw an obtuse triangle, $\triangle VMD$, with T the midpoint of $\overline{VM}$, P the midpoint of $\overline{MD}$, and $\angle M$ the obtuse angle. [Students' drawings should look like the following.]

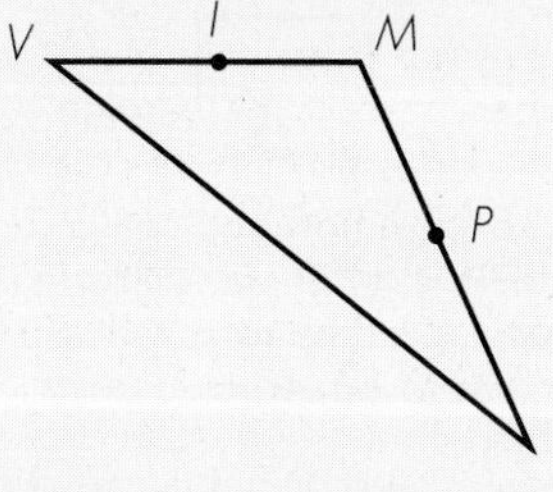

Warm-Up Activity
Draw two line segments, $\overline{AB}$ and $\overline{XP}$, on the chalkboard, each 36 inches long. Have one student locate K on $\overline{AB}$ so that K is half way from A to B. Have another student locate V on $\overline{XP}$ so that V is two-thirds of the way from X to P.

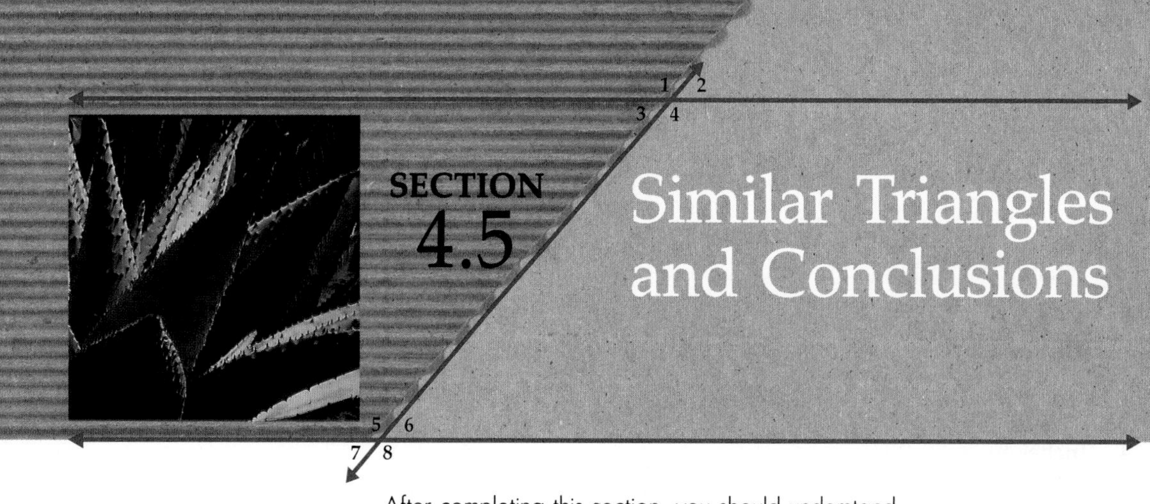

SECTION 4.5 Similar Triangles and Conclusions

After completing this section, you should understand
- ▲ how to justify conclusions involving similar triangles.
- ▲ how to solve for missing lengths, using similar triangles.

Previously, you learned how to tell if triangles were similar. You practiced mappings, wrote correspondences, solved proportions, and used scale drawings. In this section, two theorems will be introduced. These theorems, along with definitions, will be used to show similarity of triangles.

JUSTIFYING CONCLUSIONS

Often you are presented with a situation where you must analyze given information. You then make a guess, called a conjecture, that you think is correct. You have reached a conclusion. Conjectures are formed many different ways. You can form conjectures by observing patterns, by looking at examples, or by looking at all the possibilities. Sometimes conjectures are given to you. In geometry, once you have reached a conclusion, you know that the conclusion must be justified.

DISCOVERY ACTIVITY

1. Draw a large scalene triangle on a sheet of paper. Label the triangle ABC.
2. Using your ruler, find the midpoint of $\overline{AC}$. Label this point D. Do the same for $\overline{BC}$, labeling this point E.
3. Measure $\overline{DE}$ and $\overline{AB}$.
4. Find the ratio of DE to AB.
5. On the chalkboard, record the measurements and ratios of each student's triangle. Make a conjecture that appears to be true. See Theorem 4-5-1.

146

In the Discovery Activity, you may have concluded that $DE = \frac{1}{2}(AB)$, or that $DE{:}AB = 1{:}2$. This conclusion now needs to be justified (proved).

A proof is started by listing the given information, drawing a diagram, and stating the conclusion that needs to be justified.

Given: $\triangle ABC$ with midpoints D and E

Prove: $DE = \frac{1}{2}(AB)$

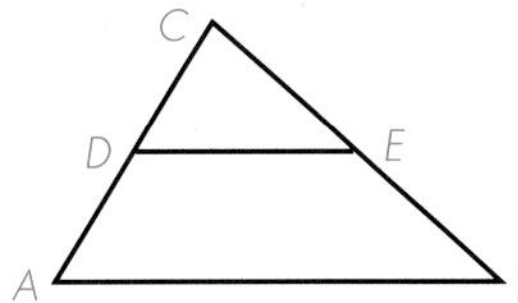

Statements	Reasons
1. $\triangle ABC$ with midpoints D and E	1. Given
2. $CD = \frac{1}{2}(CA)$ and $CE = \frac{1}{2}(CB)$	2. Definition of midpoint
3. $m\angle C = m\angle C$	3. Identity
4. $CD/CA = CE/CB = \frac{1}{2}$	4. Algebra
5. $\triangle ABC \sim \triangle DEC$	5. SAS
6. $DE/AB = \frac{1}{2}$	6. Corresponding sides of similar triangles are in equal ratio.
7. $DE = \frac{1}{2}(AB)$	7. Multiplication

THEOREM 4-5-1 **If a segment connects the midpoints of two sides of a triangle, then the length of the segment is equal to $\frac{1}{2}$ the length of the third side.**

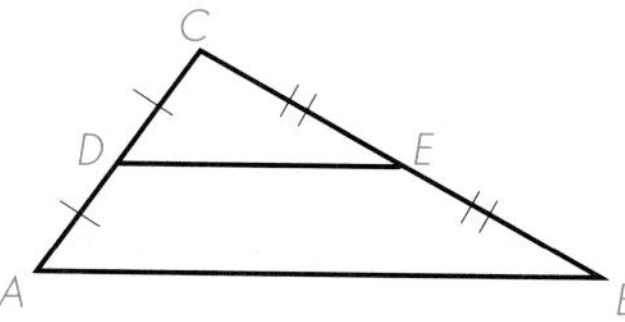

$DE = \frac{1}{2}(AB)$

Chalkboard Example
$\overline{CT} \parallel \overline{MP}$. If $AM = 16$, $AC = 12$, and $AT = 10$, find AP.

$\left[13\frac{1}{3}\right]$

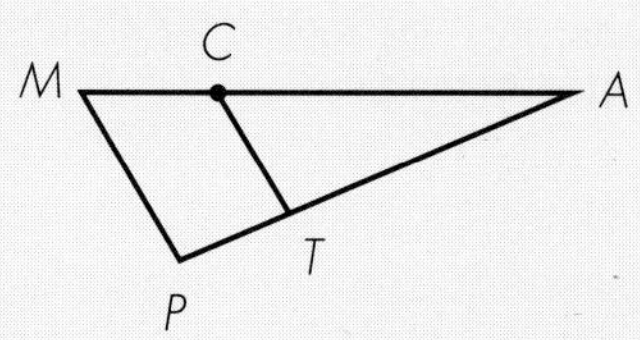

TEACHING COMMENTS

Using the Page
In the Discovery Activity (on Page 146), students look at several examples to reach the conclusion in Theorem 4-5-1. Make sure students understand each step in the proof.

Using the Page
Use the given and the diagram on this page, along with the drawings made in the Discovery Activity, to challenge students to make another conjecture about $\overline{DE}$ and $\overline{AB}$. Since $\triangle ABC \sim \triangle DEC$, ask students what they know about $m\angle CDE$ and $m\angle CAB$. [equal] If two lines cut by a transversal form corresponding angles that are equal in measure, what do they know about the lines? [parallel]

Using the Page
Have students write the paragraph proof in two-column form. Make sure students understand the reason for each step of the proof.

Example: Given: $\triangle ABC$, with $AB = 12$, $BC = 15$, $AC = 21$. D is the midpoint of $\overline{AC}$, E is the midpoint of $\overline{BC}$, and F is the midpoint of $\overline{AB}$. How long are segments $\overline{DE}$, $\overline{EF}$, and $\overline{DF}$?
$DE = \frac{1}{2}(AB) = \frac{1}{2}(12) = 6$
$EF = \frac{1}{2}(AC) = \frac{1}{2}(21) = 10.5$
$DF = \frac{1}{2}(BC) = \frac{1}{2}(15) = 7.5$

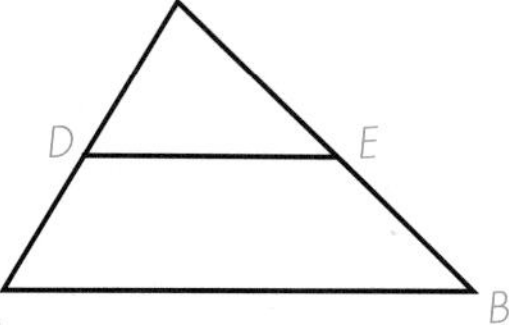

Example: Given: $\triangle ABC$, with D the midpoint of $\overline{AC}$ and $\overline{DE}$ parallel to $\overline{AB}$.
Justify: $\overline{DE}$ bisects $\overline{CB}$.

We are given $\triangle ABC$ with D the midpoint of $\overline{AC}$, and $\overline{DE}$ parallel to $\overline{AB}$. Since $\overline{DE}$ is parallel to $\overline{AB}$, $m\angle A = m\angle CDE$. $\triangle ABC \sim \triangle DEC$ by AA ($m\angle C = m\angle C$). Then the ratios of corresponding sides are equal. D is the midpoint of $\overline{AC}$, so $CD/AC = \frac{1}{2}$. Then $CE/CB = \frac{1}{2}$. Therefore, $CE = \frac{1}{2}(CB)$. So E is the midpoint of $\overline{CB}$. This means that $\overline{DE}$ bisects $\overline{CB}$.

Using the Page
In the Discovery Activity, have students draw a second parallel line and find the ratios corresponding to step 3. Students should find that the new ratios are different from the first set, but are still equal to each other.

DISCOVERY ACTIVITY

The geometric drawing tools can be used to complete this Discovery Activity.

1. On a sheet of paper, draw $\triangle ABC$. Draw a line parallel to $\overline{AB}$, intersecting $\overline{AC}$ at D and $\overline{BC}$ at E.
2. Measure the lengths of $\overline{CD}$, $\overline{DA}$, $\overline{CE}$, and $\overline{EB}$.
3. What is the ratio of CD to DA? of CE to EB? Answers will vary.
4. Are the two ratios equal? Yes
5. Compare your results to those of your classmates. Did everyone have equal ratios? Yes
6. Write a conclusion that appears to be true. See Theorem 4-5-2.

NOTEBOOK

THEOREM 4-5-2 **If a line is parallel to one side of a triangle and intersects the other sides at any points except the vertex, then the line divides the sides proportionally.**

Example: Given $\triangle ABC$ with $\overline{DE} \parallel \overline{AB}$. What is the length of $\overline{BE}$?

Let $BE = x$.
Using Theorem 4-5-2, the proportion is

$\frac{5}{4} = \frac{9}{x}$.
$5x = 36$
$x = 7.2$

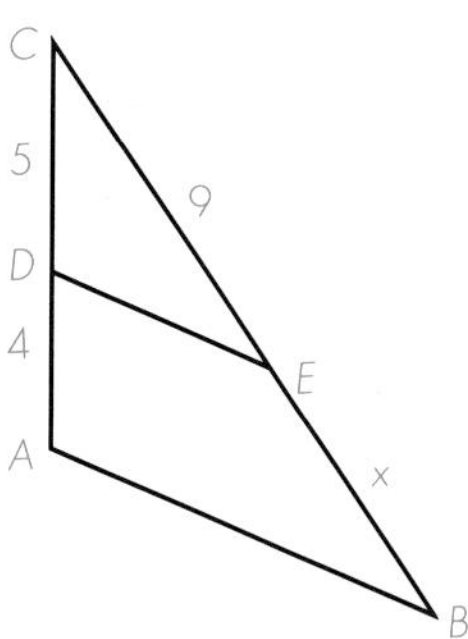

CLASS ACTIVITY

Refer to $\triangle OMN$. $\triangle OPQ \sim \triangle OMN$.

1. Find QN. 7.5
2. Find ON. 12.5

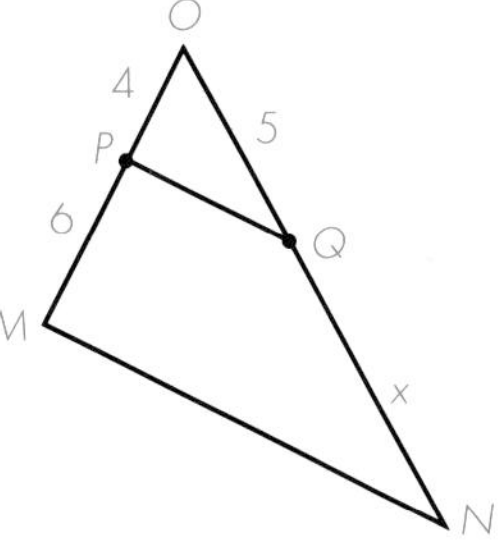

HOME ACTIVITY

Refer to $\triangle PRY$.

1. Are $\triangle PRY$ and $\triangle DEY$ similar? If so, why?
Yes; SAS
2. Solve for DP and RE. $DP = 15$; $RE = 18$

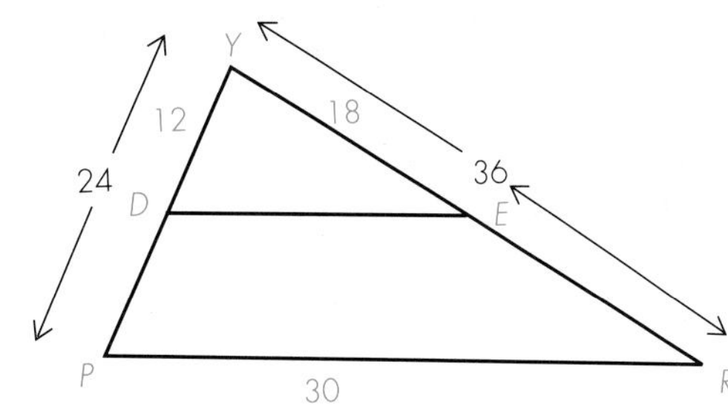

Refer to $\triangle ABC$. $AB = 8$ and $BC = 6$.

3. Are the triangles similar? If so, why?
Yes; SAS
4. Solve for DE and AD. $DE = 7.5$, $AD = 2$

Refer to $\triangle WXY$. $\triangle WXY \sim \triangle WAT$.

5. Find WX. 100
6. Find YT. 30
7. Find XA. 50
8. Find AT. 45

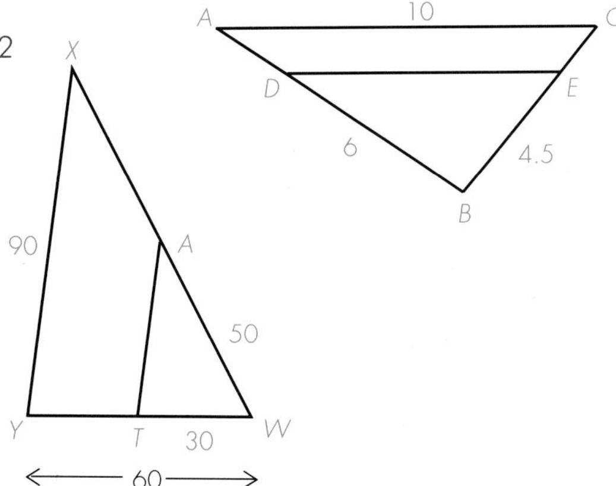

Extra Practice

1. D, E, and F are midpoints. Find DE, EF, and FD. [$DE = 11.5$, $EF = 5$, $FD = 8.5$]

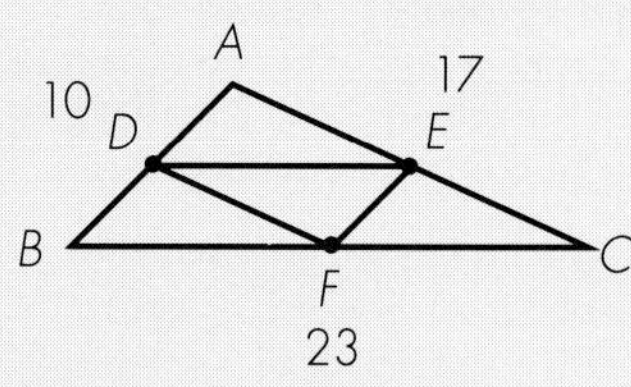

2. Refer to $\triangle KTA$. $\triangle KTA \sim \triangle KNI$. Find KI. [12.5]

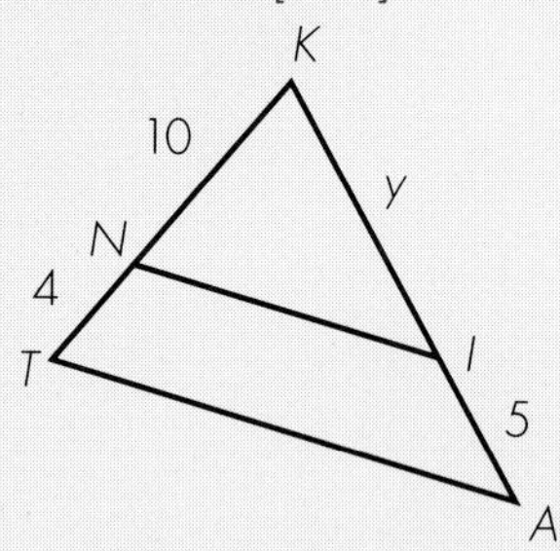

Using the Page
See Exercises 9-10. Have students find DA. Then have them find $\frac{CD}{DA}$ and $\frac{CD}{EB}$. Ask students what they found about the ratio of the corresponding segments that make up each side.

Refer to $\triangle ABC$. $AB = 38$, $BC = 54$, $AC = 24$, $CD = 16$, $CE = 36$.

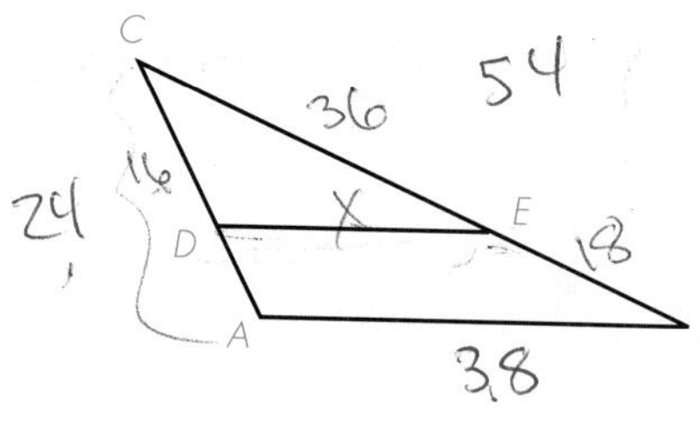

9. Are the triangles similar? If so, why?
Yes; SAS
10. Solve for DE and BE. $DE = 25\frac{1}{3}$; $BE = 18$

Refer to $\triangle ABC$.

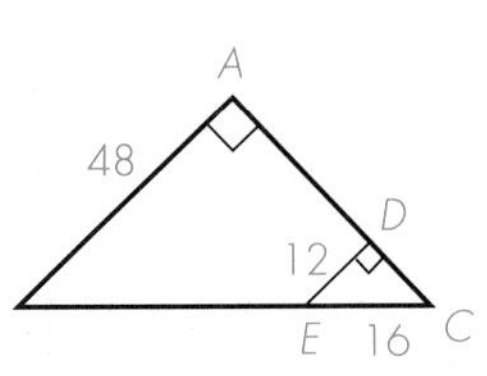

11. Are the triangles similar? If so, why?
Yes; AA
12. Solve for BC. 64

Refer to $\triangle NOP$. $\overleftrightarrow{QR} \parallel \overline{NO}$.

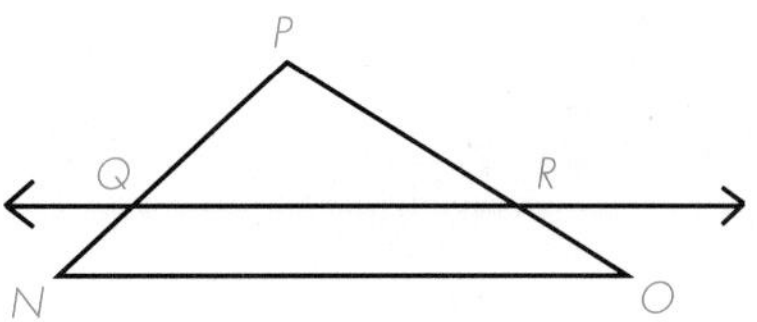

13. Is $\triangle NOP \sim \triangle QRP$? If so, why? Yes; AA
14. Write an if-then statement for Exercise 7.
See margin.

Additional Answers
14. If a line intersects two sides of a triangle and is parallel to the third side, then similar triangles are formed.

19.

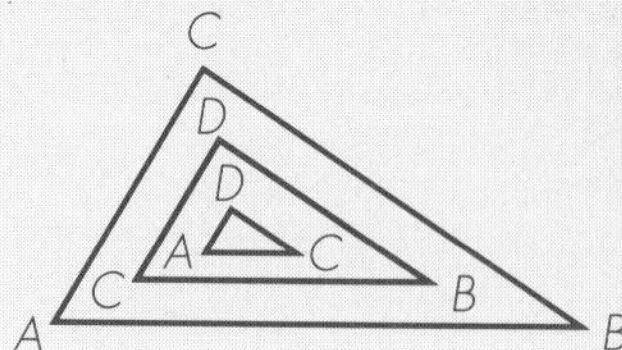

In the figure to the right, $\overleftrightarrow{AB} \parallel \overleftrightarrow{CD}$ with transversals $\overleftrightarrow{AD}$ and $\overleftrightarrow{CB}$ intersecting at O.

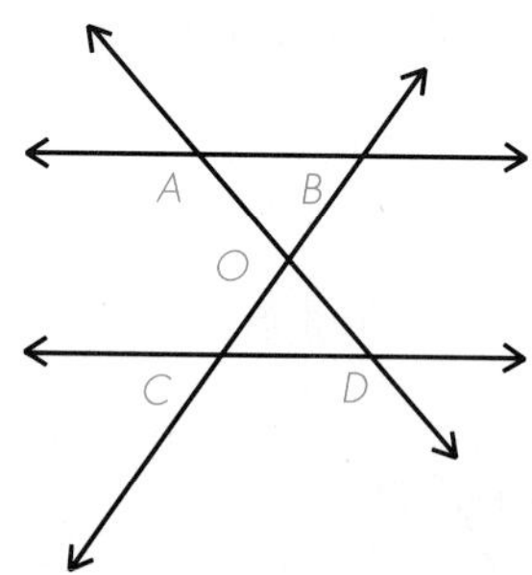

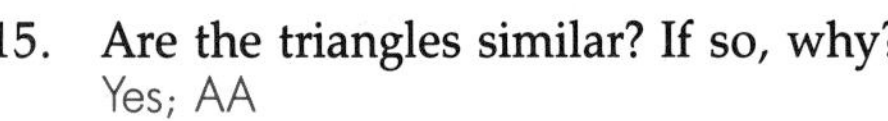

15. Are the triangles similar? If so, why?
Yes; AA

Refer to $\triangle ABC$ with $\angle C$ a right angle. $\overline{CD}$ is perpendicular to $\overline{AB}$, $m\angle A = 60°$.

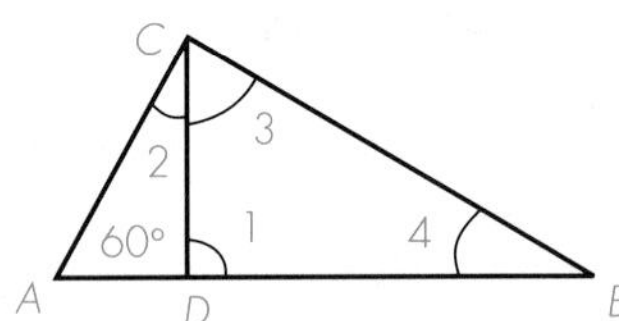

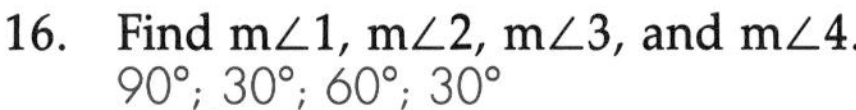

16. Find $m\angle 1$, $m\angle 2$, $m\angle 3$, and $m\angle 4$.
90°; 30°; 60°; 30°
17. How many triangles are there? List them.
3; $\triangle ABC$, $\triangle ACD$, $\triangle BCD$
18. $\triangle ABC \sim \triangle CBD \sim \triangle ACD$. Why? AA
19. Map the similar triangles. See margin.
20. Is it true that $(CD)^2 = (AD)(DB)$? Yes

Using the Page
See Exercise 20. Use the mapping from Exercise 19 to write the proportion $\frac{AD}{CD} = \frac{CD}{DB}$. Then $(CD)^2 = (AD)(DB)$.

21. Write a Logo procedure that tells the turtle to draw $\triangle XYZ$ with $XY = 70$, $YZ = 60$, and $XZ = 90$. Then tell the turtle to draw $\overline{MN} \parallel \overline{YZ}$ where M is the midpoint of $\overline{XY}$ and N is the midpoint of $\overline{XY}$. Is $\triangle XYZ \sim \triangle MYN$? If so, why? Yes; sas

Refer to $\triangle ABC$. $\triangle ABC \sim \triangle DEC$.

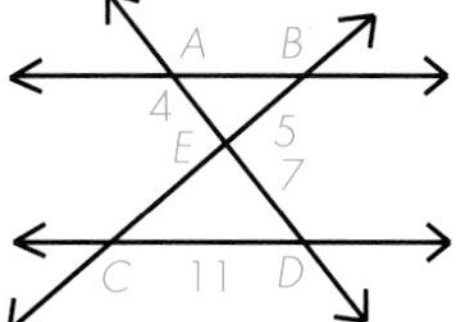

22. Find CB. 22.5

23. Find AD. 5

24. Find EB. 7.5

25. Find DE. 20

In the drawing at the right, $\overleftrightarrow{AB} \parallel \overleftrightarrow{CD}$, and transversals $\overleftrightarrow{AD}$ and $\overleftrightarrow{BC}$ intersect at E.

26. Find EC $8\frac{3}{4}$

27. Find AB. $6\frac{2}{7}$

28. In the drawing at the right, $\overleftrightarrow{AB} \parallel \overleftrightarrow{CD} \parallel \overleftrightarrow{EF}$. Find the unknowns w, x, y, and z.
$w = 3\frac{1}{3}$, $x = 6\frac{1}{2}$
$y = 8\frac{2}{3}$, $z = 8\frac{1}{3}$

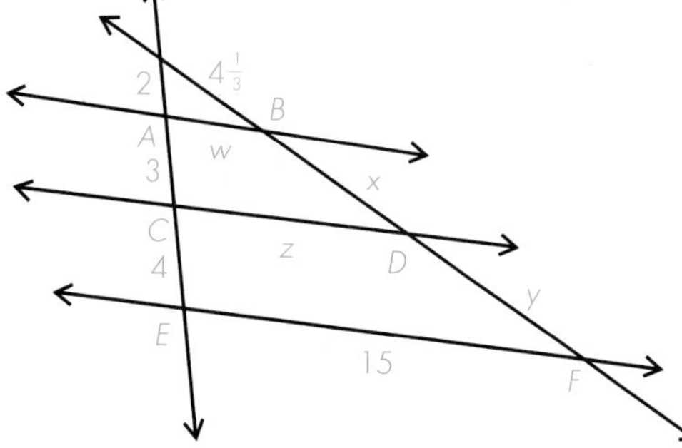

CRITICAL THINKING

This was once called the Golden Rule in arithmetic:

> Multiply the last number by the second, and divide the product by the first number.

For example, if 100 notebooks cost \$40, how much money should 15 cost?

Following the rule, $15 \cdot 40 = 600$
$\frac{600}{100} = \$6$

29. Explain why this method works. Hint: Use a proportion.
$\frac{\$40}{100} = \frac{x}{15}$, or $x = \frac{\$40 \cdot 15}{100}$

Using the Page
See Exercises 26-27. Since alternate interior angles are equal and vertical angles are equal, $\triangle AEB \sim \triangle DEC$ by AA.

FOLLOW-UP

Assessment
1. If T and C are midpoints, what is the value of $\frac{TC}{WN}$? $\left[\frac{1}{2}\right]$

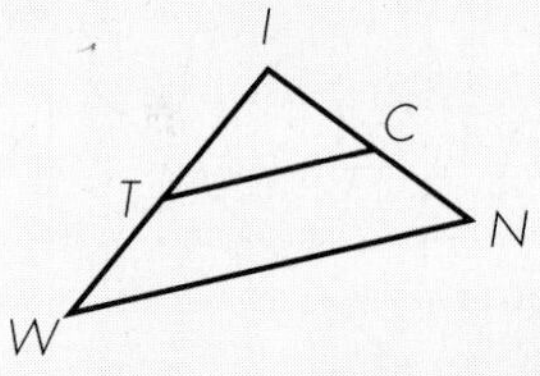

2. $\triangle XUQ \sim \triangle XKM$. Find QU. [6]

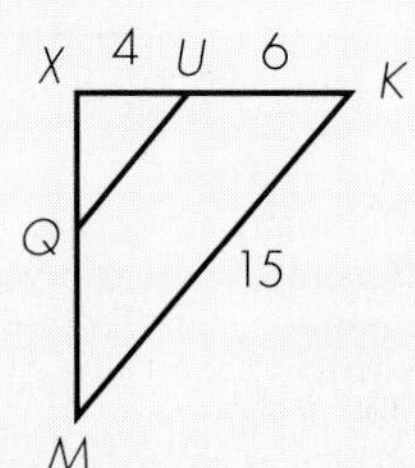

Reteaching
Students who have had difficulty with this section may benefit from Reteaching Activity 4.5.

Enrichment
For students who have mastered the material in this section, you may wish to assign Enrichment Activity 4.5.

SECTION 4.6

Resources
Reteaching 4.6
Enrichment 4.6

Objectives
After completing this section, the student should understand
▲ how to solve problems involving similar triangles.

GETTING STARTED

Quiz: Prerequisite Skills

1. $\triangle MAP \sim \triangle TOP$. Find MA. [60]

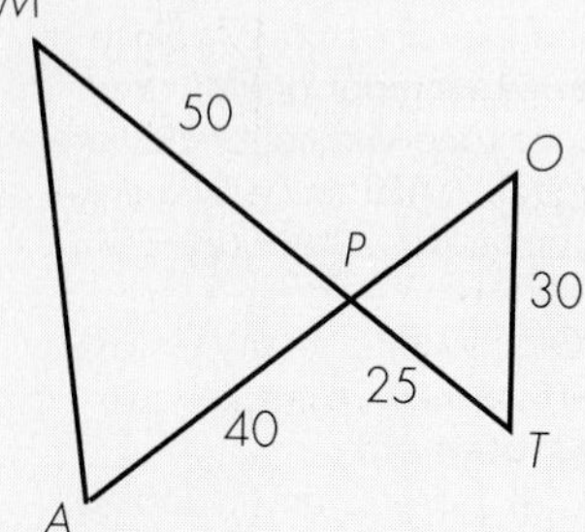

2. Find VN. [12]

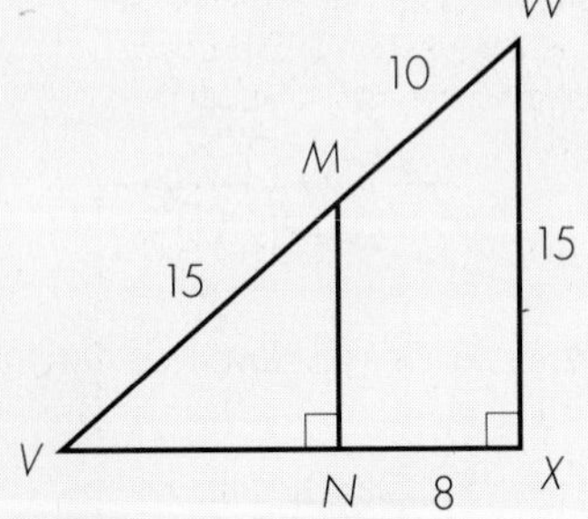

Warm-Up Activity
Choose a sunny day. Take the students outdoors. Use a yardstick or meterstick to find the height of a person (shortest in class) and the length of that person's shadow. Find the height of another person (tallest in class) and the length of this person's shadow. Compare the ratio of each person's height to the length of their shadow.

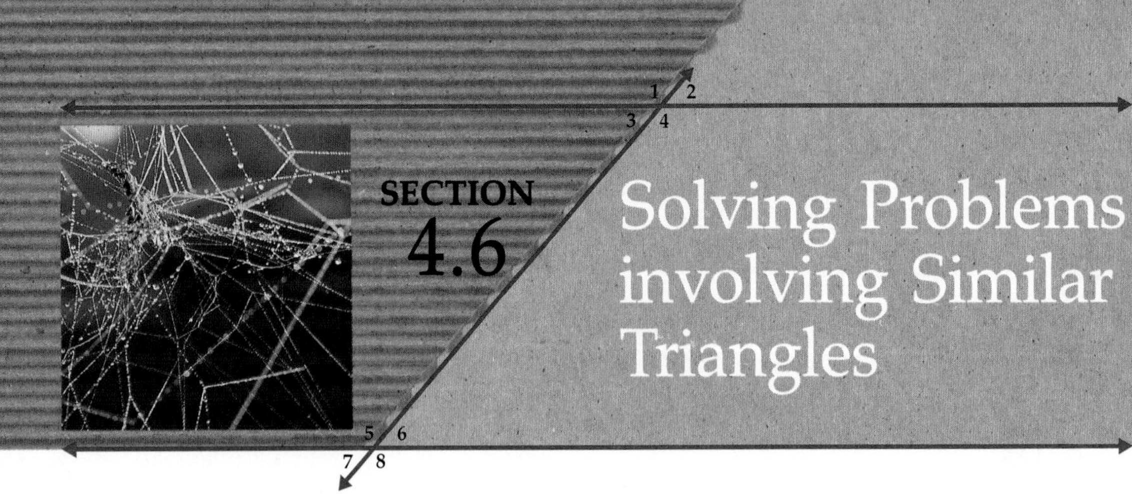

Solving Problems involving Similar Triangles

After completing this section, you should understand
▲ how to solve problems involving similar triangles.

The last section dealt with justification of similar triangles from a theoretical viewpoint. This section will deal with applications of similar triangles. It extends the theories to situations in life where they can be useful.

SOLVING PROBLEMS USING SIMILAR TRIANGLES

Some of these problems are taken from situations that may be found in your community. A couple of examples are given to show you how the people in your community might make use of similar triangles.

Example: A surveyor was hired to determine the width of a pond. She took some measurements, then made a drawing.

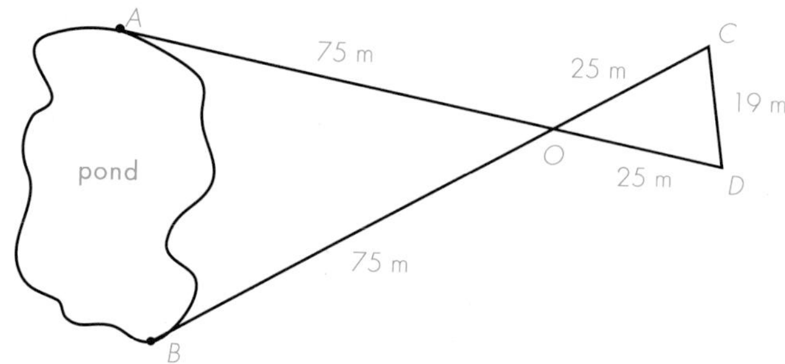

$\frac{CO}{BO} = \frac{25}{75} = \frac{1}{3}$ $\quad \frac{DO}{AO} = \frac{25}{75} = \frac{1}{3}$ $\quad$ Two ratios are equal.

The angles included between the sides are equal.

$m\angle COD = m\angle BOA$
$\triangle COD \sim \triangle BOA$ by SAS.

152

Map the triangles so that the corresponding parts are easier to see.

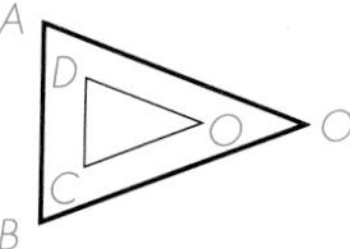

Therefore, $\frac{CD}{BA} = \frac{1}{3}$.

$\frac{19}{BA} = \frac{1}{3}$

$BA = 57$

The width of the pond is 57 meters.

Example: Some hikers noticed that a lookout tower cast a shadow that was 50 feet long. One of the hikers, who was 6 feet tall, cast a shadow that was 8 feet long. What is the height of the tower?

Both the person and the tower can be assumed to be perpendicular to the ground, thus forming right angles.

The sun beaming down forms angles that are equal, as long as the person and the object being measured are close together.

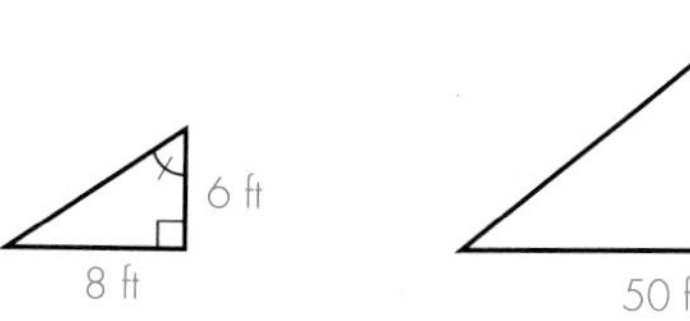

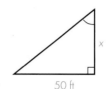

Therefore, similar triangles can be used to solve this problem.

Use the proportion $\frac{x}{6} = \frac{50}{8}$.

$$8x = 300$$

$$x = 37.5$$

The tower is 37.5 feet high.

Chalkboard Example
Continue with the Warm-Up Activity. Measure the shadow of any other person in the class. At the chalkboard, use the ratio of either of the first two people to predict the height of the third person.

TEACHING COMMENTS

Using the Page
The example on page 152 is typical of many situations where direct measurement cannot be made because of an obstacle that is in the way. Finding the distance across the base of a mountain is another typical situation. Ask students to think of some others.

Using the Page
The example on page 153 is typical of situations where vertical measurements are difficult to obtain. Ask students to think of some other situations where vertical measurement would be difficult.

Using the Page
The angle formed by the sun beaming down is the complement of the angle of depression.

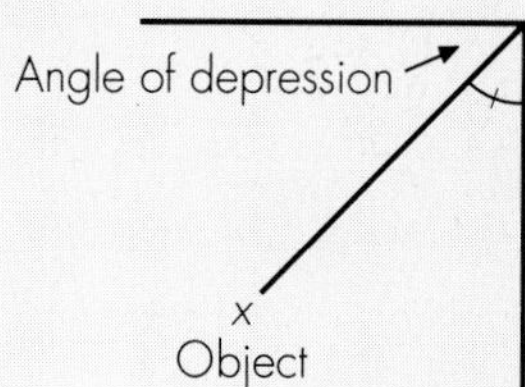

Extra Practice

1. Solve for x and y. $\triangle PAN \sim \triangle IDR$. [$x = 12$, $y = 10$]

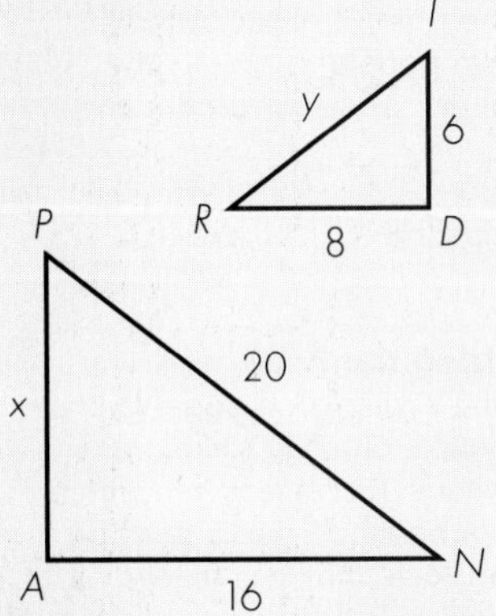

2. A brace is to be put on a trellis, as shown. $\overline{WR} \parallel \overline{FL}$. Find FW and LR. [10.8; 10.8]

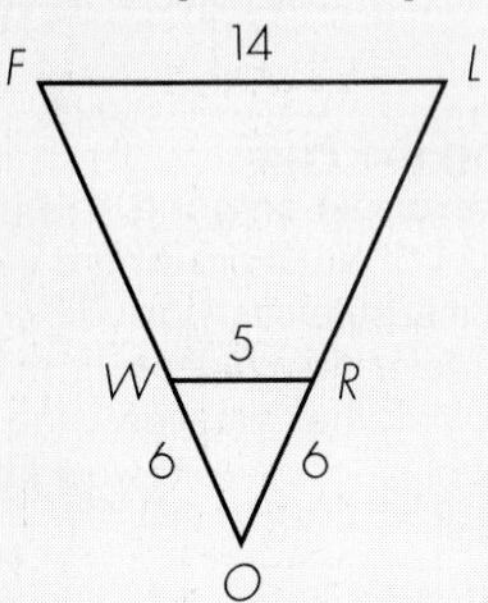

Example: A surveyor wants to measure the width of a stream. She puts a stake down at A, directly across from a lone tree at B. She measures and marks 10 meters and then 2 meters along the bank. She moves away from the bank until her second stake (C) and the tree line up. This point she finds is 9 meters from the bank. She now has enough information to find the distance across the stream.

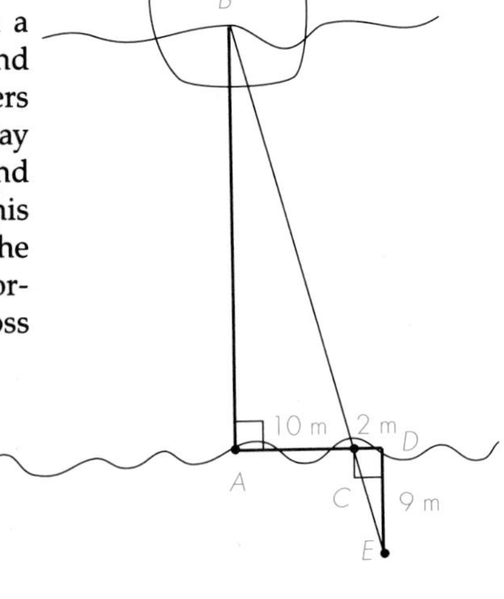

$\triangle ABC \sim \triangle DEC$ (By AA)

$\frac{AB}{DE} = \frac{AC}{DC}$

$\frac{AB}{9} = \frac{10}{2}$

$AB = 45$ m

CLASS ACTIVITY

1. Solve for x and y.
   $x = 28$, $y = 70$

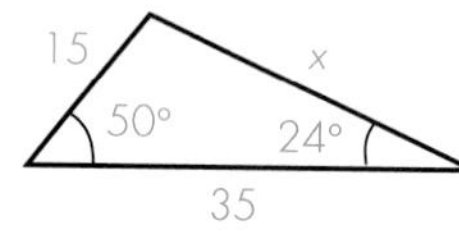

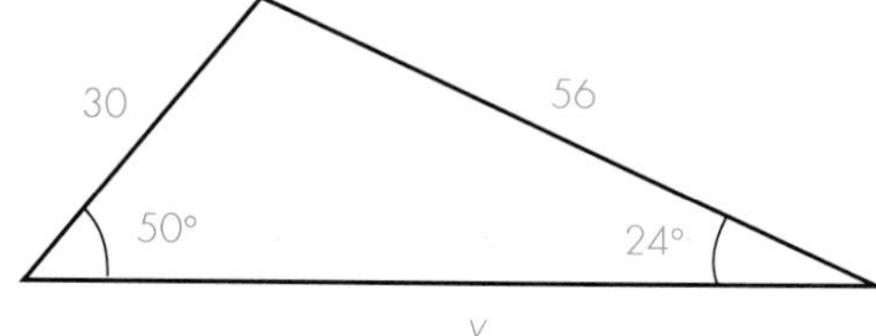

Refer to figure TAG.

2. $m\angle T =$ 50°
3. $m\angle TNA =$ 90°
4. $m\angle TAN =$ 40°
5. $m\angle GAN =$ 50°
6. Are triangles TNA, ANG, and TAG similar? If so, why?
   Yes; by AA

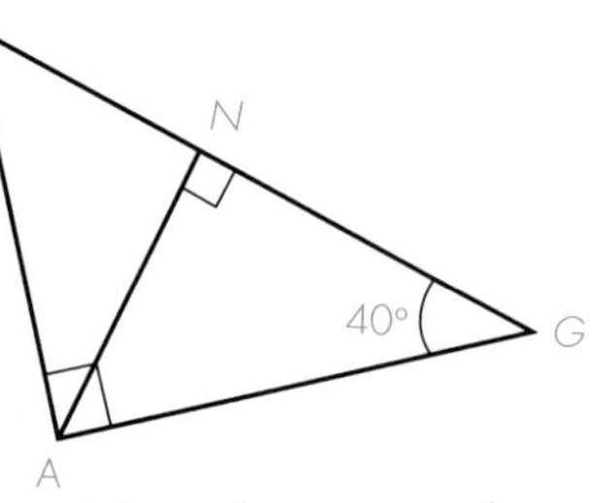

7. A person is planning to fence in a triangular area and cannot read two of the measurements on the drawing.

Given: $\triangle ABC$ with points D and E, $\overline{DE} \parallel \overline{AB}$

Solve for AC and BE. $66\frac{2}{3}'$; $26\frac{2}{3}'$

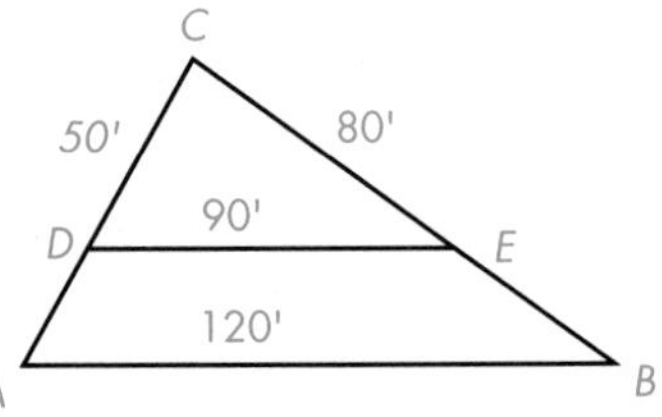

HOME ACTIVITY

1. Triangle *ABC* is projected on the screen by the overhead projector. The original triangle has sides 3, 4, and 5 inches long. The shortest side of the projected triangle is 14 inches long. What are the lengths of the other two sides of the projected triangle? $18\frac{2}{3}''$ and $23\frac{1}{3}''$

2. A surveyor needs to calculate the length of a pond. He drew the following diagram and made the measurements as indicated. What is the length of the pond? 120 ft

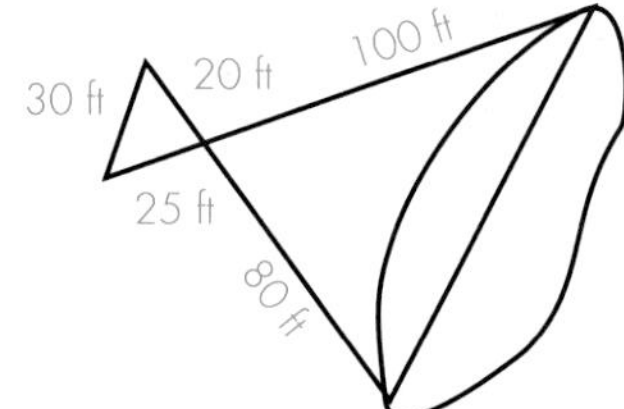

3. Photography makes use of similar figures. Solve for x and y, the dimensions of the final enlargement. $x = 15$ cm, $y = 9$ cm

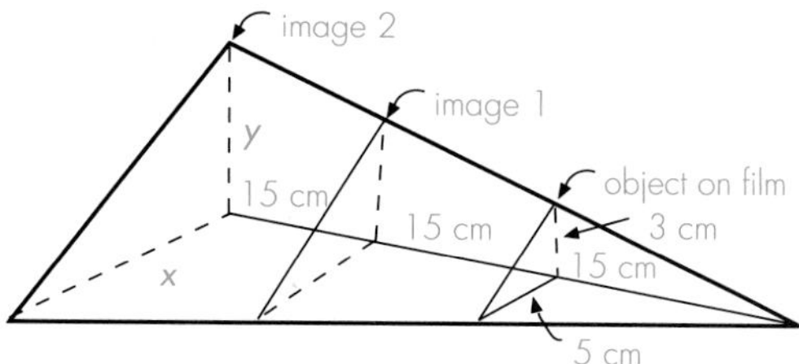

4. On a state map, city A is 5.5 inches from city Z. The scale is 1 inch to 22 miles. To the nearest mile, how far are the two cities from each other? 121 miles

Using the Page
See Exercise 1. The answer is given in terms of $\frac{1}{3}$ inches. You may want to have students write the answer to the nearest $\frac{1}{16}$ inch. Civil engineers and some other professionals use rulers marked in $\frac{1}{10}$ inches. Have students write the answers to the nearest $\frac{1}{10}$ inch.

Using the Page
Ratios were defined as one number divided by another number, where the denominator was not zero. So far, only rational numbers have been used. Point out to students that irrational numbers *can* be used for writing a ratio.

5. Solve for x and y. $x = 10$, $y = 2\sqrt{3}$

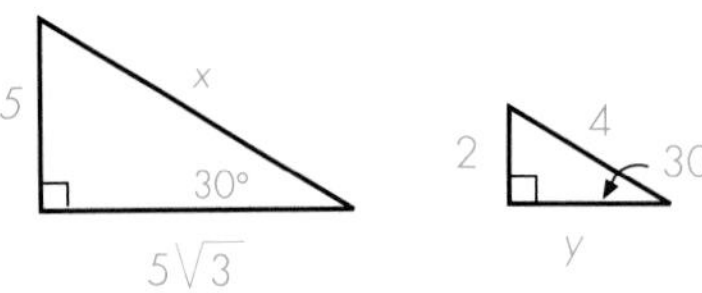

6. Find the width of the river in the figure. 40

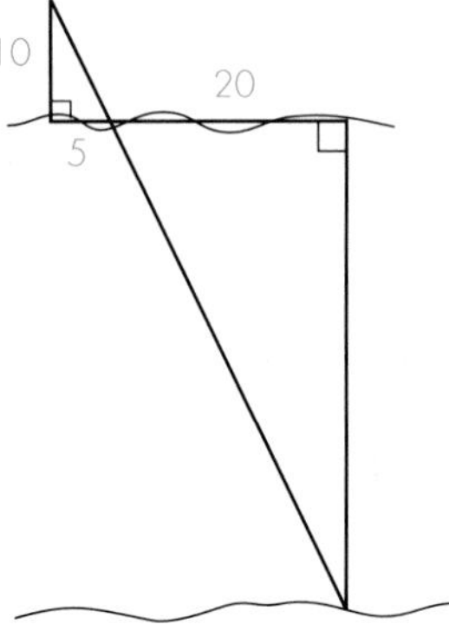

7. A student drew $\triangle ABC$ on a field, with $\overline{DE} \parallel \overline{AB}$. CA equals 25 m and CD equals 10 m. The student then had only enough time to measure CE (7 m) before it began to rain. What is the length of $\overline{CB}$? 17.5 m

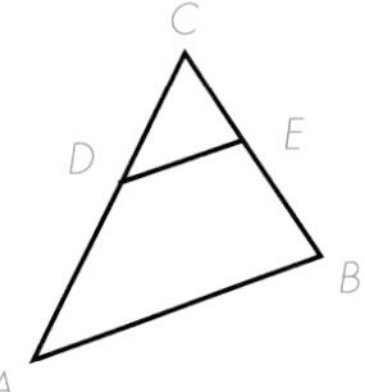

8. A group of students noticed the shadow of the flagpole was 47 feet long at the same time a person 6 feet tall cast a shadow 7 feet long. What is the height of the flagpole to the nearest foot? 40 feet

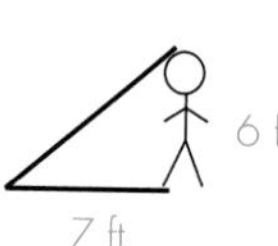

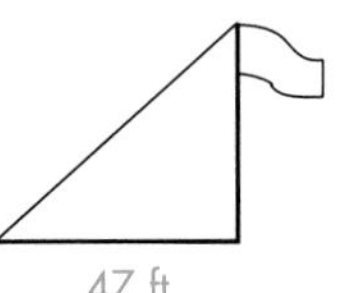

Refer to $\triangle FGH$. $m\angle H = 90°$, $HD \perp \overline{FG}$, and $m\angle F = 35°$.

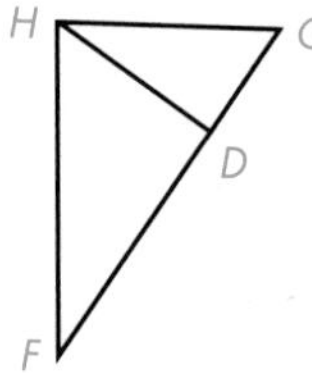

9. $m\angle G =$ 55°
10. $m\angle FDH =$ 90°
11. $m\angle FHD =$ 55°
12. $m\angle GHD =$ 35°
13. Are FGH, HGD, and FHD similar? If so, why?
Yes; AA

Refer to $\triangle ABC$.

14. Write a Logo procedure to tell the turtle to draw the figure at the right, given that $\overline{DE} \parallel \overline{FG} \parallel \overline{AB}$ with $CE = 7$, $CD = 10$, $DF = 12$, $FA = 13$, and $CB = 24.5$. Answers may vary.

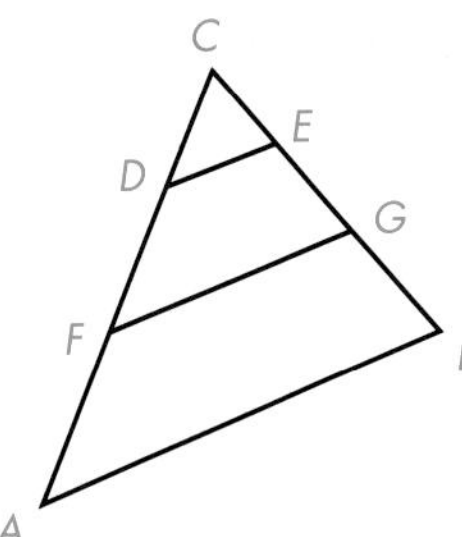

15. Are the triangles in figure ABC similar? If so, why? Yes; AA

Refer to $\triangle JKL$.

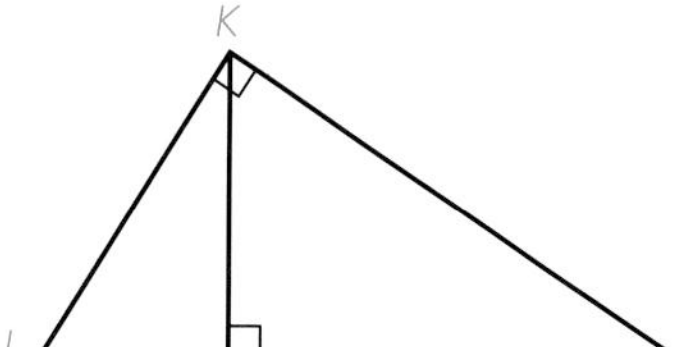

16. How many triangles are there? 3

17. Are any triangles similar? If so, why?
Yes; AA

Refer to $\triangle ABC$. A student was asked to measure the sides of this triangle, but the tape measure wasn't long enough. The only measurements that could be taken are indicated below.

Given: $\overline{DE} \parallel \overline{FG} \parallel \overline{AB}$

$CD = 18$ ft
$DF = 12$ ft
$EG = 15$ ft
$AF = 19$ ft
$DE = 20$ ft

Find the length of each segment.

18. CE 22.5 ft
19. BG 23.75 ft
20. AC 49 ft
21. CB 61.25 ft
22. FG 33.33 ft
23. AB 54.44 ft

Using the Page
See Exercises 18-22. Ask students how long the tape could possibly have been. [20 ft < length < 22.5 ft]

CRITICAL THINKING

When a beam of light strikes a mirror, it is reflected at the same angle as it hits. The diagram shows the light from the lamp striking the mirror at a 25° angle. It reflects into the girl's eye at the same angle.

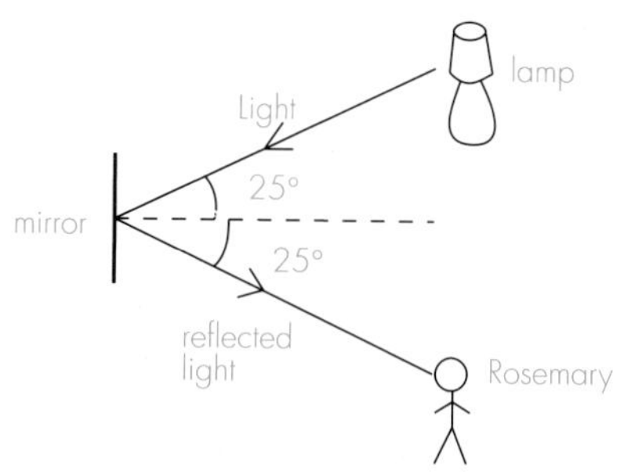

Paul knows about the reflective angles of a mirror. He decides to use this property to measure the height of a building to see how long a ladder he'll need to reach the roof. Paul has a mirror and a meter stick.

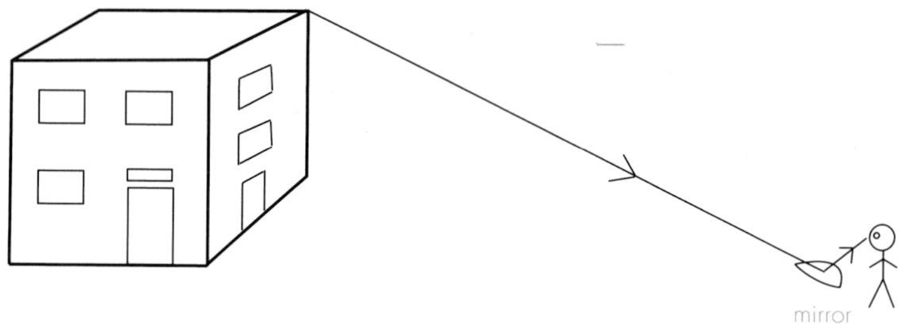

24. Draw the similar triangles in the problem. Answers will vary.

25. What measurements does Paul need to make? Height to eye, distance from feet to mirror and base of building to mirror.

26. Assign some possible measures to your drawing, with x being the height of the building. Answers will vary.

27. Write the proportion, and solve for x. Answers will vary.

FOLLOW-UP

Assessment

1. Describe a situation in which you might use similar triangles to solve a problem. [Typical response: when trying to measure through an object in the path of measurement.]

2. Find x. [60 ft]

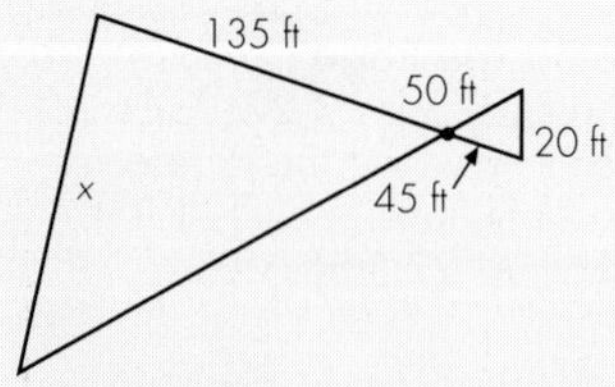

Reteaching

Students who have had difficulty with this section may benefit from Reteaching Activity 4.6.

Enrichment

For students who have mastered the material in this section, you may wish to assign Enrichment Activity 4.6.

4.1 Triangles ABC and DFE are similar.
Vertices: $A \leftrightarrow D$, $B \leftrightarrow F$, $C \leftrightarrow E$
Angles: $\angle A \leftrightarrow \angle D$, $\angle B \leftrightarrow F$, $\angle C \leftrightarrow \angle E$
Sides: $\overline{AB} \leftrightarrow \overline{DF}$, $\overline{BC} \leftrightarrow \overline{FE}$, $\overline{AC} \leftrightarrow \overline{DE}$

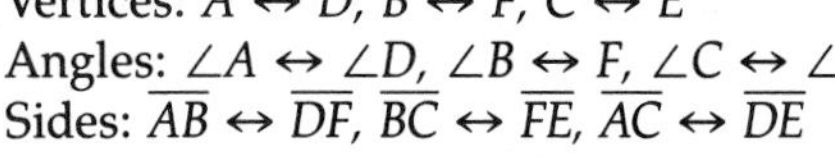

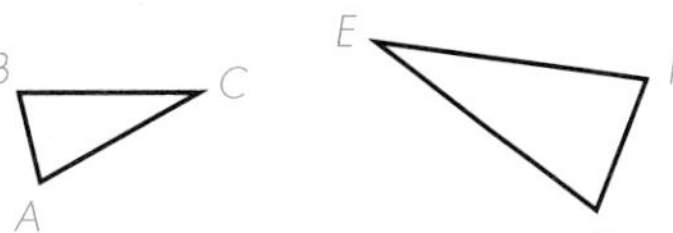

4.2 A ratio is one number divided by another number that is not zero.
$\frac{6}{7}$ $\frac{x}{3}$ $\frac{(y-5)}{13}$
A proportion is an equation consisting of two equal ratios.
$\frac{300}{25} = \frac{15}{x}$ (Multiply both sides by $25x$)
$300x = 375$
$x = 1.25$

4.3 Triangles RAP, YES, and CUT are similar.
$RAP \sim YES$ by SAS
$YES \sim CUT$ by SSS
$RAP \sim CUT$ by AA

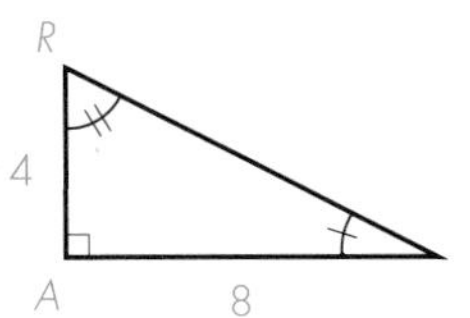

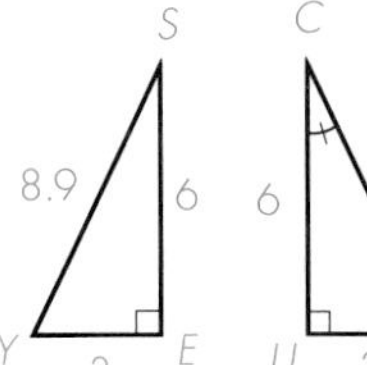

4.4 Polygons $SRTUV$ and $GHIJK$ are similar. If SR is 5 centimeters long and GH is 4 centimeters, how long is IH?
$\frac{5}{4} = \frac{2.67}{x}$ (Multiply by $4x$)
$5x = 10.68$
$x = 2.14$ cm

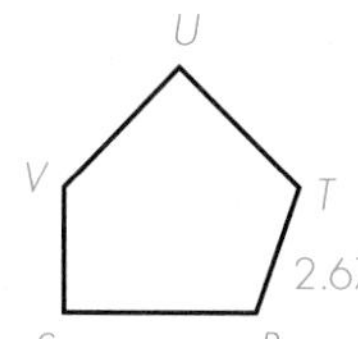

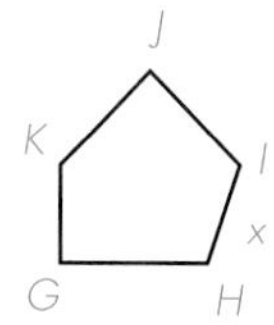

4.5 Lines YZ and PQ are parallel. What is the length of PM?
$\frac{\overline{PM}}{\overline{QM}} = \frac{\overline{YP}}{\overline{ZQ}}$ (By Theorem 4-5-2)
$\frac{PM}{7} = \frac{16}{18}$
$PM = 6\frac{2}{9}$

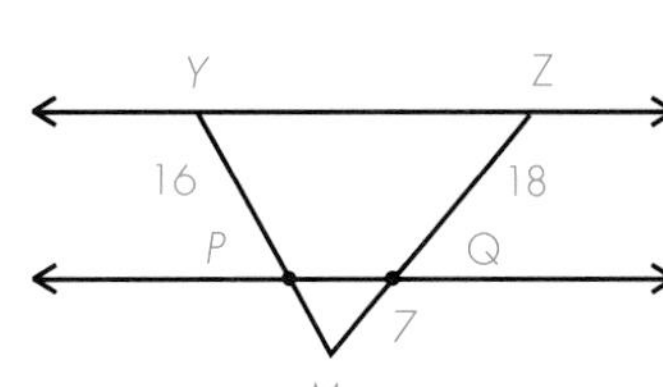

4.6 How wide is (what is the diameter of) the water tower?
$\frac{18}{45} = \frac{20}{50} = \frac{2}{5}$ and $m\angle WTA = m\angle RTE$
$\triangle WAT \sim \triangle RET$ (By SAS)
$\frac{\overline{TR}}{\overline{WT}} = \frac{\overline{ER}}{\overline{WA}}$
$\frac{20}{50} = \frac{16}{x}$
$x = 40$ ft

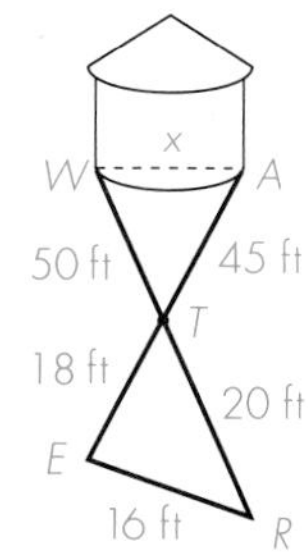

Chapter Vocabulary
conjecture
correspondence
corresponding angles
corresponding sides
corresponding vertices
enlargement factor
lowest terms
mapping
multiplier
proportion
ratio
rational number
scale
scale drawing
scale factor
similar figures
similar triangles

159

Using the Page
In this Computer Activity students learn how to combine statements to draw similar triangles and how to position them appropriately. You may wish to review the methods of drawing triangles before introducing the combined procedure.

COMPUTER ACTIVITY

Variables can be used when defining LOGO procedures. A variable in LOGO is identified by a colon preceding the variable's name. The variable :SIDE is used in the EQUITRI procedure below to represent the length of a side of the equilateral triangle.

```
TO EQUITRI
RT 30
FD :SIDE
REPEAT 2[RT 120 FD :SIDE]
END
```

To execute the procedure, you must type EQUITRI and then a number to indicate the length of a side. For example, EQUITRI 30 would assign the value 30 to :SIDE and thus tell the turtle to draw an equilateral triangle whose sides are 30 spaces long.

To draw two similar equilateral triangles, simply use the EQUITRI procedure twice.

```
TO SIMTRI1
PU FD 30 PD
EQUITRI 20
HOME
EQUITRI 10
END
```

The output is shown below.

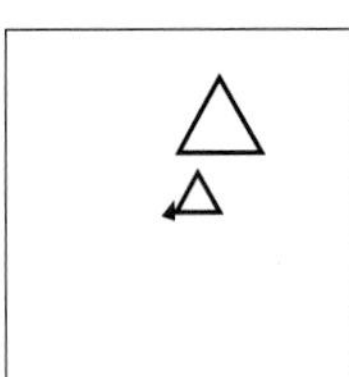

The equilateral triangle on the top has sides 20 spaces long.
The equilateral triangle on the bottom has sides 10 spaces long.

The lengths of the sides of an equilateral triangle are given in Exercises 1–3 below. Choose the length of the sides of an equilateral triangle that is similar to each given triangle. Then modify the SIMTRI1 procedure to tell the turtle to draw the triangles. Answers may vary.

1. 15 **2.** 28 **3.** 42

More than one variable can be used in a procedure. The TRI1 procedure will draw one angle and one side of an acute triangle. In the procedure, the variables :ANGLE and :SIDE are used to represent the measure of an angle and the length of a side, respectively. Notice that to find the measure of the angle the turtle will need to turn through, you will need to use the operation symbol for subtraction (−).

```
TO TRI1 :ANGLE :SIDE
RT 180 − :ANGLE
FD :SIDE
END
```

This procedure can now be used in another procedure to draw a complete triangle.

```
TO TRI3
TRI1 30 20
TRI1 65 10
TRI1 85 18
END
```

The output is shown below.

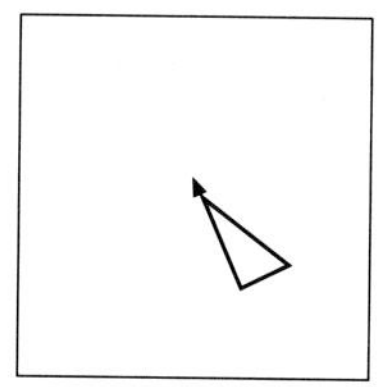

Use the TRI1 procedure in a procedure named SIMTRI2 to tell the turtle to draw the similar triangles given in Exercises 4 and 5 below.
Answers may vary.

4.

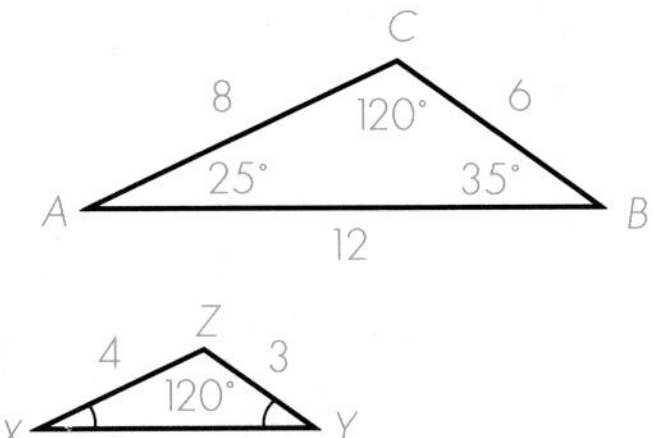

5.

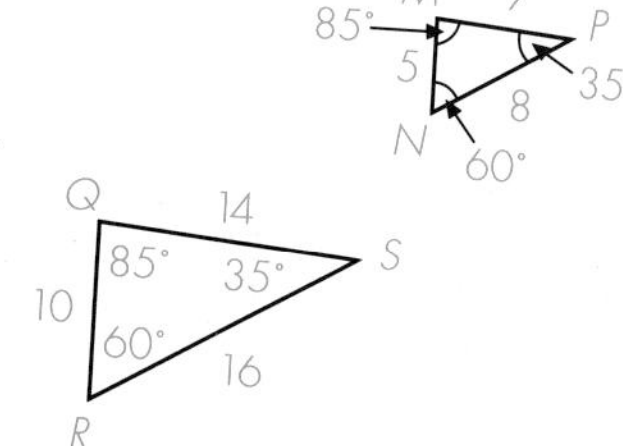

6. In $\triangle ABC$, $m\angle A = 50°$, $m\angle B = 60°$, $m\angle C = 70°$, $AB = 20$, $BC = 16$, and $AC = 18$. Find the lengths of the sides of a triangle that is similar to $\triangle ABC$. Use the SIMTRI2 procedure to tell the turtle to draw the triangles. Answers may vary.

Using the Page
Students can work individually or in groups to study the Chapter 4 Review. If you have students study in groups, they should work together to make sure all members of the group know the material in the chapter.

Informal assessment such as interviews, classroom observations, or a review of student portfolios can be used instead of paper-and-pencil tests.

Using the Page
See Exercise 6. Remind students that 7% means $\frac{7}{100}$. They can write a proportion to find the amount of tax and then add to find the total cost.

CHAPTER REVIEW

For each pair of triangles, determine why they are similar and the ratio of the sides. Map the triangles.

1.

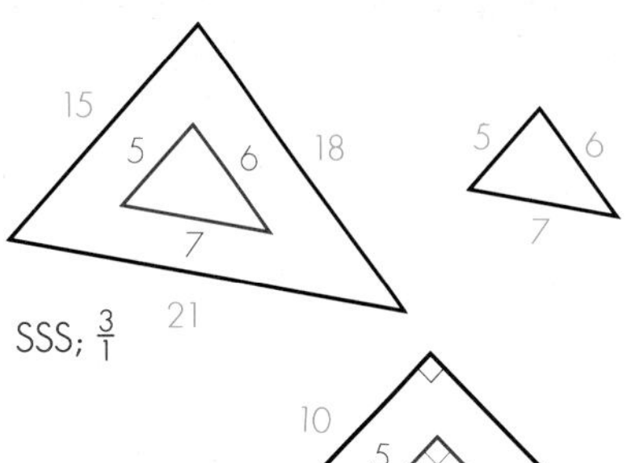

SSS; $\frac{3}{1}$

2.

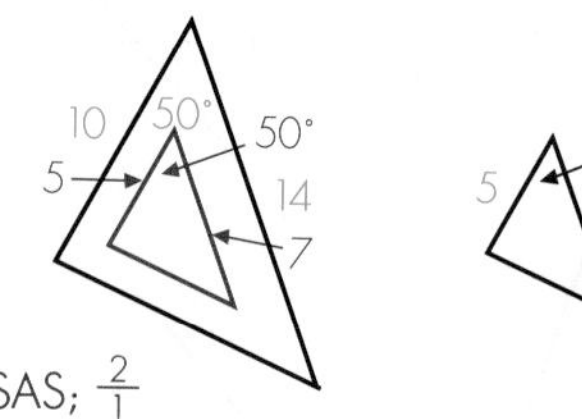

SAS; $\frac{2}{1}$

3.

10
5
10
5
10
30°
30°
20

AA; $\frac{2}{1}$

4.

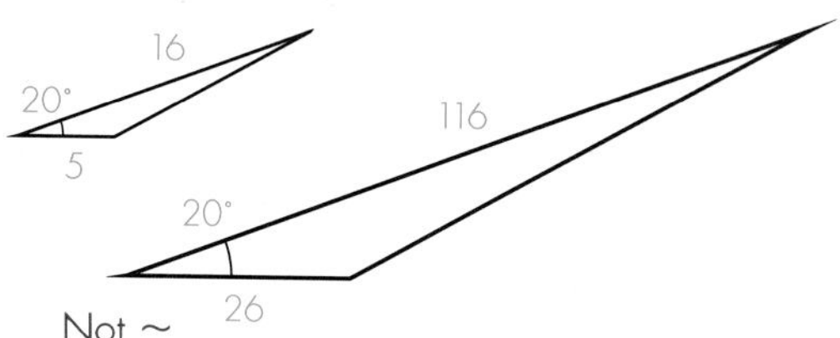

Not ~

5. In the figure to the right, $l_1 \parallel l_2$ with transversals $\overline{AD}$ and $\overline{BC}$. Are the triangles similar? If so, why? Yes; AA

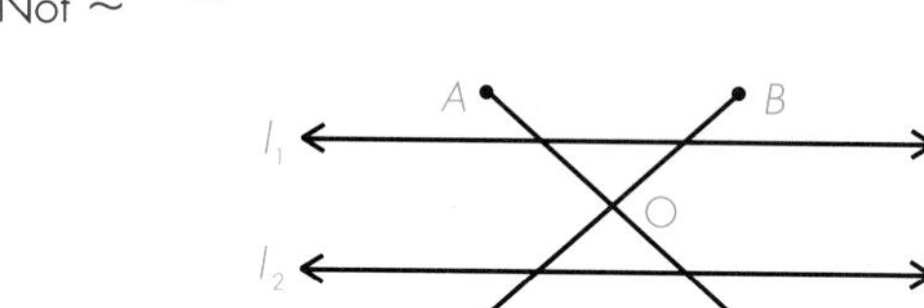

6. If 7 dictionaries cost $213.50, then what will 12 dictionaries cost at the same price per book? $366.

In $\triangle ABC$, $m\angle C = 90°$, $\overline{CD} \perp \overline{AB}$, and $m\angle A = 25°$.

7. $m\angle B =$ 65°

8. $m\angle ADC =$ 90°

9. $m\angle ACD =$ 65°

10. $m\angle BCD =$ 25°

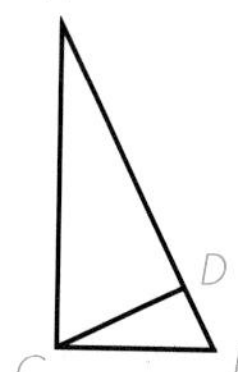

11. How many triangles are in the figure? 3

12. Are the triangles similar? If so, why?
Yes; AA

13. Refer to $\triangle QRP$. How long is ST?
15

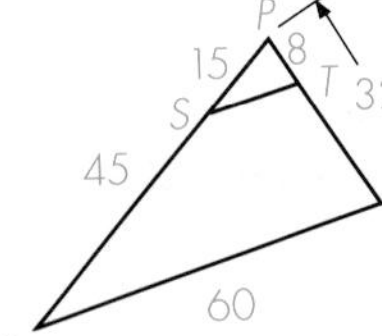

14. In the drawing at the right, the scale $\frac{1}{4}$ in. = 3 ft. What are the true measurements of the space?
9′ by 12′ by 18′

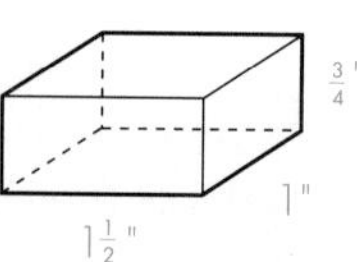

CHAPTER TEST

1. A school has a ratio of 1 teacher for every 17 students. There are 650 students in the school with an expected increase of 81 students next year. How many teachers will the school need next year? 43

2. A school has a ratio of 1 coach to every 14 players. If the school has 3 full-time and 1 half-time coach, then how many players are there? 49

3. A road map has a scale of 1 inch to 39 miles. Two cities are $5\frac{3}{4}$ inches apart on the map. How many miles are they apart? 224.25 mi

4. Solve the proportion for x. Use your calculator and round the answer to two decimal places. $\frac{(3x-4)}{12} = \frac{14}{23}$ 3.77

Which of the pairs of triangles are similar and why?

5.

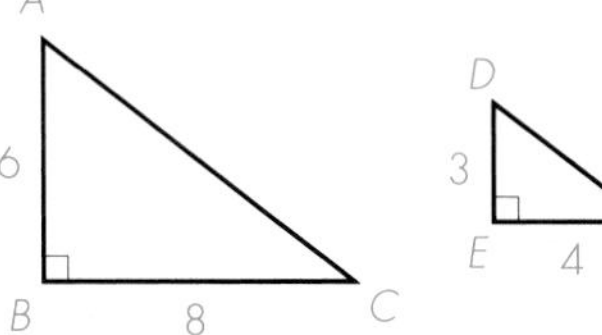

Similar; SAS

6.

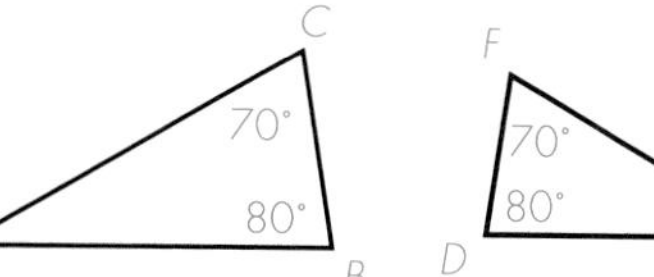

F 70° 80° D E

Similar; AA

7.

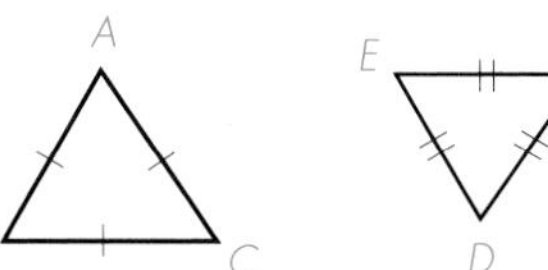

8.

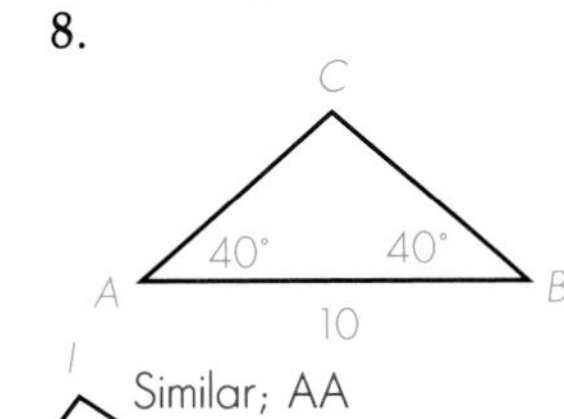

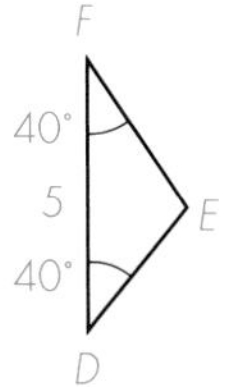

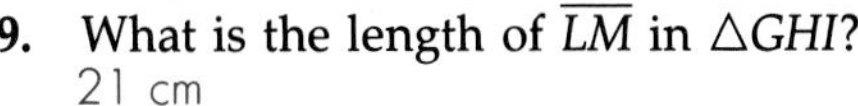

Similar; SSS

Similar; AA

9. What is the length of $\overline{LM}$ in $\triangle GHI$? 21 cm

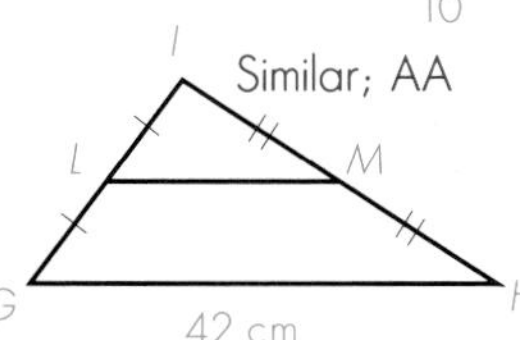

Given a right triangle ABC with C the right angle. $\overline{CD} \perp \overline{AB}$.

10. Draw the figure. See margin.
11. Map the three similar triangles. See margin.
12. Complete: $\overline{AC}/\overline{AD} = \frac{?}{?} = \frac{?}{?}$
$\overline{BC}/\overline{CD}$; $\overline{AB}/\overline{AC}$

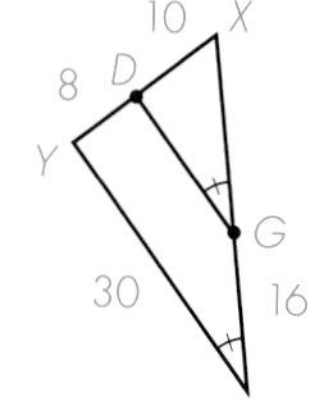

Refer to $\triangle XYZ$. $m\angle XGD = m\angle XZY$.

13. $XG =$ 20
14. $DG =$ $16\frac{2}{3}$

Assessment Resources
Achievement Tests pp. 7-8

Test Objectives
After studying this chapter, students should understand
- how to map figures.
- ratios and proportions.
- how to solve problems involving ratios and proportions.
- what similar triangles are.
- how to solve problems involving similar triangles.
- scale drawings.
- applications of scale factors.
- how to justify conclusions involving similar triangles.
- how to solve for missing lengths, using similar triangles.

Additional Answers

10.

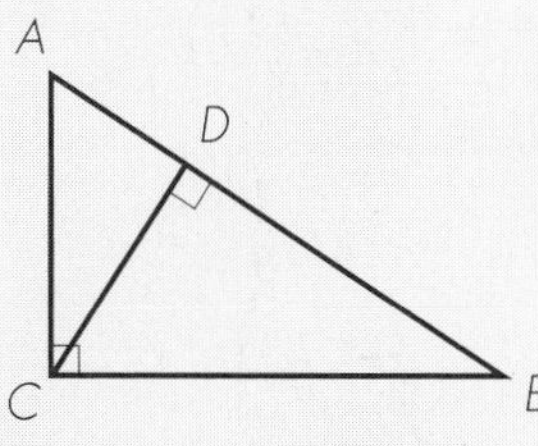

11.

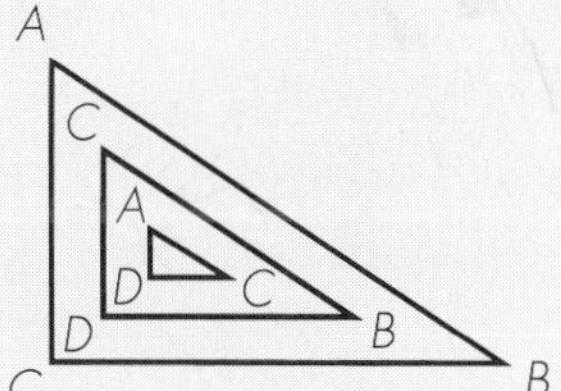

Algebra Review Objectives
Topics reviewed from algebra include
- using integers to evaluate two-variable expressions.
- solving first-degree equations in one variable.
- solving problems involving time.
- solving problems, using $d = rt$.

Example: Evaluate $x^2 + y$ using $x = 16$ and $y = -13$.
$16^2 + (-13)$
$256 - 13$
243

Example: Solve for x.

$21 - 3(2x - 7) = 15$
$21 - 6x + 21 = 15$
$-6x + 42 = 15$
$-6x = -27$
$x = 4.5$

Evaluate each expression using $x = -76$ and $y = 38$.

1. $x + y$ -38 **2.** $x - y$ -114 **3.** xy -2888

4. $\frac{x}{y}$ -2 **5.** $y - x$ 114 **6.** $|x + y|$ 38

7. $|x - y|$ 114 **8.** $\frac{y}{x}$ $-\frac{1}{2}$ **9.** $\frac{1}{x} + \frac{1}{y}$ $\frac{1}{76}$

Solve each equation for x.

10. $14x + 65 = 2x - 7$ $x = -6$ **11.** $5(2x - 6) = 3x + 12$ $x = 6$

12. $16 - 4(3x - 12) = 112$ $x = -4$ **13.** $2.5x + 3 = 3(1.5x - 5)$ $x = 9$

14. $1.6x + 6 = 2(1.8x - 9)$ $x = 12$ **15.** $2.3x + 4 = 3(1.6x - 5)$ $x = 7.6$

A student rode on her bike for 3 hours and traveled 135 miles. After a half-hour rest, she rode another 2 hours and traveled 75 miles more to her campsite.

16. If she left at 8 a.m., when did she arrive at her campsite? 1:30 p.m.

17. What was her average speed during the first three hours? 45 mph

18. What was her average speed during the last two hours? 37.5 mph

19. What was the average speed for the entire trip (including the rest time)? 38.2 mph

20. What was the average speed counting only the time she was traveling? 42 mph

Name ______________________ Date ______________

Mappings and Correspondences

When two triangles have the same shape, they can be *mapped*. To map figures means to match part of one figure with part of a second figure.

X is the largest angle of $\triangle XYZ$ and A is the largest angle of $\triangle ABC$. You can map $\angle X$ onto $\angle A$. This is called *writing a correspondence*. A double-pointed arrow is used to show a mapping or correspondence.

$$\angle X \leftrightarrow \angle A$$

Angle X corresponds to angle A.

Use the figures above to complete each statement.

1. The longest side of triangle XYZ is mapped on to the longest side of triangle ABC.

   Side $\overline{YZ}$ corresponds to side $\overline{BC}$.

2. The smallest angle of triangle XYZ is mapped on to the smallest angle of triangle ABC.

   Angle Y corresponds to angle B.

3. The vertex opposite the shortest side of triangle XYZ is mapped on to the vertex opposite the shortest side of triangle ABC.

   Y corresponds to B.

4. The middle-sized angle of triangle XYZ is mapped on to the middle-sized angle of triangle ABC.

   Angle Z corresponds to angle C.

5. Map the corresponding angles of the triangles at the right.

   $\angle N \leftrightarrow \angle E$ $\angle M \leftrightarrow \angle D$ $\angle O \leftrightarrow \angle F$

6. Map the corresponding sides of the triangles at the right.

   $\overline{NM} \leftrightarrow \overline{ED}$ $\overline{MO} \leftrightarrow \overline{DF}$ $\overline{ON} \leftrightarrow \overline{FE}$

Can each pair of triangles be mapped? Write *yes* or *no*.

7. yes

8. no

9. yes

Name ______________________ Date ______________

Pentagon Designs

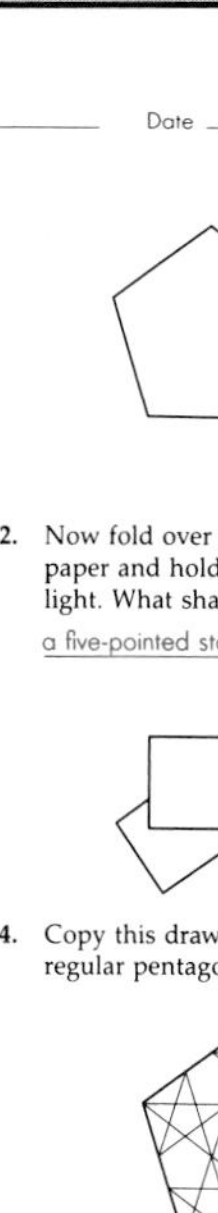

Many interesting and beautiful designs are based on the *regular pentagon*. This figure has five sides of equal length. Each of the five angles measures 72°.

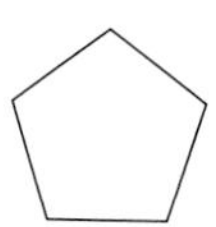

1. One way to make a pentagon is to tie a knot in a strip of paper and then press the strip flat. Use adding machine tape or other paper to make the pentagon.

2. Now fold over one end of the strip of paper and hold the knot up to a strong light. What shape do you see?

   a five-pointed star

3. Use a protractor and ruler to draw a regular pentagon. Then connect each vertex to each of the other four vertices. The drawing has been started in the figure below. What shape do you get?

   a five-pointed star

4. Copy this drawing. Start by drawing a regular pentagon with a star inside it.

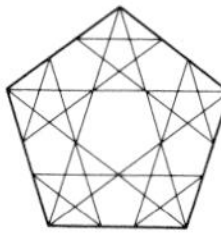

Now make the designs shown below. Students may also wish to create their own designs.

5.

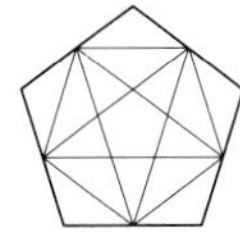

6.

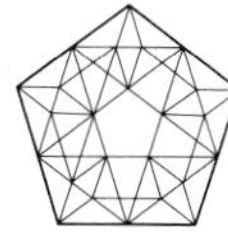

7.

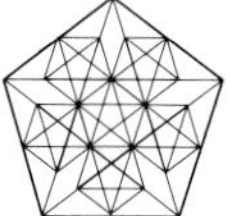

Name ______________________ Date ______________

Ratios and Proportions

A *ratio* is a comparison between two numbers. For example, the ratio of the tops of the two triangles can be written three ways.

10 to 5 $\frac{10}{5}$ 10:5

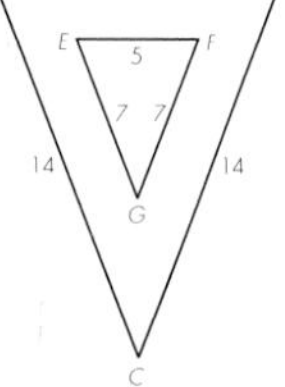

Because a ratio indicates a division, the bottom or second number can never equal zero.

Two equal ratios are called a *proportion*. If you know three numbers in a proportion, you can cross-multiply to solve for the fourth number.

$\frac{10}{5} = \frac{14}{EG}$ $10 \times EG = 5 \times 14$ $EG = \frac{5 \times 14}{10}$

Use the figure above to write or complete each ratio.

1. *BC* to *FG* equals 14 to 7
2. *BC* to *AC* equals 14: 14
3. The two long sides of triangle *EFG* FG to GE, or 7:7
4. The top of the small triangle to the top of the large triangle EF:AB, or 5:10
5. $EF:AB =$ 5:10
6. *EG* to $AC =$ 7:14
7. $\frac{EF}{FG} = \frac{5}{7}$

Complete each statement to show the cross multiplication.

8. If $\frac{2}{3} = \frac{6}{9}$, then $3 \cdot 6 = 2 \cdot$ 9
9. If $\frac{4}{8} = \frac{6}{12}$, then $8 \cdot 6 = 12 \cdot$ 4
10. If $\frac{a}{b} = \frac{3}{7}$, then $7a =$ 3b
11. If $\frac{x}{5} = \frac{12}{9}$, then $9x =$ 60

Solve each proportion for *x*.

12. $\frac{x}{5} = \frac{12}{9}$ $x = \frac{60}{9}$ or $\frac{20}{3}$
13. $\frac{x}{4} = \frac{3}{2}$ $x =$ 6
14. $\frac{x}{2} = \frac{18}{10}$ $x = \frac{18}{5}$
15. $\frac{x}{3} = \frac{10}{6}$ $x =$ 5

Name ______________________ Date ______________

The Geometric and Arithmetic Means

In some proportions, the same number is in two diagonally opposite positions. The number is called the *geometric mean*. In the proportion, *b* is the geometric mean of the two numbers *a* and *c*.

$$\frac{a}{b} = \frac{b}{c}$$

In mathematics, there are several different types of means, or averages. Another one, the *arithmetic mean*, is shown at the right. For two numbers *a* and *c*, the arithmetic mean, *d*, is one half the sum of *a* and *c*.

$$d = \frac{1}{2}(a + c)$$

Solve each problem.

1. 8 is the geometric mean of 2 and what other number? 32
2. Use the proportion for the geometric mean to solve for *b* in terms of *a* and *c*. $b = \sqrt{ac}$

Find the geometric mean and the arithmetic mean for each pair of numbers.

3. 4 and 9 — $6, 6\frac{1}{2}$
4. 3 and 12 — $6, 7\frac{1}{2}$
5. 5 and 45 — 15, 25
6. 6 and 8 — $4\sqrt{3}, 7$
7. 20 and 5 — $10, 12\frac{1}{2}$
8. 9 and 12 — $6\sqrt{3}, 10\frac{1}{2}$
9. 4 and 16 — 8, 10
10. 15 and 16 — $4\sqrt{15}, 15\frac{1}{2}$
11. What conjecture can you make about the geometric mean and the arithmetic mean of two numbers? The geometric mean is always less than the arithmetic mean.
12. In any right triangle, the altitude drawn to the hypotenuse is the geometric mean of the two segments on the hypotenuse.

    In right triangle *ABC* at the right, *h* is the geometric mean of *x* and *y*.

    Copy the figure at the right, using 10 centimeters for *x* and 6 centimeters for *y*. Use the geometric mean to find *h*.
    $2\sqrt{15}$

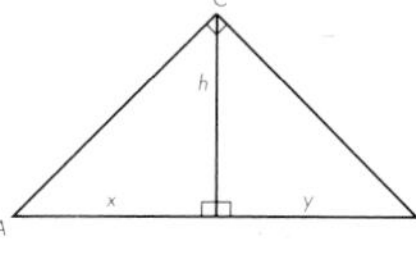

Name ______________________ Date ______________

Similar Triangles

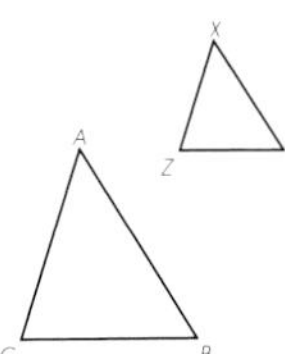

Similar triangles have the same shape. But, they do not have to be the same size.

The corresponding angles of similar triangles are congruent. Their angles have equal measures.

Corresponding sides of similar triangles are in proportion. This means ratios of pairs of corresponding sides will be equal.

Use the triangles above. For each problem, circle the letter of the correct answer.

1. Angle A
   - a. is congruent to angle Y.
   - (b.) is congruent to angle X.
   - c. corresponds to angle Z.

2. The ratio of AB to XY
   - a. is less than 1.
   - b. equals the ratio of BC to XY.
   - (c.) equals the ratio of BC to YZ.

3. The ratio of YZ to BC
   - (a.) equals the ratio of XZ to AC.
   - b. equals the ratio of XZ to AB.
   - c. equals the ratio of XZ to BC.

For each pair of triangles, solve the proportion to find the missing length.

4. $\frac{x}{5} = \frac{18}{15}$ $x =$ 6

5. $\frac{x}{4} = \frac{11}{2}$ $x =$ 22

6. $\frac{x}{50} = \frac{50}{100}$ $x =$ 25

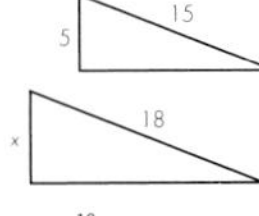

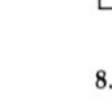

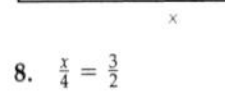

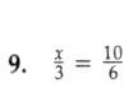

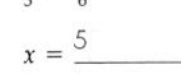

7. $\frac{x}{2} = \frac{18}{10}$ $x = \frac{18}{5}$

8. $\frac{x}{4} = \frac{3}{2}$ $x =$ 6

9. $\frac{x}{3} = \frac{10}{6}$ $x =$ 5

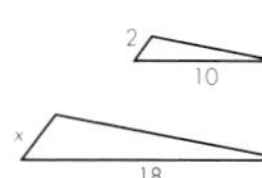

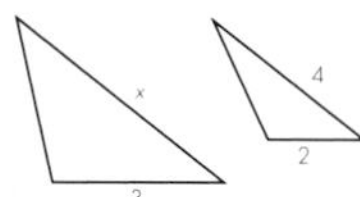

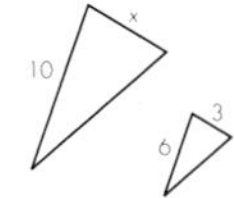

Name ______________________ Date ______________

Similar Figures and Rep-Tiles

A *rep-tile* is a figure that can be repeated to form a larger copy of itself. The large figure is similar to the small one.

In the figure at the right, the large triangle is similar to each of the four small triangles.

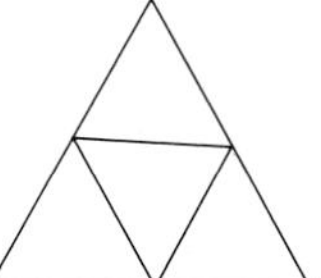

Trace four copies of each figure on a separate sheet of paper. Arrange the copies to form a large figure similar to the small one. Record your answers on these shapes.

1. 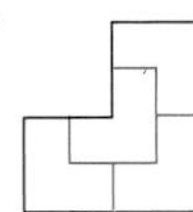

2. 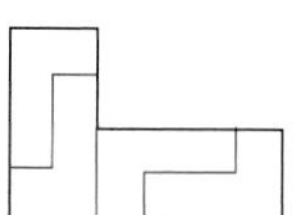

3. 

4. 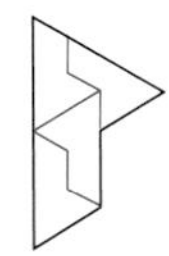

5. 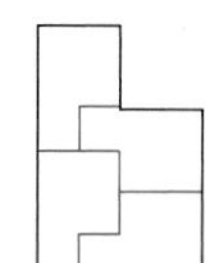

6. 

7. The figure at the right has been "rep-tiled" twice. Make a very large copy of this figure, or any other figure on this page. "Rep-tile" your figure four times.

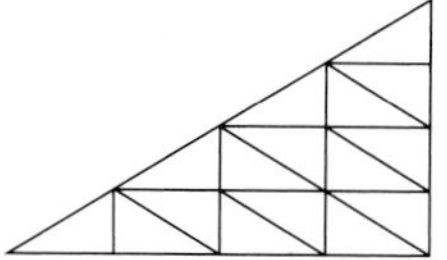

Name ______________________ Date ______________

Similar Figures and Scale Drawings

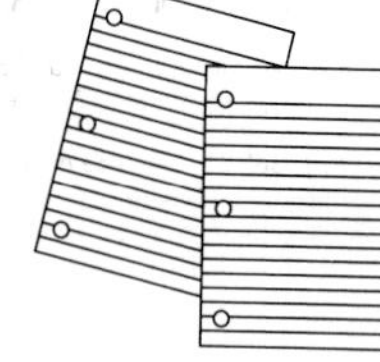

The ratios in similar figures are used to make maps and scale drawings. If a scale drawing is to be accurate, all ratios between actual lengths and corresponding lengths in the drawing must be equal.

In the drawing at the right, the width of each sheet of paper is 1 inch. The actual width of the paper is 8.5 inches. What should the length of the drawing be if the actual length of the paper is 11 inches?

$\frac{1 \text{ in. drawn}}{8.5 \text{ in. actual}} = \frac{x}{11 \text{ in. actual}}$ The scale is 1 in.:8.5 in.

$8.5x = 1 \cdot 11$

The length in the drawing should be 11 ÷ 8.5, or about 1.3 inches.

Solve the problems. The actual dimensions of the notebook paper are 8.5 inches by 11 inches.

1. A scale drawing of a sheet of notebook paper is 1 centimeter wide. What is the scale?
   1 cm:8.5 in.

2. A scale drawing of a sheet of notebook paper is 1 centimeter wide. What is the length of the drawing?
   about 1.3 cm

3. A scale drawing of a sheet of notebook paper is 10 centimeters tall. What is the scale?
   10 cm:11 in.

4. A scale drawing of a sheet of notebook paper is 10 centimeters tall. What is the width of the drawing?
   about 7.7 cm

5. In the space at the right, sketch a scale drawing of a sheet of notebook paper. Make the drawing 2 centimeters wide. What should the height of your drawing be?
   about 2.6 cm

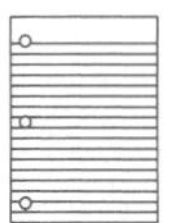

6. A scale drawing of a sheet of notebook paper has a scale of 1:10. What are the dimensions of the drawing?
   0.85 in. by 1.1 in.

7. An enlargement of a sheet of notebook paper has a scale of 5:1. What are the dimensions of the drawing?
   42.5 in. by 55 in.

Name ______________________ Date ______________

Using Scale Factors to Enlarge Drawings

Ratio and proportion can be used to enlarge or reduce a drawing, such as the one at the right, to any size.

1. A square grid has been drawn on top of the umbrellas. In unit squares, what are the dimensions of the grid?
   10 units by 8 units

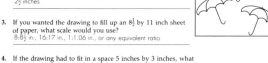

2. If you enlarged the drawing using a scale of 1 square:$\frac{1}{2}$ inch, about how wide would the top umbrella be?
   $2\frac{1}{2}$ inches

3. If you wanted the drawing to fill up an $8\frac{1}{2}$ by 11 inch sheet of paper, what scale would you use?
   8:$8\frac{1}{2}$ in., 16:17 in., 1:1.06 in., or any equivalent ratio

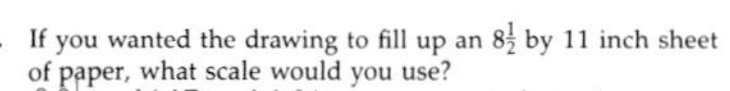

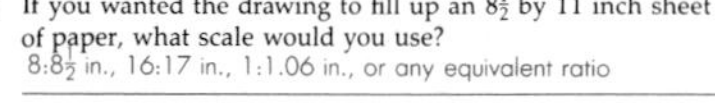

4. If the drawing had to fit in a space 5 inches by 3 inches, what scale would you use?
   8:3 in., or any equivalent

5. Use graph paper to make an enlargement of the drawing of the umbrellas. What scale did you choose?
   Answers will vary.

6. Make a reduction of the drawing of the umbrellas. What scale did you choose?
   Answers will vary.

7. When you enlarge or reduce a drawing, you can use a grid made of rectangles instead of squares. Draw a rectangular grid over the first drawing of the umbrellas. Then enlarge the drawing to four times its size. Give the dimensions of the rectangles in the small grid.
   Answers will vary.

8. Enlarge the drawing at the right to create a pattern in which each line segment is 4 inches long. The pattern can be folded to create an eight-sided solid shape called an *octahedron*. What scale did you use?
   1 in.:4.6 in., or any scale close to this

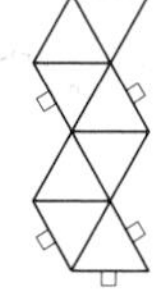

Name ______________________ Date ______________

Similar Triangles and Conclusions

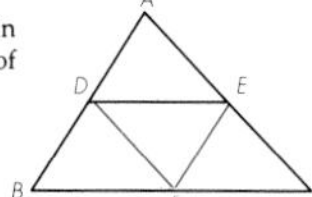

The *midpoint* of a line segment separates it into two equal pieces. In triangle *ABC*, *D* is the midpoint of side *AB*, and *E* is the midpoint of side *AC*.

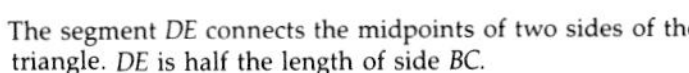

The segment *DE* connects the midpoints of two sides of the triangle. *DE* is half the length of side *BC*.

This relationship is true for any triangle. Also, *DE* is parallel to side *BC*.

Use triangle *ABC* for these problems.

1. Mark the midpoint of side *BC* with the letter *F*. Connect *D* and *F*. Side *DF* is half the length of what line segment? $\overline{AC}$

2. Connect points *E* and *F*. What line segment is twice the length of segment *EF*? $\overline{AB}$

Use the figure at the right for these problems.

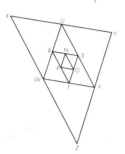

3. The midpoints of the sides of triangle *XYZ* have been connected to form triangle *UVW*. Which side of triangle *UVW* is half the length of side *XY*? $\overline{WV}$

4. Which side of triangle *XYZ* is twice the length of side *WU*? $\overline{ZY}$

5. Which side of *UVW* is half as long as side *XZ*? $\overline{UV}$

6. The midpoints of triangle *UVW* have been connected to form triangle *RST*. Complete these equations.

   a. *UV* = 2(*RT*)   b. *UW* = 2(*ST*)   c. *VW* = 2(*RS*)

7. The midpoints of triangle *RST* have been connected to form triangle *NOP*. If *XY* measures 10 inches, find the lengths of line segments *WV*, *RS*, and *PO*.
   5 in., 2.5 in., 1.25 in.

Name ______________________ Date ______________

Similar Triangles in 3-Dimensional Space

The theorems you have studied about similar triangles in 2-dimensional space are also true in 3-dimensional space.

In each proof below, provide or complete the reason for each step.

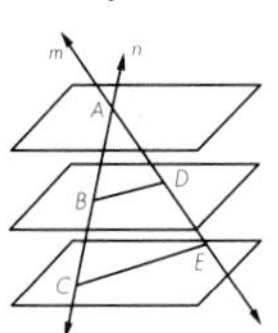

1. In the figure above, *BD* is parallel to *CE*. Prove that △*ADB* is similar to △*AEC*.

   a. By Theorem 4-5-2, *BD* ∥ *CE* implies that *AB*:*AC* = AD:AE.

   b. Because ∠*CAE* ≅ ∠*BAD* and *AB*:*AC* = *AD*:*AE*, △*ADB* is similar to △*AEC*. The definition which proves this is abbreviated SAS (or 4-3-1).

2. In the figure above, three pairs of parallel line segments are *PS* and *AD*, *SR* and *DC*, and *QR* and *BC*. Prove that triangles *EQP* and *EBA* are similar.

   a. By Theorem 4-5-2, *PS* ∥ *AD* implies that *EP*:*EA* = *ES*: ED.

   b. *SR* ∥ *DC* implies that *ES*:*ED* = *ER*:*EC*, and *QR* ∥ *BC* implies that EQ:EB = ER:EC

   c. Using substitution, *EP*:*EA* = *EQ*: EB.

   d. Because ∠*PEQ* ≅ ∠*AEB* and *EP*:*EA* = *EQ*:*EB*, the triangles *EQP* and *EBA* are similar. The definition which proves this is abbreviated SAS (or 4-3-1).

3. In the figure at the right, *NT*:*RT* = *OT*:*PT*. Prove that triangles *OTN* and *PTR* are similar.

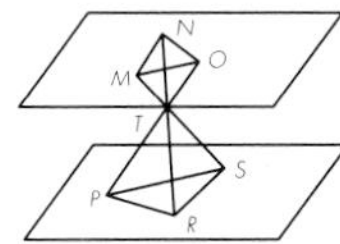

   a. ∠*OTN* ≅ ∠*PTR* because vertical angles are equal.

   b. By SAS, △*OTN* and △*PTR* are similar because ∠*OTN* ≅ ∠*PTR* and NT:RT = OT:PT

Name ______________________ Date ____________

Solving Problems Involving Similar Triangles

You can use shadows and your own height to estimate the heights of trees and buildings. You also need to pace off distances. To pace a distance, practice taking steps that are about one yard long.

In the figure at the right, the rays of the sun form two similar triangles. Using the triangles, you can write this proportion:

$$\frac{\text{your height}}{\text{your shadow}} = \frac{\text{tree's height}}{\text{tree's shadow}}$$

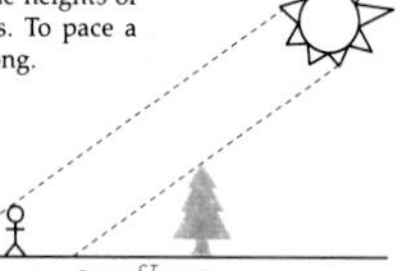

Use the two triangles to answer the following problems. Write the line segment that corresponds to each quantity.

1. Your height $\overline{ZY}$
2. The height of the tree $\overline{HT}$
3. The length of your shadow $\overline{XY}$
4. The length of the tree's shadow $\overline{ST}$

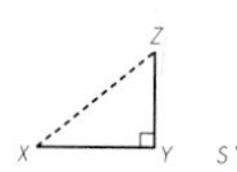

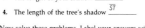

Now solve these problems. Label your answers with a unit of measurement.

5. Let's say you are 5 feet tall and your shadow is 9 feet long. The shadow of the tree is 30 feet long. Write the ratio of your height to the length of your shadow. 5:9
6. Write the ratio of the tree's height to the length of the tree's shadow. Use *HT* to stand for the tree's height. *HT*:30
7. Write a proportion with the two ratios in Exercises 5 and 6. 5:9 = *HT*:30
8. Find *HT*, the height of the tree. about 16.7 ft
9. Let's say you pace off the shadow of the tree and find that it is 10 paces long. Your shadow is 3 paces. How tall is the tree in feet? about 16.7 ft
10. How tall is the tree in yards? about 5.6 yd
11. Another tree casts a shadow that is 2 yards long. How tall is that tree in yards? about 1.1 yd
12. How tall is the second tree in feet? about 3.3 ft

Name ______________________ Date ____________

Similar Triangles and Work Problems

Two right triangles drawn as shown in the figure can be used to solve certain types of algebra problems called *work problems.* Using the figure, you can show that the following equation is true.

$$\frac{1}{x} + \frac{1}{y} = \frac{1}{z}$$

Complete these statements to prove the equation above.

1. $\triangle PAC$ is similar to $\triangle QBC$ and $\triangle ABQ$ is similar to $\triangle ACR$. Complete these proportions $\frac{z}{x} = \frac{n}{m+n}$ $\frac{z}{y} = \frac{m}{m+n}$
2. Find the sum. $\frac{n}{m+n} + \frac{m}{m+n} =$ 1
3. By substitution, $\frac{z}{x} + \frac{z}{y} =$ 1
4. Divide all three terms of the last equation by z. What new equation do you get? $\frac{1}{x} + \frac{1}{y} = \frac{1}{z}$

The equation you proved can be used to solve a work problem. If one person can complete a job in x hours and another person can complete the same job in y hours, then it will take z hours for them to complete the job if they work together.

Solve the following problems by drawing figures like the one at the top of the page. Draw line segments x and y of the given lengths. Then measure z. Segment AC can be any length.

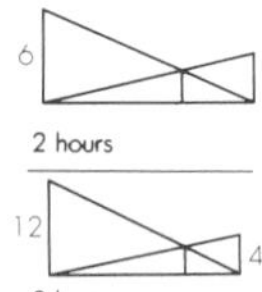

5. One person can complete a job in 6 hours and another person can complete the job in 3 hours. If they work together, how long will it take them to complete the job? 2 hours
6. Theresa can check an inventory report in 4 hours. Tom can complete the same job in 12 hours. If they work together, how long will it take them to check a report? 3 hours

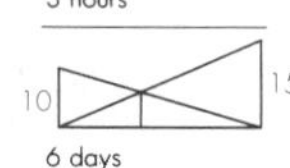

7. Arnie can paint a building in 15 days. Art takes 10 days to do the same job. If they work together, how many days will it take them to paint the building? 6 days

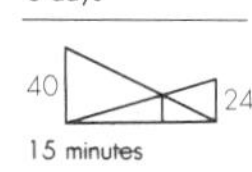

8. Sam can plant a row of tomatoes in 40 minutes. Sarah can do the same job in 24 minutes. If they work together, how long will it take them to plant each row? 15 minutes

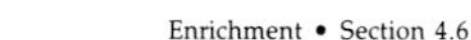

Name ______________________ Date ______________

Logo and Similar Triangles

1. Write a Logo procedure to tell the turtle to draw $\triangle ABC$ so that A is at the home position, $\overline{AB}$ is a vertical segment that is 20 units long, $m\angle A = 60$ and $m\angle B = 50$. Make a sketch of the output.

   Answers for procedure may vary.

2. Write a Logo procedure to tell the turtle to draw $\triangle XYZ$ so that X is at the home position, $\overline{XY}$ is a vertical segment that is 40 units long, $m\angle X = 60$ and $m\angle Y = 50$. Make a sketch of the output.

   Answers for procedure may vary.

3. Is $\triangle ABC \sim \triangle XYZ$? Justify your answer. yes; AA

4. Write a Logo procedure to tell the turtle to draw $\triangle DEF$ so that $\overline{DE}$ is 30 units long, $\overline{EF}$ is 25 units long, and $m\angle E = 45$. Make a sketch of the output.

   Answers may vary.

5. Write a Logo procedure to tell the turtle to draw $\triangle MNP$ so that $\overline{MN}$ is 90 units long, $\overline{NP}$ is 75 units long, and $m\angle N = 45$. Make a sketch of the output.

   Answers may vary.

6. Is $\triangle DEF \sim \triangle MNP$? Justify your answer. yes; SAS

Name________________ Class________________ Date________________

Achievement Test 4 (Chapter 4)

SIMILAR TRIANGLES

No. Correct
No. Exercises: **23**
Score
4.35 x No. Correct =

GEOMETRY FOR DECISION MAKING
James E. Elander
SOUTH-WESTERN PUBLISHING CO.

1. In a certain parking lot, the ratio of red cars to blue cars is 3:2. If there are 81 red cars, how many blue cars are there in the lot? 54
2. In a neighborhood, the ratio of brown houses to yellow houses is 3:4. If there are 24 yellow houses, how many brown houses are there in the neighborhood? 18
3. The ratio of students in the band to students in the chorus is 3:5. If there are 84 students in the band, how many are in the chorus? 140
4. The ratio of students taking mathematics to students taking science is 6:5. If 300 students are taking mathematics, how many students are taking science? 250
5. A road map has a scale of 1 inch to 35 miles. Two towns are three inches apart on the map. How many miles apart are the towns? 105 mi
6. A road map has a scale of 3 centimeters to 5 kilometers. Two towns are 35 kilometers apart. How far apart are they on the map? 21 cm
7. Solve the proportion $\frac{3x}{5} = \frac{18}{50}$ for x. 0.6
8. Solve the proportion $\frac{5x-7}{8} = \frac{23}{9}$ for x. Use a calculator and round the answer to two decimal places. 5.49

Tell whether the two triangles are similar or not. If you say they are similar, tell why.

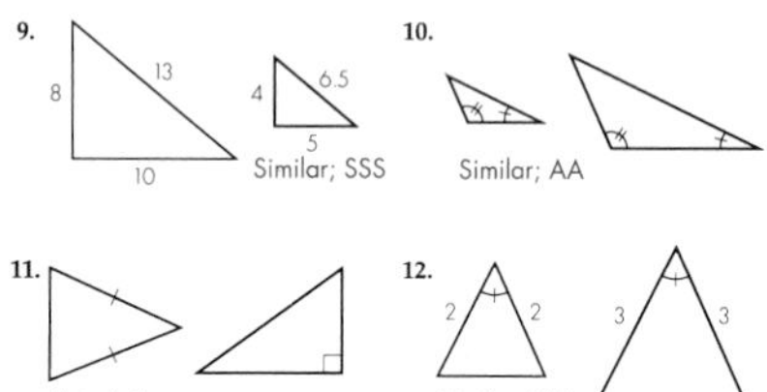

MG01AG

Tell whether or not the two pairs of triangles are similar. If you say they are similar, tell why.

13. Similar; AA

14.

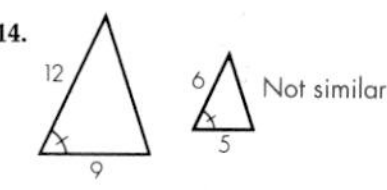

15.

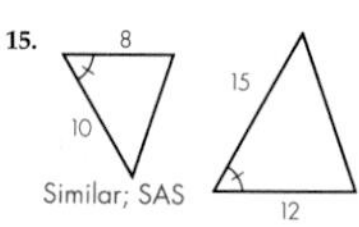

16.

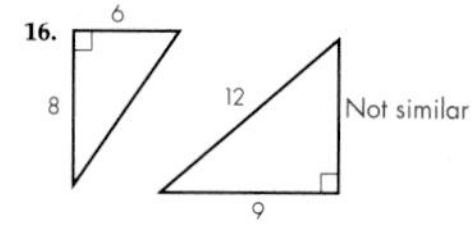

17. What is the length of $\overline{HD}$ in ΔTVM? 11.5 mm

18. Find $ST + TU + US$. 49

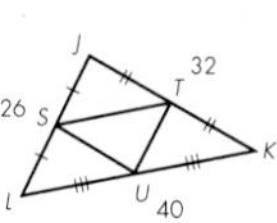

19. Map the triangles. $\Delta ABC \sim \Delta OMN$

20. Map the triangles. $\Delta JKL \sim \Delta XYZ$

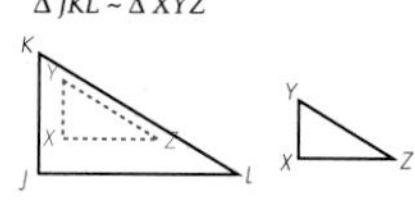

21. For the triangles in Exercise 19, write two ratios equal to AB/OM. BC/MN, AC/ON.

Refer to ΔGHI. Suppose $m\angle GMN = m\angle GHI$.

22. Find GM. $6\frac{2}{3}$

23. Find NM. $10\frac{2}{3}$

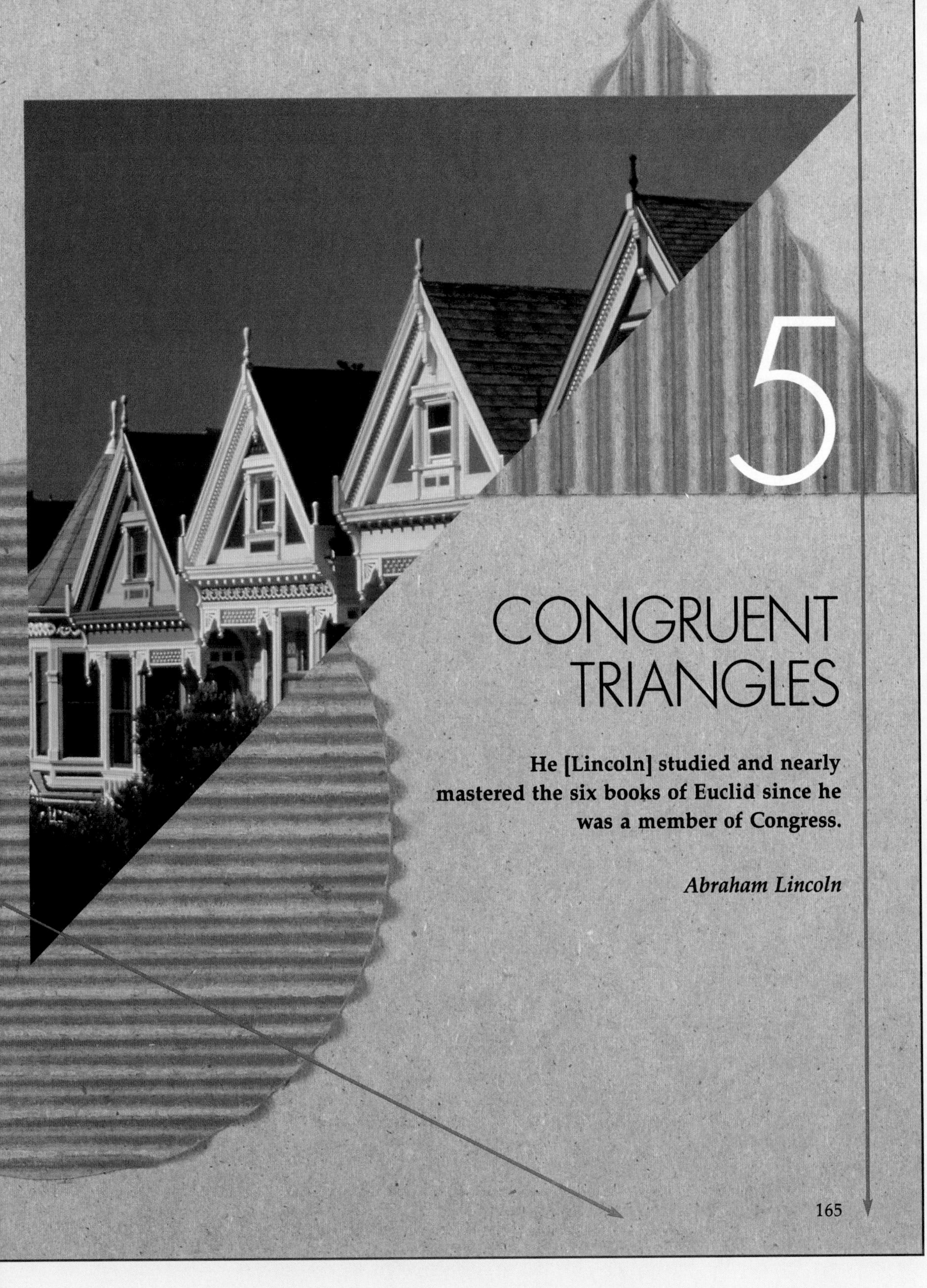

5

CONGRUENT TRIANGLES

He [Lincoln] studied and nearly mastered the six books of Euclid since he was a member of Congress.

Abraham Lincoln

165

5

CHAPTER 5 OVERVIEW

Congruent Triangles

Suggested Time: 10–12 days

The quote at the beginning of the chapter speaks of characteristics of both Lincoln and Euclid. Lincoln, having had little formal education, continued to try to improve his mind by educating himself throughout his life. Euclid's books were well-designed for such a man as this, since each new theorem was built logically on the ones that preceded it in the books. This chapter continues to build on previous knowledge. A discussion of sets and subsets leads to congruent triangles forming a subset of similar triangles. The second theorem in the chapter tells us that if a triangle has two equal sides and the included angle is bisected, then two congruent triangles are formed. The next theorem tells us that if a triangle has two equal angles, then the sides opposite these angles are equal. After looking at several proofs, students are introduced to converses of statements.

He [Lincoln] studied and nearly mastered the six books of Euclid since he was a member of Congress.

Abraham Lincoln

CHAPTER 5 OBJECTIVE MATRIX

	End of Chapter Items by Activity				Student Supplement Items by Activity		
Objectives by Number	Review	Test	Computer	Algebra Skills*	Reteaching	Enrichment	Computer
5.1.1	✓	✓			✓		
5.1.2	✓	✓			✓		
5.1.3	✓	✓			✓		
5.1.4	✓	✓		✓			
5.2.1	✓	✓			✓		✓
5.2.2	✓	✓	✓		✓		✓
5.3.1	✓	✓	✓		✓		
5.3.2	✓			✓	✓		
5.4.1	✓	✓			✓	✓	
5.4.2					✓	✓	
5.5.1				✓	✓	✓	
5.5.2		✓					

*A ✓ beside a Chapter Objective under Algebra Skills indicates that the algebra skills taught within the chapter or in previous Algebra Skills lessons are used.

CHAPTER 5 PERSPECTIVES

▲ Section 5.1

Sets and Subsets

A *set*, a *subset*, and the *universal set* are defined, and examples of each are given. Venn diagrams are used to show how sets relate to each other. After the set operations union and intersection are introduced, Venn diagrams are used to show how to solve problems about union and intersection of sets.

The Home Activity includes practice in determining members of sets, subsets of given sets, and in using Venn diagrams to solve problems. Continue to encourage students to keep notebooks for recording definitions, postulates, theorems, and examples.

▲ Section 5.2

Congruent Triangles

Through a Discovery Activity, students find that similar triangles having the same size and shape are congruent. The definition of congruent triangles uses the idea of congruent triangles being a subset of similar triangles. Students are introduced to the concept of justifying a theorem for *a general case,* rather than for a specific example.

The Home Activity includes exercises where students must decide if triangles are congruent. Exercises 5 and 6 do not include enough information for students to be able to make a decision.

▲ Section 5.3

Corresponding Parts

The concept of corresponding parts is first approached through the idea of maps having the same scale and the use of identical parts manufactured in industry. The three conditions for similar triangles are reviewed, along with the fact that for congruent triangles, the ratio of corresponding sides is one. The idea of both corresponding angles and corresponding sides being congruent is extended to other congruent figures. A theorem in this section tells us that if two angles of a triangle are equal, then the sides opposite these angles are equal.

The Home Activity emphasizes that congruent triangles must be similar, with the ratio of corresponding sides equal to one. Students then match various corresponding parts. Three justifications are included that use *corresponding parts of congruent figures* as a step in the justification.

▲ Section 5.4

The Converse of a Theorem

An if-then statement is given, with the hypothesis and the conclusion emphasized. The converse of the statement is then defined. In a Discovery Activity, students look at four statements and their converses to find that even when the truth value of a statement is known, the truth value of the converse must be examined separately. The converse of a theorem from Section 5.3 is justified, leading to a new theorem: If a triangle is isosceles, then the angles opposite the equal sides are equal.

The Home Activity provides a variety of statements for which students write converses and decide whether the converse is true or false. Some practice is given that requires knowledge of the new theorem.

▲ Section 5.5

Applications

The material in this section is two-fold. The first part of the lesson deals with how the advertising industry might present a true statement with the hope that consumers will also believe that the converse of the statement is true. The second part of the lesson deals with applications of congruent figures that students might find in their everyday lives.

The Home Activity includes exercises in which students must look at an advertisement and determine the truth of the statement and its converse. Other exercises require students to justify statements about congruent or similar figures.

SECTION 5.1

Resources
Reteaching 5.1
Enrichment 5.1
Transparency Master 5-1

Objectives
After completing this section, the student should understand
- ▲ what a set is.
- ▲ what Venn diagrams are.
- ▲ what set operations are.
- ▲ how to solve problems using sets.

Materials
compass

Vocabulary
set
well-defined
member
element
braces
subset
universal set
universe
Venn diagram
union
intersection
empty set

GETTING STARTED

Quiz: Prerequisite Skills
1. What are the even counting numbers less than 12? [2, 4, 6, 8, 10]
2. Is your teacher a student in the room? [No]
3. Is brown one of the colors of lights on a stoplight? [No]

Warm-Up Activity
Describe various items in the classroom that are all alike in some way. These items could be book titles, articles of clothing, or objects that have the same shape (such as all rectangular objects). Name one more item, and ask students if that item belongs to the collection already named.

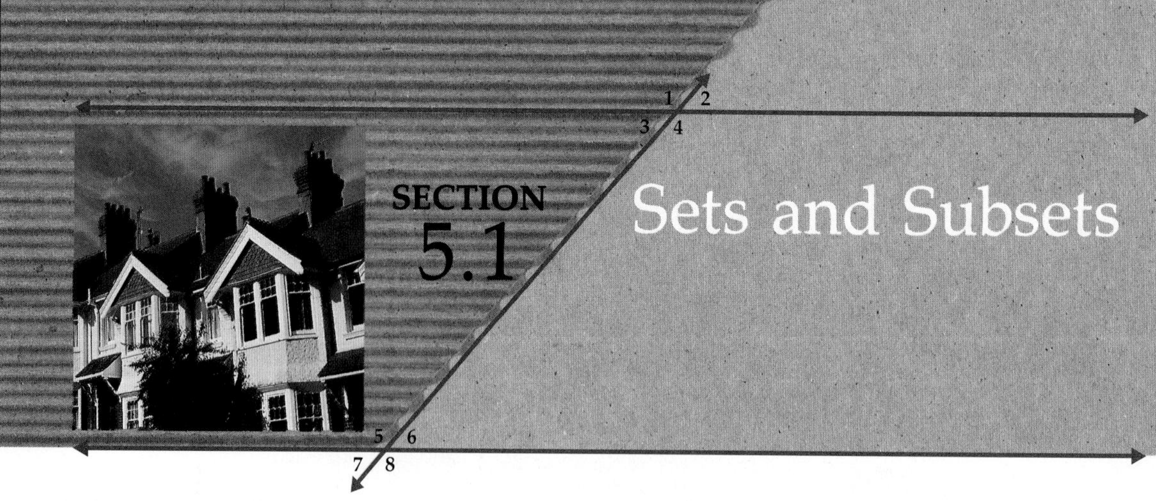

SECTION 5.1 Sets and Subsets

After completing this section, you should understand
- ▲ what a set is.
- ▲ what Venn diagrams are.
- ▲ what set operations are.
- ▲ how to solve problems using sets.

Before the study of similar triangles can be extended to congruent triangles, some terms from the study of set theory must be defined.

WHAT IS A SET?

You can tell by looking at a plate whether it belongs to a certain set of dishes.

DEFINITION 5-1-1 **A SET is a well-defined collection.**

Well-defined means that if you are given an object, you can tell whether or not it is a member, or element, of the set. Members of a set are enclosed in braces: { }. For example, $S = \{1, 2\}$ is read S is the set of numbers 1 and 2.

Example: The set of counting numbers between 1 and 10 = {2, 3, 4, 5, 6, 7, 8, 9}.

Example: $N = \{1, 3, 5, 7, \ldots\}$ = the set of odd counting numbers. The three dots (. . .) indicate that the terms continue in the given pattern, until the last member is reached. When no last member is given, the set continues forever, or to infinity.

Example: The set of all geometry students in your school is well-defined. Given a name or an ID number, you can determine whether or not a student is taking geometry.

NOTEBOOK
DEFINITION 5-1-2 **A SUBSET is any set contained in the given set.**

A is a subset of U is written as $A \subset U$.

$\{1, 2\} \subset \{1, 2, 3, 4, 5\}$

Notice that a set can be a subset of itself. It can also be a subset of a larger set, called a universal set.

NOTEBOOK
DEFINITION 5-1-3 **The UNIVERSE or UNIVERSAL SET is all the possible members in the well-defined set.**

The set of all geometry students in your school could be a universal set. Your geometry class would be a subset of the universe since your class is a part of the set of all geometry students.

CLASS ACTIVITY

Let $U = \{0, 1, 2, 3, 4, 5, 6, 7, 8, 9\}$. Write true or false about each statement.

1. $\{0, 2, 4, 6, 8\} \subset U$ True
2. $\{10\} \subset U$ False
3. $\{1, 3, 5, 7, 9\} \subset U$ True
4. The set of all even numbers $\subset U$ False
5. Describe U in words. Answers may vary. Possible answer: set of one-digit numbers.
6. Write the one-member subsets for $\{a, b, c\}$. $\{a\}, \{b\}, \{c\}$
7. Write the two-member subsets for $\{2, 4, 6, 8\}$. $\{2, 4\}, \{2, 6\}, \{2, 8\}, \{4, 6\}, \{4, 8\}, \{6, 8\}$

VENN DIAGRAMS

Venn diagrams are often used to help show how sets relate to each other. A rectangle represents the universal set for a given situation. Sets are then represented by circles within the rectangle.

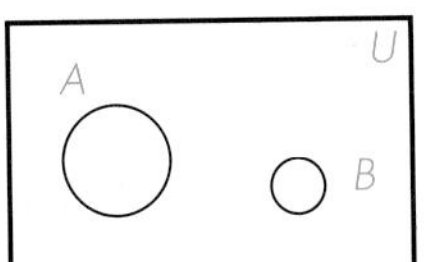

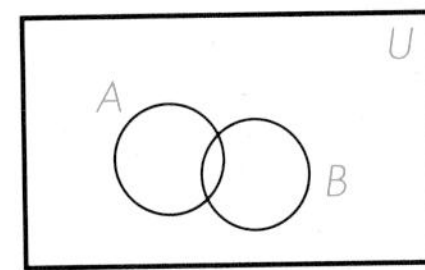

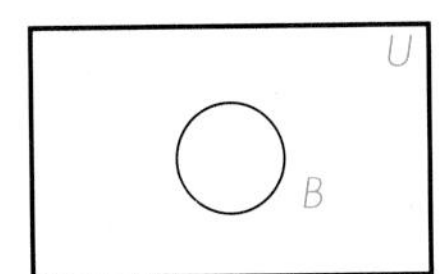

Chalkboard Example
A class was surveyed as to what types of fruit they like. Of those surveyed, 15 like bananas, 18 like oranges, and 16 like nectarines. Ten like bananas and oranges, 9 like oranges and nectarines, and 10 like bananas and nectarines. Eight like all three. Make a Venn diagram and determine how many students there are in the class. [28]

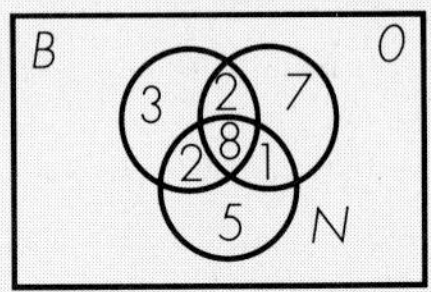

TEACHING COMMENTS

Using the Page
See the second example on page 166. Point out that enough elements must be listed so that the set is well-defined.

Using the Page
Some students may be confused by $A \subset A$. Let $A = \{1, 2, 3, 4\}$. Show students that every element of $\{1, 2, 3, 4\}$ is in $\{1, 2, 3, 4\}$.

Using the Page
Point out that a universal set can be whatever set is appropriate for a given situation. The universe may be students in a school, if the situation concerns a survey taken of a class. It may be whole numbers if the question asks how many. Rational numbers can be the universe for other situations.

Using the Page
The Venn diagrams show three different situations. In the first, sets A and B are *disjoint*—they do not have any elements in common. In the second case, sets A and B may have some elements in common. In the third case, set B is contained entirely in set A, $B \subset A$.

SET OPERATIONS

There are two operations in set theory: union and intersection.

DEFINITION 5-1-4 **The UNION of *n* sets is the set consisting of all the members in the *n* sets.**

The shaded area shows the union of two sets, *A* and *B*.

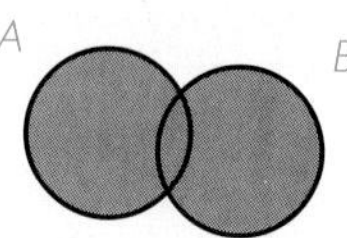

A union *B* is written as $A \cup B$.

Using the Page
Have students write their own definitions of the union of two sets. Discuss which definitions are valid and which are not. One definition for two sets might be: $A \cup B$ = all the elements in *A* or in *B* or in both.

DEFINITION 5-1-5 **The INTERSECTION of *n* sets is the set consisting of the members common to the *n* sets.**
The shaded area shows the intersection of two sets, *A* and *B*.

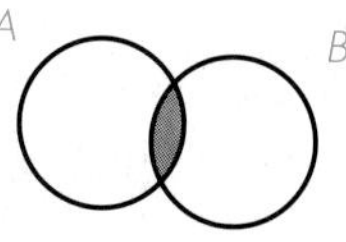

A intersection *B* is written as $A \cap B$.

Using the Page
Have students write their own definitions for intersection. Discuss which definitions are valid and which are not. Have students work in groups to write counterexamples for definitions that are not valid.

Example: Let $T = \{3, 6, 9, 12\}$ and $E = \{2, 4, 6, 8, 10, 12\}$. Draw a Venn diagram to show $T \cup E$ and to show $T \cap E$. Write the sets for $T \cup E$ and $T \cap E$.

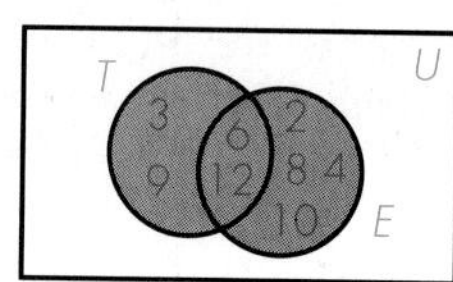

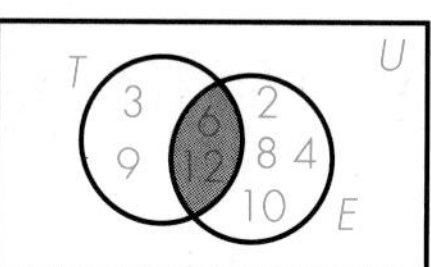

$T \cup E = \{2, 3, 4, 6, 8, 9, 10, 12\}$ — The elements are in *T* **or** *E*.

$T \cap E = \{6, 12\}$ — The elements are in *T* **and** *E*.

Example: Let $N = \{0, 1, 2\}$ and $M = \{4, 5, 6\}$. Find $N \cap M$.

There are no elements in common to both sets. The intersection is empty. This set is called the empty set.

Using the Page
You may want to mention that the empty set is also called the null set.

THEOREM 5-1-1 **The empty set is a subset of every set. The symbol is either Ø or { }.**

Example: List all of the subsets of $A = \{1, 2, 3\}$.

{1, 2, 3}	A set is a subset of itself.
{1, 2}, {1, 3}, {2, 3}	Subsets having 2 elements.
{1}, {2}, {3}	Subsets having 1 element.
{ }	Subsets having no elements, or empty set.

CLASS ACTIVITY

1. Draw a Venn diagram illustrating the following: See margin.
   U: the set, or number, of students in your school
   F: the set of freshmen *S*: the set of sophomores
   J: the set of juniors *R*: the set of seniors
2. Using Exercise 1, if there are 300 in *F*, 275 in *S*, 254 in *J*, and 199 in *R*, then how many are in *U*? 1028

A survey was taken of a group of students.
10 were taking English (*E*). 8 were taking geometry (*G*).
7 were taking history (*H*). 4 were taking *E* and *G*.
3 were taking *G* and *H*. 5 were taking *H* and *E*.
2 were taking all three subjects.

3. Draw a Venn diagram showing the information. See margin.
4. How many were taking only geometry? only English? only history? 3; 3; 1
5. How many were taking history or English? geometry or history? geometry or English? 12; 12; 14
6. How many students were interviewed? 15

Use a computer and a geometric drawing tool to complete Exercises 7–10.

7. Draw a line. Can you draw a ray on the line? Is a ray a subset of a line? Yes; yes
8. Draw an isosceles triangle. Can you make the isosceles triangle into a rectangle? Are isosceles triangles a subset of rectangles? No; no
9. Draw the union of two rays with a common endpoint. What figure is formed? An angle
10. Draw two parallel lines. Is there a point common to both lines? What set describes the intersection of these two lines? No; the empty set

Using the Page
When discussing the survey problem, emphasize that 10 were taking English means that there were a total of 10 in the circle labeled *E*.

Four were taking *E* and *G* means that there were a total of 4 in the intersection of these two circles.

Two were taking all three subjects means that 2 were in the intersection of all 3 circles.

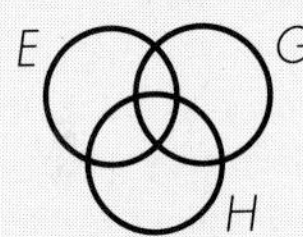

It is usually easiest to start at the intersection of all three circles, next determine the intersection of two circles at a time, and then use the total for a circle to determine how many are taking that subject only.

Additional Answers

1.

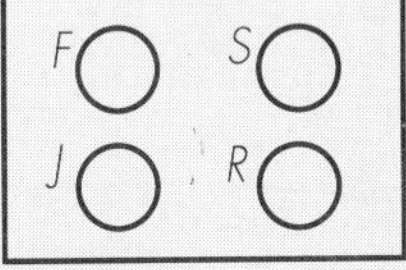

3.

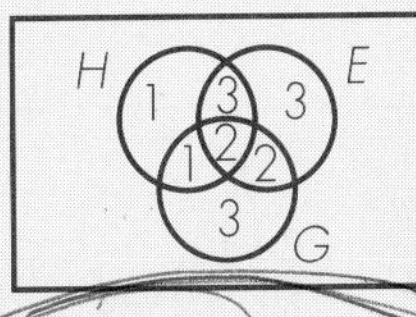

Extra Practice
Of the students surveyed, 45 like canoeing (*C*), 50 like skiing (*S*), and 65 like hiking (*H*). Thirty-two like both canoeing and skiing, 38 like both skiing and hiking, and 33 like both canoeing and hiking. Twenty-seven like all three. Five surveyed did not like any of the three. How many students were surveyed? [89]

HOME ACTIVITY

Write each set using braces, or set notation.

1. The first four counting numbers divisible by 2 and 3 {6, 12, 18, 24}
2. The first four counting numbers divisible by 2 or 3 {2, 3, 4, 6}
3. The set of vowels of the alphabet {a, e, i, o, u}
4. The set of integers whose absolute value is less than 3 {−2, −1, 0, 1, 2}
5. The set of counting numbers greater than 100 {101, 102, 103, . . .}

Write whether each set is a subset of $A = \{1, 2, 3, 4, 5, \ldots, 12\}$.

6. $B = \{2, 4, 6\}$ Yes
7. $C = \{\ \}$ Yes
8. $D = \{2, 4, 6, 8, \ldots\}$ No
9. $E = \{1, 3, 5, \ldots, 11\}$ Yes

A school's population was surveyed about extracurricular activities.

110 were out for sports (S).
95 were out for band (B).
83 were out for dramatics (D).
41 were out for D and B.
36 were out for D and S.
56 were out for B and S.
12 were out for all three.
150 were not out for any of the three.

10. Draw a Venn diagram showing the information. See margin.
11. How many students were taking only sports? only band? only dramatics? 30; 10; 18
12. How many were out for S or B? for B or D? for S or D? 149; 137; 157
13. How many students were in the school? 317

CRITICAL THINKING

14. Copy and complete the following table.

Set	Number of Elements	Subsets	Number of Subsets
$\{a\}$	1	$\{a\}, \{\ \}$	2
$\{a, b\}$	2	$\{a, b\}, \{a\}, \{b\}, \{\ \}$	4
$\{a, b, c\}$	3	See margin.	8
$\{a, b, c, d\}$	4	See margin.	16

15. How many subsets are there for a set of n elements? 2^n

Additional Answers
10.

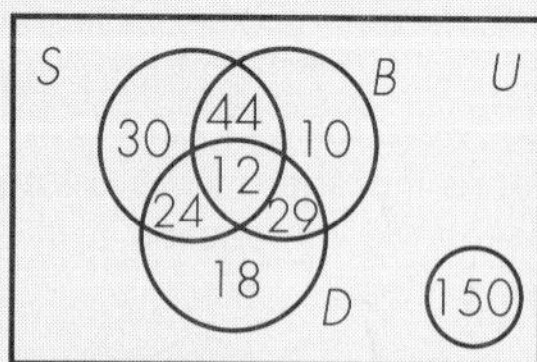

14. {a, b, c}, {a, b}, {a, c}, {b, c}, {a}, {b}, {c}, { }; {a, b, c, d}, {a, b, c}, {a, b, d}, {a, c, d}, {b, c, d}, {a, b}, {a, c}, {a, d}, {b, c}, {b, d}, {c, d}, {a}, {b}, {c}, {d}, { }

Using the Page
The Critical Thinking exercises give students a way to check to see if the correct number of subsets have been listed for a set of n elements.

FOLLOW-UP

Assessment
1. Explain why {1, 2, 10} is not a subset of {1, 2, 3, 4, 5}. [10 is not an element of {1, 2, 3, 4, 5}.]
2. Is { } a subset of {1, 2, 3, 4, 5}? [Yes]
3. Draw a Venn diagram to show $A \cap B \cap C$. [Students' drawings should look like the following.]

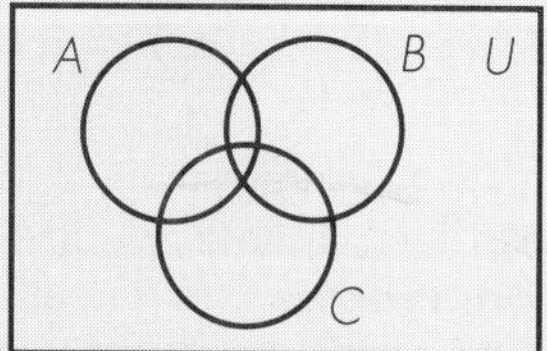

Reteaching
Students who have had difficulty with this section may benefit from Reteaching Activity 5.1.

Enrichment
For students who have mastered the material in this section, you may wish to assign Enrichment Activity 5.1.

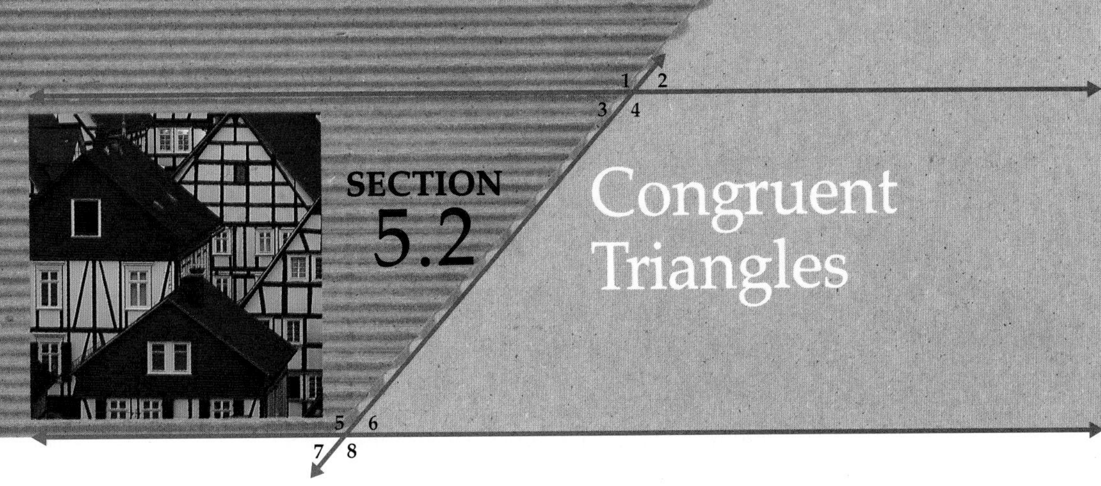

After completing this section, you should understand
- ▲ the meaning of congruent triangles.
- ▲ how to prove triangles congruent.

In Chapter 4, you solved problems involving similar triangles. You used a reversible definition for similarity, with conditions involving angle-angle (AA), side-side-side (SSS), and side-angle-side (SAS).

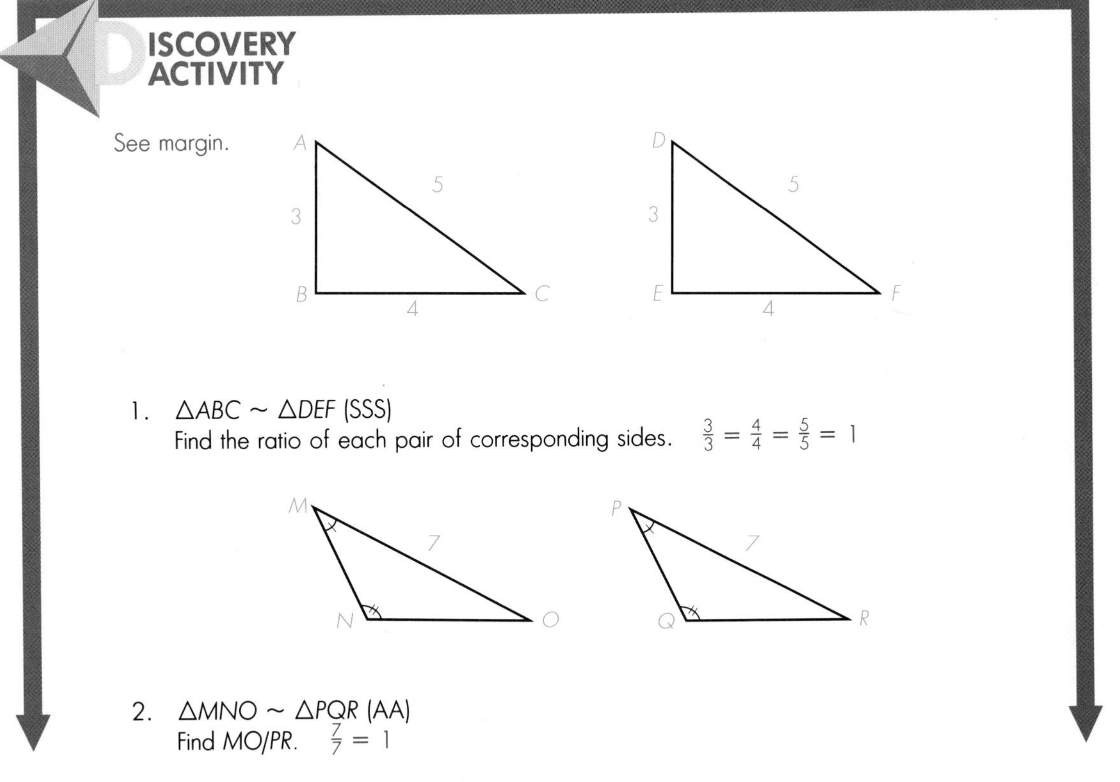

DISCOVERY ACTIVITY
See margin.
A
B
C
D
E
F
3
4
5
3
4
5
1. △ABC ~ △DEF (SSS)
Find the ratio of each pair of corresponding sides. 3/3 = 4/4 = 5/5 = 1
M
N
O
P
Q
R
7
7
2. △MNO ~ △PQR (AA)
Find MO/PR. 7/7 = 1

SECTION 5.2

Resources
Reteaching 5.2
Enrichment 5.2
Transparency Master 5-2

Objectives
After completing this section, the student should understand
- ▲ the meaning of congruent triangles.
- ▲ how to prove triangles congruent.

Materials
scissors
ruler
protractor

Vocabulary
congruent triangles
general case

GETTING STARTED

Quiz: Prerequisite Skills

1. Why is $\triangle MNO \sim \triangle VWX$?

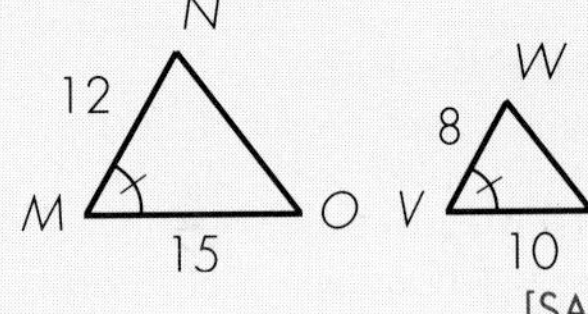

[SAS]

2. Why is $\triangle FED \sim \triangle PQR$?

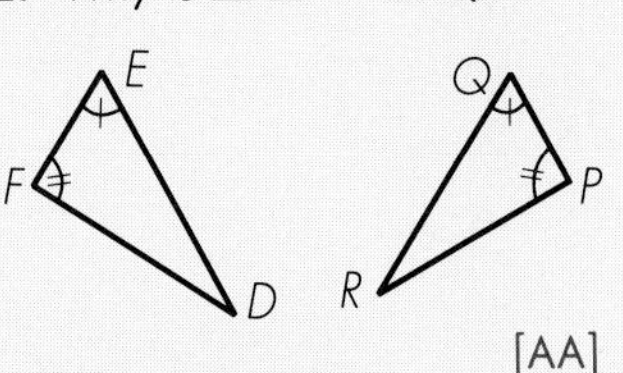

[AA]

Warm-Up Activity
Have students label the four corners of a sheet of paper with *W*, *X*, *Y*, and *Z*. Students should fold along one of the diagonals, then cut along the fold. Have students place one triangle on top of the other, and ask for a description of what they see.

Chalkboard Example
Show that $\triangle DEF \cong \triangle MNO$.

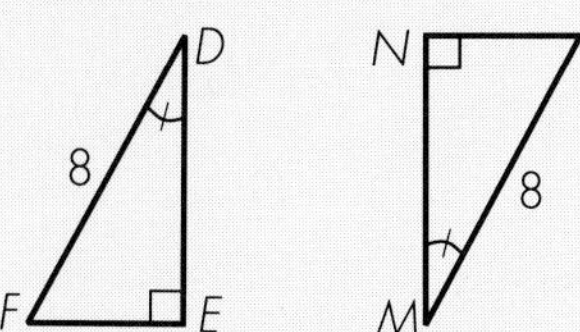

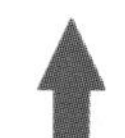

$\triangle DEF \sim \triangle MNO$ by AA, and $\frac{8}{8} = 1$.

TEACHING COMMENTS

Using the Page
A geometric drawing tool can by used to complete this Discovery Activity.

Additional Answers
See the Discovery Activity, Step 3. Some students may respond that the triangles are similar and that lengths of corresponding sides are equal. Lead students in discussion until they realize that similar means same shape, and equal corresponding sides means same size.

Using the Page
Have students write their own definitions of congruent triangles. Discuss which definitions are valid and which ones are not.

3. What appears to be true about the pairs of triangles?
Same size and shape

WHAT ARE CONGRUENT TRIANGLES?

The triangles of the Discovery Activity have not only the same shape but also the same size. In both pairs of triangles, the ratio of the lengths of the corresponding sides is one. The triangles are congruent.

DEFINITION 5-2-1 **If two triangles are similar and the ratio of corresponding sides is one, then the triangles are CONGRUENT.**

All congruent triangles are similar, but not all similar triangles are congruent. A Venn diagram can be used to display this.

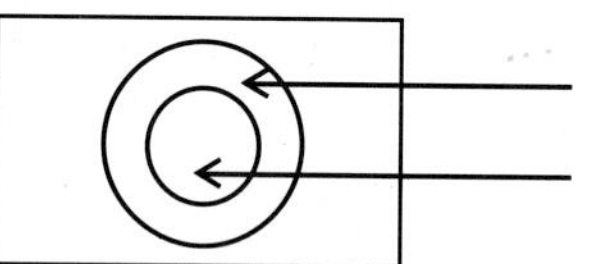

The symbol for congruency is a combination of the equal and similarity symbols, ≅.

$\triangle ABC \cong \triangle DEF$

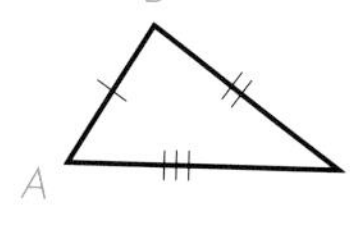

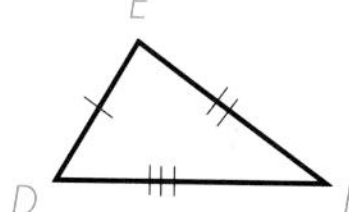

Example: Are the following triangles congruent?

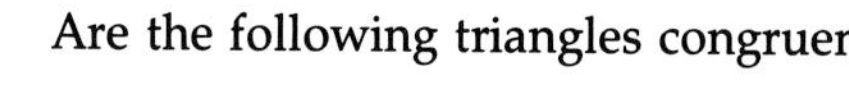

a.

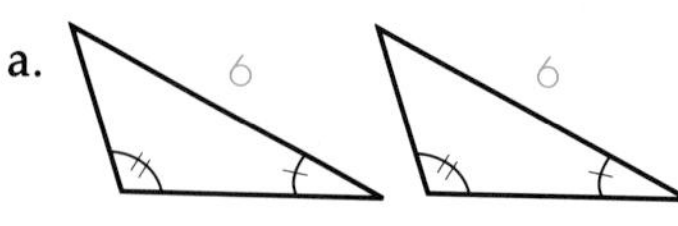

b.

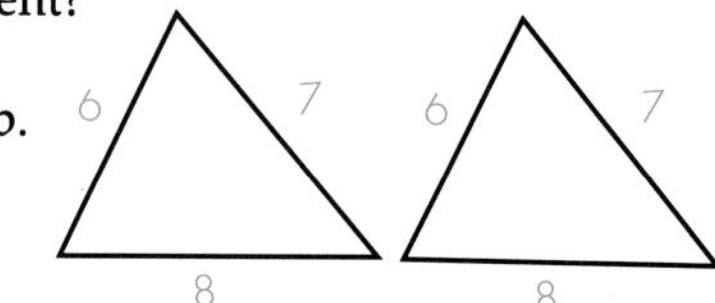

c.

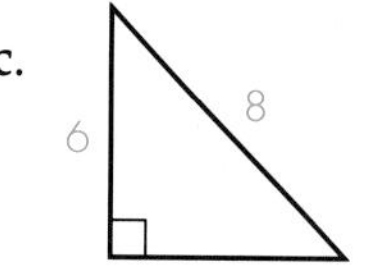

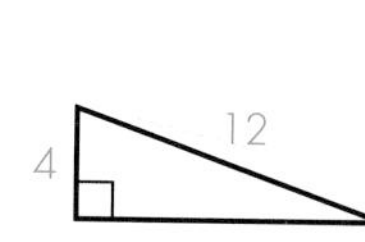

d.

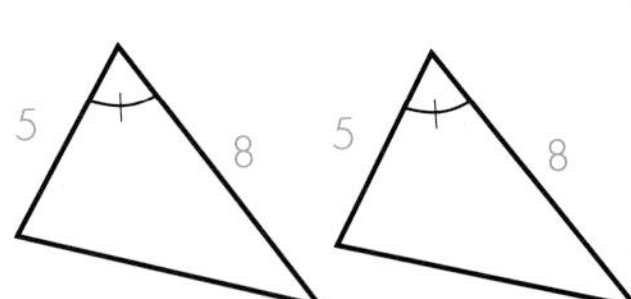

a. Yes; the triangles are similar by AA, and $\frac{6}{6} = 1$.
b. Yes; the triangles are similar by SSS, and $\frac{6}{6} = \frac{7}{7} = \frac{8}{8} = 1$.
c. No; $\frac{6}{4} \neq \frac{8}{12} \neq 1$.
d. Yes; the triangles are similar by SAS, and $\frac{5}{5} = \frac{8}{8} = 1$.

CLASS ACTIVITY

1. Use a computer and a geometric drawing tool to draw $\triangle AED$ and $\triangle BEC$ so that E is the midpoint of $\overline{AB}$, $AD = BC$, $\overline{DA} \perp \overline{AB}$, and $\overline{CB} \perp \overline{AB}$. Does $\triangle AED$ appear to be congruent to $\triangle BEC$? Justify your answer. See margin.

2. The gable end of a house is built by making $AC = BC$ and $AD = BD$. Is $\triangle ADC \cong \triangle BDC$? If so, why? Yes; $\triangle ADC \sim \triangle BDC$ by SSS; $AC/BC = AD/BD = CD/CD = 1$

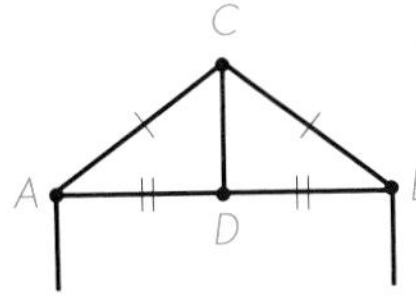

PROVING TRIANGLES CONGRUENT

When a theorem is justified, it is justified for a *general case.* For example, if the theorem is about triangles having a right angle, it is justified for *all* triangles that have a right angle.

Example: Given: $\triangle ABC$ with $AC = BC$, $\overline{CD}$ bisects $\angle ACB$.
Prove: $\triangle ACD \cong \triangle BCD$

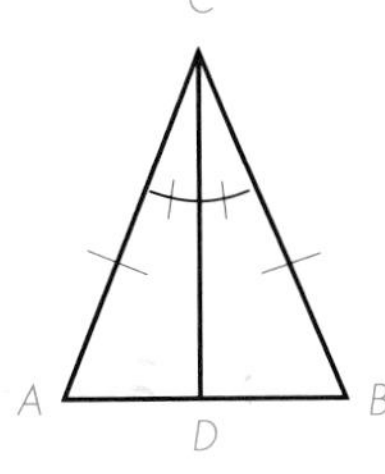

Statements	Reasons
1. $AC = BC$	1. Given
2. $\overline{CD}$ bisects $\angle ACB$.	2. Given
3. $CD = CD$	3. Identity
4. $m\angle ACD = m\angle BCD$	4. Definition of angle bisector
5. $AC/BC = CD/CD$	5. Algebra
6. $\triangle ACD \sim \triangle BCD$	6. SAS
7. $AC/BC = 1$	7. Algebra
8. $\triangle ACD \cong \triangle BCD$	8. Definition of congruent triangles

Additional Answers

1.

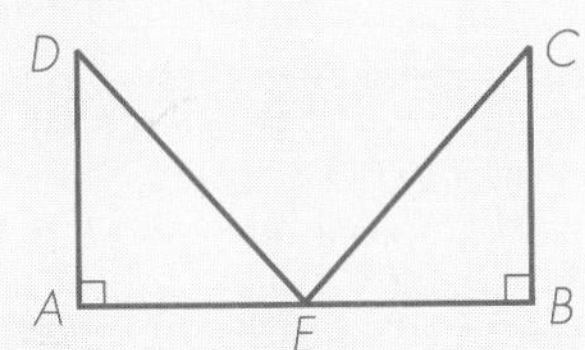

Yes; $\triangle AED \sim \triangle BEC$ by SAS; $\frac{DA}{CB} = \frac{AE}{BE} = 1$

Using the Page

Make sure that students understand what is meant by a general case. In the proof that follows, the only requirement is that two sides of a triangle are equal and that the included angle is bisected. The length of the equal sides and the size of the angle do not matter. Make sure that students understand each step of the proof.

Extra Practice

1. Is $\triangle JMK \cong \triangle LMK$? [Yes; $\triangle JMK \sim \triangle LMK$ by AA; $\frac{JK}{LK} = 1$]

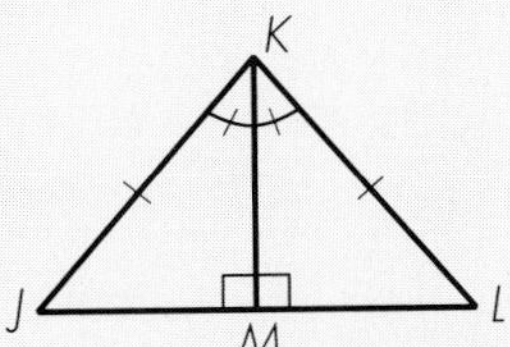

This is a general case since it applies to all triangles with two equal sides. Therefore, it will be stated as a theorem.

THEOREM 5-2-1 **If a triangle has two equal sides and the included angle is bisected, then the triangle is divided into two congruent triangles.**

CLASS ACTIVITY

1. Use a computer and a geometric drawing tool. Draw $\triangle MON$ so that $MN = MO$. Draw P on $\overline{MN}$ so that $NP = 6$ and $m\angle NMP = m\angle OMP$. Find NO. Justify your answer. See margin.

2. In $\triangle FGH$, $FG = HG$ and $\overline{GI}$ bisects $\angle FGH$. If $m\angle F = 50°$, find $m\angle H$. 50°

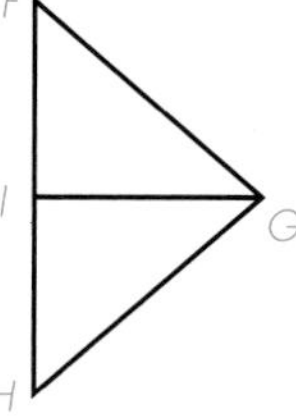

HOME ACTIVITY

State whether each pair of triangles are congruent.

1.

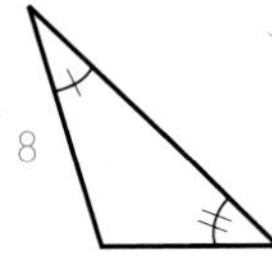

Yes

2.

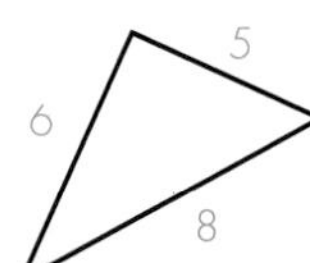

No

3.

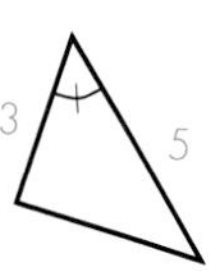

Yes

4.

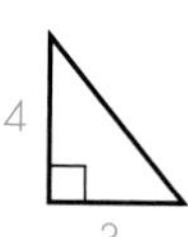

No

Additional Answers

1. $NO = 12$; since $m\angle NMP = m\angle OMP$, $\overline{MP}$ bisects $\angle NMO$. Since $MN = MO$ and the included $\angle NMO$ is bisected, by Theorem 5-2-1, $\triangle NMP \cong \triangle OMP$. So $NP = OP$, $OP = 6$, and thus $NO = 12$.

Using the Page

Students may notice that both exercises in the Class Activity make use of the fact that corresponding parts of congruent triangles are congruent. This will be stated as a definition in Section 5.3.

At this time you may wish to use Transparency Master 5-2 to illustrate that SSA is *not* a way to show congruency. Students often begin thinking that any three congruent parts of a triangle will insure congruency. This example may dispel this belief.

Refer to triangles ABC and DEF.

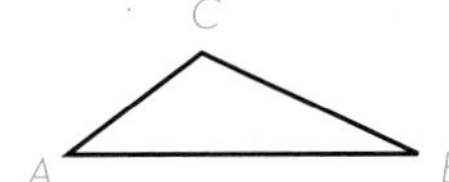

2

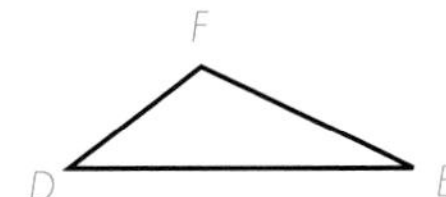

$AB = DE$, $AC = DF$, $m\angle B = m\angle E$

5. Are the triangles similar?
   Not enough information given to decide.
6. Are the triangles congruent?
   Not enough information given to decide.

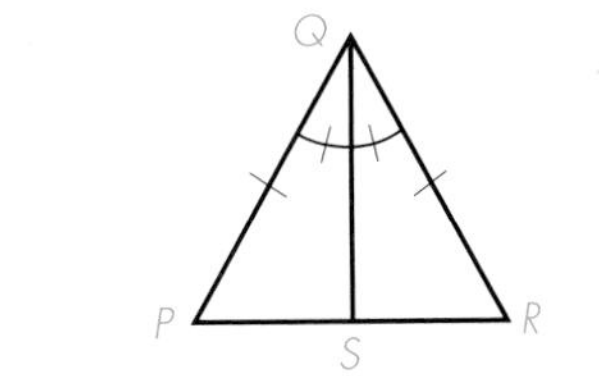

7. In $\triangle PQR$, $PQ = RQ$ and $m\angle PQS = m\angle RQS$. What do you know about $\triangle PQS$ and $\triangle RQS$? Explain your answer. $\cong$, by Theorem 5-2-1

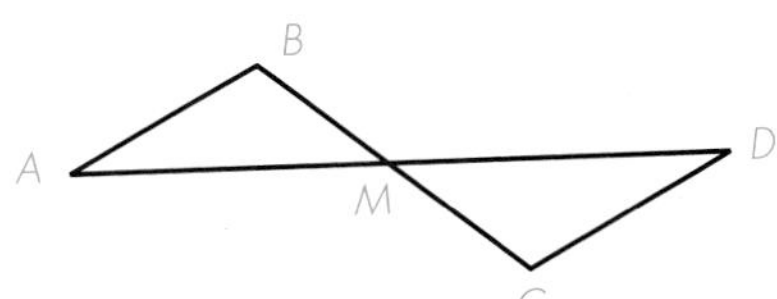

8. Given: $\triangle ABM$ and $\triangle DCM$ with M the midpoint of $\overline{AD}$ and $\overline{BC}$. Explain why $\triangle AMB \cong \triangle DMC$.
   $\triangle ABM \sim \triangle DCM$ by SAS; $AM/DM = 1$

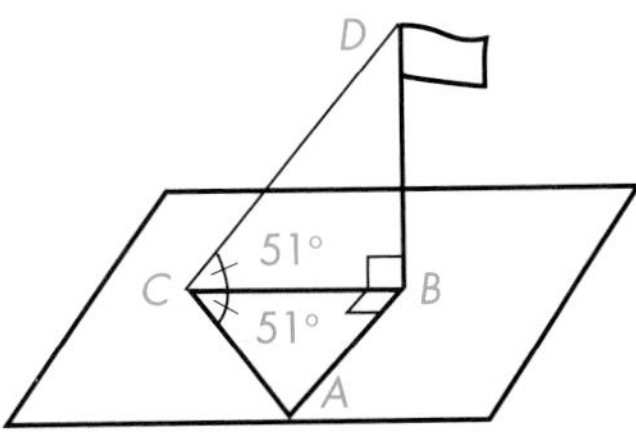

9. Determine the height of the flagpole.
   $AB = 41$ ft 41 ft

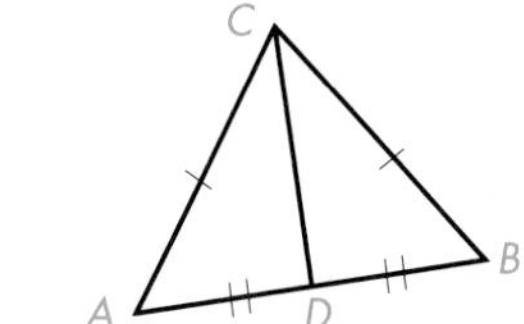

10. In $\triangle ABC$, $AC = BC$ and D is the midpoint of $\overline{AB}$. Explain why $\triangle ACD \cong \triangle BCD$.
    $CD = CD$, $\triangle ACD \sim \triangle BCD$ by SSS, $AC/BC = AD/BD = 1$

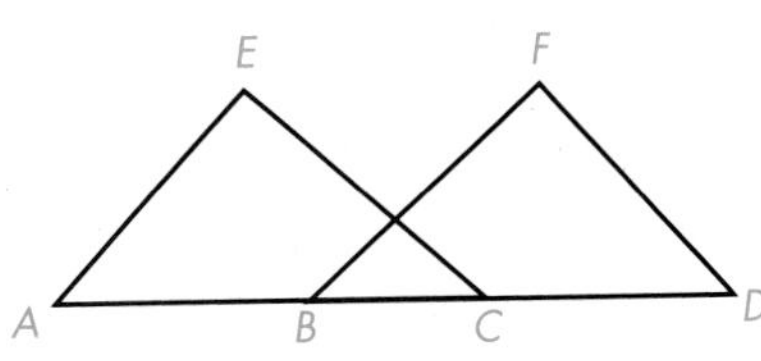

11. Given: $\triangle AEC$ and $\triangle DFB$, with $AE = DF$, $AB = CD$, and $m\angle A = m\angle D$
    Justify: $\triangle AEC \cong \triangle DFB$ (Use a paragraph proof.) See margin.

Using the Page
In Exercises 5 and 6, emphasize that enough information must be given to decide if two triangles are congruent or not. These triangles appear to be congruent, but no conclusion can be reached without knowing either that the last pair of sides are congruent or that one other pair of corresponding angles are congruent.

Using the Page
See Exercise 11. You may want to have students separate the overlapping triangles by tracing each onto a sheet of paper.

Additional Answers
11. It is given that $AB = CD$. Then $AC = BD$ by segment addition. $AE = DF$, and $m\angle A = m\angle D$. $\frac{AC}{BD} = \frac{AE}{DF}$ by algebra. Therefore, $\triangle AEC \sim \triangle DFB$ by SAS. Since $\frac{AC}{BD} = \frac{AE}{DF} = 1$, $\triangle AEC \cong \triangle DFB$.

12. In the figure at the right, $\overleftrightarrow{AB} \parallel \overleftrightarrow{CD}$, $\overleftrightarrow{AD}$ and $\overleftrightarrow{BC}$ are transversals, and $CM = BM$. Explain why $\triangle AMB \cong \triangle DMC$.
$\triangle AMB \sim \triangle DMC$ by AA; $CM/BM = 1$

13. Rectangle $ABDC$ has $\overleftrightarrow{AB} \parallel \overleftrightarrow{CD}$, $\overline{CA} \perp \overline{AB}$, $\overline{DB} \perp \overline{CD}$. Explain why $\triangle ABC \cong \triangle DCB$.
$\triangle ABC \sim \triangle DCB$ by AA (rt. angles, alt. int. angles); $CB/CB = 1$

14. In the figure at the right, $\overline{AOC}$ and $\overline{BOD}$ are line segments. Explain why $\triangle AOB \cong \triangle COD$. $\triangle AOB \sim \triangle COD$ by AA; $AO/CO = 1$

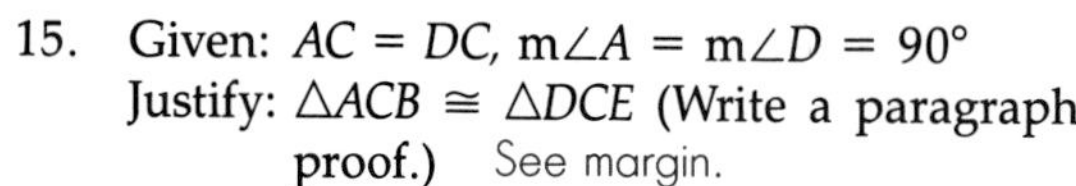

15. Given: $AC = DC$, $m\angle A = m\angle D = 90°$
Justify: $\triangle ACB \cong \triangle DCE$ (Write a paragraph proof.) See margin.

$\triangle ABC$ is equilateral. $\overline{AD}$, $\overline{BE}$, and $\overline{CF}$ are angle bisectors.

16. Is $\triangle ABD \cong \triangle ACD$? Explain your answer.
Yes; $\triangle ABD \sim \triangle ACD$ by SAS; $AB/AC = AD/AD = 1$

17. Is $\triangle BCE \cong \triangle BAE$? Explain your answer.
Yes; $\triangle BCE \sim \triangle BAE$ by SAS; $BC/BA = BE/BE = 1$

18. Is $\triangle CAF \cong \triangle BCF$? Explain your answer.
Yes; $\triangle CAF \sim \triangle BCF$ by SAS; $CA/CB = CF/CF = 1$

19. Plot the points for each triangle on the Cartesian plane. Use your ruler and protractor to determine if the triangles appear to be congruent.
Triangle 1: (0,0), (0,3), (4,0);
Triangle 2: (−1,0), (−1,4). (−4,0) Yes

CRITICAL THINKING

20. A student found an easy way to remember a certain year in history. The year is a four-digit number, such as 1988. The four digits ad up to 17. When the first two digits are turned upside down, the second two digits result. What is the year? (There are two answers.) 1961, 1691

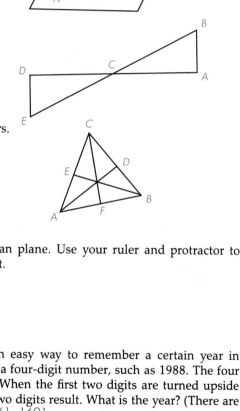

Additional Answers

15. It is given that $m\angle A = m\angle D = 90°$. $m\angle DCE = m\angle ACB$, since they are vertical angles. Then $\triangle ABC \sim \triangle DCE$ by AA. It is given that $AC = DC$, therefore $\frac{AC}{DC} = 1$. Thus $\triangle ACB \cong \triangle DCE$.

FOLLOW-UP

Assessment

1. If two triangles are not similar, can they be congruent? [No]

2. Is $\triangle PQR \cong \triangle QTR$? [Yes]

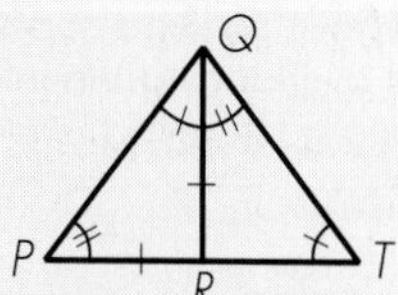

Reteaching

Students who have had difficulty with this section may benefit from Reteaching Activity 5.2.

Enrichment

For students who have mastered the material in this section, you may wish to assign Enrichment Activity 5.2.

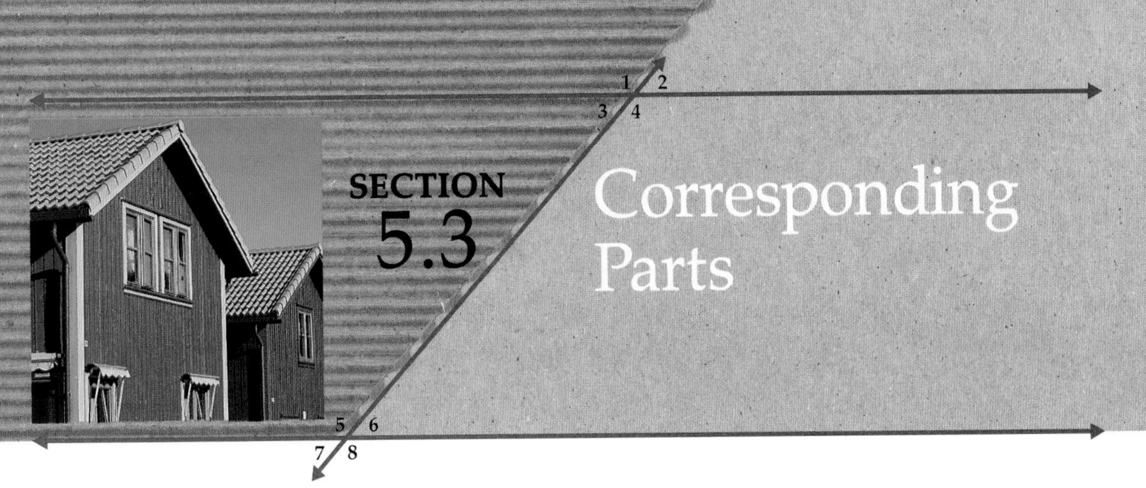

SECTION 5.3 Corresponding Parts

After completing this section, you should understand
▲ corresponding parts of congruent figures.
▲ how to determine lengths of segments and measurements of angles.

If you had two road maps of your state that had the same scale, or ratio, you could say that they were in proportion. If you measured a segment on one map, it would represent an equal length on the other map.

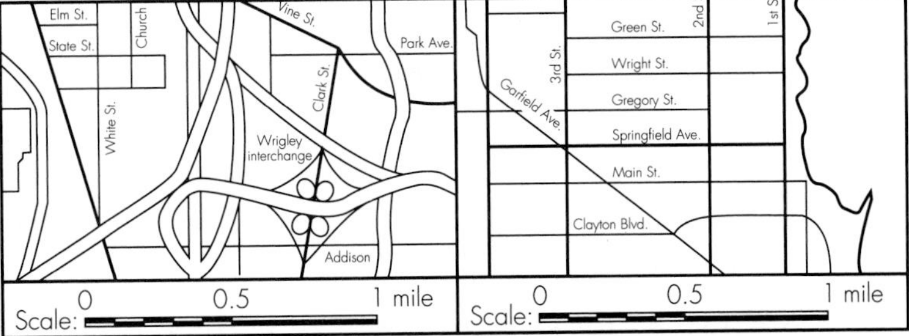

The measured segments on one map would correspond identically to the corresponding segment on the other map. The concept of corresponding, identical parts is widely used in industry. It would be very difficult (and costly!) to replace some items, such as house doors, if they did not come in standard (identical) sizes.

177

SECTION 5.3

Resources
Reteaching 5.3
Enrichment 5.3

Objectives
After completing this section, the student should understand
▲ corresponding parts of congruent figures.
▲ how to determine lengths of segments and measurements of angles.

Materials
scissors

Vocabulary
corresponding parts of congruent figures

GETTING STARTED

Quiz: Prerequisite Skills

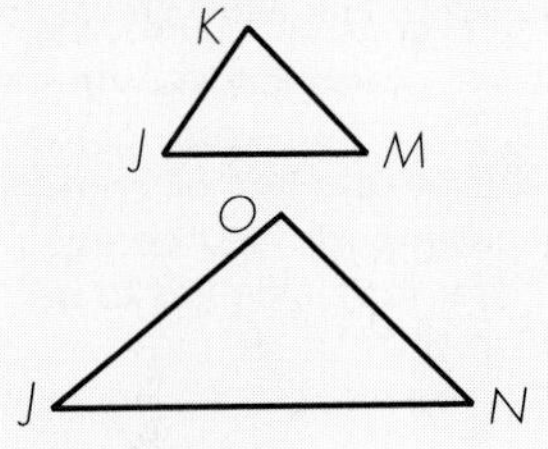

$\triangle JKM \sim \triangle NOV$

1. List the corresponding sides. [$\overline{JK}$ and $\overline{NO}$, $\overline{KM}$ and $\overline{OV}$, $\overline{MJ}$ and $\overline{VN}$]

2. List the corresponding angles. [$\angle J$ and $\angle N$, $\angle M$ and $\angle V$, $\angle K$ and $\angle O$]

Warm-Up Activity
Have students fold a sheet of paper in half, as shown.

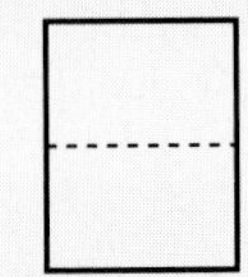

Students then cut along the fold. Have them make observations about the two rectangles.

Chalkboard Example

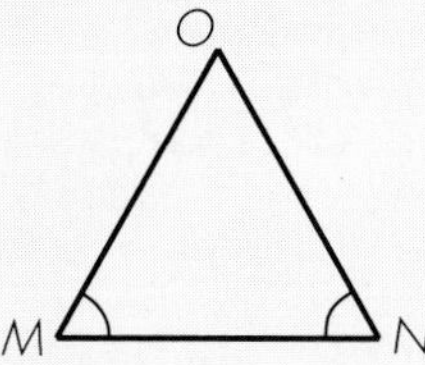

In $\triangle MON$, $m\angle M = m\angle N$. If $MO = 6$, what is NO? [6]

TEACHING COMMENTS

Using the Page
See the first of the three situations for similar triangles. When the triangles are also congruent, the situation is sometimes referred to as AAS congruency.

Using the Page
Remind students that for figures other than triangles, in order for two figures to be similar, corresponding angles must be equal, *and* corresponding sides must be in equal ratio. For all congruent figures, the ratio of corresponding sides is one. Thus, the reverse of the definition says that all corresponding parts must be equal in order for the figures to be congruent.

Using the Page
A geometric drawing tool can be used to complete the Discovery Activity.

CORRESPONDING PARTS OF CONGRUENT FIGURES

In order for triangles to be congruent, they must be similar and the ratio of the corresponding sides must be one. Recall that triangles are similar if one of three situations is present.

1. Two angles in each triangle are equal (AA).
2. The ratio of all pairs of corresponding sides are equal (SSS).
3. The ratio of two pairs of corresponding sides are equal and the included angles are equal (SAS).

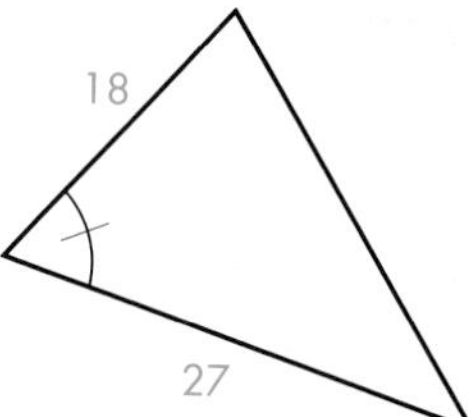

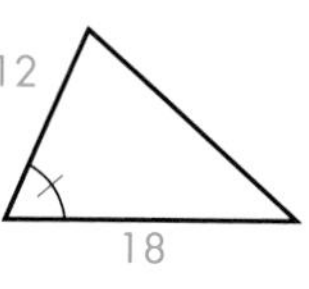

Thus, for congruent triangles, both corresponding sides and corresponding angles are equal. This can be extended to other figures, as well.

Corresponding parts of congruent figures are equal. This may be abbreviated as CPCF.

To this point, you have worked primarily with triangles. However, the definition holds for all geometric figures. Since definitions are reversible, if two figures have all corresponding parts equal, then the figures are congruent.

DISCOVERY ACTIVITY

1. Draw a 2-inch line segment on a sheet of paper, labeling the segment *AB*. See margin.
2. From *A* and from *B*, use your protractor to draw $\angle BAC$ and $\angle ABC$ so that the two angles are equal and the rays intersect at *C*.
3. Measure the lengths of $\overline{AC}$ and $\overline{BC}$.
4. What conclusion do you reach about measures of $\overline{AC}$ and $\overline{BC}$? Equal

SIDES OPPOSITE EQUAL ANGLES

The class should all have reached the same conclusion, that $AC = BC$. The following will justify the conclusion.

Given: $\triangle ABC$ with $m\angle A = m\angle B$
Justify: $AC = BC$

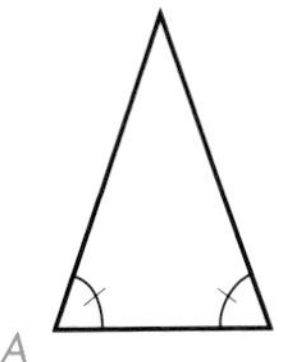

From the given, we know that $m\angle A = m\angle B$. We can draw the angle bisector from C because there is a number assigned to the measure of $\angle ACB$, and that number can be divided by 2. Let the angle bisector intersect $\overline{AB}$ at D.

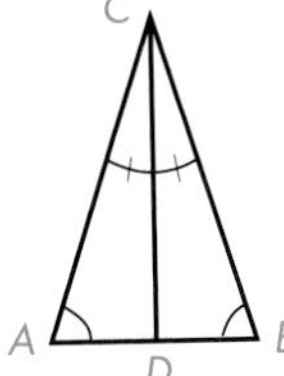

$m\angle ACD = m\angle BCD$. Then $\triangle ACD \sim \triangle BCD$ by AA. $CD = CD$, so $CD/CD = 1$. Therefore, $\triangle ACD \cong \triangle BCD$. Definition 5-3-1 tells us that the other corresponding parts are equal. So $AC = BC$.

THEOREM 5-3-1 **If a triangle has two equal angles, then the sides opposite the equal angles are equal.**

The above theorem applies to isosceles triangles. For an equiangular triangle, Theorem 5-3-1 can be applied twice.

THEOREM 5-3-2 **If a triangle is equiangular, then it is equilateral.**

CLASS ACTIVITY

1. Use a computer and a geometric drawing tool to draw an equiangular triangle. Measure each side. Repeat this activity several times, using different angle measures each time. Do your results show that Theorem 5-3-2 is true? Yes

Are the pairs of triangles congruent?

2.

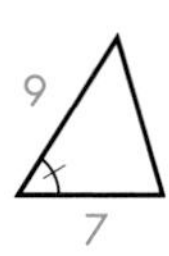

7 9

Yes

3.

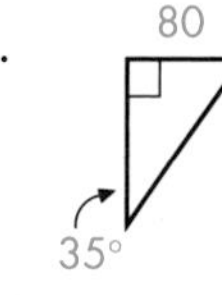

Yes

Using the Page
Make sure students understand each step of the paragraph proof. You may want to have students work in pairs to write the proof in two-column form.

Using the Page
Have students work in small groups to write a proof for Theorem 5-3-2.

Extra Practice

1. $\triangle PET \cong \triangle CAT$. If $PE = 8$ inches, what is CA? [8 inches]

2. $\triangle NOW \cong \triangle MIC$. If $m\angle I = 56°$, what is $m\angle O$? [56°]

Using the Page

See Exercise 9. Ask students if a triangle having three equal angles is isosceles. [Yes] Have students draw a Venn diagram relating isosceles triangles and equilateral triangles.

4. 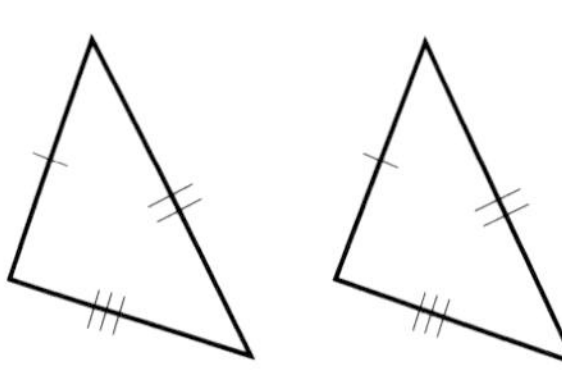

Yes

5. 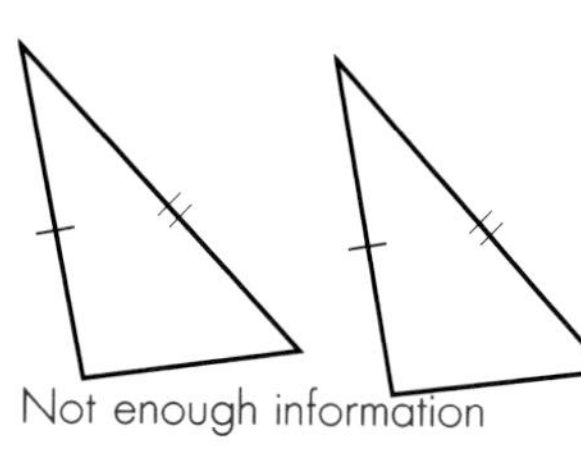

Not enough information

6. 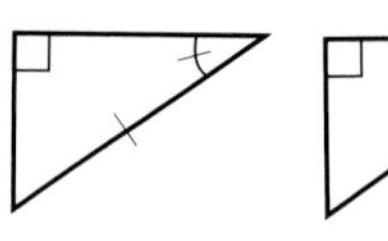

Yes

7. 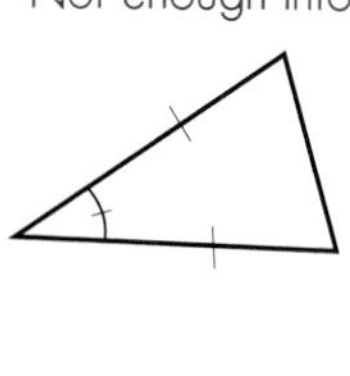

Yes

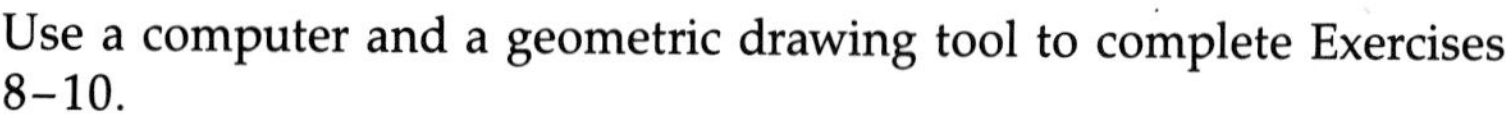

Use a computer and a geometric drawing tool to complete Exercises 8–10.

8. Draw a triangle with two equal angles. Measure each side. Repeat this activity several times, using different angle measures each time. If a triangle has two equal angles, is it isosceles? Yes

9. Draw two triangles that have equal corresponding angles. Measure each side. Repeat this activity several times, using different angle measures each time. If two triangles have equal corresponding angles, are the triangles congruent? Sometimes

10. Draw an acute triangle and a right triangle. Measure each side and each angle. Repeat this activity several times, using different angle measures and different side lengths each time. Is an acute triangle ever congruent to a right triangle? No

HOME ACTIVITY

Refer to $\triangle ABC$ and $\triangle DEF$.

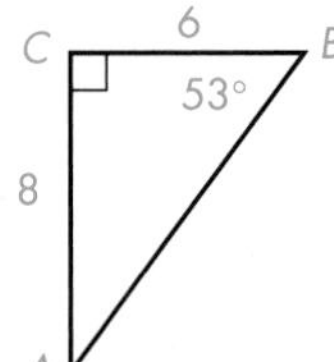

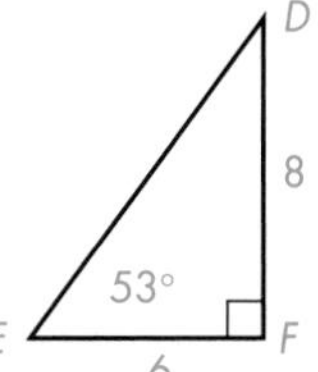

1. Why are the triangles congruent? AA, $\frac{6}{6} = \frac{8}{8} = 1$
2. Why does $AB = DE$? CPCF
3. Does $m\angle A = m\angle D$? Yes
4. Find $m\angle A$. 37°

Refer to $\triangle MNO$ and $\triangle PQR$.

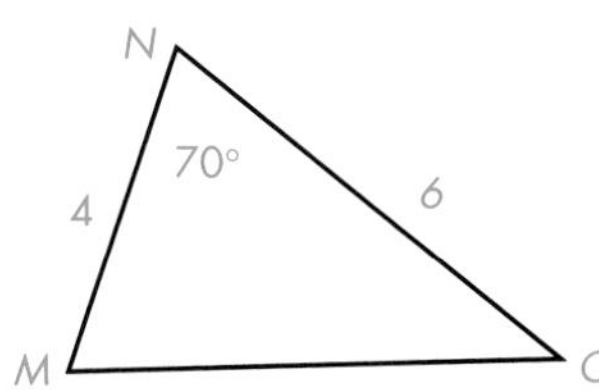

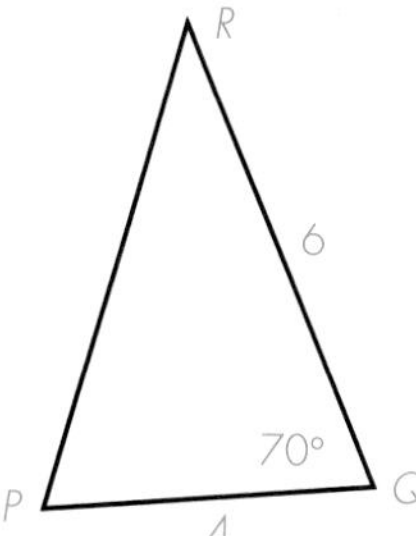

5. Are the two triangles similar? If so, why? Yes; SAS

6. Are the two triangles congruent? If so, why? Yes; $\frac{4}{4} = \frac{6}{6} = 1$

7. The length of $\overline{MO}$ is equal to the length of which side? $\overline{PR}$

8. Is $m\angle N = m\angle P$? Yes

Refer to $\triangle STU$ and $\triangle VWX$.

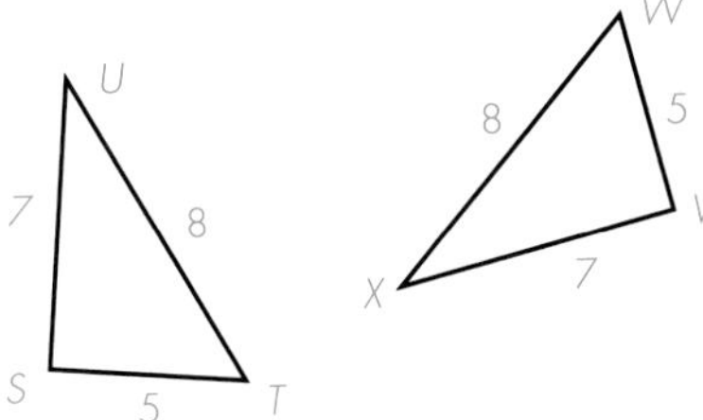

9. Are the triangles congruent? If so, why? Yes; SSS and $\frac{7}{7} = \frac{8}{8} = \frac{5}{5} = 1$

10. Is $m\angle U = m\angle W$? No

11. What angle is $\angle S$ equal to? $\angle V$

12. Classify the triangles according to the lengths of their sides. Scalene

13. If a triangle has two equal angles, then it is isosceles. Always

14. If two triangles have corresponding angles equal, the triangles are congruent. Sometimes

15. If two lines intersect, then the vertical angles are supplementary. Sometimes

16. An acute triangle and a right triangle are congruent. Never

Using the Page
See Exercise 10. Remind students that corresponding angles must be between corresponding sides. The sides with lengths 7 and 5 are not corresponding, so $\angle U$ cannot correspond to $\angle W$. Then the measures are not equal.

Using the Page
See Exercises 13-16. If the answer to an exercise is not *always*, have students add a word or phrase to the sentence so that it is always true.

17. Given: $AB = BC = CD = DA$
Justify: $m\angle A = m\angle C$

Have students provide missing reasons.

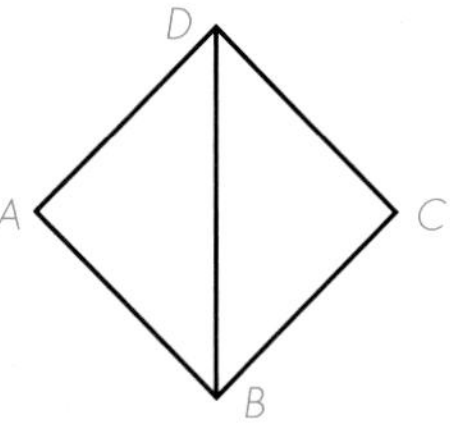
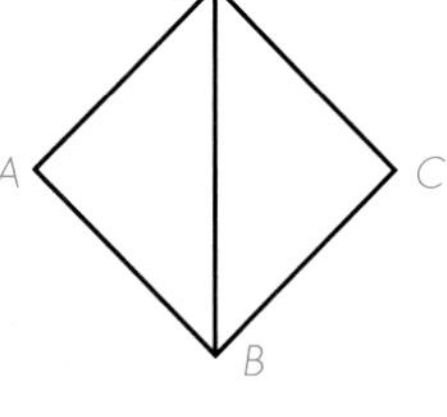

Statements	Reasons
1. $AB = BC = CD = DA$	1. Given
2. $BD = BD$	2. Identity
3. $\frac{AB}{CB} = \frac{AD}{CD} = \frac{BD}{BD}$	3. Algebra
4. $\triangle ABD \sim \triangle CBD$	4. SSS
5. $\frac{AB}{CB} = 1$	5. Algebra
6. $\triangle ABD \cong \triangle CBD$	6. Definition congruent triangles
7. $m\angle A = m\angle C$	7. CPCF

18. In $\triangle ABC$, $AC = BC$, and D, E, and F are midpoints. Write a paragraph proof to show that $m\angle DFA = m\angle EFB$. See margin.

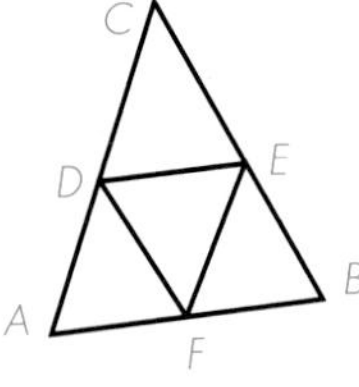

19. Given: Isosceles $\triangle ABC$ with $AC = BC$.
$\overline{CD}$ bisects $\angle ACB$. Point X is on $\overline{CD}$.
Justify: $\triangle AXB$ is isosceles.
Have students provide missing reasons.

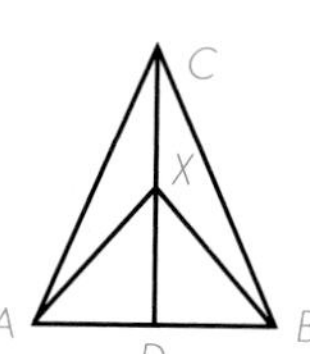

Statements	Reasons
1. Isosceles $\triangle ABC$ with $AC = BC$	1. Given
2. $\overline{CD}$ bisects $\angle ACB$.	2. Given
3. $m\angle ACD = m\angle BCD$	3. Definition angle bisector
4. $CX = CX$	4. Identity
5. $\frac{AC}{BC} = \frac{CX}{CX}$	5. Algebra
6. $\triangle ACX \sim \triangle BCX$	6. SAS
7. $\frac{CX}{CX} = \frac{AC}{BC} = 1$	7. Algebra
8. $\triangle ACX \cong \triangle BCX$	8. Definition congruent triangles
9. $AX = BX$	9. CPCF
10. $\triangle AXB$ is isosceles.	10. Definition isosceles triangle

CRITICAL THINKING

20. Without using a calculator, determine which of these four numbers is the greatest: 3^6, 6^3, 4^5, and 5^4. (Hint: Factor each number.) 4^5

Additional Answers

18. We are given that $AC = BC$. D and E are midpoints, so $AD = BE$. F is a midpoint, so $AF = BF$. $m\angle A = m\angle B$, since the base angles of an isosceles triangle are equal. $\triangle AFD \sim \triangle BFE$ by SAS. Since $\frac{AD}{BE} = 1$, $\triangle AFD \cong \triangle BFE$. Then $m\angle DFA = m\angle EFB$ by CPCF.

Using the Page

20. Students may reason as follows:
Compare 3^6 and 6^3.
$3^6 = 3^3 \cdot 3^3$ and $6^3 = 2^3 \cdot 3^3$. Since $3^3 > 2^3$, 6^3 cannot be the greatest.
Compare 3^6 and 5^4.
$3^6 = 3^3 \cdot 3^3$ and $5^4 = 5^2 \cdot 5^2$. $3^3 = 27$ and $5^2 = 25$. Since $27 > 25$, 5^4 cannot be the greatest.
Compare 3^6 and 4^5.
$3^6 = 3^3 \cdot 3^3$ and $4^5 = 2^5 \cdot 2^5$. $3^3 = 27$ and $2^5 = 32$. Since $32 > 27$, 4^5 is the greatest of the four numbers.

FOLLOW-UP

Assessment

1. $MNPQ \cong VATS$. Which side is congruent to $\overline{PQ}$?

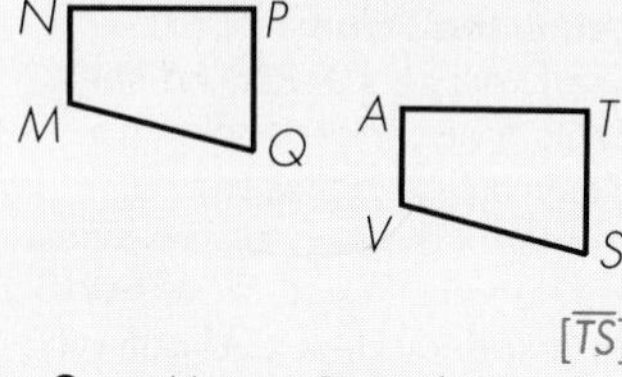

[$\overline{TS}$]

2. $\angle M \cong \angle O$. Find OP.

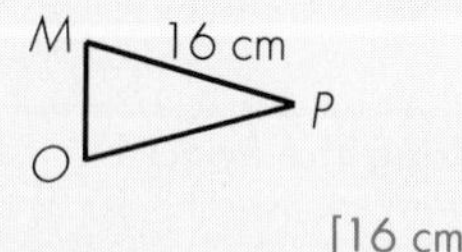

[16 cm]

Reteaching

Students who have had difficulty with this section may benefit from Reteaching Activity 5.3.

Enrichment

For students who have mastered the material in this section, you may wish to assign Enrichment Activity 5.3.

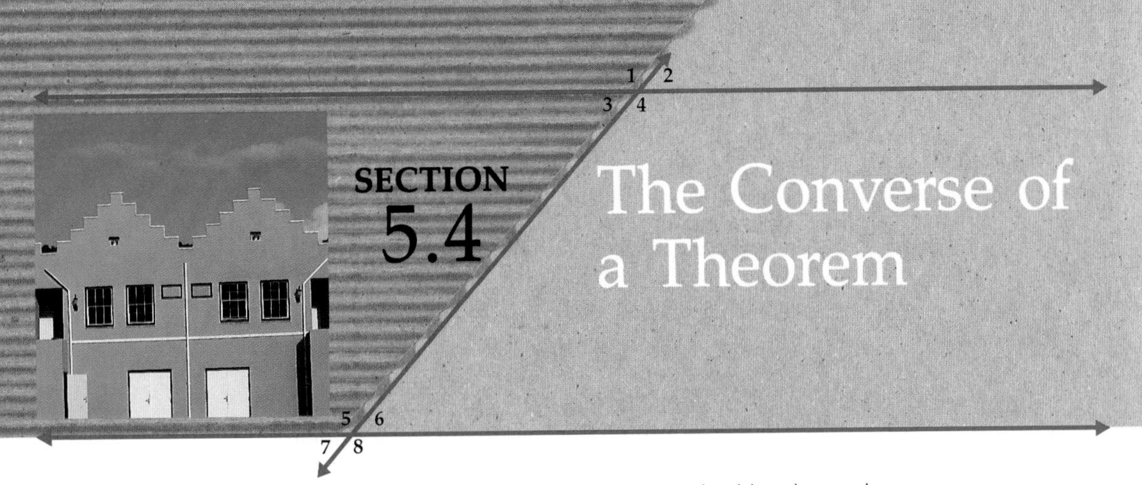

The Converse of a Theorem

After completing this section, you should understand
- ▲ the meaning of converse.
- ▲ how a statement and its converse are related.

You have used conditional statements many times so far in this course. The following statement is an example.

If two lines are parallel, then the two lines do not intersect.

Recall that the underlined part of the statement is the hypothesis, or given, and that the circled part is the conclusion.

CONVERSES

You form the converse of a conditional statement by switching the hypothesis and the conclusion.

If two lines do not intersect, then they are parallel.

The two statements are not the same. In the first, you are given the fact that two lines are parallel and conclude that the lines do not intersect. In the second, you are given the fact that two lines do not intersect and conclude that the lines are parallel.

SECTION
5.4

Resources
Reteaching 5.4
Enrichment 5.4

Objectives
After completing this section, the student should understand
- ▲ the meaning of converse.
- ▲ how a statement and its converse are related.

Vocabulary
hypothesis
conclusion
converse

GETTING STARTED

Quiz: Prerequisite Skills
Read Statement *A*. Can you conclude Statement *B* from Statement *A*?

1. *A*: $x < 9$
B: $x = 17$ [No]

2. *A*: $x \neq 7$
B: $x < 7$ or $x > 7$ [Yes]

Warm-Up Activity
Have a student make a statement. The rest of the class must be able to judge whether the statement is true or false. (Questions, orders, or exclamatory statements are not allowed.) Some examples might be: "I am the president of the U.S." [False] or "I have red hair." [True]

Chalkboard Example
Given the following statement, write its converse and state whether the converse is true or false.

If it is cold in Buenos Aires, then it is June 15th. [If it is June 15th, then it is cold in Buenos Aires; true.]

TEACHING COMMENTS

Using the Page
See page 183. You may want to have students look up the word hypothesis in a dictionary. Students should realize that the *hypotheses* are the facts that are assumed to be true and are used to justify the conclusion.

Additional Answers
4. If a ray divides an angle into two equal angles, then it is the angle bisector.
6. If two sides of a triangle are equal, then the angles opposite these sides are equal.
2. *A*: If you own a watch that you wear on your wrist, then you own a wristwatch; true.
B: If a person is an excellent skier, then that person skis in the Olympics; false.
C: If Jack owns a ring, then he owns a piece of jewelry; true.
D: If a teenager is a girl, then the teenager owns a bicycle; false.

Extra Practice
Give the converse of each statement.

1. If last month was May, then next month is July. [If next month is July, then last month was May.]
2. If I take the dog for a walk, then he is happy. [If the dog is happy, then I take him for a walk.]

DEFINITION 5-4-1 **The CONVERSE of an "If A, then B" statement is "If B, then A."**

Example: Statement: If it rains, then I carry an umbrella.
Converse: If I carry an umbrella, then it rains.

When writing the converse of a statement, you may have to change the wording of the hypothesis and conclusion slightly to make the converse read more clearly.

Example: Statement: If Jim has a dog, then he has a pet.
Converse: If he has a pet, then Jim has a dog.
Reworded converse: If Jim has a pet, then he has a dog.

The reworded converse does not change the meaning; it simply restates the converse in a way that makes it more clear.

CLASS ACTIVITY

Give the converse of each of the following statements.

1. If you are late to class, then you are tardy.
   If you are tardy, then you are late to class.
2. If you are a safe driver, then you do not have accidents.
   If you do not have accidents, then you are a safe driver.
3. If you watch the news on TV, then you are informed.
   If you are informed, then you watch the news on TV.
4. If a ray is an angle bisector, then it divides the angle into two equal angles. See margin.
5. If a triangle is isosceles, then it is equilateral.
   If a triangle is equilateral, then it is isosceles.
6. If a triangle has two equal angles, then the sides opposite these angles are equal. See margin.

DISCOVERY ACTIVITY

1. Tell whether each statement is true or false.

Statement A: If you own a wristwatch, then you own a watch that you wear on your wrist. True
Statement B: If a person skis in the Olympics, then that person is an excellent skier. True
Statement C: If Jack owns a piece of jewelry, then he owns a ring. False
Statement D: If a teenager owns a bicycle, then the teenager is a girl. False

2. Write the converse of each statement above, and tell whether it is true or false.
   See margin.

3. Copy and complete the following tables, using the results from parts 1 and 2. Entries for statement A and its converse have been done for you.

	Statement	Converse
A	true	true
B	true	false

	Statement	Converse
C	false	true
D	false	false

4. Study the tables and write a generalization that seems to be true. Even when you know that a statement is true or false, you cannot use this to tell whether the converse is true or false.

You may have discovered that even when you know that a statement is true or false, you cannot use this to tell whether the converse is true or false.

CLASS ACTIVITY

Consider each of the following statements to be true. Write the converse of each and tell whether the converse is true or false.

1. **If John goes swimming, then the temperature is above 75°F.**
   If the temperature is above 75°F, then John goes swimming; false.
2. **If Joan rides her bicycle to school, then she does not walk.**
   If Joan does not walk, then she rides her bicycle to school; true.
3. **If Maria plays tennis, then it is not snowing.**
   If it is not snowing, then Maria plays tennis; false.
4. **If it is raining, then I carry an umbrella.**
   If I carry an umbrella, then it is raining; false.

Consider each of the following statements to be false. Write the converse of each and tell whether the converse is true or false.

5. **If Alex is twelve years old, then he is a teenager.**
   If Alex is a teenager, then he is twelve years old; false.
6. **If a person is a girl, then the person owns a horse.**
   If a person owns a horse, then the person is a girl; false.
7. **If Juan owns an automobile, then he owns a convertible.**
   If Juan owns a convertible, then he owns an automobile; true.
8. **If a dog is brown, then it is a male.**
   If a dog is a male, then it is brown; false.

A SPECIAL CONVERSE

In the last section, it was proved that if a triangle has two equal angles, then the sides opposite these angles are equal (Theorem 5-3-1).

Using the Page
You may want to have some of the students work together to research and prepare a report on truth tables. Have students include values for p, q, $\sim q$, $\sim p$, $p \rightarrow q$, $q \rightarrow p$, and $p \cap q$. Students should write a sentence to illustrate each part of the truth table.

Extra Practice
1. Consider the following statement to be true. Write the converse and tell whether the converse is true or false. If Peter cooks, then he makes an omelet. [If Peter makes an omelet, then he cooks; true.]
2. Consider the following statement to be false. Write the converse and tell whether the converse is true or false. If Mona babysits, then she earns $5 per hour. [If Mona earns $5 per hour, then she babysits; false.]

Using the Page
Have students look at the diagram next to the *Given*. Point out that $\overline{CD}$ was not part of the given. It is an *auxiliary line*, drawn to help with the proof. Point out that in the proof (Step 2), you must be able to give a reason that justifies the drawing of the auxiliary line.

THE CONVERSE OF THEOREM 5-3-1 IS

If two sides of a triangle are equal, then the angles opposite the two sides are equal.

Just because Theorem 5-3-1 was proved to be true, you cannot assume its converse is true. You must prove the converse true separately.

Provide a reason for each statement in the proof.

Given: $\triangle ABC$ with $AC = BC$
Justify: $m\angle A = m\angle B$

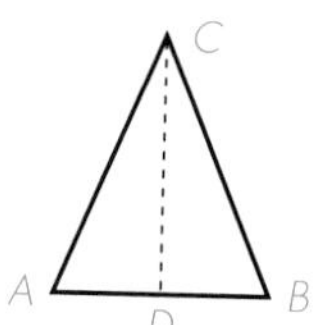

Statements	Reasons
1. $AC = BC$	1. Given
2. Draw the bisector of $\angle C$.	2. Every angle has exactly one bisector.
3. $m\angle ACD = m\angle BCD$	3. Definition of angle bisector
4. $CD = CD$	4. Identity
5. $\triangle ACD \sim \triangle BCD$	5. SAS
6. $\frac{AC}{BC} = 1$	6. Algebra
7. $\triangle ACD \cong \triangle BCD$	7. If two triangles are similar, and the ratio of corresponding sides is one, then the triangles are congruent.
8. $m\angle A = m\angle B$	8. CPCF

THEOREM 5-4-1 **If the triangle is isosceles, then the angles opposite the equal sides are equal.**

CLASS ACTIVITY

1. Write the reasons to complete the following proof.

   Given: $AB = DB$ and $AC = DC$
   Justify: $m\angle BAC = m\angle BDC$

Statements	Reasons
1. $AB = DB$ and $AC = DC$	1. Given
2. $m\angle 1 = m\angle 2$ and $m\angle 3 = m\angle 4$	2. Theorem 5-4-1
3. $m\angle 1 + m\angle 3 = m\angle 2 + m\angle 4$	3. Algebra
4. $m\angle BAC = m\angle BDC$	4. Substitution

2. Use a computer and a geometric drawing tool to draw an isosceles triangle. Measure each angle. Repeat this activity several times, using different angle measures and different side lengths each time. Do your results show that Theorem 5-4-1 is true? Yes

HOME ACTIVITY

Write the converse of each of the following statements.

1. If it is a cat, then it is a Siamese.
   If it is a Siamese, then it is a cat.
2. If it is summer, then I'll go swimming.
   If I go swimming, then it is summer.
3. If he is a judge, then he is honest.
   If he is honest, then he is a judge.
4. If you study, then you earn good grades.
   If you earn good grades, then you study.
5. If the triangle is isosceles, then it has two equal angles.
   If the triangle has two equal angles, then it is isosceles.

Consider each of the following statements to be true. Write the converse of each and tell whether the converse is true or false.

6. If I have a quarter, then I have 25 cents.
   If I have 25 cents, then I have a quarter; false.
7. If Barb is not walking, then she is riding her horse.
   If Barb is riding her horse, then she is not walking; true.
8. If clothes are washed, then they are clean.
   If clothes are clean, then they are washed; true.
9. If two lines are perpendicular, then they meet at right angles.
   If two lines meet at right angles, then they are perpendicular; true.
10. If two angles are complementary, then the sum of their measures is 90°.
    If the sum of the measures of two angles is 90°, then they are complementary; true.

Consider each of the following statements to be false. Write the converse of each and tell whether the converse is true or false.

11. If it is winter in New York, then the date is January 31.
    If the date is January 31, then it is winter in New York; true.
12. If James is not at home, then he is in school.
    If James is in school, then he is not at home; true.
13. If an animal is a dog, then it has two legs.
    If an animal has two legs, then it is a dog; false.
14. If a figure has four sides, then it is a triangle.
    If a figure is a triangle, then it has four sides; false.
15. If two angles are congruent, then they are right angles.
    If two angles are right angles, then they are congruent; true.

Extra Practice
Write the converse of each and tell whether the converse is true or false.

1. If two lines are parallel, then they do not intersect. [If two lines do not intersect, then they are parallel; false.]
2. If a figure is a square, then all four sides are equal. [If a figure has all four sides equal, then it is a square; false.]

16. Write the reasons to complete the proof.
Equilateral triangles are equiangular.
Given: $\triangle ABC$ with $AB = BC = CA$
Prove: $m\angle A = m\angle B = m\angle C$

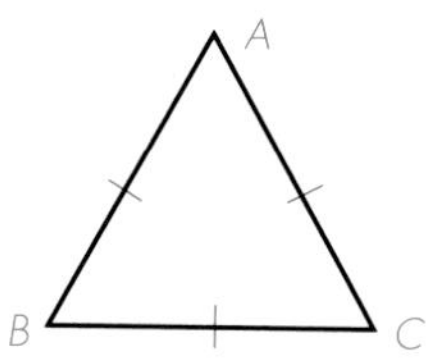

Statements	Reasons
1. $AB = CA$	1. Given
2. $m\angle B = m\angle C$	2. Theorem 5-4-1
3. $BC = CA$	3. Given
4. $m\angle A = m\angle B$	4. Theorem 5-4-1
5. $m\angle A = m\angle B = m\angle C$	5. Substitution

17. Given: $\triangle ABC$ with $AC = BC$. $\overline{CD}$ bisects $\angle ACB$.
Prove: $\overline{CD} \perp \overline{AB}$
See margin.

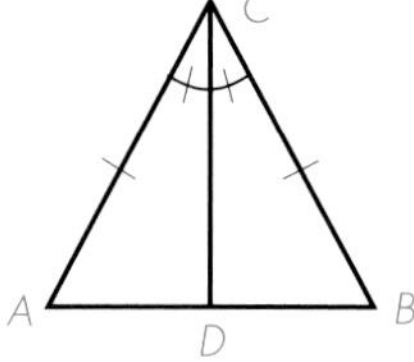

18. Given: $\triangle ABC$ with $AC = BC$. $\overline{AD}$ and $\overline{BE}$ are angle bisectors.
Prove: $AE = BD$
See margin.

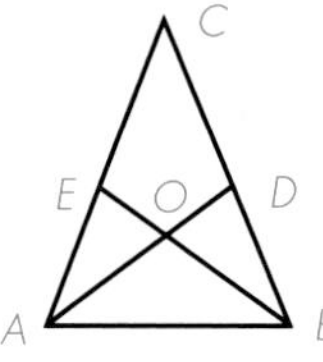

CRITICAL THINKING

19. Observe the patterns of equilateral triangles.

  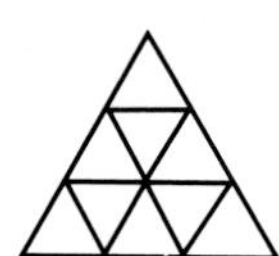 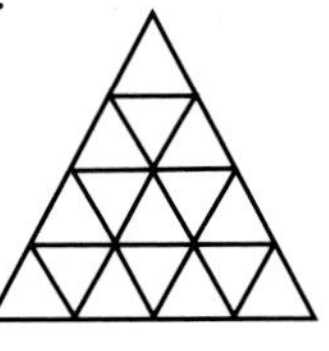

Number of rows	1	2	3	4
Number of small $\triangle$'s	1	4	9	16

Find a pattern. Predict the number of small triangles in a pattern with 15 rows. 225

Additional Answers

17. $AC = BC$ is given, and $m\angle ACD = m\angle BCD$ by the definition of an angle bisector. $CD = CD$, so $\triangle ADC \sim \triangle BDC$ by SAS. $m\angle ADC = m\angle BDC$ because corresponding angles are equal, and $m\angle ADC + m\angle BDC = 180°$ because they form a straight angle. By substitution we find both angles equal 90°, so $\overline{CD} \perp \overline{AB}$.

18. Angles opposite equal sides are equal, so $m\angle EAB = m\angle DBA$. These angles are bisected, so $m\angle EAO = m\angle DBO$, and $m\angle OAB = m\angle OBA$. $m\angle EOA = m\angle DOB$ because they are vertical angles. So by AA, $\triangle EOA \sim \triangle DOB$. Since sides opposite equal angles are equal, $AO = BO$, and $\frac{AO}{BO} = 1$. Therefore, $\triangle EOA \cong \triangle DOB$, and $AE = BD$.

FOLLOW-UP

Assessment

1. How do you form the converse of a statement? [Switch the hypothesis and the conclusion.]

2. In $\triangle UVW$, $UV = UW$. If $m\angle V = 46°$, what is $m\angle W$? [46°]

Reteaching
Students who have had difficulty with this section may benefit from Reteaching Activity 5.4.

Enrichment
For students who have mastered the material in this section, you may wish to assign Enrichment Activity 5.4.

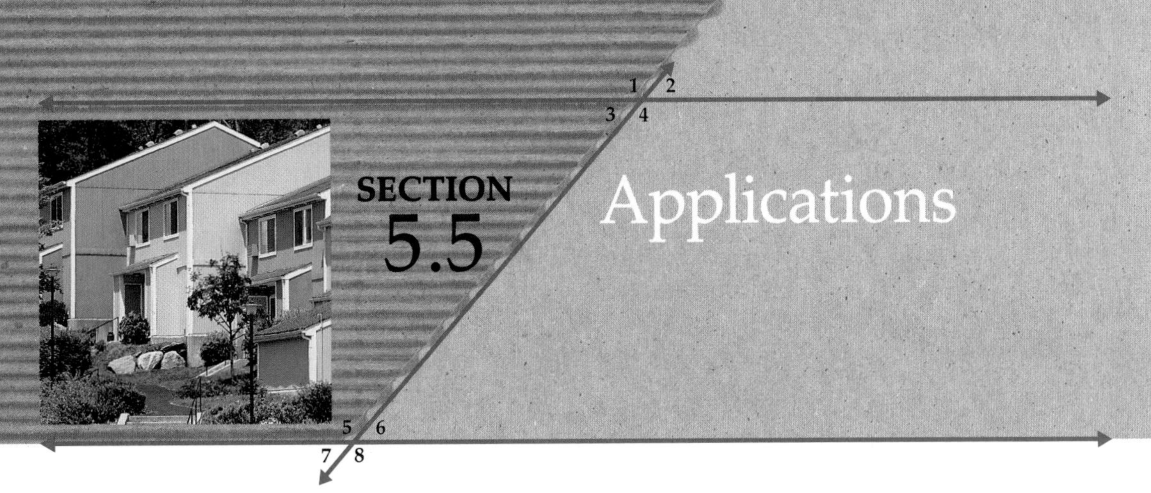

SECTION 5.5 Applications

After completing this section, you should understand
▲ applications of congruent triangles.
▲ applications of the converse.

You saw in the last section that no matter if a statement is true (or false), you must look at the converse itself to decide if it is true or false. The advertising world makes use of this. A true statement is made. Some people may believe that the converse is also true.

APPLICATIONS OF THE CONVERSE

Example: "Smart shoppers shop at store X."

In an if-then form, this statement is
If you are a smart shopper, then you shop at store X.

This can be a true statement. A smart shopper may well go to store X because the store has some good buys. The store is hoping that people will think that the converse is true: if you shop at store X, then you are a smart shopper. They are trying to appeal to a person's pride—we all like to think we are smart shoppers. But going to a particular store doesn't necessarily mean a person knows how to look for the best buy!

REPORT 5-5-1 Consult with several local merchants as to

1. the time (day of week, month, season) when they order their merchandise;
2. how they select the merchandise; and
3. how they develop advertising to attract the market.

Bring to class some examples of local advertising. When possible, write the advertisement in an if-then form. Write the converse of the statement, and check it for accuracy.

SECTION 5.5

Resources
Reteaching 5.5
Enrichment 5.5

Objectives
After completing this section, the student should understand
▲ applications of congruent triangles.
▲ applications of the converse.

Materials
merchandise stores advertisements

GETTING STARTED

Quiz: Prerequisite Skills

1. Write the converse of the statement. If you like our turkey, then you'll love our dressing. [If you love our dressing, then you'll like our turkey.]

2. $\triangle MVW = \triangle TRO$. Name the side congruent to $\overline{MW}$. [$\overline{TO}$]

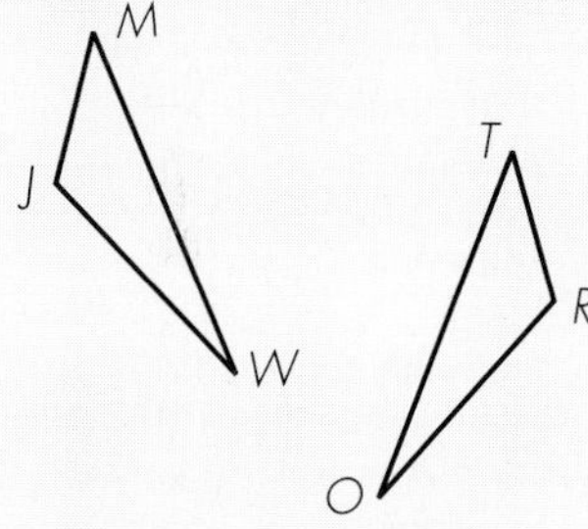

Chalkboard Example
People on the go drink Jus d'Orangé.
Write the statement in if-then form. Write the converse of the statement. What does the converse want you to think, if you assume that the converse is true? [If you are on the go, then you drink Jus d'Orangé. If you drink Jus d'Orangé, then you are on the go. People who drink Jus d'Orangé will have lots of energy.]

TEACHING COMMENTS

Using the Page
Students should find the application of converses interesting. Most advertisers will not make a false statement. But some ads are worded in such a way that someone could assume that the converse is also true and be influenced into believing something that may not be true.

Additional Answers
3. You can be as good as an NBA star just by wearing a certain brand of basketball shoes.
4. If two triangles are similar, then they are congruent. No.
5. If two angles of a triangle are equal in measure, then the sides opposite the equal angles are equal in length. Yes.
6. If the length of one side of a triangle is less than 13, then the sum of the lengths of the other two sides is 13. No.

Extra Practice
An advertisement states *Our food is the food of champion dogs.*
1. Write the statement in if-then form. [If your dog is a champion, then it eats our food.]
2. Write the converse of your statement in Exercise 1. [If your dog eats our food, it will be a champion.]
3. What do you think the ad wants you to conclude? [The type of dog food you feed your dog is enough to make it a champion.]

CLASS ACTIVITY

In a magazine directed at high school athletes, an ad shows an NBA star wearing a certain brand of basketball shoes.

1. Write your interpretation of the ad in if-then form.
   If you are an NBA star, then you wear these shoes.
2. Write the converse of the statement in Exercise 1.
   If you wear these shoes, then you are an NBA star.
3. What do you think the ad wants you to conclude? See margin.

Assume that each statement is true. Write the converse. Is the converse true? See margin.

4. If two triangles are congruent, then they are similar.
5. If two sides of a triangle are equal in length, then the angles opposite the equal sides are equal in measure.
6. If the sum of the lengths of two sides of a triangle is 13, then the length of the third side is less than 13.

APPLICATION OF CONGRUENT FIGURES

Mass production plays an important role in our society. The blueprints for houses involve drawings that are similar figures of the houses. The construction of the houses can involve thousands of congruent pieces.

Triangles are often used in construction, as triangular braces are very strong.

Example: The Williams family wants to build a fence around their backyard. They drew a diagram of two sections of fence.

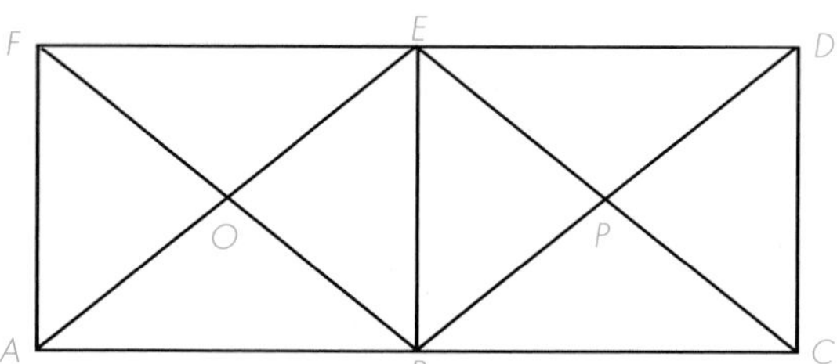

The following information is given in the diagram.

$\overline{AF} \parallel \overline{BE} \parallel \overline{CD}$
$\overline{FD} \parallel \overline{AC}$
$\overline{AE} \parallel \overline{BD}$
$\overline{FB} \parallel \overline{EC}$
$\overline{FA} \perp \overline{FD}$, $\overline{FA} \perp \overline{AC}$
$\overline{EB} \perp \overline{FD}$, $\overline{EB} \perp \overline{AC}$
$\overline{DC} \perp \overline{FD}$, $\overline{DC} \perp \overline{AC}$
$FA = EB = DC = 5$ ft
$AB = BC = FE = ED = 6$ ft

1. Find all of the triangles that are congruent to $\triangle EFA$.
   $\overline{FA} \perp \overline{FD}$, so $\angle EFA$ is a right angle. $\overline{EB} \perp \overline{FD}$, so $\angle FEB$ is a right angle. Then $m\angle EFA = m\angle FEB$. $FE = FE$. $FA = 5$ ft and $EB = 5$ ft, so $FA = EB$. Then $\triangle FEB \cong \triangle EFA$. By similar reasoning, $\triangle BAF$, $\triangle ABE$, $\triangle DEB$, $\triangle CBE$, $\triangle EDC$, and $\triangle BCD$ are congruent to $\triangle EFA$.

2. Find all of the triangles that are congruent to $\triangle FEO$.
   $\overline{FD} \parallel \overline{AC}$, so $\angle EFB \cong \angle ABF$ since they are alternate interior angles. Also, $\angle FEA \cong \angle BAE$ for the same reason. $FE = AB$. Therefore, $\triangle FEO \cong \triangle BAO$. $\overline{AE} \parallel \overline{BD}$. $\angle EAC \cong \angle DBC$ since they are corresponding angles (transversal $\overline{AC}$). $\overline{FB} \parallel \overline{EC}$. $\angle FBA \cong \angle ECA$ since they are corresponding angles (transversal $\overline{AC}$). $AB = BC$. Therefore, $\triangle BAO \cong \triangle BCP$. By use of alternate interior angles, $\triangle EDP \cong \triangle CBP$. Then $\triangle FEO \cong \triangle BAO \cong \triangle EDP \cong \triangle CBP$.

Example: The figure shows the pattern for the Jack in the Pulpit quilt. Each shape is created by cutting and sewing in a piece of fabric in a particular color. If you wanted a quilt large enough to cover a double bed, you would need to create 20 squares like this and sew them together.

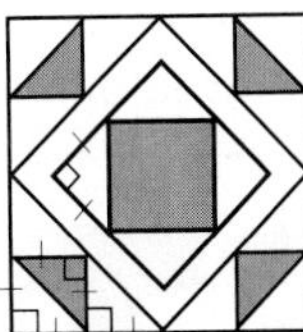

1. Are the triangular pieces congruent?

Yes, they each contain a right angle enclosed by two equal legs. Therefore, they are similar by SAS, and the ratio of two corresponding sides is one.

2. How many unshaded congruent triangles would you need to cut for the entire quilt?

There are 16 unshaded congruent triangles. You would need $16 \cdot 20 = 320$ triangles of this shape and color for your quilt.

Using the Page
Students may find it easier to find all the congruent triangles if they trace the figure at the bottom of page 190 several times, then shade in two triangles at a time as they show each pair congruent.

HOME ACTIVITY

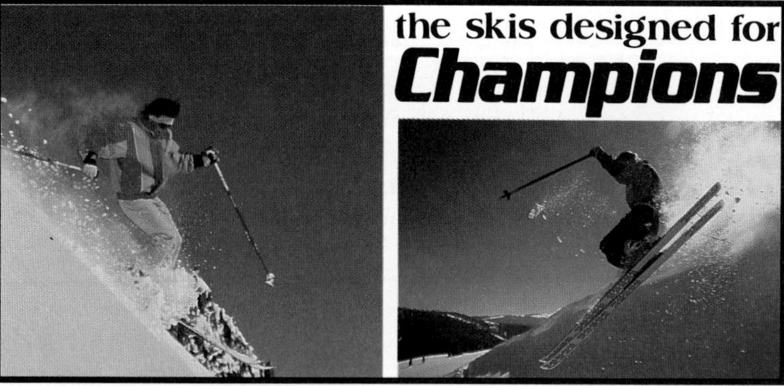

1. **Write your interpretation of the ad in if-then form.**
   If you are a champion, then the skis were designed for you.
2. **Write the converse of the statement in Exercise 1.**
   If the skis were designed for you, then you are a champion.
3. **Do you think the statement in Exercise 1 is true?** Yes
4. **Do you think the converse in Exercise 2 is true?** No

Additional Answers
For Exercises 4 and 8, some students may answer differently. For example, someone in class may be a champion skier. You may want to accept an answer as long as it can be explained reasonably.

Step out in style.

5. **Write your interpretation of this ad in if-then form.**
   If you wear these clothes, then you will be in style.
6. **Write the converse of the statement in Exercise 5.**
   If you are in style, then you wear these clothes.
7. **Do you think the statement in Exercise 5 is true?** Yes
8. **Do you think the converse in Exercise 6 is true?** Yes

The following ad was placed by a summer camp.

We promote
—Positive self-image
—Personal confidence
—Self-reliance
—Academic achievement

9. Write your interpretation of the above ad in if-then form. See margin.

10. Write the converse of the statement you wrote for Exercise 9. See margin.

11. Do you think the statement in Exercise 9 is true? Yes

12. Do you think the converse in Exercise 10 is true? No

13. In figure *ABCD*, $AB = BC = CD = DA$. $m\angle D = m\angle C = 90°$. M is the midpoint of $\overline{DC}$. Explain why $AM = BM$, and why $m\angle 1 = m\angle 2$. See margin.

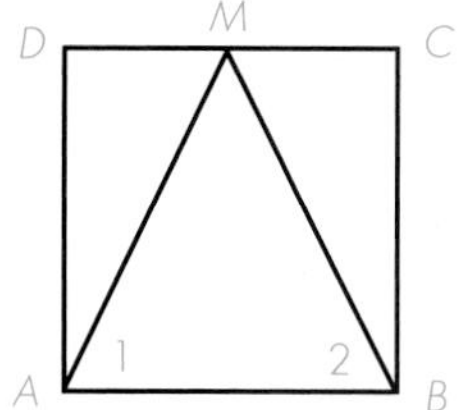

14. In this figure, $JL = LM$, $KL = LN$, and $NM = 54$ feet. What length is JK? 54 ft

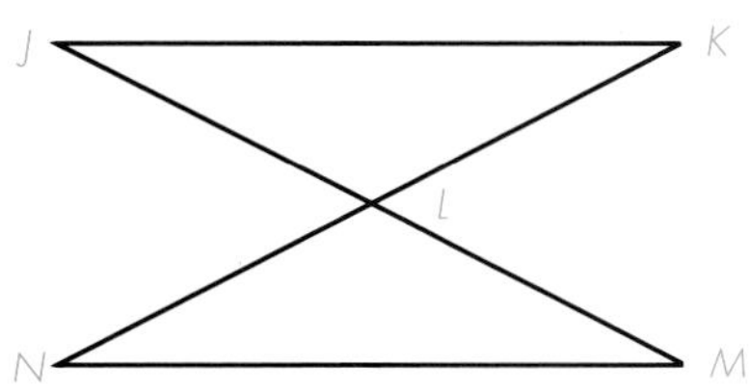

Additional Answers

9. If you attend the summer camp, then you will have a positive self-image, personal confidence, self-reliance, and academic achievement.

10. If you have a positive self-image, personal confidence, self-reliance, and academic achievement, then you attend the summer camp.

13. M is the midpoint, so $DM = CM$. $m\angle C = m\angle D = 90°$, $DA = BC$, and $\frac{DA}{CB} = 1$. Thus $\triangle MDA \cong \triangle MCB$ by SAS and ratio = 1. $MA = MB$ by CPCF. Then $m\angle 1 = m\angle 2$ by Theorem 5-4-1.

15. In this diagram of a garage roof, $AM = BM$. If $\angle PMA$ is a right angle, then how do you know that $AP = BP$? See margin.

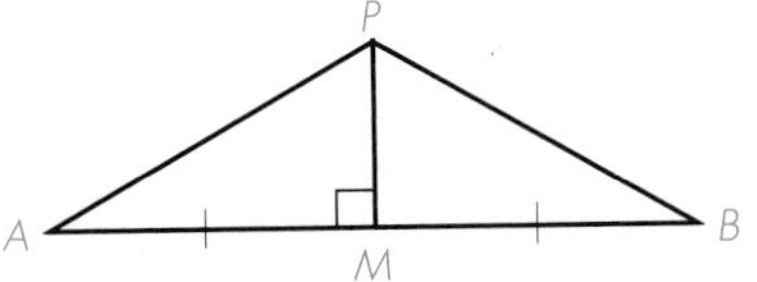

16. In this diagram of a gable roof, $AC = BC$, M is the midpoint of $\overline{AB}$. How do you know that $\overline{CM} \perp \overline{AB}$? See margin.

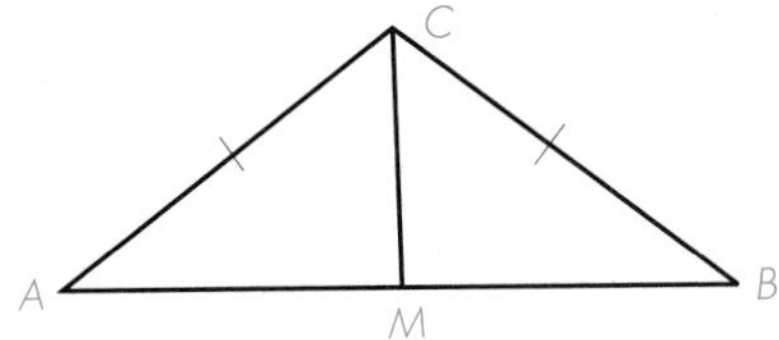

17. A designer created the following design for an ad. $AC = AE$, $AB = AD$, and $m\angle CAB = m\angle EAD$. Show that $CB = ED$. See margin.

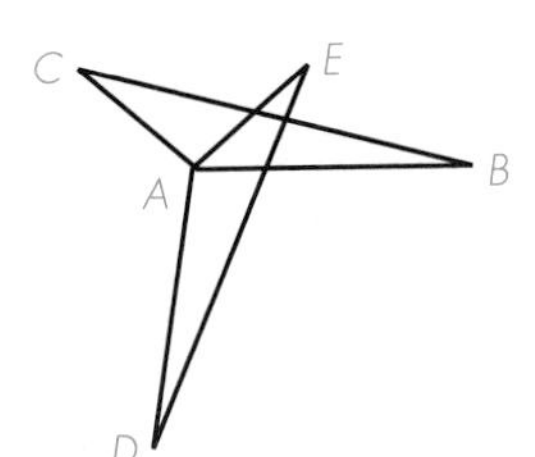

18. In this drawing of a park, $\overline{DE} \parallel \overline{AB}$. $AB = 40$ m, $AC = 25$ m, $BC = 35$ m, and $AD = 10$ m. What is the length of $\overline{CE}$? 21 m

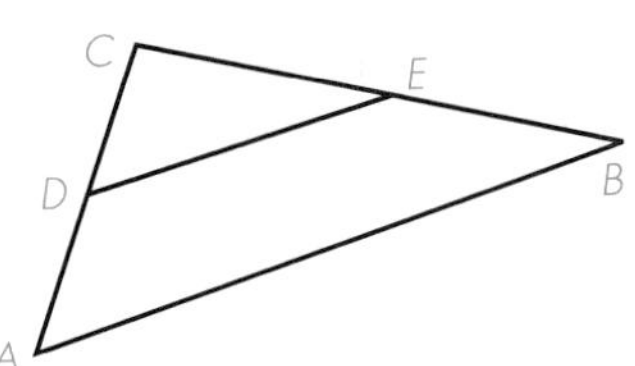

CRITICAL THINKING

19. Find a pattern in the table. Use the pattern to predict the number of diagonals in a figure having 24 sides.

Number of Sides	Number of Diagonals
3	0
4	2
5	5
6	9
7	14
8	20

$\frac{n(n-3)}{2}$; 252

Additional Answers

15. $AM = BM$; $\angle PMA$ is a right angle; so $\angle PMB$ is a right angle; $PM = PM$ and $\frac{AM}{BM} = 1$. $\triangle PMA \cong \triangle PMB$ by SAS and ratio = 1. $AP = BP$ by CPCF.

16. $AC = BC$; M is the midpoint of $\overline{AB}$, so $AM = BM$. $CM = CM$. $\frac{CM}{CM} = 1$. $\triangle ACM \cong \triangle BCM$ by SSS and ratio = 1. Then $m\angle CMA = m\angle CMB$. Since $\angle AMB$ is a straight angle, $m\angle CMA = m\angle CMB = 90°$. Then $\overline{CM} \perp \overline{AB}$.

17. $AC = AE$; $AB = AD$; $m\angle CAB = m\angle EAD$. $\frac{AC}{AE} = \frac{AB}{AD} = 1$. $\triangle CAB \cong \triangle EAD$ by SAS and ratio = 1. Then $CB = ED$ by CPCF.

FOLLOW-UP

Assessment

1. Explain why advertisers hope that people will believe that the converse of a true statement is also true. [It can influence how they think about a product or service.]

2.

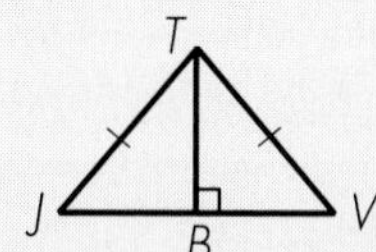

$JT = VT$, and $\overline{TB} \perp \overline{JV}$. How do you know that B is the midpoint of $\overline{JV}$? [Since $JT = VT$, $m\angle J = m\angle V$. $\overline{TB} \perp \overline{JV}$, so $\triangle TBJ$ and $\triangle TBV$ are right angles, and are therefore congruent. $\frac{JT}{TV} = 1$, so $\triangle JTB \cong \triangle VTB$ by AA and ratio = 1. Then $JB = VB$ by CPCF. Thus B is the midpoint of $\overline{JV}$ by definition of midpoint.]

Reteaching

Students who have had difficulty with this section may benefit from Reteaching Activity 5.5.

Enrichment

For students who have mastered the material in this section, you may wish to assign Enrichment Activity 5.5.

5.1 Write all the subsets of $A = \{1, 4, 9\}$. $\{1\}$, $\{4\}$, $\{9\}$, $\{1, 4\}$, $\{1, 9\}$, $\{4, 9\}$, $\{1, 4, 9\}$, $\{\ \}$

$B = \{1, 4, 7, 10, 13\}$ $C = \{2, 4, 6, 8, 10\}$
Draw Venn diagrams showing $B \cup C$ and $B \cap C$.

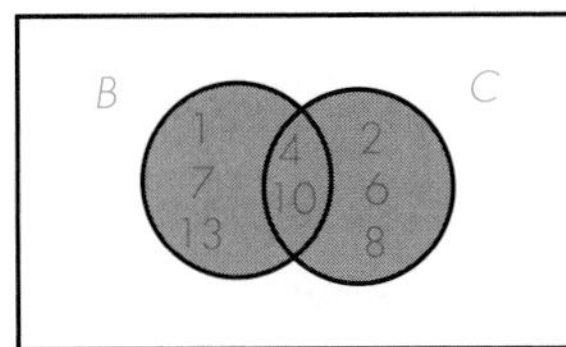

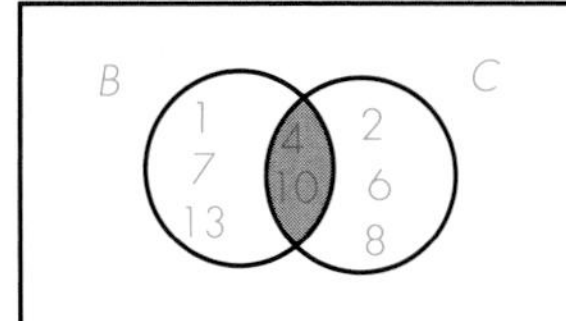

5.2 SQ bisects $\angle PQR$. Is $\triangle PQS$ congruent to $\triangle RQS$?
$PQ = QR$, $QS = QS$, and $m\angle PQS = m\angle RQS$.
$\triangle PQS \sim \triangle RQS$ (by SAS) and $\frac{SQ}{SQ} = \frac{PQ}{RQ} = 1$.
Therefore, $\triangle PQS \cong \triangle RQS$.

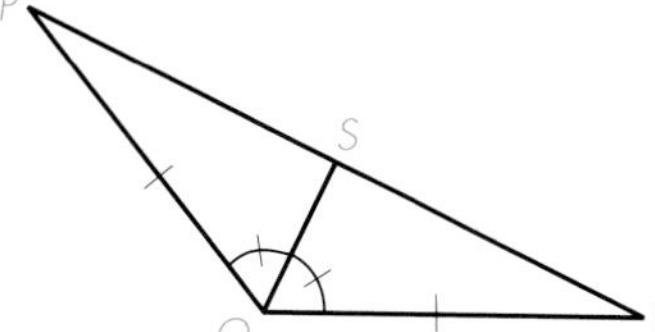

5.3 Find x in $\triangle DRY$.
$\triangle DRY \sim \triangle TON$ (by AA) and $\frac{DR}{TO} = 1$
Therefore $\triangle DRY \cong \triangle TON$, and $\frac{DY}{TN} = 1$.
$\frac{x}{10} = 1$
$x = 10$

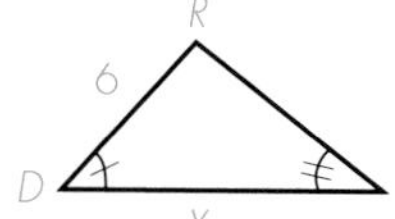

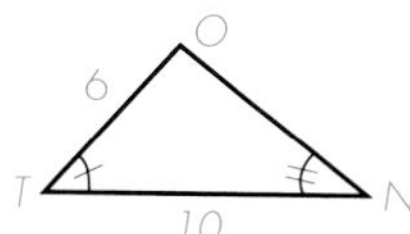

5.4 This is a true statement: If a triangle is equilateral, then it is isosceles. Is the converse true?

Converse: If a triangle is isosceles, then it is equilateral; false.

This is a true statement: If a triangle has two equal angles, then the sides opposite these angles are equal. Is the converse true?

Converse: If a triangle has two equal sides, then the angles opposite these sides are equal; true.

The truth of a statement does not determine the truth of its converse.

Chapter Vocabulary
braces
conclusion
congruent triangles
converse
corresponding parts of congruent figures
element
empty set
general case
hypothesis
intersection
member
set
subset
union
universal set
universe
Venn diagram
well-defined

195

COMPUTER ACTIVITY

Using the Page
This Computer Activity investigates similarity of right triangles. Some special methods of determining similarity in right triangles are presented. Both side relationships and angle relationships are considered.

You have learned that all right triangles contain one right angle. So we automatically know that two right triangles have one pair of angles with equal measures. Thus, to prove that two right triangles are similar, we must only prove either that another pair of angles have the same measure (AA) or that the pairs of legs of the right triangles have the same lengths (SS).

The LOGO procedure in the example below can be used to show that two right triangles are similar if one pair of the acute angles have equal measures.

Example: Write a LOGO procedure to draw two right triangles, each with one acute angle whose measure is 25. Do the two right triangles appear to be similar? Justify your answer.

First write a procedure to draw one right triangle with an acute angle with a measure of 25.

```
TO RTTRI1
BK 80
RT 90
FD 40
LT 155
FD 70
END
```

Then use the RTTRI1 procedure in a procedure to draw both of the right triangles.

```
TO RTTRI2
RTTRI1
PU HOME LT 90 FD 100 RT 90 PD
RTTRI 1
END
```

The output of RTTRI2 is shown below.

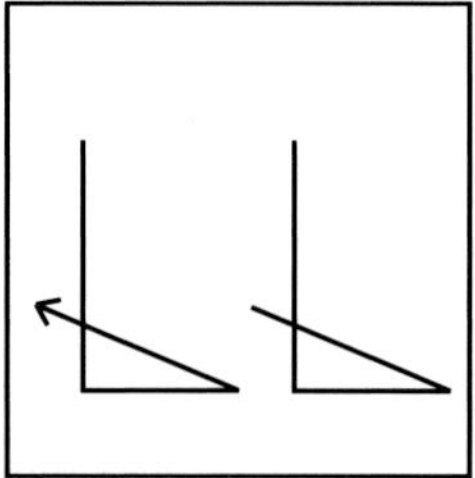

The two right triangles do appear to be similar.

196

The triangles can be proven to be congruent by AA and showing that the ratio of the corresponding sides is $\frac{40}{40}$ or one.

1. Write a LOGO procedure named RTTRI3 to draw two right triangles, each with one leg with a length of 25 spaces and one leg with a length of 35 spaces. Do the two right triangles appear to be congruent? Justify your answer.
   Answers for the procedure may vary. Yes; the triangles are similar by SAS and $\frac{25}{25} = \frac{35}{35} = 1$.

Modify the RTTRI1 procedure to tell the turtle to draw each right triangle. Then modify the RTTRI2 procedure to tell the turtle to draw a right triangle that is congruent to the given triangle.
Answers may vary.

2. 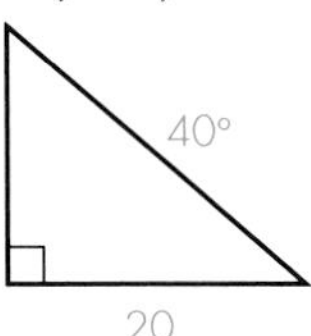

3. 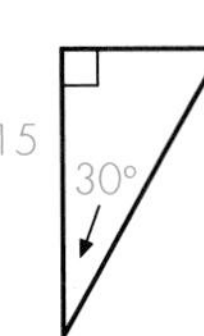

4. 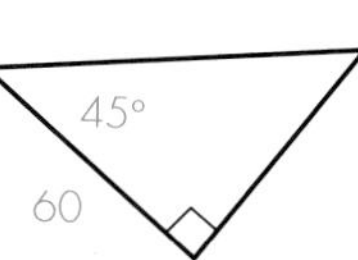

5. 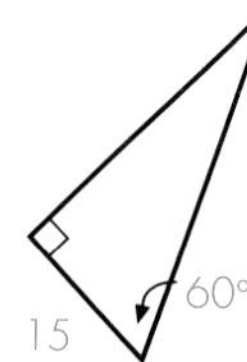

Modify the RTTRI3 procedure to tell the turtle to draw each right triangle and a right triangle that is congruent to it.

6. 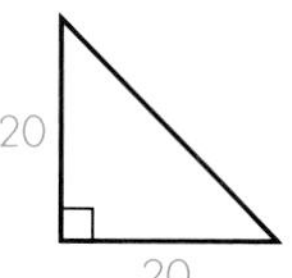

7. 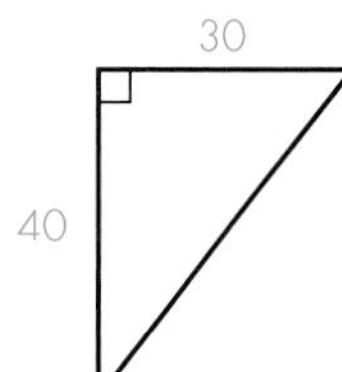

8. 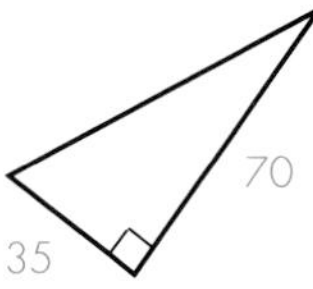

9. 

Write a LOGO procedure to tell the turtle to draw the figure. Answers may vary.

10. 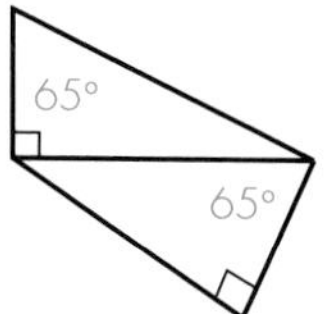

11. 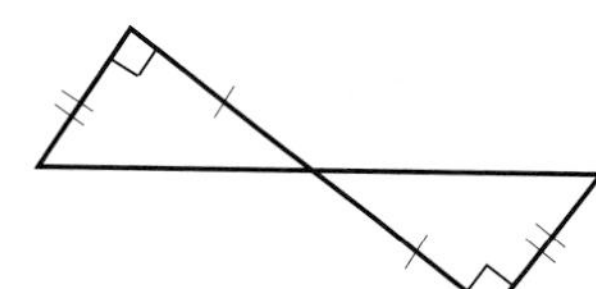

Using the Page
For Exercises 1, 6-9, and 11, students will need to experiment with the measures of the acute angles of the triangles until the legs have the indicated lengths.

Using the Page
Students can work individually or in groups to study the Chapter 5 Review. If you have students study in groups, they should work together to make sure all members of the group know the material in the chapter.

Informal assessment such as interviews, classroom observations, or a review of student portfolios can be used instead of paper-and-pencil tests.

13

1. Explain the significance of the congruence symbol.
   It is a combination of the symbols for similarity and equality.

Classify each statement as true or false.

2. All similar triangles are congruent. False
3. All congruent triangles are similar. True
4. Some similar triangles are congruent. True
5. If the ratio of the corresponding sides for similar triangles is equal to one, then the triangles are congruent. True

Refer to figures a–h.

6. Which triangles are similar? a & f; b, d, & g; c, e, & h
7. Which triangles are congruent? b & g; c & e

a. 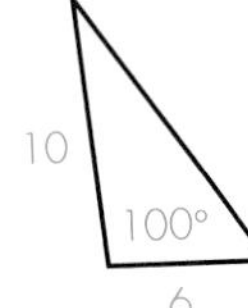

b. 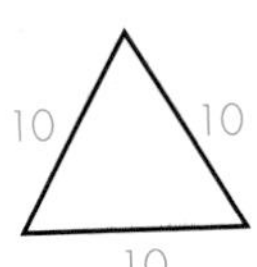

c. 

d. 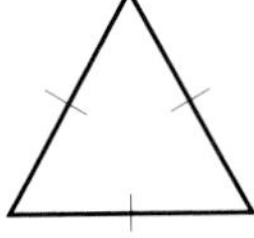

e. 

f. 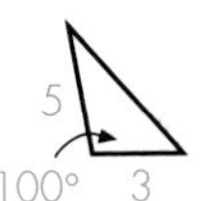

g. 

h. 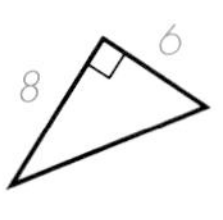

This information was collected for the after-school geometry class picnic.

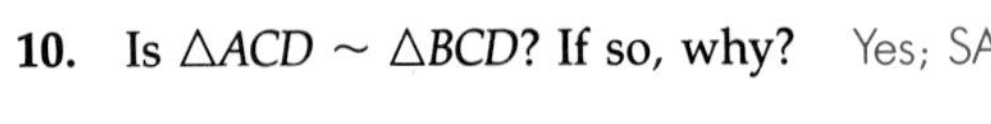

10 voted for pizza (P). 17 voted for hot dogs (H).
6 voted for P and H. 2 voted for neither P or H.

8. Draw the Venn diagram illustrating this information. See margin.
9. How many students are in this geometry class? 23

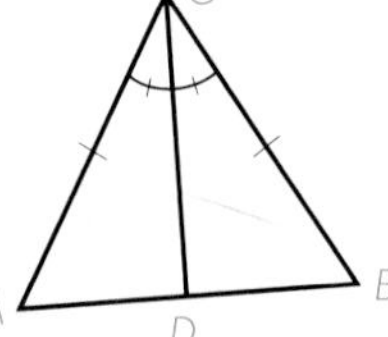

Refer to $\triangle ABC$.

10. Is $\triangle ACD \sim \triangle BCD$? If so, why? Yes; SAS

11. Is $\triangle ACD \cong \triangle BCD$? If so, why?
    Yes; $\triangle ACD \sim \triangle BCD$, $AC = BC$, so $\frac{AC}{BC} = 1$ SAS
12. Give the converse of the statement: If you want to be a better thinker, then study geometry.
    If you study geometry, then you want to be a better thinker.

Additional Answers
8.

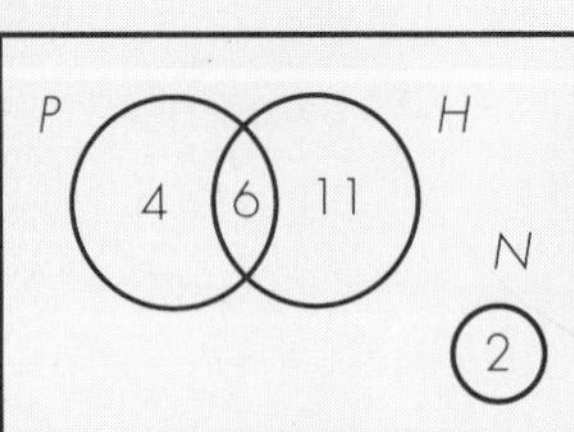

CHAPTER TEST

Let $A = \{1, 4, 9, 16, 25\}$.

1. Is $\{36\} \subset A$? No

2. List all the three-element subsets of A. See margin.

Refer to figure $ABCD$. $AB = CD = AD = BC$, $AE = CF$. $\overline{AC}$ is a straight line segment.

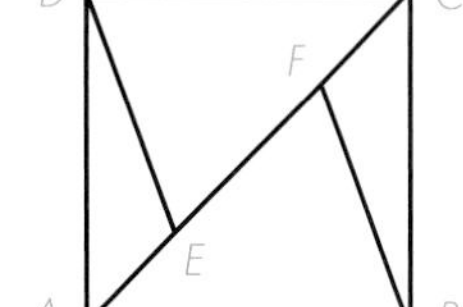

3. Is $m\angle DAC = m\angle DCA$? Explain. See margin.

4. Justify that $\triangle AED \cong \triangle CFB$. See margin.

5. Is $DE = BF$? Explain. Yes; CPCF

6. Show that $m\angle DEF = m\angle BFE$.
   $DE = BF$, so $m\angle DEF = m\angle BFE$

A car manufacturer places an ornament on the hood of the new model. The hood ornament is a triangle, △. The ad copy is

△ A SYMBOL OF QUALITY

7. Write in if-then form what you think the advertisement wants you to conclude.
   If you buy △, then you have quality.
8. Write the converse of your statement in Exercise 7. If you want quality, then you buy △.

Use the survey information for Exercises 9 and 10.

12 people like grapes (G).	15 people like nectarines (N).
5 people like both G and N.	3 people like neither G or N.

9. Draw a Venn diagram showing this information. See margin.

10. How many people were surveyed? 25

Assessment Resources
Achievement Tests pp. 9-10

Test Objectives
After studying this chapter, students should understand
- what a set is.
- what Venn diagrams are.
- what set operations are.
- how to solve problems using sets.
- the meaning of congruent triangles.
- how to prove triangles congruent.
- corresponding parts of congruent figures.
- how to determine lengths of segments and measurements of angles.
- the meaning of converse.
- applications of the converse.

Additional Answers

2. {1, 4, 9}, {1, 4, 16}, {1, 4, 25}, {1, 9, 16}, {1, 9, 25}, {1, 16, 25}, {4, 9, 16}, {4, 9, 25}, {4, 16, 25}, {9, 16, 25}

3. Yes; $AD = CD$, so the angles opposite these sides ($\triangle ADC$) are equal.

4. Draw $\overline{FD}$. $AD = CD$, $m\angle DAC = m\angle DCA$ from Exercise 3, $AE = CF$. $\frac{AE}{CF} = \frac{AD}{CD}$. So $\triangle AED \sim \triangle CFD$ by SAS. Since $\frac{AE}{CF} = \frac{AD}{CD} = 1$, $\triangle AED \cong \triangle CFD$.

9.

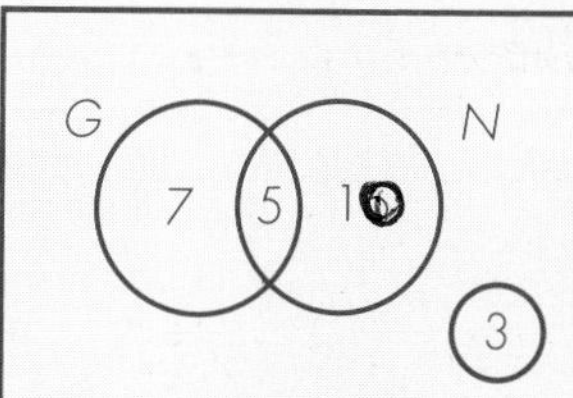

Algebra Review Objectives
Topics reviewed from algebra include

- solving a first-degree equation in one variable.
- solving a second-degree equation in one variable by using the quadratic formula.
- problem solving using guess-and-check.

ALGEBRA SKILLS

Example: First-degree equation

$$3x - 4 = 5(2x - 7)$$
$$3x - 4 = 10x - 35$$
$$3x + 31 = 10x$$
$$31 = 7x$$
$$x = 4\tfrac{3}{7}$$

Example: Second-degree equation

$2x^2 - 3x - 35 = 0$

Use the quadratic formula, $x = \frac{-b \pm \sqrt{b^2 - 4ac}}{2a}$

$$x = \frac{-(-3) \pm \sqrt{(-3)^2 - 4(2)(-35)}}{2(2)}$$
$$x = \frac{3 \pm \sqrt{9 + 280}}{4}$$
$$x = \frac{3 \pm \sqrt{289}}{4}$$
$$x = \frac{3 \pm 17}{4}$$
$$x = 5 \text{ or } x = -\tfrac{7}{2}$$

Solve each equation for x.

1. $3x - 45 = 90$ $x = 45$

2. $x^2 = 36$ $x = 6, -6$

3. $5x^2 = 125$ $x = 5, -5$

4. $17x - 5(3x - 50) = 68$ $x = -91$

5. $x^2 + x - 6 = 0$ $x = 2, -3$

6. $5x - 29 = 41$ $x = 14$

7. $2(x - 3) = 3(2x + 6)$ $x = -6$

8. $2x^2 - 2x - 12 = 0$ $x = 3, -2$

9. $x^2 + 15x + 26 = 0$ $x = -2, -13$

10. $4x + 6 = 7x - 9$ $x = 5$

11. $3x^2 - 5x - 2 = 0$ $x = 2, -\frac{1}{3}$

12. $3(2x - 7) = 4(3x + 9)$ $x = -9\frac{1}{2}$

13. $5(4x + 6) = 3(5x - 4)$ $x = -8.4$

14. $x^2 - 7x + 6 = 0$ $x = 1, 6$

15. $x^2 - 5x - 9 = 0$ $x = \frac{5 \pm \sqrt{61}}{2}$

16. $3(7x - 9) = 4(2x + 5)$ $x = 3\frac{8}{13}$

17. $3x + 17 = 5x - 9$ $x = 13$

18. $3x^2 + 6x + 4 = 0$ No real solution

19. $2x^2 - 7x + 6 = 0$ $x = 2, \frac{3}{2}$

20. $7(2x - 4) = 8(3x + 5)$ $x = -6.8$

21. A counselor asked a new student her age. The student replied that two times her age plus the square root of 36 resulted in a perfect square number. The counselor knew that the student was between 10 and 18, and that today was her birthday. How old is the new student? 15 yrs

Name ______ Date ______

Sets and Subsets

A *set* is a collection which can be defined either by listing its members, or by describing what belongs in the set. A set is *well-defined* only if for any given item you can be certain whether it belongs to the set or not.

The *union* of two sets is a new set containing all the members of both original sets, each member only once. The *intersection* of two sets is a new set containing just the members common to both sets.

Use the words *union* and *intersection* to describe the shaded part of each drawing.

1. union of the circular and rectangular regions
2. intersection of the square and triangular regions
3. all points in the union that are not in the intersection

Use the two Venn diagrams at the right. Complete each statement with one of the following: set *A*, set *Y*, union, intersection.

4. The bottom diagram shows the union of set *X* and set Y.
5. 28 and 9 are the intersection of sets *A* and *B*.
6. Set A is the set with the most members.
7. Set Y has members 12, 1, 11, 99, and 76.
8. 12 and 1 are in the intersection of sets *X* and *Y*.

Write *yes* or *no* to tell whether each set is well-defined.

9. right triangles yes
10. attractive drawings no
11. Easterners no
12. New Yorkers yes
13. successful baseball players no
14. members of baseball's Hall of Fame yes

Name ______ Date ______

Making a Precise Copy

On a Macintosh computer, an exact copy of a complicated geometric figure can be made just by choosing the copy command and then the paste command.

Before the use of computers, people used precision drawing instruments—rulers, protractors, compasses, T-squares, dividers—to construct exact copies of figures.

Two thousand years ago when the Greeks were first investigating the properties of geometric shapes, they did not even have precise rulers or protractors. So they were concerned with discovering methods for drawing precise figures using just a compass and a straightedge.

Describe three practical situations in which it would be important to make a precise copy of a figure. Accept any reasonable answers.

1. during construction of a building
2. a company mass produces a book with drawings
3. to prepare a blueprint for a machine

Use Figures 1–3 at the right for Exercises 4 and 5.

4. Make a tracing of Figure 1. Now draw the figure again using a protractor; each angle measures 108°. Why is the second drawing more accurate?
   A more accurate angle can be obtained using a protractor.
5. If you can copy a triangle, you can copy any polygon. Use Figures 2 and 3 to help you determine why this is so. Write your reason below. Now copy the two figures using a ruler and a protractor.
   All polygons can be divided into triangles.
6. Copy the four figures at the bottom of the page, using any method you wish, including a computer. Make your copies as precise as possible.

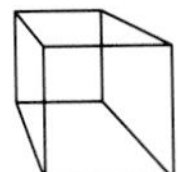
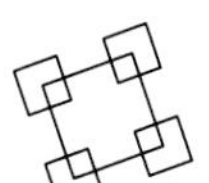

Name ______________________ Date ______________

Congruent Triangles

Recall that similar triangles have the same shape. The measurements of corresponding angles are equal.

Congruent triangles have the same shape and also the same size.

You can make congruent triangles by tracing or by folding paper.

Fold a sheet of notebook paper as shown. Cut off the top rectangle.

You will have two congruent right triangles.

1. Using a ruler, draw a triangle on an index card or sheet of paper. Make the sides any lengths you choose. Cut out your triangle and give it a name.
2. Draw four copies of your triangle as shown in the figure at the right. Notice that the left and right triangles are rotated versions of the top triangle, and that the bottom triangle is a flipped version.
3. Now choose three letters for the vertices of your triangle. For example, *GWS* (for George Washington Smith). Write the letters on all four triangles at the right. Make sure the same vertex is always labeled with the same letter.

Start with a sheet of paper that is $8\frac{1}{2}$ by 11 inches. Make the four folds as numbered at the right. Measure to the nearest eighth of an inch.

4. Cut along Fold 1. The congruent triangles have sides that measure $8\frac{1}{2}$, 11, and $13\frac{7}{8}$ inches.
5. Cut along Fold 2. One pair of congruent triangles have sides that measure $8\frac{1}{2}$, 7, and 7 inches. The second pair measures 11, 7, and 7 inches.
6. Cut along Fold 3. The congruent triangles have sides that measure $5\frac{1}{2}$, $4\frac{1}{4}$, and 7 inches.
7. Cut along Fold 4. The congruent triangles have sides that measure $5\frac{1}{2}$, $4\frac{1}{4}$, and 7 inches.

Name ______________________ Date ______________

The Five Platonic Solids

In a *regular polygon*, all the sides and all the angles are congruent. A square is a regular polygon. Figures A and B are two-dimensional polygons. B is concave and not regular; A is convex and regular.

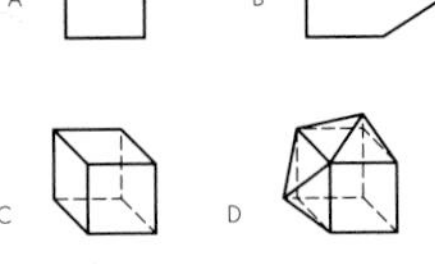

A *regular polyhedron* is a three-dimensional shape in which all the faces are regular polygons. Figure C is regular and convex. Figure D is concave and not regular.

There are only five regular convex polyhedrons. They are called the *Platonic Solids.*

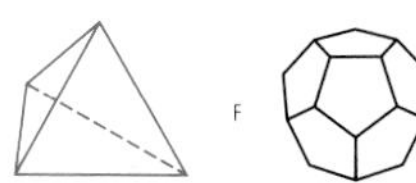

1. Figure C is one of the Platonic Solids. It is called a regular *hexahedron* because it has six congruent faces. What is another name for the hexahedron? cube
2. A second Platonic Solid is the regular *tetrahedron*. It has four faces. Each face is an equilateral triangle. Sketch a tetrahedron in the space at the right. What is another name for this shape? pyramid
3. Figure E is a regular *icosahedron*. It has 20 congruent faces. Describe one of the faces. equilateral triangle
4. Figure F is a regular *dodecahedron*. Describe one of its 12 faces. regular pentagon
5. Complete this chart to describe the characteristics of the five Platonic Solids.

	Tetrahedron	Octahedron	Icosahedron	Hexahedron	Dodecahedron
Number of Faces	4	8	20	6	12
Number of Faces Meeting at Each Vertex	3	4	5	3	3
Angle Between Edges	60°	60°	60°	90°	108°
Number of Edges	6	12	30	12	30
Number of Vertices	4	6	12	8	20

6. Build a model of the octahedron.

Name ______________________ Date ______________

Corresponding Parts

Two triangles are *congruent* if

(1) each angle of one triangle is congruent to the corresponding angle of the other triangle, and

(2) each side of one triangle is congruent to the corresponding side of the other triangle.

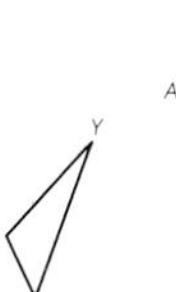

$\triangle XYZ \cong \triangle ABC$

Refer to triangles *ABC* and *XYZ* above. Write six correspondences stating which angles and which sides correspond.

1. $\angle A \leftrightarrow \angle X$ **2.** $\angle B \leftrightarrow \angle Y$ **3.** $\angle C \leftrightarrow \angle Z$

4. $\overline{AB} \leftrightarrow \overline{XY}$ **5.** $\overline{BC} \leftrightarrow \overline{YZ}$ **6.** $\overline{CA} \leftrightarrow \overline{ZX}$

Triangles *MNO* and *PQR* at the right are congruent. Write six statements to show that corresponding angles and corresponding sides are congruent.

7. $\angle M \cong \angle P$ **8.** $\angle N \cong \angle Q$ **9.** $\angle O \cong \angle R$

10. $\overline{MN} \cong \overline{PQ}$ **11.** $\overline{NO} \cong \overline{QR}$ **12.** $\overline{OM} \cong \overline{RP}$

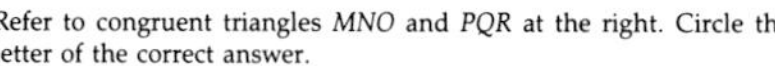

Refer to congruent triangles *MNO* and *PQR* at the right. Circle the letter of the correct answer.

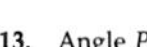

13. Angle *P*

a. is congruent to angle *N*.
(b.) is congruent to angle *M*.
c. corresponds to angle *O*.

14. The ratio of *MN* to *PQ* is

a. less than 1.
b. greater than 1.
(c.) equal to 1.

Is the pair of triangles congruent? Write *yes* or *no*.

15. yes **16.** no **17.** yes

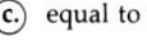

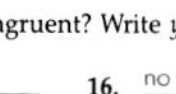

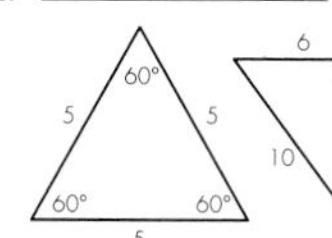

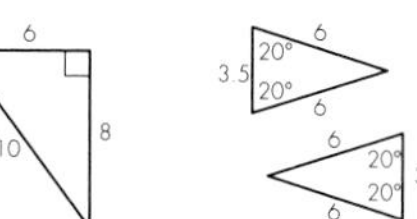

Name ______________________ Date ______________

American Pieced Quilts

The making of pieced, or "patchwork," quilts reached its height in America during the nineteenth century. Scraps left over from clothesmaking and worn garments were cut into shapes and sewed together. Many of the traditional quilt designs were based on congruent geometric shapes.

Describe each quilt in words. Tell what kinds of congruent shapes are used, and how they are arranged and colored to make the pattern. Test your descriptions by having someone who has not seen the designs try to draw them as you read your descriptions. Answers will vary.

1.

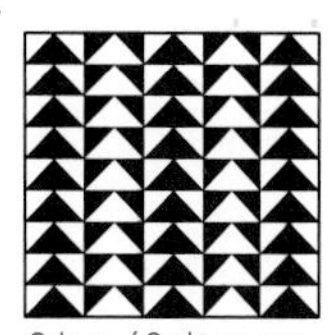

Column of 9 white rectangles each with a black isosceles triangle, whose base is the length of the rectangle, inside. The next column reverses black and white, five columns in all.

2. 3 by 4 checkerboard pattern. Each square has a contrasting octagon in its interior.

3. Checkerboard pattern with circles centered at the vertices of the squares. The checkerboard colors are reversed inside the circles

4. 5 rows of alternating black and white equilateral triangles.

5.

5 rows of alternating gray and white equilateral triangles

6. 5 rows of equilateral triangles. Gray triangles form hexagons in the pattern.

Name ________________ Date ________________

The Converse of a Theorem

Syllogism

If A, then B.
A is true.
———
B is true.

Every theorem in geometry states that *if* something is true, *then* something else is also true. The *if* part is called the *hypothesis;* it states what is given as true. The *then* part is the *conclusion.* If-then statements of this type are called *conditional statements.*

Conditional statements are used in logic arguments. For example, the logic pattern shown at the right is called a *syllogism.*

In each statement, underline the hypothesis. Circle the conclusion.

1. If the measure of an angle is between 0° and 90°, then the angle is called an acute angle.
2. If the measures of two angles total 180°, then the two angles are supplementary.
3. If two lines intersect, then the vertical angles are equal.
4. If a figure is a triangle, then the sum of the measures of the angles is 180°.

Rewrite each of the following statements as a conditional statement.

5. Obtuse angles measure between 90° and 180°.
   If an angle measures between 90° and 180°, then it is obtuse.
6. The midpoint of a line segment divides the segment into two equal parts.
   If a point divides a line segment into two equal parts, then the point is the midpoint of the segment.
7. A scalene triangle has no sides of equal length.
   If a triangle has no sides of equal length, then the triangle is scalene.
8. Perpendicular lines form right angles.
   If two lines are perpendicular, then they form right angles.

Accept each of the following statements as true. Complete each syllogism.

9. All dogs need exercise. Buster is a dog. Buster needs exercise.
10. Only careless people have accidents. I am not careless. I do not have accidents.
11. Alfred is always thirsty after he jogs. Alfred has been jogging, so he is thirsty.
12. A computer only works when it is plugged in. Sarah's computer is not plugged in, so it will not work.

Name ________________ Date ________________

Patterns in Logical Thinking

A conditional statement of the form *If A, then B* has two parts, the hypothesis, *A*, and the conclusion *B*.

Here are four logical patterns that can be made with an If-Then statement. The horizontal line segment stands for *then* or *therefore.*

Pattern 1	Pattern 2	Pattern 3	Pattern 4
If *A*, then *B*.	If *A*, then *B*.	If *A*, then *B*.	If *A*, then *B*.
A is true.	*B* is false.	*B* is true.	*A* is false.
B is true.	*A* is false.	?	?

1. Let *A* be "a figure is a triangle" and *B* be "a figure is a polygon." Write the conditional statement.
   If a figure is a triangle, then the figure is a polygon.

Can valid conclusions be drawn from Patterns 3 and 4? Use the conditional from Exercise 1 to explain your answers.

2. Pattern 3: No; A figure can be a polygon without being a triangle.
3. Pattern 4: No; If a figure is not a triangle, it may or may not be a polygon.
4. Notice that Patterns 3 and 4 do not give valid conclusions. Now try the negation of both parts of the If-Then statement. Let *A* be "a figure is not a triangle" and *B* be "a figure is not a polygon." Write the conditional statement.
   If a figure is not a triangle, then it is not a polygon.

Do the four patterns give valid conclusions when the two parts are negated? Write *yes* or *no.*

5. Pattern 1: no
6. Pattern 2: no
7. Pattern 3: yes *A* is true
8. Pattern 4: yes *B* is false

9. Explain what is logically wrong about this argument.

   If we want to find out if there is intelligent life on Mars, we will need to spend more money on the space program. So, if we spend more money on the space program, we will find out whether or not there is intelligent life on Mars.

   It is an example of Pattern 3. The conclusion does not follow from the premises.

Name ________________ Date ________________

Applications

When solving a problem that involves congruent triangles, sometimes the most difficult step is drawing the diagram.

The figures on this page are mixed up—they are not next to the appropriate exercises. First match each exercise to its figure. Write the letter of the correct figure next to the exercise number. Then find the answer to the problem. No computation is necessary.

C **1.** A rectangular gate is 30 in. wide. If a brace 50 in. long is used to steady the gate, which line segments in the figure represent the height of the gate?
$\overline{GE}$ and $\overline{AT}$

E **2.** To find the perpendicular distance across a river from point A to point B, lay off $\overline{BC}$ perpendicular to $\overline{AB}$. Then lay off $\overline{CE}$ at a right angle to $\overline{BC}$. Place a stick at point O, the midpoint of $\overline{BC}$. Then find point D on $\overline{CE}$ so that A, O, and D line up. Which segment has the same length as $\overline{AB}$?
$\overline{CD}$

B **3.** $\overline{AB}$ and $\overline{CB}$ are two rafters of a roof. They are equal in length. $\overline{DF}$ and $\overline{EG}$ are supports, perpendicular to the floor and equally distant from A and C, respectively. You measure the length of $\overline{DF}$. What is the height of the other support?
$\overline{EG}$

D **4.** To find a distance AB across a river, let $\overline{AC}$ be any convenient segment. Then construct the triangle ACD so that angles 1 and 3 are equal, and angles 2 and 4 are equal. What part of the new triangle is congruent to $\overline{AB}$?
$\overline{AD}$

A **5.** To find a distance AB across a pond, locate point O from which $\overline{AO}$ and $\overline{BO}$ can be measured. Then extend $\overline{AO}$ and $\overline{BO}$ to points D and C, respectively, so that $AO = OD$ and $BO = OC$. Which segment has the same length as $\overline{AB}$?
$\overline{CD}$

Name ______________________ Date ______________

Congruent Polygons

Polygons are *congruent* if corresponding angles have equal measure and corresponding sides have equal length. It does not matter if the congruent polygon has been rotated or flipped. The three polygons at the right are congruent.

1. Write the pairs of congruent figures in this set. b and e, d and f, a and g

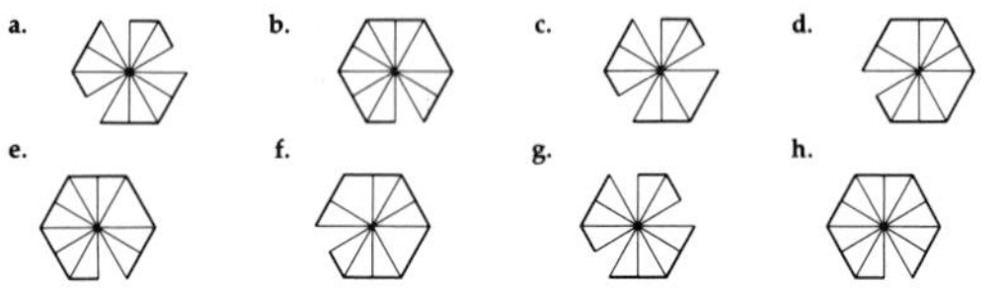

a. b. c. d.

e. f. g. h.

2. Write the pairs of congruent figures. Corresponding regions must be shaded.
a and j, b and d, c and g, e and h, f and i.

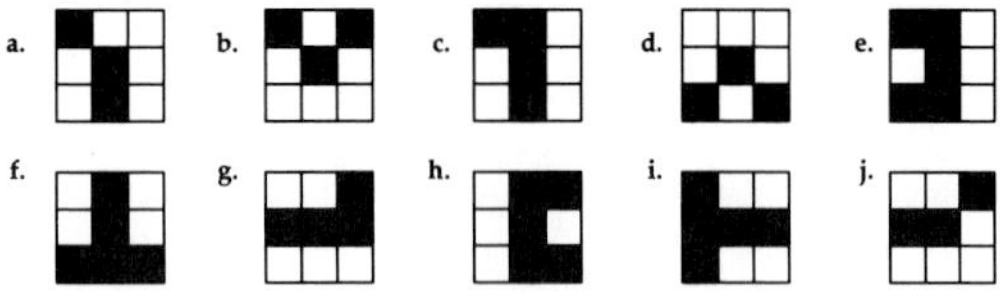

a. b. c. d. e.

f. g. h. i. j.

3. Write the pairs of congruent figures. Corresponding faces must be shaded.
a and f, b and e, c and d

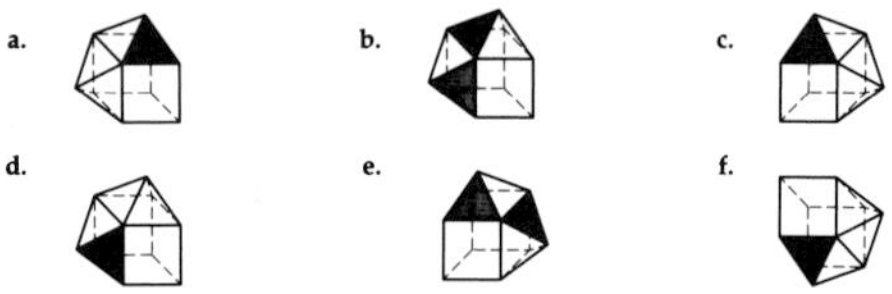

a. b. c.

d. e. f.

Name ______________________ Date ______________

Logo and Congruent Triangles

1. Draw two right triangles that can be proven to be congruent by showing that the two triangles are similar by AA and that the ratio of the corresponding sides is 1. Then write a Logo procedure to tell the turtle to draw the congruent right triangles. Answers may vary.

2. Draw two triangles that are not right triangles that can be proven to be congruent by showing that the two triangles are similar by AA and that the ratio of the corresponding sides is 1. Then write a Logo procedure to tell the turtle to draw the congruent triangles. Answers may vary.

3. Draw two right triangles that can be proven to be congruent by showing that the two triangles are similar by SAS and that the ratio of the corresponding sides is 1. Then write a Logo procedure to tell the turtle to draw the congruent right triangles. Answers may vary.

4. Draw two triangles that are not right triangles that can be proven to be congruent by showing that the two triangles are similar by SAS and that the ratio of the corresponding sides is 1. Then write a Logo procedure to tell the turtle to draw the congruent triangles. Answers may vary.

Name________________ Class________________ Date________________

Achievement Test 5 (Chapter 5)
CONGRUENT TRIANGLES

No. Correct
No. Exercises: **20**
Score
5.00 x No. Correct =

GEOMETRY FOR DECISION MAKING
James E. Elander
SOUTH-WESTERN PUBLISHING CO.

Suppose $A = \{5, 9, 11, 17\}$.

1. Is $\{5, 9\} \subset A$? Yes

2. Is $\{13\} \subset A$? No

3. List of the two-element subsets of A.
{5, 9}, {5, 11}, {5, 17}, {9, 11}, {9, 17}, {11, 17}

4. List all the three element subsets of A.
{5, 9, 11}, {5, 9, 17}, {9, 11, 17}

In the figure, $FZ = JM$, $TK = OK$, and $\angle OZM \cong \angle TJF$.

5. Is $TJ = OZ$? Yes

6. Is $FJ = MZ$? Yes

7. Justify that $\Delta TJF \cong \Delta OZM$.
$\Delta TJF \sim \Delta OZM$ by SAS, $TJ/OZ = 1$

8. Is $TF = OM$? Explain.
Yes; CPCF

In the figure, $\overline{PM} \parallel \overline{ON}$ and $PM = ON$.

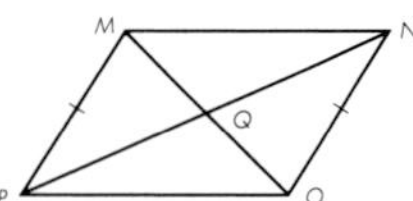

9. Is $\angle MPN \cong \angle ONP$? Explain. Yes; alternate interior angles formed by parallel lines and a transversal

10. Is $\Delta PMN \cong \Delta NOP$? Explain. Yes; SAS, $PN/PN = 1$

11. Is $MN = OP$? Explain. Yes; CPCF

12. Is $\angle MNP \cong \angle OPN$? Yes

The following advertisement appeared in a sports magazine.

People who play like pros use our tennis rackets.

13. Write the message of the advertisement in a statement of the form "If A, then B."
If you play like a pro, then you use our tennis rackets.

14. Write the converse of your answer for Exercise 13.
If you use our tennis rackets, then you will play like a pro.

15. In the figure shown, $\overline{PR} \perp \overline{RQ}$.
Show that $m\angle PRS = m\angle QRS = 45°$.
$\Delta PRS \cong \Delta QRS$ by SSS, $RS/RS = 1$.
Then $m\angle PRS = m\angle QRS$ by CPCF.
Since $m\angle PRQ = 90°$,
both $\angle PRS$ and $\angle QRS$ measure $45°$

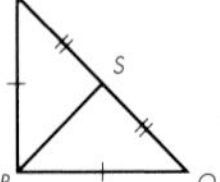

Use the survey information below for Exercises 16 and 17.

34 people like asparagus (A). 45 people like spinach (S).
25 people like both A and S. 6 people like neither A nor S.

16. Complete the Venn diagram to show the given information.

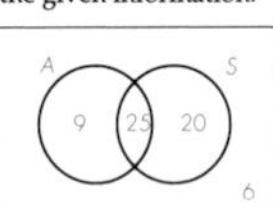

17. How many people were surveyed? 60

Use the survey information below for Exercises 18-20.

39 people like football (F). 53 people like basketball (B).
67 people like hockey (H). 21 people like F and B.
42 people like B and H. 18 people like F and H.
13 people like all three. No one dislikes all three.

18. How many people like B or H? 78

19. How many people like B only? 3

20. How many people were surveyed? 91

Tell what kind of figure each of the following is. Use symbols to name it.

1.

line segment; $\overline{BA}$ or $\overline{AB}$

2.

ray; $\overrightarrow{CA}$

3.

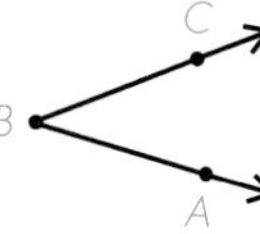

angle; $\angle ABC$ or $\angle CBA$ or $\angle B$

Measure each angle and classify it as acute, obtuse, right, or straight.

4.

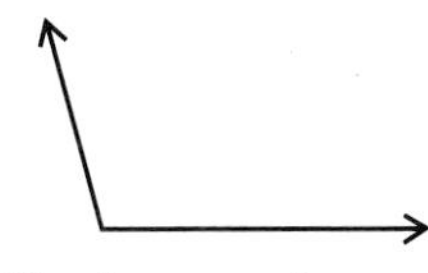

105°; obtuse angle

5.

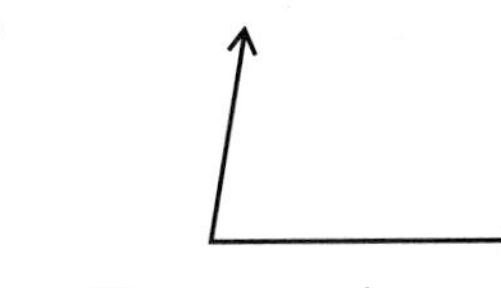

80°; acute angle

6. What is the measure of each angle formed by the bisector of a 90° angle? 45°

Each of the following pairs of angles is either complementary or supplementary. Find the measure of each angle.

7.

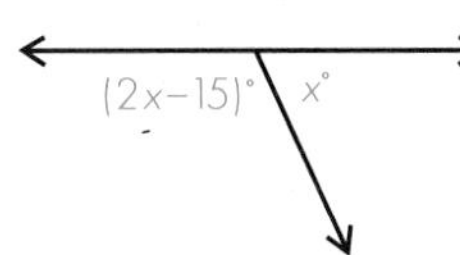

$x° = 65°$; $(2x - 15)° = 115°$

8.

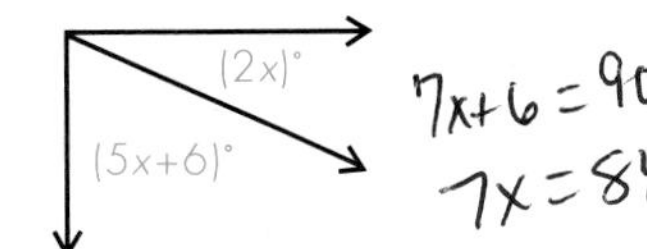

$(2x)° = 24°$; $(5x + 6)° = 66°$

Refer to the number line at the right.

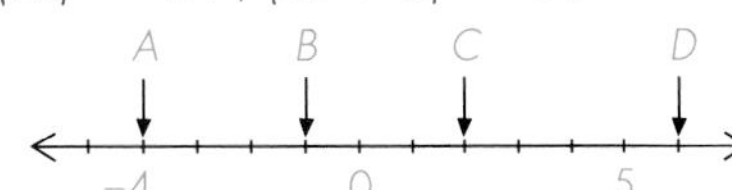

9. What are the coordinates of points *A*, *B*, *C*, and *D*? −4; −1; 2; 6

10. What are the lengths of $\overline{AC}$ and $\overline{BD}$? 6; 7

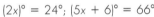

Suppose *C* is between *B* and *G*. Point *L* is the midpoint of $\overline{BC}$. $BG = 10$ and $BL = 1\frac{1}{2}$.

11. Draw a diagram to show how *B*, *L*, *C*, and *G* are situated.

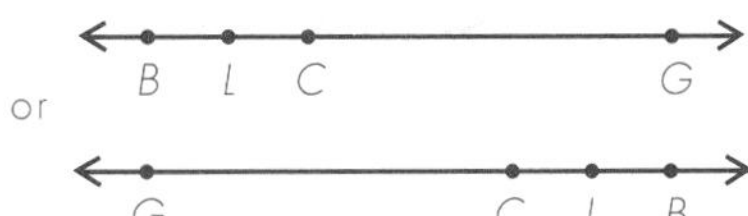

12. What is the measure of $\overline{LC}$? of $\overline{CG}$? $LC = 1\frac{1}{2}$; $CG = 7$

13. Write LOGO commands for drawing a 110° angle and its bisector. Answers will vary.

Use the diagram. Identify the following.

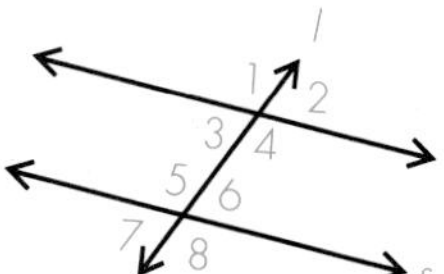

14. Corresponding angles $\angle 1, \angle 5; \angle 2, \angle 6; \angle 3, \angle 7; \angle 4, \angle 8$

15. Alternate interior angles $\angle 3, \angle 6; \angle 4, \angle 5$

16. Vertical angles $\angle 1, \angle 4; \angle 2, \angle 3; \angle 5, \angle 8; \angle 6, \angle 7$

17. Alternate exterior angles $\angle 1, \angle 8; \angle 2, \angle 7$

In the diagram, $\overleftrightarrow{ML}$ and $\overleftrightarrow{DH}$ are in plane P and do not intersect. Also, $m\angle RTM = 90°$. Identify the following.

18. Intersecting lines $\overleftrightarrow{ML}$, $\overleftrightarrow{LH}$; $\overleftrightarrow{ML}$, $\overleftrightarrow{RT}$; $\overleftrightarrow{DH}$, $\overleftrightarrow{LH}$
19. Parallel lines $\overleftrightarrow{ML}$, $\overleftrightarrow{DH}$
20. A transversal $\overleftrightarrow{LH}$
21. Skew lines $\overleftrightarrow{RS}$, $\overleftrightarrow{DH}$; $\overleftrightarrow{RS}$, $\overleftrightarrow{LH}$
22. Coplanar lines $\overleftrightarrow{RS}$, $\overleftrightarrow{ML}$; $\overleftrightarrow{ML}$, $\overleftrightarrow{DH}$, and $\overleftrightarrow{LH}$ (ML, LH, DH
23. Three collinear points M, T, L; R, T, S
24. Three noncollinear points
    R, T, M (other answers possible)

In the diagram, $r \parallel s$ and t is a transversal.

Give the degree measures of each angle.

25. $\angle 1$ 140° **26.** $\angle 2$ 40° **27.** $\angle 3$ 140°
28. $\angle 4$ 140° **29.** $\angle 5$ 140° **30.** $\angle 6$ 40°

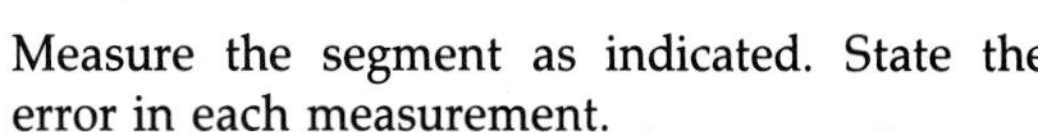

Measure the segment as indicated. State the error in each measurement.

31. to the nearest $\frac{1}{2}$ inch $2\frac{1}{2}$ in.; error: $\frac{1}{4}$ in.
32. to the nearest $\frac{1}{16}$ inch $2\frac{7}{16}$ in.; error: $\frac{1}{32}$ in.
33. Draw the closed three-dimensional figure this layout would form. rectangular prism

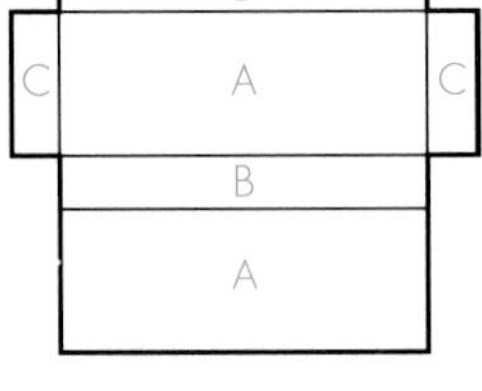

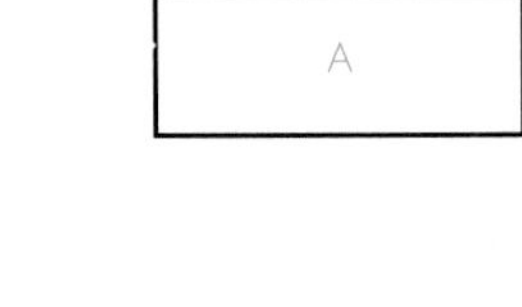

34. "If a person studies the manual carefully, that person will pass the driver's written test. Kim studied the manual carefully." If both statements are true, what can you conclude about how Kim will do on the test?
    Kim will pass the driver's written test.

Write the letter of each triangle that fits the description.

35. Acute triangle b, c, f, d
36. Scalene triangle a, d
37. Equilateral triangle c
38. Obtuse triangle none
39. Right triangle a, e
40. Isosceles triangle b, c, e, f
41. Equiangular triangle c

a. 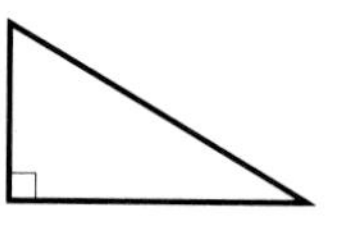

b. 

c. 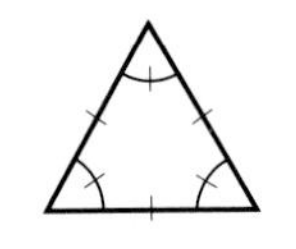

d. 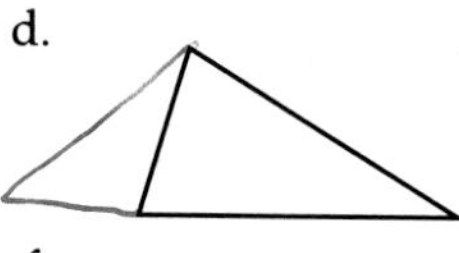

e. f. 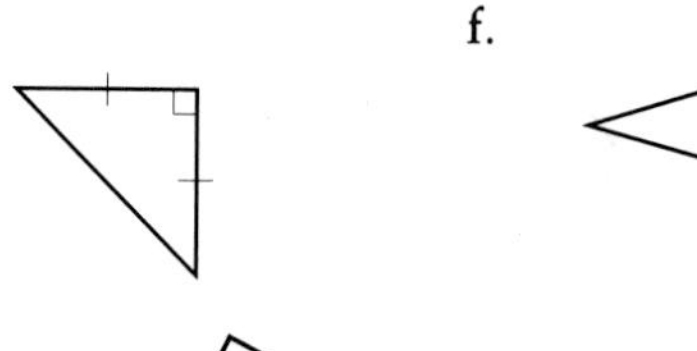

42. The length of the unknown side in the triangle is between what two numbers?
    $9 < x < 37$

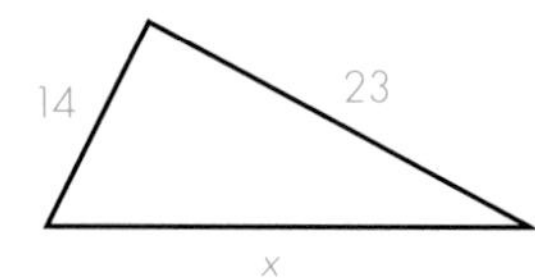

43. What is the measure of $\angle C$? 28°

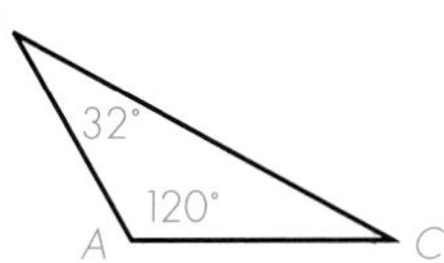

Consider the figure and the information given. Find the measure of each angle.

44. $\angle 1$ 90° **45.** $\angle 2$ 90°
46. $\angle 3$ 40° **47.** $\angle 4$ 140°
48. $\angle 5$ 130° **49.** $\angle 6$ 130°

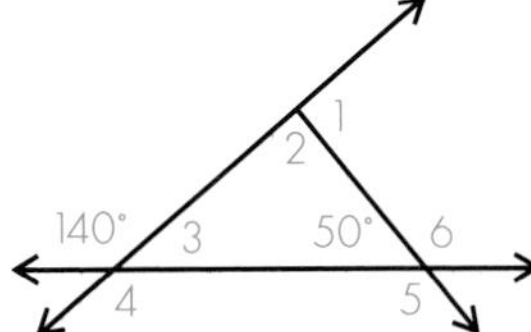

50. Plot the following points on the x-y plane: (6,1), (0,4), and (6,4). Identify the type of triangle that is formed when the points are connected by line segments.
See student drawings; right scalene triangle

51. Two cities are $4\frac{1}{4}$ in. apart on a road map. If the map uses a scale of 1 in. = 30 mi, what is the actual distance between the cities?
127.5 mi

52. A factory wants to have a ratio of 2 managers for every 15 trainees. If 90 people are in the training program, how many managers will it need? 12 managers

Identify the pairs of similar triangles. Tell how you know they are similar.

53.

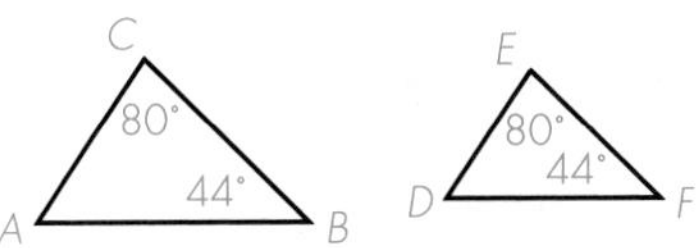

E, 80°, 44°, D, F

Similar; AA

54.

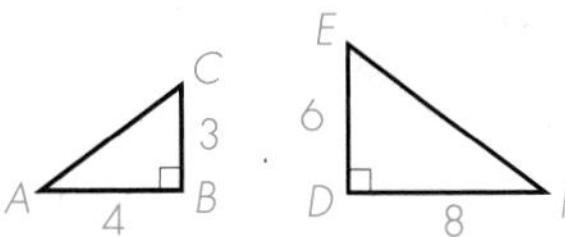

Similar; SAS

55.

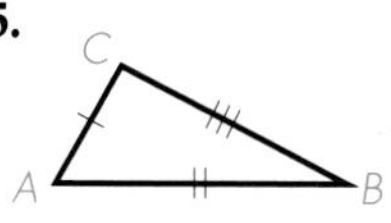

F, D, E

Similar SSS

56.

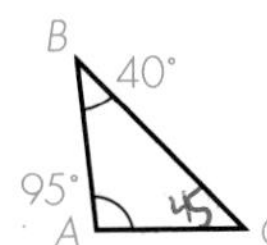

F, E, 55°, 40°, D

Not similar

57. What is the length of $\overline{RS}$ in $\triangle ABC$? 14 m

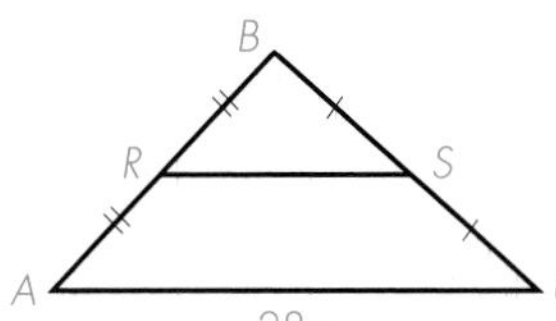

Refer to the diagram. $m\angle SQR = m\angle SVT$. Find the length of each segment.

58. $\overline{SV}$ = 18
59. $\overline{TV}$ = 15

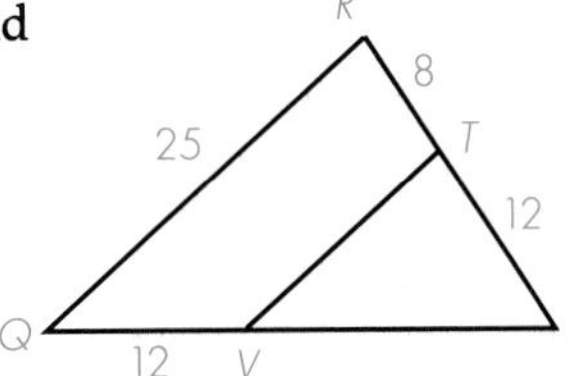

Classify each statement as true or false.

60. No similar triangles are congruent. False

61. If two triangles are congruent, then the ratio of two corresponding sides is equal to 1. True

Refer to figures a – f.

62. Which triangles are similar?
a, c, e; b, d, f

63. Which triangles are congruent? a, e; b, f

a.

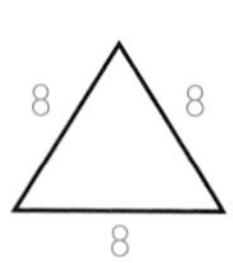

b.

c.

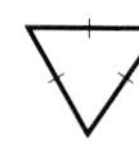

d.

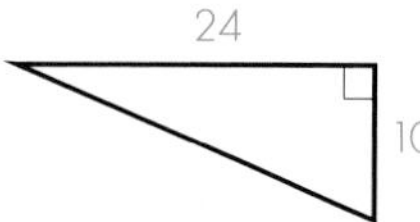

e.

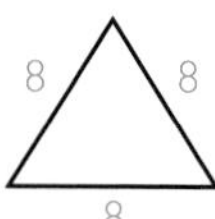

f.

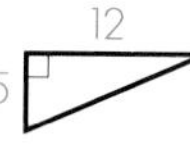

Refer to $\triangle ABC$.

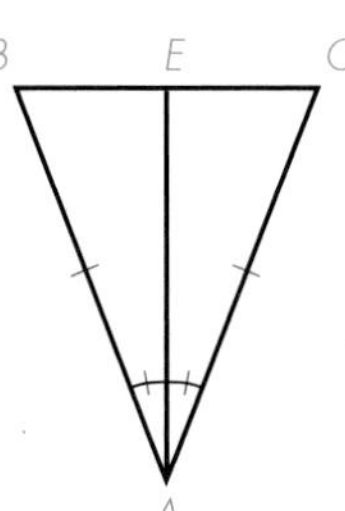

64. Is $\triangle ACE \sim \triangle ABE$? If so, why? Yes; SAS

65. Is $\triangle ACE \cong \triangle ABE$? If so, why?
Yes; If a triangle has two equal sides and the included angle is bisected, then the triangle is divided into congruent triangles.

The members of a geometry class were asked what after-school clubs they belong to.

12 are in the computer club
8 are in the photography club
5 are members of both clubs
6 are in neither club

66. Draw a Venn diagram illustrating this information.

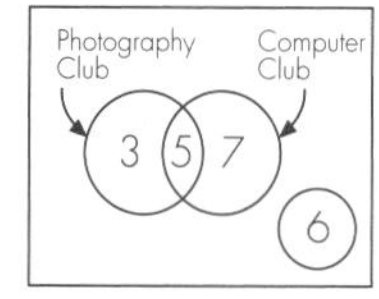

67. How many students are in this geometry class? 21 students

68. In the figure at the right, $m\angle 1 = m\angle 2$, $PA = PB$, and $PC = PD$.
Justify: $\overline{PB}$ bisects $\angle APD$

m∠PAB = m∠1, since the angles are opposite equal sides in △APB. m∠PCD = m∠2 (same reason). So △PAB ~ △PCD (AA). m∠APB = m∠CPD, since they are correspond parts of similar triangles. So $\overline{PB}$ bisects ∠APD, by the definition of bisector.

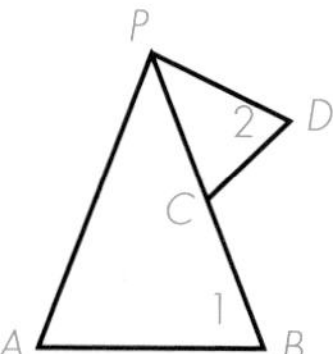

Give the converse of each of the following statements.

69. If you read the newspaper, then you are informed.
If you are informed, then you read the newspaper.

70. If you got into the theatre, then you had a ticket. If you had a ticket, then you got into the theatre.

71. If a person has visited London, then the person has been to England.
If the person has been to England, then the person has visited London.

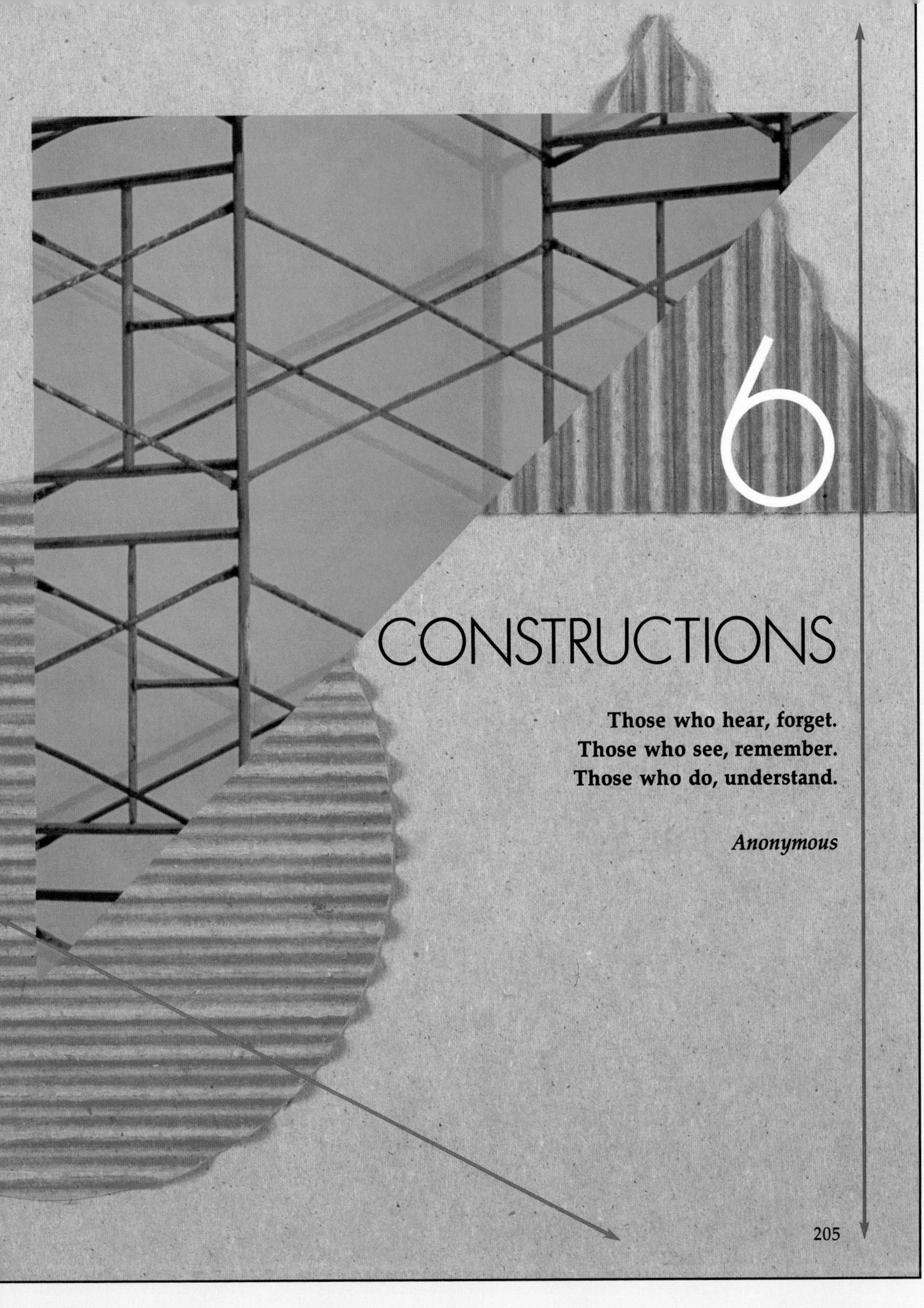

6

CONSTRUCTIONS

Those who hear, forget.
Those who see, remember.
Those who do, understand.

Anonymous

205

CHAPTER 6 OVERVIEW

Constructions

Suggested Time: 12–14 days

As the quote at the beginning of this chapter indicates, Chapter 6 will provide a wealth of understanding for students as they use the basic tools of geometry to learn to do many basic constructions.

These constructions include copying and bisecting line segments and angles, and constructing parallel and perpendicular lines. The first theorem in this chapter tells us that if two lines are intersected by a transversal so that the alternate interior angles are equal, then the lines are parallel. This is the converse of an earlier theorem.

Those who hear, forget.
Those who see, remember.
Those who do, understand.

Anonymous

CHAPTER 6 OBJECTIVE MATRIX

	End of Chapter Items by Activity				Student Supplement Items by Activity		
Objectives by Number	**Review**	**Test**	**Computer**	**Algebra Skills**	**Reteaching**	**Enrichment**	**Computer**
6.1.1	✔	✔				✔	
6.1.2	✔	✔				✔	
6.1.3	✔	✔			✔		
6.1.4	✔	✔			✔		
6.2.1					✔		
6.2.2	✔	✔			✔		
6.3.1	✔	✔			✔		
6.4.1	✔	✔	✔		✔	✔	✔
6.4.2	✔	✔			✔	✔	
6.4.3	✔	✔			✔	✔	
6.4.4	✔	✔					
6.5.1	✔	✔			✔		
6.6.1	✔	✔		✔	✔	✔	

*A ✔ beside a Chapter Objective under Algebra Skills indicates that algebra skills taught within the chapter or in previous Algebra Skills lessons are used.

CHAPTER 6 PERSPECTIVES

▲ Section 6.1

Bisection of Line Segments and Angles

Students first learn to copy a line segment and an angle and then to bisect each, using only a compass and a straightedge. The construction of bisecting a line segment is justified by means of a paragraph proof.

The Home Activity provides practice in copying and bisecting line segments and angles. One exercise introduces using the compass as a walking tool in order to show segment addition.

6

▲ Section 6.2

Perpendiculars

Students learn to draw a perpendicular to a given line through a given point. The given point is on the line for the first construction but is not on the line for the second construction. Both constructions are related through Discovery Activities to the construction for finding the midpoint of a line segment. A justification for one of the constructions is given; in the justification, students see that *by construction* is a valid reason for a step in a proof.

The Home Activity includes exercises in which students draw line segments of a designated length and locate the given point through specific directions. They then use this diagram in constructing perpendiculars.

▲ Section 6.3

Constructing a Line Parallel to Another Line

Students are first referred back to Postulate 3-1-1, regarding the number of parallel lines through a point not on the line. Two other possibilities are listed, introducing the idea of non-Euclidean geometry. A report on non-Euclidean geometry is suggested for groups of students to prepare for extra credit. The work in this book assumes that Postulate 3-1-1 holds true. Justification for copying angles is given, then used in the construction of parallel lines.

The Home Activity includes practice in constructing a line parallel to a given line through a given point. A couple of exercises involve constructing a line through a vertex of a triangle that is parallel to the side opposite the vertex.

▲ Section 6.4

Constructions Involving Triangles

Concurrent lines are defined, as are a *median*, an *altitude*, an *angle bisector*, and the *perpendicular bisector of a side of a triangle*. Through a series of Discovery Activities, students find that the three medians of a triangle are concurrent, as are the three altitudes, the three perpendicular bisectors of the sides, and the three angle bisectors. A project is suggested that allows students to see that the point of concurrency of the medians of a triangle is the triangle's center of gravity.

The Home Activity involves practice finding the various points of concurrency of triangles.

▲ Section 6.5

Dividing a Line Segment into *n* Equal Parts

Prior to this section, students could only bisect a line segment, or a portion of the line segment. Students are now introduced to a technique that will allow them to divide a line segment into any number of congruent segments. This technique involves copying angles to construct parallel lines. A justification of the construction is given. A Discovery Activity shows how line segments can be divided by using tools other than geometric tools.

The Home Activity includes practice in dividing line segments into no more than six equal parts. A few of the exercises allow students to try the method of the Discovery Activity.

▲ Section 6.6

Theorems from Constructions

The Triangle Midpoint Theorem extends previous work. Students now find that the line segment containing the midpoints of two sides of a triangle is not only half the length of the third side, but is parallel to the third side as well. Through a Discovery Activity, students find that the point of concurrency of the perpendicular bisectors of a triangle is the center of a circle, called the circumcircle. Through another Discovery Activity, they find that the angle bisectors are concurrent at a point that is the center of the in-circle of the triangle.

The Home Activity includes practice finding these points of concurrency.

SECTION
6.1

Resources
Reteaching 6.1
Enrichment 6.1

Objectives
After completing this section, the student should understand
- ▲ how to copy a line segment using the tools of geometry.
- ▲ how to copy an angle using the tools of geometry.
- ▲ how to bisect a line segment.
- ▲ how to bisect an angle.

Materials
compass
straightedge
ruler
protractor
library resources

GETTING STARTED

Quiz: Prerequisite Skills
1. If a line segment is bisected, then it is divided into how many equal parts? [2]
2. A ray that divides an angle into two equal angles is called what? [angle bisector]

Warm-Up Activity
Have student use a ruler to bisect a line segment and a protractor to bisect an angle. Have students exchange papers and check the drawings.

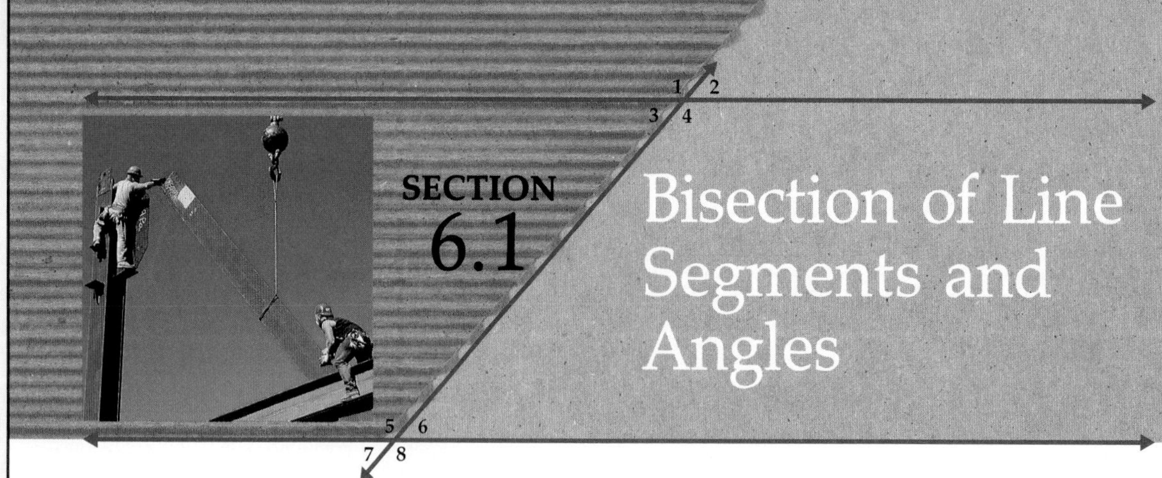

SECTION 6.1 Bisection of Line Segments and Angles

After completing this section, you should understand
- ▲ how to copy a line segment using the tools of geometry.
- ▲ how to copy an angle using the tools of geometry.
- ▲ how to bisect a line segment.
- ▲ how to bisect an angle.

In Chapter 1, you learned that the tools for geometry include a compass and a straightedge (or ruler). In this chapter, you will learn how to use these tools and how to solve some problems using these tools.

COPYING A LINE SEGMENT

Given a segment, $\overline{AB}$, copy the segment onto line l.

A B l

Place a point on l.

C l

Place the point of your compass on point A. Open the setting on the compass until the pencil point is on point B. Move the point of the compass over to the point on line l. Make an arc. You have just copied a line segment.

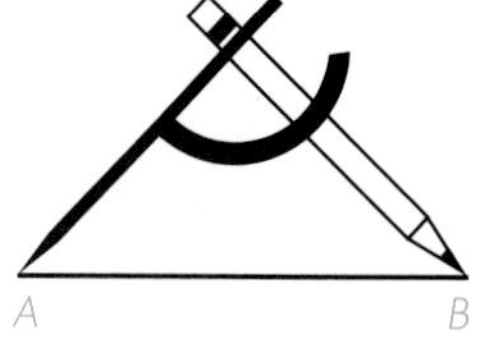

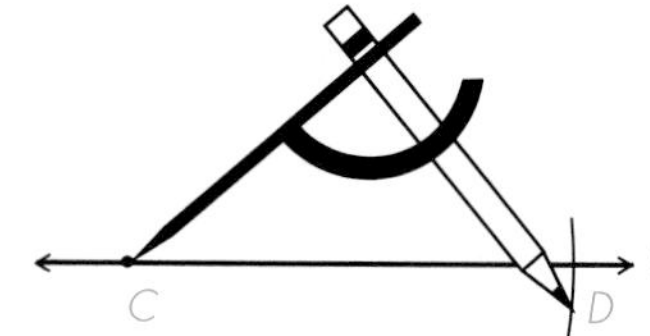

COPYING AN ANGLE

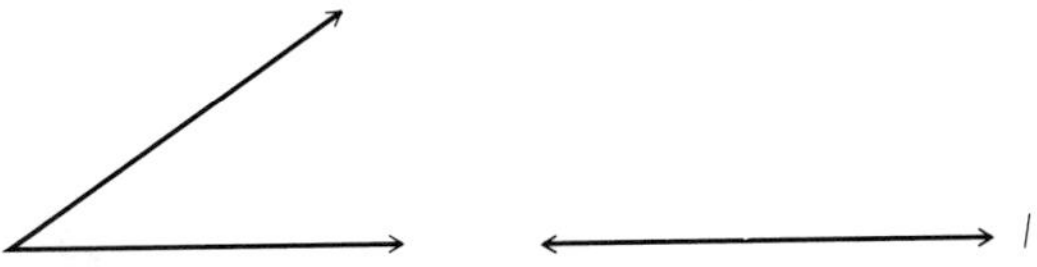

Copy the angle. Place the point of the compass on the vertex of the angle. Mark an arc on each ray of the angle.

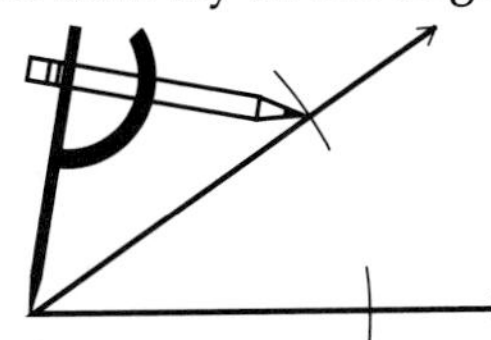

Mark a point on line *l*. Using the same setting on the compass, mark one arc on line *l*, and another above it.

Place the point of your compass where an arc intersects one ray of the angle. Open the setting until the point of the pencil touches the intersection of the arc and the other ray.

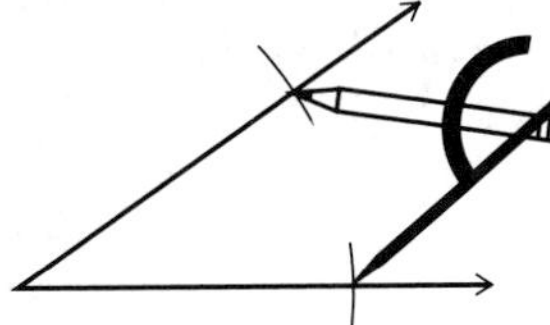

Keeping this setting, place the point of the compass at the point on line *l* where the arc intersects *l*. Make a small arc.

Complete the angle.

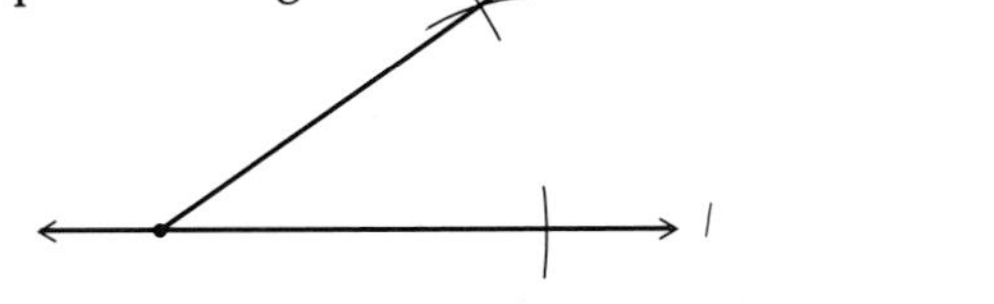

BISECTING A LINE SEGMENT

With an unruled straightedge, you can't measure a segment in order to find the midpoint. You must bisect the segment in order to find the midpoint.

Chalkboard Example
Bisect the line segment, then bisect each half.

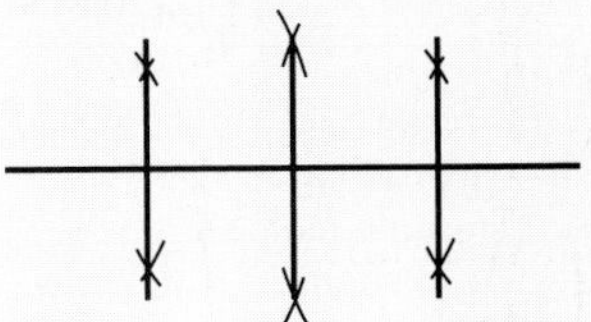

TEACHING COMMENTS

Using the Page
You may want to have students use a notebook or several sheets of paper as padding so that the compass point does not leave a hole in their desk.

Using the Page
Before they begin the constructions, make sure students understand how to change settings on their compasses and how to adjust the lead or pencil.

Using the Page
The directions for copying an angle indicate that a mark should be made above the line. There may be instances where the arc should be drawn below the line so that the angle can be copied below the line.

Using the Page
An alternate method for bisecting a segment is to draw one large arc from each endpoint. Connect the points of intersection of the two arcs.

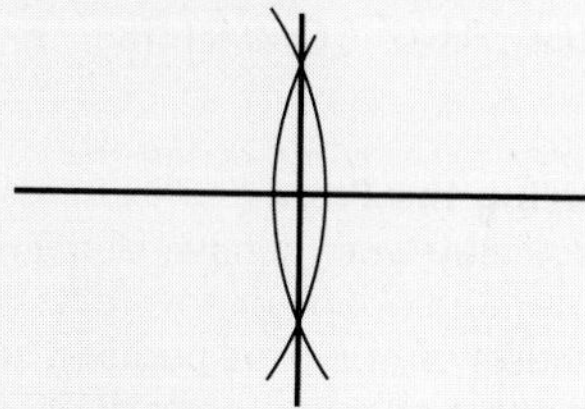

Bisect $\overline{AB}$.

To bisect a segment, open your compass to a setting that is just a little more than half the length of $\overline{AB}$. Mark two arcs from A and two from B, as shown below.

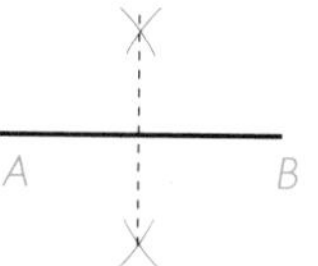

Connect the points where the arcs intersect. This segment crosses $\overline{AB}$ at its midpoint, thus bisecting it.

Show that M is really the midpoint of $\overline{AB}$.

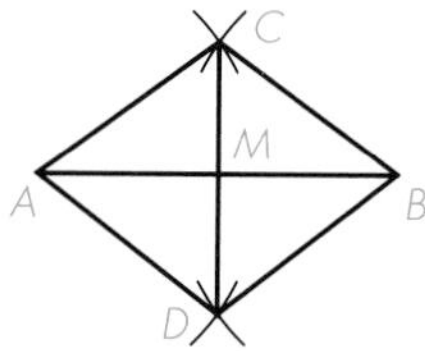

Using the Page
Make sure students understand the steps in the paragraph proof that justifies the construction for bisecting a line segment.

$\overline{AC}$, $\overline{BC}$, $\overline{AD}$, and $\overline{BD}$ were constructed so that they were all the same length. Then $\triangle ACD \cong \triangle BCD$, by SSS and the ratio of corresponding sides is 1. This means $\angle ACM \cong \angle BCM$, and since $AC = BC$ and $CM = CM$, $\triangle ACM \cong \triangle BCM$. Then $AM = BM$ by CPCF. By the definition of midpoint, the construction is justified.

CLASS ACTIVITY

1. Use a computer and a geometric drawing tool to draw any line segment. Use the principles from the construction for bisecting a line segment to bisect the segment. Answers may vary.

BISECTING AN ANGLE

Place the point of your compass on the vertex of the angle. Mark an arc on both rays of the angle.

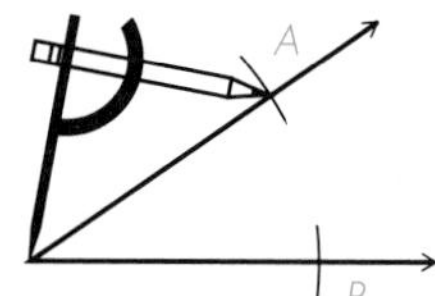

Using the same setting, place the point on A, and mark an arc in the center. Do the same from B. Label the vertex C and the intersecting arcs D. $\overrightarrow{CD}$ is the angle bisector.

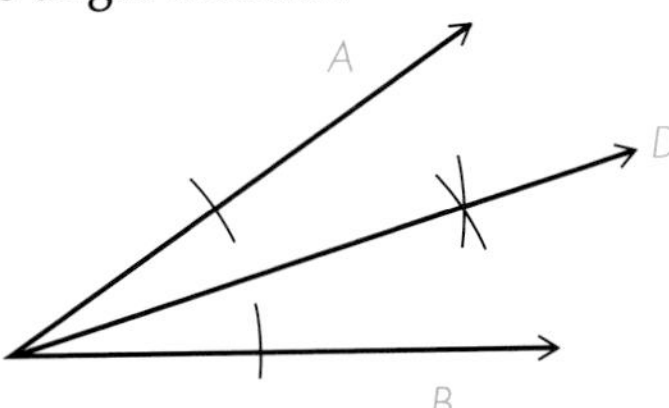

REPORT 6-1-1 The compass we use today can hold a setting until it is changed. Euclid used a collapsible compass. He lost the setting every time he picked the compass up to move it! Research how Euclid could bisect See margin.

1. a line segment.
2. an angle.

Suggested reference:
Cajori, F. *A History of Elementary Mathematics*. New York: Macmillan Publishing Co., 1905.

CLASS ACTIVITY

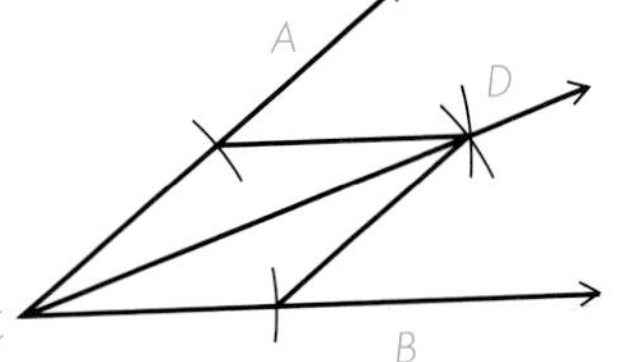

1. Show that $\overline{CD}$ is really the bisector of $\angle ACB$. (Hint: Show that $\triangle CAD \sim \triangle CBD$.) See margin.

2. Use a computer and a geometric drawing tool to draw any angle. Use the principles from the construction for bisecting an angle to bisect the angle. Answers may vary.

Copy each line or angle. Then use a compass and straightedge to bisect each.

3. 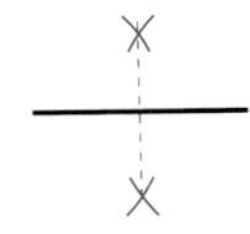

4. 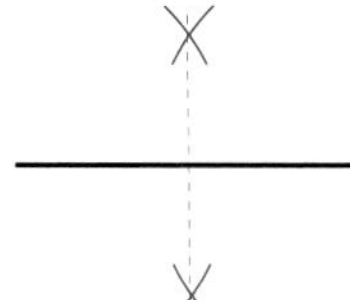

5. 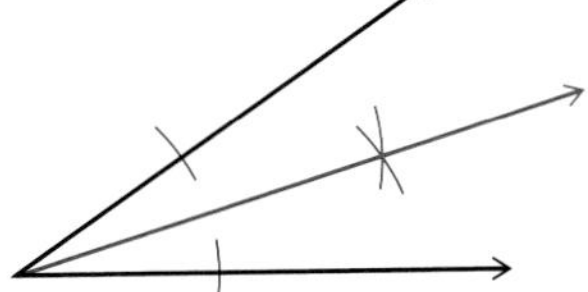

6. 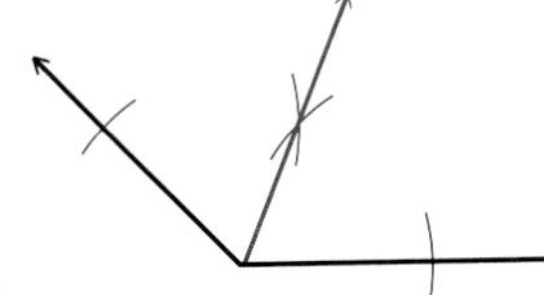

Using the Page

The directions at the top of the page indicate that the same setting should be used. Point out that the construction is still valid if the compass setting is changed. Emphasize that the setting must be such that the two arcs will intersect in a point D.

Using the Page

Some people prefer to draw one continuous arc that passes through A and B when bisecting an angle.

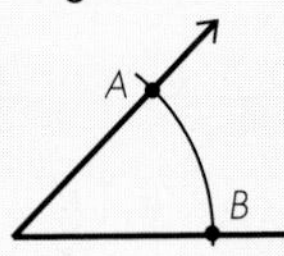

The remainder of the construction is the same as before.

Additional Answers

1. $\overline{AC}$ and $\overline{CB}$ were constructed so that they are the same length. $\overline{AD}$ and $\overline{BD}$ were constructed so that they are the same length. So $\triangle CAD \cong \triangle CBD$ by SSS and the ratio of the corresponding sides is 1. This means $m\angle ACD \cong m\angle BCD$ by CPCF. By definition of angle bisector, the construction is justified.

Extra Practice

1. Bisect the line segment.

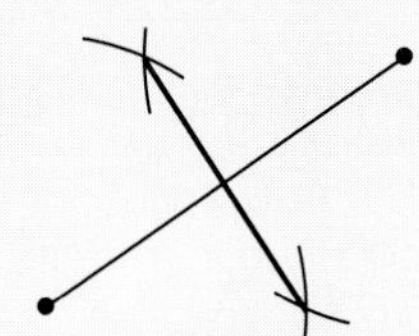

2. Bisect the angle.

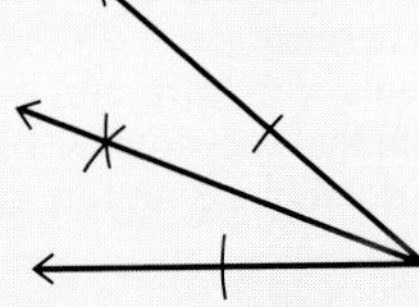

Additional Answers

3. Draw a point on a line l_2. Label it M. From M, mark an arc on l_2 for the length of $\overline{AB}$. Place the compass point on the intersection of this arc and l_2. Mark off an arc for the length of $\overline{CD}$. Label the intersection as point N.

4. Scaled-down sketch

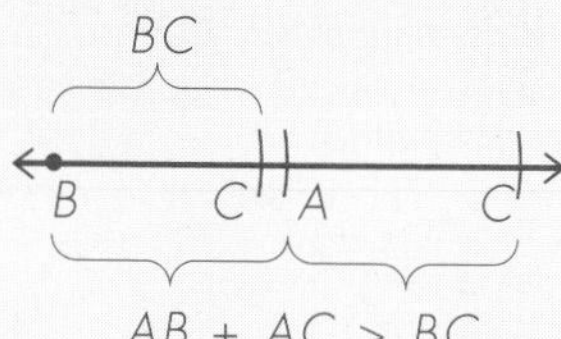

9.

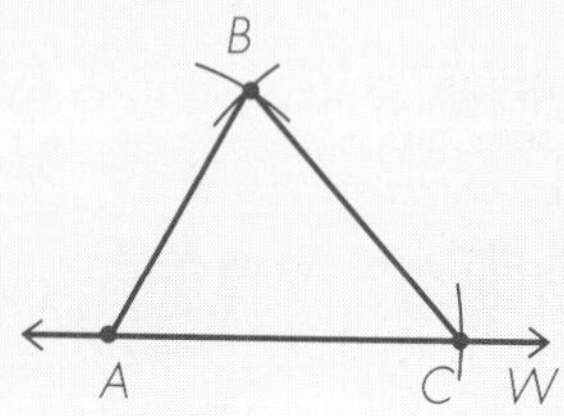

FOLLOW-UP

Assessment

1. Bisect the line segment.

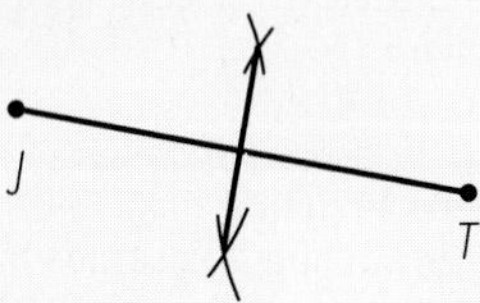

2. Bisect the angle.

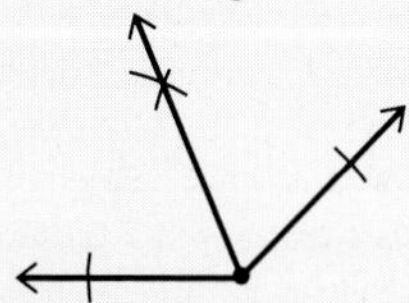

Reteaching

Students who have had difficulty with this section may benefit from Reteaching Activity 6.1

Enrichment

For students who have mastered the material in this section, you may wish to assign Enrichment Activity 6.1.

HOME ACTIVITY

1. Draw a $6\frac{1}{2}$-inch line segment. Bisect it. See students' constructions.

2. Draw a 5-inch line segment. By construction, divide the segment into four equal segments. Bisect segment, then bisect each congruent segment.

3. On l_1, look at line segments $\overline{AB}$ and $\overline{CD}$. Construct $\overline{MN}$ on a line so that $MN = AB + CD$. (HINT: Use your compass as a walking tool) See margin.

4. Use your compass to show that $AB + AC > BC$ in $\triangle ABC$. See margin.

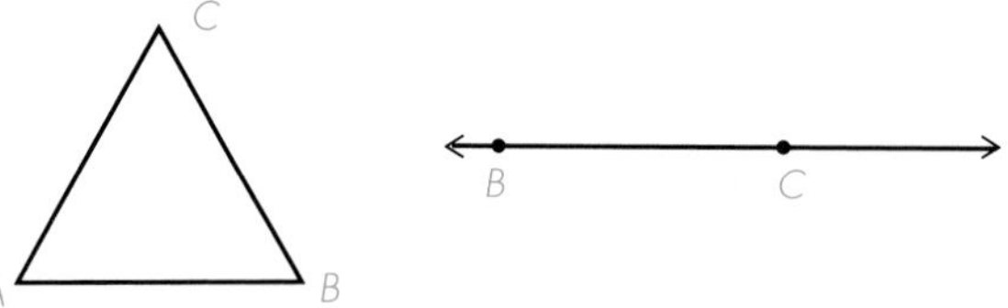

Copy each angle. Construct the angle bisector for each. Label the bisecting ray $\overrightarrow{CM}$.

5. 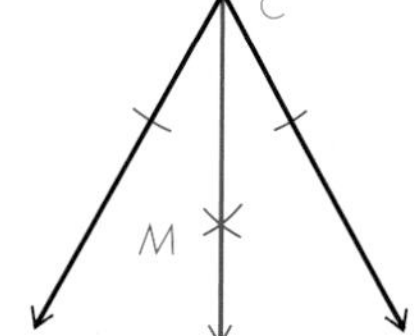

6. 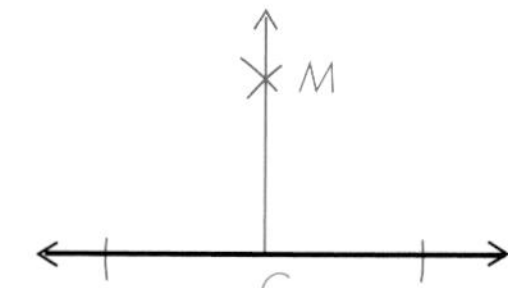

7. 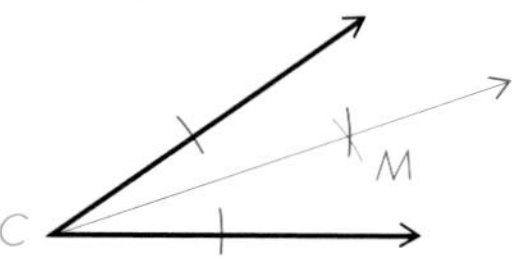

8. Draw a 60° angle. Use your compass and straightedge to divide the angle into four equal angles. Bisect angle, then bisect each congruent angle.

9. Copy the three segments. Copy $\overline{AC}$ onto line w, then use the other two segments to construct a triangle that has sides with lengths of the given segments. See margin.

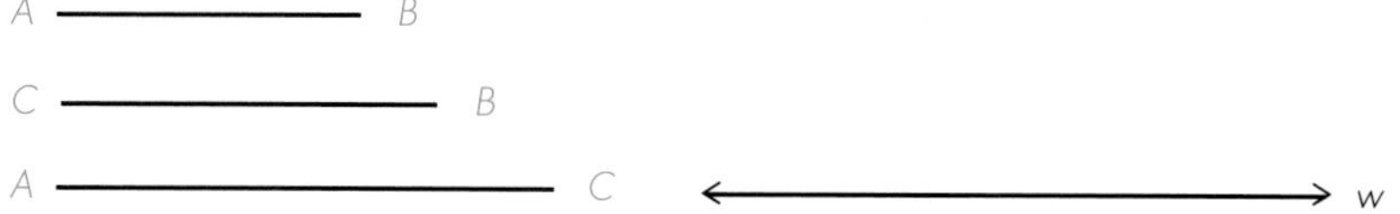

10. Construct a triangle so that $AB = 4$ cm, $AC = 6$ cm, and $m\angle A = 50°$. (Use your protractor to draw $\angle A$.) See students' constructions.

CRITICAL THINKING

11. How many points are needed to divide a line segment into 100 equal segments? Try a simpler case first, and look for a pattern. 99

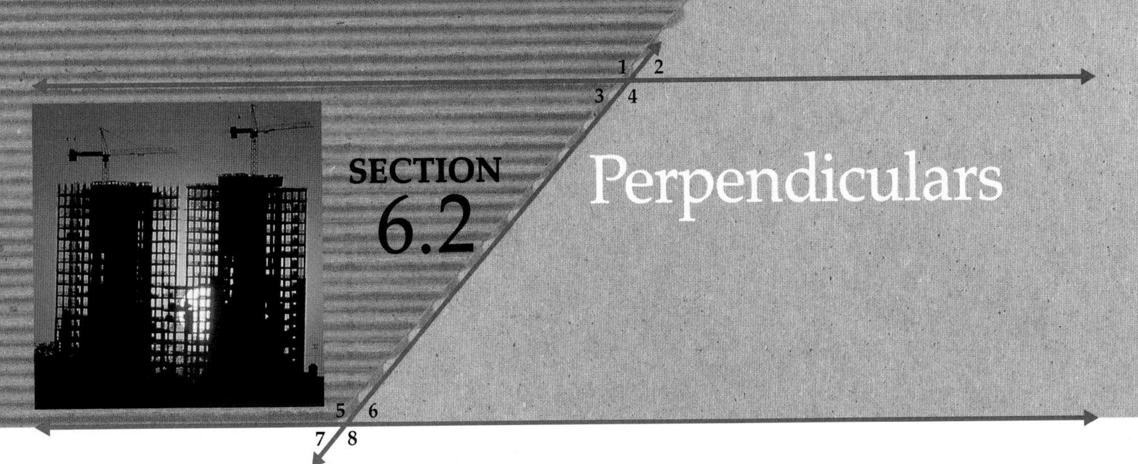

SECTION 6.2 Perpendiculars

After completing this section, you should understand
- ▲ how to construct a perpendicular to a line from a point on the line.
- ▲ how to construct a perpendicular to a line from a point not on the line.

Recall that perpendicular lines form a 90° angle, or a right angle. The symbol for perpendicularity is $\perp$.

PERPENDICULARS TO A LINE FROM A POINT ON THE LINE

Given line $\overleftrightarrow{AB}$ and point C on $\overleftrightarrow{AB}$, construct a line through C that is perpendicular to $\overleftrightarrow{AB}$.

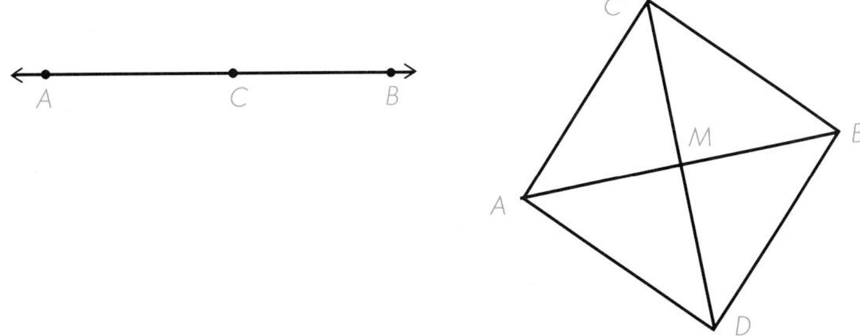

Recall that in the last section, we showed that $\triangle ACM \cong \triangle BCM$, where $AC = BC = AD = BD$. Then by CPCF, $\angle CMA \cong \angle CMB$. Since $m\angle AMB = 180°$, both $\angle CMA$ and $\angle CMB$ are right angles. Then $\overline{CM} \perp \overline{AB}$. Thus, the bisector constructed was perpendicular to the segment $\overline{AB}$.

In given line $\overleftrightarrow{AB}$, C is not the midpoint of $\overline{AB}$. If C were a midpoint, you could construct a perpendicular at C as you did in the last section.

SECTION 6.2

Resources
Reteaching 6.2
Enrichment 6.2

Objectives
After completing this section, the student should understand
- ▲ how to construct a perpendicular to a line from a point on the line.
- ▲ how to construct a perpendicular to a line from a point not on the line.

Materials
compass
straightedge
ruler

Vocabulary
perpendicular lines
by construction

GETTING STARTED

Quiz: Prerequisite Skills

1. What term describes two lines that meet to form a 90° angle? [perpendicular]

2. $\angle NOK$ is a straight angle. $m\angle PON = m\angle POK$. What is the measure of $\angle PON$? [90°]

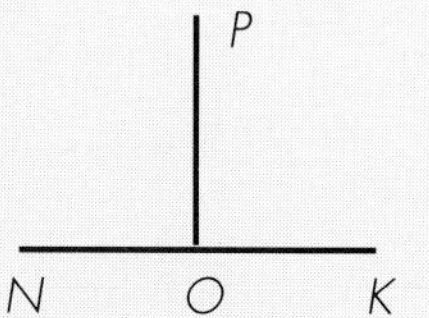

Warm-Up Activity
Have students draw a line segment on a sheet of paper. They should then take a sheet of ruled paper and align one of the lines on it with the segment drawn. Next, have them draw a line along the edge of the ruled paper so that it passes through the first segment drawn. Have students measure the angle formed to see how close the measure is to 90°.

Chalkboard Example
Construct a perpendicular to the line from point *L*.

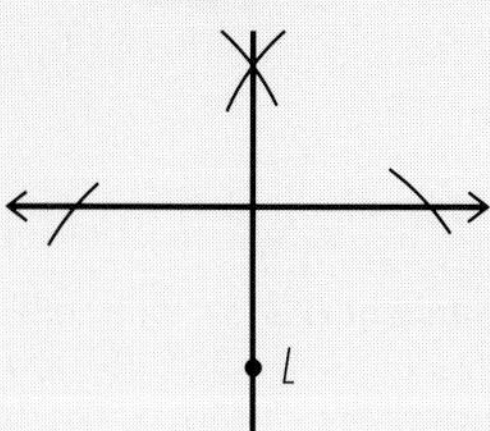

TEACHING COMMENTS

Using the Page
See Page 211. You might mention that the symbol for perpendicularity on a diagram is ⊾.

Using the Page
Make sure students understand that in Section 6.1 they were given the endpoints of a line segment and constructed the perpendicular bisector to find the midpoint of the segment. For this construction, students use a given point C and the same compass setting to construct two endpoints so that C is the midpoint of a segment. Then the perpendicular can be constructed through point C.

Using the Page
Arcs were made above and below point C. Only one set of intersecting arcs is really needed, since that intersection and point C would determine a line. The second set of arcs helps as a check of the alignment.

DISCOVERY ACTIVITY

1. Copy line $\overleftrightarrow{AB}$ on page 211 onto a sheet of paper.
2. Open the setting on your compass so that it is less than *AC* or *BC*. Without changing the setting, place the point of the compass on C and make a mark to the left and to the right of C. Label the points of intersection *D* and *E*.
3. What do you know about *CD* and *CE*? Equal
4. What name can be given to point C? Midpoint for $\overline{DE}$

You may have found that by using the same compass setting, you can make *C* the midpoint of a line segment. Now you can construct a perpendicular at *C*.

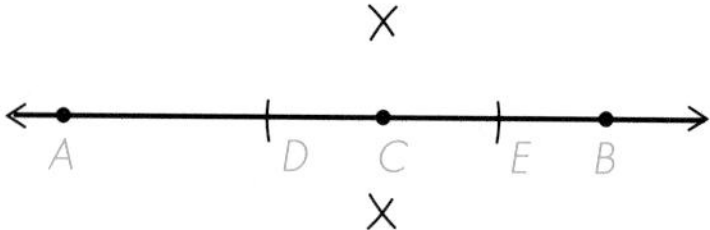

Open the compass setting so that it is slightly more than *CD* or *CE*. From *D*, make a mark above the line and below the line. Do the same from *E*.

Connect the intersections of the arcs to form the perpendicular to $\overleftrightarrow{AB}$ through *C*.

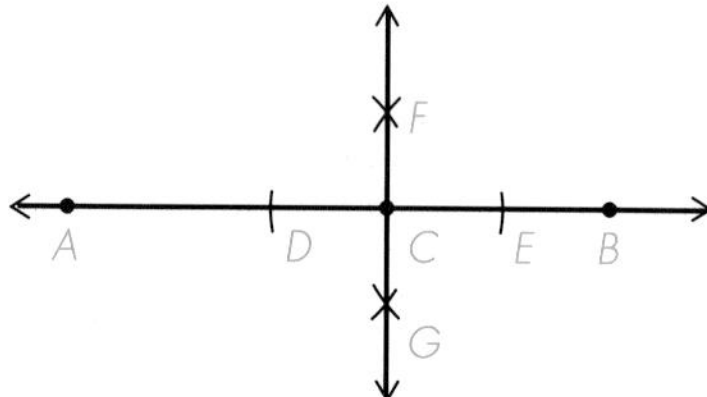

To justify that $\overleftrightarrow{FG} \perp \overleftrightarrow{AB}$, you can show that $\angle BCF$ is a right angle.

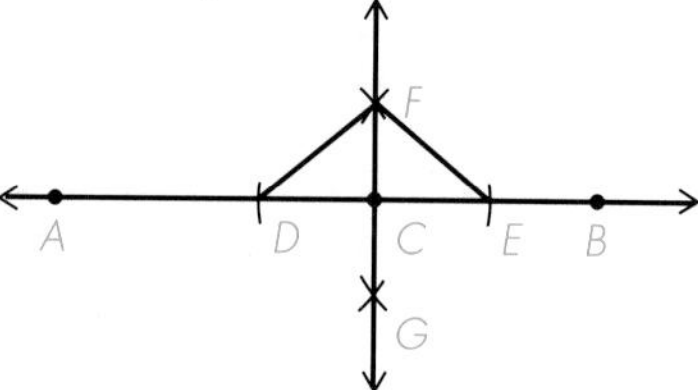

Connect F to D and F to E. You are able to do this because two points determine a line. $EF = DF$ since these were marked off without changing the setting on the compass. This reasoning is called "by construction."

Steps	Reasons
1. $CF = CF$	1. Identity
2. $CD = CE$	2. By construction
3. $DF = EF$	3. By construction
4. $\triangle DEF$ is isosceles.	4. Definition of isosceles triangle
5. $m\angle CDF = m\angle CEF$	5. Angles opposite the equal sides of an isosceles triangle are equal.
6. $CF/CF = CD/CE = 1$	6. Algebra
7. $\triangle DFC \sim \triangle EFC$	7. SAS
8. $\triangle DFC \cong \triangle EFC$	8. Definition of congruent triangles
9. $m\angle DCF = m\angle ECF$	9. CPCF
10. $m\angle DCE = 180°$	10. Definition of straight angle
11. $m\angle ECF + m\angle DCF = 180°$	11. Supplementary angles
12. $2(m\angle ECF) = 180°$	12. Substitution
13. $m\angle ECF = 90°$	13. Algebra
14. $m\angle BCF = 90°$	14. Same angle
15. $\angle BCF$ is a right angle.	15. Definition of right angle

Using the Page
You might prefer to include connecting F to D and F to E as the first step in the justification.

Using the Page
Remind students that straight angles can be assumed from a diagram, but that right angles cannot be assumed.

Extra Practice
Draw a perpendicular to the line through point C.

1.

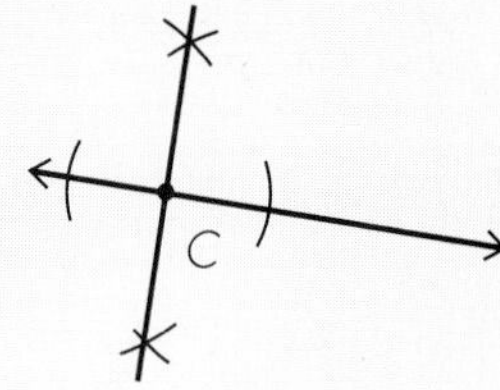

2.

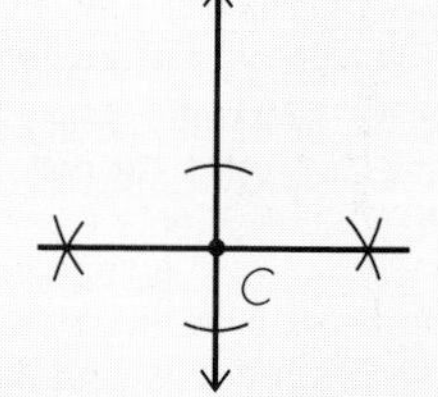

CLASS ACTIVITY

1. Use a computer and a geometric drawing tool to draw any line with point C between points A and B. Use the principles from the construction of a perpendicular to a line from a point on the line to draw a perpendicular to the line through point C.
Answers may vary.

Additional Answers

2. scaled-down sketch

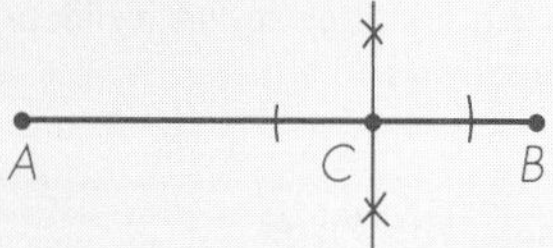

3. scaled-down sketch

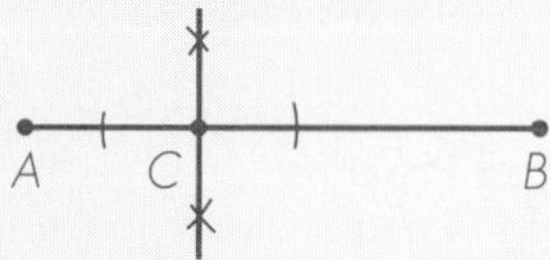

2. scaled-down sketch

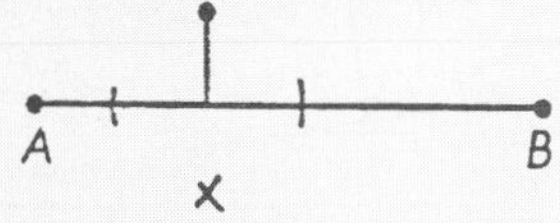

Using the Page

Have students work in small groups to write the justification for Question 5 of the Discovery Activity.

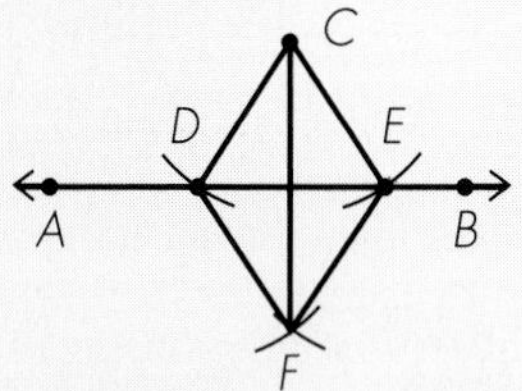

$CD = CE$ by construction. $DF = EF$ by construction. $CF = CF$. $\frac{CD}{CE} = 1$. Then $\triangle CDF \cong \triangle CEF$ by SSS, ratio = 1. $m\angle DCF = m\angle ECF$ by CPCF. $CJ = CJ$. Then $\triangle CDJ \cong \triangle CEJ$ by SAS, ratio = 1. $m\angle CJD = m\angle CJE$ by CPCF. Then each angle is a right angle, since $\angle DJE$ is a straight angle. Then $\overleftrightarrow{CF} \perp \overleftrightarrow{AB}$.]

2. Draw a 3-inch line segment, $\overline{AB}$, on your paper. Mark point C 1 inch from B. Construct the perpendicular to $\overline{AB}$ at C. See margin.

3. Draw a 12-centimeter line segment, $\overline{AB}$, on your paper. Mark a point C 5 centimeters from A. Construct the perpendicular to $\overline{AB}$ at C. See margin.

PERPENDICULARS TO A LINE FROM A POINT NOT ON THE LINE

The next exercise shows how to construct a perpendicular to a line from a point not on the line.

DISCOVERY ACTIVITY

1. Copy line $\overleftrightarrow{AB}$ onto a sheet of paper.
2. Place the point of your compass on point C and open the setting so that you can make two arcs on $\overleftrightarrow{AB}$. Label the intersections of the arcs and $\overleftrightarrow{AB}$ D and E.
3. What do you know about CD and CE? Equal
4. Keeping the same setting, place the point of your compass on D and make an arc below $\overleftrightarrow{AB}$. Do the same from E. Label the intersection of the arcs F.
5. Connect C and F. What appears to be true? $\overleftrightarrow{CF} \perp \overleftrightarrow{AB}$

The justification for $\overleftrightarrow{CF} \perp \overleftrightarrow{AB}$ in the Discovery Activity is similar to the proof for a perpendicular from a point on the line. Try the justification on your own.

CLASS ACTIVITY

1. Use a computer and a geometric drawing tool to draw any line. Draw points A and B on the line and point C above the line and between points A and B. Use the principles from the construction of a perpendicular to a line from a point not on the line to draw a perpendicular to the line through point C. Answers may vary.
2. Draw a 5-inch segment, $\overline{AB}$, and a point C 2 inches from A and 2 inches above $\overline{AB}$. Construct the perpendicular from C to the line segment. See margin.

3. Draw a 10-centimeter segment and mark a point below the segment. Construct a perpendicular to the segment from the point. See students' constructions.

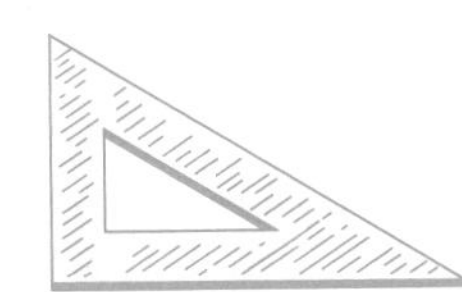

HOME ACTIVITY

Draw the perpendicular at or from *C* to *AB*. See margin.

1. A C B

2. A C B

3. C A B

4. 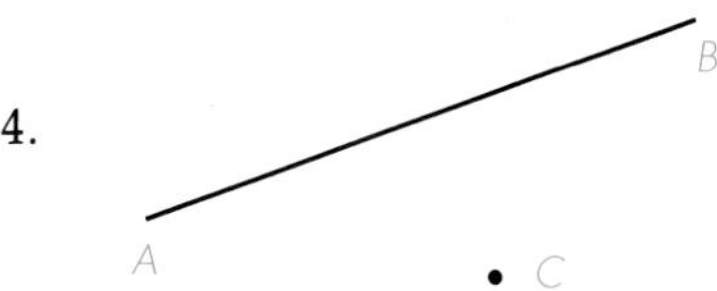

See students' constructions.

5. Draw $\overline{AB}$, a 6-inch line segment, and mark *C* 1.5 inches from *A*. Construct the perpendicular to $\overline{AB}$ from *C*. See margin.
6. Draw $\overline{AB}$, a 9-centimeter line segment, and mark *C* 3.5 centimeters from *A*. Construct the perpendicular to $\overline{AB}$ from *C*. See margin.
7. Draw $\overline{AB}$, a 60-millimeter line segment, and mark *C* 20 millimeters from *A* and 15 millimeters below $\overline{AB}$. Construct the perpendicular to $\overline{AB}$ from *C*. See margin.
8. Draw a 5-inch line segment $\overline{AB}$, and mark a point *D* 1 inch above the segment and 3 inches from *A*. Construct the perpendicular to $\overline{AB}$ from *D*. See margin.
9. Draw $\overline{MN}$, a 7-inch line segment, and mark *P* $3\frac{1}{4}$ inches from *N*. Construct the perpendicular to $\overline{MN}$ from *P*. See margin.
10. Draw a 5-centimeter line segment $\overline{XY}$, and mark a point *Z* 2.4 centimeters from *Y* and 1.8 centimeters above $\overline{XY}$. Construct the perpendicular to $\overline{XY}$ from *Z*. See margin.
11. Draw a $4\frac{1}{2}$-inch line segment $\overline{PQ}$, and mark a point *R* $1\frac{1}{4}$ inches from *Q*. Construct the perpendicular to the segment from *R*. See margin.
12. Draw a 3.2-centimeter line segment $\overline{HI}$, and mark a point *J* 1.6 centimeters from *H* and 2.1 centimeters above $\overline{HI}$. Construct the perpendicular to $\overline{HI}$ from *J*. See margin.
13. Draw a 43-millimeter line segment $\overline{DE}$, and mark a point *F* 26 millimeters from *E*. Construct the perpendicular to $\overline{DE}$ from *F*. See margin.
14. Draw a $3\frac{1}{4}$-inch line segment $\overline{GK}$, and mark a point *T* $1\frac{1}{2}$ inches from *K* and $\frac{3}{4}$ inch below $\overline{GK}$. Construct the perpendicular to $\overline{GK}$ from *T*. See margin.
15. Draw an angle and pick a point between the two rays. From that point construct perpendiculars to the two rays. Answers may vary.

Additional Answers

1.
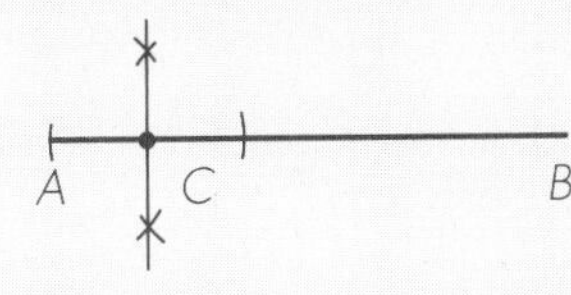

2.
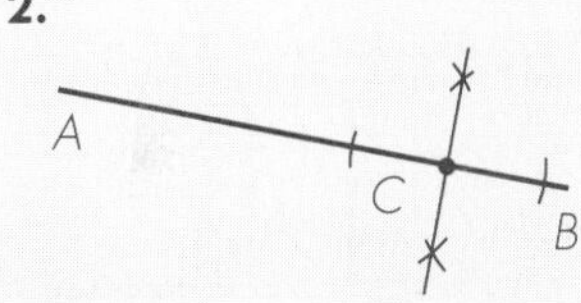

3.
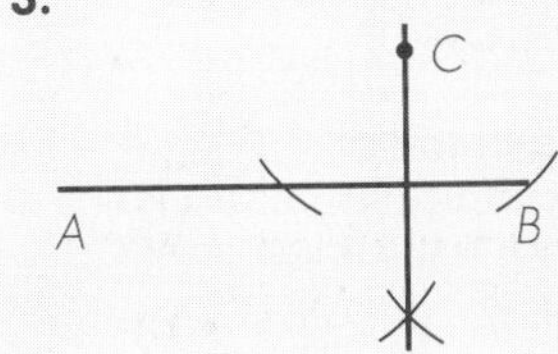

4.
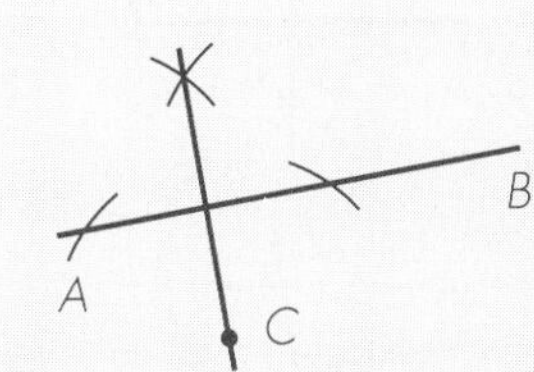

16. A pipeline is to be constructed from the oil well to the river. By construction, indicate the path that is perpendicular to the river. See margin.

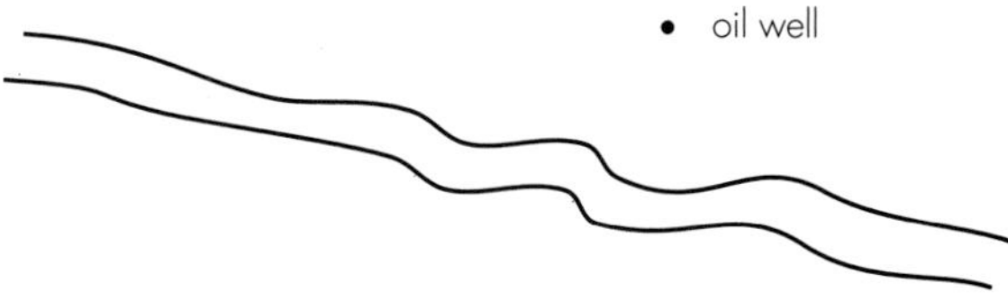

17. A surveyor knows line $\overleftrightarrow{AB}$ on a map is a north-south line. Show, by construction, how to determine an east-west line through city P.

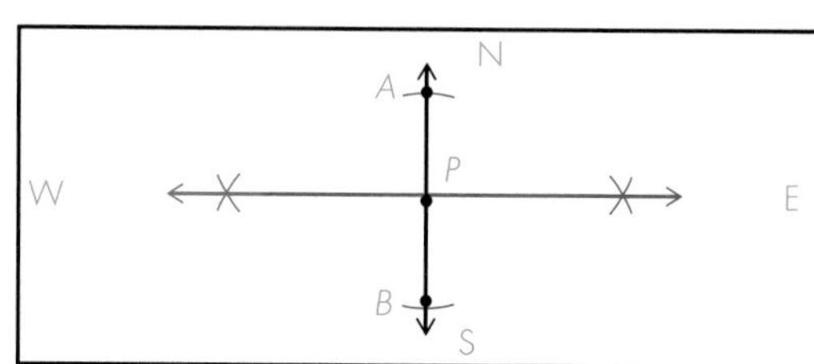

18. Perform the following directions: See margin.
   a. Draw line segment $\overline{AB}$ 5 inches in length.
   b. Mark point C, 2 inches from A, and D, 2 inches from B.
   c. From C and D, construct perpendiculars.
   d. From C, mark a point E, 1 inch above C on the perpendicular.
   e. From E, construct a perpendicular to $\overline{CE}$. Call it $\overline{GH}$.
   f. Extend $\overline{GH}$ to intersect the perpendicular at D. Label the point M.
   g. Darken 1.5 inches from C through E, and do the same from D through M. Now darken $\overline{EM}$.
   h. What does the darkened figure remind you of? Answers may vary.
19. How would you construct an angle of 45°? All you are given to start with is a line on your paper. Construct perpendicular lines, then bisect the 90° angle.

CRITICAL THINKING

20. Which printed capital letters in the alphabet consist of perpendicular line segments? Form one of these letters by construction.
   E, F, H, I, L, T; constructions may vary.

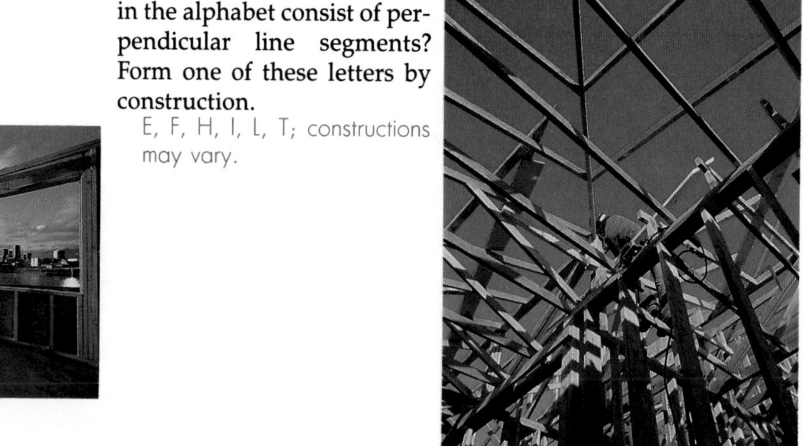

Additional Answers

16.

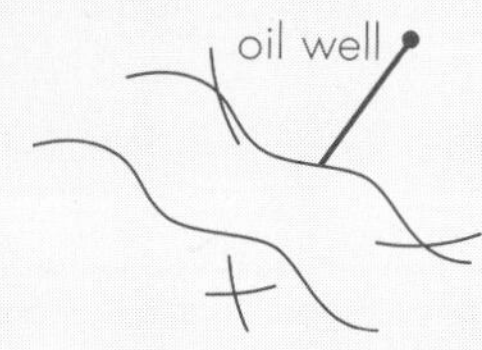

18. scaled-down sketch

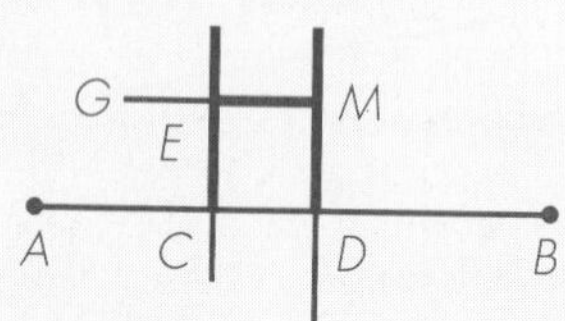

FOLLOW-UP

Assessment
Construct the perpendicular to each line through point C.

1.

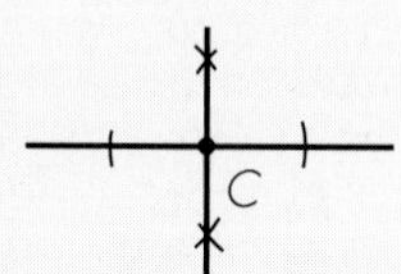

2.

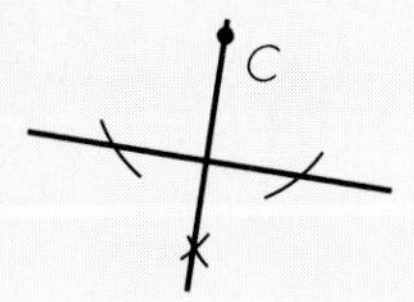

Reteaching
Students who have had difficulty with this section may benefit from Reteaching Activity 6.2.

Enrichment
For students who have mastered the material in this section, you may wish to assign Enrichment Activity 6.2.

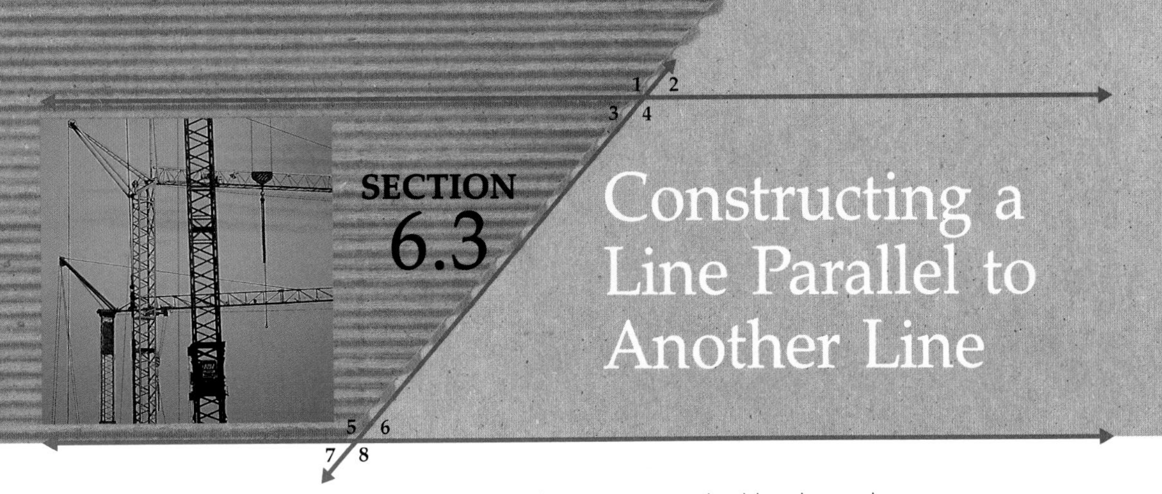

SECTION 6.3 Constructing a Line Parallel to Another Line

After completing this section, you should understand

▲ how to construct a line parallel to another line.

In the blueprint above, you can find many examples of parallel lines. How were they constructed?

NUMBER OF PARALLEL LINES THROUGH A POINT

Recall Theorem 3-1-1: if the figure is a triangle, then the sum of the angles is 180°. The proof of this theorem was based on the assumption (Postulate 3-1-1) that, through a point not on a line, there is only one line parallel to a given line.

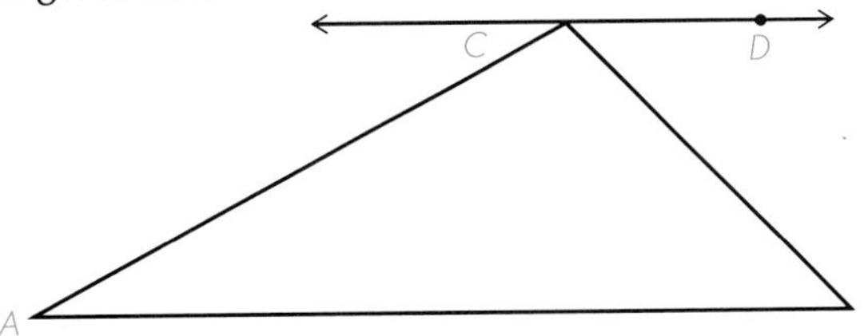

Mathematicians in the past have investigated other possibilities that could be true if Postulate 3-1-1 did not hold. The other possibilities are

1. there are many lines through a point parallel to the given line; and
2. there are no lines through a point parallel to the given line.

These possibilities may be true in non-Euclidean geometry, which plays an important role in navigating a ship or plane. For the work in this book, Postulate 3-1-1 will hold; we assume there is exactly one line parallel to a given line through a point not on the line.

SECTION 6.3

Resources
Reteaching 6.3
Enrichment 6.3
Transparency Master 6-1

Objectives
After completing this section, the student should understand
▲ how to construct a line parallel to another line.

Materials
compass
straightedge
ruler
library resources

GETTING STARTED

Quiz: Prerequisite Skills

1. According to Postulate 3-1-1, how many lines are there through a point not on a line that are parallel to the line? [one]

2. If two lines are parallel and are intersected by a transversal, what do you know about the alternate interior angles formed? [equal]

Warm-Up Activity
Have students draw lines on each side of a ruler, a box, or any other item that would yield parallel lines. Then have them draw a transversal and make observations about the figures they have drawn.

REPORT 6-3-1 Mathematicians developed non-Euclidean geometry to deal with the possibilities of either no parallel lines or more than one parallel line through a point. Use the references below to write a brief report on non-Euclidean geometry.

Suggested sources:
Bergamini, David. *Mathematics.* New York: Time-Life Books, 1963.

Chalkboard Example
Construct a line through *P* parallel to $\overleftrightarrow{AB}$.

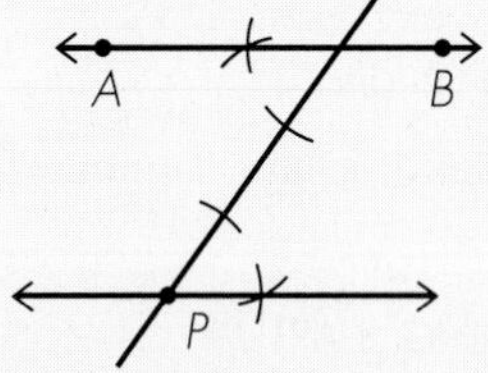

TEACHING COMMENTS

Using the Page
You can use a globe to demonstrate another fact of non-Euclidean geometry: parallel lines can intersect. On the globe, parallel lines are defined to be great circles—circles that are formed by a plane passing through the center of the earth and intersecting the surface of the earth. The lines of longitude are parallel because they are all great circles. The two lines shown intersect at the north and the south poles.

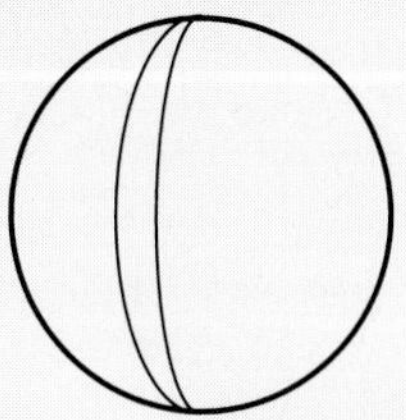

CONSTRUCTING A LINE PARALLEL TO A GIVEN LINE

To construct a parallel line, you first need to recall how to copy an angle.

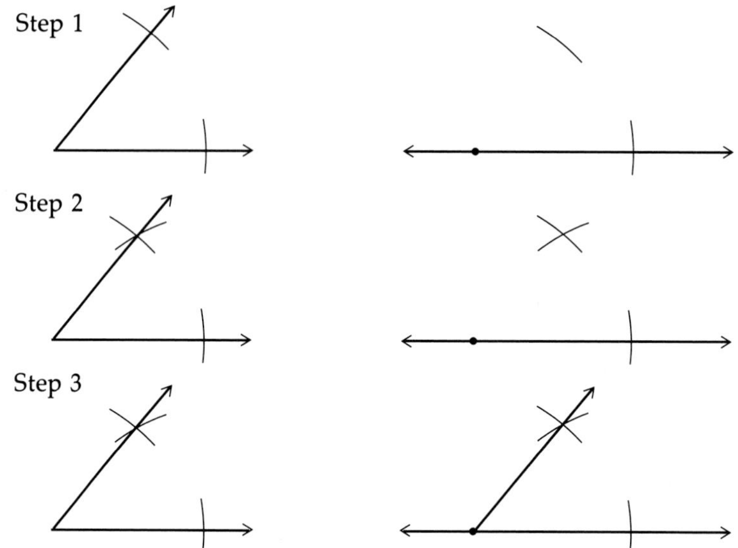

JUSTIFYING THE METHOD FOR COPYING AN ANGLE

Here is the justification of the method of copying an angle.

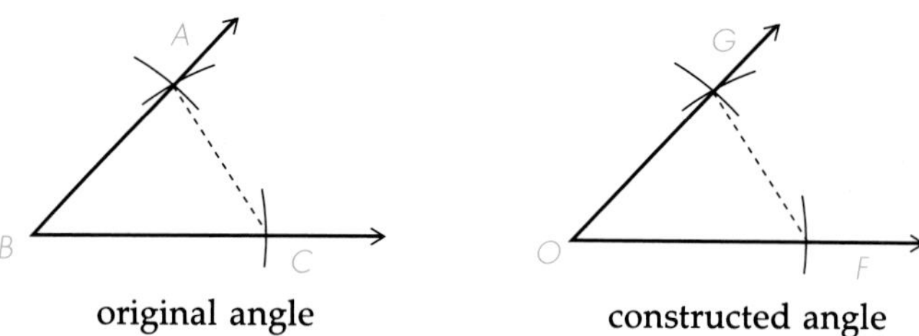

original angle constructed angle

Steps	Reasons
1. Draw $\overline{AC}$, $\overline{GF}$.	1. Two points determine a line.
2. $BC = OF$, $BA = OG$, $AC = GF$	2. By construction
3. $BC/OF = BA/OG = AC/GF$	3. Algebra
4. $\triangle ABC \sim \triangle GOF$	4. SSS
5. $BC/OF = BA/OG = AC/GF = 1$	5. Algebra
6. $\triangle ABC \cong \triangle GOF$	6. Definition of congruent triangles
7. $\angle ABC \cong \angle GOF$	7. CPCF

Using the Page
See Step 1 in the proof. Remind students that when auxiliary lines are drawn, they must be justified with a reason.

DISCOVERY ACTIVITY

1. On a sheet of paper, draw line $\overleftrightarrow{AB}$. Choose a point above the line, and label the point P.
2. Choose a point on $\overleftrightarrow{AB}$. Label the point F. Draw $\overleftrightarrow{FP}$.
3. Copy $\angle PFB$ to form $\angle FPN$. $\angle FPN$ is to be the alternate interior angle for $\angle PFB$. See margin.
4. What appears to be true about $\overleftrightarrow{NP}$ and $\overleftrightarrow{AB}$? They are parallel.
5. Make a second copy of line $\overleftrightarrow{AB}$ and point P. Choose a different point F on $\overleftrightarrow{AB}$.
6. Copy $\angle PFB$ as before. Does your answer to question 4 change? No

Using the Page
Make sure students copy the angle correctly in step 2 of the Discovery Activity. This step involves copying the angle in a different orientation than the original angle.

Use Transparency Master 6-1 to show students that equal corresponding angles also means that the lines are parallel. Given $m\angle 1 = m\angle 3$. But $m\angle 1 = m\angle 2$, since they are vertical angles. Thus $m\angle 2 = m\angle 3$. Since $\angle 2$ and $\angle 3$ are alternate interior angles, the lines are parallel.

Additional Answers

3.

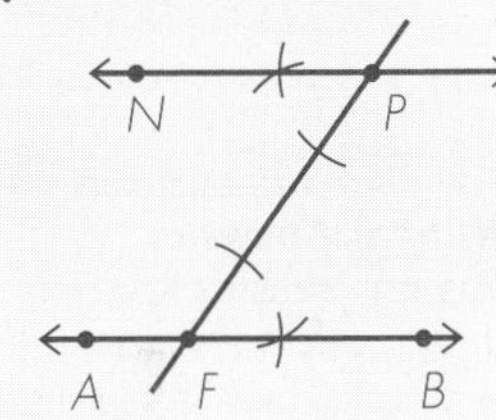

If two lines are intersected by a transversal so that the alternate interior angles are equal, then the two lines appear to be parallel. This is indeed the case.

JUSTIFYING THE CONSTRUCTION

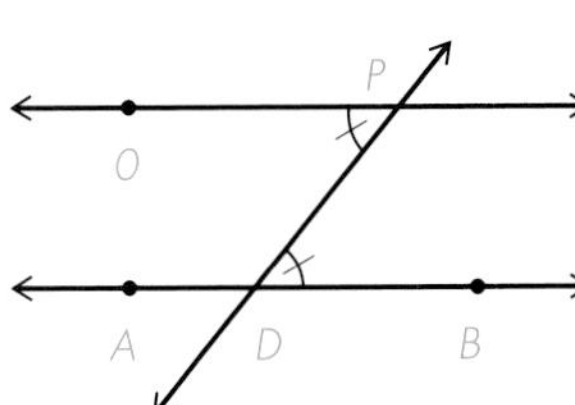

Recall Theorem 2-3-2: if two parallel lines are intersected by a transversal, then the alternate interior angles are equal. In the Discovery Activity, you constructed alternate interior angles to be equal. Your conclusion is that the lines are parallel. This is the converse of Theorem 2-3-2, so we must justify this conclusion separately.

Given: Lines $\overleftrightarrow{AB}$ and $\overleftrightarrow{OP}$ with transversal so that $\angle OPD \cong \angle PDB$.

Justify: Lines $\overleftrightarrow{AB}$ and $\overleftrightarrow{OP}$ are parallel.

Using the Page

Students should realize that $\overleftrightarrow{OPF}$ is a straight line. The (false) assumption that $\overleftrightarrow{OP}$ and $\overleftrightarrow{AB}$ intersect leads to formation of ∠3, and, therefore, the triangle. Students should realize that an angle of a triangle cannot have zero measurement.

Extra Practice

Draw a line segment $1\frac{1}{2}$ inches long. Label it $\overline{TW}$. Mark a point C, $\frac{1}{2}$ inch from W and $\frac{3}{4}$ inch below $\overline{TW}$. Construct a line parallel to $\overline{TW}$ through C.

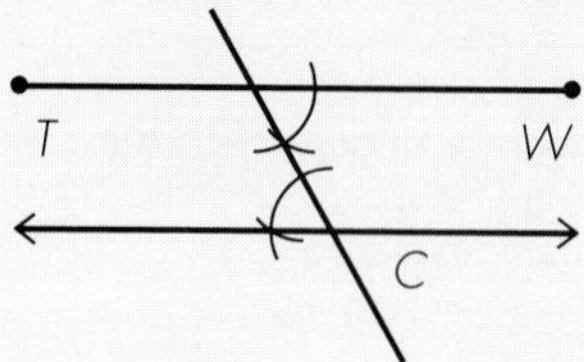

Additional Answers

2. Possible construction

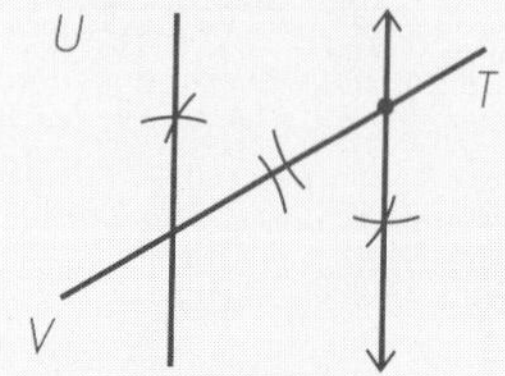

3. Possible construction

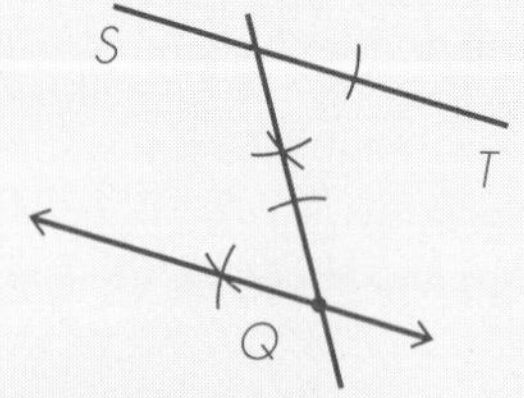

4. Possible construction scaled-down sketch

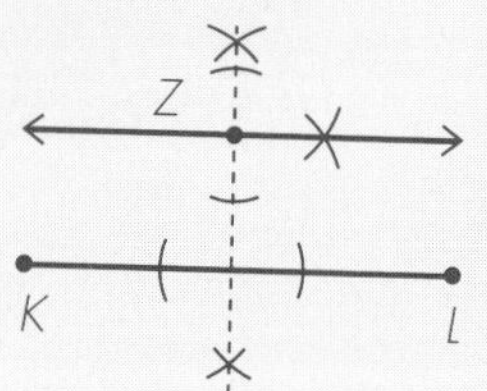

Indirect reasoning will be used for this justification.

Only two possibilities exist: either $\overleftrightarrow{OP}$ is parallel to $\overleftrightarrow{AB}$ or $\overleftrightarrow{OP}$ is not parallel to $\overleftrightarrow{AB}$. Assume that $\overleftrightarrow{OP}$ is not parallel to $\overleftrightarrow{AB}$. If the lines are not parallel, then $\overleftrightarrow{AB}$ and $\overleftrightarrow{OP}$ intersect. This means a triangle is formed, as indicated in the figure below.

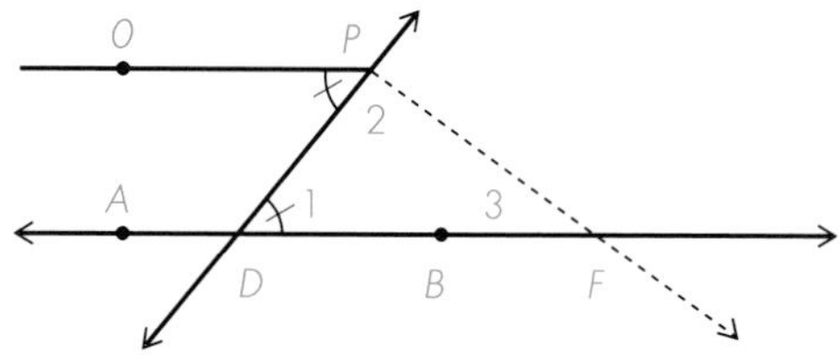

It was given that $\angle OPD \cong \angle PDB$. Because $\overleftrightarrow{OP}$ is a line, $m\angle 1 + m\angle 2 = 180°$. In $\triangle DPF$, $m\angle 1 + m\angle 2 + m\angle 3 = 180°$. But this means that $m\angle 3 = 0°$. This is impossible. Since the assumption led to a contradiction, the assumption must be false, or invalid. This means that $\overleftrightarrow{OP} \parallel \overleftrightarrow{AB}$.

THEOREM 6-3-1 **If two lines are intersected by a transversal so that the alternate interior angles are equal, then the lines are parallel.**

Thus, the method presented in the Discovery Activity has been justified as a way of constructing parallel lines.

CLASS ACTIVITY

1. Use a computer and a geometric drawing tool to draw any two lines intersected by a transversal so that the alternate interior angles are equal. What appears to be true about the lines? Repeat this activity several times. Do your findings support Theorem 6-3-1? They are parallel; Yes.

Construct a line parallel to the given line through the given point. See margin.

2.

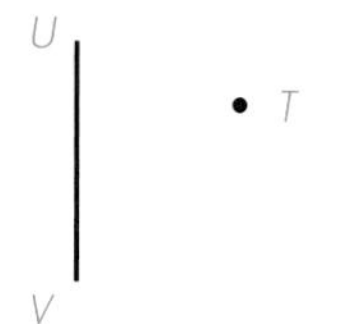

3.

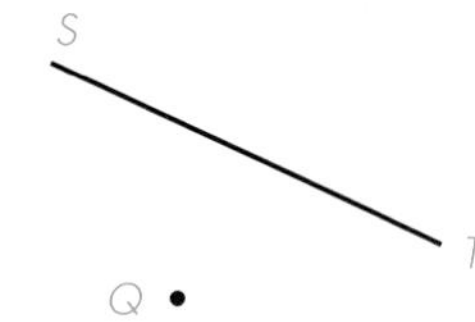

4. Draw a line segment 15 centimeters long. Label it $\overline{KL}$. Mark a point Z, 6 centimeters from K and 3 centimeters above $\overline{KL}$. Construct a line parallel to $\overline{KL}$ through Z. See margin.

5. Draw a 4-inch line segment. Choose two points, *A* and *B*, so that *A* is 1 inch above the line segment and *B* is 2 inches below the line segment. Construct a line through *A* and one through *B* that are parallel to the segment. See margin.

Additional Answers

5. Possible construction scaled down sketch

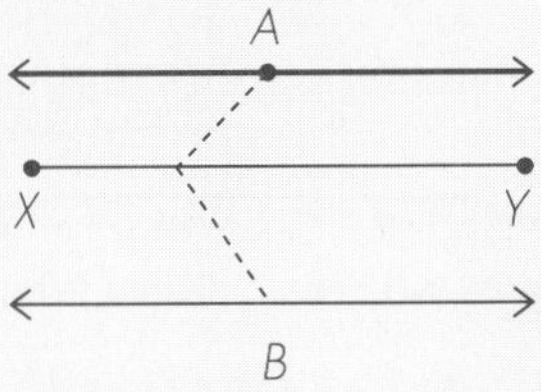

A

X Y

B

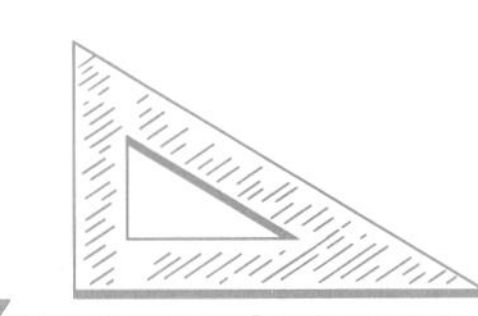

HOME ACTIVITY

Construct a line parallel to the given line through the given point(s). Possible constructions drawn.

1.

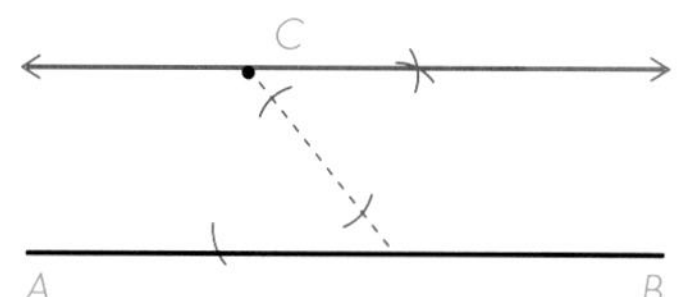

2.

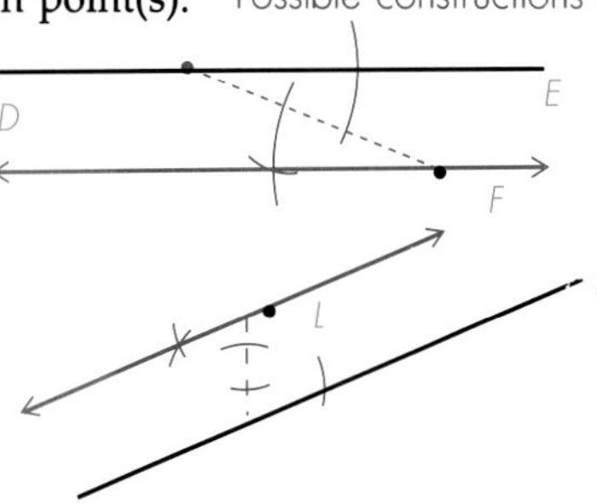

3.

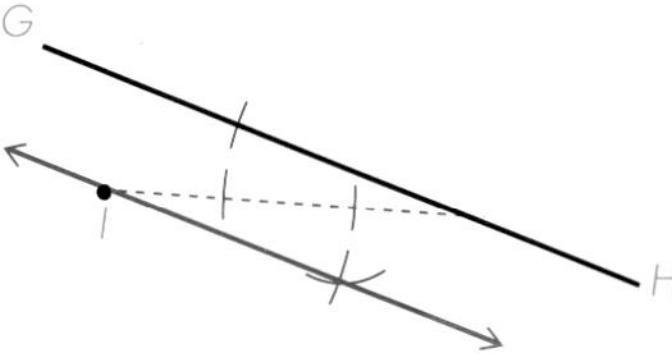

4.

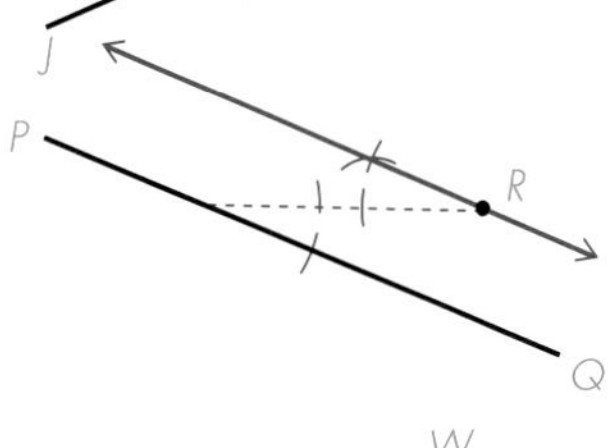

5.

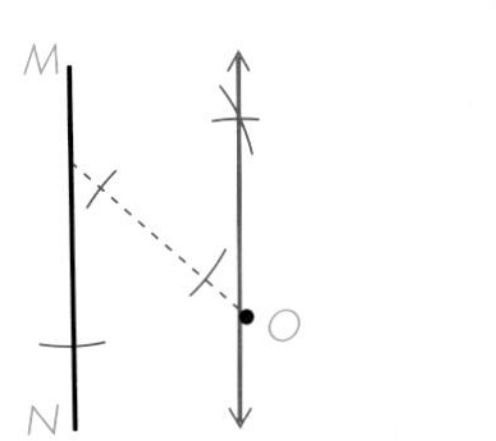

6.

7.

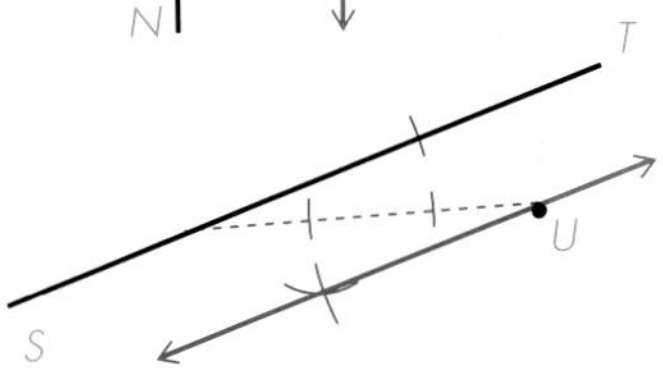

T

U

S

8.

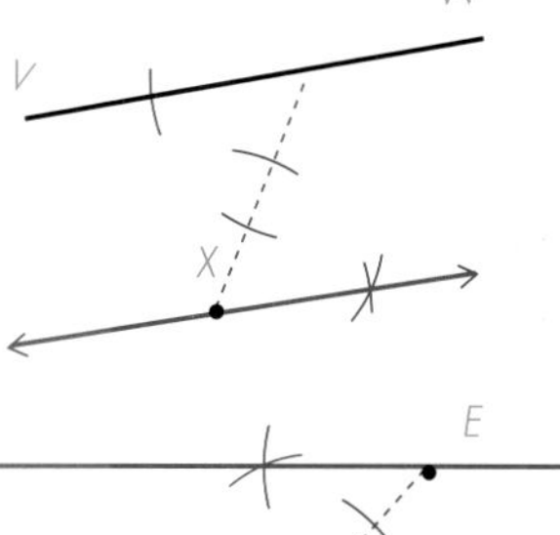

9.

11.

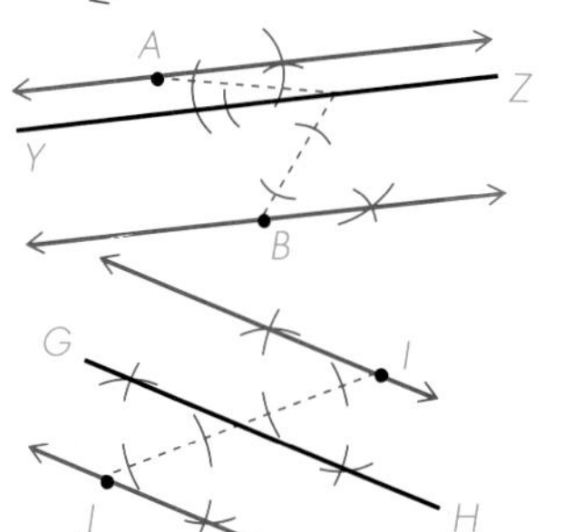

10.

12.

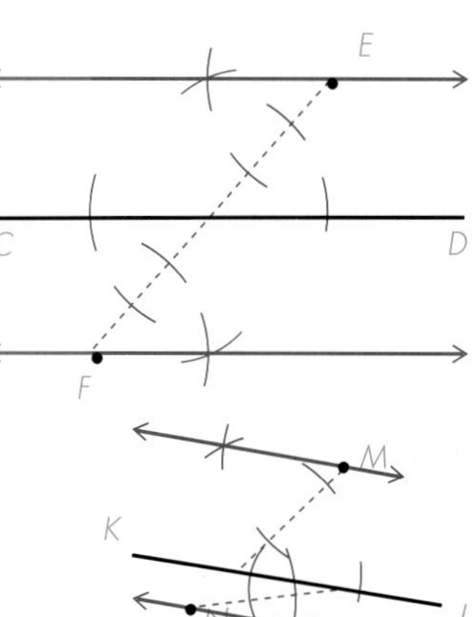

Additional Answers
For Exercises 13-16, scaled down sketches are shown.

13.

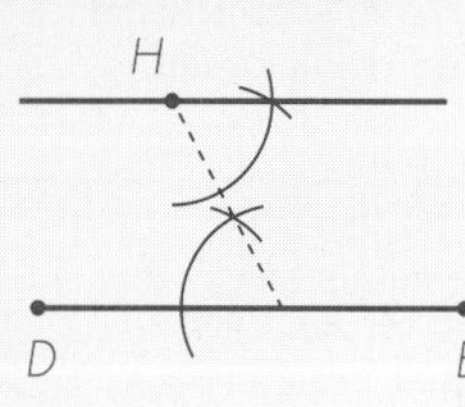

14.

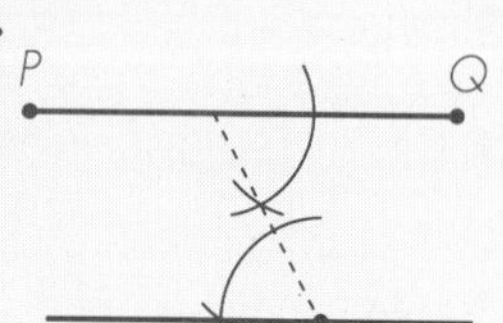

15.

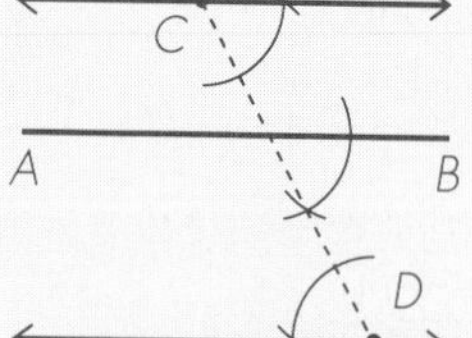

16.

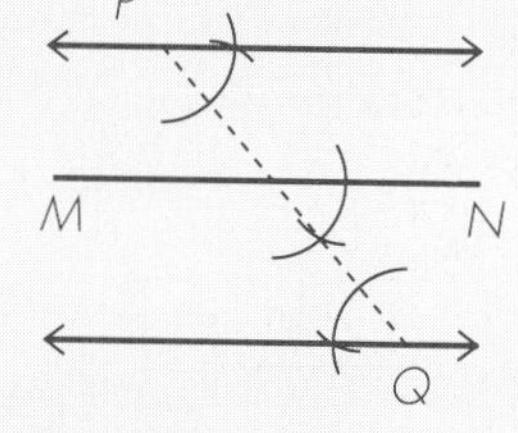

FOLLOW-UP

Assessment

1. What two angles, formed by two lines and a transversal, are used to construct parallel lines? [alternate interior]

2. Construct a line through K parallel to GH.

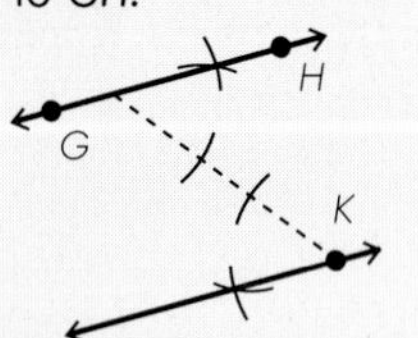

Reteaching
Students who have had difficulty with this section may benefit from Reteaching Activity 6.3.

Enrichment
For students who have mastered the material in this section, you may wish to assign Enrichment Activity 6.3.

13. Draw a line segment $\overline{DE}$ 4.5 inches long. Choose a point H so that H is 2.25 inches from D and 1.25 inches above $\overline{DE}$. Construct the line through H that is parallel to $\overline{DE}$. See margin.

14. Draw a line segment $\overline{PQ}$, 130 millimeters long. Choose a point T that is 64 millimeters from Q and 57 millimeters below $\overline{PQ}$. Construct the line through T that is parallel to $\overline{PQ}$. See margin.

15. Draw a line segment $\overline{AB}$, 6 inches long. Choose a point C 2 inches from A and $1\frac{1}{2}$ inches above $\overline{AB}$, and a point D $1\frac{1}{2}$ inches from B and 1 inch below $\overline{AB}$. Draw a line through each of these points that is parallel to $\overline{AB}$. See margin.

16. Draw a line segment $\overline{MN}$, 180 millimeters long. Choose a point P 53 millimeters from M and 26 millimeters above $\overline{MN}$, and a point Q, 74 millimeters from N and 36 millimeters below $\overline{MN}$. Draw a line segment through each of these points that is parallel to $\overline{MN}$. See margin.

17. Draw an acute triangle and label it $\triangle SWC$. Construct the line through C parallel to $\overline{SW}$. Answers may vary.

18. Construct a right angle, then draw a right triangle ACB, with C the vertex of the right angle. Construct a line through A parallel to $\overline{BC}$, and a line through B parallel to AC, and a line through C parallel to $\overline{AB}$. Extend the constructed lines until they intersect. How many triangles are formed? 5

Draw an acute triangle. Bisect each angle and extend the bisectors until each intersects the opposite side.

19. What do you observe about the three angle bisectors? They intersect at one point.

20. How many triangles are there in the final figure? 16

CRITICAL THINKING

21. Use indirect reasoning to solve the following: Sara, Bob, and Cheryl are three bowlers with different averages. If only one of the following statements is true, then who has the highest average? Sara

Bob has the highest average.

Sara does not have the highest average.

Cheryl does not have the lowest average.

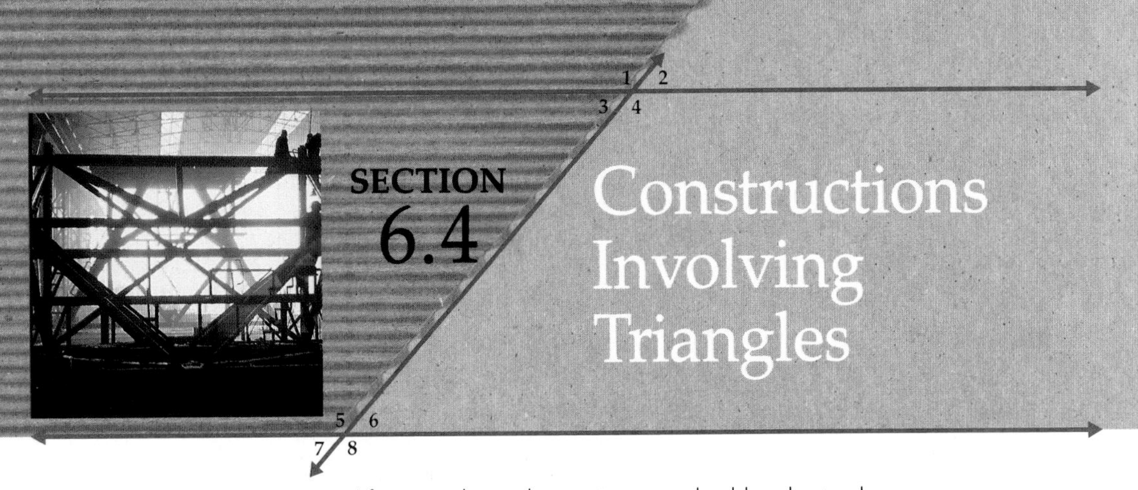

After completing this section, you should understand
- ▲ the medians of a triangle.
- ▲ the altitudes of a triangle.
- ▲ the angle bisectors of a triangle.
- ▲ the perpendicular bisectors of the sides of a triangle.

So far in this chapter, you have learned how to bisect line segments, how to bisect angles, how to construct perpendiculars, and how to construct a line parallel to a given line.

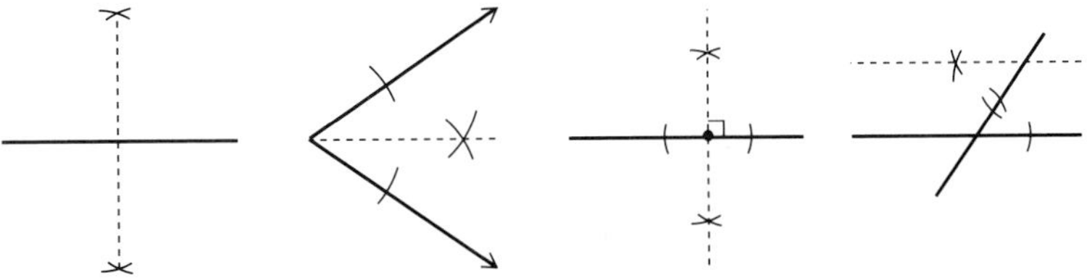

This section incorporates these constructions into applications with triangles. Some new terminology will be introduced.

MEDIANS OF A TRIANGLE

You may already be familiar with medians because of the median of a highway. The median divides a highway into two equal portions.

The median of a triangle divides a side of the triangle into two equal portions.

SECTION 6.4

Resources
Reteaching 6.4
Enrichment 6.4
Transparency Master 6-2

Objectives
After completing this section, the student should understand
- ▲ the medians of a triangle.
- ▲ the altitudes of a triangle.
- ▲ the angle bisectors of a triangle.
- ▲ the perpendicular bisectors of the sides of a triangle.

Materials
ruler
protractor
compass
straightedge
cardboard
nail

Vocabulary
median
concurrent lines
centroid
center of gravity
perpendicular bisector of a side
angle bisector in a triangle
altitude

GETTING STARTED

Quiz: Prerequisite Skills
See students' constructions.

1. Draw a line segment, then construct the perpendicular bisector.

2. Draw an angle, then bisect it.

3. Draw a line and a point not on the line. Construct a perpendicular from the point to the line.

Warm-Up Activity
Bring in photos of gymnasts who are balancing on uneven parallel bars, rings, and so on. Lead students into a discussion on how a gymnast must use the center of gravity of his/her body in order to balance.

Chalkboard Example
Find the centroid of the following triangle.

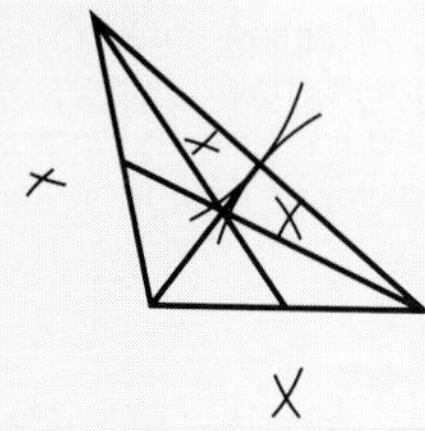

TEACHING COMMENTS

Using the Page
The diagram on this page shows the median from C to $\overline{AB}$. Remind students that $\overline{CA}$ and $\overline{CB}$ are the sides *adjacent* to ∠C.

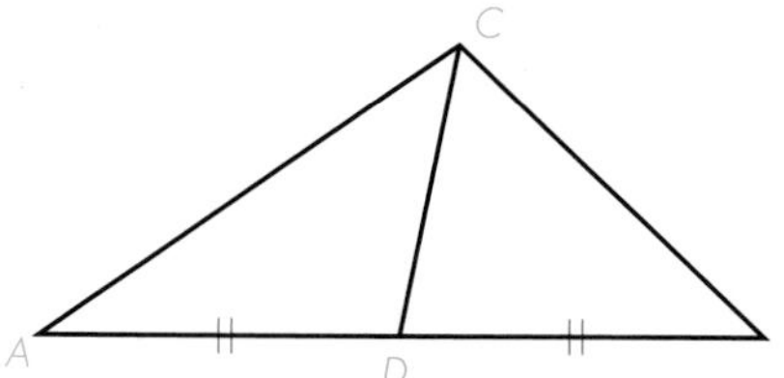

$\overline{CD}$ is one of the medians of $\triangle ABC$. $\overline{CD}$ divides $\overline{AB}$ into two equal portions or lengths.

DEFINITION 6-4-1 **A MEDIAN of a triangle is a line segment joining the vertex of an angle and the midpoint of the opposite side.**

To draw a median, you must locate the midpoint of a side of the triangle. Connect this point to the vertex opposite it.

DISCOVERY ACTIVITY

1. Use a ruler and a protractor to draw $\triangle ABC$ on a sheet of paper. In $\triangle ABC$, let $AB = 3$ inches, $m\angle A = 40°$, and $m\angle B = 60°$.
2. By construction, locate the midpoint of $\overline{AB}$, $\overline{AC}$, and $\overline{BC}$. Label the midpoints D, E, and F, respectively.
3. Connect points A and F, B and E, and C and D. What name is given to $\overline{AF}$, $\overline{BE}$, and $\overline{CD}$? Medians
4. How many medians does a triangle have? 3
5. Write a statement that appears to be true about the medians that you constructed. They intersect at one point.

You may have found that the medians all intersect at the same point. Lines that do this are called concurrent lines.

DEFINITION 6-4-2 **CONCURRENT lines are lines that intersect at the same point.**

The point at which the lines intersect is the point of concurrency. The point of concurrency for the medians of a triangle is the centroid.

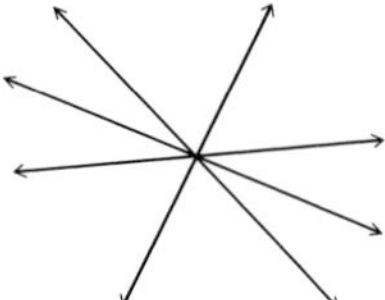

DEFINITION 6-4-3 **The point where the medians of a triangle intersect is the CENTROID, or the center of gravity, of the triangle.**

The definition mentions an important fact about the centroid—it is the center of gravity of a triangle. This idea is explored in the following project.

PROJECT 6-4-1 Cut a large triangle from a piece of cardboard. Draw the medians, and label the point where the medians intersect M. Using the point of a nail or a pencil, try to support the triangle by placing the point on M. Does the triangle balance? Yes

Move the nail or point away from M and try to balance the triangle. Does the triangle balance? No

Try this experiment with triangles of different shapes, and write a conclusion. Present your experiment and findings to the class.

You may have found that you could balance the triangle at its centroid but not anywhere else. The triangle balances at its center of gravity.

PERPENDICULAR BISECTORS OF THE SIDES OF A TRIANGLE

The three perpendicular bisectors of the sides of a triangle also have a special property, as you will discover.

DISCOVERY ACTIVITY

1. Use a ruler and a protractor to draw $\triangle ABC$ on a sheet of paper. Let $AB = 10$ cm, $m\angle A = 40°$, and $AC = 12$ cm.

2. Construct the perpendicular bisectors of each side of the triangle. Label the midpoints of the sides D, E, and F. Extend the perpendicular bisectors. What do you observe? They intersect at one point.

You may have discovered that the perpendicular bisectors of the sides of a triangle are concurrent.

Using the Page
Emphasize that the center of gravity is that point where an object will balance.

Transparency Master 6-2 can be used to show students by measuring that the centroid C is $\frac{2}{3}$ of the distance from a vertex to the side opposite the vertex.

Using the Page
You may want to have students find the centroid before they cut the triangle out. If students have trouble balancing the triangle at the centroid, you might have them use the eraser-end of the pencil, rather than the point.

ANGLE BISECTORS OF A TRIANGLE

Next, you will investigate the angle bisectors of a triangle.

Using the Page
You may want to have students repeat the first Discovery Activity, using a right triangle or an obtuse triangle the second time.

DISCOVERY ACTIVITY

1. On a sheet of paper, draw a large acute triangle.
2. Construct the angle bisector for each angle.
3. Extend each angle bisector until it intersects the opposite side.
4. Write a conclusion that appears to be true about the bisectors.
   They are concurrent.

You may have observed that the three angle bisectors are concurrent.

DEFINITION 6-4-4 **An ANGLE BISECTOR of a triangle is a line segment from a vertex of a triangle to the point where the angle bisector of that angle intersects the opposite side.**

ALTITUDES OF A TRIANGLE

The altitude of a triangle also involves a perpendicular segment.

DISCOVERY ACTIVITY

1. On a sheet of paper, draw a large acute triangle.
2. From each vertex, construct a line segment that is perpendicular to the opposite side.
3. Write a conclusion that appears to be true about the altitudes.
   They are concurrent.

You may have discovered that the three perpendiculars from the vertices to the opposite sides are concurrent. These line segments are called altitudes.

NOTEBOOK

DEFINITION 6-4-5 **An ALTITUDE of a triangle is a perpendicular line segment from a vertex of the triangle to the opposite side or to the line determined by the opposite side.**

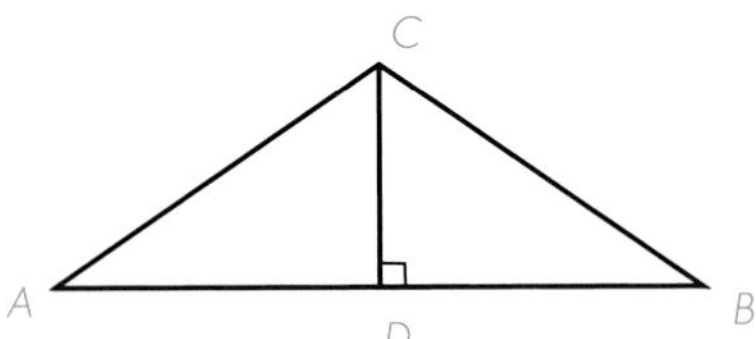

$\overline{CD}$ is the altitude from C.

You used an acute triangle in the Discovery Activity. Observe what happens when the triangle is obtuse.

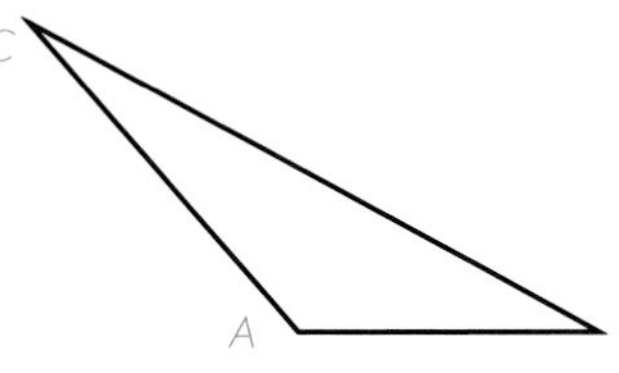

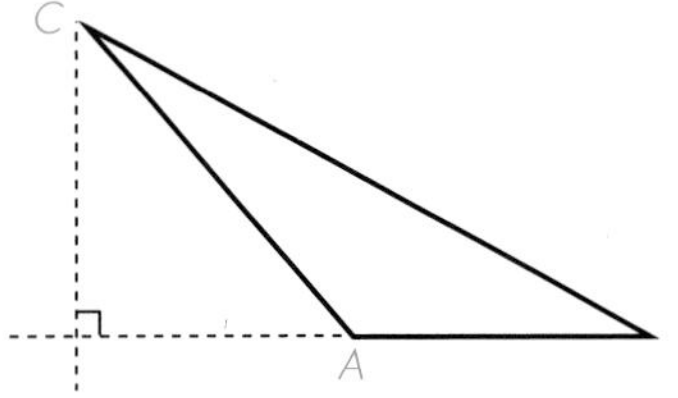

In order to draw the altitude from C, you must extend $\overline{AB}$.

You may have noticed that in each construction, the line segments intersect at a common point. Only one, the centroid, or center of gravity, was discussed. The usefulness of the other points of concurrency will be discussed in Section 6.6.

Using the Page
Have students draw an obtuse triangle, as shown, and actually do the construction for the altitude from C to the extension of $\overline{AC}$.

CLASS ACTIVITY

Copy each triangle four times. For each triangle, construct the medians, the perpendicular bisectors of the sides, the angle bisectors, and the altitudes. Label each example. See students' constructions.

1.

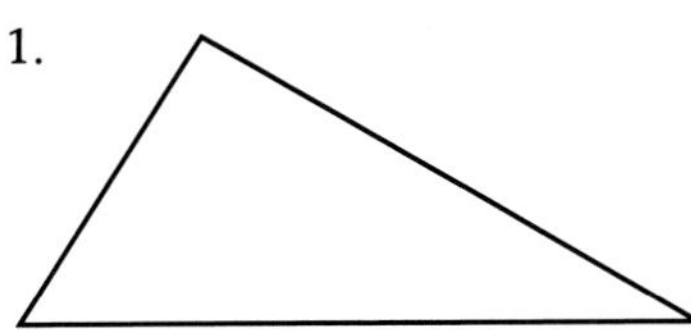

2.

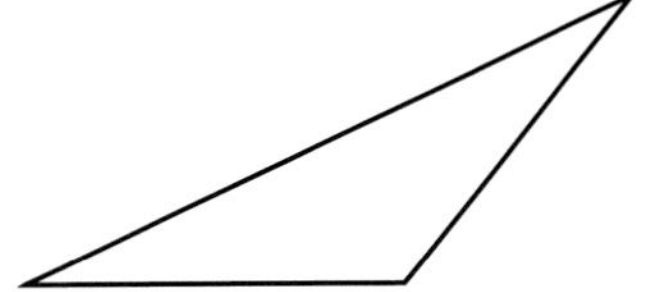

3. Draw $\triangle ABC$ with $AB = 4$ inches, $BC = 3.5$ inches, and $m\angle B = 50°$. Construct the medians and label the centroid M.
See margin.
4. Construct $\triangle DEF$ with $DE = 3$ inches, $DF = 2.5$ inches, and $EF = 4.5$ inches. Construct the perpendicular bisectors of the sides. Label the point of concurrency P. Classify the triangle according

Additional Answers
3. scaled-down sketch

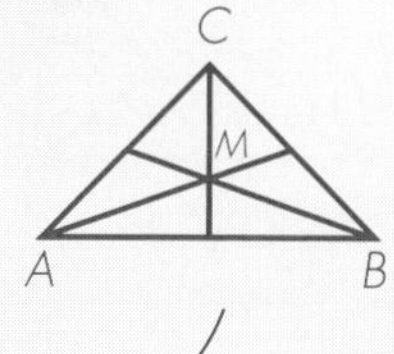

Extra Practice

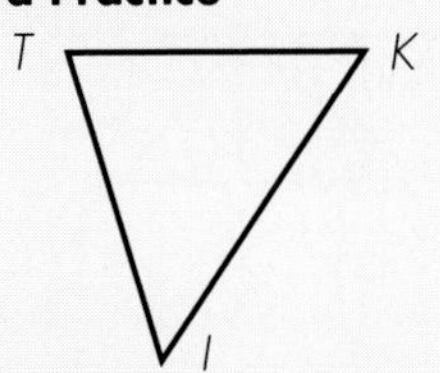

Copy $\triangle TIK$ for each exercise. See students' constructions.

1. Construct the medians.
2. Construct the perpendicular bisectors of the sides.
3. Construct the angle bisectors.
4. Construct the altitudes.

Additional Answers

For Exercises 4-6, scaled-down sketches are shown.

4.

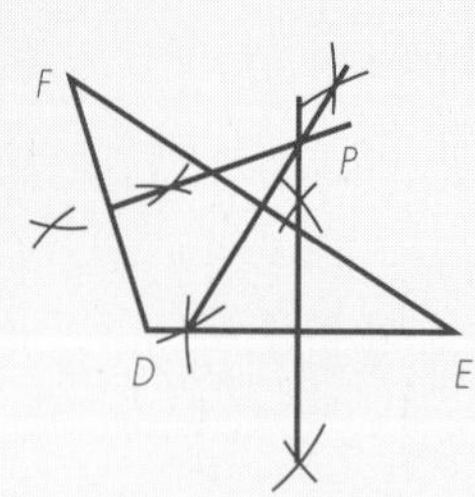

5.

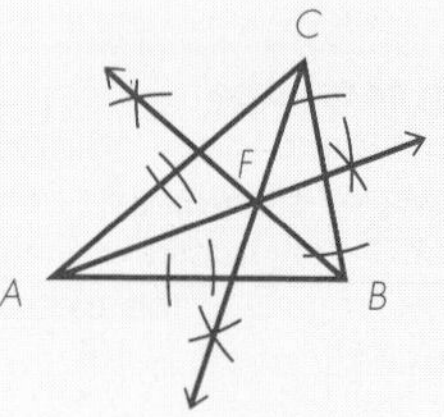

6.

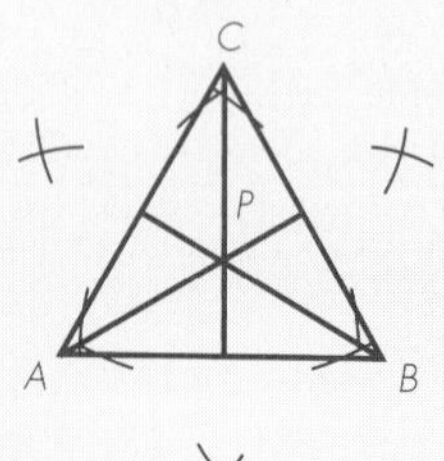

For Exercises 3-5, scaled-down sketches are shown.

3.

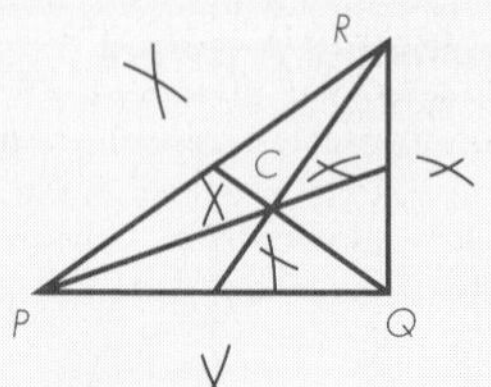

4.

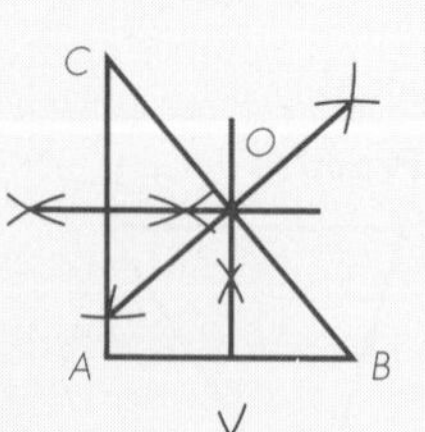

5.

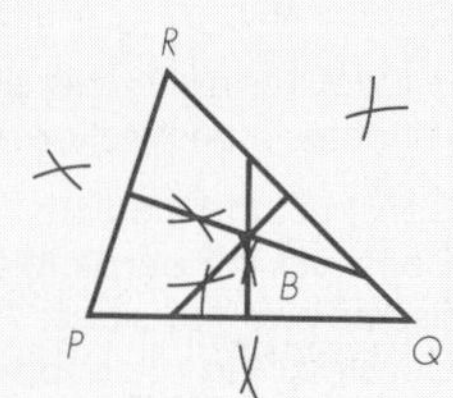

to its sides and its angles.
Scalene, obtuse; see margin for construction.

5. Construct $\triangle ABC$, given that AB = 12.5 cm, BC = 10 cm, and AC = 15 cm. Construct the three angle bisectors and label the point of concurrency F. Classify the triangle according to its sides and its angles. Scalene, acute; see margin for construction.

6. Construct $\triangle ABC$ with AB = 4 inches, BC = 4 inches, and CB = 4 inches. Construct the three altitudes. Label the point of concurrency P. Classify the triangle according to its sides and its angles. Equilateral, equiangular, acute; see margin for construction.

Use a computer and a geometric drawing tool to complete Exercises 7 and 8.

7. Draw any triangle and the three medians of the triangle. Repeat this activity several times. Is the centroid ever outside the triangle? Explain your answer. No; each segment is drawn from the midpoint of a side to the angle opposite the side.

8. Draw any triangle and the three altitudes of the triangle. Repeat this activity several times. Do the altitudes ever intersect outside the triangle? Explain your answer. Yes; sometimes you must extend a side in order to construct the perpendicular.

HOME ACTIVITY

Copy each triangle four times. For each triangle, construct the medians, the perpendicular bisectors of the sides, the angle bisectors, and the altitudes. Label each example. See students' constructions.

1.

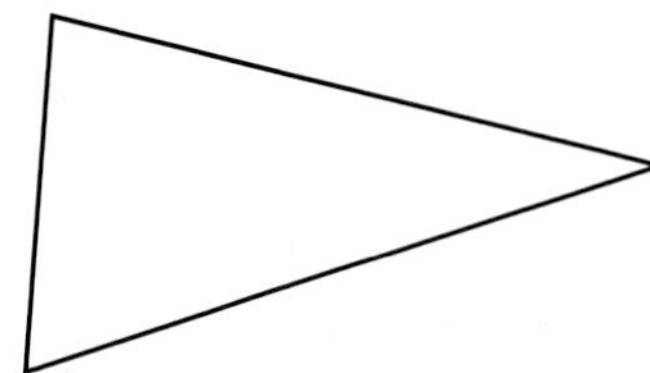

2.

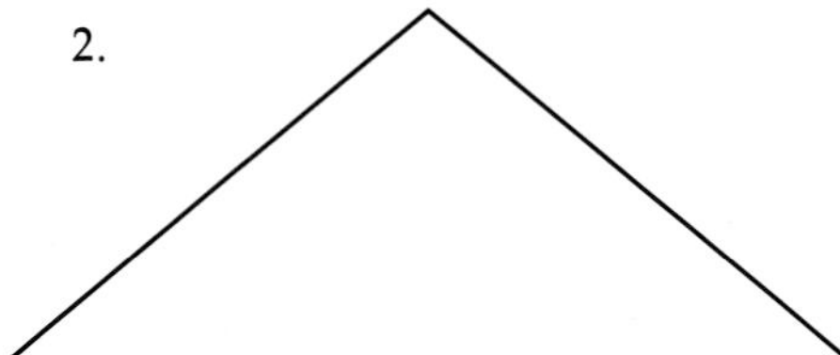

3. Construct $\triangle PQR$ with PQ = 12 cm, QR = 9 cm, and PR = 15 cm. Construct the medians and label the centroid C. See margin.

4. Construct $\triangle ABC$ with AB = 3 inches, BC = 5 inches, and AC = 4 inches. Construct the perpendicular bisectors of the sides. Label the point of concurrency O. Classify $\triangle ABC$ according to its sides and angles. Scalene, right; see margin for construction.

5. Draw $\triangle PQR$ with PQ = 14 cm, $m\angle P = 80°$, and $m\angle Q = 40°$. Construct the perpendicular bisectors of the sides. Label the point of concurrency B. See margin.

6. Construct an equilateral triangle with sides measuring 2 inches. Construct the three medians, then construct the three perpendicular bisectors of the sides. What do you observe?
They are the same.

7. Construct $\triangle ABC$, with $AB = 1$ inch, and $AC = BC = 2$ inches. Construct the median from C, and construct the perpendicular bisector of $\overline{AB}$. What do you observe? They are the same.

8. Draw $\triangle DHK$ with $DK = 8$ cm, $m\angle D = 100°$, and $m\angle K = 50°$. Construct the angle bisectors. Label the point of concurrency P. See margin.

9. Draw $\triangle THM$ with $TM = 10$ cm, $m\angle T = 110°$, and $m\angle M = 35°$. Construct the three altitudes. Label the point of concurrency K. See margin.

10. Construct $\triangle ABC$, with $AB = 3$ inches, $CA = 4$ inches, and $BC = 5$ inches. Construct the three altitudes, and label the point of concurrency P. What do you observe about the point P?
It is the same as point A.

11. Construct an isosceles right triangle. Construct the three perpendicular bisectors of the sides. Where is the point of concurrency? On the hypotenuse

12. Draw a large equilateral triangle. Construct each of the following: the three medians, the three angle bisectors, the three altitudes, and the three perpendicular bisectors of the sides. What do you observe? The points of concurrency are all the same.

CRITICAL THINKING

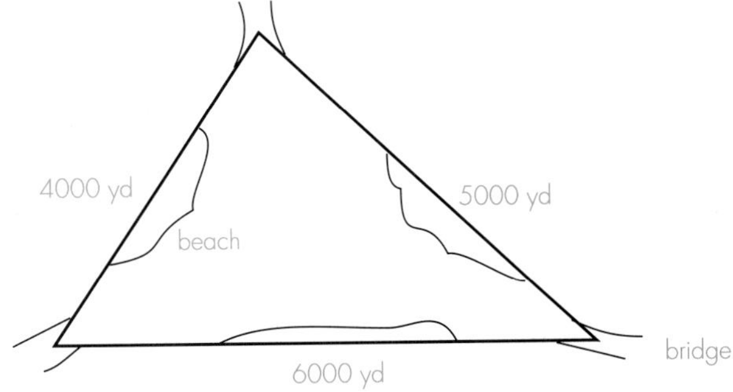

You live on a triangular island that has a beach on each side and a bridge at each vertex. The measurements of the island are 4000 yards by 6000 yards by 5000 yards.

13. You plan to build a cabin equidistant from the three beaches. Would you use the medians, the altitudes, the angle bisectors, or the perpendicular bisectors of the sides to help you decide where to build? Angle bisectors (incenter)

14. If you were to build your cabin equidistant from each bridge, would you use the medians, the altitudes, the angle bisectors, or the perpendicular bisectors of the sides to help you find the spot to build? Perpendicular bisectors (centroid) (orthocenter)

Additional Answers
For Exercises 8 and 9, scaled-down sketches are shown.

8.

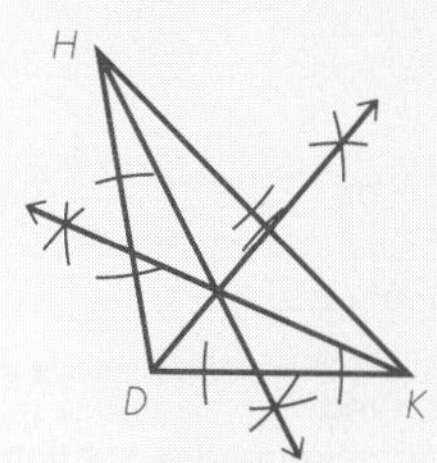

9.

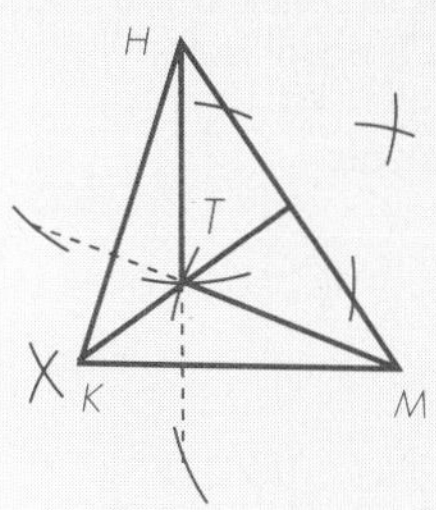

FOLLOW-UP

Assessment
1. What name is given to lines intersecting in one point? [concurrent]
2. Copy $\triangle TAL$ four times. Construct the three medians, the three angle bisectors, the three perpendicular bisectors of the sides, and the three altitudes. [See students' constructions.]

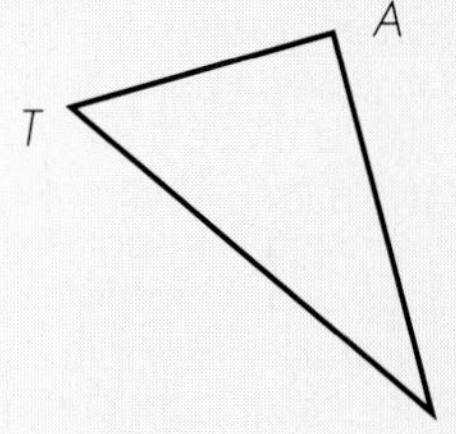

Reteaching
Students who have had difficulty with this section may benefit from Reteaching Activity 6.4.

Enrichment
For students who have mastered the material in this section, you may wish to assign Enrichment Activity 6.4.

SECTION 6.5

Resources
Reteaching 6.5
Enrichment 6.5

Objectives
After completing this section, the student should understand
▲ how to divide a line segment into *n* equal parts.

Materials
compass
straightedge
ruler

GETTING STARTED

Quiz: Prerequisite Skills

1. Bisect the segment.

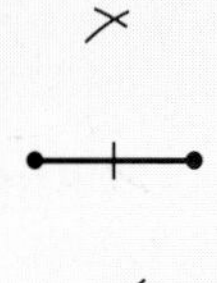

2. Copy the angle.

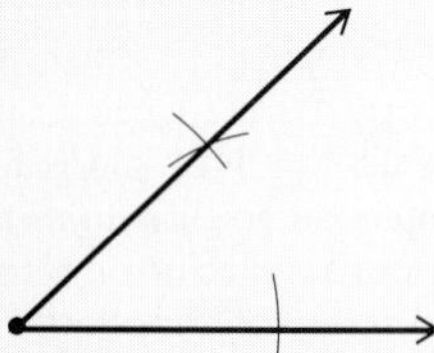

Warm-Up Activity
Have students draw a line segment 3 inches long and then use a ruler to divide it into three equal parts. Then have students draw a line segment 4 inches long and try to divide it into three equal parts with their rulers. Have them discuss any difficulties they encounter.

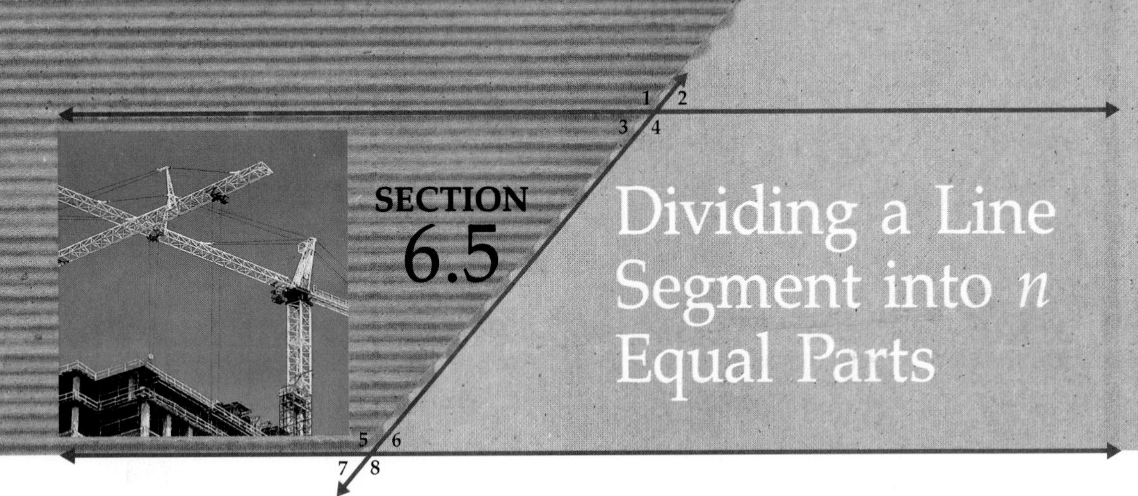

SECTION 6.5 Dividing a Line Segment into *n* Equal Parts

After completing this section, you should understand
▲ how to divide a line segment into *n* equal parts.

Section 6.1 not only introduced you to the tools and basic constructions of geometry, but also suggested a challenging problem. In that section, you learned to bisect a line segment.

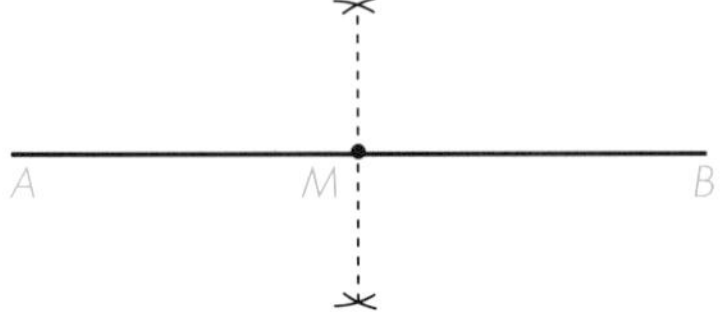

Applying the method of bisecting a line segment to $\overline{AM}$ and $\overline{MB}$ would result in four equal line segments. The process could continue, resulting in a sequence of equal line segments.

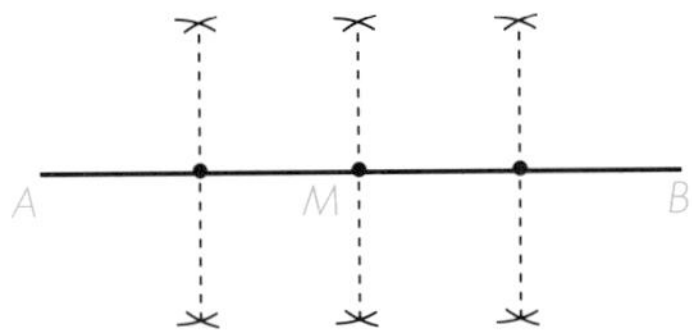

This method will work for dividing a line segment into 2^x congruent parts. However, a new method of dividing the segment is needed for a number of equal parts that is not a power of two.

DIVIDING A LINE SEGMENT INTO THREE EQUAL PARTS

The object is to find two points on a line segment that will divide it into three equal parts.

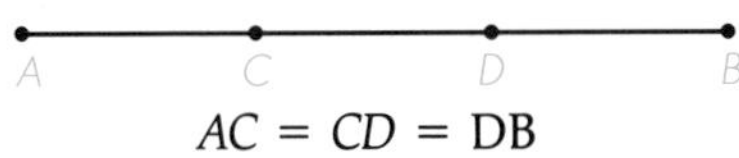

$AC = CD = DB$

Divide $\overline{AB}$, 10 centimeters long, into three equal parts.

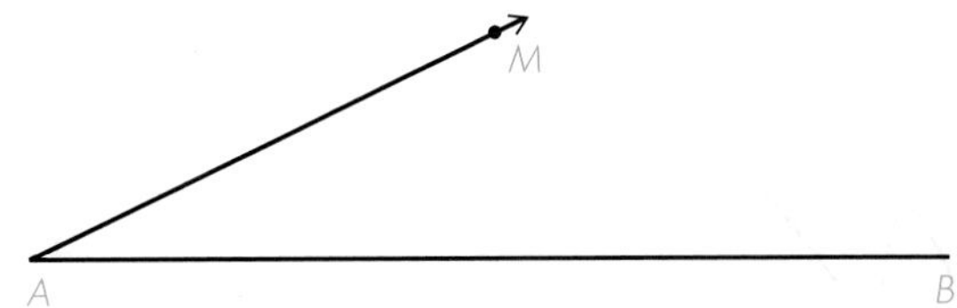

Draw a ray, $\overrightarrow{AM}$, where M is not on $\overline{AB}$.

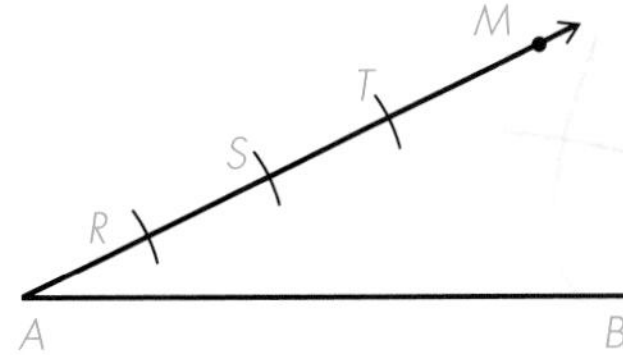

Open a compass setting to about 1.5 centimeters. With the point of the compass on A, mark an arc that intersects $\overrightarrow{AM}$. Label the intersection of the arc and $\overrightarrow{AM}$ point R. Now place the point of the compass on R, and make an arc on $\overrightarrow{AM}$, labeling the intersection S. Repeat the process, with the third point labeled T. Since the setting on the compass was not changed, $AR = RS = ST$.

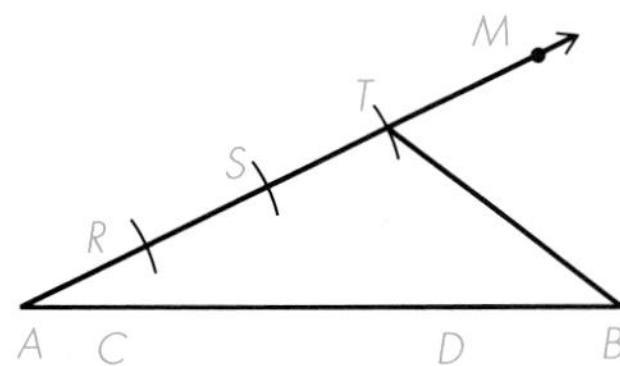

Connect points T and B with a line segment.

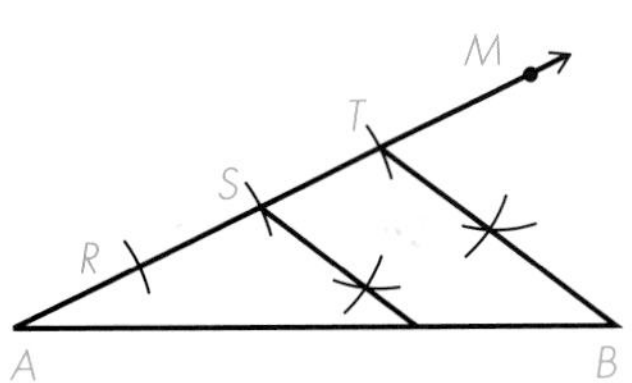

Construct a line through point S parallel to $\overline{TB}$. (See Section 6.3 if necessary.)

M
T
S
R
A
B

Chalkboard Example
Draw two line segments of equal length. Divide the first segment into four equal parts by the method of Section 6.1. Then divide the second segment into four equal parts by the method of this section.

TEACHING COMMENTS

Using the Page
See Page 230. Have students estimate where the points of intersection would be to divide the four equal segments of $\overline{AB}$ each in half. Have them count to verify that this would result in eight equal parts.

Using the Page
For this construction, we make use of the fact that if corresponding angles are equal, then two lines cut by a transversal are parallel. Students use $\angle STB$ to construct the corresponding angle with vertex S.

Using the Page
Make sure students understand each step of the proof. You may want to have students work in pairs and write a paragraph proof.

Construct a line through point R parallel to $\overline{TB}$.

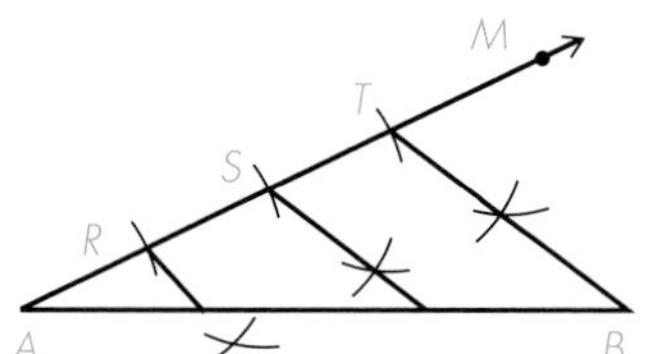

C and D divide $\overline{AB}$ into three equal parts, or line segments.

Using the same method, you can divide a line segment into any number of equal parts, or n equal parts.

The construction method used should be justified. Refer to the completed construction, and supply the reasons for each step in the proof.

Steps	Reasons
1. $\angle A \cong \angle A$	1. Identity
2. $\angle ARC \cong \angle ASD \cong \angle ATB$	2. By construction
3. $\triangle ARC \sim \triangle ASD \sim \triangle ATB$	3. AA
4. $AR/AS = AC/AD$, and $AR/AT = AC/AB$	4. Definition of similar triangles
5. $AR = RS = ST$	5. By construction
6. $AS = 2(AR)$, and $AT = 3(AR)$	6. Algebra
7. $AR/AS = AC/AD = \frac{1}{2}$, and $AR/AT = AC/AB = \frac{1}{3}$	7. Substitution
8. $AD = 2(AC)$, and $AB = 3(AC)$	8. Algebra
9. $AD = AC + CD$	9. Segment addition
10. $2(AC) = AC + CD$	10. Substitution
11. $AC = CD$	11. Algebra
12. $AB = AD + DB$	12. Segment addition
13. $3(AC) = 2(AC) + DB$	13. Substitution
14. $AC = DB$	14. Algebra
15. $AC = CD = DB$	15. Substitution

PHYSICAL MODEL OF DIVIDING A SEGMENT INTO EQUAL PARTS

Construction workers or carpenters may need to divide a piece of wood into congruent segments. These people probably wouldn't carry a compass, but they can use a method that is similar to the one in this section for dividing the piece of wood.

DISCOVERY ACTIVITY

1. On a sheet of paper, draw a line segment, $\overline{AB}$, 4 inches long. This will be the carpenter's piece of wood, a two by four, 8 feet long.
2. Place your ruler so that it is on A. Select a point M that is not on $\overline{AB}$, and draw $\overrightarrow{AM}$.
3. Use the width of your ruler, rather than your compass, to mark points R, S, and T on $\overrightarrow{AM}$ so that $AR = RS = ST$.
4. Draw $\overline{TB}$.
5. Now, use your ruler to draw parallels to $\overline{TB}$ through points S and R. Remember, the carpenter is using another board, rather than a ruler. The carpenter would mark on the second board two points, X and Y, when $\overline{TB}$ was drawn. Then the carpenter would slide the board to S so that X and Y that lie on $\overrightarrow{AM}$ are still on $\overrightarrow{AM}$, and draw the next line through S. Do the same with your ruler. Label the points of intersection with $\overline{AB}$ C and D.

This method is probably faster than using a compass to construct the parallel line segments but may not be quite as accurate.

CLASS ACTIVITY

1. Use a computer and a geometric drawing tool to draw any line segment. Use the principles from the construction for dividing a line segment into n equal parts to divide the line segment into three equal parts. Answers may vary.
2. Draw a line segment 3 inches long. Bisect the segment using the method of this section. See margin.
3. Draw a line segment 10 centimeters long. Divide the line segment into three equal parts. See margin.
4. Draw a line segment 8 centimeters long. Divide the line segment into five equal parts. See margin.

Using the Page

You may want to have the students try the Discovery Activity with a piece of wood and a ruler, or with two pieces of wood. Slats that are used for holding plastic over windows may be small enough for classroom use.

Additional Answers

For Exercises 2-4, scaled-down sketches are shown.

2.

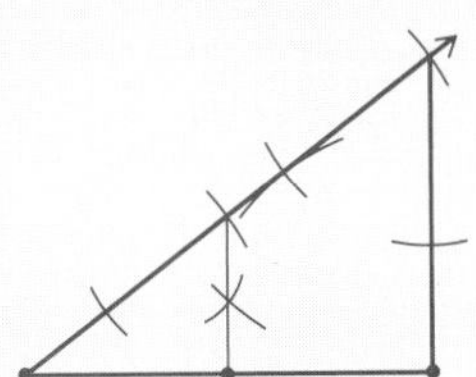

3.

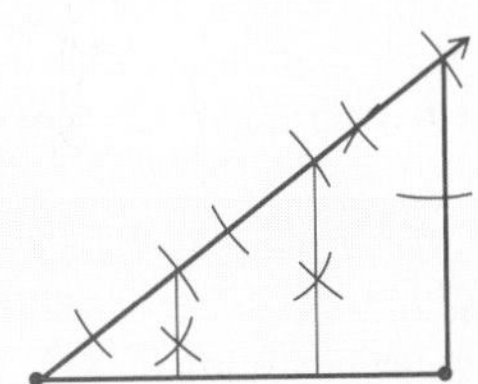

4.

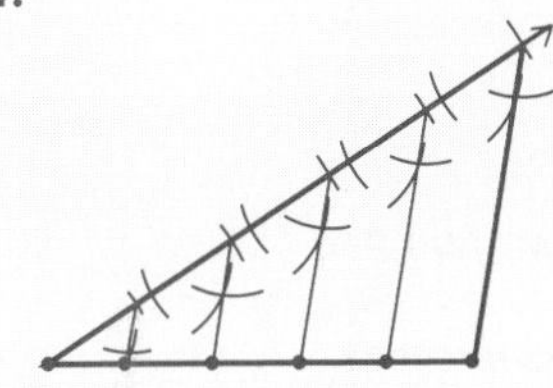

Extra Practice

1. Divide the line segment into six equal parts.

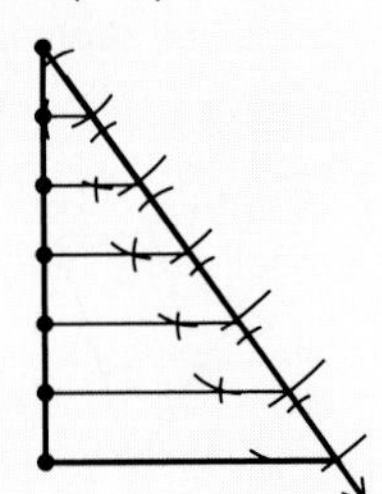

Using the Page
As students complete the exercises in the Home Activity, suggest that they estimate the location of the division points before they begin the constructions.

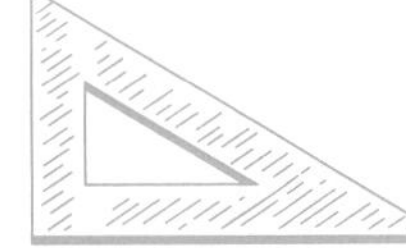

HOME ACTIVITY

See students' constructions.

1. Draw a line segment 7 centimeters long. Bisect the segment using the method of this section.
2. Draw a line segment 5 inches long. Bisect the segment using the method of this section.
3. Draw a line segment 8 centimeters long. Divide the segment into three equal parts.
4. Draw a line segment $3\frac{1}{2}$ inches long. Divide the segment into three equal parts.
5. Draw a line segment 9 centimeters long. Divide the segment into four equal parts using the method of this section.
6. Draw a line segment $4\frac{3}{4}$ inches long. Divide the segment into four equal parts using the method of this section.
7. Draw a line segment 12 centimeters long. Divide the segment into five equal parts.
8. Draw a line segment $4\frac{1}{2}$ inches long. Divide the segment into five equal parts.
9. Draw a line segment 15 centimeters long. Divide the segment into six equal parts.
10. Draw a line segment 7 inches long. Divide the segment into six equal parts.
11. Draw a line segment 4 inches long. Bisect the segment using the method in the Discovery Activity.
12. Draw a line segment 8 centimeters long. Divide the segment into three equal parts using the method of this section.
13. Draw a line segment 5 inches long. Divide the segment into four equal parts using the method in the Discovery Activity.
14. Draw a line segment 12 centimeters long. Divide the segment into five equal parts using the method in the Discovery Activity.

CRITICAL THINKING

Copy the angle to the right. Construct a line segment connecting the rays of the angle at points equidistant from *R*. Trisect (divide into three equal parts) the line segment you constructed. Draw line segments to *R* from the points that trisect the line.

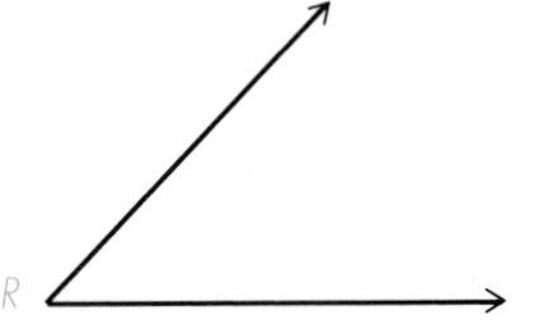

15. Do these connecting segments trisect the angle? No
16. Would your answer to Exercise 15 be the same for any angle? Yes

FOLLOW-UP

Assessment
1. To construct the parallel line, you use the fact that if lines are parallel, ________ angles are equal. [corresponding]
2. Divide the line segment into three equal parts.

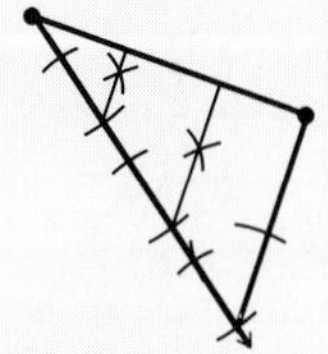

Reteaching
Students who have had difficulty with this section may benefit from Reteaching Activity 6.5.

Enrichment
For students who have mastered the material in this section, you may wish to assign Enrichment Activity 6.5.

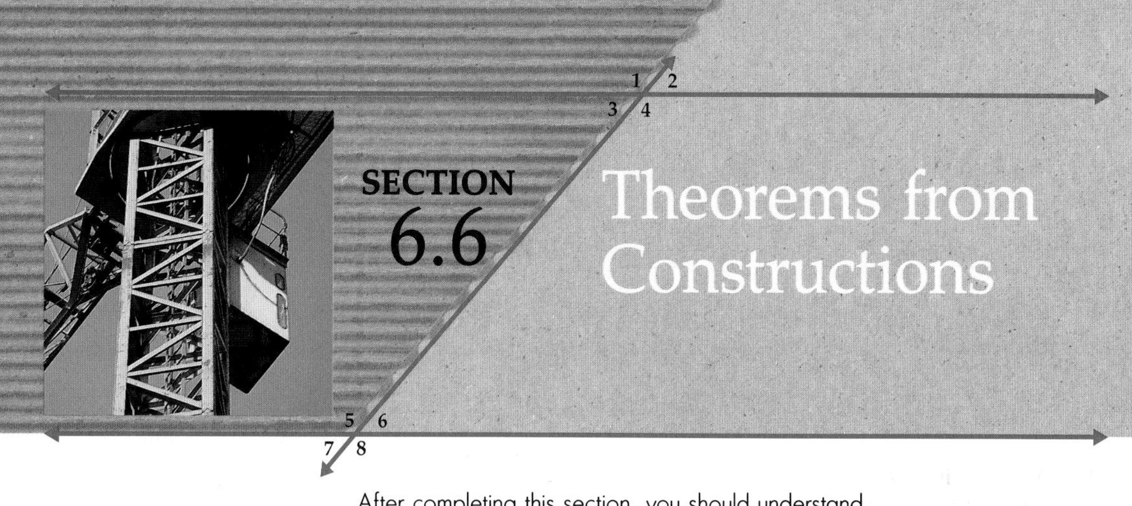

SECTION 6.6 Theorems from Constructions

After completing this section, you should understand
▲ how constructions lead to conclusions.

In Section 6.4, you learned that the medians of a triangle are concurrent at the centroid, the center of gravity of the triangle. The other constructions that you did lead to some important theorems about triangles.

THE TRIANGLE MIDPOINT THEOREM

For $\triangle ABC$, construct the bisectors of $\overline{AC}$ and $\overline{BC}$. Label the midpoints E and F. Draw $\overline{EF}$.

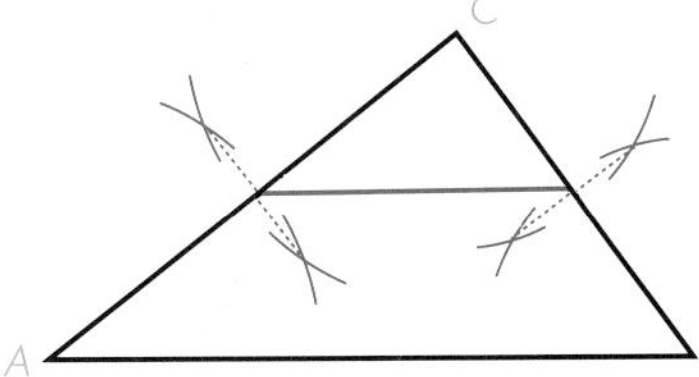

$\overline{EF}$ appears to be parallel to $\overline{AB}$. Measuring their lengths, EF appears to be half of AB.

THEOREM 6-6-1 **If the midpoints of two sides of a triangle are joined, then the line segment determined or formed is parallel to the third side and is equal to $\frac{1}{2}$ its length.**

The justification of this theorem is in two parts.

Given: $\triangle ABC$ with E and F midpoints by construction
Justify: a. $EF = \frac{1}{2}(AB)$
b. $\overline{EF} \parallel \overline{AB}$

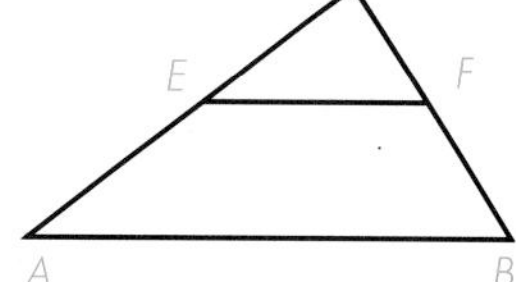

SECTION 6.6

Resources
Reteaching 6.6
Enrichment 6.6

Objectives
After completing this section, the student should understand
▲ how constructions lead to conclusions.

Materials
compass
straightedge
ruler
protractor

Vocabulary
circle
center
circumcircle
circumcenter
in-circle
incenter

GETTING STARTED

Quiz: Prerequisite Skills

1. In how many points do the perpendicular bisectors of the sides of a triangle intersect? the angle bisectors? the altitudes? [1; 1; 1]

2. Bisect the angle.

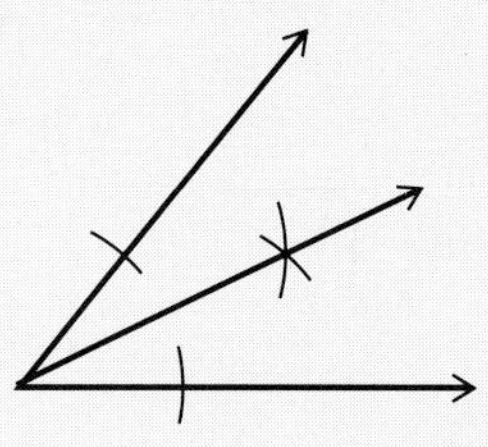

3. Construct the perpendicular to $\overline{AB}$ from P.

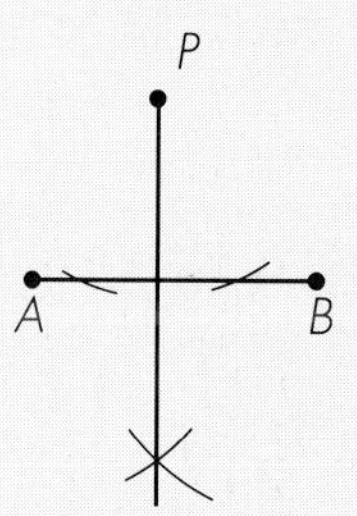

Warm-Up Activity
Have students draw a large triangle and then construct the perpendicular bisectors of the sides. Label the point of concurrency T. Place a circular object (coin, jar lid, or similar item) so that it is centered over T. Have students make observations about the distance from the edge of the circular object to the three vertices of the triangle.

Chalkboard Example

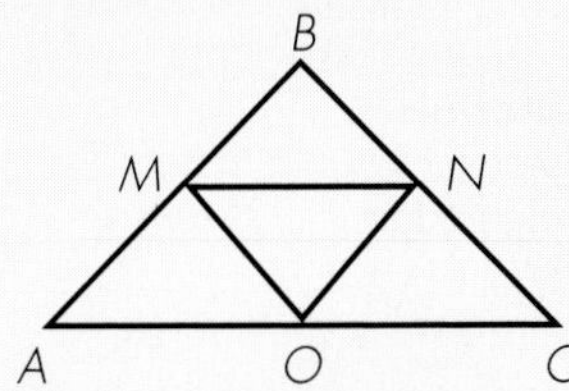

$AB = 10$, $BC = 12$, $AC = 14$. Find $MN + NO + MO$ if M, N, and O are midpoints. [18]

TEACHING COMMENTS

Using the Page
Ask the students why they can draw $\overline{EF}$ on Page 235. [Two points determine a line.]

Using the Page
It can be proven that if corresponding angles are equal, then the lines are parallel. Then, $\angle A \cong \angle CEF$ would be enough for $\overline{EF} \parallel \overline{AB}$.

Extra Practice
1.

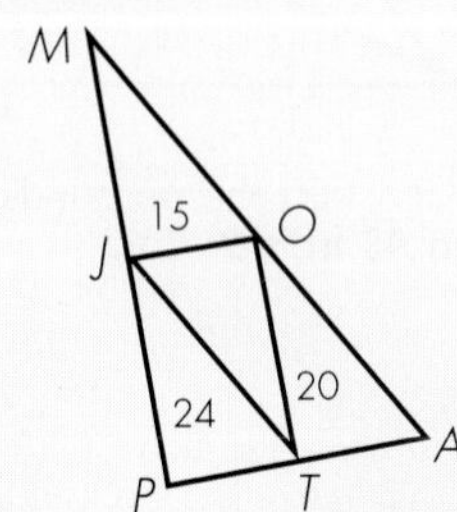

Find $MA + AP + MP$. Points J, O, and T are midpoints. [118]

Steps	Reasons
Part a	
1. $EC = \frac{1}{2}(AC)$	1. Midpoint divides a segment into two equal parts.
2. $FC = \frac{1}{2}(BC)$	2. Same as reason 1
3. $\angle C \cong \angle C$	3. Identity
4. $\triangle ABC \sim \triangle EFC$	4. SAS
5. $EF = \frac{1}{2}(AB)$	5. Sides of similar triangles are in equal ratio.
Part b	
6. $\angle A \cong \angle CEF$	6. Corresponding angles of similar triangles are equal.
7. $\overline{EF} \parallel \overline{AB}$	7. If two lines are intersected by a transversal so that the corresponding angles are equal, then the lines are parallel.

CLASS ACTIVITY

1. Use a computer and a geometric drawing tool to draw any $\triangle ABC$. Find the midpoint of $\overline{AC}$ and label it E. Find the midpoint of $\overline{BC}$ and label it F. Draw line segment $\overline{EF}$. Find the measures of $\overline{AB}$, $\overline{EF}$, $\angle CEF$, and $\angle CAB$. Repeat this activity several times. How does the measure of $\overline{EF}$ relate to the measure of $\overline{AB}$? What do the measures of $\angle CEF$ and $\angle CAB$ tell you about $\overline{EF}$ and $\overline{AB}$? Do your findings support Theorem 6-6-1? $EF = \frac{1}{2}AB$; $\overline{EF} \parallel \overline{AB}$; yes

Find the length of the line segment.

2. $\overline{PQ}$ 3. $\overline{XY}$

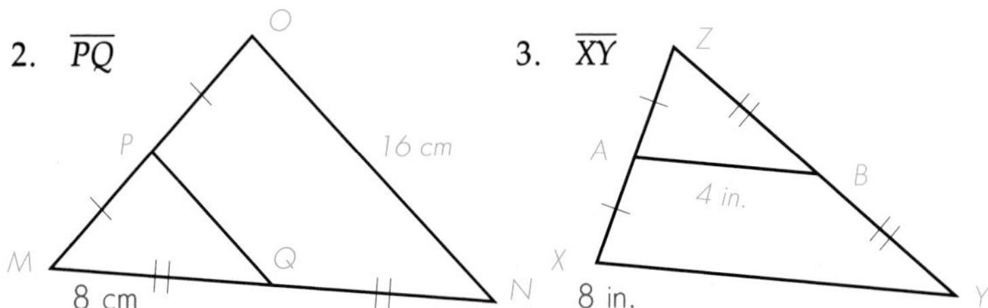

PERPENDICULAR BISECTORS OF TRIANGLE SIDES

Recall that the perpendicular bisectors of the sides of a triangle are concurrent.

DISCOVERY ACTIVITY

1. On a sheet of paper, draw a large acute triangle, *ABC*.
2. Construct the perpendicular bisectors of all three sides, and label the midpoints *D*, *E*, and *F*. Label the point of concurrency O.
3. Place the point of your compass on O. Open the setting on your compass until the point of the pencil is on *A*. Rotate the compass until the pencil point is back on *A*. What do you observe? The arc passes through *B* and *C*.

You may have found that the arc passes through all three vertices.

Using the Page
Have students repeat the Discovery Activity using an obtuse or a right triangle. Students should find that the arc still passes through all three vertices.

NOTEBOOK
DEFINITION 6-6-1 **A CIRCLE is the set of all points on a plane equidistant from a point called the center.**

In the Discovery Activity, *O* was the center of the circle. This is justified below.

Using the Page
To justify that O is the center of the circle, it is necessary to show that the radii are all of equal length. This is accomplished by showing that two radii at a time are equal, using the fact that the center is on the perpendicular bisector of the segment containing two points on the circle.

Given: $\overline{MN}$, with constructed perpendicular bisector, *K* the midpoint of $\overline{MN}$, and *P* any point besides *K* on the perpendicular bisector.
Justify: $PM = PN$

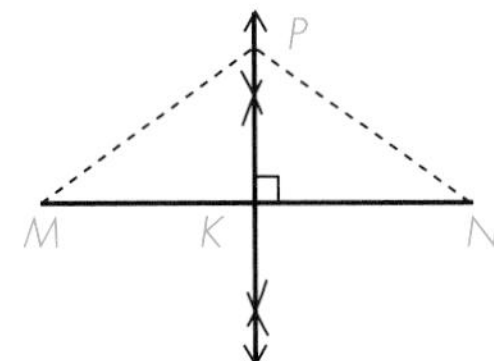

Steps	Reasons
1. Draw $\overline{PM}$, $\overline{PN}$.	1. Two points determine a line.
2. $MK = NK$	2. Midpoint divides a segment into two equal parts.
3. $PK = PK$	3. Identity
4. $m\angle MKP = m\angle NKP$	4. By construction (perpendicular bisector)
5. $\triangle MKP \sim \triangle NKP$	5. SAS
6. $MK/NK = 1$	6. Algebra
7. $\triangle MKP \cong \triangle NKP$	7. Similar triangles and ratio of sides = 1
8. $MP = NP$	8. CPCF

THEOREM 6-6-2 **If a point is on the perpendicular bisector, then it is equidistant from the endpoints of the segment.**

Thus, in the Discovery Activity, $OA = OB = OC$, and O is the center of the circle. The circle goes around the outside of the triangle, and is called the circumcircle. Point O is the circumcenter of the triangle.

Extra Practice
Find by construction the circumcenter of each triangle.

1.

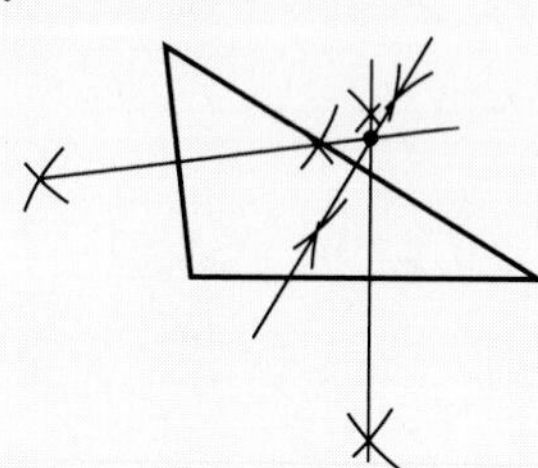

2.

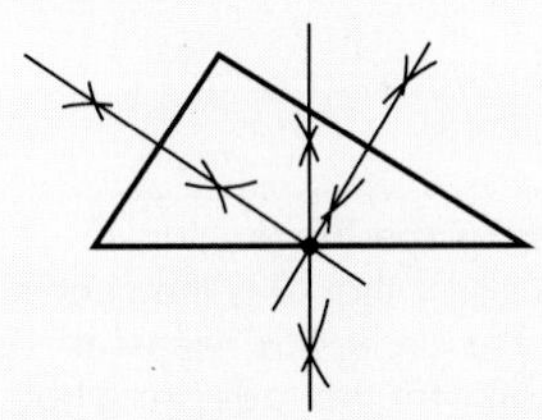

CLASS ACTIVITY

1. Copy $\triangle DEF$. Construct the circumcenter.
   Construct by using ⊥ bisectors of sides

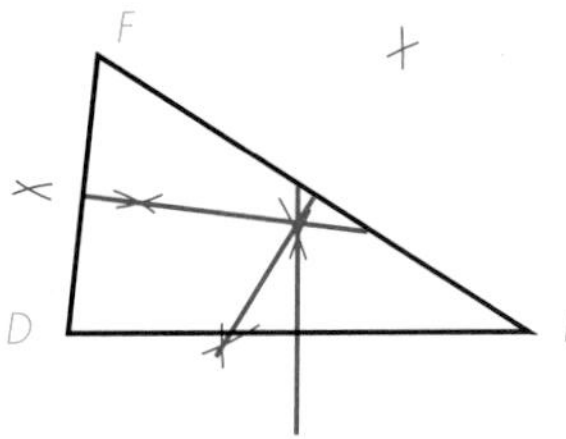

THE ANGLE BISECTORS OF A TRIANGLE

Recall that the angle bisectors of a triangle are also concurrent.

DISCOVERY ACTIVITY

1. On a sheet of paper, draw a large triangle. Bisect the angles. Label the point of concurrency of the angle bisectors T.
2. From T, construct the perpendiculars to the sides of the triangle. Label the points of intersection of the perpendiculars and the sides X, Y, and Z. Measure TX, TY, and TZ. What do you observe? They are equal.
3. Place the point of your compass on T and the pencil point on X. Construct a circle. What do you observe? Circle passes through Y and Z also.

Using the Page
After locating point T in the Discovery Activity, you may want to use the circular object as described in the Warm-Up Activity. Have students center it over point T and compare the distance from the edge of the circular object to the sides of the triangle.

You may have found that the circle you constructed passes through X, Y, and Z. The circle is inside the triangle and so is called the in-circle.

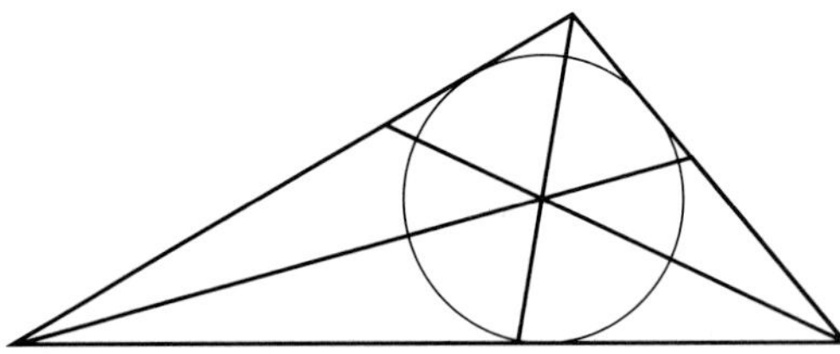

THEOREM 6-6-3 **If a point is on the angle bisector, then the point is equidistant from the sides of the angle.**

Using the Page
In Theorem 6-6-3, emphasize that when distance from a point to a line is measured, the distance is defined to be the length of the perpendicular from the point to the line.

Given: $\angle ABC$ with $\overrightarrow{BT}$ the angle bisector, $\overline{TD} \perp \overrightarrow{BC}$, and $\overline{TE} \perp \overrightarrow{BA}$

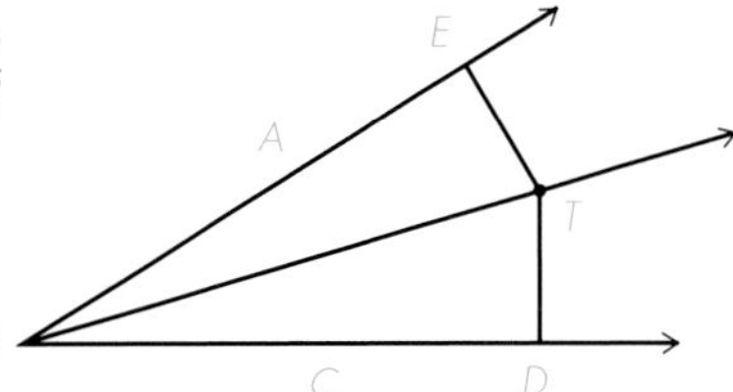

Justify: $TD = TE$

Supply the missing reasons.

Steps	Reasons
1. $\angle TBD \cong \angle TBE$	1. Definition of angle bisector
2. $BT = BT$	2. Identity
3. $\overline{TD} \perp \overline{BC}$, $\overline{TE} \perp \overline{BA}$	3. Given
4. $\angle TEB$ and $\angle TDB$ are right angles.	4. Definition of perpendicular lines
5. $\angle TEB \cong \angle TDB$	5. All right angles are equal.
6. $\triangle BTD \sim \triangle BTE$	6. AA
7. $BT/BT = 1$	7. Algebra
8. $\triangle BTD \cong \triangle BTE$	8. Triangles similar, ratio of sides = 1
9. $TD = TE$	9. CPCF

CLASS ACTIVITY

1. Construct the in-circle of a triangle whose sides measure 3 centimeters, 5 centimeters, and 6 centimeters. See margin.

2. Construct the in-circle of a right triangle whose sides measure 3 inches, 4 inches, and 5 inches. See margin.

HOME ACTIVITY

1. Draw an obtuse triangle with two sides measuring 8 cm and 6 cm, and the included angle equal to 110°. Construct the perpendicular bisectors of the sides. Label the circumcenter O. Construct the circumcircle. See margin.

2. Draw an acute triangle ABC, with $m\angle A = 60°$, $AB = 3$ inches, and $m\angle B = 40°$. Construct the angle bisectors. Label the incenter V. Construct the in-circle. See margin.

3. D and E are midpoints of two sides of $\triangle MNO$. If $MN = 32$ mm, what is DE? 16 mm

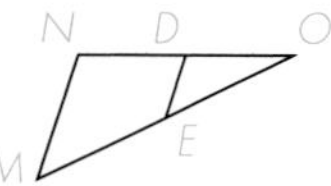

Extra Practice

1. Locate the incenter of the triangle, and label it V.

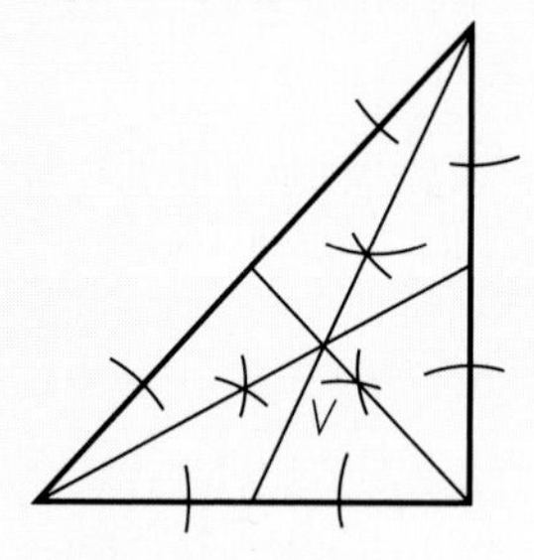

Additional Answers

1.

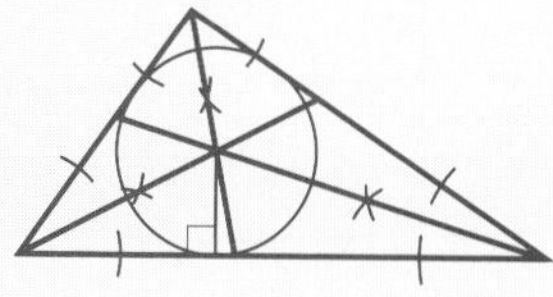

2.

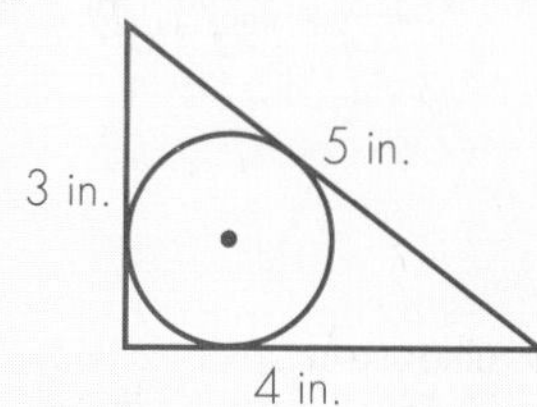

For Exercises 1 and 2, scaled-down sketches are shown.

1.

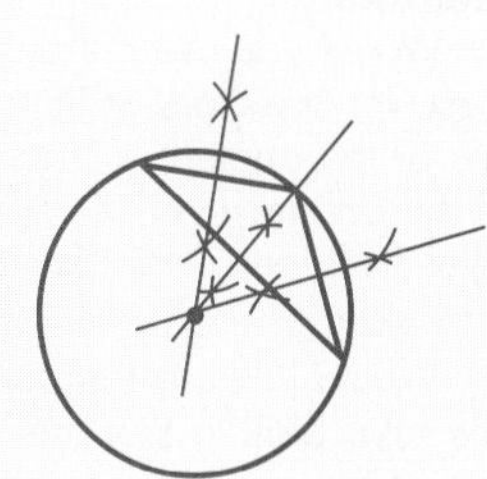

2.

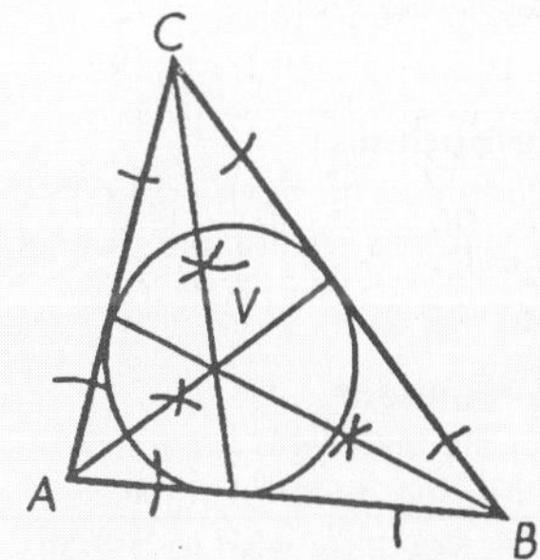

Additional Answers
For Exercises 6-9, scaled-down sketches are shown.

6.

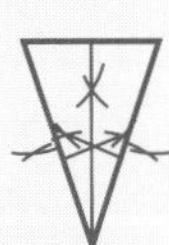

7.

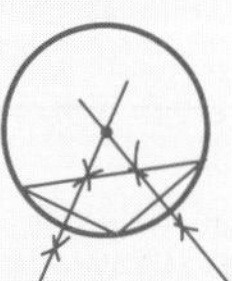

8.

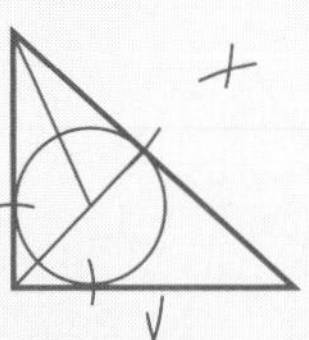

9.

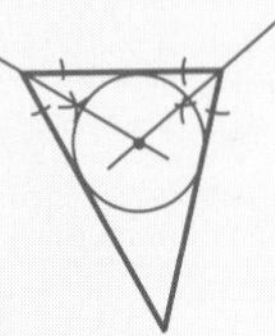

13. Choose a point C, and draw $\triangle ABC$. Construct the midpoints of $\overline{AC}$ and $\overline{BC}$, labeling the midpoints D and E. Measure $\overline{DE}$. $AB = 2(DE)$.

FOLLOW-UP

Assessment

1. Would you use the perpendicular bisectors of the sides or the angle bisectors to find the circumcenter of a triangle? [perpendicular bisectors of the sides]

2. Would you use the perpendicular bisectors of the sides or the angle bisectors to find the incenter of a triangle? [angle bisectors]

Reteaching
Students who have had difficulty with this section may benefit from Reteaching Activity 6.6.

Enrichment
For students who have mastered the material in this section, you may wish to assign Enrichment Activity 6.6.

In $\triangle ABC$, D, E, and F are midpoints.

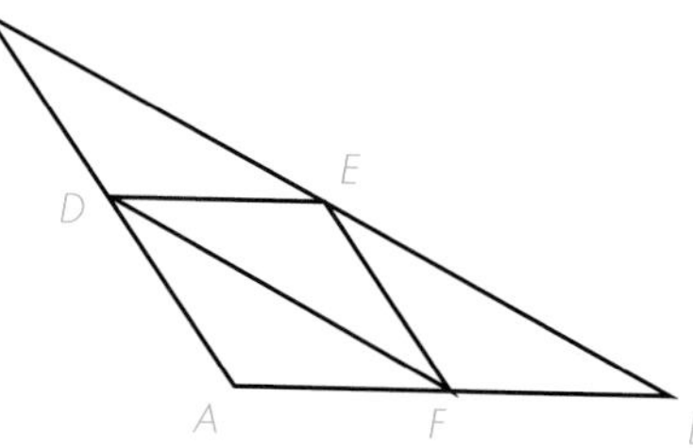

4. If $AC = 8$, $AB = 10$, and $BC = 15$, what does $DF + DE + EF$ equal? $16\frac{1}{2}$

5. If $DE = 6$, $DF = 8$, and $EF = 4$, what does $AB + AC + BC$ equal? 36

6. Construct a triangle with sides of length 3 inches, 3 inches, and 2 inches. Construct the circumcenter of the triangle. See margin.

7. Draw a triangle with sides of length 3 inches and 4 inches, and the included angle equal to 115°. Construct the circumcircle of the triangle. See margin.

8. Construct a triangle with sides of length 8 cm, 10 cm, and 12 cm. Construct the in-circle of the triangle. See margin.

9. Draw a triangle with sides of length 4 inches and $5\frac{1}{2}$ inches, and included angle equal to 75°. Construct the incenter of the triangle. See margin.

10. Construct an equilateral triangle with sides of length 4 inches. Construct the circumcenter and the incenter. What do you observe? They are the same point.

11. Draw an isosceles triangle with two of the angles measuring 70° and the base equal to 6 centimeters. Construct the three medians, labeling the centroid *M*. Construct the three altitudes, labeling the point of concurrency *A*. Construct the three angle bisectors, labeling the incenter *B*. What do you observe about *M*, *A*, and *B*?
They all lie on the median/altitude/angle bisector from the vertex to the base.

12. Construct an isosceles right triangle, *ABC*, with $\angle C$ the right angle. Construct the median from $\angle C$ to $\overline{AB}$. Label the point *O* where the median intersects $\overline{AB}$. Measure *OA*, *OB*, and *OC*. What do you observe? They are equal.

CRITICAL THINKING

13. A surveyor was asked to determine the distance from *A* to *B*. The direct line is blocked by the building, as indicated in the figure. Using a theorem from this section, devise a method to help the surveyor. See margin.

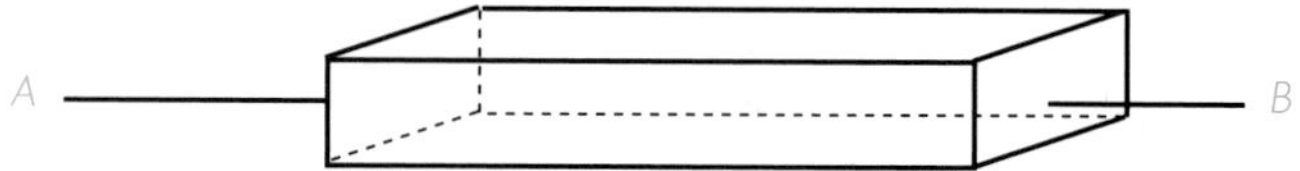

Chapter Vocabulary
altitude
angle bisector in a triangle
by construction
center
center of gravity
centroid
circle
circumcenter
circumcircle
concurrent lines
incenter
in-circle
median
perpendicular bisector of a side
perpendicular lines

6.1 Bisect the line segment.

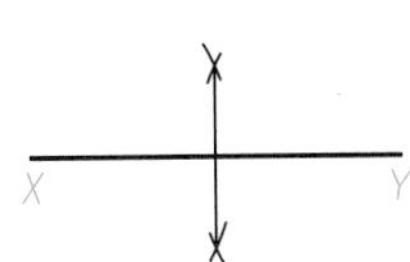

Bisect the angle.

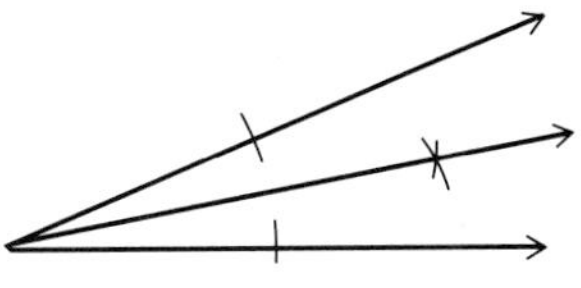

6.2 Construct a perpendicular to AB from C.

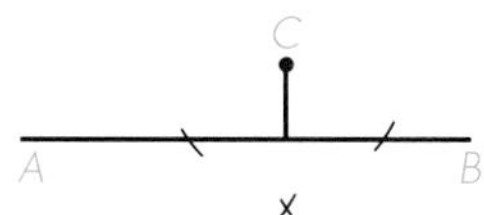

6.3 Construct a line parallel to EF through D.

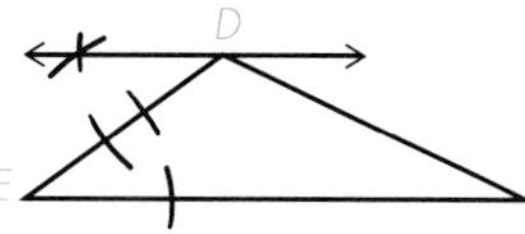

6.4 Find the incenter R and the circumcenter S of GHI.

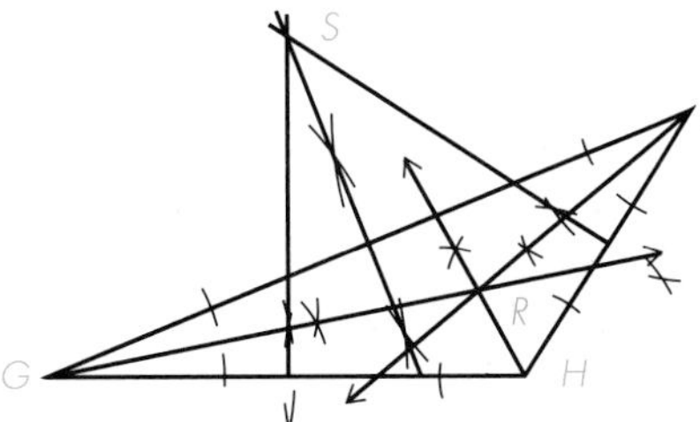

6.5 Divide JK into three equal parts.

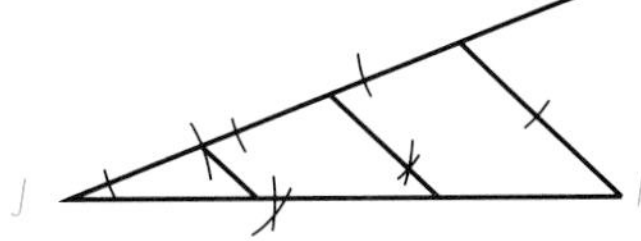

6.6 NP is the bisector of $\angle MNO$. Find x. Because NP is an angle bisector, $MP = OP$. So $x = 20$ cm.

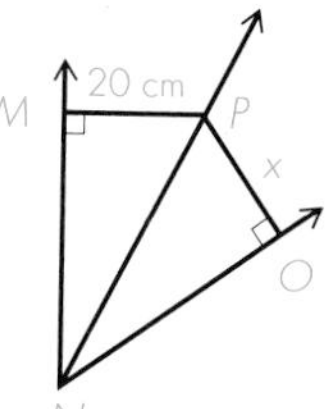

241

COMPUTER ACTIVITY

Using the Page
This Computer Activity shows students how to draw medians in any right triangle. This procedure is easily extended to non-right triangles. Students must expect to do some computation to accomplish this task. They must learn to identify the midpoint of a line segment given its endpoints on the coordinate plane. Perhaps some time should be spent calculating midpoints before moving to the computer.

Additional Answers

1. ```
   TO MEDIANS
     SETXY 10 0
     SETXY 0 10
     SETXY 0 0
     PU SETXY 5 0 PD
     SETXY 0 10
     PU SETXY 5 5 PD
     SETXY 0 0
     PU SETXY 0 5 PD
     SETXY 10 0
   END
   ```
2. ```
   TO MEDIANS
     PU SETXY 5 5 PD
     SETXY 15 5
     SETXY 5 20
     SETXY 5 5
     PU SETXY 10 5 PD
     SETXY 5 20
     PU SETXY 10 12.5 PD
     SETXY 5 5
     PU SETXY 5 12.5 PD
     SETXY 15 5
   END
   ```
3. ```
   TO MEDIANS
     PU SETXY 0 (−6) PD
     SETXY 8 (−6)
     SETXY 0 2
     SETXY 0 (−6)
     PU SETXY 4 (−6) PD
     SETXY 0 2
     PU SETXY 4 (−2) PD
     SETXY 0 (−6)
     PU SETXY 0 (−2) PD
     SETXY 8 (−6)
   END
   ```

The three vertices of a right triangle have been graphed on the x-y plane at the right. In order to draw the medians of this triangle, you must find the coordinates of the midpoint of each side.

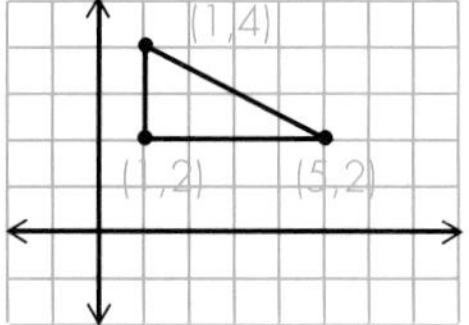

Consider the side whose endpoints have coordinates of (1, 2) and (5, 2). We can count 4 units between these points, so it follows that the midpoint would divide the segment into two segments that are each 2 units long. So the midpoint must have coordinates (3, 2). Notice that these coordinates can be computed by adding the x-coordinates $(1 + 5 = 6)$ and then dividing the sum by 2 $(6 \div 2 = 3)$, and by adding the y-coordinates $(2 + 2 = 4)$ and then dividing the sum by 2 $(4 \div 2 = 2)$. The general rule for finding the coordinates of the midpoint of a line segment is given below.

The coordinates of the midpoint of the line segment whose endpoints have coordinates (x_1, y_1) and (x_2, y_2) are

$$\left(\frac{x_1 + x_2}{2}, \frac{y_1 + y_2}{2}\right).$$

You can use LOGO and this rule to tell the turtle to draw triangles and their medians on the x-y plane.

Example: The coordinates of the three vertices of a triangle are (10, 5), (10, 15), and (30, 5). Write a procedure named MEDIANS that tells the turtle to draw the triangle and its three medians.

First, compute the coordinates of the midpoint of each side.

The side whose endpoints have coordinates of (10, 5) and (10, 15) has a midpoint whose coordinates are $\left(\frac{10 + 10}{2}, \frac{5 + 15}{2}\right)$, or (10, 10). Thus, the first median will start at (10, 10) and go to (30, 5). The side whose endpoints have coordinates of (10, 15) and (30, 5) has a midpoint whose coordinates are $\left(\frac{10 + 30}{2}, \frac{15 + 5}{2}\right)$, or (20, 10). Thus, the second median will start at (20, 10) and go to (10, 5). The side whose endpoints have coordinates (10, 5) and (30, 5) has a midpoint with coordinates $\left(\frac{10 + 30}{2}, \frac{5 + 5}{2}\right)$ or (20, 5). Thus, the third median will start at (20, 5) and go to (10, 15).
   ```

Now, write the procedure.

```
TO MEDIANS
PU SETXY 10 5 PD
SETXY 10 15              First, draw the triangle.
SETXY 30 5
SETXY 10 5
PU SETXY 10 10 PD        Draw the first median.
SETXY 30 5
PU SETXY 20 10 PD        Draw the second median.
SETXY 10 5
PU SETXY 20 5 PD         Draw the third median.
SETXY 10 15
END
```

The output is shown below.

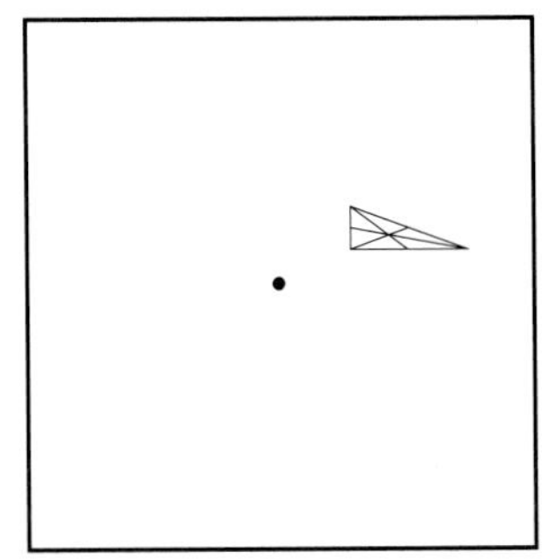

The coordinates of the three vertices of a triangle are given below. Modify the MEDIANS procedure to tell the turtle to draw each triangle and its three medians. See margin.

1. (0, 0), (10, 0), (0, 10)
2. (5, 5), (15, 5), (5, 20)
3. (0, −6), (8, −6), (0, 2)
4. (−6, −4), (−6, 6), (0, −4)
5. (−25, −25), (−25, −10), (−5, −25)
6. (−19, 18), (−19, 12), (−13, 18)
7. Modify the MEDIANS procedure to tell the turtle to draw a triangle of your choice and its three medians. Answers may vary.

Additional Answers

4. TO MEDIANS
```
PU SETXY −6 (−4) PD
SETXY −6 6
SETXY 0 (−4)
SETXY −6 (−4)
PU SETXY −6 1 PD
SETXY 0 (−4)
PU SETXY −3 1 PD
SETXY −6 (−4)
PU SETXY −3 (−4) PD
SETXY −6 6
END
```

5. TO MEDIANS
```
PU SETXY −25 (−25) PD
SETXY −25 (−10)
SETXY −5 (−25)
SETXY −25 (−25)
PU SETXY −25 (−17.5)
PD
SETXY −5 (−25)
PU SETXY −15 (−17.5)
PD
SETXY −25 (−25)
PU SETXY −15 (−25) PD
SETXY −25 (−10)
END
```

6. TO MEDIANS
```
PU SETXY −19 18 PD
SETXY −19 12
SETXY −13 18
SETXY −19 18
PU SETXY −19 15 PD
SETXY −13 18
PU SETXY −16 15 PD
SETXY −19 18
PU SETXY −16 18 PD
SETXY −19 12
END
```

1. Copy $\overline{AB}$. Bisect $\overline{AB}$, and label the midpoint M. See margin.

A B

2. Draw an acute angle, $\angle ABC$, and copy the angle onto $\overline{EF}$ with the vertex at E. Answers will vary.
3. Copy the angle and bisect it.

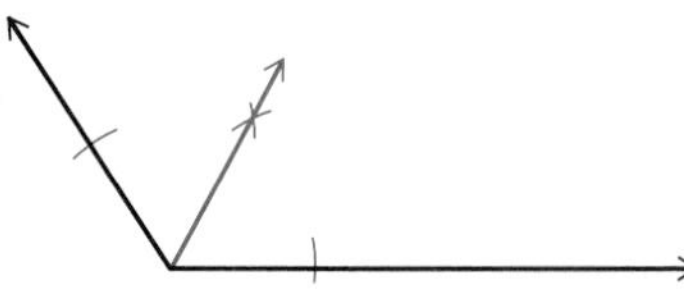

4. Copy the figure, and construct a line perpendicular to $\overleftrightarrow{AB}$ through C.

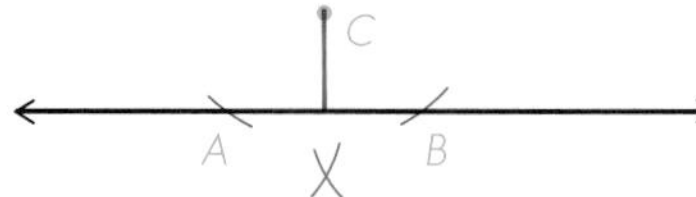

5. Copy the figure from Exercise 4. Construct a line parallel to $\overleftrightarrow{AB}$ through C. See margin.
6. Draw a line segment 4 inches long. By construction, divide the segment into three equal parts. See margin.
7. Draw a large obtuse triangle ABC, with $\angle C$ the obtuse angle. Construct the angle bisector of $\angle C$, the median from B, and the altitude from A. See margin.
8. What is the measure of $\overline{KL}$ in $\triangle ABC$? K and L are midpoints. 31 cm

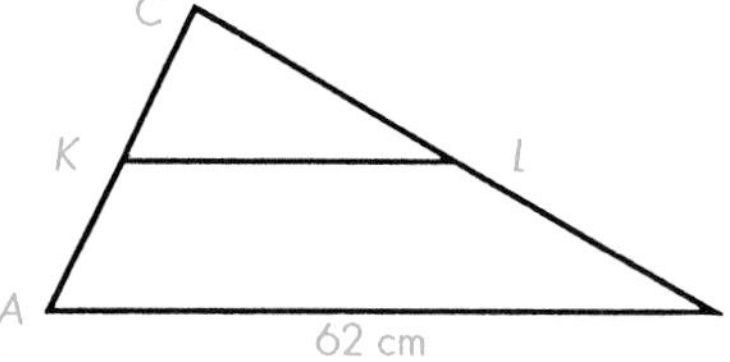

9. Construct the circumcenter for the triangle. See margin.

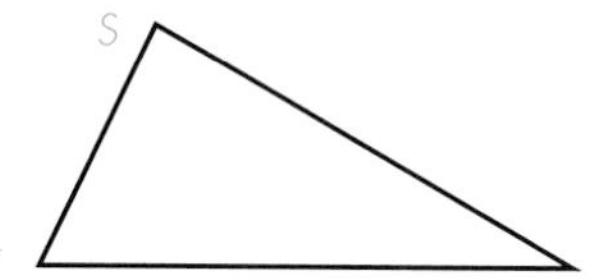

10. Construct the incenter for the triangle.

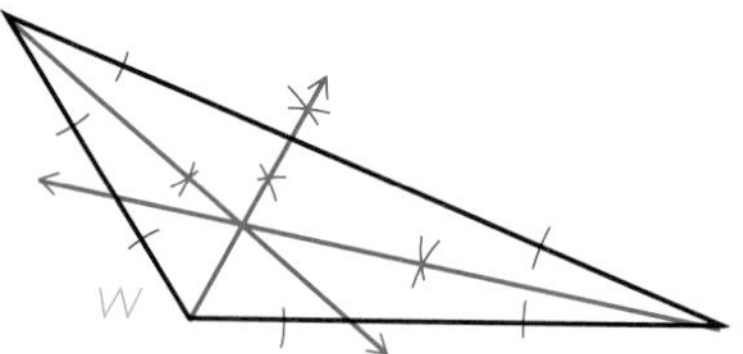

Using the Page

Students can work individually or in groups to study the Chapter 6 Review. If you have students study in groups, they should work together to make sure all members of the group know the material in the chapter.

Informal assessment such as interviews, classroom observations, or a review of student portfolios can be used instead of paper-and-pencil tests.

Additional Answers

For Exercises 1, 5, 6, 7, and 9, scaled-down sketches are shown.

1.

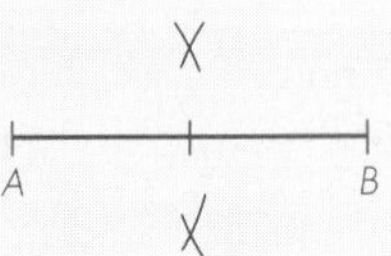

5.

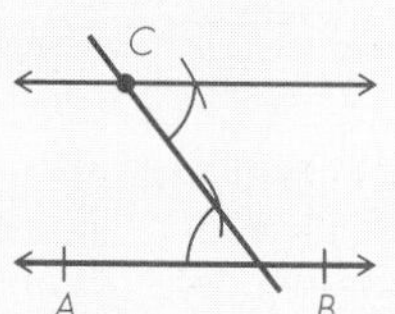

6.

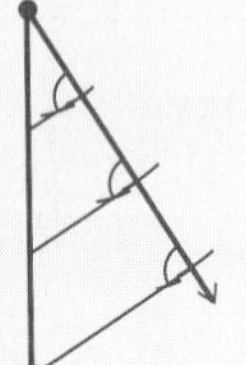

7.

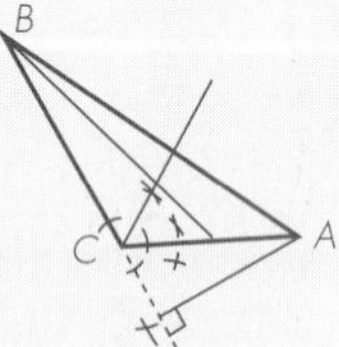

9.

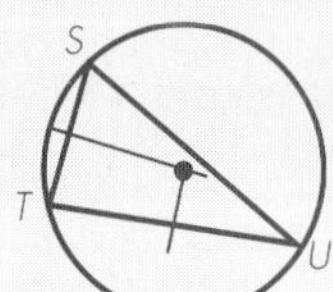

CHAPTER TEST

1. Draw a line segment $2\frac{3}{4}$ inches long. Bisect the segment. See margin.
2. Copy the figure. Construct the line perpendicular to $\overline{MN}$ through W.

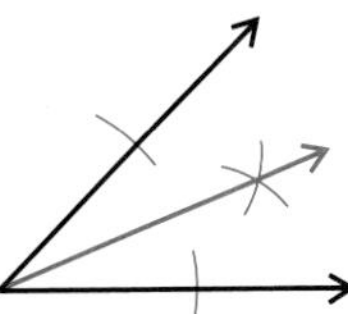

3. Copy the figure from Exercise 2. Construct the line parallel to $\overline{MN}$ through W. See margin.
4. Draw a line segment 9 centimeters long. By construction, divide the segment into three equal parts. See margin.
5. Copy and then bisect the angle.

6. Draw an obtuse angle measuring 105°. Copy the angle by construction onto $\overline{VW}$ with the vertex at V. See students' constructions.
7. Copy $\triangle HGI$. Construct the angle bisector for $\angle H$. From G, construct the median. From I, construct the altitude. Construct the perpendicular bisector of $\overline{GH}$.

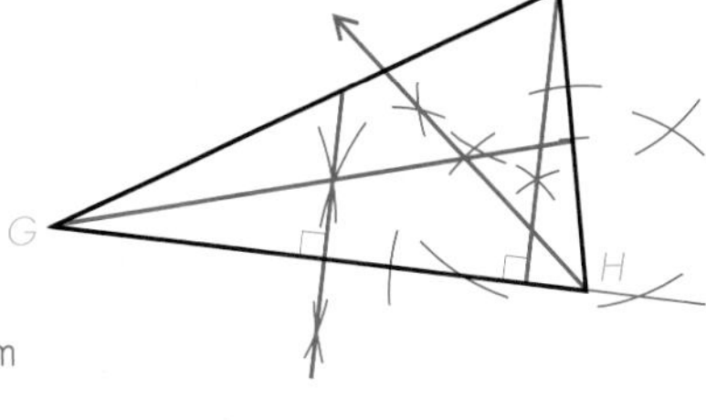

8. X and W are midpoints. Find the length of $\overline{VT}$. 31.2 cm

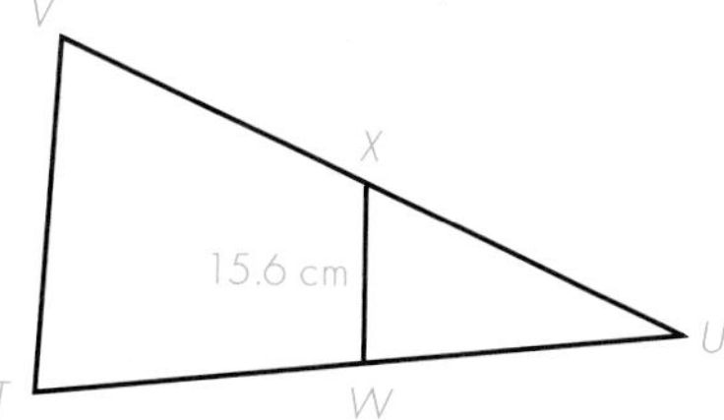

9. Construct the circumcenter for the triangle. See margin.

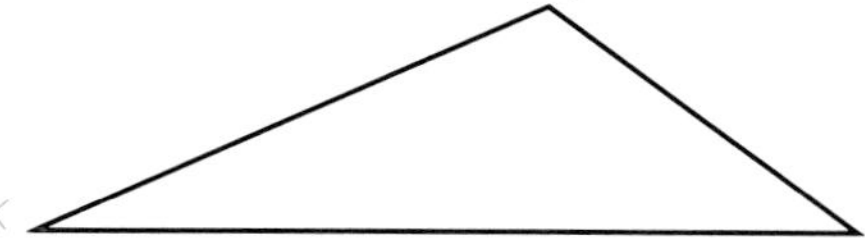

10. Construct the incenter for the triangle.

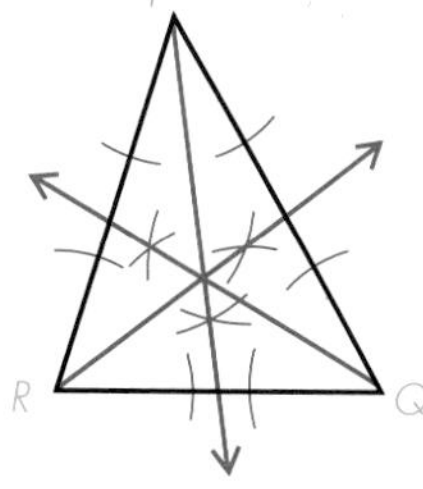

Assessment Resources
Achievement Tests pp. 11-12

Test Objectives
After completing this chapter, students should understand

- how to copy a line segment using the tools of geometry.
- how to copy an angle using the tools of geometry.
- how to bisect a line segment.
- how to bisect an angle.
- how to construct a perpendicular to a line from a point not on the line.
- how to construct a line parallel to another line.
- the medians of a triangle.
- the altitudes of a triangle.
- the angle bisectors of a triangle.
- the perpendicular bisectors of the sides of a triangle.
- how to divide a line segment into n equal parts.
- how constructions lead to conclusions.

Additional Answers
Scaled-down sketches are given for Exercises 1, 3, 4, and 9.

1.

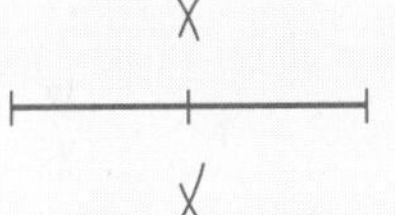

3.

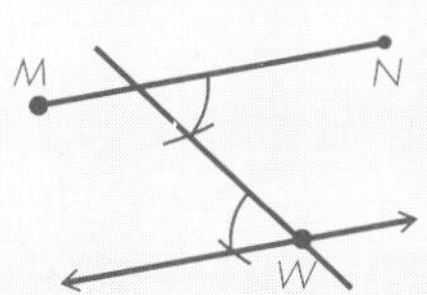

4.

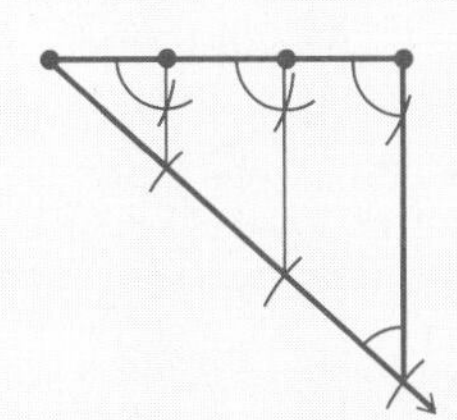

9.

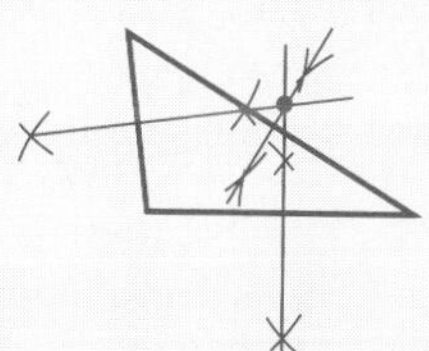

Algebra Review Objectives
Topics reviewed from algebra include

- finding exact values of square roots.
- estimating square roots, using a calculator.
- estimating square roots, using the table on Page 562.
- solving second-degree equations.

ALGEBRA SKILLS

The square roots of 9 are 3 and −3. Because 9 is a perfect square, $\sqrt{9}$ is a rational number. The square roots of numbers that are not perfect squares are irrational numbers: the roots have a decimal that is nonterminating or nonrepeating. Irrational roots may be expressed as decimal approximations or in square-root form.

Example: $\sqrt{5} = ?$ Using a calculator, enter 5, and press the square-root key $\boxed{\sqrt{\ }}$.

$5 \boxed{\sqrt{\ }} \approx 2.236068$

Example: $\sqrt{500} = ?$ Factor the number, looking for perfect squares.
$\sqrt{500} = \sqrt{5 \cdot 100} = 10\sqrt{5}$
You can leave the answer in square-root form, or use the table of square roots on page 562. Locate 5 in the leftmost column, headed n. The column headed $\sqrt{n}$ gives 2.236 for the approximate value of $\sqrt{5}$.
$\pm 10\sqrt{5} \approx 10(\pm 2.236) = \pm 22.36$

Example: $x^2 = 16$
$\sqrt{x^2} = \pm\sqrt{16}$ Take the square root of both sides of the equation.
$x = 4$ or $x = -4$

Example: Reformulate the left side as a square. Take the square root of both sides.

$$x^2 + 8x + 16 = 169$$
$$(x + 4)^2 = 169$$
$$x + 4 = \pm 13$$
$$x = 9 \text{ or } x = -17$$

Find the square roots to two decimal places, using the table on page 562.

1. $\sqrt{7}$ 2.65
2. $\pm\sqrt{36}$ 6
3. $\pm\sqrt{90}$ ±9.49
4. $\pm\sqrt{300}$ ±17.32
5. $\sqrt{30}$ 5.48
6. $\sqrt{(9)^2}$ 9
7. $\sqrt{(-5)^2}$ 5
8. $\sqrt{-36}$ No real answer

Solve for x. If x is not a perfect square, leave the answer in square-root form.

9. $x^2 = 64$ — $x = 8$ or $x = -8$
10. $2x^2 = 8$ — $x = 2$ or $x = -2$
11. $(x - 4)^2 = 125$ — $x = 4 + 5\sqrt{5}$ or $x = 4 - 5\sqrt{5}$
12. $x^2 = (3)^2 + (4)^2$ — $x = 5$ or $x = -5$
13. $(x + 3)^2 = 56$ — $x = 2\sqrt{14} - 3$ or $x = -2\sqrt{14} - 3$
14. $x^2 - 4x + 4 = 100$ — $x = 12$ or $x = -8$
15. $3x^2 + 6x + 3 = 48$ — $x = 3$ or $x = -5$

Solve for x, to the nearest two decimal places. Use the table of square roots or your calculator.

16. $x^2 = 3$ — $x \approx \pm 1.73$
17. $5x^2 = 180$ — $x = \pm 6$
18. $(x + 2)^2 = 33$ — $x \approx 3.74$ or $x \approx -7.74$
19. $(2x)^2 = (20)^2 + (15)^2$ — $x = \pm 12.5$
20. $x^2 = 2(17)^2$ — $x \approx \pm 24.04$
21. $x^2 - 6x + 9 = 32$ — $x \approx 8.66$ or $x \approx -2.66$

Name ______________________ Date ____________

Bisection of Line Segments and Angles

The steps for bisecting an *angle* are illustrated in the top row of figures; the steps for bisecting a *segment* are shown in the second row. In each step, the new arcs or lines are shown heavier than the rest of the drawing.

1. 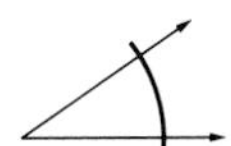
2. 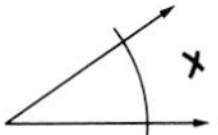
3. 

1. 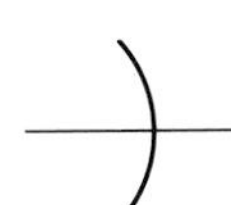
2. 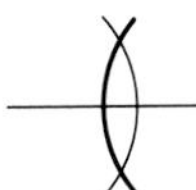
3. 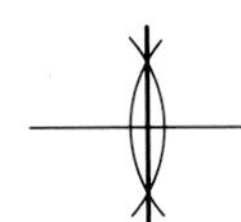

---

1. a. Bisect angle *ACE* shown below. Choose a point on the angle bisector and call it *O*. What are the measurements of angles *ACO* and *OCE*?
   15°, 15°

   b. Bisect segment *AB*. Show that the two parts are equal by using two hatch marks on each part. Mark the midpoint *M*.

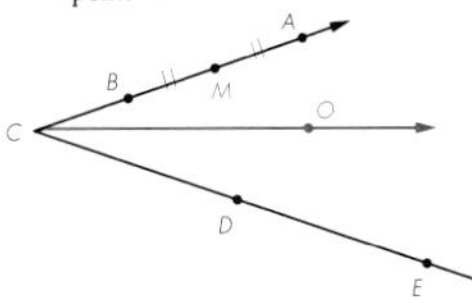

2. a. Bisect segment *YZ* shown below. Label the bisecting segment *AB*, with *A* above segment *YZ*. Draw segment *YA*.

   b. Now bisect angle *XYA*. (You will need to extend ray *YX* to do this.) Find the measurements of angles *XYA* and *AYZ*.
   Answers will vary, but their sum must be 90°.

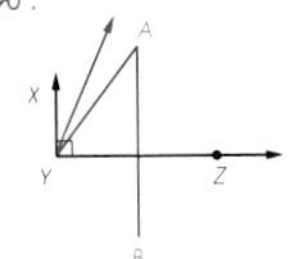

3. Make a careful tracing of the four-sided shape at the right. Construct the bisector of each of the four interior angles. What figure is formed by the four angle bisectors?
   rectangle

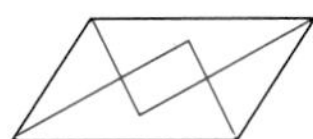

Name ______________________ Date ____________

Star Polygons

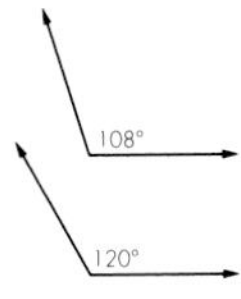

A *star polygon*, also called a *stellated* polygon, is formed by extending the sides of a convex polygon to create a star-like figure. A pentagon leads to a five-pointed star, a hexagon to a six-pointed star, and so on. Any convex polygon with no right angles and more than four sides can be stellated.

Pentagonal stars formed from regular pentagons use angles of 72° and 108°; hexagonal stars formed from regular hexagons use angles of 60° and 120°.

1. Connect the dots in order to make a five-pointed star.
2. Complete the figure so that the fifth line segment touches the circle in just one point.
3. Draw a five-pointed star inside the pentagon by connecting the vertices.

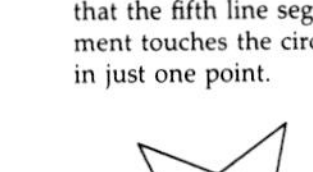

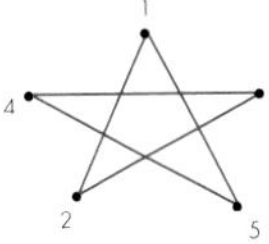

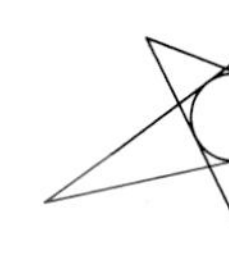

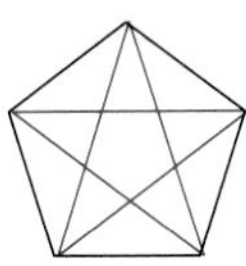

The figures in the first row are based on the pentagon; those in the second row are based on the hexagon. Start with a regular pentagon or hexagon. Use a ruler to complete an identical figure.

4. 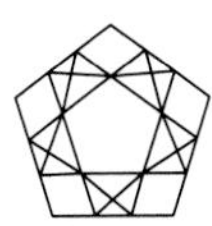
5. 
6. 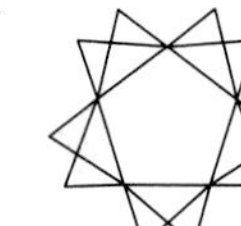
7. 
8.  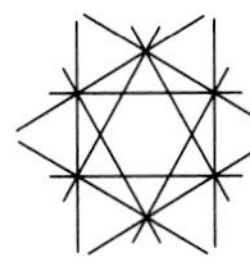 
9. 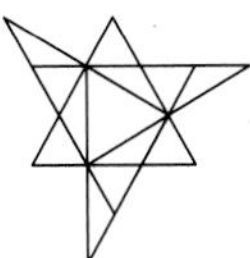

Name ______________________ Date ____________

Perpendiculars

In the top row of figures, the steps for drawing a perpendicular to a line from a given point on that line are shown. In the second row, the steps for a similar construction when the given point is not on the line are shown.

1. 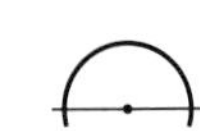
2. 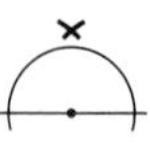
3. 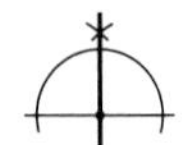

1. 
2. 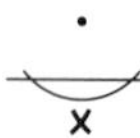
3. 

---

1. a. A construction has been started in Figure 1. Label the point marked with the dot as the point *O*. Label the two points where the arcs meet the line with the letters *N* and *W*.
   b. Now finish constructing a perpendicular to line *b* at the point *O*. Make sure the opening of your compass is wider than the length *NO*.

2. a. In Figure 2, label the vertical line *m* and the horizontal line *n*. Mark their intersection point *F*. Label the two points where the arcs meet line *n* with the letters *A* and *B*. Label the point where the arcs meet line *m* with the letter *C*.
   b. Draw *CA* and *CB*. How do their lengths compare?
   They are equal.
   c. Mark a new point, *R*, anywhere above line *n*, but not on line *m*. Construct a perpendicular to line *n* from point *R*. Label it line *s*.
   d. Construct a perpendicular to line *m* from point *R*. Label it line *t*. Write two pairs of parallel lines in the figure.
   s ∥ m, t ∥ n

3. Construct the drawing in Figure 3. Start by drawing side *AC*. Construct a perpendicular to side *AC* from point *B*. Complete right triangle *ABC* with side *AB* any length you wish. Finally, construct a perpendicular to side *AB* from point *C*.

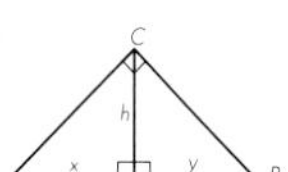

Name ______________________ Date ____________

Constructing Regular *N*-Gons

There are many ways to construct regular polygons using just a compass and straightedge. The drawings on this page involve only circles and perpendicular lines.

In each figure, the heavy line segments show the beginning of a regular *n*-gon. Construct each design without using a protractor. Complete the *n*-gon and write its name. (Hint: For Exercise 4, the name of the figure starts with the letters *do*.)

1. regular octagon

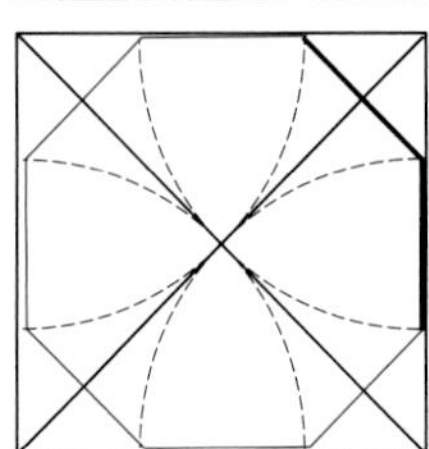

2. regular octagon

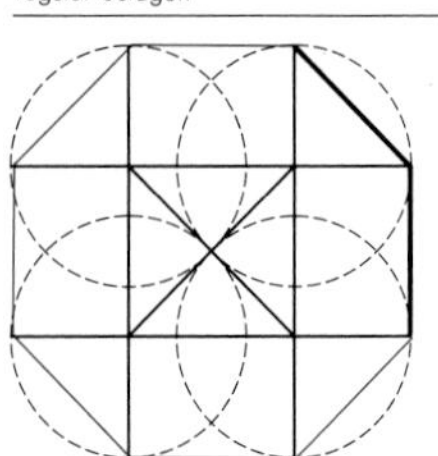

3. regular hexagon

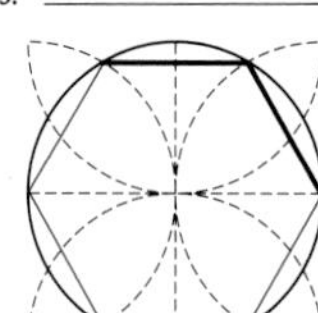

4. regular dodecagon

5. regular octagon

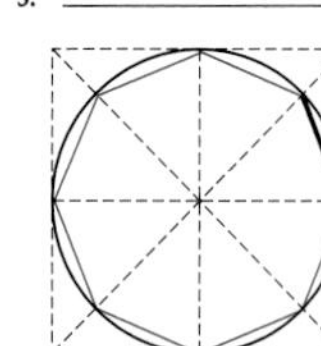

Name ______________________ Date ______________

Constructing a Line Parallel to Another Line

Here is one way to construct parallel lines. It involves copying an angle.

1. 2. 3.

The drawing at the right shows a second way to make parallel lines. Paper-folding is yet another way. Describe how to fold a sheet of paper

1. to make two parallel lines. Fold the paper in half horizontally. Unfold and then fold the bottom half again horizontally into halves. Or, fold into thirds.

2. to make three parallel lines. Fold the paper into quarters by folding it in half twice.

Use the figures at the top of the page for Exercises 3–6.

3. In Step 1 of the construction, label the horizontal line *m* and the point above line *m* with the letter *P*. A line has been drawn through point *P*. Where this line meets *m*, mark the point *Q*. Now two arcs are drawn from *P* and *Q* with the same compass opening.

4. In Step 2, a smaller compass opening is used to draw two new arcs. Label the top intersection point *O* and the lower intersection point *R*.

5. In Step 3, a line is drawn through *P* and *O*. This is the line parallel to line *m* through the point *P*.

6. Mark another point on the slanted line above point *P*. Label this point *S*. Which two angles are congruent? $\angle SPO \cong \angle PQR$

The three figures below all use parallel lines. Make copies of the figures on a separate sheet of paper.

7. 8. 9.

Name ______________________ Date ______________

The Golden Rectangle

The Golden Ratio $\frac{1+\sqrt{5}}{2}$

Once you have learned to construct perpendicular and parallel lines, you can construct rectangles. Of all rectangles, the *Golden Rectangle* is thought to have the most pleasing proportions. It is often used by artists and has many unusual properties. A rectangle is a Golden Rectangle if it has a length-to-width ratio equal to the expression at the right.

1. Compute the value of the Golden Ratio to the nearest thousandth. 1.618

2. Construct a rectangle measuring 100 mm by 162 mm. How closely does this approximate a Golden Rectangle? within 0.2 mm

3. The large rectangle in Figure 1 is a Golden Rectangle. Explain how you know that all the smaller, interior rectangles are also Golden Rectangles. Corresponding acute angles of the triangles are congruent. The right triangles are similar. Thus, the rectangles are similar and their sides are in equal ratio.

4. Make a construction like that shown in Figure 1. Make the sides of the interior rectangles equidistant.

5. Figure 2 shows "whirling squares" inside a Golden Rectangle. When you cut a square from one end of a Golden Rectangle, the smaller rectangle left over is also a Golden Rectangle. Make a large copy of Figure 2. Use 10 inches as the width of your rectangle.

6. By drawing a quarter circle inside each square of Figure 2 you can make the spiral shown in Figure 3. Make this spiral in the rectangle you drew in Exercise 5.

7. Choose one or more of the designs at the bottom of the page to construct. Or create a new design of your own that uses the Golden Rectangle or the spiral.

1

2

3

Name ______________________ Date ______________

Constructions Involving Triangles

Altitude
Connects a vertex with the opposite side; must be perpendicular to the side

Median
Connects a vertex with the opposite midpoint

Angle Bisector
Divides an angle of the triangle into two equal parts

For each triangle in Exercises 1–12, write whether the thick line is an *altitude, median,* or *angle bisector.* For some triangles, there is more than one answer.

1. median
2. altitude
3. angle bisector
4. altitude, median, angle bisector
5. angle bisector
6. altitude, median, angle bisector
7. median
8. altitude
9. altitude
10. median
11. median
12. median

13. In the space at the right, construct a right triangle *ABC* with the right angle labeled point *C*. Which two sides of the triangle are also altitudes?
sides *AC* and *BC*

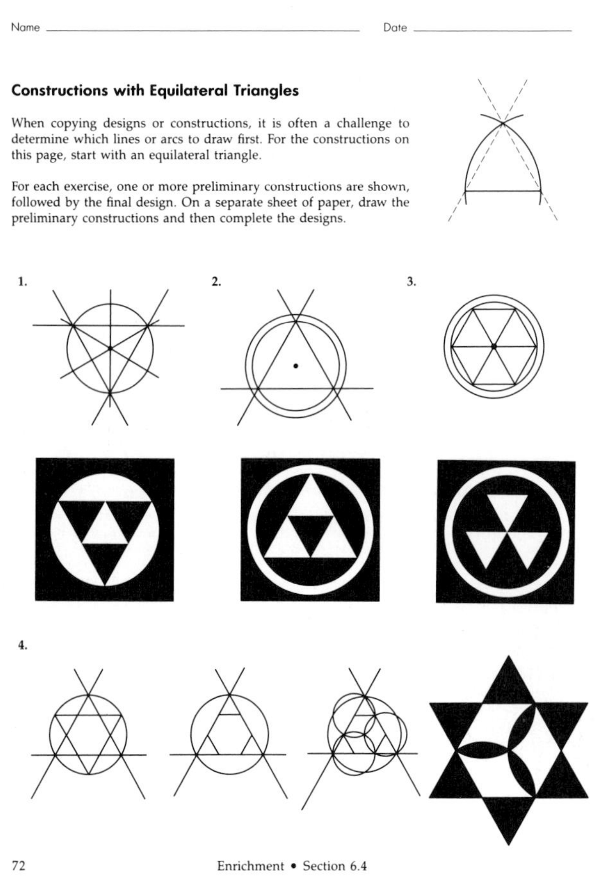

Name ______________________ Date ______________

Constructions with Equilateral Triangles

When copying designs or constructions, it is often a challenge to determine which lines or arcs to draw first. For the constructions on this page, start with an equilateral triangle.

For each exercise, one or more preliminary constructions are shown, followed by the final design. On a separate sheet of paper, draw the preliminary constructions and then complete the designs.

1. 2. 3.

4.

Name ______________________ Date ____________

Dividing a Line Segment into *n* Equal Parts

Segment AC in the drawing is 5 cm long. The parallel lines are spaced equally, 1 cm apart. Segments AC and AB are transversals that intersect the five parallel lines.

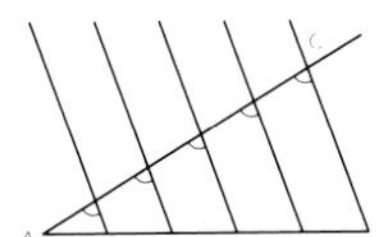

If parallel lines cut off equal segments on one transversal, they cut off equal segments on every transversal.

This property of parallel lines can be used to divide a line segment into any number of equal segments.

A sheet of notebook paper can be used to divide a line segment into equal segments. Follow these steps to divide segment MN into five equal parts.

1. Lay a sheet of notebook paper over MN. Line up one edge of the paper with point M.
2. Hold the paper at M and turn it until the fifth line of the paper touches point N.
3. On the notebook paper, mark six dots at M, N, and the four equally-spaced points in between.
4. Fold the notebook paper just above the dots. Then use it to mark the division points on segment MN.
5. On another sheet of paper, draw a right triangle. Divide each side into four equal segments. Then make the designs shown below. Create a fourth design of your own and record it on the blank triangle at the right.

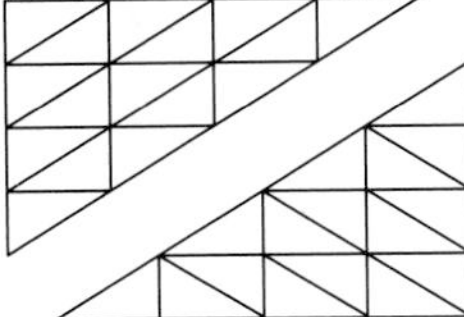

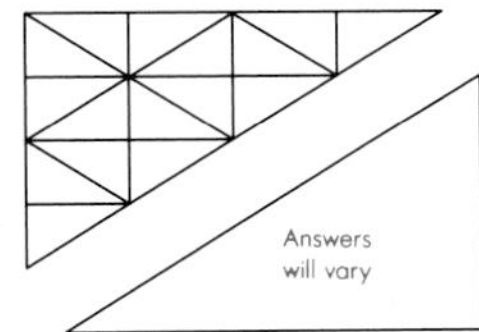

Name ______________________ Date ____________

Woven Figures

Woven figures are more challenging to construct than they may first appear.

1. Copy this construction.

2. In the space below, draw a figure which interweaves a square and a circle.

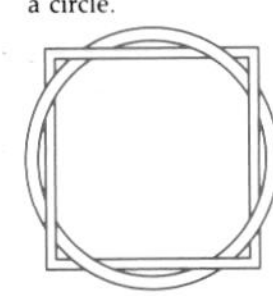

3. Explain why the woven figures are not drawings of two-dimensional objects.
No part of a two-dimensional object can be on top of or underneath another part of the object.

4. Copy this figure.

5. Copy this figure.

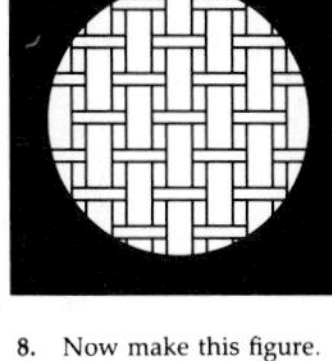

6. Compare and contrast the two figures you made for Exercises 4 and 5.
Possible responses: Figure 4 uses no circles. In Figure 5, the horizontal strips are narrower than the vertical strips. In Figure 4, the strips are on a black background.

7. Copy this prism.

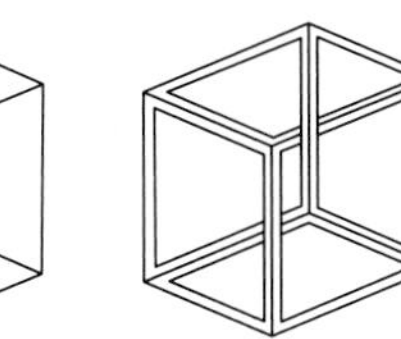

8. Now make this figure.

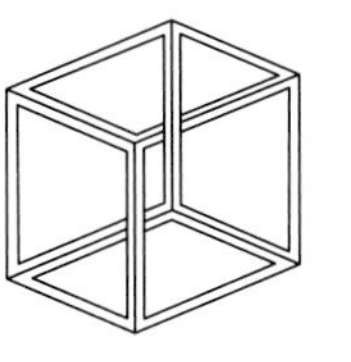

9. Describe what is "wrong" with the figure in Exercise 8. Why is it called an *impossible* figure?
Two "back" vertices are connected by a segment which is "on top" of a segment connecting two "front" vertices. The figure could never be created with real-life rigid objects.

Name ______________________ Date ____________

Theorems from Constructions

Theorem A	*Theorem B*	*Theorem C*
Midpoints of Two Sides of a Triangle	**Point on Perpendicular Bisector**	**Point on Angle Bisector**
If the midpoints of two sides of a triangle are joined, then the line segment formed is parallel to the third side and its length is one-half that of the third side.	If a point is on the perpendicular bisector of a line segment, then it is equidistant from the endpoints of the segment.	If a point is on the bisector of an angle, then the point is equidistant from the sides of the angle.

Each of Exercises 1–4 is an illustration of one of the three theorems above. Write *A*, *B*, or *C* to indicate the correct theorem.

1. Segment *CD* is the perpendicular bisector of segment *AB*. This theorem tells us that the lengths of the segments *PA* and *PB* must be equal. B

2. *M* is the midpoint of side *XZ* and *N* is the midpoint of side *YZ*. This theorem tells us that the length of segment *XY* is twice the length of segment *MN*. A

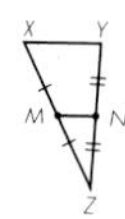

3. Segment *AD* is perpendicular to segment *CB*. *CD* and *DB* are equal. This theorem tells us that *AC* = *AB* and that *OC* = *OB*. So, triangles *ABC* and *OBC* are both isosceles triangles. B

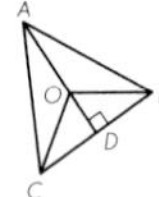

4. The angles *PRS* and *QRS* have equal measurements. Thus, *SR* is the angle bisector of *R*. This theorem tells us that *SQ* equals *SP*. C

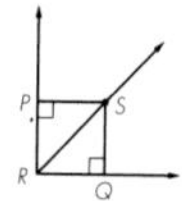

5. Another theorem you have studied in this chapter states the following: If two lines are intersected by a transversal so that the alternate interior angles are equal, then the lines are parallel.

The following statement is an example of the theorem. Complete the statement below.

Lines *m* and n are intersected by the transversal labeled *t*. The alternate interior angles *x* and *y* are equal. Therefore, we know that lines m and n are parallel.

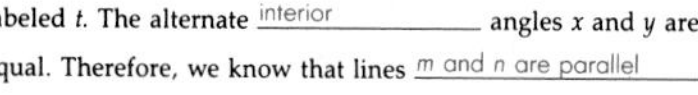

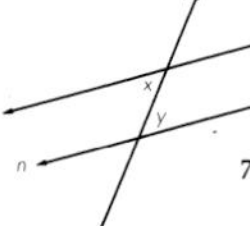

Name ______________________ Date ____________

Dissection Puzzles

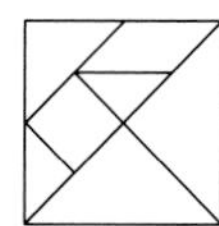

One little-known type of recreational mathematics consists of *dissection puzzles*. In these puzzles, the goal is to transform one figure into another by making a given number of straight cuts and then rearranging the pieces.

The seven tangram pieces, shown at the right, are a form of dissection puzzle. The seven pieces can be used to make many different figures.

Constructed from wood and brightly painted, dissection puzzles make great gifts!

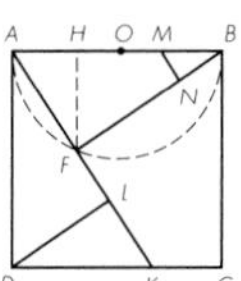

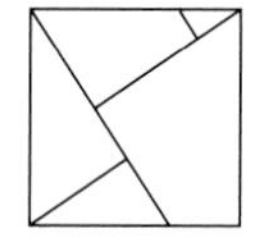

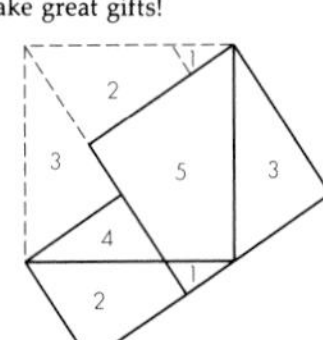

1. Make a copy of the drawing above. Point *O* is the midpoint of segment *AB*. $AH = \frac{1}{3}AB$ and *MB* = *KC*. Lines that look perpendicular are perpendicular.

2. Segment *HF* is only used for the drawing. Cut the square into the five pieces shown.

3. The dissection forms two squares, one having twice the area of the other. Which pieces make up the two new squares? 2 and 4; 1, 3, and 5

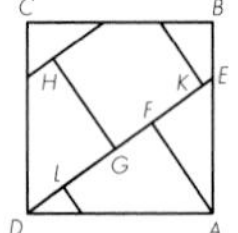

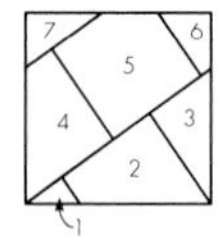

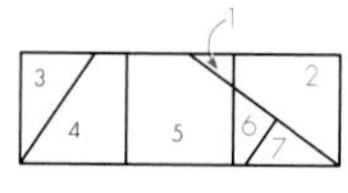

4. The goal in the dissection above is to make three congruent squares. *AE* is half the length of a diagonal. *GH* = *GK* = *FL* = *AF*

5. Segment *CH* is only used for the drawing. Cut the square into the seven pieces shown.

6. Finish numbering the diagram to show how the seven pieces make three congruent squares.

Name ______________________ Date ______________

Logo and Medians

The coordinates of two of the three vertices of a triangle are given in each exercise below. Choose the coordinates of the third vertex so that the coordinates of the midpoints of each side will be integers. Then modify the MEDIANS procedure from page 243 in the textbook to tell the turtle to draw each triangle and its three medians. Answers may vary.

1. (0, 0), (4, 0)

2. (0, 3), (2, 5)

3. (3, −1), (5, 1)

4. (−6, −1), (−2, −5)

5. (−6, 6), (−2, 6)

6. (2, −5), (4, −3)

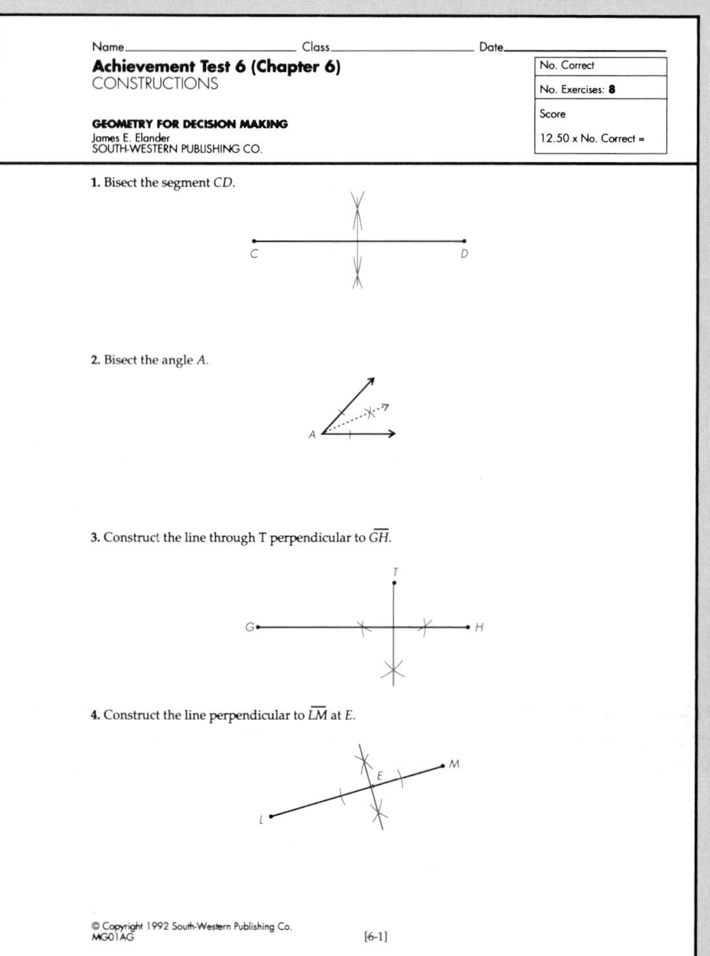

Name__________ Class__________ Date__________

Achievement Test 6 (Chapter 6)
CONSTRUCTIONS

No. Correct
No. Exercises: **8**
Score
12.50 x No. Correct =

GEOMETRY FOR DECISION MAKING
James E. Elander
SOUTH-WESTERN PUBLISHING CO.

1. Bisect the segment *CD*.

2. Bisect the angle *A*.

3. Construct the line through T perpendicular to $\overline{GH}$.

4. Construct the line perpendicular to $\overline{LM}$ at *E*.

MG01AG

5. Construct the line through *T* parallel to $\overleftrightarrow{GH}$.

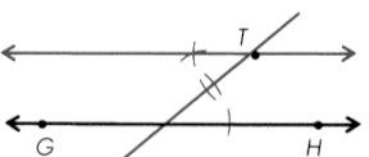

6. Draw a line segment 3 inches long. By construction, divide segment *AB* into three equal parts.

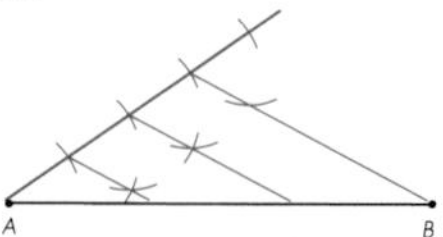

7. Construct an angle congruent to $\angle A$ that has $\overrightarrow{EC}$ as a side.

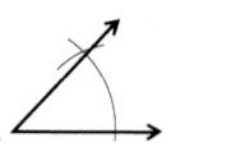

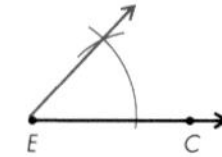

8. In Δ *PQR*, *G* and *M* are midpoints. What is the length of $\overline{RQ}$? 32.6 m

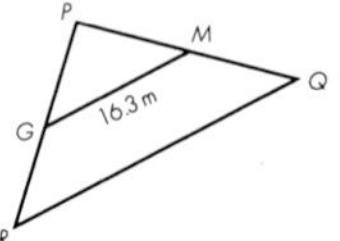

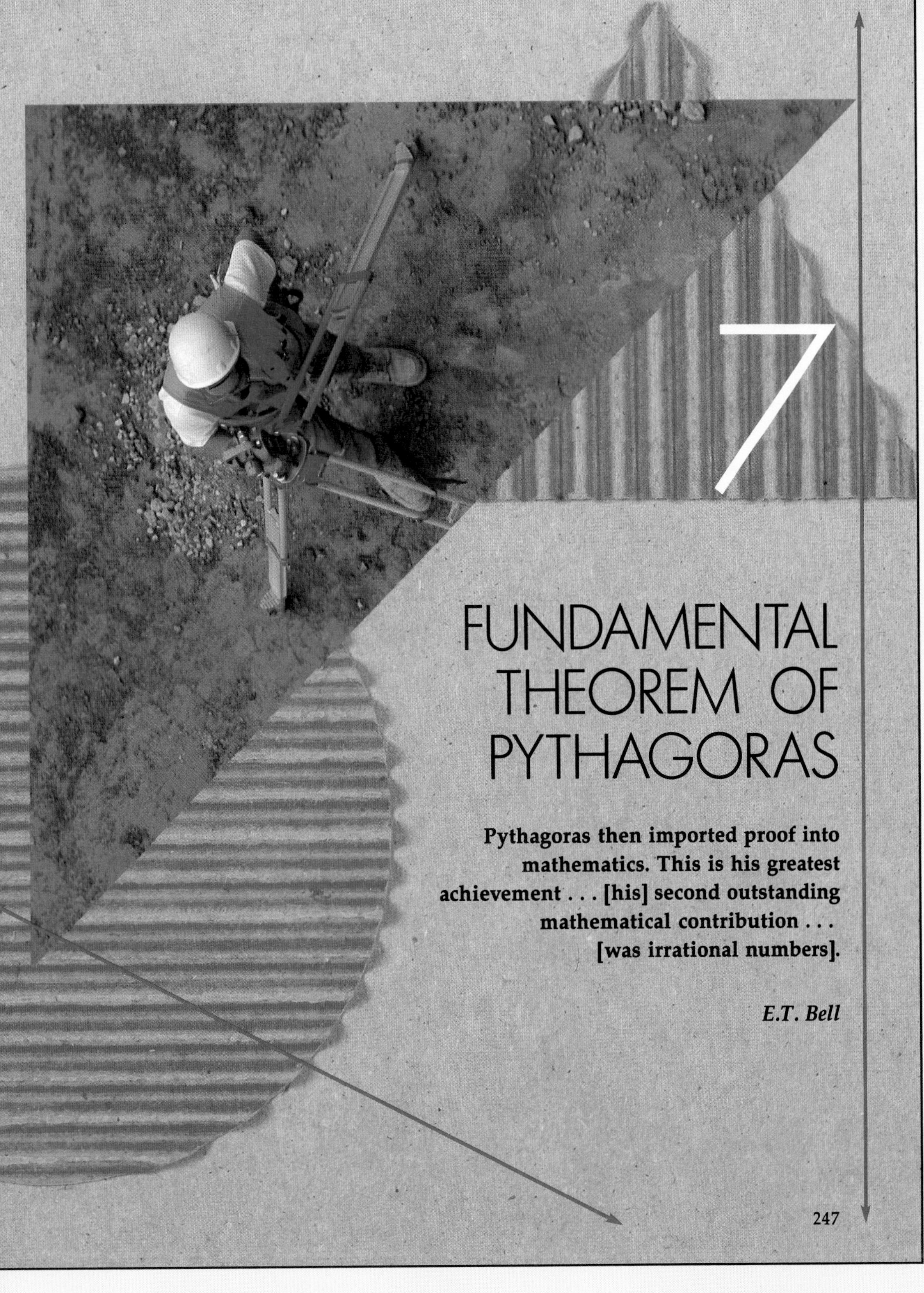

7

FUNDAMENTAL THEOREM OF PYTHAGORAS

Pythagoras then imported proof into mathematics. This is his greatest achievement . . . [his] second outstanding mathematical contribution . . . [was irrational numbers].

E.T. Bell

247

CHAPTER 7 OVERVIEW

Fundamental Theorem of Pythagoras

Suggest Time: 12–16 days

The Pythagorean Theorem is rightly considered to be one of the most important, if not *the* most important theorem, in geometry. This is reflected in the quotation by E.T. Bell that appears at the beginning of the chapter. It also accounts for the appearance of the Pythagorean Theorem relatively early in the book. The author believes that students should become acquainted with the theorem as early in the school year as is convenient and then should revisit the theorem several times during the remainder of the course.

The opening page shows a surveyor. This can be related to the discussion of indirect measurement in Section 7.3 in which two surveyors, without trigonometry tables, are able to find the distance between two landmarks on opposite sides of a pond.

> Pythagoras then imported proof into mathematics. This is his greatest achievement . . . . [His] second outstanding mathematical contribution . . . [was irrational numbers].
>
> *E.T. Bell*

CHAPTER 7 OBJECTIVE MATRIX

	End of Chapter Items by Activity				Student Supplement Items by Activity		
Objectives by Number	**Review**	**Test**	**Computer**	**Algebra Skills**	**Reteaching**	**Enrichment**	**Computer**
7.1.1	✓	✓	✓	✓	✓	✓	✓
7.1.2	✓	✓		✓	✓		✓
7.2.1	✓	✓		✓	✓	✓	
7.2.2	✓	✓		✓	✓	✓	
7.3.1	✓	✓		✓	✓		
7.3.2	✓	✓		✓	✓	✓	
7.4.1	✓	✓		✓	✓		
7.4.2		✓		✓	✓		
7.4.3	✓	✓		✓			
7.5.1			✓	✓	✓	✓	✓
7.5.2	✓	✓		✓	✓	✓	✓

*A ✓ beside a Chapter Objective under Algebra Skills indicates that algebra skills taught within the chapter or in previous Algebra Skills lessons are used.

CHAPTER 7 PERSPECTIVES

▲ Section 7.1

The Pythagorean Theorem

In the Discovery Activity, students are led, through direct measurement, to see the Pythagorean relationship that exists among the three sides of a right triangle. Since the first two triangles, *ABC* and *PQR*, have sides with integral lengths, the missing side can be found either by measuring the sides with a ruler or simply by *counting* the units on the hypotenuse. (The complete data for the first of these triangles are listed in the table on page 249 for students to study as a model.) Only in the last two triangles is it necessary to use measurement, rather than counting, to record the lengths of the sides.

After the Discovery Activity has been completed, the relationship that it reveals is proved as the Pythagorean Theorem. The theorem and its proof should be discussed in class for a number of reasons. The first reason is the importance of the theorem itself in the history of mathematics. Secondly, the details of the proof involve important geometric concepts such as similarity that deserve being reviewed. Finally, the proof itself is a model of mathematical reasoning that involves both algebraic and geometric concepts.

The converse of the theorem is also stated but is not proved. (See the Lesson Notes for this section for the outline of a proof of the converse.) Although the proof of the Pythagorean Theorem and its converse is important, some of your students may have difficulty following the abstractions in the proof, both geometric and algebraic. Encourage these students to accept the theorems as postulates and stress the use to the concept to solve problems. After students have had ample experience in the use to the theorem, a review of the proof may give them more insight into the relationships involved.

7

▲ Section 7.2

The Isosceles Right Triangle

This section studies the simplest of right triangles, the *isosceles right triangle*. After a proof that each acute angle of an isosceles right triangle has 45° as its degree measure, students are led to the discovery that the length of the hypotenuse is related to the length of each leg by the formula $c = s\sqrt{2}$.

If students have difficulty with the special relationships involved in dealing with isosceles right triangles, stress that isosceles right triangle relationships can always be treated as regular right triangle relationships in which the *a* and the *b* are equal.

▲ Section 7.3

The 30°-60° Right Triangle

This section completes the two-section coverage of the "special" right triangles, the 45°-45° right triangle of the previous section and the 30°-60° right triangle of this section.

Unlike most triangles, these two special triangles can be analyzed using simple techniques to develop relationships among all the sides and angles. As a consequence, these special triangles have the advantage of allowing students to explore the field of indirect measurement without getting into trigonometry.

Most students have little difficulty remembering that the side opposite the 30° angle is $1/2c$. If students have difficulty with the side opposite the 60° angle, emphasize exercises that deal only with the relationship between the hypotenuse and the side opposite the 30° angle. Once that relationship has been thoroughly internalized, suggest that students use the standard Pythagorean Theorem to find the third side of the triangle—the one opposite the 60° angle.

▲ Section 7.4

Angle-Side Relationships

This section considers the question of how the angles of a triangle are related to the sides. Theorems 7-4-1 and 7-4-2 state the *inequality* relationships that hold. (Unequal sides are opposite unequal angles and conversely.) It is worth noting that using available geometric tools, one must settle for conclusions that are stated as inequalities rather than as equalities. This fact may be compared to the exact *equality* relationships that can be stated in the 45°-45° and the 30°-60° right triangles.

▲ Section 7.5

Pythagorean Theorem and Descartes

This section continues the study of coordinate geometry that was begun in Section 3.6. The main point here is to use the Pythagorean Theorem to introduce the Distance Formula. The Distance Formula is exploited more thoroughly in Chapter 13, along with other formulas from coordinate geometry.

When using the Distance Formula, students should realize that it usually is a matter of personal preference which point is called (x_1, y_1) and which is called (x_2, y_2). Once the choice is made, however, the students must be consistent and not choose x_1 from one point and y_1 from the other point.

SECTION 7.1

Resources
Reteaching 7.1
Enrichment 7.1

Objectives
After completing this section, the student should understand
- ▲ the Pythagorean Theorem.
- ▲ squares and square roots of numbers.

Materials
centimeter ruler
protractor
compass
calculator

Vocabulary
hypotenuse
leg of a right triangle
Pythagorean Theorem

GETTING STARTED

Quiz: Prerequisite Skills

1. How many degrees are there in the three angles of a triangle? [180]

2. In a triangle, what is the largest possible number of:
a. acute angles? [3]
b. obtuse angles? [1]
c. right angles? [1]

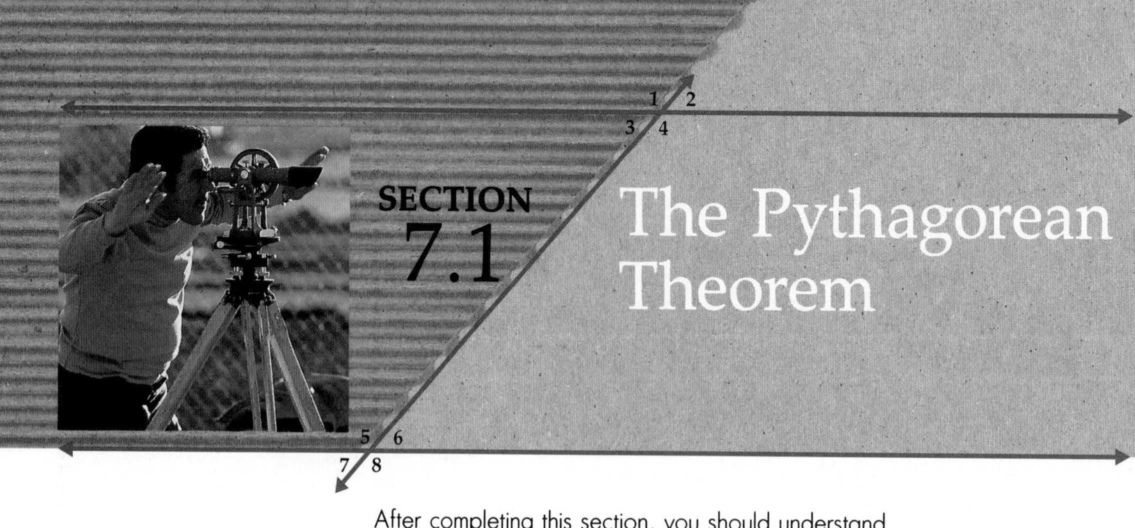

SECTION 7.1 The Pythagorean Theorem

After completing this section, you should understand
- ▲ the Pythagorean Theorem.
- ▲ squares and square roots of numbers.

You are familiar with the three types of triangles shown below.

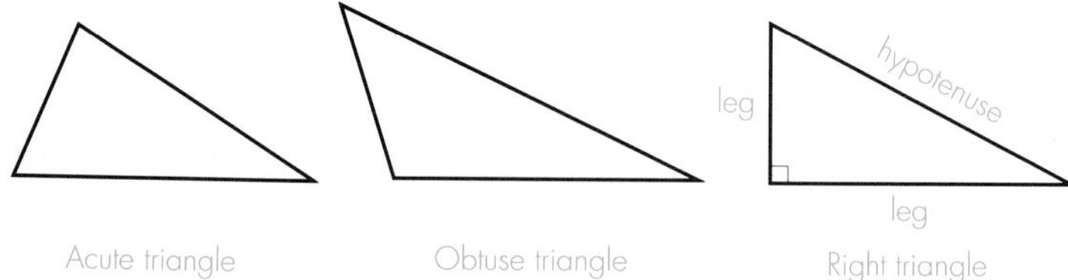

RIGHT TRIANGLES

Notice that there are special names for the three sides of the right triangle. The side opposite the right angle is the *hypotenuse*. The other two sides are the *legs* of the right triangle.

DEFINITION 7-1-1: **The HYPOTENUSE of a right triangle is the side opposite the right angle.**

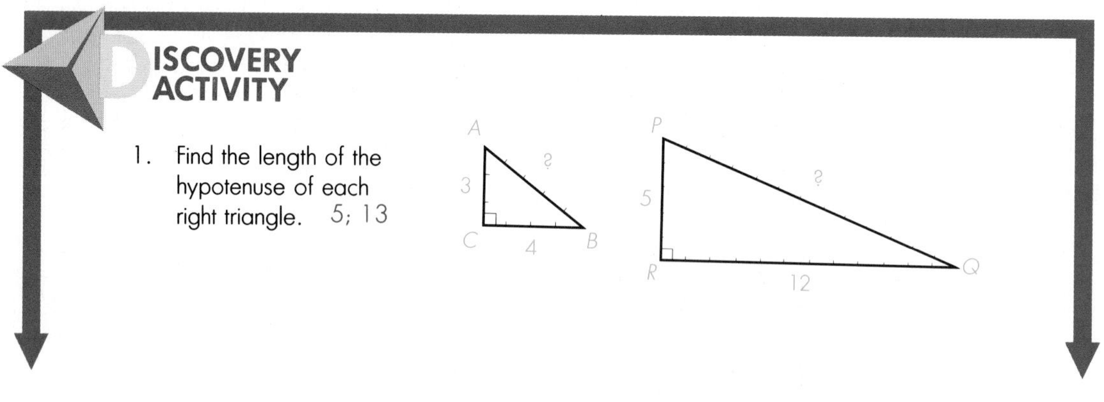

1. Find the length of the hypotenuse of each right triangle. 5; 13

2. Measure the sides of each right triangle to the nearest tenth of a centimeter.

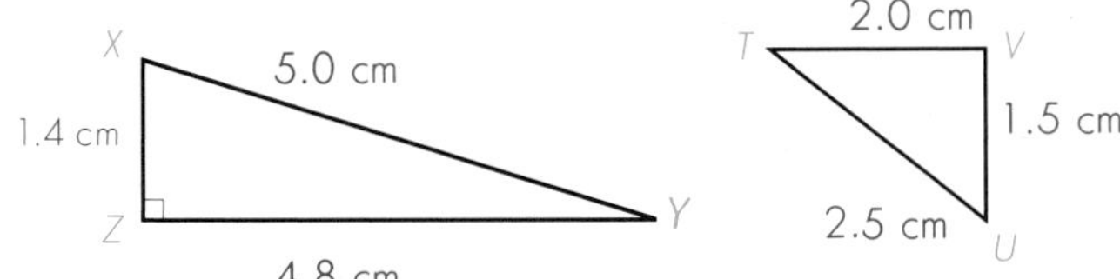

3. Copy and complete the following table for the triangles in Exercises 1 and 2. The entries for $\triangle ABC$ have been done for you.

	Length of first leg	Length of second leg	Length of hypotenuse	Square of first leg	Square of second leg	Square of hypotenuse
$\triangle ABC$	3	4	5	9	16	25
$\triangle PQR$	5	12	13	25	144	169
$\triangle XYZ$	1.4 cm	4.8 cm	5.0 cm	1.96	23.04	25
$\triangle TUV$	2.0 cm	1.5 cm	2.5 cm	4	2.25	6.25

4. Study the table in Exercise 3. What do you notice about how the squares of the three sides of each right triangle are related? Write a generalization that seems to be true. $(1\text{st leg})^2 + (2\text{nd leg})^2 = (\text{hyp.})^2$
5. Construct a triangle with sides of lengths 100 mm, 96 mm, and 28 mm. Use the methods of Section 6.4.
6. Square the length of each side of the triangle you constructed. How are the squares related? $96^2 + 28^2 = 100^2$
7. Measure the largest angle of the triangle you constructed. What kind of angle is it? Right angle
8. Write a generalization based on your observations concerning the triangle you constructed. If the sum of the squares of the lengths of two sides of a triangle equals the square of the length of the third side, then the triangle is a right triangle.

Based upon your observations in Exercises 1–4 of the Discovery Activity, you might conclude that the following statement seems to be true: If a triangle is a right triangle, then the sum of the squares of its two legs is equal to the square of the hypotenuse. Based upon your observations in Exercises 5–8 of the Discovery Activity, you might conclude that the *converse* of the above statement also seems to be true: If the sum of the squares of the lengths of two sides of a triangle is equal to the square of the length of the third side, then the triangle is a right triangle.

CLASS ACTIVITY

Each of the following is the set of lengths of three sides of a triangle. Is the triangle a right triangle?

1. 6, 8, 10 Yes
2. 7, 10, 12 No
3. 26, 10, 24 Yes

TEACHING COMMENTS

Using the Page

Point out to students that in the first exercise of the Discovery Activity, they need merely *count* the number of units in the hypotenuse. Only in Exercise 2 are the lengths of the sides measured. Students should be reassured that since the measurements are to be found to the nearest tenth of a centimeter, entries in the table for Exercise 3 may vary. As a result, the generalization of Exercise 4 may not be exact for many students.

Extra Practice
Each of the following is the set of lengths of three sides of a triangle. Is the triangle a right triangle?

1. 7, 8, 9 [No]
2. 16, 30, 34 [Yes]
3. 10, 15, 20 [No]

Each of the following triangles is a right triangle. Find the missing length. Choose from the following: 12, 12.5, 14, 15, 15.5, 16.

4. 15

5. 12.5

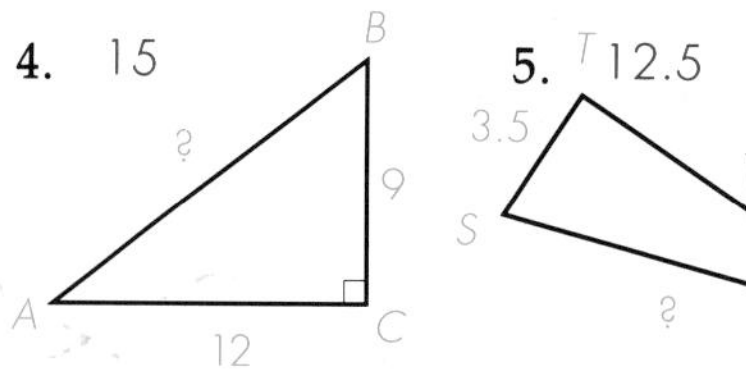

6. 12

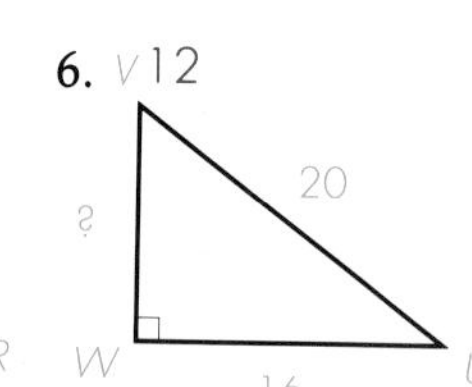

THE PYTHAGOREAN THEOREM

In the figure at the right, a and b are the legs of the right triangle ABC. The hypotenuse is c. The right angle is $\angle C$. Earlier in this section you learned how the three sides of the triangle are related. You saw that the following statement was true: $a^2 + b^2 = c^2$. You can prove that the statement is true.

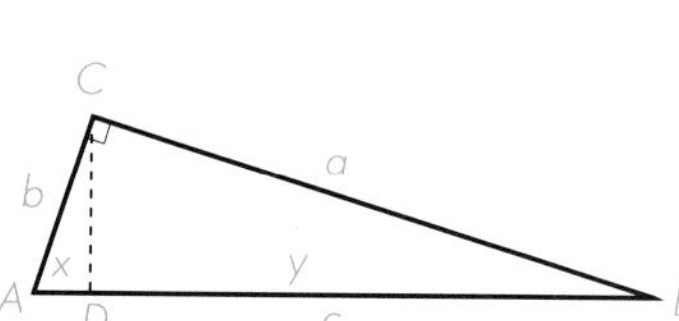

Given: Right triangle ABC with $m\angle C = 90°$ and $\overline{CD} \perp \overline{AB}$

Justify: $a^2 + b^2 = c^2$

First, show that $\triangle ABC \sim \triangle ACD$ and $\triangle ABC \sim \triangle CBD$. Have students supply the missing reasons.

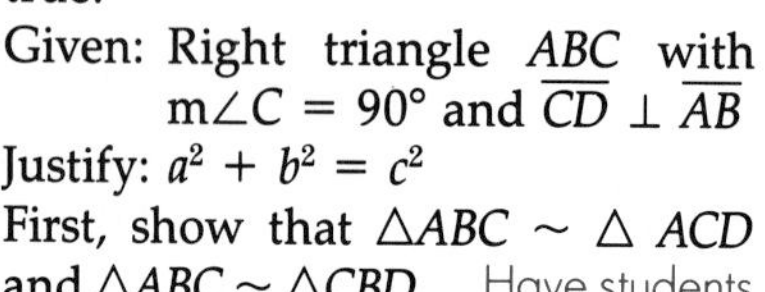

Step	Reason
1. In triangles ABC and ACD, $m\angle A = m\angle A$.	1. Identical angles
2. In triangles ABC and ACD, $m\angle C = m\angle ADC$.	2. All right angles are equal.
3. $\triangle ABC \sim \triangle ACD$	3. AA
4. In triangles ABC and CBD, $m\angle B = m\angle B$.	4. Identical angles
5. In triangles ABC and CBD, $m\angle C = m\angle CDB$.	5. All right angles are equal.
6. $\triangle ABC \sim \triangle CBD$	6. AA
7. $\frac{b}{x} = \frac{c}{b}$; $\frac{a}{y} = \frac{c}{a}$	7. If two figures are similar, then the corresponding sides are in equal ratio.
8. $b^2 = cx$; $a^2 = cy$	8. Property of algebra
9. $b^2 + a^2 = cx + cy$	9. Property of algebra
10. $b^2 + a^2 = c(x + y)$	10. Property of algebra
11. $x + y = c$	11. From diagram
12. $b^2 + a^2 = c \cdot c, = c^2$	12. Substitution; step 10

THEOREM 7-1-1 **Pythagorean Theorem: If a right triangle has sides of lengths a, b, and c, where c is the hypotenuse, then $a^2 + b^2 = c^2$.**

The converse of the Pythagorean Theorem is also true. Its proof is not given.

THEOREM 7-1-2 **Converse of the Pythagorean Theorem: If a triangle has three sides of lengths a, b, and c, such that $a^2 + b^2 = c^2$, then the triangle is a right triangle.**

When you know the lengths of two sides of a right triangle, you can use the Pythagorean Theorem to find the third side. When you need to find the square root of a number to find the missing length, use a calculator or the table of squares and square roots on page 562.

Example:

$$c^2 = a^2 + b^2$$
$$= 3^2 + 5^2$$
$$= 9 + 25, \text{ or } 34$$
$$c = \sqrt{34}, \text{ or } 5.831$$

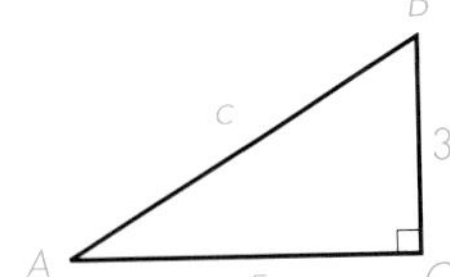

CLASS ACTIVITY

Find the missing length. Use a calculator or the table of squares and square roots on page 562. Round lengths to the nearest tenth.

1. 39

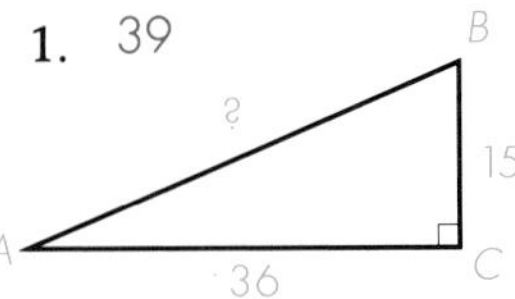

2. 7.1

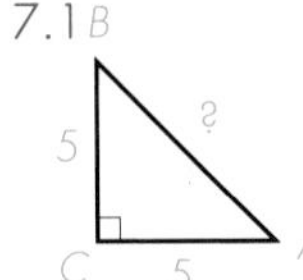

3. 22.4

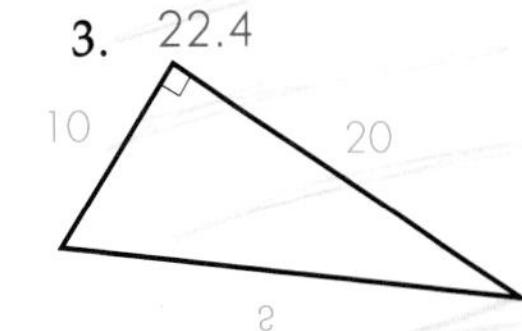

Copy and complete the following table. The lengths of the legs of right triangle ABC are a and b. The length of the hypotenuse is c. Use a calculator or the table of squares and square roots on page 562. Round lengths to the nearest tenth.

	a	b	c	$a^2 + b^2 = c^2$
4.	8	6	10	$8^2 + 6^2 = c^2$
5.	3.6	1.5	3.9	$3.6^2 + 1.5^2 = c^2$
6.	10	12	15.6	$10^2 + 12^2 = c^2$
7.	8	8.9	12	$8^2 + b^2 = 144$

Using the Page

The "property of algebra" in step 8 of the proof of Theorem 7-1-1 is that of "multiplying each side of an equation by the same number." Here, that number is bx for the first equation and ay for the second equation. (A "property of proportions" could also be used here.) In step 9 the "property of algebra" is that of "adding the corresponding sides of two equations." In step 10, the "property of algebra" is the distributive property of multiplication with respect to addition.

Using the Page

Proof of the converse of the Pythagorean Theorem:

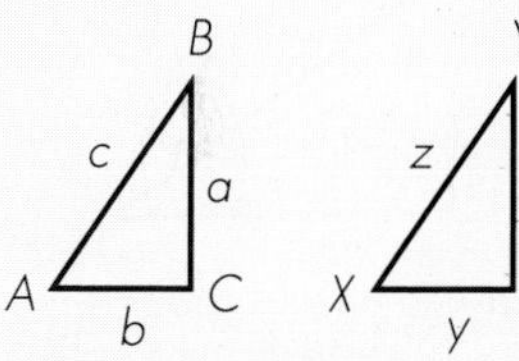

In $\triangle ABC$, $c^2 = a^2 + b^2$. To prove that $m\angle C = 90°$, construct a second triangle, $\triangle XYZ$, such that $x = a$, $y = b$, and $m\angle Z = 90°$. Then by the Pythagorean Theorem, $z^2 = x^2 + y^2$. By construction, $x = a$ and $y = \angle b$, so $x^2 + y^2 = a^2 + b^2 = c^2$. Thus, $z^2 = c^2$ and $z = c$. Then $\triangle ABC$ and $\triangle XYZ$ are congruent by SSS. Thus, $m\angle C = m\angle Z = 90°$.

HOME ACTIVITY

Each of the following is the set of lengths of three sides of a triangle. Is the triangle a right triangle?

1. 20, 48, 52 Yes
2. 0.6, 0.8, 1 Yes
3. 4, 5, 6 No
4. 27, 364, 365 Yes

Find the missing length. Use a calculator for squares and square roots. Round answers to the nearest tenth.

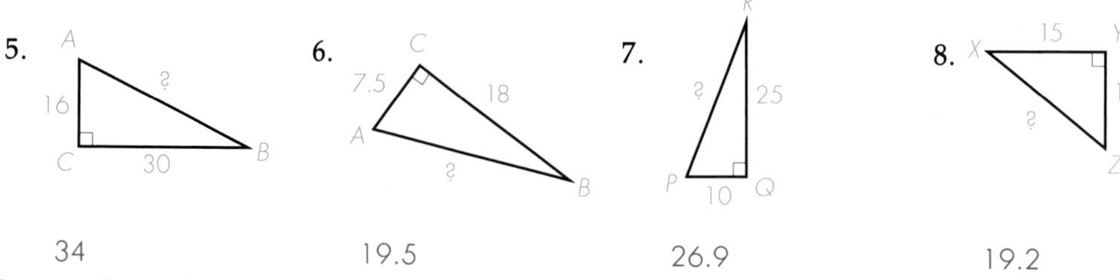

34 19.5 26.9 19.2

Copy and complete the following table. The lengths of the legs of right triangle ABC are a and b. The length of the hypotenuse is c. Use a calculator or the table of squares and square roots on page 562. Round lengths to the nearest tenth.

	a	b	c	$a^2 + b^2 = c^2$
9.	8	15	17	$8^2 + 15^2 = c^2$
10.	24	7	25	$24^2 + 7^2 = c^2$
11.	6	7	9.2	$6^2 + 7^2 = c^2$
12.	15	13.2	20	$15^2 + b^2 = 20^2$

13. A classroom measures 25 ft by 30 ft. To the nearest tenth of a foot how far apart are diagonally opposite corners of the floor? 39.1 ft
14. A regulation baseball diamond has the shape of a square. The distance from home plate to first base is 90 ft. Find the distance between first base and third base to the nearest tenth of a foot. 127.3 ft
15. A brace is needed to reinforce the gate shown below. To the nearest hundredth of a foot, what length of lumber must be bought? 8.94 ft

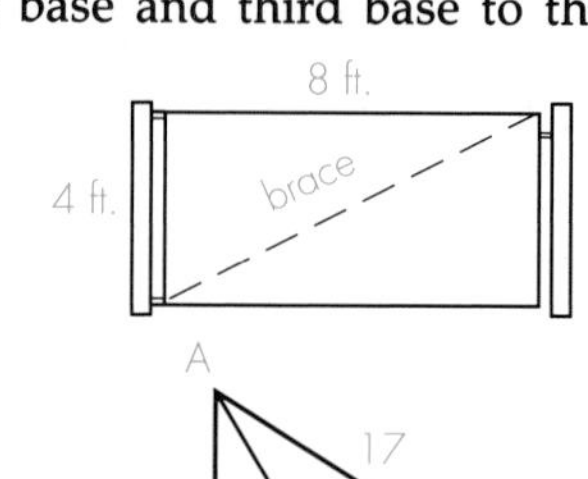

16. In the figure, find the length of $\overline{AB}$. (Hint: Find the length of another segment first.)
$AB = 10$

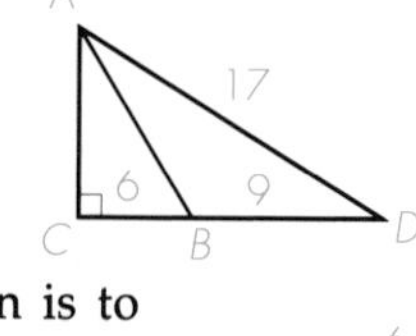

CRITICAL THINKING The piece of luggage shown is to carry an umbrella. Will an umbrella 30 in. long fit inside? No

FOLLOW-UP

Assessment

1. Explain why each figure is not a right triangle.

a.

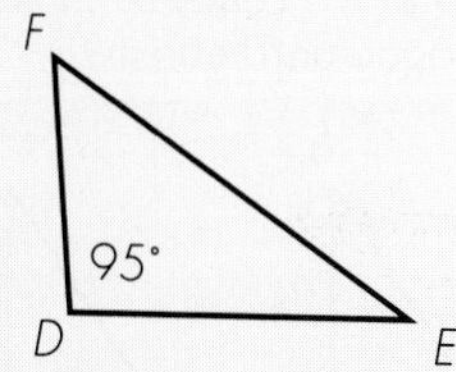

[Since one angle is obtuse, neither of the remaining angles can be right.]

b.

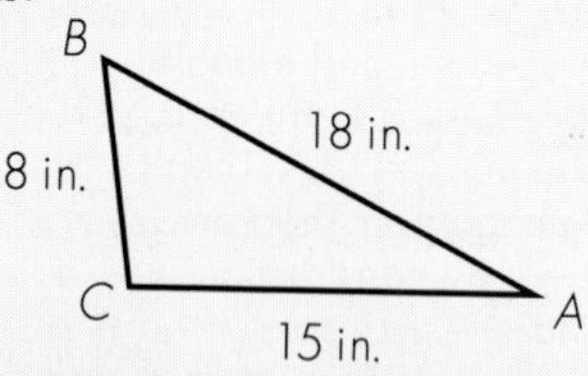

[Since the relationship $a^2 + b^2 = c^2$ does not hold for the three sides, the triangle is not a right triangle.]

2. Find the missing length in right triangle ABC. [28]

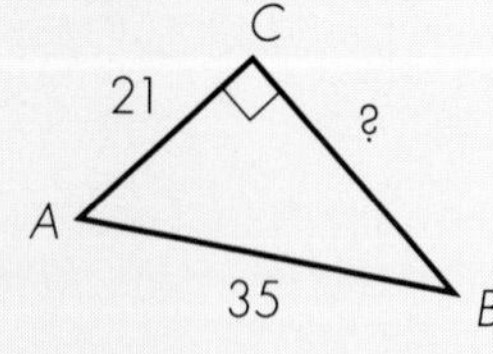

Reteaching
Students who have had difficulty with this section may benefit from Reteaching Activity 7.1.

Enrichment
For students who have mastered the material in this section, you may wish to assign Enrichment Activity 7.1.

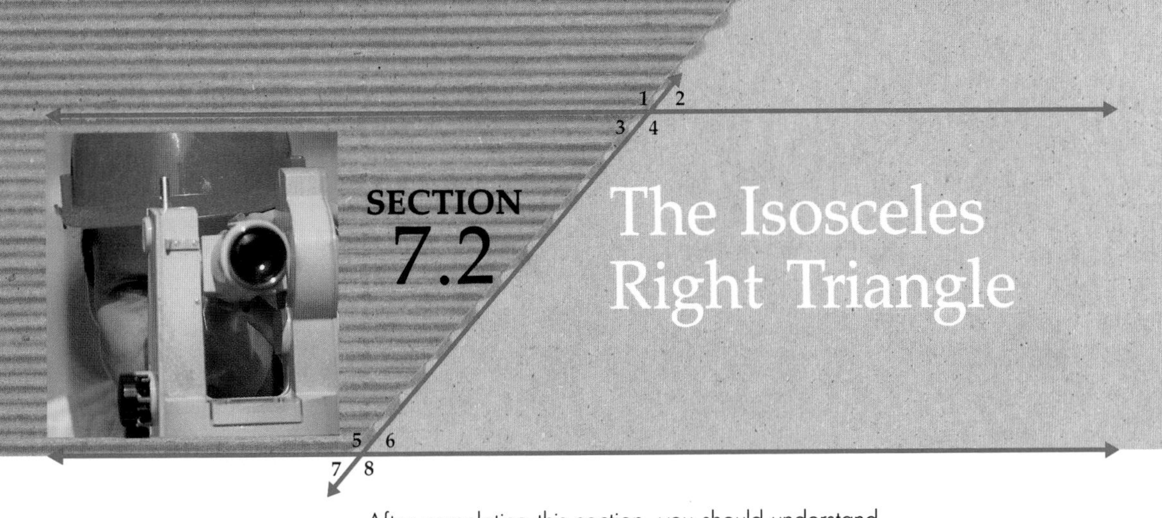

The Isosceles Right Triangle

After completing this section, you should understand
▲ the properties of isosceles right triangles.
▲ how to solve problems using isosceles right triangles.

The three triangles below are all isosceles. However, just one is both isosceles and right.

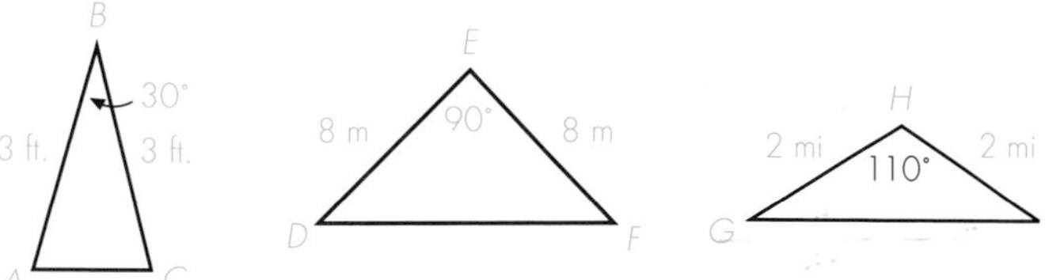

The triangle in the middle, $\triangle DEF$, is an *isosceles right* triangle.

Examples: $\triangle RST$ is a *right* triangle but is not isosceles.

$\triangle UVW$ is an *isosceles* triangle but is not right.

$\triangle XYZ$ is a triangle that is both *right* and *isosceles.*

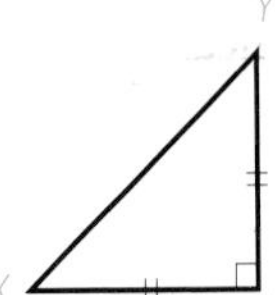

SECTION
7.2

Resources
Reteaching 7.2
Enrichment 7.2

Objectives
After completing this section, the student should understand
▲ the properties of isosceles right triangles.
▲ how to solve problems using isosceles right triangles.

Materials
centimeter ruler
calculator
compass [for warm-up]
protractor [for warm-up]

Vocabulary
isosceles right triangle

GETTING STARTED

Quiz: Prerequisite Skills

1. What term is used to describe a triangle with two sides that have the same length? [isosceles]

2. Triangle *ABC* is a right triangle.

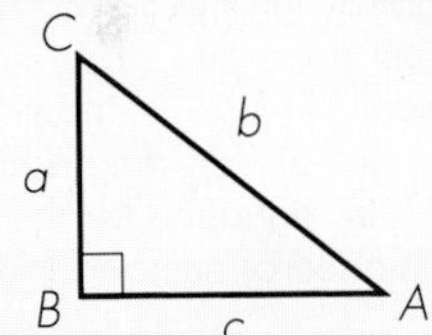

What is the formula that relates the lengths of the three sides [$a^2 + b^2 = c^2$]

Warm-Up Activity
Have students construct the perpendicular bisector of line segment $\overline{AD}$ that is 6 cm long. Label the midpoint of $\overline{AD}$ as C. Then lay off the length AC on the perpendicular bisector as BC. Draw $\overline{AB}$. An isosceles right triangle is formed. Have the students measure angles A and B. They should obtain 45° as the reading for both angle measures.

TEACHING COMMENTS

Using the Page
The "property of algebra" in step 4 of Theorem 7-2-1 is the property of "combining like (or similar) terms." The "property of algebra" of step 5 is that of "subtracting the same number from each side of an equation." In step 6, the property is that of "dividing each side of the equation by the same (nonzero) number."

THE 45°-45° RIGHT TRIANGLE

You can prove that each of the acute angles of an isosceles right triangle has an angle measure of 45°. Provide a reason for each step in the proof.
Have students provide a reason for each missing step in the proof.

Given: Triangle ABC with $AC = BC$
Justify: $m\angle A = 45°$ and $m\angle B = 45°$

Step	Reason
1. $m\angle A + m\angle B + 90° = 180°$	1. If a figure is a triangle, then the sum of the angles is 180°.
2. $m\angle A = m\angle B$	2. If a triangle is isosceles, then the angles opposite the equal sides are equal.
3. $m\angle A + m\angle A + 90° = 180°$	3. Substitution
4. $2m\angle A + 90° = 180°$	4. Property of algebra
5. $2m\angle A = 180° - 90°$, or $90°$	5. Property of algebra
6. $m\angle A = 90° \div 2$, or $45°$	6. Property of algebra
7. $m\angle B = 45°$	7. Substitution (step 2)

THEOREM 7-2-1 **If a triangle is an isosceles right triangle, then the acute angles are each 45°.**

DISCOVERY ACTIVITY

1. Use a centimeter ruler to measure the hypotenuse of each isosceles right triangle below. Measure to the nearest tenth of a centimeter.

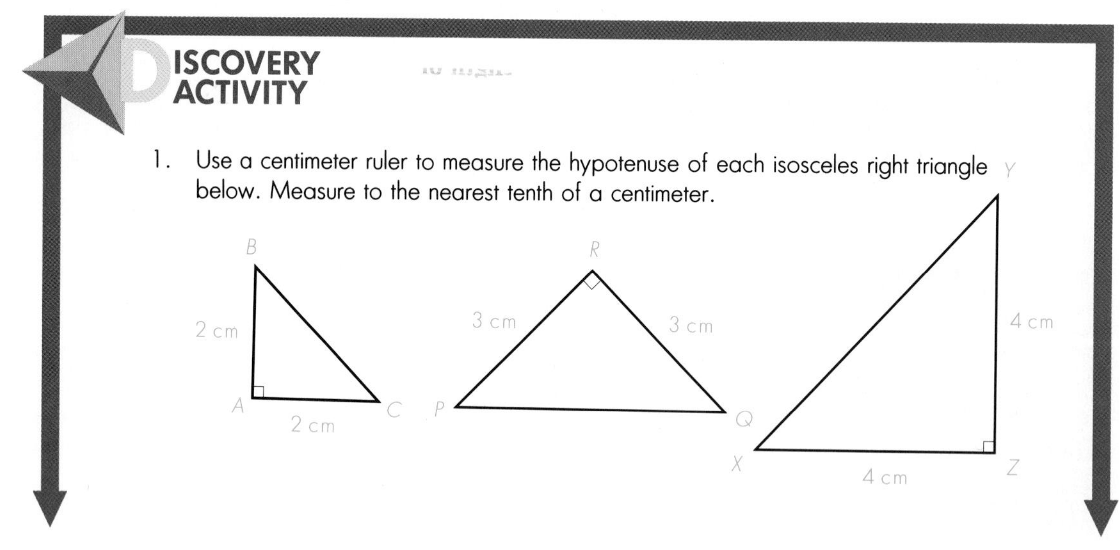

2. Copy and complete the following table. Use the results of Exercise 1.

	Length of leg	Length of hypotenuse	Hypotenuse ÷ leg
△ABC	2 cm	2.8 cm	1.4
△PQR	3 cm	4.2 cm	1.4
△XYZ	4 cm	5.6 cm	1.4

3. Study the table and write a generalization that seems to be true.

From the table, you might conclude that the ratio $\frac{\text{hypotenuse}}{\text{leg}}$ has a value of approximately 1.4. If this is so, then the relationship can also be written as *hypotenuse* ≈ 1.4 × *leg*. (The symbol ≈ means *is approximately equal to*.) You can use algebra to show that this relationship *is* true for all isosceles right triangles.

Here is the proof. The three sides of right triangle *ABC* are s, s, and c, where c is the length of the hypotenuse. By the Pythagorean Theorem, $s^2 + s^2 = c^2$. Solve for c in terms of s.

$$s^2 + s^2 = c^2$$
$$2s^2 = c^2$$
$$c^2 = 2s^2$$
$$c = s\sqrt{2}$$

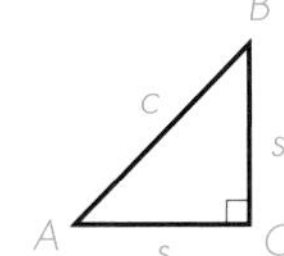

(Take the square root of each side.)

Use a calculator or the table of squares and square roots on page 562 to find the value of $\sqrt{2}$: $\sqrt{2} \approx 1.414$, or about 1.4. Thus, in right triangle *ABC*, $c \approx 1.414s$.

NOTEBOOK

THEOREM 7-2-2 **The length of the hypotenuse of an isosceles right triangle is $\sqrt{2}$ (about 1.414) times the length of either leg. In symbols this is written as $c = s\sqrt{2}$, or $c \approx 1.414s$.**

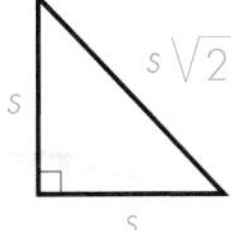

The formula of Theorem 7-2-2 is written as $c = s\sqrt{2}$ rather than as $c = \sqrt{2s}$ to stress that only the 2 (not 2*s*) is under the radical symbol.

Example: 1. To find the length of the hypotenuse of isosceles right triangle *DAN, multiply* the length of a leg by $\sqrt{2}$, or 1.414.

$$n = 2.6\sqrt{2}$$
$$\approx 2.6(1.414)$$
$$\approx 3.6764, \text{ or about } 3.7$$

Using the Page
The ratio "hypotenuse ÷ leg" is a trigonometric ratio that is not covered in Chapter 14 *Trigonometry*. The ratio is the *secant* (or *cosecant*) of 45°. This ratio is the reciprocal of the *sine* (or *cosine*) of 45°, ratios that are covered in Chapter 14. You might mention, without emphasis, that the ratio "leg ÷ hypotenuse" will appear later in the course.

Example: 2. To find the length of one leg of isosceles right triangle *KIM*, *divide* the length of the hypotenuse by $\sqrt{2}$, or 1.414.

$$s = 8.9 \div \sqrt{2}$$
$$\approx 8.9 \div 1.414$$
$$\approx 6.29\text{, or about 6.3 in.}$$

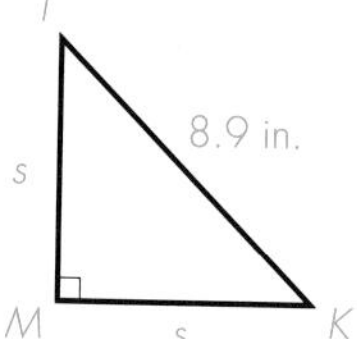

CLASS ACTIVITY

In this activity, use a calculator or the table of squares and square roots on page 562. Answer to the nearest tenth.
Find the length of the hypotenuse of each isosceles right triangle.

1. P, 4.7, ?, 45°, A, L — 6.6

2. 11.7

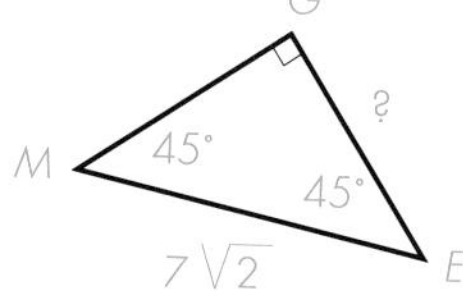

Find the length of one leg of each isosceles right triangle.

3. A, 11.6 ft, ?, R, Y — 8.2

4. G, ?, M, 45°, 45°, $7\sqrt{2}$, E — 7

You can use the Pythagorean Theorem to find the missing length of a side of an isosceles right triangle. Study the examples shown.

Example:

$$a^2 + b^2 = c$$
$$11^2 + 11^2 = c^2$$
$$121 + 121 = c^2$$
$$242 = c^2$$
$$\sqrt{242} = c$$
$$c = \sqrt{242}$$
$$c \approx 15.6 \text{ ft}$$

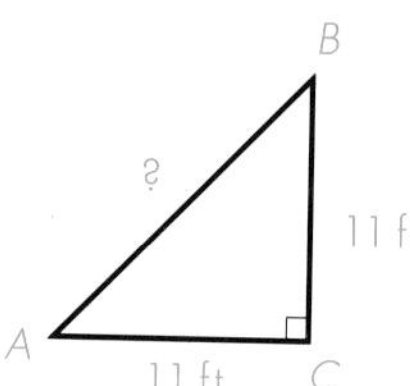

Extra Practice

1. Find the length of the hypotenuse. [$3\sqrt{2}$, or about 4.2]

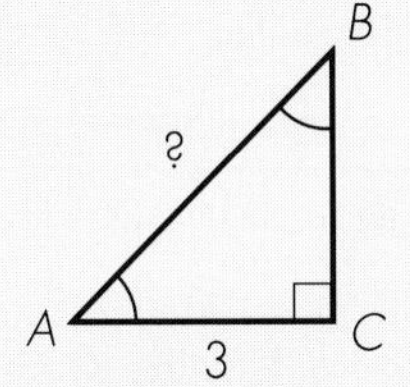

2. Find the length of one leg of the isosceles triangle. [9.9]

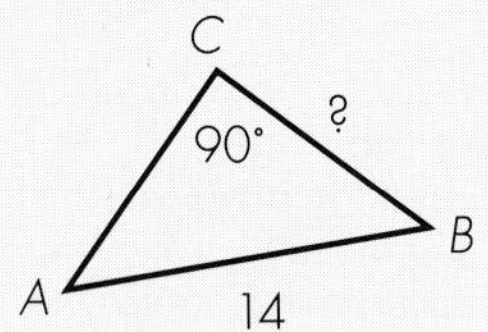

Example:

$$a^2 + b^2 = c^2$$
$$s^2 + s^2 = (5\sqrt{2})^2$$
$$2s^2 = 5^2 \cdot 2$$
$$2s^2 = 50$$
$$s^2 = 25$$
$$s = 5$$

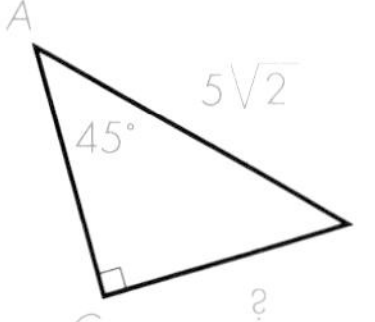

Use the Pythagorean Theorem to find the missing length of a side.

5. 7.1

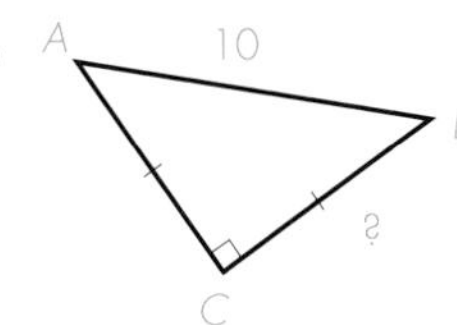

6. 15

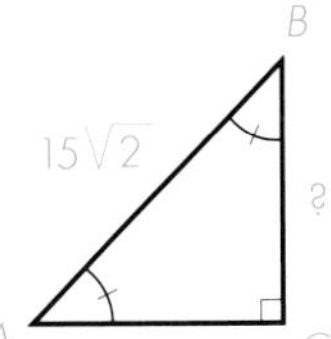

Extra Practice
Use the Pythagorean Theorem to find the missing length of a side.

1.
B
?
10√2
C
A
[10]

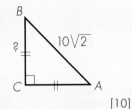

2.

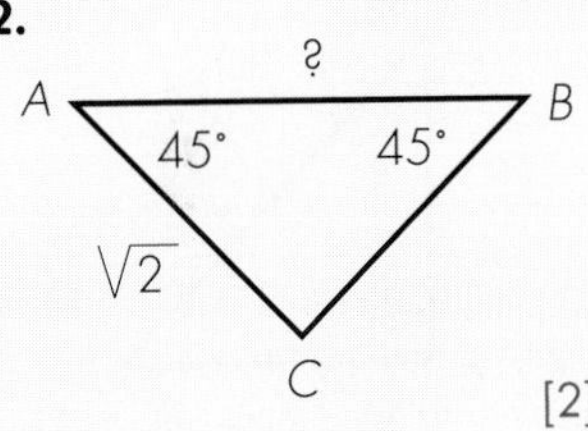

[2]

HOME ACTIVITY

In these exercises, use a calculator for squares and square roots or use the table on page 562. Find the length of the hypotenuse of each isosceles right triangle.

1. 17.0

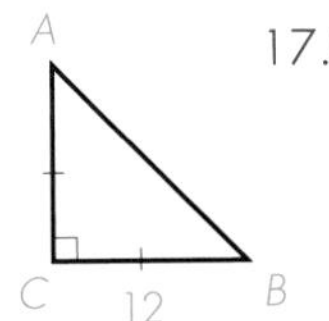

2. 12.7

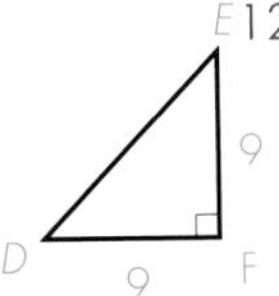

3. 58.0

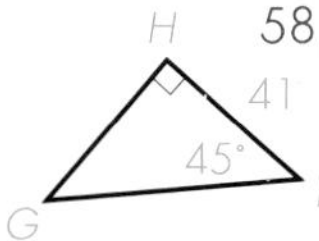

4. 10

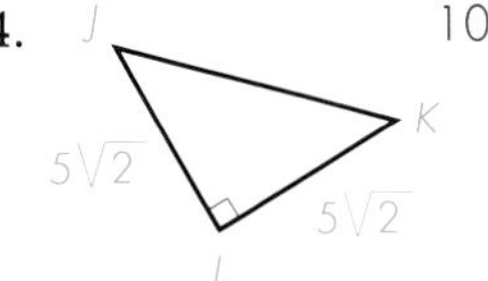

Find the length of one leg of each isosceles right triangle.

5. 12.7

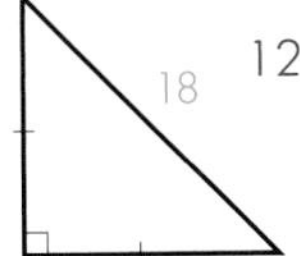

6. 18.5

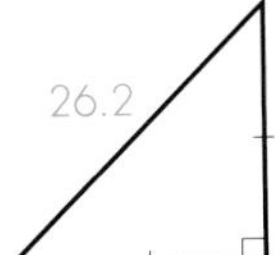

7. 70.7

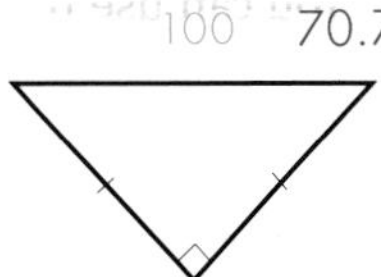

8. 200.0

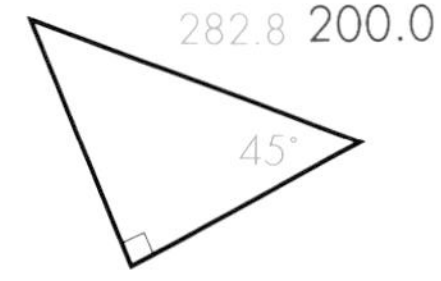

Use the Pythagorean Theorem to find the missing length of a side. Answer to the nearest tenth.

9. 18.4

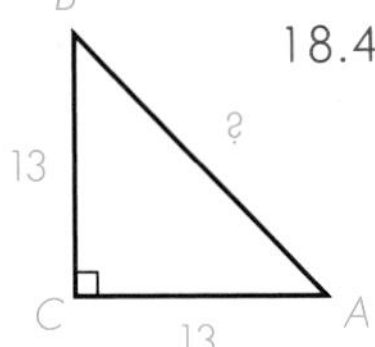

10. 18

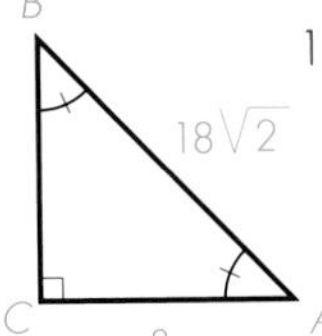

11. 70.7

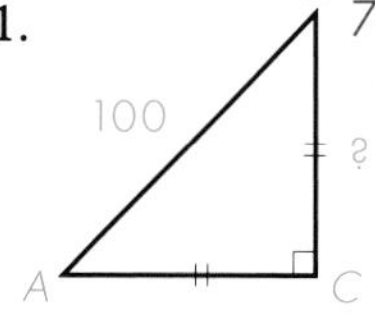

12. 36

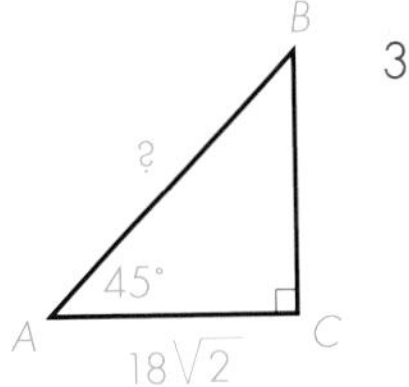

FOLLOW-UP

Assessment

1. Which triangles are *necessarily* isosceles right triangles? Justify your answer.

a.

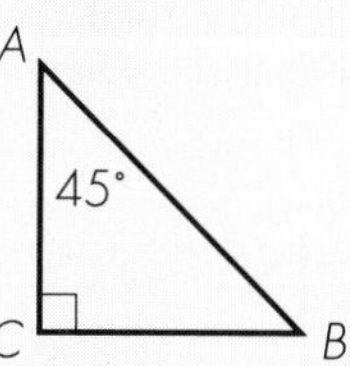

[180° − 90° − 45° = 45°; The right triangle is isosceles by Theorem 5-3-1.]

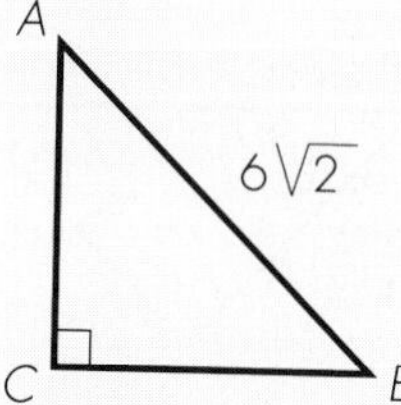

[Not necessarily isosceles; $\triangle ABC$ is isosceles if $AC = BC = 6$; otherwise it is not.

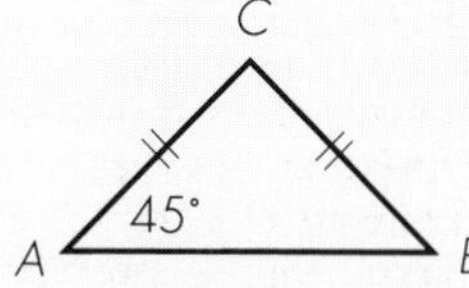

[By Theorem 5-4-1, $m\angle B = 45°$; $180° - 45° - 45° = 90°$. Thus, the isosceles triangle is also a right triangle.]

2. Find the length of the missing side. [about 11.3]

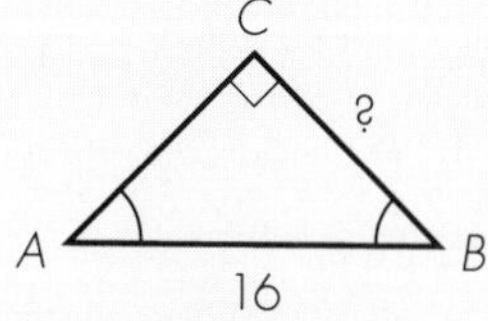

Reteaching

Students who have had difficulty with this section may benefit from Reteaching Activity 7.2.

Enrichment

For students who have mastered the material in this section, you may wish to assign Enrichment Activity 7.2.

Copy and complete the table below. Answer to the nearest tenth.

	Length of one leg of an isosceles right triangle	Length of hypotenuse
13.	1.0	$1\sqrt{2}$, or 1.4
14.	2	$2\sqrt{2}$, or 2.8
15.	3	$3\sqrt{2}$, or 4.2
16.	5.0	$5\sqrt{2}$, or 7.071
17.	10.0	$10\sqrt{2}$, or 14.1

18. Write a Logo procedure named ISOSRTTRI to draw an isosceles right triangle. Use variables in the procedure so that the lengths of the sides can be different each time the procedure is called. Answers may vary.

19. The roof of a shed forms a right angle at its peak. About how many feet long is the beam that runs from point A to point B? About 17 ft

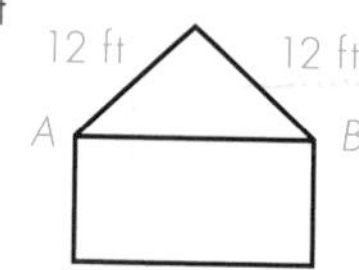

20. Alice jogged from A to C and then on to B. Returning she jogged from B to A. About how much shorter was the return trip? 58.6 ft

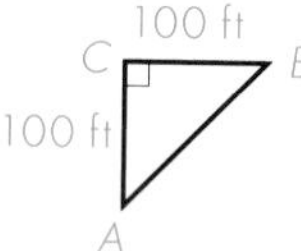

CRITICAL THINKING

In the basement of a building, two water pipes are parallel and 25 ft apart. If a furnace were not in the way, Mr. Karpov could run a connecting pipe from point A on water pipe 1 to the point on water pipe 2 that is closest to A. Instead, he must use a 45° elbow pipe at A to make the connection elsewhere on pipe 2. To the nearest tenth of a foot how long will the connection pipe be? About 35.4 ft

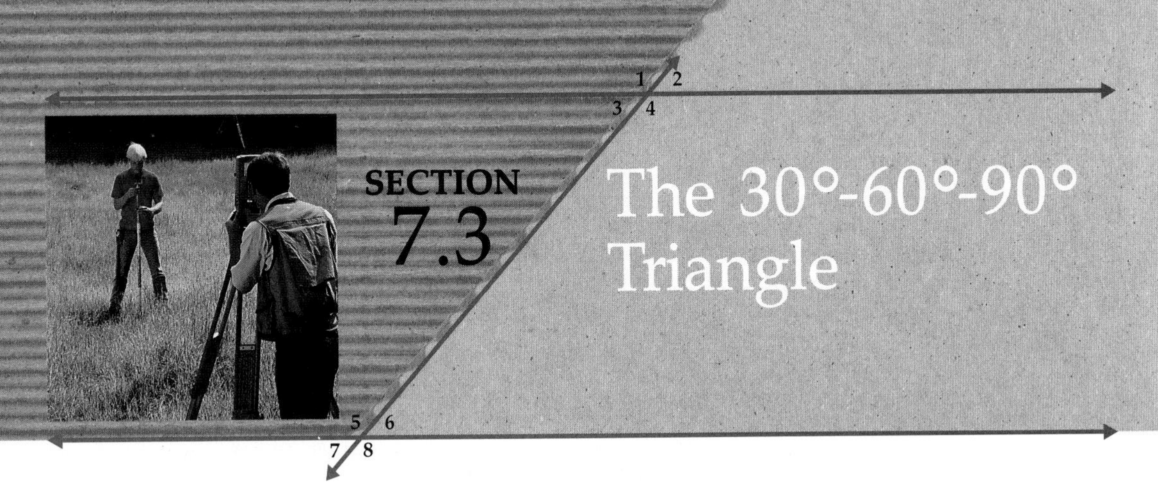

SECTION 7.3

The 30°-60°-90° Triangle

After completing this section, you should understand
- ▲ theorems about 30°-60° right triangles.
- ▲ how to solve problems using 30°-60° right triangles.

There are two special triangles that you will often encounter in geometry and later in trigonometry. You already know about the first such triangle, the isosceles right triangle. The second such special triangle is the *30°–60° right triangle.* A few 30°-60° right triangles are shown below.

LEG OPPOSITE THE 30° ANGLE

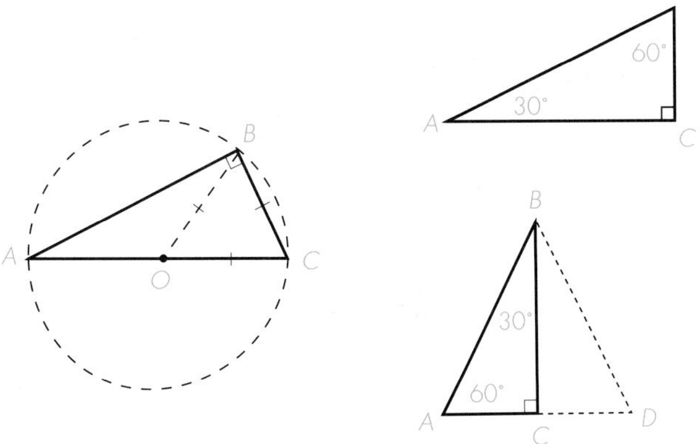

For most right triangles, it is necessary to use trigonometry to find how the sides of a triangle are related to each other and to the angles. For 30°-60° right triangles, you do not need trigonometry to find these relationships.

SECTION 7.3

Resources
Reteaching 7.3
Enrichment 7.3

Objectives
After completing this section, the student should understand
- ▲ theorems about 30°-60° right triangles.
- ▲ how to solve problems using 30°-60° right triangles.

Materials
centimeter ruler
protractor
calculator

Vocabulary
deductive reasoning
equiangular
equilateral
indirect measurement
inductive reasoning
30°-60° right triangle

GETTING STARTED

Warm-Up Activity
This paper-folding activity is an informal way of generating a 30°-60° right triangle. Have students draw or construct an equilateral triangle and label *ABD*. (Use equilateral triangle *ABD* near the bottom of page 259 as a model.) Next, have them cut out the triangle and fold it in such a way that points *A* and *D* coincide, forming two new congruent triangles *ACB* and *DCB* (see model). By direct measurement, students can confirm that the two triangles are 30°-60° right triangles.

DISCOVERY ACTIVITY

1. On a separate piece of paper, draw line segment $\overline{AC}$ 4 cm long.
2. Use a protractor and a ruler to draw $\angle ACX$ equal to 90°.
3. Use your protractor and a ruler to draw $\angle CAY$ equal to 30°. Use B to label the intersection of $\overrightarrow{CX}$ and $\overrightarrow{AY}$.
4. Measure $\overline{AB}$ to the nearest tenth of a centimeter. 2.3 cm
5. Measure $\overline{CB}$ to the nearest tenth of centimeter. 4.6 cm
6. Find the ratio $CB : AB$. 1:2 or $\frac{1}{2}$
7. Repeat steps 1–6 using a 6-cm line segment instead of a 4-cm segment. See margin
8. Based on your experiments, state a generalization that appears to be true. See margin

It appears as though the shorter leg of a 30°-60° right triangle has half the length of the hypotenuse. This is a conclusion based upon *inductive reasoning,* that is, reasoning using the results of a number of experiments or observations.

It is possible to show that the conclusion is correct using *deductive reasoning,* that is, reasoning based upon undefined terms, defined terms, postulates, and previously justified theorems.

The following proof is based upon the fact that an equilateral triangle can be divided into two 30°-60° right triangles.

Equilateral triangle ABC has three equal sides, by definition of *equilateral.* It is also *equiangular,* so that the measure of each angle is 60°. (The proof that an equilateral triangle is also equiangular is asked for in the HOME ACTIVITY.)

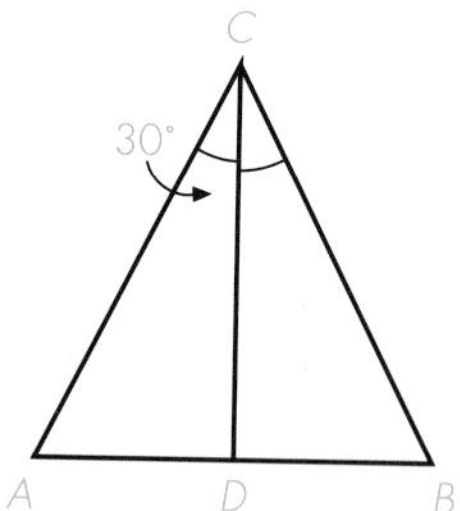

TEACHING COMMENTS

Additional Answers

7. In Step 4, AB is about 6.8 cm. In step 5, CB is about 3.4 cm. In step 6, the ratio is 1:2, or $\frac{1}{2}$.

8. It appears that in a 30°-60° right triangle, the length of the leg opposite the 30° angle is half the length of the hypotenuse.

In the figure, $\overline{CD}$ bisects angle C. Therefore, $m\angle ACD = 30°$ (half of 60°). You can show that $\triangle ACD \cong \triangle BCD$ by SAS. Thus $m\angle ADC$ and $m\angle BDC$ are each 90° by CPCF. So, $\triangle ACD$ is a 30°-60° right triangle.

Also, $AD = BD$ by CPCF. From this you can see that $AD = \frac{1}{2}AB$

Since $AB = AC$, you can use substitution to conclude that $AD = \frac{1}{2}AC$

THEOREM 7-3-1 **If a triangle is a 30°-60° right triangle, then the length of the leg opposite the 30° angle is half the length of the hypotenuse.**

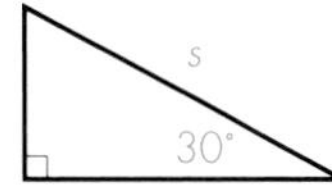

Examples:

1. In $\triangle XYZ$, $\overline{YZ}$ is opposite the 30° angle. $YZ = x = \frac{1}{2}(20)$, or 10

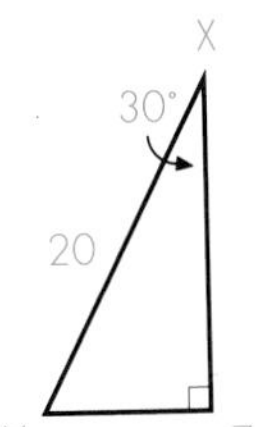

2. In $\triangle NOP$, $\overline{OP}$ is opposite the 30° angle. $OP = 24 = \frac{1}{2}p$, so $p = 2 \cdot 24 = 48$

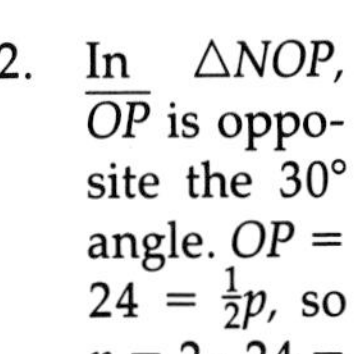
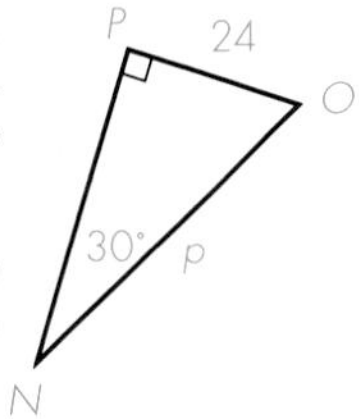

CLASS ACTIVITY

Find the missing lengths.

1. 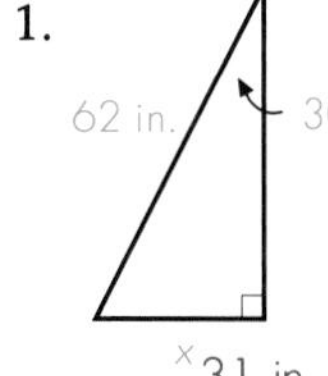

2. 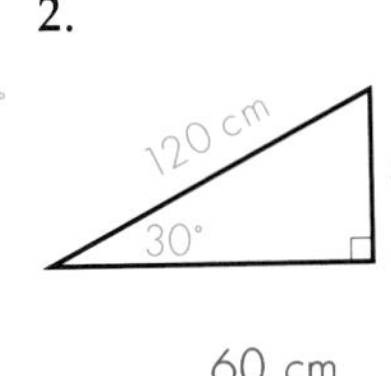

3. 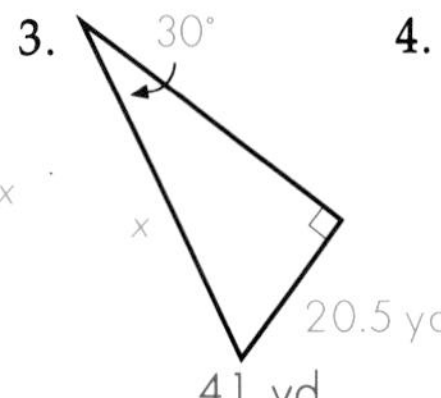

4. 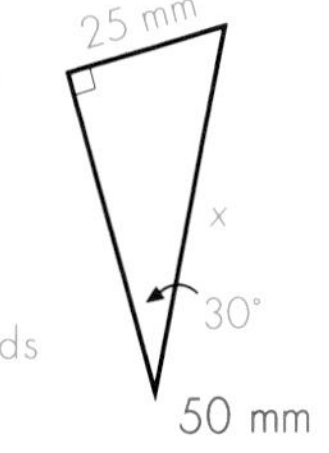

LEG OPPOSITE THE 60° ANGLE

You know that the length of the leg opposite the 30° angle of a 30°-60° right triangle is half the length of the hypotenuse. You can now find out how the leg opposite the 60° angle is related to the hypotenuse. Use the Pythagorean Theorem and algebra. In right triangle ABC, $m\angle A = 30°$ and $m\angle B = 60°$. So, $a = \frac{1}{2}c$, or $c = 2a$. Substitute $2a$ for c in the Pythagorean Theorem.

$$a^2 + b^2 = c^2$$
$$a^2 + b^2 = (2a)^2$$
$$a^2 + b^2 = 4a^2$$

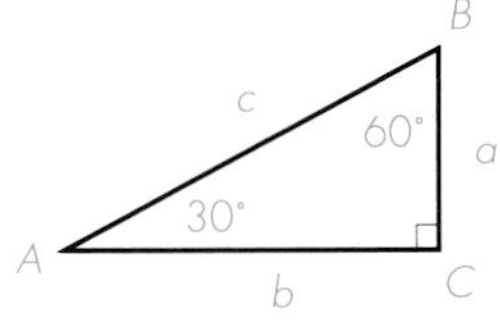

Using the Page

In the proof of Theorem 7-3-1, $\overline{CD}$ is constructed as an angle bisector. Some students will conclude that the segment "automatically" bisects $\overline{AB}$ and is perpendicular to it. Inform students that when a line is constructed, it is geometrically impossible to specify more than two conditions on it. Attempting to impose more than two conditions on a line is an example of *overdefinition*. Here, the two conditions are that the line include point C and that it bisect ∠C. You might point out that an alternative proof of Theorem 7-3-1 is possible that is based upon some other set of two conditions on line segment $\overline{CD}$. For example, the line segment could have been constructed as a bisector of $\overline{AB}$ instead of as a bisector of ∠C.

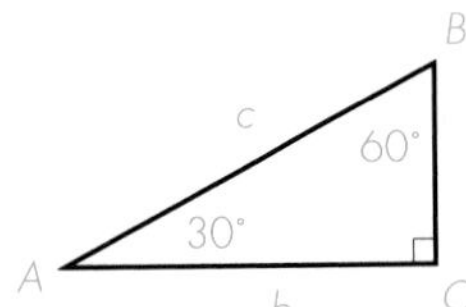

Next, solve for b.

$$a^2 + b^2 - a^2 = 4a^2 - a^2$$
$$b^2 = 3a^2$$

Take the square root of each side of this equation.

$$b = a\sqrt{3}$$

Now substitute $\frac{1}{2}c$ for a.

$$b = \left(\tfrac{1}{2}c\right)\sqrt{3}, \text{ or } c\,\tfrac{\sqrt{3}}{2}.$$

NOTEBOOK

THEOREM 7-3-2 **If a triangle is a 30°-60° right triangle, then the length of the leg opposite the 60° angle is $\frac{\sqrt{3}}{2}$ times the length of the hypotenuse.**

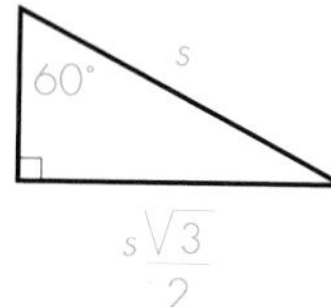

Examples: Find the missing length.

1. $b = c\,\frac{\sqrt{3}}{2}$
   $= (20)\frac{\sqrt{3}}{2}$
   $= 10\sqrt{3}$
   $\approx 10 \cdot 1.732$
   $= 17.32$ in.

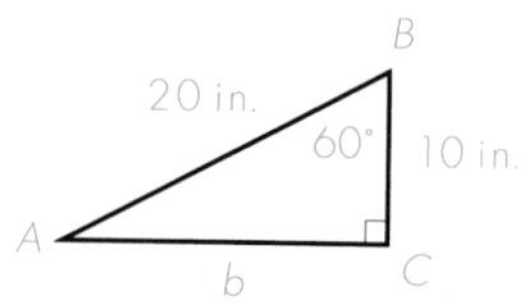

2. 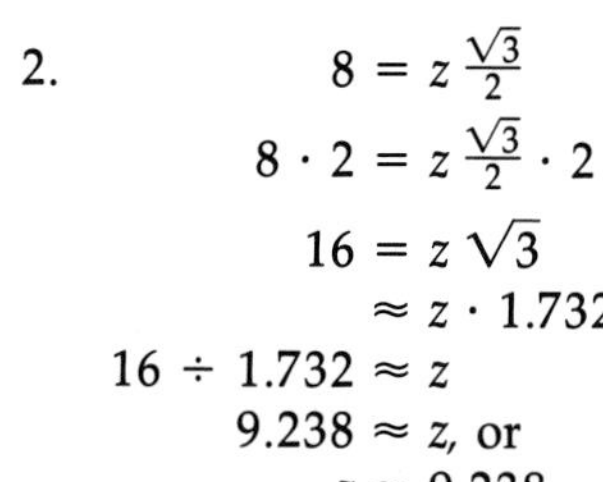

$$8 = z\,\frac{\sqrt{3}}{2}$$
$$8 \cdot 2 = z\,\frac{\sqrt{3}}{2} \cdot 2$$
$$16 = z\sqrt{3}$$
$$\approx z \cdot 1.732$$
$$16 \div 1.732 \approx z$$
$$9.238 \approx z, \text{ or}$$
$$z \approx 9.238$$

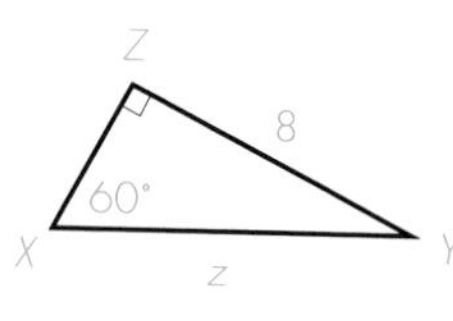

Extra Practice
Find the value of x. Answer to the nearest tenth. Use 1.732 for $\sqrt{3}$.

1.

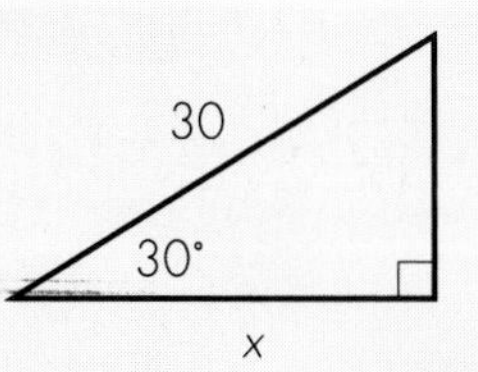

[26.0]

2.

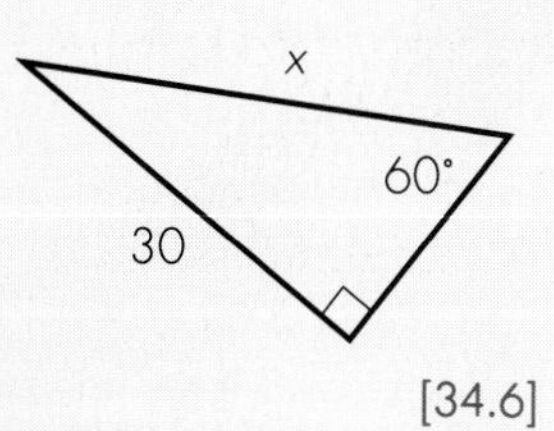

[34.6]

CLASS ACTIVITY

Find the value of x. Answer to the nearest tenth. Use 1.732 for $\sqrt{3}$.

1. 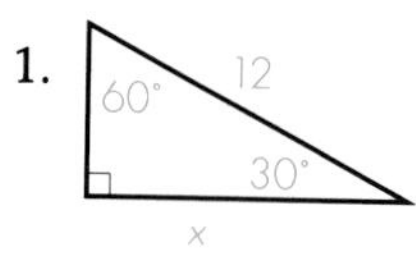

10.4

2. 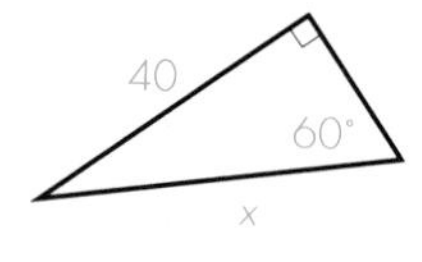

46.2

3. 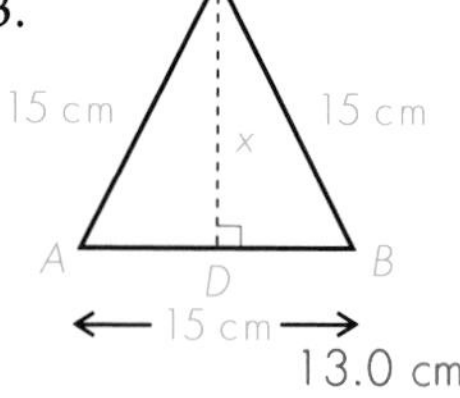

13.0 cm

INDIRECT MEASUREMENT

The relationships among the three sides of a 30°-60° right triangle depend on the shape of the triangle but not on its size. They are illustrated in the triangles below.

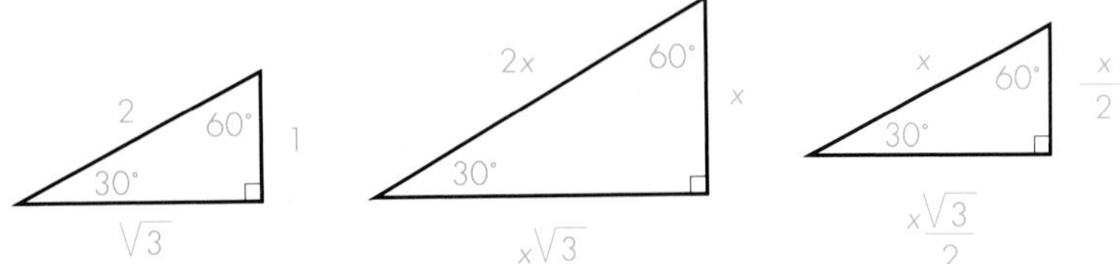

These relationships are the same for all such triangles, large and small. For this reason, you can use 30°-60° right triangles to measure distances indirectly.

Example: Alicia and Susan were surveying a property and needed to find the distance between an oak tree and a large rock on opposite sides of a pond. They paced off several feet along their side of the pond, forming a right angle at *C*. Next they found a point *B* such that $m\angle ABC = 60°$. The distance *BC* was found to measure 116 ft. They found *AB* and *AC* as follows.

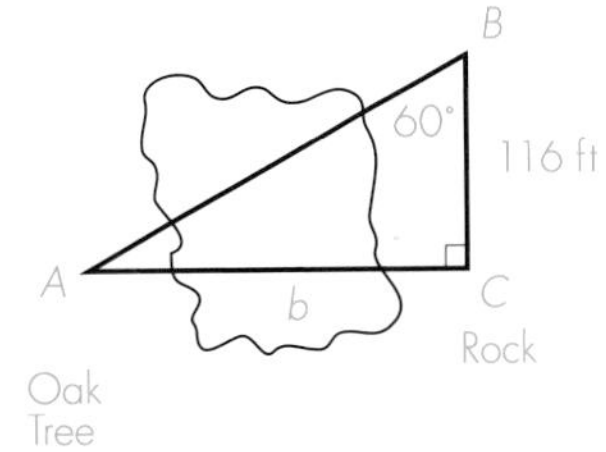

$\frac{1}{2}AB = 116$, so $AB = 232$

$b = 232 \cdot \frac{\sqrt{3}}{2}$ or $116\sqrt{3}$

$b \approx 116\ (1.732)$, or about 201 ft

CLASS ACTIVITY

Find the distance across the pond. Answer to the nearest whole unit. Use 1.732 for $\sqrt{3}$.

1. 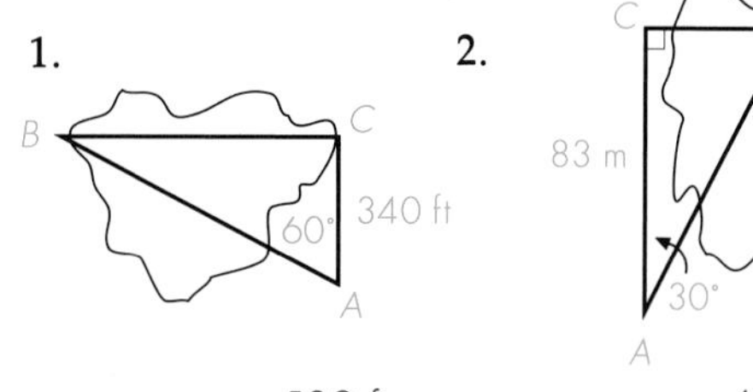

589 ft

2. 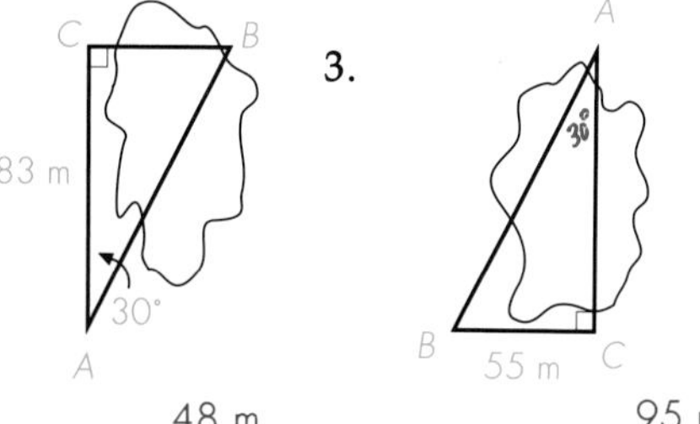

48 m

3. 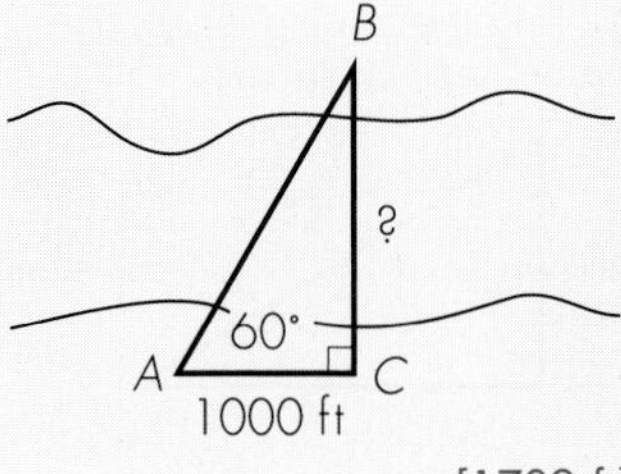

95 m

Using the Page

An alternate solution to the Example is to use the "basic" 30°-60° right triangle with sides of 1, $\sqrt{3}$, and 2 (see the left figure at the top of the page) and to solve the proportion $b:116 = \sqrt{3}:1$.

Extra Practice

Find the distance across the river from *C* to *B*. Answer to the nearest whole unit. Use 1.732 for $\sqrt{3}$.

1.

[1732 ft]

2. 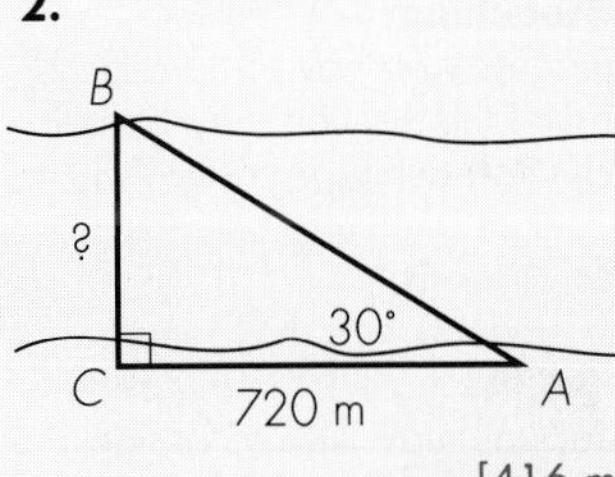

[416 m]

HOME ACTIVITY

Find the missing lengths. Answer to the nearest tenth. Use 1.732 for $\sqrt{3}$.

1. 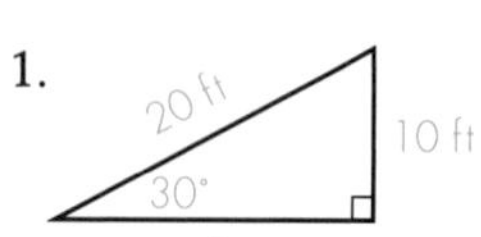

2. 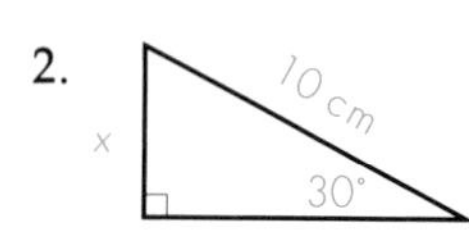

3. 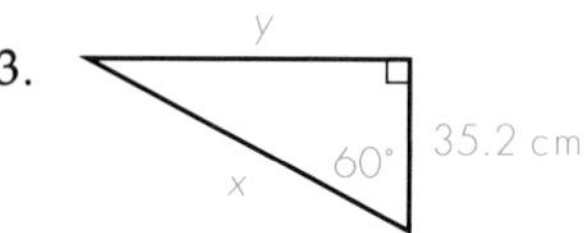

x = 17.3 ft   x = 5.0 cm   x = 70.4 cm, y = 61.0 cm

Find the missing measure. Answer to the nearest whole unit. Use 1.732 for $\sqrt{3}$.

4. 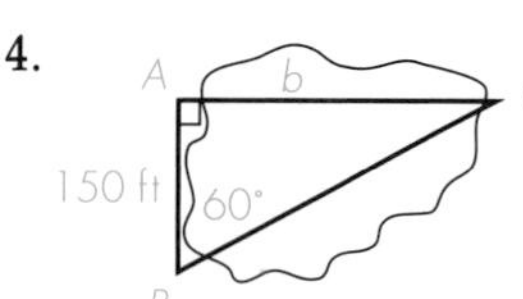

5. 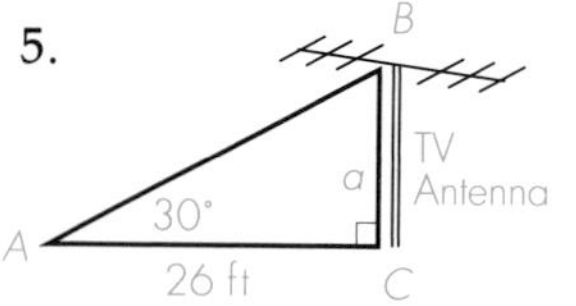

6. 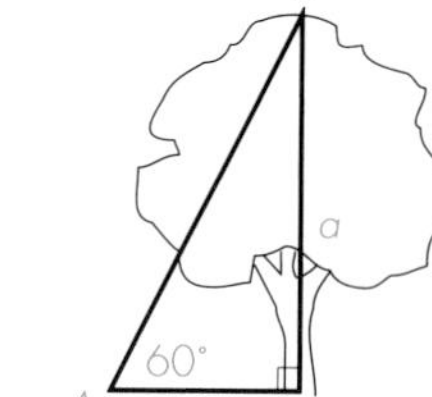

b = 260 ft   a = 15 ft   a = 21 ft

7. Write a Logo procedure named RTTRI3060 to draw a 30°-60° right triangle. Use variables in the procedure so that the lengths of the sides can be different each time the procedure is called. Answers may vary.

8. In the figure, $\overline{XY} \parallel \overline{UV}$. Copy the figure. Then bisect angles *ABY* and *BAV*. Call the intersection point of the bisectors *C*. First, find *x* and *y*. Then find m∠*ABC*. Explain how you arrived at your answers. See margin.

9. Prove that if a triangle is equilateral then it is also equiangular. See margin.

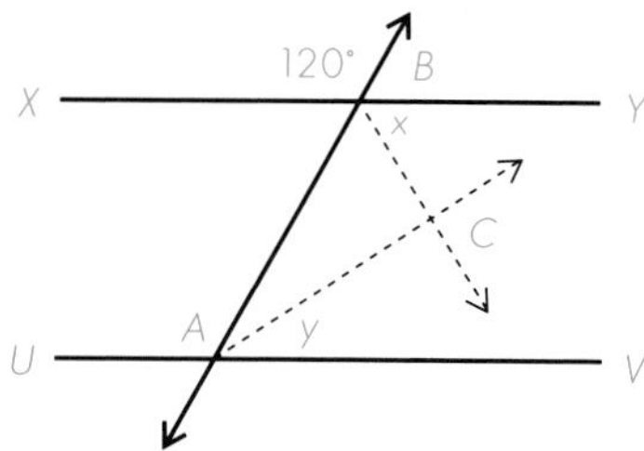

CRITICAL THINKING

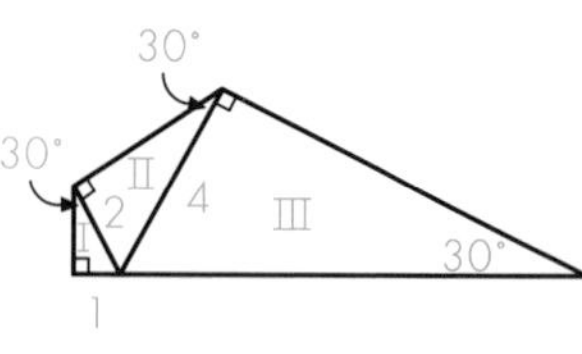

10. Copy the figure at the left and continue the pattern until you have drawn a total of five connected 30°-60° right triangles. Find the length of the last hypotenuse to be drawn when the total number of triangles is:

1. 1 — 2
2. 2 — 4
3. 3 — 8
4. 4 — 16
5. 5 — 32
6. 10 — 1024
7. *n* — 2^n

Additional Answers

8. m∠*ACB* = 90°; x = 8 cm; y = 13.9 cm. m∠*ABY* = 120° (vertical angles are equal), and m∠*BAV* = 60° (interior angles on the same side of transversal are supplementary). Therefore, the two angle bisectors meet to form a 30°-60° right triangle. To find x and y, use the theorems about 30°-60° triangles.

9. Suppose that △*ABC* is equilateral. Since an equilateral triangle is also isosceles, m∠*A* = m∠*B*. Similarly, m∠*B* = m∠*C*. Thus, all three angles have the same measure.

FOLLOW-UP

Assessment

1. If the shortest side of a 30°-60° right triangle is 1 in. long, what are the lengths of the other two sides? [$\sqrt{3}$ in., or about 1.732 in.; 2 in.]

2. Find the missing length to the nearest whole unit.

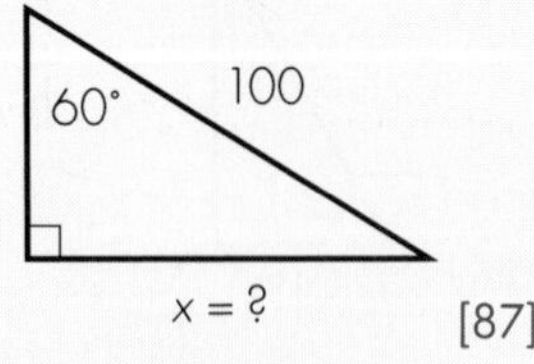

[87]

Reteaching

Students who have had difficulty with this section may benefit from Reteaching Activity 7.3.

Enrichment

For students who have mastered the material in this section, you may wish to assign Enrichment Activity 7.3.

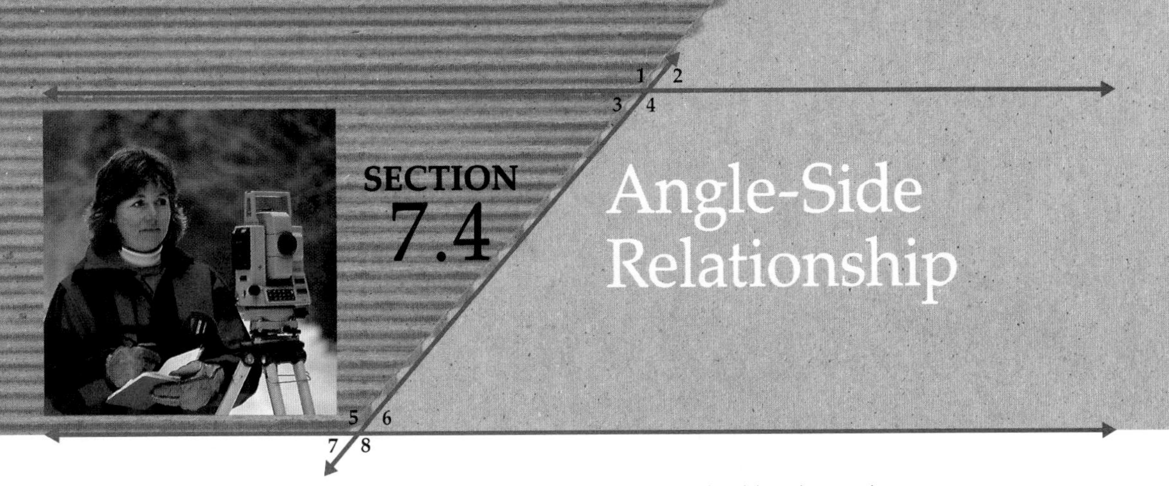

SECTION 7.4 Angle-Side Relationship

After completing this section, you should understand
▲ that the side opposite the greatest angle of a triangle is the longest side.
▲ that the angle opposite the longest side of a triangle is the greatest angle.
▲ applications of triangle inequalities.

Surveyors use triangles and indirect measurement to help them establish the boundaries of large pieces of land. Very often the sides of the a triangle are not of equal length nor the angles of equal measure. In such cases, how are the sides related to the angles? In this section, you will learn one way in which they are related.

UNEQUAL ANGLES IN A TRIANGLE

In Section 3.2, you learned that the sum of two sides of a triangle is greater than the third side. In this section you will learn about another inequality involving triangles.

DISCOVERY ACTIVITY

1. On a separate piece of paper, draw a triangle such as the one at the right in which $m\angle C > m\angle B$.
2. Cut out the triangle and fold it along PM so that vertices B and C coincide as shown in the second diagram. Notice that M is the midpoint of $\overline{BC}$.
3. How is AC related to the sum of AP and PC?

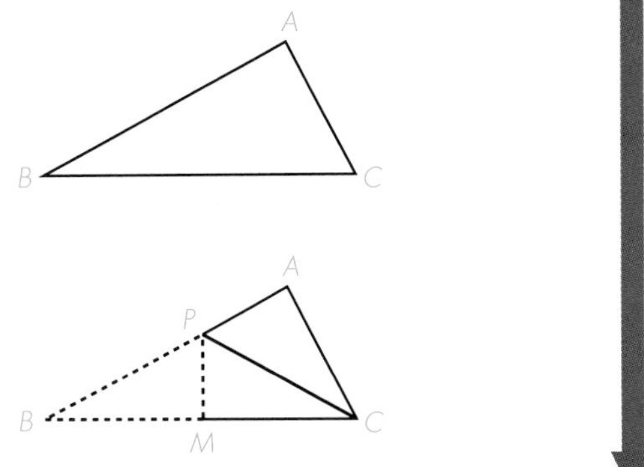

SECTION 7.4

Resources
Reteaching 7.4
Enrichment 7.4
Transparency Master 7-1

Objectives
After completing this section, the student should understand
▲ that the side opposite the greatest angle of a triangle is the longest side.
▲ that the angle opposite the longest side of a triangle is the greatest angle.
▲ applications of triangle inequalities.

Materials
scissors

GETTING STARTED

Quiz: Prerequisite Skills

1. Write three inequalities that involve the sides of $\triangle ABC$. [$AB + AC > BC$, $AB + BC > AC$, $AC + BC > AB$]

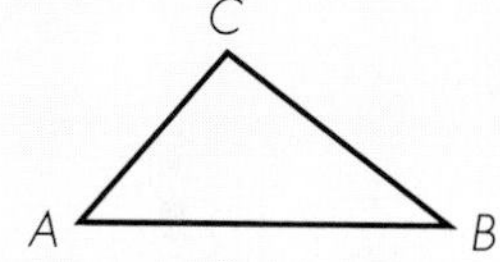

2. How is angle measure w related to angle measures x, y, and z? [$w + x = 180°$; $w = y + z$]

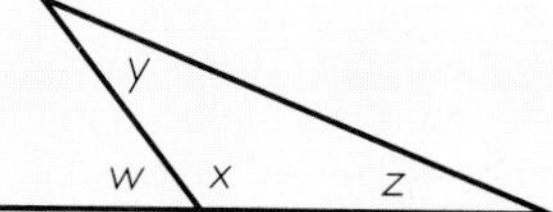

(Hint: Review Theorem 3-2-1.) $AP + PC > AC$

4. How is the sum $AP + PC$ related to AB? $AP + PC = AB$

5. What can you conclude about the relationship between AB and AC? $AB > AC$

From the Discovery Activity, it appears that you can make the following conjecture: If in a triangle, one angle is greater than the other, then the side opposite the greater angle is larger than the side opposite the other angle.

The Discovery Activity also suggests a method of proving the conjecture.

Given: $\triangle ABC$ with
$m\angle C > m\angle B$

Justify: $AB > AC$

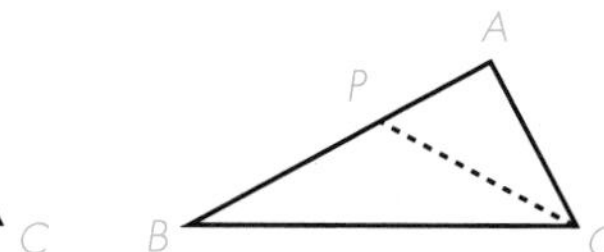

In $\triangle ABC$, copy $\angle B$ with C as the vertex and $\overline{BC}$ as one of its two sides. The other side of the new angle intersects $\overline{AB}$ in point P. Fill in the reasons in the proof.

Steps	Reasons
1. $PC = PB$	1. If a triangle has two equal angles, then the sides opposite the equal angles are equal.
2. $AP + PC > AC$	2. The sum of the lengths of two sides of a triangle is greater than the length of the third side.
3. $AP + PB > AC$, or $AB > AC$	3. Substitution

In the above proof, it was assumed that point P lies in the interior of $\angle BCA$.

THEOREM 7-4-1 **If a triangle has one angle greater than another angle, then the side opposite the greater angle is longer than the side opposite the other angle.**

TEACHING COMMENTS

Using the Page
In the proof of Theorem 7-4-1, it is noted that "it was assumed that point P lies in the interior of $\angle BCA$." In a more rigorous course, this fact would follow as a corollary of a theorem such as the following:

If an angle is convex, then any chord of the angle, except for the endpoints of the chord, is inside the angle.

Here a *chord of an angle* is a line segment that has its endpoints on different sides of the angle.

Transparency Master 7-1 can be used to demonstrate a paper-folding activity that justifies Theorem 7-4-1.

Example: In isosceles triangle ABC, $m\angle ABC = 30°$. How are the lengths of the three sides related?

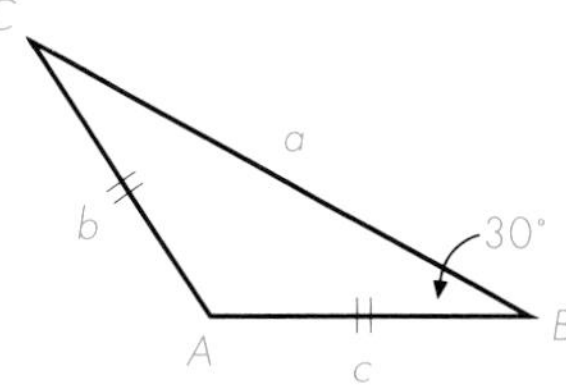

Since $\triangle ABC$ is isosceles, $b = c$. Also, $m\angle B = m\angle C$, since *in an isosceles triangle the angles opposite the equal sides are equal* (Theorem 5-4-1). So, $m\angle C = 30°$. Since the sum of the interior angles of the triangle is 180°, $m\angle A = 120°$.

Thus, $m\angle A > m\angle B$. From Theorem 7-4-1 it follows that $a > b$. Finally, since $b = c$, $a > c$.

CLASS ACTIVITY

For each triangle ABC, write $a < b$, $a = b$, or $a > b$ if sufficient information is provided to justify one of these conclusions. Otherwise write "insufficient information." (The triangles are *not* necessarily drawn to show their proper shapes.)

1. C a > b; angles 70° at A, 50° at B

2. 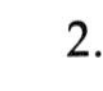 

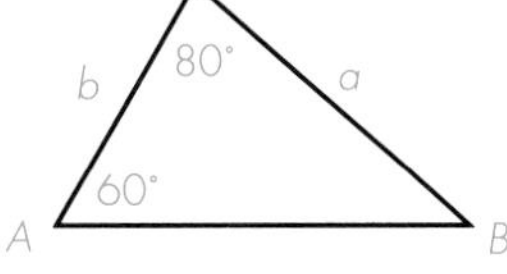

3. 

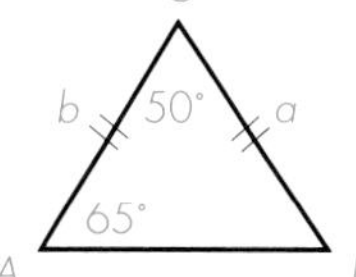

4. C a < b; angle 44° at A

5. C a < b
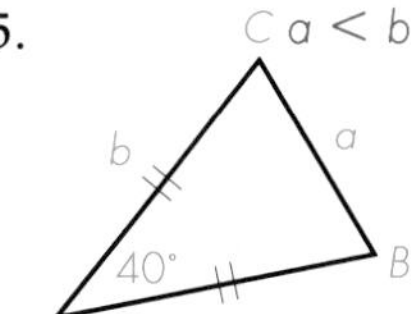

6. C Insuf. Infor.
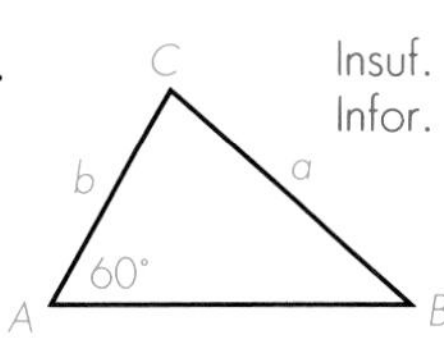

7. Use Theorem 7-4-1 to show that in a right triangle, the hypotenuse must always be the longest side. (Hint: Can either of the non-right angles be equal to 90°? greater than 90°? Why?)

UNEQUAL SIDES IN A TRIANGLE

In Chapter 5 you learned that the converse of a true theorem is not necessarily true. Here is Theorem 7-4-1 and its converse.

Theorem: If a triangle has one angle greater than another angle, then the side opposite the greater angle is longer than the side opposite the other angle.

Extra Practice

For each triangle ABC, write $a < b$, $a = b$, $a > b$, or "insufficient information."

1.

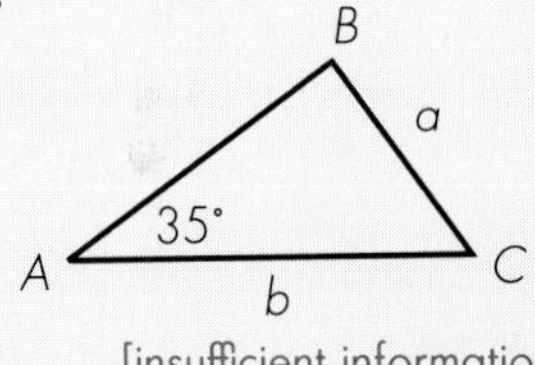

[insufficient information]

2.

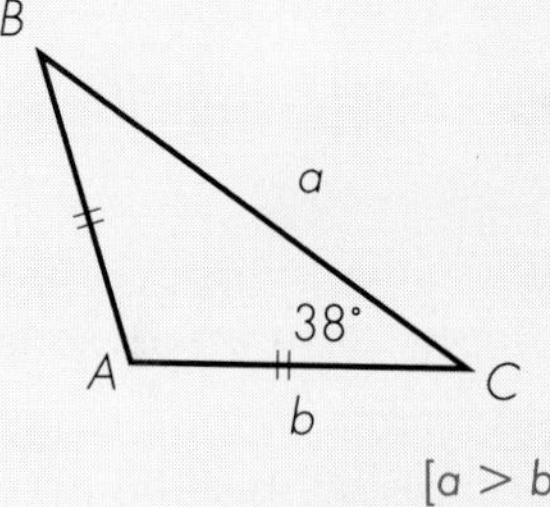

[$a > b$]

Additional Answers

7. Let $\angle A$ and $\angle B$ be the two non-right angles. Then in $m\angle A + m\angle B + 90° = 180°$. Thus $m\angle A + m\angle B = 90°$. Since $m\angle A$ and $m\angle B$ are each positive, $m\angle A < 90$ and $m\angle B < 90$. By Theorem 7-4-1, the hypotenuse is the longest side.

Converse: If a triangle has one side longer than another side, then the angle opposite the longer side is greater than the angle opposite the other side.

Is the converse of the theorem true? The next Discovery Activity will help you answer this question.

DISCOVERY ACTIVITY

1. On a separate piece of paper, draw a scalene triangle such as the one at the right in which $AC > AB$.

2. Cut out the triangle and fold the left part of the triangle in such a way that side $\overline{AB}$ coincides with a portion of side $\overline{AC}$.

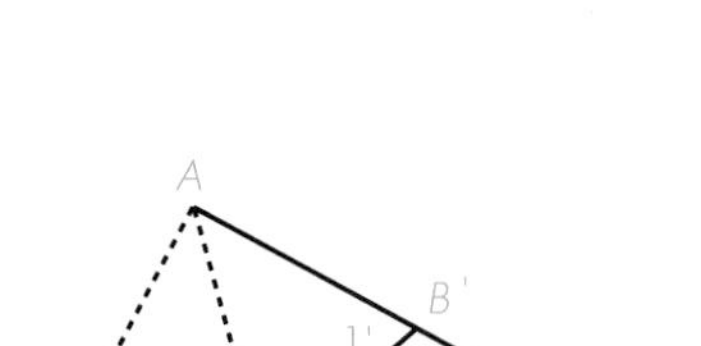

3. Unfold the triangle and inspect the fold along line segment $\overline{AP}$. What does this line segment bisect? $\angle A$

4. What two triangles are congruent? Why? $\triangle BAP \cong \triangle B'AP$; SAS

5. How is $\angle 1$ related to $\angle 1'$? How do you know? $m\angle 1 = m\angle 1'$; CPCF

6. How is $\angle 1'$ related to $\angle 2$? How do you know? (Hint: Review Theorem 3-5-1) $m\angle 1' > m\angle 2$; An exterior angle of a triangle is equal to the sum of the two remote interior angles. Thus, it is greater than either.

7. What can you conclude about the relationship between $\angle 1$ and $\angle 2$? $m\angle 1 > m\angle 2$

From the above Discovery Activity, it appears that the converse of Theorem 7-4-1 *is* true.

THEOREM 7-4-2 **If a triangle has one side longer than another side, then the angle opposite the longer side is greater than the angle opposite the other side.**

The proof of Theorem 7-4-2 is not given.

Using the Page
The proof of Theorem 7-4-2 is not given. However, like the proof of Theorem 7-4-1, such a proof would be based upon the approach taken in the Discovery Activity that precedes it. The main difference between the Activity and a formal proof is in the way in which line segment $\overline{AP}$ is created. In the Discovery Activity, $\overline{AP}$ becomes the bisector of $\angle A$ indirectly, as the result of a paper-folding process. In the proof, on the other hand, $\overline{AP}$ would be formed directly as the result of the construction of the bisector of $\angle A$.

CLASS ACTIVITY

For each triangle *ABC*, write m∠*A* < m∠*B*, m∠*A* = m∠*B*, or m∠*A* > m∠*B* if sufficient information is provided to justify one of these conclusions. Otherwise write "insufficient information." (The triangles are *not* necessarily drawn to show their proper shapes.)

1.

m∠A > m∠B

2.

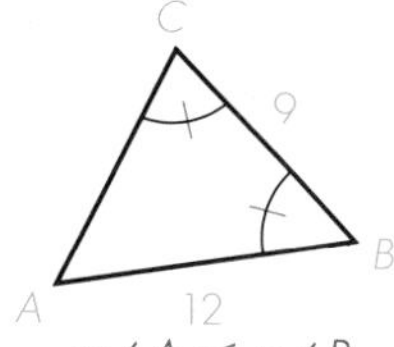

m∠A < m∠B

3.

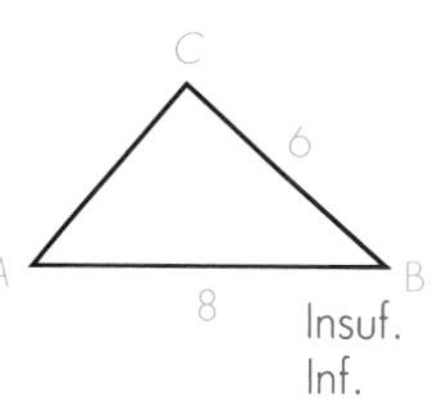

Insuf. Inf.

4. List the angles of △*FGH* from least to greatest.
∠G, ∠F, ∠H

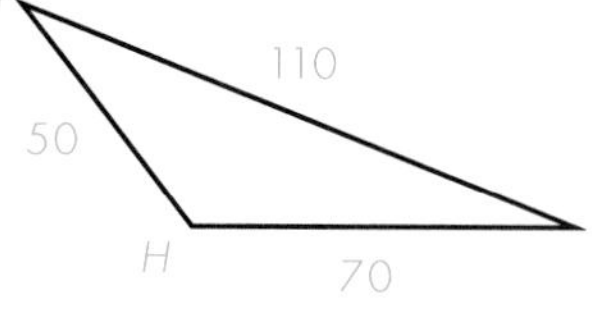

5. List the sides of △*RST* from shortest to longest.
$\overline{RT}$, $\overline{ST}$, $\overline{RS}$

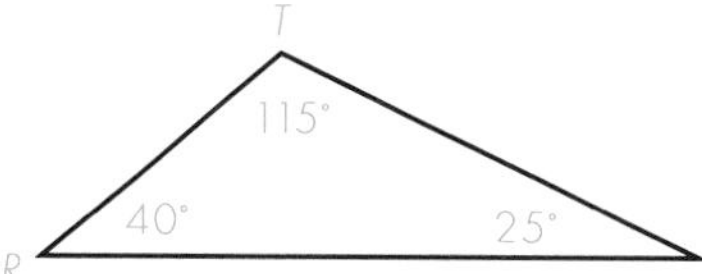

HOME ACTIVITY

For each triangle, write *a* < *b*, *a* = *b*, or *a* > *b* if sufficient information is provided to justify one of these conclusions. Otherwise write "insufficient information." (The triangles are *not* necessarily drawn to show their proper shapes.)

1. a > b

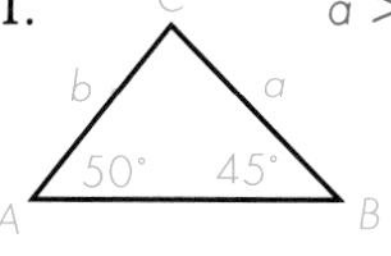

2. a < b

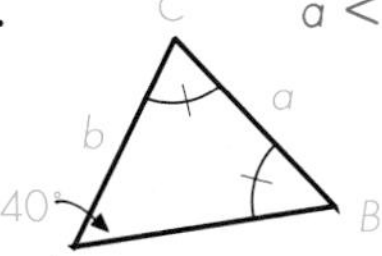

3. Insuf. Inf.

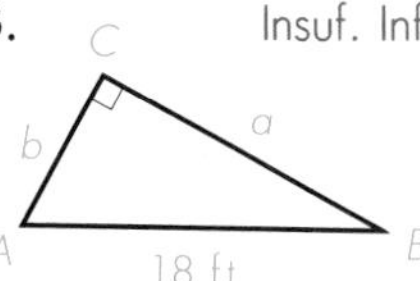

4. a < b

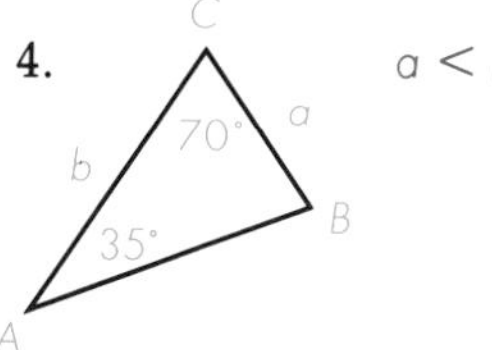

For each triangle, write m∠*A* < m∠*B*, m∠*A* = m∠*B*, or m∠*A* > m∠*B* if sufficient information is provided to justify one of these conclusions. Otherwise write "insufficient information." (The triangles are *not* necessarily drawn to show their proper shapes.)

5. Insuf. Inf.

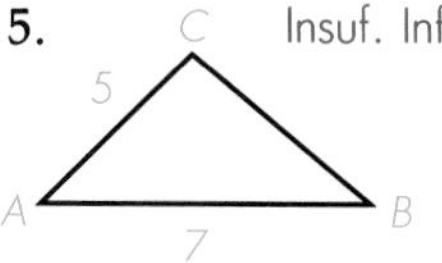

6. m∠A = m∠B

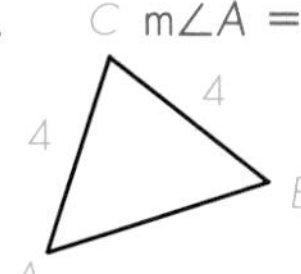

7. m∠A > m∠B

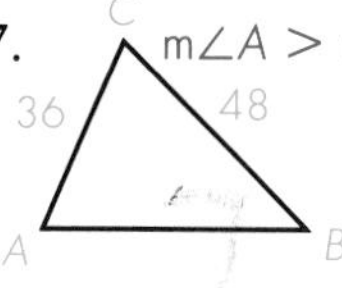

8. m∠A > m∠B

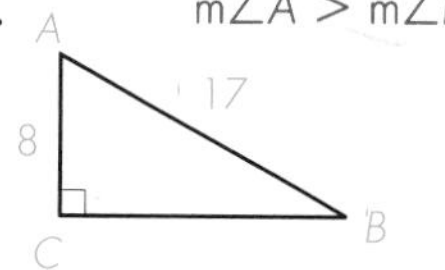

Extra Practice

1. Write m∠A < m∠B, m∠A = m∠B, m∠A > m∠B, or "insufficient information." [m∠A < m∠B]

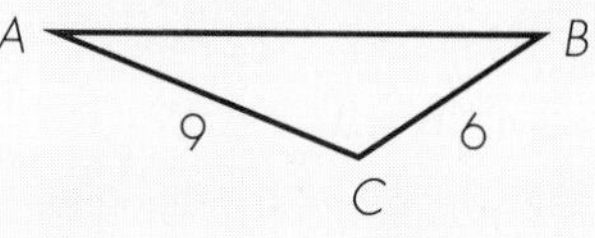

2. List the angles of △*RST* from least to greatest. [∠T, ∠R, ∠S]

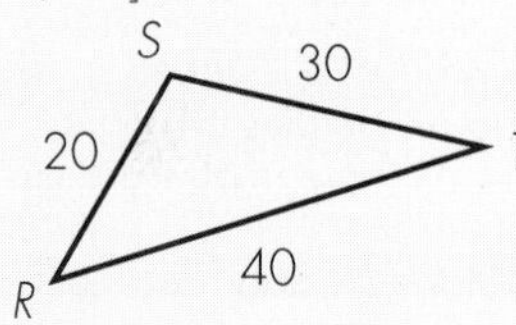

Additional Answers

11. Since m$\angle 2 <$ m$\angle 1$, $PM < PO$, by Theorem 7-4-1. But $PN < PM$, and so $PN < PO$. So by Theorem 7-4-2, m$\angle 2 <$ m$\angle 3$.

12. By Theorem 3-5-1, m$\angle 2$ = m$\angle 4$ + m$\angle 3$. Since angle measures are positive, m$\angle 2 >$ m$\angle 3$. It is given that m$\angle 1 >$ m$\angle 2$. So m$\angle 1 >$ m$\angle 3$. By Theorem 7-4-1, it follows that $WY > WX$.

13. Since $AC > BC$, it follows from Theorem 7-4-2 that m$\angle B >$ m$\angle A$. Thus, $\frac{1}{2}$m$\angle B > \frac{1}{2}$m$\angle A$. Therefore, by Theorem 7-4-1, $AD > BD$.

14. Since the triangle is scalene, no two sides are equal. Theorem 7-4-2 can be used for each pair of sides to show that no two angles are equal; that is, all three angles are unequal.

15. Since $\angle 1$ is an exterior angle of $\triangle ADC$, m$\angle 1 = x +$ m$\angle A$. Thus, m$\angle 1 > x$. Thus, $BC > BD$ by theorem 7-4-1.

FOLLOW-UP

Assessment

1. If three angles of a triangle are all unequal, what can you conclude about the lengths of the three sides? [They are all unequal.]

2. If at least two sides of a triangle are equal, what can you conclude about the three angles? [At least two angles are equal.]

Reteaching

Students who have had difficulty with this section may benefit from Reteaching Activity 7.4.

Enrichment

For students who have mastered the material in this section, you may wish to assign Enrichment Activity 7.4.

9. List the sides of $\triangle PQR$ from shortest to longest. $\overline{RQ}$, $\overline{PR}$, $\overline{PQ}$

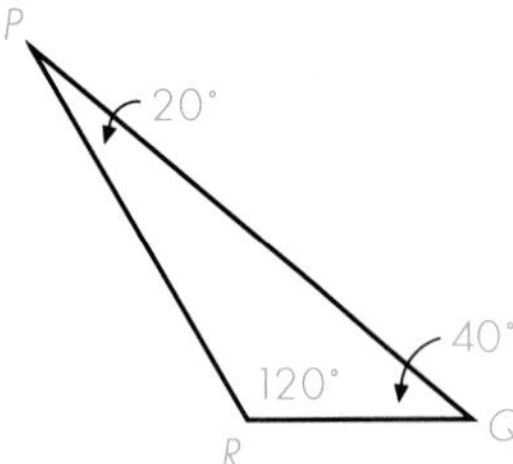

10. List the angles of $\triangle WXY$ from least to greatest. $\angle X$, $\angle W$, $\angle Y$

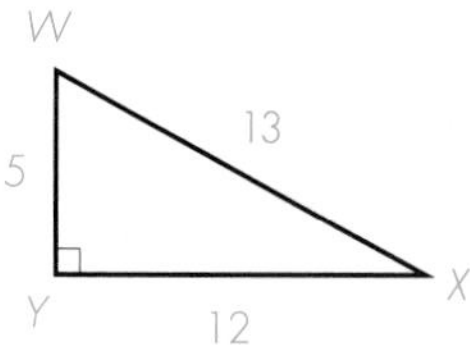

In Exercises 11–14, be prepared to give reasons for your answers.

11. Given: $\triangle MPO$ with
m$\angle 2 <$ m$\angle 1$ and
$PN < PM$
What can you conclude about m$\angle 2$ and m$\angle 3$? See margin.

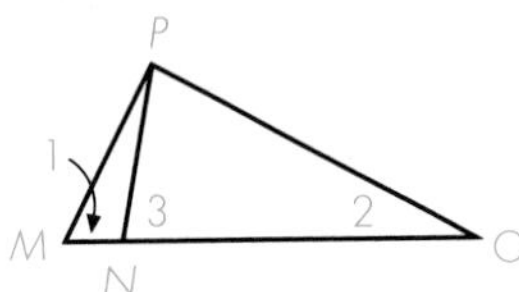

12. Given: $\triangle WXY$ with
m$\angle 1 >$ m$\angle 2$
What can you conclude about m$\angle 1$ and m$\angle 3$? about WY and WX? (Hint: Review Theorem 3-5-1.) See margin.

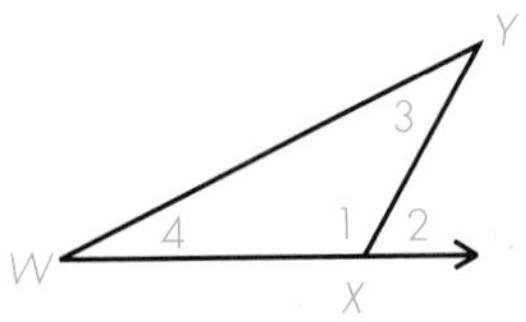

13. Given: $\triangle ABC$ with
$AC > BC$
The bisectors of $\angle A$ and $\angle B$ meet at D. What can you conclude about BD and AD? See margin.

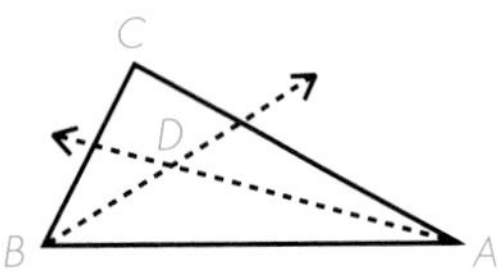

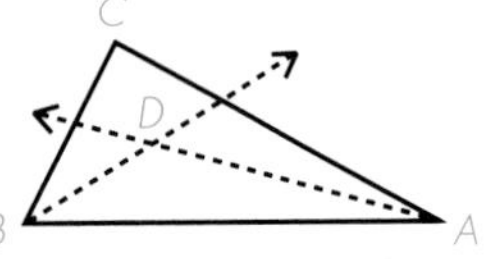

14. Given: $\triangle DEF$ is a scalene triangle. What can you conclude about the three angles? See margin.

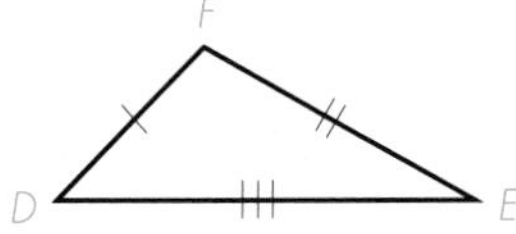

CRITICAL THINKING

15. In $\triangle ABC$, the bisector of $\angle C$ intersects $\overline{AB}$ in point D. Show that $BC > BD$. See margin.

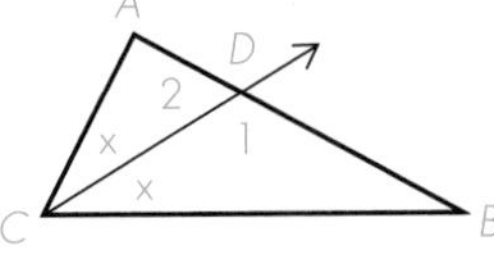

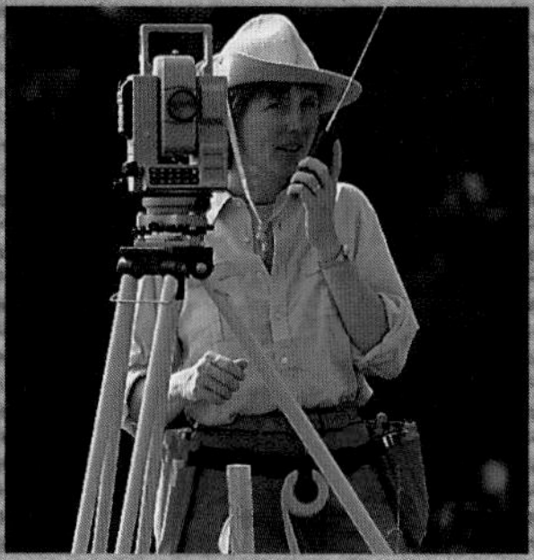

SECTION 7.5 Pythagorean Theorem and Descartes

After completing this section, you should understand
- ▲ how the Pythagorean Theorem is applied to the coordinate plane.
- ▲ the distance formula.

You have seen how surveyors can use a 30°-60° right triangle to locate landmarks on property. There are other ways in which landmarks can be located. For example you can give the location of each point in relation to a pre-selected reference point.

The point of reference is called the **origin.** The origin is the intersection of two perpendicular lines of reference, called the **x-axis** and the **y-axis.**

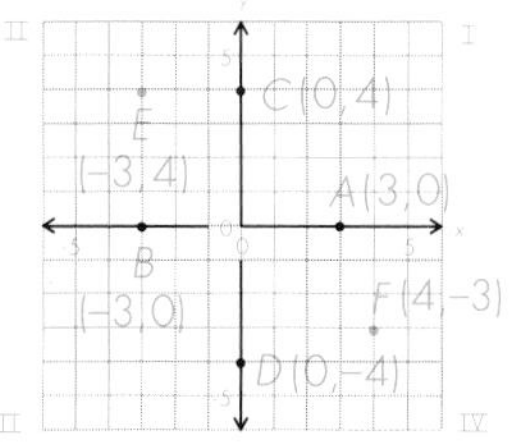

The x- and y-axes divide the plane into four quadrants (I, II, III and IV). A point is represented by a pair of numbers. The first number indicates how far the point is to the right or left of the origin, the second number tells how far it is above or below the origin. For example, point A is (3, 0) and is 3 units to the right of the origin. Point D is $(0, -4)$ and is 4 units below the origin. The pairs for points E and F use the same numbers (-3 and 4) but in a different order. For this reason, a pair such as $(4, -3)$ is called an **ordered pair.** The first number is the x-coordinate and the second number is the y-coordinate. The x- and y-axes are called the **coordinate axes.** The plane that they define is called the **coordinate plane.**

SECTION 7.5

Resources
Reteaching 7.5
Enrichment 7.5
Transparency Masters 7-2 and 7-3

Objectives
After completing this section, the student should understand
- ▲ how the Pythagorean Theorem is applied to the coordinate plane.
- ▲ the Distance Formula.

Materials
library resources
graph paper

Vocabulary
coordinate axes
coordinate geometry
coordinate plane
Distance Formula
ordered pair
origin
quadrant
x-coordinate
y-coordinate

GETTING STARTED

Quiz: Prerequisite Skills

1. Refer to the number line below.

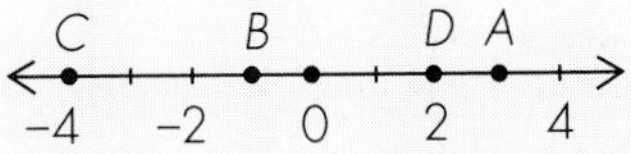

Find the following lengths.

a. OA [3] b. OC [4] c. AD [1] e. BD [3]

2. Simplify.

a. $|-2|$ [2]
b. $|4 - (-2)|$ [6]
c. $|-2 - 4|$ [6]
d. $|-x|^2$ [x^2]

TEACHING COMMENTS

Using the Page
Students who prepare a report on René Descartes will find that he lived from 1596 until 1650, a period of time that included such notable events as the deaths of Shakespeare and Cervantes (both in 1616), the landing of the Pilgrims in the New World (1620), the accession of Louis XIV (1643), and the beheading of Charles I (1649). Prominent mathematicians of the same era were Pierre de Fermat (1601-1665) and Blaise Pascal (1623-1662). Descartes's major works are the *Discourse on the Method of Reasoning Well and Seeking Truth in the Sciences* (1637) and, as an appendix to this treatise, *La Géométrie*, the work in which he introduced analytic geometry to the world.

The man who is credited with introducing coordinates into geometry is René Descartes.

REPORT 7-5-1 Prepare a report on René Descartes that includes the following information.

1. The time in history when Descartes lived
2. The important mathematical works that Descartes wrote
3. The names of other important mathematicians of Descartes's era
4. Some major world events during Descartes's lifetime

Sources:
Bell, Eric Temple, *Men of Mathematics*. New York: Simon & Schuster, 1937.
Boyer, Carl B., *A History of Mathematics*. Princeton, NJ: Princeton University Press, 1985.

The study of geometry by means of coordinates is called **coordinate geometry.** The Discovery Activity shows how coordinate geometry can be used to examine distance.

DISCOVERY ACTIVITY

In the figure, $\overline{AB}$ and $\overline{CD}$ are horizontal, $\overline{EF}$ and $\overline{GH}$ are vertical.

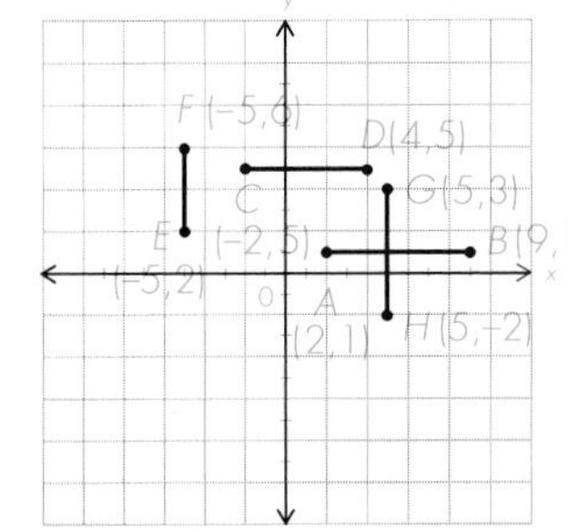

1. Count squares to find AB and CD. 7; 6
2. Count squares to find EF and GH. 4; 5
3. How could you use the coordinates to find the lengths of the segments?
Horizontal: subtract x-coordinates; vertical: subtract y-coordinates. Use the positive difference.

The Discovery Activity suggests the following definition.

NOTEBOOK

DEFINITION 7-5-1 **The LENGTH of the horizontal line segment with endpoints (x_1, y) and (x_2, y) is $|x_2 - x_1|$. For the vertical line segment with endpoints (x, y_1) and (x, y_2), the length is $|y_2 - y_1|$.**

The absolute value symbol (| |) means that you use the positive values of the differences in the definition.

Additional Answers
6. $PQ = \sqrt{18}$, $QR = \sqrt{32}$, $PR = \sqrt{50}$, $(\sqrt{18})^2 + (\sqrt{32})^2 = (\sqrt{50})^2$. By the converse of the Pythagorean Theorem (Theorem 7-1-2), the triangle is a right triangle.

NONHORIZONTAL AND NONVERTICAL LINE SEGMENTS

You can use the Pythagorean Theorem to find a formula for the length of a line segment that is neither horizontal nor vertical.

In $\triangle ABD$,

$$d^2 = a^2 + b^2$$
$$d^2 = (|x_2 - x_1|)^2 + (|y_2 - y_1|)^2$$
$$d^2 = (x_2 - x_1)^2 + (y_2 - y_1)^2$$

So $d = \sqrt{(x_2 - x_1)^2 + (y_2 - y_1)^2}$

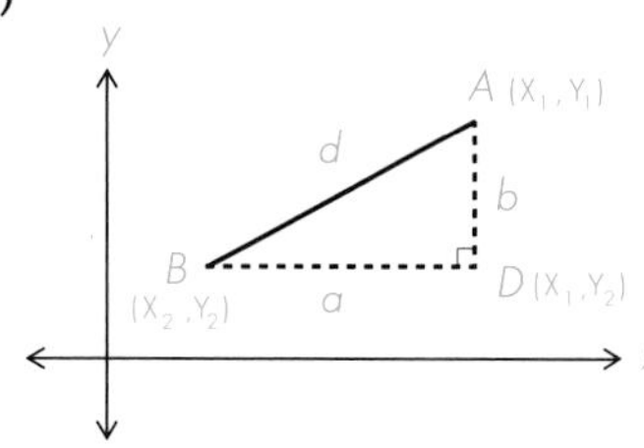

THEOREM 7-5-1 **The distance between points $A(x_1, y_1)$ and $B(x_2, y_2)$ is**

$$d = \sqrt{(x_2 - x_1)^2 + (y_2 - y_1)^2} \quad \textbf{(The Distance Formula)}$$

CLASS ACTIVITY

Mark the points on a coordinate plane. Then find the length of the segment connecting the two points.

1. A (3, 6) B (6, 2) 5
2. C (1, 2) D (1, −1) 3
3. E (−5, −1) F (−1, 4) $\sqrt{41}$, or 6.4
4. M (−4, 5) N (−2, 5) 2

HOME ACTIVITY

Mark the points on a coordinate plane. Then find the length of the segment connecting the two points.

1. A (−3, 3) B (5, −3) 10
2. C (4, 4) D (4, 1) 3
3. E (−2, −7) F (6, 8) 17
4. G (0, −4) H (8, −5) $\sqrt{65}$, or 8.1
5. $\triangle ABC$ has vertices $A(0, 1)$, $B(3, 7)$, and $C(6, 1)$. Show that the triangle is isosceles. $AB = \sqrt{45}$, $BC = \sqrt{45}$
6. $\triangle PQR$ has vertices $P(1, 1)$, $Q(4, 4)$, and $R(8, 0)$. Show that the triangle is a right triangle. (Hint: Review Theorem 7-1-2.) See margin.

CRITICAL THINKING

7. Which is larger, $\sqrt{10} + \sqrt{17}$ or $\sqrt{53}$? Find the answer without using tables or a calculator. (Hint: From algebra, you know that $(a + b)^2 = a^2 + 2ab + b^2$.) See margin.

Extra Practice
Mark the points on a coordinate plane. Then find the length of the segment connecting the two points.

1. $A(2, 4)$ $B(-3, 4)$ [5]
2. $C(-4, 1)$ $D(3, 0)$ [$\sqrt{50}$, or 7.1]

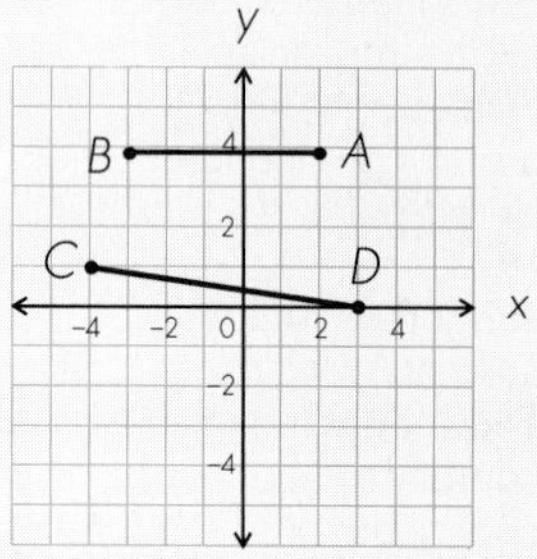

Additional Answers
6. See page 272.
7. Square both numbers to see which has the larger square. $(\sqrt{10} + \sqrt{17})^2 = 10 + 2\sqrt{170} + 17$ and $(\sqrt{53})^2 = 53$. $10 + 2\sqrt{170} + 17 = 27 + 2\sqrt{170}$. Note that $13^2 = 169$. Therefore, $27 + 2\sqrt{170} > 27 + 2\sqrt{169} = 27 + 2(13) = 53$. Since $\sqrt{10} + \sqrt{17}$ has a larger square than $\sqrt{53}$, $\sqrt{10} + \sqrt{17}$ is the larger number.

FOLLOW-UP

Assessment
1. Find the distance between the points $A(5, 2)$ and $B(-4, 2)$. [9]
2. Find the distance between the points $C(8, -2)$ and $D(5, 2)$. [5]

Reteaching
Students who have had difficulty with this section may benefit from Reteaching Activity 7.5.

Enrichment
For students who have mastered the material in this section, you may wish to assign Enrichment Activity 7.5.

Chapter Vocabulary
coordinate axes
coordinate geometry
coordinate plane
deductive reasoning
Distance Formula
equiangular
equilateral
hypotenuse
indirect measurement
inductive reasoning
isosceles right triangle
leg of a right triangle
ordered pair
origin
Pythagorean Theorem
quadrant
30°-60° right triangle
x-coordinate
y-coordinate

7.1 The hypotenuse of a right triangle is the side opposite the right triangle. The other two sides are the legs of the right triangle. In $\triangle ABC$, the right angle is $\angle C$, so the hypotenuse is $\overline{AB}$. The legs are $\overline{AC}$ and $\overline{BC}$.

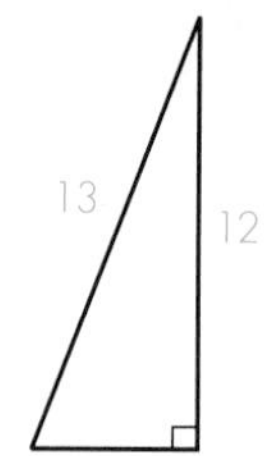

If a right triangle has sides of lengths a, b, and c, where c is the hypotenuse, then $a^2 + b^2 = c^2$. In right triangle ABC, $5^2 + 12^2 = 13^2$.

If a triangle has three sides of lengths a, b, and c, such that $a^2 + b^2 = c^2$, then the triangle is a right triangle. For $\triangle DEF$, since $8^2 + 6^2 = 10^2$, you know that $\angle DEF$ is a right triangle.

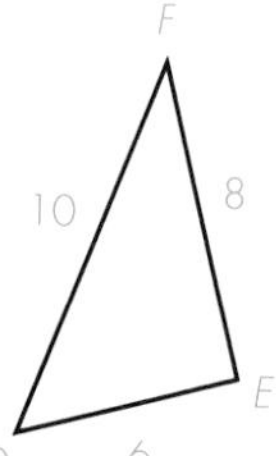

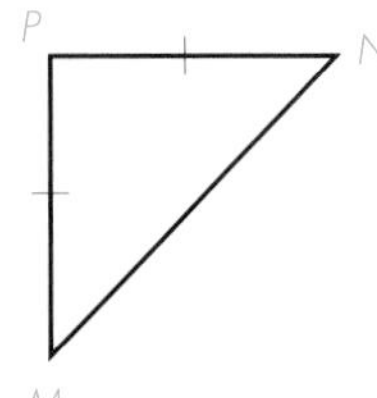

7.2 In an isosceles right triangle, the acute angles are each 45°. In the isosceles right triangle MNP, the acute angles are $\angle M$ and $\angle N$. $m\angle M = 45°$ and $m\angle N = 45°$.

The length of the hypotenuse of an isosceles right triangle is $\sqrt{2}$ (about 1.414) times the length of either leg. $\triangle PQR$ has legs of length 7, so the length of the hypotenuse is: $QR = 7\sqrt{2}$, or about 9.899

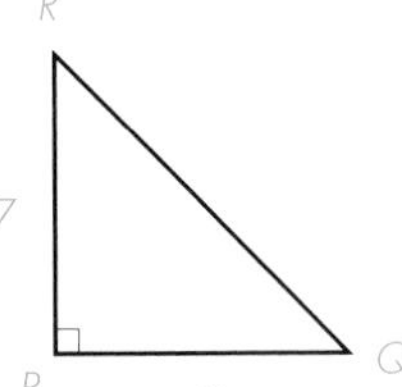

7.3 In a 30° − 60° right triangle, the length of the leg opposite the 30° angle is half the length of the hypotenuse. The length of the leg opposite the 60° angle is $\frac{\sqrt{3}}{2}$ times the length of the hypotenuse. In $\triangle XYZ$

$$YZ = \tfrac{1}{2} \cdot 13 = 6\tfrac{1}{2}$$

$$XZ = \tfrac{\sqrt{3}}{2} \cdot 13, \text{ or about } 11.258$$

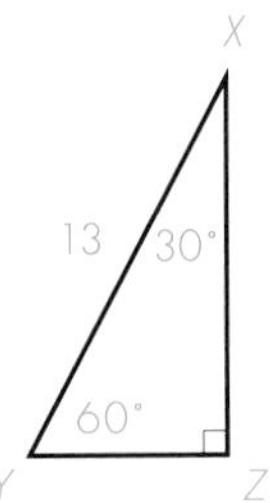

274

7.4 If two angles of a triangle are unequal, then the side opposite the greater angle is longer than the side opposite the other angle. In $\triangle ABC$, $m\angle C > m\angle A$. Therefore $AB > BC$.

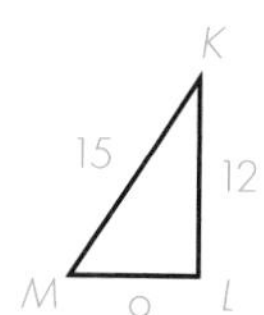

If two sides of a triangle are unequal, then the angle opposite the longer side is greater than the angle opposite the other side. In $\triangle KLM$, $KL > LM$. Therefore $m\angle M > m\angle K$.

7.5 In a coordinate plane, the length of a horizontal segment equals the absolute value of the difference of the x-coordinates of the endpoints. If the segment is vertical, its length equals the absolute value of the difference of the y-coordinates of the endpoints.

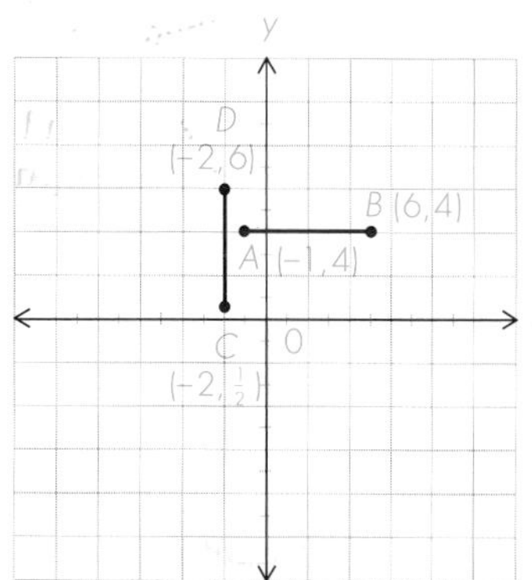

$\overline{AB}$ is horizontal, so

$$AB = |6 - (-1)| = |7| = 7$$

$\overline{CD}$ is vertical, so

$$CD = |\tfrac{1}{2} - 6| = |-5\tfrac{1}{2}| = 5\tfrac{1}{2}$$

For a segment that is neither horizontal nor vertical and whose endpoints are (x_1, y_1) and (x_2, y_2), the length is the distance d between the endpoints:

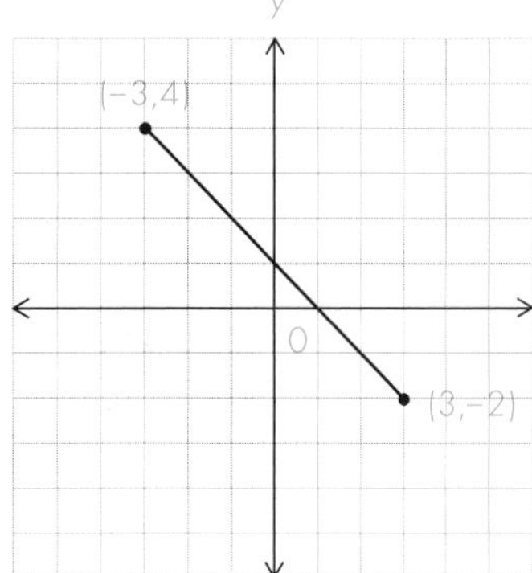

$$d = \sqrt{(x_2 - x_1)^2 + (y_2 - y_1)^2}$$

If the endpoints are (−3, 4) and (3, −2), the distance is:

$$d = \sqrt{(3-(-3))^2 + (-2-4)^2}$$

$$d = \sqrt{36 + 36}$$

$d = \sqrt{72}$, or about 8.49

Using the Page
This Computer Activity investigates Pythagorean Triples, three positive whole numbers a, b, and c, that satisfy the equation $a^2 + b^2 = c^2$. Students use a simple computer program to generate these triples and then use the triples to draw right triangles.

Although this lesson is designed to be done using a computer, most of the activity can be performed using a calculator if the computer is not available. Exercises 1-10 are easily computed using a hand calculator. Problem 11 can be modified to permit students to draw the triangle on graph paper.

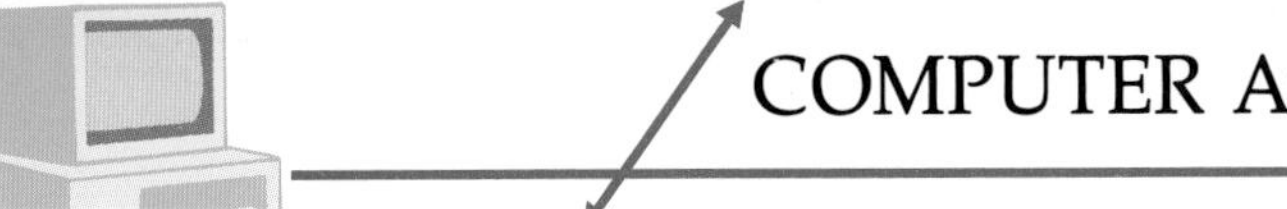

COMPUTER ACTIVITY

Three positive whole numbers *a*, *b*, and *c* that satisfy the equation $a^2 + b^2 = c^2$ are called a Pythagorean Triple. The ordered triple (a, b, c) is used to represent a Pythagorean Triple, where c is the largest number. The table below shows several Pythagorean Triples, where *a* is odd and *n* tells what line in the list the triple is on.

n	a	b	c
1	3	4	5
2	5	12	13
3	7	24	25
4	9	40	41
.	.	.	.
.	.	.	.
.	.	.	.
n	$2n + 1$	$2n^2 + 2n$	$2n^2 + 2n + 1$

The *n*th Pythagorean Triple in the list is $(2n + 1, 2n^2 + 2n, 2n^2 + 2n + 1)$.

You can write a Logo procedure to find these Pythagorean Triples. In order to do this, you will need to use the operation symbol + for addition, * for multiplication, and ↑ for exponentiation.

For example, $2n^2 + 2n$ is written as 2 * *n* ↑ 2 + 2 * *n*. You will also need to use the OUTPUT command to print the triples. For example, OUTPUT 2 * :*N* + 1 will print the result of 2 * :*N* + 1 after a value is given for :*N*.

Example: Write a procedure named PYTHTRIPLE to find the *n*th Pythagorean Triple in the table above.

```
TO PYTHTRIPLE :N
  OUTPUT 2 * :N + 1
  OUTPUT 2 * :N ↑ 2 + 2 * :N
  OUTPUT 2 * :N ↑ 2 + 2 * :N + 1
END
```

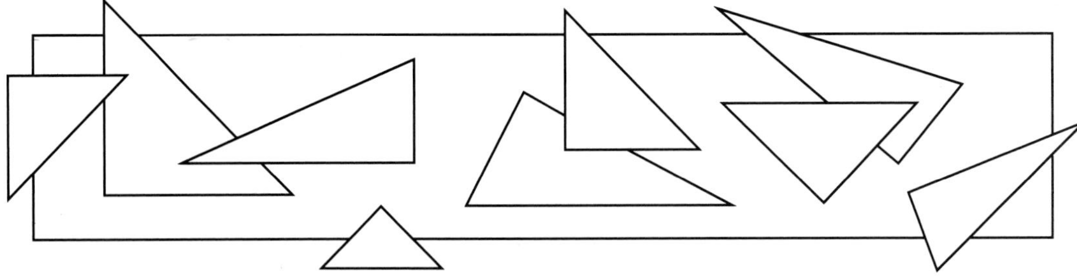

If :N is 5, the output will be:
11
60
61

Use the output from the PYTHTRIPLE procedure to help you fill in the chart below. Use the values in the n column for $:N$ in the procedure.

	n	a	b	c
1.	6	____	____	____
2.	7	____	____	____
3.	8	____	____	____
4.	9	____	____	____
5.	10	____	____	____
6.	11	____	____	____
7.	12	____	____	____
8.	13	____	____	____
9.	14	____	____	____
10.	15	____	____	____

11. Write a Logo procedure to draw any right triangle whose sides have lengths $2n + 1$, $2n^2 + 2n$, and $2n^2 + 2n + 1$, where n is a positive whole number. Then execute the procedure to draw right triangles whose sides have the lengths given by the Pythagorean Triples from Exercises 1 and 2.

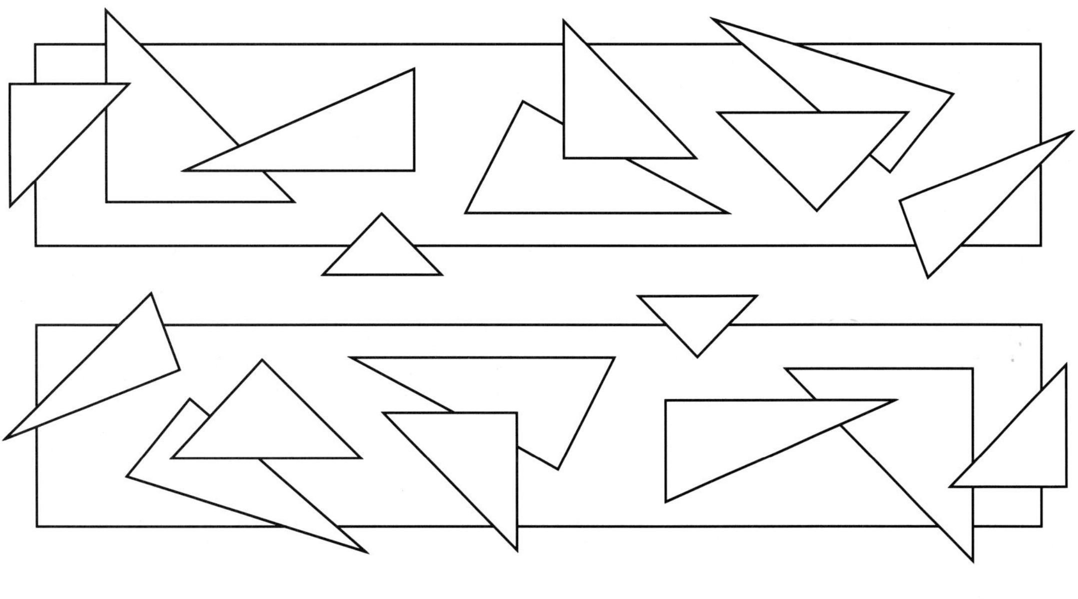

Using the Page
Students can work individually or in groups to study the Chapter 7 Review. If you have students study in groups, they should work together to make sure all members of the group know the material in the chapter.
Informal assessment such as interviews, classroom observations, or a review of student portfolios can be used instead of paper-and-pencil tests.

Assessment Resources
Achievement Tests pp. 13-14

Each of the following is the set of lengths of three sides of a triangle. Is the triangle a right triangle?

1. 7, 8, 9 No
2. 16, 30, 34 Yes
3. 11, 60, 61 Yes
4. 9, 10, 12 No

Find the missing lengths to the nearest tenth. Use a calculator or the table of squares and square roots on page 562.

5. 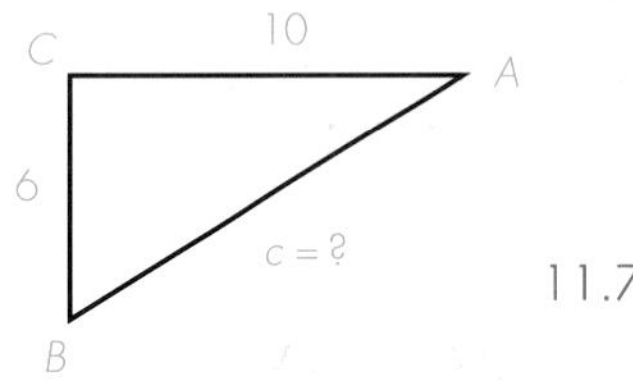

11.7

6. 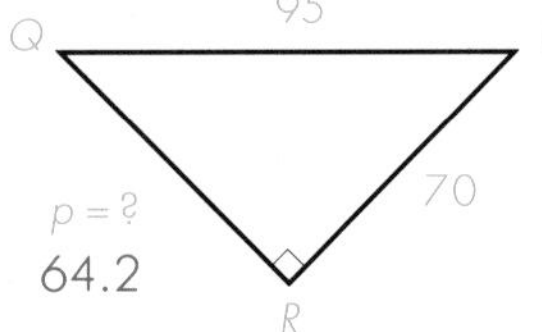

64.2

7. 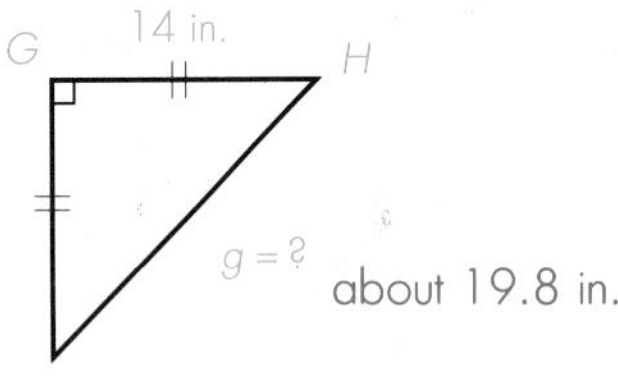

about 19.8 in.

8. 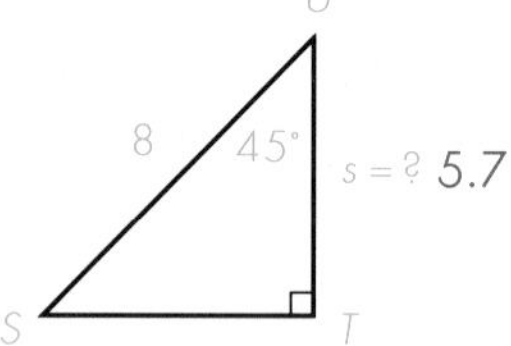

5.7

9. 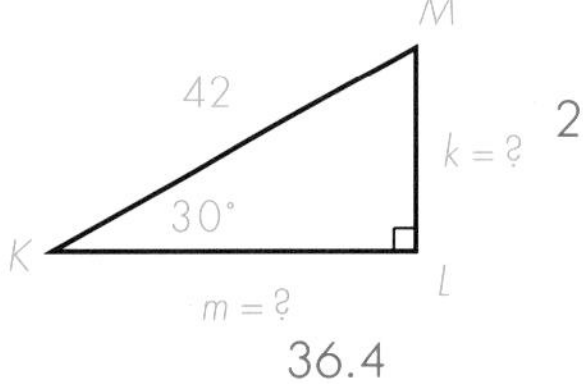

21
36.4

10. 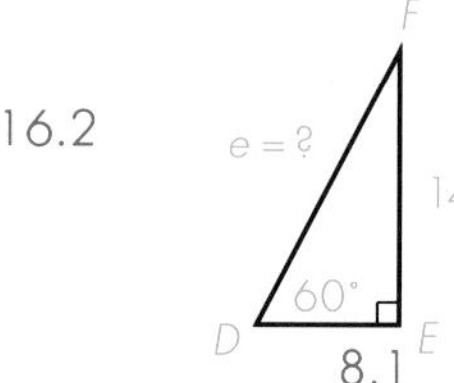

16.2
8.1

11. List the lengths of the sides of $\triangle XYZ$ from shortest to longest. x, z, y

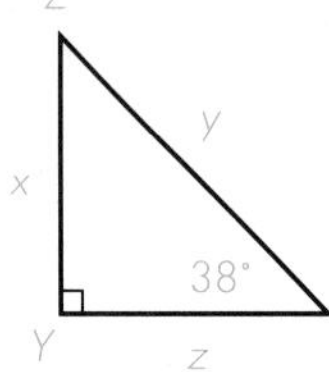

12. Write $m\angle A < m\angle B$, $m\angle A = m\angle B$, $m\angle A > m\angle B$, or "insufficient information."
Insufficient information

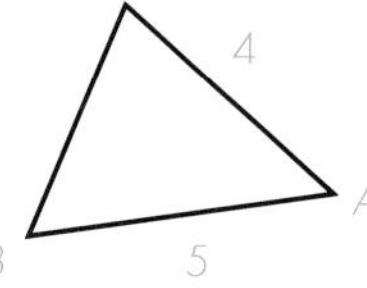

Find the length of the line segment that connects the two points.

13. 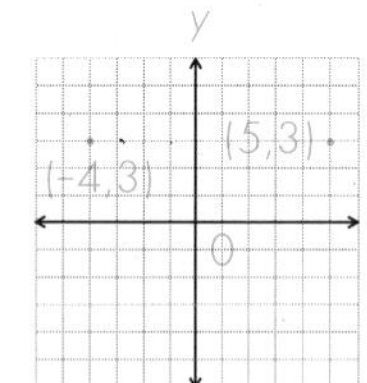

9

14. $\sqrt{65} \approx 8.1$
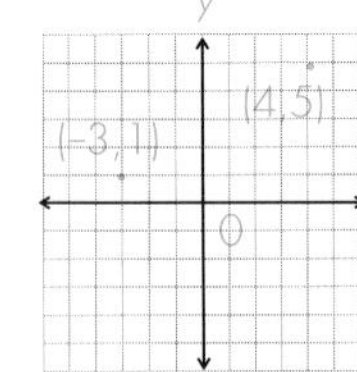

15. (−1, −3) and (−1, 4) 7

16. (−5, 2) and (3, −6) 11.3

CHAPTER TEST

You may use a calculator or the table of square and square roots on page 562. Round decimal answers to the nearest tenth.

1. The numbers 13, 84, and 85 are the lengths of the three sides of a triangle. Is the triangle a right triangle? Yes

2. Find the missing length. 19.5

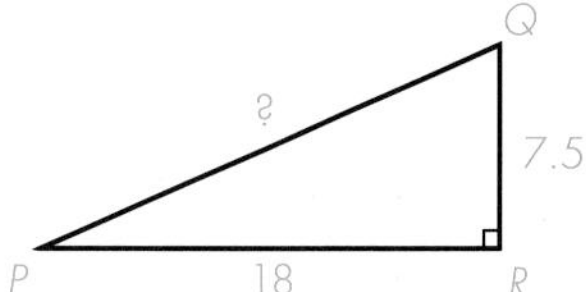

3. A window measures 4 ft by 6 ft. How far is it from one corner to the diagonally opposite corner? 7.2 ft

4. When an 18-in. candle just fits in the bottom of a rectangular box 10 in. wide. How long is the box? 15.0 in.

Find the missing lengths.

5. $17\sqrt{2}$, or about 24.0

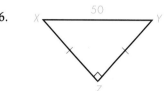

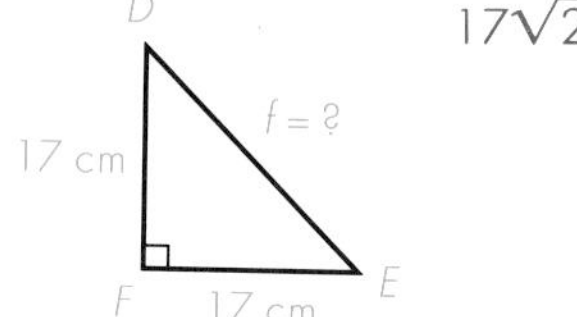

6. 35.4

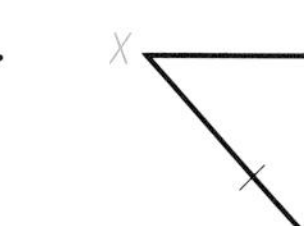

7.

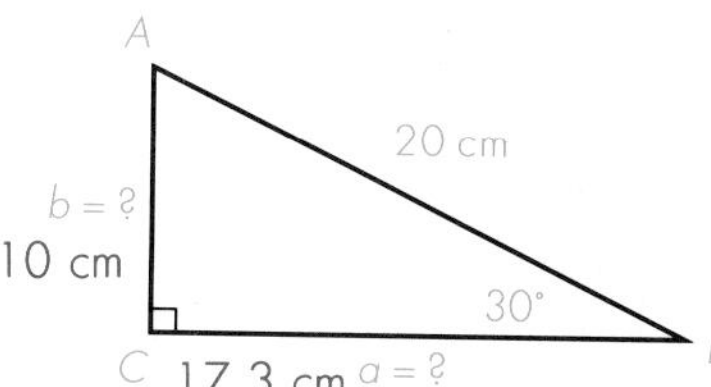

8. 114.3 ft

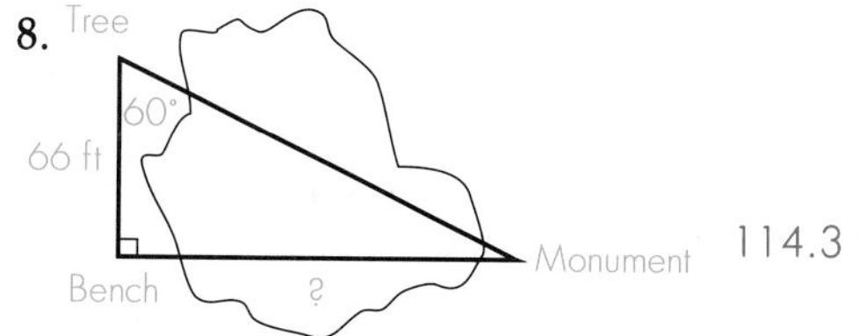

9. List the angles of $\triangle ABC$ from least to greatest. $\angle B$, $\angle A$, $\angle C$

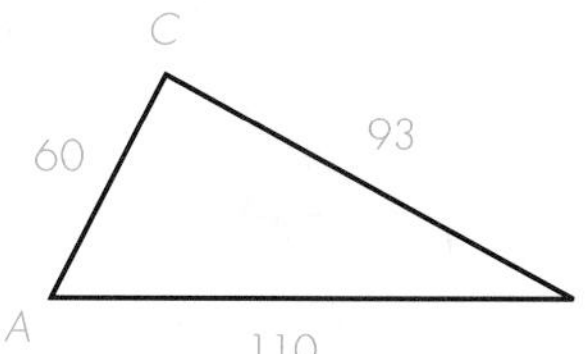

10. Write $x < y$, $x = y$, $x > y$, or "insufficient information." $x < y$

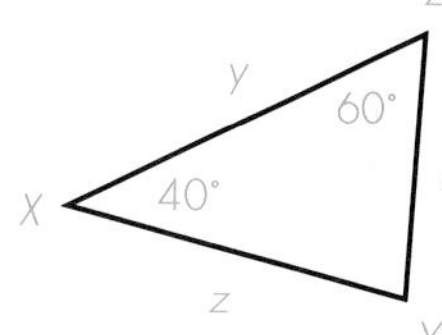

Find the length of the line segment that connects the two points.

11. (2, 4) and (2, −2) 6

12. (−3, 4) and (3, 2) $\sqrt{40} \approx 6.3$

13. (2, −6) and (−8, −6) 10

14. (−1, −1) and (3, 7) $\sqrt{80} \approx 8.9$

Test Objectives

After studying this chapter, students should understand

- the Pythagorean Theorem.
- squares and square roots of numbers.
- the properties of isosceles right triangles.
- how to solve problems using isosceles right triangles.
- theorems about 30°-60° right triangles.
- how to solve problems using 30°-60° right triangles.
- that the side opposite the greatest angle of a triangle is the longest side.
- that the angle opposite the longest side of a triangle is the greatest angle.
- how the Pythagorean Theorem is applied to the coordinate plane.
- the distance formula.

Algebra Review Objectives
Topics reviewed from algebra include
- solving linear equations in one variable.
- using linear equations to find the missing part of a geometric figure.
- problem solving using linear equations in one variable.

ALGEBRA SKILLS

Example: Solve for x.

$$\begin{aligned}6(8x - 3) &= 5x + 3(5 - x) - 10\\ 48x - 18 &= 5x + 15 - 3x - 10\\ 48x - 18 &= 2x + 5\\ 48x - 18 - 2x &= 2x + 5 - 2x\\ 46x - 18 &= 5\\ 46x - 18 + 18 &= 5 + 18\\ 46x &= 23\\ \frac{46x}{46} &= \frac{23}{46}\\ x &= \frac{1}{2}\end{aligned}$$

Example: Solve for x.

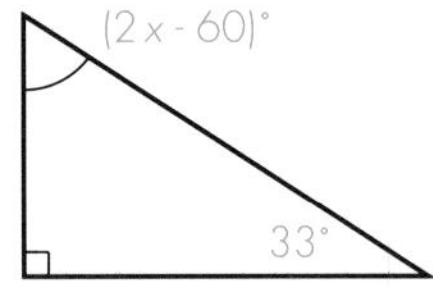

$$\begin{aligned}(2x - 60) + 33 + 90 &= 180\\ 2x + 63 &= 180\\ 2x &= 117\\ x &= 58\tfrac{1}{2}\end{aligned}$$

1. Solve for x.

a. $x - 14 = 27$ 41

b. $x + 33 = 81$ 48

c. $4x = 34$ $8\frac{1}{2}$

d. $-5x = 35$ -7

e. $3x - 1 = 50$ 17

f. $53 = 14x + 7$ $3\frac{2}{7}$

g. $\frac{x}{5} = \frac{17}{6}$ $14\frac{1}{6}$

h. $\frac{x - 1}{8} = \frac{1 + x}{5}$ $-4\frac{1}{3}$

i. $0.6x + 1.2 = 4.8$ 6

j. $5x - 17 = 28 - 3x$ $5\frac{5}{8}$

k. $5(x + 2) = 7 - x$ $-\frac{1}{2}$

l. $19 - 4x = \frac{1}{2}(13 - 6x)$ $12\frac{1}{2}$

m. $-6(x - 1) - 8 = 15x + 1$ $-\frac{1}{7}$

n. $-(7x + 1) = x - 12(2 + x) + 3$ -5

2. Find the value of the unknown angles.

a. 44.7°

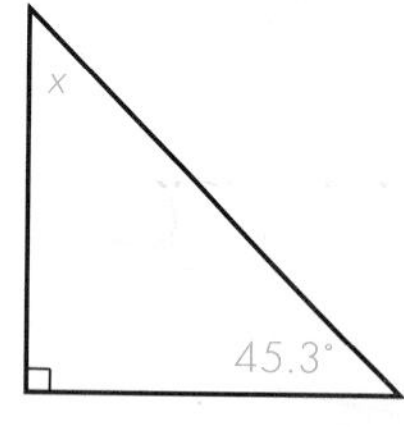

b. 37.5° 52.5°

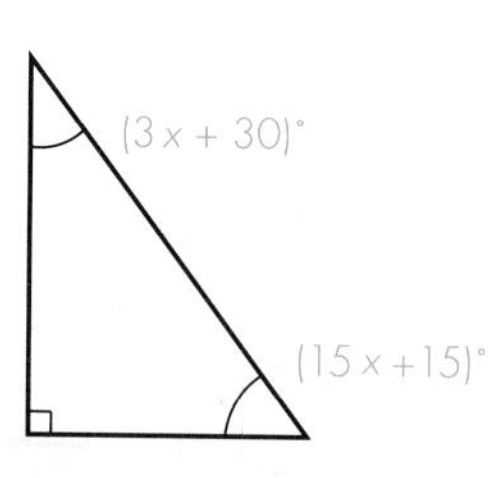

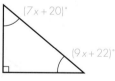

c. 54° 36°

d. 41° 49°

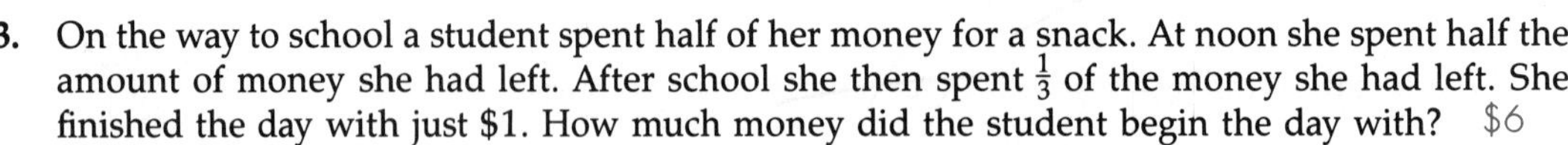

3. On the way to school a student spent half of her money for a snack. At noon she spent half the amount of money she had left. After school she then spent $\frac{1}{3}$ of the money she had left. She finished the day with just \$1. How much money did the student begin the day with? \$6

280

Name ______________________ Date ______________

The Pythagorean Theorem

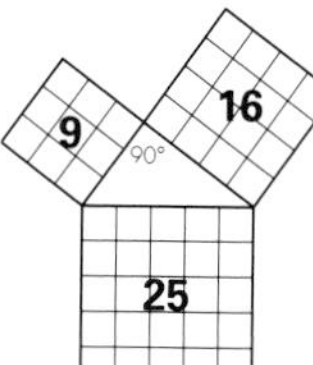

One of the most well-known theorems in mathematics is the Pythagorean Theorem. It states that if squares are drawn on the three sides of a right triangle, the square on the longest side of the triangle (called the hypotenuse) will be equal in size to the other two sides put together.

The drawing shows one example of the Pythagorean Theorem.

$a^2 + b^2 = c^2 \quad 3^2 + 4^2 = 5^2$

If you know two sides of a right triangle, you can use the theorem to find the third side. Pretend you do not know the length of the 4-unit side.

$$3^2 + b^2 = 5^2$$
$$b^2 = 5^2 - 3^2$$
$$b^2 = 25 - 9$$
$$b^2 = 16$$
$$b = 4$$

Draw a square on paper or an index card. Draw a diagonal to divide your square in half. Cut along the diagonal and you will have a triangle the same shape as triangle *ABC*.

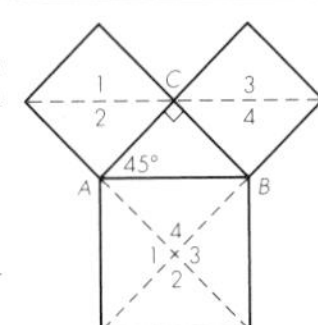

1. Use your triangle to help you copy the figure at the right. Which side is the hypotenuse?
side AB

2. One square has been divided into two triangles with a dotted line segment. Use dotted lines to show the Pythagorean Theorem. Start by drawing a diagonal in the largest square.

Sketch each triangle and estimate the length of the side *x*. Then use the Pythagorean Theorem to see how close your estimate is. Estimates will vary.

3. Estimate: ______
Computed: 15

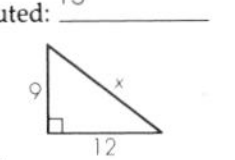

4. Estimate: ______
Computed: 20

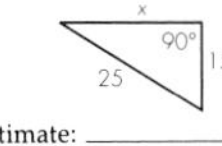

5. Estimate: ______
Computed: 5

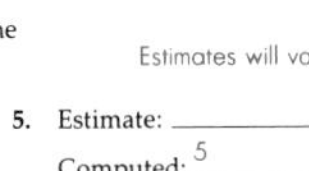

6. Estimate: ______
Computed: 24

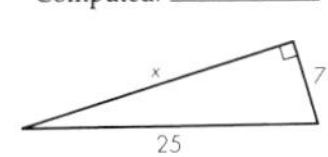

7. Estimate: ______
Computed: 8

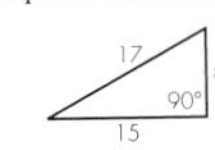

8. Estimate: ______
Computed: 34

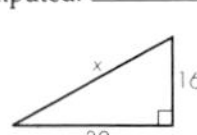

Name ______________________ Date ______________

The Pythagorean Theorem and Similar Triangles

Recall that two triangles are similar if two angles of one triangle have the same measure as two corresponding angles of the other triangle. Also, if two triangles are similar, then the corresponding sides are proportional.

Combine what you know about similar triangles with the Pythagorean Theorem to solve these problems.

In the figure at the right, angles *ABC* and *ADE* have the same measure. Follow these steps to find the length of segment *ED*.

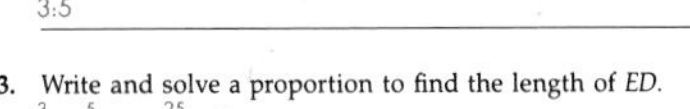

1. Use the Pythagorean Theorem to find the length of segment *BC*.
5 units

2. The ratio of *AB* to *AD* is 3:5. What is the ratio of the sides *BC* to *ED*?
3:5

3. Write and solve a proportion to find the length of *ED*.
$\frac{3}{5} = \frac{5}{x}$; $x = \frac{25}{3}$ units

Follow these steps to find *x* in the figure at the right.

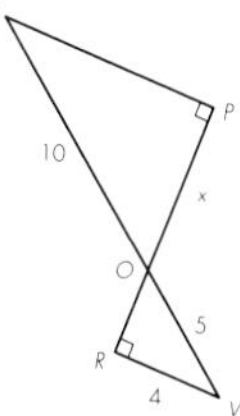

4. Give two pairs of angles to show that the two triangles are similar.
m∠TPO = m∠WRO, m∠POT = m∠ROW

5. Use the Pythagorean Theorem to find the length of side *RO*.
3 units

6. Write and solve a proportion to find *x*.
$\frac{5}{10} = \frac{3}{x}$; x = 6 units

Find *x* in each figure.

7. $x = \frac{36}{5}$ units

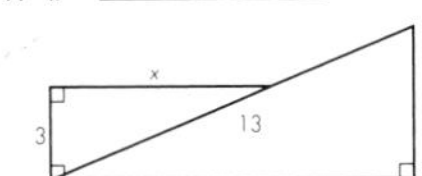

8. $x = \frac{24}{5}$ units

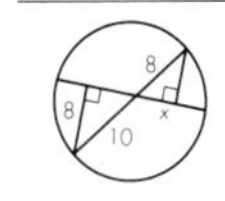

Name ______________________ Date ______________

The Isosceles Right Triangle

When you draw a diagonal in a square, you create two *isosceles right triangles.* The diagonal cuts two opposite angles of the square in half. So, two of the angles in the isosceles right triangle measure half of 90°, or 45° each.

Two sides of the isosceles right triangle are the same length. To find the length of the hypotenuse, we can use the Pythagorean Theorem.

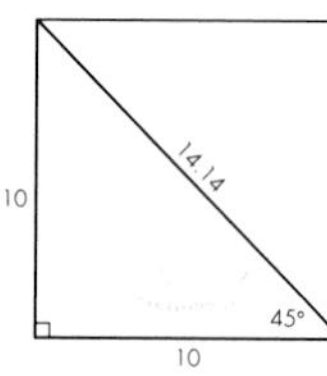

$c^2 = a^2 + b^2$ $\qquad c^2 = 10^2 + 10^2$ $\qquad c^2 = 200$

Notice that 200 equals 2 times 100. So, c equals $\sqrt{2}$ times 10.

The hypotenuse of an isosceles right triangle *always* equals $\sqrt{2}$ times the length of each shorter side. $\sqrt{2}$ is about 1.414. A rough estimate can be made using 1.5.

For each problem, circle the letter of the only answer that is always mathematically true.

1. An isosceles triangle
   a. always has one right angle.
   (b.) always has two sides of the same length.
   c. always has three sides of the same length.

2. The hypotenuse of a right triangle
   a. is equal to the square root of 2.
   b. is the shortest of the three sides.
   (c.) is opposite the 90° angle.

Find the length of side x in each isosceles right triangle. Round to the nearest hundredth, if necessary.

3. 12 units

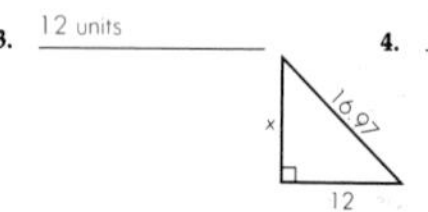

4. 15 units

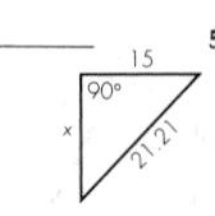

5. 16.97 units

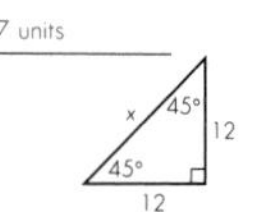

6. 25.45 units

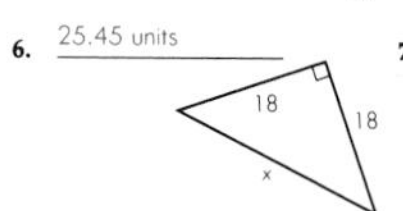

7. 21.21 units

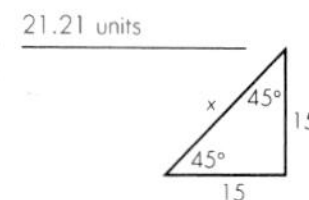

8. 30 units

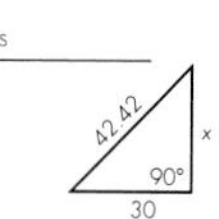

Name ______________________ Date ______________

Tangrams and Right Isosceles Triangles

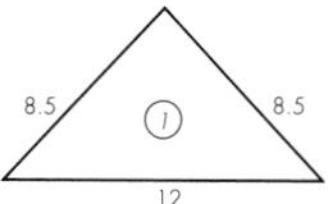

One of the oldest puzzles is the Chinese game of *tangrams.* It has been a popular Oriental pastime for several thousand years. The seven tangram pieces can be put together to make a wide variety of shapes.

1. Construct the seven tangram pieces shown at the right and below. Use heavy paper. Use centimeters for the lengths.

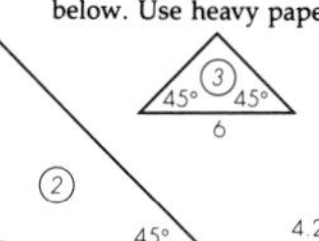

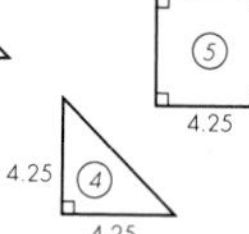

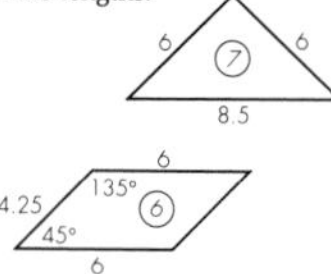

2. Assemble the seven tangram pieces to make one large square. Sketch your answer in the square at the right. See solid lines.
   What is the length of each side of the large square?
   12 units

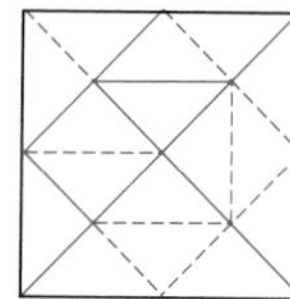

3. The seven pieces are called *tans.* Each of the five largest tans can be divided into isosceles right triangles congruent to the two smallest tans. Make sketches to show how this division can be done. See dotted lines at right.
   Altogether, how many identical isosceles right triangles are there in the seven tangram pieces?
   16

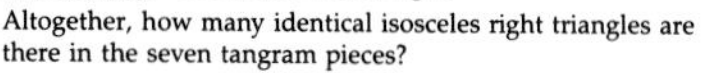

4. A *convex polygon* has no outside angles less than 180 degrees. There are 13 possible convex polygons that can be made with the tangram pieces. Of these shapes, one shape is a triangle, six shapes have four sides, two shapes have five sides, and four shapes have six sides.

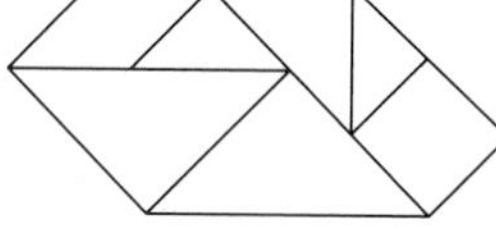

The square above is one of the 13 possibilities. Another one is shown at the right. How many of the 11 other shapes can you find? Answers will vary. Here are two possibilities.

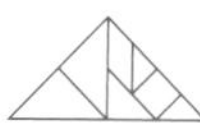

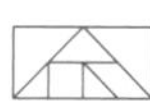

Name ______________________ Date ______________

The 30°-60°-90° Triangle

Recall that an equilateral triangle has three sides of the same length. You can think of a 30°-60° right triangle as one half of an equilateral triangle. One side of the 30°-60° right triangle is half as long as the hypotenuse. To find the length of the other side, we can use the Pythagorean Theorem.

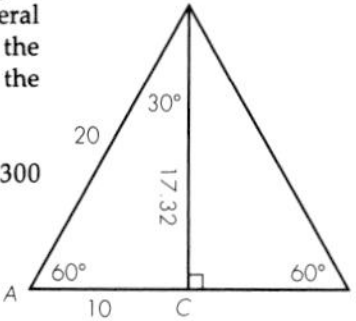

$b^2 = c^2 - a^2$ $\qquad b^2 = 20^2 - 10^2$ $\qquad b^2 = 300$

Notice that 300 equals 3 times 100. So, b equals $\sqrt{3}$ times 10.

The hypotenuse of a 30°-60° right triangle *always* equals $\sqrt{3}$ times the length of the shortest side. $\sqrt{3}$ is about 1.732. A rough estimate can be made using 1.75.

1. Copy triangle *ACB* in the figure above. Label the three vertices and mark the measurements of the three angles. How are the two smaller angles related?
m∠ABC is one half of m∠BAC.

2. Make another copy of triangle *ACB*. Mark the measurements of the three sides. How is the shortest side related to the longest side?
It is one half as long.

3. Draw an equilateral triangle with three 2-inch sides. Fold it in half to make two 30°-60° right triangles. How long is the shortest side of each 30°-60° right triangle?
1 inch

4. Copy these two triangles. Use the numbers 1, 2, $\sqrt{2}$, and $\sqrt{3}$ to label the six sides.
m∠C = 60°
m∠F = 45°
$\overline{AC} \cong \overline{DF}$

Find the length of side x to the nearest hundredth, if necessary, and the measure of angle O in each 30°-60° right triangle.

5. 7.5 units; 30°

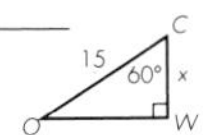

6. 20 units; 60°

7. 25.98 units; 30°

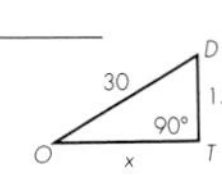

8. 10.39 units; 90°

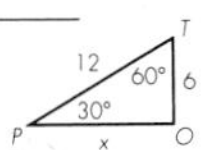

Name ______________________ Date ______________

Tessellations with 30°-60°-90° Triangles

The large shape at the right is a *regular hexagon*. It has six sides of the same length. The interior of the hexagon is completely covered with 30°-60°-90° triangles. A geometrical design that completely covers a plane area is called a *tessellation*.

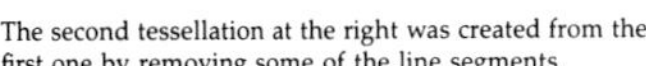

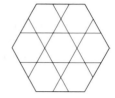

The second tessellation at the right was created from the first one by removing some of the line segments.

1. Describe the shapes that meet at each vertex of the tessellation.
2 equilateral triangles and 2 regular hexagons

2. This tessellation is described with the symbols $3^2 \cdot 6^2$. Explain what you think the symbols mean.
2 three-sided figures and 2 six-sided figures at each vertex

3. If the shortest side in each small 30°-60°-90° triangle is 1 centimeter, find the lengths of the sides of the hexagons and triangles in the second tessellation.
2 cm

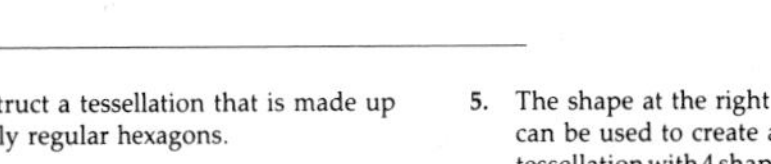

4. Construct a tessellation that is made up of only regular hexagons.
   a. Use a number with an exponent to describe this tessellation.
   6^3
   b. With a different color, draw line segments to connect the centers of the hexagons. Describe the tessellation formed by the connected colored lines.
   Equilateral triangles are formed.
   c. Use a number with an exponent to describe the new tessellation.
   3^6

5. The shape at the right can be used to create a tessellation with 4 shapes meeting at each vertex.

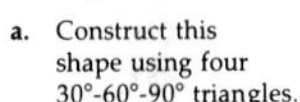

   a. Construct this shape using four 30°-60°-90° triangles.
   b. Use a number with an exponent to describe this tessellation.
   4^4
   c. Connect the centers of the shapes that make up the tessellation. What do you notice?
   The result is the same tessellation translated half a shape over and half a shape down.

Name ______________________ Date ______________

Angle-Side Relationship

Here are three of the theorems you've studied that involve triangle inequalities.

1. The sum of the lengths of two sides of a triangle is greater than the length of the third side. $a + b > c$

2. If a triangle has one angle with measure greater than another angle, then the side opposite the greater angle is longer than the side opposite the other angle. If $m\angle A > m\angle B$, then $a > b$.

3. If a triangle has a side longer than another side, then the measure of the angle opposite the longer side is greater than the measure of the other angle. If $a > b$, then $m\angle A > m\angle B$.

Use the figure above. For each problem, circle the letter of the right answer.

1. Theorem 1 tells us that
   - (a.) BC is less than the sum of AB and AC.
   - b. BC is greater than the sum of AB and AC.
   - c. BC is the longest side.

2. Theorem 2 tells us that
   - a. the greater the angle, the shorter the opposite side.
   - b. angle A is always larger than angle B.
   - (c.) knowing something about two angles in a triangle gives us information about the sides opposite those angles.

Use the figure at the right. Complete the symbolic statements of the three theorems.

3. Theorem 1: In $\triangle XYZ$, $x + y > z$, $x + z > y$, and $z + y > x$.

4. Theorem 2: In $\triangle XYZ$, $m\angle X > m\angle Y$ implies that $x > y$, $m\angle Z > m\angle Y$ implies that $z > y$, and $m\angle X > m\angle Z$ implies that $x > z$.

5. Theorem 3: In $\triangle XYZ$, $x > y$ implies $m\angle X > m\angle Y$, $z > y$ implies $m\angle Z > m\angle Y$, and $x > z$ implies $m\angle X > m\angle Z$.

Name ______________________ Date ______________

The Hinge Theorem

Similar to the two Triangle Inequalities in Chapter 3 and the Angle-Side Relationships in this section, the Hinge Theorem tells us when one line segment must be longer than another.

> **THE HINGE THEOREM**
>
> If two sides of one triangle are congruent, respectively, to two sides of a second triangle, and the included angle of the first triangle is larger than the included angle of the second, then the third side of the first triangle is longer than the third side of the second.

1. The drawing at the right shows two sticks connected with a hinge at the point marked B. The other ends of the sticks, A and C, are joined with a rubber band.

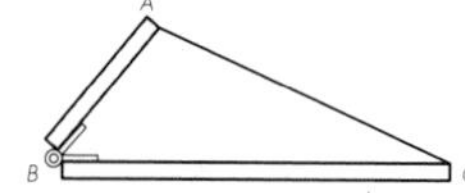

Which of the following is a true statement based on the Hinge Theorem? b

   - a. Two sticks of the same lengths as AB and BC would stretch a rubber band to the same length as AC.
   - b. As the hinge is opened wider, the rubber band is stretched longer.
   - c. The length of the stretched rubber band is greater than the length of either stick.

2. Use the two figures at the right. Which of the statements below is the Hinge Theorem? c

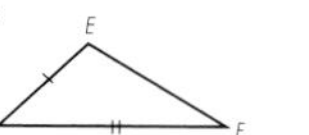

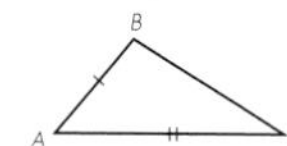

   - a. Given $\triangle ABC$ and $\triangle DEF$, with $BC > EF$. If $m\angle A > m\angle D$, then $AB = DE$ and $AC = DF$.
   - b. Given $\triangle ABC$ and $\triangle DEF$, with $AB = DE$ and $AC = DF$. If $BC > EF$, then $m\angle A > m\angle D$.
   - c. Given $\triangle ABC$ and $\triangle DEF$, with $AB = DE$ and $AC = DF$. If $m\angle A > m\angle D$, then $BC > EF$.

3. Write a symbolic statement of the Converse Hinge Theorem.
   Given $\triangle ABC$ and $\triangle DEF$, with $AB = DE$ and $AC = DF$. If $BC > EF$, then $m\angle A > m\angle D$.

> **THE CONVERSE HINGE THEOREM**
>
> If two sides of one triangle are congruent, respectively, to two sides of a second triangle, and the third side of the first triangle is longer than the third side of the second, then the included angle of the first triangle is larger than the included angle of the second.

Name ______________________ Date ____________

Pythagorean Theorem and Descartes

The Pythagorean Theorem has many different uses. One of them is a way to find the distance between two points that are labeled with ordered pairs.

An ordered pair is two numbers that give you the location of a point. In the figure, point A is labeled (2, 1) because it is 2 units to the right of the original, and 1 unit up.

Once you know that side BC is 4 units and side AC is 3 units, you can use the Pythagorean Theorem to find the length d.

$d^2 = 4^2 + 3^2$
$d^2 = 16 + 9$
$d^2 = 25$
$d = 5$

(5, 5) B a d A C (2, 1) b (5, 1) (0, 0)

A B C D

Use the number lines above and to the right for these problems.

E F G H

1. Copy the horizontal number line above. Start with point A and number the line from 0 to 9. Then find the lengths of these segments.

   AB = 3 AC = 5
   BD = 5 BC = 2

2. Copy the vertical number line at the right. Start with point E and number the line from 0 to 8. Then find the lengths of these segments.

   EH = 8 FG = 5
   FH = 6 EG = 7

3. Mark two points on the horizontal number line. Label them x_1 and x_2. Write a formula to find the length of the segment between the points.

   $l = |x_2 - x_1|$

4. Mark two points on the vertical number line. Label them y_1 and y_2. Write a formula to find the length of the segment between the points.

   $l = |y_2 - y_1|$

5. Copy the triangle at the top of this page. Label the points A and B with the ordered pairs (x_1, y_1) and (x_2, y_2). Now use the Pythagorean Theorem to express the square of the length d as the sum of two other squares.

   $d^2 = (x_2 - x_1)^2 + (y_2 - y_1)^2$

Name ____________________ Date ____________

The Distance Formula in 3-Dimensional Space

The location of a point in three-dimensional space can be described with three numbers—an *ordered triple.* Some of the vertices of the rectangular solid at the right have been marked with ordered triples.

(0,5,8) (12,5,8) (0,0,8) (12,5,0) (0,0,0) (12,0,0)

Use the rectangular solid at the right above for Exercises 1–2.

(3,4,12)

1. How many vertices does the solid have?
8

2. Write ordered triples for the vertices that are not labeled.
(0, 5, 0) and (12, 0, 8)

3. In the space at the right, sketch a rectangular solid with opposite vertices (0, 0, 0) and (3, 4, 12).

(0,0,0)

4. Compare the dimensions of this solid with the one drawn at the top of the page.
$\frac{1}{4}$ the length, $\frac{4}{5}$ the depth, $\frac{3}{2}$ the height

5. Write ordered triples for the vertices that are not labeled.
(0, 0, 12), (0, 4, 12), (0, 4, 0), (3, 0, 0), (3, 4, 0), (3, 0, 12)

6. Use the diagram below to show that the square of the length of the diagonal, *d*, is equal to the sum of the squares of the three dimensions (length, width, height).

Start by writing x^2 and d^2 as sums of squares. Then use substitution to express the square of *d* as a sum of squares.
$x^2 = w^2 + l^2$, $d^2 = x^2 + h^2$, so $d^2 = w^2 + l^2 + h^2$

h d h w x w l

7. Write a formula for the distance between two points in three-dimensional space. Use the triples (x_1, y_1, z_1) and (x_2, y_2, z_2) in your formula.

HINT: Recall that the distance formula for 2-dimensional space is
$d = \sqrt{(x_1 - x_2)^2 + (y_1 - y_2)^2}$.
$d = \sqrt{(x_1 - x_2)^2 + (y_1 - y_2)^2 + (z_1 - z_2)^2}$

Name ______________________ Date ______________

Pythagorean Triples

The procedure below can be used to help you determine whether an ordered triple is a Pythagorean triple.

```
TO TRIPLE :A :B :C
OUTPUT :A ↑ 2 + :B ↑ 2
OUTPUT :C ↑ 2
END
```

If the ordered triple is a Pythagorean triple, the results of both OUTPUT commands will be the same.

Use the TRIPLE procedure shown above to determine whether each ordered triple is a Pythagorean triple. Write *yes* or *no*.

1. (10, 24, 26) yes **2.** (2, 3, 4) no **3.** (12, 35, 37) yes

4. (30, 74, 75) no **5.** (40, 75, 85) yes **6.** (4, 5, 6) no

7. (26, 28, 38) no **8.** (24, 45, 51) yes **9.** (14, 48, 50) yes

The SQRT command is used in Logo to find the square root of a number. This command can be used in a procedure to find the third number of a Pythagorean triple when the other two numbers are known. For example, the procedure below can be used to find :C given :A and :B.

```
TO FINDTRIP :A :B
OUTPUT SQRT (:A ↑ 2 + :B ↑ 2)
END
```

In Exercises 10–13, :A and :B are given. Use the FINDTRIP procedure to find :C.

10. :A = 36, :B = 48 60 **11.** :A = 8, :B = 15 17

12. :A = 18, :B = 24 30 **13.** :A = 16, :B = 30 34

14. Write a procedure to find :A given :B and :C.

```
TO FINDTRIP :B :C
OUTPUT SQRT (:C ↑ 2 − :B ↑ 2)
END
```

15. Write a procedure to find :B given :A and :C.

```
TO FINDTRIP :A :C
OUTPUT SQRT (:C ↑ 2 − A ↑ 2)
END
```

Name______________ Class______________ Date______________

Achievement Test 7 (Chapter 7)
FUNDAMENTAL THEOREM OF PYTHAGORAS

No. Correct
No. Exercises: **25**
Score
4.00 x No. Correct =

GEOMETRY FOR DECISION MAKING
James E. Elander
SOUTH-WESTERN PUBLISHING CO.

You may use a calculator or the table of squares and square roots on page 562. Round decimal answers to the nearest tenth.

1. The numbers 15, 91, and 92 are the lengths of the three sides of a triangle. Is the triangle a right triangle? No

2. The numbers 9, 40, and 41 are the lengths of the three sides of a triangle. Is the triangle a right triangle? Yes

Find the missing length.

3. 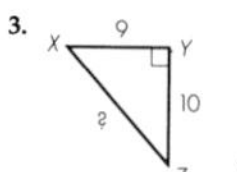
$\sqrt{181}$, or 13.5

4. 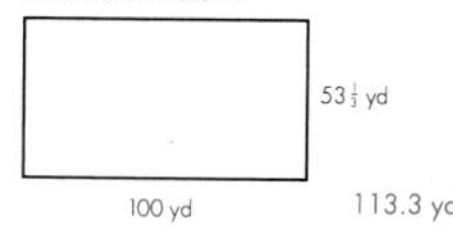
$\sqrt{507}$, or 22.5

5. What is the greatest straight-line distance that a football player can run on the football field shown below?

53 1/3 yd
100 yd

113.3 yd.

6. On the way to school, Kerry sometimes takes a short cut instead of turning at the corner. How much distance does she save if she takes the short cut?

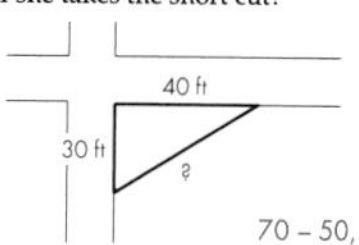
70 – 50, or 20 feet.

Find the missing lengths.

7. 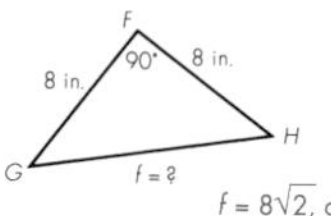
$f = 8\sqrt{2}$, or 11.3 in

8. 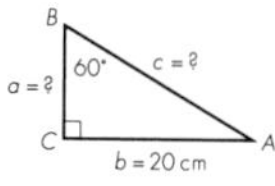
$a \approx 11.5$ cm; $c \approx 23.1$ cm

9. d = ? 30° 60 ft e = ?
e = 30 ft;
d = $30\sqrt{3}$, or 52.0 ft

10. 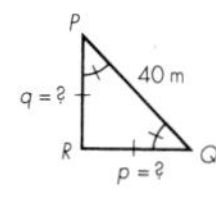
$p = q = 20\sqrt{2}$, or 28.3 m

MG01AG

11. In triangle *STU*, $m\angle TUS = 90°$. Find the length of median $\overline{UV}$.

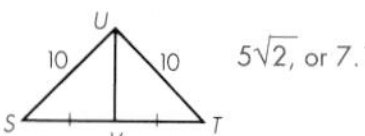
$5\sqrt{2}$, or 7.1

In Items 12-15, the triangles are not necessarily drawn to show their proper shapes.

12. List the angles of Δ *ABC* from least to greatest.

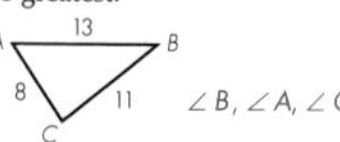
∠B, ∠A, ∠C

13. List the sides of Δ *XYZ* from shortest to longest.

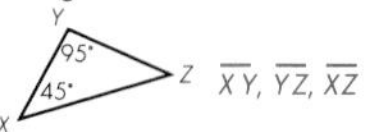
$\overline{XY}$, $\overline{YZ}$, $\overline{XZ}$

14. Write $a < b$, $a = b$, $a > b$, or "insufficient information."

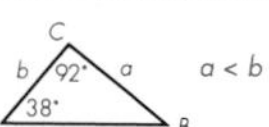
$a < b$

15. Write $m\angle X < m\angle Y$, $m\angle X = m\angle Y$, $m\angle X > m\angle Y$, or "insufficient information."

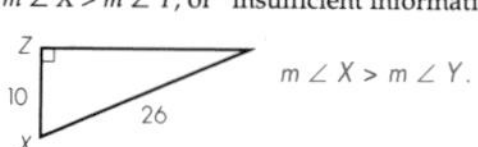
$m\angle X > m\angle Y$.

Find the length of the line segment that connects the two points.

16. (1, 4) (8, 4) — 7

17. (–2, 5) (–2, –1) — 6

18. (7, 5) (1, 1) — $\sqrt{52}$, or 7.2

19. 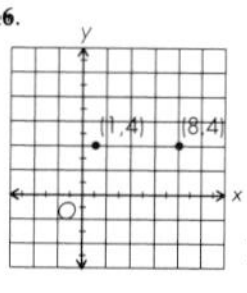
$\sqrt{52}$, or 7.2

20. 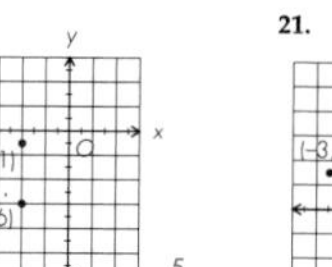
5

21. 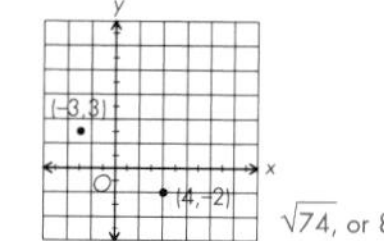
$\sqrt{74}$, or 8.6

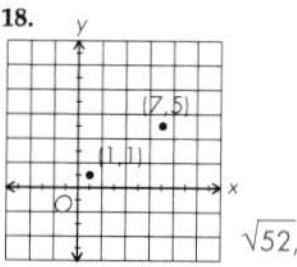
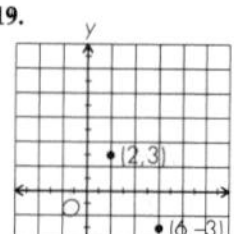

22. (0, 0) and (5, 9) $\sqrt{106}$, or 10.3

23. (5,1) and (–6, -3) $\sqrt{137}$, or 11.7

24. (–4, 8) and (–4, 11) 19

25. (7, –1) and (15, –2) $\sqrt{65}$, or 8.1

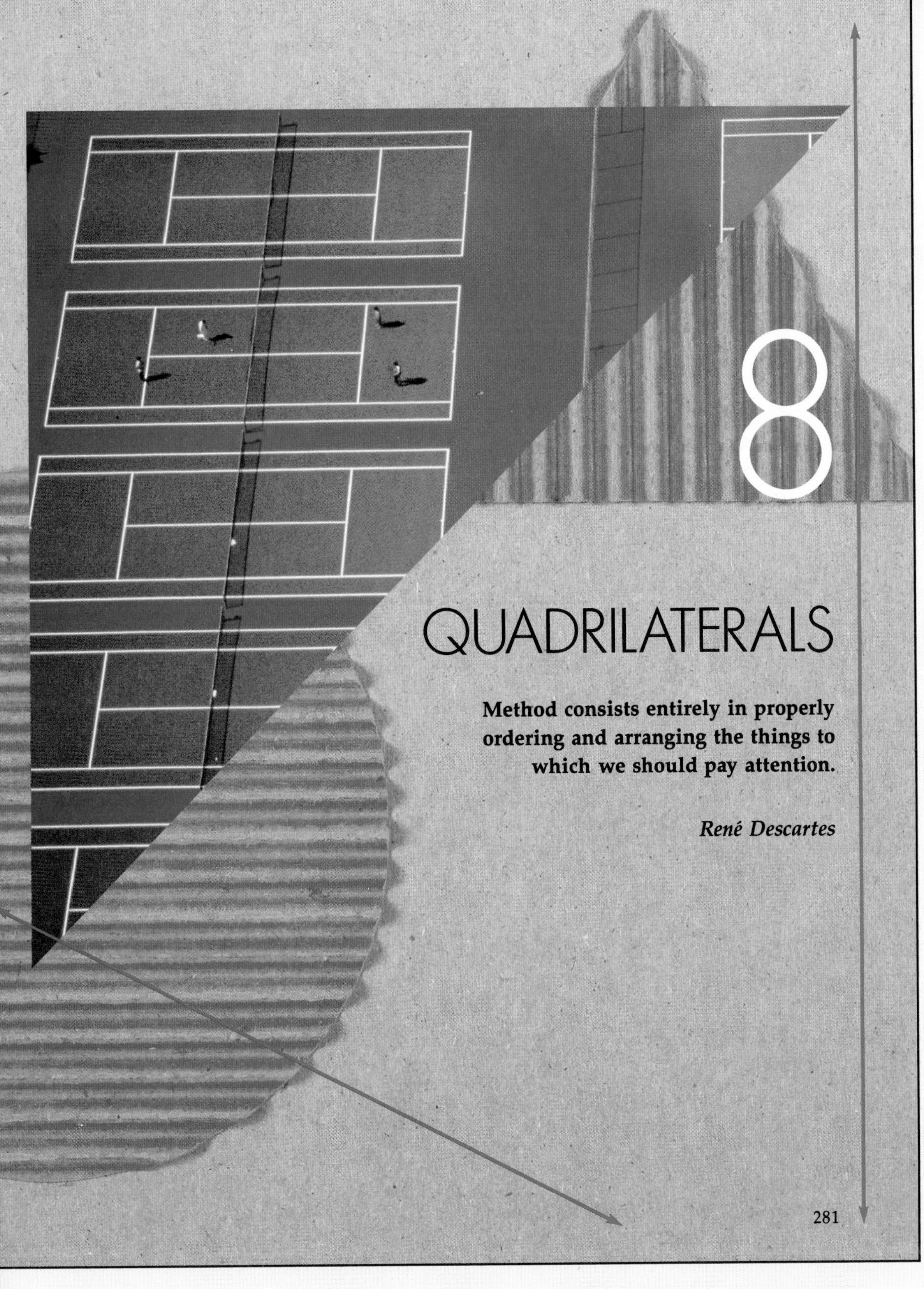

8

QUADRILATERALS

Method consists entirely in properly ordering and arranging the things to which we should pay attention.

René Descartes

281

CHAPTER 8 OVERVIEW

Quadrilaterals

Suggested Time: 10–12 days

The definition of a triangle is used as the basis for the discussion of quadrilaterals. Students will explore and compare the different properties that distinguish types of quadrilaterals and become familiar with the parts of a quadrilateral. The first theorem in the chapter states that the sum of the measures of the interior angles of a quadrilateral is 360°. This theorem provides a basic tool for students to use in finding angle measures, parallel sides, and, in Chapters 9 and 10, the perimeter and area of quadrilaterals, other polygons, and circles.

Method consists entirely in properly ordering and arranging the things to which we should pay attention.

René Descartes

CHAPTER 8 OBJECTIVE MATRIX

	End of Chapter Items by Activity				Student Supplement Items by Activity		
Objectives by Number	**Review**	**Test**	**Computer**	**Algebra Skills**	**Reteaching**	**Enrichment**	**Computer**
8.1.1	✓	✓			✓	✓	✓
8.1.2		✓			✓		
8.2.1	✓	✓	✓	✓	✓		✓
8.2.2	✓	✓	✓	✓	✓	✓	✓
8.3.1	✓	✓	✓	✓	✓	✓	✓
8.3.2	✓	✓	✓	✓	✓	✓	✓
8.4.1	✓	✓	✓		✓	✓	✓
8.5.1	✓	✓		✓	✓	✓	✓
8.5.2	✓	✓		✓	✓	✓	✓

*A ✓ beside a Chapter Objective under Algebra Skills indicates that algebra skills taught within the chapter or in previous Algebra Skills lessons are used.

CHAPTER 8 PERSPECTIVES

▲ Section 8.1

Properties of Quadrilaterals

Through a Discovery Activity, students determine properties common to different types of quadrilaterals. The term *quadrilateral* is defined, and students discover the difference between convex and concave figures. A second Discovery Activity helps students conclude that any quadrilateral can be divided into two triangles, so that the sum of its interior angle measures is 360°. Another experiment shows them that by physically adding a diagonal to a quadrilateral, they are constructing a rigid figure formed from two triangles.

The Home Activity reinforces the terms introduced in the section. In Critical Thinking, students are asked to find a relationship between the number of sides of a polygon and the number of triangles formed by its diagonals, in order to predict the sum of its interior angle measures.

▲ Section 8.2

Trapezoids

Trapezoids and isosceles trapezoids are defined. Through a Discovery Activity, the class finds that two pairs of angles in an isosceles trapezoid are equal. A two-column proof of this theorem follows. In a second Discovery Activity, students are asked to find relationships between the median of a trapezoid and its bases. The theorem involving the median of a trapezoid is justified in a two-column proof.

The Home Activity provides reinforcement of the two trapezoid theorems and also asks students to write their own proof showing that the diagonals of an isosceles trapezoid are equal in length. In Critical Thinking, students are asked to find a relationship between the number of sides of a polygon and its diagonals.

▲ Section 8.3

Parallelograms

Parallelograms are defined as quadrilaterals with two pairs of parallel sides. In a Discovery Activity, students draw conclusions about the measures of opposite sides and angles of parallelograms. These conclusions are justified in a two-column proof. The diagonals of a parallelogram are explored, and it is proven that they bisect each other.

The Home Activity reinforces the theorems involving both trapezoids and parallelograms, and includes an additional proof that adjacent angles of a parallelogram are supplementary. The proof shows only that two pairs of angles are supplementary. You may want to discuss in class the reasons why the statement is true for all four angles. Critical Thinking provides a practical application of the use of quadrilaterals. Have students color and display their work.

▲ Section 8.4

Rectangles

This section reviews the quadrilaterals studied thus far. Continue to encourage students to add definitions, theorems, and examples to their notebooks. The Discovery Activity leads to the definition of a rectangle. Have students discuss why the definition need not state that all four angles of a rectangle are right angles. Other Class Activity exercises ask students to prove that the diagonals of a rectangle are equal in length and form isosceles triangles within the rectangle. The suggested group project provides an excellent opportunity for students to work cooperatively to find real-world applications for the figures they have studied. Those with an interest in architecture or photography may be encouraged to extend the project to include research on the Golden Rectangle (Critical Thinking) or a photographic display of local landmarks.

▲ Section 8.5

Squares and Rhombuses

The diagram at the beginning of the section provides an opportunity for students to relate the properties of different quadrilaterals and to discuss the fact that each succeeding figure contains all the properties of the one before it. You may also want to discuss why the converse is not true. The first Discovery Activity gives a clear illustration of a figure with four equal sides that is not a square. In the second Discovery Activity, students learn that the diagonals of a rhombus are perpendicular, another property commonly attributed only to a square. The third Discovery Activity leads students to complete the list of properties of a square.

The Home Activity provides a table that students may copy into their notebooks to use as a reference on the properties of quadrilaterals.

SECTION 8.1

Resources
Reteaching 8.1
Enrichment 8.1

Objectives
After completing this section, the student should understand
- ▲ the properties that determine a quadrilateral.
- ▲ the difference between convex and concave quadrilaterals.

Materials
ruler
protractor
cardboard strips
paper fasteners

Vocabulary
quadrilateral
diagonal
convex
concave

GETTING STARTED

Quiz: Prerequisite Skills

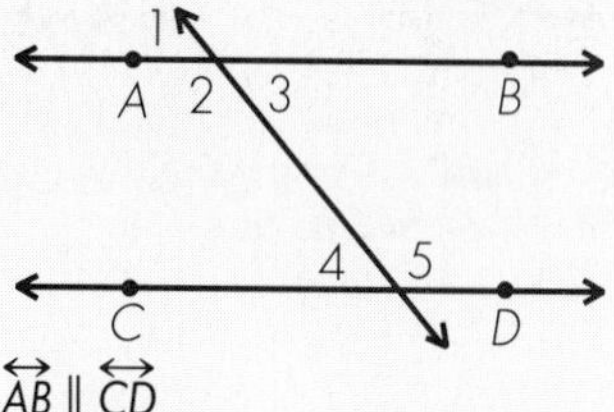

$\overleftrightarrow{AB} \parallel \overleftrightarrow{CD}$

1. Which pairs of angles have equal measures? Explain how you know. [m∠1 = m∠3, vertical angles; m∠1 = m∠4, corresponding angles; m∠2 = m∠5, alternate interior angles]

2. What is the sum of m∠2 and m∠3? Why? [180°; straight line]

TEACHING COMMENTS

Using the Page
Have students try to form a figure that contains four points, three of which are collinear. They should conclude that the figure will be a triangle.

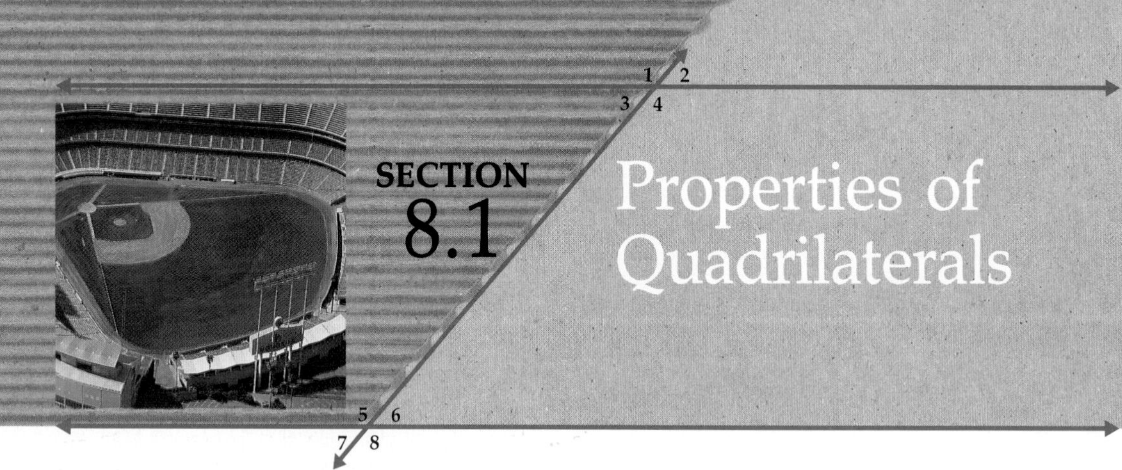

SECTION 8.1 Properties of Quadrilaterals

After completing this section, you should understand
- ▲ the properties that determine a quadrilateral.
- ▲ the difference between convex and concave quadrilaterals.

In Chapter 3 you found that a triangle is a figure formed by the line segments that connect three noncollinear, coplanar points. If you connect four coplanar points in order, you will form another kind of figure.

Examples:

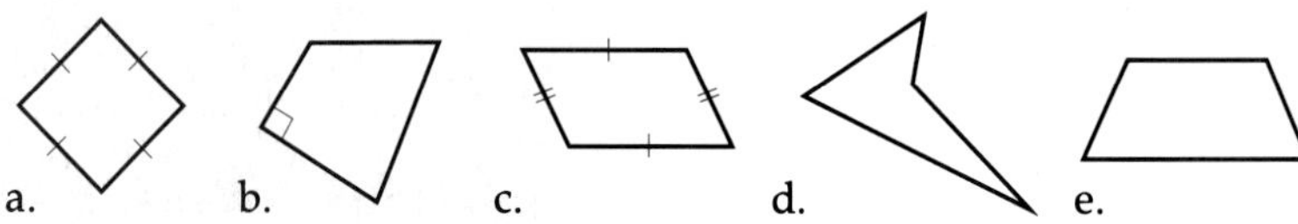

DISCOVERY ACTIVITY

1. Copy and complete the table for figures a–e shown above.

Figure	a	b	c	d	e
Number of sides	4	4	4	4	4
Number of vertices	4	4	4	4	4
At least one pair of opposite sides parallel	√		√		√
opposite sides congruent	√		√		
Opposite sides parallel and congruent	√		√		
No parallel sides		√		√	
Number of collinear points	2	2	2	2	2

2. Use the information in the table to write a description of the figures that will be true for all of them. The figures all have 4 sides and 4 vertices or angles. None of the figures has more than 2 collinear points.

You may have discovered that the figures all have four sides and four vertices.

DEFINITION 8-1-1 **A QUADRILATERAL is a closed, plane, four-sided figure.**

A quadrilateral consists of four line segments that intersect only at their endpoints. It is a *closed* figure because the line segments are connected consecutively. It is a *plane* figure because all four points are on the same plane.

$\overline{JK}$ and $\overline{KL}$ are adjacent sides.
$\overline{KL}$ and $\overline{JM}$ are opposite sides.
K and L are adjacent vertices.
J and L are opposite vertices.
$\angle L$ and $\angle M$ are adjacent angles.
$\angle J$ and $\angle L$ are opposite angles.

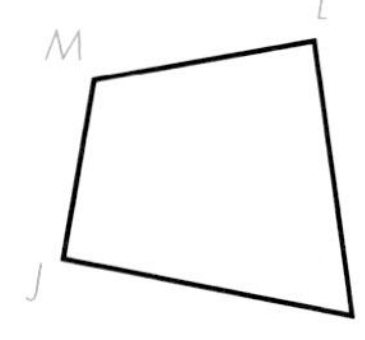

Two additional line segments, the diagonals, are also associated with a quadrilateral.

NOTEBOOK DEFINITION 8-1-2 **A DIAGONAL is a line segment determined by two non-adjacent vertices.**

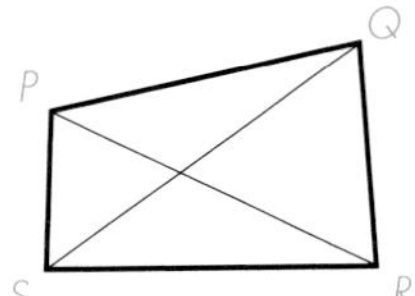

Example: $\overline{PR}$ and $\overline{QS}$ are diagonals of $PQRS$.

CLASS ACTIVITY

Refer to figure $GHJK$.

1. Name the pairs of opposite sides and angles.
2. List the pairs of adjacent sides and angles. See margin.
3. Name the diagonals.
   GJ and KH

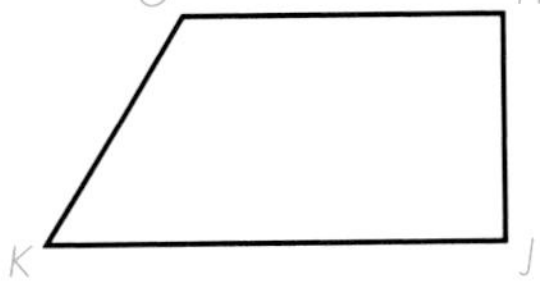

1. $\overline{GH}$ and $\overline{KJ}$, $\overline{GK}$ and $\overline{HJ}$; $\angle G$ and $\angle J$, $\angle H$ and $\angle K$

Is the figure a quadrilateral? If not, explain why.

4. Yes
5. No; line segments intersect at point that is not an endpoint.
6. No; triangle

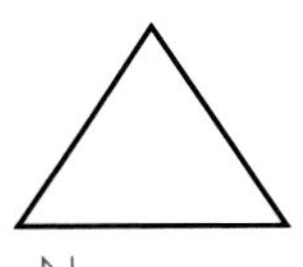

7. Yes

Using the Page
Students may write descriptions covering some but not all points of similarity between the figures. Help them find all the attributes common to the figures.

Using the Page
During the discussion of Definition 8-1-1, discuss what is meant by a closed figure. Have students draw examples of open and closed figures on the chalkboard. Then discuss what is meant by a plane figure. Have students draw examples of plane and space figures.

Using the Page
During the discussion of Definition 8-1-2, ask whether a triangle can have a diagonal. [No]

Additional Answers
2. GH and HJ, HJ and JK, JK and KG, KG and GH; $\angle G$ and $\angle H$, $\angle H$ and $\angle J$, $\angle J$ and $\angle K$, $\angle K$ and $\angle G$

Extra Practice
Refer to figure $ABCD$.

1. Name the diagonals. [$\overline{AC}$, $\overline{BD}$]

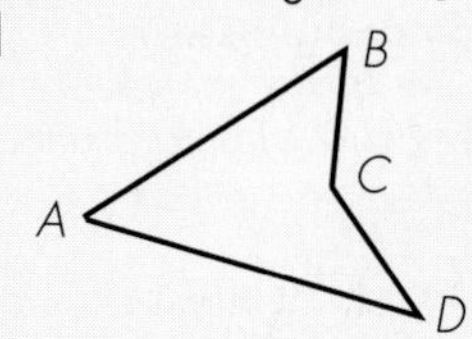

2. Name pairs of adjacent sides. [$\overline{AB}$, $\overline{BC}$; $\overline{BC}$, $\overline{CD}$; $\overline{CD}$, $\overline{DA}$; $\overline{DA}$, $\overline{AB}$]

Using the Page
Have students look up non-mathematical definitions for *convex* and *concave*. If possible, display a convex mirror. Ask students to determine the least number of sides a figure must have in order to be concave. [4]

Using the Page
Have students draw several concave quadrilaterals and explain why the figures have an interior angle measure of 360°.

TYPES OF QUADRILATERALS

Compare quadrilateral *ABCD* with quadrilateral *EFGH*. Notice that in *ABCD*, the measure of each angle is less than 180°. In *EFGH*, one angle measures more than 180°. The diagonals $\overline{AC}$ and $\overline{BD}$ in *ABCD* differ from diagonal $\overline{EG}$ in *EFGH* by being inside the figure; $\overline{EG}$ is outside.

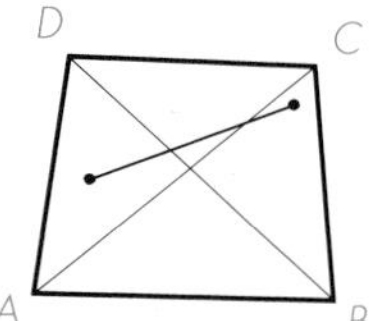

DEFINITION 8-1-3 **A CONVEX quadrilateral is a quadrilateral with the measure of each interior angle less than 180°.**

ABCD is convex because each of its angles is less than 180°, and its diagonals lie inside the figure. In fact, any line segment connecting any two points inside a convex quadrilateral will lie inside the figure.

DEFINITION 8-1-4 **A CONCAVE quadrilateral is a quadrilateral with one interior angle whose measure is greater than 180°.**

EFGH is concave because the measure of interior ∠*H* is greater than 180°, and one of its diagonals, $\overline{EG}$, lies outside the figure. In fact if any line segment connecting any two points inside a quadrilateral intersects the sides of the figure, then the figure is concave.

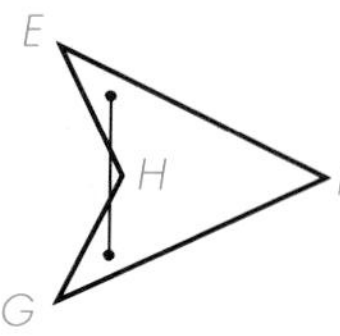

DISCOVERY ACTIVITY

Recall that if a figure is a triangle, then the sum of the measures of its angles is 180° (Theorem 3-1-1).

1. Draw a quadrilateral *ABCD*.
2. Draw the diagonal $\overline{AC}$.
3. How many triangles have you formed? 2
4. What is the sum of the angles in △*ABC*? 180°
5. Write a conclusion about the sum of the measures of the interior angles of quadrilateral ABCD. Give a reason for your conclusion. See margin.
6. Do you think your conclusion will be true for any quadrilateral? Why? See margin.

Additional Answers
5. The sum of the interior angles is 360°; the diagonal of a quadrilateral forms two triangles, 2 · 180° = 360°.
6. Yes; all quadrilaterals may be divided into two triangles by their diagonals.

THEOREM 8-1-1 **The sum of the measures of the interior angles of a quadrilateral is 360°.**

CLASS ACTIVITY

Tell whether the figure is convex or concave. Find the sum of the measures of the interior angles.

1. 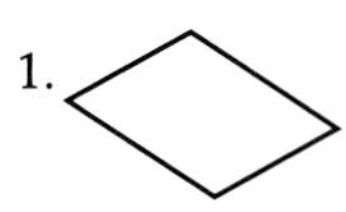
Convex; 360°

2. Concave; 360°

3. 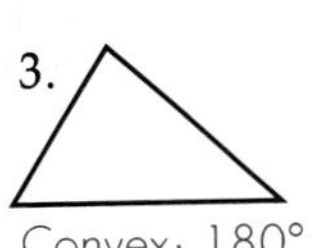
Convex; 180°

4. 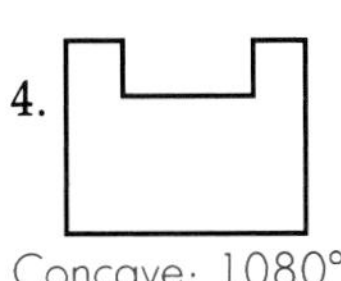
Concave; 1080°

A SPECIAL PROPERTY OF TRIANGLES

The physical use of diagonals is an important factor in architecture. In the construction of buildings, bridges, and other structures, the diagonal provides support for four-sided figures. Try this experiment.

1. Use four cardboard strips and paper fasteners to construct a convex quadrilateral fastened at the vertices.

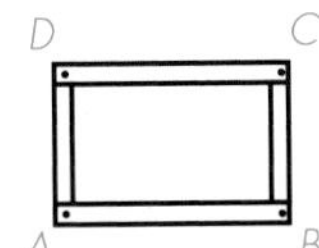

2. Hold the quadrilateral by the sides $\overline{AD}$ and $\overline{BC}$ and move your hands up and down. What happens to the quadrilateral?
3. Cut another cardboard strip to fit from A to C. Attach it at A and C. Repeat step 2.

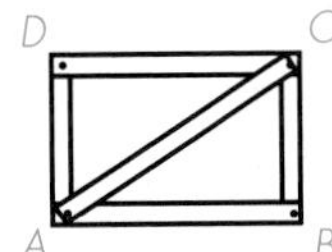

4. What do you observe?
The diagonal makes the figure rigid.

THEOREM 8-1-2 **A triangle is a rigid or non-flexible figure.**

Once the diagonal is attached, the quadrilateral becomes inflexible or rigid. Notice that the diagonal converts the quadrilateral into two triangles. This is why the triangle is called a rigid figure.

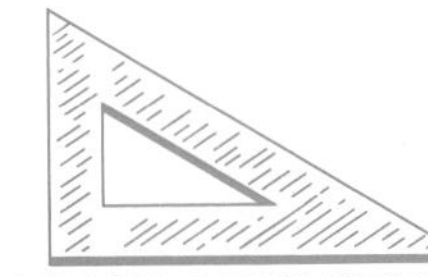

HOME ACTIVITY

Refer to figure *LMNO*.

1. Name the diagonals. $\overline{LN}$ and $\overline{MO}$
2. Name the pairs of opposite sides. $\overline{LO}$ and $\overline{MN}$, $\overline{LM}$ and $\overline{MO}$
3. Name the pairs of opposite angles. ∠L and ∠N, ∠M and ∠O

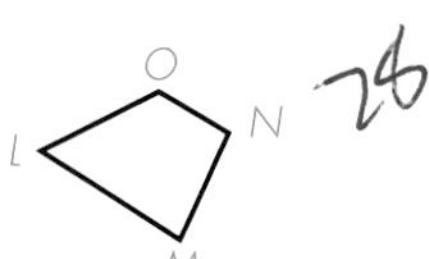

Extra Practice
Tell whether the figure is convex or concave. Find the sum of the measures of the interior angles.

1. 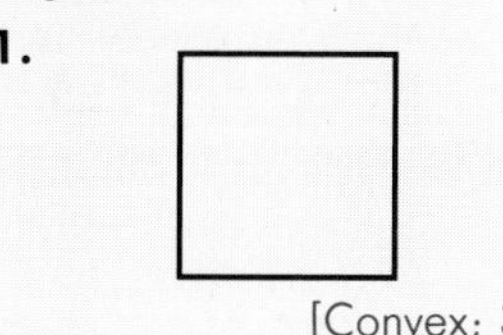
[Convex; 360°]

2. 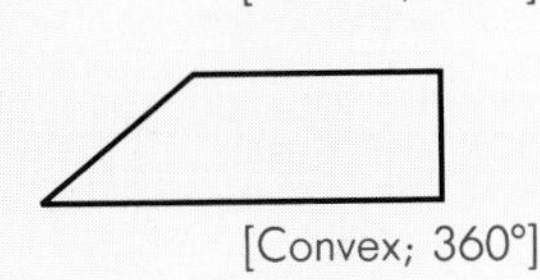
[Convex; 360°]

3. 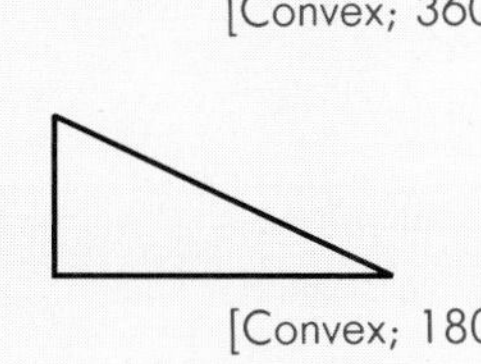
[Convex; 180°]

4. 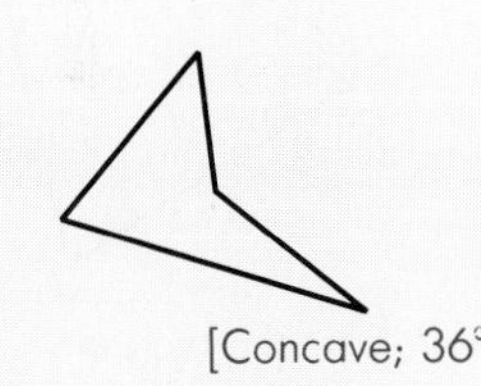
[Concave; 36°]

Using the Page
Have students experiment with quadrilaterals that have four sides of unequal length as well as parallelograms and isosceles trapezoids. You may wish to have them bring to class pictures of structures that use quadrilaterals divided by diagonals.

Refer to figure *QRST*.

4. Name the diagonals. $\overline{RT}$ and $\overline{QS}$

5. Name the adjacent sides.
$\overline{QR}$ and $\overline{RS}$, $\overline{RS}$ and $\overline{ST}$, $\overline{ST}$ and $\overline{TQ}$, $\overline{TQ}$ and $\overline{QR}$

6. Name the adjacent angles.
$\angle Q$ and $\angle R$, $\angle R$ and $\angle S$, $\angle S$ and $\angle T$, $\angle T$ and $\angle Q$

Which of the figures below are

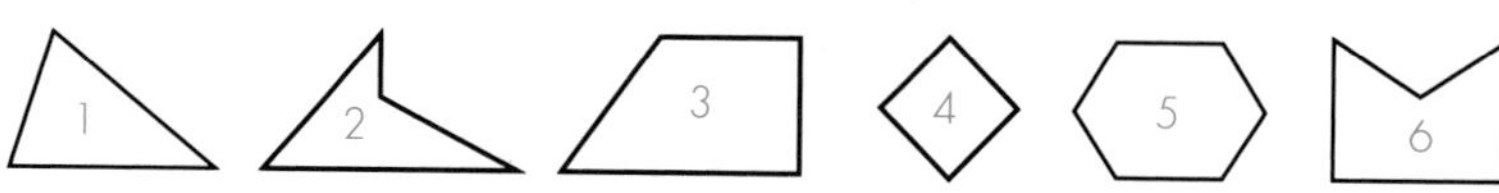

7. quadrilaterals? 2, 3, 4

8. triangles? 1

9. concave? 2, 6

10. convex? 1, 3, 4, 5

11. not quadrilaterals? 1, 5, 6

Find the sum of the measures of the interior angles.

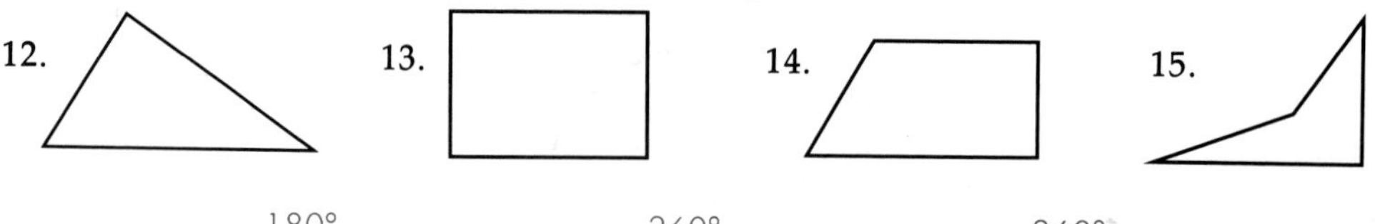

12. 180°

13. 360°

14. 360°

15. 360°

Write Logo commands that tell the turtle to draw each figure described below.

16. Convex quadrilateral
Answers may vary.

17. Concave quadrilateral
Answers may vary.

CRITICAL THINKING

18. Copy and complete the table.

Number of Sides	Sum of Interior Angles	Name of Polygon	Number of Sides	Sum of Interior Angles	Name of Polygon
3	180°	triangle	8	1080° (6 × 180°)	octagon
4	360° (2 × 180°)	quadrilateral	10	1440° (8 × 180°)	decagon
5	540° (3 × 180°)	pentagon	12	1800° (10 × 180°)	dodecagon
6	720° (4 × 180°)	hexagon	n	$(n - 2) \times 180°$	*n*-gon

Using the Page
18. Have students draw four-, five-, and six-sided polygons and list the number of diagonals to find the pattern.

FOLLOW-UP

Assessment
Refer to figure *GHJK*.

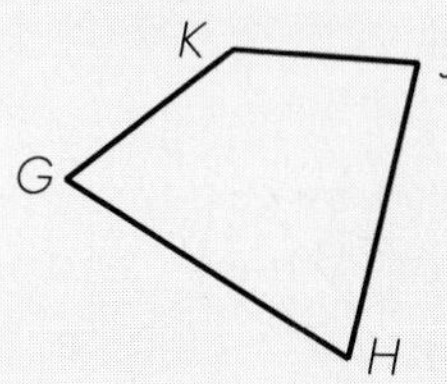

1. Name the diagonals. [$\overline{GJ}$, $\overline{KH}$]

2. Name the two pairs of opposite sides. [$\overline{GH}$, $\overline{KJ}$; $\overline{GK}$, $\overline{JH}$]

3. Name two pairs of opposite angles. [$\angle G$, $\angle J$; $\angle H$, $\angle K$]

Find the sum of the measures of the interior angles.

4.

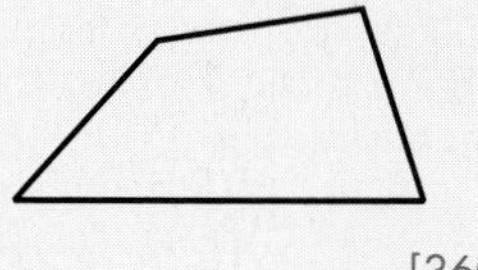

[360°]

5.

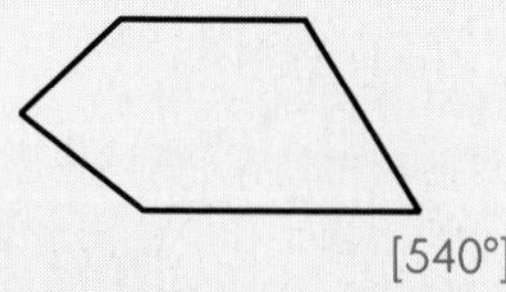

[540°]

Reteaching
Students who have had difficulty with this section may benefit from Reteaching Activity 8.1.

Enrichment
For students who have mastered the material in this section, you may wish to assign Enrichment Activity 8.1.

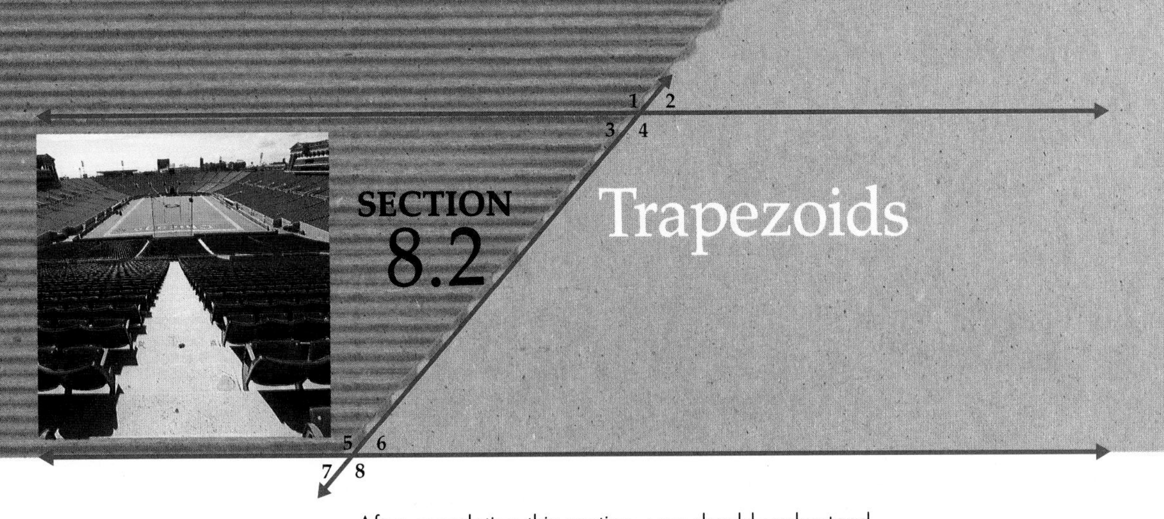

Trapezoids

After completing this section, you should understand
▲ the properties of trapezoids and isosceles trapezoids.
▲ how to apply the theorems about isosceles trapezoids and medians of trapezoids.

A trapezoid is a member of the set of quadrilaterals. This means that any property of a quadrilateral also applies to a trapezoid: It is a four-sided closed figure, and the sum of its interior angles is 360°. One more condition is necessary for a quadrilateral to be a trapezoid—one pair of sides must be parallel.

A TRAPEZOID is a quadrilateral with one and only one pair of parallel sides.

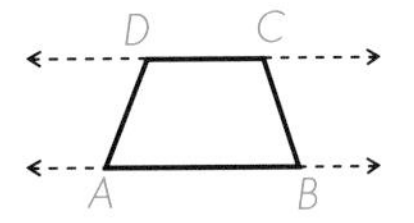

The parallel sides are called **BASES.** A special class of trapezoids has non-parallel sides that are equal in length. These sides are called **LEGS.**

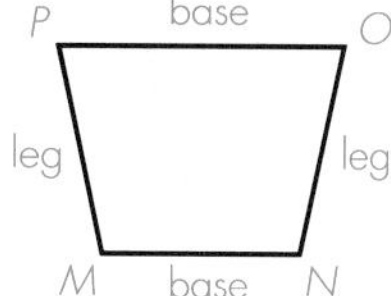

An ISOSCELES TRAPEZOID is a trapezoid with two non-parallel equal sides.

DISCOVERY ACTIVITY

1. On graph paper, draw several isosceles trapezoids.
2. Use a protractor to measure the interior angles of each trapezoid.

SECTION
8.2

Resources
Reteaching 8.2
Enrichment 8.2
Transparency Master 8-1

Objectives
After completing this section, the student should understand
▲ the properties of trapezoids and isosceles trapezoids.
▲ how to apply the theorems about isosceles trapezoids and medians of trapezoids.

Materials
ruler
protractor
compass

Vocabulary
trapezoid
isosceles trapezoid
median
base
leg

GETTING STARTED

Chalkboard Example
Draw an irregular quadrilateral and discuss the fact that no two sides are parallel. Have students experiment to see how many different kinds of quadrilaterals they can draw with just two parallel sides.

TEACHING COMMENTS

Using the Page
During the discussion of Definition 8-2-2, ask the class how they might relate what they know about isosceles triangles to the isosceles trapezoid.

3. What do you observe about the measures of the angles in each figure? Compare your observations with your classmates'. Two pairs of angles are always equal.

Using the Page
Go through each step of the proof of Theorem 8-2-1 to be sure students understand the reasons. Point out that step 9 justifies the first part of the proof and step 11 justifies the second part.

ISOSCELES TRAPEZOIDS

You may have discovered that an isosceles trapezoid has two pairs of angles with equal measures. You can use what you know about similar and isosceles triangles to prove that the measures of the base angles of an isosceles trapezoid are equal.

Given: Trapezoid $ABCD$ with $AD = BC$
Justify: $m\angle DAB = m\angle CBA$ and $m\angle ADC = m\angle BCD$

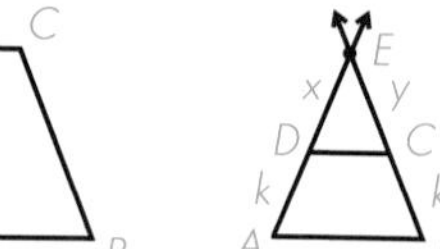

Have students supply the missing reasons.

Steps	Reasons
1. $AD = BC$	1. Given
2. Extend $\overline{AD}$ and $\overline{BC}$ until they intersect at E.	2. Two non-parallel lines intersect at one point.
3. $\overline{DC} \parallel \overline{AB}$, so $m\angle DAB = m\angle EDC$ and $m\angle CBA = m\angle ECD$.	3. Corresponding angles have equal measures.
4. $\triangle ABE \sim \triangle DCE$	4. AA
5. $\frac{x}{(x+k)} = \frac{y}{(y+k)}$	5. Corresponding sides are in equal ratio.
6. $x(y+k) = y(x+k)$ $xy + xk = yx + yk$ $xk = yk$ $x = y$	6. Algebra
7. $\triangle EDC$ is isosceles.	7. Two equal sides
8. $m\angle EDC = m\angle ECD$	8. Angles opposite equal sides are equal.
9. $m\angle DAB = m\angle CBA$	9. Substitution
10. $m\angle ADC + m\angle EDC = 180°$ and $m\angle BCD + m\angle ECD = 180°$	10. Supplementary angles
11. $m\angle ADC = m\angle BCD$	11. Supplementary to equal angles

THEOREM 8-2-1 **In an isosceles trapezoid, the measures of each pair of base angles are equal.**

CLASS ACTIVITY

Find the missing angle measures for each isosceles trapezoid.

1. D C A B 70°

m∠B = 70°
m∠C = 110°
m∠D = 110°

2. 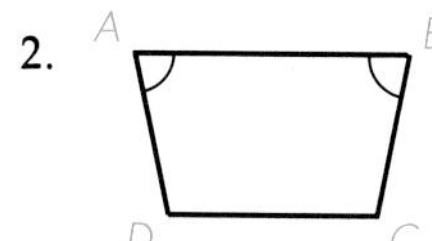

m∠A = (6x + 50)°
m∠B = (7x + 45)°
m∠A = 80° m∠C = 100°
m∠B = 80° m∠D = 100°

3. P O M N 65° 43°

m∠MPN = 72°
m∠NPO = 43°
m∠PON = 115°
m∠ONP = 22°

4. T S Q R 24° 38°

m∠SRQ = 62°
m∠QSR = 80°
m∠QST = 38°
m∠STQ = 118°

5. **Is a trapezoid concave or convex? Why?** Convex; all angles are less than 180°.

DEFINITION 8-2-3 **The MEDIAN of a trapezoid is the line segment joining the midpoints of the two non-parallel sides.**

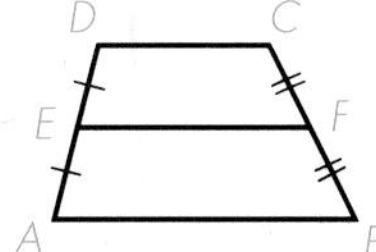

DISCOVERY ACTIVITY

1. On graph paper, draw several *ABCD* trapezoids with *AB* ∥ *DC*.
2. Use a compass to locate the midpoints of the two non-parallel sides for each figure. Label the midpoints *E* and *F*.
3. Draw the medians.
4. Measure to find the length of the median and two bases for each figure. *EF* = ?
   AB = ? *DC* = ? *AB* + *DC* = ?
5. How does the length of a median compare to the sum of the lengths of the two bases? It is half as long.
6. Write a conclusion stating what you have discovered.

Extra Practice

Find the missing angle measures for each isosceles trapezoid.

1.

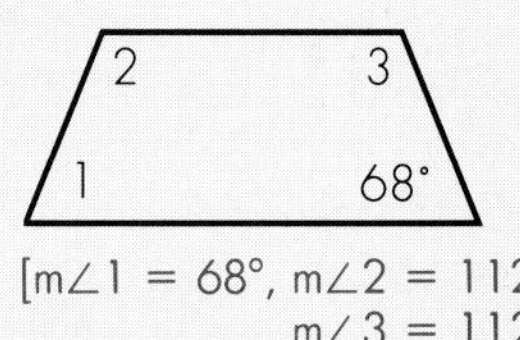

[m∠1 = 68°, m∠2 = 112°, m∠3 = 112°]

2.

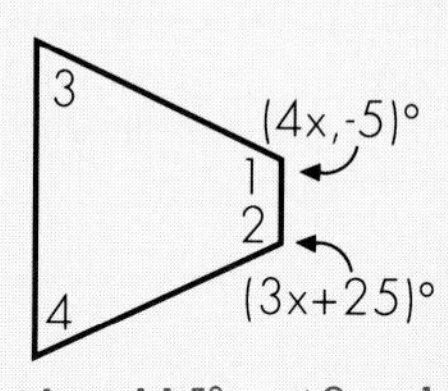

[m∠1 = 115°, m∠2 = 115°, m∠3 = 65°, m∠4 = 65°]

3.

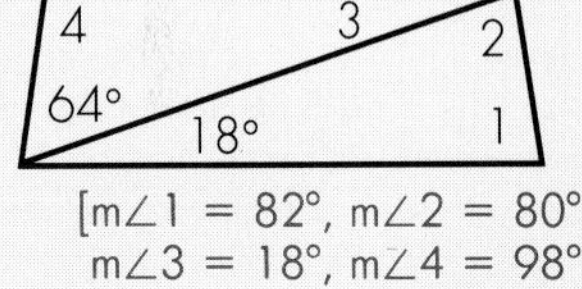

[m∠1 = 82°, m∠2 = 80°, m∠3 = 18°, m∠4 = 98°]

Using the Page

During the discussion of definition 8-2-3, ask students to explain how the median of a trapezoid compares to the median of a triangle.

Using the Page

During the Discovery Activity, have a student explain how to use a compass to find the midpoint of a line segment. Remind students to measure carefully.

Using the Page
Before beginning the proof, ask students to suggest a strategy and to tell what information they will need to complete the proof.

MEDIANS OF TRAPEZOIDS

You may have found that the length of the median of a trapezoid is one half the sum of the lengths of the bases. You may also have noticed that the median of a trapezoid appears to be parallel to its bases.

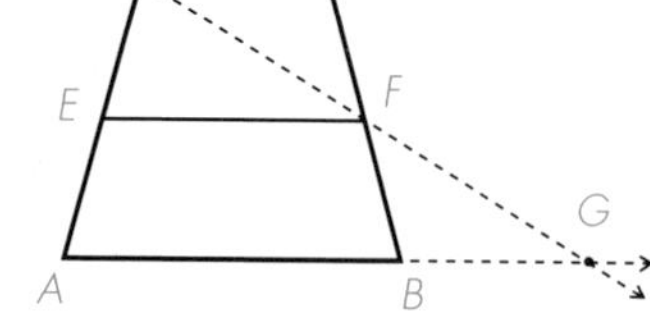

Given: Trapezoid $ABCD$ with $\overline{AB} \parallel \overline{DC}$
E is the midpoint of $\overline{AD}$.
F is the midpoint of $\overline{BC}$.
Justify: $EF = \frac{1}{2}(AB + DC)$;
$\overline{EF} \parallel \overline{AB}$ and $\overline{EF} \parallel \overline{DC}$

Have students supply the missing reasons.

Steps	Reasons
1. $\overline{AB} \parallel \overline{DC}$; E and F are midpoints of $\overline{AD}$ and $\overline{BC}$.	1. Given
2. Draw the median $\overline{EF}$.	2. Two points determine a line.
3. Extend $\overline{AB}$ and draw $\overleftrightarrow{DF}$ to intersect $\overline{AB}$ at G.	3. Two non-parallel lines intersect at one point.
4. $m\angle BFG = m\angle CFD$	4. Vertical angles are equal.
5. $m\angle FBG = m\angle FCD$	5. Alternate interior angles are equal.
6. $CF = FB$	6. Definition of midpoint
7. $\triangle BFG \cong \triangle CFD$	7. ASA
8. $\overline{DC} \cong \overline{BG}$ and $\overline{DF} \cong \overline{FG}$	8. CPCF
9. For $\triangle ADG$, $\overline{EF} \parallel \overline{AG}$, and $EF = \frac{1}{2}AG$.	9. Line joining midpoints of a triangle is parallel to the third side and equal to $\frac{1}{2}$ its length.
10. $AG = AB + BG$	10. Definition of line segment
11. $AG = AB + DC$	11. Substitution
12. $EF = \frac{1}{2}(AB + DC)$	12. Substitution

NOTEBOOK **THEOREM 8-2-2** **The median of a trapezoid is parallel to the bases and equal to one-half the sum of their lengths.**

Using the Page
Students who have difficulty understanding Theorem 8-2-2 might be encouraged to draw several trapezoids on graph paper, draw and measure each median, then compare to find how it relates to the sum of the bases. You wish to use Transparency Master 8-1 to show that the theorem is true of all types of trapezoids.

CLASS ACTIVITY

Name the bases, legs, and median.

1. Bases: $\overline{AB}$, $\overline{DC}$; median: $\overline{EF}$; legs: $\overline{AD}$, $\overline{BC}$

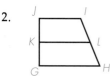

2. Bases: $\overline{GH}$, $\overline{JI}$; median: $\overline{KL}$; legs: $\overline{GJ}$, $\overline{HI}$

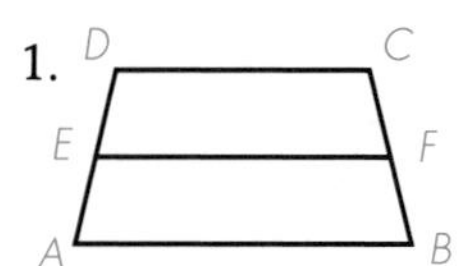

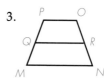

3. Bases: $\overline{MN}$, $\overline{PO}$; median: $\overline{QR}$; legs: $\overline{MP}$, $\overline{NO}$

Find the missing lengths.

4. 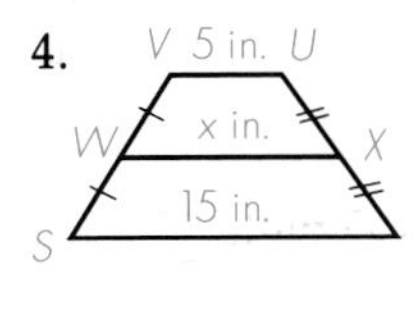

$x = 10$ in.

5. 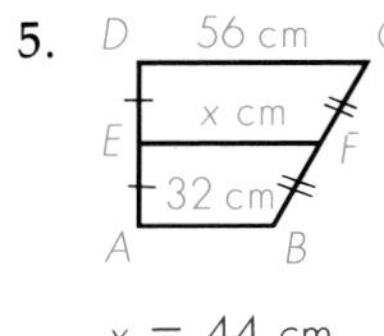

$x = 44$ cm

6. 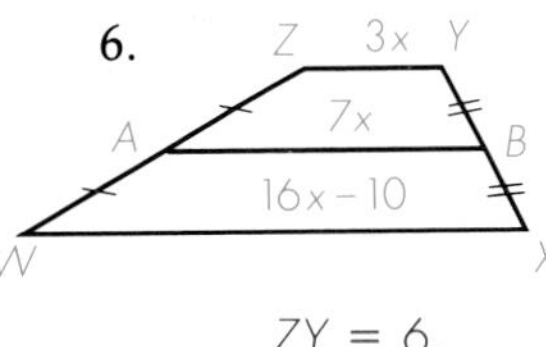

$ZY = 6$, $WX = 22$, $AB = 14$

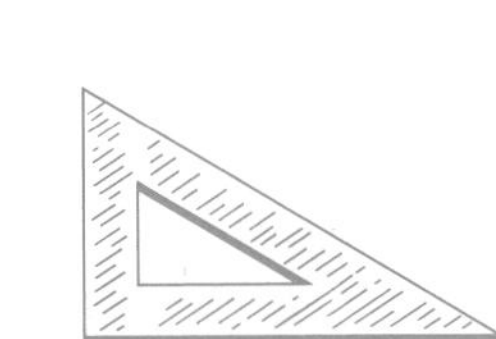

HOME ACTIVITY

40

Is the figure a trapezoid? If not, explain why.

1.  No; triangle
2.  Yes
3. 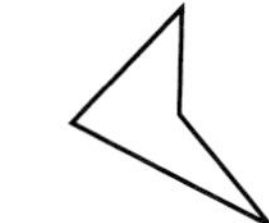 No; concave
4.  No; parallelogram
5. 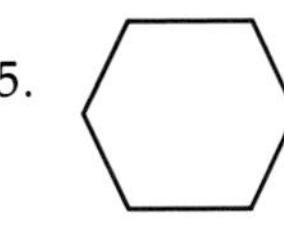 No; hexagon

Name the bases, legs, and median for each figure. The points between the vertices are the midpoints of the sides. See margin.

6. 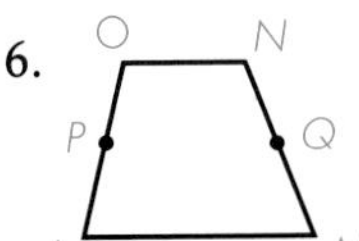

7. 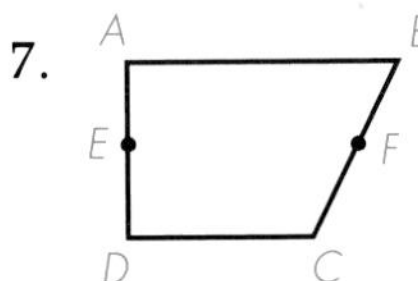

8. 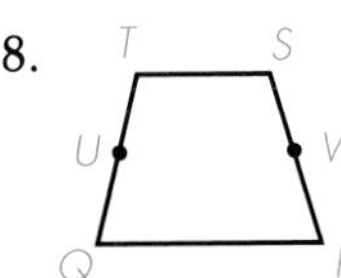

Use pencil and paper to find the measurement of the numbered angles in each trapezoid shown below. Then write Logo commands to draw each trapezoid to check your answers.

9. 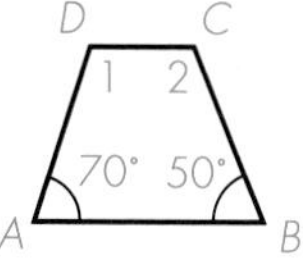

$m\angle 1 = 110°$
$m\angle 2 = 130°$

10. 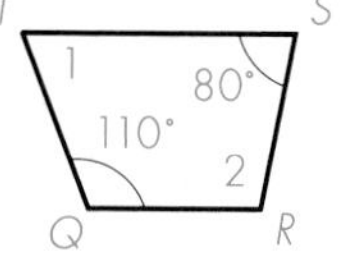

$m\angle 1 = 70°$
$m\angle 2 = 100°$

11. 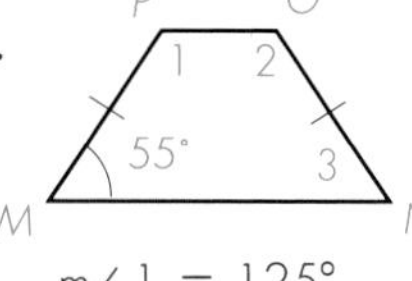

$m\angle 1 = 125°$
$m\angle 2 = 125°$
$m\angle 3 = 55°$

Find the measures of the numbered angles.

12. 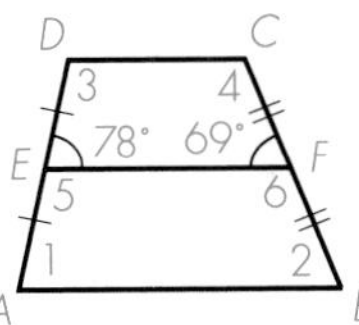

See margin.

13. 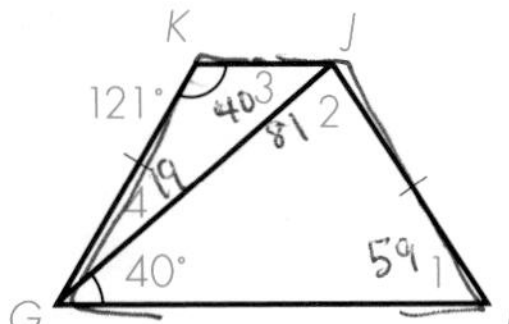

14. 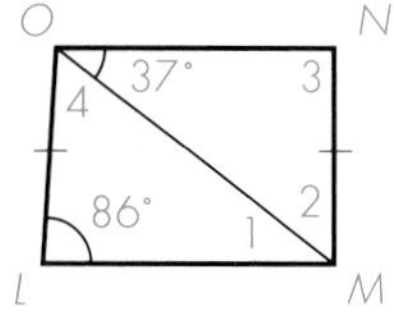

Extra Practice
Find the missing lengths.

1. 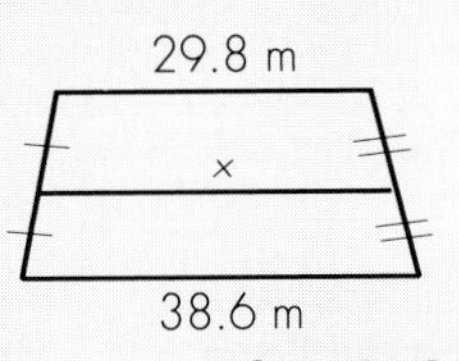

[x = 34.2 cm]

2. 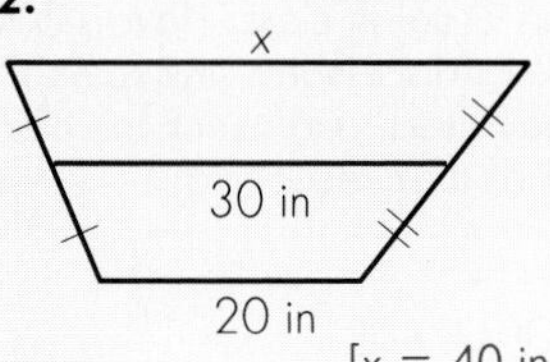

[x = 40 in.]

3. 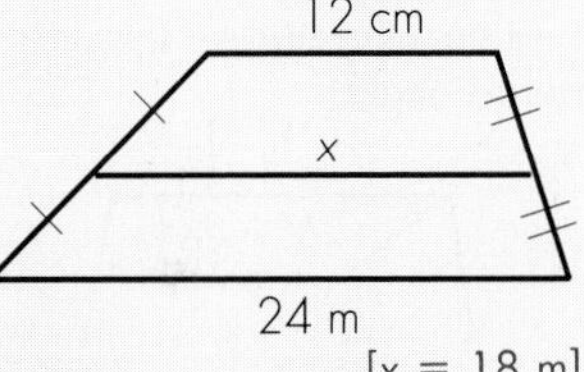

[x = 18 m]

Additional Answers

6. Bases: $\overline{LM}$, $\overline{ON}$; legs: $\overline{LO}$, $\overline{MN}$; median: $\overline{PQ}$

7. Bases: $\overline{DC}$, $\overline{AB}$; legs: $\overline{DA}$, $\overline{CB}$; median: $\overline{EF}$

8. Bases: $\overline{QR}$, $\overline{TS}$; legs: $\overline{QT}$, $\overline{RS}$; median: $\overline{UV}$

12. $m\angle 1 = 78°$, $m\angle 2 = 69°$, $m\angle 3 = 102°$, $m\angle 4 = 111°$, $m\angle 5 = 102°$, $m\angle 6 = 111°$

13. $m\angle 1 = 59°$, $m\angle 2 = 81°$, $m\angle 3 = 40°$, $m\angle 4 = 19°$

14. $m\angle 1 = 37°$, $m\angle 2 = 49°$, $m\angle 3 = 94°$, $m\angle 4 = 57°$

Using the Page

18. You may wish to review this proof in class. Have students discuss the strategies they used, and check the clarity of their reasoning.

FOLLOW-UP

Assessment

Find the measures of the numbered angles.

1.

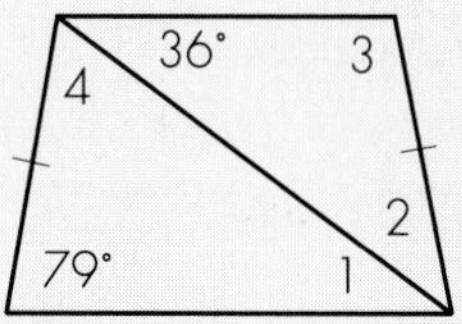

[m∠1 = 36°, m∠2 = 43°, m∠3 = 101°, m∠4 = 65°]

2.

[m∠1 = 81°, m∠2 = 70°, m∠3 = 99°, m∠4 = 110°, m∠5 = 81°, m∠6 = 70°]

Find the length.

3.

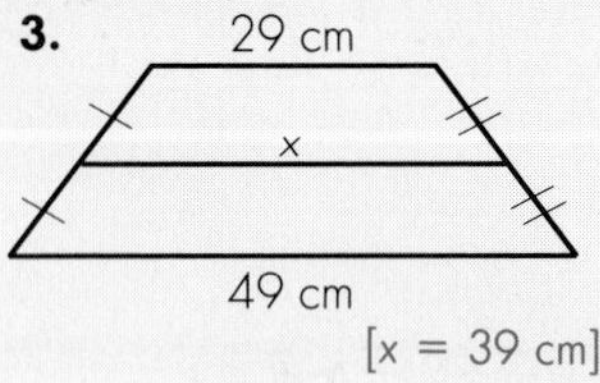

[x = 39 cm]

Reteaching

Students who have had difficulty with this section may benefit from Reteaching Activity 8.2.

Enrichment

For students who have mastered the material in this section, you may wish to assign Enrichment Activity 8.2.

Find the missing lengths.

15.
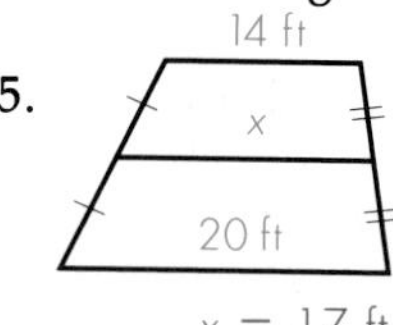

x = 17 ft

16.
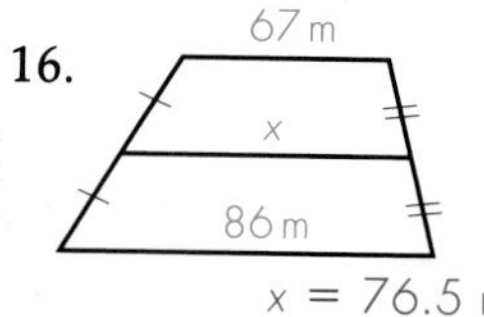

x = 76.5 m

17.
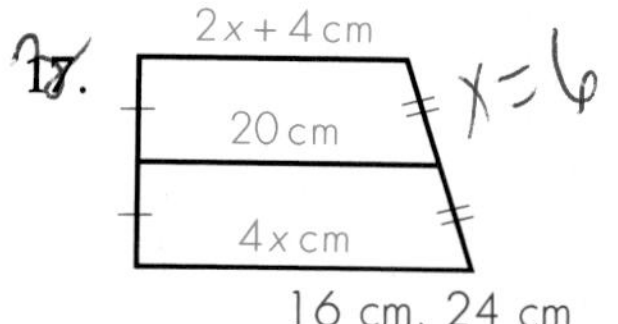

16 cm, 24 cm

18. Write reasons to complete the following proof. Then write a sentence summarizing the result.

Given: $ABCD$ is an isosceles trapezoid with $\overline{AB} \parallel \overline{DC}$.
Justify: $AC = BD$

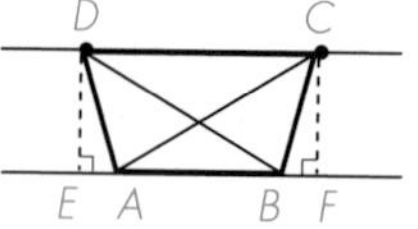

Steps	Reasons
a. Draw perpendiculars from D to $\overleftrightarrow{AB}$ and C to $\overleftrightarrow{AB}$, intersecting $\overleftrightarrow{AB}$ at E and F.	a. Construction
b. $DE = CF$	b. Points on parallel lines are equidistant.
c. $DA = CB$	c. Definition of isosceles trapezoid
d. $\triangle DEA \cong \triangle CFB$	d. Hypotenuse-leg
e. $m\angle DAE = m\angle CBF$	e. CPCF
f. $m\angle DAB = m\angle CBA$	f. Supplements of equal angles
g. $\triangle DAB \cong \triangle CBA$	g. SAS
h. $AC = BD$	h. CPCF

The diagonals of an isosceles trapezoid are congruent.

Refer to isosceles trapezoid *MNOP*.

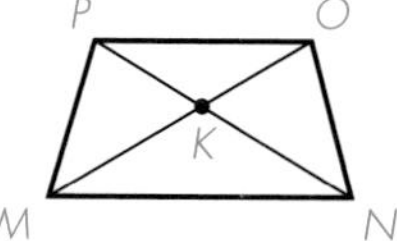

19. Name the pairs of similar triangles you can find. Explain how you know.
By AA, $\triangle PKO \sim \triangle MKN$, $\triangle PKM \sim \triangle OKN$, $\triangle PMN \sim \triangle ONM$, $\triangle PMO \sim \triangle ONP$

20. Name the pairs of congruent triangles.

CRITICAL THINKING

Draw each polygon and its diagonals. Copy and complete the following table. Can you predict the number of diagonals in a polygon with 12 sides? 54 diagonals

Polygon	Number of Sides	Number of Diagonals
Triangle	3	0
Quadrilateral	4	2
Pentagon	5	5
Hexagon	6	9
Heptagon	7	14
Octagon	8	20
Nonagon	9	27
Decagon	10	35

n + add the # of previous side, 1 less than

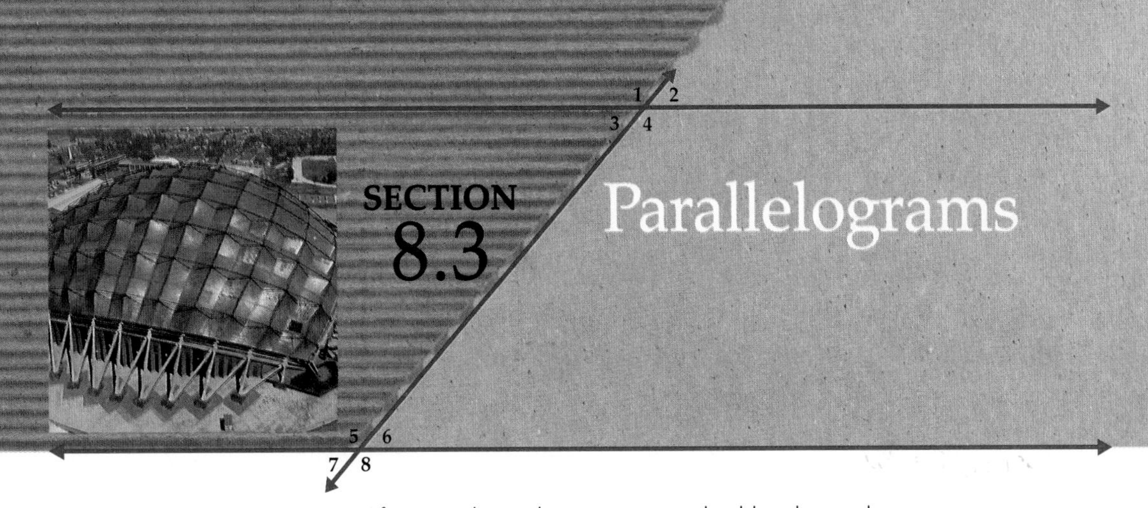

Parallelograms

After completing this section, you should understand
▲ the properties of a parallelogram.
▲ how to apply the theorems about parallelograms.

Quadrilaterals and trapezoids are all four-sided figures, but trapezoids have the additional condition that one pair of sides is parallel. Parallelograms are another subset of quadrilaterals.

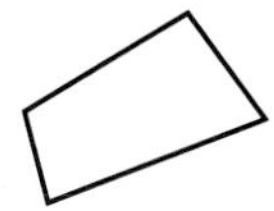
no parallel sides

1 pair of parallel sides

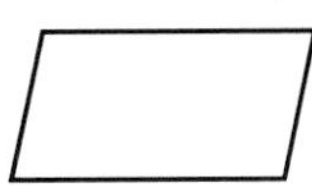
2 pairs of parallel sides

DEFINITION 8-3-1 **A PARALLELOGRAM is a quadrilateral with both pairs of opposite sides parallel.**
$\overline{AB} \parallel \overline{DC}$
$\overline{AD} \parallel \overline{BC}$

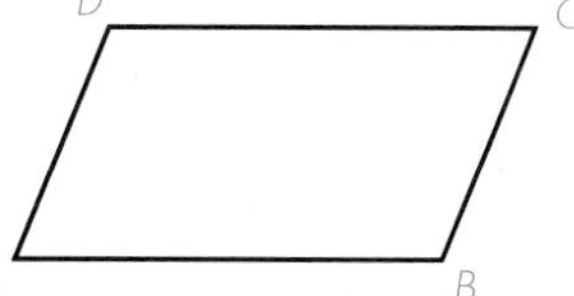

DISCOVERY ACTIVITY

1. Draw a pair of parallel lines.
2. Draw another pair of parallel lines (transversals) that intersect both of the original lines. Label the intersecting points *ABCD*.
3. What do you know about figure *ABCD*? It is a parallelogram.
4. Measure each pair of opposite sides. Then measure each pair of opposite angles. What do you observe? Opposite sides and opposite angles have equal measures.
5. Draw another parallelogram with sides and angles of a different measure. Repeat step 4. Are your conclusions the same? Compare your observations with other students'.

SECTION
8.3

Resources
Reteaching 8.3
Enrichment 8.3

Objectives
After completing this section, the student should understand
▲ the properties of a parallelogram.
▲ how to apply the theorems about parallelograms.

Materials
ruler
protractor
colored chalk (optional)

Vocabulary
parallelogram

GETTING STARTED

Quiz: Prerequisite Skills
Find the sum of the measures of the interior angles.

1. [360°]

2. [180°]

Warm-Up Activity
Have students construct and cut out pairs of congruent equilateral triangles, isosceles triangles, and 30°-60° right triangles. Then have them use each pair of triangles to form as many quadrilaterals as they can. Ask for an informal description of each figure. [Quadrilaterals with no parallel sides, with two pairs of parallel sides, with four equal sides]

TEACHING COMMENTS

Using the Page
During the Discovery Activity (page 293), you may wish to have students work in small groups, assigning a different-sized parallelogram to each student in a group. Then have each group compare the results and draw a conclusion.

Using the Page
During the discussion of the proof of Theorem 8-3-1, you may wish to draw the diagram on the chalkboard and, as the proof progresses, mark the corresponding angles in different-colored chalks. This helps students understand how the angles of the congruent triangles are related to the angles of the parallelogram.

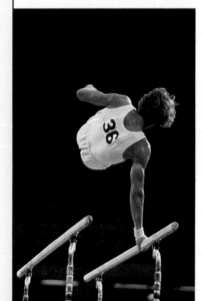

You may have found that the measures of opposite sides and opposite angles are equal for the parallelograms you drew. Do you think that this property will be true for all parallelograms?

Given: ▱$ABCD$
Justify: $m\angle A = m\angle C$, $m\angle B = m\angle D$, $AB = DC$, $AD = BC$

Have students supply the missing reasons.

Steps	Reasons
1. Draw diagonals $\overline{AC}$ and $\overline{DB}$.	1. Two points determine a line.
2. $\overline{AB} \parallel \overline{DC}$ and $\overline{AD} \parallel \overline{BC}$	2. Definition of parallelogram
3. $m\angle CAB = m\angle DCA$ and $m\angle DAC = m\angle ACB$	3. Alternate interior angles are equal.
4. $AC = AC$	4. Identity
5. $\triangle ACD \cong \triangle CAB$	5. ASA
6. $AB = DC$ and $AD = BC$	6. CPCF
7. $m\angle ADC = m\angle CBA$	7. CPCF
8. $m\angle CAB = m\angle DCA$ $+\ m\angle DAC = m\angle ACB$ $m\angle DAB = m\angle DCB$	8. Algebra
9. $m\angle A = m\angle C$	9. Substitution

THEOREM 8-3-1 **If a quadrilateral is a parallelogram, then the opposite sides are equal in length and the measures of the opposite angles are equal.**

CLASS ACTIVITY

Find the following measures for ▱$ABCD$.

1. DC 15
2. BC 10
3. $m\angle B$ 110°
4. $m\angle C$ 70°
5. $m\angle D$ 110°

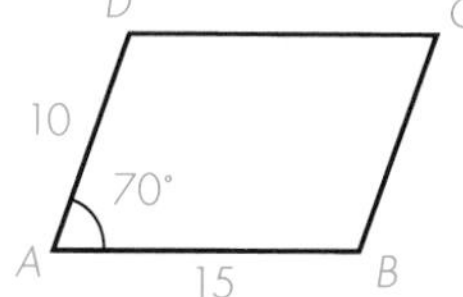

Find the following measures for ▱$PQRS$.

6. SP 42
7. SR 26
8. $m\angle P$ 120°
9. $m\angle Q$ 60°
10. $m\angle R$ 120°
11. $m\angle S$ 60°

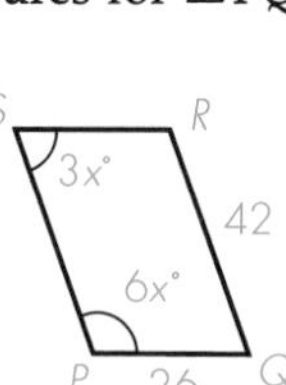

Find the angle measures.

12. 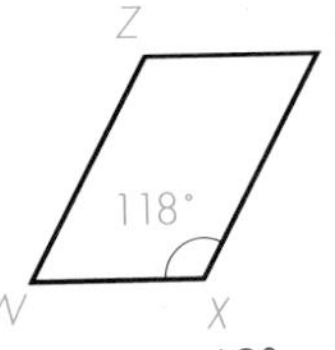

$m\angle W =$ 62°
$m\angle Y =$ 62°
$m\angle Z =$ 118°

13. 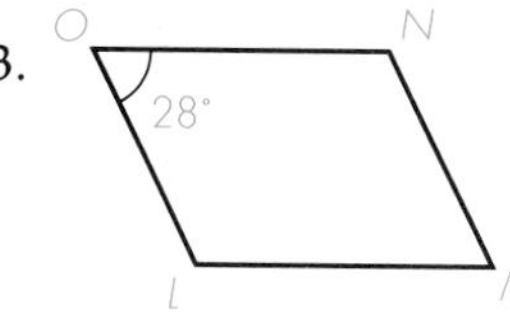

$m\angle L =$ 152°
$m\angle M =$ 28°
$m\angle N =$ 152°

14. 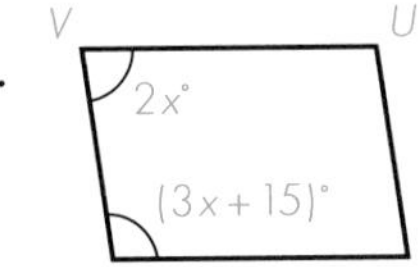

$m\angle S =$ 114°
$m\angle T =$ 66°
$m\angle U =$ 114°
$m\angle V =$ 66°

15. 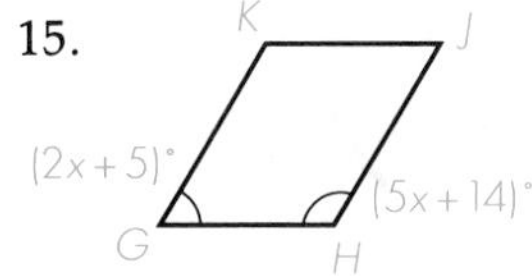

$m\angle G =$ 51°
$m\angle H =$ 129°
$m\angle J =$ 51°
$m\angle K =$ 129°

DISCOVERY ACTIVITY

1. Draw a parallelogram and label it *ABCD*, with diagonals $\overline{AC}$ and $\overline{BD}$.
2. Label the point where $\overline{AC}$ and $\overline{BD}$ intersect O.
3. Find the lengths of $\overline{AO}$ and $\overline{OC}$. Then find the lengths of $\overline{DO}$ and $\overline{OB}$. What do you observe? Lengths are equal.
4. Draw another parallelogram of a different size. Repeat steps 2 and 3. What do you find? Same relationship
5. Write a conclusion about the diagonals of a parallelogram.

Using the Page
Again, you may wish to group students and assign a different-sized parallelogram to each student in a group. Have each group compare results and draw a conclusion about the relationship between the diagonals of a parallelogram.

You may have discovered that the diagonals of a parallelogram bisect each other. This conclusion can be justified for all parallelograms.

Given: ▱*ABCD* with diagonals $\overline{AC}$ and $\overline{BD}$ intersecting at *O*

Justify: $AO = OC$, $DO = OB$
Have students supply missing reasons.

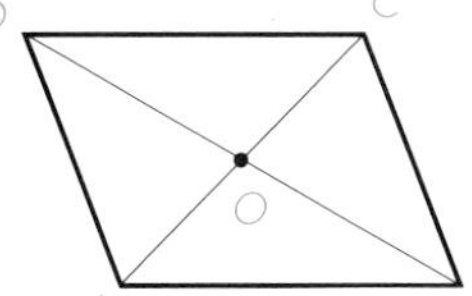

Using the Page
During discussion of the proof, ask the class to determine how many pairs of congruent triangles are formed by one diagonal of a parallelogram. [1] Then have them find the number of pairs of congruent triangles formed by two diagonals. [4]

Additional Answers

5. m∠B = 57°
m∠C = 123°
m∠D = 157°

6. m∠U = 115°
m∠UVT = 28°
m∠SVT = 37°
m∠STV = 28°

7. m∠L = 42°
m∠M = 138°
m∠N = 42°
m∠O = 138°

8. m∠SRP = 72°
m∠PTQ = 64°
m∠PQS = 44°
m∠QSR = 44°
m∠STP = 116°
m∠SPR = 36°
m∠SQR = 28°
m∠PRQ = 36°

Extra Practice

Find each measure for ▱$PQRX$.

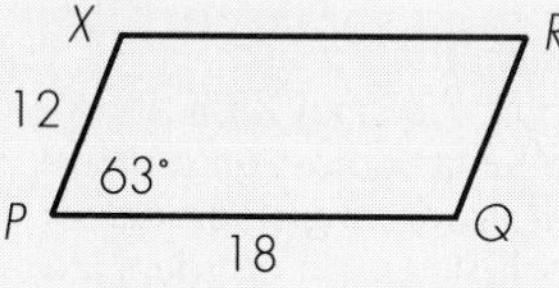

1. XR [18] **2.** QR [12]
3. m∠Q [117°]
4. m∠R [63°]

Find each length for ▱$ABCD$.

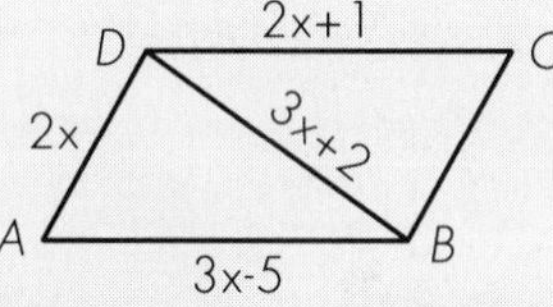

5. AB [13] **6.** CD [13]
7. AD [12] **8.** BD [20]

Find each angle measure for ▱$GHJK$.

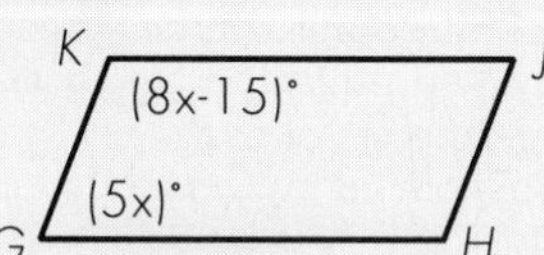

9. m∠G [75°]
10. m∠H [105°]
11. m∠J [75°]
12. m∠K [105°]

Steps	Reasons
1. $\overline{AB} \parallel \overline{DC}$	1. Definition of parallelogram
2. m∠ODC = m∠OBA and m∠OCD = m∠OAB	2. Alternate interior angles are equal.
3. $AB = DC$	3. Opposite sides of parallelogram are equal.
4. $\triangle AOB \cong \triangle COD$	4. ASA
5. $AO = OC$ and $DO = OB$	5. CPCF

THEOREM 8-3-2 **If the quadrilateral is a parallelogram, then the diagonals bisect each other.**

CLASS ACTIVITY

Find the missing lengths for each parallelogram.

1.

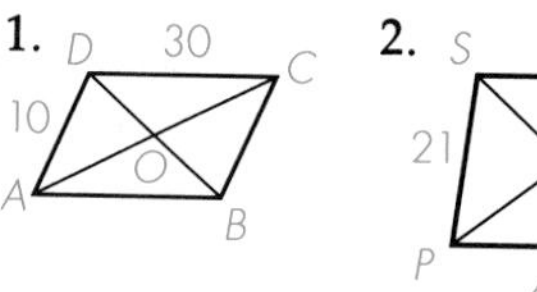

2.

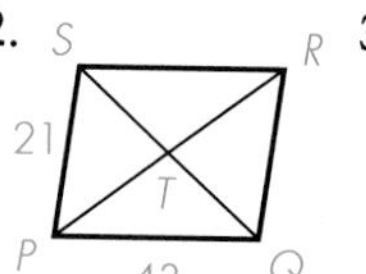

3.

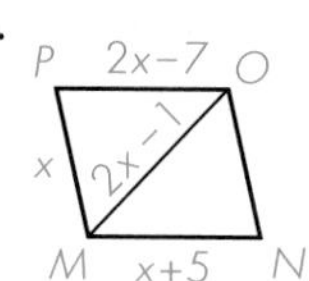

4.

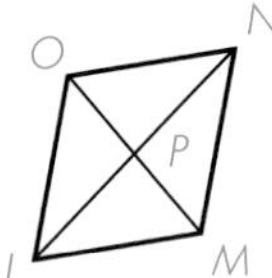

1.	2.	3.	4.
AC = 35 DO = 14	ST = 23 QR = 21	MN = 17	OP = 24
BD = 28 AB = 30	PT = 25 SR = 43	PO = 17	PN = 28
AO = 17.5 BC = 10	SQ = 46 TQ = 23	MP = 12	OM = 48
OB = 14 OC = 17.5	PR = 50 TR = 25	MO = 23	LN = 56

Find the missing angle measures for each parallelogram.

5.

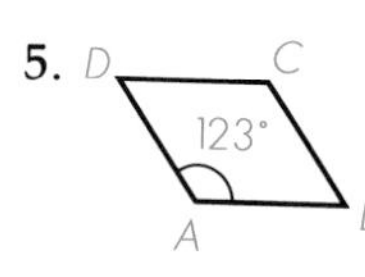

6.

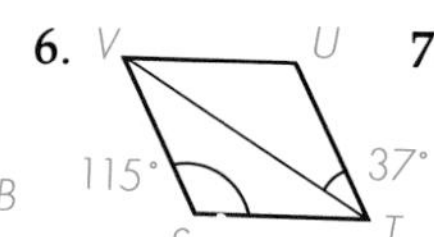

7.

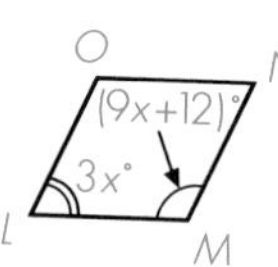

8.

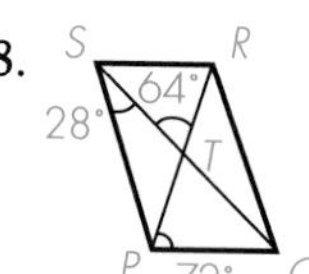

5.	6.	7.	8.	
m∠B =	m∠U =	m∠L =	m∠SRP =	m∠STP =
m∠C =	m∠UVT =	m∠M =	m∠PTQ =	m∠SPR =
m∠D =	m∠SVT =	m∠N =	m∠PQS =	m∠SQR =
See margin.	m∠STV =	m∠O =	m∠QSR =	m∠PRQ =

Is the quadrilateral a parallelogram? Give a reason for your answer.

9.

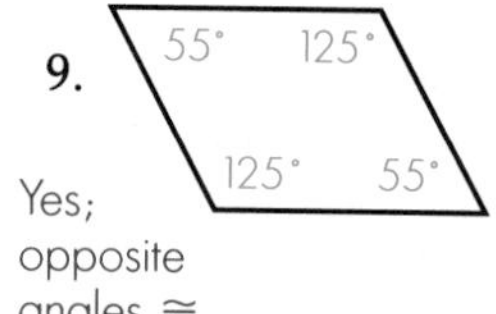

Yes; opposite angles ≅

10.

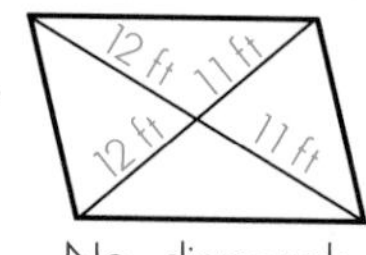

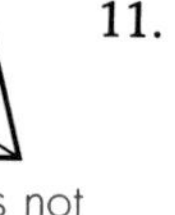

No; diagonals not bisected

11.

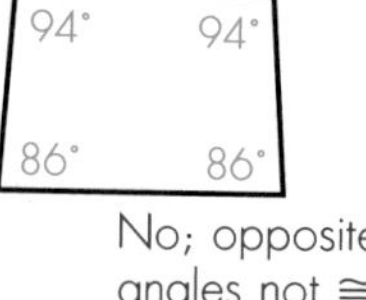

No; opposite angles not ≅

12. 1.3 m, 2.6 m, 1.3 m, 2.5 m

No; opposite sides not ≅

13.

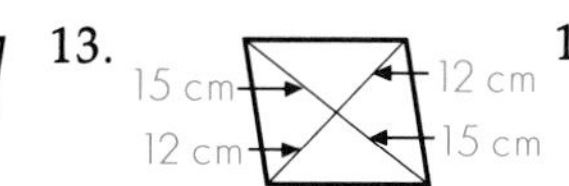

Yes; diagonals bisected

14.

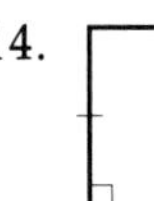

No; opposite sides not ≅

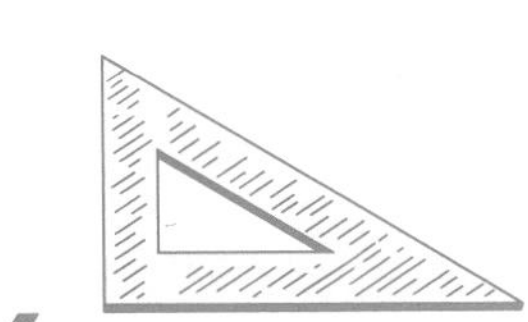

HOME ACTIVITY

Find the angle measures for each parallelogram.

1.

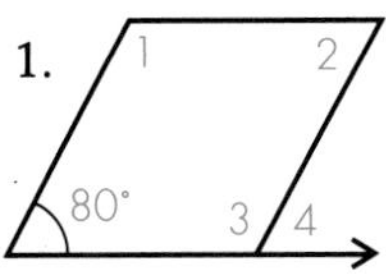

m∠ 1 = 100° m∠3 = 100°
m∠ 2 = 80° m∠4 = 80°

2.

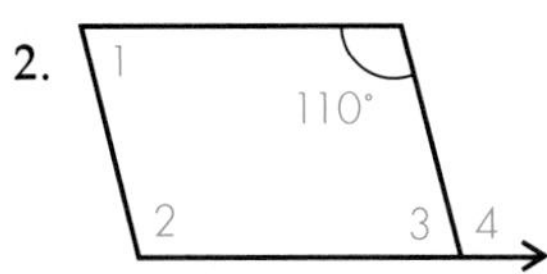

m∠1 = 70° m∠3 = 70°
m∠2 = 110° m∠4 = 110°

3.

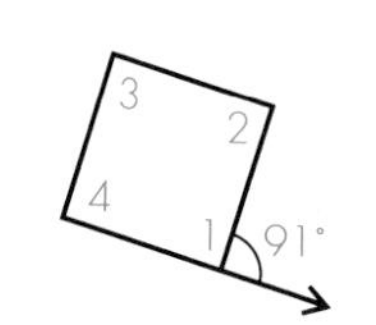

m∠1 = 89° m∠3 = 89°
m∠2 = 91° m∠4 = 91°

4.

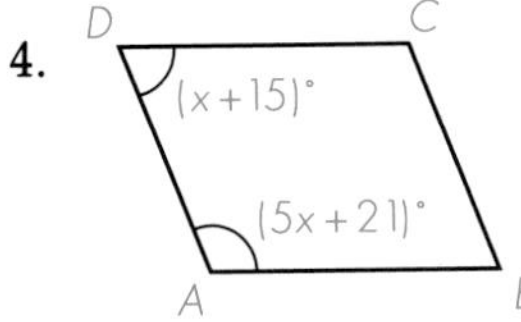

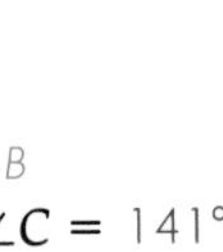

m∠A = 141° m∠C = 141°
m∠B = 39° m∠D = 39°

5.

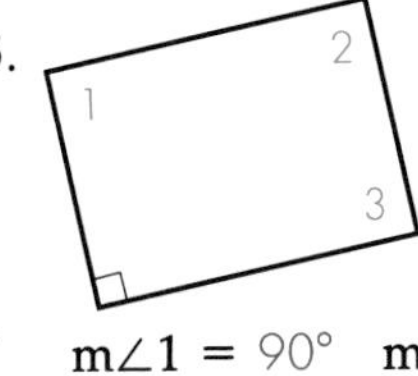

m∠1 = 90° m∠3 = 90°
m∠2 = 90°

6.

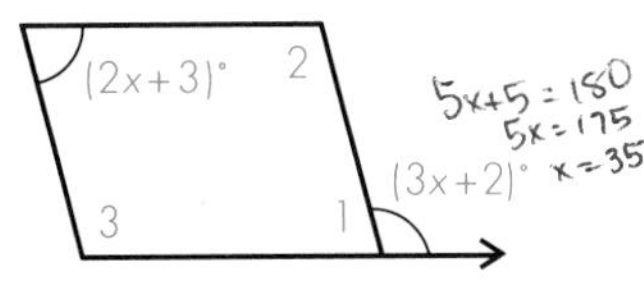

m∠1 = 73° m∠3 = 107°
m∠2 = 107°

Find the missing measures for each parallelogram.

7.

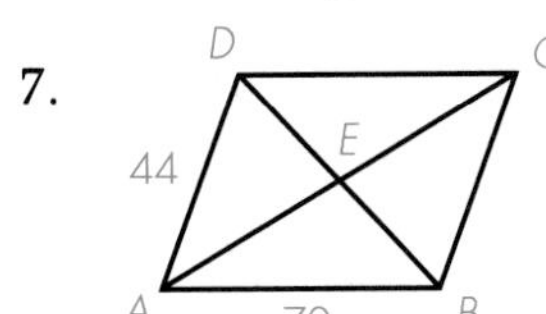

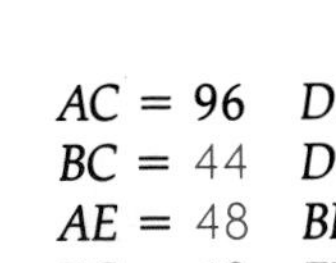

AC = 96 DB = 64
BC = 44 DC = 70
AE = 48 BE = 32
EC = 48 ED = 32

8.

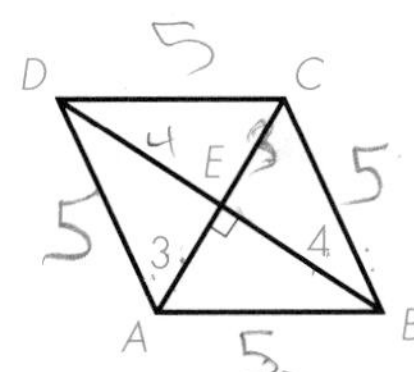

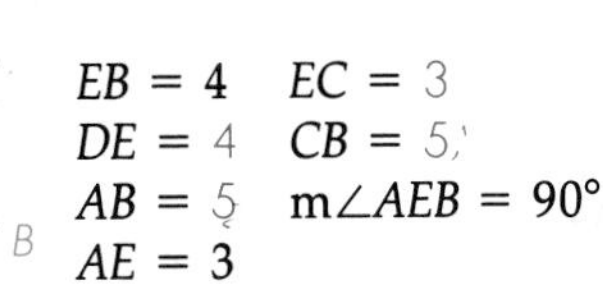

EB = 4 EC = 3
DE = 4 CB = 5
AB = 5 m∠AEB = 90°
AE = 3

Write Logo commands that tell the turtle to draw each parallelogram described below.

9. ▱ABCD with m∠ABC = 80°, AB = 75, BC = 50 Answers may vary.

10. ▱HIJK with m∠IJK = 40°, IJ = 40, JK = 110
Answers may vary.

11. The seats in a stadium are supported by the structure outlined in the picture. $\overline{WX} \parallel \overline{ZY}$, $\overline{ZB} \parallel \overline{YX}$, $\overline{WZ} \parallel \overline{AY}$, m∠ZWX = 52°, and m∠YXW = 33°. What is the measure of ∠ZCA? 85°

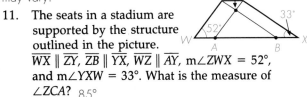

Extra Practice

Is the quadrilateral a parallelogram? Give a reason for your answer.

1.

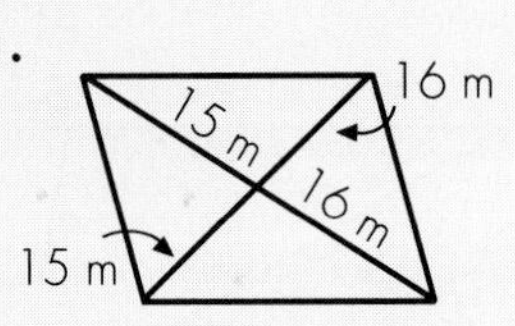

[No; diagonals are not bisected.]

2.

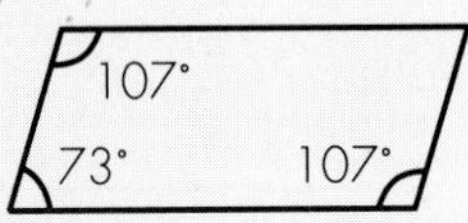

[Yes; opposite angles are congruent.]

3.

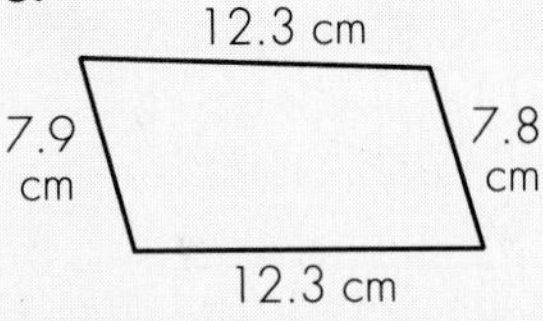

[No; opposite sides are not congruent.]

Find the missing measures for ▱ABCD.

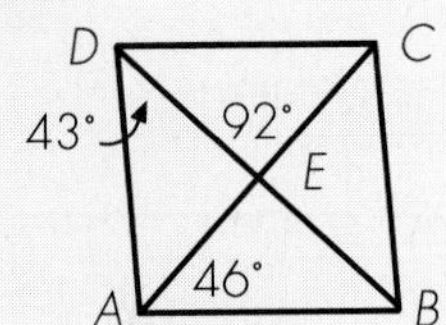

4. m∠DCA [46°]
5. m∠CDA [42°]
6. m∠AEB [92°]
7. m∠AED [88°]
8. m∠DAC [49°]
9. m∠DBC [43°]
10. For ▱PSRT, PQ = 16 and QS = 19. Find the length of each diagonal.

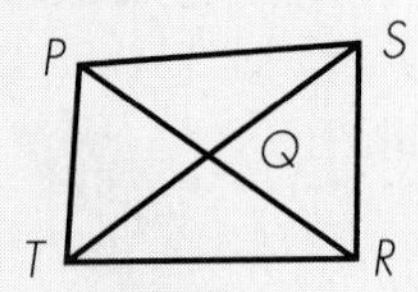

[PR = 32, ST = 38]

FOLLOW-UP

Assessment
Is the quadrilateral a parallelogram? Give a reason for your answer.

1.

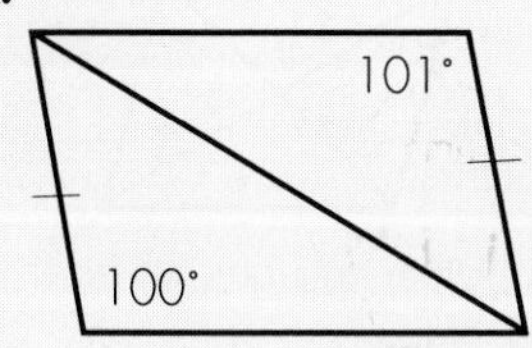

[No; opposite angles are not congruent.]

2.

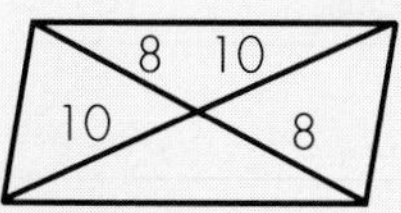

[Yes; diagonals are bisected.]

Find the angle measures for each parallelogram.

3.

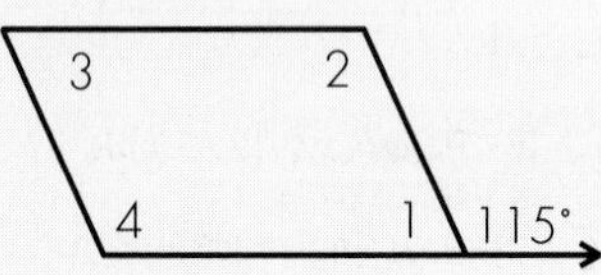

[m∠1 = 65°, m∠2 = 115°, m∠3 = 65°, m∠4 = 115°]

4.

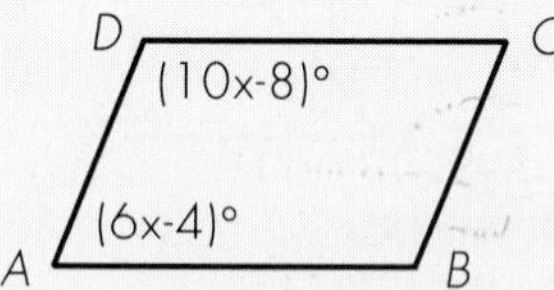

[m∠A = 68°, m∠B = 112°, m∠C = 68°, m∠D = 112°]

Reteaching
Students who have had difficulty with this section may benefit from Reteaching Activity 8.3.

Enrichment
For students who have mastered the material in this section, you may wish to assign Enrichment Activity 8.3.

Is the quadrilateral a parallelogram? Give a reason for your answer.

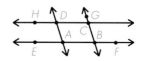

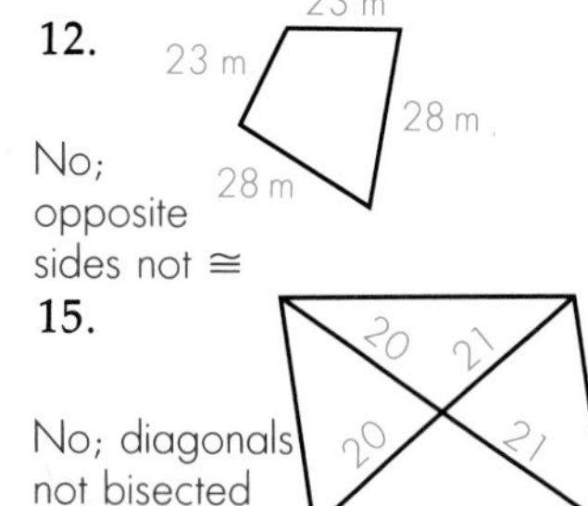

12. No; opposite sides not ≅

13. 15 a a 15 — Yes; diagonals bisected

14. 75° 105° 105° 75° — Yes; opposite angles ≅

15. No; diagonals not bisected

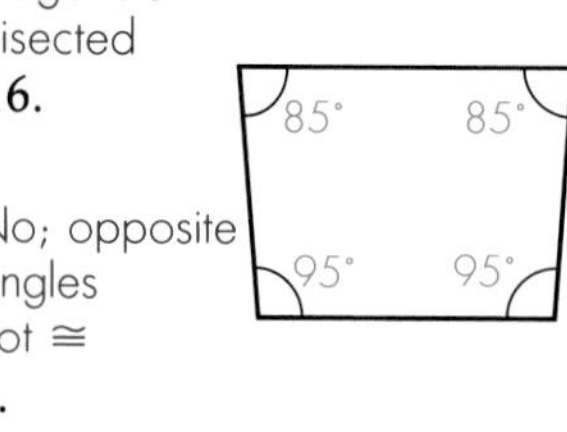

16. No; opposite angles not ≅

17. 3.2 m 4.8 m 4.8 m 3.2 m — Yes; opposite sides ≅

18. Write a reason for each step.

Given: ▱*ABCD*
Justify: The adjacent angles of a parallelogram are supplementary.

H D G A C B E F

Steps	Reasons
a. $\overline{AB} \parallel \overline{DC}$ and $\overline{AD} \parallel \overline{BC}$	a. Opposite sides of ▱ are ∥.
b. m∠*DAE* = m∠*CBA*	b. Corresponding angles
c. m∠*DAE* + m∠*DAB* = 180°	c. Linear pair of angles suppl.
d. m∠*CBA* + m∠*DAB* = 180°	d. Substitution
e. m∠*DCG* = m∠*ABC*	e. Corresponding angles
f. m∠*DCG* + m∠*DCB* = 180°	f. Linear pair of angles suppl.
g. m∠*ABC* + m∠*DCB* = 180°	g. Substitution

h. Do you think this will be true for all four angles of the parallelogram? Why? Yes; m∠*C* = m∠*A* and m∠*D* = m∠*B*

CRITICAL THINKING

19. Tessellations are patterns made by using congruent figures placed side by side so that there are no gaps or overlapping parts in the pattern. Any two figures have only one side in common.

a. Construct and cut out an equilateral triangle. Then cut out 11 copies of the triangle.
b. Use the triangles to make a tessellation. What is the sum of the measures of the angles that meet at any one point? 360°
c. Draw and cut out a quadrilateral. Label the angles *A, B, C,* and *D*. Make 11 copies of the quadrilateral.
d. Use the quadrilaterals to make a tessellation. Be sure that there are no gaps or overlapping parts in the pattern. What angles meet at any one point? Angles *A, B, C,* and *D*
e. What is the sum of the measures of the angles that meet at any one point? 360°

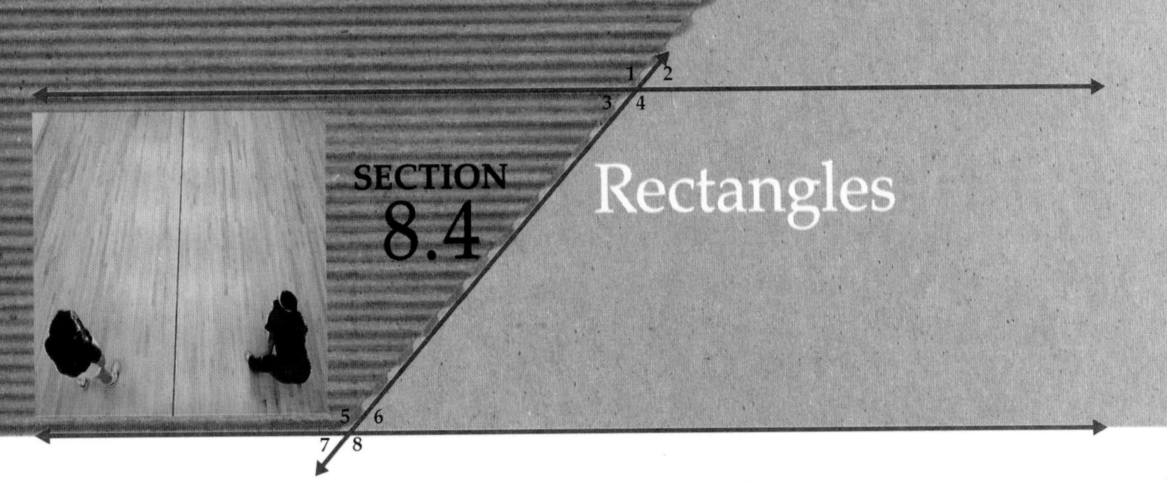

Rectangles

After completing this section, you should understand
▲ the properties of a rectangle.

The diagram below shows the set of quadrilaterals and the subsets of figures you have investigated thus far in this chapter. Also included, as a subset of parallelograms, is another figure, the rectangle.

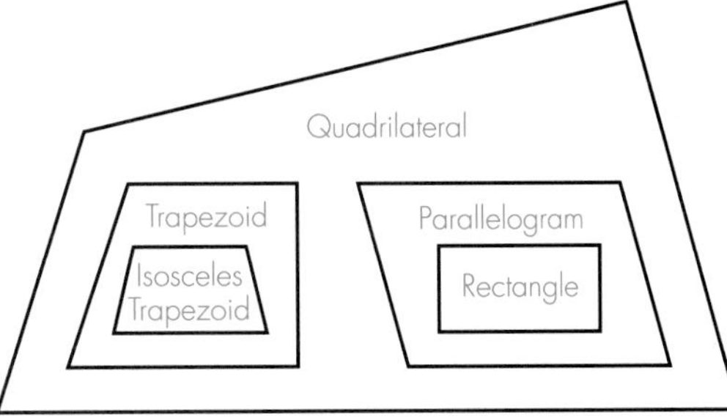

You have discovered that the following properties are true.

A **quadrilateral** has four sides and four angles.
The sum of the measures of the angles is 360°.

A **trapezoid** has one pair of parallel sides.
The median is parallel to both bases and equal to one-half the sum of the lengths of the bases.

A **parallelogram** has two pairs of parallel sides.
Both pairs of opposite angles have equal measures.
The diagonals bisect each other.

Rectangles possess several new properties in addition to the ones listed above.

SECTION
8.4

Resources
Reteaching 8.4
Enrichment 8.4

Objectives
After completing this section, the student should understand
▲ the properties of a rectangle.

Materials
ruler
protractor
colored chalk
library resources

Vocabulary
rectangle

GETTING STARTED

Quiz: Prerequisite Skills
In quadrilateral *ABCD*, $\overline{AB} \parallel \overline{CD}$, and m∠*A* = 90°.

1. What is m∠*D*? [90°]

2. Is *ABCD* a parallelogram? Explain your answer. [You can't tell without more information.]

In ▱*QRST*, *QR* = 12 inches, and *RS* = 8 inches.

3. What is *ST*? [12 in.]

4. What is *TQ*? [8 in.]

Chalkboard Example
Draw the diagram shown on page 299. As students identify the properties that are true for each figure, list them on the board next to that figure. Ask a volunteer to draw each figure, showing special properties with colored chalk.

TEACHING COMMENTS

Using the Page
After the Discovery Activity, ask students to draw some pairs of diagonals that are congruent and that bisect each other. Have them join the endpoints and describe the quadrilaterals formed. Then have them draw some pairs of diagonals that are congruent but do not bisect each other. Have them join the endpoints and describe the quadrilaterals.

Additional Answers
1. Opposite angles are equal in measure, so m∠A = 90°, so m∠A + m∠C = 180°. Thus, m∠B = m∠D = 180°, since the sum of the measures of all four angles is 360°. m∠B = m∠D, so the measure of each must be 90°.

Using the Page
Ask what other pairs of triangles could have been used in the proof. [△ACD and △CAB] Discuss whether the diagonals of any parallelogram are necessarily equal in length. Have students provide counterexamples.

DISCOVERY ACTIVITY

1. Draw a line and mark on it two points 6 centimeters apart. Label the line segment $\overline{AB}$.
2. Use a protractor to construct 90° angles at *A* and *B*. The rays of each angle should be on the same side of $\overline{AB}$.
3. Measure 2 centimeters up on the ray at point *A* and mark point *D*. Do the same on the ray at point *B* and mark point *C*.
4. Draw *DC*, then draw diagonals $\overline{AC}$ and $\overline{BD}$.
5. Does quadrilateral *ABCD* appear to be a parallelogram? Yes
6. Measure the four sides and the four angles. What do you find? Opposite sides equal, opposite angles equal
7. Measure the diagonals. What do you find? Diagonals equal

You may have discovered two new properties in the quadrilateral *ABCD* you constructed: the four angles have equal measures, and the diagonals are equal in length.

DEFINITION 8-4-1 **A RECTANGLE is a parallelogram with one right angle.**

CLASS ACTIVITY

1. Why do you think the definition of a rectangle does not state that all four of the angles are right angles? Use your knowledge of the properties of a parallelogram to answer. See margin.
2. Justify that the diagonals of a rectangle are always equal in length. Complete the reasons for the steps.
   Given: Rectangle *ABCD* with diagonals $\overline{AC}$ and $\overline{BD}$
   Justify: *AC* = *BD*

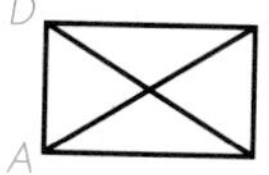

Steps	Reasons
a. *AD* = *BC*	a. Opposite sides of rectangle are equal.
b. *AB* = *AB*	b. Identity
c. m∠*ABC* = m∠*BAD*	c. Measures of angles of rectangle are 9(
d. △*DAB* ≅ △*CBA*	d. SAS
e. *AC* = *BD*	e. CPCF

THEOREM 8-4-1 **If a quadrilateral is a rectangle, then the diagonals are equal in length and all four angles are right angles.**

Example: *ABCD* is a rectangle. $AB = 12$, $BC = 5$, and $AC = 13$. Find the following measures.

$AD = ?$ 5 $DC = ?$ 12 $BD = ?$ 13 $m\angle ADC = ?$ 90°

Example: *LMNO* is a rectangle. $LM = 4$ and $MN = 3$. Find *LN*.
Hint: What kind of a right triangle is $\triangle LMN$? 3-4-5 right △
$LN = ?$ 5

CLASS ACTIVITY

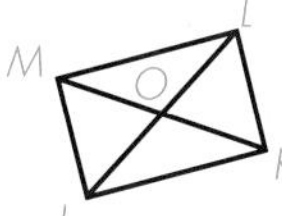

JKLM is a rectangle.

1. What kind of triangle is $\triangle JOK$? Why? Isosceles; $JO = OK$
2. What kind of a triangle is $\triangle MJK$? Why? Right △; $m\angle J = 90°$
3. If $MO + OL = 30$, what is the measure of *MK*? 30
4. Name all pairs of congruent segments in rectangle *WXYZ*.

4. $\overline{WX} \cong \overline{ZY}$, $\overline{ZW} \cong \overline{YX}$, $\overline{WY} \cong \overline{ZX}$, $\overline{WO} \cong \overline{OY}$, $\overline{ZO} \cong \overline{OX}$, $\overline{ZO} \cong \overline{OW}$, $\overline{YO} \cong \overline{OX}$, $\overline{WO} \cong \overline{OX}$, $\overline{YO} \cong \overline{OZ}$

Study the information given for figures 5–7. Which of the parallelograms would be rectangles? (Assume that the information is correct even though the figure may not be accurately drawn.) 7

5. 

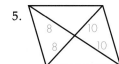

6. 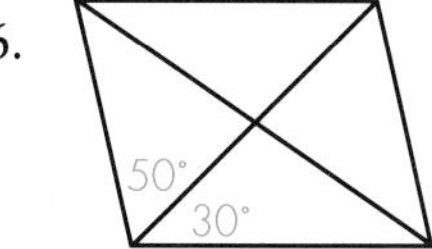

7. 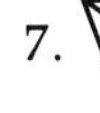 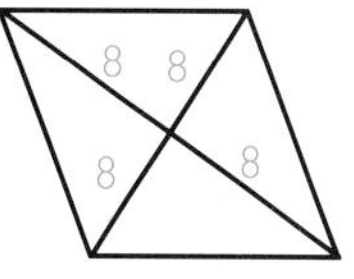

Use a computer and a geometric drawing tool to complete Exercises 8–10.

8. Draw a parallelogram. Measure each angle. Repeat this activity several times, using different angles each time. Is every parallelogram a rectangle? No
9. Draw a quadrilateral with one right angle. Measure the other three angles. Repeat this activity several times, using different angles each time. If a quadrilateral has one right angle, is it a rectangle? No
10. Draw a parallelogram and its diagonals. Measure each diagonal. Repeat this activity several times, using different angle measures each time. If the diagonals of a parallelogram are equal in length, is it a rectangle? Yes

Using the Page
Before beginning the Class Activity, you may want to review the different types of right triangles and the theorems that explain how to find the lengths of the sides [Theorems 7-1-1, 7-2-2, 7-3-1, and 7-3-2].

Extra Practice
ABCD is a rectangle. $AO = 12.5$, $AB = 20$, and $BC = 15$. Find the measure.

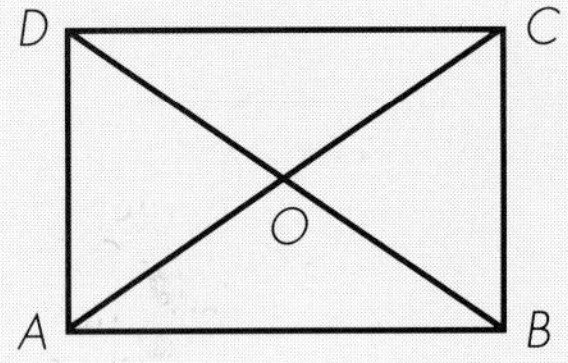

1. *AC* [25] **2.** *OD* [12.5]
3. *AD* [15]
4. $m\angle ABC$ [90°]

For rectangle *MNPQ*, $QN = 30$ and $m\angle MQN = 60°$. Find the measure.

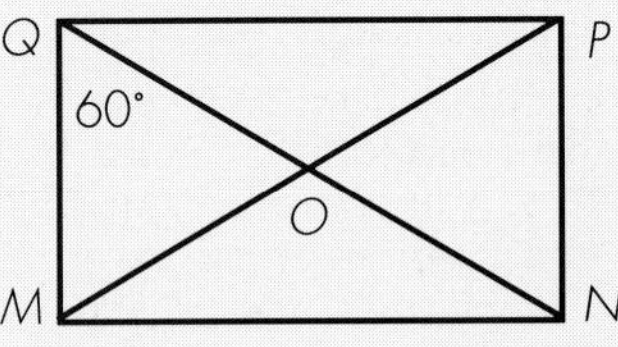

5. $m\angle QNM$ [30°]
6. *QM* [15]
7. *MN* [$15\sqrt{3}$]
8. *OP* [15]
9. $m\angle NQP$ [30°]
10. $m\angle QNP$ [60°]

Using the Page
You may wish to do Exercise 11 with the class and have a volunteer write the proof on the chalkboard as the class determines the steps and reasons.

Using the Page
For the project, you may wish to have students gather library books on architecture and find illustrations of well-known buildings and bridges that incorporate various quadrilaterals in their construction. The library may also be a source for pictures of buildings in your community.

11. Write steps and reasons to justify the following.

Given: *WXYZ* is a rectangle.

Justify: $\triangle WXO$ is isosceles.
See margin.

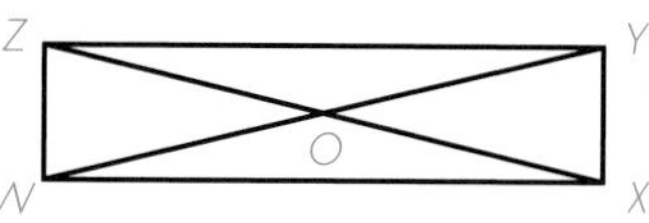

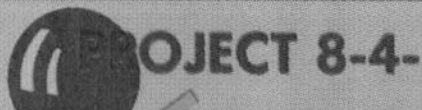

PROJECT 8-4-1

Rectangles and triangles are the geometric building blocks of architecture and construction. As you travel from home to school, take note of these quadrilaterals as you see them used in homes, playgrounds, office buildings, sports fields, bridges, and other structures.

Work with a group to create a display or collage illustrating the use of triangles, rectangles, trapezoids, and parallelograms in the architecture of your community. Use your own photographs and sketches, or pictures cut from local newspapers and magazines.

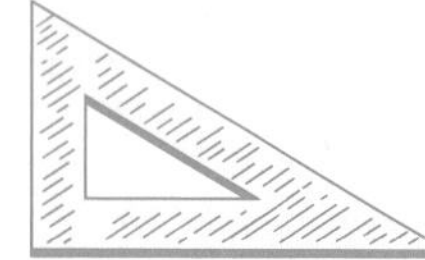

HOME ACTIVITY

ABCD is a rectangle. Find the length of the segments

1. *AC* 30
2. *DC* 24
3. *DB* 30
4. *BC* 18
5. *DO* 15
6. *CO* 15

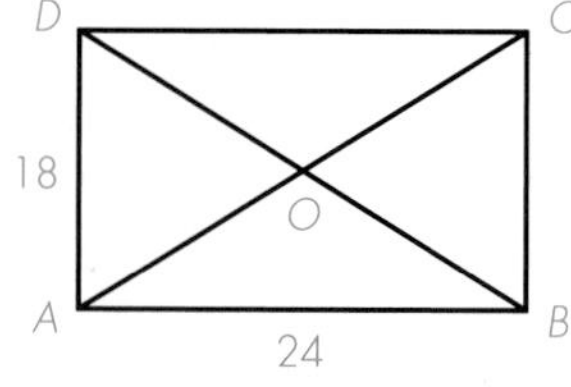

LMNO is a rectangle.

7. What kind of triangle is $\triangle OLN$? 3-4-5 right triangle
8. What is the length of $\overline{LN}$? 50
9. What kind of triangle is $\triangle LPM$? Isoceles

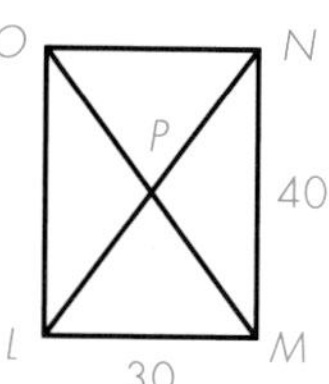

10. List the pairs of congruent triangles in rectangle *PQRS*.
$\triangle SPQ \cong \triangle RQP$, $\triangle SPR \cong \triangle RQS$, $\triangle STR \cong \triangle PTQ$, $\triangle STP \cong \triangle RTQ$, $\triangle PSR \cong PQR$, $\triangle SPQ \cong SRQ$

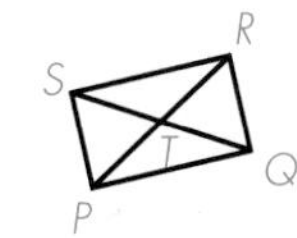

11. Study the information given for figures a–d. Which of the parallelograms would be rectangles? (Assume that the information is correct even though the figure may not be drawn accurately.) b and d

a. 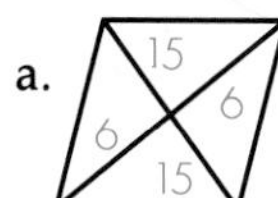

b. 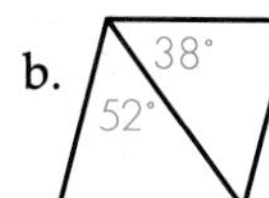

c. 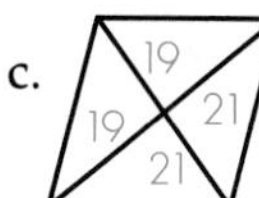

d. 

12. $WY = ZX$ by diagonals of rectangles are equal; $WV = ZX$ by opposite sides of parallelogram are equal; $WV = WY$ by substitution; $\triangle WYV$ is isosceles by definition; it has two equal sides.

12. Write the steps and reasons to justify the following.

Given: *WXYZ* is a rectangle.
WVXZ is a parallelogram.
Justify: $\triangle WYV$ is isosceles.

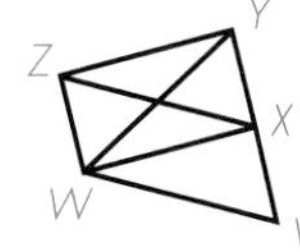

13. *ABCD* and *EFGH* are two rectangles. Find the sum of the measures of angles 1, 2, 3, and 4. (Hint: Use what you know about parallel lines and transversals.)
360°; $m\angle 2 + m\angle 3 = 180°$; $m\angle 1 + m\angle 4 = 180°$

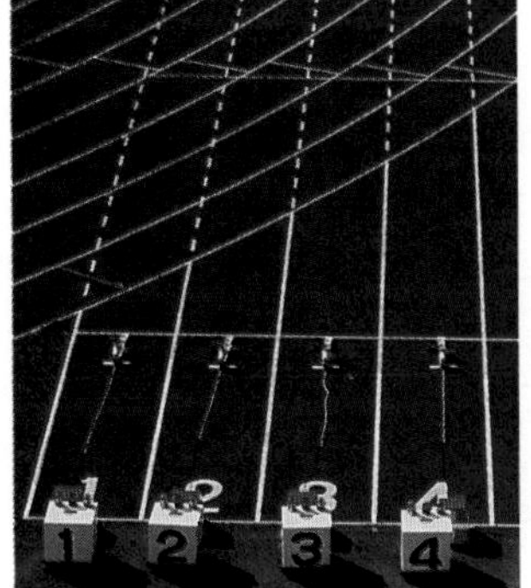

The ancient Greeks considered the Golden Rectangle to be one of the most beautifully proportioned geometric forms, and used it in much of their architecture. All of the figures below are Golden Rectangles. What do they have in common?

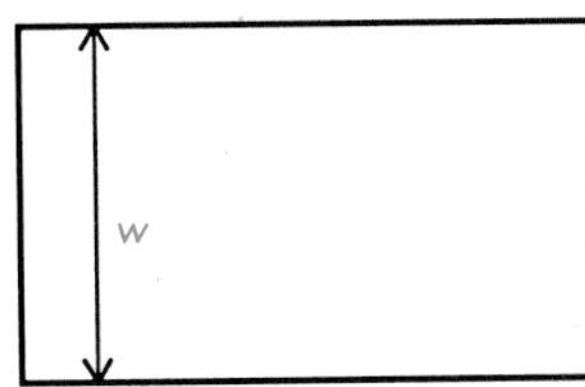

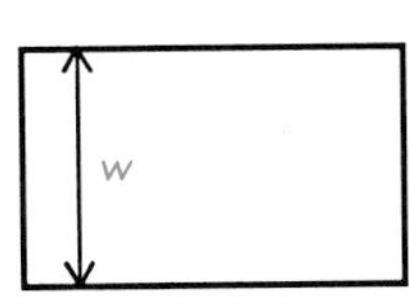

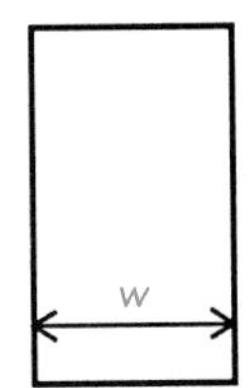

Each is about 0.6

Measure the length and width of each rectangle in millimeters. Write the ratio of width (w) to length (l) as a decimal rounded to hundredths. What do you observe about the decimal ratio?
The ratio 0.61803 . . . is called the golden ratio.

FOLLOW-UP

Assessment

DEFG is a rectangle. Find the length of the segment.

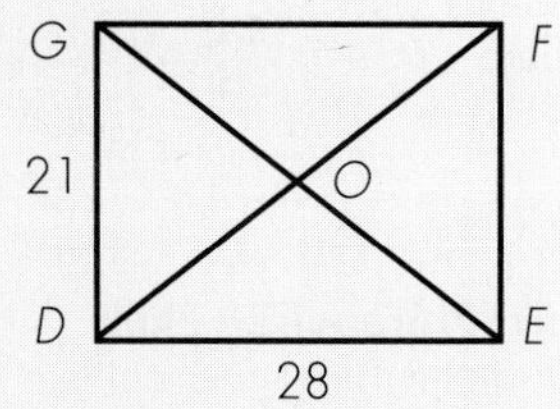

1. *DF* [35] **2.** *GF* [28]
3. *GE* [35] **4.** *OF* [17.5]
5. Is the parallelogram a rectangle? Explain.

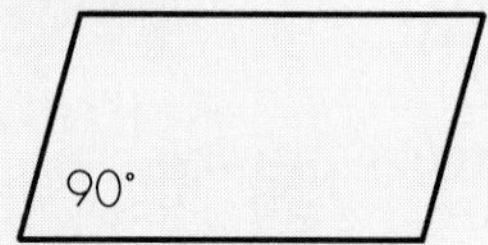

[Yes; all four angles measure 90°.]

6. *PQRS* is a parallelogram. $PQ = 4x - 3$, $PR = 3x - 2$, $SR = 2x + 25$, and $SQ = 2x + 12$. Show that *PQRS* is a rectangle. [$x = 14$; $PR = SQ = 40$; diagonals are congruent.]

Reteaching

Students who have had difficulty with this section may benefit from Reteaching Activity 8.4.

Enrichment

For students who have mastered the material in this section, you may wish to assign Enrichment Activity 8.4.

SECTION 8.5

Resources
Reteaching 8.5
Enrichment 8.5
Transparency Master 8-2

Objectives
After completing this section, the student should understand
▲ the properties of a rhombus.
▲ the properties of a square.

Materials
ruler
protractor
cardboard strips
paper fasteners

Vocabulary
rhombus
square

GETTING STARTED

Quiz: Prerequisite Skills
Tell whether the statement is true or false.

1. Every rectangle is a parallelogram. [True]

2. If a quadrilateral has one pair of parallel sides and one right angle, it is a rectangle. [False]

Warm-Up Activity
This paper-folding activity may serve as a basis for discussion of the properties of a square and its diagonals. Students fold the top right corner of the short side of a sheet of notebook paper diagonally down to the lower left corner. They mark the lower edge and cut off the extra paper. Then they fold the resulting square so there are two diagonals.

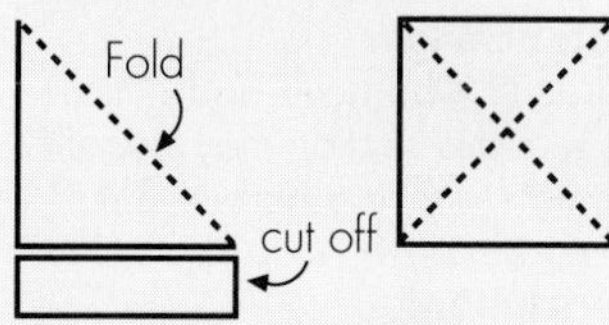

Ask for an informal description of the figure and its diagonals.

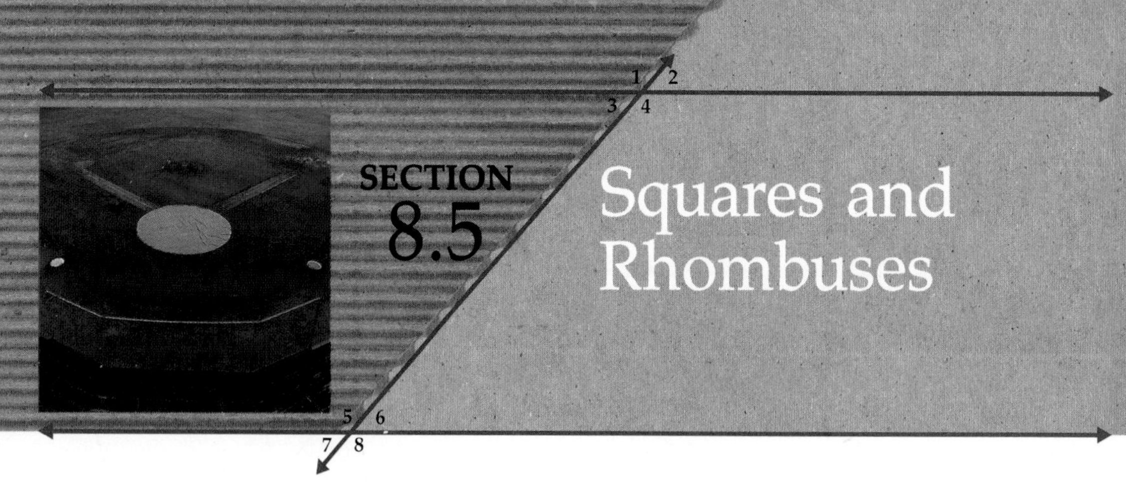

SECTION 8.5 Squares and Rhombuses

After completing this section, you should understand
▲ the properties of a rhombus.
▲ the properties of a square.

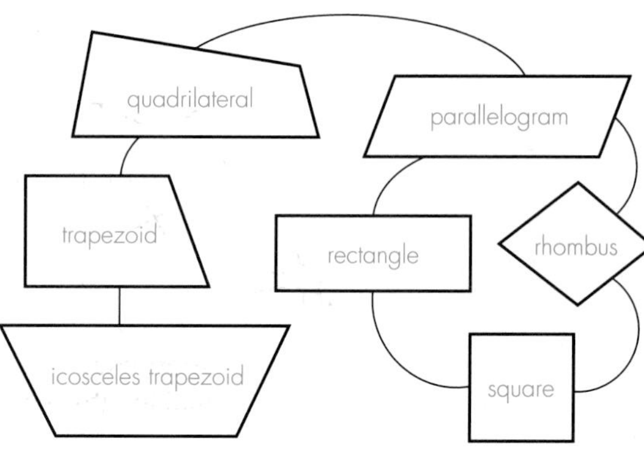

The diagram shows two new figures to investigate—rhombus and square. You can tell from the diagram that a rhombus will have all of the properties of a parallelogram, and a square will have properties common to both the rectangle and rhombus.

DISCOVERY ACTIVITY

1. Use four strips of cardboard equal in length and fasten them at the endpoints with brads to form figure *ABCD*. Be sure the figure is flexible.

2. Place the figure on a piece of paper, holding it down on $\overline{AB}$. Move point C back and forth as far as it will go in each direction.

3. Notice the position of the opposite sides as you move C back and forth. In every position, except when $\overline{DC}$ is on $\overline{AB}$, the figure forms a rhombus. Why is a rhombus a parallelogram? opposite sides equal in length

4. Now hold a pencil point at C, still holding the figure firmly on AB. As you move point C, use the pencil to trace the path of point C. What do you observe about the line you traced? Point C makes an arc or semicircle around point B.

This activity may have shown you that the sides of a rhombus are always parallel as well as equal in length.

DEFINITION 8-5-1 **A RHOMBUS is a parallelogram with sides of equal length.**

CLASS ACTIVITY

ABCD is a rhombus.

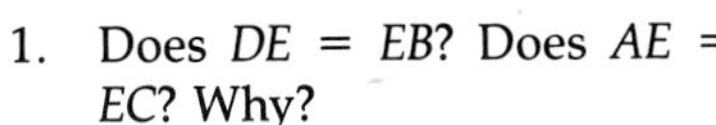

1. Does *DE* = *EB*? Does *AE* = *EC*? Why?
   Yes; since a rhombus is a parallelogram, the diagonals bisect each other.

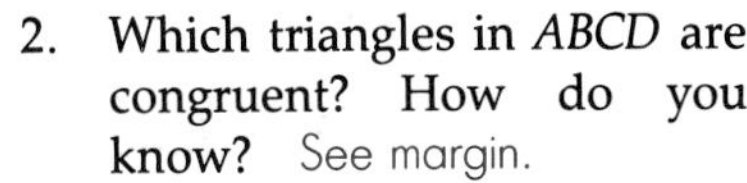

2. Which triangles in *ABCD* are congruent? How do you know? See margin.

From the definition you know that a rhombus is a parallelogram with four sides of equal length. There is another property that distinguishes the rhombus.

DISCOVERY ACTIVITY

1. Draw three different rhombuses and label each one *ABCD*.
2. Draw diagonals $\overline{AC}$ and $\overline{BD}$. Label the intersection O.
3. With a protractor, measure the four angles formed at the intersection of the diagonals. Do this for each figure and record your measurements. All 90°
4. Write a conclusion to summarize your results.

You may have discovered that, since all four angles at the intersection of the diagonals are right angles, the diagonals of a rhombus are perpendicular to each other.

TEACHING COMMENTS

Using the Page
As students do the Discovery Activity, discuss with them the appearance of the figure as it changes shape. Do the sides always look parallel? Equal in length? How do they know that the sides are equal? [Cardboard strips are of equal length.] This activity should provide a clear counterexample to the common misconception that the only figure with four sides of equal length is a square.

Additional Answers
2. $\triangle ADB \cong \triangle DCB$, $\triangle ADC \cong \triangle ABC$, $\triangle ADE \cong \triangle CBE$, $\triangle DEC \cong \triangle BEA$; by SSS

Using the Page
During the Discovery Activity, you may wish to have students try to form a rhombus with diagonals that bisect each other but do not intersect at an angle of 90°. They should conclude that this will result in a parallelogram or, if the diagonals are congruent, in a rectangle.

Using the Page
In discussing Theorems 8-5-1 and 8-5-2, ask students to determine the one property of a rhombus that makes it different from a rectangle. [Perpendicular diagonals] Again, since students often think that the only figure with perpendicular diagonals is a square, this theorem gives a counterexample in the rhombus.

Using the Page
Go through each step of the proof to be sure students understand the reasons. Have them explain why, if m∠AOD = 90°, the other three angles around point O must each also measure 90°.

Extra Practice
ABCD is a rhombus. Find the measure.

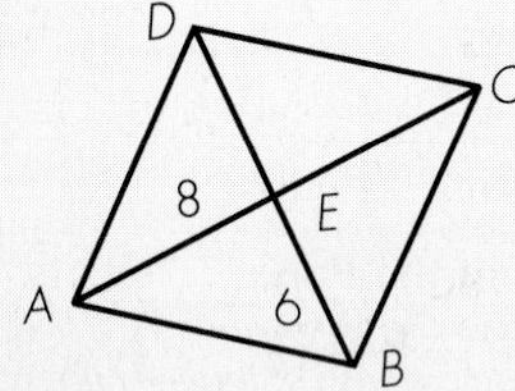

1. *AC* [16] **2.** *BD* [12]
3. *AD* [10] **4.** *DC* [10]
5. For rhombus *PQRS*, what is the value of x? [27.5]

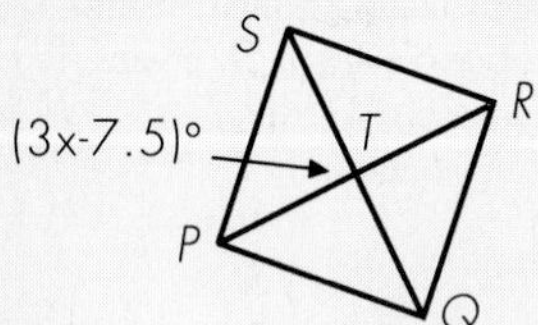

NOTEBOOK **THEOREM 8-5-1** **If a parallelogram is a rhombus, then the diagonals are perpendicular.**

Given: *ABCD* is a rhombus with diagonals $\overline{AC}$ and $\overline{BD}$ intersecting at *O*.

Justify: $\overline{AC} \perp \overline{BD}$ and m∠*AOD* = 90°

Have students supply the missing reasons.

Steps	Reasons
1. *AD* = *DC* = *CB* = *BA*	1. Definition of rhombus
2. *AO* = *OC* and *DO* = *OB*	2. Diagonals of parallelogram bisect each other.
3. △AOD ≅ △AOB	3. SSS
4. ∠AOD ≅ ∠AOB	4. CPCF
5. m∠AOD + m∠AOB = 180°	5. Linear pair of angles = 180°
6. 2(m∠AOD) = 180° m∠AOD = 90°	6. Substitution and algebra
7. m∠AOD = m∠COB and m∠AOB = m∠DOC	7. Vertical angles
8. $\overline{AC} \perp \overline{BD}$	8. Definition of perpendicular lines

NOTEBOOK **THEOREM 8-5-2** **If a figure is a rhombus, then opposite sides are parallel, opposite sides are equal in length, opposite angles have equal measures, the diagonals bisect each other, and the diagonals are perpendicular.**

SQUARES

From the tree diagram at the beginning of this section you can see that a square is both a rectangle and a rhombus.

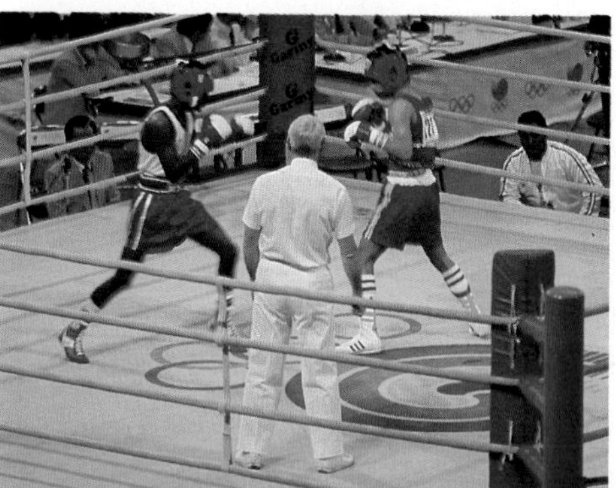

NOTEBOOK **DEFINITION 8-5-2** **A SQUARE is a rectangle with sides that are equal in length.**

DISCOVERY ACTIVITY

1. Construct two squares, each with sides of a different length. Recall that since a square is a rectangle, each of its angles measures 90°.
2. Draw the diagonals for each square.
3. With a protractor, measure the angles formed by the intersection of the diagonals. What do you observe? They are ⊥.

You may have found that the diagonals of a square are perpendicular.

THEOREM 8-5-3 **If a figure is a square, then opposite sides are parallel, opposite sides are equal in length, there are four right angles, the diagonals are equal in length, the diagonals bisect each other, and the diagonals are perpendicular.**

CLASS ACTIVITY

Given: Square *ABCD* with diagonals $\overline{AC}$ and $\overline{BD}$ intersecting at *O*; *AB* = 5 cm

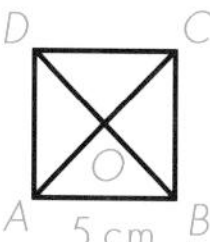

Find the following measures. Use a calculator when necessary.

1. $AD = ?$ 5 cm
2. $BC = ?$ 5 cm
3. $CD = ?$ 5 cm
4. $AC = ?$ 7.07 cm
5. $BD = ?$ 7.07 cm
6. $OA = ?$ 3.54 cm
7. $OD = ?$ 3.54 cm
8. $OC = ?$ 3.54 cm

Name the quadrilateral described by each set of diagonals.

9. 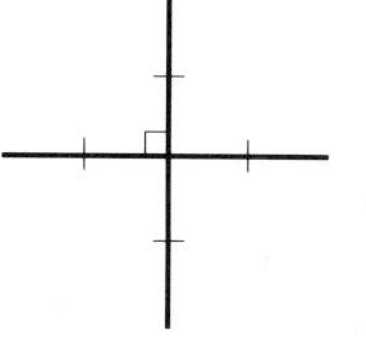 Parallelogram

10. 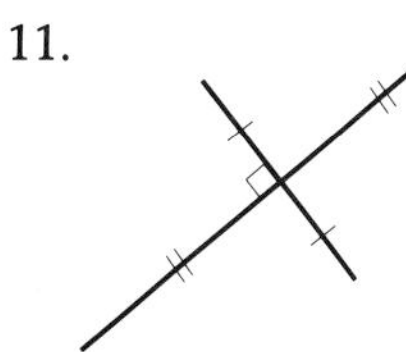 Square

11. Rhombus

12. Use a computer and a geometric drawing tool to draw a quadrilateral that is not a parallelogram. Find and connect the midpoints of adjacent sides. What kind of quadrilateral has been formed? Parallelogram

Using the Page
During the Discovery Activity, you may wish to have students measure the diagonals of the square as well. Discuss the reasons why a square is both a rhombus and a rectangle.

Using the Page.
During the discussion of Theorem 8-5-3, you may use Transparency Master 8-2 and the diagram shown on page 304 to clarify the relationships between the quadrilaterals.

Extra Practice
Given: Square *ABCD* with side *AB* = 12 and diagonals $\overline{AC}$ and $\overline{BD}$ intersecting at *E*. Find the measure.

1. *BC* [12]
2. *DC* [12]
3. *AC* [$12\sqrt{2}$]
4. *AE* [$6\sqrt{2}$]
5. $m\angle AEB$ [90°]
6. *DE* [$6\sqrt{2}$]

Name the quadrilateral described by each set of diagonals.

7. 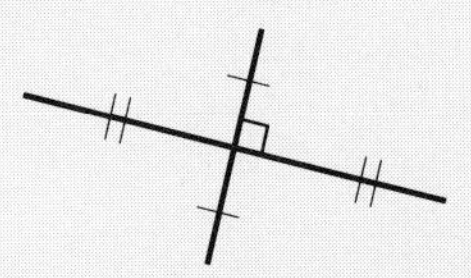

[Rhombus]

8. 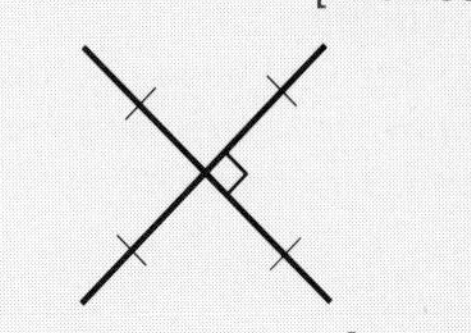

[Square]

Additional Answers

6. The number of small squares on each new checkerboard is being added in each case; this results in the sum of consecutive squares: $1^2 + 2^2 + 3^3$ . . . . It may be helpful for students to draw a diagram for each array of squares.

FOLLOW-UP

Assessment

Refer to rhombus *PQRS*.

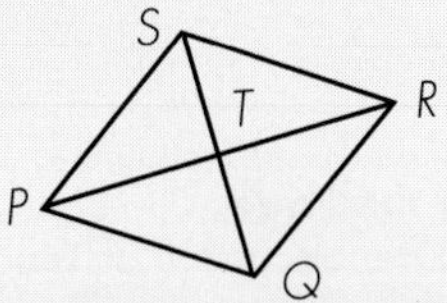

1. If $PT = 12$, what is PR? [24]

2. If $SR = 25$, what is PS? [25]

3. What is $m\angle PTS$? [90°]

4. If *PQRS* is a square, what is $m\angle PQR$? [90°]

Name the quadrilateral described by each set of diagonals.

5.

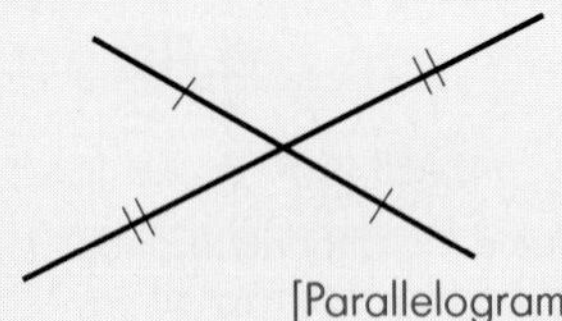

[Parallelogram]

6.

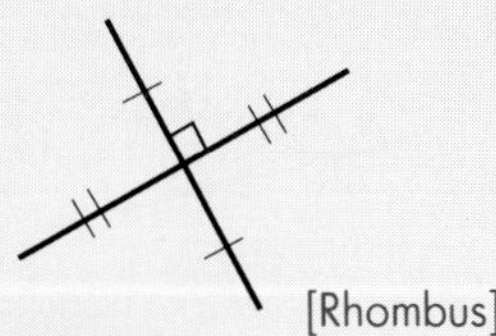

[Rhombus]

Reteaching

Students who have had difficulty with this section may benefit from Reteaching Activity 8.5.

Enrichment

For students who have mastered the material in this section, you may wish to assign Enrichment Activity 8.5.

HOME ACTIVITY

1. Copy and complete the following table. Mark an *X* in the column under each figure if the property is true for that figure.

Properties of Quadrilaterals					
Property	Parallelogram	Rectangle	Square	Rhombus	Trapezoid
Both pairs of opposite sides equal.	X	X	X	X	
All sides are equal.			X	X	
Both pairs of opposite sides parallel.	X	X	X	X	
Each pair of opposite angles equal.	X	X	X	X	
All angles right.		X	X		
Diagonals bisect each other.	X	X	X	X	
Diagonals are equal.		X	X		
Diagonals are perpendicular.			X	X	
Adjacent angles are supplementary.	X	X	X	X	
Two sides are parallel.	X	X	X	X	X
Sum of measures of angles = 360°.	X	X	X	X	X
Diagonals bisect opposite angles.			X	X	

Name the quadrilateral described by each set of diagonals.

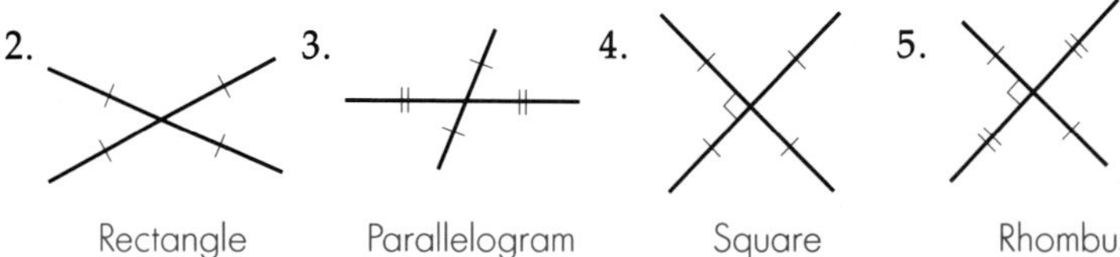

2. Rectangle
3. Parallelogram
4. Square
5. Rhombus

CRITICAL THINKING

6. The game of checkers is played on a square game board that is divided into 64 small squares, 8 on a side. How many squares are there on a checkerboard?

The total cannot be 64, since you can immediately see 65 squares—64 small squares and 1 large square—the checkerboard itself.
On a board with only one square, the number of squares is 1.
On a board that is a 2 by 2 array of squares, the total number of squares is 5: $1 + 4$.
On a 3 by 3 board, the total number of squares is 13: $1 + 4 + 9$.
What is being added in each case? See margin.
Use inductive reasoning to predict the total number of squares for the 8 by 8 checkerboard. Show your work.
$1^2 + 2^2 + 3^2 + 4^2 + 5^2 + 6^2 + 7^2 + 8^2 = 204$

Chapter Vocabulary
base
concave
convex
diagonal
isosceles trapezoid
leg
median
parallelogram
quadrilateral
rectangle
rhombus
square
trapezoid

8.1 A quadrilateral is a closed, plane, four-sided figure. It may be convex or concave. The sum of the measures of the interior angles of a quadrilateral as 360°. What is the measure of $\angle D$?

$$50 + 108 + 92 + x = 360$$
$$250 + x = 360$$
$$x = 110 \qquad m\angle D = 110°$$

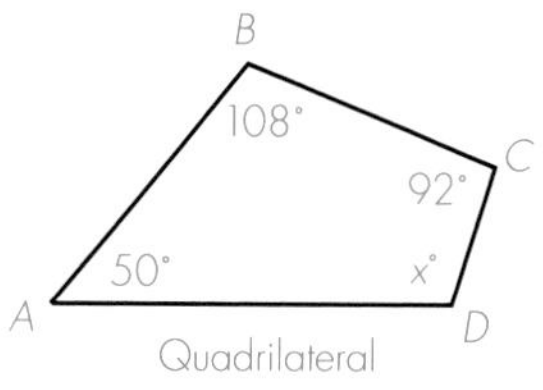

Quadrilateral

8.2 A trapezoid is a quadrilateral with one pair of parallel sides. An isosceles trapezoid has equal non-parallel sides and equal base angles. The median of a trapezoid joins the midpoints of the non-parallel sides, is parallel to the bases, and has a length that is half the sum of the lengths of the two bases. What is the length of JG?

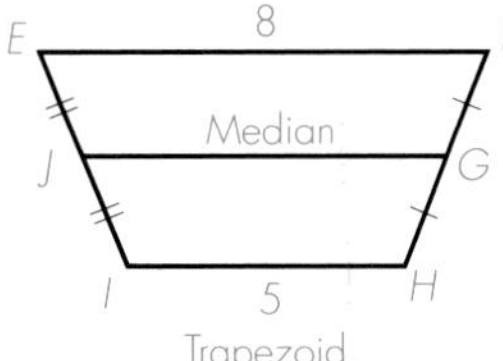

Trapezoid

$$JG = \tfrac{1}{2}(8 + 5)$$
$$JG = 6.5$$

8.3 A parallelogram is a quadrilateral with two pairs of opposite sides parallel. Opposite sides and opposite angles of a parallelogram are equal. What is the measure of $\angle L$?

$$m\angle L = m\angle N = 122°$$
$$m\angle L = 122°$$

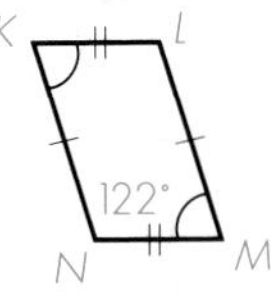

Parallelogram

8.4 A rectangle is a parallelogram with one right angle. A rectangle has diagonals equal in length. The diagonals bisect each other. $SQ = 13$. What is the length of TR?

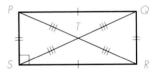

$$PR = SQ = 13$$
$$TR = \tfrac{1}{2}PR = \tfrac{1}{2}(13) = 6.5$$

Rectangle

8.5 A rhombus is a parallelogram with all four sides equal in length. In a rhombus the diagonals intersect at right angles and bisect each other. A square is a rectangle with all four sides equal in length. The diagonals of a square are equal and bisect each other at right angles. If $UV = 5$ and $AC = 8$, what are the lengths of VW and ED?

$$VW = UV = 5 \qquad BD = AC = 8$$
$$VW = 5 \qquad ED = \tfrac{1}{2}(BD) = \tfrac{1}{2}(8) = 4$$

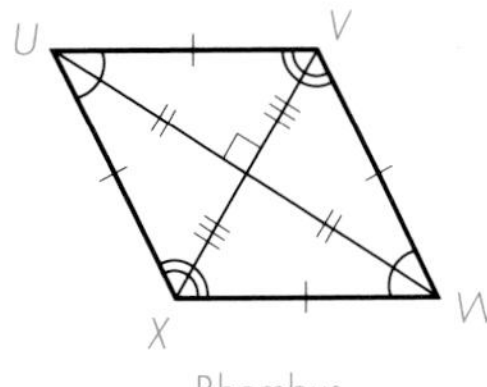

Rhombus

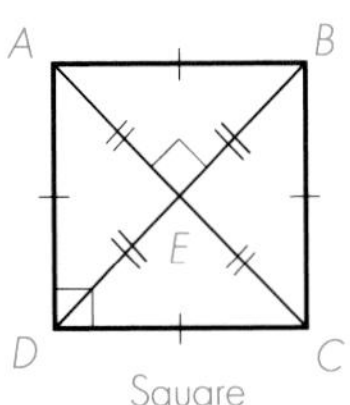

Square

309

COMPUTER ACTIVITY

You can use Logo to draw a parallelogram on the x−y plane when given just the coordinates of three of the vertices.

Example: The coordinates of three of the vertices of ▱*ABCD* are *A*(0,0), *B*(2,2), and *C*(5,2). Find the coordinates of vertex *D*. Then write a procedure using the procedure XYPLANE from page 118 and the command SETXY to draw the parallelogram.

First, graph the coordinates for the given vertices on the x−y plane. Then, use the fact that the opposite sides of a parallelogram are equal and the distance formula to find the coordinates of *D*. $\overline{BC}$ and $\overline{AD}$ are opposite sides of the parallelogram, so $BC = AD$. $BC = |5 - 2| = 3$ so $AD = 3$. Thus, the coordinates of *D* must be (3,0).

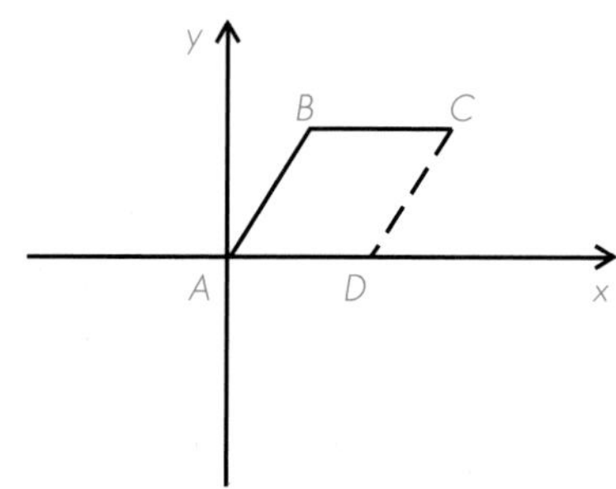

The Logo procedure below tells the turtle to draw this parallelogram.

```
TO PARA
  XYPLANE
  SETXY 2 2
  SETXY 5 2
  SETXY 3 0
  SETXY 0 0
END
```

Modify the procedure PARA from the example above to tell the turtle to draw each parallelogram. See margin.

1.

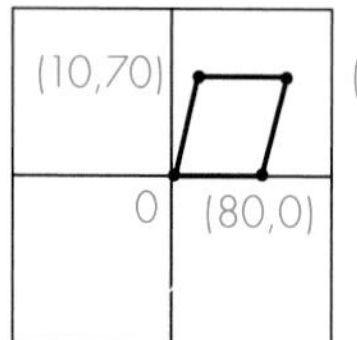

2.

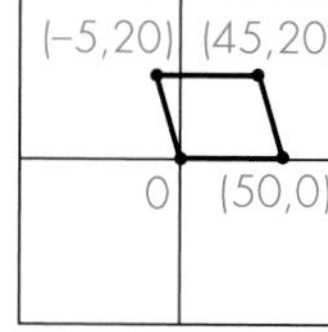

3.

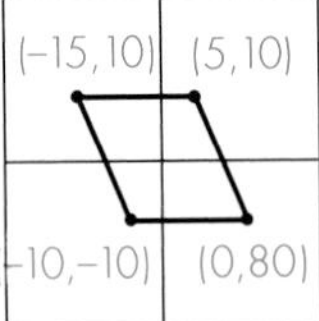

4.

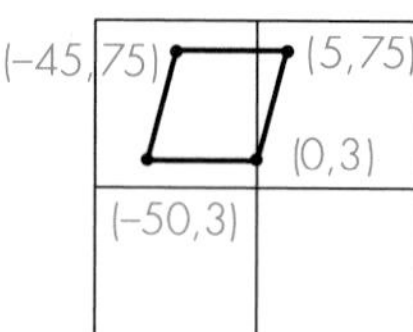

Using the Page

This Computer Activity suggests a way for students to construct parallelogram trapezoids, and rectangles on their computers when only three vertices are known.

For parallelograms the students graph the coordinates of the three known vertices, compute the length of the one side, and from that determine the coordinates of the fourth vertex.

Similar techniques are applied to trapezoids and rectangles.

Additional Answers

1. TO PARA
   XYPLANE
   SETXY 80 0
   SETXY 90 70
   SETXY 10 70
   SETXY 0 0
   END
2. TO PARA
   XYPLANE
   SETXY 50 0
   SETXY 45 20
   SETXY −5 20
   SETXY 0 0
   END
3. TO PARA
   XYPLANE
   PU SETXY 10 (−10) PD
   SETXY 5 10
   SETXY −15 10
   SETXY −10 (-10)
   END
4. TO PARA
   XYPLANE
   PU SETXY 0 3 PD
   SETXY 5 75
   SETXY −45 75
   SETXY −50 3
   SETXY 0 3
   END

The coordinates of three of the vertices of ▱*ABCD* are given in each exercise below. Find the coordinates of vertex *D*. Then modify the procedure PARA to tell the turtle to draw each parallelogram. See margin.

5. *A*(0,0), *B*(6,12), *C*(15,12) (9,0)

6. *A*(2,4), *B*(7,8), *C*(10,8) (5,4)

7. *A*(5,−1), *B*(7,12), *C*(13,12) (11,−1)

8. *A*(−4,2), *B*(−6,−3), *C*(1,−3) (3,2)

Use the procedure XYPLANE in procedures that tell the turtle to draw each figure. Give each procedure the indicated name. Answers may vary.

9. TRAPEZOID1

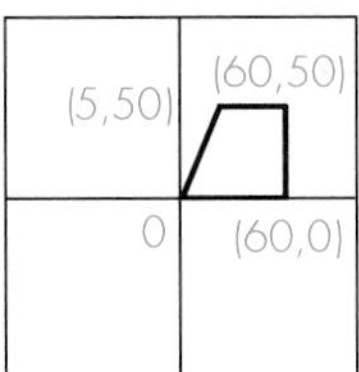

10. RECTANGLE1

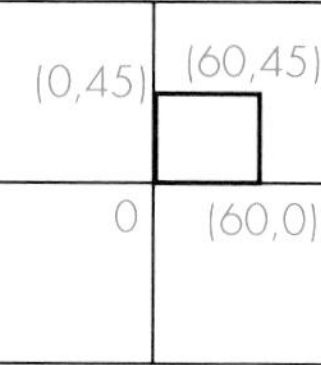

11. TRAPEZOID2

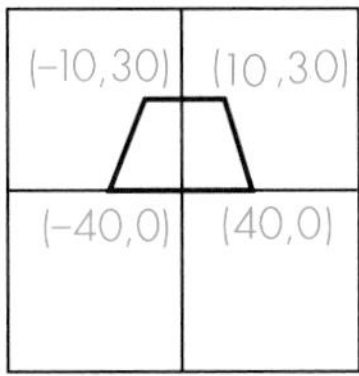

12. RECTANGLE2

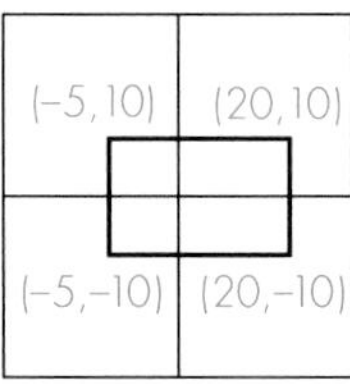

The coordinates of three of the vertices of the specified quadrilateral are given below. Find the coordinates of the fourth vertex. Then modify the indicated procedure from Exercises 9–12 to tell the turtle to draw each quadrilateral.

13. (0,4), (6,2), (2,2); TRAPEZOID1 Answers may vary.

14. (3,5), (8,5), (8,2); RECTANGLE1 (3,2)

15. (−3,2), (4,7), (6,7); TRAPEZOID2 Answers may vary.

16. (−4,−1), (−4,2), (3,−1); RECTANGLE2 (3,2)

Additional Answers

5.
```
TO PARA
XYPLANE
SETXY 6 12
SETXY 15 12
SETXY 9 0
SETXY 0 0
END
```

6.
```
TO PARA
XYPLANE
PU SETXY 2 4 PD
SETXY 7 8
SETXY 10 8
SETXY 5 4
SETXY 2 4
END
```

7.
```
TO PARA
XYPLANE
PU SETXY 5 (−1) PD
SETXY 7 12
SETXY 13 12
SETXY 11 (−1)
SETXY 5 (−1)
END
```

8.
```
TO PAA
XYPLANE
PU SETXY −4 2 PD
SETXY −6 (−3)
SETXY 1 (−3)
SETXY 3 2
SETXY −4 2
END
```

Using the Page
Students can work individually or in groups to study the review. If you have students study in groups, they should work together to make sure all members of the group know the material in the chapter.

Informal assessment such as interview, classroom observations, or a review of student portfolios can be used instead of paper-and-pencil tests.

Additional Answers
4. Yes; diagonals bisect each other.
5. Yes; opposite angles are congruent and supplementary.
6. No; diagonals are not bisected.
7. No; both pairs of opposite sides are not congruent.

Refer to quadrilateral *PQRS*.

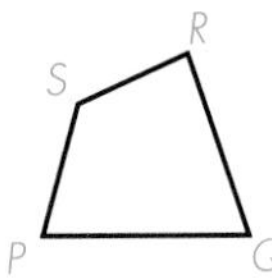

1. Name one pair of opposite sides. $\overline{SR}$ and $\overline{PQ}$ or $\overline{SP}$ and $\overline{RQ}$
2. Name one pair of adjacent angles. $\angle S$ and $\angle R$
3. $m\angle P + m\angle Q + m\angle R + m\angle S = ?$ 360°

Is the figure a parallelogram? Give a reason for your answer. See margin.

4. 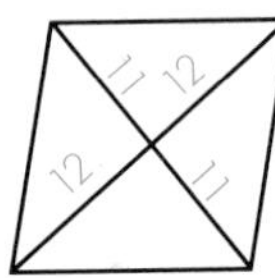

5. 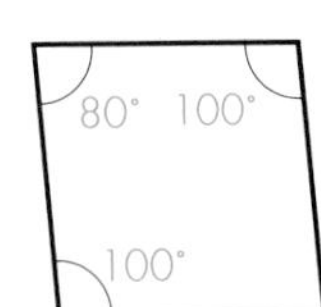

6. 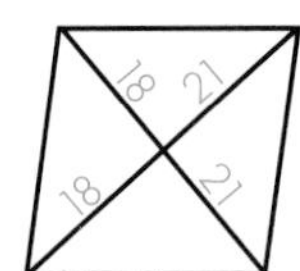

7. 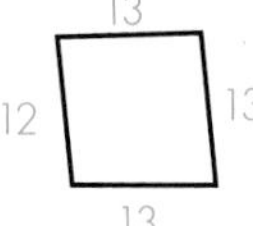

$\overline{EF}$ is the median of trapezoid *ABCD*. *BC* = 18. Find the following measures.

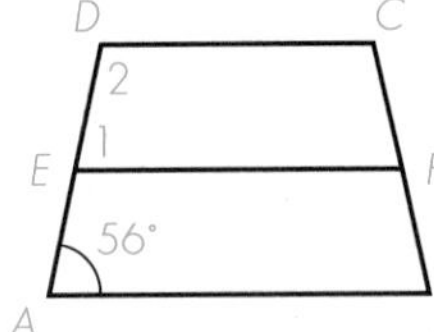

8. BF = ? 9
9. $m\angle 1 = ?$ 56°
10. $m\angle 2 = ?$ 124°
11. Do you know that *AD* = 18? Why or why not? No; can't tell whether *ABCD* is isosceles.

JKLM is a parallelogram. Find the following measures.

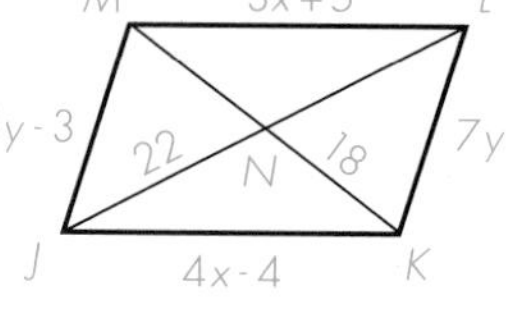

12. *JK* 32
13. *ML* 32
14. *KL* 21
15. *MJ* 21
16. *NL* 22
17. *NM* 18

18. Name four pairs of congruent triangles in rhombus *GHJK*.
$\triangle GHK \cong \triangle JHK$, $\triangle GKJ \cong \triangle GHJ$, $\triangle GOK \cong \triangle JOH$, $\triangle GOH \cong \triangle JOK$

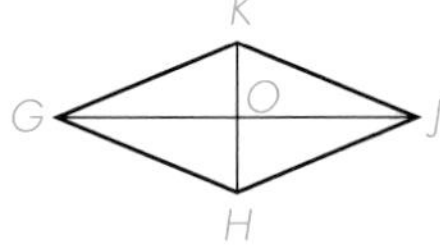

Find the indicated angle measures.

19.  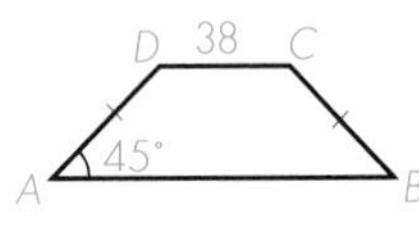

$m\angle B = ?$ 45°
$m\angle C = ?$ 135°
$m\angle D = ?$ 135°

20. 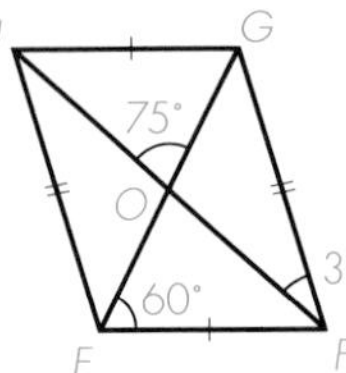

$m\angle HGE = ?$ 60°
$m\angle GHF = ?$ 45°
$m\angle EOH = ?$ 105°
$m\angle HFE = ?$ 45°

21. 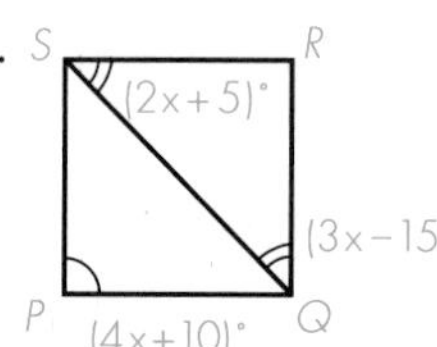 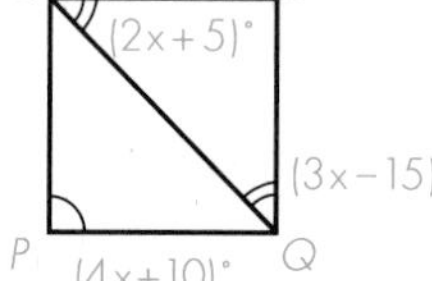

$m\angle RSQ = ?$ 45°
$m\angle SQP = ?$ 45°
$m\angle SPQ = ?$ 90°
$m\angle SRQ = ?$ 90°

22. 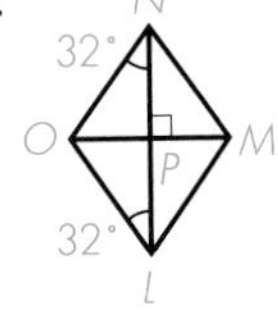 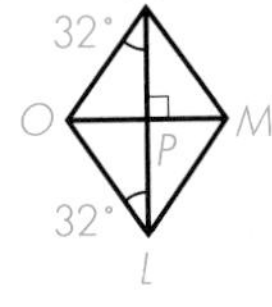

$m\angle NOP = ?$ 58°
$m\angle POL = ?$ 58°
$m\angle NML = ?$ 116°
$m\angle OLM = ?$ 64°

23. A rectangular garden measures 30 feet by 40 feet. How many feet of garden hose are needed to reach from one corner to the opposite corner? 50 ft

312

CHAPTER 8 TEST

Which of the following figures are

1. quadrilaterals? 1, 4, 5
2. parallelograms? 4, 5
3. rectangles? 4, 5
4. trapezoids? 1
5. squares? 5
6. none of these? 2, 3

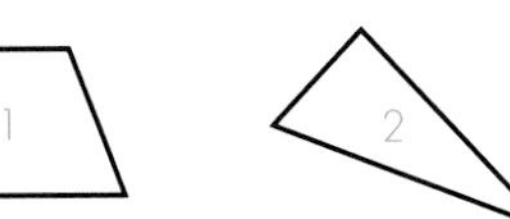

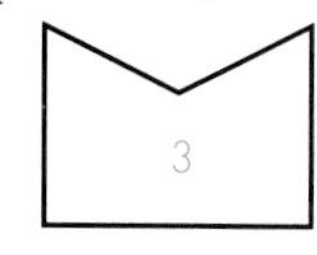

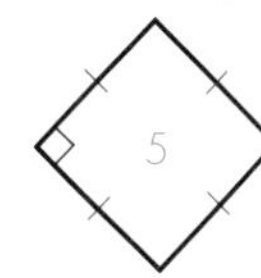

ABCD is a trapezoid.

7. Name one pair of opposite sides and one pair of adjacent sides. See margin.
8. Name one pair of opposite angles and one pair of adjacent angles. See margin.
9. If *ABCD* is an isosceles trapezoid, what is true for $m\angle A$ and $m\angle B$? measures are equal
10. $m\angle A + m\angle D = ?$ 180°
11. If $\overline{EF}$ is the median, what is true for $\overline{AE}$ and $\overline{ED}$? $\cong$
12. If $AB = 15$ and $DC = 12$, what is the length of $\overline{EF}$? 13.5

Is the figure a parallelogram? Give a reason for your answer. See margin.

13. 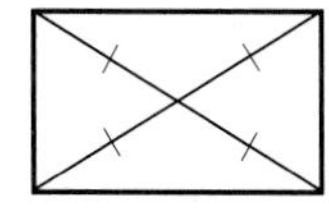

14. 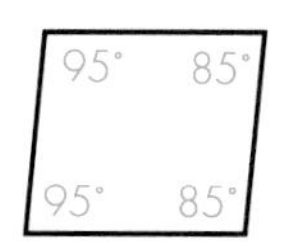

15. 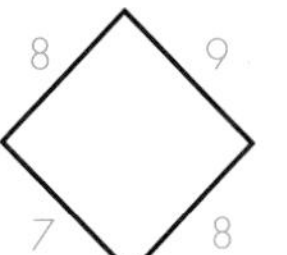

16. 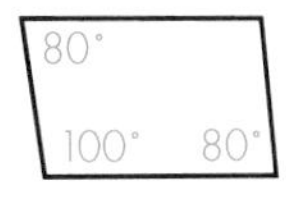

Find the indicated lengths or angle measures for each figure.

17. Rectangle

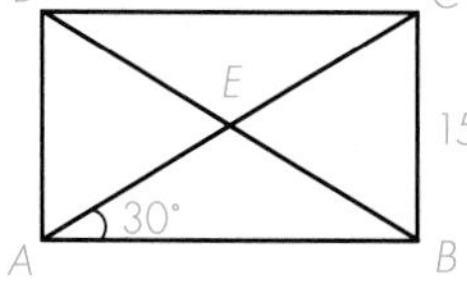

AC	30	*DB*	30
AB	$15\sqrt{3}$	*DC*	$15\sqrt{3}$
DE	15	*AE*	15

18. Isosceles trapezoid

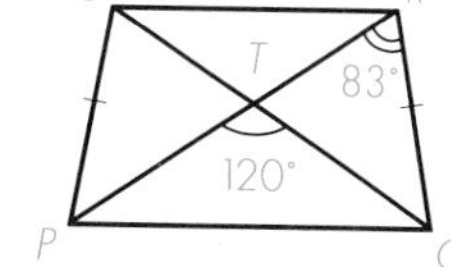

$m\angle RST$	30°	$m\angle SRT$	30°	$m\angle TSP$	37°
$m\angle RPQ$	30°	$m\angle SQP$	30°	$m\angle STP$	60°
$m\angle RTQ$	60°	$m\angle TQR$	37°	$m\angle TPS$	83°
$m\angle STR$	120°				

19. Parallelogram

PO	34	*NO*	20
QN	22	*PN*	44
		MQ	18
		MO	36

20. Rhombus

GH	10
GO	8
OH	6
$m\angle KOG$	90°

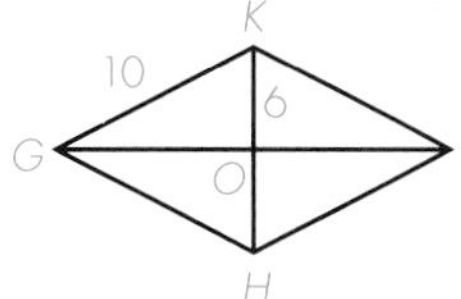

Assessment Resources
Achievement Tests pp. 15–16

Test Objectives
After studying this chapter, students should understand
- the properties that determine a quadrilateral.
- the properties of trapezoids and isosceles trapezoids.
- how to apply the theorems about isosceles trapezoids and medians of trapezoids.
- the properties of a parallelogram.
- how to apply the theorems about parallelograms.
- the properties of a rectangle.
- the properties of a rhombus.
- the properties of a square.

Additional Answers

7. Opposite: $\overline{DC}$ and $\overline{AB}$, or $\overline{AD}$ and $\overline{BC}$; adjacent: $\overline{AB}$ and $\overline{BC}$, $\overline{BC}$ and $\overline{CD}$, $\overline{CD}$ and $\overline{DA}$, or $\overline{DA}$ and $\overline{AB}$

8. Opposite: $\angle A$ and $\angle C$, or $\angle B$ and $\angle D$; adjacent: $\angle A$ and $\angle B$, $\angle B$ and $\angle C$, $\angle C$ and $\angle D$, or $\angle D$ and $\angle A$

13. Yes; diagonals bisect each other.

14. No; opposite angles are not congruent.

15. No; opposite sides are congruent.

16. Yes; opposite angles are congruent and supplementary.

Algebra Review Objectives
Topics reviewed from algebra include
- simplifying expressions containing radicals.
- rationalizing the denominator of an expression containing a radical.
- carrying out operations on expressions containing radicals.
- graphing equations and inequalities on the number line.
- identifying figures graphed on the number line.
- problem solving using variables.

Additional Answers

9. [Ray]

0 2

10. [Ray]

11. [Point]

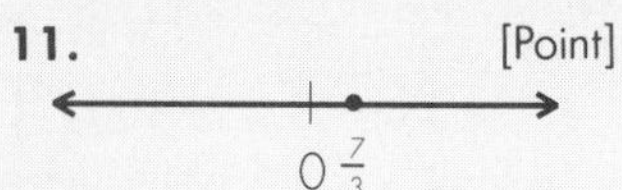

12. [Line segment]

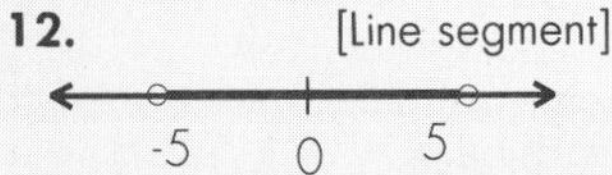

13. [Line segment]

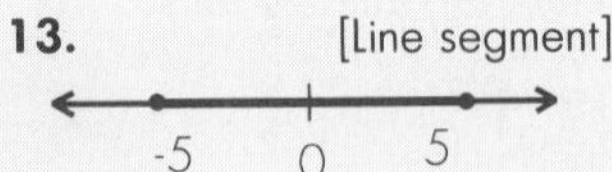

14. [Point]

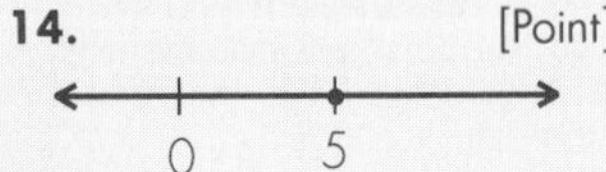

15. [Line segment]

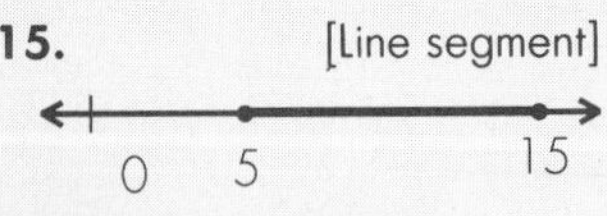

16. [Line]

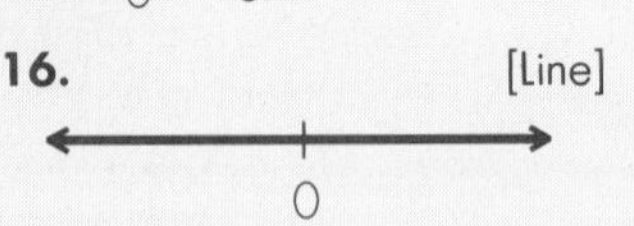

18. They have opposite signs; the product is negative, so one factor must be positive and one negative.

ALGEBRA SKILLS

To add or subtract expressions containing radicals, first simplify the expression, then combine like terms.

Example: Find $x + y$ where $x = 4 + \sqrt{12}$ and $y = 2 - \sqrt{27}$.
First simplify terms. $x = 4 + \sqrt{12} = 4 + \sqrt{4 \cdot 3} = 4 + 2\sqrt{3}$
$y = 2 - \sqrt{27} = 2 - \sqrt{9 \cdot 3} = 2 - 3\sqrt{3}$
So, $x + y = 4 + 2\sqrt{3} + 2 - 3\sqrt{3} = 6 - \sqrt{3}$

Multiply radical expressions the same way you multiply binomials.

Example:
$$\begin{aligned} xy &= (4 + 2\sqrt{3})(2 - 3\sqrt{3}) \\ &= 4(2 - 3\sqrt{3}) + 2\sqrt{3}(2 - 3\sqrt{3}) \\ &= 8 - 12\sqrt{3} + 4\sqrt{3} - 6 \cdot 3 \\ &= 8 - 8\sqrt{3} - 18 \\ &= -10 - 8\sqrt{3} \end{aligned}$$

When you divide radical expressions, you may have to rationalize the denominator—that is, rewrite the denominator without the radical. Use the idea that $(a - b)(a + b) = a^2 - b^2$.

Example: $x/y = \frac{4 + 2\sqrt{3}}{2 - 3\sqrt{3}} \cdot \frac{2 + 3\sqrt{3}}{2 + 3\sqrt{3}} = \frac{8 + 4\sqrt{3} + 12\sqrt{3} + 18}{4 - 27}$

$x/y = \frac{26 + 16\sqrt{3}}{-23}$

Let $x = 1 + \sqrt{8}$ and $y = 3 - \sqrt{18}$. Evaluate the following expressions.

1. $x + y$ $4 - \sqrt{2}$ **2.** $x - y$ $-2 + 5\sqrt{2}$ **3.** xy $-9 + 3\sqrt{2}$ **4.** $y - x$ $2 - 5\sqrt{2}$
5. x/y $\frac{5 + 3\sqrt{2}}{-3}$ **6.** y/x $\frac{15 - 9\sqrt{2}}{-7}$ **7.** $(x - y)^2$ $54 - 20\sqrt{2}$ **8.** $(x + y)^2$ $18 - 8\sqrt{2}$

The following expressions represent geometric figures when they are graphed on the number line. Draw the graph and name the figure. See margin.

9. $x \geq 2$
10. $3x \leq 12$
11. $3(2x - 6) = -4(3x - 6)$
12. $x \leq 5$ and $x \geq -5$
13. $|x| \leq 5$
14. $|x - 5| = 0$
15. $|x - 10| \leq 5$
16. $|x| \leq |x|$

17. In a school of 850 students, 62% registered for math classes and 175 signed up for music classes. How many more students took math than music? 352 students

Suppose you begin with a number. You subtract 2 from your number. Then you subtract 8 from the original number. You multiply the two differences and get a product of -9.

18. What do you know about the signs of the two numbers you multiplied? Why? See margin.
19. Try to find the original number by guessing.
20. Solve for the original number using algebra. Let n = the number. $(n - 2)(n - 8) = -9$; $n = 5$

314

Name ______________________ Date ____________

Properties of Quadrilaterals

A *quadrilateral* is a closed 2-dimensional figure with four straight sides.

These shapes are quadrilaterals. These shapes are not quadrilaterals.

Is not closed

Is not 2-dimensional

Does not have four sides

Does not have straight sides

Write the letter of the definition for each word.

1. quadrant d
2. quadruped c
3. quadrilateral e
4. quadruple a
5. quadricentennial b

a. to multiply by four
b. the four hundredth anniversary of an event
c. an animal with four feet
d. one quarter of a circle
e. a geometric shape with four straight sides

Give a reason why each shape is not a quadrilateral.

6. Does not have straight sides
7. Does not have four sides
8. Is not closed
9. Does not have four sides
10. Does not have straight sides
11. Is not 2-dimensional
12. *ABC* with each side of length 3 inches
Does not have four sides
13. *ABCD* with $AB = 10$ inches, and $BC = CD = DA = 1$ inch
Is not closed
14. The cube with vertices *ABCDEFGH*
Is not 2-dimensional

Name ______________________ Date ____________

Inscribed and Circumscribed Polygons

A When each vertex of a polygon is a point on a circle, the polygon is *inscribed* in the circle. We also say the circle is *circumscribed* about the polygon.

B The side of a polygon is *tangent* to a circle if there is exactly one point of intersection. When every side of a polygon is tangent to a circle, the circle is *inscribed* in the polygon. We also say that the polygon is *circumscribed* about the circle.

Make a sketch of each figure described in Exercises 1–4.

1. A circle inscribed in a triangle.
2. A quadrilateral circumscribed about a circle.
3. A square inscribed in a circle.
4. A circle inscribed in a square.
5. Compare the lengths of $\overline{OB}$ and $\overline{OD}$. Then compare the lengths of $\overline{PA}$ and $\overline{PC}$. What do you notice?
Each pair of lengths is equal.
6. Copy the drawing and complete it to make a five-pointed star. Label the other three points *Q*, *M*, and *N*. Find the point where $\overline{QN}$ is tangent to the circle. Label this point *E*. Write three pairs of congruent line segments.
$\overline{QB} \cong \overline{QE}$, $\overline{MC} \cong \overline{MD}$, $\overline{NE} \cong \overline{NA}$
7. A quadrilateral is circumscribed about a circle. Side *EC* measures 17 units. What is the length of the segment labeled *y*?
7 units
8. Side *AE* measures 9 units and side *BC* measures 12 units. Find the length of side *AB*.
4 units

Name ______________________ Date ______________

Trapezoids

A *trapezoid* is a quadrilateral with one pair of parallel sides. The other two sides are not parallel.

These shapes are not trapezoids.

Have more than one pair of parallel sides

Does not have a pair of parallel sides

Has more than four sides

Use the figure at the right. The two horizontal lines are parallel.

1. The angles labeled *a* and *b* are vertical angles. Is *a* greater than, less than, or equal to *b*? equal to
2. The angles labeled *b* and *f* are supplementary. What is sum of *b* and *f*? 180°
3. What is the sum of *c* and *e*? 180°
4. What is the sum of the angles of a trapezoid? 360°

Use the figure at the right. $\overline{TR}$ is parallel to $\overline{PA}$.

5. The sum of the angles of a triangle is 180°. What is the measure of angle *PRA*? 90°
6. Because *TRAP* is an isosceles trapezoid, angles *T* and *TRA* have equal measures. Angle *T* measures 115°. Write the measures of the four angles of the trapezoid.
$m\angle T = m\angle TRA = 115°$; $m\angle APT = m\angle A = 65°$

The angle labeled *x* measures 30°, 45°, or 60°. Find *x*.

7.

8.

9. Lines *p*, *q*, and *r* are parallel.

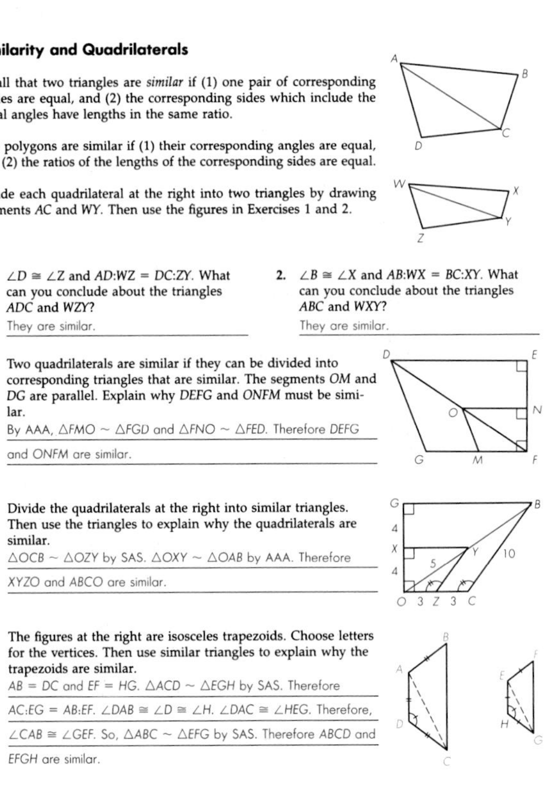

Name ______________________ Date ______________

Similarity and Quadrilaterals

Recall that two triangles are *similar* if (1) one pair of corresponding angles are equal, and (2) the corresponding sides which include the equal angles have lengths in the same ratio.

Two polygons are similar if (1) their corresponding angles are equal, and (2) the ratios of the lengths of the corresponding sides are equal.

Divide each quadrilateral at the right into two triangles by drawing segments *AC* and *WY*. Then use the figures in Exercises 1 and 2.

1. $\angle D \cong \angle Z$ and $AD{:}WZ = DC{:}ZY$. What can you conclude about the triangles *ADC* and *WZY*?
They are similar.

2. $\angle B \cong \angle X$ and $AB{:}WX = BC{:}XY$. What can you conclude about the triangles *ABC* and *WXY*?
They are similar.

3. Two quadrilaterals are similar if they can be divided into corresponding triangles that are similar. The segments *OM* and *DG* are parallel. Explain why *DEFG* and *ONFM* must be similar.
By AAA, $\triangle FMO \sim \triangle FGD$ and $\triangle FNO \sim \triangle FED$. Therefore *DEFG* and *ONFM* are similar.

4. Divide the quadrilaterals at the right into similar triangles. Then use the triangles to explain why the quadrilaterals are similar.
$\triangle OCB \sim \triangle OZY$ by SAS. $\triangle OXY \sim \triangle OAB$ by AAA. Therefore *XYZO* and *ABCO* are similar.

5. The figures at the right are isosceles trapezoids. Choose letters for the vertices. Then use similar triangles to explain why the trapezoids are similar.
$AB = DC$ and $EF = HG$. $\triangle ACD \sim \triangle EGH$ by SAS. Therefore $AC{:}EG = AB{:}EF$. $\angle DAB \cong \angle D \cong \angle H$. $\angle DAC \cong \angle HEG$. Therefore, $\angle CAB \cong \angle GEF$. So, $\triangle ABC \sim \triangle EFG$ by SAS. Therefore *ABCD* and *EFGH* are similar.

Name ______________________ Date ______________

Parallelograms

A *parallelogram* has two pairs of parallel sides. Since the edges of a ruler are parallel, you can use a ruler to draw a parallelogram.

Match each description to one of the figures below.

1. a trapezoid b
2. not a closed figure g
3. a parallelogram with no right angles a
4. the only 3-dimensional shape d
5. a parallelogram with right angles h
6. the only concave figure f
7. a pentagon c
8. quadrilateral that is not a trapezoid or a parallelogram e

Use a ruler to draw these parallelograms in the spaces below.

9. *ABCD* with angle *A* less than 90°. Write the measurements of angles *B*, *C*, and *D*.
Answers will vary.

10. *ABCD* with angle *A* greater than 90°. Find the sum of the measures of adjacent angles *C* and *D*.
180°

11. *ABCD* with angle *A* equal to 90° and *AB* not equal to *BC*. What is another name for this shape?
rectangle

Name ______________________ Date ______________

The Parallelogram Rule for Adding Vectors

There is an important application of parallelograms in science. If an object is being pulled in the direction of $\overrightarrow{AB}$ with a force of 50 pounds, and in the direction of $\overrightarrow{AC}$ with a force of 100 pounds, the object will actually move in the direction of $\overrightarrow{AD}$ as if pulled by a force of strength *x*.

The arrows in the figure which represent the forces are called *vectors.* The vector *AD* is the *resultant* of the two force vectors *AB* and *AC*.

To find the resultant of two vectors, you construct a parallelogram and draw its diagonal.

You will need a protractor and a centimeter ruler. For each exercise, trace the appropriate diagram. Then draw a parallelogram and its diagonal. Measure the length of the diagonal. Use ratio and proportion to find the answer to the problem. Student answers will vary.

1. A plane is traveling 60 km/h in a direction making an angle of 60° to the wind. The wind is blowing due west at 30 km/h. What is the resultant velocity and direction of the plane?
80 km/h; 40° N of W

2. A sailboat is heading east at 30 km/h while the wind is blowing it northward at 8 km/h. Find the resultant velocity and direction of the sailboat.
30 km/h; 18° N of E

3. Two tugboats pull on a ship. They pull at an angle of 30° to each other. One pulls with a force of 600 Newtons. The other pulls with 750 Newtons of force. What is the resultant force on the ship?
1300 Newtons

Draw diagrams to help solve Exercises 4 and 5.

4. A motorboat has a speed of 15 km/h. It crosses a river whose current has a speed of 3 km/h. In order to cross the river at right angles, the boat should be pointed in which direction?
11° into the current

5. A steamer is being propelled east at the rate of 15 mph. The wind is driving it north at 5 mph. Find the velocity of the steamer. What direction is it traveling?
16 mph; 18° N of E

Name ______________________ Date ____________

Rectangles

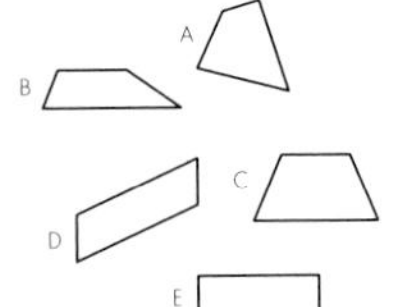

The five figures at the right are all quadrilaterals; each of them has four sides. The *rectangle*, Figure E, has four right angles, and opposite sides are parallel and congruent. In Figures A–D, assume that lines that appear to be parallel are parallel.
If a parallelogram has one right angle, then it is a rectangle.

Refer to the figures above. Write the letters of all the figures that match each description.

1. quadrilaterals A, B, C, D, E
2. parallelograms D, E
3. rectangles E
4. convex polygons A, B, C, D, E
5. trapezoids B, C
6. open figures none
7. Complete this chart to describe the characteristics of the given quadrilaterals. Write *yes* or *no* in each box.

	Parallelograms	Trapezoids	Isosceles Trapezoids	Rectangles
Are there two pairs of parallel sides?	yes	no	no	yes
Does each pair of adjacent angles have a sum of 180°?	yes	no	no	yes
Is at least one pair of sides congruent?	yes	no	yes	yes
Is the sum of the measures of the angles 360°?	yes	yes	yes	yes
Can all four sides be different lengths?	no	yes	no	no

Find the measure of angle ABC and the length of segment AB.

8. $\overline{FD}$ is parallel to $\overline{AB}$.
   m$\angle ABC$ = 56° AB = 4

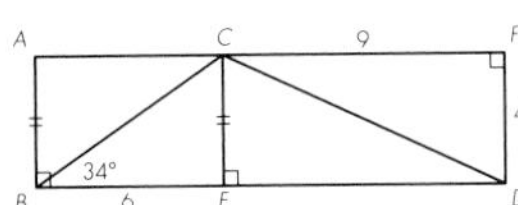

9. XYZ is an equilateral triangle and XD = CY = 3.
   m$\angle ABC$ = 90°
   AB = 9

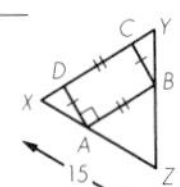

Name ______________________ Date ____________

Polyhedrons, Prisms, and Parallelepipeds

Any solid figure formed by portions of plane surfaces is called a *polyhedron*.

When two planes intersect a polyhedron, the figure they form is called a *prism*. If the planes are not parallel, the figure is called a *truncated* prism.

If the lateral sides of a prism are right angles to its bases, it is a *right* prism. If the angles are not right angles, the prism is an *oblique* prism.

A *parallelepiped* is a prism whose bases are parallelograms.

Write the letter of the figure at the right that matches each description. Use each letter only once.

1. right triangular prism F
2. right parallelepiped C
3. truncated prism D
4. rectangular oblique parallelepiped E
5. polyhedron that is not a prism A
6. right prism with five-sided bases B

Write the letters of all the figures that match each description.

7. All the faces are parallelograms. C, E
8. All the faces are quadrilaterals. C, D, E
9. All the faces are rectangles. C
10. There are more than six faces. A, B
11. The lateral faces are rectangles. B, C, F
12. The lateral faces are parallelograms. B, C, E, F
13. Draw a right prism with trapezoidal bases.
14. Draw a truncated cube.

15. Draw a polyhedron with only four faces.

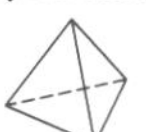

Name ____________ Date ____________

Squares and Rhombuses

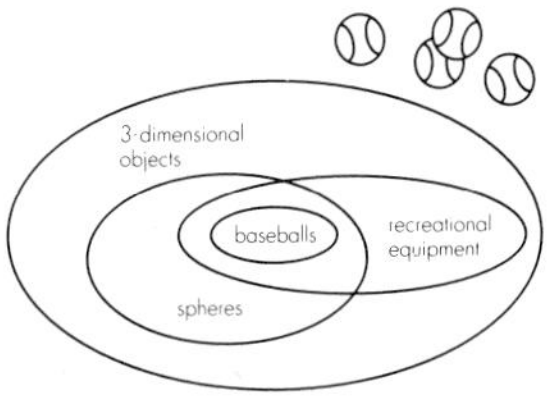

An object can have many names, depending on the various sets it belongs to. The Venn diagram at the right shows the relationships among various sets of objects.

All baseballs are spheres. But not all spheres are baseballs. (An orange is a sphere.) Some recreational equipment are baseballs; some aren't. (A tennis racket is not.)

3-dimensional objects
baseballs
recreational equipment
spheres

1. Choose *all, none,* or *some* to complete this statement.
   Some of the rectangles are also rhombuses.

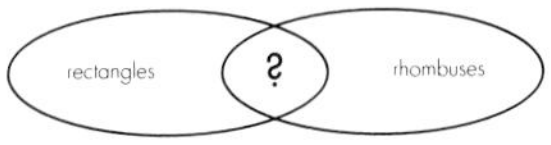

2. Choose *squares, trapezoids,* or *quadrilaterals.*
   Squares are figures that are both rectangles and rhombuses.

rectangles ? rhombuses

3. Choose *all, none,* or *some* to complete this statement.
   None of the trapezoids are also parallelograms.

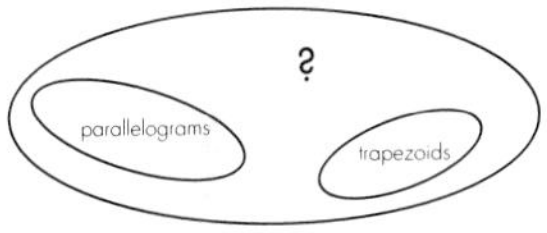

4. Choose *squares, rectangles,* or *quadrilaterals.* Parallelograms and trapezoids are members of the set of quadrilaterals.

?
parallelograms
trapezoids

5. Choose *all, none,* or *some* to complete this statement.
   All of the squares are also parallelograms.

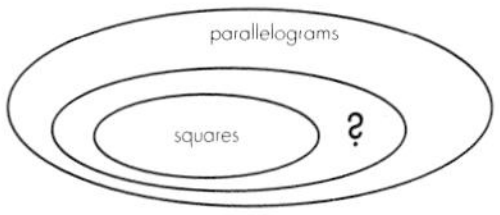

6. Choose *squares, rectangles,* or *trapezoids.*
   The rectangles are a subset of the parallelograms that include the set of squares.

parallelograms
squares ?

Name ______________________ Date ____________

Shapes Formed By Angle Bisectors

You have already learned how to bisect an angle with a compass.

1.

2.

3.

Use the figures at the right for these exercises.

1. Bisect the four angles of the parallelogram. What shape is formed by the angle bisectors?
   a rectangle
2. Draw four more parallelograms on a separate sheet of paper. Sketch the angle bisectors. What shape is formed by the bisectors?
   a rectangle
3. Complete this statement: The angle bisectors of the angles of a parallelogram form a
   rectangle
4. Bisect the angles of the rectangle. What shape is formed?
   a square
5. Complete this statement: The bisectors of the angles of a rectangle form a
   square
6. Bisect the angles of the trapezoid. Describe the figure that is formed.
   a quadrilateral with two opposite right angles
7. Bisect the base angles of the isosceles triangle. What figure is formed by the bisectors and the base?
   another isosceles triangle
8. What happens when you bisect the angles of a square?
   The angle bisectors are the diagonals of the square.

Name ______________________ Date ______________

Logo and Quadrilaterals

Draw each quadrilateral specified below on the *x-y* plane, indicating the coordinates of each vertex. Then write a LOGO procedure to tell the turtle to draw each quadrilateral. Answers may vary.

1. square

2. rectangle

3. parallelogram

4. rhombus

5. trapezoid

6. a quadrilateral that is not a parallelogram or a trapezoid

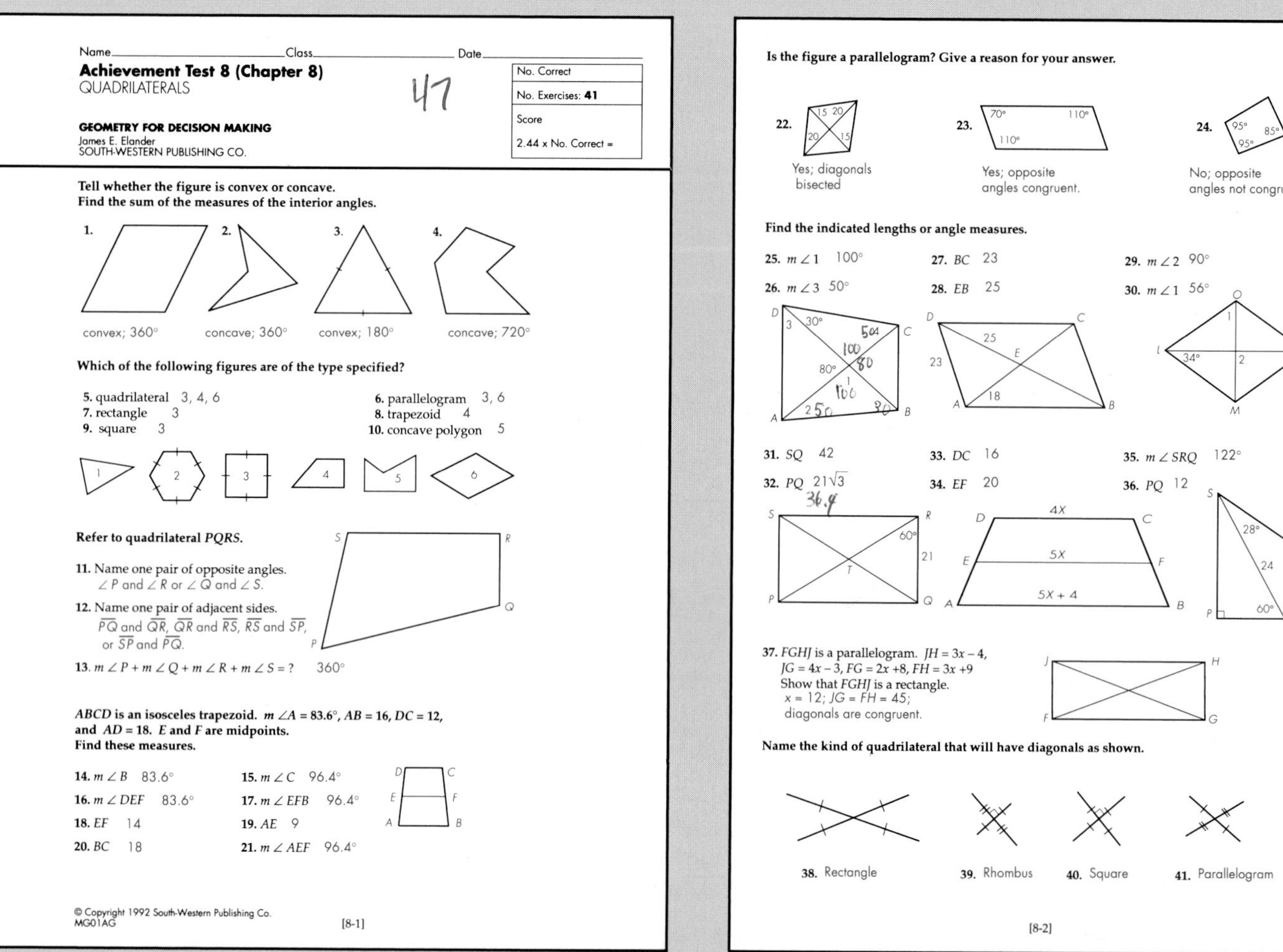

Name__________ Class__________ Date__________

Achievement Test 8 (Chapter 8)
QUADRILATERALS

47

No. Correct
No. Exercises: **41**
Score
2.44 x No. Correct =

GEOMETRY FOR DECISION MAKING
James E. Elander
SOUTH-WESTERN PUBLISHING CO.

Tell whether the figure is convex or concave. Find the sum of the measures of the interior angles.

1. convex; 360°
2. concave; 360°
3. convex; 180°
4. concave; 720°

Which of the following figures are of the type specified?

5. quadrilateral 3, 4, 6
6. parallelogram 3, 6
7. rectangle 3
8. trapezoid 4
9. square 3
10. concave polygon 5

Refer to quadrilateral *PQRS*.

11. Name one pair of opposite angles.
$\angle P$ and $\angle R$ or $\angle Q$ and $\angle S$.
12. Name one pair of adjacent sides.
$\overline{PQ}$ and $\overline{QR}$, $\overline{QR}$ and $\overline{RS}$, $\overline{RS}$ and $\overline{SP}$, or $\overline{SP}$ and $\overline{PQ}$.
13. $m\angle P + m\angle Q + m\angle R + m\angle S = ?$ 360°

***ABCD* is an isosceles trapezoid. $m\angle A = 83.6°$, $AB = 16$, $DC = 12$, and $AD = 18$. *E* and *F* are midpoints. Find these measures.**

14. $m\angle B$ 83.6°
15. $m\angle C$ 96.4°
16. $m\angle DEF$ 83.6°
17. $m\angle EFB$ 96.4°
18. EF 14
19. AE 9
20. BC 18
21. $m\angle AEF$ 96.4°

MG01AG

[8-1]

Is the figure a parallelogram? Give a reason for your answer.

22. Yes; diagonals bisected
23. Yes; opposite angles congruent.
24. No; opposite angles not congruent

Find the indicated lengths or angle measures.

25. $m\angle 1$ 100°
26. $m\angle 3$ 50°
27. BC 23
28. EB 25
29. $m\angle 2$ 90°
30. $m\angle 1$ 56°
31. SQ 42
32. PQ $21\sqrt{3}$ 36.4
33. DC 16
34. EF 20
35. $m\angle SRQ$ 122°
36. PQ 12

37. *FGHJ* is a parallelogram. $JH = 3x - 4$, $JG = 4x - 3$, $FG = 2x + 8$, $FH = 3x + 9$
Show that *FGHJ* is a rectangle.
$x = 12$; $JG = FH = 45$;
diagonals are congruent.

Name the kind of quadrilateral that will have diagonals as shown.

38. Rectangle
39. Rhombus
40. Square
41. Parallelogram

[8-2]

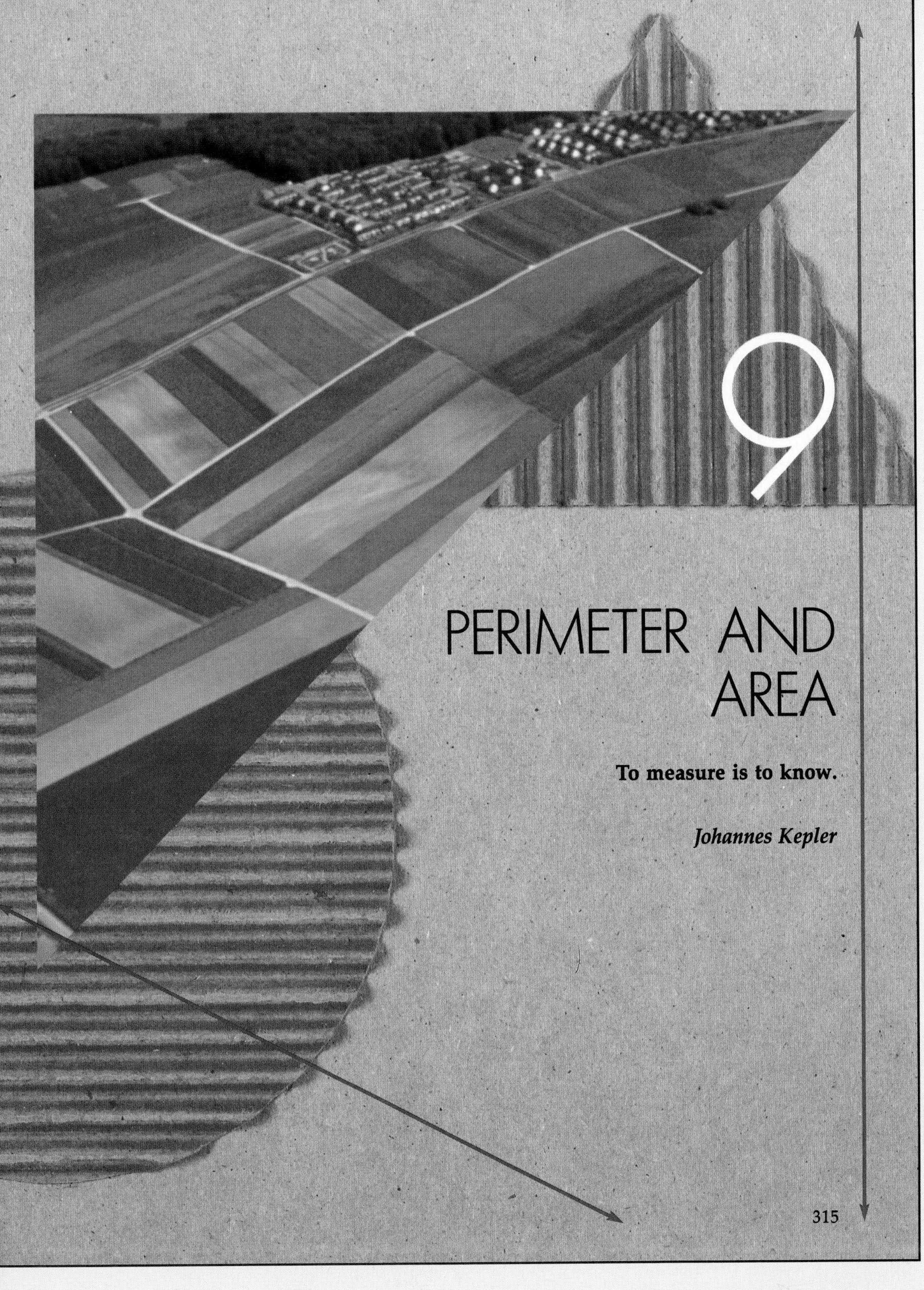

9

PERIMETER AND AREA

To measure is to know.

Johannes Kepler

9

315

CHAPTER 9 OVERVIEW

Perimeter and Area

Suggested Time: 10–12 days

To measure is to know.

Johannes Kepler

Kepler's statement is especially appropriate for this chapter since, for most people, mathematical perception of their surroundings depends on the measurement of space, whether it is estimation of distance or of the size of an object. The concepts of perimeter and area should be familiar to most students. The area theorems in this chapter are all based on the postulate for finding the area of a rectangle. Students are encouraged to derive the areas of various polygons in different ways, by applying the formulas for area of a triangle and area of a parallelogram. Another key concept in the chapter is the relationship between the perimeters and areas of similar figures.

CHAPTER 9 OBJECTIVE MATRIX

	End of Chapter Items by Activity				Student Supplement Items by Activity		
Objectives by Number	Review	Test	Computer	Algebra Skills	Reteaching	Enrichment	Computer
9.1.1	✓	✓	✓		✓	✓	✓
9.1.2	✓	✓	✓	✓	✓	✓	
9.2.1	✓	✓	✓	✓	✓		✓
9.2.2	✓	✓		✓	✓		✓
9.3.1	✓	✓		✓	✓	✓	
9.4.1	✓	✓		✓	✓	✓	
9.5.1	✓	✓			✓	✓	
9.5.2	✓	✓		✓	✓		✓

*A ✓ beside a Chapter Objective under Algebra Skills indicates that algebra skills taught within the chapter or in previous Algebra Skills lessons are used.

CHAPTER 9 PERSPECTIVES

▲ Section 9.1

Perimeter and Area

The aerial photographs of farmland that accompany this chapter illustrate the uses we make of familiar geometric figures in dividing up our environment. You may wish to discuss, as the chapter progresses, some of the reasons why these figures are used in specific instances; for example, fields are shaped a certain way to allow circular irrigation or rectangular plowing. The terms area and perimeter are defined. Remind students that the numbers assigned for the area or perimeter of a figure are positive numbers.

In a Discovery Activity students find that the area of a figure is the number of square units contained in its interior and observe that, for a rectangle, this number is equal to the product of length and width. The exercises in the Home Activity provide practice in finding area and perimeter, and the Critical Thinking problem begins the discussion of the relationship between perimeter and area.

▲ Section 9.2

Perimeter and Area of a Triangle

In the Discovery Activity, the class may find the relationship between the area of a rectangle and a right triangle by doubling the size of the triangle to form the rectangle. Or you may wish to have them use a geometric drawing tool to construct a triangle and an altitude to one of its sides. Students discover that the area of a triangle is one-half the product of its base and its height. The properties of isosceles right triangles, 30°-60° right triangles, and equilateral triangles are used to provide the missing information needed to find area. Finding the perimeter of a triangle will be a review for most students.

The Home Activities provide practical applications for area and perimeter, as well as an exercise asking students to write a Logo procedure for drawing a triangle with a specific area. The Critical Thinking involves finding different measurements for a room, given one perimeter.

▲ Section 9.3

Perimeter and Area of a Parallelogram

In the Discovery Activity, the class finds that the formula for the area of a parallelogram is the same as the formula for the area of a rectangle. You may wish to have students use a geometric drawing tool to construct a parallelogram. In discussing the fact that the area of any type of parallelogram can be found by the same formula, it may be helpful to draw several parallelograms, rhombuses, rectangles, and squares on the chalkboard, and have students identify several sets of altitudes and bases for each. The statement that the area of a rhombus is equal to one-half the product of its diagonals is first shown to be true for a unit square, then justified in a two-column proof for any rhombus. A mapmaking project affords the opportunity for students to work cooperatively on a practical application of perimeter and area. In the Home Activity, students are given other application problems, and the Critical Thinking problem continues the discussion of the relationship between perimeter and area.

▲ Section 9.4

Perimeter and Area of a Trapezoid

This section illustrates the various possibilities for triangulation in finding the area of a trapezoid. In the Discovery Activity, students see that the figure can be divided in two ways. They find that the area of the two triangles that form the trapezoid can be combined to find the area of the whole figure. This division is used as the basis for the two-column proof of the area theorem for trapezoids. In a Class Activity, students show that the area of a trapezoid can also be found by doubling the figure to form a parallelogram. The Critical Thinking problem provides a third way to find the area. These exercises in combining areas will be useful when students are asked to find the areas of many-sided polygons.

▲ Section 9.5

Applications

Since there are usually several ways to divide a nonparallel quadrilateral into figures for which the area formulas are known, it is important that students understand that they can use whatever configurations are easiest for them. The report on units of measure provides an opportunity for cooperative work as well as a forum for those students with dramatic talent. You may wish to review similarity before beginning the discussion of the practical application of the relationship between the corresponding sides and the areas of similar figures. In the Discovery Activity, the students use ratios and proportions to compare sides and areas of similar figures, and find that the square of the ratio of the corresponding sides of similar figures is the same as the ratio of their areas.

SECTION 9.1

Resources
Reteaching 9.1
Enrichment 9.1

Objectives
After completing this section, the student should understand
- ▲ the definitions of perimeter and area.
- ▲ how to find the perimeter and area of a rectangle.

Materials
graph paper

Vocabulary
perimeter
area

GETTING STARTED

Warm-Up Activity
On a piece of graph paper, have students trace the outline of one or two objects, such as a belt buckle, a watch, a button, or a comb. Have them consider how they might find the amount of material needed to cover the surface of each object. Ask for informal descriptions of their strategies.

TEACHING COMMENTS

Using the Page
In the Discovery Activity, students may find only one point of similarity among the figures. Help them conclude that the figures are closed, straight-sided plane figures.

Additional Answers
2. Possible answer: walk, counting steps as feet or paces as yards.

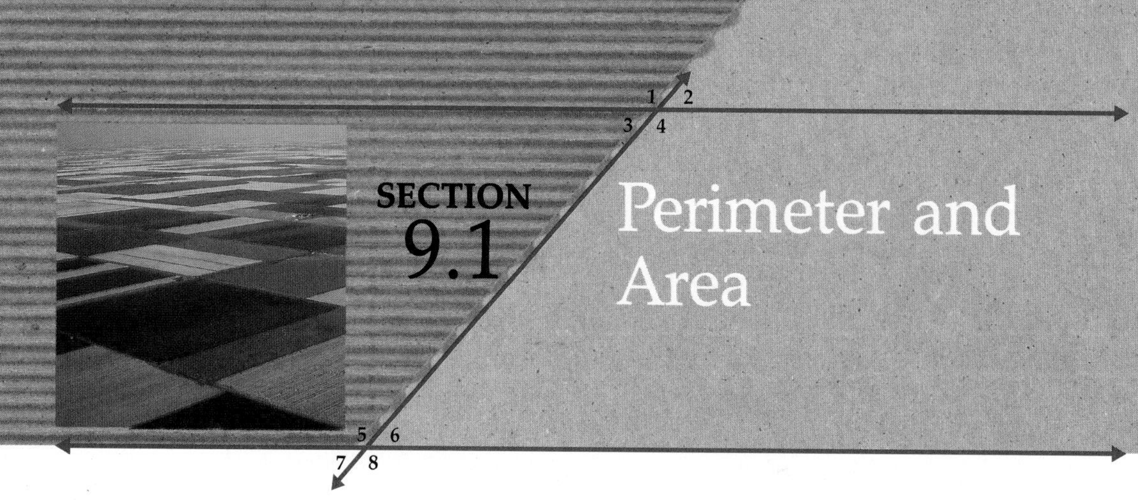

SECTION 9.1 Perimeter and Area

After completing this section, you should understand
- ▲ the definitions of perimeter and area.
- ▲ how to find the perimeter and area of a rectangle.

In Chapter 8 you investigated the properties that define different geometric figures. Since all of these figures are part of the structure of the natural and artificial world we live in, it is important to understand how to find the amount of space they occupy.

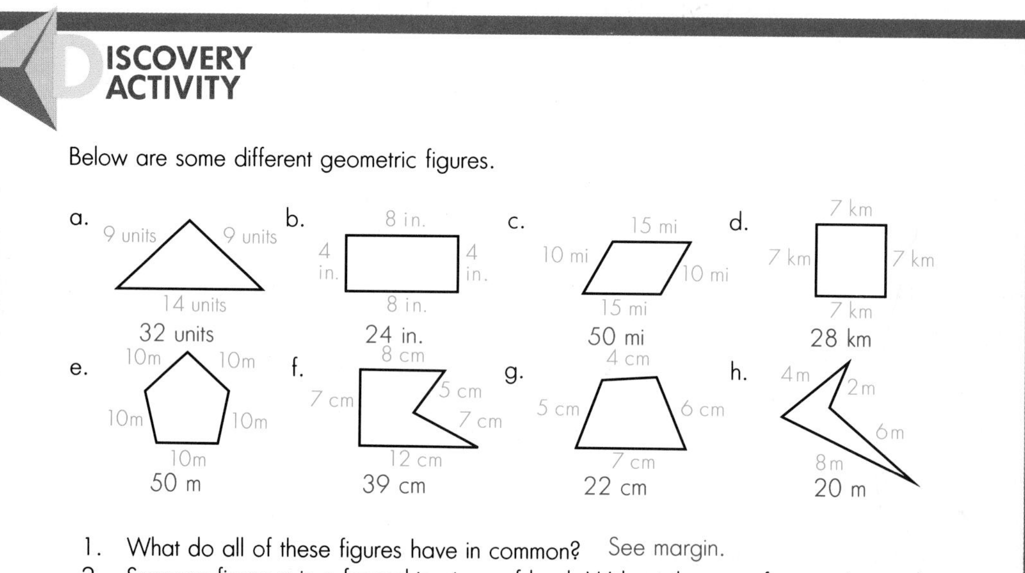

Below are some different geometric figures.

1. What do all of these figures have in common? See margin.
2. Suppose figure a is a fenced-in piece of land. Without the use of measuring tools, how could you find the distance around it? See margin.
3. How can you find the distance around each figure using the measurements given? Find this distance for figures a–h. Add sides.

You may have discovered that you can find the distance around any polygon by adding the lengths of the sides.

DEFINITION 9-1-1 **The PERIMETER of a figure is the distance around the figure or the sum of the lengths of all the sides.**

The word perimeter comes from the Greek words for distance, *meter,* and around, *peri.*

Example: How far do you have to run when you hit a home run in a baseball game?

Perimeter = 90 ft + 90 ft + 90 ft + 90 ft = 360 ft

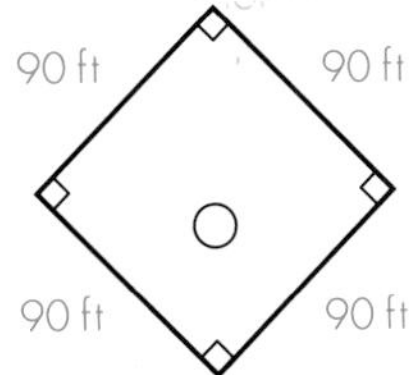

Example: What is the perimeter (P) of a basketball court?

P = 84 ft + 50 ft + 84 ft + 50 ft = 268 ft

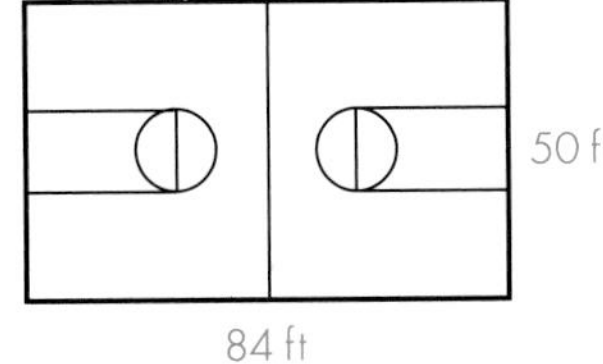

AREA

Any closed plane figure has an interior region. It is possible to assign a number called *area* to describe this interior region. The shaded regions are the interiors of these polygons.

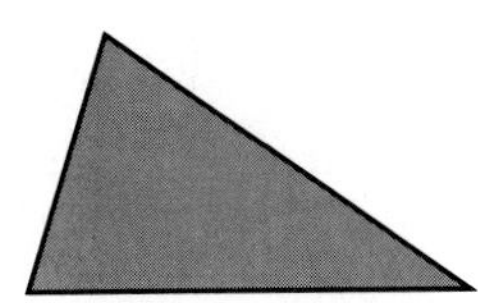

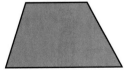

Using the Page
During the discussion of Definition 9-1-1, ask students for their own examples of times when they needed a perimeter measurement. Discuss the methods they used to find it. You may wish to have the class do some actual measuring to find the perimeter of several objects in the classroom or of the room itself.

Using the Page
During the Discovery Activity, some students may need to draw additional rectangles and count the number of squares before they can make the connection between the lengths of the sides and the area. For other students, this activity will be a review.

DISCOVERY ACTIVITY

1. Count the number of squares in the interior of each figure.

a. 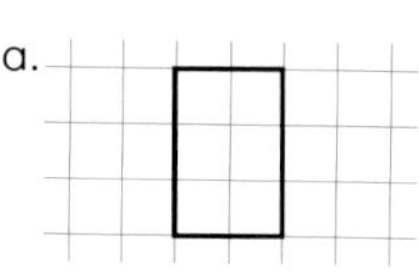 6

b.  18

c. 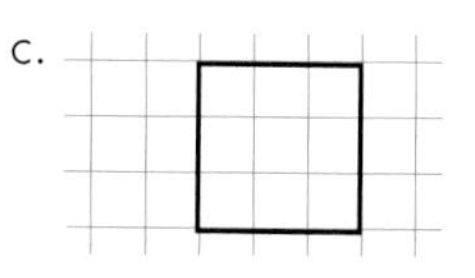 9

d. 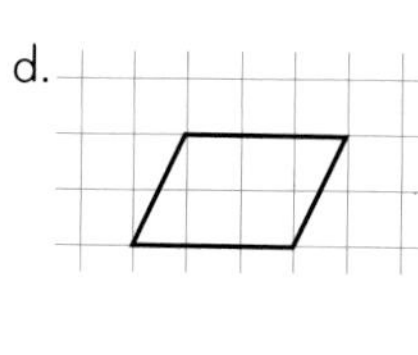 6

e. 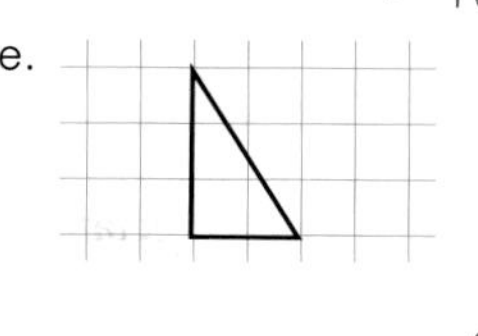 3

f. 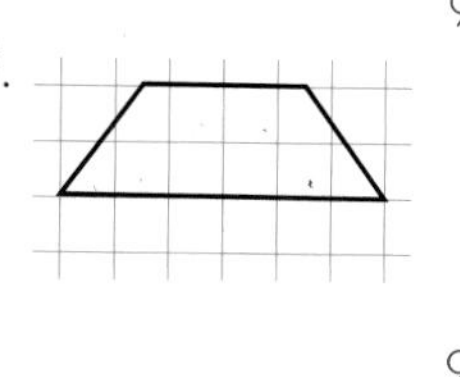 9

2. In figures a–c, it is easy to count squares. How did you count squares for figures d–f? Answers may vary.

3. There is a pattern to the number of squares you counted in figures a–c. Copy and complete the following table.

Figure	Number of Interior Squares	Measure of Sides
a	6 squares	3 by 2
b	18 squares	6 by 3
c	9 squares	3 by 3

4. What do you observe? The number of squares is equal to the length times the width.

You may have found that multiplying the length times the width of a rectangular figure gives the number of square units found in the interior. This number is the *area* of the figure.

DEFINITION 9-1-2 **The AREA of a plane figure is the number of square units contained in the interior.**

The figures a–c above are all rectangles. The table you completed shows a pattern that can be stated as a formula.

POSTULATE 9-1-1 **If the figure is a rectangle, then the number assigned for the area is the length times the width.**

Using the Page
After the discussion of Postulate 9-1-1, you may wish to have students find several rectangles with the same areas but different lengths and widths.

Area is always expressed in square units. If the unit of measure is feet, the area is expressed as feet times feet, or square feet.

Example: What is the area of a basketball court?

The court is a rectangle.

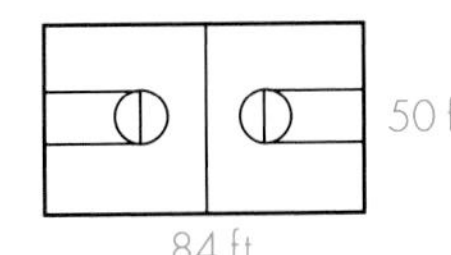

So, $A = lw$

$A = 84 \text{ ft} \times 50 \text{ ft} = 4200 \text{ sq ft}$

CLASS ACTIVITY

Find the area of the rectangle.

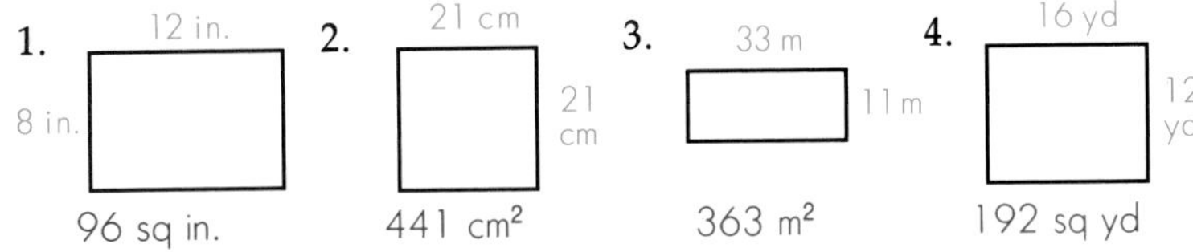

96 sq in. 441 cm² 363 m² 192 sq yd

Find the length and width of each rectangle.

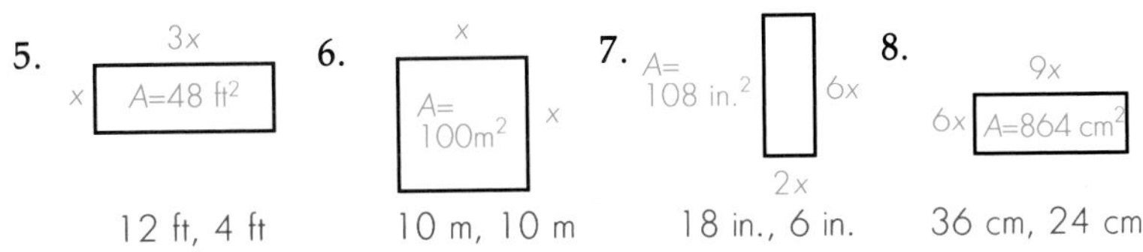

12 ft, 4 ft 10 m, 10 m 18 in., 6 in. 36 cm, 24 cm

Find the area of the shaded region for each rectangle.

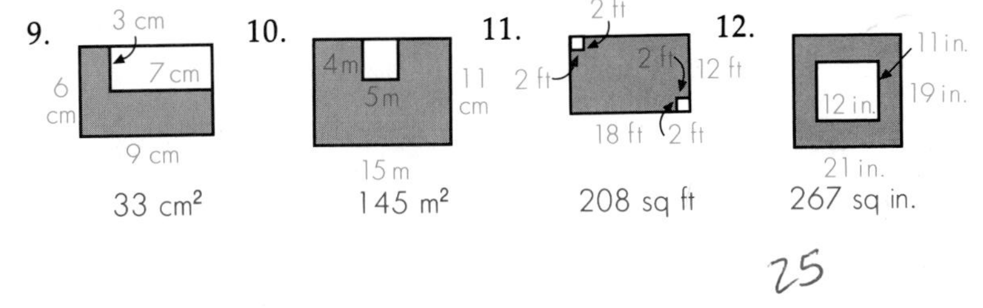

33 cm² 145 m² 208 sq ft 267 sq in.

25

HOME ACTIVITY

Find the perimeter.

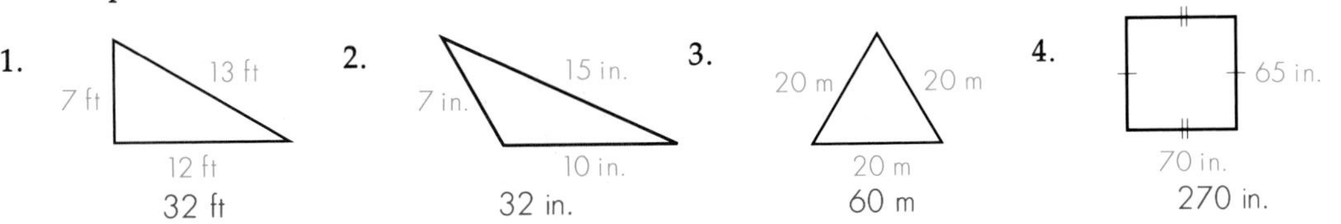

32 ft 32 in. 60 m 270 in.

Extra Practice

1. What is the area of a baseball diamond? Is the diamond a rectangle? [8100 sq ft; yes]

Find the area and perimeter.

2.

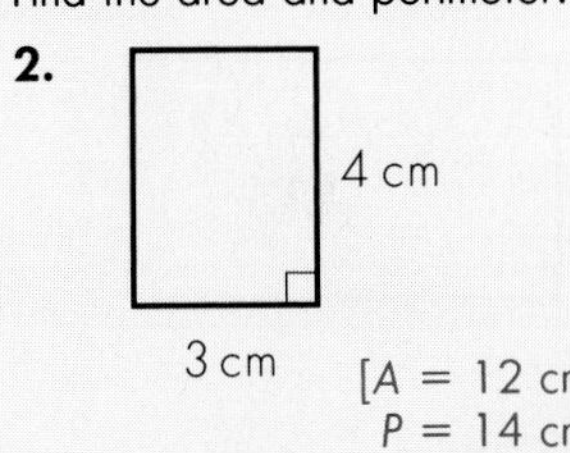

[$A = 12$ cm²; $P = 14$ cm]

3.

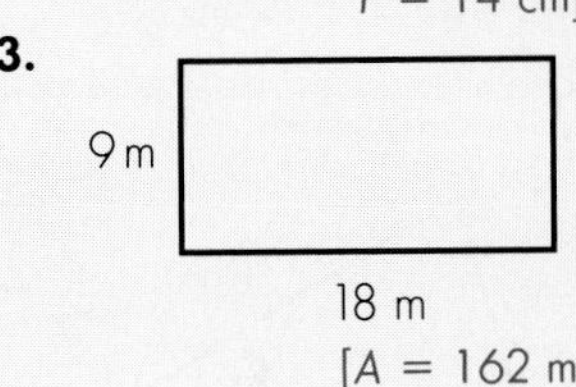

[$A = 162$ m²; $P = 54$ m]

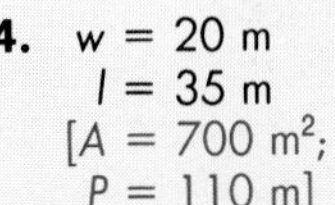

4. w = 20 m
l = 35 m
[$A = 700$ m²; $P = 110$ m]

5. w = 6 yd
l = 18 yd
[$A = 108$ sq yd; $P = 48$ yd]

Find the area of the shaded region.

6.

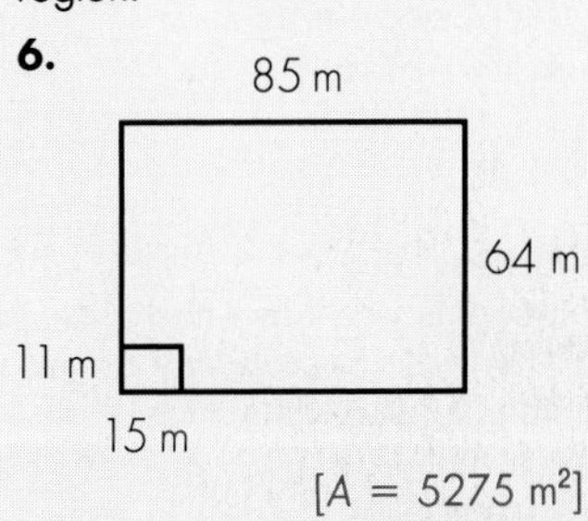

[$A = 5275$ m²]

Write a LOGO procedure to tell the turtle to draw a rectangle that has each perimeter. Then modify the procedure to draw another rectangle with the same perimeter. Answers may vary.

5. 20 6. 100 7. 14

Find the area of the rectangle.

8. 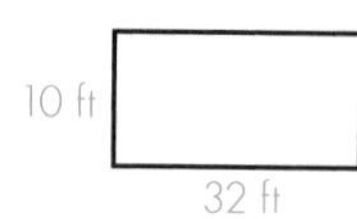

320 sq ft

9. 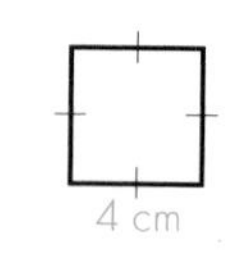

16 cm²

10. 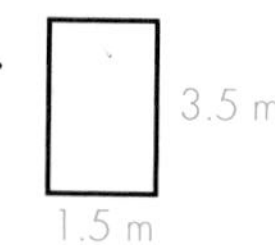

5.25 m²

Find the length and width of each rectangle.

11. 3x, A=735 cm², 5x
35 cm, 21 cm

12. 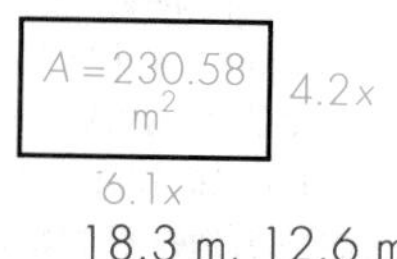

18.3 m, 12.6 m

13. A = 225 sq in., x
15 in.

Find the area of the shaded region of each rectangle.

14. 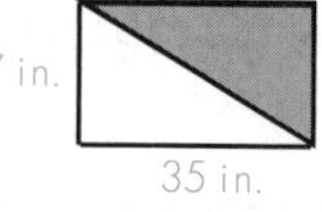

472.5 sq in.

15. 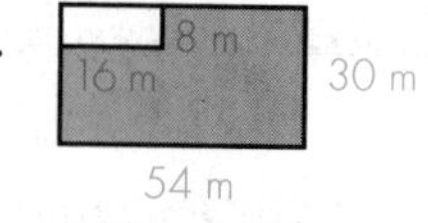

1492 m²

16. 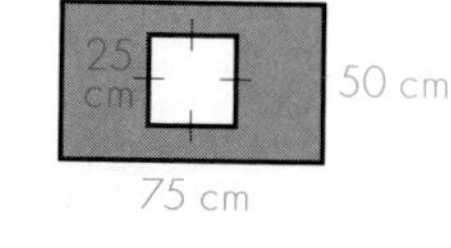

3125 cm²

A rectangular backyard measures 150 ft by 75 ft.

The owner of the property wants to sod the yard and then fence it.

17. Sod is sold for $1.55 per sq yd. How much will it cost to sod the yard? $1937.50

18. Fencing costs $3.75 per foot. How much will it cost to fence the yard? $1687.50

CRITICAL THINKING

A rectangle has a length of 20 units and a width of 15 units.

19. What is the area? What is the perimeter? 300 sq units; 70 units
20. What is the area if you double the length and width? What is the perimeter? 1200 sq units; 140 units
21. What happens to the area and perimeter when the dimensions of the rectangle are doubled? 4*A*; 2*P*
22. What do you think will happen to the area and perimeter if the length and width are halved? $A = \frac{1}{4}$ original area, $P = \frac{1}{2}$ original perimeter

Additional Answers
21. *A* quadruples, *P* doubles.

FOLLOW-UP

Assessment
Find the missing measures.

1.

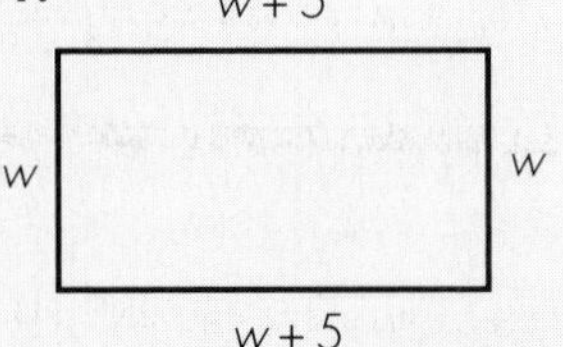

P = 38 ft
[W = 7 ft]

2.

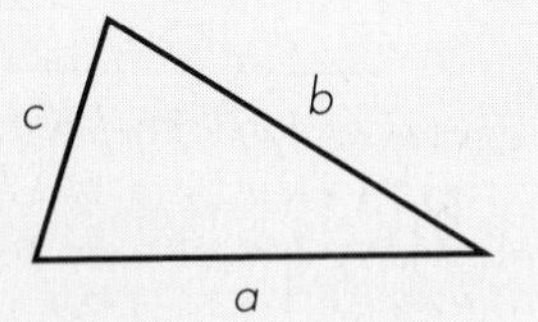

b = 2c
a = 2c + 3
P = 48 m

[a = 21 m, b = 18 m, c = 9 m]

Find the area of the rectangle.

3. w = 6 m
l = 9 m
[A = 54 m²]

4. w = 6 ft
l = 36 ft
[A = 216 sq ft]

Reteaching
Students who have had difficulty with this section may benefit from Reteaching Activity 9.1.

Enrichment
For students who have mastered the material in this section, you may wish to assign Enrichment Activity 9.1.

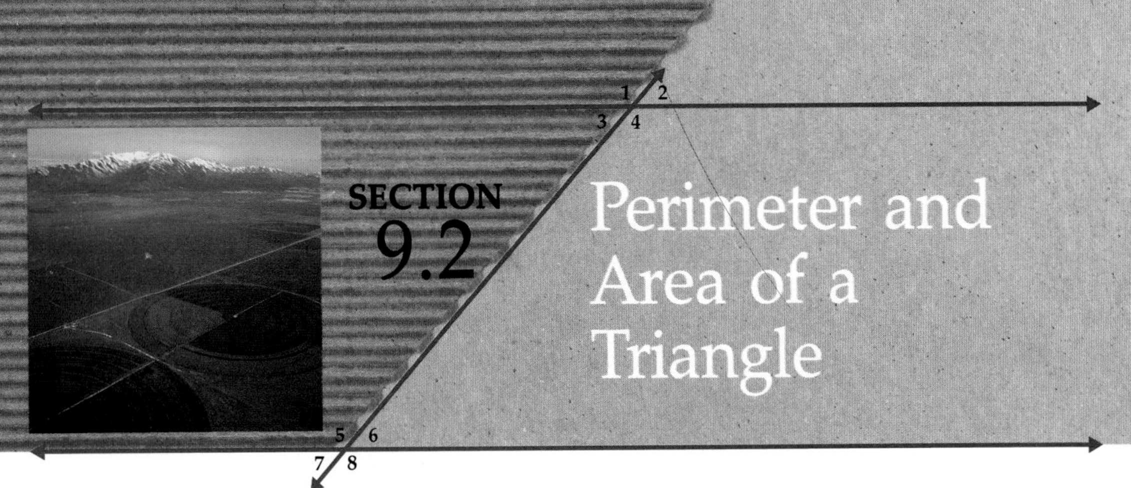

Section 9.2 Perimeter and Area of a Triangle

After completing this section, you should understand
▲ how to find the perimeter and area of a triangle.
▲ how to solve problems involving the area of a triangle.

Very often, the shape of a piece of land or of a building is made up of a group of triangles. In Section 9.1, you found the area of some triangular figures by counting the number of squares contained in the figure. This section will investigate an easier way to find the area of a triangle.

Discovery Activity

1. On graph paper, draw a triangle with sides of 3 and 4 units. See margin.
2. What is the length of the third side? How do you know? 5; right triangle, so $3^2 + 4^2 = 5^2$
3. Notice that you cannot make an exact count of the number of squares in the interior of the triangle. Instead, make the triangle into a rectangle.

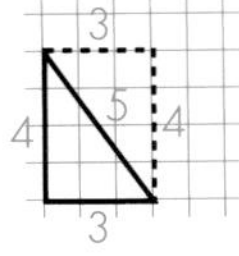

4. What is the area of the rectangle? 12 sq units
5. How many congruent triangles does the rectangle contain? 2
6. What is the area of one of the triangles? 6 sq units

SECTION 9.2

Resources
Reteaching 9.2
Enrichment 9.2

Objectives
After completing this section, the student should understand
▲ how to find the perimeter and area of a triangle.
▲ how to solve problems involving the area of a triangle.

Materials
graph paper
ruler
meter stick
string

GETTING STARTED

Quiz: Prerequisite Skills
Find the missing measures.

1.

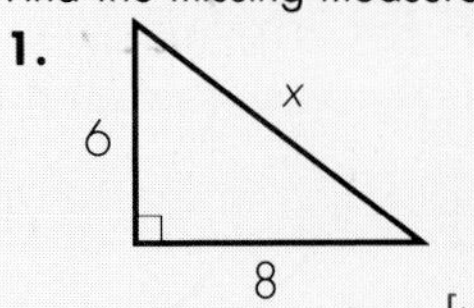

[x = 10]

2.

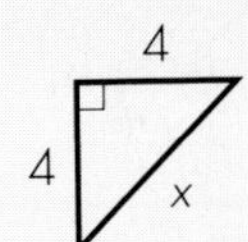

[$x = 4\sqrt{2}$]

3.

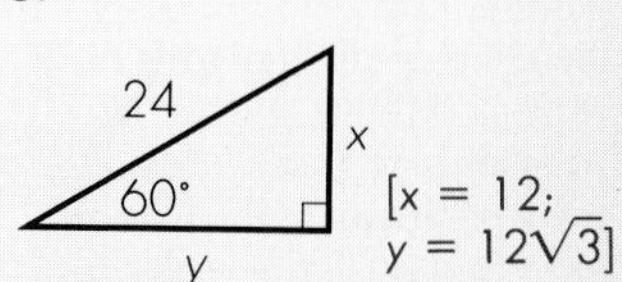

[$x = 12$; $y = 12\sqrt{3}$]

TEACHING COMMENTS

Using the Page
You may wish to extend the Discovery Activity by having students use a geometric drawing tool to draw any $\triangle ABC$. Then have them draw $\overline{AD}$ so that $\overline{AD} \perp \overline{BC}$. Ask students to measure $\overline{AD}$ and $\overline{BC}$ and find the area of $\triangle ABC$. Students should repeat this activity at least four times. Have them describe the relationships among the length of $\overline{AD}$, the length of $\overline{BC}$, and the area of $\triangle ABC$. They should discover that $\frac{1}{2} \cdot AD \cdot BC$ = area of $\triangle ABC$.

Using the Page
During the discussion of Theorem 9-2-1, remind the class that the area of any figure is always a positive number. Ask volunteers to draw additional pairs of bases and altitudes for triangles similar to those pictured.

7. What is the relationship between the area of the rectangle and the area of a right triangle? The area of a right triangle is $\frac{1}{2}$ the area of the rectangle.

8. Now suppose that you want to find the area of a triangle that is not a right triangle. On graph paper, draw an acute triangle and an obtuse triangle, each with sides having lengths a, b, and c.

9. Draw congruent triangles with sides a, b, and c next to the original triangles so that they form parallelograms.

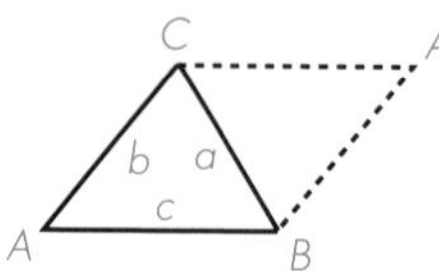

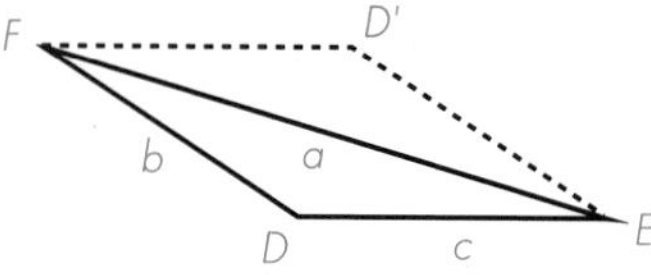

10. In each figure, draw an altitude to base, forming triangles ($\triangle CHA$ and $\triangle D'HE$). Slide the triangles to the left or right, so that they transform the parallelograms into rectangles.

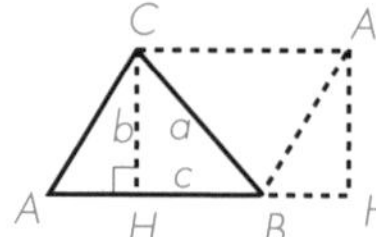

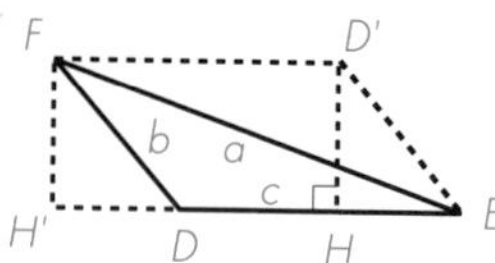

11. What is the area of the rectangle? $A = lw$ or bh

12. What can you conclude about the area of the triangle?
It is $\frac{1}{2}$ the area of the rectangle.

The rectangles you formed are made up of the two original triangles. Since the area of a rectangle is length times width or base times height, you may have concluded that the area of the triangle is $\frac{1}{2}bh$.

THEOREM 9-2-1 **If the figure is a triangle, then the number assigned to the area is $\frac{1}{2}$ the base times the height.**

Example: What is the area of $\triangle NMP$?

$A = \frac{1}{2} \cdot b \cdot h$
$\overline{PO}$ is the altitude or height. So,
$A = \frac{1}{2}(18 \cdot 4) = 20 \text{ cm}^2$

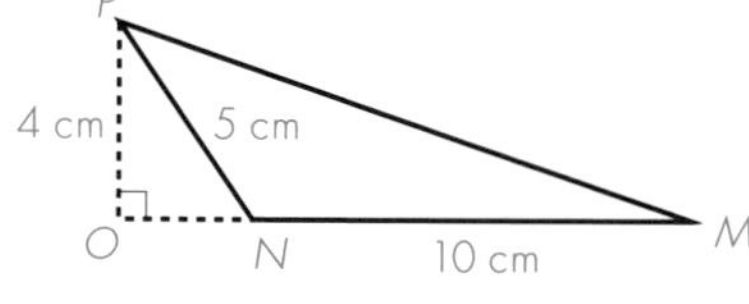

Example: $(\triangle PQR) = 272 \text{ cm}^2$. $PQ = 34$ cm. What is the length of the altitude $\overline{RS}$?

$$(\triangle PQR) = \tfrac{1}{2}bh$$
$$272 = \tfrac{1}{2} \cdot 34 \cdot (RS)$$
$$272 = 17\ (RS)$$
$$16 \text{ cm} = RS$$

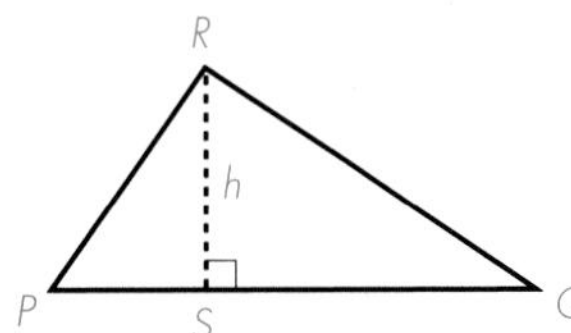

CLASS ACTIVITY

Find the area.

1. 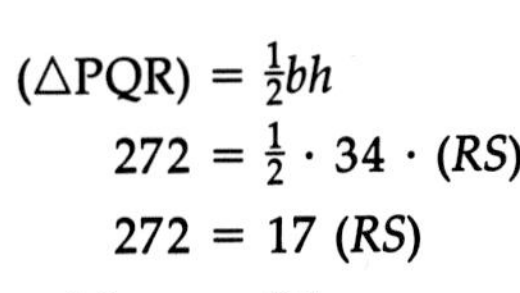

800 sq ft

2. 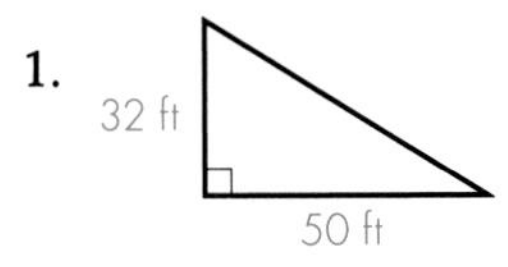

445.5 m²

3. 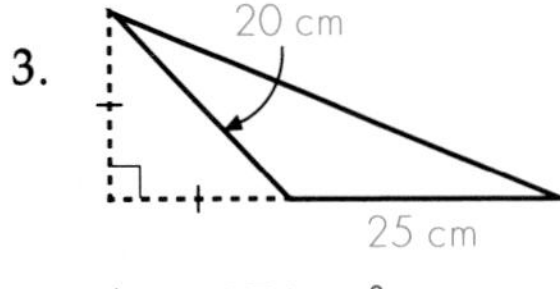

about 176 cm²

4. The diagram shows a plan for a large garden. The triangular area is to be planted with roses. What is the area of the rest of the garden? 2232 sq ft

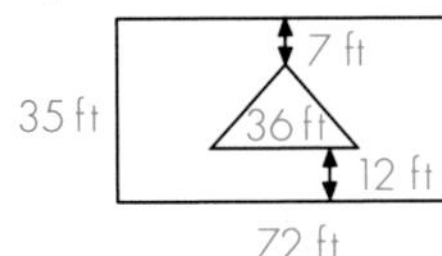

Find the missing altitude or base for each triangle.

5. 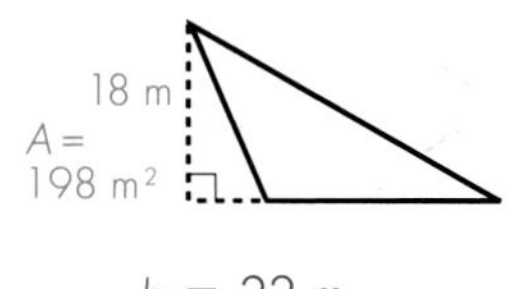

$b = 22$ m

6. 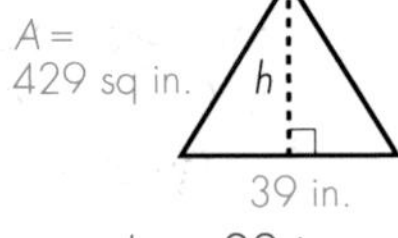

$h = 22$ in.

7. 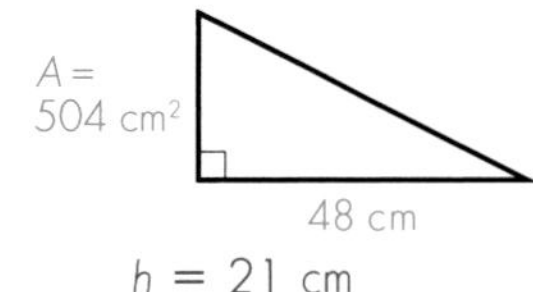

$h = 21$ cm

AREAS OF SPECIAL TRIANGLES

In Chapter 7, you studied several triangles with special properties.

Extra Practice

Find the area.

1. 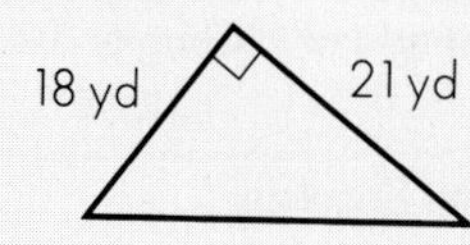

[$A = 189$ sq yd]

2. 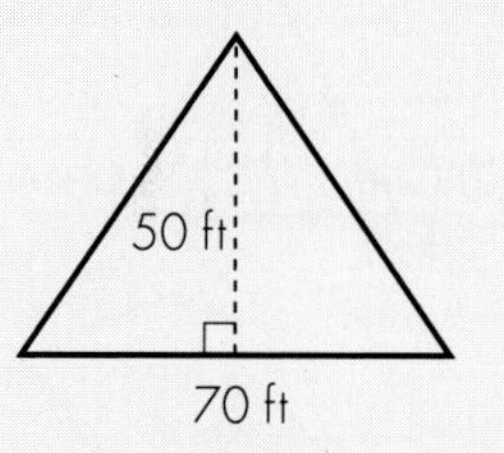

[$A = 1750$ sq ft]

Find the missing altitude or base.

3. 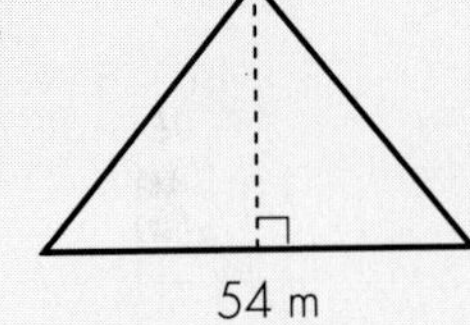

$A = 864 \text{ m}^2$
[$h = 32$ m]

4. 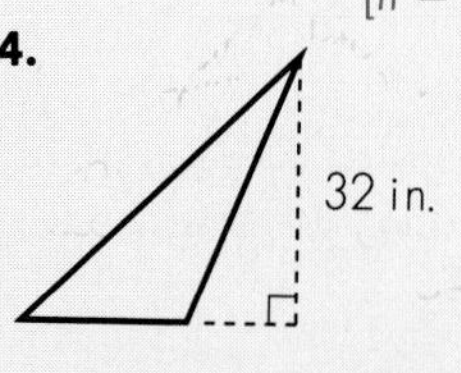

$A = 336$ sq in.
[$b = 21$ in.]

Using the Page
During the discussion of the examples of different types of triangles, you may wish to have volunteers draw the diagrams on the chalkboard and use colored chalk to outline the angles or sides that make each type of triangle unique. Encourage students to record these properties in their notebooks if they have not already done so. This will provide a quick reference that will be useful in later sections of this chapter and in Chapter 10.

Extra Practice
Find the area.

1.

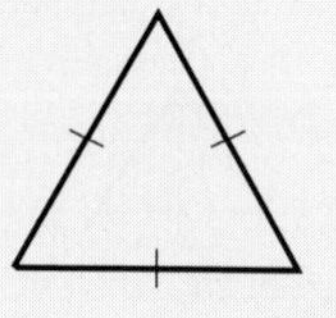

[$A = 64\sqrt{3}$ cm^2]

2.

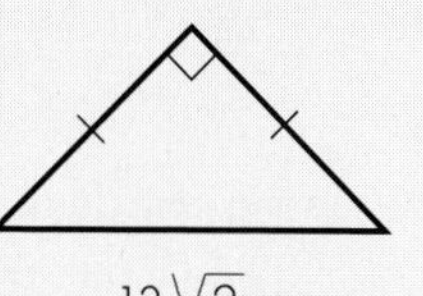

$13\sqrt{2}$ m

[$A = 84.5$ m^2]

3.

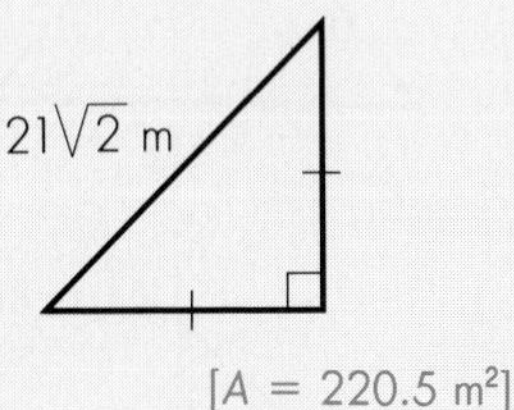

[$A = 220.5$ m^2]

You can use the properties of these triangles to find the area or a missing base or altitude measurement.

Example: What is the area of $\triangle ABC$?

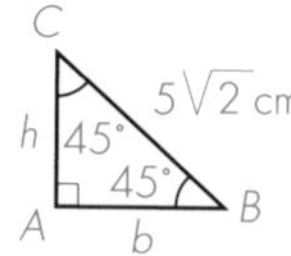

You need to find the lengths of the base and altitude. Since $m\angle B = m\angle C = 45°$, you know that $\triangle ABC$ is an isosceles right triangle. The hypotenuse $BC = 5\sqrt{2}$, so $AB = 5$ cm and $AC = 5$ cm.
Area $\triangle ABC = \frac{1}{2}(5 \cdot 5) = \frac{1}{2} \cdot 25 = 12.5$ cm^2

Example: What is the area of $\triangle LMN$?

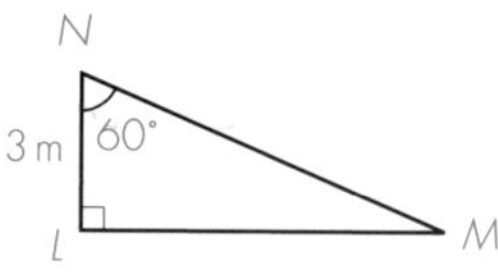

Altitude $\overline{NL} = 3$ m. What is the length of the base?
$\triangle LMN$ is a 30°–60° right triangle, so $LM = NL\sqrt{3}$ or $3\sqrt{3}$ m.
Area $\triangle LMN = \frac{1}{2}(3\sqrt{3} \cdot 3)$ or about 7.79 m^2

CLASS ACTIVITY

Find the area.

1. $12\sqrt{2}$ in.

   72 sq in.

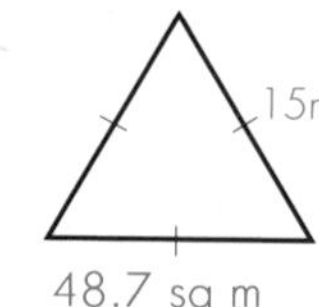

2. 15m

   48.7 sq m

3. 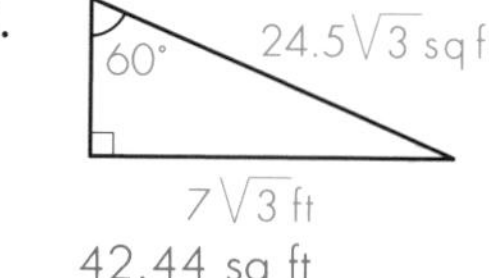

   42.44 sq ft

4. A farmer's field is fenced off in the shape of an isosceles right triangle. The lengths of the sides are shown. How much fence did it take to enclose the field? What is the area of the field?
   130 yd; $A = 722$ sq yd

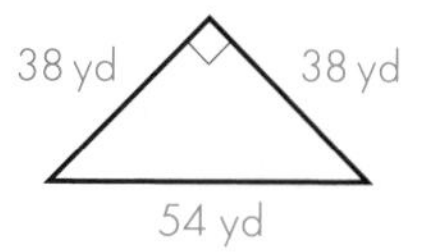

5. Place a meter stick or a long piece of string diagonally across your desktop. Estimate the area in square inches of one of the triangles formed. Then measure the base and altitude and find the area. Was your estimate reasonable? Answers will vary.

THE PERIMETER OF A TRIANGLE

You learned in Section 9.1 how to find the distance around an area. The perimeter of an area is the sum of the lengths of the sides.

The perimeter of a figure should be expressed in an appropriate unit of measure.

Example: The perimeter of a field is 300 yards.
Since 1 yd = 3 ft, it is also true that P = 900 ft.
Since 1 ft = 12 in., the perimeter can also be expressed as 10,800 in., but this measurement would probably not be useful.

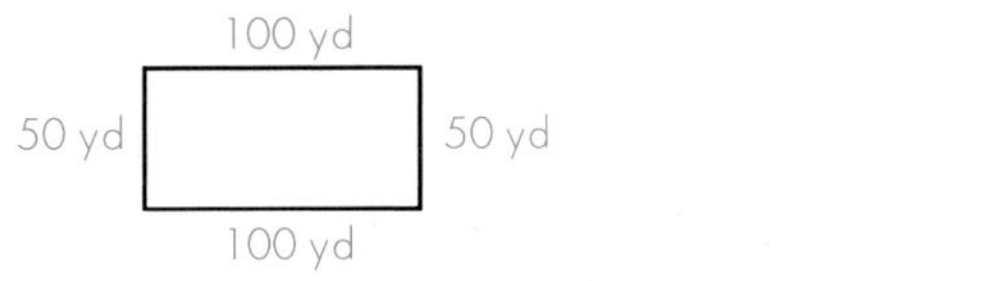

Using the Page
You may wish to extend the discussion of appropriate units of measure to area as well. For example, have students suppose there is a large area of land to be fenced in, fertilized, and planted as a garden. Ask the class to determine how they would choose the most useful unit of measure to find perimeter and area. On what conditions would they base their decisions? [Possible answers: unit of length fencing is sold in, unit of area amounts of fertilizer and seed are based upon, and so on]

CLASS ACTIVITY

Find the perimeter.

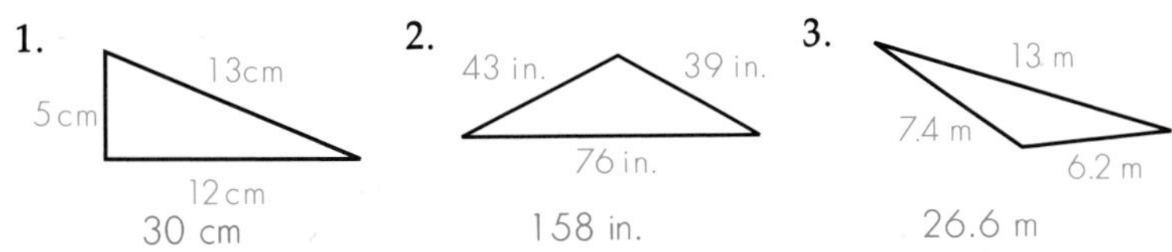

30 cm   158 in.   26.6 m

Find the missing measurements.

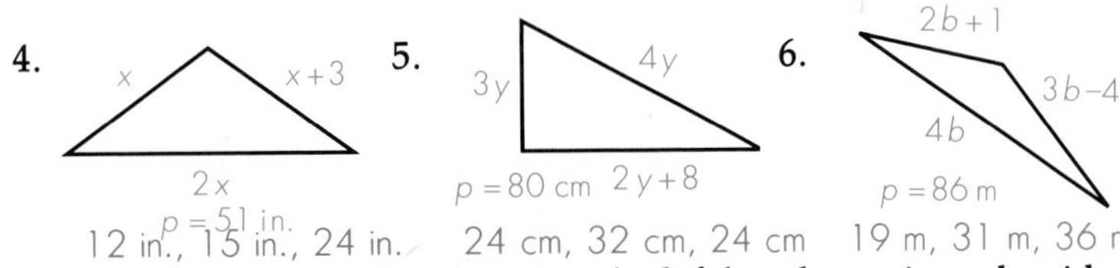

12 in., 15 in., 24 in.   24 cm, 32 cm, 24 cm   19 m, 31 m, 36 m

7. A pyramid-shaped sculpture at the left has three triangular sides. All of the edges, including the base, are to be outlined with reflective tape to illuminate the sculpture at night. How many feet of tape will be needed? 45 ft

8. The perimeter of a triangle is 150 meters. What is the perimeter in centimeters? 15,000 cm

Refer to $\triangle ABC$.

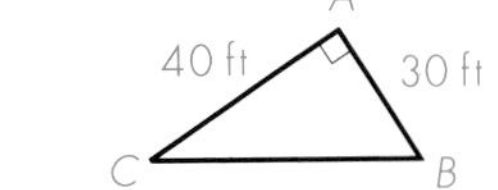

9. What is the length of $\overline{BC}$? 50 ft

10. What is the perimeter of $\triangle ABC$? 120 ft

Extra Practice
Find the area.

1.

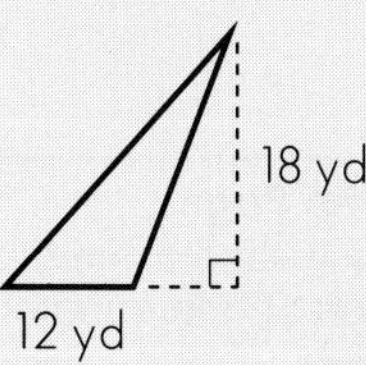

[A = 108 sq yd]

2.

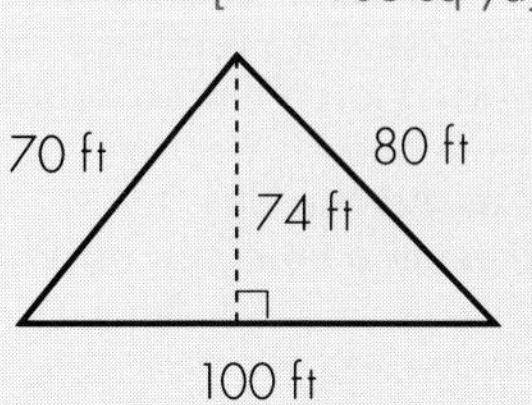

[A = 3700 sq ft]

3.

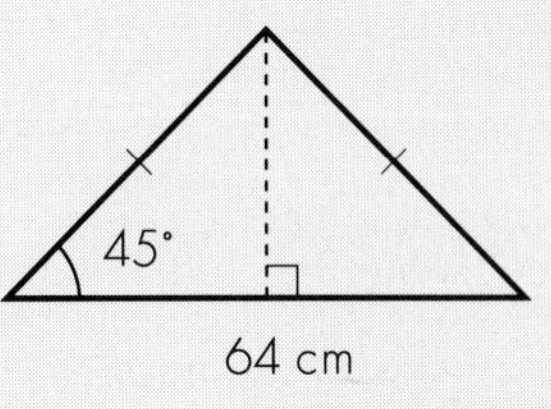

[A = 1024 cm^2]

4.

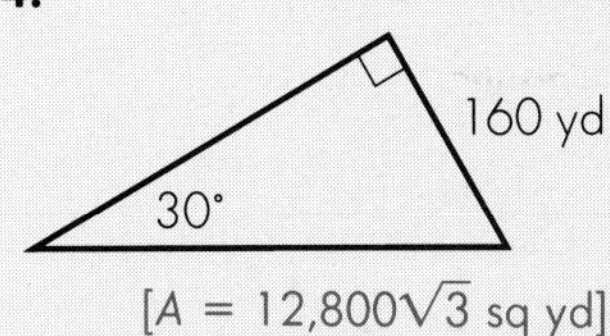

[$A = 12{,}800\sqrt{3}$ sq yd]

HOME ACTIVITY

Find the area.

1.

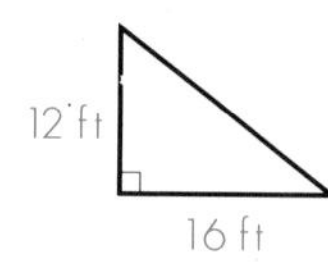

96 sq ft

2.

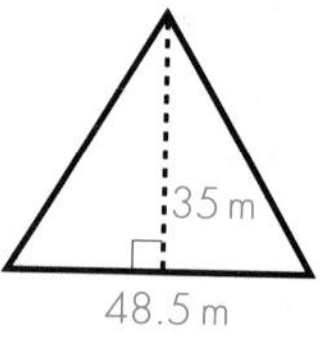

848.75 m^2

3.

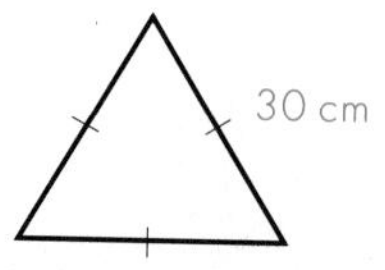

389.7 cm^2

Find the missing altitude or base.

4.

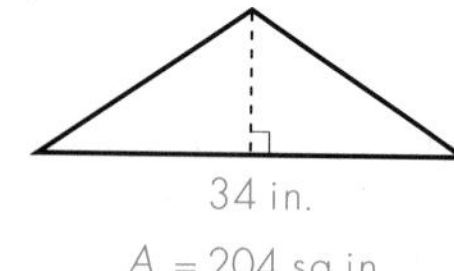

A = 204 sq in.

h = 12 in.

5.

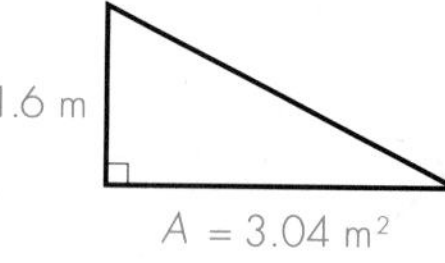

A = 3.04 m^2

b = 3.8 m

6.

A = 212.4 sq ft 35.4 ft

h = 12 ft

Find the area. The triangles are isosceles right triangles.

7.

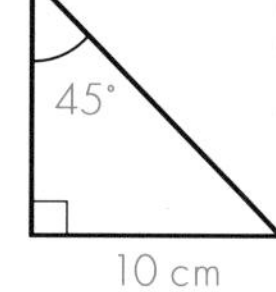

50 cm^2

8.

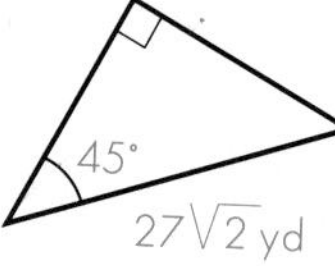

364.5 sq yd

Find the area. The triangles are 30°–60° right triangles.

9.

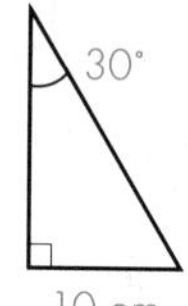

86.6 cm^2

10.

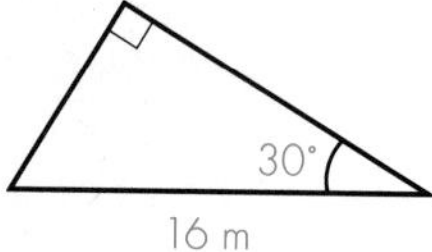

110.9 m^2

Find the perimeter.

11.

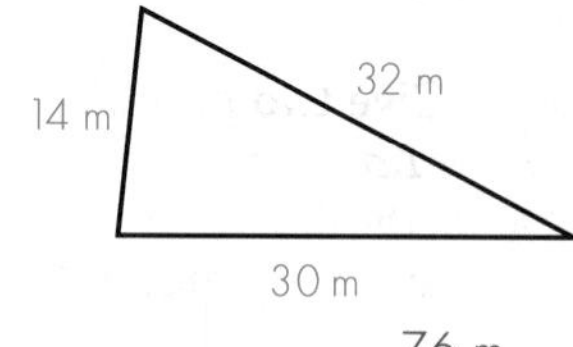

76 m

12.

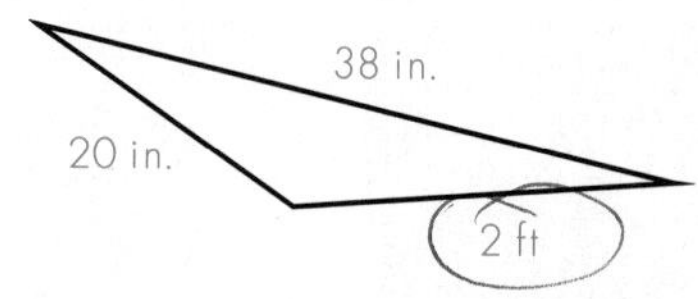

82 in. or 6 ft 10 in.

Write a LOGO procedure to tell the turtle to draw a triangle that has each area. Then modify the procedure to draw another triangle with the same area. Answers may vary.

13. 24 14. 72 15. 144

16. Find the perimeter and area of *PQRS*. P = 190 ft; A = 1800 sq ft

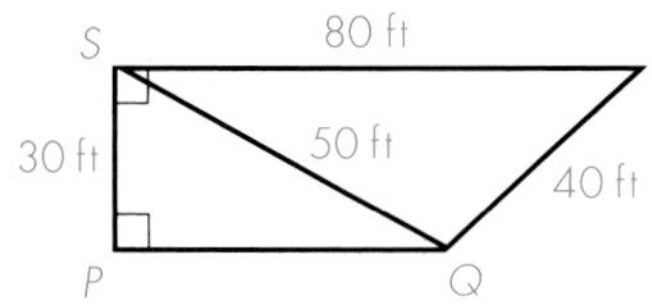

The diagram shows a floor plan for a log cabin.

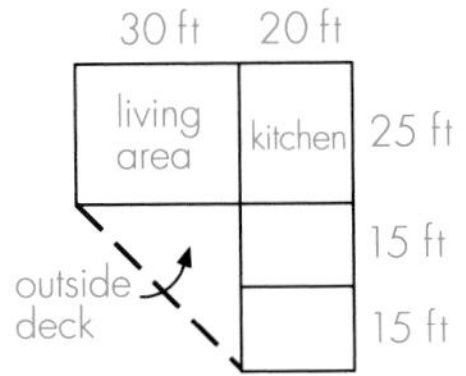

17. What is the total indoor and outdoor deck area of the cabin? 1850 sq ft 1050
18. What is the area of the deck? 450 sq ft
19. The drawing shows a driveway leading to a 3-car garage. What are the area and perimeter of the driveway? A = 1350 sq ft; P = 210 ft

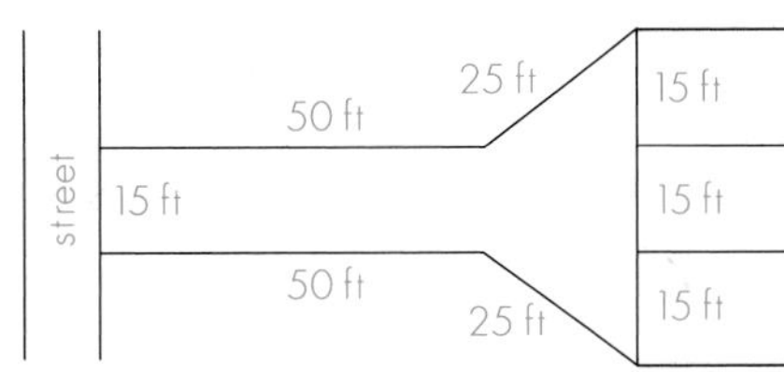

CRITICAL THINKING

Lauren said that her room had a perimeter of 528 inches.

20. Do you think this a reasonable size for a room? Yes
21. What would be a more meaningful expression of perimeter? 44ft
22. Can you tell the shape of the room from knowing only the perimeter? No
23. Assume that the room is rectangular. Give two possible sets of measurements for the length and width. Answers may vary. Possible answers: 10 ft by 12 ft; 8 ft by 14 ft; 9 ft by 13 ft
24. What are the measurements if the room is a square? 11 ft by 11 ft

Using the Page
Students can use the x-y plane to draw the triangles in Exercises 13-15.

FOLLOW-UP

Assessment
Find the area.

1.
26 m
30°

[$A = 169\sqrt{3}$ m^2]

2.

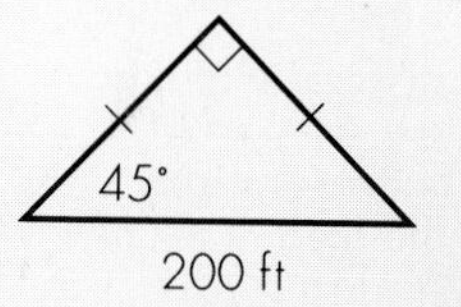

[A = 10,000 sq ft]

Find the perimeter.

3.

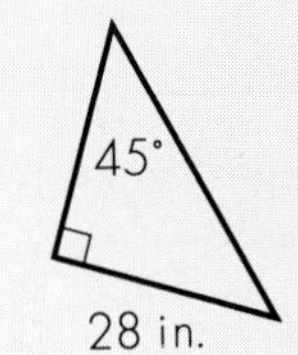

[$P = (56 + 28\sqrt{2})$ sq in.]

Reteaching
Students who have had difficulty with this section may benefit from Reteaching Activity 9.2.

Enrichment
For students who have mastered the material in this section, you may wish to assign Enrichment Activity 9.2.

SECTION 9.3

Resources
Reteaching 9.3
Enrichment 9.3
Transparency Master 9-1

Objectives
After completing this section, the student should understand how to find the perimeter and area of a parallelogram.

Materials
graph paper
compass
protractor
measuring tools

GETTING STARTED

Quiz: Prerequisite Skills
Find the area of the rectangle.

1. $l = 10.8$ m
$w = 6.5$ m
[$A = 70.2$ m^2]

2. $l = 41$ cm
$w = 32$ cm
[$A = 1312$ cm^2]

Find the area of the triangle.

3. $b = 38$ in.
$h = 10.5$ in.
[$A = 199.5$ sq in.]

4. $b = 46$ mm
$h = 21$ mm
[$A = 483$ mm^2]

Warm-Up Activity
Have students draw several parallelograms on graph paper. Then ask them to describe how they might find the area of each figure. Ask for informal estimates of the areas.

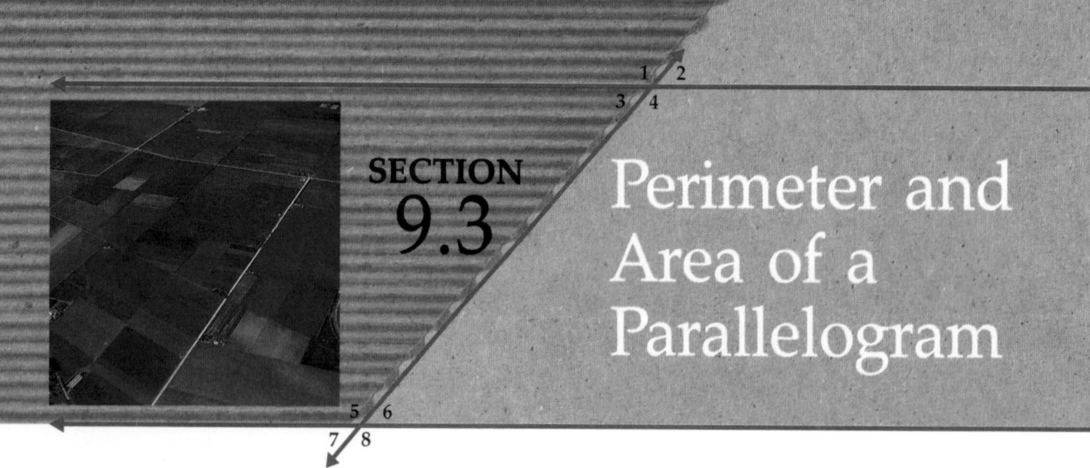

SECTION 9.3

Perimeter and Area of a Parallelogram

After completing this section, you should understand
▲ how to find the perimeter and area of a parallelogram.

In learning how to find the area of a triangle, you saw that two cases were considered: the right triangle, in which the altitude was one of the sides of the triangle, and the non-right triangle, in which you had to determine the altitude.

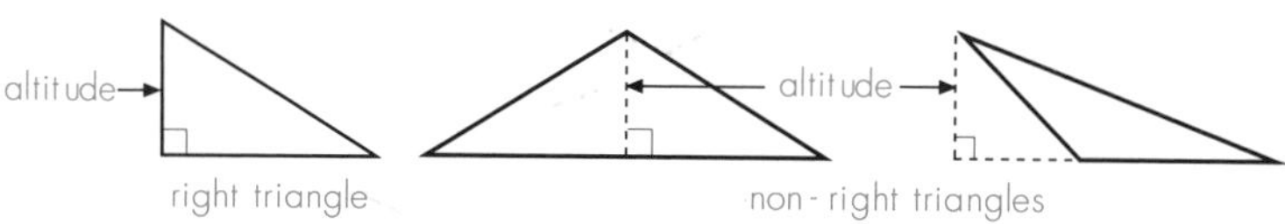

Now you will investigate finding the area of any parallelogram. Recall that the set of parallelograms contains several different figures.

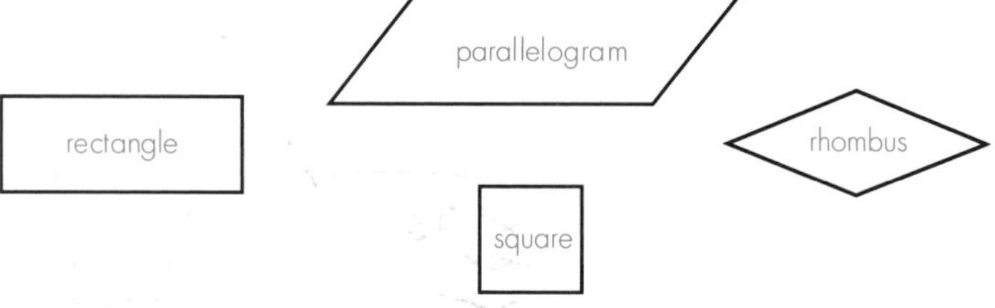

So a method for finding the area of a parallelogram will also give a method for finding the area of a rectangle, rhombus, or square.

DISCOVERY ACTIVITY

1. On graph paper, draw a large parallelogram. Label it *ABCD*.

2. Using a compass or protractor, construct a line segment perpendicular to $\overline{AB}$ from point D. Label the point of intersection E.

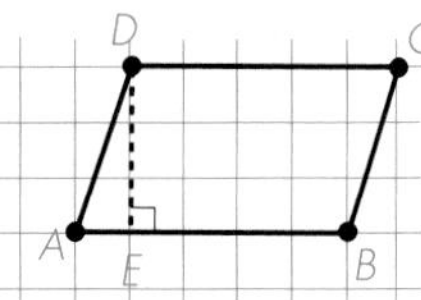

3. Cut off $\triangle ADE$ and place it at the other side of the parallelogram, so that $\overline{AD}$ lies on $\overline{BC}$.
4. What kind of figure have you formed? How can you find the area?
   Rectangle; $A = lw$
5. Write a conclusion telling what you have learned about the area of a parallelogram.

You may have discovered from the activity that the area of a parallelogram is the same as the area of a rectangle with congruent base and altitude.

NOTEBOOK

THEOREM 9-3-1 **If the figure is a parallelogram, then the area is the length of the base times the height.**

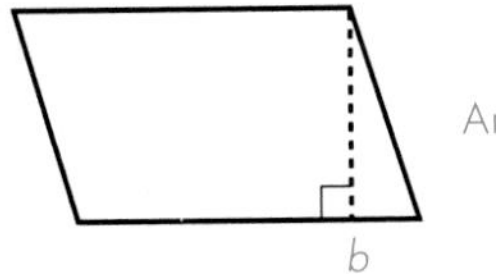

Area = bh square units

Notice that when you refer to the area of parallelograms, you use the terms *base* and *height* instead of *length* and *width*. A parallelogram has two pairs of parallel bases. Once you have chosen a base, the corresponding height or altitude is a line segment perpendicular to that base, with its endpoint on the opposite side.

Example: What is the area of *GHJK?*

$A = bh$
$A = 18 \text{ cm} \cdot 9 \text{ cm} = 162 \text{ cm}^2$

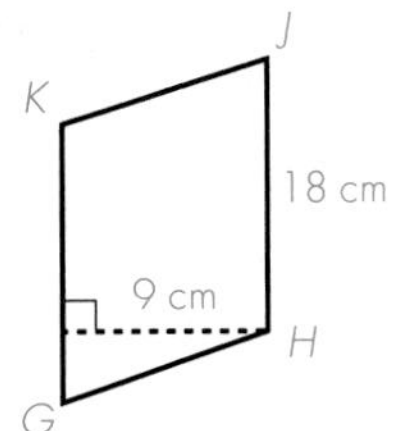

TEACHING COMMENTS

Using the Page
You may wish to extend the Discovery Activity by having students use a geometric drawing tool to draw any $\square ABCD$ that is not a rectangle. Then have them draw $\overline{DE}$ so that $\overline{DE} \perp \overline{AB}$. Ask students to measure DE and AB and find the area of $\square ABCD$. Students should repeat this activity at least four times. Have them describe the relationships among the length of $\overline{DE}$, the length of $\overline{AB}$, and the area of $\square ABCD$. They should discover that $DE \cdot AB$ = area of $\square ABCD$.

Using the Page
During the discussion of Theorem 9-3-1, you may wish to have students identify alternate pairs of bases and heights for the parallelogram shown.

Extra Practice
Find the area of each parallelogram.

1.

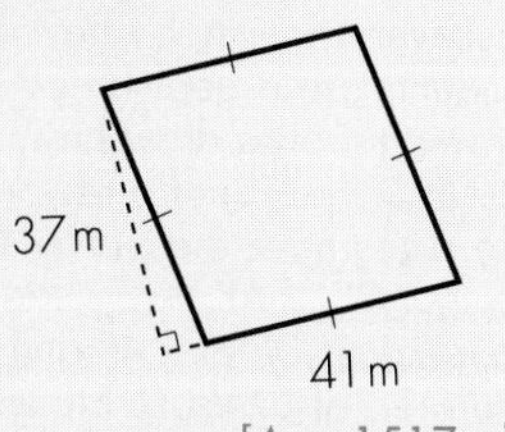

[A = 1517 m²]

2.

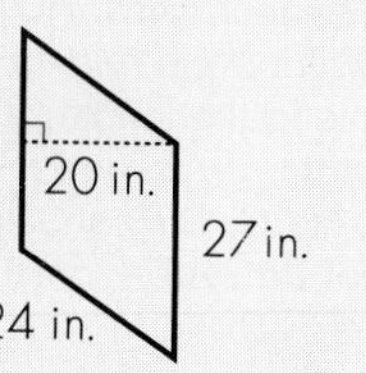

[A = 540 sq in.]

3.

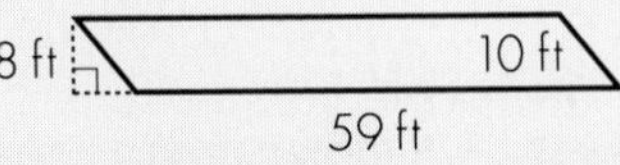

[A = 472 sq ft]

4. The area of rhombus *ABCD* is 1800 m². *AB* = 45 m. What is the height of *ABCD*? [40 m]

5. Square *PQRS* has a side of 13 cm. What is the area? [169 cm²]

CLASS ACTIVITY

Find the area of each parallelogram.

1. 100 ft — 10,000 sq ft
2. 56 yd, 94 yd — 5264 sq yd
3. 91 cm, 116 cm, 310 cm — 28,210 cm²

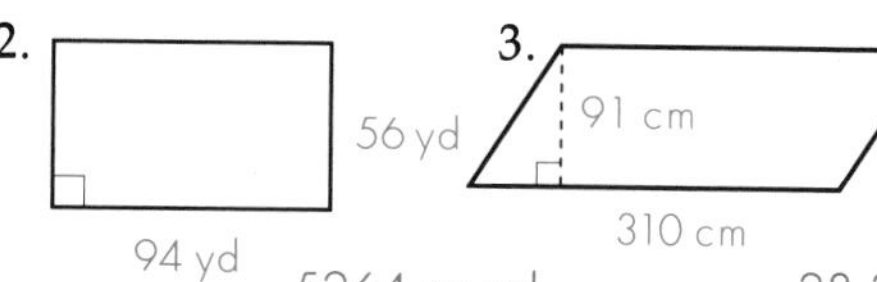

4. $b = 415$ m, $h = 75$ m — $A = 31{,}125$ m²
5. $b = 12$ in., $h = 6.5$ in. — $A = 78$ sq in.
6. $s = 90$ ft — $A = 8100$ sq ft
7. $b = 37$ m, $h = 21$ m — $A = 777$ m²
8. $b = 50$ cm, $h = 60$ cm — $A = 3000$ cm²
9. $s = 13.5$ in. — $A = 182.25$ sq in.

Find the missing base or height of each parallelogram.

10.

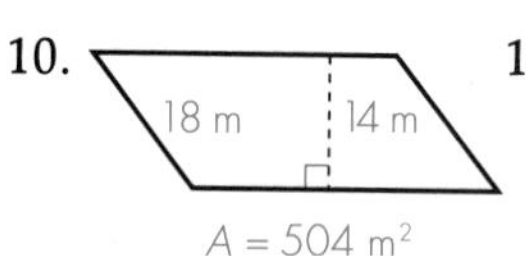

$b = 36$ m

11. A = 225 cm² — $s = 15$ cm

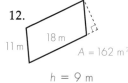

12. 11 m, 18 m, A = 162 m² — $h = 9$ m

13. *PQRS* is a rhombus with an area of 896 sq in. *PQ* = 32 in. What is the height of *PQRS*? 28 in.

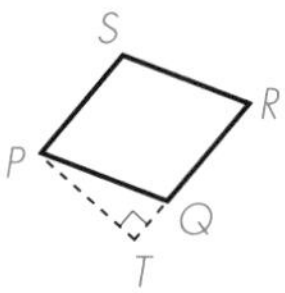

14. *ABCD* is a rectangle. *TBVD* is a parallelogram. $\overline{AT} = \overline{BC}$ and $\overline{DC} = 3\overline{AT}$. Find the area of the shaded region. 150 m

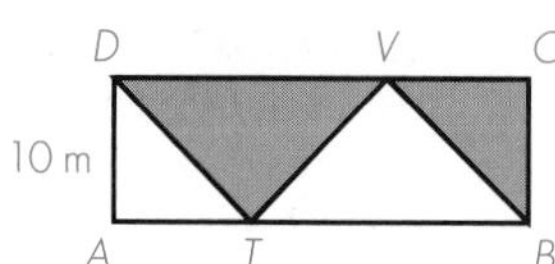

AREA OF A RHOMBUS

Recall that a rhombus is a quadrilateral with four congruent sides, and diagonals that are perpendicular and bisect each other. The area of a rhombus can be found by multiplying the lengths of the diagonals and dividing by 2.

To show that this method works, first try it for a square with sides equal to 1 unit.

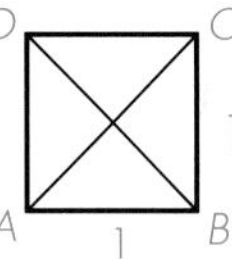

From the Pythagorean Theorem, $c^2 = a^2 + b^2$,

so $AC^2 = 1^2 + 1^2 = 2$,
$AC = \sqrt{2}$.
Since $AC = BD$, $DB = \sqrt{2}$.

$\frac{1}{2}(AC \cdot BD) = \frac{1}{2} \cdot \sqrt{2} \cdot \sqrt{2} = 1$

From the theorem of the area of a parallelogram, you know that $A = bh$, so for square $ABCD$, $= 1 \cdot 1 = 1$.

Both methods give the same answer for the area of a square. Now we justify the method for any rhombus.

Given: Rhombus $ABCD$ with diagonals d_1 and d_2
$DB = d_1$, $AC = d_2$
Justify: Area $ABCD = \frac{1}{2}d_1d_2$

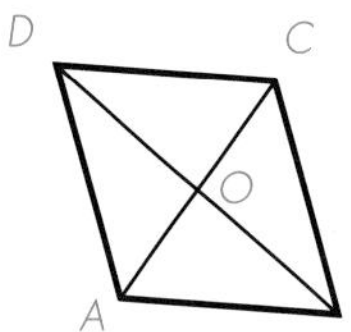

Supply the missing reasons.

Steps	Reasons
1. $A(ABCD) = A(\triangle ADC) + A(\triangle ABC)$	1. Addition
2. $DO = OB = \frac{1}{2}d_1$	2. Diagonals of rhombus bisect each other.
3. $d_1 \perp d_2$, so $\frac{1}{2}d_1$ is an altitude of both $\triangle ADC$ and $\triangle ABC$.	3. Diagonals of rhombus are perpendicular.
4. $A(\triangle ADC) = \frac{1}{2}d_2 \cdot \frac{1}{2}d_1 = \frac{1}{4}d_2d_1$ $A(\triangle ABC) = \frac{1}{2}d_2 \cdot \frac{1}{2}d_1 = \frac{1}{4}d_2d_1$	4. Triangle Area theorem
5. $A(ABCD) = \frac{1}{4}d_1d_2 + \frac{1}{4}d_1d_2$ $= \frac{1}{2}d_1d_2$	5. Algebra

Using the Page
Before beginning the discussion of the proof for the area of a rhombus, go through the example that uses a unit square. You may wish to use the square shown on Transparency Master 9-1 to illustrate both the square and the rhombus. Explain that a square of any size could have been used, but that a unit square allows the simplest possible numbers.

Using the Page
During the discussion of the proof for the area of a rhombus, have a volunteer outline the strategy used in the proof.

PERIMETER OF A PARALLELOGRAM

If quadrilateral $ABCD$ is a parallelogram with $AB = 16$ m and $BC = 12$ m, you know that $\overline{DC}$ and $\overline{AD}$ are equal to the lengths of the opposite sides.

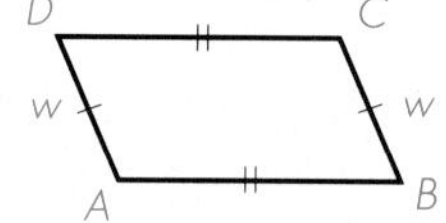

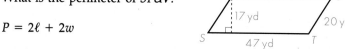

Then from the definition of perimeter,

$P = \ell + w + \ell + w$, or $P = 2\ell + 2w$.

THEOREM 9-3-2 **If the figure is a parallelogram, then the perimeter is twice the sum of two adjacent sides.**

Example: What is the perimeter of $STUV$?

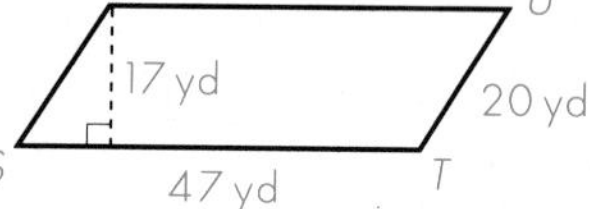

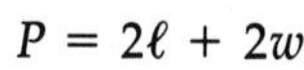

$P = 2\ell + 2w$

$P = 2(47 \text{ yd}) + 2(20 \text{ yd}) = 134$ yd

Example: $ABCD$ is a rhombus with $AB = 80$ in.

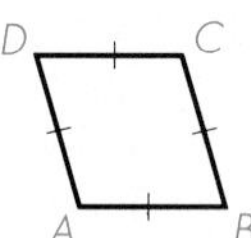

What is the perimeter?

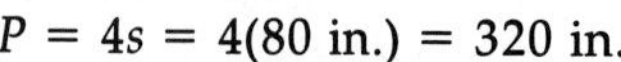

$P = 4s = 4(80 \text{ in.}) = 320$ in.

CLASS ACTIVITY

Find the perimeter of the parallelogram.

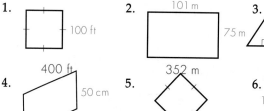

1.

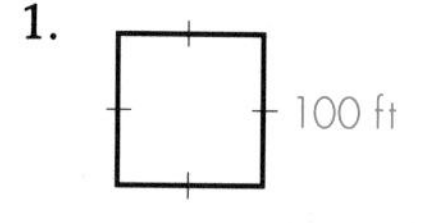

400 ft

2. 101 m, 75 m

352 m

3.

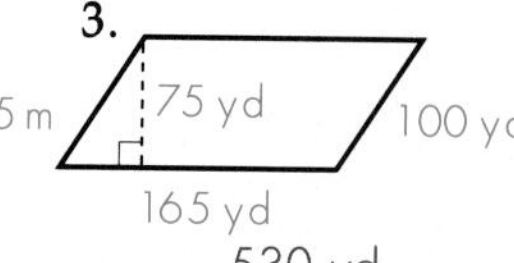

530 yd

4. 50 cm, 75 cm

250 cm

5. 90 ft

360 ft

6.

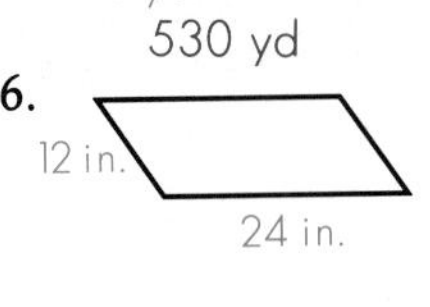

72 in.

Find the missing length or width of the parallelogram.

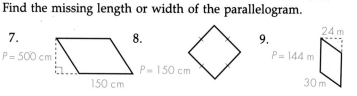

7. $P = 500$ cm, 150 cm

$w = 100$ cm

8. $P = 150$ cm

$s = 37.5$ m

9.

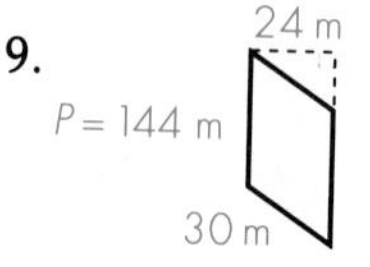

$\ell = 42$ m

Using the Page

During the discussion of Theorem 9-3-2, have students list the quadrilaterals for which this theorem will be true. [Parallelogram, rectangle, rhombus, square]

Extra Practice

Find the perimeter of the parallelogram.

1.

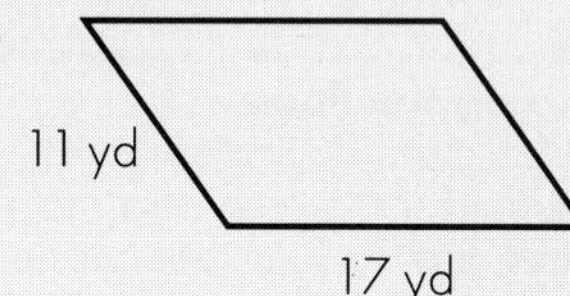

[$P = 56$ yd]

2.

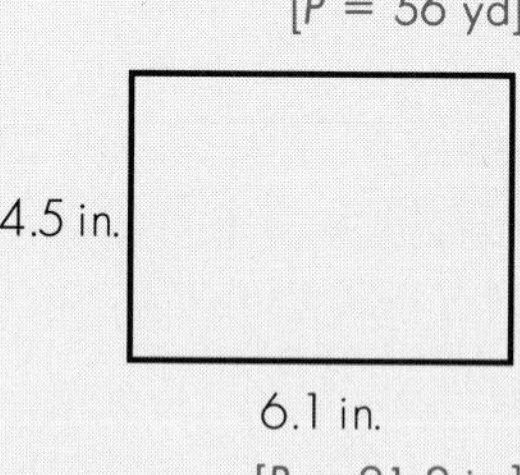

[$P = 21.2$ in.]

Find the missing length or width for each parallelogram.

3.

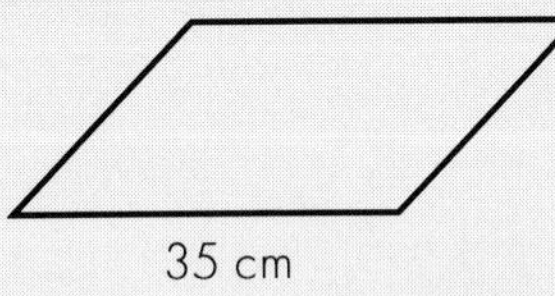

$P = 120$ in.
[$w = 25$ in.]

4.

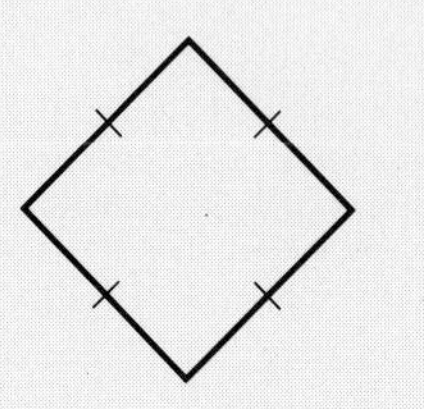

$P = 325$ m
[l, $w = 81.25$ m]

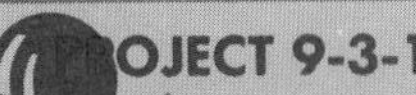

PROJECT 9-3-1

Work with a partner or a small group.

1. Decide on a subject for your map. Choose several blocks in an area of your community that you know well, one level of your school building, a small park, or another accessible location.
2. Decide on an appropriate unit of measure.
3. Measure the area by pacing it off or by using a measuring tool. Be sure to find the perimeters of all structures or objects you plan to include in the map.
4. Decide on a size for your finished map. Choose a scale.
5. Transfer the data to the map. Check the accuracy of measurements on the map by using the scale. For example, a scale might be 1 cm = 10 m or $\frac{1}{2}$ in. = 10 yd.
6. Title the map. Identify what you have included (label streets, buildings, room, and so on), and make a key.
7. Use the map to write several problems involving perimeter and area.
8. Display your map and exchange sets of problems with other groups. Solve the problems using the maps.

Using the Page
For convenience in the map-making project, you may wish to form several small groups of students who live near each other, and give each group a different part of the community to map.

Using the Page
If possible, provide students with tape measures or meter sticks to use in measuring distances. When students have chosen a scale for their maps, encourage them to use calculators to transfer their data to the maps and to check accuracy. If the class decides to map several adjacent areas of the community or the school, you may wish to display the maps in their relative positions.

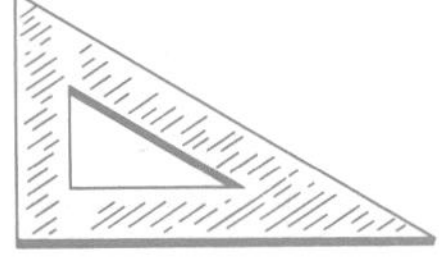

HOME ACTIVITY

Find the area of each parallelogram.

1. 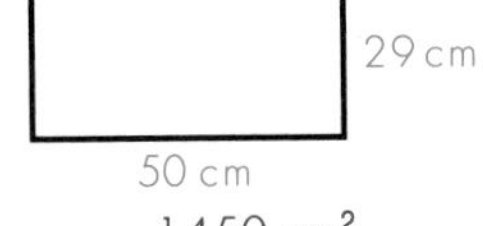

1450 cm²

2. 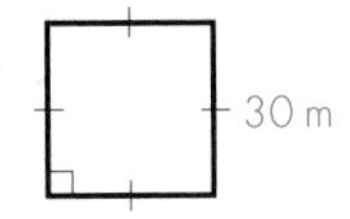

900 m²

3. 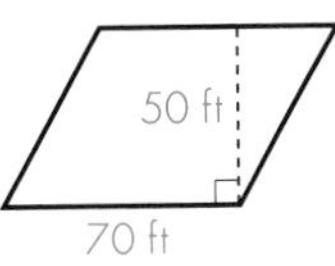

3500 sq ft

4. 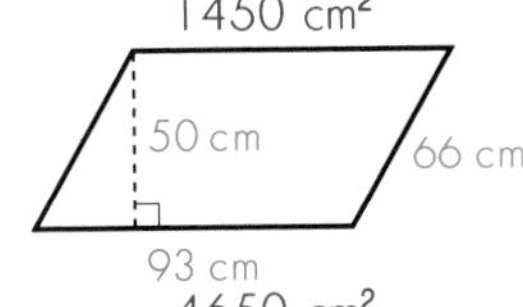

4650 cm²

5. 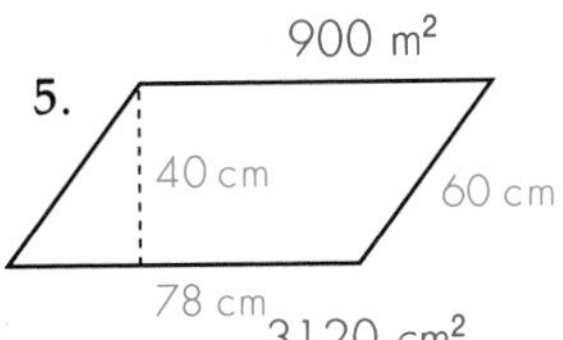

3120 cm²

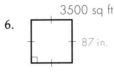

6. 87 in.
7569 sq in.

7. $b = 31$ cm, $h = 25$ cm   A = 775 cm²

8. $s = 47$ yd   A = 2209 sq yd

Find the missing measures.

9. 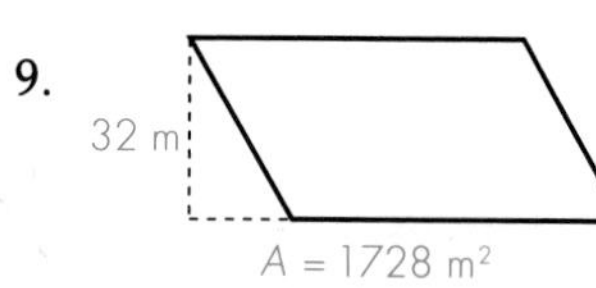

A = 1728 m²
b = 54 m

10. A = 5776 sq ft
s = 76 ft

11. 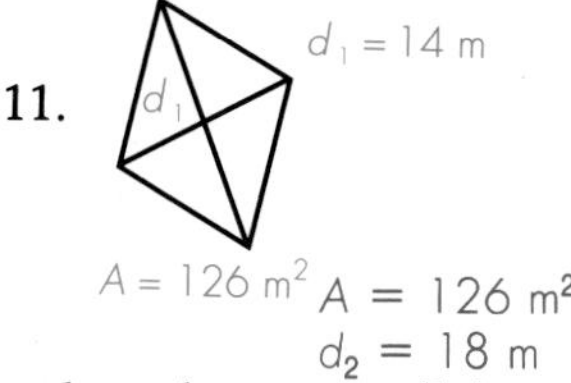

A = 126 m²
d_2 = 18 m

Write a LOGO procedure to tell the turtle to draw a parallelogram that has each area. Then modify the procedure to draw another parallelogram with the same area. Answers may vary. See margin.

12. 48
13. 36
14. 160

Find the area of the parallelogram.

15. 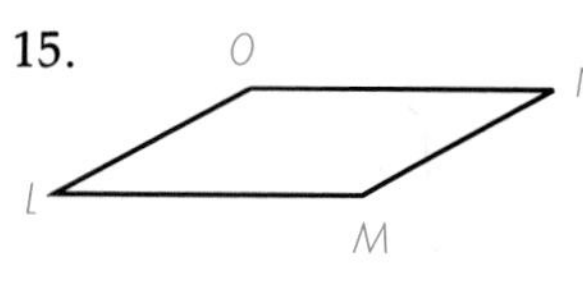

m∠L = 30°
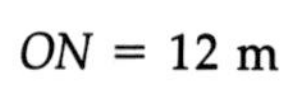
ON = 12 m
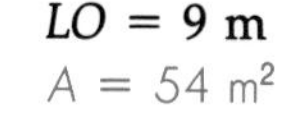
LO = 9 m
A = 54 m²

16. Z Y Q X

m∠QXY = 60°
ZY = 125 mm
YX = 80 mm
A = $5000\sqrt{3}$ mm²

Find the area of the shaded region.

17. *DEFG* and *HJKL* are parallelograms.

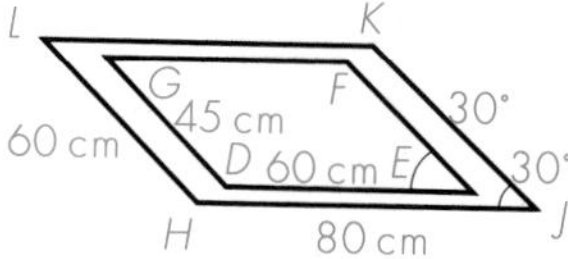

1050 cm²

Use the diagonals of the rhombus to find the area.

18. 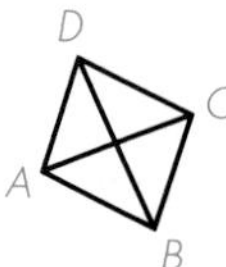

AC = 27 m *BD* = 32 m
432 m²

19. 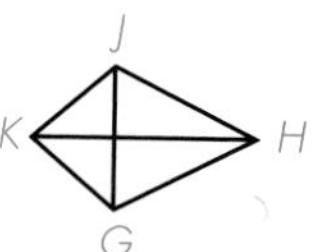

GJ = 15 ft *KH* = 28 ft
210 sq ft

Write a LOGO procedure to tell the turtle to draw a rhombus for each of the following areas. Then modify the procedure to draw another rhombus with the same area. Answers may vary.

20. 16
21. 28
22. 99

Using the Page
Students can use the x-y plane to draw the parallelograms in Exercises 12-14 and the rhombuses in Exercises 20-22.

Extra Practice
Find the area of the parallelogram.

1. 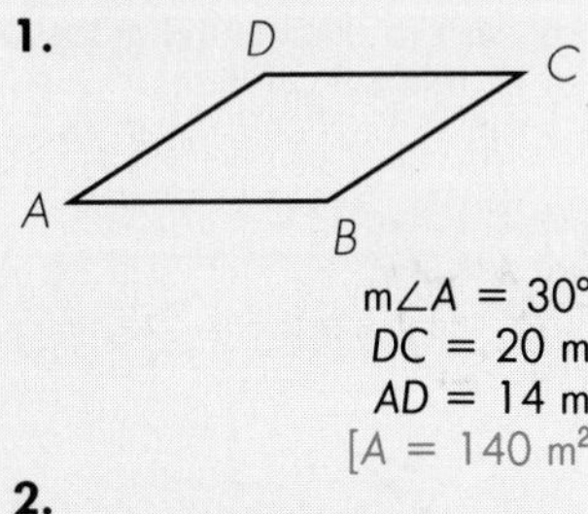

m∠A = 30°
DC = 20 m
AD = 14 m
[A = 140 m²]

2. 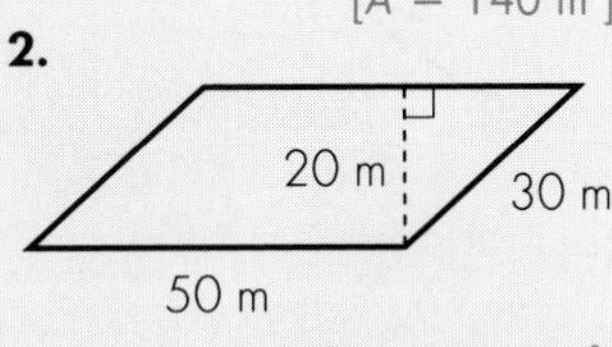

[A = 1000 m²]

Find the area of the rhombus.

3. 

MO = 12 yd
PN = 17 yd
[A = 102 sq yd]

4. 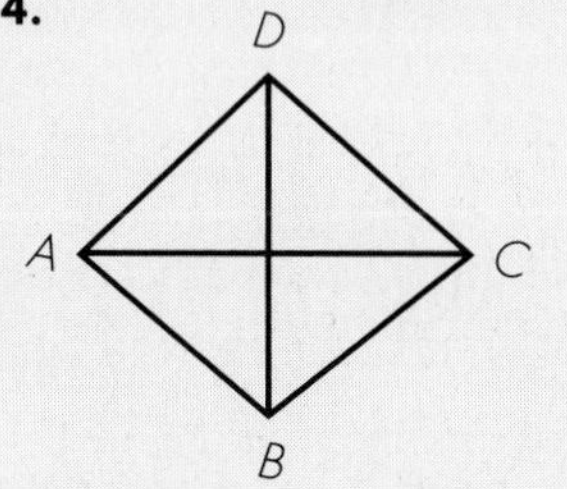

AC = 18.3 cm
BD = 15.5 cm
[A = 141.825 cm²]

Find the perimeter of each parallelogram.

23. 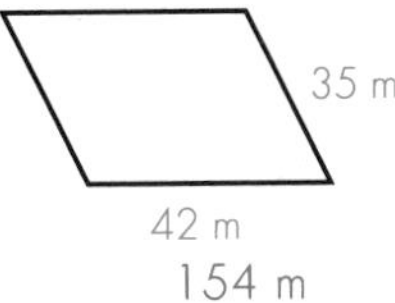

154 m

24. 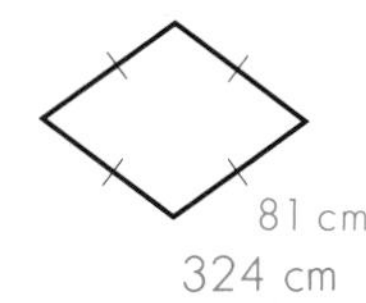

324 cm

25. 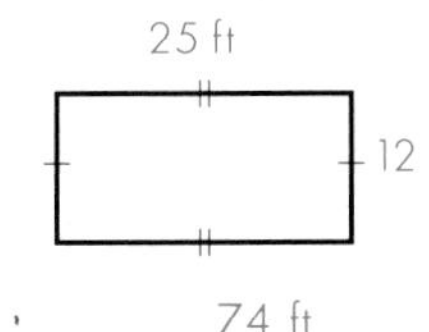

74 ft

26. The plan shows a classroom 30 feet wide by 40 feet long. If carpet costs $6.45 per square foot, what would be the cost of carpeting the whole room? $7740

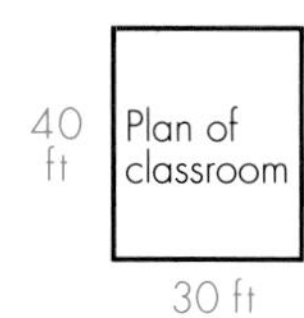

27. The drawing shows a room in a house.
Find the area of the ceiling and the area of the four walls. 396 sq ft; 640 sq ft

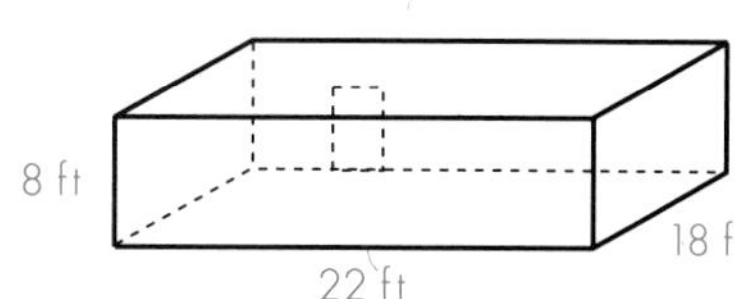

28. One gallon of paint covers 425 square feet. If you paint the walls and ceiling in the same color with two coats of paint, how many gallons will you need to buy? 5 gal.

29. If the paint costs $18.75 per gallon, how much will you spend? $93.75

CRITICAL THINKING

30. Terry has 120 feet of fencing that he wants to put up around his garden. Copy and complete the table.

Length (ft)	Width (ft)	Perimeter (ft)	Area
50	10	120	500 sq ft
45	15	120	675 sq ft
40	20	120	800 sq ft
35	25	120	875 sq ft
30	30	120	900 sq ft

31. What will be the shape and dimensions of the greatest area Terry can enclose? Greatest area is a 30 by 30 square.

FOLLOW-UP

Assessment

1. Find the perimeter and area of ▱*ABCD*.

m∠$A = 30°$, $AB = 27$ in., $BC = 12$ in.
[$P = 78$ in.; $A = 162$ sq in.]

2. Find the measure of *QS* for rhombus *PQRS*.

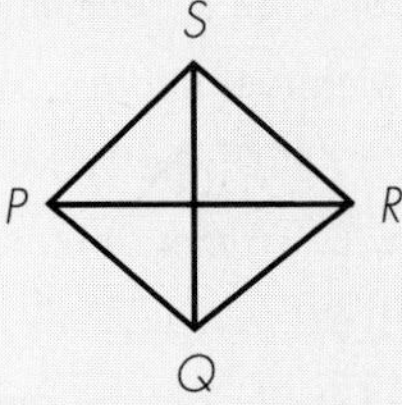

$A = 189$ m^2, $PR = 21$ m
[$QS = 18$ m]

Reteaching
Students who have had difficulty with this section may benefit from Reteaching Activity 9.3.

Enrichment
For students who have mastered the material in this section, you may wish to assign Enrichment Activity 9.3.

SECTION 9.4

Resources
Reteaching 9.4
Enrichment 9.4
Transparency Masters 9-1, 9-2

Objectives
After completing this section, the student should understand
▲ how to find the perimeter and area of a trapezoid.

Materials
graph paper
ruler
dot paper

GETTING STARTED

Warm-Up Activity
Have students draw several trapezoids on graph paper. Challenge them to develop a formula for finding the area of a trapezoid.

TEACHING COMMENTS

Using the Page
You may wish to extend the Discovery Activity by having students use a geometric drawing tool to draw any trapezoid *ABCD*. Then have them draw $\overline{DE}$ so that $\overline{DE} \perp \overline{AB}$ and $\overline{DE} \perp \overline{DC}$. Ask students to measure $\overline{DE}$, $\overline{DC}$, and $\overline{AB}$ and find the area of trapezoid *ABCD*. Students should repeat this activity at least four times. Have them describe the relationships among the length of $\overline{DE}$ the length of $\overline{DC}$, the length of $\overline{AB}$, and the area of trapezoid *ABCD*. They should discover that $\frac{1}{2} \cdot DE \cdot (DC + AB) =$ area of trapezoid *ABCD*.

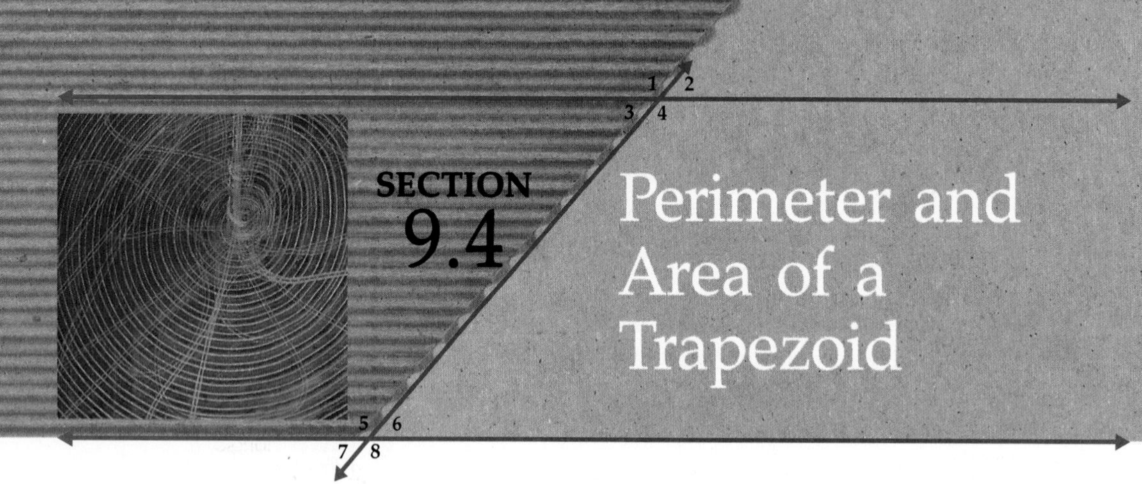

SECTION 9.4
Perimeter and Area of a Trapezoid

After completing this section, you should understand
▲ how to find the perimeter and area of a trapezoid.

In Chapter 8, you learned that there are three general types of quadrilaterals.

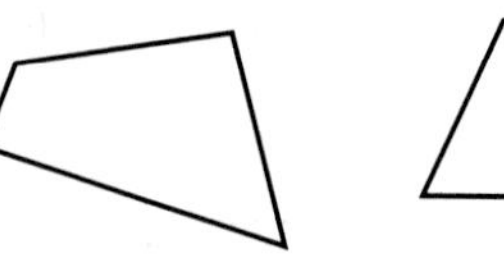
quadrilateral— no sides parallel

trapezoid—one pair of sides parallel

parallelogram—two pairs of sides parallel

The method used to find the area of a trapezoid takes advantage of the area theorems you have already investigated, and uses algebra to combine them.

DISCOVERY ACTIVITY

1. Draw a trapezoid and label it *ABCD*. See margin.
2. From points *D* and *C*, draw or construct perpendiculars to $\overline{AB}$. Label the intersections of the perpendiculars with the base *E* and *F*.
3. What familiar figures have you formed? Rectangle, 2 right triangles
4. Describe how you would find the area of *ABCD*. Add areas of rectangle and triangles.
5. Draw another trapezoid and label it *GHJK*. How could you divide the trapezoid into two triangles? Draw diagonals.
6. How would you find the area of trapezoid *GHJK*? Add areas of 2 triangles.

You may have found that the area of a trapezoid is equal to the sum of the areas of the two triangles it contains.

Given: Trapezoid $GHJK$ with diagonal $\overline{GJ}$, $\overline{GH} \parallel \overline{KJ}$

Justify: Area $GHJK = \frac{1}{2}h(b_1 + b_2)$

Supply the missing reasons.

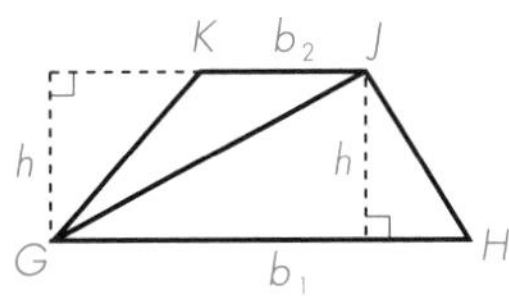

Steps	Reasons
1. $A(\triangle GHJ) = \frac{1}{2}b_1h$	1. Area of triangle
2. $A(\triangle JKG) = \frac{1}{2}b_2h$	2. Area of triangle
3. $A(GHJK) = A(\triangle GHJ) + A(\triangle JKG)$	3. Addition
4. $A(GHJK) = \frac{1}{2}b_1h + \frac{1}{2}b_2h$	4. Substitution
5. $A(GHJK) = \frac{1}{2}h(b_1 + b_2)$	5. Algebra

THEOREM 9-4-1 **If the figure is a trapezoid, then the area is $\frac{1}{2}$ the height times the sum of the bases.**

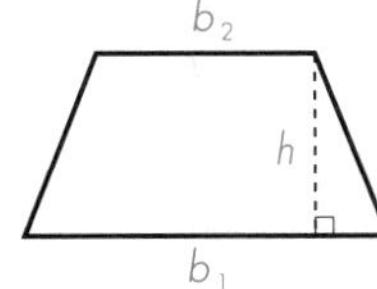

$A = \frac{1}{2}h(b_1 + b_2)$ square units

Recall that the height is the length of the altitude, or perpendicular, between the two parallel bases or sides.

Example: A city park is in the shape of a trapezoid. What is its area?

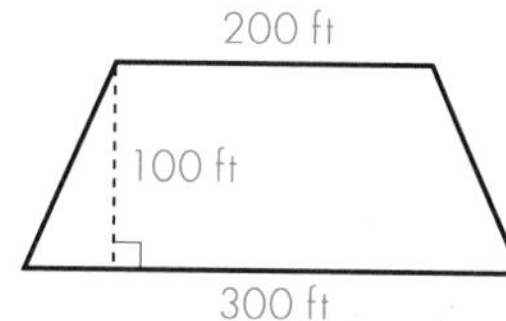

$A = \frac{1}{2}h(b_1 + b_2)$

$h = 100$ ft, $b_1 = 300$ ft, $b_2 = 200$ ft

$A = \frac{1}{2} \cdot 100(300 + 200)$

$= 50 \cdot 500 = 25{,}000$ sq ft

Using the Page

As you go through each step of the proof of Theorem 9-4-1, be sure students understand that two partial areas are being added to get the whole area. You may wish to use the trapezoids shown on Transparency Master 9-1 to illustrate that the height of the trapezoid can be any perpendicular drawn from one base to the opposite base.

Using the Page
Go through the reasoning in the example carefully, so that students can see that they are using information drawn from Chapters 7 and 8 as well as from Chapter 9. It may be helpful to do the exercise orally and have students give a reason for each step.

Example: *ABCD* is an isosceles trapezoid. What is the area?

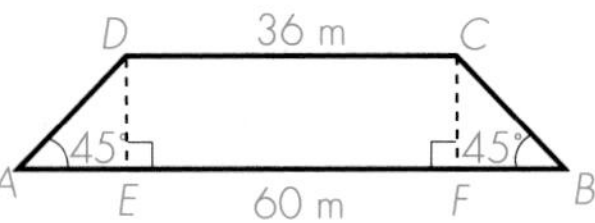

1. What information is missing? Height
2. Use what you know about 45° right triangles to find the height. Draw perpendiculars $\overline{DE}$ and $\overline{CF}$ from points *D* and *C* to $\overline{AB}$. What is figure *EFCD?* Rectangle
3. What is the length of $\overline{EF}$? 36 m
   Find the length of $\overline{AE}$ and $\overline{FB}$. 24 m
4. What do you know about $\overline{AE}$ and $\overline{DE}$ in $\triangle ADE$ and $\overline{BF}$ and $\overline{CF}$ in $\triangle BCE$? Why? They have equal lengths; 45° right $\triangle$
5. So, if $AE = BF$ and $AE + BF = 24$ m, then $AE = 12$ m and $BF = 12$ m. Since $AE = DE$ and $BF = CF$, $DE = 12$ m and $CF = 12$ m. Thus, the height of the trapezoid is 12 m.
6. Find the area. $\frac{1}{2} \cdot 12\ (36 + 60) = 576$ sq m

Extra Practice
Find the area of the trapezoid.

1.

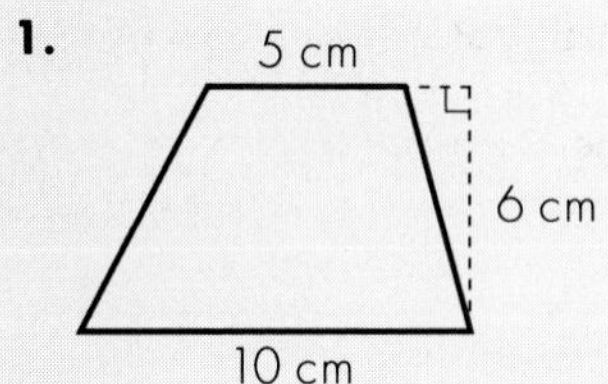

[$A = 45$ cm^2]

2.

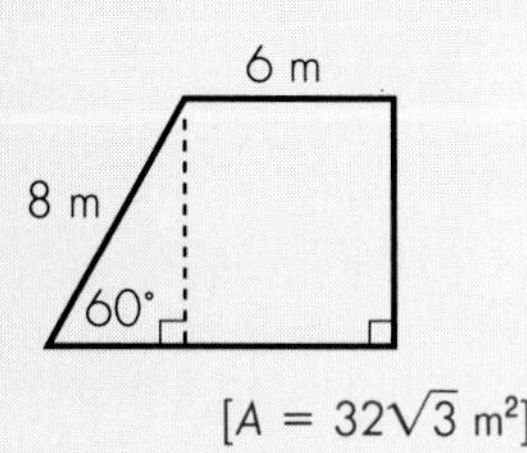

[$A = 32\sqrt{3}$ m^2]

CLASS ACTIVITY

Find the area of the trapezoid.

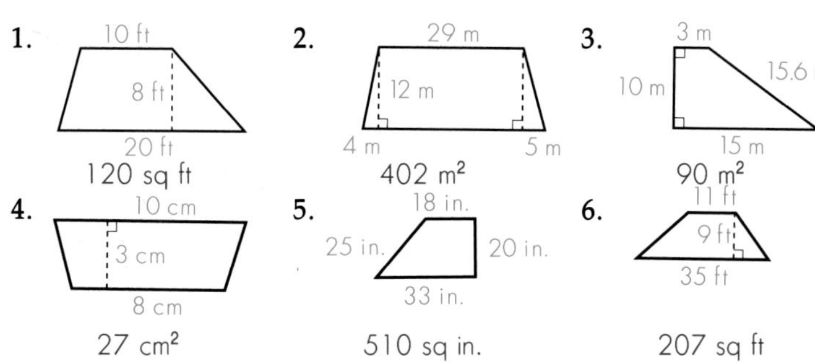

7. *MNOP* is a trapezoid. Find the area.

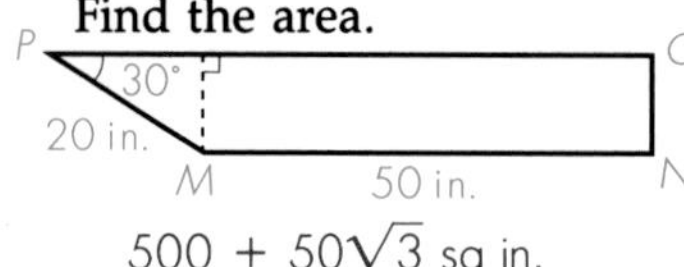

$500 + 50\sqrt{3}$ sq in.

8. Find the area of trapezoid *ABCD.*

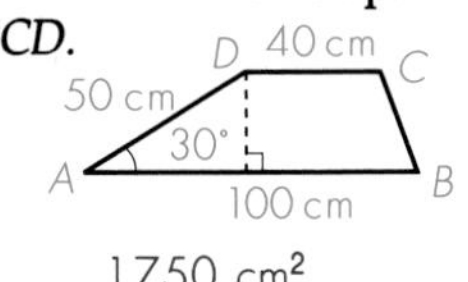

1750 cm^2

9. Find the area of isosceles trapezoid *STWX*.

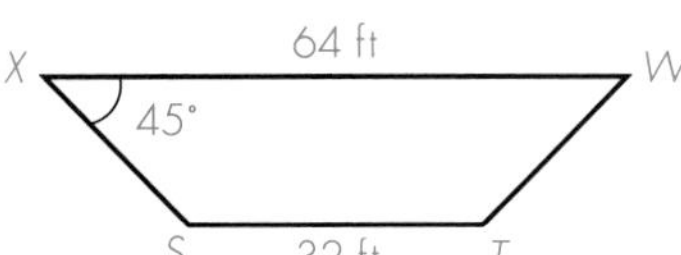

768 sq ft

10. Find the area of trapezoid *GHJK*.

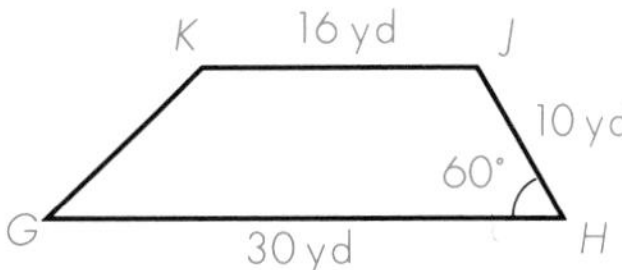

$115\sqrt{3}$ sq yd

Find the missing base or height for each trapezoid.

11. 24 m, 35 m, $A = 936$ m² — $b = 43$ m

12. 27 cm, 51 cm, $A = 741$ cm² — $h = 19$ cm

13. 30 m, 45°, 45°, 46 m, $A = 304$ m² — $h = 8$ m

There is another way to derive the formula for the area of a trapezoid. Trapezoid *ABCD* can be seen as one-half of parallelogram *AGHD*.

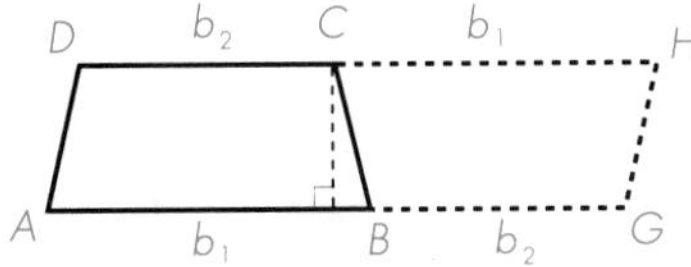

14. Since *AGHD* is a parallelogram, *AG* = *DH* = ? + ? $b_1 + b_2$
15. Area of *AGHD* = ? $(b_1 + b_2)h$
16. Since $ABCD = \frac{1}{2}AGHD$, Area *ABCD* = ? $\frac{1}{2}h(b_1 + b_2)$

PERIMETER OF A TRAPEZOID

You have seen that to find the perimeter of any figure, you add the lengths of the sides. For a trapezoid,

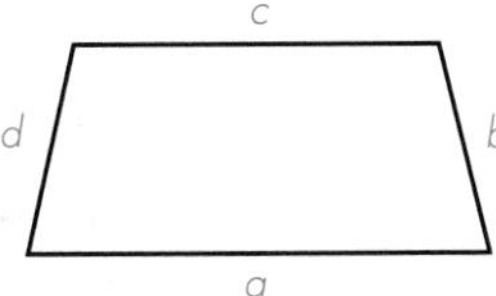

Perimeter $(P) = a + b + c + d$.

Example: Find the perimeter of isosceles trapezoid *ABCD*.

43 cm, 35 cm, 68 cm

$P = 68 \text{ cm} + 35 \text{ cm} + 43 \text{ cm} + 35 \text{ cm}$

$P = 181 \text{ cm}$

Extra Practice

Find the area of the trapezoid.

1.

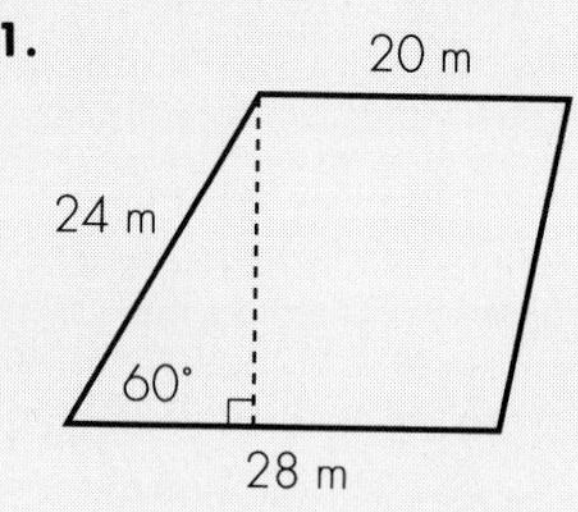

$[A = 288\sqrt{3} \text{ m}^2]$

2.

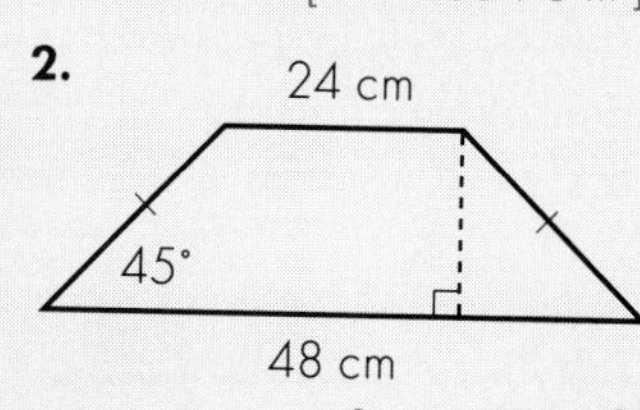

$[A = 432 \text{ cm}^2]$

Using the Page

In discussing this alternate method for deriving the area of a trapezoid, it may be helpful to some students to copy the diagram on graph paper, making sure that ▱*AGHD* is actually twice the size of trapezoid *ABCD*. Be sure that students understand which are the corresponding bases.

CLASS ACTIVITY

Find the perimeter and the area of each trapezoid.

1. 25 ft; 15 ft; 12 ft; 13 ft; 39 ft

$P = 92$ ft, $A = 384$ sq ft

2. 22 in.; 9 in.; 15 in.; 20 in.; 18 in.; 12 in.

$P = 105$ in., $A = 306$ sq in.

3. 45 cm; 25 cm; 20 cm; 41.2 cm; 15 cm; 36 cm

$P = 207.2$ cm, $A = 1410\ cm^2$

4. *ABCD* is an isosceles trapezoid with $DC = \frac{2}{3}AB$. What is the perimeter? the area?

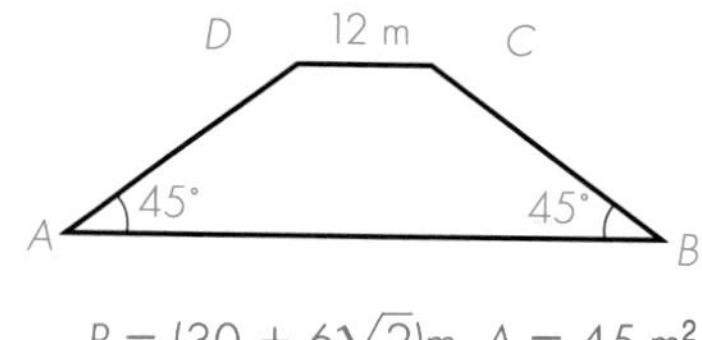

$P = (30 + 6\sqrt{2})$m, $A = 45\ m^2$

5. *LMNO* is a trapezoid with $LM = 2ON$. What is the perimeter? the area?

L 48 cm M; 30°; 13 cm; 15 cm; O; N

$P = 100$ cm, $A = 270\ cm^2$

CONSTRUCTION AND AREA

On dot paper, draw a square with an area of 1 square unit, and a triangle and a trapezoid, both with areas of $1\frac{1}{2}$ square units. Draw the figures below. Find the area of each figure without using any of the area formulas.

$A = 3$ sq units, $B = 1\frac{1}{2}$ sq units, $C = 8$ sq units, $D = 7$ sq units, $E = 6$ sq units, $F = 10$ sq units, $G = 12$ sq units

Using the Page

After students have completed the construction and area activity, you may wish to extend it by introducing them to Pick's Theorem (discovered in 1899). Use the plain dot section of Transparency Master 9-2 and explain that the dots are called lattice points and a polygon drawn with its vertices at lattice points is called a lattice polygon. Assume that each small lattice square has a side of one unit. Draw several figures similar to the following on the transparency.

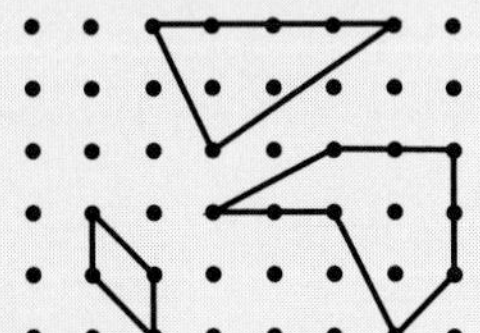

Pick's Theorem states that the area of any lattice polygon is equal to $L + \frac{1}{2}B - 1$.
L is the number of points inside the polygon, and B is the number of points on the border of the polygon. Have students find out if the theorem is true for each of the polygons shown. Then have them try it for the figures in their texts.

HOME ACTIVITY

Find the perimeter and the area of the trapezoid.

1. 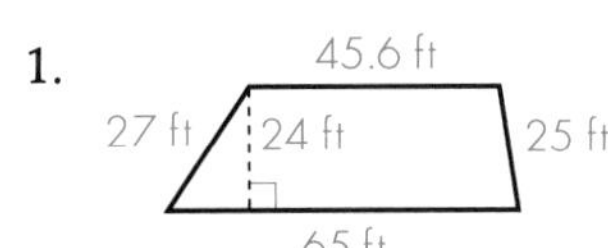

$P = 162.6$ ft, $A = 1327.2$ sq ft

2. 100 m, 80 m, 59.1 m, 75 m, 200 m

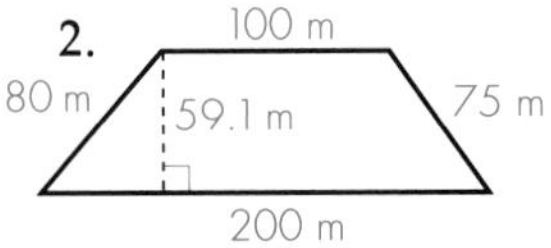

$P = 455$ m, $A = 8865$ m^2

3. 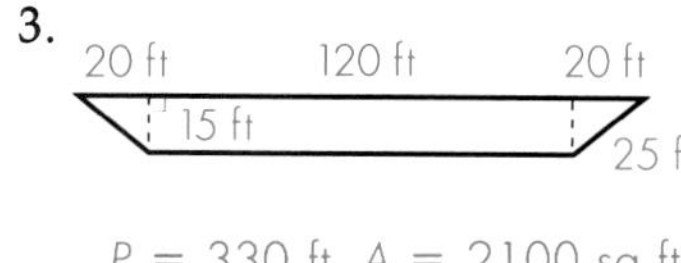

$P = 330$ ft, $A = 2100$ sq ft

Find the missing measure of the trapezoid.

4. 16 m, 40 m

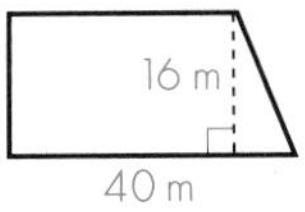

$A = 568$ m^2

$b = 31$ m

5. 84 cm, 65 cm

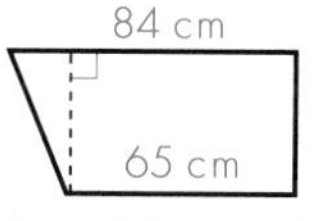

$A = 3948.5$ cm^2

$h = 53$ cm

6. 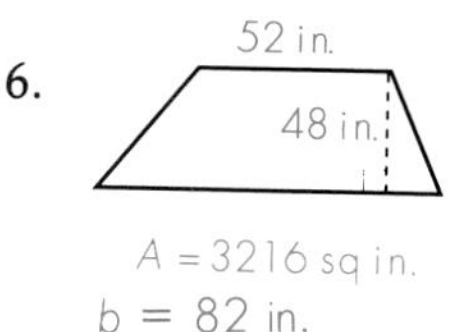

$A = 3216$ sq in.

$b = 82$ in.

Find the area of each trapezoid.

7. D 27 m C, 45°, A 65 m B

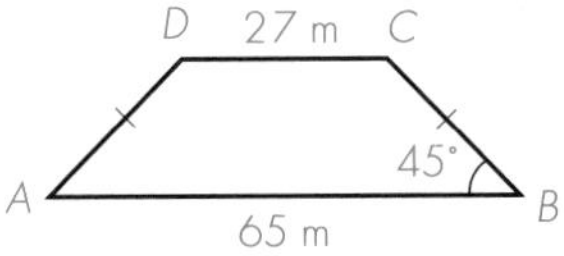

$A = 874$ m^2

8. 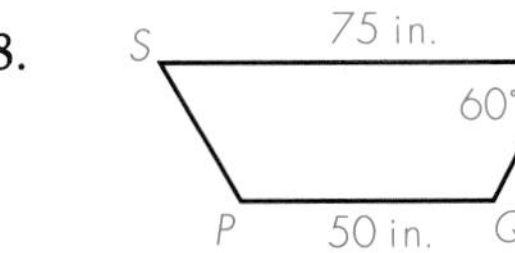

$A = 1375$ sq in.

9. $RN = \frac{1}{2}LM$, $NM = 32$ m

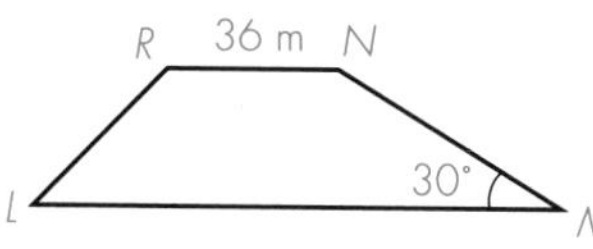

$A = 864$ m^2

10. $KJ = \frac{3}{5}GH$

K, J, 45°, G, 100 cm, H

$A = 1600$ cm^2

11. $\overline{EF}$ is the median of trapezoid *ABCD*. $DG = 10$ m, $AB = 36$ m, and $DC = 25$ m. What is the area of trapezoid *EFCD*? 138.75 m^2

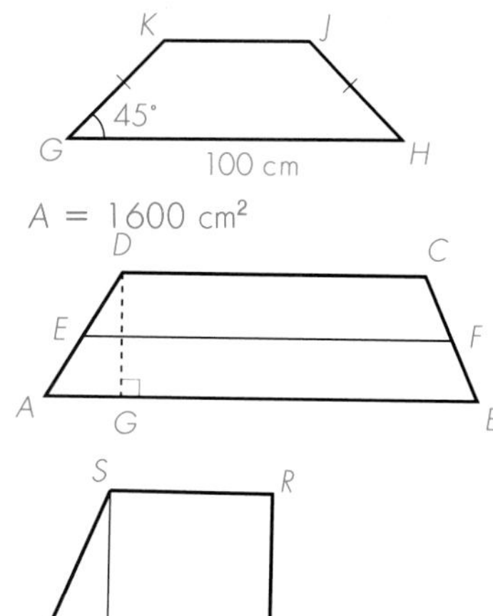

12. *TQRS* is a trapezoid with an area of 180 cm^2. *PQRS* is a square with $PQ = 12$ cm. Find the length of $\overline{SR}$ and $\overline{TQ}$. $SR = 12$ cm, $TQ = 18$ cm

Additional Answers
23.

Steps	Reasons
1. $A(ABCD) = A(\text{I}) + A(\text{II}) + A(\text{III})$	**1.** Addition
2. $(ABCD) = \frac{1}{2}ah + \frac{1}{2}ch + \frac{1}{2}bh$	**2.** Definition of area for triangles and rectangles
3. $A(ABCD) = \frac{1}{2}ah + \frac{1}{2}ch + \frac{1}{2}bh + \frac{1}{2}bh = \frac{1}{2}h(a + c + b + b)$	**3.** Algebra
4. $a + b + c = AB$ and $b = CD$	**4.** Identity
5. $A(ABCD) = \frac{1}{2}h(AB + CD)$	**5.** Substitution

Assessment
Find the area of the trapezoid.

1.

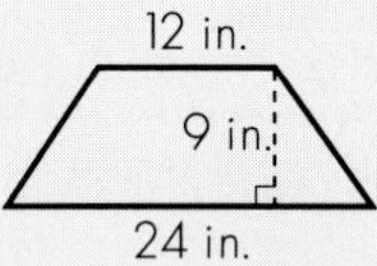

[A = 162 sq in.]

2.

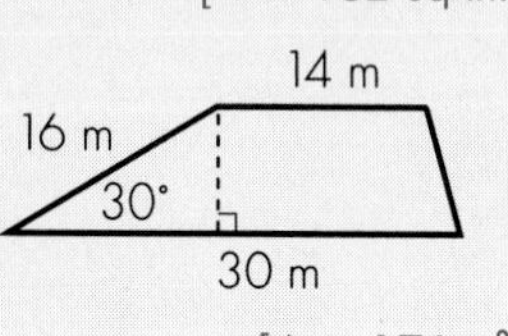

[A = 176 m²]

Reteaching
Students who have had difficulty with this section may benefit from Reteaching Activity 9.4.

Enrichment
For students who have mastered the material in this section, you may wish to assign Enrichment Activity 9.4.

Write a LOGO procedure to tell the turtle to draw a trapezoid that has each area. Then modify the procedure to draw another trapezoid with the same area. Answers may vary. See margin.

13. 64 14. 210 15. 150

A marching band planned a new formation to use during the football season. They will march in two lines from points A and B at 45° angles. When the first person in each line reaches the center line in the field, the lines will turn and march toward each other. The band will form the outline of a trapezoid.

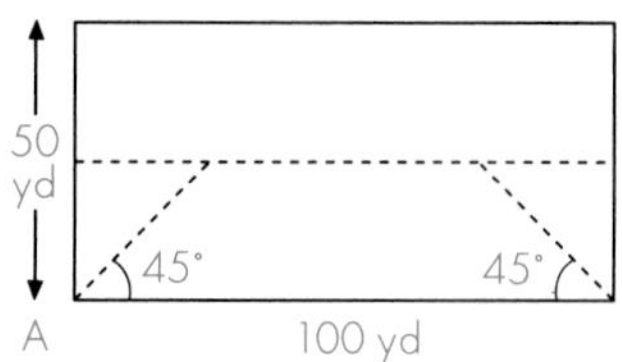

16. What is the area of the trapezoid? 1875 sq yd
17. If the band members stand 1 yard apart and there is a person on points A and B, how many players are in the band? 121

The diagram shows a city park.

18. What is the perimeter of trapezoid $ADFG$? 682 ft
19. What is the perimeter of the entire park? 785.5 ft
20. What is the area of the trapezoid? 20,000 sq ft
21. What is the area of the entire park? 25,745.75 sq ft
22. If the scale for the drawing were 1 inch = 200 feet, then how many inches long would $\overline{AD}$ be on the drawing? $1\frac{1}{2}$ in.

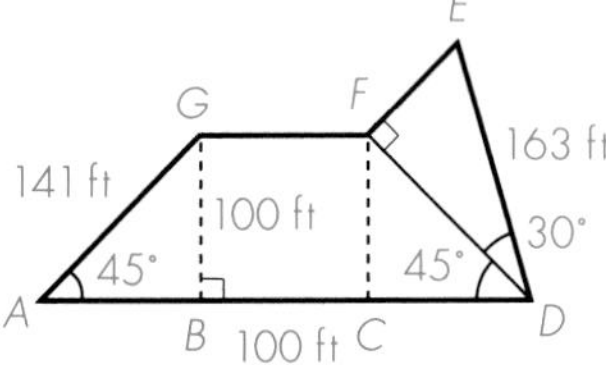

CRITICAL THINKING

23. In the Discovery Activity on page 336, you found that a trapezoid can be divided into two right triangles and a rectangle. Show that the area of trapezoid $ABCD$ is equal to the combined areas of triangles I and II and rectangle III. Write a two-column proof. See margin.

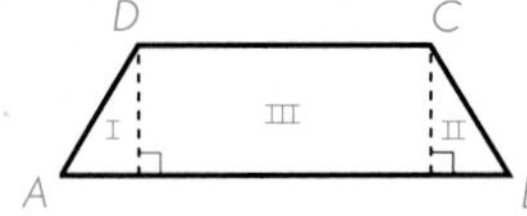

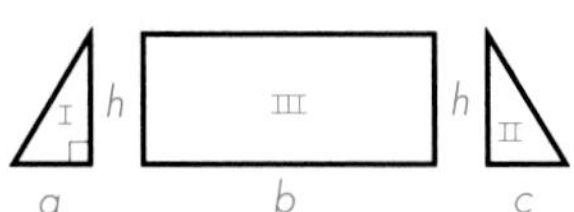

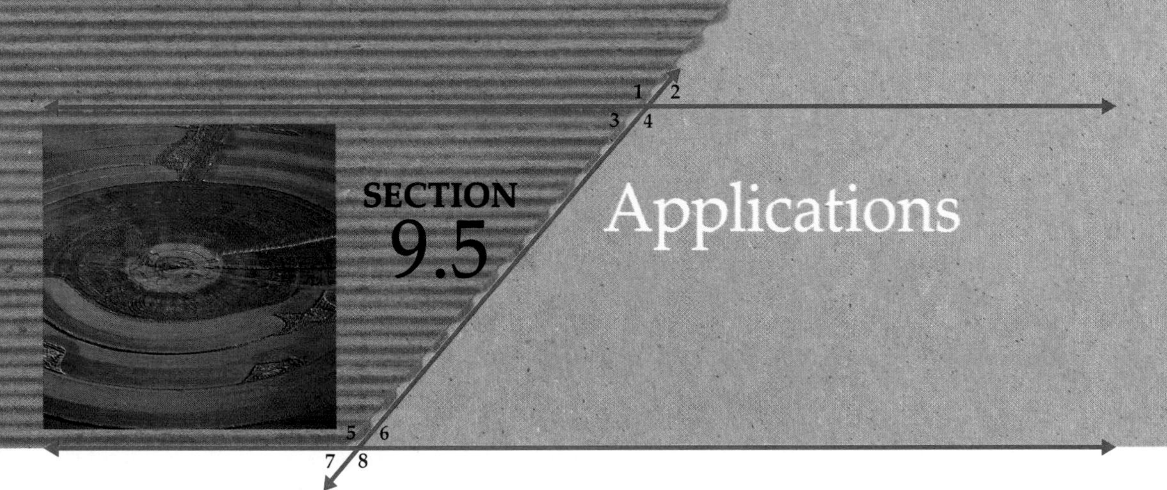

After completing this section, you should understand
- ▲ how to find the area of a quadrilateral with no parallel sides.
- ▲ the relationship between area and perimeter of similar figures.

In this chapter, you have investigated methods of finding area for several different kinds of quadrilaterals. Often it is necessary to find the areas of quadrilaterals that do not have a specific formula.

Look at figures 1–3.

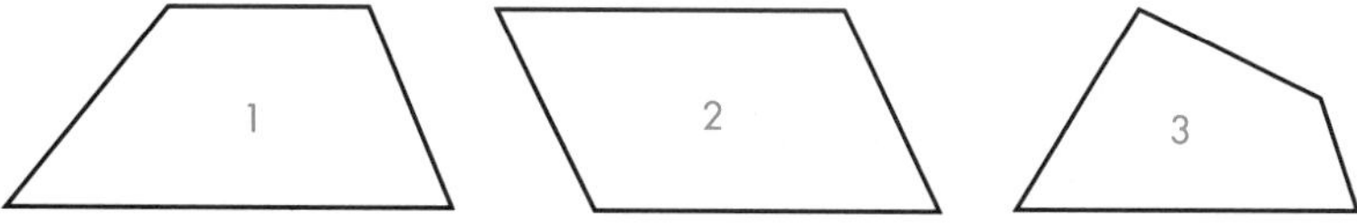

The areas of figures 1 and 2 can be found by formula; figure 3 has no parallel sides, so it is not a parallelogram and has no specific area formula. To find the area of quadrilaterals with no parallel sides, partition the figure into other figures for which you know a formula for the area.

AREA OF QUADRILATERALS WITH NO PARALLEL SIDES

Example: Find the area of quadrilateral *ABCD*. You can partition *ABCD* into a triangle and a trapezoid by drawing a line segment from *C*, parallel to $\overline{AB}$, and intersecting $\overline{AD}$ at *E*.

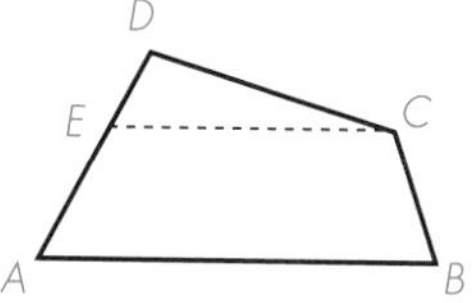

SECTION 9.5

Resources
Reteaching 9.5
Enrichment 9.5
Transparency Master 9-2

Objectives
After completing this section, the student should understand
- ▲ how to find the area of a quadrilateral with no parallel sides.
- ▲ the relationship between area and perimeter of similar figures.

Materials
ruler
graph paper
library resources

Vocabulary
similar

GETTING STARTED

Warm-Up Activity
Have students cut a large square from graph paper and fold it in half to make two congruent triangles. Challenge them to construct by folding and/or drawing, three smaller right triangles, one small square, and one small trapezoid within one half of the large square.
One solution:

TEACHING COMMENTS

Using the Page
Quadrilateral *ABCD* has been reproduced on Transparency Master 9-2 and may be used to illustrate this example.

The area of quadrilateral *ABCD* thus becomes the sum of the areas of trapezoid *ABCE* and triangle *DEC*.

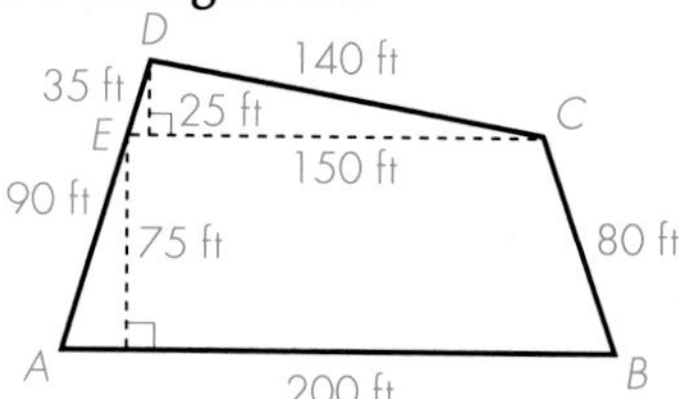

To find the areas of trapezoid *ABCE* and $\triangle DEC$ you need to know the altitudes of triangle and trapezoid and the lengths of $\overline{AB}$ and $\overline{EC}$ (bases). In actual situations, these lengths are usually obtained by measurement. Here, the measurements have been given to you.

$$
\begin{aligned}
A(ABCD) &= A(ABCE) + A(\triangle DEC) \\
&= \tfrac{1}{2}h(b_1 + b_2) + \tfrac{1}{2}bh \\
&= \tfrac{1}{2} \cdot 75 \text{ ft } (150 \text{ ft} + 200 \text{ ft}) + \tfrac{1}{2}(150 \text{ ft} \cdot 25 \text{ ft}) \\
&= 13{,}125 \text{ sq ft} + 1875 \text{ sq ft} \\
&= 15{,}000 \text{ sq ft}
\end{aligned}
$$

Using the Page
You may wish to have students divide quadrilateral *ABCD* in alternate ways to show that the area will be the same no matter how the figure is divided.

Since there is no formula for finding the area of quadrilaterals with no parallel sides, you must partition the figure into triangles, trapezoids, or parallelograms, and find the sum of the combined areas.

To find the perimeter of quadrilateral *ABCD,* find the sum of the lengths of the sides.
$P(ABCD) = 200 \text{ ft} + 125 \text{ ft} + 140 \text{ ft} + 80 \text{ ft} = 545 \text{ ft}$

CLASS ACTIVITY

Find the area and perimeter of each quadrilateral.

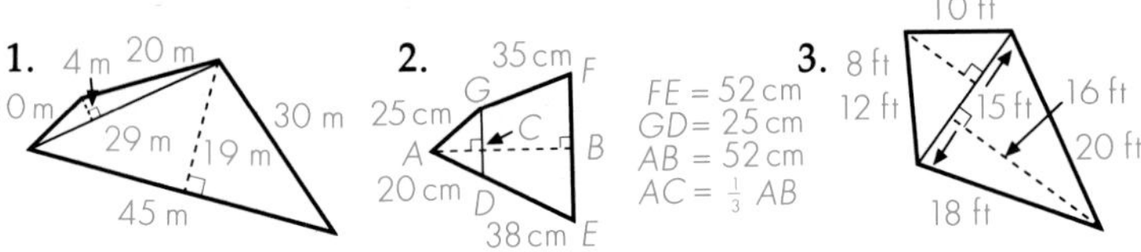

1. $A = 485.5 \text{ m}^2$, $P = 105$ m
2. $A = 1551 \text{ cm}^2$, $P = 170$ cm
3. $A = 180$ sq ft, $P = 60$ ft
4. The garden shown at the right is made up of flower beds; a reflecting pond, and lawn. To buy grass seed and fertilizer for the lawn, you need to know its area. Partition the space and find the area of the lawn. 1312.5 sq ft

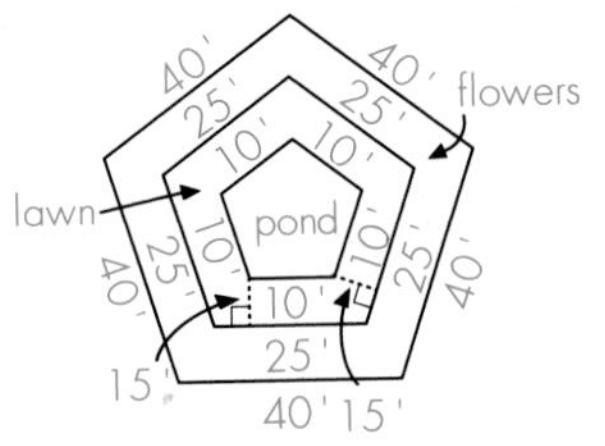

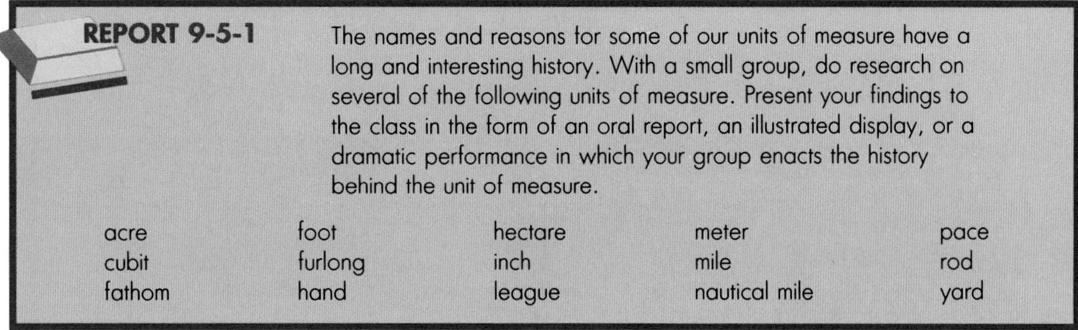

REPORT 9-5-1 The names and reasons for some of our units of measure have a long and interesting history. With a small group, do research on several of the following units of measure. Present your findings to the class in the form of an oral report, an illustrated display, or a dramatic performance in which your group enacts the history behind the unit of measure.

acre	foot	hectare	meter	pace
cubit	furlong	inch	mile	rod
fathom	hand	league	nautical mile	yard

Using the Page
The report on units of measure provides a good opportunity for small groups that might like to write a sketch or dramatization of the history of some older units of measure. You may wish to have students consult their history teachers for ideas.

RELATIONSHIPS BETWEEN SIDES, AREA, AND PERIMETER

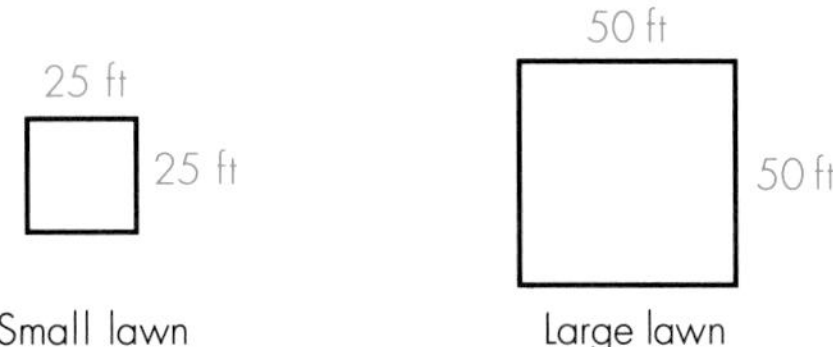

Suppose that you have a job mowing the smaller lawn. You charge $4 for the job. What would you charge to mow the larger lawn?

a. $8 b. $10 c. $12 (d.) $16

Before you decide on an amount, compare the areas of the two lawns. Both yards are squares, so the area is ℓw or s^2.

A(small lawn) = 25 · 25 = 625 square feet
A(large lawn) = 50 · 50 = 2500 square feet

The area of the large lawn is four times that of the small lawn, so you should charge $16.

You can use the concept of ratio and proportion (from Chapter 4) to solve problems like this one.

$$\frac{\text{Area}_1}{\text{Area}_2} = \frac{\text{Cost}_1}{\text{Cost}_2} = \frac{625}{2500} = \frac{\$4}{C_2}$$

$$625 \cdot C_2 = 2500 \cdot 4$$
$$625C_2 = 10{,}000$$
$$C_2 = \$16$$

DEFINITION 9-5-1 **Two figures are SIMILAR if the corresponding angles are equal in measure and the corresponding sides are in equal ratio.**

Notice that this definition of similar figures is valid for all similar figures, not just for triangles.

Using the Page
During the discussion of Definition 9-5-1, you may wish to have students construct several sets of similar figures on graph paper and map the correspondences between them.

Using the Page
You may wish to extend the Discovery Activity by having students explore the ratios of corresponding sides and areas for several sets of similar triangles, parallelograms, or trapezoids. You may wish to use the plain dot half of Transparency Master 9-2 to illustrate these figures for students to copy.

Additional Answers
5.

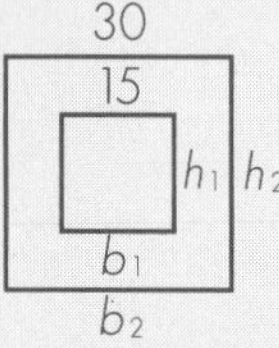

Extra Practice
1. The ratio of the perimeters of two similar figures is $\frac{2}{7}$. What is the ratio of the areas? $\left[\frac{4}{49}\right]$
2. The ratio of the area of two similar figures is $\frac{64}{25}$. What is the ratio of the perimeters? $\left[\frac{8}{5}\right]$

DISCOVERY ACTIVITY

1. Draw two squares, one with sides of 15 units and one with sides of 30 units.
2. Are the two figures similar? Yes
3. Are the corresponding angles equal? Yes; 90°
4. Are corresponding sides in equal ratio? What is the ratio? Yes; $\frac{1}{2}$
5. Map the figures. See margin.
6. What is the area of each square? 225 square units, 900 square units
7. Write the ratio of the areas, $\frac{A_1}{A_2}$. $\frac{A_1}{A_2} = \frac{225}{900}$
8. In place of A_1 and A_2, write the formula for the area of any parallelogram, b_1h_1 and b_2h_2.
   So, $\frac{A_1}{A_2} = \frac{b_1h_1}{b_2h_2} = \frac{225}{900}$
9. Rewrite the proportion.
   $$\frac{b_1}{b_2} \cdot \frac{h_1}{h_2} = \frac{225}{900}$$
   $$\frac{1}{2} \cdot \frac{1}{2} = \frac{225}{900}$$
   $$\left(\frac{1}{2}\right)^2 = \frac{1}{4}$$
10. What does the proportion tell you about the ratios of corresponding sides and areas of similar figures? Write a sentence stating your conclusion.

You may have discovered that the square of the ratio of the corresponding sides of similar figures is the same as the ratio of their areas.

POSTULATE 9-5-2 **If two similar figures have a ratio of a/b for their corresponding sides, then**
a. the ratio of their perimeters is a/b, and
b. the ratio of their areas is $(a/b)^2$.

Example: Figure A and figure B are similar.

$\frac{\text{Perimeter A}}{\text{Perimeter B}} = \frac{6}{18} = \frac{1}{3}$ Fig. A

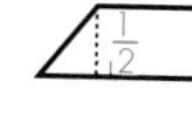

$\frac{\text{Area A}}{\text{Area B}} = \frac{1}{9} = \left(\frac{1}{3}\right)^2$ Fig. B

CLASS ACTIVITY

Each pair of figures is similar. Find the ratios of the perimeters and areas. Figures are not necessarily drawn to scale.

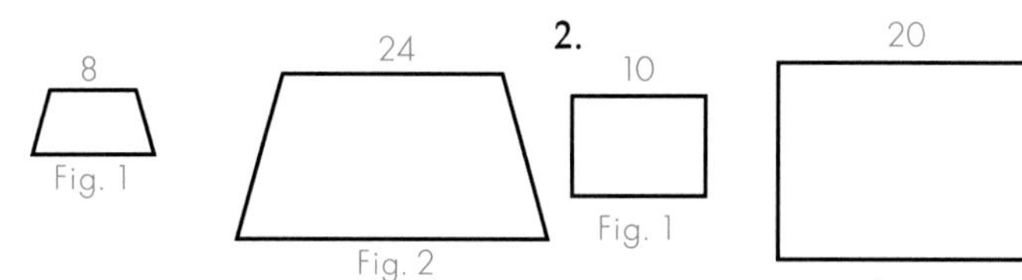

1. $\frac{P_1}{P_2} = \frac{1}{3}$ $\frac{A_1}{A_2} = \frac{1}{9}$

2. $\frac{P_1}{P_2} = \frac{1}{2}$ $\frac{A_1}{A_2} = \frac{1}{4}$

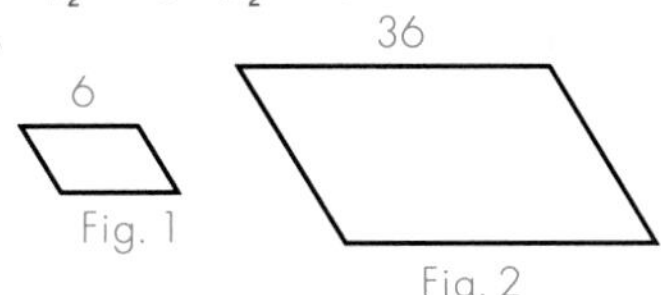

4. Fig. 1 6; Fig. 2 9

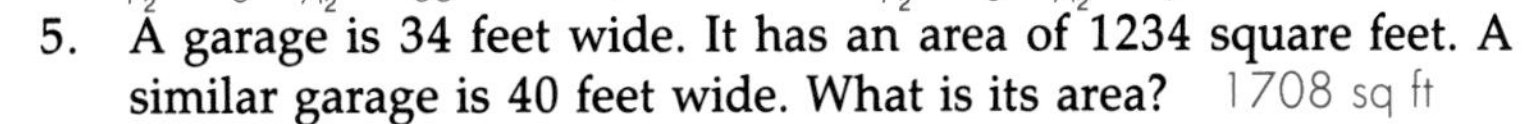

3. $\frac{P_1}{P_2} = \frac{1}{6}$ $\frac{A_1}{A_2} = \frac{1}{36}$

4. $\frac{P_1}{P_2} = \frac{2}{3}$ $\frac{A_1}{A_2} = \frac{4}{9}$

5. **A garage is 34 feet wide. It has an area of 1234 square feet. A similar garage is 40 feet wide. What is its area?** 1708 sq ft

HOME ACTIVITY

Find the area and perimeter of each figure.

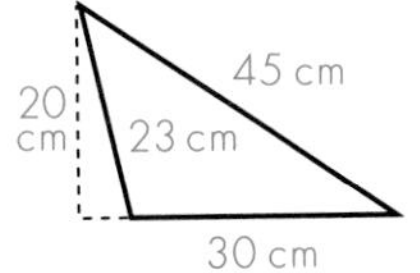

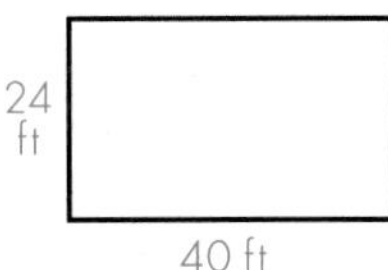
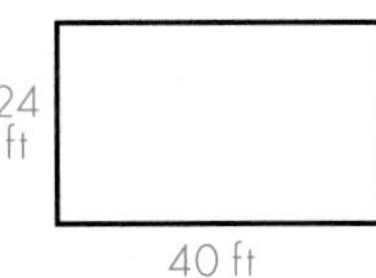

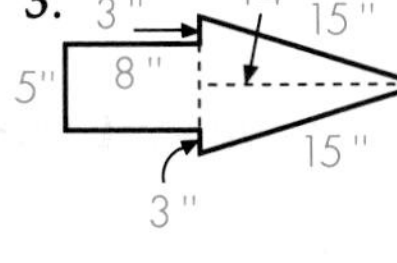

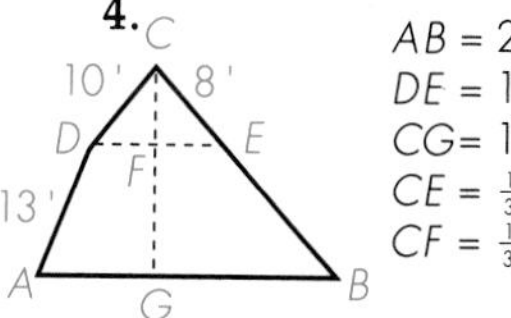

1. $A = 300$ cm², $P = 98$ cm
2. $A = 960$ sq ft, $P = 128$ ft
3. $A = 117$ sq in., $P = 57$ in.
4. $A = 310\frac{1}{3}$ sq ft, $P = 75$ ft

The ratio of the area of the figure shown to the area of a similar figure is given. Write a Logo procedure to tell the turtle to draw the figure given and the similar figure. Answers may vary. See margin.

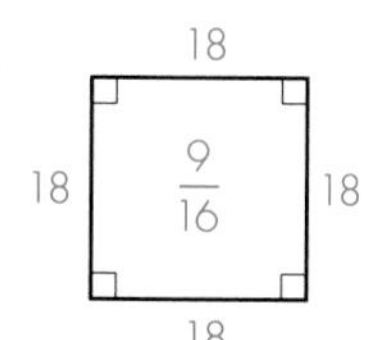

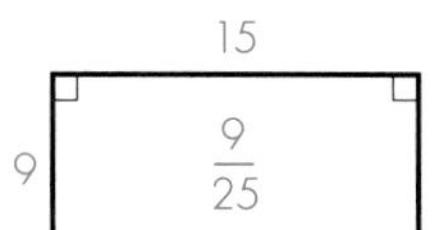

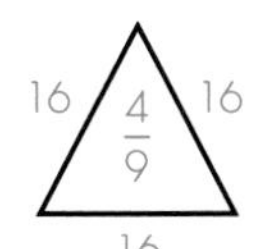

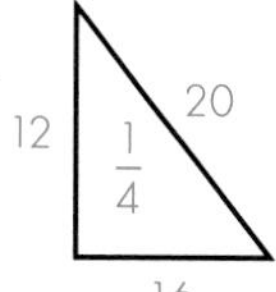

Extra Practice

Find the area and perimeter of each figure.

1.

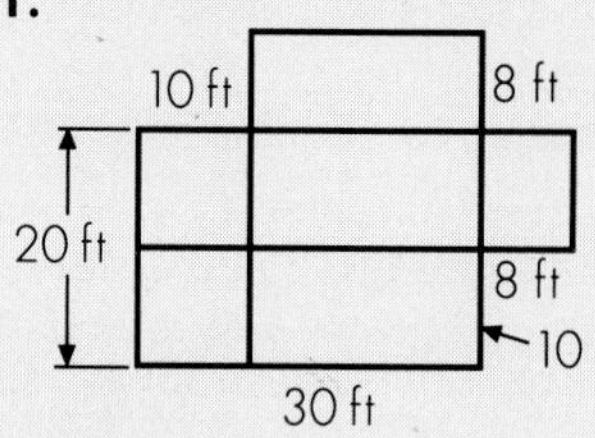

[$A = 840$ sq ft; $P = 132$ ft]

2.

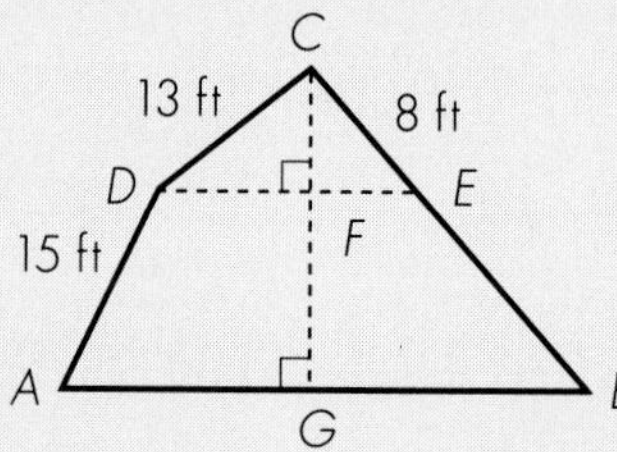

$AB = 36$ ft $CF = \frac{1}{3}CG$
$CE = \frac{1}{3}CB$ $DE = 18$ ft
$CG = 21$ ft
[$A = 441$ sq ft; $P = 88$ ft]

3. XYZ is a right triangle with $ZK \perp XY$. Find the ratio $\frac{A(\triangle YZK)}{A(\triangle ZXK)}$.

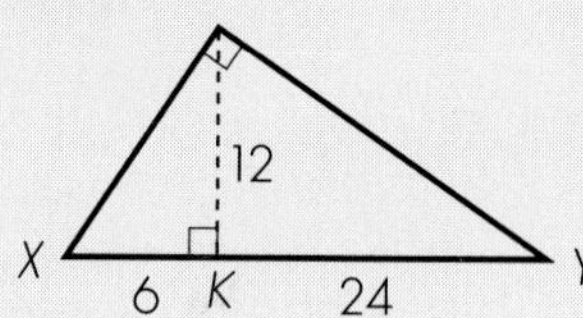

Explain your answer.
[$\triangle YZK \sim \triangle ZXK$ because corresponding sides have equal ratios: $\frac{6}{12} = \frac{12}{24} = \frac{1}{2}$. So $\frac{A(\triangle YZK)}{A(\triangle ZXK)} = \frac{1}{4}$.]

Using the Page

Students can use the x-y plane to draw the similar figures in Exercises 5-8.

9. X, Y, and Z are the midpoints of the sides of $\triangle QRS$. What is the ratio of the area of $\triangle XYZ$ to the area of $\triangle QRS$? Explain your answer.
$\overline{ZY} = \frac{1}{2}\overline{QR}$, $\overline{ZX} = \frac{1}{2}\overline{SR}$, $\overline{YX} = \frac{1}{2}\overline{SQ}$; $\overline{ZY}/\overline{QR} = \overline{ZX}/\overline{SR} = \overline{YX}/\overline{SQ} = \frac{1}{2}$, so $\frac{A(\triangle XYZ)}{A(\triangle QRS)} = \left(\frac{1}{2}\right)^2 = \frac{1}{4}$.

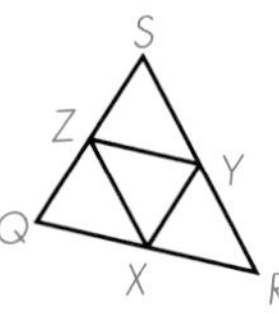

CRITICAL THINKING

The figures below are made up of arrays of dots and represent geometric numbers. Find each number and record it.

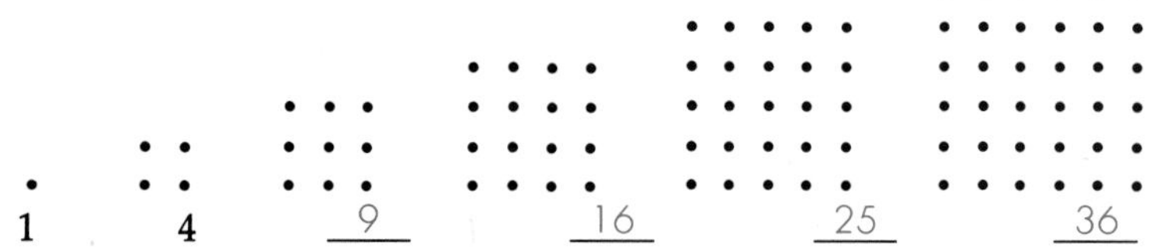

1 4 9 16 25 36

10. What geometric number is represented by each of these numbers? Square

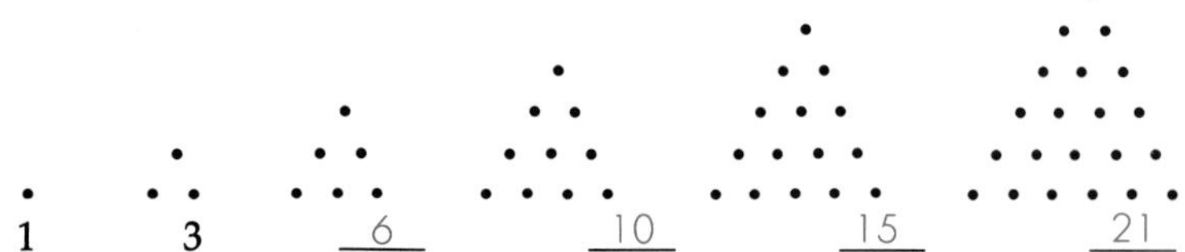

1 3 6 10 15 21

11. What geometric number is represented by each of these numbers? Triangle

The numbers associated with the square arrays of dots are called square numbers. The ones found in the triangular arrays are called triangular numbers. You can use an abbreviation to represent a square or triangular number. For example, S_3 means the third square number, a square with three dots on each side. T_6 means the sixth triangular number, a triangle with six dots on a side.

There are several interesting patterns involving these geometric numbers. Copy and complete the table.

n	1	2	3	4	5	6	7	8	9	10	11	12	13	14
S_n	1	4	9	16	25	36	49	64	81	100	121	144	169	196
T_n	1	3	6	10	15	21	28	36	45	55	66	78	91	105

12. Find the sum of any two consecutive triangular numbers. What do you notice about the sum? It is a square number.

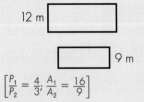

FOLLOW-UP

Assessment
1. Find the ratios of the perimeters and areas for the similar figures.

$\left[\frac{P_1}{P_2} = \frac{4}{3}; \frac{A_1}{A_2} = \frac{16}{9}\right]$

Reteaching
Students who have had difficulty with this section may benefit from Reteaching Activity 9.5.

Enrichment
For students who have mastered the material in this section, you may wish to assign Enrichment Activity 9.5.

CHAPTER SUMMARY

Chapter Vocabulary
area
perimeter
similar

9.1 Find the perimeter of figure *ABCD*.
$P = 3.5 \text{ m} + 3.5 \text{ m} + 2.5 \text{ m} + 2 \text{ m}$
$P = 11.5 \text{ m}$

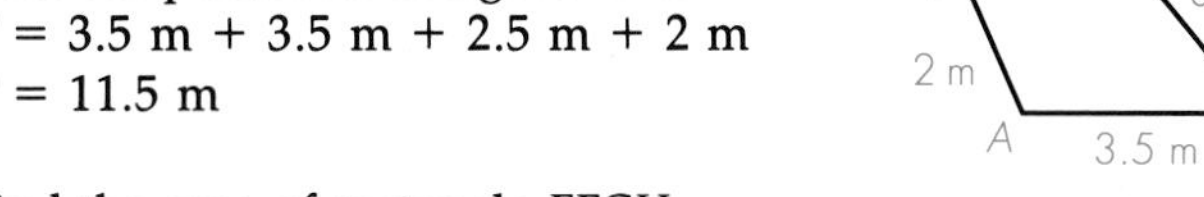

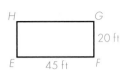

Find the area of rectangle *EFGH*.

$A = 45 \text{ ft} \cdot 20 \text{ ft}$
$A = 900 \text{ sq ft}$

9.2 Find the area of $\triangle JKL$.

$A = \frac{1}{2}(3 \text{ cm} \cdot 5 \text{ cm})$
$A = 7.5 \text{ cm}^2$

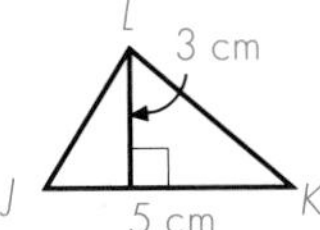

9.3 Find the perimeter and the area of parallelogram *MNOP*.

$P = 2(9 \text{ mi} + 10 \text{ mi})$
$P = 38 \text{ mi}$
$A = 6.5 \text{ mi} \cdot 10 \text{ mi}$
$A = 65 \text{ sq mi}$

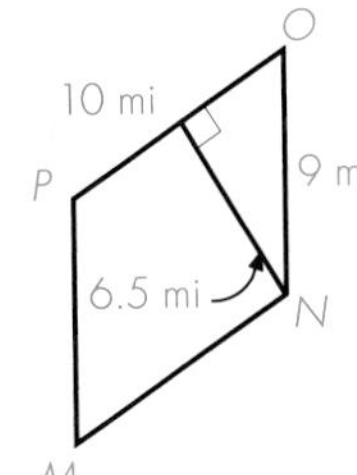

9.4 Find the area of trapezoid *QRST*.

$A = \frac{1}{2}(22 \text{ mm})(52 \text{ mm} + 70 \text{ mm})$

$A = 1342 \text{ mm}^2$

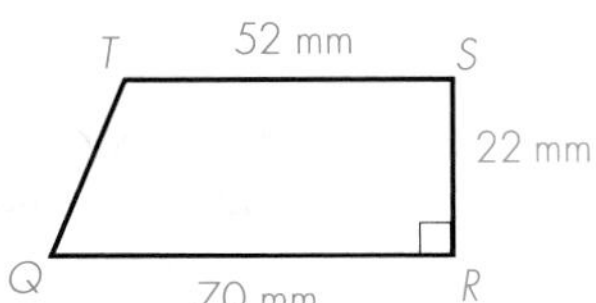

9.5 What is the ratio between the perimeter of $\triangle UVW$ and that of $\triangle XYZ$?

The ratio between corresponding sides of $\triangle UVW$ and $\triangle XYZ$ is 1:2, so the two triangles are similar. By Postulate 9-5-1, the ratio between the perimeters is also 1:2.

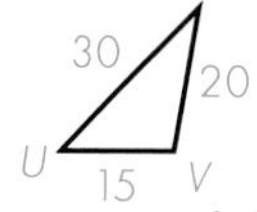

What is the ratio of the area of $\triangle UVW$ to that of $\triangle XYZ$?

By Postulate 9-5-1, square the ratio of the sides of similar figures to get the ratio of the areas.

$\frac{A(\triangle UVW)}{A(\triangle XYZ)} = \left(\frac{1}{2}\right)^2 = \frac{1}{4}$

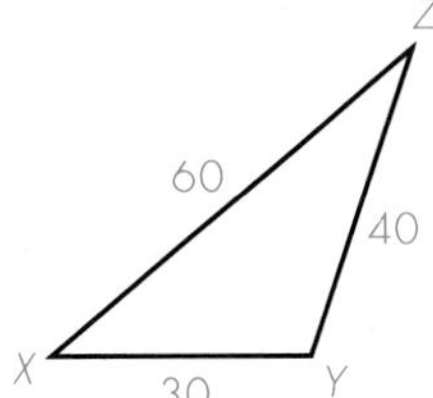

349

Using the Page

The method suggested for solving Heron's Formula using a calculator may have to be modified if students do not have access to calculators with parentheses, storage, and recall keys. It may be necessary for them to combine the calculator with paper-and-pencil. For example, first they compute and record the value of s. Then they compute and record the values for $(s - a)$, $(s - b)$, and $(s - c)$. Finally they compute $s(s - a)(s - b)(s - c)$ and find the square root of the product.

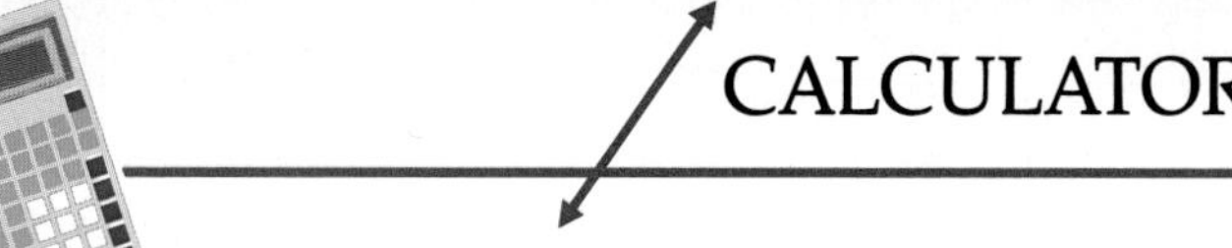

CALCULATOR ACTIVITY

Carla wants to plant a garden in a triangular plot. She knows that the lengths of the three sides of the plot are 9 feet, 10 feet, and 13 feet. Carla wants to find the area of her plot.

To find the area, Carla could make a scale drawing of her plot, find the height of the triangle, and then find the area. Or she could use a formula developed in the first century A.D. by the Greek mathematician Heron of Alexandria.

Heron's formula states that if a triangle has sides with lengths a, b, and c, then the area of the triangle is

$$\sqrt{s(s - a)(s - b)(s - c)}$$

$$\text{where } s = \tfrac{1}{2}(a + b + c).$$

A calculator can be very useful when using Heron's formula to find an area. The Parentheses keys, which are labeled (and), can be used to group numbers together just as in the formula. For example, to calculate s for Carla's garden plot, press the keys shown below.

Press:	(	9	+	1	0	+	1	3	)	÷ 2	=
Display:		9		19			32				16

Another key that would be very useful in calculating area using Heron's formula is the Store key, which is labeled STO. The value for s that was calculated above can be stored in the memory of the calculator by pressing the Store key after pressing the [=] key. The value of s can then be retrieved from memory each time it is needed by pressing the Recall key, which is labeled RCL.

To calculate the radicand to find the area of Carla's garden plot using Heron's formula, press the keys shown below. Notice that the [RCL] key is pressed in place of s, and the [(] and [)] keys are pressed in place of the parentheses.

Formula: $s \quad (\; s \; - \; a \;) \quad (\; s \; - \; b \;) \quad (\; s \; - \; c \;)$

Press: [RCL] [×] [(] [RCL] [−] [9] [)] [×] [(] [RCL] [−] [1] [0] [)] [×] [(] [RCL] [−] [1] [3] [)] [=]

Display: 16   7 112   6 672   3 2016

350

To find the area of Carla's garden plot, we must now find the square root of 2016. To do this, press the Square Root key, which is labeled $\sqrt{\ }$ or $\sqrt{x}$, after pressing the [=] key.

Press: [$\sqrt{\ }$]

Display: 44.899889

Rounded to the nearest hundredth, the area of Carla's triangular garden plot is 44.90 square feet.

Use a calculator and Heron's formula to calculate the area of each triangle with side lengths as given below. Round each answer to the nearest hundredth.

1. 3 in., 7 in., 8 in. 10.39 sq in.
2. 4 m, 11 m, 12 m 21.93 m^2
3. 15 ft, 15 ft, 15 ft 97.43 sq ft
4. 8 cm, 9 cm, 10 cm 34.20 cm^2
5. 6 yd, 10 yd, 12 yd 29.93 sq yd
6. 7 in., 11 in., 16 in. 31.94 sq in.
7. 5 m, 9 m, 13 m 16.07 m^2
8. 6 mi, 8 mi, 10 mi 24 sq mi
9. 9 ft, 17 ft, 21 ft 74.41 sq ft
10. 5 cm, 12 cm, 13 cm 30 cm^2
11. Mark wants to paint the front of his triangular house including the door. To do this, he needs to find the area of the front of the house. Use a calculator and Heron's formula to find this area. 642.98 sq ft

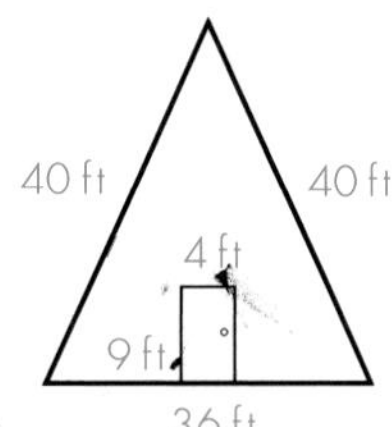

12. Suppose Mark decides not to paint the door of his house. Find the area of the surface he wants to paint. 606.98 sq ft

Using the Page
It may be helpful to have students work in pairs to calculate the areas of the triangles in Exercises 1-10. They can compare answers and, in case of disagreement, work together to find the error. An additional exercise follows.
Find the area of quadrilateral *PQRS*.

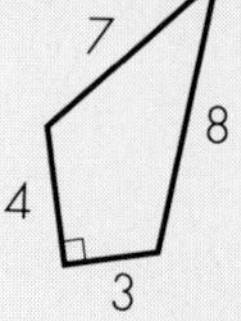

[$A \approx 23.320508$]

Using the Page

Students can work individually or in groups to review. If you have students study in groups, they should work together to make sure all members of the group know the material in the chapter.

Informal assessment such as interviews, classroom observations, or a review of student portfolios can be used instead of paper-and-pencil tests.

Find the perimeter and area.

1. 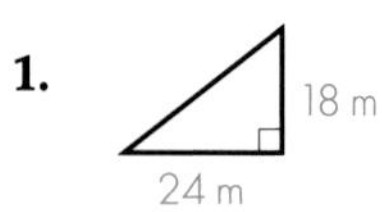

P = 72m
A = 216 m^2

2. 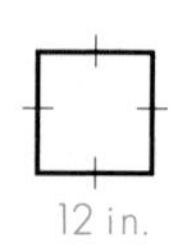

P = 48 in.
A = 144 sq in.

3. 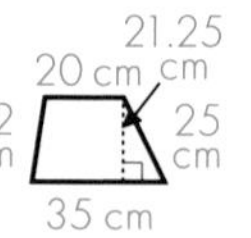

P = 102 cm
A = 584.4 cm^2

4. 7' 7.5' 8'
P = 31 ft
A = 56 sq ft

5. 40" 22.2" 30° 25"
P = 87.2 in.
A = 250 sq in.

6. 15 m 10 m 6 m
P = 54 m
A = 144 m^2

7. 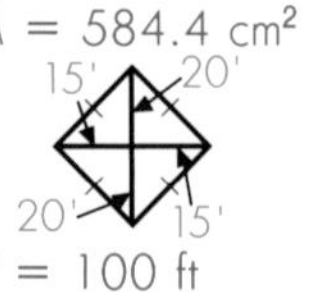

P = 100 ft
A = 600 sq ft

8. 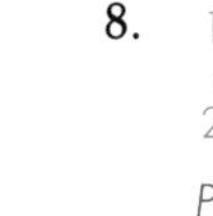
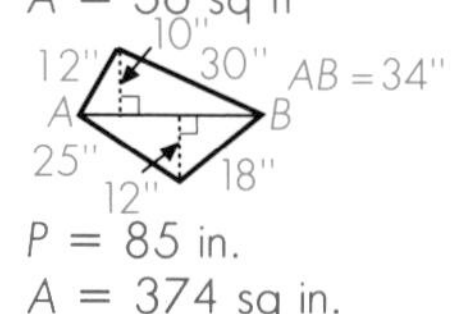

P = 85 in.
A = 374 sq in.

9. Find the area and the perimeter of trapezoid *PQRS*. A = 108 sq ft; P = 36 + 12$\sqrt{2}$ ft

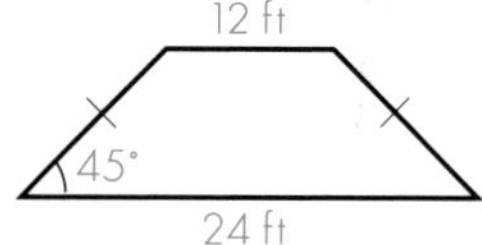

The area of each figure is given. Find the missing base and/or height.

10. A = 72 cm^2
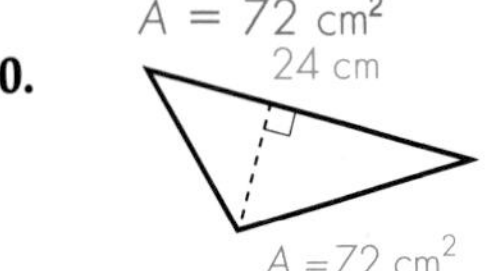

h = 6 cm

11. A = 125 cm^2
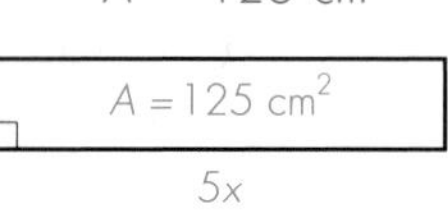

h = 5 cm, b = 25 cm

12. A = 135 sq yd
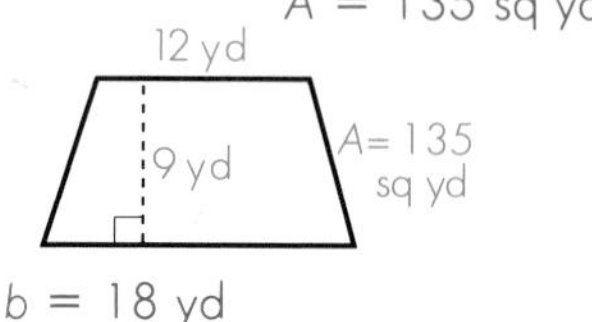

b = 18 yd

Find the area of the shaded region.

13. 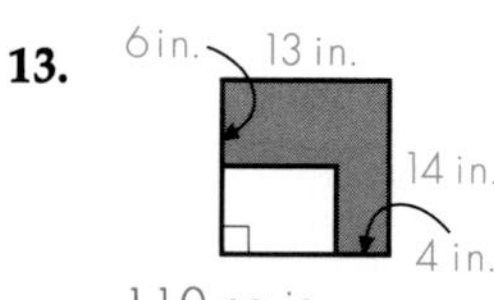

110 sq in.

14. 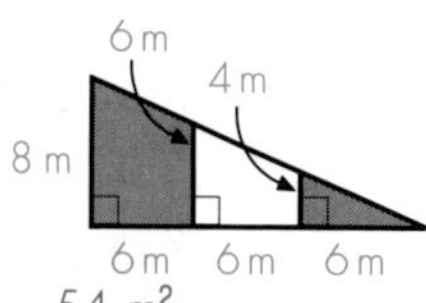

54 m^2

15. 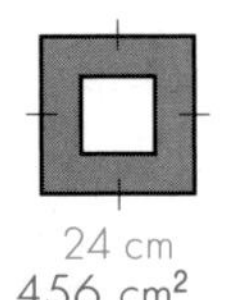

456 cm^2

16. 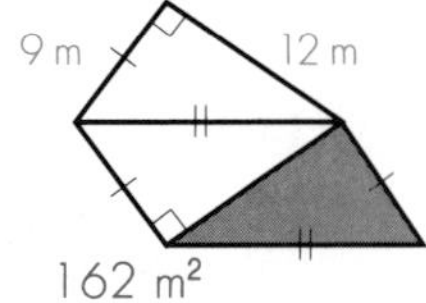

162 m^2

Find the ratios of the perimeters and area for each pair of similar figures.

17. 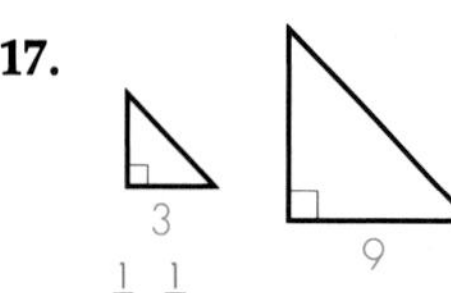

$\frac{1}{3}$, $\frac{1}{9}$

18. 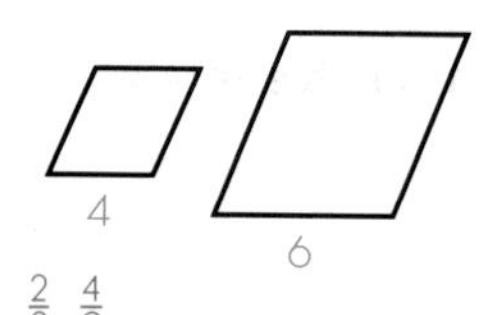

$\frac{2}{3}$, $\frac{4}{9}$

19. 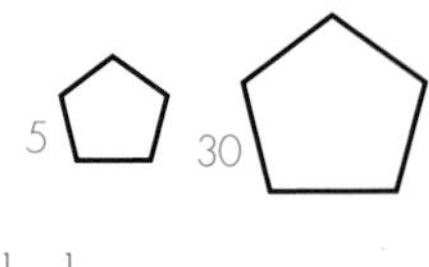

$\frac{1}{6}$, $\frac{1}{36}$

20. Ellen wants to rope off a section of a gym and use it as an exercise area. She has a 100-meter length of rope. What are the dimensions of the largest area she can enclose? 25 m by 25 m

21. The bottom of a swimming pool measures 15 meters by 20 meters. It is to be tiled with rectangular tiles that are 50 centimeters by 25 centimeters. How many tiles are needed? 2400 tiles

Find the area and perimeter.

1.
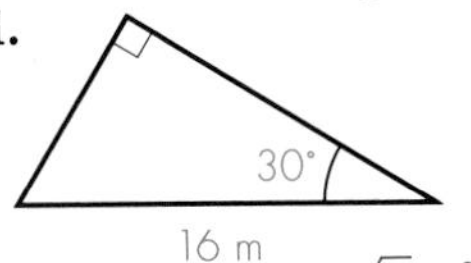

$A = 32\sqrt{3}$ m², $P = 24 + 8\sqrt{3}$ m

2.
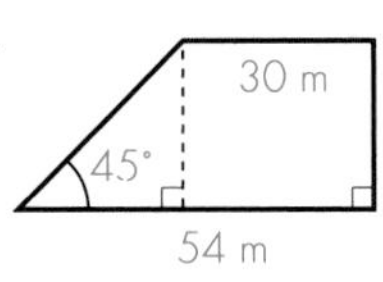

$A = 1008$ m², $P = 108 + 24\sqrt{2}$ m

3.
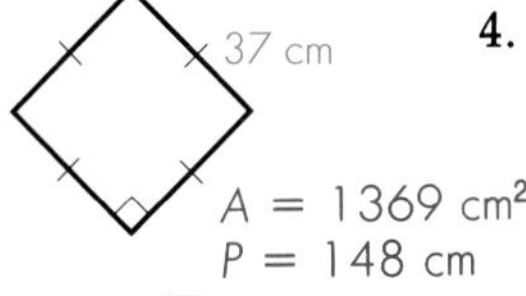

$A = 1369$ cm²
$P = 148$ cm

4.
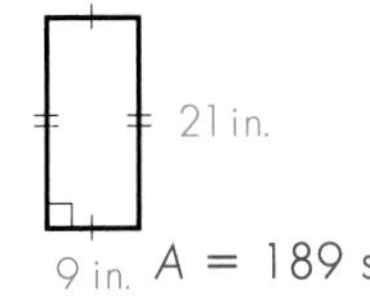

$A = 189$ sq in.
$P = 60$ in.

5.

$A = 675$ cm³, $P = 112$ cm

6.
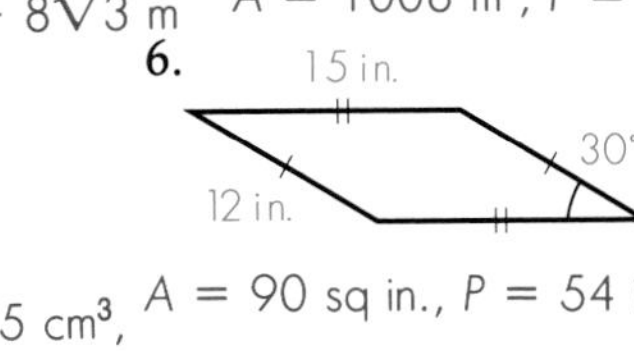

$A = 90$ sq in., $P = 54$ in.

7.
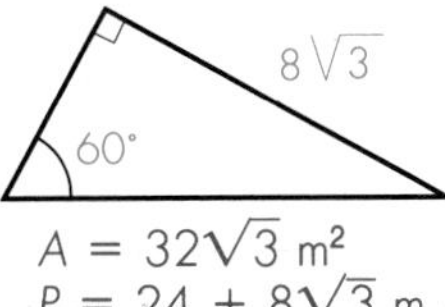

$A = 32\sqrt{3}$ m²
$P = 24 + 8\sqrt{3}$ m

8.
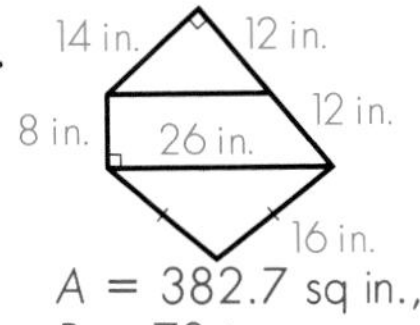

$A = 382.7$ sq in., $P = 78$ in.

9. $\overline{JK}$ is the median of trapezoid *LMNO*. *NP* = 42 cm, *LM* = 120 cm, and *ON* = 60 cm. What is the area of trapezoid *JKNO*? 1575 cm²

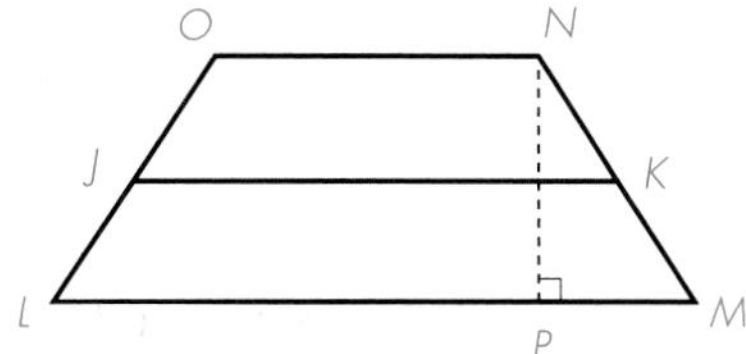

The area of the figures is given. Find the missing measure.

10. $A = 169$ m²
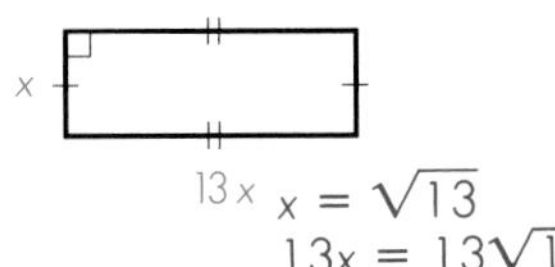

$x = \sqrt{13}$
$13x = 13\sqrt{13}$

11. $A = 45$ cm²
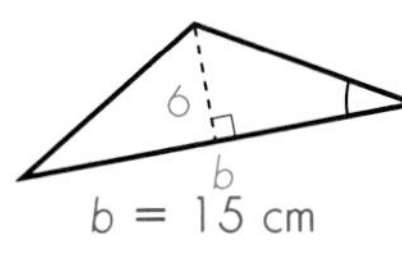

$b = 15$ cm

12. $A = 300$ sq yd
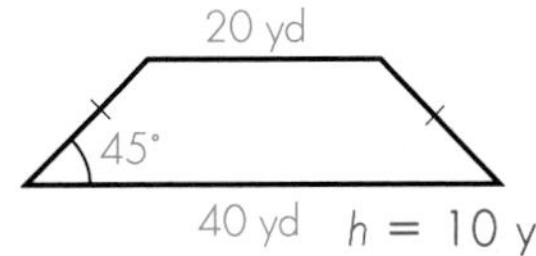

$h = 10$ yd

Find the area of the shaded region.

13.
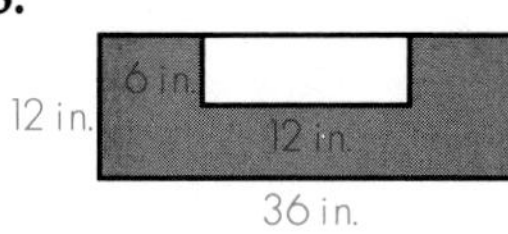

360 sq in.

14.
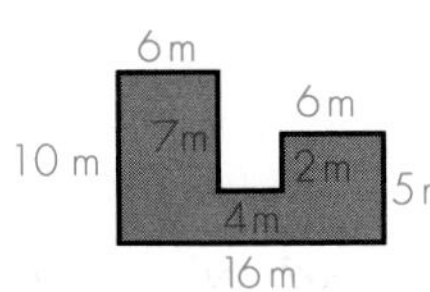

102 m²

15.
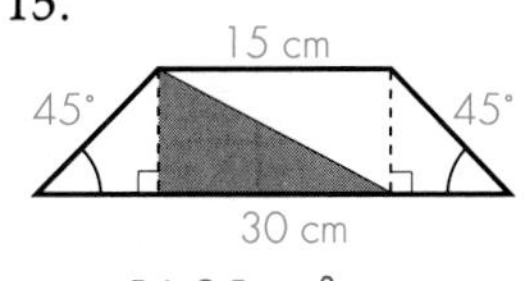

56.25 cm²

Find the ratios of the perimeters and areas for the following similar figures.

16.
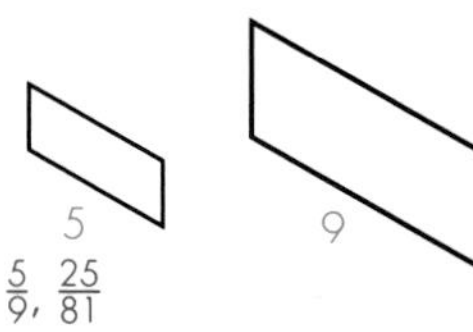

$\frac{5}{9}, \frac{25}{81}$

17.
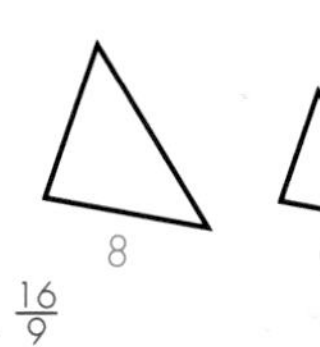

$\frac{4}{3}, \frac{16}{9}$

18.
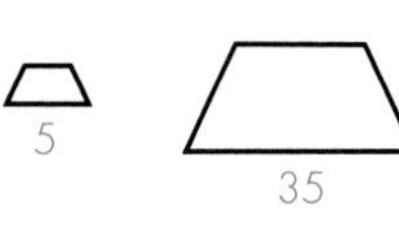

$\frac{1}{7}, \frac{1}{49}$

The area of a yard is 10,800 square feet. A similar yard has a side of 50 feet that corresponds to a side of 60 feet in the larger yard.

19. What is the area of the smaller yard? 7500 sq ft
20. What is the ratio of the perimeters, smaller to larger? $\frac{5}{6}$

Assessment Resources
Achievement Tests pp. 17-18

Test Objectives
After studying this chapter, students should understand
- the definitions of perimeter and area.
- how to find the perimeter and area of a rectangle.
- how to find the perimeter and area of a triangle.
- how to find the perimeter and area of a parallelogram.
- how to find the perimeter and area of a trapezoid.
- how to find the area of a quadrilateral with no parallel sides.
- the relationship between area and perimeter of similar figures.

Algebra Review Objectives
Topics reviewed from algebra include
- using the addition, multiplication, and division properties of equations to solve problems.

Several properties of equations are useful in solving geometry problems.

Addition Property: If $x = y$, then $x + a = y + a$.
You can add the same number to each side of an equation.

Multiplication Property: If $x = y$, then $x(a) = y(a)$.
You can multiply both sides of an equation by the same number.

Division Property: If $x = y$, then $x/a = y/a$, $a \neq 0$.
You can divide both sides of an equation by the same number.

Solve the following equations. Round decimals to the nearest tenth.

1. $\frac{13}{x} = \frac{144}{69}$ $x = 6.2$
2. $37(12x - 14) = -7(15x - 40)$ $x = 1.5$
3. $3 - 4(3x - 6) = 4x - 11$ $x = 2.4$
4. $3 - 4(\frac{3}{8}x - 6) = \frac{3}{4}x - 9$ $x = 16$
5. $2x^2 - 5x + 3 = 2(x - 1)(x - 2)$ $x = 1$
6. $5 - (2(x + 5) + 1) = x - 3$ $x = -1$
7. $(2x - 7)(3x + 15) = 0$ $x = \frac{7}{2}$ or -5
8. $x^2 + 5x = -6$ $x = -2$ or -3

9. The side of an equilateral triangle is 2 units longer than the side of a square. The perimeter of the square is greater than the perimeter of the triangle. The length of the side of the square is an integer. The side of the square must be at least how long? Are there an infinite number of solutions? $s > 6$; yes

Given: Square $ABCD$. $\overline{HA} = \overline{EB} = \overline{FC} = \overline{GD}$.

$\triangle AEH \cong \triangle BFE \cong \triangle CGF \cong \triangle DHG$.
$A(ABCD)A(EFGH) = A(\triangle AEH) + A(\triangle BFE) + A(\triangle CGF) + A(\triangle DHG)$.

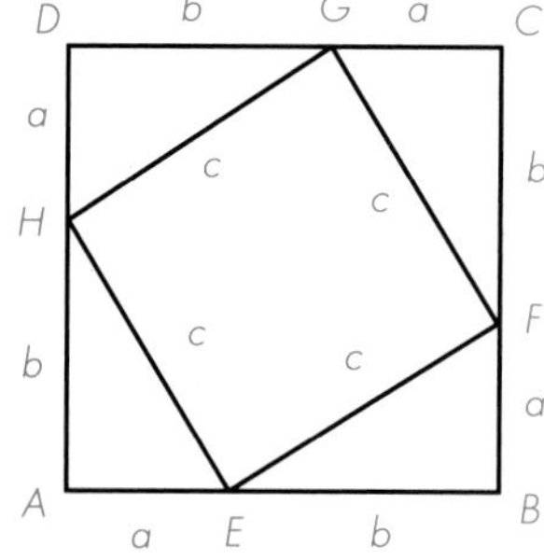

Complete the following.

10. $A(ABCD) = (a + b)^2 =$ $a^2 + 2ab + b^2$

11. Area of each triangle = $\frac{1}{2}ab$

12. $A(EFGH) =$ c^2

13. Substitute the values for the areas that you obtained in Exercises 10–12 in the equation for Area ABCD above. Solve the equation so that c^2 is on one side and all other terms are on the other side. $a^2 + b^2 = c^2$

14. What theorem does this equation represent? Pythagorean theorem

Name ______________________ Date ____________

Perimeter and Area

To find the perimeter of a rectangle, you can use the formula $P = 2\ell + 2w$, where ℓ is the length and w is the width.

To find the area of a rectangle, you can use the formula $A = \ell w$.

Find the perimeter and area of the rectangle at the right.

$P = 2\ell + 2w$
$= 2(16) + 2(12)$
$= 32 + 24$
$= 56$

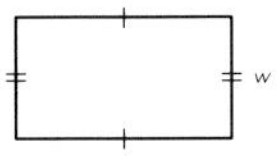

$A = \ell w$
$= 16 \times 12$
$= 192$

The area is 192 m².

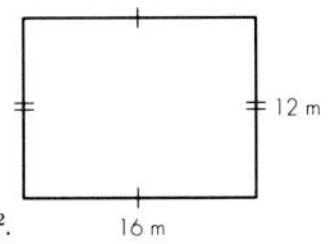

The perimeter is 56 m.

Find the perimeter and area of each rectangle.

1.

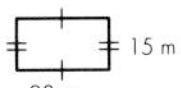

perimeter: 86 m
area: 420 m²

2.

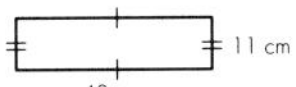

perimeter: 108 cm
area: 473 cm²

3.

perimeter: 68 in.
area: 289 in.²

4.

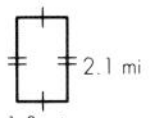

perimeter: 6.8 mi
area: 2.73 mi²

5.

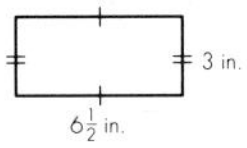

$6\frac{1}{2}$ in.

perimeter: 19 in.
area: $19\frac{1}{2}$ in.²

6.

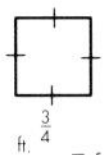

$1\frac{3}{4}$ ft.

perimeter: 7 ft
area: $3\frac{1}{16}$ ft²

Find the length and width of each rectangle.

7.

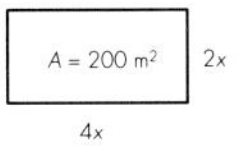

length: 20 m
width: 10 m

8.

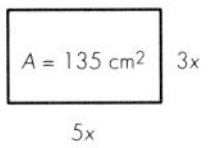

length: 15 cm
width: 9 cm

9.

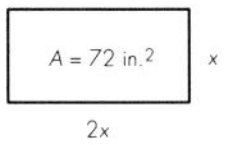

length: 12 in.
width: 6 in.

Name ______________________ Date ____________

Geometry and Algebra

Many formulas in algebra can be related to the area formulas for rectangles and squares. For example, the Distributive Property of Multiplication over Addition states that $a(b + c) = ab + ac$. In the rectangle at the right, notice that the area of the large rectangle can be stated as the sum of the areas of the two smaller rectangles.

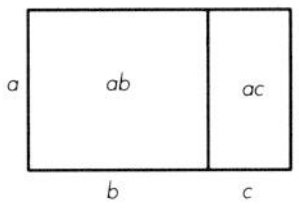

For each figure, write the area as the sum of the areas of the four smaller rectangles. Combine like terms if possible.

1.

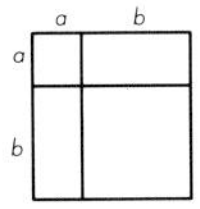

$(a + b)(a + b) =$ $a^2 + 2ab + b^2$

2.

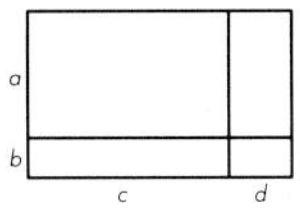

$(a + b)(c + d) =$ $ac + ad + bc + bd$

3. The first figure below represents the area $c^2 - b^2$. Think of cutting the figure apart and rearranging the pieces as shown in the second figure. Write the length and width of the resulting rectangle. Then write the area as the product of the length and width.

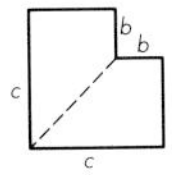

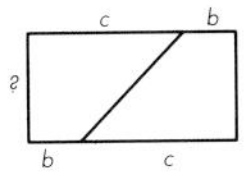

length: (c + b)
width: (c − b)
area: (c + b)(c − b)

In the figure at the right, the shaded region is a square with sides of length $x - y$. One expression for the area of this region is $(x - y)^2$. Write an algebraic expression for each area.

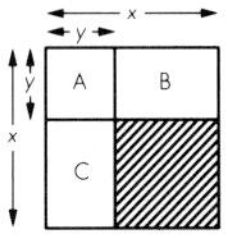

4. area A y^2

5. area B $(x - y)y$

6. area C $(x - y)y$

7. Use the answers for Exercises 4–6 and the following equation to write an algebraic expression for $(x - y)^2$. You will need to simplify the right side of the equation.

$(x - y)^2 = x^2 -$ area A $-$ area B $-$ area C $x^2 - 2xy + y^2$

Name ______________________ Date __________

Area and Perimeter of a Triangle

To find the area of a triangle, you can use the formula shown below, where b and h are the lengths of the base and height, respectively. Think of a triangle as being half of a rectangle.

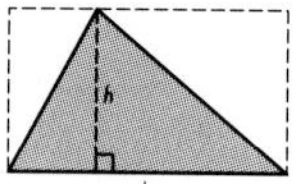

$A = \frac{1}{2}bh$

To find the perimeter of a triangle, add the lengths of the sides.

Find the area and perimeter of the triangle below.

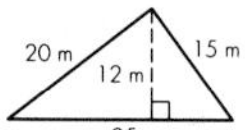

$A = \frac{1}{2}bh$
$= \frac{1}{2} \times 25 \times 12$

The area is 150 m².

P = 25 + 15 + 20
= 60

The perimeter is 60 m.

Find the area.

1. 900 ft²

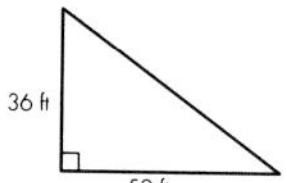

2. 525 cm²

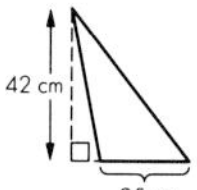

3. 3.5 $2\sqrt{3}$ m²

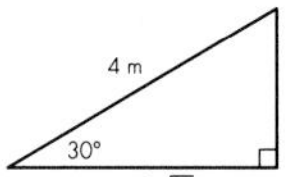

Find the height.

4. $A = 40$ m²

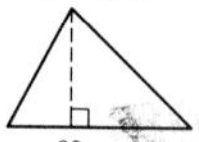

~~10 m~~ 2.9

5. $A = 150$ mm²

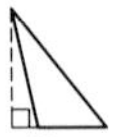

25 mm

6. $A = 60$ cm²

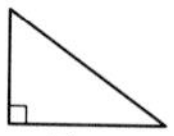

10 cm

Find the length of each side.

7. P = 29 cm

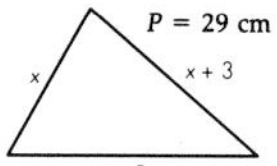

7 cm, 10 cm, 12 cm

8. P = 52 mm

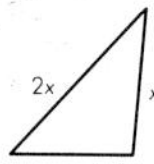

12 mm, 16 mm, 24 mm

9. P = 140 ft

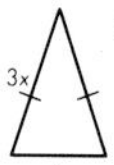

20 ft, 60 ft, 60 ft

Name ______________________ Date __________

3-Dimensional Grids

Sometimes designers or architects use a 3-dimensional grid to draw the interior of a room. Think of the grid as the interior of a huge box, with each square representing one square foot of area. The grid is drawn in perspective, that is, certain lines converge at a single point. In the grid shown below, the room is 15 feet wide by 8 feet high by 8 feet deep.

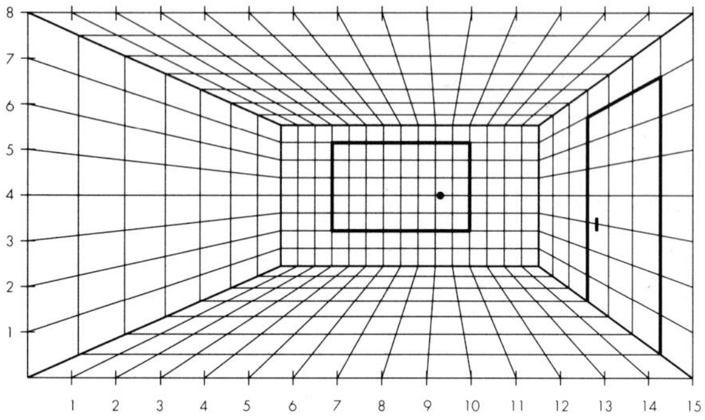

For Exercises 1–6, find the following.

1. area of the floor 120 ft²

2. perimeter of the floor 46 ft

3. area of the window 40 ft²

4. perimeter of the window 26 ft

5. area of the left wall 64 ft²

6. perimeter of the left wall 32 ft

7. Draw a window on the left wall.

8. Trace the grid above on another sheet of paper. Use the new grid to draw a simple design of a room interior.

$40 = \frac{1}{2} \cdot 28 \cdot h$

Name ______________________ Date ______________

Perimeter and Area of a Parallelogram

To find the area of a parallelogram, use this formula:

$A = bh$, where b and h are the lengths of the base and height, respectively

To find the area of a rhombus, use this formula:

$A = \frac{1}{2}d_1d_2$, where d_1 and d_2 are the lengths of the diagonals

To find the perimeter of a parallelogram or a rhombus, add the lengths of the sides.

$P = a + b + a + b$
$= 2a + 2b$

$P = c + c + c + c$
$= 4c$

Find the area of the parallelogram.

1. 48 cm, 70 cm — 3,360 cm²
2. 64 m, 22 m — 1,408 m²
3. 91 m, 75 m — 6,825 m²

Use the length of the diagonals of the rhombus to find the area.

4. $MP = 43$ m, $QN = 38$ m — 817 m²
5. $RT = 20$ ft, $WS = 35$ ft — 350 ft²

Find the perimeter.

6. 15 in., 27 in. — 84 in.
7. 6.4 m, 3.9 m — 20.6 m
8. $6\frac{1}{4}$ ft — 25 ft

Name ______________________ Date ______________

Pentominoes

Think of arranging five equal squares in all possible ways so that each square shares at least one side with another square. The twelve possible arrangements are shown below. These arrangements are sometimes called **pentominoes.**

These pentominoes can be arranged in many different ways. Since each pentomino has an area of 5 square units, the total area of all twelve pentominoes is 60 square units.

Copy each pentomino on centimeter graph paper. Then cut out the pentominoes and rearrange them to cover the rectangular grid below. The area of the rectangle is 60 cm². NOTE: Pieces can be flipped.

Name ______________________ Date ______________

Perimeter and Area of a Trapezoid

To find the area of a trapezoid, use the formula $A = \frac{1}{2}h(b_1 + b_2)$, where h is the length of the altitude and b_1 and b_2 are the lengths of the bases.

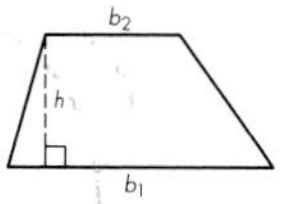

To find the perimeter of a trapezoid, add the lengths of the sides.

Find the area and perimeter of this trapezoid.

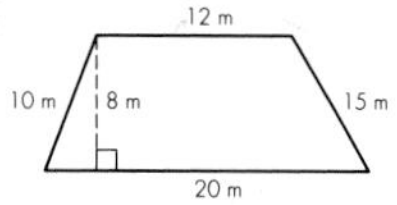

$A = \frac{1}{2} \times h(b_1 + b_2)$

$A = \frac{1}{2} \times 8(20 + 12)$

$= 128$

The area is 128 m².

$P = 10 + 20 + 15 + 12$

$P = 57$

The perimeter is 57 m.

Find the area and perimeter.

1.

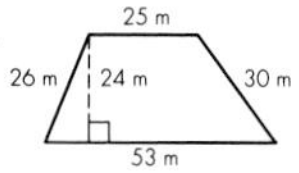

area: 936 m²

perimeter: 134 m

2.

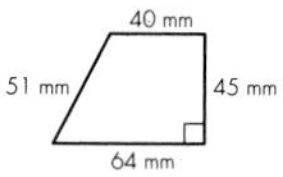

area: 2,340 mm²

perimeter: 200 mm

3.

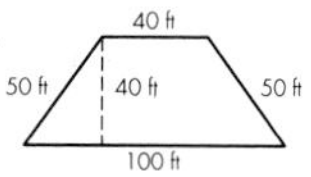

area: 2,800 ft²

perimeter: 240 ft

Find the missing measure.

4.

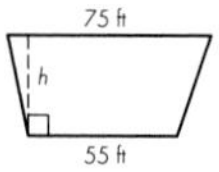

$A = 2{,}600$ ft²

$h =$ 40 ft

5.

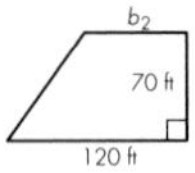

$A = 7{,}000$ ft²

$b_2 =$ 80 ft

6.

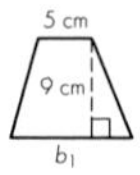

$A = 54$ cm²

$b_1 =$ 7 cm

Find the area.

7.

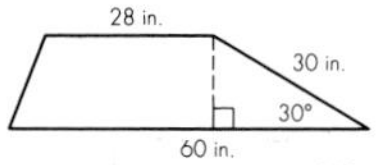

660 in.²

8.

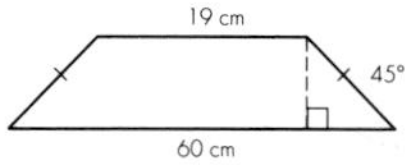

216 cm²

869.75

Name ______________________ Date ______________

The Case of the Disappearing Squares

Puzzles like the ones on this page have been in existence for many years. Your parents may have tried to solve the same puzzles when they were in school. See if you can explain the unusual results. It may be helpful to copy the patterns on graph paper, cut the patterns apart, and rearrange the pieces as shown.

1. The first square has an area of 121 square units. When the pieces are rearranged as in the second figure, two squares are missing.

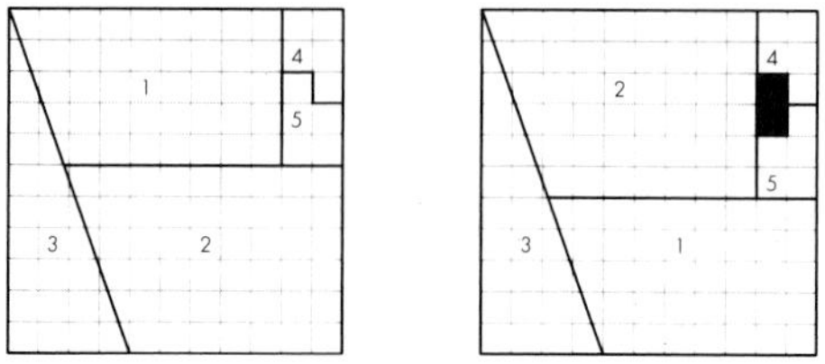

Explanation: When the pieces are rearranged, the figure looks like a square, but it is not a square. The grid lines of regions 1 and 2 do not align with those of region 3.

2. The square has an area of 64 square units. When the pieces are rearranged, the rectangle has an area of 65 square units.

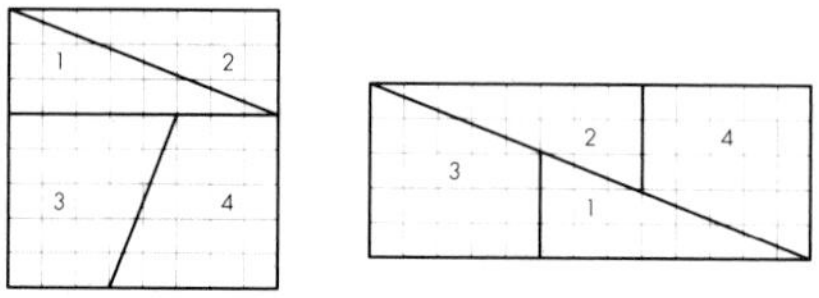

Explanation: When the pieces are rearranged, the diagonal has a slight bend in it. There is actually a slight gap with an area of 1 square unit.

Name ______________________ Date ______________

Applications

To find the area of a quadrilateral with no parallel sides, you can partition the figure into separate regions for which you can find the areas.

In quadrilateral *ABCD*, you can draw a line through point *D* that is parallel to $\overline{AB}$. The line intersects $\overline{BC}$ at a point *E*. Now you can find the areas of $\triangle DEC$ and trapezoid *ABED*. NOTE: There are other ways to partition the figure.

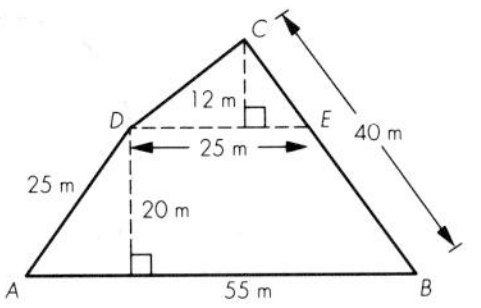

Find the area of quadrilateral *ABCD*.

Area of $\triangle DEC = \frac{1}{2} \times 25 \times 12 = 150$ Area of trapezoid $ABED = \frac{1}{2} \times 20(55 + 25) = 800$

So the total area is 150 m² + 800 m², or 950 m².

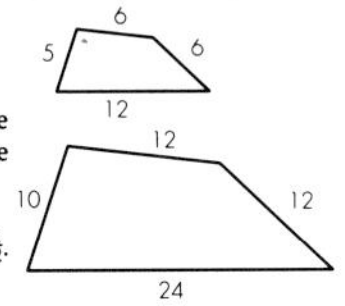

When two figures are similar, the ratios of corresponding sides are equal. If the ratio of corresponding sides is $\frac{a}{b}$, then the ratio of the perimeters of the figures is also $\frac{a}{b}$ and the ratio of the areas is $\left(\frac{a}{b}\right)^2$.

For the similar figures at the right, the ratio of corresponding sides is $\frac{1}{2}$. So the ratio of the perimeters is $\frac{1}{2}$ and the ratio of the areas is $\frac{1}{4}$.

Find the area.

1. $RT = 26$ cm, $UV = 14$ cm, $WS = 40$ cm — 702 cm²

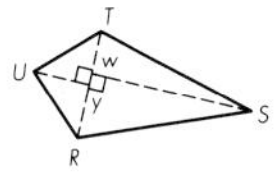

2. 1185 m²

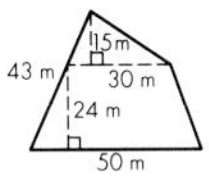

The given figures are similar. Find the ratio of the perimeters and the ratio of the areas.

3. perimeters: $\frac{2}{1}$ areas: $\frac{4}{1}$

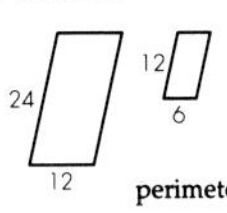

4. perimeters: $\frac{3}{2}$ areas: $\frac{9}{4}$

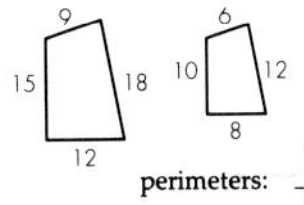

Name ______________________ Date ______________

More Proofs of the Pythagorean Theorem

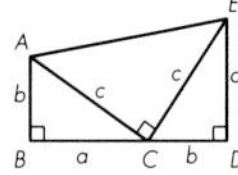

1. President James Garfield discovered a proof of the Pythagorean Theorem in 1876. At that time, he was a member of the House of Representatives. The proof involves finding the area of *ABDE* in two different ways, first using three triangles and then as a trapezoid.

   a. What is the area of $\triangle ABC$? $\frac{1}{2}ab$

   b. What is the area of $\triangle DEC$? $\frac{1}{2}ab$

   c. What is the area of $\triangle ACE$? $\frac{1}{2}c^2$

   d. Write the area of *ABDE* as the sum of the areas of three triangles. Then simplify.
   $\frac{1}{2}ab + \frac{1}{2}ab + \frac{1}{2}c^2 = ab + \frac{1}{2}c^2$

   e. Write the area of *ABDE* by thinking of *ABDE* as trapezoid where $\overline{AB}$ and $\overline{DE}$ are the parallel bases.
   $\frac{1}{2}(a + b)(a + b) = \frac{1}{2}a^2 + ab + \frac{1}{2}b^2$

   f. In the space below, use answers for **d** and **e** to show that $c^2 = a^2 + b^2$.

   $$ab + \frac{1}{2}c^2 = \frac{1}{2}a^2 + ab + \frac{1}{2}b^2$$
   $$\frac{1}{2}c^2 = \frac{1}{2}a^2 + \frac{1}{2}b^2$$
   $$c^2 = a^2 + b^2$$

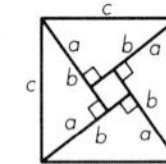

2. A Hindu mathematician named Bhâskara discovered the following proof of the Pythagorean Theorem in the twelfth century. The proof involves four congruent triangles and one square.

   a. What is the area of each triangle? $\frac{1}{2}ab$

   b. What is the length of each side of the large square? c

   c. What is the length of each side of the small square? $b - a$

   d. In the space below, write an equation for c^2, the area of the large square, in terms of the areas of the four triangles and the small square. Then simplify to show that $c^2 = a^2 + b^2$.

   $$c^2 = \frac{1}{2}ab + \frac{1}{2}ab + \frac{1}{2}ab + \frac{1}{2}ab + (b - a)^2$$
   $$= 2ab + b^2 - 2ab + a^2$$
   $$= a^2 + b^2$$

Name ______________________ Date ____________

Heron's Formula

1. Heron's Formula, $A = \sqrt{s(s-a)(s-b)(s-c)}$, can be used to calculate the area of a triangle if you only know the lengths of the three sides. However, it is easier to use the formula $A = \frac{1}{2}bh$ with a certain type of triangle, even when only the lengths of the sides are given. Name the type of triangle. right triangle

Use a calculator and the formula $A = \frac{1}{2}bh$ or Heron's Formula to calculate the area of each triangle below. Round each answer to the nearest hundredth, if necessary.

2. 24 m²

3. 101.82 in.²

4. 43.63 cm²

5. 30 ft²

6. 120 m²

7. 38.94 cm²

8. 88.61 yd²

9. 54 in.²

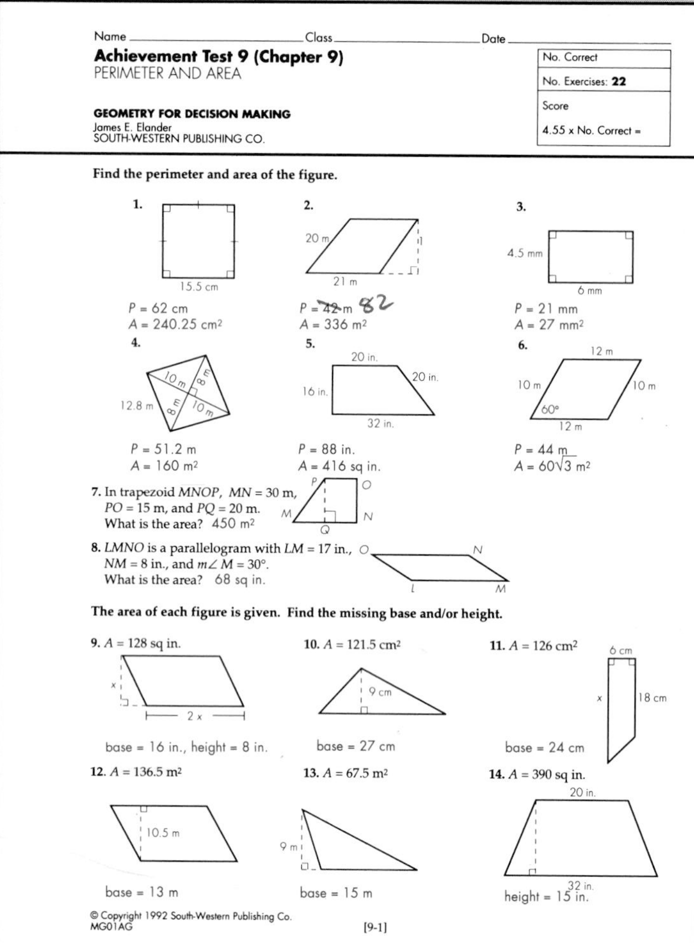

Name ____________ Class ____________ Date ____________

Achievement Test 9 (Chapter 9)
PERIMETER AND AREA

No. Correct
No. Exercises: **22**
Score
4.55 x No. Correct =

GEOMETRY FOR DECISION MAKING
James E. Elander
SOUTH-WESTERN PUBLISHING CO.

Find the perimeter and area of the figure.

1. 15.5 cm
P = 62 cm
A = 240.25 cm²

2. 20 m, 11, 21 m
P = ~~42~~ m 82
A = 336 m²

3. 4.5 mm, 6 mm
P = 21 mm
A = 27 mm²

4. 10 m, 8 m, 12.8 m
P = 51.2 m
A = 160 m²

5. 20 in., 20 in., 16 in., 32 in.
P = 88 in.
A = 416 sq in.

6. 12 m, 10 m, 10 m, 60°, 12 m
P = 44 m
A = $60\sqrt{3}$ m²

7. In trapezoid *MNOP*, *MN* = 30 m, *PO* = 15 m, and *PQ* = 20 m. What is the area? 450 m²

8. *LMNO* is a parallelogram with *LM* = 17 in., *NM* = 8 in., and $m\angle M$ = 30°. What is the area? 68 sq in.

The area of each figure is given. Find the missing base and/or height.

9. *A* = 128 sq in. (x, 2x)
base = 16 in., height = 8 in.

10. *A* = 121.5 cm² (9 cm)
base = 27 cm

11. *A* = 126 cm² (6 cm, x, 18 cm)
base = 24 cm

12. *A* = 136.5 m² (10.5 m)
base = 13 m

13. *A* = 67.5 m² (9 m)
base = 15 m

14. *A* = 390 sq in. (20 in., 32 in.)
height = 15 in.

MG01AG

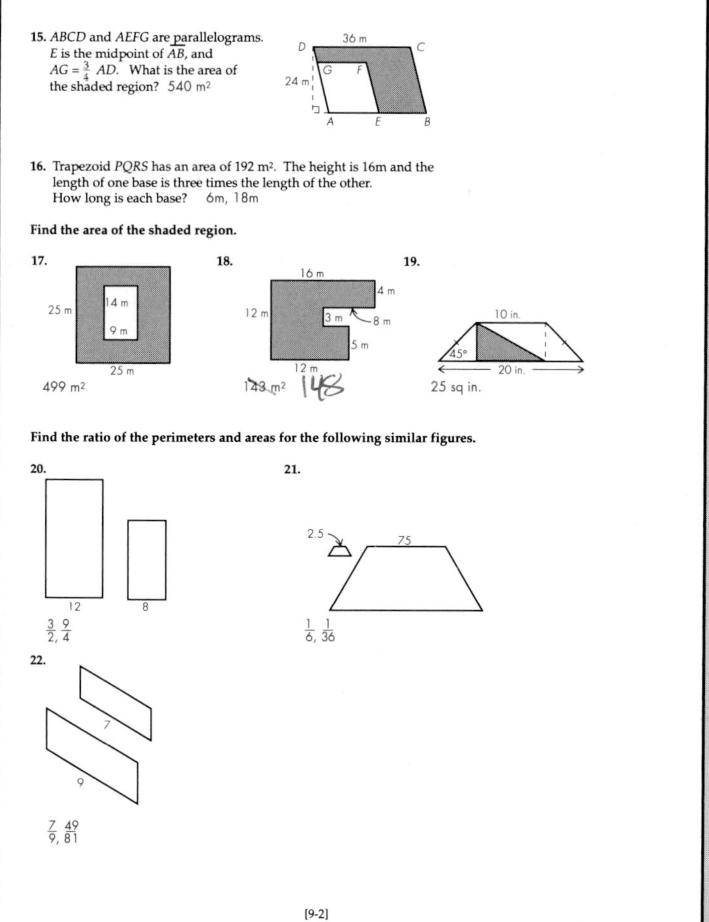

15. *ABCD* and *AEFG* are parallelograms. *E* is the midpoint of $\overline{AB}$, and $AG = \frac{3}{4}AD$. What is the area of the shaded region? 540 m²

16. Trapezoid *PQRS* has an area of 192 m². The height is 16m and the length of one base is three times the length of the other. How long is each base? 6m, 18m

Find the area of the shaded region.

17. 25 m, 14 m, 9 m, 25 m
499 m²

18. 16 m, 4 m, 12 m, 3 m, 8 m, 5 m, 12 m
~~143~~ m² 148

19. 10 in., 45°, 20 in.
25 sq in.

Find the ratio of the perimeters and areas for the following similar figures.

20. 12, 8
$\frac{3}{2}, \frac{9}{4}$

21. 2.5, 75
$\frac{1}{6}, \frac{1}{36}$

22. 7, 9
$\frac{7}{9}, \frac{49}{81}$

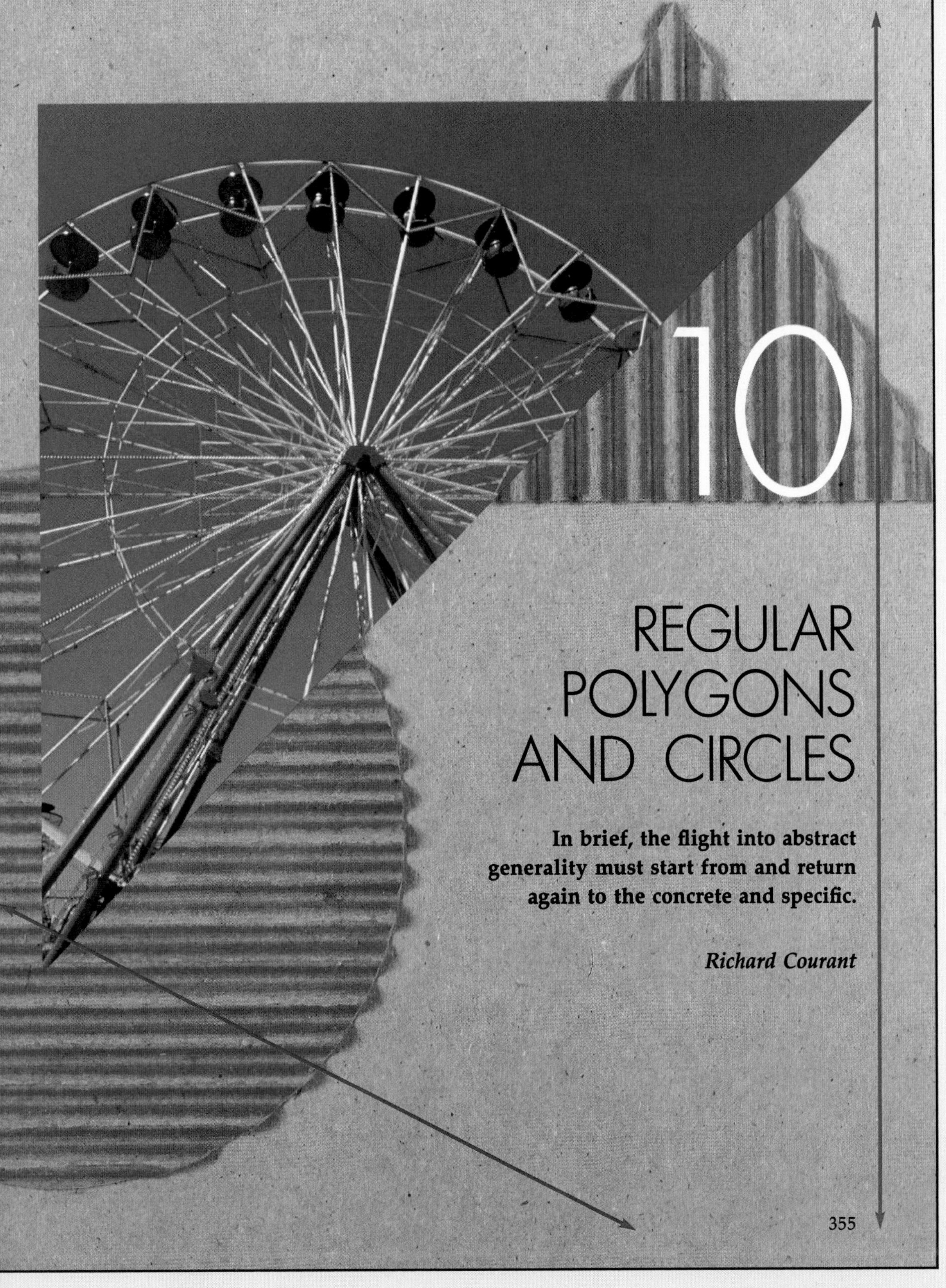

10

REGULAR POLYGONS AND CIRCLES

In brief, the flight into abstract generality must start from and return again to the concrete and specific.

Richard Courant

355

CHAPTER 10 OVERVIEW

Regular Polygons and Circles

Suggested Time: 12–14 days

In Chapters 8 and 9, students learned to make generalizations about the properties of quadrilaterals and came to understand how to find area and perimeter. In this chapter, this knowledge is extended to finding the area of regular polygons and to finding the circumference and area of a circle. Students learn that, as the number of sides of a regular polygon increases, the perimeter and area of the polygon begin to approach the circumference and area of a circle. By extension, they find that they can use the properties of angles of polygons to solve problems involving area of circles.

In brief, the flight into abstract generality must start from and return again to the concrete and specific.

Richard Courant

CHAPTER 10 OBJECTIVE MATRIX

	End of Chapter Items by Activity				Student Supplement Items by Activity		
Objectives by Number	**Review**	**Test**	**Computer**	**Algebra Skills***	**Reteaching**	**Enrichment**	**Computer**
10.1.1	✓	✓	✓		✓	✓	
10.1.2	✓	✓	✓	✓			
10.1.3				✓			
10.2.1	✓			✓	✓	✓	
10.3.1	✓	✓	✓				
10.3.2	✓	✓		✓			
10.3.3	✓	✓		✓	✓	✓	
10.4.1	✓	✓			✓		
10.4.2	✓	✓		✓	✓		
10.4.3	✓	✓				✓	✓
10.5.1	✓	✓			✓		
10.5.2	✓	✓		✓			
10.6.1	✓	✓		✓	✓	✓	

*A ✓ beside a Chapter Objective under Algebra Skills indicates that algebra skills taught within the chapter or in previous Algebra Skills lessons are used.

CHAPTER 10 PERSPECTIVES

▲ Section 10.1

Regular Polygons

In the Class Activity, students extend their knowledge of geometric figures by drawing and naming polygons with increasing numbers of sides. The term *regular polygon* is defined. The Discovery Activity helps the class conclude that, since the diagonals of a polygon divide it into triangles, it is possible to calculate the measures of the interior angles if the number of sides (and thus the number of interior triangles) is known. Theorem 10-1-1 provides a formula for this. If students have difficulty with these concepts, have them cut out various polygons, draw the diagonals from one vertex, and cut the polygons along the diagonals. Then have them count the number of triangles and multiply that number by 180°.

10

▲ Section 10.2

Area of Regular Polygons

In the Discovery Activity, students divide a pentagon into triangles using the angle bisectors and then use the additive property of area to determine their own formula for finding the area of a pentagon. In finding a general formula, the relationship between the perimeter of a pentagon and the bases of its interior triangles becomes important. In the Class and Home Activities, students are called upon to use all of the properties of triangles they have learned—the Pythagorean Theorem, properties of 30°-60° and 45° right triangles—in order to find the areas of regular polygons. Have students especially investigate regular octagons and hexagons, stressing the way these figures can be related to equilateral, 30°-60° right triangles (hexagons) and 45°-right triangles (octagons). Students having difficulty with the concepts can fold or cut out octagons and hexagons to verify the concepts.

▲ Section 10.3

Circumference and Area of a Circle

Several important terms associated with circles are defined for students. The first Discovery Activity enables students to find out for themselves that the relationship between the diameter of a circle and its circumference is a constant—the irrational number π. In the second Discovery Activity, students find the area of a circumscribed hexagon in an attempt to approximate the area of a circle. They see that the more sides a polygon has, the closer its perimeter comes to being the circumference of a circle with a radius equal to the height of the polygon. Since the circumference is related to the radius by the constant π, it should be no surprise to students to find that the area is also related to this constant. Students may prepare a report on one of several famous mathematicians of antiquity, with special emphasis on the history of π. You may wish to ask interested students to do research on modern uses of π.

▲ Section 10.4

Angles and Circles

Several new terms associated with circles are introduced: *chord*, *secant*, *tangent*, *arc*, and *central angle*. The study of angles and their special relationships to circles is approached through the Discovery Activity, in which students explore the various ways in which a pair of intersecting lines can intersect a circle. As each case is explored, one particular property of the angles of a triangle is used repeatedly: the measure of an exterior angle of a triangle is equal to the sum of the measures of the two remote interior angles. For students having difficulty with these new concepts, review the Exterior Angle Theorem and have them calculate the measure of the exterior angles of various triangles given the measures of the angles of the triangles. The proofs of the theorems in this section lead students to understand how the measures of inscribed, interior, and exterior angles and the arcs they intersect on a circle are related.

▲ Section 10.5

Chords and Secants

The emphasis is on the relationship between pairs of line segments that intersect circles. In the first Discovery Activity, students find that the products of the segment lengths of two intersecting chords are equal. This theorem and Theorem 10-5-2, which states that the perpendicular bisector of a chord passes through the center of the circle, provide students with two important methods for solving problems involving circles. The second Discovery Activity helps students find the similar relationship between the square of the length of a tangent and the product of the length of a secant and its external segment. On their own, students are asked to justify similar statements about the lengths of two secants and of two tangents. For students having difficulty with the proofs, stress its application. Internalization of the concept by application thereof will often make the proof more acceptable to them.

▲ Section 10.6

Applications Involving Circles

The connection between polygons and circles is developed by having students inscribe several polygons in a circle. The link between the diameter and radius of a circle and the diagonals and height of a regular polygon provides the basis for solving problems involving both types of figures. Properties of triangles, properties of quadrilaterals, and formulas for area and perimeter are now all available for application.

SECTION 10.1

Resources
Reteaching 10.1
Enrichment 10.1

Objectives
After completing this section, the student should understand
- ▲ the definition of a regular polygon.
- ▲ how to find the measure of the interior angles of a polygon.
- ▲ how to find the perimeter of a polygon.

Materials
protractor

Vocabulary
polygon
regular polygon

GETTING STARTED

Quiz: Prerequisite Skills
Draw a triangle *ABC*.

1. Identify an altitude and a base.

2. Is there a diagonal? [No]

3. Find the sum of the measures of the interior angles. [180°]

Draw a parallelogram *ABCD*.

4. Identify two pairs of bases and heights.

5. Name the diagonals. [*AC* and *BD*]

6. Find the sum of the measures of the interior angles. [360°]

Chalkboard Activity
Draw this figure on the chalkboard. Have students determine how many triangles, parallelograms, and trapezoids are in the figure.

[13 triangles, 13 parallelograms, 18 trapezoids]

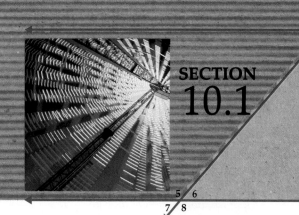

SECTION 10.1 Regular Polygons

After completing this section, you should understand
- ▲ the definition of a regular polygon.
- ▲ how to find the measure of the interior angles of a polygon.
- ▲ how to find the perimeter of a polygon.

In Chapters 8 and 9 you investigated the properties of polygons with three and four sides—triangles and quadrilaterals. A polygon can have any number of sides, it can be convex or concave, and it can be regular or nonregular.

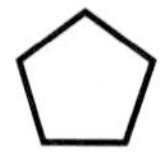
convex, regular

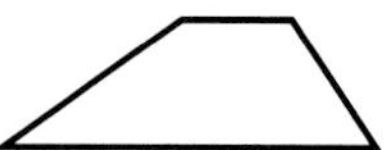
convex, nonregular

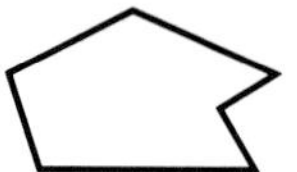
concave, nonregular

You can classify a polygon by the number of its sides.

CLASS ACTIVITY

The number of sides of a polygon is given under *n*. Match each figure with its name.

	n	Name	
1.	3	Triangle	e
2.	4	Quadrilateral	a
3.	5	Pentagon	c
4.	6	Hexagon	g
5.	7	Heptagon	d
6.	8	Octagon	h
7.	9	Nonagon	f
8.	10	Decagon	b

a. b. c. d. e. f. g. h.

In general, the term *n*-gon is used to describe a polygon with *n* sides.

DEFINITION 10-1-1 **A POLYGON is a closed plane figure consisting of line segments.**

Recall that the points making up a polygon are coplanar, and that the line segments (or sides) determined by these points do not intersect.

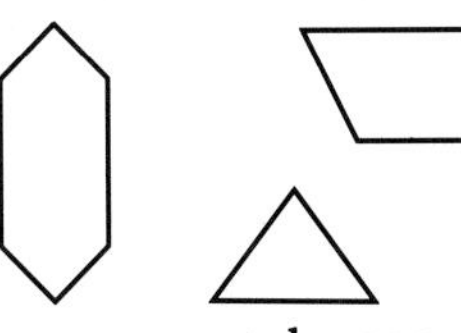

polygons

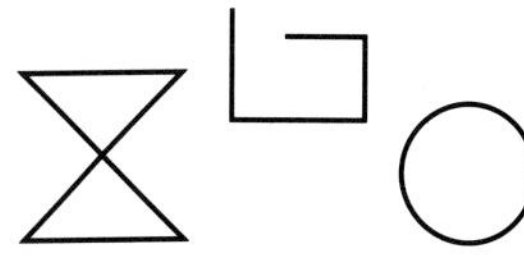

not polygons

You have already worked with several kinds of polygons that have sides of equal length and angles of equal measure.

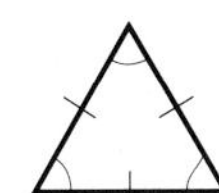

equilateral triangle

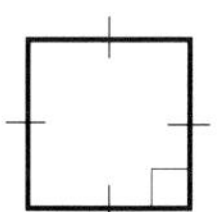

square

DEFINITION 10-1-2 **A REGULAR POLYGON is a polygon that is equilateral and equiangular.**

Example:

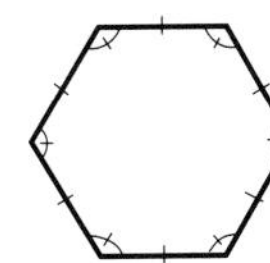

regular hexagon

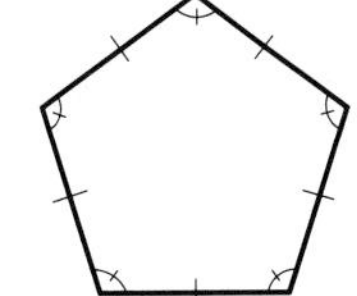

regular pentagon

CLASS ACTIVITY

Identify each polygon by the number of sides. Tell whether it is convex or concave, regular or nonregular.

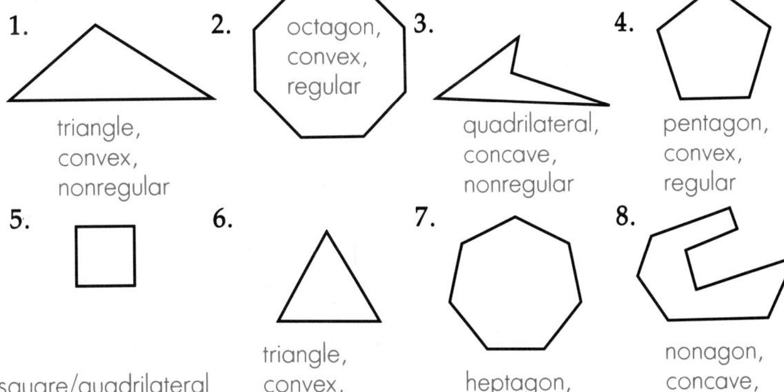

TEACHING COMMENTS

Using the Page
During the discussion of Definition 10-1-1, be sure students understand the line segments that make up a polygon intersect only at the vertices of the polygon. Then ask them to explain how the definition of a polygon differs from that of a quadrilateral. [Quadrilateral has four sides and four vertices; polygon can have any number.]

Using the Page
During the discussion of Definition 10-1-2, ask students to name the equilateral and equiangular three-sided and four-sided polygons. [equilateral triangle, square]

DISCOVERY ACTIVITY

1. Trace this regular pentagon. Notice that all of the sides are of equal length and all of the angles have equal measures.

2. Draw all the diagonals from point A. How many triangles did you form? 3

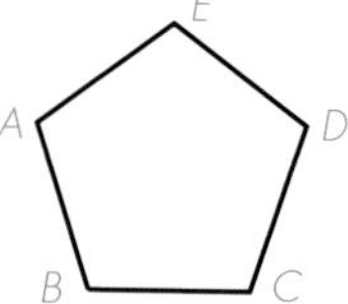

3. What is the sum of the measures of the angles of the triangles you formed? $3 \cdot 180° = 540°$

4. Do you think that the sum of the measures of the angles in the triangles is equal to the sum of the measures of the angles of the pentagon? Why? See margin.

5. How can you find the measure of each angle of the pentagon? How do you know? $540° \div 5 = 108°$; all angles have equal measure.

6. Repeat steps 1–5 for a regular hexagon. What measure do you get for each angle of the hexagon? 120°

You may have discovered that it is possible to calculate the measures of the interior angles of a regular polygon if you know the number of sides.

For a regular pentagon, $n = 5$ Sum of angle measures =	For a regular hexagon, $n = 6$ Sum of angle measures =
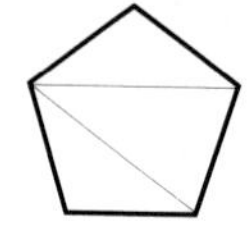 $3 \cdot 180° = 540°$	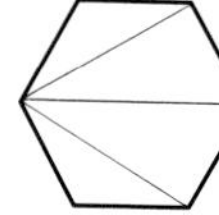 $4 \cdot 180° = 720°$
Measure of each angle =	Measure of each angle =
$\frac{3 \cdot 180°}{5} = \frac{540°}{5} = 108°$	$\frac{4 \cdot 180°}{6} = \frac{720°}{6} = 120°$

Or in general, for a regular polygon with n sides, the measure of each angle of the polygon is $\frac{(n-2)180°}{n}$. See margin.

THEOREM 10-1-1 **Each angle of a regular n-gon is equal to $(n - 2)$ times 180° divided by n, where n is the number of sides in the n-gon.**

Using the Page
During the Discovery Activity, observe as students draw the diagonals in each figure. Help them to understand that they are trying to find the least number of triangles that make up the polygon. They should see that the easiest way to draw the correct number of diagonals is to draw them from only one vertex of the polygon.

Additional Answers
4. Yes; $m\angle A = m\angle EAD + m\angle DAC + m\angle CAB$; $m\angle C = m\angle BCA + m\angle ACD$; $m\angle D = m\angle CDA + m\angle ADE$; $m\angle B = m\angle B$; $m\angle E = m\angle E$

Using the Page
Students can discover the formula $\frac{(n-2)180°}{n}$ with the help of a geometric drawing tool. First, have them draw several different regular polygons. For each polygon, have students find the measure of one angle. Then ask them to generalize by writing an expression for the measure of each angle of a regular polygon with n sides.

Using the Page
In the discussion of Theorem 10-1-1, be sure the class understands how the theorem can be used to find both the sum of all the interior angles of the polygon and the measure of an individual interior angle. Then ask students to find the least number of degrees possible for the measure of an interior angle of a regular polygon, and have them explain their answers. [60°]

Example: Find the measure of an interior angle of a regular 15-gon.

$$\frac{(n-2)180°}{n} = \frac{(15-2)180°}{15} = \frac{13 \cdot 180°}{15} = 156°$$

Example: What is the sum of the angle measures of a 30-gon?

$$(n-2)180° = (30-2)180° = 28 \cdot 180° = 5040°$$

CLASS ACTIVITY

Find the sum of the angle measures for each polygon.

1. Hexagon 720°
2. 12-gon 1800°
3. 20-gon 3240°
4. 42-gon 7200°

Find the measure of an interior angle and an exterior angle for each regular polygon.

5. Heptagon 128.6°; 51.4°
6. 18-gon 160°; 20°
7. Nonagon 140°; 40°
8. 24-gon 165°; 25°
9. The measure of an interior angle of a regular polygon is 156°. How many sides does the polygon have? 15

PERIMETER OF A POLYGON

In Chapter 9 you learned that the perimeter of any quadrilateral is the distance around it or the sum of the lengths of the sides.

The same is true for any polygon.

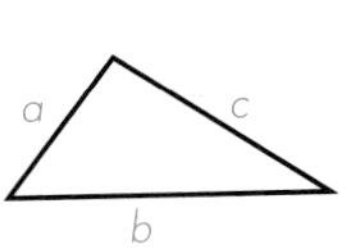

$P = a + b + c$

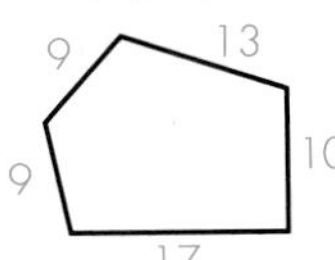

$P = 58$

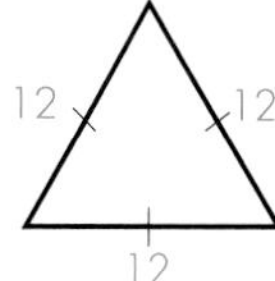

$P = 36$

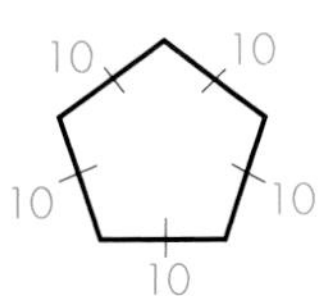

$P = 50$

CLASS ACTIVITY

Use a calculator to find the perimeter.

1. 8 cm, 12 cm — 40 cm
2. 8 yd — 64 yd
3. 5 in., 8 in. — 23 in.
4. 14 ft — 70 ft
5. Use a calculator to find the perimeter of a regular 15-gon with a side measuring 13 meters. 195 m
6. The perimeter of a regular decagon is 325.8 ft. Use a calculator to find the measure of one side. 32.58 ft

Using the Page
After students have completed the Class Activity, you may wish to review the definition of an exterior angle of a triangle, and have students extend this definition to include the exterior angle of a polygon.

Extra Practice
Find the sum of the angle measures for each polygon.

1. Pentagon [540°] **2.** 16-gon [2520°]

Find the measure of an interior angle for each regular polygon.

3. 11-gon [147.27°] **4.** 22-gon [163.63°]

5. The measure of an interior angle of a regular polygon is 162°. How many sides does the polygon contain? [20]

Find the measure of an exterior angle for each regular polygon.

6. Hexagon [60°] **7.** Nonagon [40°]

Find the perimeter of each regular polygon with the given side.

7. 15-gon, $s = 12$ cm [$P = 180$ cm]

8. 20-gon, $s = 11$ in. [$P = 220$ in.]

9. Pentagon, $s = 3.6$ m [$P = 18$ m]

10. The perimeter of a regular 12-gon is 50.4 cm. What is the length of one side? [4.2 cm]

HOME ACTIVITY

Identify each polygon by the number of its sides. Tell whether the polygon is convex or concave, regular or nonregular.

1. 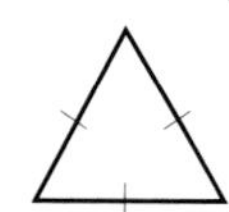
triangle, convex, regular

2. 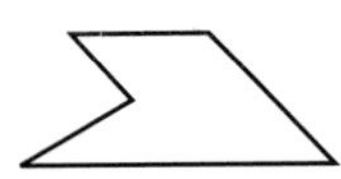
pentagon, concave, nonregular

3. 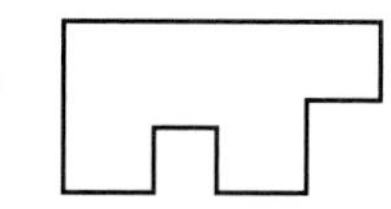
decagon, concave, nonregular

4. 
hexagon, convex, regular

Find the sum of the angle measures for each polygon.

5. Square
360°
6. Triangle
180°
7. 12-gon
1800°
8. 100-gon
17,640°
9. Octagon
1080°

Write a LOGO procedure to tell the turtle to draw each regular polygon.
Answers may vary.

10. Pentagon
11. Hexagon
12. Octagon

Find the measure of an exterior angle for each regular polygon.

13. Square
90°
14. Hexagon
60°
15. 28-gon
12.9°
16. Nonagon
40°

The measure of an angle of a regular polygon is given. Find the number of sides.

17. 144°
$n = 10$
18. 156°
$n = 15$
19. 165.6°
$n = 25$

20. Find the perimeter of polygon *ABCD*. 106 ft
21. What is the are of ∠BDC? 180 sq ft
22. What is the area of *ABCD*? 564 sq ft

9 ft C
D
32 ft
A 24 ft B

CRITICAL THINKING

Suppose you could program the LOGO turtle to move one foot and then turn right one second, repeating the routine until the turtle arrived at its starting point.

23. What regular polygon would it draw? 1,296,000-gon
24. How many miles would it travel to form the polygon? 245.5 mi

Using the Page
Encourage students to check their work in the Home Activity with a calculator, if they have one available.

FOLLOW-UP

Assessment
1. A regular polygon has an interior angle of 135°. How many sides does it have? [8]
2. What is the sum of the angle measures of a polygon with 18 sides? [2880°]
3. The perimeter of a regular 15-gon is 342 m. What is the length of one side? [22.8 m]

Reteaching
Students who have had difficulty with this section may benefit from Reteaching Activity 10.1.

Enrichment
For students who have mastered the material in this section, you may wish to assign Enrichment Activity 10.1.

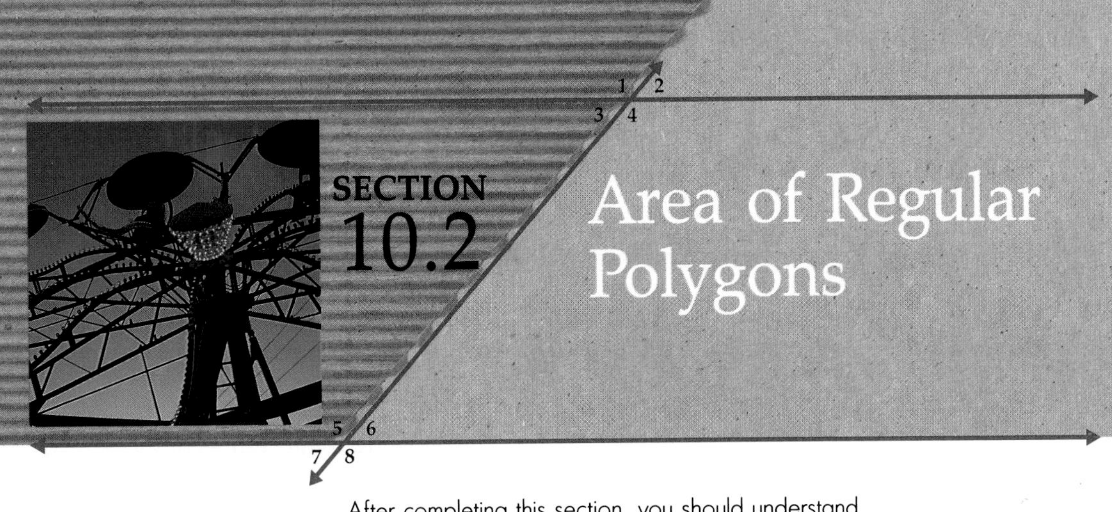

Area of Regular Polygons

After completing this section, you should understand
▲ how to find the area of a regular polygon.

How could you find the area of the side of the Ferris wheel shown in the photograph? What geometric figures does it contain? In the last chapter, you investigated finding the area of quadrilaterals with no parallel sides and no specific area formula. To find the area you could

1. count the number of square units of measure contained in the figure; or

2. divide the figure into triangles, parallelograms, or trapezoids, find those areas, and combine them.

For many-sided polygons such as the Ferris wheel, these methods may be inexact or impractical.

DISCOVERY ACTIVITY

1. Trace this regular pentagon.
2. Use a protractor or compass to construct the bisector of each angle. Extend the bisectors until they intersect in the center of the pentagon. Label the point of intersection O.
3. You have divided the pentagon into five triangles. Are the triangles congruent? How do you know?
4. How could you write a formula for finding the area of a regular pentagon? $5 \cdot \frac{1}{2}bh$

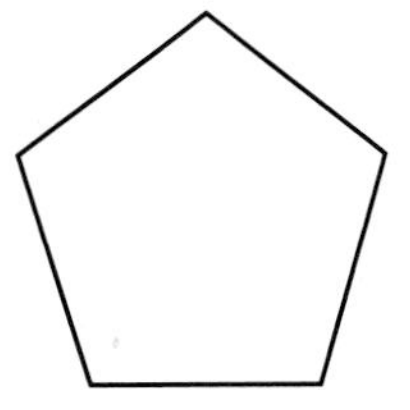

Yes; by AA and ratio of corresponding sides = 1.

361

SECTION 10.2

Resources
Reteaching 10.2
Enrichment 10.2
Transparency Master 10-1

Objectives
After completing this section, the student should understand
▲ how to find the area of a regular polygon.

Materials
protractor
compass
ruler

GETTING STARTED

Quiz: Prerequisite Skills
Find the area.

1.

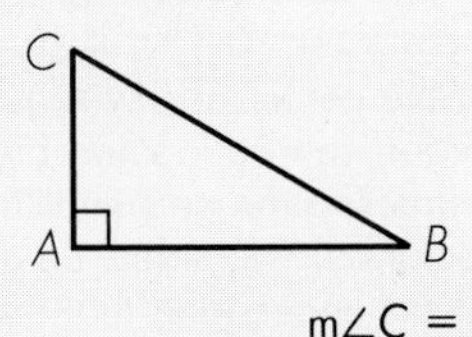

m∠C = 60°
BC = 30 in.
[$A = 225\sqrt{3}$ sq in.]

2.

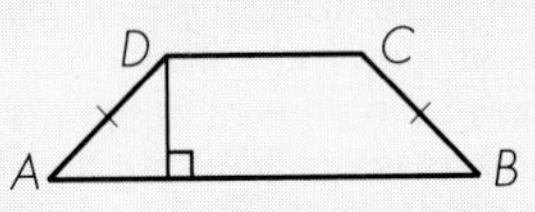

AB = 18 cm
DC = 12 cm
m∠A = 45°
[$A = 45$ cm^2]

3.

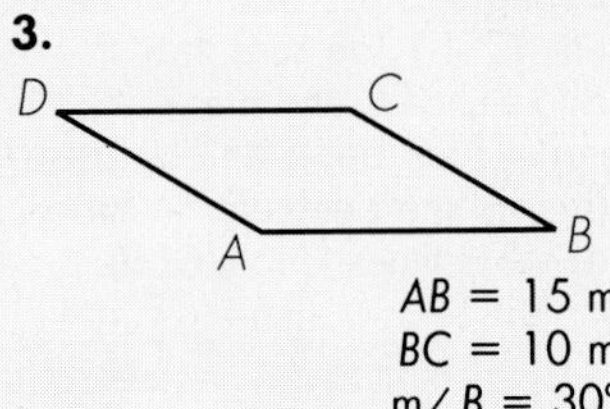

AB = 15 m
BC = 10 m
m∠B = 30°
[$A = 75$ m^2]

5. Trace this regular hexagon and repeat steps 2 and 3. How could you write a formula for finding the area for a regular hexagon?
$6 \cdot \frac{1}{2}bh$

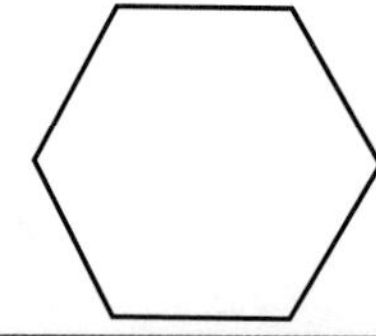

You may have discovered that the area of a regular pentagon or hexagon is equal to the sum of the areas of the congruent triangles contained in them.

When you drew the angle bisectors from the vertices of the hexagon, congruent triangles were formed. Can you explain why the six triangles are congruent? Where do the bisectors intersect?
SAS; center of hexagon

The line segments connecting the vertices of a regular polygon with the center bisect the angles and form congruent triangles. You can see that the area of a regular polygon is a multiple of the area of one of these triangles.

The regular heptagon has perpendicular bisectors of the seven sides. Note that the bisectors intersect at point O. If you put the point of a compass on point O and the pencil on a vertex of the heptagon and draw a circle, the vertices of the heptagon will all be on the circle.

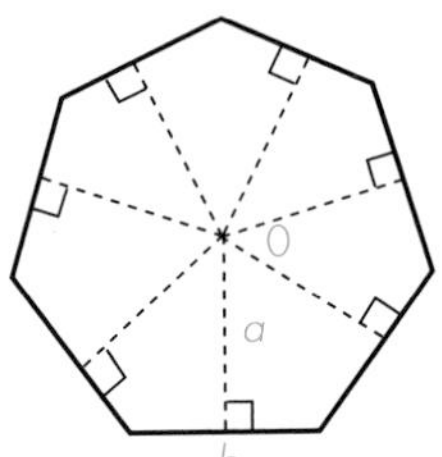

The center of any regular polygon can be located by constructing the perpendicular bisectors of two sides. The point where the bisectors intersect is the center. The length of the bisector, from a base to the center, is the *apothem* of the polygon.

For a regular heptagon, $A = 7 \cdot \frac{1}{2}ba$. You can express the area of the heptagon in terms of its perimeter.

$P = 7b$

$A = \frac{1}{2}(7ba)$

$A = \frac{1}{2}Pa$ (by substitution)

THEOREM 10-2-1 **The area of a regular polygon is $\frac{1}{2}Pa$, where P is the perimeter of the polygon and a is the length of the apothem.**

Example: What is the area of a regular pentagon with a side of 4 meters and an apothem of 2.75 meters?

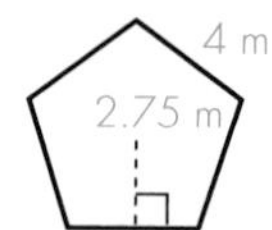

$P = 5 \times 4 = 20$ m, $a = 2.75$ m $\quad A = \frac{1}{2}Pa = \frac{1}{2} \times 20 \times 2.75 = 27.5\text{ m}^2$

TEACHING COMMENTS

Using the Page
You may wish to have students trace the heptagon shown on the page and have them construct the perpendicular bisectors of two of the sides. Then have them use a compass to draw the circle that contains the vertices of the heptagon.

Using the Page
You may wish to use Transparency Master 10-1 to illustrate how the center of any regular polygon can be located by constructing the perpendicular bisectors of two sides. Be sure students understand that the height of the polygon is the length of the perpendicular bisector, and not the length of the angle bisector. Some students may find it easier to think of the height of the polygon as the altitude of one of the triangles formed by the angle bisectors.

Using the Page
During the discussion of Theorem 10-2-1, be sure that students understand the connection between one base (b) and the perimeter (number of sides times length of b).

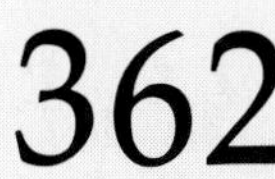

Example: What is the area of this regular polygon?

Since the polygon is a hexagon, $m\angle DEF$ is 120° and $m\angle DEO$ is 60°.

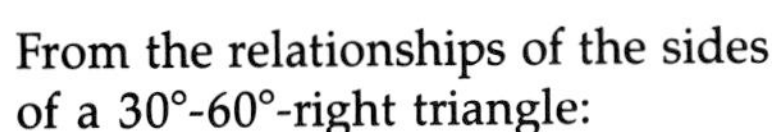

From the relationships of the sides of a 30°-60°-right triangle:

$DE = 5$ in., $OD = 5\sqrt{3}$ in. and
$CD = 10$ in.
$P = 6 \cdot 10 = 60$

$$A = \frac{1}{2}Pa = \frac{1}{2} \cdot 60 \cdot 5\sqrt{3} = 150\sqrt{3} \text{ sq in.}$$

Example: What is the area of a regular hexagon with a side of 8 centimeters?

Since a side of the hexagon is 8 cm, you can find the perimeter: $6 \cdot 8 = 48$ cm.

$m\angle ABC = 120°$, so $m\angle ABO = 60°$.

$AB = 4$ cm, so $AO = 4\sqrt{3}$ cm (height).

$$A = \frac{1}{2}Pa = \frac{1}{2} \cdot 48 \cdot 4\sqrt{3} = 96\sqrt{3} \text{ cm}^2$$

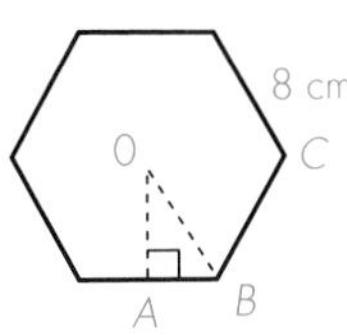

CLASS ACTIVITY

Use a calculator to find the area of each regular polygon.

1.

$A = 120 \text{ cm}^2$

2.
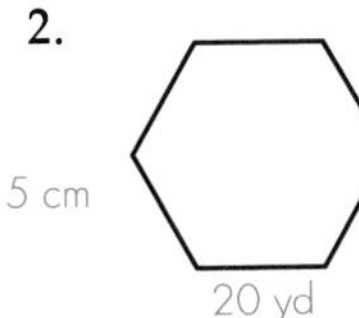

$A = 600\sqrt{3}$ sq yd

3.
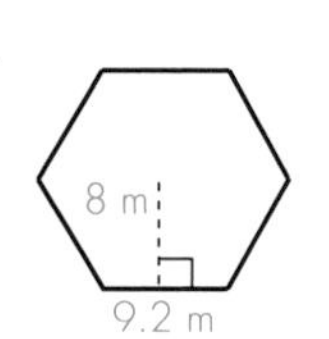

$A = 220.8 \text{ m}^2$

4.
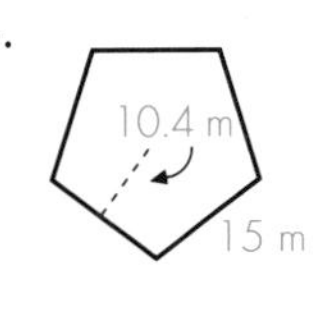

$A = 390 \text{ m}^2$

Use a calculator to solve the problems.

5. The top of a circus tent is made up of a regular hexagon with sides of 50 feet. What is its area? $3750\sqrt{3}$ sq ft

6. A regular polygon has three sides, each measuring 10 feet. The apothem is 8.7 feet. What is the area? 43.5 sq ft

7. Find the area of a regular hexagon with a side of 10 meters and an apothem of 8.7 meters. 261 m^2

Using the Page

You may wish to use Transparency Master 10-1 to illustrate the examples for the class. Discuss the different ways students can use what they know about the properties of 30°-60° and 45° right triangles and equilateral triangles to help them find missing information in problems. Ask students to explain how they could find the base and height of a regular hexagon for which they know only the length of an angle bisector. [Use properties of 30°-60° right triangle.]

Extra Practice

Use a calculator to find the area of each regular polygon.

1.
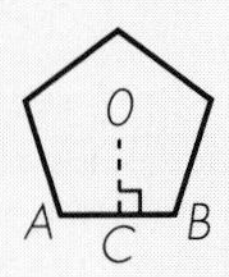

$OC = 3.75$ cm
$AB = 6$ cm
$[A = 56.25 \text{ cm}^2]$

2.
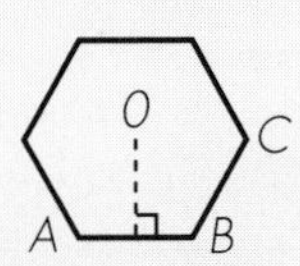

$AB = 32$ mm
$m\angle ABC = 120°$
$[A = 1536\sqrt{3} \text{ mm}^2]$

3.
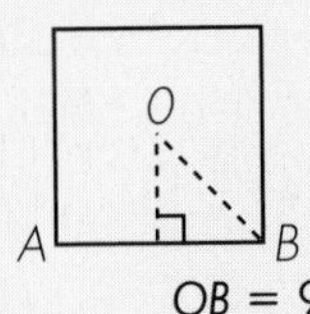

$OB = 9\sqrt{2}$ in.
[A = 324 sq in.]

4.
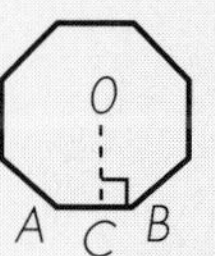

$OC = 10.8$ ft
$AB = 9$ ft
[A = 388.8 sq ft]

Show that the angle bisectors of a regular hexagon divide the hexagon into six equilateral triangles.

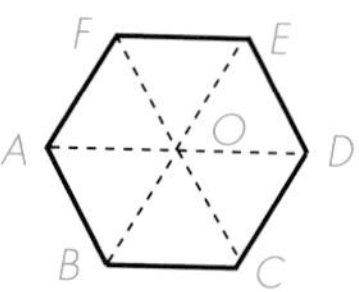

8. Find the measure of an angle of a regular hexagon. 120°

9. What is the measure of $\angle OAB$? of $\angle OBA$? Why?
Both 60°; angle was bisected.

10. What is the measure of each of the six angles around point O? How do you know? 60°; 360° ÷ 6 = 60°

11. Are the six triangles congruent? Why?
Yes; AA or SAS and $\frac{\text{base}}{\text{base}} = 1$.

12. Are the six triangles equilateral? Why? Yes; each angle of a triangle measures 60°, so all sides are congruent.

HOME ACTIVITY

Using the Page
Encourage students to check their work in the Home Activity with a calculator, if they have one available.

Find the area of each regular polygon. See margin.

1. $8\sqrt{3}$ cm; 16 cm

$64\sqrt{3}$ cm²

2. 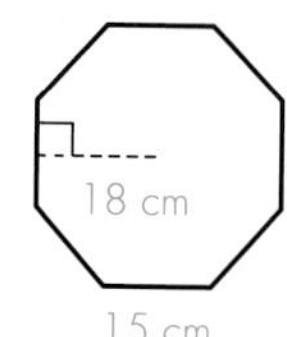

1080 cm²

3. 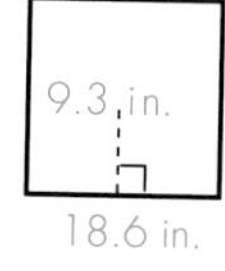

345.96 sq in.

4. 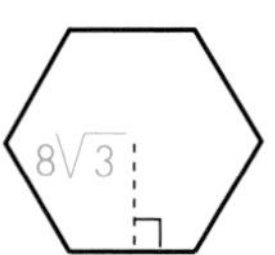

$384\sqrt{3}$ cm²

5. 

110 sq in.

6. 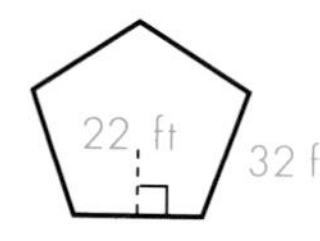

1760 sq ft

7. Copy and complete the table.

Figure	Number of Sides	Length of Side	Apothem	Area
Hexagon	6	1 m	0.87 m	2.61 m
Heptagon	7	1 m	1.04 m	3.64 m
Octagon	8	1 m	1.21 m	4.84 m
Nonagon	9	1 m	1.37 m	6.17 m
Decagon	10	1 m	1.54 m	7.70 m

8. What do you observe about the areas of the polygons as the number of sides increases?
Area also increases.
9. What is the perimeter of each polygon in Exercise 7? 6 m, 7 m, 8 m, 9 m, 10 m

One of the most famous buildings in the United States is the Pentagon in Arlington, Virginia, where the Defense Department is located. This tremendous office complex has 17.5 miles of corridors. Each side of the building is 921 feet long, and the apothem of the pentagon is 633.8 feet.

10. What is the perimeter of the Pentagon?
4605 ft
11. What is the area (in square feet) of the Pentagon? What is the area in square yards?
about 1,459,000 sq ft; about 162,000 sq yd
12. How many square yards are there in a football field? about 5500 sq yd
13. How many football fields can fit into the area of the Pentagon? about 30

A band shell is to be built on a 40-foot by 50-foot piece of land in an amusement park. The shell will have either 10 sides, each 10 feet long, or 12 sides, each 12 feet long, depending on how it fits on the piece of land.

14. What is the area of a 10-sided band shell with 10-foot sides and a height of 15.4 feet? 770 sq ft
15. What is the area of a 12-sided band shell with 12-foot sides and a height of 22.4 feet?
about 1600 sq ft
16. Which shell will fit on a 40-foot by 50-foot piece of land? 10-sided shell

17. Write a LOGO procedure using the procedure XYPLANE from page 118 and the command SETXY to plot and connect the points $A(2, 0)$, $B(4, 0)$, $C(6, 2)$, $D(4, 4)$, $E(2, 4)$, and $F(0, 2)$. Show the output on a piece of graph paper. What is the name of the polygon that is drawn? Is the polygon regular? Explain.
Hexagon; no; the sides are not equal in length.

Use a ruler, compass, and protractor to do the following construction.

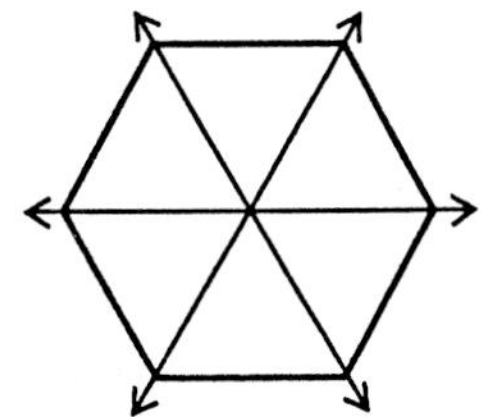

18. Construct a regular hexagon. Draw a circle and mark the center.
Since a hexagon has 6 sides, divide 360° by 6.
Draw 6 angles, each 60°, with the vertices around the point at the center of the circle.
Connect the points where the rays of the angles intersect the circle.

Extra Practice
Use a calculator to find the area of the shaded region of each regular polygon.

1.

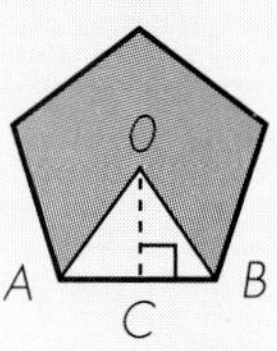

$OC = 24$ cm
$OA = 30$ cm
[$A = 1728$ cm^2]

2.

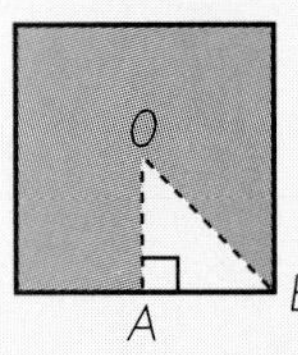

$OB = 3\sqrt{2}$ cm
[$A = 31.5$ cm^2]

3.

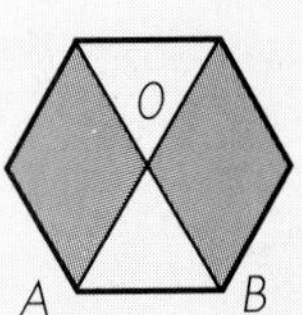

$AB = 15$ m
[$A = 225\sqrt{3}$ m^2]

19. Construct a regular nonagon. Follow the steps above. What angle measure did you use? $360° \div 9 = 40°$

20. Construct a regular octagon. What angle measure did you use? 45°

21. Construct a regular pentagon. What angle measure did you use? 72°

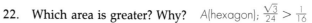

CRITICAL THINKING

Compare the areas of a square and a hexagon with the same perimeter. Give reasons for the following steps.

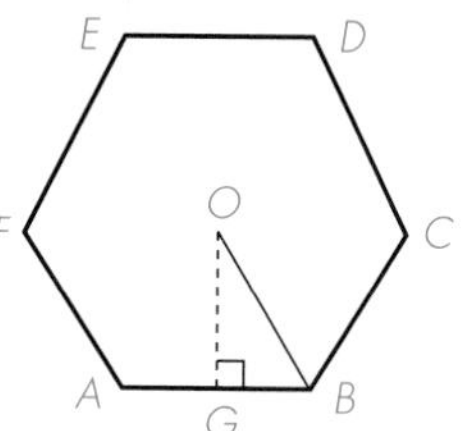

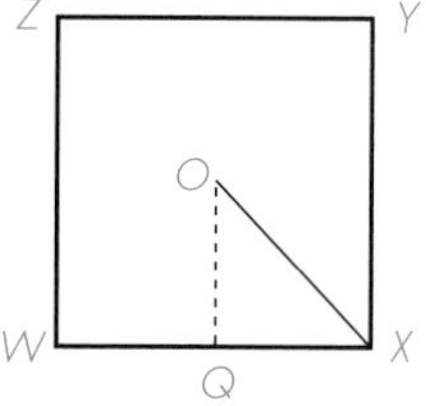

Hexagon	Square
$P(ABCDEF) = P$	$P(WXYZ) = 4P$
$AB = \frac{P}{6}$ Why? Side is $\frac{1}{6}P$	$WX = \frac{P}{4}$ Why? Side is $\frac{1}{4}P$.
$OG = \frac{P\sqrt{3}}{12}$	$OQ = \frac{P}{8}$
$A = \frac{1}{2}Ph$	$A = \frac{1}{2}Ph$
$A = \frac{1}{2}\left(P\frac{\sqrt{3}}{12}\right) \cdot P$	$A = \frac{1}{2}\left(\frac{P}{8}\right) \cdot P$
$A = \frac{\sqrt{3}}{24} \cdot P \cdot P$	$A = \frac{1}{16} \cdot P \cdot P$

22. Which area is greater? Why? A(hexagon); $\frac{\sqrt{3}}{24} > \frac{1}{16}$

Additional Answers

22. $OG = \frac{s\sqrt{3}}{2}$ because $\triangle OGB$ is a 30°-60° right triangle. $OQ = \frac{s}{2}$ because $\triangle OQX$ is a 45° right triangle.

FOLLOW-UP

Assessment
Find the area of each regular polygon.

1.

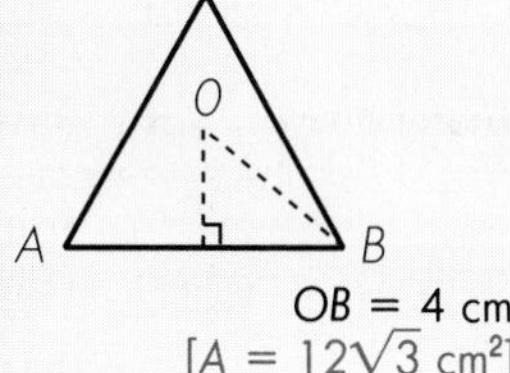

$OB = 4$ cm
$[A = 12\sqrt{3}\ \text{cm}^2]$

2.

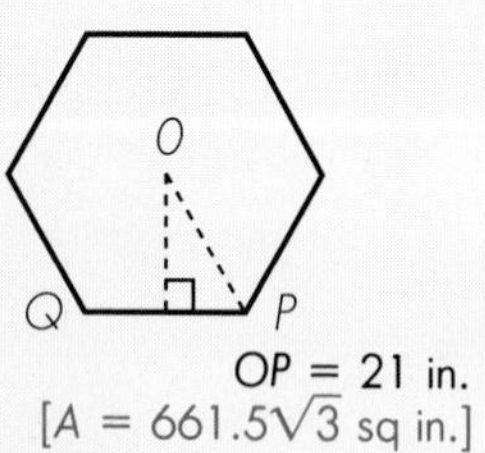

$OP = 21$ in.
$[A = 661.5\sqrt{3}$ sq in.$]$

Reteaching
Students who have had difficulty with this section may benefit from Reteaching Activity 10.2.

Enrichment
For students who have mastered the material in this section, you may wish to assign Enrichment Activity 10.2.

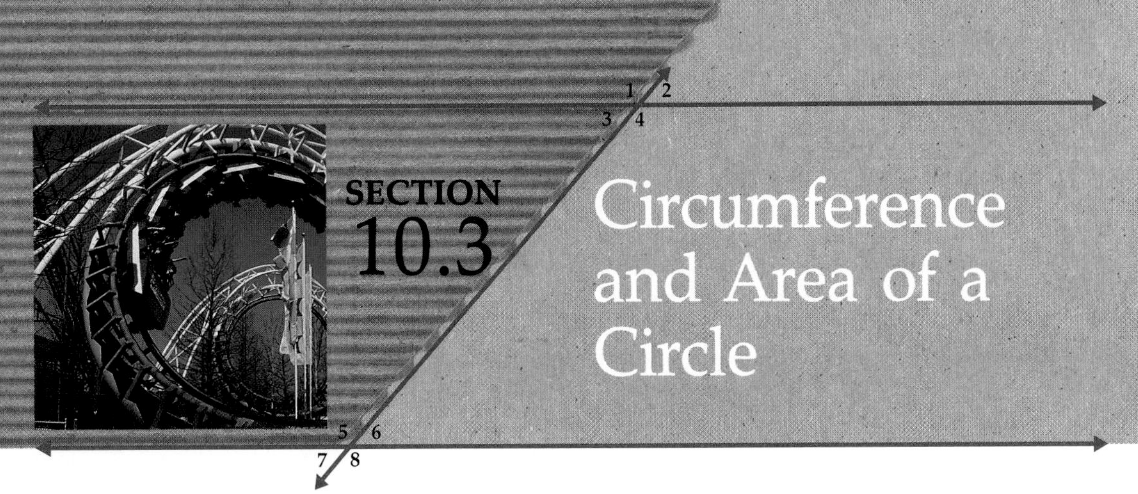

SECTION 10.3 Circumference and Area of a Circle

After completing this section, you should understand
- ▲ the definition of a circle and its parts.
- ▲ how to find the circumference of a circle.
- ▲ how to find the area of a circle.

The problem of finding an accurate method for determining the circumference and area of a circle has occupied mathematicians, engineers, and other scientists for thousands of years. Recall that a circle is the set of all points on a plane that are equidistant from a point called the center.

DEFINITION 10-3-1 **The RADIUS of a circle is a line segment whose endpoints are the center of a circle and a point on the circle.**

The *diameter* is the distance across a circle through the center. It is twice the length of the radius. The *circumference* of a circle is the distance around the circle.

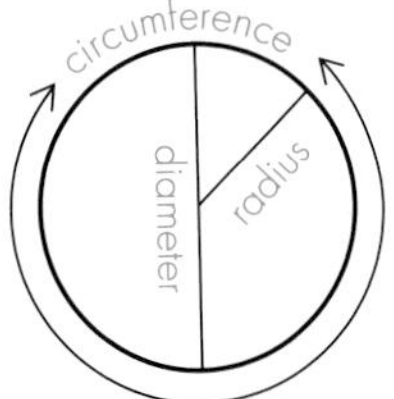

Use a compass to draw three circles, each one with a greater radius than the last.

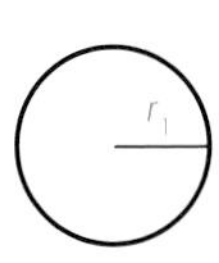

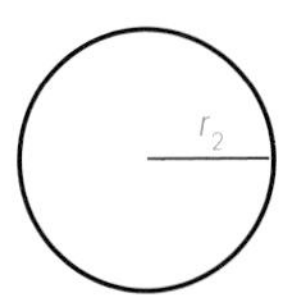

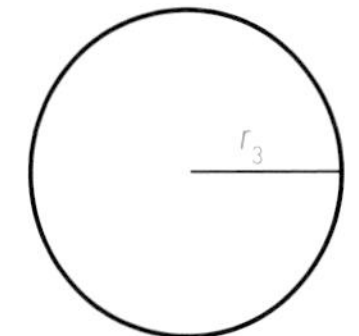

You can see that the circles get larger as you increase the length of the radius. And as the circles get larger, the circumference, or distance around the circle, also gets larger.

SECTION 10.3

Resources
Reteaching 10.3
Enrichment 10.3
Transparency Master 10-2

Objectives
After completing this section, the student should understand
- ▲ the definition of a circle and its parts.
- ▲ how to find the circumfer-ence of a circle.
- ▲ how to find the area of a circle.

Materials
compass
circular objects
tape
ruler
tape measure
library resources

Vocabulary
radius
diameter
circumference
pi

GETTING STARTED

Quiz: Prerequisite Skills
Use a calculator to express each fraction as a decimal rounded to the nearest hundredth.

1. $\frac{5}{8}$ [0.63] **2.** $\frac{7}{12}$ [0.58]
3. $\frac{38}{7}$ [5.43] **4.** $\frac{47}{100}$ [0.47]
5. $\frac{35}{8}$ [4.38] **6.** $\frac{25}{9}$ [2.78]
7. $\frac{6}{5}$ [1.20] **8.** $\frac{21}{11}$ [1.91]

Chalkboard Activity
Draw several polygons, both regular and nonregular, on the chalkboard, and have students explain how to find the perimeters. Then draw a circle and display several circular objects, such as a jar lid, a button, a bangle bracelet, or a wheel. Ask students to explain how they would find the distance around each circle.

TEACHING COMMENTS

Using the Page
During the Discovery Activity, remind students that the more accurately they measure, the more consistent the ratio between circumference and diameter will be. You may wish to have students work in small groups to make and record measurements.

DISCOVERY ACTIVITY

1. Collect a set of different-sized circular objects, such as bicycle or wagon wheels, bangle bracelets, plastic or metal rings. Use a small piece of tape to mark a point on the circumference of each circle.

2. Determine the circumference of each circle by rolling it through one revolution, marking the points where the taped point begins and ends.

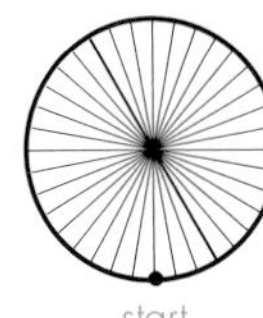

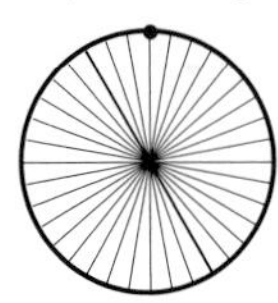
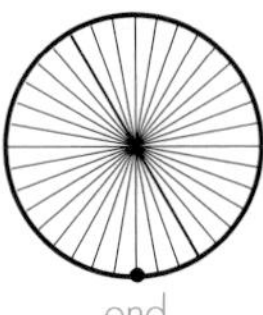

3. Measure and record the distance between the points.

4. Measure the diameter, or distance across the circle through the center.

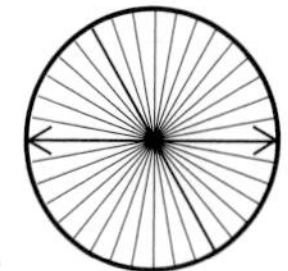

5. Find the ratio of the circumference to the diameter. (distance from step 3 ÷ diameter)

6. Repeat the activity for several other circular objects. Record the ratios you find.

7. What do you observe about the ratio of circumference to diameter (C/d)?
It is about 3:1.

You may have discovered that the circumference of a circle is approximately three times the diameter. The more accurate your measurements, the closer the value of the ratio will be to 3.14 . . . . The ratio of circumference to diameter, or circumference to two times the radius, is a *constant.* This means that this ratio will be the same for any circle.

This constant is an irrational number that has been given a symbol, π, **pi.** The value of π is only approximate—to 10 decimal places, $\pi \approx 3.1415926535$.

POSTULATE 10-3-1 **The circumference of a circle is π times the diameter (d).**

$$C = \pi d \approx 3.14d$$

Since the diameter is twice the radius, $C = \pi(2r) = 2\pi r$.

Using the Page
During the discussion of π, you may wish to have students look up the term *constant* and explain its mathematical definition. You may also wish to review these definitions of *irrational number*: a number that cannot be represented as the quotient of two integers $\left(\text{there are no integers } m \text{ and } n \text{ such that } \pi = \frac{m}{n}\right)$; a number that is represented by an infinite nonrepeating decimal.

Example: What is the circumference of a 26-inch bicycle wheel?

$C = \pi d \quad d = 26$ in.

$C = 26\pi \approx 26(3.14) \approx 81.64$ in.

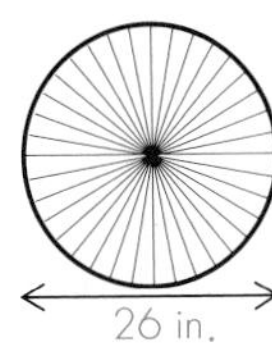

Example: What is the circumference of a circle with a radius of 12 meters?

$C = 2\pi r \quad C = 2\pi \cdot 12 = 24\pi \approx 24 \cdot 3.14 \approx 75.36$ m

CLASS ACTIVITY

For each diameter find the length of the radius.

1. 32.4 cm
   16.2 cm
2. 12 in.
   6 in.
3. 15.32 cm
   7.66 cm
4. 8.5 m
   4.25 m

For each radius find the length of the diameter.

5. 3.75 ft
   7.5 ft
6. 7 yd
   14 yd
7. 32.1 mm
   64.2 mm
8. $18\frac{1}{2}$ ft
   37 ft
9. You can use the [π] key on a calculator to help you find the circumference of a circle. Press the [π] key on a calculator. What number is shown on the display? 3.1415927

Use a calculator to find the circumference. Round your answer to the nearest hundredth.

10. $d = 2$ in.
    6.28 in.
11. $r = 2$ ft
    12.57 ft
12. $d = 8$ cm
    25.13 cm
13. $r = 50$ yd
    314.16 yd
14. π is the ratio of the __________ to the __________.
    circumference, diameter

Extra Practice
Use a calculator to find the circumference to the nearest tenth. Use the calculator's value for π.

1. $d = 120$ m
2. $r = 99.4$ in.
[C = 377.0 m]
[C = 624.5 in.]
3. $d = 10$ m
4. $d = 1000$ cm
[C = 31.4 m]
[C = 3141.6 cm]
5. $d = 3.2$ in.
6. $r = 16$ cm
[C = 10.1 in.]
[C = 100.5 cm]
7. $d = 4.5$ ft
[C = 14.1 ft]

THE AREA OF A CIRCLE

It is possible to determine the area of a circle by counting the number of square units in the interior, but this is a difficult and inaccurate method. The area of a circle is also related to its diameter and radius and to the constant, π. You can see that, as the length of the diameter increases, so does the area.

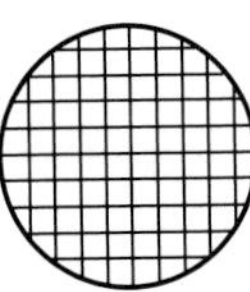
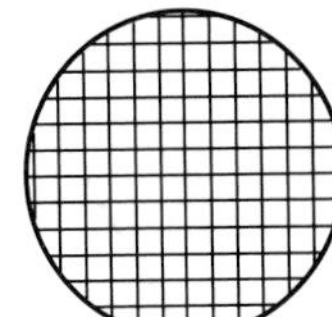
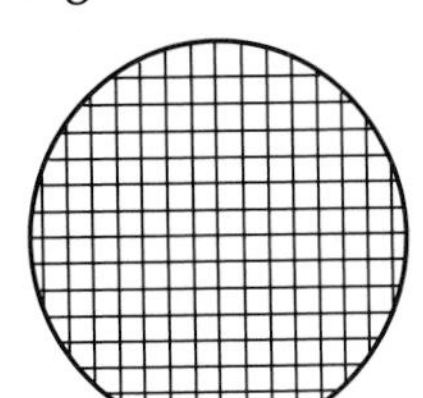

DISCOVERY ACTIVITY

1. Construct a circle with a 1-inch radius. Mark the center O. Label point A on the circle.

2. Place the point of a compass on A and the pencil tip on O and mark off equal segments around the circle. Label the points on the circle.

3. Connect the points. What polygon is inscribed in the circle? Is it a regular polygon? Hexagon; yes
4. What is the length of $\overline{OA}$? of $\overline{AB}$? Why? See margin.

5. What is the perimeter of the hexagon? What is the area? 6 in.; 2.6 sq in.

6. How is the circumference of the circle related to the perimeter of the hexagon? Use the formula for circumference, $C = \pi d$, to explain. $C = \pi d > P$ (hexagon)

7. How do you think the area of the circle is related to the area of the hexagon? A (circle) $> A$ (hexagon)
8. How might you get a closer estimate of the circle's area? Use a figure with more sides and find its area.

Additional Answers
4. 1 in.; 1 in.; OA is the radius; since figure is regular hexagon, the sides are equal to $\overline{OA}$.

You may have discovered that the perimeter and area of the hexagon are close to, but less than, the circumference and area of the circle.

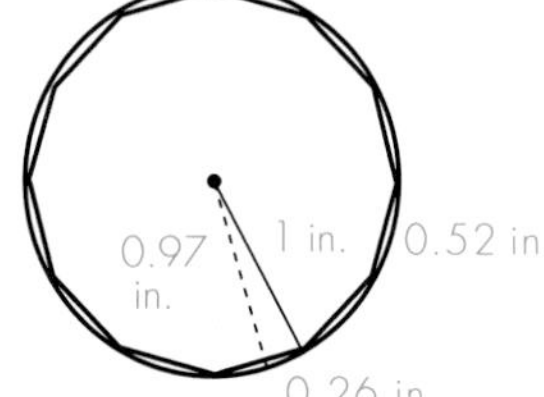

Now find the area of a regular 12-gon inscribed in a circle with a 1-inch radius.

$$A = \frac{1}{2}Ph = \frac{1}{2}(0.97)6.24 = 3.03 \text{ sq in.}$$

Notice that as the number of sides increases,
- the height of the polygon approaches the length of the radius;
- the perimeter of the polygon approaches the circumference of the circle;
- the area of the polygon approaches the area of the circle.

$$h \to r$$
$$P \to C = 2\pi r$$
$$A = \frac{1}{2}Ph \to \frac{1}{2} \cdot 2\pi r \cdot r \text{ or } \pi r^2$$

Using the Page
For students who have difficulty understanding how the perimeter of the inscribed polygon approaches the circumference of the circle, you may wish to use Transparency Master 10-2 to illustrate additional many-sided polygons inscribed in circles with a 1-inch radius, and have students repeat the exercise.

THEOREM 10-3-1 **The area of a circle is the radius squared times the number π.**

Example: A circle has a diameter of 4 feet. What is the area?

$$A = \pi r^2 \approx 3.14(2)^2 \approx 12.56 \text{ sq ft}$$

Example: A circle has an area of 625π. What is the radius?

$$\begin{aligned} A &= \pi r^2 \\ 625\pi &= \pi r^2 \\ 625 &= r^2 \\ 25 &= r \end{aligned}$$

REPORT 10-3-1 Prepare and present an oral report on one of the following mathematicians of ancient Greece. In your report, pay particular attention to the mathematician's contributions to geometry and to the history of π.

Plato
Pythagoras
Euclid
Hippocrates of Chios
Archimedes

Using the Page
The report provides opportunity for cooperative work in research, writing, and presenting. You may wish to have interested students extend their research to modern-day interpretations and uses of π.

CLASS ACTIVITY

Use a calculator to find the area. Round your answer to the nearest hundredth.

1. 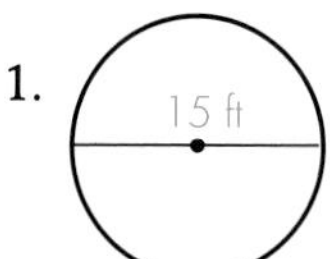

2. 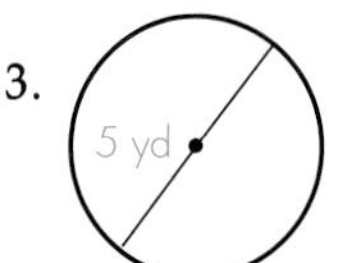

3. 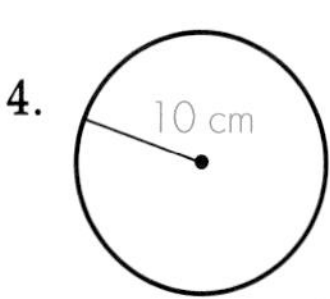

4. 10 cm

A = 176.71 sq ft A = 6.16 sq in. A = 19.63 sq yd A = 314.16 cm^2

Find the area of each circle with the given measure. Use $\pi \approx 3.14$.

5. $d = 4$ ft 12.56 sq ft
6. $C = 6\pi$ cm 28.26 cm^2
7. $r = 4$ m 50.24 m^2
8. $C = 10\pi$ in. 78.5 sq in.
9. $d = 100$ m 7850 m^2
10. $r = 12$ yd 452.16 sq yd
11. $d = 1.5$ km 1.77 km^2
12. $C = 12\pi$ m 113.04 m^2

Find the diameter of each circle.

13. $A = 11{,}304 \text{ cm}^2$ d = 120 cm
14. $A = 628$ sq ft d = 28.28 ft
15. $A = 3140 \text{ m}^2$ d = 63.24 m
16. Use a calculator to find the area of the shaded region. The centers of the circles are O and A, the length of radius $\overline{AB}$ is 8 cm, and $\overline{OB} = 2\overline{AB}$. 603.19 cm^2

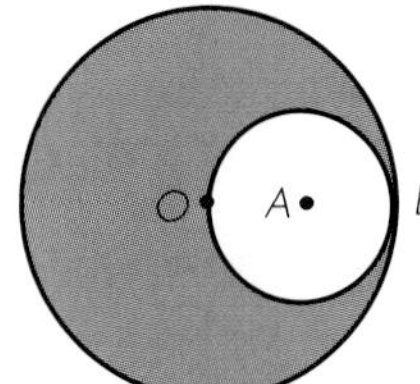

Extra Practice
Use a calculator to find the area. Round the answer to the nearest hundredth.

1. $d = 9.2$ ft [$A = 66.48$ sq ft]
2. $r = 1.2$ in. [$A = 4.52$ sq in.]
3. $r = 15$ cm [$A = 706.86$ cm^2]
4. $d = 21.5$ m [$A = 363.05$ m^2]

Find the diameter of each circle.

5. $A = 153.94$ sq in. [$d = 14$ in.]
6. $A = 36.3168 \text{ cm}^2$ [$d = 6.8$ cm]

Find the area of the shaded region of the circle.

7.

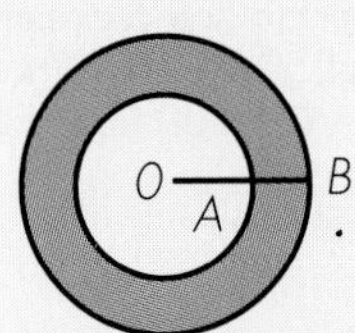

$OB = 12$ cm
$OA = \frac{2}{3}OB$
[$A = 251.3 \text{ cm}^2$]

HOME ACTIVITY

Copy and complete the table. Use $\pi \approx 3.14$. See margin.

	Diameter	Radius	Circumference	Area
1.	16 in.	8 in.	50.24 in.	200.96 sq in.
2.	60 cm	30 cm	188.4 cm	2826 cm^2
3.	42 yd	21 yd	131.88 yd	1384.74 sq yd
4.	11.84 m	5.92 m	37.18 m	109.9 m^2
5.	20 in.	10 in.	62.8 in.	314 sq in.
6.	10 ft	5 ft	31.4 ft	78.5 sq ft

A circle has a radius of 2 centimeters.

7. Find the area in terms of π. 4π = 12.56
8. Find the circumference in terms of π. 4π = 12.56

An equilateral triangle circumscribes a circle with a diameter of 5.8 feet. The side of the triangle is 10 feet.

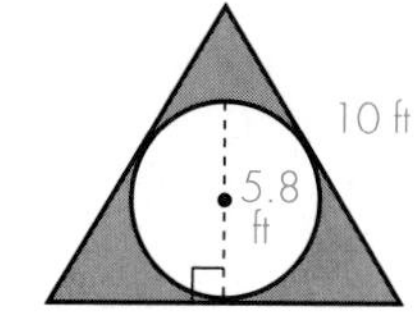

9. What is the area of the triangle? $25\sqrt{3}$ sq ft or 43.3 sq ft
10. What is the area of the circle? Use $\pi \approx 3.14$. 26.4 sq ft
11. What is the area of the shaded region? 16.9 sq ft

CRITICAL THINKING

Two circles have radii of 3 centimeters and 5 centimeters. Use $\pi \approx 3.14$.

12. What is the circumference of the smaller circle? 18.84 cm
13. What is the circumference of the larger circle? 31.4 cm
14. What is the ratio of the radii, smaller to larger? 0.6
15. What is the ratio of the circumferences, smaller to larger? 0.6
16. Find the area of the smaller circle. 28.26 cm^2
17. Find the area of the larger circle. 78.5 cm^2
18. What do you think the ratio of the areas, smaller to larger, will be? Why? 0.36; ratio of radii squared
19. Write a conclusion about what you found. See margin.

Using the Page
Encourage students to check their work in the Home Activity with a calculator, if they have one available.

Additional Answers
8. The ratio of the radii to two circles is proportional to the ratio of their circumferences. The square of the ratio of two radii is proportional to the areas of the circles.

FOLLOW-UP

Assessment
1. Two circles have radii of 12 inches and 9 inches. How much greater is the circumference of the larger circle? [18.84 in.]
2. A circle has a circumference of 50.24 cm. What is the area? [A = 200.96 cm^2]

Reteaching
Students who have had difficulty with this section may benefit from Reteaching Activity 10.3.

Enrichment
For students who have mastered the material in this section, you may wish to assign Enrichment Activity 10.3.

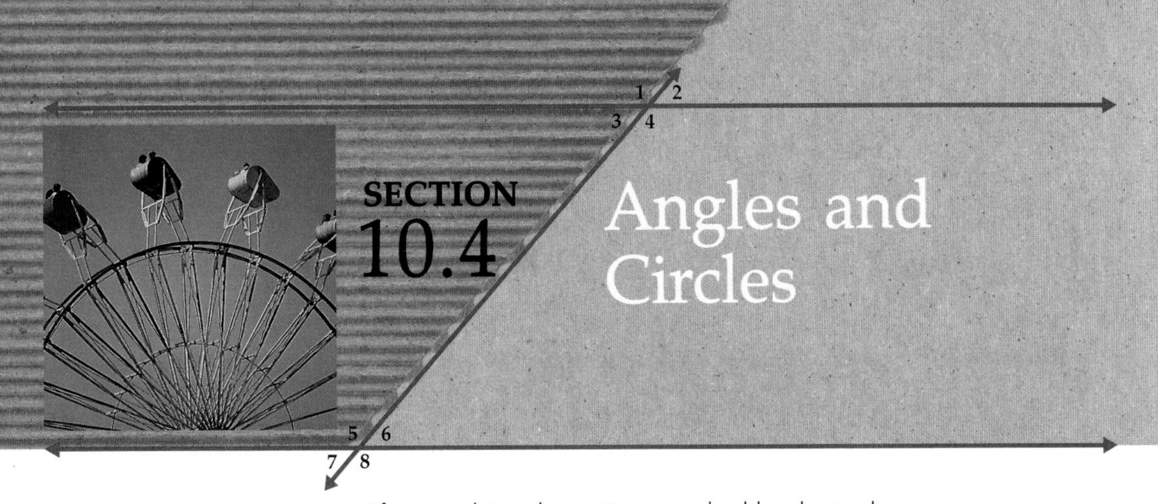

SECTION 10.4 Angles and Circles

After completing this section, you should understand

▲ how to identify central and inscribed angles and their intersected arcs.
▲ how to find the measures of angles intersecting circles.
▲ how to identify a tangent to a circle.

Before beginning an investigation of angles and circles, it will be helpful to define some more terms associated with a circle.

CHORD: a line segment whose endpoints are two points on the circle

SECANT: a line that intersects the circle at two points

TANGENT: a line that intersects the circle at one point

ARC: a part of the circumference of the circle

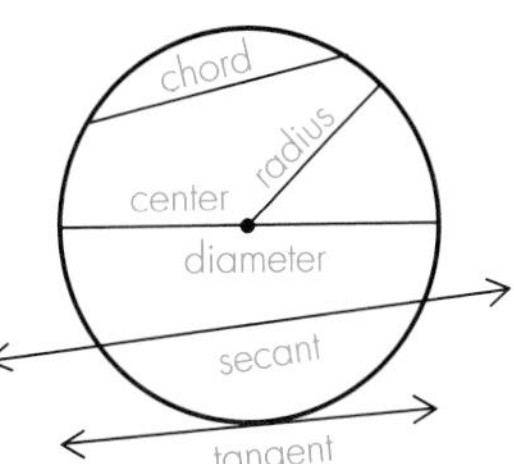

DISCOVERY ACTIVITY

1. Use a compass to draw six circles on a sheet of paper.

2. Look at the pair of intersecting lines shown at the right. How many ways can you find for these lines to intersect a circle? Think of placing point X at different places in, on, or outside of the circle.

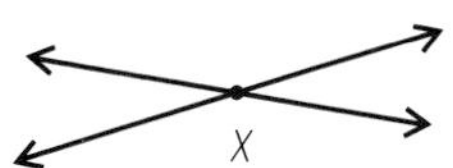

3. Draw as many cases as you can find.

SECTION
10.4

Resources
Reteaching 10.4
Enrichment 10.4

Objectives
After completing this section, the student should understand
▲ how to identify central and inscribed angles and their intersected arcs.
▲ how to find the measures of angles intersecting circles.
▲ how to identify a tangent to a circle.

Materials
ruler
compass
protractor

Vocabulary
chord
secant
tangent
arc
central angle
inscribed angle

GETTING STARTED

Warm-Up Activity
Have students use a compass to construct two circles, one with a radius of 2 inches and one with a radius of 4 inches, and then cut out the circles. Have them fold each circle in fourths. Ask how they would compare the arcs formed by one-fourth of each circle. Then have them measure one of the angles at the center of each circle.

Ask what they observe about the measures of the two angles. [They are equal.] Does the size of the circles make a difference in the size of the angle at the center? [No.]

TEACHING COMMENTS

Using the Page
After the Discovery Activity, on page 373, you may wish to have students come to the board to draw the different ways the lines can intersect the circle. Have the class decide if there are any additional cases.

Using the Page
During the discussion of Definition 10-4-2, point out to students that there are many central angles that can intersect $\widehat{AB}$; the angle that has the same measure as $\widehat{AB}$ is the angle that intersects $\widehat{AB}$ at A and B.

Using the Page
In the discussion of inscribed angles, point out that the angle has its vertex on the circle, and its rays must intersect the circle at two points.

Using the Page
You may wish to draw the diagram for Theorem 10-4-1 on the chalkboard and outline the different angles with colored chalk as you proceed through the proof.

You may have discovered that there are five ways for these lines to intersect a circle.

INTERSECTION AT THE CENTER

DEFINITION 10-4-1 **A CENTRAL ANGLE is an angle whose vertex is at the center of the circle.**

You can measure a central angle with a protractor. Since the rays of the angle intersect the circle and the complete circle contains 360°, the arc intersected by the angle contains the same number of degrees as the central angle.

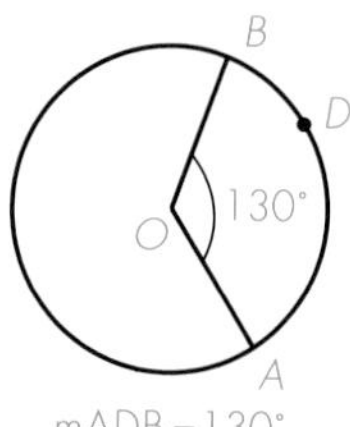

mADB = 130°

DEFINITION 10-4-2 **The ARC of the circle has the same measure as the central angle that intersects the arc. The symbol for arc is $\widehat{AB}$ or $\widehat{ADB}$.**

INTERSECTION ON THE CIRCLE

Angle ABC intersects $\widehat{AC}$ of the circle, with the vertex B on the circle. This angle is called an inscribed angle.

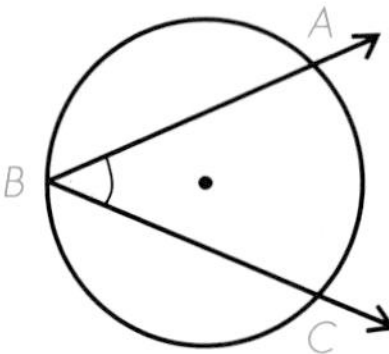

DEFINITION 10-4-3 **An INSCRIBED ANGLE is an angle whose vertex is on the circle and whose rays intersect the circle in two other points.**

It is possible to find the measure of $\angle ABC$ if you know the measure of the intersected arc AC. Have students supply the missing reasons.

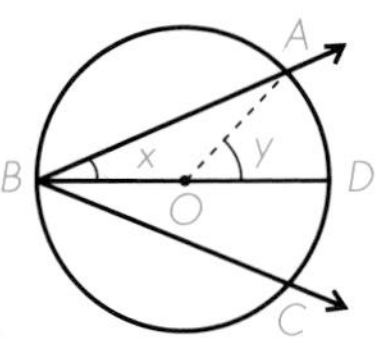

Steps	Reasons
1. Draw $\overleftrightarrow{BO}$ and mark the point where $\overleftrightarrow{BO}$ intersects the circle O.	1. Two points determine a line.
2. Draw $\overline{OA}$ to form $\triangle ABO$. $\triangle ABO$ is isosceles.	2. $\overline{OB} = \overline{OA}$; radii of circle.
3. $m\angle ABO = m\angle BAO$	3. Angles opposite congruent sides of isosceles triangle

Let $\angle y = \angle AOD$ and $\angle x = \angle ABO$.

4. $m\angle y = m\angle x + m\angle BAO$	4. Exterior angle
5. $m\angle y = 2(m\angle x)$ and $\frac{1}{2}(m\angle y) = m\angle x$	5. Substitution, algebra
6. $m\angle y = m\widehat{AD}$	6. Arc has measure of central angle.
7. $m\angle x = \frac{1}{2}(m\widehat{AD})$	7. Substitution

Repeat steps 2–7 to show that $m\angle CBO = \frac{1}{2}(m\widehat{CD})$.

8. $m\angle ABD + m\angle CBD = \frac{1}{2}(m\widehat{AD}) + \frac{1}{2}(m\widehat{CD})$ and $m\angle ABC = \frac{1}{2}(m\widehat{ADC})$	8. Algebra

NOTEBOOK

THEOREM 10-4-1 **The measure of an inscribed angle is $\frac{1}{2}$ the number of degrees in the intersected arc.**

INTERSECTION INSIDE THE CIRCLE

To find the measure of $\angle AED$, follow these steps.

1. Draw chord $\overline{CA}$.
2. Does $m\angle x = m\angle CAB + m\angle ACD$? Why? — Yes; exterior angles
3. $m\angle ACD = \frac{1}{2}(m\widehat{AD})$ and $m\angle CAB = \frac{1}{2}(m\widehat{BC})$. Why? — Inscribed angles
4. $m\angle x = \frac{1}{2}(m\widehat{AD} + m\widehat{BC})$

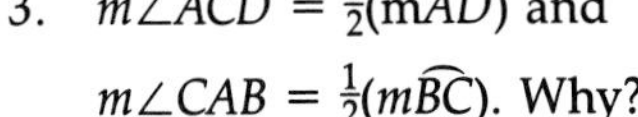

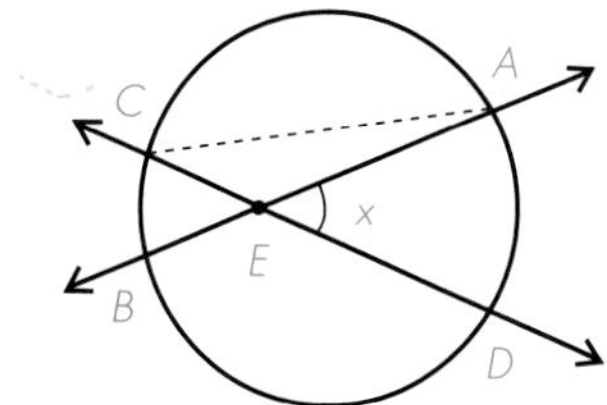

NOTEBOOK

THEOREM 10-4-2 **If the vertex of the angle is within the circle, then the measure of the angle is $\frac{1}{2}$ the sum of the degrees in the intersected arcs.** This theorem is true even for the case where the vertex of the angle is at the center of the circle.

Example: Find $m\angle x$.

$$m\angle x = \frac{1}{2}(m\widehat{AD} + m\widehat{BC})$$

$$m\angle x = \frac{1}{2}(60° + 80°) = 70°$$

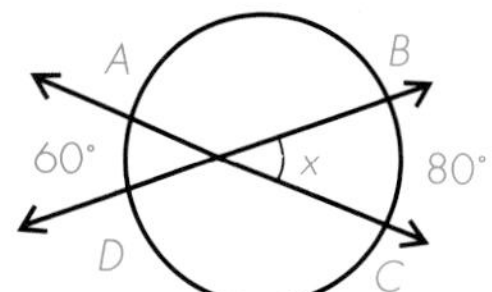

Using the Page
After discussing the example, you may wish to have students find the measure of the angle that is supplementary to $\angle x$. [110°] Then have them find the sum of the measures of $\widehat{AB}$ and $\widehat{DC}$. [220°]

Example: Find $m\angle PQR$.
The measure of all the arcs is 360°. Solve for x to find the measure of each arc.

$$x + (2x + 5) + (3x + 20) + (5x + 5) = 360°$$
$$11x + 30 = 360°$$
$$11x = 330°$$
$$x = 30°$$

5x+5 R x Q T P 2x+5 S 3x+20

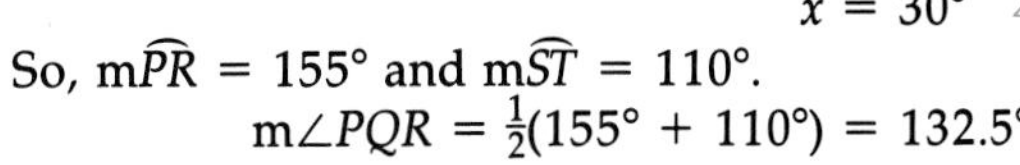

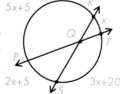

So, $m\widehat{PR} = 155°$ and $m\widehat{ST} = 110°$.

$$m\angle PQR = \frac{1}{2}(155° + 110°) = 132.5°$$

Additional Answers
1. Radii: $\overline{OA}$, $\overline{OB}$, $\overline{OC}$, $\overline{OH}$, $\overline{OG}$; diameters: $\overline{GH}$, $\overline{BC}$; chords: $\overline{AF}$, $\overline{GH}$, $\overline{BC}$; secant: $\overline{DE}$; tangent: $\overleftrightarrow{BJ}$; central angles: $\angle COH$, $\angle COA$, $\angle AOG$, $\angle GOB$, $\angle BOH$; inscribed angles: $\angle JBC$, $\angle CBK$
2. Radii: $\overline{OX}$, $\overline{OA}$, $\overline{OY}$, $\overline{OZ}$; diameters: $\overline{XY}$, $\overline{AZ}$; chords: $\overline{AB}$, $\overline{XY}$, $\overline{AZ}$; secants: $\overline{PQ}$, $\overline{DE}$; tangent: $\overleftrightarrow{CR}$; central angles: $\angle XOA$, $\angle AOY$, $\angle YOZ$, $\angle ZOX$; inscribed angle: $\angle ABZ$

Extra Practice
1. In circle O, $m\angle PRS = 50°$ and $m\widehat{QT} = 40°$. Find $m\widehat{PS}$.

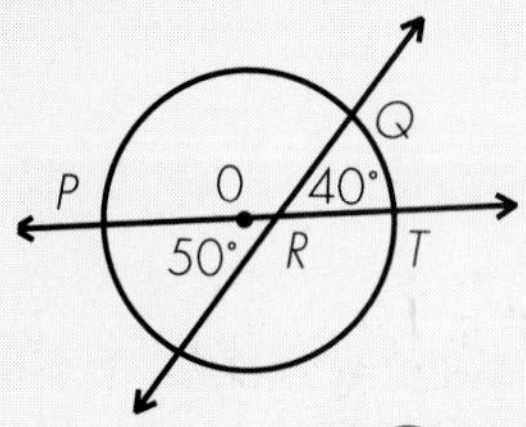

[$m\widehat{PS} = 60°$]

2. $m\widehat{XL} = x$, $m\widehat{XM} = 5x + 20$, $m\widehat{MZ} = 3x + 10$, and $m\widehat{ZL} = 7x + 10$. Find $m\widehat{MYZ}$.

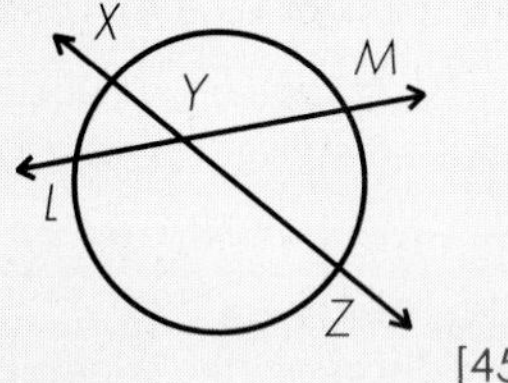

[45°]

CLASS ACTIVITY

For each circle, name the radii, diameters, chords, secants, tangents, central angles, and inscribed angles.

1.

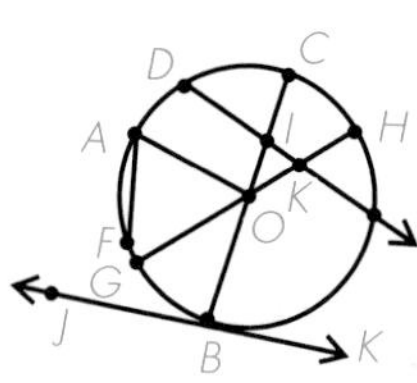

See margin.

2.

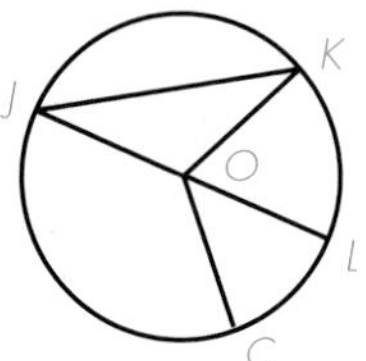

See margin.

In circle O, $m\widehat{KL} = 65°$, $m\widehat{CL} = 48°$, and $\overline{JL}$ is a diameter. Find the measures of

3. $\angle KOL$ 65°
4. $\angle KJL$ 32.5°
5. $\angle LOC$ 48°
6. $\angle JOK$ 115°

ANGLES OUTSIDE A CIRCLE

In the circle to the right, the two lines intersect so that the vertex of the angle is outside the circle and the rays intersect the circle. Find $m\angle x$.

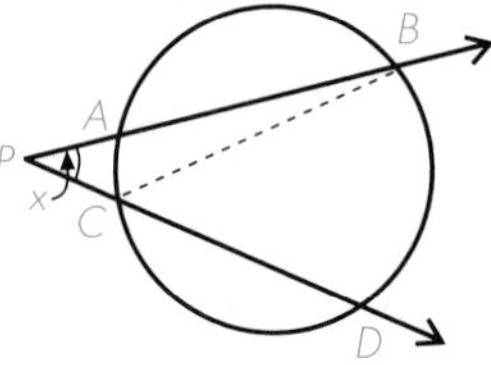

Have students supply the missing reasons.

Steps	Reasons
1. Draw $\overline{CB}$.	1. Two points determine a line.
2. $m\angle DCB = m\angle ABC + m\angle CPB$	2. Exterior angle
3. $m\angle CPB = m\angle DCB - m\angle ABC$	3. Algebra
4. $m\angle DCB = \frac{1}{2}(m\widehat{BD})$ and $m\angle ABC = \frac{1}{2}(m\widehat{AC})$	4. Inscribed angles
5. $m\angle CPB = \frac{1}{2}(m\widehat{BD} - m\widehat{AC})$	5. Substitution

THEOREM 10-4-3 **If the vertex of the angle is outside the circle and the rays intersect the circle, then the measure of the angle is $\frac{1}{2}$ the difference of the measure in degrees of the two intersected arcs.**

Example: $m\widehat{RT} = 100°$ and $m\widehat{QS} = 30°$. Find $m\angle P$.

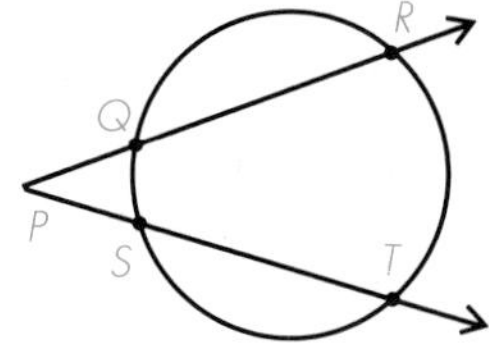

$m\angle P = \frac{1}{2}(m\widehat{RT} - m\widehat{QS})$

$m\angle P = \frac{1}{2}(100° - 30°) = 35°$

TANGENTS TO A CIRCLE

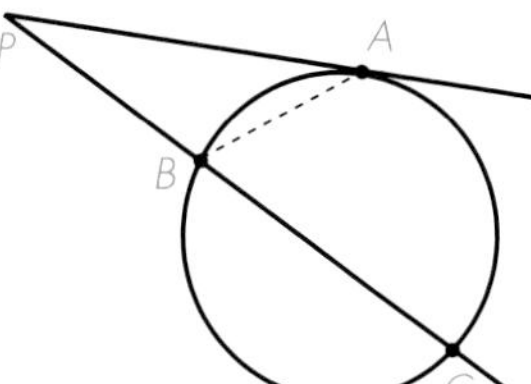

Two lines intersect so that the vertex of an angle is outside the circle; $\overleftrightarrow{PC}$ passes through the circle, and $\overleftrightarrow{PA}$ is tangent to the circle at point *A*.

This case is another case of an exterior angle. $m\angle APB = m\angle ABC - m\angle PAB$, so $m\angle APB = \frac{1}{2}(m\widehat{AC} - m\widehat{AB})$.

Example: $m\widehat{TQ} = 40°$ and $m\widehat{TR} = 130°$. Find $m\angle P$.

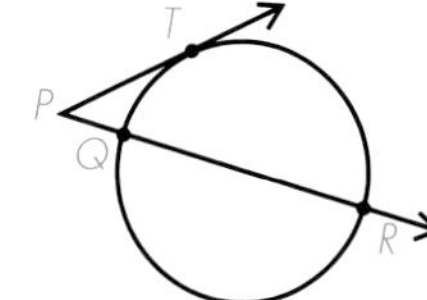

$$m\angle P = \tfrac{1}{2}(m\widehat{TR} - m\widehat{TQ})$$

$$m\angle P = \tfrac{1}{2}(130° - 40°) = 45°$$

Notice that $\overleftrightarrow{PT}$ intersects, or is tangent to, the circle at only one point.

DEFINITION 10-4-4 **If a line intersects a circle at one point, then the line is TANGENT to the circle.**

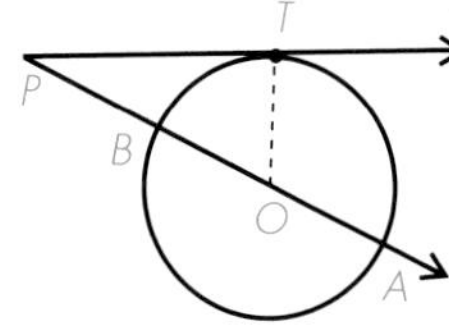

What is the measure of the angle formed by a tangent to a circle and a line through the center of the circle? Given circle *O* with tangent $\overleftrightarrow{PT}$, find the measure of $\angle PTO$.

Have students supply missing reasons.

Steps	Reasons
1. $m\angle TOA = m\angle OPT + m\angle PTO$	1. Exterior angle
2. $m\angle PTO = m\angle TOA - m\angle OPT$	2. Algebra
3. $m\angle TOA = m\widehat{TA}$	3. Central angle
4. $m\angle OPT = \frac{1}{2}(m\widehat{TA} - m\widehat{TB})$	4. Angle outside circle is $\frac{1}{2}$ difference of arcs.
5. $m\angle PTO = m\widehat{TA} - \frac{1}{2}(m\widehat{TA} - m\widehat{TB})$ $= \frac{1}{2}(m\widehat{TA}) + \frac{1}{2}(m\widehat{TB})$ $= \frac{1}{2}(m\widehat{TA} + m\widehat{TB})$	5. Substitution and algebra
6. $m\widehat{TA} + m\widehat{TB} = 180°$	6. $\overline{AB}$ is a diagonal.
7. $m\angle PTO = \frac{1}{2}(180°) = 90°$	7. Substitution

THEOREM 10-4-4 **A tangent to a circle forms a 90° angle with the radius of the circle at the point of intersection.**

This theorem tells you that the tangent and the radius are perpendicular.

Using the Page
In discussing Definition 10-4-4, be sure students understand what it means for a line to intersect a circle at only one point. It may be helpful to use a physical example of this—perhaps a jar lid balanced on a desk top.

Using the Page
In the discussion of the proof for Theorem 10-4-4, it may be helpful to draw the diagram on the chalkboard and outline each set of related angles and arcs in colored chalk, so that students understand which substitutions are possible.

Extra Practice

1. $m\widehat{QR} = 95°$ and $m\angle S = 40°$. Find $m\widehat{TP}$.

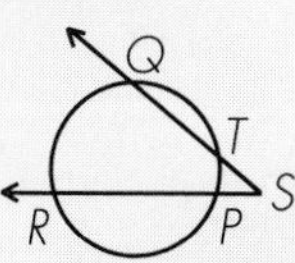

[15°]

2. $m\widehat{AB} = 40°$. O is the center of the circle. Find the measure of each numbered angle.

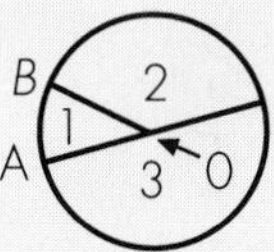

[$m\angle 1 = 40°$, $m\angle 2 = 140°$, $m\angle 3 = 180°$]

FOLLOW-UP

Assessment

1. In circle O, $m\angle JMK = 38°$ and $m\widehat{NL} = 60°$. Find $m\widehat{JM}$.

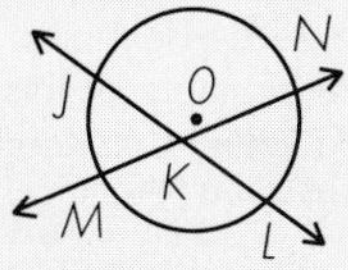

[16°]

2. $m\widehat{QR} = 90°$, $m\widehat{RT} = 100°$, and $m\widehat{ST} = 110°$. Find the measure of each numbered angle.

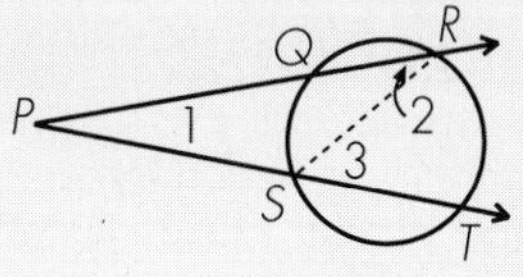

[$m\angle 1 = 20°$, $m\angle 2 = 30°$, $m\angle 3 = 50°$]

Reteaching

Students who have had difficulty with this section may benefit from Reteaching Activity 10.4.

Enrichment

For students who have mastered the material in this section, you may wish to assign Enrichment Activity 10.4.

CLASS ACTIVITY

1. Use a computer and a geometric drawing tool to draw any circle. Label a point on the circle and draw a radius to the point. Then draw a line that is perpendicular to the radius at the point. Repeat this activity several times, using a circle with a different radius each time. What seems to be true about the line? Do you think it is possible to draw a line through a point on a circle so that the line also passes through the interior of the circle? Do your results show that Theorem 10-4-4 is true?
   It is tangent to the circle; no; yes.
2. Draw an angle inscribed in a semicircle. Find the measure of the angle. Repeat this activity several times, using a circle with a different radius each time. If an angle is inscribed in a semicircle, is the measure of the angle 90°? Justify your answer.
   Yes; intersected arc is 180°.

Find the measure of each angle.
$m\widehat{AB} = 100°$, $m\widehat{AD} = 80°$

3. $m\angle 1$ 80° 4. $m\angle 2$ 100°
5. $m\angle 3$ 80° 6. $m\angle 4$ 100°

$m\widehat{BD} = 170°$, $m\widehat{DF} = 10°$, $m\widehat{CF} = 100°$

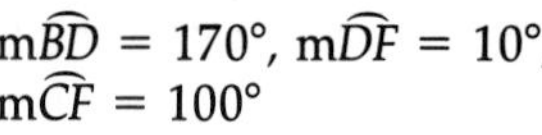

7. $m\angle 1$ 90° 8. $m\angle 2$ 90°
9. $m\angle 3$ 45° 10. $m\angle 4$ 45°
11. $m\angle 5$ 135°

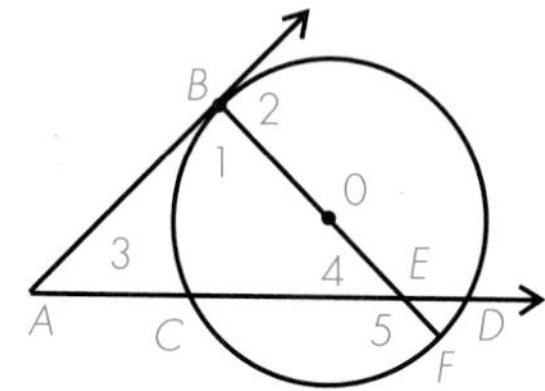

HOME ACTIVITY

Identify the following for circle *O*.

1. Radius — $\overline{OB}$, $\overline{OA}$, $\overline{OC}$
2. Diameter — $\overline{AC}$
3. Center — O
4. Central angle — $\angle AOB$, $\angle BOC$, $\angle AOC$
5. Tangent — $\overleftrightarrow{PT}$
6. Inscribed angle — $\angle TAC$, $\angle TCA$, $\angle ATC$
7. Chord — $\overline{TC}$, $\overline{TA}$, $\overline{AC}$
8. Arc — $\widehat{AT}$, $\widehat{TC}$, $\widehat{CB}$, $\widehat{BA}$

In circle *O*, $m\widehat{TA} = 140°$ and $m\widehat{AB} = 120°$. Find the measure of each angle.

9. $m\angle 1$ 70° 10. $m\angle 2$ 140° 11. $m\angle 3$ 20°
12. $m\angle 4$ 20° 13. $m\angle 5$ 20° 14. $m\angle 6$ 70°
15. $m\angle 7$ 50° 16. $m\angle 8$ 40° 17. $m\angle 9$ 110°

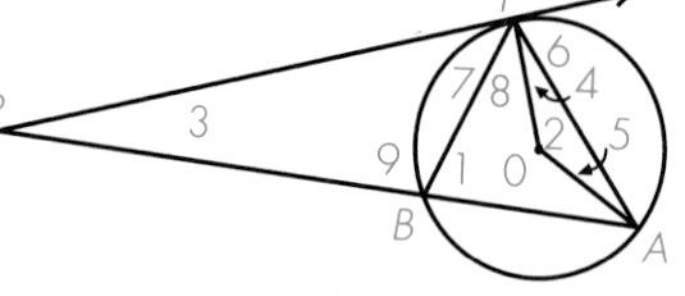

CRITICAL THINKING

18. From point P outside a circle, draw $\overrightarrow{PA}$ and $\overrightarrow{PB}$ tangent to the circle. Show that the ray from *P* through the center of the circle bisects $\angle APB$.

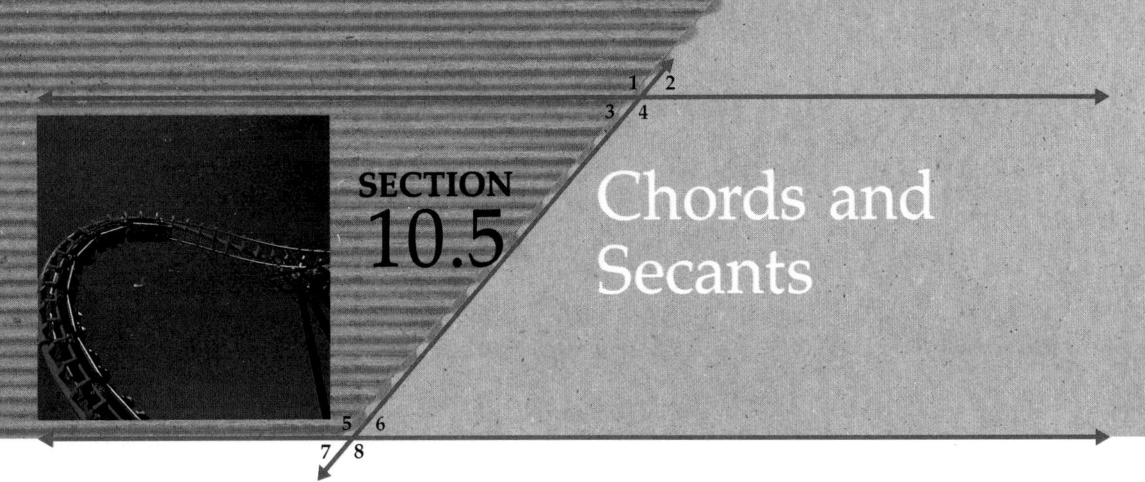

SECTION 10.5 Chords and Secants

After completing this section, you should understand

- ▲ the definitions of a chord and a secant.
- ▲ how to use the theorems about chords and secants to solve problems.

In Section 10.4, chords were briefly discussed as you investigated finding the measures of angles that intersect circles. Several properties associated with these lines are useful in solving problems about circles.

DISCOVERY ACTIVITY

1. Use a compass to draw a large circle. Label the center O. See margin.
2. Mark four points on the circle and label them *A*, *B*, *C*, and *D*.
3. Draw $\overline{AC}$ and $\overline{BD}$. Your figure should resemble this one. Label the intersection of $\overline{AC}$ and $\overline{BD}$, *E*. *E* is not the center of the circle.

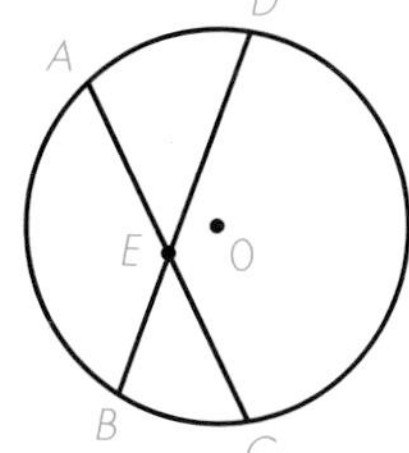

4. Measure (in centimeters to the nearest tenth) $\overline{AE}$, $\overline{EC}$, $\overline{BE}$, and $\overline{ED}$. Copy and complete the following.
   AE = ________ *EC* = ________ Product of *AE* and *EC* = ________
   BE = ________ *ED* = ________ Product of *BE* and *ED* = ________
5. What do you notice about the products? They are equal.
6. Repeat steps 1–4 with a different circle and different segments. What can you conclude? Do you think you will always get this result?

SECTION 10.5

Resources
Reteaching 10.5
Enrichment 10.5

Objectives
After completing this section, the student should understand
- ▲ the definitions of a chord and a secant.
- ▲ how to use the theorems about chords and secants to solve problems.

Materials
compass
ruler
protractor

Vocabulary
chord
secant
length of a tangent

GETTING STARTED

Quiz: Prerequisite Skills
Solve each proportion for x.

1. $\frac{x}{45} = \frac{16}{15}$ [48]

2. $\frac{15}{24} = \frac{30}{x}$ [48]

3. $\frac{x}{84} = \frac{3}{14}$ [18]

4. $\frac{6}{21} = \frac{x}{70}$ [20]

Warm-Up Activity
Have students construct a circle with a radius of 4 inches and cut it out. Have students fold the circle in on itself several times so that the edge of the circle meets the center point each time. Have them measure the distance between the center and the fold lines. Ask for an informal description of what they observe. [The distances from the fold lines to the center are equal.]

379

TEACHING COMMENTS

Using the Page
During the discussion of Definition 10-5-1, ask the class to decide whether every diameter is a chord and whether any radius is a chord. [Yes; No.]

Using the Page
As you go through the proof of Theorem 10-5-1, be sure that students understand the correspondence ratios between the two similar triangles in step 5. Point out that, to justify the theorem, you need only show that two of the ratios are equal.

Using the Page
When you reach step 10 of the proof of Theorem 10-5-1, have students explain why there can be only one perpendicular bisector of a line.

You may have discovered that when two chords intersect, the product of the segment lengths of one chord is equal to the product of the segment lengths of the other chord.

DEFINITION 10-5-1 **A CHORD of a circle is a line segment whose endpoints are two points on the circle.**

Given: A circle with chords $\overline{AC}$ and $\overline{BD}$ intersecting at E
Justify: $AE \cdot CE = BE \cdot DE$
Have students supply the missing reasons.

Steps	Reasons
1. Draw chords $\overline{AD}$ and $\overline{BC}$.	1. Two points determine a line.
2. $m\angle AED = m\angle BEC$	2. Vertical angles
3. $m\angle ADB = m\angle ACB$	3. Inscribed angles intersecting same arc have equal measure.
4. $\triangle ADE \sim \triangle BCE$ Map the triangles.	4. AA
5. $AD/CB = DE/CE = AE/BE$	5. Writing correspondences
6. $AE \cdot CE = BE \cdot DE$	6. Algebra

THEOREM 10-5-1 **If two chords intersect, then the product of the segment lengths of one chord is equal to the product of the segment lengths of the other chord.**

Chords have another property that you can use in solving problems.

Given: Circle O with chord $\overline{AB}$; $\overleftrightarrow{CD}$ is the perpendicular bisector of $\overline{AB}$.
Justify: $\overleftrightarrow{CD}$ passes through center O. Have students supply the missing reasons.

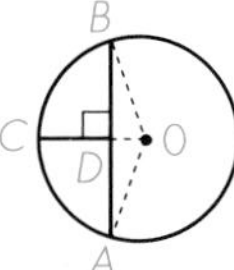

Steps	Reasons
1. $AD = DB$	1. Definition of bisector
2. Draw $\overline{OA}$, $\overline{OD}$, and $\overline{OB}$.	2. Two points determine a line.
3. $OA = OB$	3. Radii of same circle
4. $m\angle OAD = m\angle OBD$	4. Base angles of isosceles triangle are equal.
5. $\triangle AOD \cong \triangle BOD$	5. SAS and $OD/OD = 1$
6. $m\angle ADO = m\angle BDO$	6. CPCF
7. $m\angle ADO + m\angle BDO = 180°$	7. Supplementary angles
8. $m\angle ADO = m\angle BDO = 90°$	8. Substitution and algebra
9. $\overline{DO}$ is perpendicular bisector of $\overline{AB}$.	9. Definition of perpendicular bisector
10. $\overleftrightarrow{DO}$ is the same line as $\overleftrightarrow{CD}$.	10. Segment has only one perpendicular bisector.
11. $\overleftrightarrow{CD}$ passes through center O.	11. Point O is on $\overleftrightarrow{CO}$.

THEOREM 10-5-2 **The perpendicular bisector of a chord of a circle passes through the center of the circle.**

Example: In circle O, $\overline{AB}$ is a diameter, $\overline{CD} \perp \overline{AB}$, $AE = 5$, and $EB = 15$. Find DE and CE.

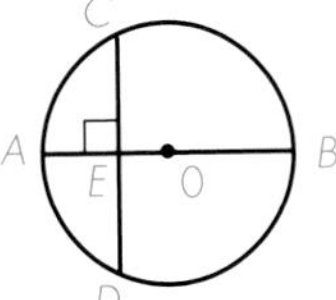

Since diameter $\overline{AB}$ is perpendicular to $\overline{CD}$, $DE = CE$.

$$DE \cdot CE = AE \cdot EB$$
$$(DE)^2 = 5 \cdot 15 = 75$$
$$DE = 5\sqrt{3} = CE$$

DISCOVERY ACTIVITY

1. Use a compass to draw a large circle. Label the center O. See margin.
2. Label a point on the circle T and draw $\overline{OT}$.
3. At point T, construct a perpendicular to $\overline{OT}$.
4. Label a point P on the perpendicular, about 6 centimeters from T.
5. Label another point on the circle B, so that $\overline{BP}$ intersects the circle at point A as shown.
6. Measure in centimeters the lengths of $\overline{PT}$, $\overline{PA}$, and $\overline{PB}$.

   PT = ________ PA = ________ PB = ________
   $(PT)^2$ = ________ $PA \cdot PB$ = ________
7. What do you observe? Do you think this will always be true? Try steps 1–6 for a different-size circle and tangent.

Using the Page
The geometric drawing tools can be used to complete this Discovery Activity.

You may have discovered that the square of the length of the tangent equals the product of the lengths of the secant and its external segment.

DEFINITION 10-5-2 **A SECANT is a line segment that intersects a circle in two points and has one endpoint on the circle and the other endpoint outside the circle.**

DEFINITION 10-5-3 **The LENGTH OF A TANGENT is the length of a segment from a point outside a circle to the point of intersection of the tangent with the circle.**

Using the Page
In discussing Definitions 10-5-2 and 10-5-3, be sure that students understand which line segment is the tangent and which is the secant with its external segment.

Using the Page

As you go through the proof of Theorem 10-5-3, be sure students do not confuse this theorem with Theorem 10-5-1, which involves the product of the segment lengths of two intersecting chords.

Given: Circle with tangent $\overleftrightarrow{PT}$ and secant $\overline{PB}$

Justify: $(PT)^2 = PB \cdot PA$

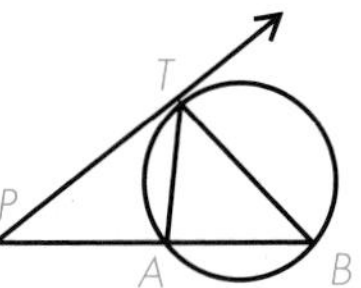

Steps	Reasons
1. Draw $\overline{AT}$ and $\overline{BT}$.	1. Two points determine a line
2. $m\angle TBA = m\angle PTA$	2. Inscribed angles intersecting the same arc are equal.
3. $m\angle P = m\angle P$	3. Identity
4. $\triangle TPB \sim \triangle APT$	4. AA
Map the triangles.	
5. $PT/AP = PB/PT = BT/TA$	5. Writing correspondences
6. $(PT)^2 = PB \cdot PA$	6. Algebra

Given a tangent and a secant from a point, then the length of the tangent squared is equal to the product of the lengths of the secant and its external segment.

Extra Practice

1. Find the length of $\overline{EC}$.

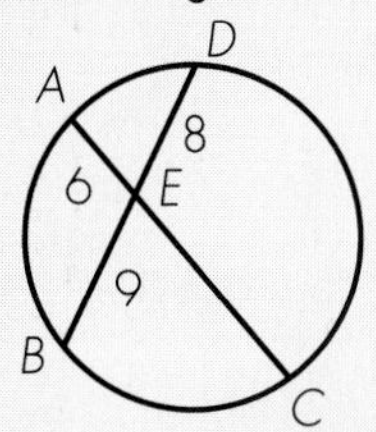

[EC = 12]

2. AC = 4 and CD = 5. Find AB.

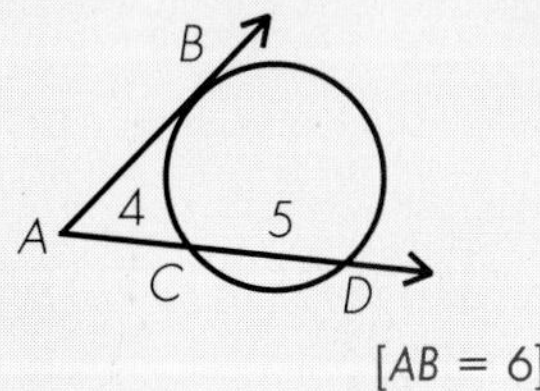

[AB = 6]

CLASS ACTIVITY

Find the value of *x*.

1. 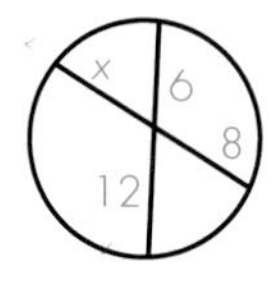

$x = 9$

2. 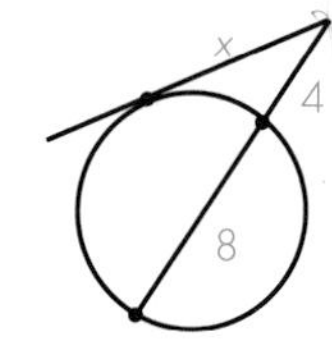

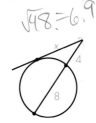

$x = 4\sqrt{3}$

3. 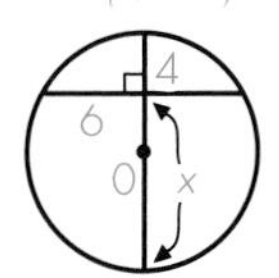

$x = 9$

4. In circle *O*, *BC* = 6 and *AC* = 12. What is the length of *DC*? 24

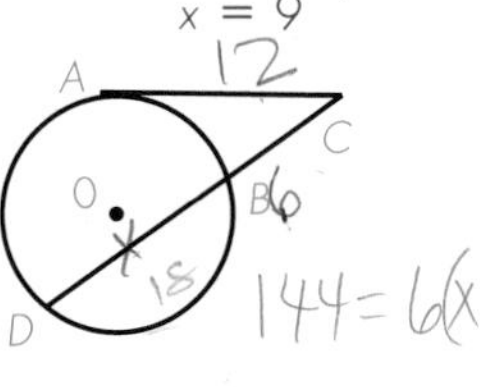

5. Use a computer and a geometric drawing tool to draw any circle and two secants that intersect outside the circle. Label the point at which the two secants intersect *A*. Label the points at which one of the secants intersects the circle *B* and *C* and the points at which the other secant intersects the circle *D* and *E*. Find the lengths of $\overline{AC}$, $\overline{AB}$, $\overline{AE}$, and $\overline{AD}$. Repeat this activity several times, using a circle with a different radius and secants of different lengths each time. What do you notice about the product of *AC* and *AB* and the product of *AE* and *AD* in each case?
The products are equal.

6. Use the figure you drew in Exercise 5 and draw $\overline{DC}$ and $\overline{BE}$. Write reasons to show that $AC \cdot AB = AE \cdot AD$.

Steps	Reasons
a. Draw $\overline{DC}$ and $\overline{BE}$.	a. Two points determine a line
b. $m\angle BCD = m\angle DEB$	b. Interior angles intersect same arc.
c. $m\angle A = m\angle A$	c. Identity
d. $\triangle AEB \sim \triangle ACD$	d. AA

Map the triangles.

e. $AE/AC = AB/AD = BE/DC$	e. Writing correspondences
f. $AC \cdot AB = AE \cdot AD$	f. Algebra

Write a conclusion to summarize the result of the proof. See margin.

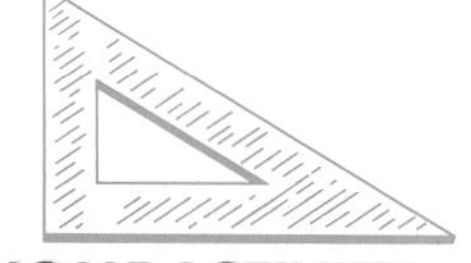

HOME ACTIVITY

For circle O, classify the following.

1. $\overleftrightarrow{PT}$ Tangent
2. $\overline{AB}$ Chord
3. $\overline{PB}$ Secant
4. $\angle PBT$ Inscribed angle
5. O Center
6. $\overset{\frown}{AB}$ Arc
7. $\overline{TO}$ Radius
8. $\angle TOB$ Central angle

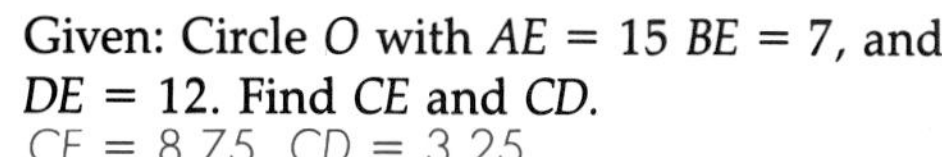

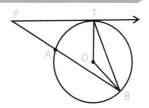

9. Given: Circle O with $AC = 24$, $DX = 14$, and $BX = 10$. Find CX. $CX = 10$ or 14

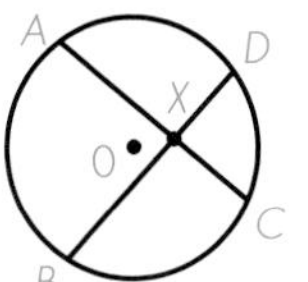

10. Given: Circle O with $AE = 15$ $BE = 7$, and $DE = 12$. Find CE and CD.
$CE = 8.75$, $CD = 3.25$

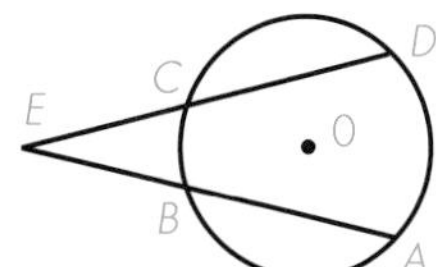

11. Given: Circle O with $BC = 4$ and $DC = 8$. Find AC.

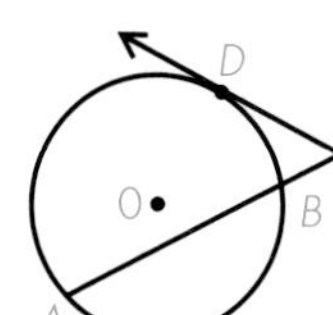

$AC = 16$

12. Given: Circle O with radius $\overline{OC}$, $AB = 12$, and $OD = 8$. Find the length of the radius.

$r = 10$

Additional Answers

6. If two secants intersect outside a circle and each one intersects the circle at two points, then the product of the lengths of one secant and its external segment equals the product of the lengths of the other secant and its external segment.

The diagram shows a section of a broken wheel. $\overline{AB} \perp \overline{DE}$, $AF = BF = 6$ cm, $EF = 2$ cm, and $\overline{ED}$ is a diameter.

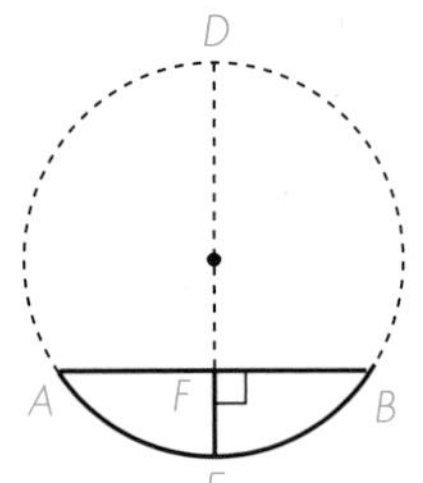

13. What would have been the length of $\overline{ED}$?
20 cm
14. What was the original circumference of the wheel? 62.8 cm
15. What was the area of the wheel? 314 cm^2

Given: Circle O with $PT = 12$, $PF = 18$, $m\angle TOA = 90°$, $m\widehat{TF} = 168°$. $\overleftrightarrow{PT}$ is an tangent.

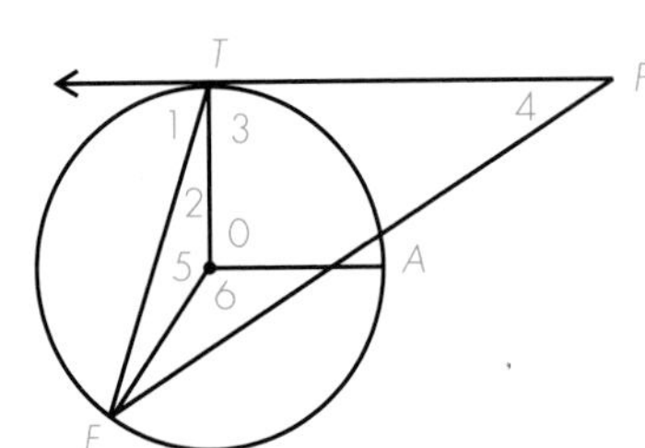

16. Find the measures of the numbered angles.
See margin.
17. Find PA. 8

18. Given: Circle O with tangents $\overleftrightarrow{PQ}$ and $\overleftrightarrow{PR}$. $PQ = 6$ cm, $m\angle RPO = 20°$. Find $m\angle QRP$ and the length of $\overline{PR}$.
$m\angle QPR = 70°$, $PR = 6$ cm

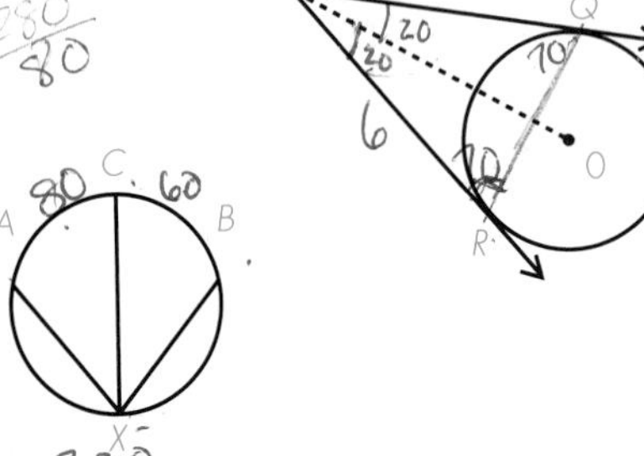

Given: Circle O with $m\widehat{CB} = 60$ and $m\widehat{AXB} = 220°$. Find the measure.

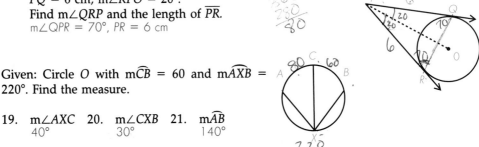

19. $m\angle AXC$
40°
20. $m\angle CXB$
30°
21. $m\widehat{AB}$
140°

CRITICAL THINKING

22. I'm tired of mowing the lawn outside my house, so I bought a sheep to eat the grass. The sheep is tied to one corner of the house. If it takes about 4000 square feet of grass to feed an adult sheep, will mine get enough to eat where she is?
Yes; the reachable area = 4101.63 sq ft

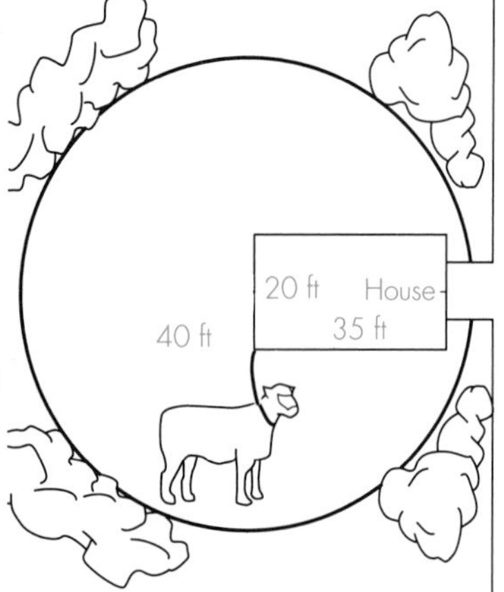

Additional Answers
16. $m\angle 1 = 84°$; $m\angle 2 = 6°$; $m\angle 3 = 90°$; $m\angle 4 = 39°$; $m\angle 5 = 168°$; $m\angle 6 = 102°$

FOLLOW-UP

Assessment
1. Find x.

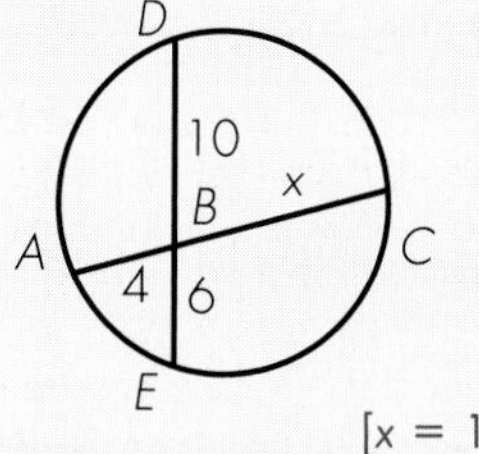

[x = 15]

2. Find x.

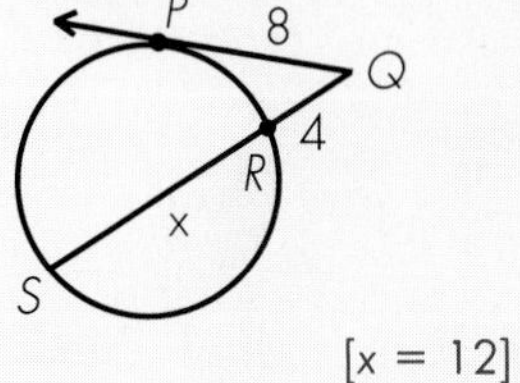

[x = 12]

Reteaching
Students who have had difficulty with this section may benefit from Reteaching Activity 10.5.

Enrichment
For students who have mastered the material in this section, you may wish to assign Enrichment Activity 10.5.

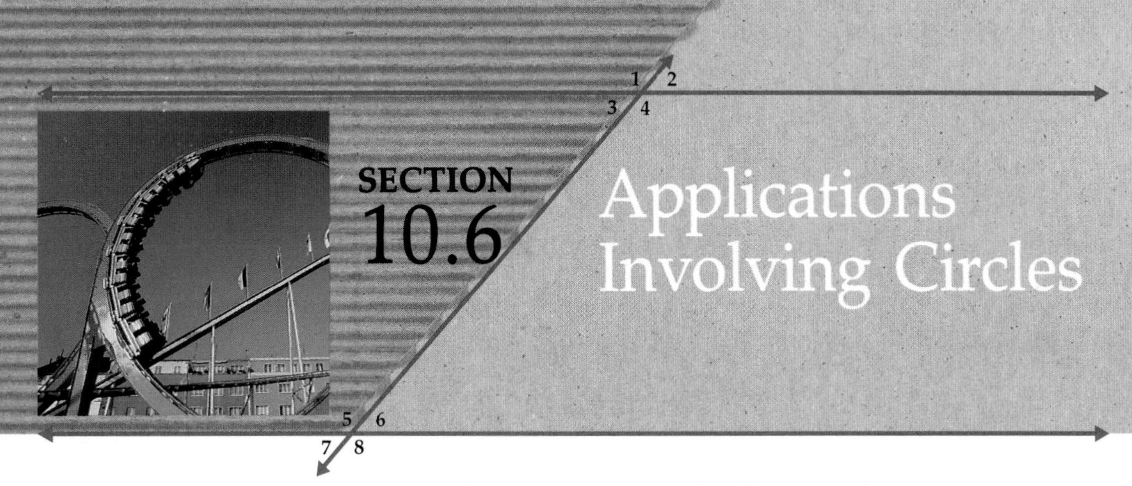

SECTION 10.6 Applications Involving Circles

After completing this section, you should understand
▲ how to solve problems using theorems about regular polygons and circles.

From your investigation of the properties of regular polygons and circles, you can see that there is a special relationship between these figures.

DISCOVERY ACTIVITY

1. Draw two circles, each with a radius of 8 centimeters. Cut them out.
2. Fold one circle in half three times.
3. Open the circle. Draw a line connecting the adjacent endpoints of the fold lines. What figure have you formed? Octagon
4. Fold the other circle in half, then fold the sides in so that they meet in a cone shape.

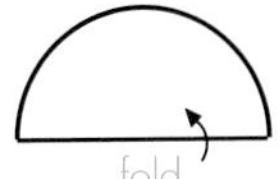

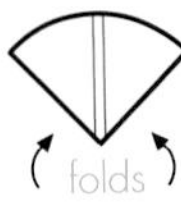

5. Open the circle and draw chords to connect adjacent endpoints. What figure have you formed? Square
6. Cut out several more circles and try to form a hexagon, an equilateral triangle, and a pentagon.
7. What do these figures all have in common? Vertices on circle

SECTION 10.6

Resources
Reteaching 10.6
Enrichment 10.6
Transparency Master 10-2

Objectives
After completing this section, the student should understand
▲ how to solve problems using theorems about regular polygons and circles.

Materials
compass
ruler

Vocabulary
inscribed
circumscribed

GETTING STARTED

Warm-Up Activity
Have students trace a circular object such as a jar lid on a piece of paper. On the chalkboard draw a carpenter's square and explain what it is. Ask them to explore how they could locate the center of their circle using only this tool. [Place the 90° angle on the circumference; mark the points where the sides of the tool intersect the circle. This is a diagonal. Repeat. The intersection of the two diagonals is the center of the circle.]

TEACHING COMMENTS

Using the Page
After the Discovery Activity, you may wish to use Transparency Master 10-2 to illustrate the inscribed polygons and to show a circumscribed polygon.

You may have found that each of the polygons you drew inside a circle has its vertices on the circumference of the circle. The polygon is **inscribed** in the circle. The circle is **circumscribed** about the polygon.

When a circle is inscribed in a polygon, each side of the polygon is tangent to the circle.

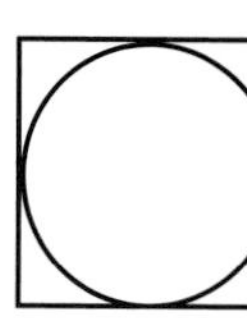
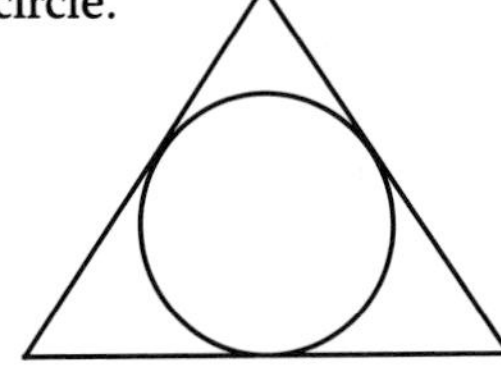
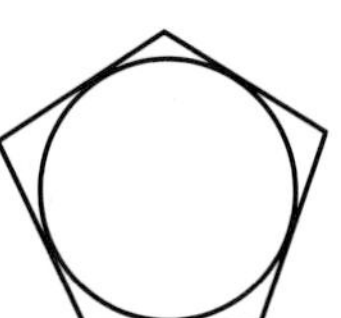

Example: $\triangle ABC$ is equilateral and inscribed in circle O. $AB = 8$ in. What is the radius of the circle?

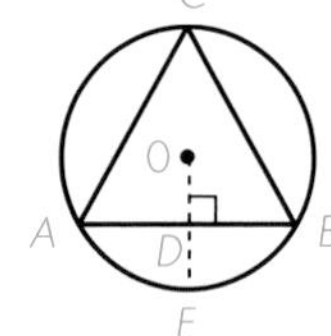

Radius $\overline{OE}$ is perpendicular to chord $\overline{AB}$ at D. $AD = DB$. Why?
Radius is ⊥ bisector of chord.
Since $AB = 8$ in., $AD = 4$ in. What is OD? $\frac{4}{\sqrt{3}}$

$$(AO)^2 = (AD)^2 + (OD)^2$$
$$(AO)^2 = 4^2 + \left(\frac{4}{\sqrt{3}}\right)^2$$
$$(AO)^2 = 16 + 5\tfrac{1}{3} = 21\tfrac{1}{3}$$
$AO = 4.6$ in. The radius of the circle is 4.6 in.

Extra Practice

1. A square plug is used to check the diameter of a hole drilled in a piece of metal. What length side should the square have to test a hole with a diameter of 4.32 cm? Round to the nearest hundredth.
[3.05 cm]

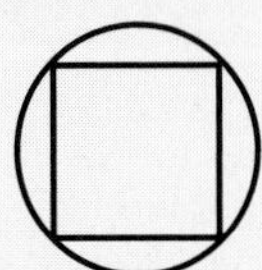

2. A square field measures 250 meters on a side. The diagrams show two types of irrigation systems. Which system will irrigate more land?

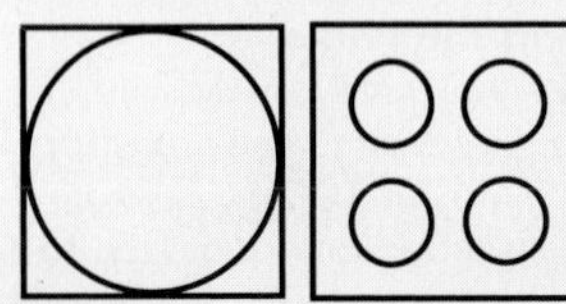

[Both will irrigate the same amount—49,062.5 m²]

CLASS ACTIVITY

1. Use a computer and a geometric drawing tool to draw any circle. Draw two diameters of the circle and label them $\overline{AC}$ and $\overline{BD}$. Draw $\overline{AB}$, $\overline{BC}$, $\overline{CD}$, and $\overline{DA}$. Repeat this activity several times, using a circle with a different diameter each time. Describe figure $ABCD$. Rectangle

2. $\triangle ABC$ is equilateral and inscribed in a circle. Show that the angles of $\triangle ABC$ divide the circle into three arcs with the same measure. Give a reason for your answer.

$m\angle A = m\angle B = m\angle C = 60°$; so $m\widehat{AC} = m\widehat{CB} = m\widehat{BA} = 120°$; inscribed angles are equal.

3. A circular clock face is inscribed in a square frame. The side of the square measures 8 feet. What is the area of the part of the square that is not used for the face of the clock? 13.76 sq ft

AREA OF A SECTOR

You can use the formula for the area of a circle to find the area of a part of the circle called a sector.

$\overset{\frown}{POQ}$ and $\angle POQ$ are the boundaries of sector POQ. You know that $m\angle POQ = m\overset{\frown}{POQ}$, because $\angle POQ$ is a central angle. You can use this ratio to find the area of sector POQ.

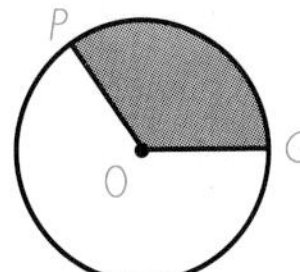

$$\frac{\text{Area(sector } POQ)}{\text{Area(circle } O)} = \frac{m\angle POQ}{360°}$$

Example: $\overline{AB}$ is a chord of circle O. Radius $OA = 10$ cm and $m\angle AOB = 90°$. What is the area of the shaded region?

Area(circle O) $= \pi r^2 = 100\pi$ cm^2
Area($\triangle AOB$) $= \frac{1}{2}bh = \frac{1}{2}(10 \cdot 10) = 50$ cm^2

$$\frac{\text{Area(sector } AOB)}{100\pi} = \frac{90°}{360°}$$

Area(sector AOB) $= \frac{1}{4} \cdot 100\pi = 25\pi \approx 78.5$ cm^2
Area(shaded region) $=$ Area(sector AOB) $-$ Area($\triangle AOB$)
$= 78.5 - 50 = 28.5$ cm^2

CLASS ACTIVITY

1. A circular theater is divided into six congruent sectors. The diameter of the theater is 60 feet. Two of the sectors are used for audience seating. What is the area of the space taken up by seats? 942 sq ft

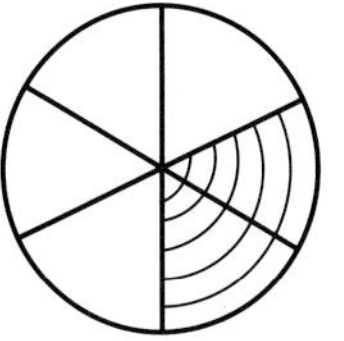

Find the area of the shaded region. Use $\pi \approx 3.14$.

2. 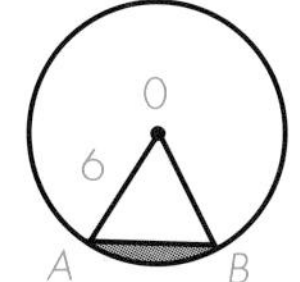

$r = 6$ ft
$m\angle AOB = 60°$
$A = 3.25$ sq ft

3. 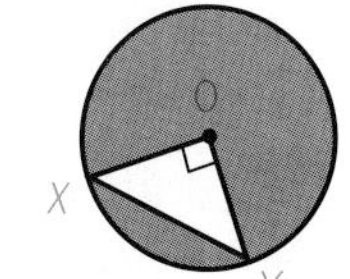

$r = 12$ in.
$m\angle XOY = 90°$
$A = 380.16$ sq in.

4. 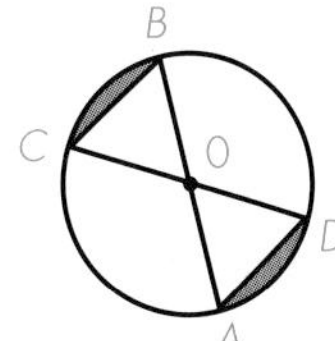

$AB = 36$ cm
$m\angle AOD = 60°$
$A = 58.53$ cm^2

Extra Practice

1. A circle with a radius of 12 centimeters is divided into nine congruent parts. What is the area of one of the parts? [$A = 50.24$ cm^2]

2. $XZ = ZY = 12$; $XO = 15$. Find OZ.

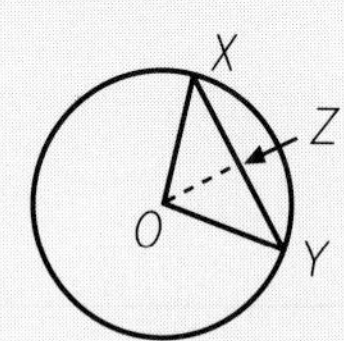

[$OZ = 9$]

3. $\overline{AB} \perp \overline{CD}$; $AB = 8$, $AC = 10$. Find CD.

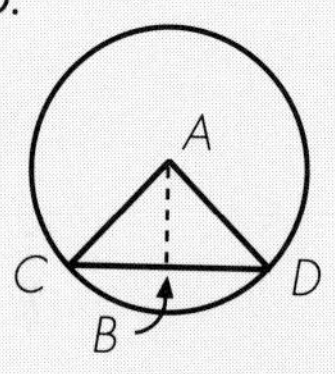

[$CD = 12$]

Using the Page
During the discussion of circles and similarity, ask students to consider whether all circles are similar. Have the give a reason for their answers. [Yes; the circumference of every circle is a multiple of the constant π.]

CIRCLES AND SIMILARITY

In Chapter 9, you learned that for two similar polygons with sides S_1 and S_2, perimeters P_1 and P_2, and areas A_1 and A_2,

$\frac{S_1}{S_2} = \frac{P_1}{P_2}$ and $(\frac{S_1}{S_2})^2 = \frac{A_1}{A_2}$.

Do you think that similar ratios will also be true for circles? Try it for two circles, one with a radius of 4 feet and one with a radius of 8 feet.

1. Find the circumference of each circle.

   Use $\pi \approx 3.14$.
   $C_1 = 25.12$ ft; $C_2 = 50.24$ ft
2. What is the ratio of C_1/C_2? Is this ratio equal to the ratio of the radii of the two circles?
   0.5 or $\frac{1}{2}$; yes
3. Predict the ratio of the areas of circle_1 and circle_2.
   Answers may vary. Possible answer: The ratio will be 0.5, or the area will double.
4. Find the area of each circle.
   $A_1 = 50.24$ sq ft; $A_2 = 200.96$ sq ft
5. What is the ratio $\frac{A_1}{A_2}$? Does your answer match your prediction? 0.25 or $\frac{1}{4}$

So, for two circles with radii r and R,

$$\frac{\text{Circumference (circle } r)}{\text{Circumference(circle } R)} = \frac{r}{R} \text{ and } \frac{\text{Area(circle } r)}{\text{Area(circle } R)} = \left(\frac{r}{R}\right)^2$$

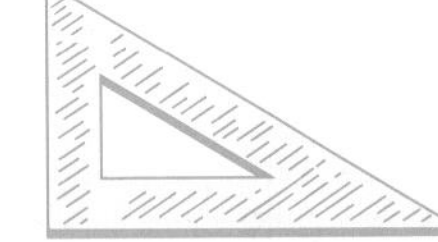

HOME ACTIVITY

Using the Page
Encourage students to check their work in the Home Activity with a calculator, if they have one available.

Find the area of the shaded region. Use $\pi = 3.14$.

1. 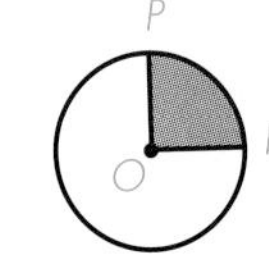

   $OP = 12$ cm
   $m\angle POR = 90°$
   $A = 113.04 \text{ cm}^2$

2. 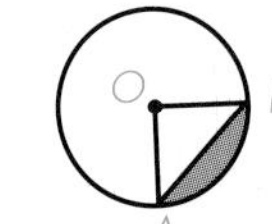

   $OA = 10$m
   $m\angle BOA = 90°$
   $A = 28.5\text{m}^2$

3. 24 in.
   24 in.
   $A = 123.84$ sq in.

4. A B D C
   $AB = 8$m
   $AD = 6$m
   $A = 30.5\text{m}^2$

5. Two circles have radii of 8 centimeters and 12 centimeters. What is the ratio of their areas? $\frac{4}{9}$

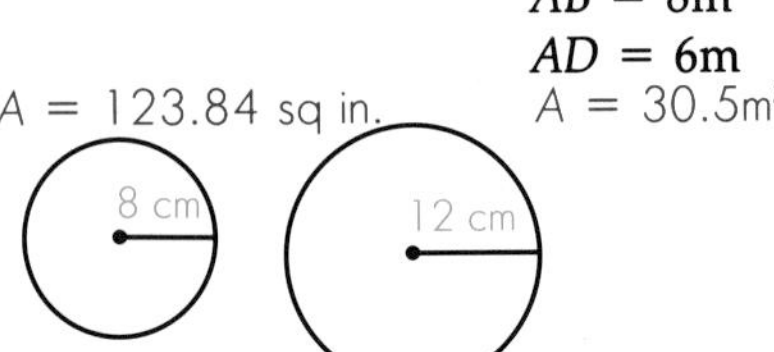

Find the value of x for each circle.

6. 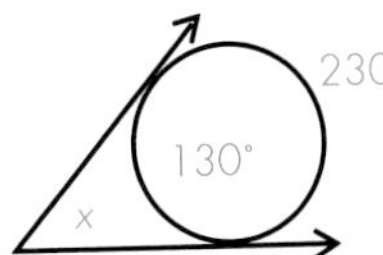

7. 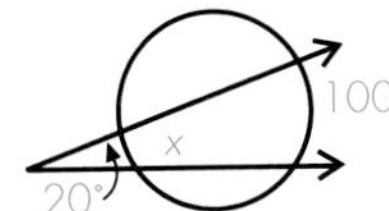

8. 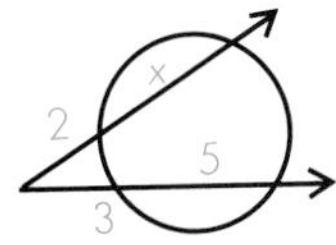

9. 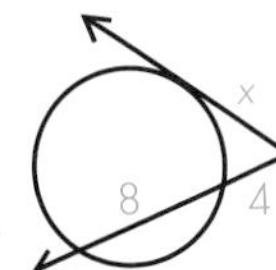

$x = 50°$ $x = 60°$ $x = 10$ $x = 6.9$

10. A circular theater building has a stage that is 4 meters from the center of the circle. The radius of the building is 20 meters. How wide is the front edge of the stage? 39.2m

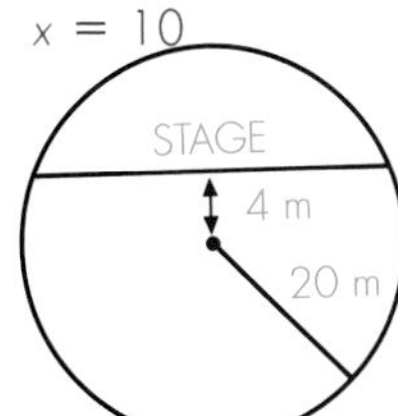

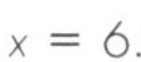

11. A Ferris wheel is made up of a regular hexagon divided in triangles and surrounded by a circular frame. The wheel has a radius of 50 feet. How many feet of lights would it take to outline the circumference and the spokes of the wheel? 614 ft

12. Quadrilateral $PQRX$ is inscribed in circle O. Show that $\angle P$ is supplementary to $\angle R$ and $\angle Q$ is supplementary to $\angle X$. $m\angle P = \frac{1}{2} m\widehat{QRX}$; $m\angle R = \frac{1}{2} m\widehat{QPX}$; $m\widehat{QRX} + m\widehat{QPX} = 360°$; $\frac{1}{2}(m\widehat{QRX} + m\widehat{QPX}) = 180°$; $m\angle P + m\angle R = \frac{1}{2}(m\widehat{QRX} + m\widehat{QPX}) = 180°$; same steps for $\angle X$ and $\angle Q$

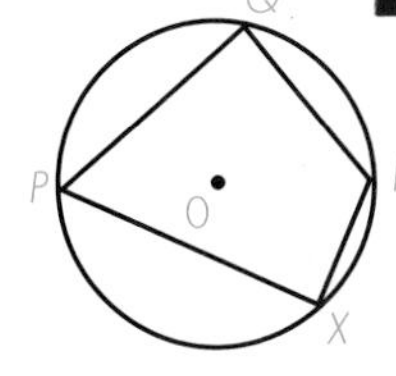

CRITICAL THINKING

In Montana there is an unusual natural stone formation in the shape of a tall pillar, called Pompey's Pillar. Assume that the formation has a circular base. A surveyor stands 1000 feet from the base of the pillar and finds a tangent to the pillar is 1500 feet.

13. How can you use this information to find the diameter of the pillar? See margin.
14. What is the diameter? 1250 ft

Additional Answers

12. $m\angle P = \frac{1}{2}(m\widehat{QRX})$; $m\angle R = \frac{1}{2}(m\widehat{QPX})$; $m\widehat{QRX} + m\widehat{QPX} = 360°$; $\frac{1}{2}(m\widehat{QRX} + m\widehat{QPX}) = 180°$; $m\angle P + m\angle R = \frac{1}{2}(m\widehat{QRX} + m\widehat{QPX}) = 180°$; same steps for $\angle X$ and $\angle Q$.

13.

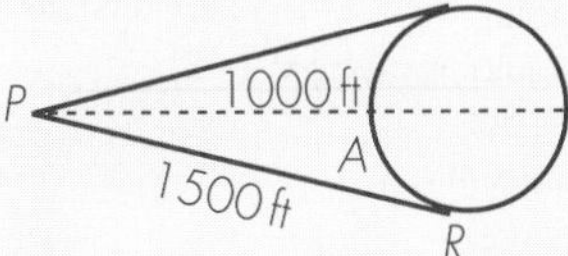

$PA(PA + \text{diameter of pillar}) = (PR)^2$

FOLLOW-UP

Assessment

1. In circle O, Z is the midpoint of $\overline{XY}$. $OZ = XZ$; $OX = 2\sqrt{2}$. Find XY.

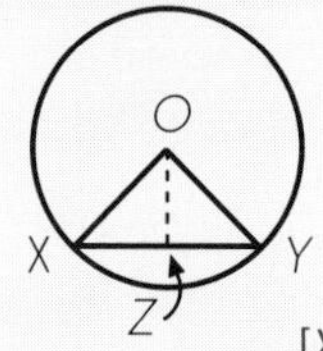

[$XY = 4$]

2. $\overline{AB}$ is a chord of circle O. Radius $OB = 4$ cm and $m\angle AOB = 60°$. Find the area of the shaded region.

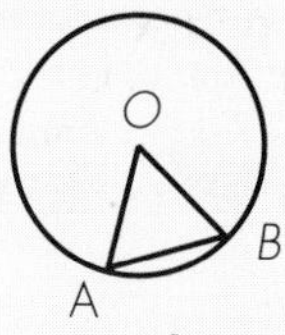

[$A = 1.44\ \text{cm}^2$]

3. Two circles have radii of 6 centimeters and 9 centimeters. What is the ratio of their areas? $\left[\frac{4}{9}\right]$

Reteaching

Students who have had difficulty with this section may benefit from Reteaching Activity 10.6.

Enrichment

For students who have mastered the material in this section, you may wish to assign Enrichment Activity 10.6.

Chapter Vocabulary

arc
central angle
chord
circumference
circumscribed
diameter
inscribed
inscribed angle
length of a tangent
pi
polygon
radius
regular polygon
secant
tangent

10.1 *ABCDE* is a regular polygon. Its sides have equal lengths and its angles have equal measures. Where n is the number of sides, the measure of $\angle A$ is

$$m\angle A = \frac{(n-2)180°}{n}$$
$$m\angle A = \frac{3 \cdot 180°}{5} = 108°$$

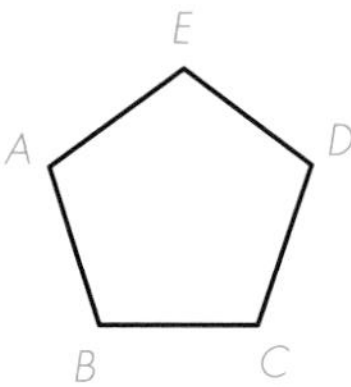

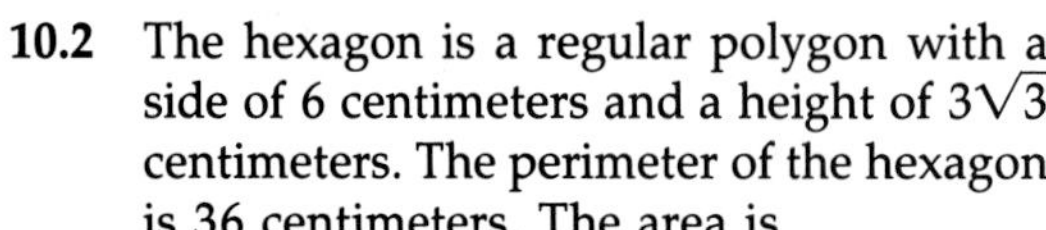

10.2 The hexagon is a regular polygon with a side of 6 centimeters and a height of $3\sqrt{3}$ centimeters. The perimeter of the hexagon is 36 centimeters. The area is

$$A = \tfrac{1}{2}Ph$$
$$A = \tfrac{1}{2} \cdot 36 \cdot 3\sqrt{3} = 54\sqrt{3} \text{ cm}^2$$

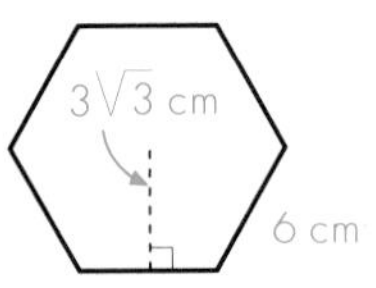

10.3 Circle O has a diameter of 20 inches and a radius of 10 inches. The circumference of the circle is
$C = \pi d$ or $C = 2\pi r$ $\quad \pi \approx 3.14\ldots$
$C \approx 3.14 \cdot 20 \approx 62.8$ in.
The area of the circle is
$A = \pi r^2$
$A = \pi(10)^2 = \pi \cdot 100 \approx 314$ sq in.

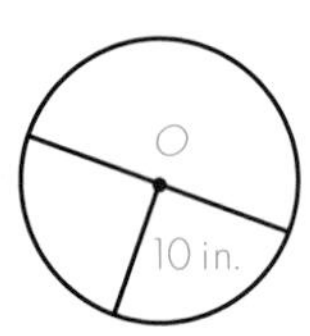

10.4 A central angle has its vertex at the center of a circle. For circle O, $m\widehat{AB} = 70°$ and $m\widehat{DC} = 110°$.

$\angle AOB$ is a central angle, so
$m\angle AOB = m\widehat{AB} = 70°$.

An inscribed angle has its vertex on the circle.

$\angle DBC$ is an inscribed angle, so
$m\angle DBC = \frac{1}{2}(m\widehat{DC}) = \frac{1}{2} \cdot 110° = 55°$.

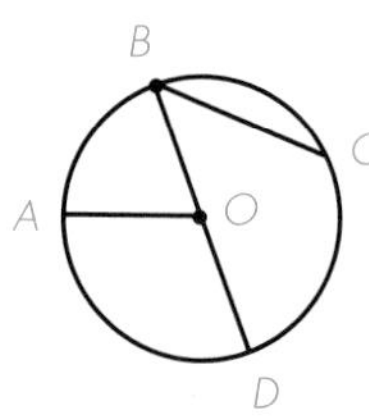

For circle O, $m\widehat{QR} = 85°$ and $m\widehat{PS} = 75°$. $\overline{PR}$ and $\overline{QS}$ are chords that intersect inside the circle.

$$m\angle PTS = \tfrac{1}{2}(m\widehat{QR} + m\widehat{PS})$$
$$m\angle PTS = \tfrac{1}{2}(85° + 75°) = 80°$$

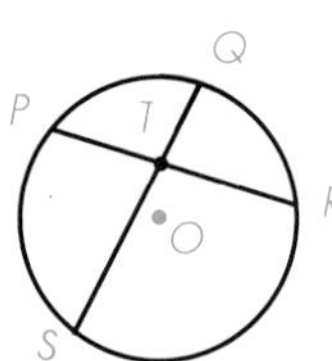

A secant is a line segment that intersects a circle in two points. $\overline{XZ}$ and $\overline{XT}$ are secants. $m\widehat{ZT} = 130°$ and $m\widehat{YS} = 40°$.

$$m\angle X = \tfrac{1}{2}(m\widehat{ZT} - m\widehat{YS})$$
$$m\angle X = \tfrac{1}{2}(130° - 40°) = 45°$$

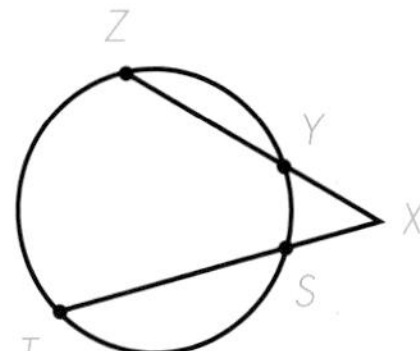

A tangent is a line that intersects a circle in one point. Line $\overleftrightarrow{AB}$ is tangent to circle O at point B. Secant $\overline{AD}$ intersects circle O at C and D. Radius $\overline{OB}$ is perpendicular to tangent $\overleftrightarrow{AB}$.

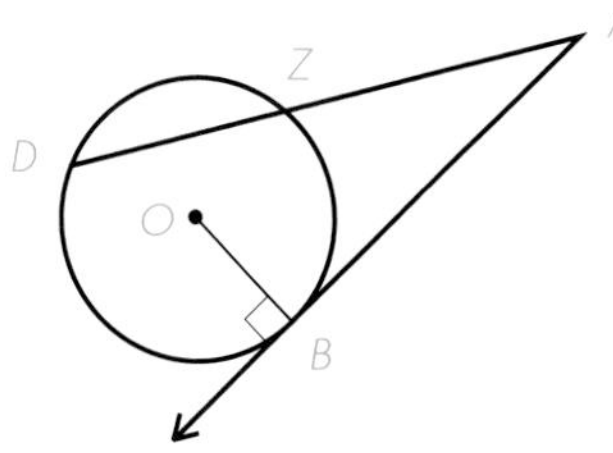

$m\widehat{BD} = 140°$ and $m\widehat{BC} = 80°$.

$$m\angle A = \tfrac{1}{2}(m\widehat{BD} - m\widehat{BC})$$
$$m\angle A = \tfrac{1}{2}(140° - 80°) = 30°$$

10.5 A chord of a circle has its endpoints on the circle. Chords $\overline{AB}$ and $\overline{CD}$ intersect at point E. Find AE.

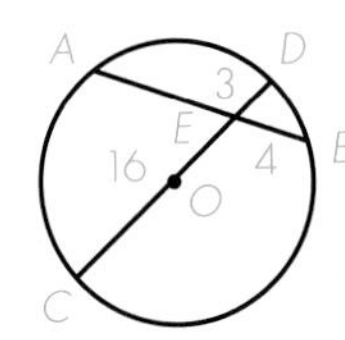

$$AE \cdot EB = CE \cdot ED$$
$$AE \cdot 4 = 16 \cdot 3$$
$$AE = 12$$

The perpendicular bisector of chord $\overline{AB}$ passes through the center of the circle. $\overline{OD} = 10$. Find AB.

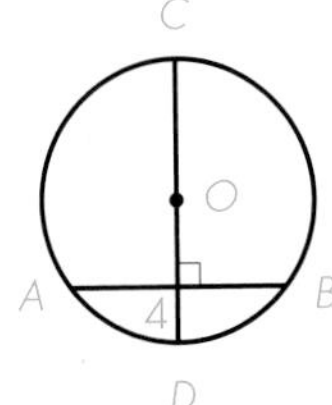

$$(AO)^2 = (OE)^2 + (AE)^2$$
$$10^2 = 6^2 + (AE)^2$$
$$8 = AE$$
$$AB = 2(AE) = 16$$

$\overleftrightarrow{PQ}$ is a tangent and $\overline{PS}$ is a secant of the circle. $PS = 25$ and $PR = 10$. Find PQ.

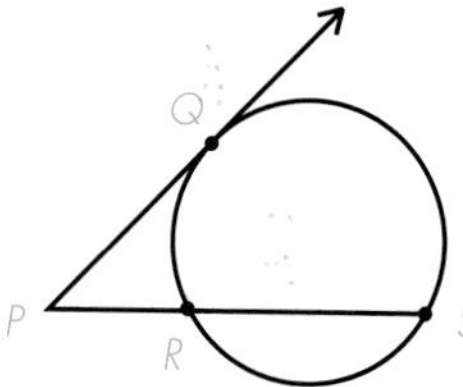

$$(PQ)^2 = PS \cdot PR$$
$$(PQ)^2 = 25 \cdot 10$$
$$PQ = 15.8$$

Using the Page
Use the computer to illustrate the Total Turtle Trip Theorem. Have students compute the exterior angle for different *n*-gons. Then discuss the procedure used to draw a circle. Have students work individually or in small groups at the computer to complete the activity.

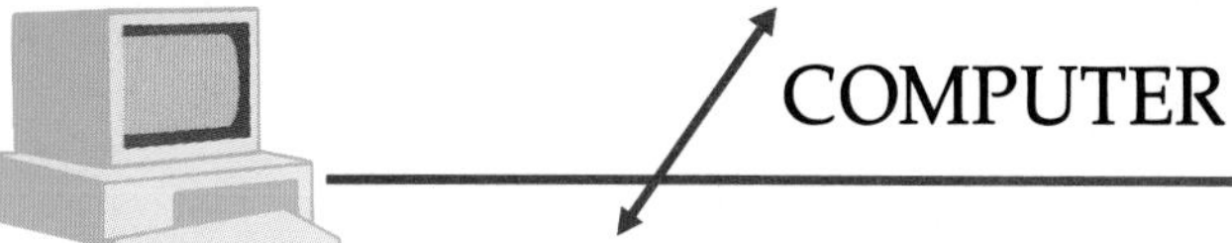

COMPUTER ACTIVITY

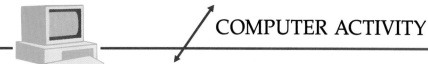

To draw a regular polygon using Logo, the turtle must turn through the exterior angles of the polygon. Since you know that the sum of the exterior angles for any polygon is 360°, the measure of each exterior angle of a regular n-gon is $360 \div n$. So, to draw a regular n-gon, the turtle must make n turns of $(360 \div n)°$ each. In Logo, this is called the Total Turtle Trip Theorem, or the TTT Theorem.

Example: Use the TTT Theorem to write a procedure to draw a regular pentagon.

The turtle must make five turns to draw a pentagon. So the measure of the angle at each turn is $(360 \div 5)°$ or 72°. The measure of the angle at each turn can be computed in the procedure by the computer.

```
TO PENTAGON
  RT 90
  REPEAT 5 [FD 50 LT 360/5]
END
```

The output is shown below.

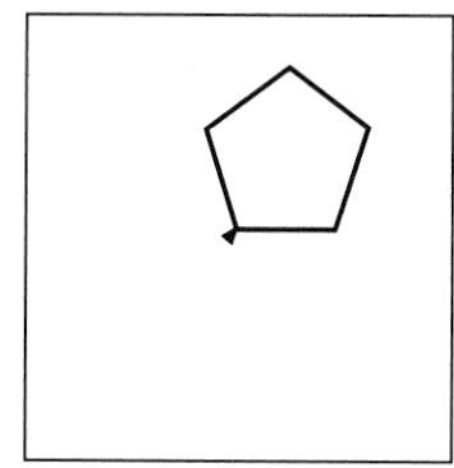

Use the TTT Theorem to write a procedure to tell the turtle to draw each polygon.
Answers may vary.

1. Octagon
2. Nonagon
3. Decagon
4. 15-gon
5. 20-gon
6. 30-gon

7. What figure does the 30-gon look like? A circle

You have already discovered that as the number of sides of a regular polygon increases, the perimeter of the polygon approaches the circumference of the circle. This information along with the results of Exercises 1–7 indicates that a circle can be drawn in Logo.

The TTT Theorem tells us that to draw a regular polygon, the turtle turns through 360° to return to its starting position. This is also true when the turtle draws a circle. The basic command used to draw a circle using Logo is REPEAT X[FD Y RT Z], where $X \cdot Z = 360$. When using this command, be sure that as X gets larger, Y gets smaller so that the circle will fit on the screen.

Example: Find X so that the command REPEAT X[FD 2 RT 5] can be used in a procedure to tell the turtle to draw a circle. Then write the procedure.

We know that $X \cdot Z = 360$. In the command REPEAT X[FD 2 RT 5], $Z = 5$. So, to find X, substitute 5 for Z and solve.

$$X \cdot 5 = 360$$
$$X = 72$$

So the command is REPEAT 72[FD 2 RT 5]. The procedure to draw the circle is given below.

```
TO CIRCLE
  RT 90
  REPEAT 72[FD 2 RT 5]
END
```

Find the unknown value so that each command can be used in a procedure to tell the turtle to draw a circle. Then write the procedure. See margin.

8. REPEAT X[FD 5 RT 12] **9.** REPEAT X[FD 8 RT 36] **10.** REPEAT X[FD 7 RT 18]

11. REPEAT 90[FD 2 RT Z] **12.** REPEAT 24[FD 2 RT Z] **13.** REPEAT 180[FD 1 RT Z]

Write a Logo procedure to draw a circle in each position on the screen. Answers may vary.

14. Upper right-hand corner **15.** Upper left-hand corner

16. Lower right-hand corner **17.** Lower left-hand corner

Write procedures that tell the turtle to draw each figure. Answers may vary.

18.

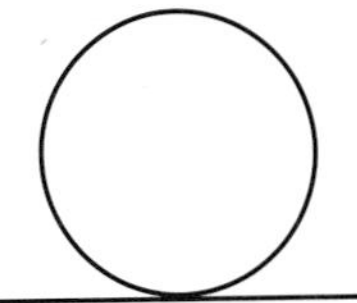

19.

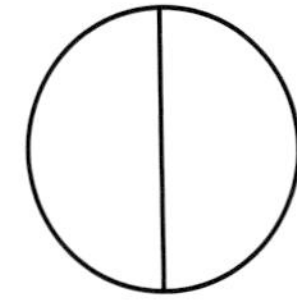

Additional Answers

8. TO CIRCLE
RT 90
REPEAT 30 [FD 5 RT 12]
END

9. TO CIRCLE
RT 90
REPEAT 10 [FD 8 RT 36]
END

10. TO CIRCLE
RT 90
REPEAT 20 [FD 7 RT 18]
END

11. TO CIRCLE
RT 90
REPEAT 90 [FD 2 RT 4]
END

12. TO CIRCLE
RT 90
REPEAT 24 [FD 2 RT 15]
END

13. TO CIRCLE
RT 90
REPEAT 180 [FD 1 RT 2]
END

Using the Page
Students can work individually or in groups to review Chapter 10. If you have students study in groups, they should work together to make sure all members of the group know the material in the chapter.

Informal assessment such as interviews, classroom observations, or a review of student portfolios can be used instead of paper-and-pencil tests.

CHAPTER REVIEW

Find the measure of an interior angle for each regular polygon.

1. Hexagon 120° **2.** Decagon 144° **3.** 21-gon 162.9° **4.** 18-gon 160°

5. The measure of an interior angle of a regular polygon is 150°. How many sides does the polygon have? 12

6. A regular pentagon has a side of 20 centimeters and an apothem of 14 centimeters. What is its area? 700 cm²

Find the circumference and area. Use $\pi = 3.14$.

7.

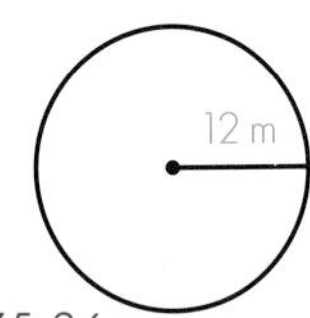

C = 75.36 m;
A = 452.16 m²

8.

C = 56.52 in.;
A = 254.34 sq in.

9.

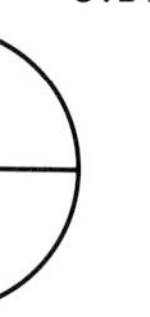

C = 31.4 ft;
A = 78.5 sq ft

10.

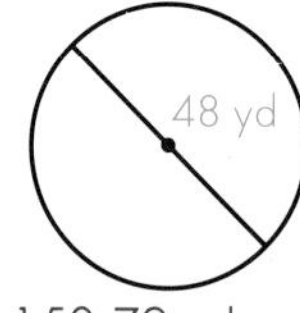

C = 150.72 yd;
A = 1808.64 sq yd

11. How many revolutions will a 26-inch bicycle wheel make in traveling a distance of one mile? (1 mi = 5280 ft) 776 revolutions

In circle O, $m\angle A = 40°$, $m\angle BDA = 20°$, and $\overline{CE}$ is a diameter. Find the measure.

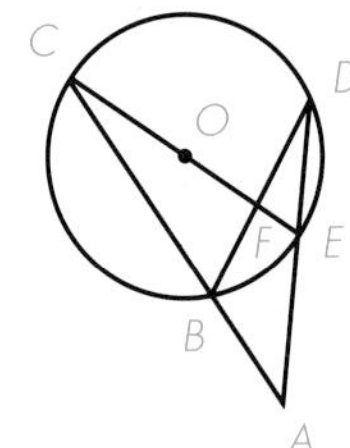

12. $m\angle ACE$ 20° **13.** $m\angle CBD$ 60° **14.** $m\angle DBA$ 120°

15. $m\widehat{BE}$ 40° **16.** $m\widehat{CB}$ 140° **17.** $m\widehat{CD}$ 120°

18. $m\angle CFD$ 80° **19.** $m\widehat{DE}$ 60° **20.** $m\angle DFE$ 100°

Find the value of x for each circle.

21.

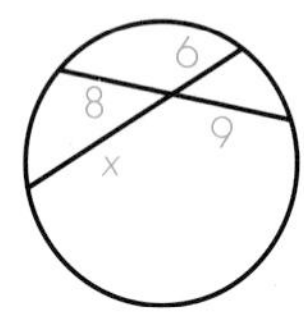

x = 12

22.

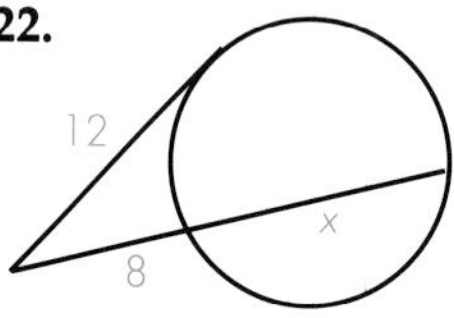

x = 10

23.

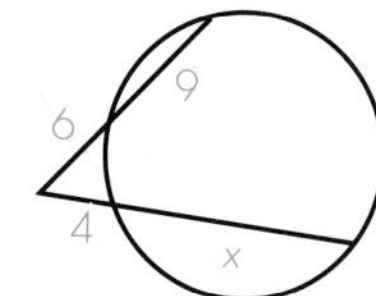

x = 18.5

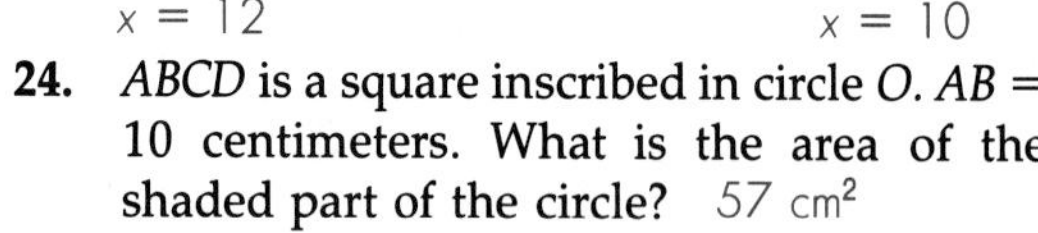

24. $ABCD$ is a square inscribed in circle O. AB = 10 centimeters. What is the area of the shaded part of the circle? 57 cm²

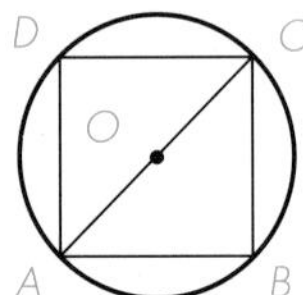

CHAPTER TEST

Find the measure of an interior angle for each regular polygon.

1. Octagon 135° **2.** Nonagon 140° **3.** 28-gon 167.14° **4.** 40-gon 171°

Find the measure of an exterior angle for each regular polygon.

5. Hexagon 60° **6.** Pentagon 72° **7.** Decagon 36° **8.** 20-gon 18°

9. The sum of the angle measures for a regular polygon is 7200°. How many sides does the polygon have? 42

Find the circumference and area of each circle. Use $\pi \approx 3.14$.

10.

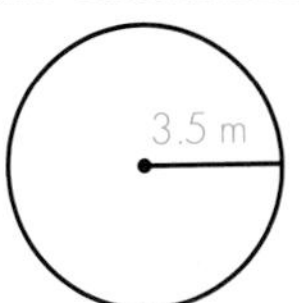

C = 21.98 m; A = 38.47 m²

11.

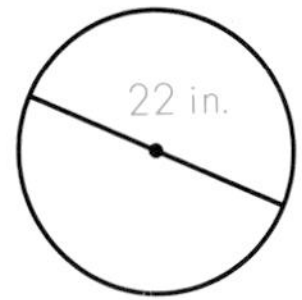

C = 69.08 in.; A = 379.94 sq in.

12.

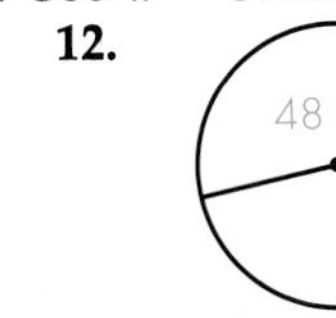

C = 150.72 yd; A = 1808.64 sq yd

13.

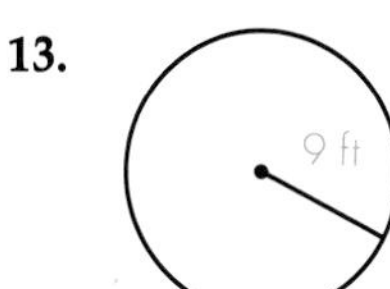

C = 56.52 ft; A = 254.34 sq ft

A circular jogging path has a circumference of one mile.

14. What is the area in square miles of the space enclosed by the jogging path? Use $\pi \approx 3.14$. 0.08 sq mi

15. What is the length in feet of a path from one side of the circle to the other, through the center? (1 mi = 5280 ft) 1681.5 ft

In circle O, $m\widehat{TS} = 110°$, $m\angle TVU = 50°$, $m\angle QOR = 60°$, and $\overline{TR}$ is a diameter.

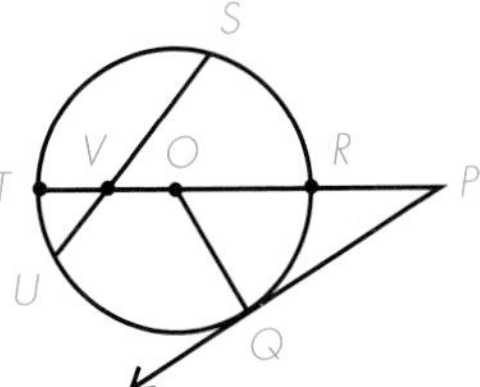

Find these measures.

16. $m\widehat{SR}$ 70° **17.** $m\widehat{TU}$ 30° **18.** $m\angle TPQ$ 30°

19. $m\widehat{QR}$ 60° **20.** $m\widehat{UQ}$ 90° **21.** $m\angle TVS$ 130°

22. $m\angle OQP$ 90° **23.** $m\angle TOQ$ 120° **24.** $m\angle SVR$ 50°

25. $ABCD$ is a rectangle inscribed in circle O. AD = 18 cm and AB = 24 cm. What is the area of the shaded region? 274.5 cm²

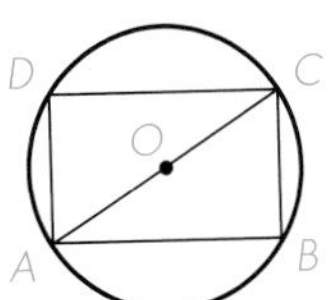

Find the value of x.

26.

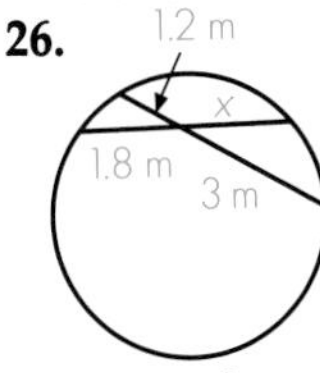

x = 2 m

27.

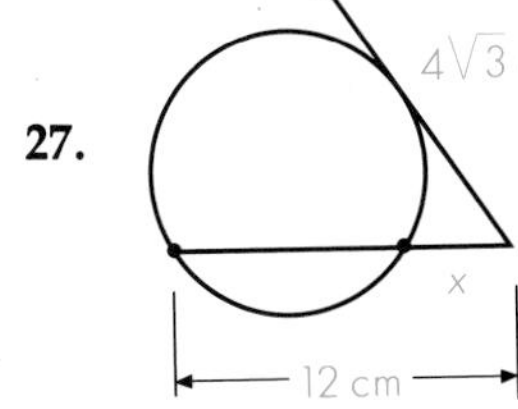

x = 4 cm

28.

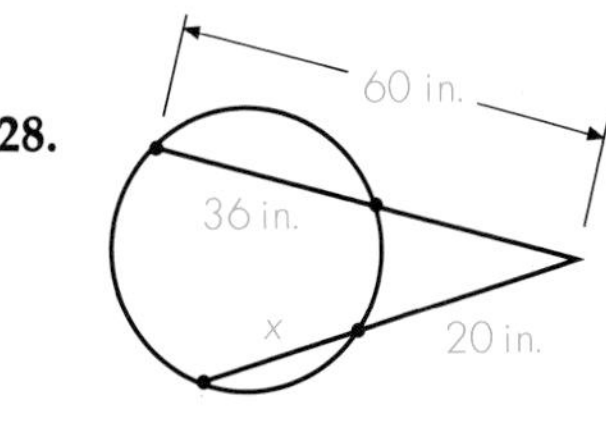

x = 52 in.

Assessment Resources
Achievement Tests pp. 19-20

Test Objectives
After studying this chapter, students should understand
- the definition of a regular polygon.
- how to find the measure of the interior angles of a polygon.
- how to find the area of a regular polygon.
- the definition of a circle and its parts.
- how to find the circumference of a circle.
- how to find the area of a circle.
- how to identify central and inscribed angles and their intersected arcs.
- how to find the measures of angles intersecting circles.
- how to identify a tangent to a circle.
- the definitions of a chord and a secant.
- how to use the theorems about chords and secants to solve problems.
- how to solve problems using theorems about regular polygons and circles.

395

Algebra Review Objectives
Topics reviewed from algebra include

- evaluating functions by substituting values for the variables.
- solving equations with one variable.

ALGEBRA SKILLS

The formulas you use to find the perimeter, area, and circumference of regular polygons and circles are called functions. These functions show the relationships between measures.

A **relation** is a set of ordered pairs.

$$R = \{(2, 4), (4, 8), (8, 16), (16, 32)\}$$

A **function** is a relation in which the first term in each ordered pair is different.

Example: $P = \{s, 4s\} = \{(1, 4), (2, 8), (3, 12), (4, 16)\}$

Function P shows the relation between the first term in the ordered pair (the length of a side of a square) and the second term in the ordered pair (the perimeter of the square).

Example: $Q = \{(1, 3), (2, 6), (1, 9), (3, 12)\}$ Is Q a function?

Two of the ordered pairs in relation Q have the same first term, so Q is not a function.

To evaluate a function, substitute values for the variables and perform the operations.

Function $S = \{4\pi r^2\}$, where $\pi \approx 3.14$.

1. Copy and complete the table.

Value of r		Value of S
1	$S = 4(3.14)1^2$	12.56
2	$S = 4(3.14)2^2$	50.24
4		200.96
8		803.84
16		3215.36
0.5		3.14
0.001		0.0000126

2. If the value of function S is 1234, what is the positive value of r to hundredths? 9.91

3. Function $x = \{x^2 + 1\}$. Find the value of f(x) for $x = \{-1, 0, 2, 3, 5\}$. $f(x) = \{2, 1, 5, 10, 26\}$

4. Function $\left\{(x, y)\ y = \frac{-5}{3}x + 2\right\}$. Find the value of y for $x = \{-2, -1, 0, 1, 2\}$.
$y = \left\{\frac{16}{3}, \frac{11}{3}, 2, \frac{1}{3}, -\frac{4}{3}\right\}$

Solve the equations.

5. $(y + 4) - (y - 6) = 5(y - 8)$
$y = 10$

6. $16 - 4(2x + 1) = 6x + (1 - 3x)$
$x = 1$

7. $\frac{3}{16} = \frac{x}{(x - 13)}$
$x = -3$

8. $\frac{x^2}{3} + \frac{7x}{3} + 4 = 0$
$x = -3, x = -4$

Name ______________________ Date ______________

Regular Polygons

Polygons A, B, and C are *concave*. It is possible to find two points inside the polygon such that the line segment joining the points is not entirely inside the polygon.

Polygons D, E, and F are *convex*. Any non-tangent line that intersects the polygon intersects it in exactly two points. Each interior angle measures less than 180°.

Polygons G, H, and I are *non-regular*. Either the sides are not all congruent, or the angles are not all congruent, or both.

Polygons J, K, and L are *regular*. The sides are congruent and the interior angles are also congruent. Regular polygons are both *equilateral* (equal sides) and *equiangular* (equal angles).

A B C D E F G H I J K L

Use the figures above for Exercises 1–6.

1. Which of polygons D through L are concave? G
2. Which of polygons A through C and G through L are convex? H, I, J, K, L
3. Which of polygons A through F and J through L are non-regular? A, B, C, D, E
4. Which of polygons A through I are regular? F
5. Which of polygons A through L are quadrilaterals? D, H, J
6. Which pairs of polygons appear to be congruent? B and G, E and I, F and L

Circle the letter of the best description for each polygon.

7.
   a. non-regular hexagon
   b. regular quadrilateral
   c. (circled) non-regular pentagon
   d. regular triangle

8.
   a. concave pentagon
   b. (circled) convex hexagon
   c. convex octagon
   d. concave triangle

9.
   a. (circled) regular convex quadrilateral
   b. non-regular concave quadrilateral
   c. regular convex pentagon
   d. regular concave hexagon

Name ______________________ Date ______________

Spirals

A *spiral* is a curve traced by a point that moves around a fixed point, called the pole, from which the point continually moves away. There are different types of spirals. The one at the right is called an *Archimedean* spiral. The distance from its pole increases in arithmetic sequence. For this spiral, the distance increases one-half centimeter for every 45° of rotation. The groove in a phonograph record is in the shape of an Archimedean spiral.

Spirals occur often in nature. Some examples are the webs of some spiders, elephant tusks, heads of daisies, the chambered nautilus, and the shapes of many galaxies.

1. 2. 3.

The two spirals at the right are distortions of the spiral at the left. On the lines below, describe how the two distortions were made.

1. Figure 2: Compressed or shrunk about 50% in width
2. Figure 3: Compressed or shrunk about 50% in height

The spirals below are called *Baraville* spirals. They are created by drawing nested regular polygons and then shading sections as shown. Each successive polygon is rotated a certain number of degrees. Give the number of degrees of rotation for each polygon below. Then copy the spirals on separate sheets of paper. Try creating a Baraville spiral in a set of nested octagons.

3. Degrees rotated: 45°
4. Degrees rotated: 30°

Name ______________________ Date ______________

Area of Regular Polygons

The perimeter of a regular polygon is given by the formula $p = ns$, where n is the number of sides and s is the length of each side.

The perpendicular distance from a side of a regular polygon to its center is called the *apothem*. The figure at the right shows the five apothems for pentagon *PENTA*.

One formula for the area of a regular polygon is $A = pa \div 2$, where p is the perimeter and a is the length of an apothem.

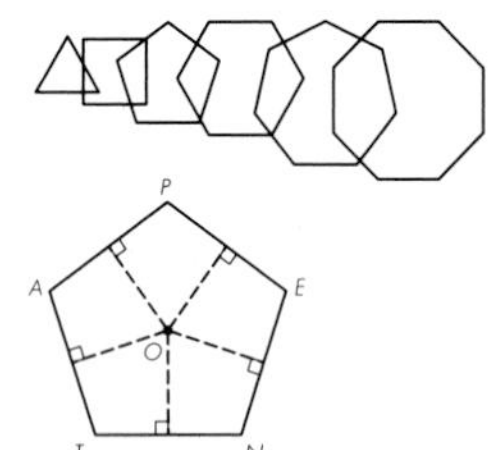

---

1. Figure 1 shows the first six regular polygons. If all the sides are 2 inches in length, what are the six perimeters?

   a. equilateral triangle 6 in.
   b. square 8 in.
   c. pentagon 10 in.
   d. hexagon 12 in.
   e. septagon 14 in.
   f. octagon 16 in.

1.

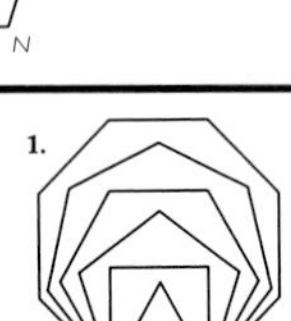

2. In a few sentences, describe how Figure 2 differs from Figure 1.
   In Figure 1, all the polygons have a common side. In Figure 2, the centers of all the polygons coincide.

2.

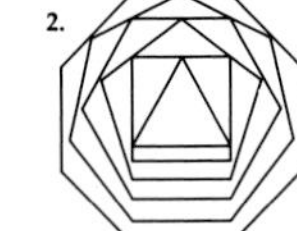

3. In Figure 3, the dashed apothems are each perpendicular to their respective sides. All the sides have the same length, s. Match each polygon with its area formula.

   a. triangle II
   b. square V
   c. pentagon III
   d. hexagon I
   e. septagon IV
   f. octagon VI

   I. $A = 6s(BC) \div 2$
   II. $A = 3s(AB) \div 2$
   III. $A = 5s(JK) \div 2$
   IV. $A = 7s(ED) \div 2$
   V. $A = 4s(FG) \div 2$
   VI. $A = 8s(GH) \div 2$

3.

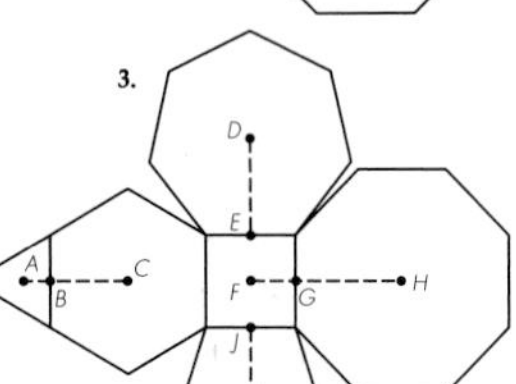

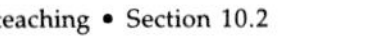

Name ______________________ Date ______________

Creating Equal Areas

The figures at the right show a method for dividing right triangles into sections with equal areas. In each figure, the nine sections of the triangle have the same area.

To make this division, start by dividing the two legs of the right triangle into three equal lengths. (See Section 6.5.) Then, in Figure 1, a semicircle is drawn on the longer leg of the triangle. Perpendicular segments are drawn to the semicircle from the two points which divide the longer leg into thirds. From the right-hand vertex, two arcs are drawn through the points where the perpendicular segments and the semicircle intersect. The triangle is then divided as shown.

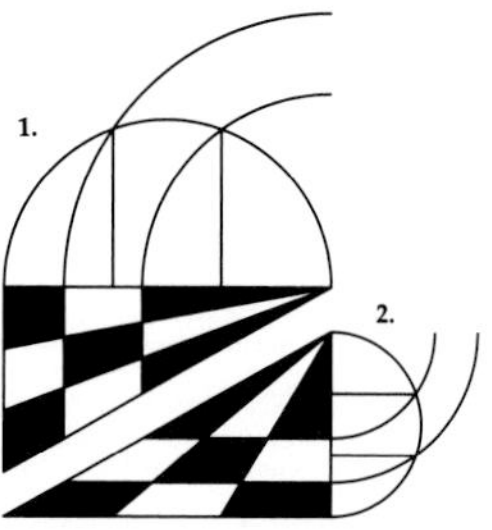

In Figure 2, the construction is similar. However, here the semicircle is drawn on the shorter leg of the triangle.

1. On a sheet of graph paper, draw a right triangle with legs that measure 9 cm and 12 cm. Then copy the construction shown in Figure 1. Find the areas of the nine regions to show that the areas are equal.
   The area of each region is 6 cm².

2. Draw another right triangle with legs that measure 9 cm and 12 cm. Then copy the construction shown in Figure 2. Find the areas of the nine regions.
   The area of each region is 6 cm².

The regular polygons below have also been divided into sections with equal areas. To create these divisions, start by making the triangular constructions shown to the left of each polygon. Complete the constructions on separate sheets of paper. Then try the same method for a regular octagon.

3.

4.

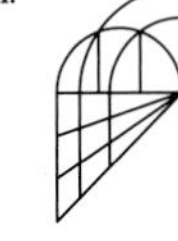

Name ______________________ Date ______________

Circumference and Area of a Circle

The four figures at the right can help you visualize that the area of a circle equals a little more than 3 times r squared, where r is the length of the radius of the circle.

1. $4r^2$ **2.** About $3r^2$ **3.** About $3r^2$ **4.** About $3r^2$

In Figure 1, each small square has sides of length r, the same as the radius of the circle in Figure 4.

In Figure 2, a corner has been cut off of each small square from Figure 1.

In Figure 3, the four pieces from Figure 2 have been rearranged to make a regular hexagon.

Figure 4 shows the same hexagon with an inscribed circle.

To remember the formula for the area of a circle, think of this joke:

First Student: The area equals pi r squared. Second Student: That can't be right—pie are round!

---

1. Fill in each blank with one of the words in the box.

radius	**diameter**	**circumference**	**area**

The length of a radius of a circle is one-half the length of a diameter. This means that in a circle with a diameter of 8 inches, the radius would be 4 inches. It is helpful to know the length of a radius of a circle because then you can substitute this value into the formula πr^2 to get the area of the circle. Or, you can substitute the length of the radius for r in the formula $2\pi r$ to get the circumference.

Think about how you would find the area of the shaded region in each figure. Then write the letter of the figure that matches each description.

C **2.** Find the area of the circle. Then subtract the area of the triangle.

D **3.** Subtract the area of the square from the area of the circle.

B **4.** Find the area of two triangles. Then subtract the area of a circle.

A **5.** Subtract the area of the circle from the area of the square.

Name ______________________ Date ______________

Estimating Areas

A square grid can be used to estimate the area of regions with curved or irregular boundaries. Start by shading the squares that contain any part of the boundary. This is called the *border area.*

The squares inside the border region make up the *inner area;* the squares in both the border and the inner area make up the *outer area.*

The actual area of the region is an amount somewhere between the inner and outer areas. One estimate is the average of the two—the sum of the inner and outer areas divided by 2.

Border area = 32 sq units
Inner area = 37 sq units
Outer area = 69 sq units
Average = 53 sq units

Use the method described above to estimate the area of each ellipse. Accept reasonable answers; possible answers given.

1. 33 sq units **2.** 61 sq units **3.** 27 sq units

4. 27 sq units **5.** 18 sq units **6.** 20 sq units

Another method for finding area is Pick's Formula, $A = (b \div 2) + i - 1$, where b is the number of grid intersection points on the boundary and i is the number of grid intersection points in the interior. This formula works only for polygons with vertices on the intersection points of a square grid, such as the polygon at the right.

7. Use Pick's Formula to find the area of the polygon. $100\frac{1}{2}$ sq units

8. Use the method described at the top of this page. Possible answer: 99 sq units

9. Compare the answers to Exercises 7 and 8. Answers will vary.

Name ______________________ Date ______________

Angles and Circles

The angles in the circles below are *central angles.* The vertex of a central angle is the center of its circle.

The measure of a central angle is the same as the measure of the intercepted arc.

For example, a central angle that intercepts one-quarter of a circle measures one-fourth of 360°, or 90°. The intercepted arc also has a measure of 90°.

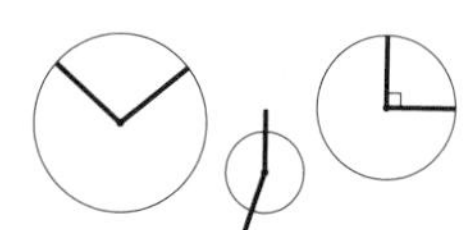

The angles in the circles below are *inscribed angles.* The vertex of an inscribed angle is a point on the circle, and its rays intersect the circle.

The measure of an inscribed angle is one-half that of its intercepted arc. Notice that the measure is also one-half that of a central angle that intercepts the same arc.

For example, an inscribed angle that intercepts a quarter-circle measures one-half of 90°, or 45°.

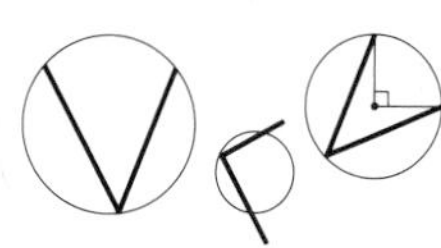

For each figure, write whether the angle shown in heavy lines is *central, inscribed,* or *neither.*

1. inscribed
2. central
3. neither
4. central

For each figure, write the measure of the arc shown with a heavy curve.

5. 105°
6. 180°
7. 110°
8. 180°

Name ______________________ Date ______________

Spheres and Tangents

The sphere at the right is tangent to the plane at just one point, the point at the very bottom of the sphere.

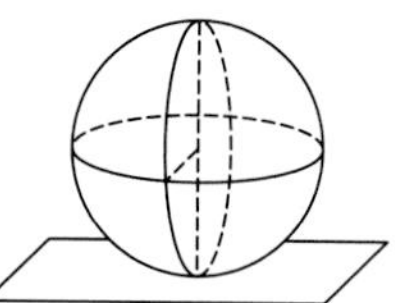

For each figure, describe the set of points that are tangent to the sphere.

1. Figure 1: 4 points, one in the center of each face of the pyramid

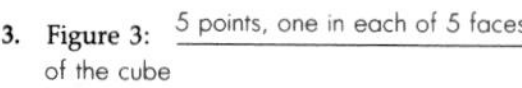

2. Figure 2: 1 point at the bottom of the sphere, plus the circle at the "equator" of the sphere

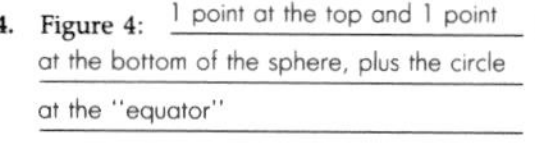

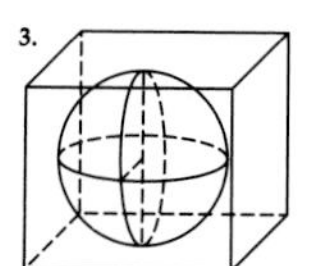

3. Figure 3: 5 points, one in each of 5 faces of the cube

4. Figure 4: 1 point at the top and 1 point at the bottom of the sphere, plus the circle at the "equator"

5. In the space at the right, sketch a sphere inside a cube. Make each face of the cube tangent to the sphere. How many points of tangency are there? 6

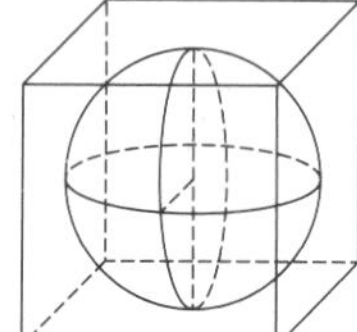

6. If the radius of the sphere drawn for Exercise 5 were 6 inches, what be the dimensions of the cube? 12 in. wide, 12 in. high, 12 in. deep

Name ______________________ Date ______________

Chords and Secants

Knowing the definitions of the following terms is important when solving problems involving circles.

arc A part of the circumference of a circle.
area of a circle The number of square units that cover the interior of a circle, equal to πr^2, where r is the length of the radius.
center of a circle The point equidistant from all points on the circle. A circle and its center are named with the same letter.
central angle An angle whose vertex is at the center of a circle.
chord A line segment whose endpoints are on a circle.
circumference The distance around a circle, equal to $2\pi r$, where r is the radius.
diameter A chord passing through the center of a circle. Its length is twice that of a radius of the circle.
inscribed angle An angle whose vertex is on the circle and whose rays intersect the circle.
pi A constant value represented by the Greek letter π. It is approximately equal to 3.14.
radius A line segment whose endpoints are the center of a circle and a point on the circle.
secant A line or line segment that intersects a circle in two points.
tangent A line or line segment that intersects a circle in exactly one point. It is also perpendicular to the radius of the circle.

Sketch each of the following. Label all points and dimensions as given.

1. Circle B with radius BC and inscribed angle ACD. B is the midpoint of diameter CD.

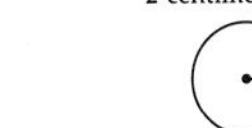

2. A circle with diameter MN and tangent NT
3. A circle with chord AB and secant CD
4. Circle O with central angle TOP and radii OR and OS
5. A circle with circumference approximately equal to 2 times 3.14 times 3 inches
6. A circle with an area approximately equal to 3.14 times the square of 2 centimeters

Write *true* or *false* for each statement.

7. An arc is a part of the circumference of the circle. true
8. A tangent passes through the center of the circle. false
9. A diameter is a chord. true
10. A radius is a chord. false

Name ______________________ Date ______________

Polar Coordinates

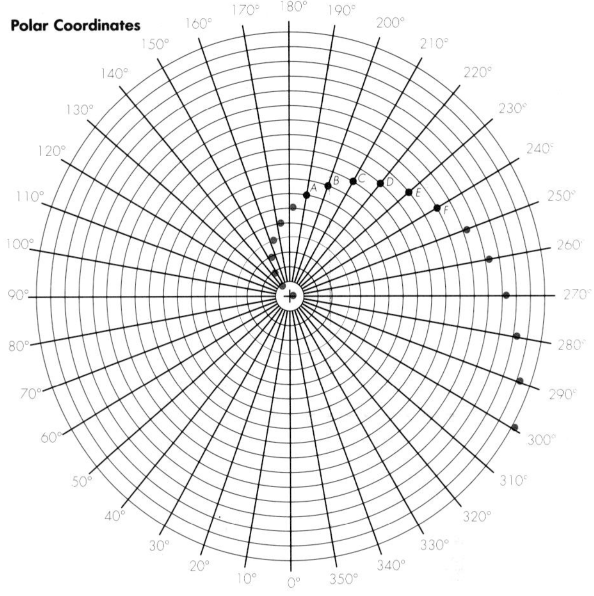

1. The figure above shows a polar coordinate system. In a polar system, the location of a point is fixed by a distance, r, and an angle, Θ. Start at the bottom of the figure and number the rays clockwise from 0° to 350°.
2. The radius of each circle is 1 unit larger than the radius of the circle inside it. Write the coordinates of the points A, B, and C in the form (r, Θ).
(7, 190°); (8, 200°); (9, 210°)
3. The points A through F are part of a spiral. Complete the spiral. Each point is one unit farther away from the center, and 10° farther in a clockwise direction.
4. Graph the equation $r = 4$. What shape do you get?
a circle

Name ______________________ Date ____________

Applications Involving Circles

When solving a word problem involving circles, it is important to be able to draw the correct diagram and then interpret it to help you solve the problem.

The figures on this page are mixed up—they are not next to the appropriate exercises. First match each exercise to its figure. Write the letter of the correct figure next to the exercise number. Then find the answer to the problem. You will not have to do any computation.

C ____ **1.** The twelve spokes of a wheel are spaced so that the points at which they are attached to the rim divide the rim into equal arcs. Write an angle formed by two adjacent spokes?
$\angle AWB$ or $\angle BWC$

D ____ **2a.** An inexperienced archer creates a target so that he will have the same chance of hitting the bulls eye as he does of hitting the outside rings. Which part of the drawing is the radius of the bulls eye?
$\overline{GJ}$

b. Describe how you could use areas to prove that his chances of hitting the bulls eye are equal to his chances of hitting either of the outside rings.
Prove that these three areas are equal: the area of the bulls eye; the area of the largest circle minus the area of the middle-sized circle; the area of the middle-sized circle minus the area of the bulls eye

A ____ **3.** This figure shows the method of construction of one form of *mansard* roof. Which chords in the diagram represent the roof?
$\overline{PQ}$, $\overline{QR}$, $\overline{RS}$, and $\overline{ST}$

B ____ **4.** The circumference of a log is 5 feet. Which part of the figure shows a cross section of the largest square beam that can be cut from the log?
square $WXYZ$

a. Q R S P V U T

b. W X Z Y

c. A B C D W

d. J K L G

Name ______________________ Date ____________

Proving Theorems and Deriving Formulas

1. Use Figure 1. Prove that the area of a triangle is equal to one-half the product of its perimeter and the radius of the inscribed circle. HINT: One radius is shown in Figure 1. You will need to draw two more radii. What three smaller triangles make up the area of triangle ABC?
area $\triangle ABC$ = area $\triangle AOB$ + area $\triangle BOC$ + area $\triangle COA$
$= \frac{1}{2}(OE)(AB) + \frac{1}{2}(OF)(BC) + \frac{1}{2}(OD)(CA)$
$= \frac{1}{2}$(radius of circle) $(AB + BC + CA)$
$= \frac{1}{2}$(radius of circle) (perimeter of $\triangle$)

2. Use Figure 2. Prove that the sum of the areas of the two shaded regions is equal to the area of right triangle DEF. $\overline{DF}$ and $\overline{DE}$ are the diameters of their respective circles. HINT: Start with the Pythagorean Theorem. Then use a formula for the area of a circle that gives the area in terms of the diameter.
$FD^2 + ED^2 = FE^2$; $A = \pi(\frac{d}{2})^2 = \frac{\pi d^2}{4}$; Area of a semicircle is $\frac{1}{2} \cdot \frac{\pi d^2}{4}$, or $\frac{\pi d^2}{8}$. $\frac{\pi}{8}(FD^2) + \frac{\pi}{8}(ED^2) = \frac{\pi}{8}(FE)^2$; $(a + b) + (c + d) = x + b + d$; $a + c = x$

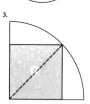

3. Use Figure 3. The square is inscribed in a quarter-circle. Derive a formula for the total area of the unshaded regions.
Area of unshaded regions = area of quarter-circle— area of shaded square; $A = \frac{1}{4}\pi r^2 - \frac{1}{2}r^2 = \frac{1}{4}r^2(\pi - 2)$, or $A = \frac{r^2}{4}(\pi - 2)$

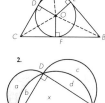

4. Use Figure 4. A *segment* of a circle is a region bounded by an arc of the circle and the chord of the arc. Describe a method for finding the area of a segment of a circle.
Find the area of the sector of the circle, then subtract the area of the triangle.

1. A D E O C F B

2. D c a b d x F E

3. r

4.

Name ____________________ Date ____________

Logo and Circles

Two circles are said to be *internally tangent* if they intersect in exactly one point and the center of the smaller circle is inside the larger circle. Two circles are said to be *externally tangent* if they intersect in exactly one point and the center of the smaller circle is outside the larger circle. Logo procedures can be written to draw circles that are internally tangent and circles that are externally tangent.

The procedure to tell the turtle to draw two circles that are internally tangent at the home position is given below.

```
TO INTANGENT
REPEAT 40[FD 5 LT 9]     Draw the small circle.
REPEAT 40[FD 10 LT 9]    Draw the large circle.
END
```

The procedure to tell the turtle to draw two circles that are externally tangent at the home position is given below.

```
TO EXTANGENT
REPEAT 40[FD 5 LT 9]     The turtle turns left.
REPEAT 40[FD 5 RT 9]     The turtle turns right.
END
```

Write procedures that tell the turtle to draw each figure. Answers may vary.

1.

2.

3.

4.

Name________________ Class________________ Date________________

Achievement Test 10 (Chapter 10)
REGULAR POLYGONS AND CIRCLES

No. Correct
No. Exercises: **27**
Score
3.70 x No. Correct =

GEOMETRY FOR DECISION MAKING
James E. Elander
SOUTH-WESTERN PUBLISHING CO.

Find the measure of an interior angle and an exterior angle for each regular polygon.

1. Heptagon $128\frac{4}{7}°$; $51\frac{3}{7}°$
2. 12-gon 150°; 30°
3. Octagon 135°; 45°
4. 20-gon 162°; 18°
5. The measure of an interior angle of a regular polygon is 160°. How many sides does it have? 18
6. An equilateral triangle has a side of 20 m. What is its area? $100\sqrt{3}$ m²

Find the circumference and area. Use $\pi \approx 3.14$. Round your answers to the nearest tenth.

7. 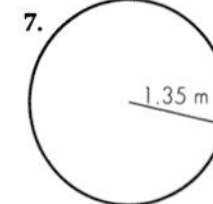

C = 8.5 m
A = 5.7 m²

8. 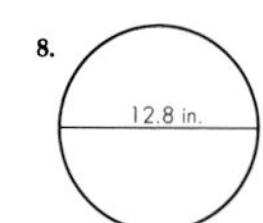

C = 40.2 in.
A = 128.7 sq in

9. 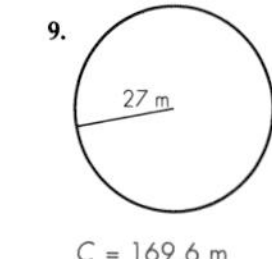

C = 169.6 m
A = 2289.1 m²

10. 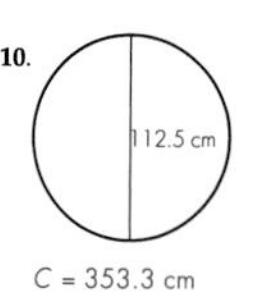

C = 353.3 cm
A = 9935.2 cm²

11. A square with sides of 5 cm is inscribed in a circle. The diagonal *AC* is a diameter of the circle. To the nearest hundredth, what is the area of the shaded part of the circle? 14.25 cm²

Find the indicated measures.

12. 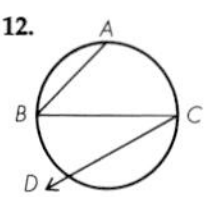

$m\overset{\frown}{AC} = 90°$
$m\angle BCD = 29°$
$m\angle ABC =$ ___ 45°
$m\overset{\frown}{BD} =$ ___ 58°

13. 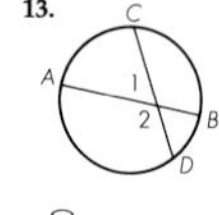

$m\overset{\frown}{AC} = 80°$
$m\overset{\frown}{CB} = 65°$
$m\overset{\frown}{AD} = 175°$
$m\angle 1 =$ ___ 60°
$m\angle 2 =$ ___ 120°

14. 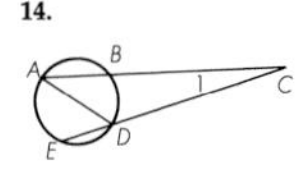

$m\overset{\frown}{AB} = 105°$
$m\overset{\frown}{AE} = 100°$
$m\overset{\frown}{ED} = 85°$
$m\angle 1 =$ ___ 15°
$m\angle 2 =$ ___ 35°
$m\angle 3 =$ ___ 50°

15. 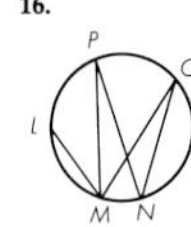

$m\overset{\frown}{ADC} = 220°$
$m\angle D =$ ___ 70°
$m\angle B =$ ___ 110°

16. 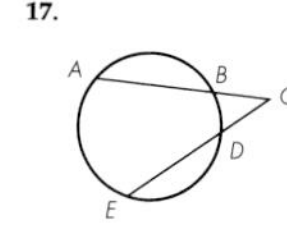

$m\overset{\frown}{LP} = m\overset{\frown}{PO} = 70°$
$m\angle LMP =$ ___ 35°
$m\angle LMO =$ ___ 70°
$m\angle PNO =$ ___ 35°

17.

$m\angle C = 40°$
$m\overset{\frown}{BD} = x$
$m\overset{\frown}{AE} = 2x + 8$
$m\overset{\frown}{BD} =$ ___ 72°
$m\overset{\frown}{AE} =$ ___ 152°

In circle *O*, $m\overset{\frown}{CB} = 60°$, $m\overset{\frown}{AH} = m\overset{\frown}{BE} = 50°$, $m\overset{\frown}{HG} = 30°$, and $\overline{AB}$ is a diameter. Find these measures.

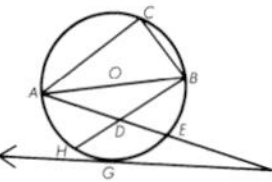

18. $m\angle CAB =$ ___ 30°
19. $m\angle ADH =$ ___ 50°
20. $m\angle EFG =$ ___ 15°
21. $m\angle ACB =$ ___ 90°
22. $m\angle CBA =$ ___ 60°
23. $m\angle HDE =$ ___ 130°

24. In the circle O, *AB* = 32 in. and *OC* = 12 in. What is the radius? 20 in.

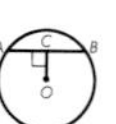

25. In the circle, $m\overset{\frown}{DB} = x$, $m\overset{\frown}{AD} = 3x + 5$, $m\overset{\frown}{BC} = 8x + 23$, and $m\overset{\frown}{AC} = 2x + 10$. Find $m\angle CEB$. 140.5°

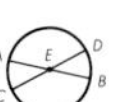

26. In circle *O*, chord $\overline{XY}$ is 12 cm and radius $\overline{OY}$ is 10 cm. How far is chord $\overline{XY}$ from the center of the circle? 8 cm

27. Circle *A* has a radius of 9 m. Circle *B* has a radius of 36 m. How many times greater is the area of circle *B* than the area of circle *A*? 16 times

In the diagram, $l \parallel m$ and t is a transversal. Give the degree measure of each angle.

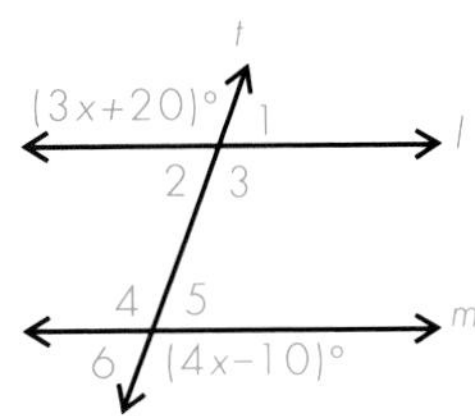

1. $\angle 1$ 70° **2.** $\angle 2$ 70° **3.** $\angle 3$ 110°
4. $\angle 4$ 110° **5.** $\angle 5$ 70° **6.** $\angle 6$ 70°

7. Measure the angle to the nearest degree. State the error in the measurement.
68°; error: $\frac{1}{2}$°

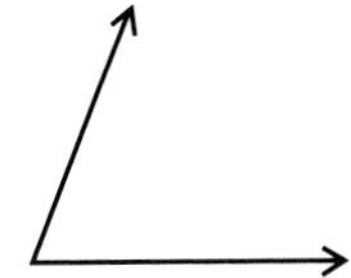

Use the information in the diagram. Find the measure of each angle.

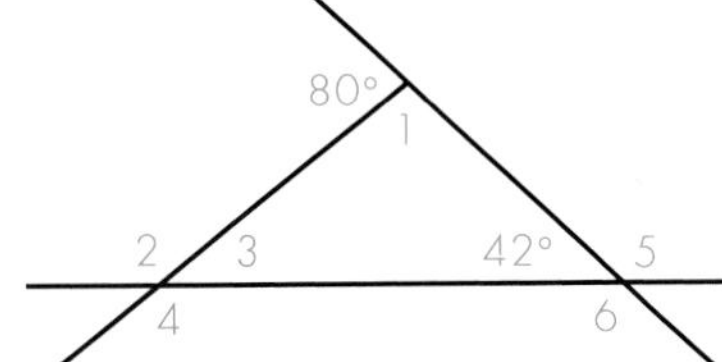

8. $\angle 1$ 100° **9.** $\angle 2$ 142°
10. $\angle 3$ 38° **11.** $\angle 4$ 142°
12. $\angle 5$ 138° **13.** $\angle 6$ 138°

Which of the pairs of triangles are similar and why?

14.

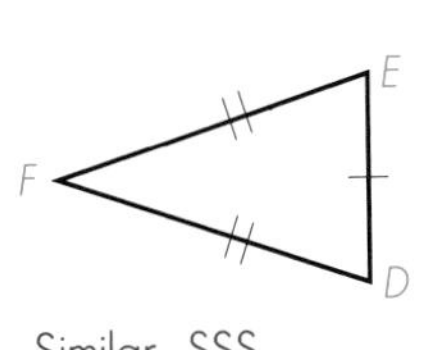

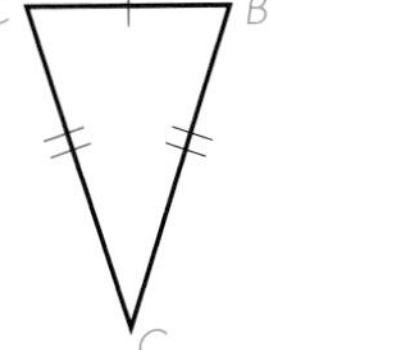

Similar; SSS

15.

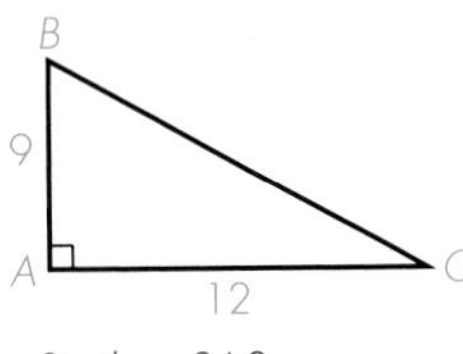

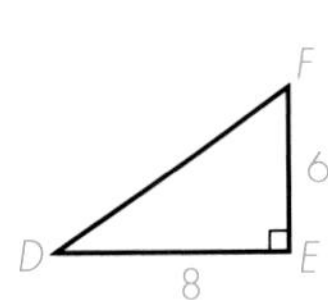

Similar; SAS

In the diagram, $m\angle BAC = m\angle DEC$. Find each length.

16. AE 7
17. DE 15

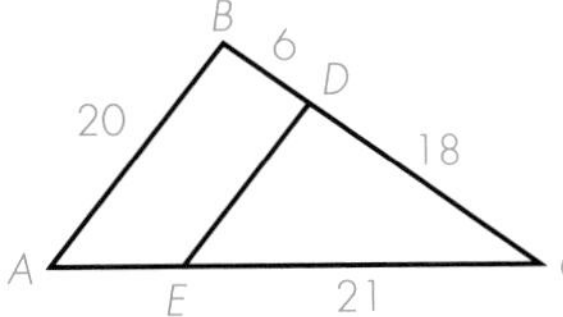

Refer to figures a–f.

18. Which triangles are congruent? c and e; b and f

19. Which triangles are similar? a, c, e; b, d, f

a. 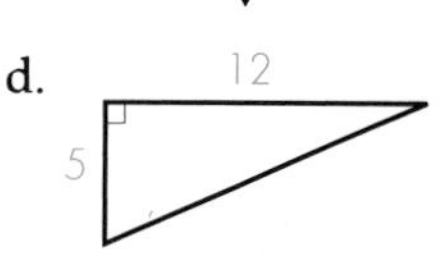

b. 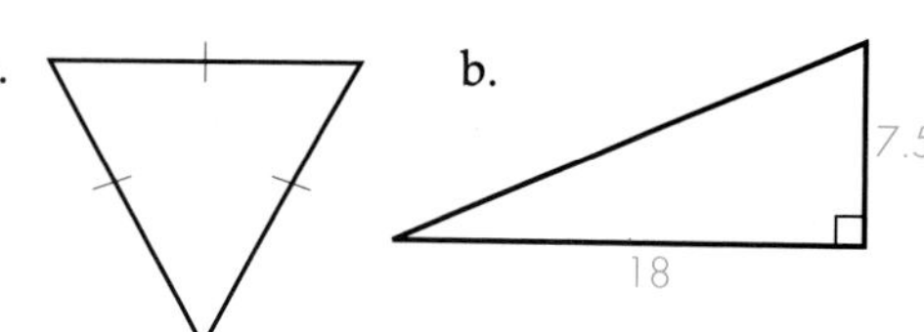

c. 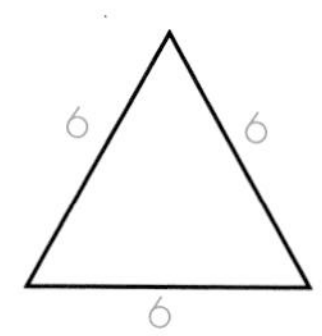

d. 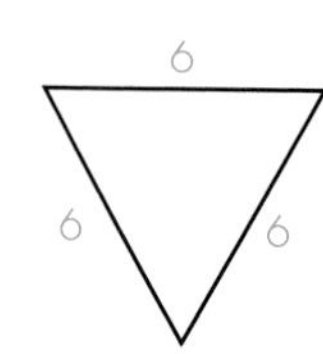

e. 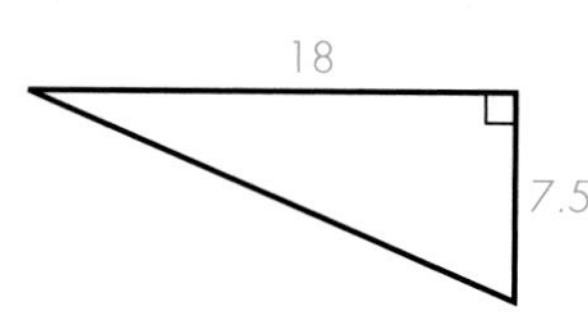

f.

Write the converse of each statement.

20. If you passed the driving test, then you will get a driver's license.
If you will get a driver's license, then you passed the driving test.

21. If you have been to Paris, then you have been to the capital of France.
If you have been to the capital of France, then you have been to Paris.

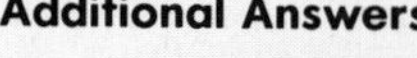

Additional Answers

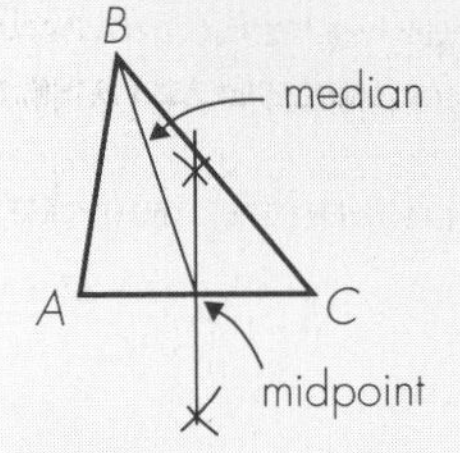

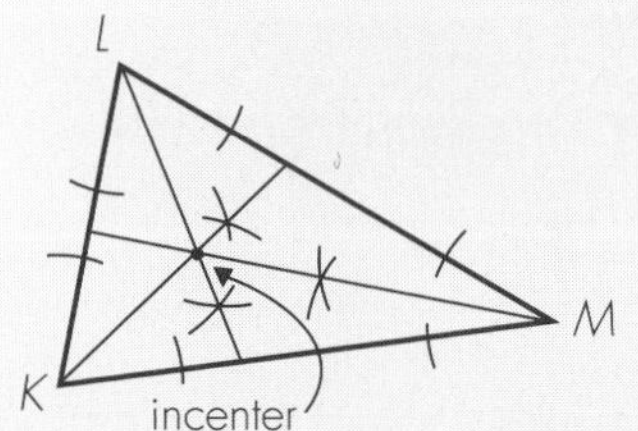

22. Draw and label an acute triangle ABC. Construct the median from B to $\overline{AC}$. See margin.

23. In $\triangle DEF$, G and H are midpoints. What is the length of $\overline{DF}$? 24.2 cm

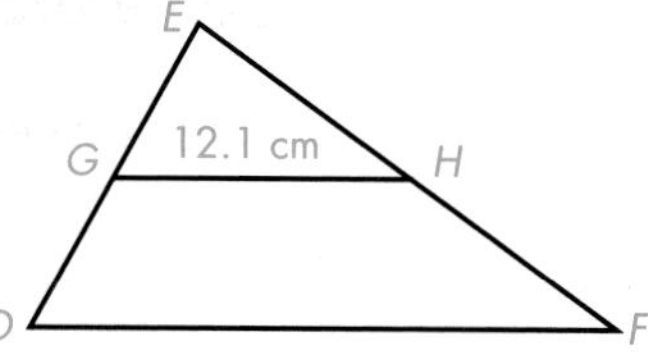

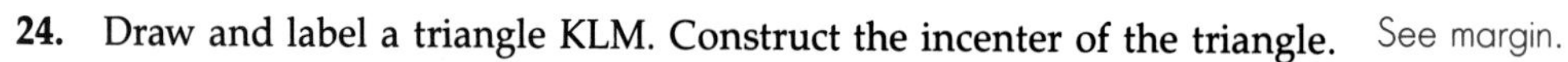

24. Draw and label a triangle KLM. Construct the incenter of the triangle. See margin.

Find the missing lengths to the nearest tenth. Use a calculator or the table of squares and square roots on page 562.

25.

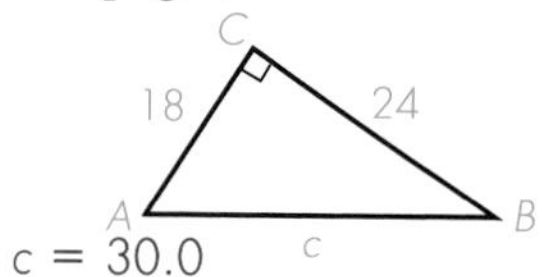

c = 30.0

26.

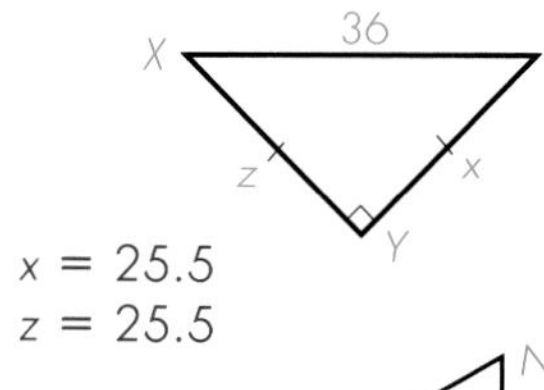

x = 25.5
z = 25.5

27.

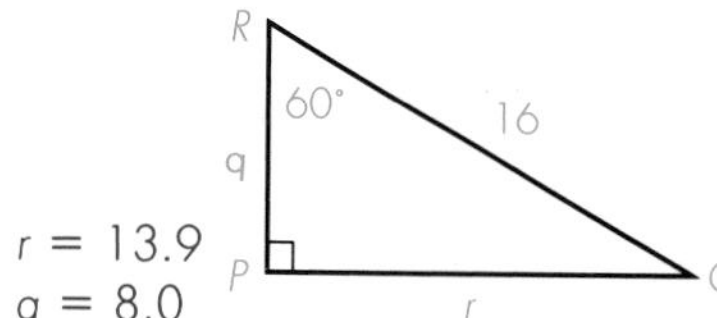

r = 13.9
q = 8.0

28.

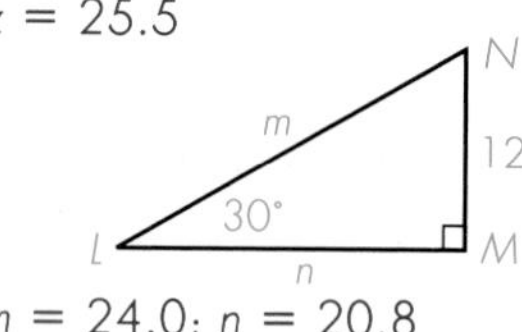

m = 24.0; n = 20.8

29. List the lengths of the sides of $\triangle PQR$ from least to greatest.

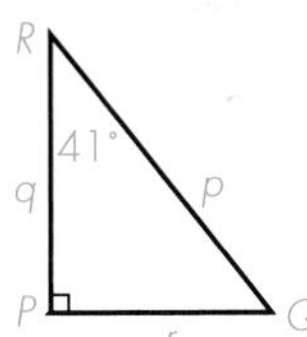

r, q, p

30. Write $m\angle A < m\angle B$, $m\angle A = m\angle B$, $m\angle A > m\angle B$, or "insufficient information."

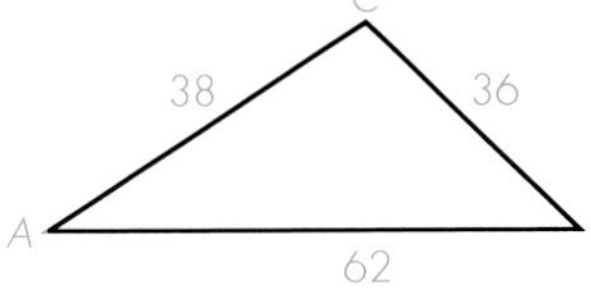

$m\angle A < m\angle B$

Find the length of the line segment that connects the two points.

31.

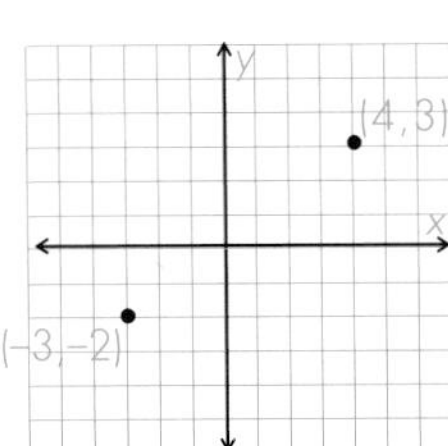

$\sqrt{74} \approx 8.6$

32.

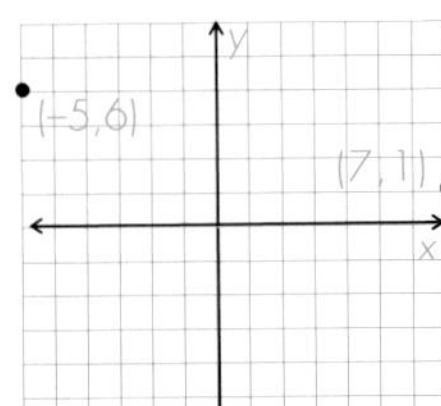

13

Refer to quadrilateral *ABCD*.

D, C, A, B

33. Name one pair of opposite sides.
$\overline{AB}$ and $\overline{DC}$ or $\overline{AD}$ and $\overline{BC}$

34. Name one pair of adjacent angles.
$\angle A$ and $\angle B$, $\angle B$ and $\angle C$, $\angle C$ and $\angle D$, or $\angle D$ and $\angle A$

35. $m\angle A + m\angle B + m\angle C + m\angle D = ?$ 360°

Is the figure a parallelogram? Give a reason for your answer.

36.

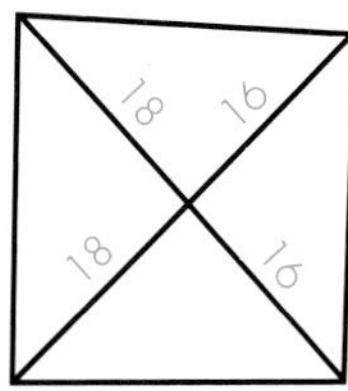

No; diagonals do not bisect each other.

37.

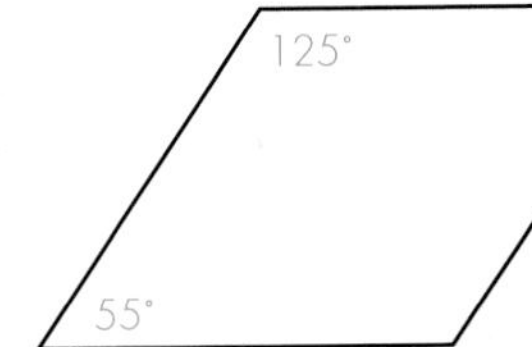

Yes; opposite angles are equal in measure.

Figure *MNOP* is an isosceles trapezoid. Find the measures of the angles.

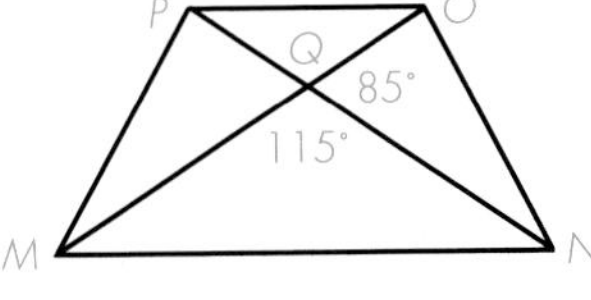

38. $\angle OMN$ 32.5° **39.** $\angle PQM$ 65° **40.** $\angle MPO$ 117.5°

Figure LMNO is a rhombus. Find the lengths of the line segments.

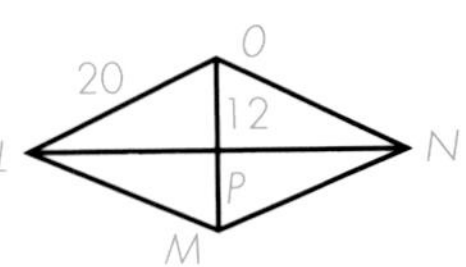

41. $\overline{LM}$ 20 **42.** $\overline{PM}$ 12 **43.** $\overline{LP}$ 16

Find the height of the figure.

44.

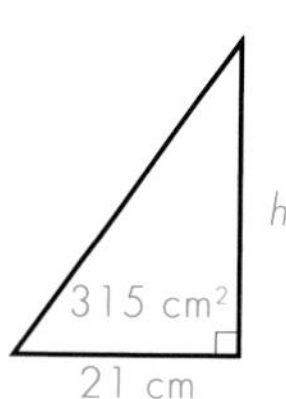

$h = 30$ cm

45.

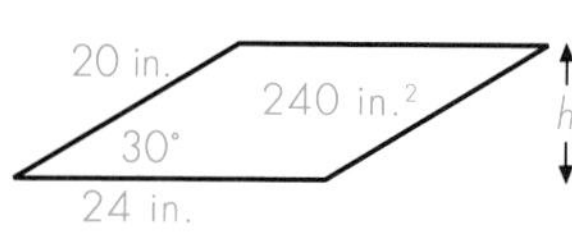

$h = 10$ in.

46. A rectangular plot of land has a length of 300 feet and an area of 45,000 square feet. The length of a similar plot is 500 feet. What is the area? 125,000 ft²

Find the measure of an interior angle for each regular polygon.

47. hexagon 120°

48. 20-gon 162°

49. decagon 144°

50. 50-gon 172.8°

51. A regular octagon has a side of 5 cm and an apothem of 6 cm. What is the area? 120 cm²

Find the circumference and area. Use $\pi \approx 3.14$. Express answers to the nearest tenth.

52.

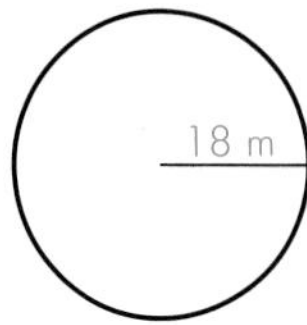

C = 113.0 m; A = 1017.4 m²

53.

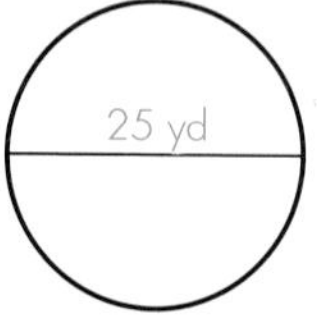

C = 78.5 yd; A = 490.6 yd²

In circle O, $m\widehat{CF} = 115°$, $m\angle CGE = 60°$, $m\angle DOA = 70°$, and $\overline{CB}$ is a diameter. Find each measure.

54. $m\widehat{FB}$ 65°

55. $m\widehat{CE}$ 55°

56. $m\angle CAD$ 20°

57. $m\widehat{ED}$ 55°

58. $m\angle ODA$ 90°

59. $m\angle FGB$ 60°

Find the value of x.

60.

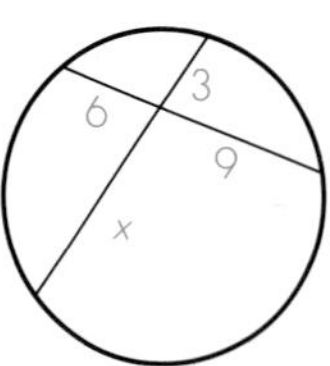

$x = 18$

61.

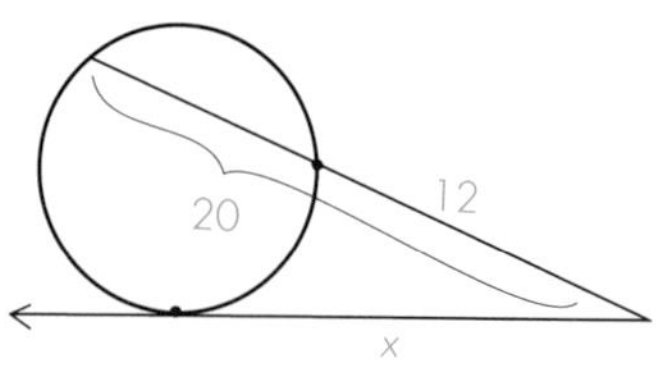

$x = \sqrt{240} \approx 15.5$

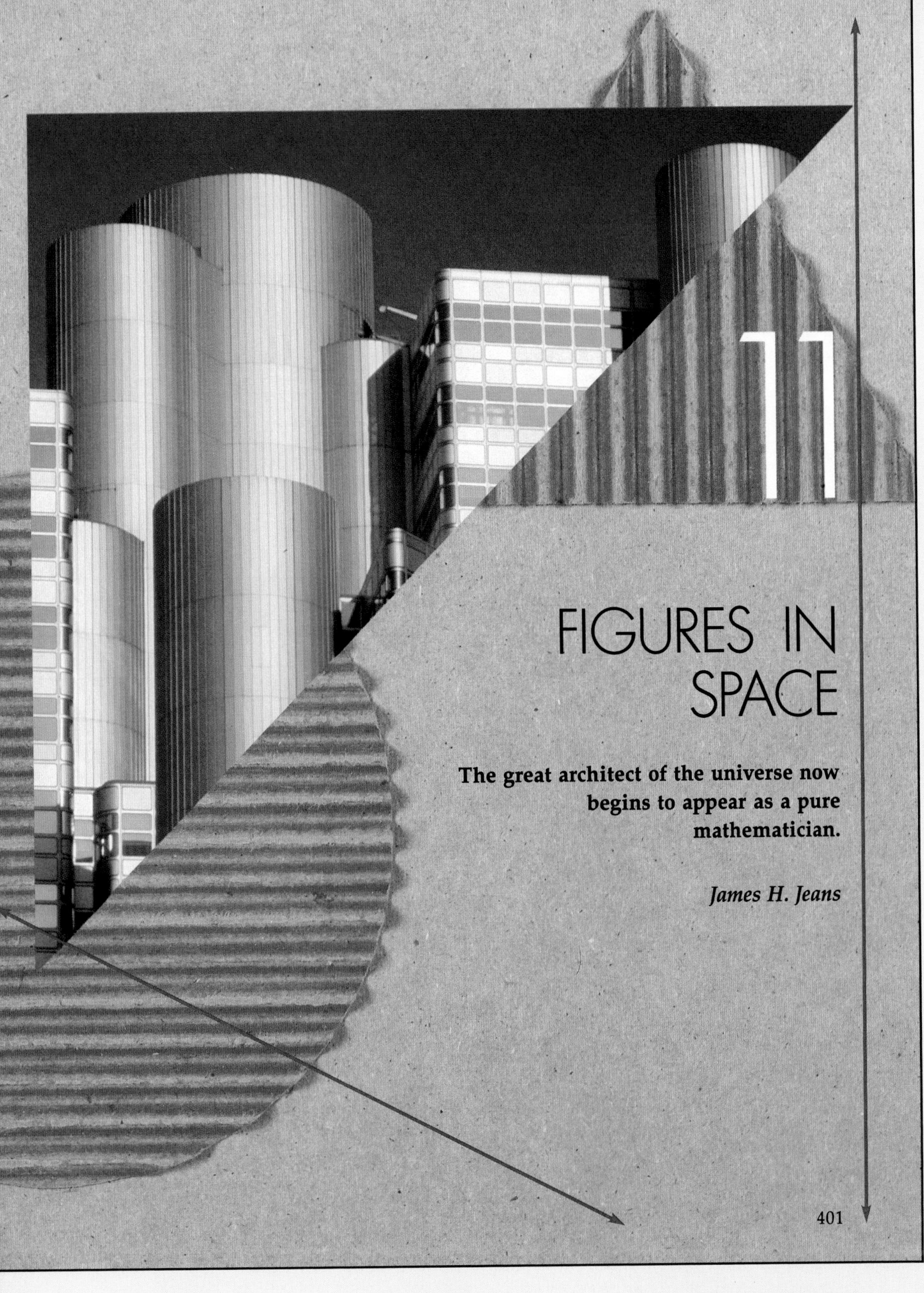

11

FIGURES IN SPACE

The great architect of the universe now begins to appear as a pure mathematician.

James H. Jeans

401

11

CHAPTER 11 OVERVIEW

Figures in Space

Suggested Time: 12–14 days

The comment of James H. Jeans that appears at the beginning of this chapter is especially relevant to the polyhedrons and other three-dimensional figures of this chapter. Simple underlying relationships exist among the basic solid figures such as the prisms, cylinders, pyramids, cones, and spheres. In particular, the relationship between the volume of a prism and a pyramid of the same base and height is a simple one; their ratio is 3:1. The same ratio holds for the volume of a circular cylinder and a cone of the same base and height. These relationships and many others are developed in this chapter.

The great architect of the universe now begins to appear as a pure mathematician.

James H. Jeans

CHAPTER 11 OBJECTIVE MATRIX

	End of Chapter Items by Activity				Student Supplement Items by Activity		
Objectives by Number	**Review**	**Test**	**Computer**	**Algebra Skills***	**Reteaching**	**Enrichment**	**Computer**
11.1.1	✓	✓	✓		✓		✓
11.1.2	✓	✓		✓			
11.2.1	✓	✓	✓		✓		
11.2.2	✓	✓		✓	✓	✓	
11.3.1	✓	✓	✓		✓		✓
11.3.2	✓	✓			✓		
11.3.3	✓	✓		✓			
11.3.4	✓	✓		✓			
11.4.1	✓	✓	✓	✓	✓		
11.4.2	✓	✓		✓	✓		
11.5.1	✓	✓					
11.5.2							
11.6.1	✓	✓		✓	✓	✓	
11.6.2	✓	✓		✓	✓	✓	

*A ✓ beside a Chapter Objective under Algebra Skills indicates that algebra skills taught within the chapter or in previous Algebra Skills lessons are used.

CHAPTER 11 PERSPECTIVES

▲ Section 11.1

Area of Prisms and Cylinders

The chapter begins with an examination of the surface area of prisms and cylinders. The Discovery Activity helps students to familiarize themselves with the properties of prisms. This leads to a definition of *prism*. The *cylinder* is then defined. Finally, area formulas are developed for both the prism and the right circular cylinder.

▲ Section 11.2

Volume of Prisms and Cylinders

The section begins in a concrete way by examining cubical blocks and assembling them into right rectangular prisms of increasing size. This process leads to the basic formula for the volume of a right rectangular prism, $V = lwh$. Then *Cavalieri's Principle* is used to find volumes of prisms that are neither right nor rectangular.

Cavalieri's Principle also is used to develop the volume formula for a circular cylinder.

▲ Section 11.3

Area of Pyramids and Cones

The section begins by considering the class of objects of which prisms are a subclass, the class (or set) of *polyhedrons.* This opens the way for an introduction to another kind of polyhedron, the *pyramid.*

In the Discovery Activity, students are led to find the formula for the lateral area of a pyramid. The formula for the total area of a pyramid is then presented.

Attention then moves to *cones,* in particular to *right circular cones.* The formula for the lateral area of a cone is developed somewhat along the lines of the lateral area of a regular pyramid.

▲ Section 11.4

Volume of Pyramids and Cones

In this section, students are led to see that the volume of a pyramid is related in a simple way to the volume of a prism that has the same height and a base of the same area. Specifically, three such pyramids have the same volume as one such prism.

An experiment describes how to bring three identical pyramids together to form what appears to be a prism.

The derivation of the volume of a cone is based upon the observation that a cone can be thought of as a "pyramid" with a very large number of lateral faces. Reasoning in this way, you can easily persuade students that the formula for the volume of a right circular cone is essentially the same as the formula for the volume of a pyramid that has the same height as the cone and that has a base with the same area as the cone's base. For simplicity, Theorem 11-4-3 is stated for a right circular cone, but the theorem would apply equally well to a cone that is not a right cone.

▲ Section 11.5

Applications

In earlier sections, the author preferred to concentrate on the mathematical aspects of the area and volume formulas for the various types of polyhedrons. This section is devoted to applications of prisms, cylinders, pyramids, and cones.

One aspect of the section is worth special comment; namely, the use of the techniques of *dimensional analysis* in solving problems. When students are faced with calculations that involve numbers that represent units of measure, they often do not know "when to multiply and when to divide." Students who are familiar with the methods of dimensional analysis are more likely to reason their way through such problems.

▲ Section 11.6

Area and Volume of a Sphere

The last section of this chapter is devoted to the area and volume of a sphere. It may not be immediately obvious that a relationship exists between the sphere and the figures studied earlier in the chapter. Nevertheless, some relationships that can be found are brought out in the student text.

The first relationship is the one that Archimedes found between a sphere and a cylinder in which the sphere is inscribed.

The second relationship that is drawn between the sphere and one of the other solids of the chapter is found in the proof of the formula for the area of a sphere (Theorem 11-6-2). Here the sphere is visualized as a bundle of an infinite number of tightly packed pyramids.

SECTION 11.1

Resources
Reteaching 11.1
Enrichment 11.1

Objectives
After completing this section, the student should understand
- ▲ what a prism and a cylinder are.
- ▲ how to find the area of prisms and cylinders.

Materials
3-by-5 in. cards
ruler
tape
scissors
calculator

Vocabulary
polygonal region
prism
base
edge
lateral edge
base edge
lateral face
vertex
right prism
oblique prism
cylinder
circular cylinder
axis
right circular cylinder
oblique circular cylinder
total surface area
lateral area
height
altitude

GETTING STARTED

Quiz: Prerequisite Skills
A rectangular screen door is 8 ft high and 3.5 ft wide.

1. What is its area? [28 ft^2]

2. What is its perimeter? [23 ft]

A circular window has a 20-inch diameter.

3. What is its area? [$100\pi \approx 314$ in.^{2}]

4. What is its circumference? [$20\pi \approx 62.8$ in.]

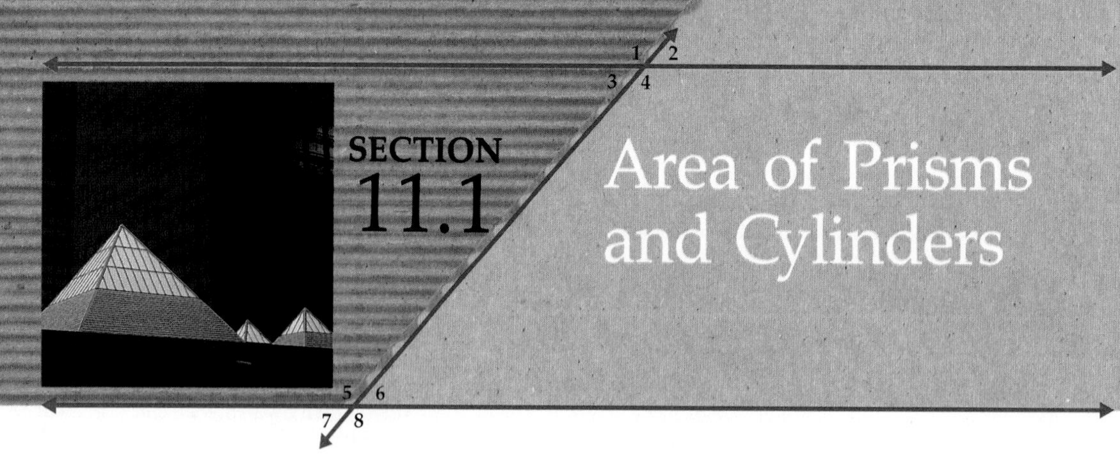

Area of Prisms and Cylinders

After completing this section, you should understand
- ▲ what a prism and a cylinder are.
- ▲ how to find the area of prisms and cylinders.

The figures at the right all have the shape of a *prism.*

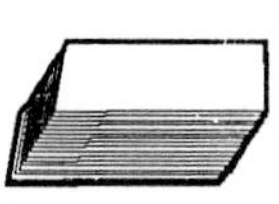

DISCOVERY ACTIVITY

1. On a 3 by 5 card, draw four lines parallel to the short ends of the card. Fold the card and tape the ends to form the open-ended, three-dimensional figure shown in the second diagram. Put it on a sheet of paper and trace around the bottom to make two pentagonal ends. Cut these out and tape them to top and bottom to make the closed figure shown in the third diagram.

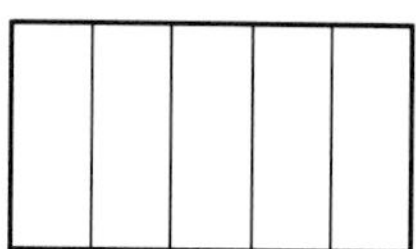
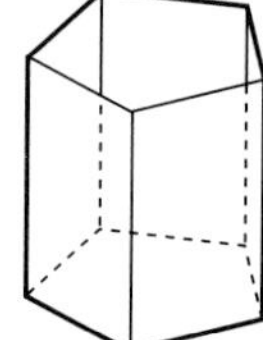
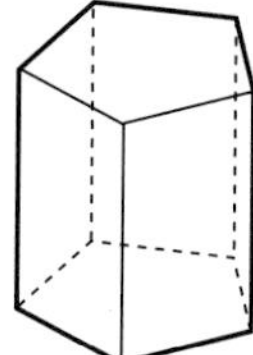

2. Take more 3 by 5 cards and make similar models in which the top and bottom are triangular, square, and so on.
3. Are the top and bottom ends parts of parallel planes? Yes
4. Are the ends congruent? Yes
5. What can you say about the vertical edges (the lines where you folded or taped)? They are parallel.

The top, bottom, and sides of the model you made in the Discovery Activity are *polygonal regions* and the three-dimensional figures themselves are *prisms.*

NOTEBOOK

DEFINITION 11-1-1: A POLYGONAL REGION is a polygon together with all the points inside the polygon.

NOTEBOOK

DEFINITION 11-1-2: Suppose two congruent polygons are situated in parallel planes so that all the line segments joining corresponding vertices are parallel. The union of the two congruent polygonal regions and all the line segments joining corresponding points of the polygons is a PRISM.

Any three-dimensional figure that can be formed in this way is a prism. The parallel polygonal regions in the definition are called the *bases* of the prism. Line segments joining corresponding vertices of the bases are called **lateral edges.** The sides of the bases are called **base edges**. Any two corresponding base edges and the lateral edges that connect them determine a parallelogram-shaped region called a **lateral face** of the prism. The point where two base edges intersect a lateral edge is called a **vertex** (plural: vertices).

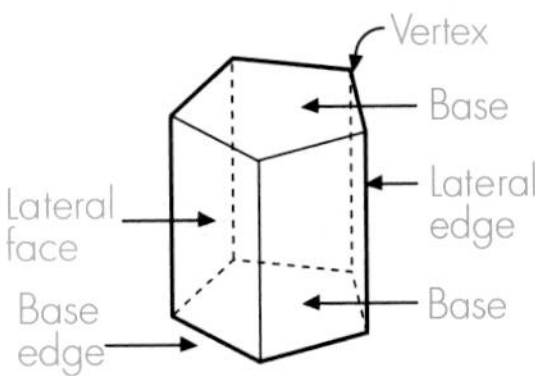

If the lateral edges of a prism are perpendicular to the bases, the prism is a **right prism.** Otherwise, it is an **oblique** prism. A prism is named according to the shape of its bases.

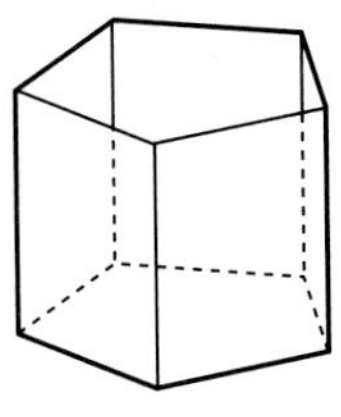

Right pentagonal prism

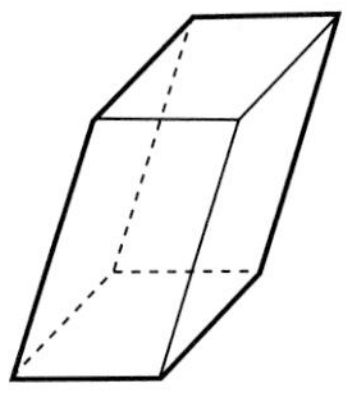
Oblique square prism

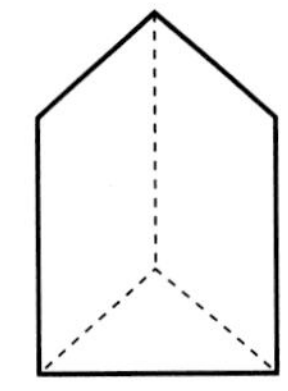
Right triangular prism

Example: The figure shown is an oblique triangular prism.

Bases: $\triangle FGH$ and $\triangle F'G'H'$

Lateral faces: $\square FGG'F'$, $\square GHH'G'$, and $\square HFF'H'$

Edges: $\overline{FG}$, $\overline{GH}$, $\overline{FH}$, $\overline{F'G'}$, $\overline{G'H'}$, $\overline{H'F'}$, $\overline{FF'}$, $\overline{GG'}$, and $\overline{HH'}$

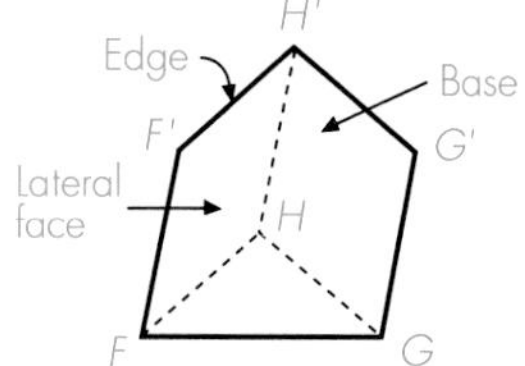

TEACHING COMMENTS

Using the Page

After the discussion of Definition 11-1-2, have students describe a prism in their own words. Ask students to list objects that are shaped like prisms. [Possible examples might be a book, a classroom, and a cereal box, which are shaped like rectangular prisms, or a peaked roof, which is shaped like a triangular prism. Some students may mention the Pentagon as an example of a pentagonal prism.]

CLASS ACTIVITY

Identify each edge, base, and lateral face. See margin.

1. 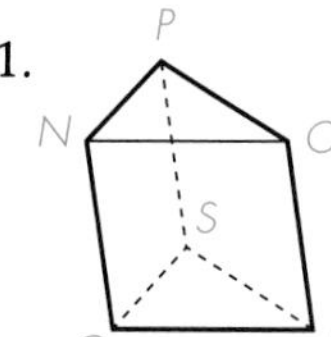

2. 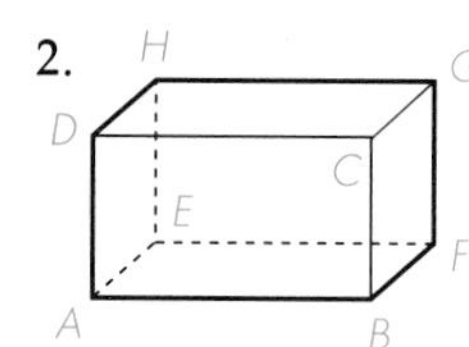

3. 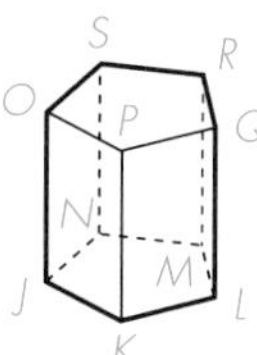

Carefully draw each figure.

4. An oblique rectangular prism
See margin.

5. A right hexagonal prism
See margin.

CYLINDERS

Cylinders are formed very much like prisms, but their bases are curved regions.

DEFINITION 11-1-3: **A CYLINDER is three-dimensional figure consisting of two congruent curved regions in parallel planes and the line segments joining corresponding points on the curves that determine the regions. Segments joining corresponding points of the curves are parallel.**

If the curved regions are circular, the cylinder is a **circular cylinder.** The segment joining the centers of the bases is called the **axis.** If the axis is perpendicular to the bases, the cylinder is a **right** circular cylinder. Otherwise, it is an **oblique** circular cylinder.

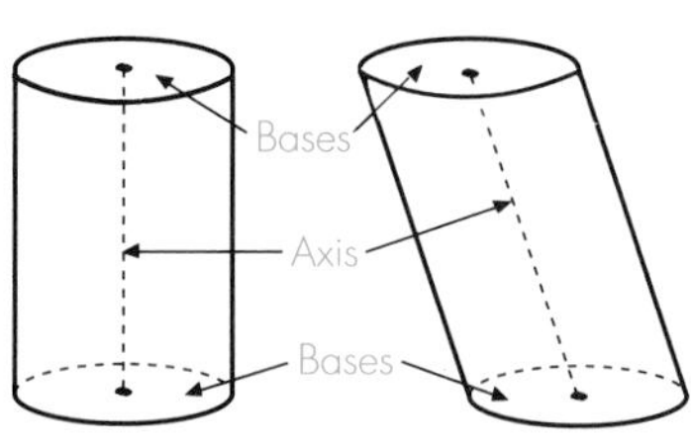

SURFACE AREA OF A RIGHT PRISM

With enough information you can find the **total surface area** of a right prism. Just find the total area of all lateral faces (the **lateral area**) and the area of the two bases.

Example: Find the total surface area of the right triangular prism. There are 3 rectangles and 2 right triangles.

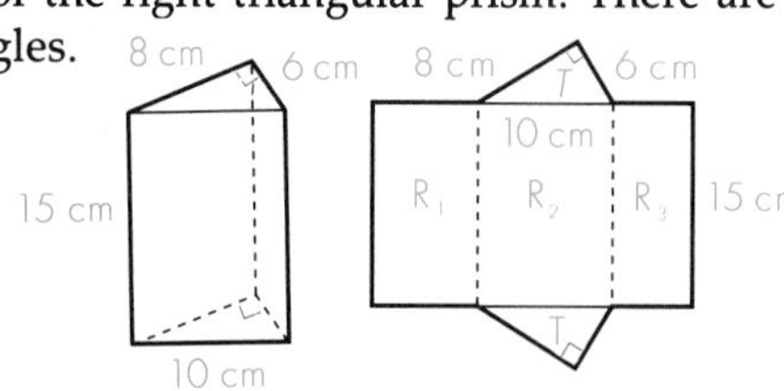

Additional Answers

1. Edges: $\overline{PN}$, $\overline{NO}$, $\overline{OP}$, $\overline{PS}$, $\overline{NQ}$, $\overline{OR}$, $\overline{SQ}$, $\overline{QR}$, $\overline{RS}$
Bases: △PNO, △SQR
Lateral faces: ▱NORQ, ▱OPRS, ▱PNQS

2. Edges: $\overline{CD}$, $\overline{DH}$, $\overline{HG}$, $\overline{GC}$, $\overline{CB}$, $\overline{DA}$, $\overline{HE}$, $\overline{GF}$, $\overline{AB}$, $\overline{BF}$, $\overline{FE}$, $\overline{EA}$
Bases: ▭ABFE, ▭DCGH
Lateral faces: ▭ABCD, ▭BFGC, ▭EFGH, ▭ADHE

3. Edges: $\overline{PQ}$, $\overline{QR}$, $\overline{RS}$, $\overline{SO}$, $\overline{OP}$, $\overline{PK}$, $\overline{QL}$, $\overline{RM}$, $\overline{SN}$, $\overline{OJ}$, $\overline{KL}$, $\overline{LM}$, $\overline{MN}$, $\overline{NJ}$, $\overline{JK}$
Bases: Pentagon OPQRS, Pentagon JKLMN
Lateral faces: ▭OPJK, ▭PQLK, ▭QRML, ▭RSNM, ▭SOJN

4.

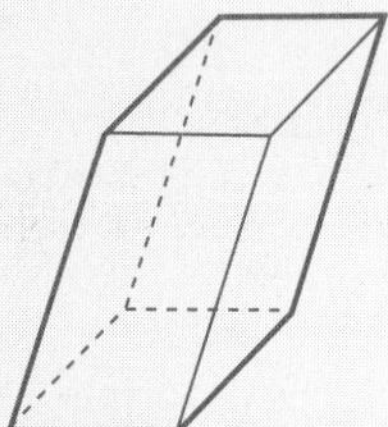

5.

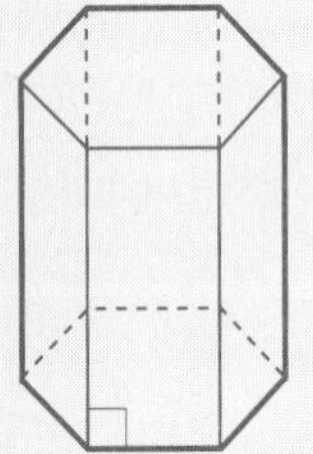

Area of bases = 2T = $2(\frac{1}{2}bh) = 2 \times \frac{1}{2} \times 8 \times 6$, or 48 cm^2
Lateral area = $R_1 + R_2 + R_3$
= (15 × 8) + (15 × 10) + (15 × 6)
= 15(8 + 10 + 6) (Use the property ab + ac + ad = a(b + c + d).)
= 15(24), or 360 cm^2
Total surface area = 48 + 360, or 408 cm^2

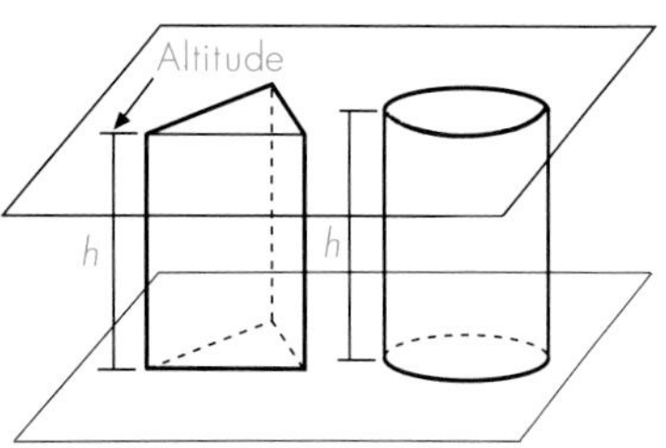

In the example, a shortcut was used to find the lateral area:
(15 × 8) + (15 × 10) + (15 × 6) = 15(8 + 10 + 6)
Notice that 15 is the distance between the planes that contain the bases. Also, (8 + 10 + 6) is the **perimeter** of the base. The distance between the planes containing the bases is called the **height** of the prism or cylinder.

NOTEBOOK **THEOREM 11-1-1** **If the height of a prism is *h* and each base has perimeter *p* and area *B*, the formula for the total surface area TA is TA = 2*B* + *ph*.**

SURFACE AREA OF A RIGHT CIRCULAR CYLINDER

The lateral area of a right circular cylinder is found in the same way as the lateral area of a right prism.

Lateral area = circumference of base × height of cylinder
= $2\pi r \times h$, or $2\pi rh$
Area of bases = 2 × (area of each base)
= $2 \times \pi r^2$, or $2\pi r^2$

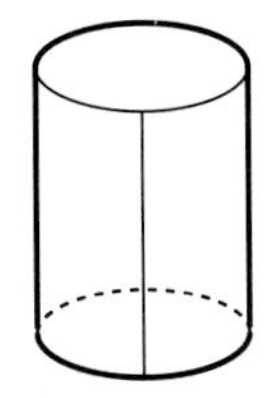

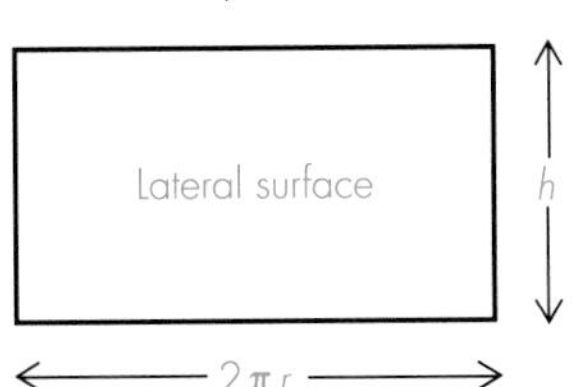

The total surface area of a right circular cylinder is found in the same way as the total surface area of a right prism:
Total area = 2 × (area of base) + lateral area

NOTEBOOK **THEOREM 11-1-2** **If the height of a right circular cylinder is *h* and the radius of each circular base is *r*, then the formula for the total surface area is: Total area = $2\pi r^2 + 2\pi rh$**

CLASS ACTIVITY

Refer to the trapezoidal prism at the right. The height of the trapezoid is 1.5. Use a calculator to find the indicated region.

B 5 C
2 1.5 3.1
A 8 D
9

Using the Page
Have students compare the formulas given in Theorems 11-1-1 and 11-1-2:
$TA = 2B + ph$
$TA = 2\pi r^2 + 2\pi rh$
In the formula for total surface area of a cylinder, ask them to explain why πr^2 replaces B and $2\pi r$ replaces p. [The base of a cylinder is a circle with area πr^2 and its perimeter, or circumference, is $2\pi r$.]

Using the Page
Some students may note that $2\pi r^2 + 2\pi rh$ can be expressed as $2\pi r(r + h)$, which can simplify the computation of the total surface area of a cylinder.

Extra Practice
1. Identify each edge, base, and lateral face of the right triangular prism.

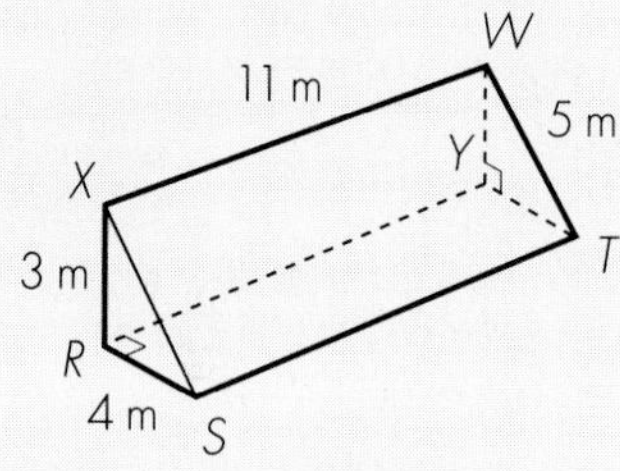

[Edges: $\overline{RX}$, $\overline{XS}$, $\overline{RS}$, $\overline{YW}$, $\overline{WT}$, $\overline{YT}$, $\overline{RY}$, $\overline{ST}$, $\overline{XW}$; Bases: △RXS and △YWT; Lateral faces: Rectangles RYTS, RYWX, XWTS]
2. Find the total surface area of the prism in Exercise 1. [144 m^2]
3. Find the total surface area of a cylinder with a height of 8 feet and a base radius of 2 feet. [$40\pi \approx 126$ ft^2]

Additional Answers
(for page 406)
1. Lateral face MORP is a rectangle, as are the other faces. Right; Edges: $\overline{MN}$, $\overline{NO}$, $\overline{OM}$, $\overline{MP}$, $\overline{NQ}$, $\overline{OR}$, $\overline{PQ}$, $\overline{QR}$, $\overline{RP}$; Bases: △PQR, △MNO; Lateral faces: ▭MORP, ▭ONQR, ▭NMPQ

Additional Answers

2. Right; Edges: $\overline{AB}$, $\overline{BC}$, $\overline{CD}$, $\overline{DE}$, $\overline{EF}$, $\overline{FG}$, $\overline{GH}$, $\overline{HA}$, $\overline{AI}$, $\overline{BJ}$, $\overline{CK}$, $\overline{DL}$, $\overline{EM}$, $\overline{FN}$, $\overline{GO}$, $\overline{HP}$, $\overline{IJ}$, $\overline{JK}$, $\overline{KL}$, $\overline{LM}$, $\overline{MN}$, $\overline{NO}$, $\overline{OP}$, $\overline{PI}$; Bases: Octagon *ABCDEFGH*, Octagon *IJKLMNOP*; Lateral faces: Rectangles *ABJI*, *BCKJ*, *CDLK*, *DEML*, *EFNM*, *FGON*, *GHPO*, *HAIP*

3. No right angle is shown. Oblique; Edges: $\overline{WX}$, $\overline{XY}$, $\overline{YZ}$, $\overline{ZW}$, $\overline{WP}$, $\overline{XQ}$, $\overline{YR}$, $\overline{ZS}$, $\overline{PQ}$, $\overline{QR}$, $\overline{RS}$, $\overline{SP}$; Bases: Quadrilaterals *WXYZ* and *PQRS*; Lateral faces: Parallelograms *PWZS*, *SZYR*, *RYXQ*, *QXWP*

4. Right; Edges: $\overline{DE}$, $\overline{EF}$, $\overline{FG}$, $\overline{GD}$, $\overline{DH}$, $\overline{EI}$, $\overline{FJ}$, $\overline{GK}$, $\overline{HI}$, $\overline{IJ}$, $\overline{JK}$, $\overline{KH}$; Bases: Rectangles *DEFG* (given) and *HIJK*: Lateral faces: Rectangles *DHKG*, *GKJF*, *FJIE*, *EIHD*. Answers for bases and lateral faces may vary.

5. Lat: 220
Tot: 276

6. Lat: $48\pi \approx 151$ cm^2
Tot: $66\pi \approx 208$ cm^2

7. Lat: 600
Tot: 660

8. Lat: 64 in.^{2}
Tot: 96 in.^{2}

9. Lat: 4368 cm^2
Tot: 5688 cm^2

10. Lat: $960\pi \approx 3016$ m^2
Tot: $1248\pi \approx 3921$ m^2

11. Lat: about 305 in.^{2}
Tot: about 316 in.^{2}

12. Lat: $216\pi \approx 679$ m^2
Tot: $378\pi \approx 1188$ m^2

FOLLOW-UP

Assessment

1. Find the total surface area of a prism with each edge 10-in. long. [600 in.^{2}]

2. A well is shaped like a cylinder with bases 1 meter in diameter. Find the lateral area of the well if it is 8 meters deep. [$8\pi \approx 25$ m^2]

Reteaching

Students who have had difficulty with this section may benefit from Reteaching Activity 11.1.

Enrichment

For students who have mastered the material in this section, you may wish to assign Enrichment Activity 11.1.

1. Find the area of each trapezoidal base. (Hint: Review Section 9.5.) 9.75
2. Find the perimeter of trapezoid *ABCD*. 18.1
3. Find the lateral area of the prism. 162.9
4. Find the total area of the prism. 182.4

Refer to the right circular cylinder at the right. The radius of each base is 2.2 in.

5. Find the area of each circular base. $4.84\pi \approx 15.2$ in.^{2}
6. Find the circumference of each base. $4.4\pi \approx 13.8$ in.
7. Find the lateral area of the cylinder. $22\pi \approx 69.1$ in.^{2}
8. Find the total area of the cylinder. $31.68\pi \approx 99.5$ in.^{2}

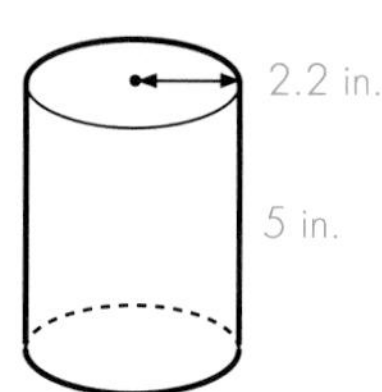

HOME ACTIVITY

Indicate whether the prism is *right* or *oblique*. Then identify each edge, base, and lateral face. See margin.

1. 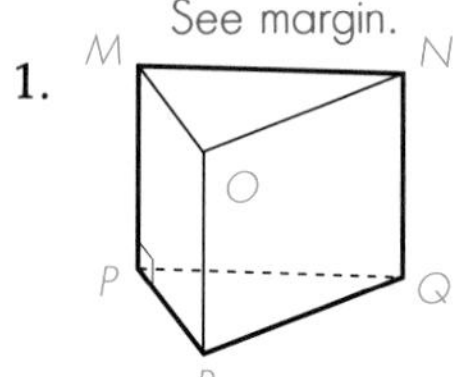

2. 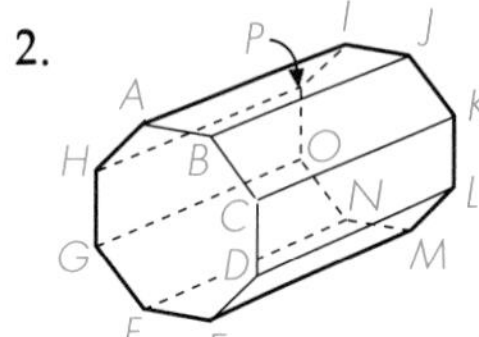

3. 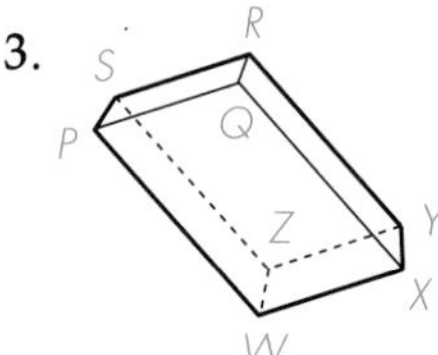

4. 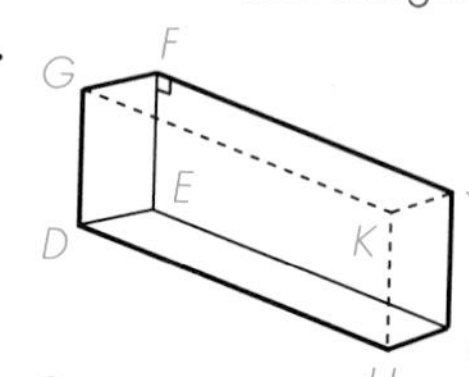

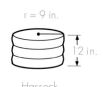

Find the lateral area and total area for each prism or right circular cylinder. See margin.

5.

6.

7. 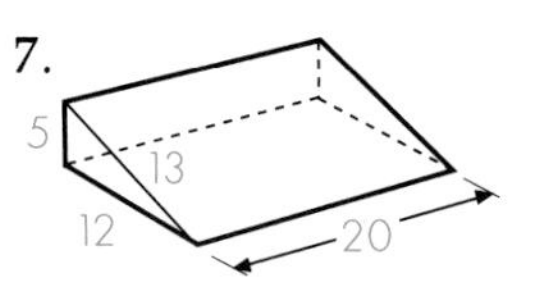

8. 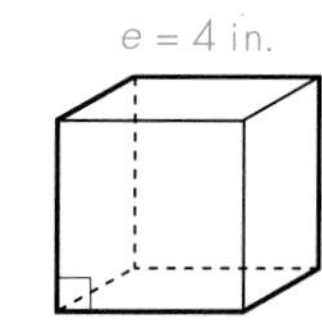

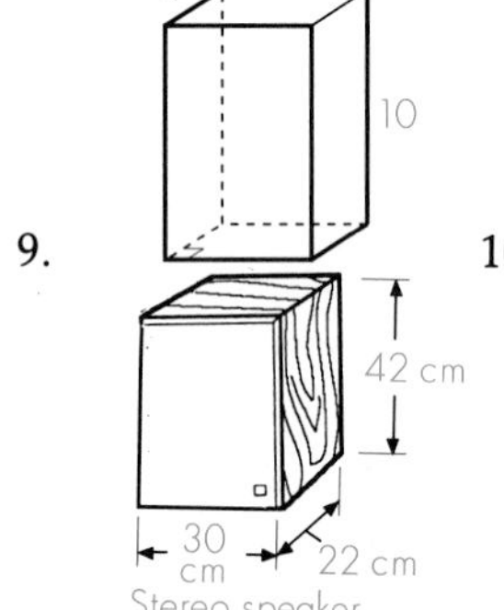

9.

Stereo speaker

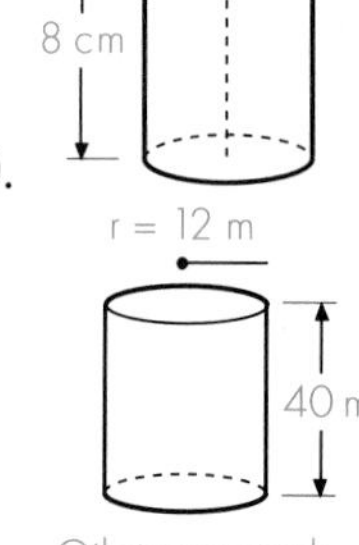

10.

Oil storage tank

11.

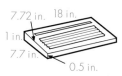

Computer keyboard

12.

Hassock

CRITICAL THINKING

13. Which cylinder has the larger lateral area? Same: 16π
14. Which cylinder has the larger total area? The cylinder on the left: $48\pi > 24\pi$

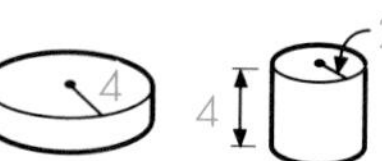

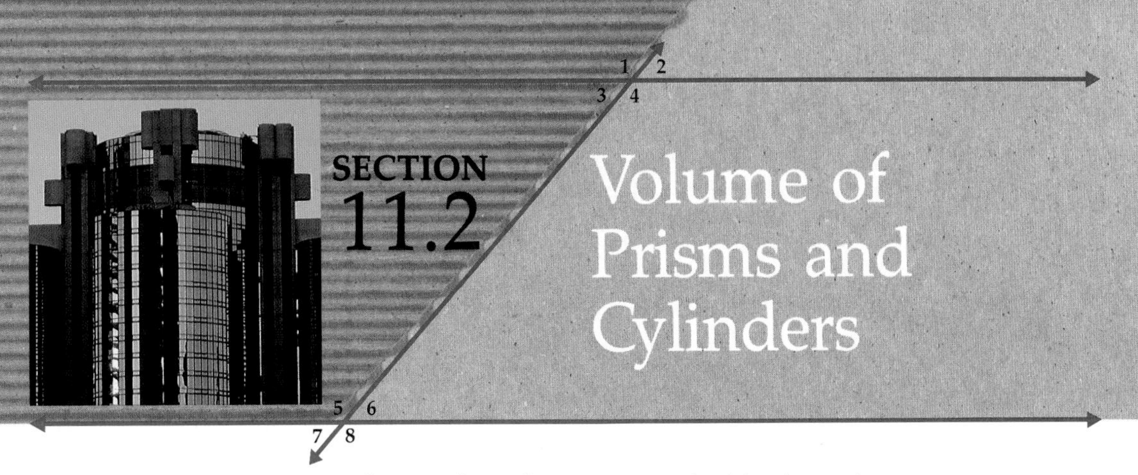

After completing this section, you should understand
▲ what volume is.
▲ how to find the volume of prisms and cylinders.

You are familiar with right rectangular prisms such as those shown here.

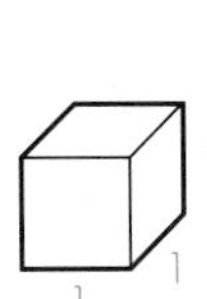

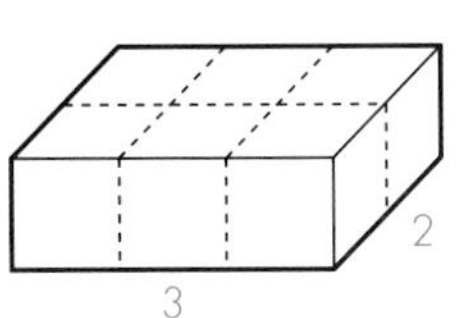

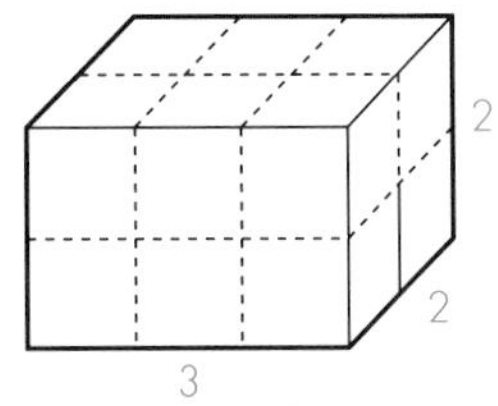

These prisms occupy space. How do you tell how much space they occupy? You can think of the larger prisms being filled with **unit cubes,** like the one on the left.

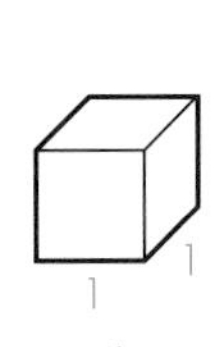

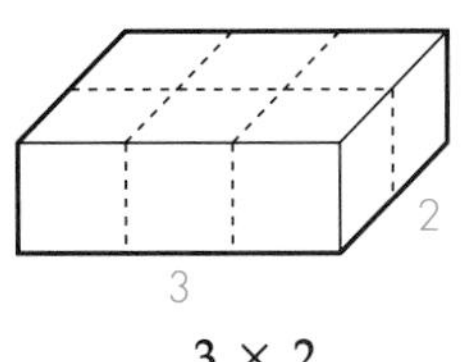

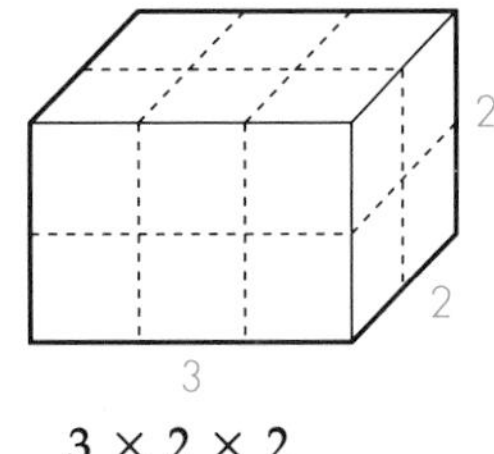

1 $\quad 3 \times 2 \quad 3 \times 2 \times 2$

The prism on the left is a **cube** because all twelve of its edges are equal. It is a **unit cube** because each edge measures one unit of length. If the unit of length is the centimeter, then the space occupied by the unit cube is 1 **cubic centimeter.** In the figures just above, the prisms occupy a space of 1 cubic centimeter (1 cm^3), 6 cubic centimeters (6 cm^3), and 12 cubic centimeters (12 cm^3), respectively. These numbers are the respective **volumes** of the three figures.

SECTION
11.2

Resources
Reteaching 11.2
Enrichment 11.2
Transparency Master 11-1

Objectives
After completing this section, the student should understand
▲ what volume is.
▲ how to find the volume of prisms and cylinders.

Materials
stack of paper or index cards
calculator

Vocabulary
cube
unit cube
cubic centimeter
volume
Cavalieri's Principle

GETTING STARTED

Quiz: Prerequisite Skills

1. How can you tell that the prism below is a right rectangular prism? [The bases are rectangles, and the lateral edges are perpendicular to the bases.]

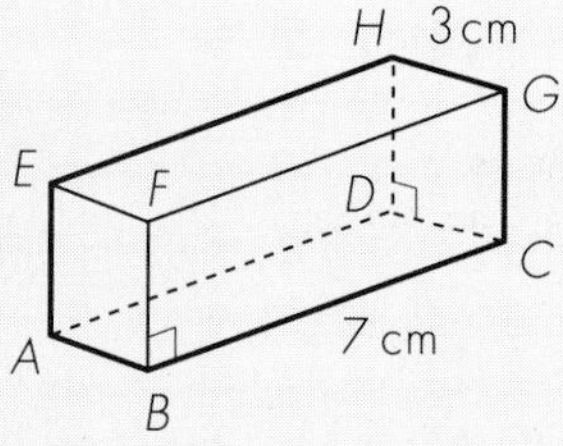

2. Find the length of $\overline{HE}$. [7 cm]

3. Find the area of rectangle $ABCD$. [21 cm^2]

4. Find the total area of the prism. [122 cm^2]

TEACHING COMMENTS

Using the Page
Point out the distinction between a *unit cube* and *cubic unit*. A unit cube is a special prism. A cubic unit is the amount of space occupied by a unit cube.

There are cubic units other then the cubic centimeter, such as the cubic meter (m^3), the cubic foot (ft^3), and so on.

NOTEBOOK

DEFINITION 11-2-1: **The VOLUME of a figure is the number of cubic units that the figure contains.**

You can find how many unit cubes will fit inside a right rectangular prism if you know its *length, width,* and *height.*

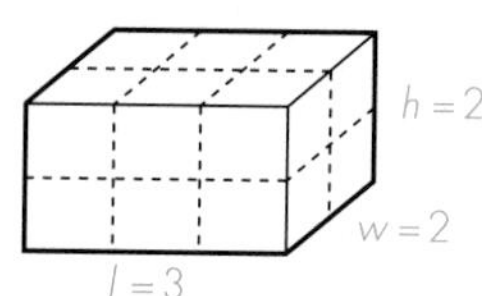

volume = number of cubes
= 3 × 2 × 2
= 12 cubic units

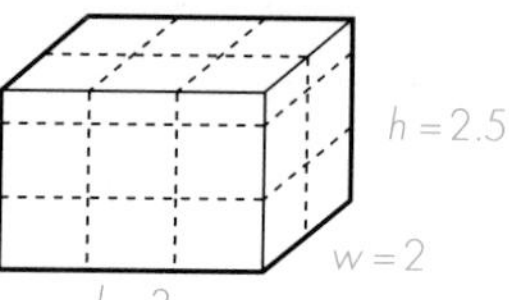

volume = number of cubes
= 3 × 2 × 2.5
= 15 cubic units

This suggests the following postulate.

POSTULATE 11-2-1 **The volume of a right rectangular prism is the product of its length, width, and height: $V = lwh$.**

Examples: Find the volume of each right rectangular prism.

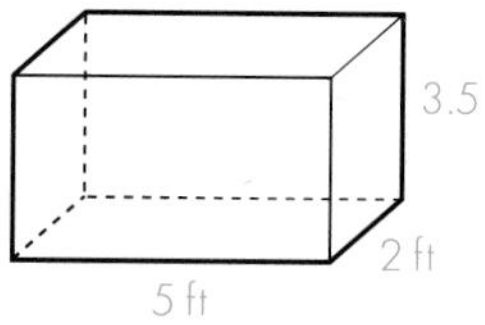

$V = lwh$
$= 5 \times 2 \times 3.5$
$= 35\ ft^3$

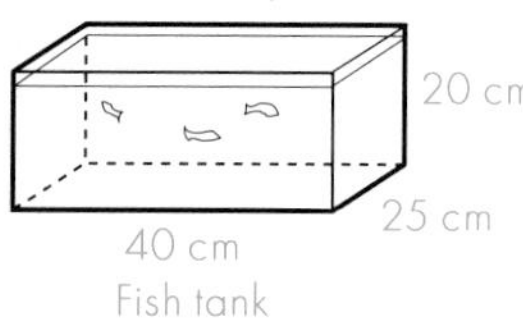

Fish tank

$V = lwh$
$= 40 \times 25 \times 20$
$= 20{,}000\ cm^3$

SOLIDS WITH EQUAL VOLUMES

At the right are shown two prisms that have the same height and the same dimensions for the base. How do the volumes compare? The next Discovery Activity will help you to answer this question.

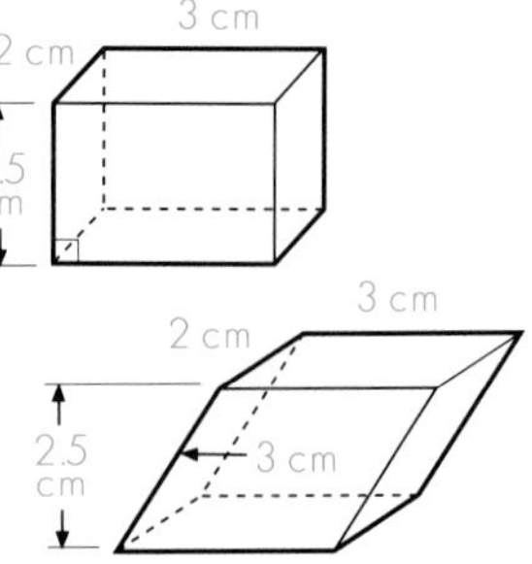

This Activity may be done either in actuality or in your imagination.

1. Open a ream (500 sheets) of copier paper and place the ream flat on top of a table, so that the ream retains its original shape as a right rectangular prism.

2. Gently tap one end of the ream so that the ream slants, as shown in the second figure.

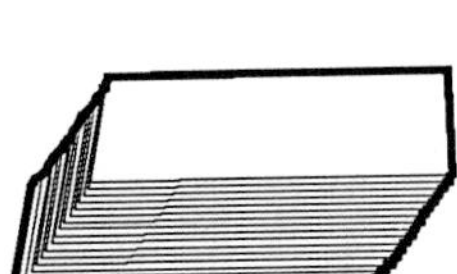

3. How does changing the shape of the ream affect the space it occupies? In other words, how does it affect its volume? Volume does not change.

Using the Page
The Discovery Activity can also be illustrated using a pack of cards or a stack of books.

As a result of the Discovery Activity, you should see that the volume of the ream of paper will not change. This illustrates an important idea known as **Cavalieri's Principle.**

POSTULATE 11-2-2 **(Cavalieri's Principle) If two solid figures have equal heights and bases of the same area, and if every plane parallel to the bases always cuts off two cross-sections of equal area, then the two solids have equal volume.**

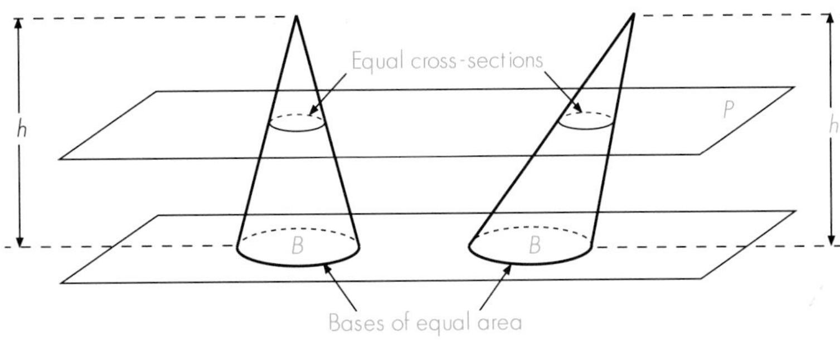

Cavalieri's Principle can be used to compare two prisms that have equal heights and bases of the same area.

Using the Page
As a two-dimensional analogy to Cavalieri's Principle, recall that two triangles with equal bases and altitudes have the same area. Also, a rectangle and parallelogram have the same area if their bases have the same length and their heights are equal.

Transparency 11-1 can be used to help students see how Cavalieri's Principle leads to the formulas stated in Theorems 11-2-1 and 11-2-2.

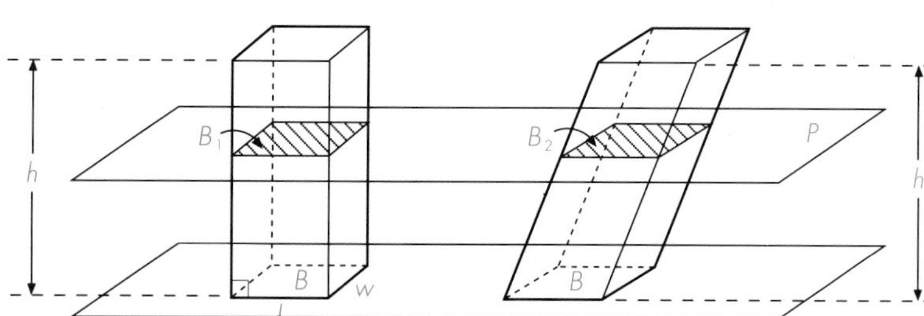

In the figure above, P is any plane that is parallel to the bases of the prisms. The plane cuts each prism in a cross-section. The left prism is a right rectangular prism with dimensions *l*, *w*, and *h*. The area of the base of each prism is *B*. It can be shown that the area B_1 of the left cross-section is equal to *B*, the area of the base of the prism at the left. Similarly, the area B_2 can be shown to equal *B*, the area of the base of the prism at the right. Therefore, $B_1 = B_2$. Thus, the two prisms obey Cavalieri's Principle. The same is true even when the bases are shapes other than rectangles.

Using the Page

Point out that in some of the exercises in this chapter, the area of a base is given. At other times, the area must be computed.

NOTEBOOK

THEOREM 11-2-1 **The volume of a prism is the product of the area *B* of one of its bases and the height *h* of the prism: $V = Bh$.**

Examples: Find the volume of each prism.

1.

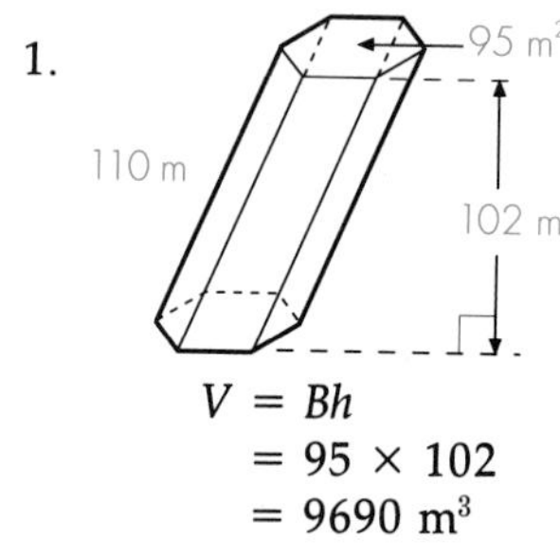

$V = Bh$
$= 95 \times 102$
$= 9690 \text{ m}^3$

2.

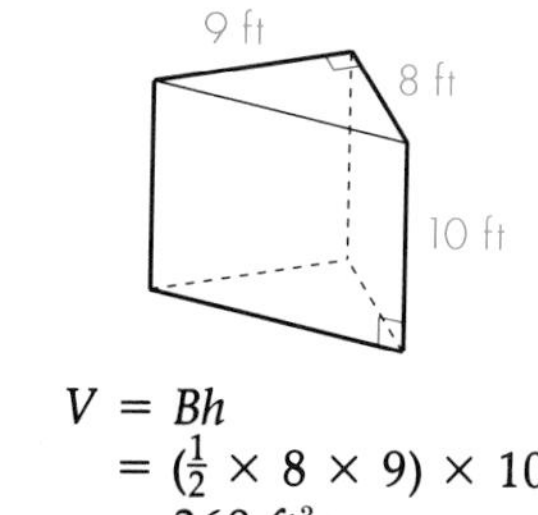

$V = Bh$
$= (\frac{1}{2} \times 8 \times 9) \times 10$
$= 360 \text{ ft}^3$

CLASS ACTIVITY

Find the volume of each prism.

1.

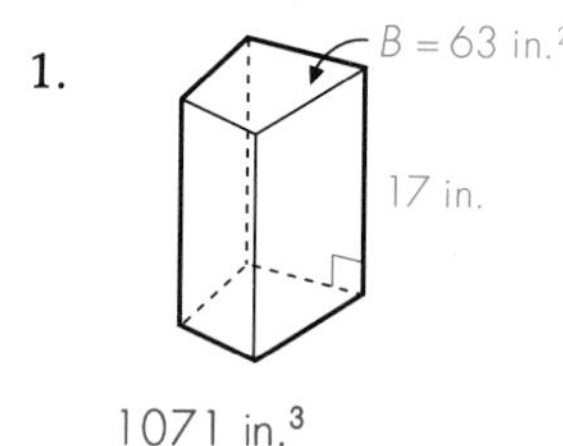

1071 in.^{3}

2.

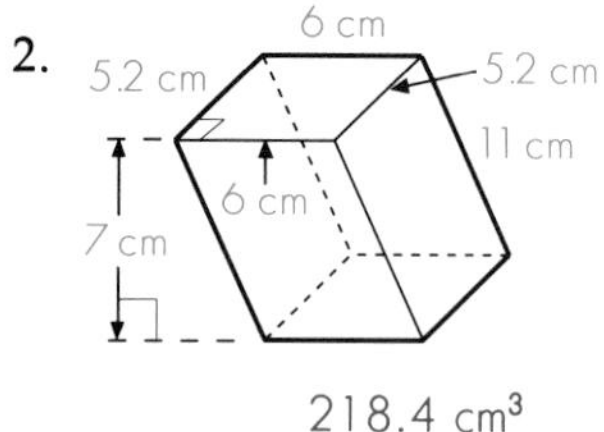

218.4 cm^3

Extra Practice

1. Find the volume of a prism with a base area of 8.5 cm^2 and a height of 4 cm. [34 cm^3]
2. Find the volume of the right rectangular prism.

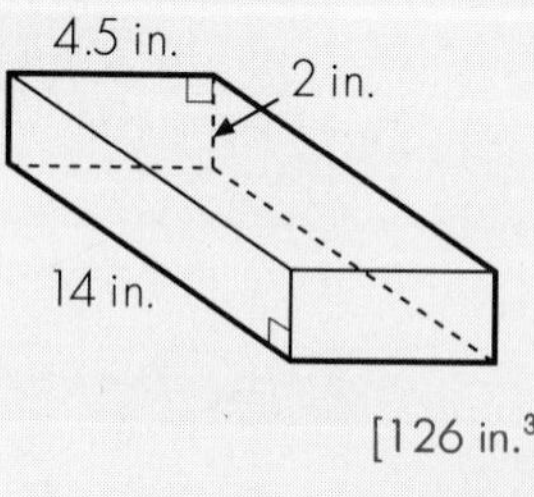

[126 in.^{3}]

THE VOLUME OF A CYLINDER

Two cylinders that have the same height and bases of equal area can be studied in the same manner as the two prisms on the previous page.

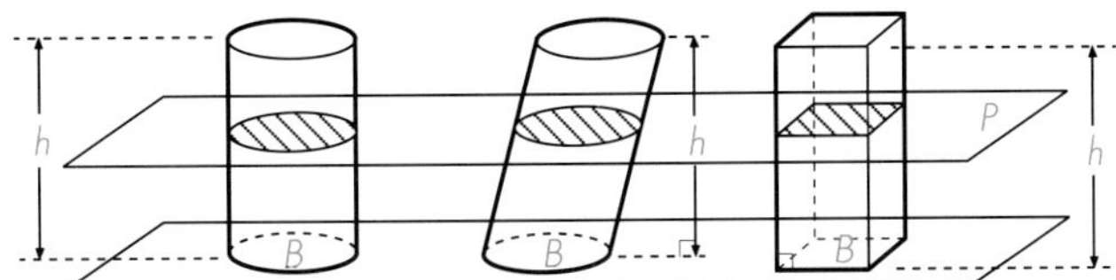

In the diagram above, the prism and the two cylinders all have bases with the same area *B*. They also have the same height *h*. It can be shown that in plane P, which is parallel to the bases of the two cylinders, all the cross-sections of the two cylinders have the same area. So, the cylinders obey Cavalieri's Principle. This leads to the following theorem.

THEOREM 11-2-2 **The volume of a circular cylinder is the product of the area *B* of one of its bases and the height *h* of the cylinder: $V = Bh$.**

Example: Find the volume of the circular cylinder.

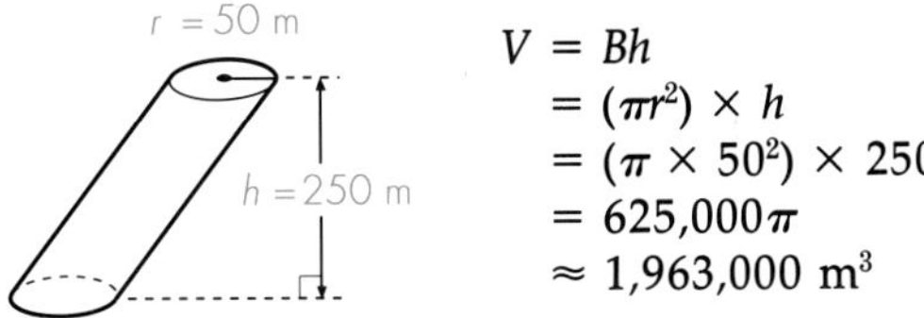

$$\begin{aligned} V &= Bh \\ &= (\pi r^2) \times h \\ &= (\pi \times 50^2) \times 250 \\ &= 625{,}000\pi \\ &\approx 1{,}963{,}000 \text{ m}^3 \end{aligned}$$

To the nearest thousand cubic meters, V is 1,963,000 m^3.

CLASS ACTIVITY

Use a calculator to find the volume of each cylinder. Answer to the nearest whole unit.

1. 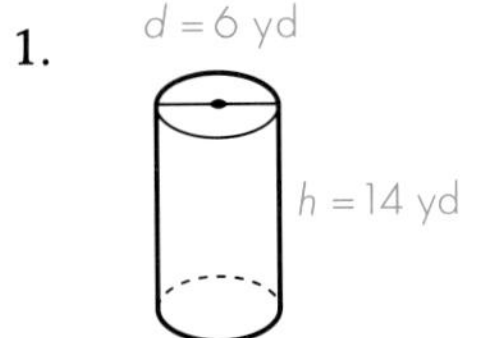

about 396 yd^3

2. 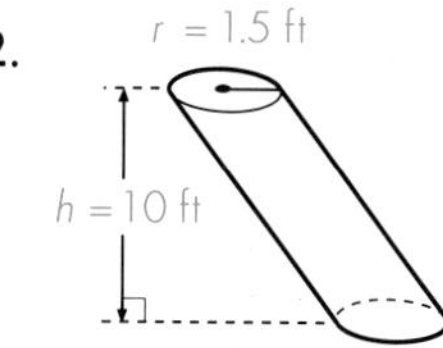

about 71 ft^3

Extra Practice
Find the volume of each cylinder. Answer to the nearest whole unit.

1. 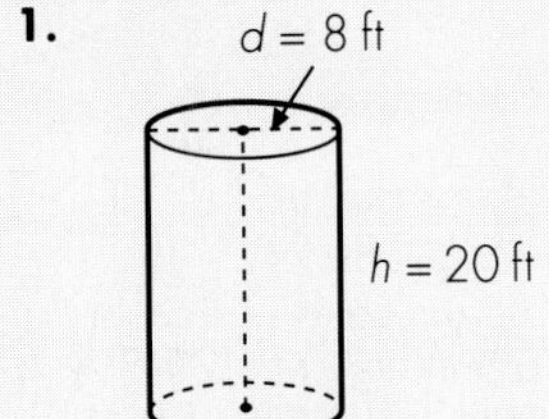

[$320\pi \approx 1005$ ft^3]

2. 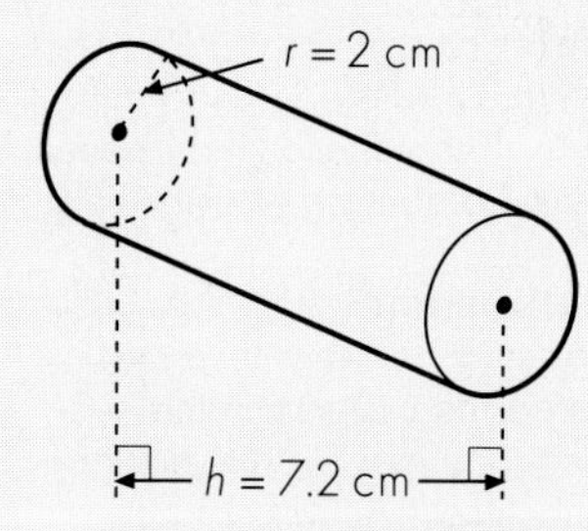

[$28.8\pi \approx 90$ cm^3]

HOME ACTIVITY

In this activity, use your calculator as needed. Find the volume of each prism. Answer to the nearest whole unit.

1. 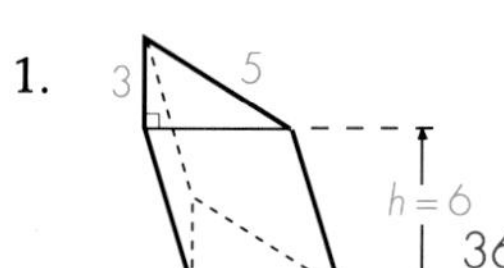

36

2. e = 6
216

3. 1520 mm³
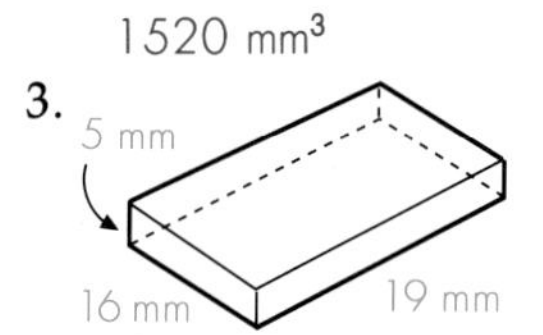

4. $10\sqrt{3} \approx 17$ in.³
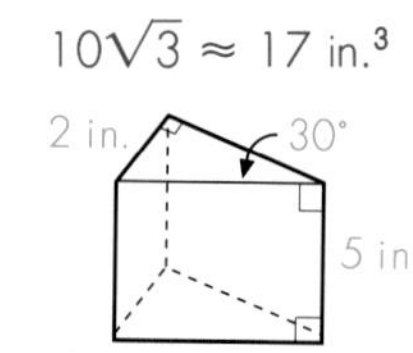

Find the volume of each cylinder.

5. 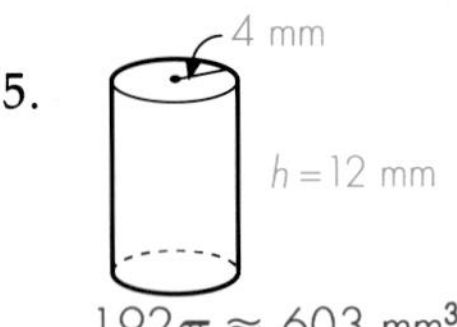

$192\pi \approx 603$ mm³

6. h = 13 in.
B = 60 in.²
780 in.³

7. 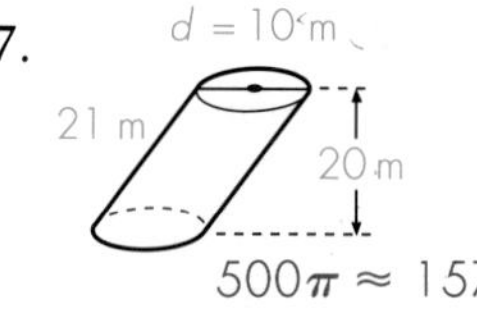

$500\pi \approx 1570$ m³

8. 96 ft³
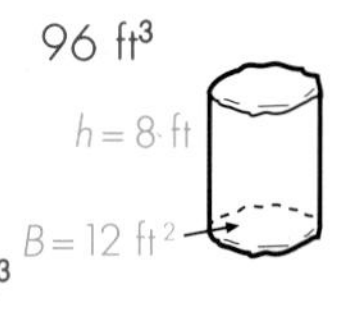

9. Find the volume of the gasoline storage tank at the right. Give your answer in gallons. (1 ft³ = 7.5 gal)
about 442,740 gal

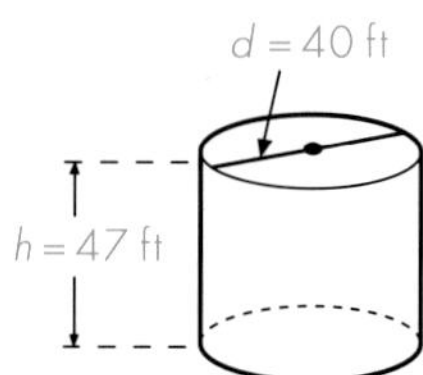

10. Find the volume of the barn. 2880 yd³

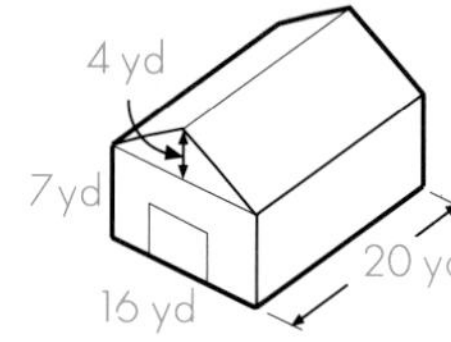

11. A basketball court has a volume of 126,000 ft³. The floor dimensions are 84 ft by 50 ft. How high is the ceiling? 30 ft

12. The length of a diameter of an automobile cylinder is called the *bore.* The distance the piston moves in the cylinder is the *stroke.* The engine capacity of a car is the combined volume of all its cylinders. Find the engine capacity of a 6-cylinder engine if the bore is 3.88 in. and the stroke is 3.25 in.
about 73.4π, or 230 in.³

Stroke
Bore

CRITICAL THINKING

13. A cylinder can be modeled using a rectangular piece of paper in two ways, as shown below. Which cylinder has the greater volume?
The second has the greater volume.

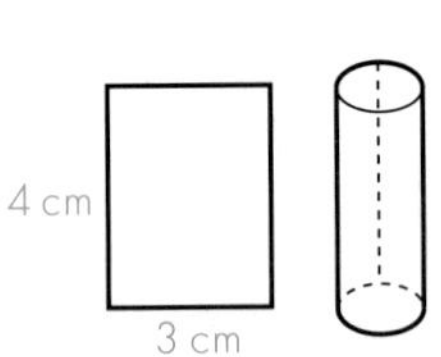

First cylinder

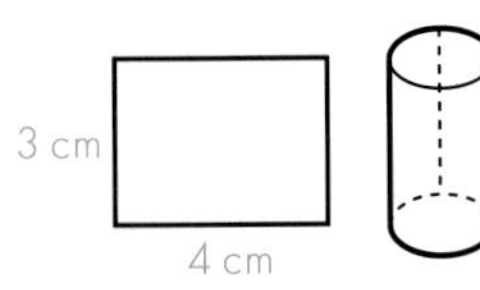

Second cylinder

Using the Page
For Exercise 1 of the Home Activity, note that more information than necessary is given. For Exercise 4, students may need to review the relation between lengths of sides of a 30°-60°-90° triangle.

FOLLOW-UP

Assessment
1. Find the volume of the prism.

5 cm
20 cm
12 cm

[600 cm³]

2. A cylindrical container is 2.5 in. high. The radius of a base is 5 in. Find the volume to the nearest whole unit. [$62.5\pi \approx 196$ in.³]

Reteaching
Students who have had difficulty with this section may benefit from Reteaching Activity 11.2.

Enrichment
For students who have mastered the material in this section, you may wish to assign Enrichment Activity 11.2.

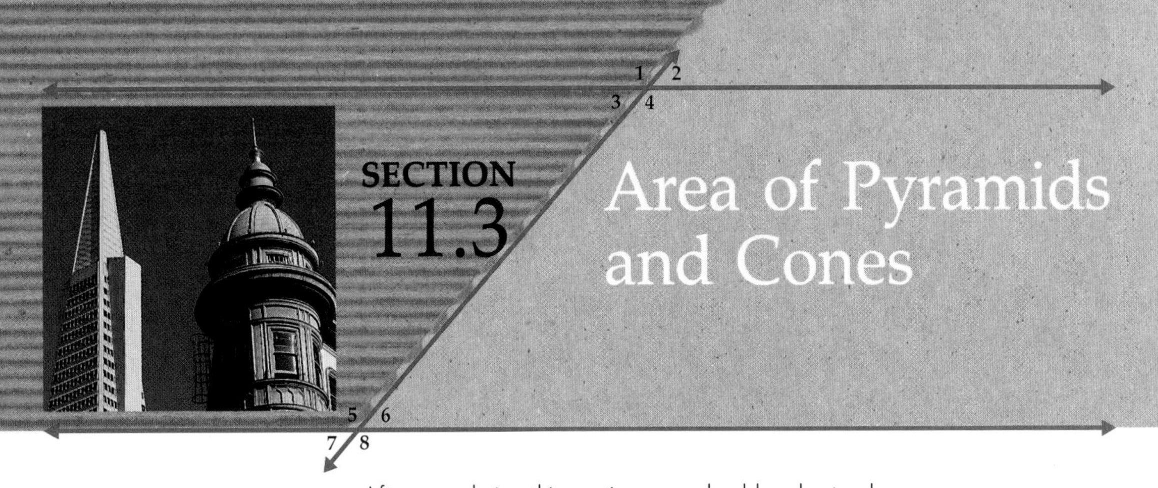

Area of Pyramids and Cones

After completing this section, you should understand
- ▲ what a pyramid is.
- ▲ what a cone is.
- ▲ how to calculate the surface area of a pyramid.
- ▲ how to calculate the surface area of a cone.

A closed, three-dimensional figure made up entirely of polygonal regions is called a **polyhedron.** Each of these figures is a polyhedron.

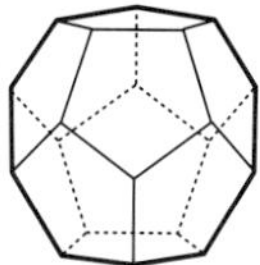
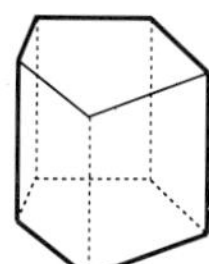
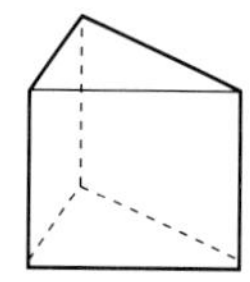

You have already studied one special kind of polyhedron, namely prisms. The figures that follow are examples of another special kind of polyhedron.

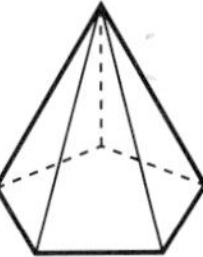
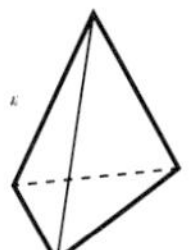
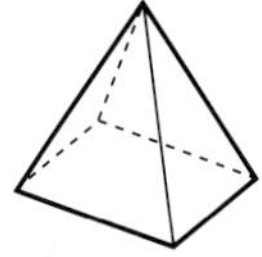

This kind of polyhedron is called a **pyramid.**

PYRAMIDS

To make a pyramid, start with a polygonal region and a point P not in the plane of the region. Then connect that point to the vertices of the region by means of line segments.

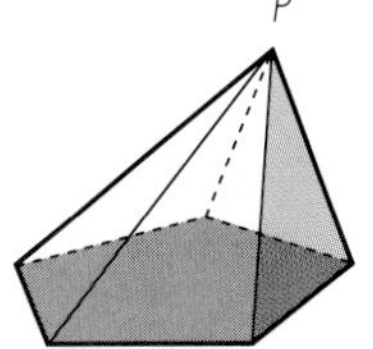

SECTION 11.3

Resources
Reteaching 11.3
Enrichment 11.3
Transparency Master 11-2

Objectives
After completing this section, the student should understand
- ▲ what a pyramid is.
- ▲ what a cone is.
- ▲ how to calculate the surface area of a pyramid.
- ▲ how to calculate the surface area of a cone.

Materials
calculator

Vocabulary
polyhedron
pyramid
base
vertex
lateral edge
lateral face
regular pyramid
slant height
lateral area
triangular pyramid
tetrahedron
quadrangular pyramid
hexagonal pyramid
square pyramid
cone
height
oblique cone
right cone
axis
sector of a circle
circular cone

GETTING STARTED

Quiz: Prerequisite Skills

1. Find the perimeter of a regular pentagon if a side is 5.2 in. long. [26 in.]

The radius of a circle is 4 cm.

2. Find the area of the circle to the nearest unit. [$16\pi \approx 50$ cm^2]

3. Find the circumference of the circle to the nearest unit. [$8\pi \approx 25$ cm]

Each side of the polygonal region and the line segments connecting its endpoints to P determine a triangular region. All of these triangular regions together with the polygonal region you started with form a pyramid. You can also think of the pyramid in the manner described in this definition.

DEFINITION 11-3-1: A PYRAMID is a polyhedron formed by a polygonal region and all the line segments connecting a point P not in the plane of the region to the points of the polygon that determine the region.

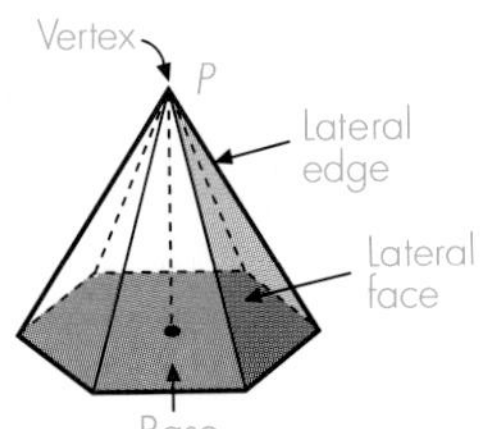

The polygonal region mentioned in the definition is the *base* of the pyramid. Point P is the *vertex* of the pyramid. Segments joining the vertex P to the vertices of the base are called *lateral edges.* Each triangular region determined by two lateral edges and an edge of the base is a *lateral face.* The perpendicular distance from the vertex P to the plane of the base is the *height* of the pyramid. The pyramid is called by the kind of polygon it has as its base.

TEACHING COMMENTS

Using the Page
The Discovery Activity provides an informal proof of the formula for the lateral area of a regular pyramid, which is given in Theorem 11-3-1. Students need to recognize that $5 \times \frac{1}{2}se$ is equivalent to $\frac{1}{2} \times 5e \times s$, and that $5e$ can be replaced by p. It may be useful to start the discussion of this activity by giving e and s specific values, say $e = 4$ and $s = 6$. Then students can compare $5 \times \left(\frac{1}{2} \times 4 \times 6\right)$ with $\frac{1}{2} \times (5 \times 4) \times 6$, where (5×4) represents the perimeter.

DISCOVERY ACTIVITY

Consider a pyramid whose base is a regular pentagon. Suppose all the lateral faces are congruent isosceles triangles. Let e be the length of each side of the base. Let s be the height to one of the triangular lateral faces.

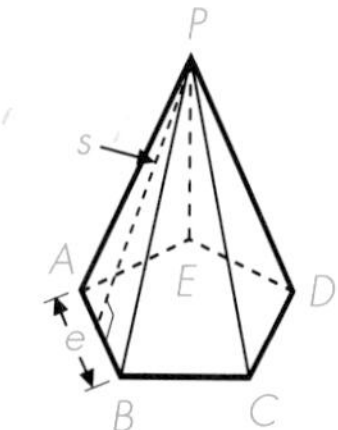

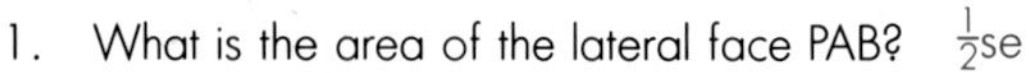

1. What is the area of the lateral face PAB? $\frac{1}{2}se$

2. What is the area of each of the other lateral faces? $\frac{1}{2}se$

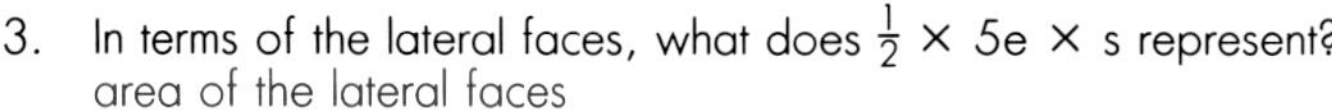

3. In terms of the lateral faces, what does $\frac{1}{2} \times 5e \times s$ represent? area of the lateral faces
4. In terms of the base, what does 5e represent? perimeter of the base

In the Discovery Activity, the pyramid was a special kind of pyramid called a *regular* pyramid.

A **regular pyramid** is a pyramid that has a regular polygonal region as its base and lateral edges that are all of equal length. The height s of each lateral face of a regular pyramid is called the **slant height** of the pyramid.

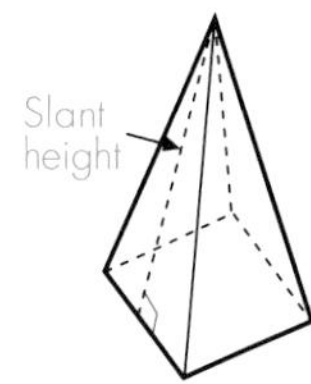

From the Discovery Activity, you may have discovered how the slant height of a pyramid and the perimeter of the base can be used to express the **lateral area** of the pyramid.

THEOREM 11-3-1 **If p is the perimeter of the base of a regular pyramid and s is the slant height, the lateral area LA is given by the formula $LA = \frac{1}{2}ps$.**

Recall that a pyramid is identified by the shape of its base, as illustrated below.

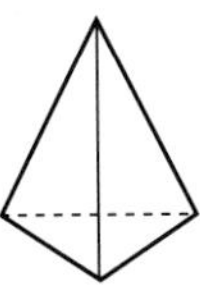

Triangular pyramid **(Tetrahedron)**

Regular quadrangular pyramid

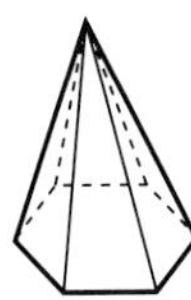

Hexagonal pyramid

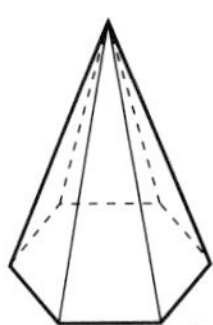

Regular hexagonal pyramid

Example: Find the lateral area of the regular square pyramid below.

$$\text{Lateral area} = \tfrac{1}{2}ps$$
$$= \tfrac{1}{2}(4 \times 7) \times 10$$
$$= 140 \text{ mm}^2$$

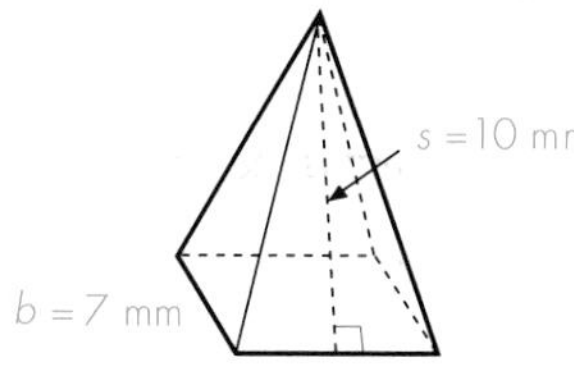

The **total area** of a pyramid is the sum of its lateral area and the area of the base.

THEOREM 11-3-2 **If B is the area of the base of a regular pyramid, p the perimeter of the base, and s the slant height of the pyramid, then the total area (TA) is given by the formula TA = B + $\frac{1}{2}$ps.**

Using the Page

To emphasize the two conditions that define a regular pyramid, have students model pyramids that are not regular pyramids. They can use D-Stix, or pencils of equal length, to illustrate pyramids that meet the second but not the first condition (equal edges, irregular base). They can try to draw sketches of pyramids that meet the first condition but not the second (regular base, unequal edges.) Students can also sketch or model pyramids that meet neither of the conditions.

Transparency 11-2 provides a visual summary of figures in space that are investigated in this chapter.

Example: Find the total area of the regular tetrahedron at the right.

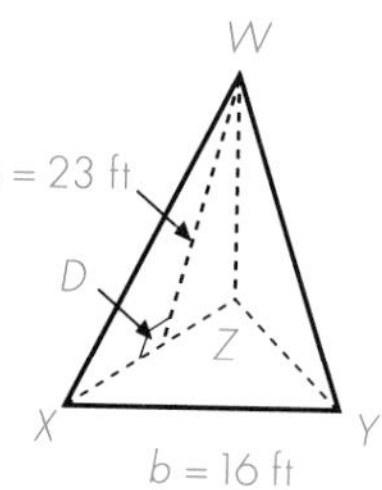

The base *XYZ* is an equilateral triangle. The altitude *ZD* of the triangular base divides the triangle into two congruent 30°-60° right triangles (see Section 7.3). So $XD = 8$ and $ZD = 8\sqrt{3}$.

Area of base of pyramid $= \frac{1}{2} \times 16 \times 8\sqrt{3} = 64\sqrt{3}$

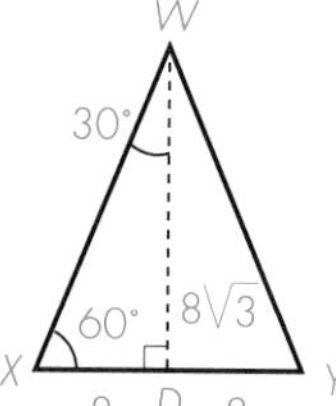

Total area of pyramid $= B + \frac{1}{2}ps$
$= 64\sqrt{3} + \frac{1}{2} \times (3 \times 16) \times 23$
$= 64\sqrt{3} + 552$
$\approx 64 \times 1.732 + 552$
$\approx 110.8 + 552$, or about 663 ft^2

CLASS ACTIVITY

Find the total area of each regular pyramid.

1. 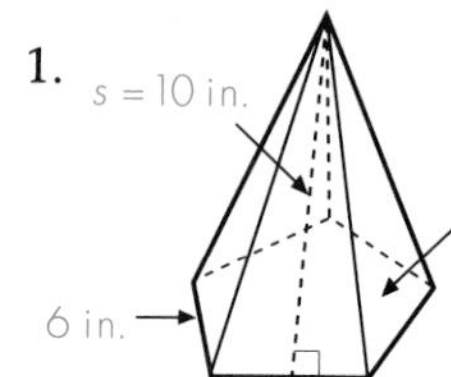

211.9 $in.^2$

Area of pentagonal base: 61.9 in.

2. 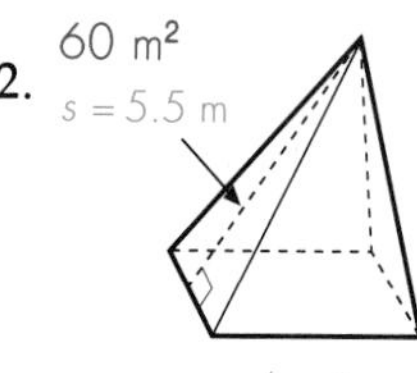

60 m^2

Extra Practice
Find the total area of each regular pyramid.

1.

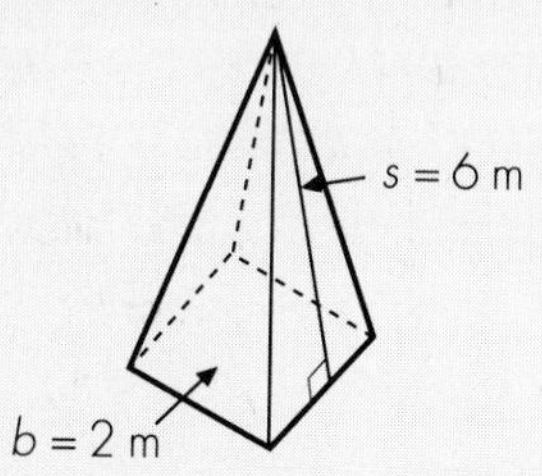

square pyramid

[28 m^2]

2.

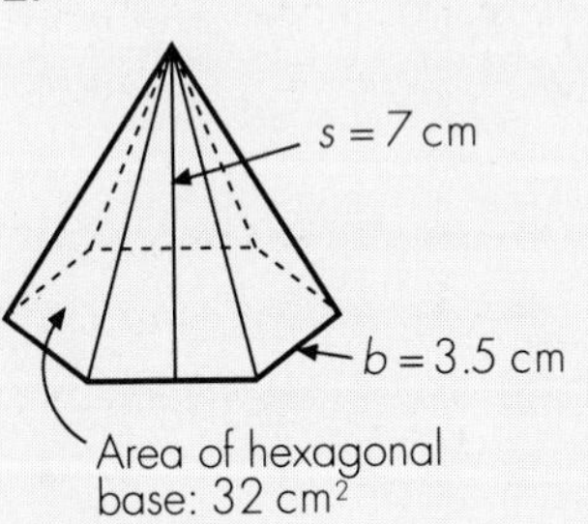

[105.5 cm^2]

CONES

Each figure below has a vertex not in plane K and consists of a closed region and all the segments joining the vertex to points on the boundary of the region.

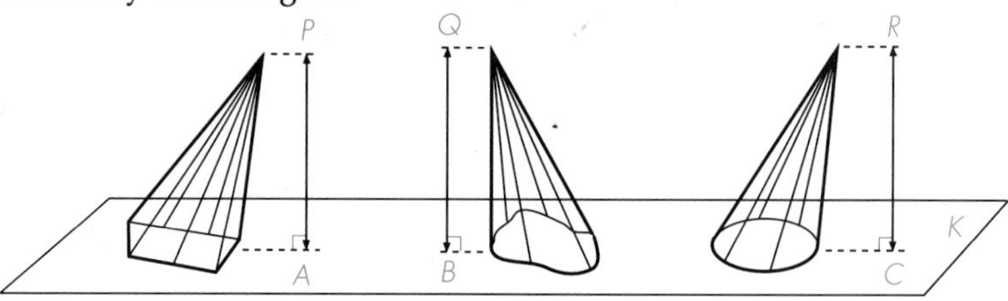

In the left figure, the region in plane K is polygonal. The figure is a pyramid. In the second figure, the boundary of the region in plane K is an irregular closed curve. Because the boundary is curved, the figure is a **cone.** Thus, a cone is similar to a pyramid except that the boundary of its base is curved. In the third figure, the region in plane K is circular. The cone is a **circular cone.**

DEFINITION 11-3-2 **A cone** is a figure formed by a closed curve region and all the line segments joining points on the boundary of the region to a point not in the plane of the region. This point is called the **vertex** of the cone. The perpendicular distance from the vertex to the plane of the base is the **height** of the cone. Two kinds of circular cones are shown at the right, an *oblique cone* and a *right cone.* The segment joining the vertex to the center of the base is called the **axis** of the cone. In a right cone, the axis is perpendicular to the base. The right cone has both a height and a **slant height,** which is the length of a line segment that joins the vertex to the circle.

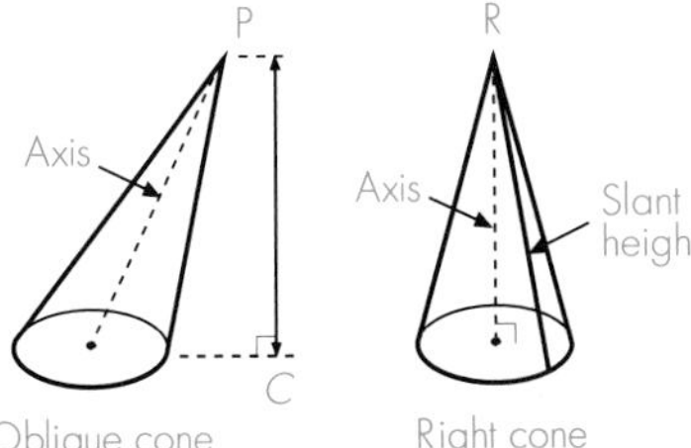

The pie-shaped figure at the left below is called a *sector* of a circle. It can be shown that the area of the sector is $\frac{1}{2}$ the product of the arc length *AB* and the radius *s* of the circle from which the sector was cut. When the sector is folded so that line segments *AP* and *BP* coincide, a right circular cone is formed. The lateral area of the cone is equal to the area of the sector.

Lateral area = Area of sector

$$= \frac{1}{2} \times \text{arc length} \times \text{radius}$$
$$= \frac{1}{2} \times 2\pi r \times s$$
$$= \pi r s$$

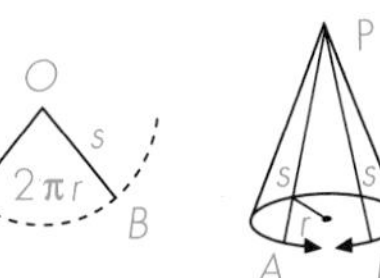

Notice that the circumference of the cone's base equals the arc length of the circle from which the sector was cut.

THEOREM 11-3-3: The lateral area LA of a right circular cone with a base of radius *r* and slant height *s* is given by LA = πrs. *r* is the radius of the base and *s* is the slant.

THEOREM 11-3-4: The total area TA of a right circular cone is the sum of the area of the base and the lateral area. If *r* is the radius of the base and *s* is the slant height, the total area is given by the formula TA = $\pi r^2 + \pi rs$.

Example: Find the total area.

Total area $= \pi r^2 + \pi rs$
$= \pi \times 3^2 + \pi \times 3 \times 5$
$= 9\pi + 15\pi$
$= 24\pi$
$\approx 24 \times 3.14$, or about 75 mm^2

r = 3 mm
s = 5 mm

Using the Page

In Theorem 11-3-3, the formula $LA = \pi rs$ is given for the lateral area of a right circular cone. It may be useful to point out that $LA = \frac{1}{2}Cs$ is an alternate formula, where C is the circumference of the base of the cone. Association with the formula $A = \frac{1}{2}bh$ for the area of a triangle may help students recall the derivation of $LA = \pi rs$, in case they forget the formula.

Using the Page

Point out the similarity between the formula $TA = \pi r^2 + \pi rs$, given in Theorem 11-3-4, and the formula $TA = B + \frac{1}{2}ps$ for the regular pyramid given in Theorem 11-3-2. By comparing formulas, students may be better able to remember them.

Using the Page

You may wish to point out the common factor, πr, in the formula $TA = \pi r^2 + \pi rs$. The students can then compute the total area in the example on the next page (page 418) by using the formula $TA = \pi r(r + s)$.

$TA = \pi r(r + s)$
$TA = \pi \times 3 \times (3 + 5)$
$= \pi \times 3 \times 8$
$= 24\pi$

CLASS ACTIVITY

Use a calculator to find the total area of each right circular cone.

1. $176\pi \approx 553$ cm^2

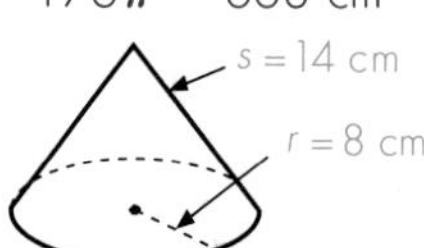

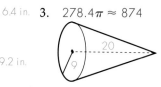

2. 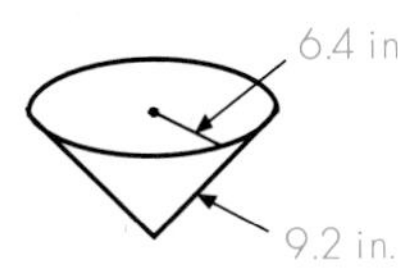

$99.84\pi \approx 313$ in.^{2}

3. $278.4\pi \approx 874$

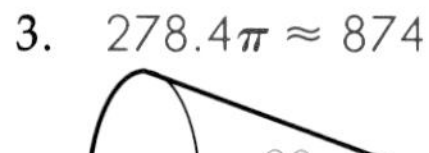

Hint: Use $a^2 + b^2 = c^2$.

HOME ACTIVITY

Find the total area of each regular pyramid or right circular cone.

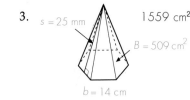

1. 480 in.^{2}

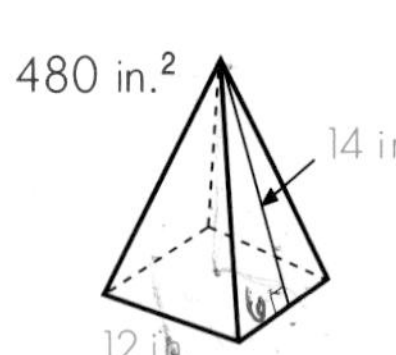

2. $450\pi \approx 1413$ ft^2

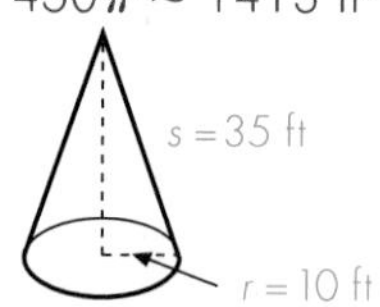

3. 1559 cm^2

s = 25 mm

B = 509 cm^2

b = 14 cm

4. $36\pi \approx 113$

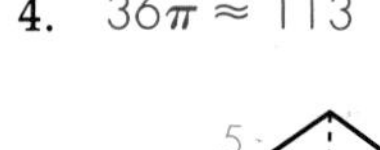
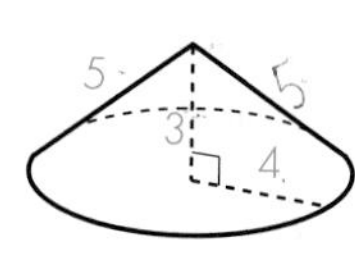

5. $96\pi \approx 302$ ft^2

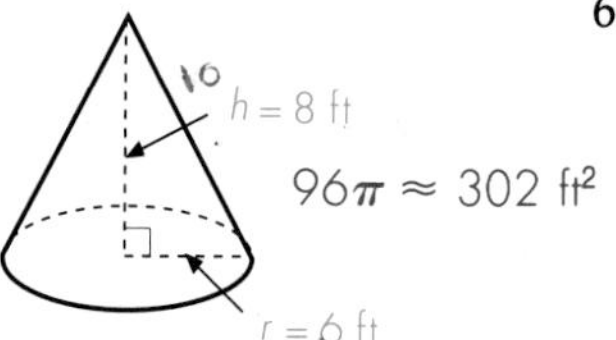

6. about 268 m^2

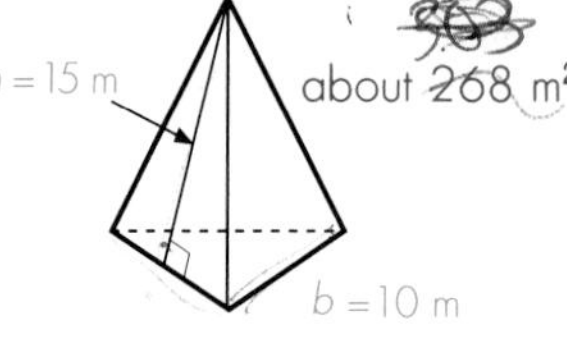

7. The top of the Washington Monument in Washington, D.C. consists of a regular square pyramid with a height of 55 ft. The length of a side of the base of the pyramid is about 34.4 ft. Find the lateral area of the pyramid. about 3965 ft^2

8. The roof of a house has the shape of a right circular cone. Its height is 24 ft and the diameter of the base is 20 ft. Shingles for the roof cost $24 per bundle and 3 bundles are needed for every 100 ft^2. Find the cost of the shingles. $600 for 25 whole bundles to cover 817 ft^2

CRITICAL THINKING

9. A 5-12-13 right triangle is rotated about the hypotenuse to form the double cone shown at the right. Find the area of the common base of the two cones.
$\left(\frac{60}{13}\right)^2\pi$, or about 67

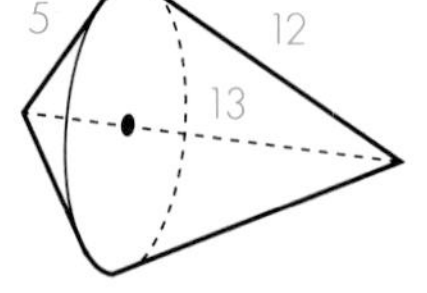

FOLLOW-UP

Assessment
Find the total area of each right circular cone or regular pyramid.

1.

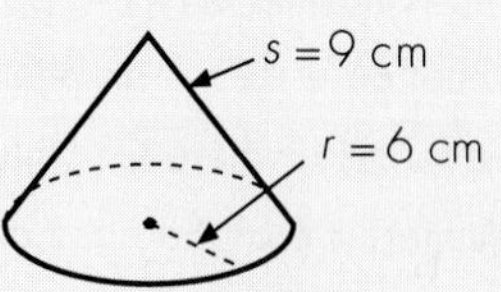

[$90\pi \approx 283$ cm^2]

2.

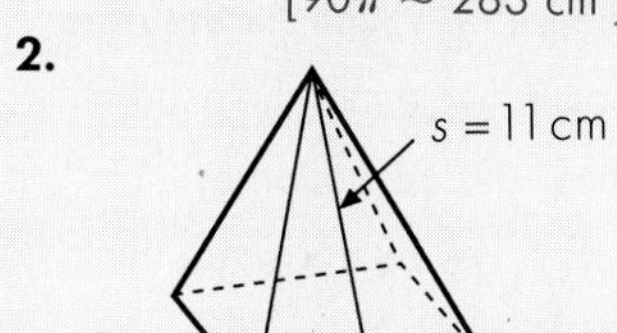

[279 cm^2]

3.

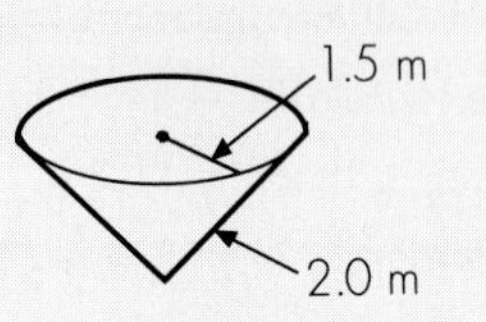

[$5.25\pi \approx 16.5$ m^2]

Reteaching
Students who have had difficulty with this section may benefit from Reteaching Activity 11.3.

Enrichment
For students who have mastered the material in this section, you may wish to assign Enrichment Activity 11.3.

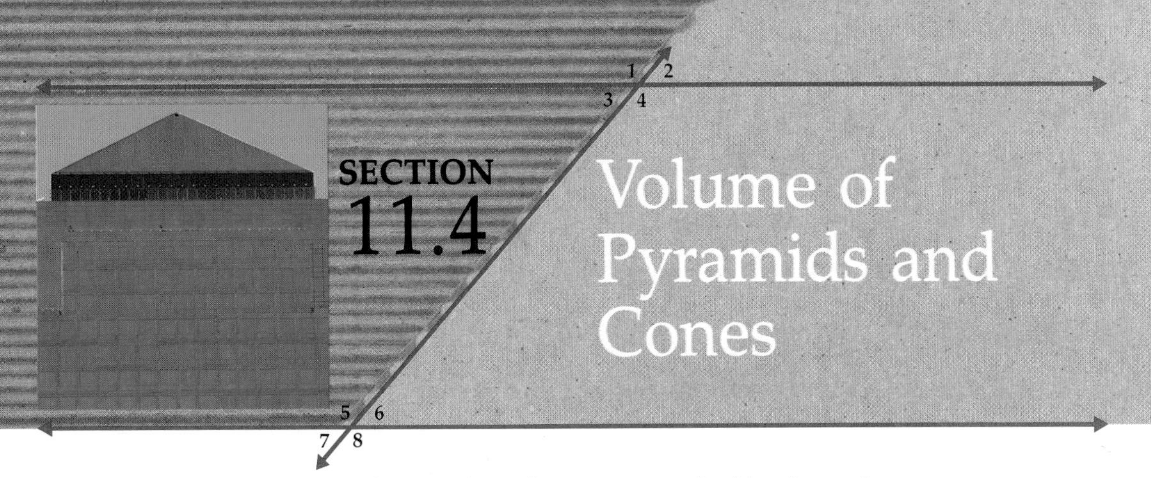

SECTION 11.4 Volume of Pyramids and Cones

After completing this section, you should understand
- ▲ how to find the volume of a pyramid.
- ▲ how to find the volume of a cone.

You know how to calculate the total area and the volume of a right prism and a right circular cylinder. You also know how to calculate the total area of a regular pyramid and a right circular cone. Now you will learn how to calculate the volume of both a pyramid and a cone.

MAKING A PRISM FROM THREE PYRAMIDS

In the diagram below, three pyramids of the same volume and shape are colored red, white, and blue, respectively. From their initial positions, the two outside pyramids approach the middle pyramid, as shown.

Initial position

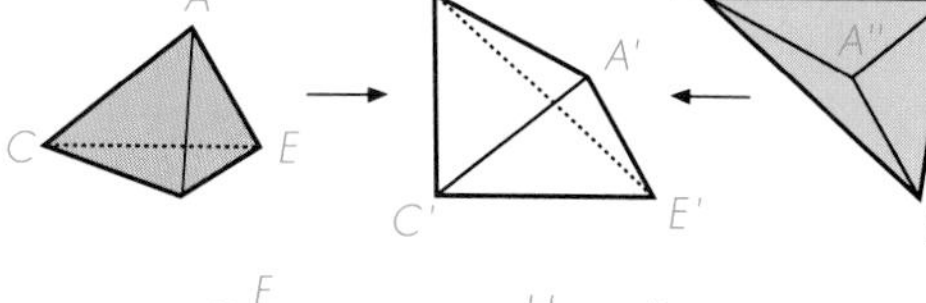

Pyramids come together

Polyhedron is formed

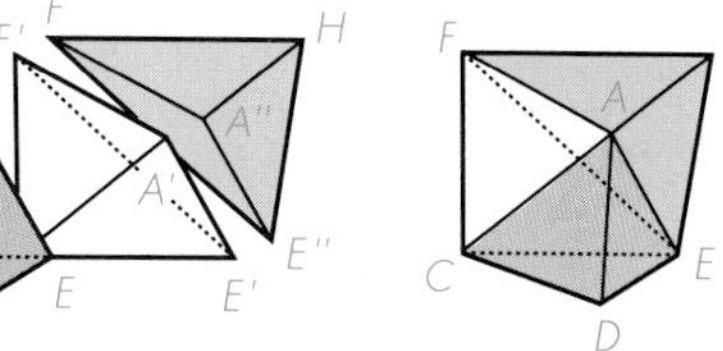

Notice how the outside pyramids fit with the middle pyramid.

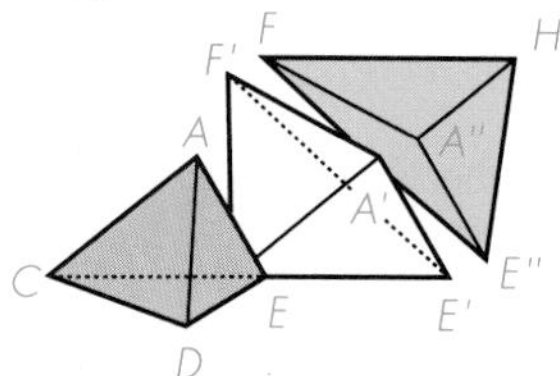

SECTION 11.4

Resources
Reteaching 11.4
Enrichment 11.4

Objectives
After completing this section, the student should understand
- ▲ how to find the volume of a pyramid.
- ▲ how to find the volume of a cone.

Materials
calculator
cutouts and tape
(See margin notes for page 420.)

GETTING STARTED

Quiz: Prerequisite Skills

1. Find the volume of the right triangular prism.

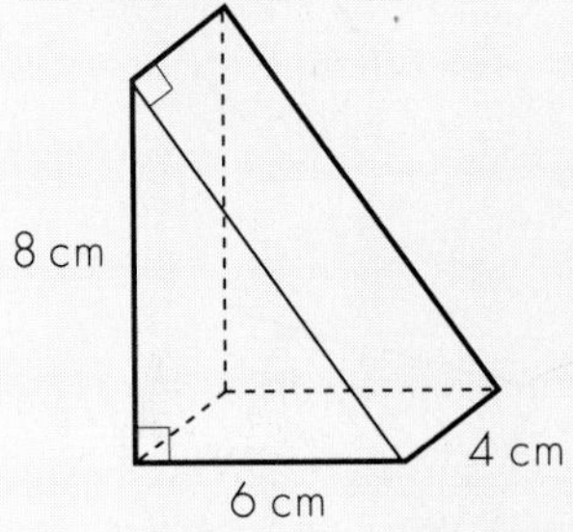

[96 cm^3]

2. Find the volume of the right circular cone.

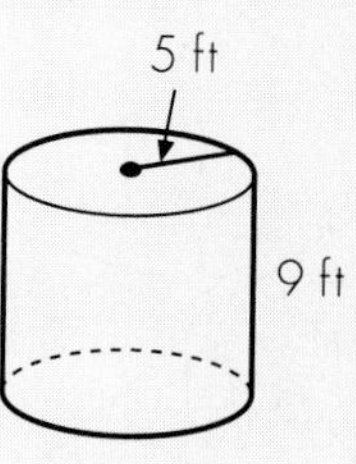

[$225\pi \approx 707$ ft^3]

TEACHING COMMENTS

Using the Page
As a group activity, students can make models of three pyramids that fit together to form a triangular prism as described. The figures below show possible patterns for such pyramids. Cutouts from the patterns can be scaled to whatever size is desirable. These pyramids should fit together to form a right prism with triangular bases 4, 5, and 6 units on the sides and a height of 4 units.

Students should fold and tape the outside triangles backward so that the letters and numbers are visible on the outside of each pyramid.

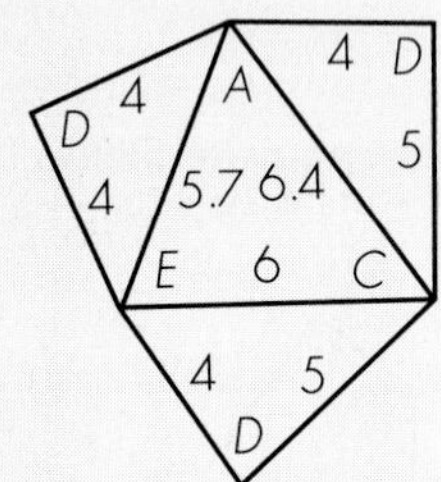

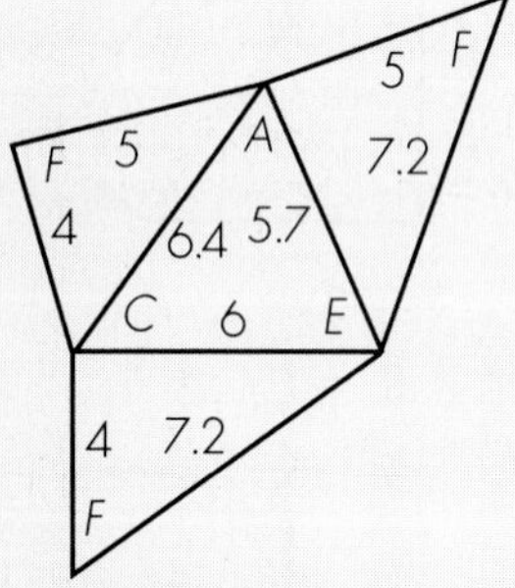

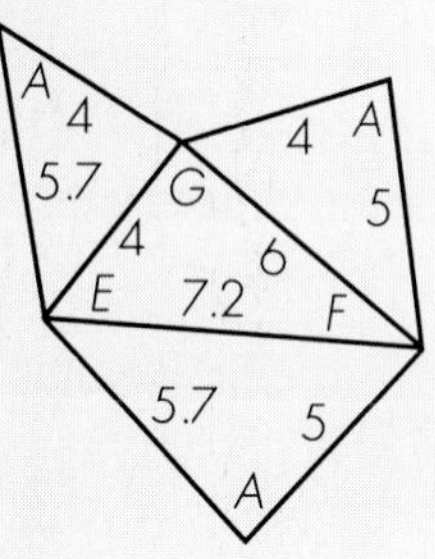

Face ACE of the red pyramid exactly fits onto face $A'C'E'$ of the white pyramid.

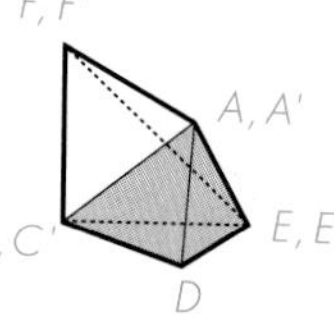

Face $A''E''F$ of the blue pyramid exactly fits onto face $A'E'F'$ of the white pyramid.

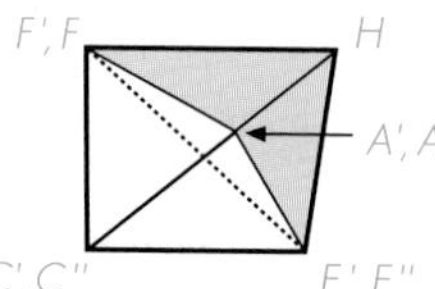

The 3 pyramids come together to form a polyhedron. Is the polyhedron a prism? The Discovery Activity below will deal with that question.

DISCOVERY ACTIVITY

Instead of assembling three pyramids into one solid figure (hoping that it is a triangular prism), do the reverse. Start with a triangular prism and partition it into pyramids as shown below.

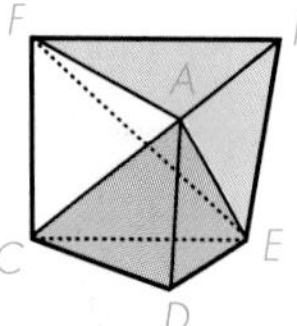

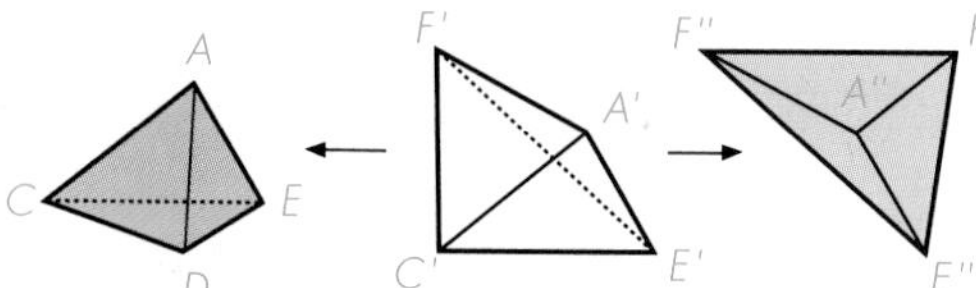

1. What kind of quadrilateral is $CEHF$, the back face of the prism? parallelogram
2. Are triangles CEF and EFH congruent? How do you know? Yes; ASA (or SSS)
3. Taking A as the vertex of pyramids $ACEF$ and $AEHF$, do you think that the two pyramids have the same height? the same volume? Yes; Yes
4. Are triangles CDE and AHF congruent? How do you know? Yes; SSS

5. Do you think that pyramid *EAHF*, with vertex *E*, has the same height as pyramid *ACDE*, with vertex *A*? Do the two pyramids have the same volume? Yes; Yes

6. Do all 3 pyramids have the same volume? Yes

Using the Page
If the students have made models of pyramids *ACEF*, *AEGF*, and *ACDE*, the models can be used as an aid in following the discussion in the Discovery Activity.

As a result of the Discovery Activity, you may have been able to conclude that a triangular prism can be partitioned into three pyramids of the same volume. Here is a summary of the proof that this is so.

The back face of the prism, quadrilateral *CEHF*, is a parallelogram, since the lateral faces of all prisms are parallelograms. Therefore, diagonal *EF* divides the parallelogram into two congruent triangles *CEF* and *EHF*. Two pyramids, *ACEF* and *AEHF*, then have congruent bases. They also have the same altitude (not shown) drawn from vertex *A* to the plane of their bases. It can be shown that if two pyramids have the same height and bases with the same area, then they have the same volume. Therefore, pyramids *ACEH* and *AEHF* have the same volume. A similar argument can be used to show that pyramids *EAHF* and *ACDE* have the same volume. Since all three pyramids have the same volume *V*, you can conclude that:

$3 \times$ volume of 1 pyramid = volume of triangular prism

or volume of 1 pyramid = $\frac{1}{3}$ volume of triangular prism

NOTEBOOK

THEOREM 11-4-1 **The volume of a triangular pyramid is equal to one-third of the product of the area of the base and the height. If B is the area of the base and h is the height, then the volume is given by the formula $V = \frac{1}{3} Bh$.**

Using the Page
Theorem 11-4-1 is sometimes presented to students without explanation because of the difficulty in visualizing the three pyramids of equal volume.

Any pyramid can be partitioned into triangular pyramids, as illustrated. Therefore, Theorem 11-4-1 can be extended to cover all pyramids.

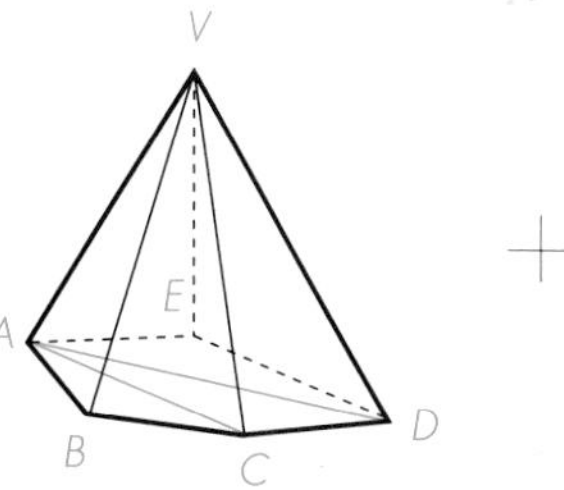

NOTEBOOK

THEOREM 11-4-2 **The volume of any pyramid is equal to one-third of the product of the area of the base and the height. The formula is $V = \frac{1}{3} Bh$.**

Examples: Find the volume of each pyramid.

1.

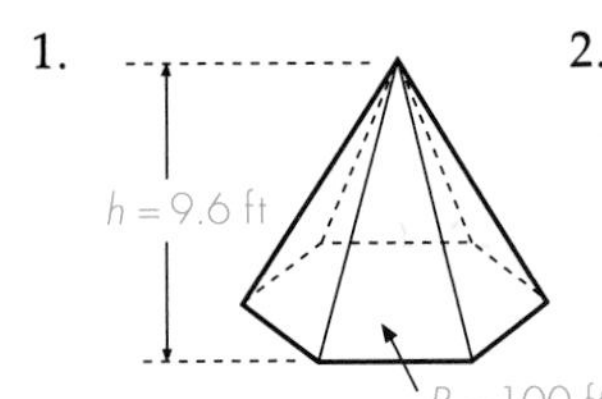

2.

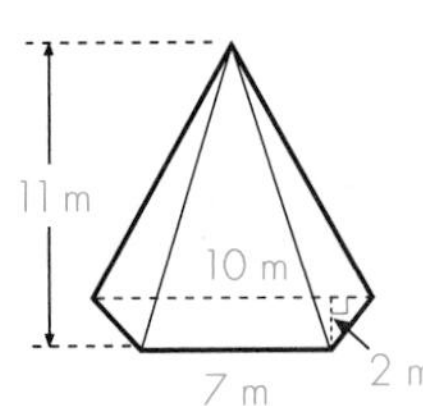

$V = \frac{1}{3}\,Bh$

$= \frac{1}{3} \times 100 \times 9.6$

$= 320 \text{ ft}^3$

The base is a trapezoid. Use $\frac{1}{2}(a + b)h$ for B.

$V = \frac{1}{3}\,Bh$

$= \frac{1}{3}\left[\frac{1}{2}(10 + 7) \times 2\right] \times 11$

$= \frac{1}{3}\,(17) \times 11$

$= \frac{187}{3}$

$\approx 62.3 \text{ m}^3$

Extra Practice

Find the volume of each pyramid.

1.

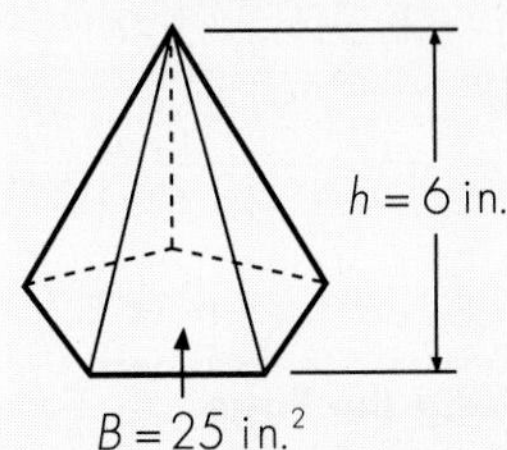

[50 in.³]

2.

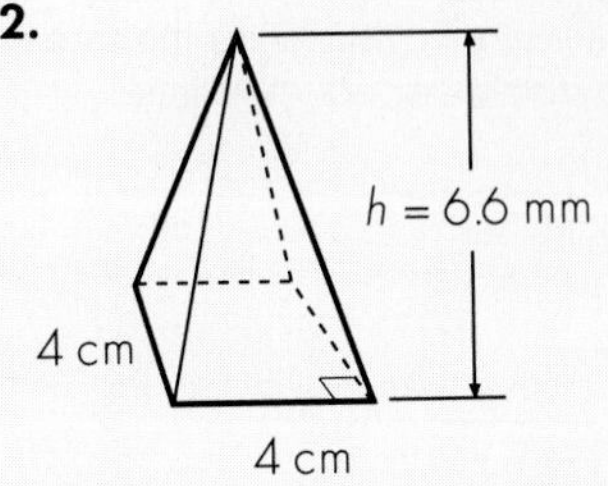

[35.2 cm³]

3.

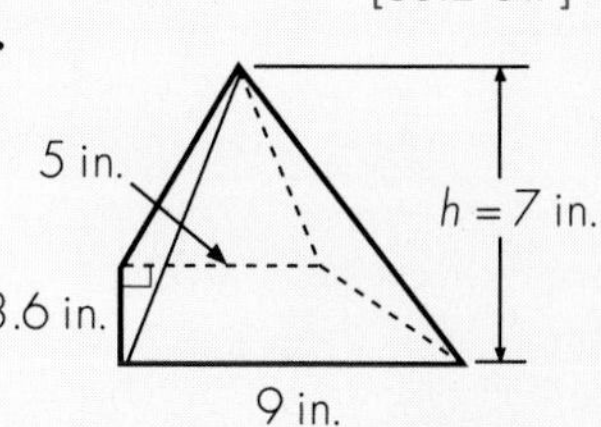

[58.8 in.³]

CLASS ACTIVITY

Find the volume of each pyramid.

1.

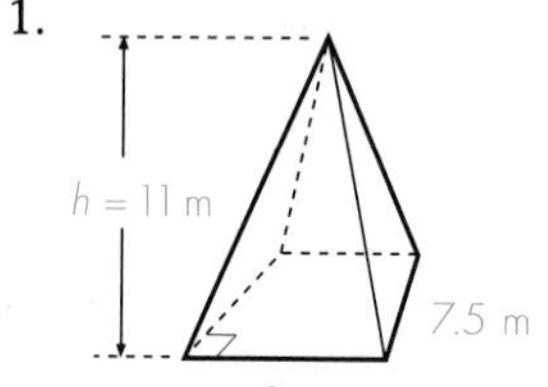

220 cm³

2.

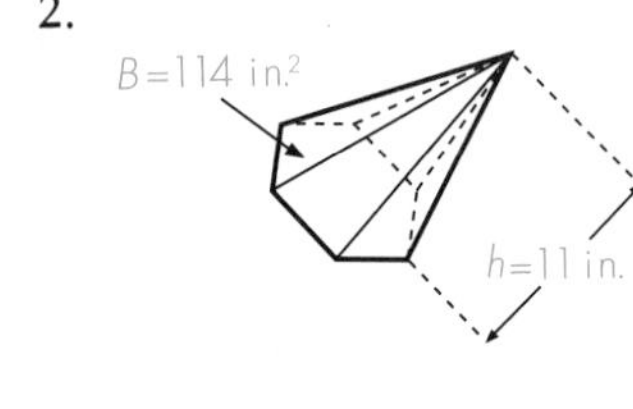

418 in.³

THE VOLUME OF A CONE

If you select equally spaced points around the circular base of a right cone, you can put a regular polygon inside the base of the cone. By connecting the vertices of this polygon to the vertex of the cone you can fit a regular pyramid inside the cone. The more sides the regular polygon has, the closer its area is to that of the circular base of the cone. Also, the closer the volume of the regular pyramid is to the volume of the cone. The height h of the pyramids and cone stay the same.

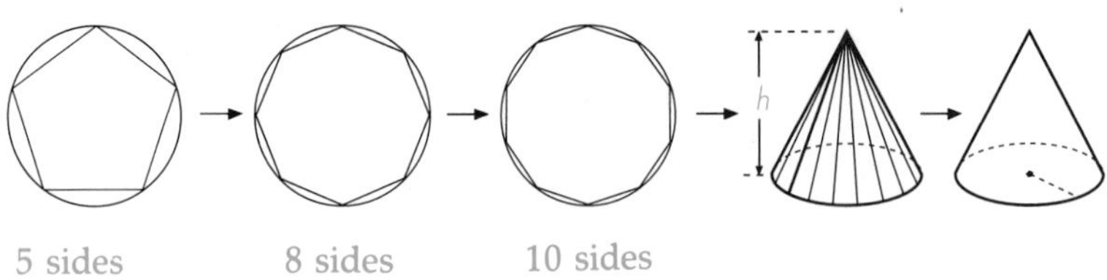

Since the volume of each pyramid is $V = \frac{1}{3}$(area of polygon)h, these volumes are approaching closer and closer to $\frac{1}{3}Bh$, where B is the area of the base of the cone. This thinking leads to the following theorem.

THEOREM 11-4-3: The volume of a right circular is equal to one-third the product of the area of the base and the height of the cone: $V = \frac{1}{3}Bh = \frac{1}{3}\pi r^2h$.

Example: Find the volume of the right circular cone below.

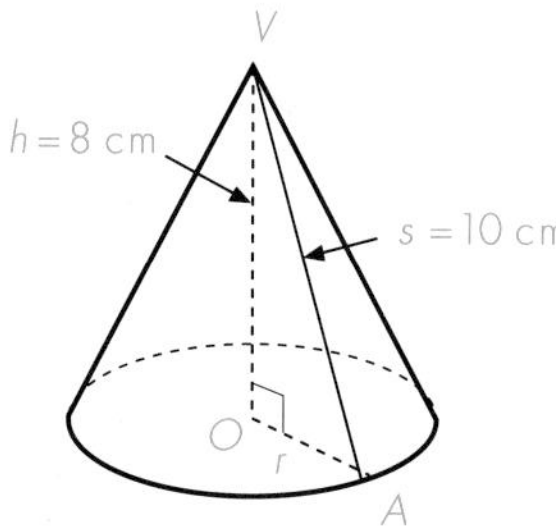

You need to find the radius of the base so you can use the formula for the volume of a cone. Use the Pythagorean Theorem.

$$\begin{aligned} r^2 + 8^2 &= 10^2 \\ r^2 + 64 &= 100 \\ r^2 &= 36 \\ r &= 6 \end{aligned}$$

Next, use Theorem 11-4-3.

$$\begin{aligned} V &= \tfrac{1}{3}\pi r^2 h \\ &= \tfrac{1}{3}\pi \times 6^2 \times 8 \\ &= 96\pi \approx 301 \text{ cm}^3 \end{aligned}$$

Using the Page
Point out that the volume of a cone is $\frac{1}{3}$ the volume of a cylinder with the same base area and height.

Extra Practice
Find the volume of the right circular cone.

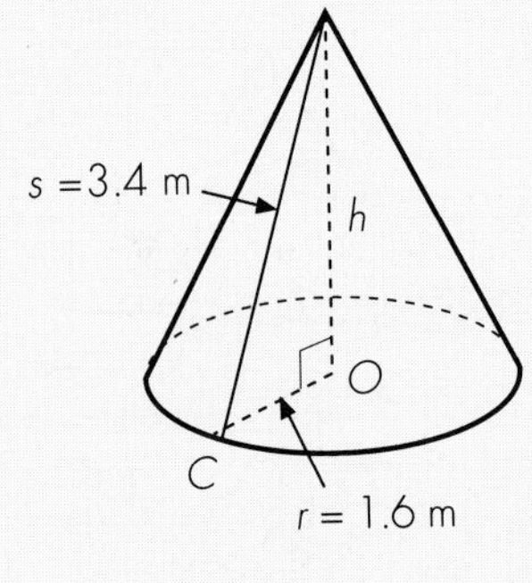

[$2.56\pi \approx 8$ m^3]

CLASS ACTIVITY

Use a calculator to find the volume of each cone.

1. $1536\pi \approx 4823$ in.^{3}

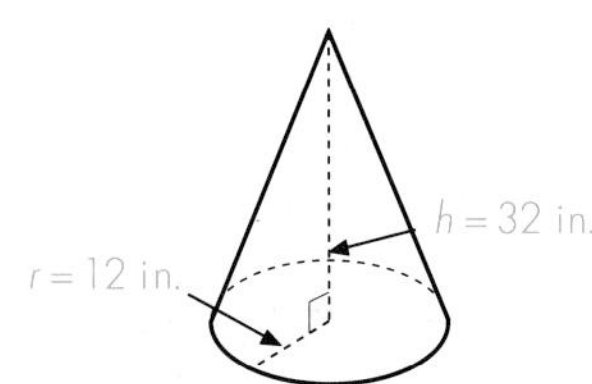

2. $12.5\pi \approx 39.3$ cm^3

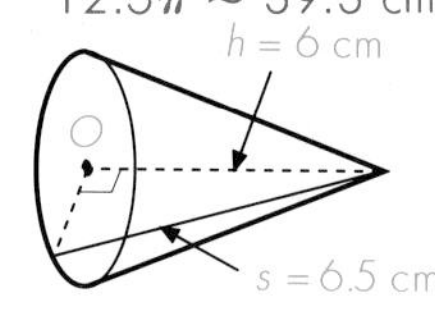

HOME ACTIVITY

Find the volume of each pyramid.

1. 391 ft^3

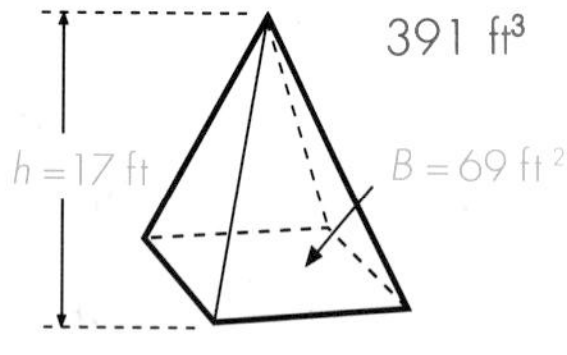

2. 11.5 m^3

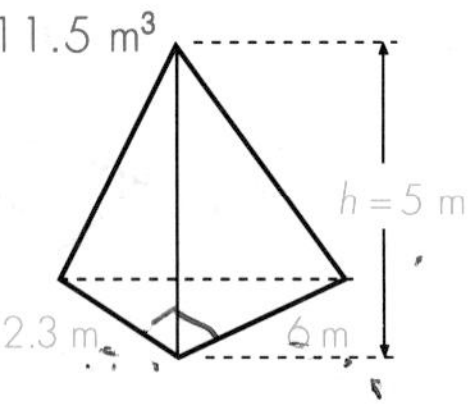

3. 37.3 in.^{3}

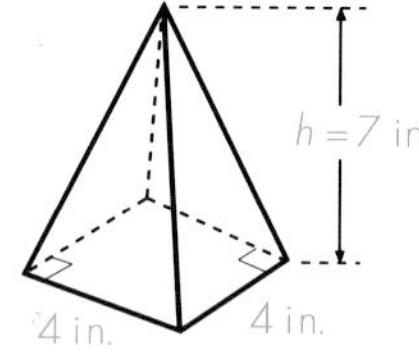

Find the volume of each cone.

4. 2000 m^3

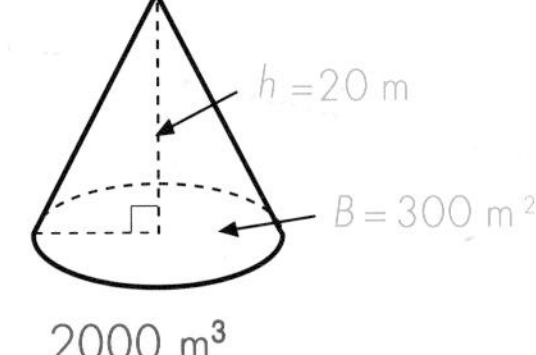

5. 48.3 km

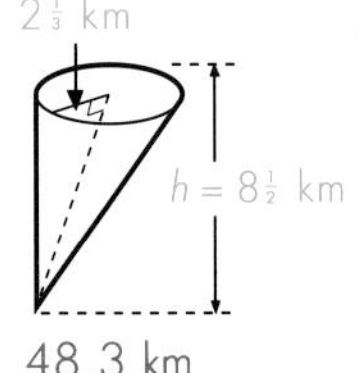

6. 1005 m^3

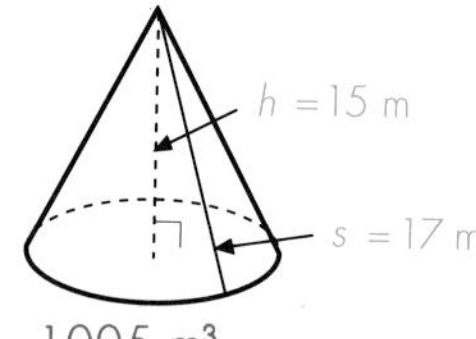

7. The Great Pyramid of Egypt has a square base that is 230 m long. Its height is 147 m. Find the volume. 2,592,100 m^3

8. A conical container is used to collect rain water. Its height is 4 ft and the diameter of the base measures 7 ft. How many gallons of water will it collect? (1 ft^3 = 7.5 gal) $\frac{49}{3}\pi \approx 51.3$ ft^3; 385 gal

9. Alex and Pierre pitched a tent that had the shape of a pyramid. The rectangular base measured 2.3 m by 2.6 m. The volume was 4.2 m^3. Find the height. about 2.1 m

CRITICAL THINKING

10. A pile of grain has the shape of a right circular cone with a height of 3 ft. The base has a circumference of 12.5 ft. Find the volume of the pile. $\frac{39.1}{\pi} \approx 12.4$ ft^3

FOLLOW-UP

Assessment
Find the volume of each pyramid or cone.

1.

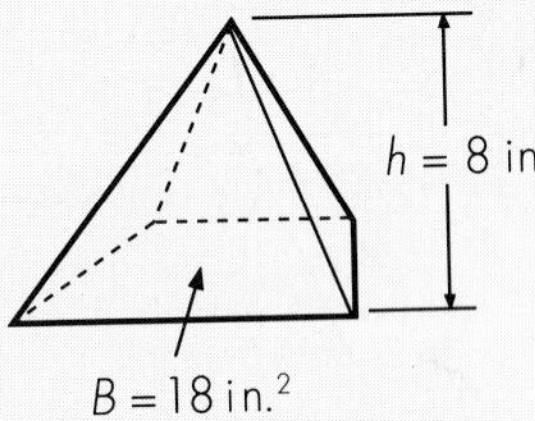

[48 in.^{3}]

2.

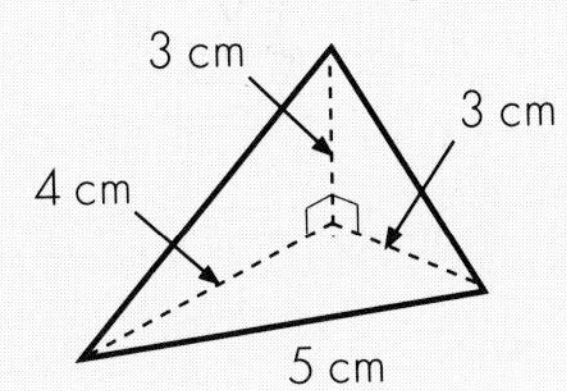

[6 cm^3]

3.

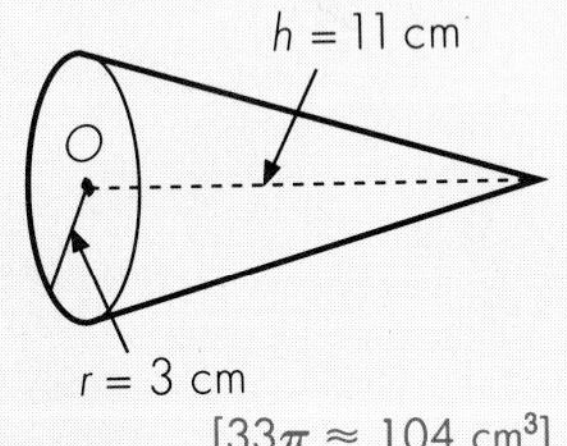

[$33\pi \approx 104$ cm^3]

Reteaching
Students who have had difficulty with this section may benefit from Reteaching Activity 11.4.

Enrichment
For students who have mastered the material in this section, you may wish to assign Enrichment Activity 11.4.

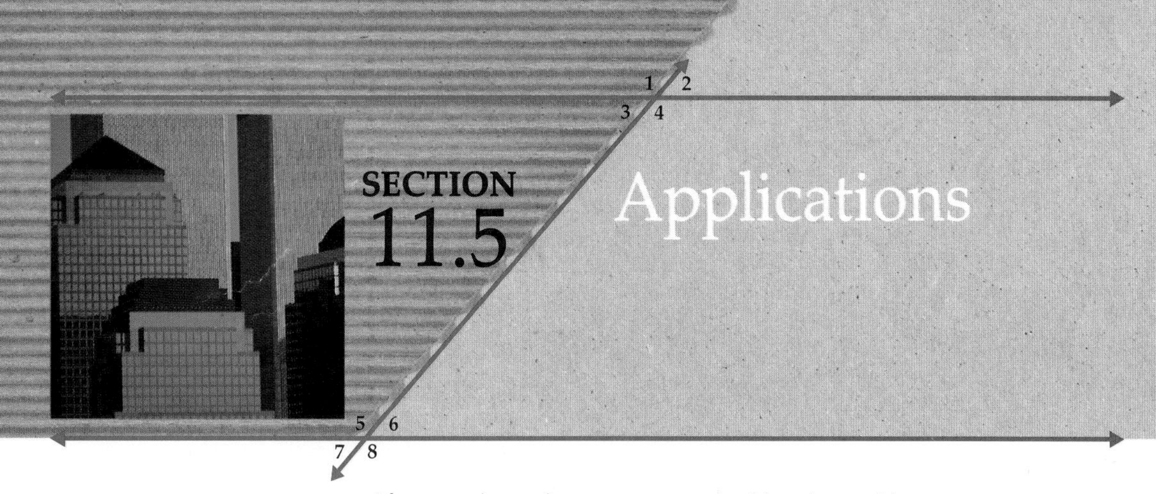

SECTION 11.5 Applications

After completing this section, you should understand how
- ▲ prisms and cylinders are used in everyday life.
- ▲ pyramids and cones are used in everyday life.

You have studied prisms, pyramids, cylinders, and cones in the earlier sections of this chapter. In this section, you will increase your familiarity with these figures and their applications.

Examples of prisms can be found everywhere around you, in the form of buildings, commercial products, and decorative features in your surroundings. One such example is the *swimming pool.*

Example: Most swimming pools have both a shallow end and a deep end. The bottoms of the two sections are connected by a gradual slope that allows the bather to choose the water depth that he or she prefers. One such pool is shown below.

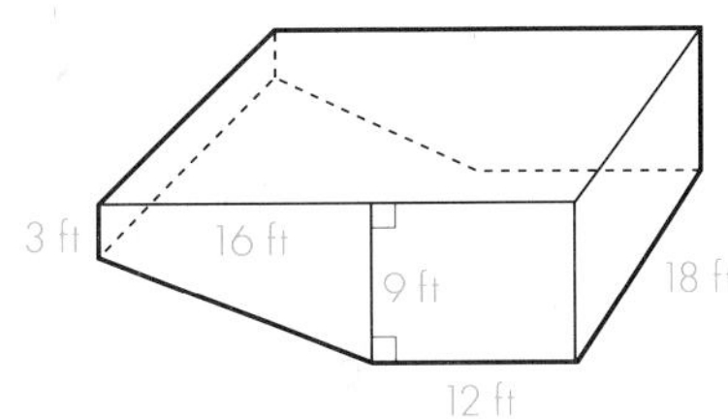

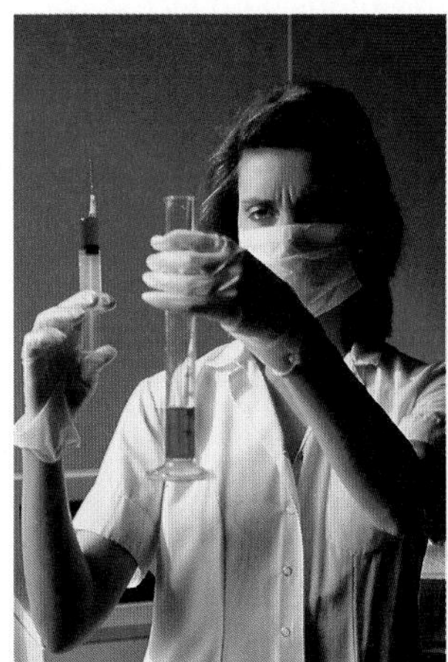

1. What kind of polygon is the shape of the two bases?
2. Find the volume of water that the pool can hold.
3. When the pool is being filled, water flows in at the rate of 67 gal/min. How long does it take to fill up the pool? There are 7.5 gallons in 1 ft^3. This can be written as 7.5 gal/ft^3.

To answer question 1, think of the front side as a base of a prism.

SECTION 11.5

Resources
Reteaching 11.5
Enrichment 11.5

Objectives
After completing this section, the student should understand
- ▲ how prisms and cylinders are used in everyday life.
- ▲ how pyramids and cones are used in everyday life.

Vocabulary
conversion factor

GETTING STARTED

Chalkboard Example
A pipe has a diameter of 14 in. If water flows through the pipe at a rate of 12 in. per second, how many gallons flow through the pipe in 1 minute? [1 gal = 231 $in.^3$]

The volume of water that flows through the pipe in 1 second is the volume of a cylinder with a height of 12 in. and a diameter of 14 in. First find this volume in cubic inches.

$V = \pi \times 7^2 \times 12$
$= 588\pi \approx 1847$ $in.^3$

Now find the number of gallons in 1847 $in.^3$.

$$V = \frac{1847}{231} \approx 8 \text{ gal}$$

Finally, find the number of gallons that flow through the pipe in 1 minute. 8 gallons in 1 second is equivalent to 8 × 60 gallons in one minute.

$8 \times 60 = 480$

About 480 gallons of water flow through the pipe in 1 minute.

TEACHING COMMENTS

Using the Page
It is often convenient to treat units of measurement (labels) as if they were algebraic quantities. We may do this without thinking about it. For example, it is understood that

$96 \text{ ft}^2 + 108 \text{ ft}^2 = 204 \text{ ft}^2$
and
$204 \text{ ft}^2 \times 18 \text{ ft} = 3672 \text{ ft}^3$.

This method of thinking is particularly useful in converting from one unit of measurement to another. In the Chalkboard Example, to convert in.³ to gal, it would be possible to multiply by $\frac{1 \text{ gal}}{231 \text{ in.}^3}$, since 1 gal = 231 in^3.

$$\frac{1847 \text{ in.}^3}{1} \times \frac{1 \text{ gal}}{231 \text{ in.}^3} = 8 \text{ gal}$$

The example on this page uses the rate 67 gal/min as a factor in order to find a number of minutes when a number of gallons is known. The form

$$\frac{1 \text{ min}}{67 \text{ gal}}$$

is chosen, since the label "gal" is to be eliminated.

The pool is a prism with two quadrilaterals as its bases. Next consider how to find the volume of this prism. The base can be separated into a trapezoid and a rectangle, so let B_1 = the area of the trapezoid and B_2 = the area of the rectangle. So

$$\begin{aligned} V &= Bh \\ &= (B_1 + B_2)h \\ &= \left[\tfrac{1}{2}(3 + 9) \times 16 + (9 \times 12)\right] \times 18 \\ &= (96 + 108) \times 18 \\ &= 204 \times 18 \\ &= 3672 \text{ ft}^3 \end{aligned}$$

To find how long it will take to fill the pool, rewrite 3672 ft³ as $\frac{3672 \text{ ft}^3}{1}$ and 7.5 gal/ft³ as $\frac{7.5 \text{ gal}}{1 \text{ ft}^3}$. Then multiply.

$$\frac{3672 \text{ ft}^3}{1} \times \frac{7.5 \text{ gal}}{1 \text{ ft}^3} \times = \frac{27{,}540 \text{ gal}}{1 \times 1}, \text{ or } 27{,}540 \text{ gal}$$

The rate of flow of water can be written in the following ways.

$$67 \text{ gal/min} \qquad \frac{67 \text{ gal}}{1 \text{ min}} \qquad \frac{1 \text{ min}}{67 \text{ gal}}$$

Use the third of these ways together with the number of gallons to find the number of minutes required to fill the pool.

$$\frac{27{,}540 \text{ gal}}{1} \times \frac{1 \text{ min}}{67 \text{ gal}} = \frac{411 \text{ min}}{1}, \text{ or } 411 \text{ min}$$

It takes about 411 min to fill the pool. The number of hours to fill the pool is 411 ÷ 60, or about 6.9 h.

In the example, the expression $\frac{7.5 \text{ gal}}{1 \text{ ft}^3}$ is called a *conversion factor,* because it can be used to convert cubic feet to gallons (and vice versa). Notice that conversion factors and similar expressions (such as water-flow rates) can be written and multiplied as if they were fractions. This technique helps you organize your calculations so that your final answer will be automatically expressed in the correct kind of units.

CLASS ACTIVITY

A swimming pool has the shape of a rectangular prism 25 ft long, 12 ft wide, and 7 ft deep. It costs 72 cents to run 1000 gal of water into the pool.

1. Find the volume of the pool. 2100 ft³
2. Find the cost of filling the pool. $11.34

SHAPES OTHER THAN PRISMS

Sometimes large storage structures are cylindrical in shape. Examples are farm silos and gasoline storage tanks.

Example: The county plans to paint the outside of a 65 ft-high cylindrical water tank that has a diameter of 6 ft.

One can of paint covers 425 ft^2 and costs \$15.56. The county officials estimate that the labor cost is 5 times the cost of the paint. How much is the total estimated cost of painting the tower?

Since the diameter is 6 ft, the radius of the tank is 3 ft.

$$\begin{aligned}\text{Lateral area of tower} &= 2\pi rh \\ &= 2 \times \pi \times 3 \times 65 \\ &= 390\pi \\ &\approx 390 \times 3.14 \\ &\approx 1225 \text{ ft}^3\end{aligned}$$

$$\begin{aligned}\text{Number of cans of paint} &= \frac{1225 \text{ ft}^3}{1} \times \frac{1 \text{ can of paint}}{425 \text{ ft}^3} \\ &= \frac{1225 \times 1 \text{ cans of paint}}{1 \times 425} \\ &= 2.88 \text{ cans of paint}\end{aligned}$$

The actual number of cans of paint needed is the next higher whole number. There will be 3 cans of paint needed.

$$\begin{aligned}\text{Cost of paint} &= 3 \times 15.56, \text{ or } \$46.68 \\ \text{Cost of labor} &= 5 \times \text{cost of paint} \\ &= 5 \times 46.68, \text{ or } \$233.40 \\ \text{Total cost} &= \text{cost of paint} + \text{cost of labor} \\ &= 46.68 + 233.40 \\ &= 280.08, \text{ or about } \$280\end{aligned}$$

Example: The Great Pyramid of Egypt is about 480 ft high and has a square base with a side that is about 755 ft long. Suppose that a modern building (with the shape of a square prism) has the same height as the Great Pyramid and the same dimensions. Assume that each floor of a building is 10 ft high and that a city block is a square that measures 300 ft on each side.

a. How many floors would the building have?
b. How many city blocks would the building occupy?

a. Number of floors $= 480 \text{ ft} \times \frac{1 \text{ floor}}{10 \text{ ft}}$

$$= \frac{480 \times 1 \text{ floor}}{10}, \text{ or } 48 \text{ floors}$$

Extra Practice

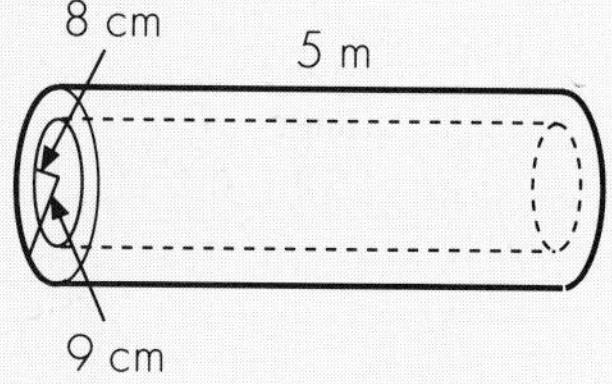

A metal pipe is 5 meters long. It has an inner radius of 8 cm and an outer radius of 9 cm. To the nearest pound, how heavy is the pipe if 4 cm^3 of the metal weighs 1 ounce? (Recall that 1 meter = 100 cm and 1 pound = 16 ounces.) [417 lb]

b. Number of blocks occupied by Great Pyramid =

$$\frac{1 \text{ city block}}{300 \times 300 \text{ ft}^2} \times \frac{755 \times 755 \text{ ft}^2}{\text{Base of Pyramid}} =$$

$$\frac{755 \times 755 \times 1}{300 \times 300} \times \frac{\text{city blocks}}{\text{Base of Pyramid}} = 6.33 \text{ blocks/Pyramid base}$$

or slightly more than 6 city blocks in area.

Example: A lumber company plans to plant new trees on the side of a mountain that has the shape of a right circular cone. The mountain is 3000 ft high and has a base with a circumference of 18 mi. (1 mi = 5280 ft) Each tree is to have 100 ft² of space. To the nearest hundred thousand, how many trees can be planted?
First, find the radius of the base.

$$C = 2\pi r$$
$$18 = 2\pi r$$
$$\frac{18}{2\pi} = r$$
$$2.86 \approx r$$
$$r \approx 2.86 \text{ mi}$$
$$\approx 2.86 \text{ mi} \times \frac{5280 \text{ ft}}{1 \text{ mi}} \approx 15{,}100 \text{ ft}$$

Next, find the area of the side of the mountain. The slant height is found by using the Pythagorean Theorem.

$$r^2 + h^2 = s^2$$
$$15{,}100^2 + 3000^2 = s^2$$
$$237{,}010{,}000 = s^2$$

s
h
r

Take the square root of each side of the equation.
$15{,}395 \approx s$, or $s \approx 15{,}395$ ft

$$\text{Lateral area of right circular cone} = \pi r s$$
$$\approx \pi \times 15{,}100 \times 15{,}395$$
$$\approx 730{,}308{,}000 \text{ ft}^2$$

Number of trees $= 730{,}308{,}000 \text{ ft}^2 \times \frac{1 \text{ tree}}{100 \text{ ft}^2} \approx 7{,}303{,}080$, or about 7,300,000 trees.

Extra Practice
An aquarium 30 inches long and 25 inches wide is filled 20 inches high with water.

1. Suppose that each fish needs about 1000 cubic inches of water. At most, how many fish should the aquarium hold? [15 fish]
2. The water in the aquarium is changed twice a week. About how many gallons per week are used for the aquarium? (1 gal = 231 in.³) [130 gal]

CLASS ACTIVITY

Use a calculator for Exercises 1 and 2.

1. Find the volume of a large, cylindrical can of flour that is 30 cm high and that has a base with a diameter of 24 cm. about 13,572 cm³
2. A conical scoop is used to empty the can of flour of Exercise 1. The height of the cone is 12 cm and the diameter of the base is 12 cm. How many scoops are required? about 30 scoops

HOME ACTIVITY

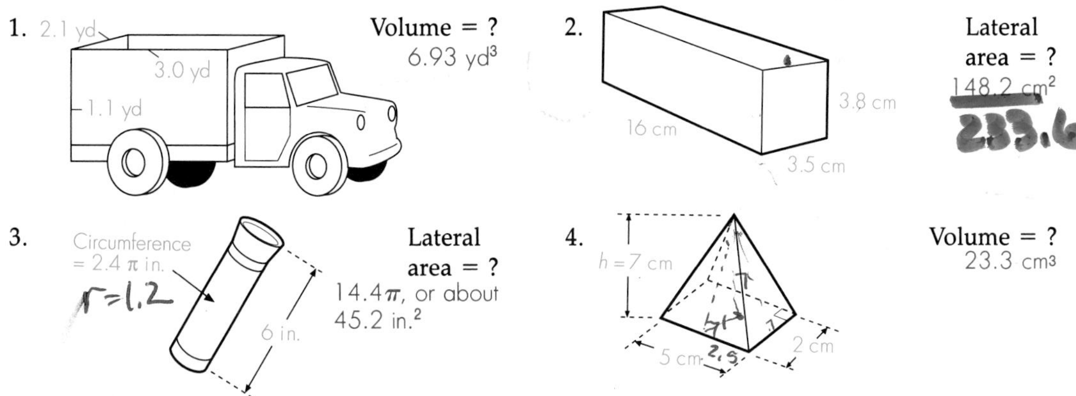

3. Lateral area = ? 14.4π, or about 45.2 in.^{2} (Circumference = 2.4π in.; 6 in.)

4. Volume = ? 23.3 cm3 ($h = 7$ cm; 5 cm; 2 cm)

5. A pencil factory is ordering paint for a batch of pencils that it is shipping. To prepare the order, the factory needs to know the lateral area of each of its pencils. Assume that a pencil is a regular hexagonal prism with each side of the base 0.125 in. long. Each lateral edge is 7 in. long. Find the lateral area of a pencil. 5.25 in.^{2}

6. The inside radius of the graduated cylinder at the right is 1.5 cm. How much will the level of liquid rise if 20 cm^3 of additional liquid are poured in? about 2.8 cm

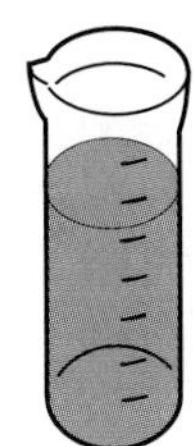

7. A large Christmas tree is to be erected in a local community. The height is to be 30 ft and the diameter of the base is to be 15 ft. There is to be an average of one ornament for each square foot of the tree's surface. Assuming that the tree is a right circular cone, find the approximate number of ornaments required. about 232π, or 728 ornaments

8. In banks, coins are separated by type and then wrapped into cylindrical stacks. Assume that a quarter is 0.0675 in. thick and has a diameter of 0.95 in. What is the area of a paper coin wrapper for a roll of 10 dollars worth of quarters? Ignore the extra paper that will be needed to overlap the first layer and the ends of the roll. 2.565π, or about 8.1 in.^{2}

FOLLOW-UP

Assessment

1. A square sandbox 6 ft on a side is to be filled 8 in. high with sand. The sand costs $3 for a 50-lb bag. If 1 cubic foot of sand weighs 75 lb, how much will it cost to fill the box? [$108]

2. A cylindrical water tank with a 3-ft diameter and a height of 6 ft is one-fifth full.

a. How many cubic feet of water must be added to fill the tank? [$10.8\pi \approx 34$ ft^3]

b. How long will it take to fill the tank if water flows in at the rate of 51 gal/min? (7.5 gal = 1 ft^3) [5 min]

9. The botanical laboratory shown has the shape of one-half of a right circular cylinder. Except for several semicircular frames, the entire structure is translucent in order to provide the plants with the largest possible amount of exposure to light. Find the total area of the structure that allows the passage of light. Ignore the thickness of the frames. 718.8π, or about 2257 ft^2

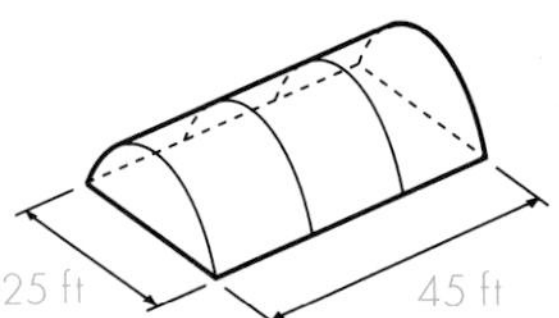

10. The triangular prism shows the original design of the roof of a house. To save materials, the builder changed the design. How many square feet of area are saved by constructing the second roof? (Include the ends.) 1896.8–1744.5, or about 152 ft^2

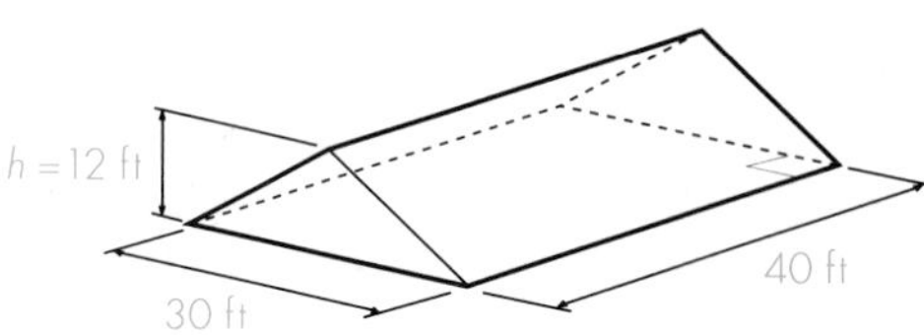

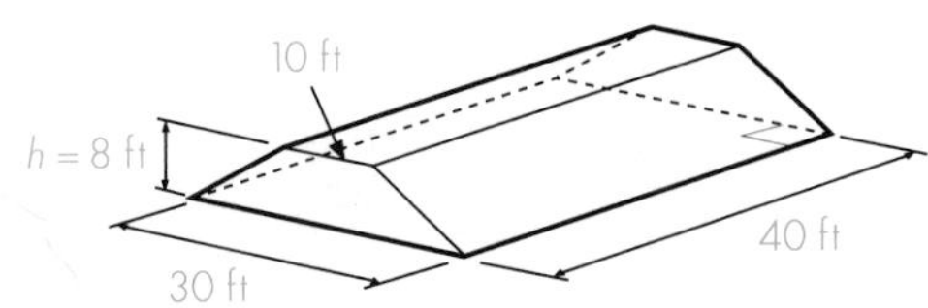

CRITICAL THINKING

11. Two squares have been removed from opposite ends of a chess board, as shown. Roberto is attempting to cover the remaining squares with a set of dominoes. Each domino can cover two adjacent squares. Can Roberto cover all the squares? Justify your answer.

Each domino covers a black square and a white square. Therefore, the total number of covered black squares has to equal the total number of covered white squares. But this cannot happen since the two removed squares are of the same color. The number of black squares in the modified board is no longer equal to the total number of white squares. As a result, the squares cannot be covered by dominoes.

Reteaching
Students who have had difficulty with this section may benefit from Reteaching Activity 11.5.

Enrichment
For students who have mastered the material in this section, you may wish to assign Enrichment Activity 11.5.

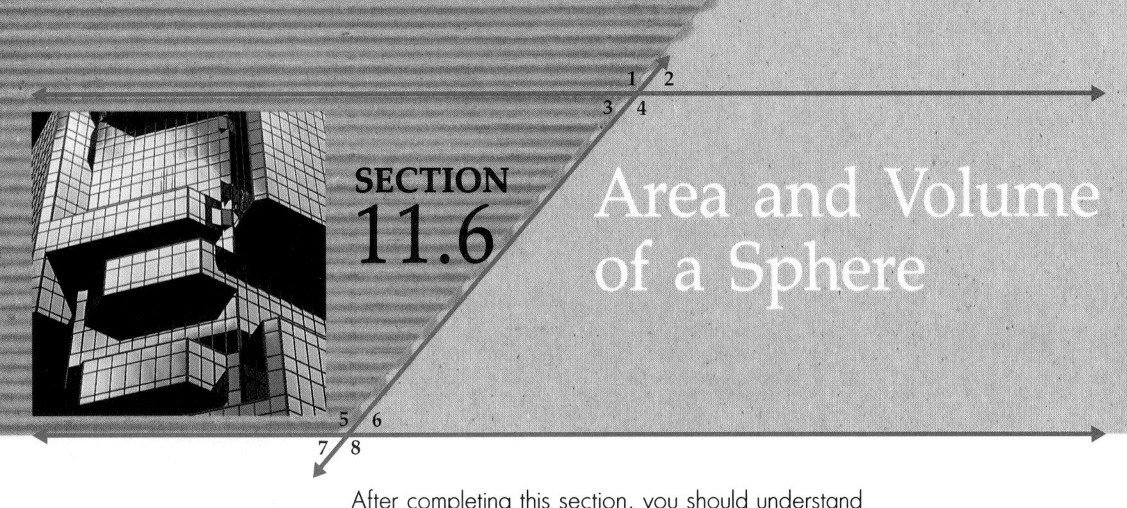

SECTION 11.6 Area and Volume of a Sphere

After completing this section, you should understand
- ▲ how to find the volume of a sphere.
- ▲ how to find the area of a sphere.

In a *plane,* a circle is the set of points equidistant from a point called the center. In *space,* the set of such points is a **sphere.**

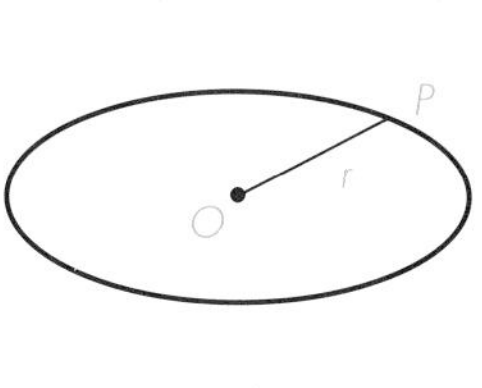

Circle

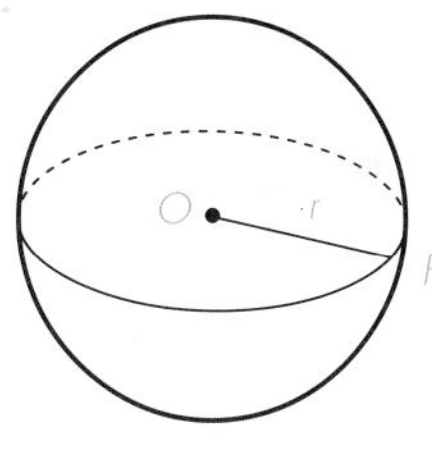

Sphere

DEFINITION 11-6-1 **A sphere** is the set of all points equidistant from a point called the **center.** The distance from any point of the sphere to the center is the **radius** of the sphere.

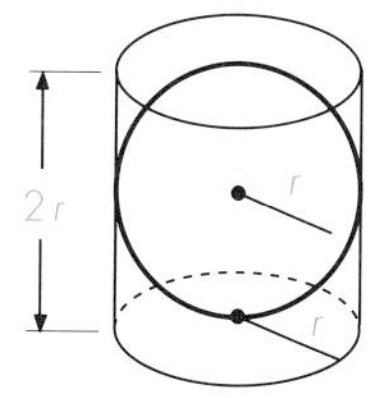

At the right, a sphere is inscribed in a cylinder. The radius of the base of the cylinder is r, the same as the radius of the sphere. The height of the cylinder is the same as the diameter of the sphere, $2r$. Over 2200 years ago, Archimedes, the greatest mathematician of his time, discovered a relationship between the above two figures. He regarded the relationship as so important that he requested that the figures be etched on his tomb. This is the relationship he discovered: $\frac{\text{Volume of inscribed sphere}}{\text{Volume of cylinder}} = \frac{2}{3}$

SECTION 11.6

Resources
Reteaching 11.6
Enrichment 11.6

Objectives
After completing this section, the student should understand
- ▲ how to find the volume of a sphere.
- ▲ how to find the area of a sphere.

Materials
library resources
calculator

Vocabulary
Archimedes
center of sphere
radius of sphere
sphere

GETTING STARTED

Quiz: Prerequisite Skills
Find the lateral area and the volume of each figure.

1.

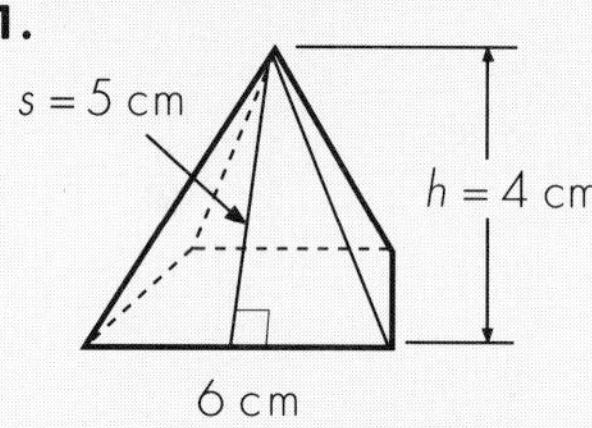

regular square pyramid
[LA: 60 cm^2; Vol: 48 cm^3]

2.

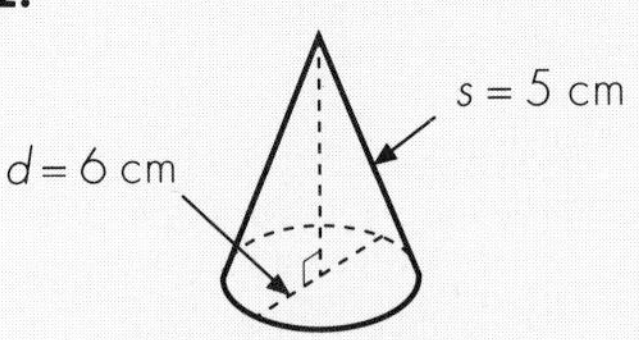

right circular cone
[LA: $15\pi \approx 47$ cm^2
Vol.: $12\pi \approx 38$ cm^3]

REPORT 11-6-1 Prepare a report on Archimedes explaining and illustrating the following:

1. the time in history when he lived
2. his mathematical accomplishments

Sources:

Bell, Eric T. *Men of Mathematics.* New York: Simon and Schuster, 1937.
Eves, Howard. *An Introduction to the History of Mathematics.* New York: Holt, Rinehart, and Winston, 1976.
Kline, Morris. *Mathematical Thought from Ancient to Modern Times.* New York: Oxford University Press, 1972.

You can use the relationship that Archimedes found to derive a formula for the volume of a sphere.

$$\frac{\text{Volume of inscribed sphere}}{\text{Volume of cylinder}} = \frac{2}{3}$$

$$\begin{aligned}\text{Volume of sphere} &= \tfrac{2}{3}\ \text{Volume of cylinder}\\ &= \tfrac{2}{3} \times \pi r^2 h\\ &= \tfrac{2}{3} \times \pi r^2 \times 2r\\ &= \tfrac{2}{3} \times 2 \times \pi r^2 \times r, \text{ or } \tfrac{4}{3}\ \pi r^3\end{aligned}$$

NOTEBOOK

THEOREM 11-6-1 **The formula for the volume of a sphere is Volume of sphere = $\frac{4}{3}\ \pi r^3$, where r is the radius of the sphere.**

Example: Find the volume of the sphere of radius 5.

$$\begin{aligned}V &= \tfrac{4}{3}\ \pi r^3\\ &= \tfrac{4}{3} \times \pi \times 5^3\\ &= \tfrac{4}{3} \times \pi \times 125\\ &= \tfrac{500}{3}\ \pi\\ &\approx 167 \times 3.14, \text{ or about } 523\ \text{mm}^3\end{aligned}$$

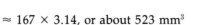

TEACHING COMMENTS

Using the Page
After a discussion of the example, have students find the volumes in the exercises below. Then have the students compare the volumes of two spheres when one radius is 2 times the other. They should notice that when one radius is multiplied by 2, the volume is multiplied by 8.

Extra Practice
Find the volume of each sphere.

1. $r = 3$ cm
[Volume: $36\pi \approx 113\ \text{cm}^3$]
2. $r = 6$ cm
[Volume: $288\pi \approx 905\ \text{cm}^3$]

AREA OF A SPHERE

The following Discovery Activity will help you find the formula for the area of a sphere.

In the figure at the right, points A, B, C, and D are four points of a sphere. Similarly, X, Y, and Z are three points of the sphere. The points in each set are connected by line segments as shown to create quadrilateral $ABCD$ and $\triangle XYZ$. Then seven other line segments are formed by connecting each vertex of the two polygons to point O, the center of the sphere.

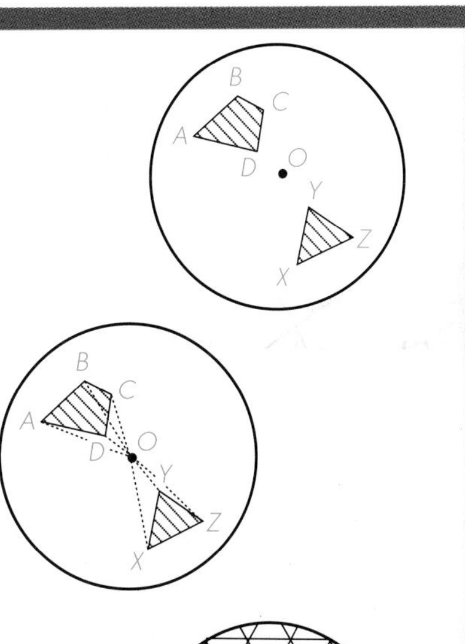

1. What kind of solid figure is polyhedron $OXYZ$? polyhedron $OABCD$? How do you know? pyramid; pyramid; def. of pyramid

2. Use your answer to question 1 to tell the formula for the volume of each of the two figures. $\frac{1}{3}$ Bh

Suppose that instead of using just two polygons, you had 1000 very small polygons and thus 1000 pyramids packing the inside of the sphere.

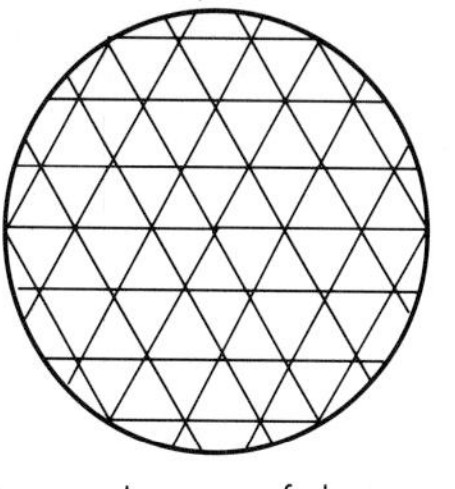

3. Would the surface of the entire sphere have more area than the total area of the bases of the pyramids? less area? the same area? More area

4. Would the sphere have more volume than the total volume of the pyramids? less volume? the same volume? More volume

5. What would be the answers to Questions 3 and 4 if the number of small pyramids were increased to 10,000? to 1,000,000? Same answers

In the Discovery Activity, you may have discovered something that happens as the number of pyramids packing the sphere increases and their size decreases. The total area of their bases approximates the area of the sphere more closely. Also, their total volume approximates the volume of the sphere more closely.

Using the Page
The Discovery Activity helps students to visualize the relationship between the total volume of the packed pyramids, as their number grows larger, and the volume of the sphere. They also see that their total base areas approach the surface area of the sphere. This leads to a development of the formula for the surface area of a sphere, which is given in Theorem 11-6-2 on page 434.

Using the Page
After a discussion of the example, have students find the surface areas in the exercises below. Then have the students compare the surface areas of two spheres when one radius is 2 times the other. They should notice that when the radius is multiplied by 2, the surface area is multiplied by 4.

Extra Practice
Find the surface area of each sphere.

1. $r = 3$ cm

$[36\pi \approx 113 \text{ cm}^2]$

2. $r = 6$ cm

$[144\pi \approx 452 \text{ cm}^2]$

You can use these ideas to find a formula for the volume of a sphere.

The volume of a small pyramid with height h and base of B_1 is:

$\frac{1}{3} B_1 h$

Suppose that the total number of pyramids is n.
If you add up the volumes of all the pyramids, then you get the following result.

$\frac{1}{3} B_1 h + \frac{1}{3} B_2 h + \frac{1}{3} B_3 h + \ldots + \frac{1}{3} B_n h$

The three dots (. . .) mean "and so on."
If the number of pyramids is very large, then you can make three reasonable assumptions:

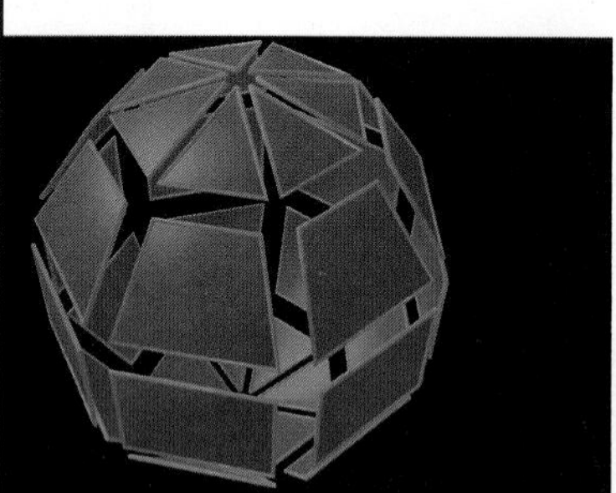

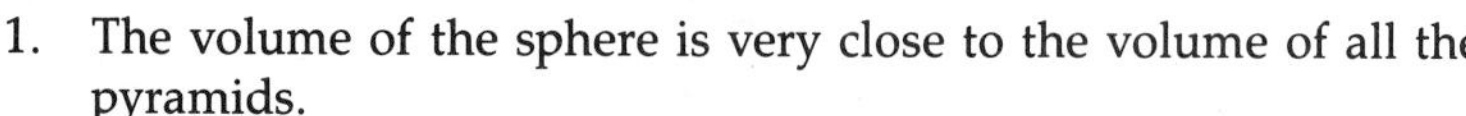

1. The volume of the sphere is very close to the volume of all the pyramids.
2. The height of each pyramid is very close to the radius of the sphere.
3. The area of the sphere is very close to the sum of all the base areas.

Volume of sphere $\approx \frac{1}{3} B_1 h + \frac{1}{3} B_2 h + \frac{1}{3} B_3 h + \ldots + \frac{1}{3} B_n h$ (Assumpt. 1)

$\frac{4}{3} \pi r^3 \approx \frac{1}{3} h(B_1 + B_2 + B_3 + \ldots + B_n)$ (Algebra Prop.)

$\frac{4}{3} \pi r^3 \approx \frac{1}{3} r(B_1 + B_2 + B_3 + \ldots + B_n)$ (Assumpt. 2)

$\frac{4}{3} \pi r^3 \approx \frac{1}{3} r \times$ (total of base areas) (Assumpt. 3)

Next, multiply each side by 3 and divide each side by r.

$4 \pi r^2 \approx$ total of base areas

This result suggests the following theorem.

THEOREM 11-6-2 **The formula for the surface area A of a sphere of radius r is $A = 4 \pi r^2$.**

Example: Find the area of the sphere at the right.

$A = 4\pi r^2$

$= 4 \times \pi \times 5^2$

$= 100 \pi$

$\approx 100 \times 3.14$, or about 314 mm^2

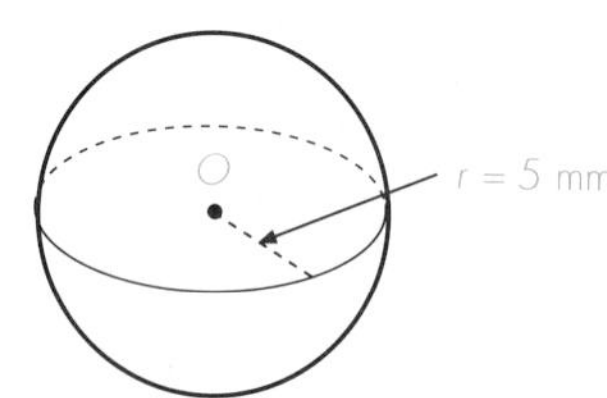

CLASS ACTIVITY

Use a calculator to find the volume and area of each sphere.

1. 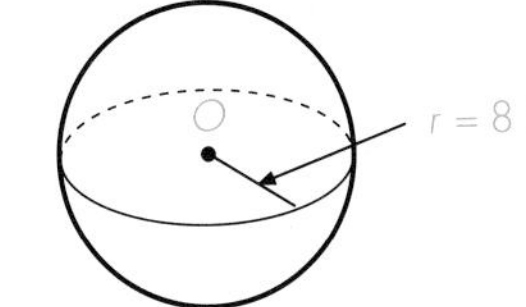

$\frac{2048}{3}\pi$, or 2144 in.^{3}; 804 in.^{2}

2. d = 6.4 mm

137 m^3; 129 m^2

HOME ACTIVITY

Find the volume and area of each sphere. See margin.

1. 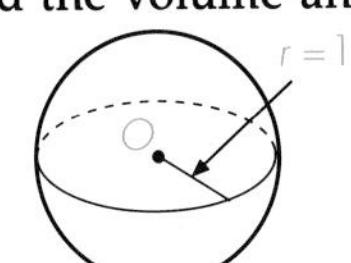

2. 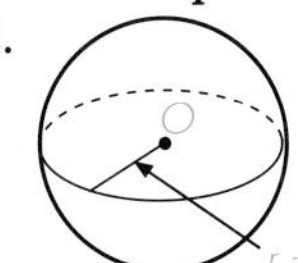

r = 6.2 cm

3. 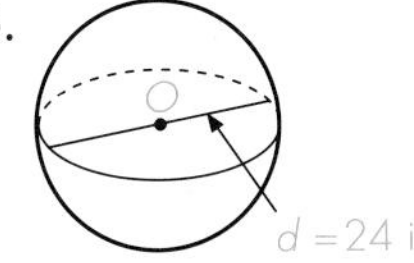

4. C = 8π ft

5. Baseball: $r = 1.4$ in. 11.5 in.^{3}; 24.6 in.

6. Golf ball: $d = 4.2$ cm 38.8 cm^3; 55.4 cm^2

7. Tennis ball: $r = 1.26$ in. 8.38 in.^{3}; 20.0 in.^{2}

8. The moon: $r = 1080$ mi 5,280,000,000 mi^3; 14,700,000 mi^2

9. The radius of the earth is about 7900 mi. About 70% of its surface is water. If the world's population is 5,400,000,000, what is the average number of people for each square mile of land? about 23 people/mi^2

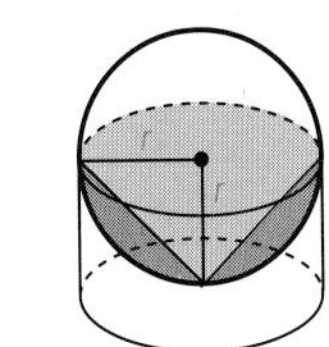

10. If a plane passes through a sphere at its center, then the cross-section that is formed is called a **great circle** (see diagram at the right). What is the relationship between the area of a great circle of a sphere and the surface area of the sphere? Area of sphere = 4 × (area of great circle)

CRITICAL THINKING

11. A great circle of a sphere divides a circle into two **hemispheres** of equal volume. The cylinder, cone, and hemisphere at the right all have radius r; r is also the height of the cylinder and cone. Show that the volume of the hemisphere is the average of the volume of the cylinder and cone.

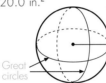

$$\frac{\frac{1}{3}\pi r^3 + \pi r^3}{2} = \frac{2}{3}\pi r^3 = \frac{1}{2}\left(\frac{4}{3}\pi r^3\right)$$

Additional Answers

1. 4188 yd^3; 1257 yd^2
2. 998 cm^3; 483 cm^2
3. 7240 in^3; 1810 in^2
4. 268 ft^3; 201 ft^2

FOLLOW-UP

Assessment

Find the volume and area of each sphere.

1. 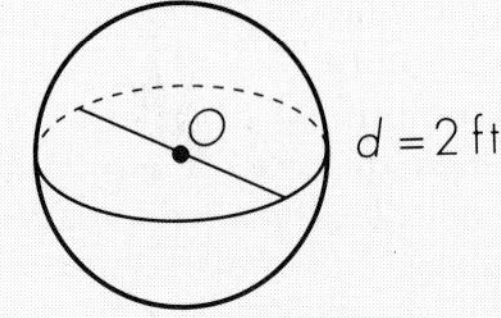

$\left[\frac{4}{3}\pi \approx 4.19 \text{ ft}^3; 4\pi \approx 12.6 \text{ ft}^2\right]$

2. 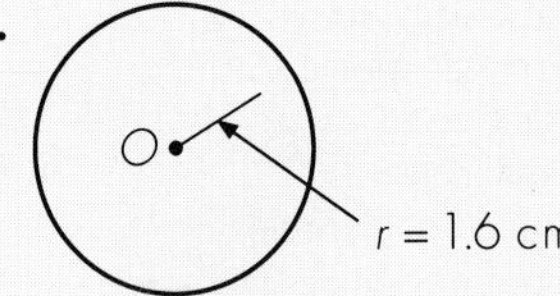

$\left[\frac{16.384}{3}\pi \approx 17.1 \text{ cm}^3; 10.24\pi \approx 32.2 \text{ cm}^2\right]$

3. The planet Mercury has a diameter of about 3100 mi. Find the surface area to the nearest hundred thousand square miles. [30,200,000 mi^2]

Reteaching

Students who have had difficulty with this section may benefit from Reteaching Activity 11.6.

Enrichment

For students who have mastered the material in this section, you may wish to assign Enrichment Activity 11.6.

CHAPTER SUMMARY

Chapter Vocabulary
altitude
Archimedes
area
axis
base
base edge
Cavalieri's Principle
center of sphere
circular cone
circular cylinder
cone
conversion factor
cube
cubic centimeter
cylinder
edge
height
hexagonal pyramid
lateral area
lateral edge
lateral face
oblique circular cylinder
oblique cone
oblique prism
perimeter
polygonal region
polyhedron
prism
pyramid
quadrangular pyramid
radius of sphere
regular pyramid
right circular cylinder
right cone
right prism
sector of a circle
slant height
sphere
square pyramid
tetrahedron
total surface area
triangular pyramid
vertex
volume

11.1 A **prism** is a closed figure in space. Its bases are two congruent polygonal regions in parallel planes, and the segments joining corresponding vertices are parallel segments called lateral edges. Two corresponding sides of the bases and the lateral edges that connect them determine a parallelogram-shaped region called a lateral face. Each point where a base edge intersects a lateral edge is a vertex of the prism. If the lateral edges are perpendicular to the bases, the prism is a right prism. Otherwise it is oblique. In the figure, *ABCDE* is the top base of the prism. $\overline{EJ}$ is a lateral edge. ▱*AEJF* is a lateral face. Points *F* and *D* are two vertices. The prism shown is a right prism.

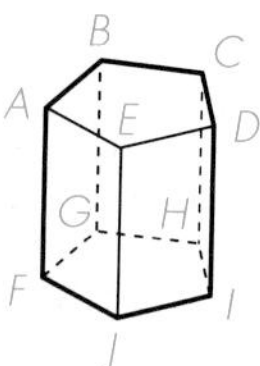

The height of a prism is the distance between the planes of the bases. If a right prism has bases of area *B* and perimeter *p*, then the lateral area *LA* and total area *TA* of the prism are given by the formula $LA = ph$ and $TA = 2B + ph$.

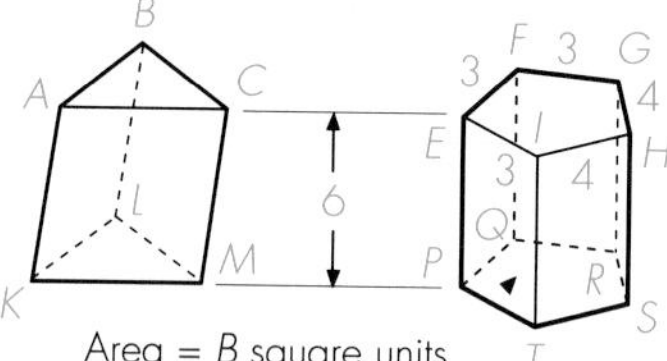

The bases of a **cylinder** are congruent curved regions that lie in parallel planes. The segments joining corresponding points on the boundries of these regions are all parallel. If the axis of a circular cylinder is perpendicular to the bases, the cylinder is a right cylinder. Otherwise it is oblique. The height of a cylinder is the distance between the planes of the bases. In the figure, the left cylinder has regions 1 and 2 as bases. The second cylinder is a right cylinder. Both cylinders have a height of 9.

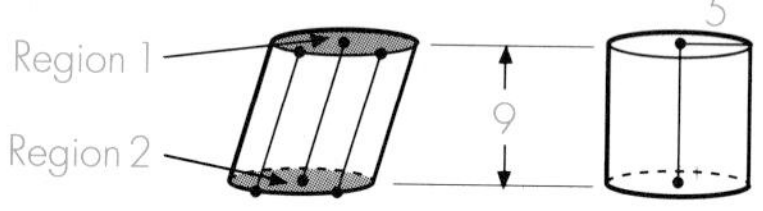

11.2 The volume of a space figure is the number of cubic units it contains. A right rectangular prism of length *l*, width *w*, and height *h* has volume $V = lwh$. The volume of a prism whose bases have area B and whose height is *h* is $V = Bh$. The volume of a cylinder whose bases have area *B* and whose height is *h* is $V = Bh$.

$V = 120 \text{ cm}^3$ $V = 78 \text{ m}^3$ $V = 52\pi \approx 163 \text{ m}^3$

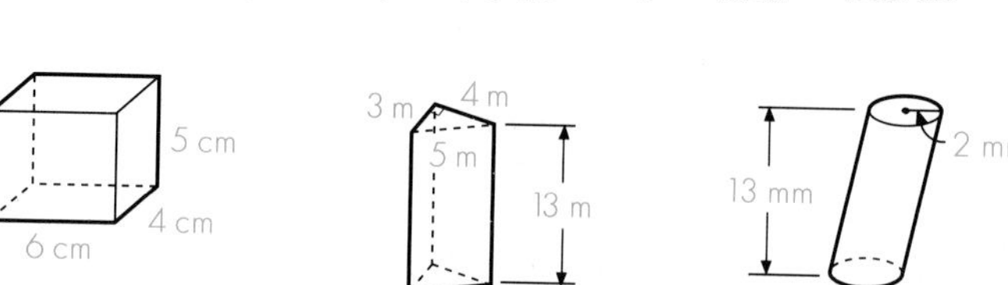

436

11.3 A pyramid has a polygonal region for its base. All the points of the polygon are joined to a point P (the vertex) not in the plane of the base. The segments joining the vertices of the base to P are lateral edges of the pyramid. The lateral faces of a pyramid are triangular regions. The height of a pyramid is the distance from the vertex of the pyramid to the plane of the base. If the base of a pyramid is a regular polygonal region and all the lateral edges are of equal length, the pyramid is a regular pyramid. The slant height of a regular pyramid is the height of a lateral face.

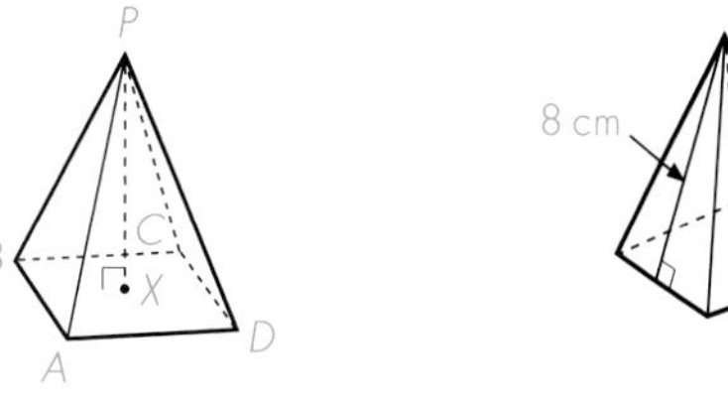

Cones are formed much the same way as pyramids, except the base is a region with a curved boundary. The line segment joining the vertex to the center of a circular cone is the axis of the cone. If the axis is perpendicular to base, the cone is a right cone. The height of a cone is the distance from the vertex to the plane of the base. For a right circular cone, the length of a segment from the vertex to a point on the circle is the slant height.

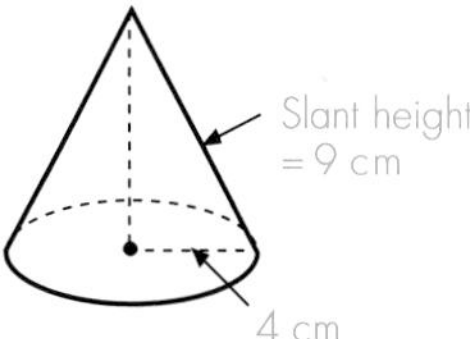

For a right circular cone whose base has radius r and whose slant height is s, $LA = \pi rs$ and $TA = \pi r^2 + \pi rs$. For the right circular cone in the diagram, $LA = \pi \times 4 \times 9 = 36\pi$ (about 113 cm^2) and $TA = \pi \times 4^2 + \pi \times 4 \times 9 = 52\pi$ (about 163 cm^2).

11.4 If a pyramid has a base of area B and a height of h, its volume is $V = \frac{1}{3}Bh$. If a cone has a height of h and a base of radius r, its volume is $V = \frac{1}{3}\pi r^2 h$. In the diagram, the pyramid is a square pyramid. So $V = \frac{1}{3} \times 25 \times 10 = 83.3$ cm^3. For the circular cone, $V = \frac{1}{3} \times \pi \times 3^2 \times 10 = 30\pi$, or about 94 cm^3.

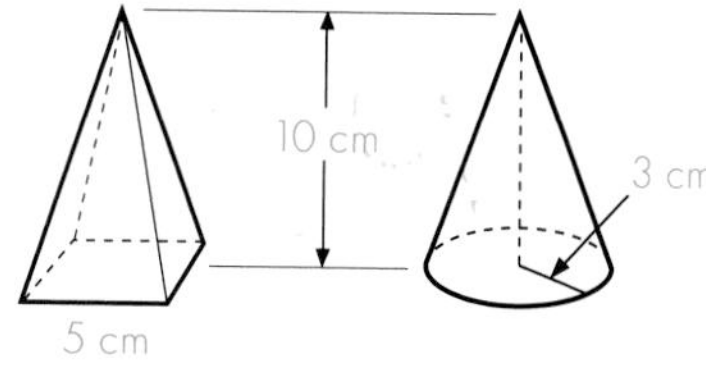

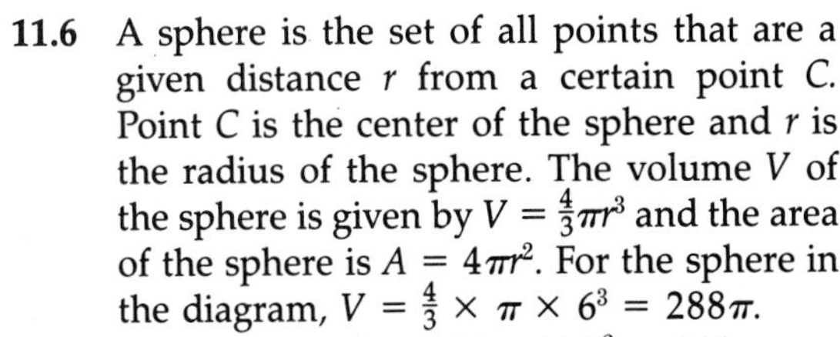

11.6 A sphere is the set of all points that are a given distance r from a certain point C. Point C is the center of the sphere and r is the radius of the sphere. The volume V of the sphere is given by $V = \frac{4}{3}\pi r^3$ and the area of the sphere is $A = 4\pi r^2$. For the sphere in the diagram, $V = \frac{4}{3} \times \pi \times 6^3 = 288\pi$.

$A = 4 \times \pi \times 6^2 = 144\pi$.

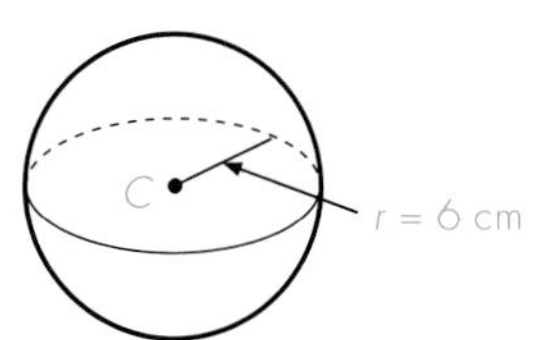

Using the Page
This Computer Activity permits students to draw right rectangular prisms and to determine their volumes. It then extends the process to right triangular prisms. When volume relationships of pyramids are introduced, students may use the programs to internalize formulas studied in the text.

COMPUTER ACTIVITY

The simplest way to draw three-dimensional figures with the LOGO turtle is to use the x-y plane and the SETXY command.

Example: Write a procedure using the command SETXY to tell the turtle to draw a right rectangular prism whose length is 40 units, width is 20 units, and height is 15 units. Then write a procedure to compute the volume of the prism.

The RECPRISM procedure below tells the turtle to draw the right rectangular prism.

```
TO RECPRISM
  SETXY 40 0
  SETXY 50 10
  SETXY 50 20
  SETXY 10 20
  SETXY 0 10
  SETXY 40 10
  SETXY 50 20
  PU SETXY 40 10 PD
  SETXY 40 0
  PU HOME PD
  SETXY 0 10
  END
```

Since this is a three-dimensional figure, the width in the figure drawn by the turtle will take perspective into account. The output is shown below.

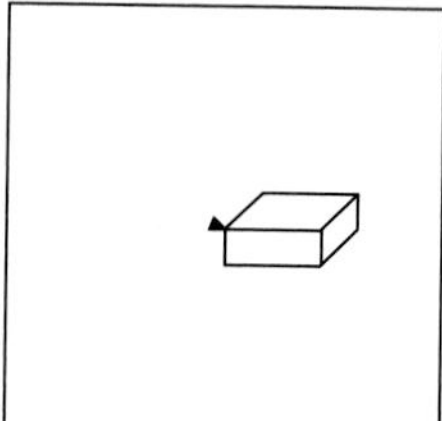

The procedure to find the volume of a right rectangular prism is given below.

```
TO VOLRECPRI :LENGTH :WIDTH :HEIGHT
  OUTPUT :LENGTH * :WIDTH * :HEIGHT
  END
```

To find the volume of the right rectangular prism whose length, width, and height are given above, simply type VOLRECPRI 40 20 15. The output will be 12000.

Modify the RECPRISM procedure to tell the turtle to draw each right rectangular prism described in Exercises 1 through 5. Then use the VOLRECPRI procedure to compute the volume of each figure.
Answers for procedures may vary.

1. length = 25, width = 15, height = 20 7500
2. length = 30, width = 18, height = 25 13500
3. length = 48, width = 13, height = 36 22464
4. length = 86, width = 9, height = 55 42570
5. length = 70, width = 29, height = 61 123830

Write procedures to compute the volume of each figure described in Exercises 6 through 8. Give each procedure the indicated name. See margin.

6. a right triangular prism; VOLTRIPRI
7. a triangular pyramid; VOLTRIPYR
8. a quadrangular pyramid; VOLQUADPYR

Use the VOLTRIPRI procedure from Exercise 6 to compute the volume of each right triangular prism.

9.

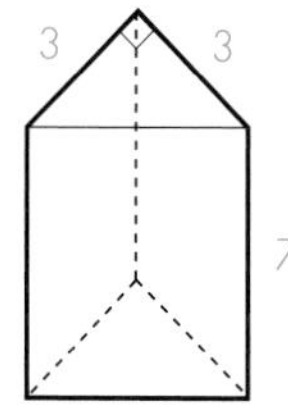

31.5

10.

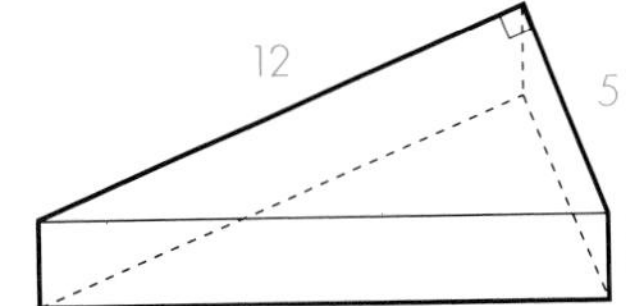

60

Use the VOLTRIPYR procedure from Exercise 7 to compute the volume of each triangular pyramid.

11.

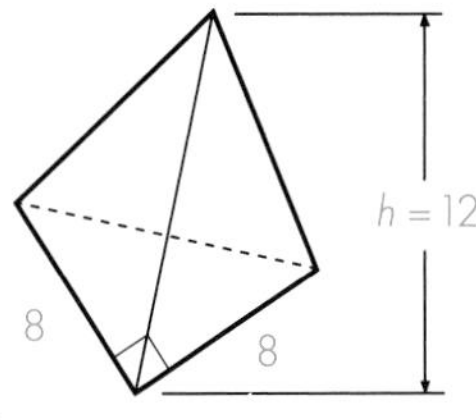

128

12.

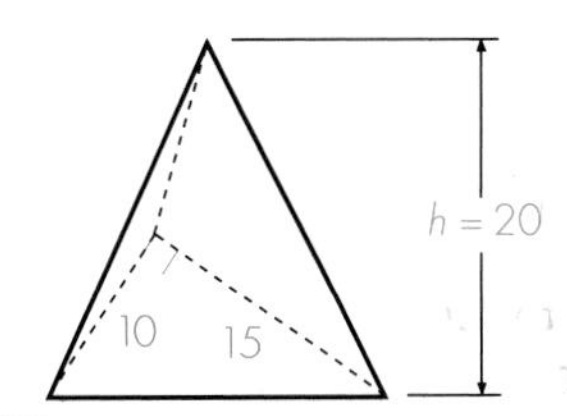

500

Use the VOLQUADPYR procedure from Exercise 8 to compute the volume of each regular square pyramid.

13.

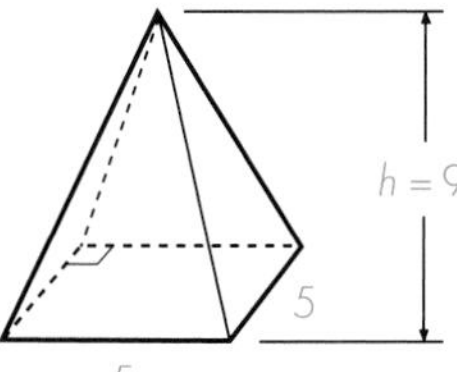

75

14.

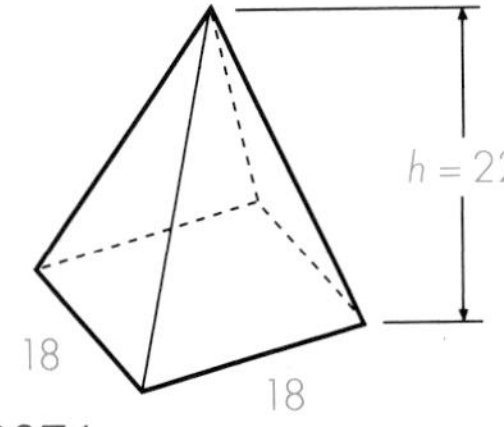

2376

Using the Page
Students can work individually or in groups to study the Chapter 11 Review. If you have students study in groups, they should work together to make sure all members of the group know the material in the chapter.

Informal assessment such as interviews, classroom observations, or a review of student portfolios can be used instead of paper-and-pencil tests.

Additional Answers
2. $24\pi \approx 75.4$ in.^{2}; $32\pi \approx 101$ in.^{2}

Find the lateral area and total area for each prism or right circular cylinder.

1. 68 m^2; 94 m^2

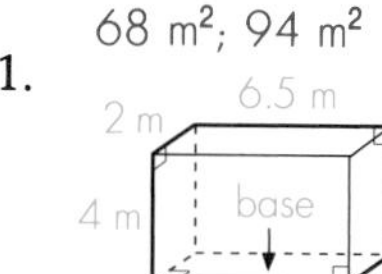

2. 75.4 in.^{2}; 101 in.^{2}

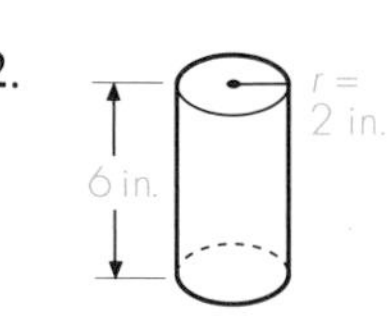

3. 36 ft^2; 54 ft^2

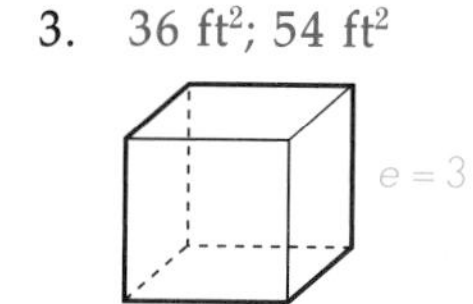

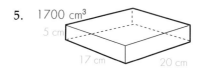

4. 1800; 2040

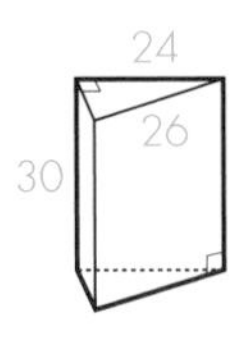

Find the volume of each prism or cylinder.

5. 1700 cm^3

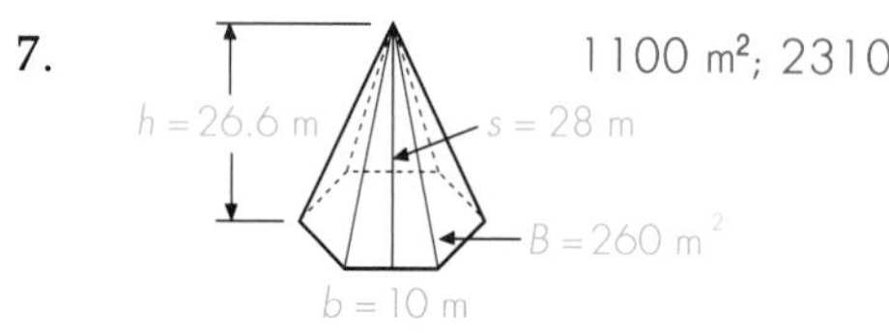

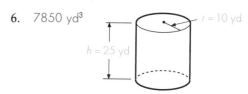

6. 7850 yd^3

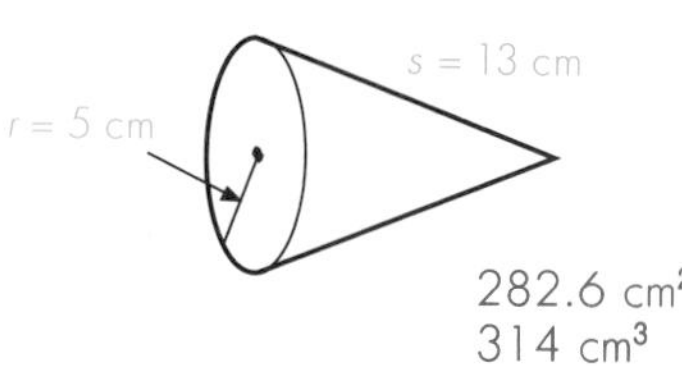

Find the total area and volume of each regular pyramid or right circular cone.

7. 1100 m^2; 2310 m^3

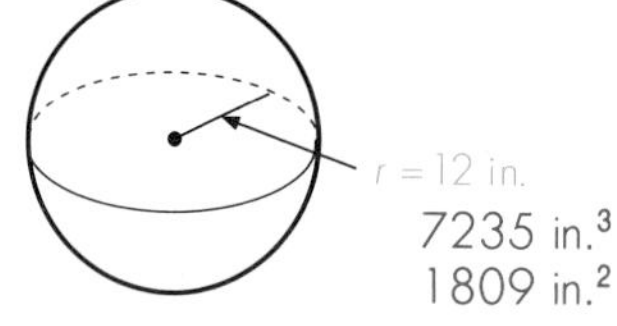

8. 282.6 cm^2 314 cm^3

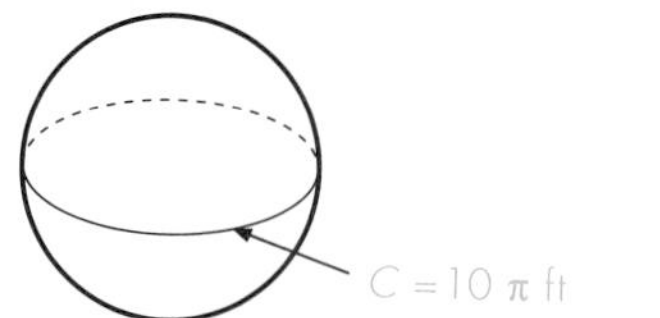

Find the volume and area of each sphere.

9. r = 12 in. 7235 in.^{3} 1809 in.^{2}

10. C = 10π ft 524 ft^3; 314 ft^2

A company is considering buying a water tank. One type is cylindrical, the other spherical. The radius of the sphere is 10 ft. The radius of a base of the cylinder is 8 ft and its height is 30 ft.

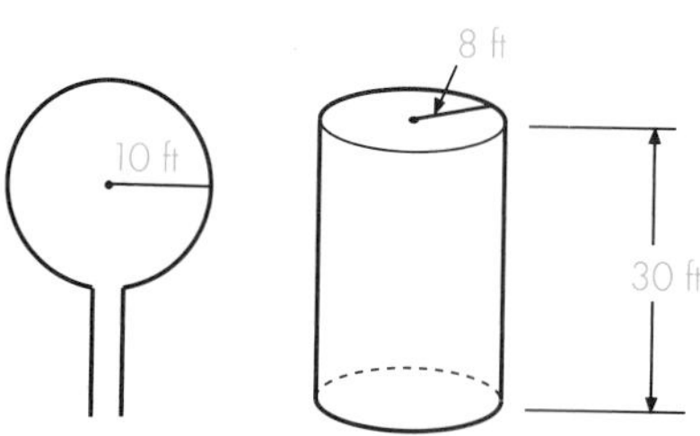

11. Which tank holds more? How many cubic feet more? cylindrical; 1843 ft^3

12. How many more gallons does the larger tank hold? (1 ft^3 = 7.5 gal) 13,822 gal

CHAPTER TEST

Find the lateral area and total area for each prism or right circular cylinder.

1.

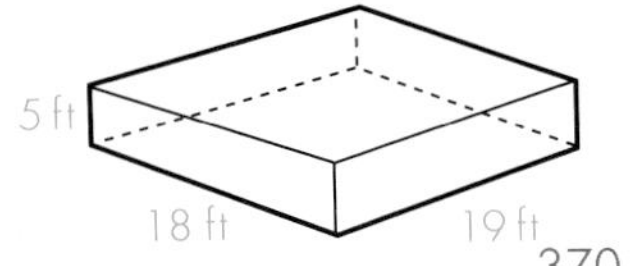

370 ft^2; 1054 ft^2

2.

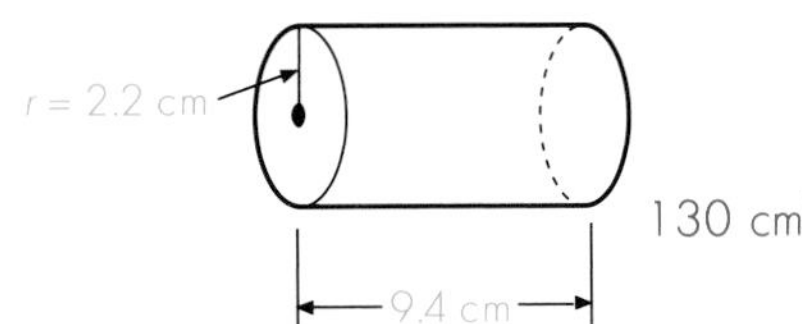

130 cm^2

Find the volume of each prism or cylinder.

3. 216 in.^{3}

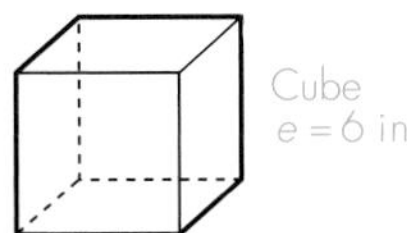

4. 56 m^3

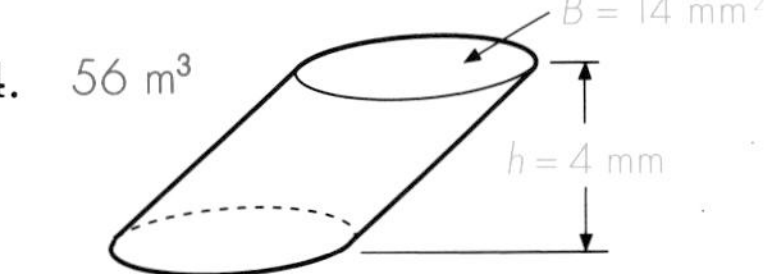

Find the total area and volume of each regular pyramid or right circular cone.

5. 120 cm^2; 75.6 cm^3

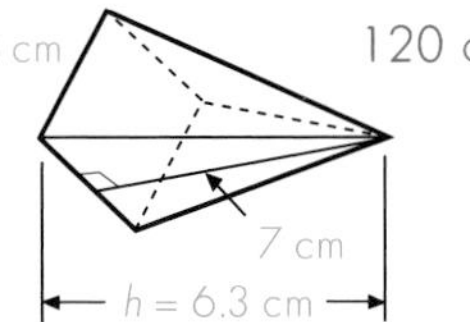

6.

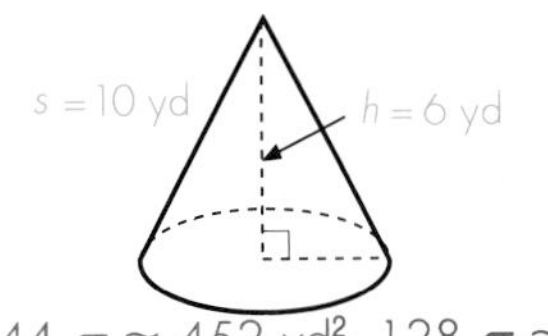

$144\pi \approx 452$ yd^2; $128\pi \approx 402$ yd^3

Find the volume and area of each sphere.

7. $\frac{16,384}{3}\pi \approx 17,149$ mm^3; $1024\pi \approx 3215$ mm^2

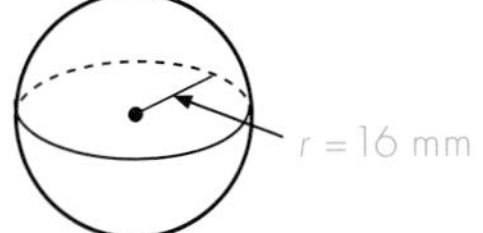

8. $167\pi \approx 524$ mm^3; $100\pi \approx 314$ mm^2

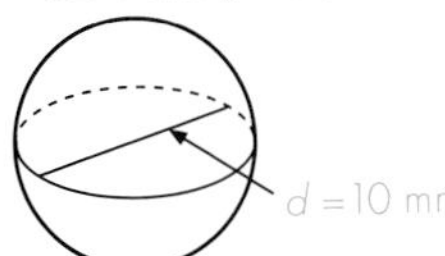

9. The Vertical Assembly Building at the John F. Kennedy Space Center in Florida has the shape of a rectangular prism. The height is 525 ft, and the dimensions of the floor area are 716 ft by 518 ft. Find the lateral area of the building. 1,295,700 ft^2

10. A cylindrical swimming pool has a diameter of 50 ft and a height of 8 ft. If the pool is being filled at the rate of 60 gal/min, how long will it take to fill the pool? (7.5 gal = 1 ft^3) 1964 min, or about 33 h

Additional Answers

2. $41.36\pi \approx 130$ cm^2; $51.04\pi \approx 160$ cm^2

Assessment Resources

Achievement Tests pp. 21-22

Test Objectives

After studying this chapter, students should understand

- how to find the area of prisms and cylinders.
- how to find the volume of prisms and cylinders.
- how to calculate the surface of a pyramid.
- how to calculate the surface area of a cone.
- how to find the volume of a pyramid.
- how to find the volume of a cone.
- how to find the area of a sphere.
- how to find the volume of a sphere.
- how prisms and cylinders are used in everyday life.

441

ALGEBRA REVIEW

Study these examples.

In the coordinate plane, find the set of points that satisfy the given condition. Then describe the location of the points.

1. The x-coordinate is always equal to the y-coordinate.

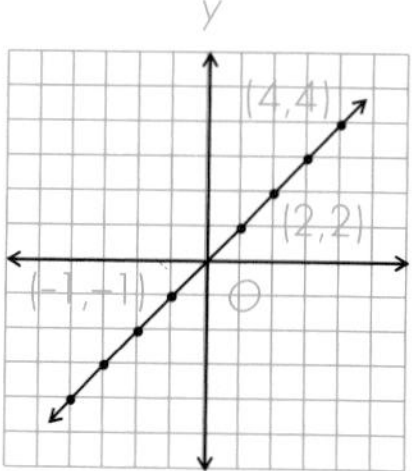

These are all the points on the line that bisects the angle formed by the positive x- and y-axes and the vertical angle.

2. The y-coordinate is always equal to $2\frac{1}{2}$.

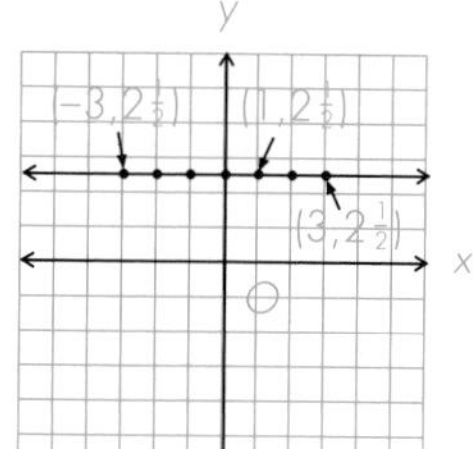

These are all the points on the horizontal line $2\frac{1}{2}$ units above the x-axis.

Example: Find and describe the set of all points that satisfy *both* of these two following conditions.

a. The sum of the coordinates is 7.

b. The x-coordinate is 6.

a y b (1,6) (6,5) (3,4) (5,2) (6,1) x O

The set consists of a single point at (6, 1).

Follow Examples 1 and 2 in solving these problems. In the coordinate plane, find the set of points that satisfy the given condition. Then describe the location of the points. See margin.

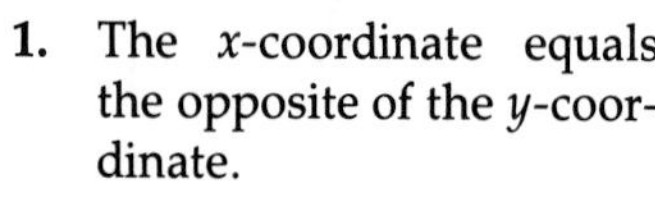

1. The x-coordinate equals the opposite of the y-coordinate.
2. The y-coordinate equals twice the x-coordinate.
3. The y-coordinate is the absolute value of the x-coordinate.

Follow Example 3 in solving these problems. Find and describe the set of all points that satisfy both of the given conditions. See margin.

4. Condition a: The y-coordinate is 2 greater than the x-coordinate. Condition b: The sum of the x- and y-coordinates is 5.
5. Condition a: The x-coordinate is 4. Condition b: The absolute value of the y-coordinate is 2.
6. Condition a: The absolute value of the x-coordinate is 3. Condition b: The absolute value of the y-coordinate is 3.

Algebra Review Objectives
Topics reviewed from algebra include
- graphing in the coordinate plane and describing the location of a set of points that satisfy one condition or two conditions.

Additional Answers

1. These are all the points on the line bisecting the angle formed by the positive x-axis and negative y-axis and its vertical angle.

2. These are all the points on the line determined by (0, 0) and (1, 2).

3. These are all the points on a V-shaped figure formed by the rays that bisect the two angles formed by the x-axis and the positive y-axis.

4. One point: $\left(1\frac{1}{2}, 3\frac{1}{2}\right)$

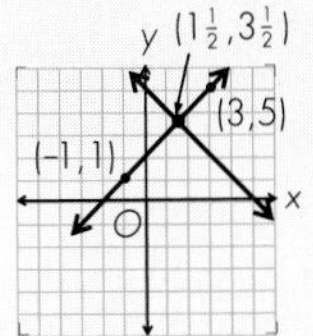

5. Two points: (4, 2) and (4, −2)

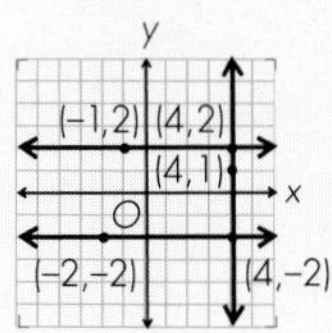

6. Four points: (3, 3), (−3, 3), (−3, −3) and (3, −3)

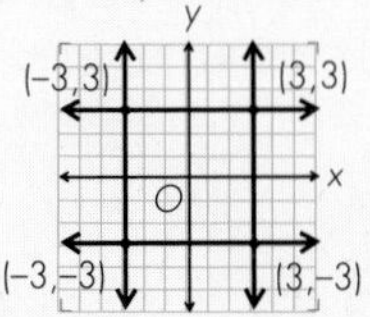

Name ______________________ Date ______________

Area of Prisms and Cylinders

A *prism* must have two congruent and parallel polygons for faces, called the *bases.* The other faces are congruent to each other, but are not necessarily parallel. They are called *lateral faces.*

Like a prism, the two bases of a *cylinder* must also be congruent and parallel. But these bases are curved figures rather than polygons. Usually the bases are circles, but can be any curve.

Prisms and cylinders are either *right* or *oblique.* They are right if the lateral faces or lateral surface form right angles with the bases. Otherwise they are oblique and look slanted.

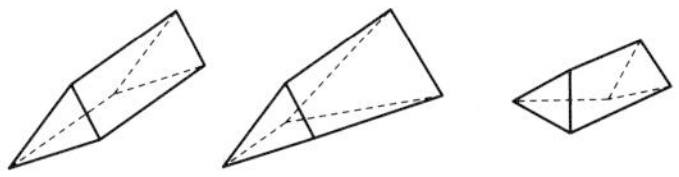

Prism
The bases are triangles.

Not a Prism
The triangular bases are not congruent.

Not a Prism
The bases are not parallel.

Cylinder
The bases are curved figures.

Circular Cylinder
The bases are circles.

Not a Cylinder
The bases are not congruent.

Write the letter of the most accurate name for each figure.

1. c
- **a.** right rectangular prism
- **b.** oblique circular cylinder
- **c.** right circular cylinder
- **d.** triangular prism

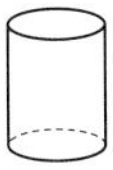

2. b
- **a.** right circular prism
- **b.** right rectangular prism
- **c.** oblique triangular prism
- **d.** oblique cylinder

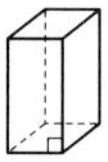

3. a
- **a.** right triangular prism
- **b.** right circular prism
- **c.** right triangular cylinder
- **d.** oblique triangular prism

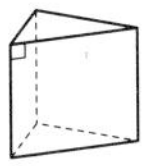

4. d
- **a.** right pentagonal prism
- **b.** right circular cylinder
- **c.** oblique hexagonal prism
- **d.** oblique pentagonal prism

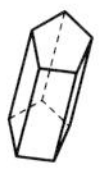

5. c
- **a.** right circular cylinder
- **b.** oblique circular prism
- **c.** oblique circular cylinder
- **d.** right pentagonal prism

6. d
- **a.** right triangular prism
- **b.** right rectangular prism
- **c.** right circular prism
- **d.** right hexagonal prism

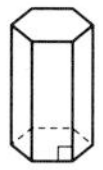

Name ______________________ Date ______________

From a Different Point of View

When you look at a 3-dimensional object, your point of view determines what you see. At the right are two views of a cup—a side view and a top view.

1. Draw the top view for each object.

a. A cube

b. A pyramid with a regular hexagon for a base

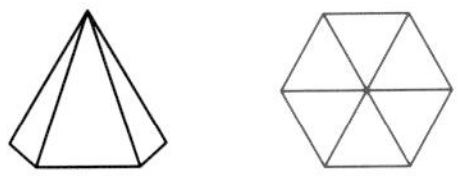

2. Draw two side views for each object.

a. A cube

b. A pyramid with a square for a base

Write the letter of the figure that shows the given shape from a different point of view.

3. b **a.** **b.** **c.**

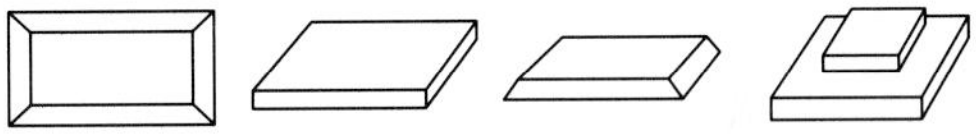

4. a **a.** **b.** **c.**

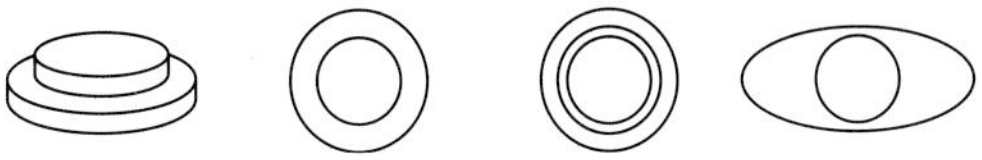

5. b **a.** **b.** **c.**

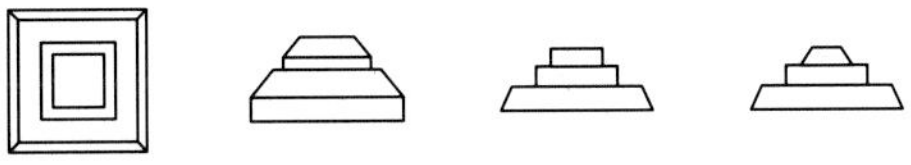

Name ______________________ Date ______________

Volume of Prisms and Cylinders

The *height* of a prism or cylinder is the perpendicular distance between the two bases. The top row of figures shows two hexagonal right prisms and a right circular cylinder. The height is labeled *h*.

The first two figures in the bottom row are an oblique triangular prism and an oblique circular cylinder. Note that the height is the perpendicular distance between the bases, not the distance along an edge.

The last figure is a right rectangular prism. The height can be the length of any of the edges. Choose two opposite bases. Then the height is the distance between them.

Shade in the two bases of each figure. Then write the height.

1. height = 10 cm
2. height = 10 cm
3. height = 15 cm
4. height = 8 cm
5. height = 8 cm
6. height = 10 cm
7. height = Answers will vary.
8. height = Answers will vary.
9. height = Answers will vary.

Name ______________________ Date ______________

Puzzles with Hidden Cubes

In the figures on this page, various shapes have been constructed by gluing cubes together. For each figure, use your visual imagination to answer the questions.

1. How many cubes
   a. are there in all? 13
   b. have just 1 face glued? 0
   c. have just 2 faces glued? 3
   d. have just 3 faces glued? 7
   e. have just 4 faces glued? 2
   f. have just 5 faces glued? 1

2. How many cubes
   a. are there in all? 17
   b. have just 1 face glued? 0
   c. have just 2 faces glued? 1
   d. have just 3 faces glued? 8
   e. have just 4 faces glued? 6
   f. have just 5 faces glued? 2

3. How many cubes
   a. are there in all? 14
   b. have just 1 face glued? 5
   c. have just 2 faces glued? 0
   d. have just 3 faces glued? 8
   e. have just 4 faces glued? 0
   f. have just 5 faces glued? 1

Name ______________________ Date ______________

Area of Pyramids and Cones

Pyramids and *cones* are formed by connecting each point of a polygon or curve to a point outside the plane of the figure. The point is called the *vertex*.

If the vertex is directly over the center of the base, the figure is a *right pyramid* or a *right cone*. If the vertex is not over the center of the base, the figure is *oblique*. The top row of figures are right; the second row are oblique. The *height, h,* is the perpendicular distance from the vertex to the plane of the base.

Right regular pyramids (those with regular polygons for bases) and right circular cones also have *slant height, s.* This is the height of each triangular face of the pyramid. For the cone, it is the distance from the vertex to any point on the circle. In the bottom row of figures, the slant heights are shown with heavy lines.

Write the letter of the figure that best matches each description. Use each figure only once.

1. oblique triangular pyramid D
2. right trapezoidal pyramid B
3. right circular cone H
4. right square pyramid F
5. right pentagonal pyramid A
6. oblique circular cone E
7. right regular hexagonal pyramid G
8. right non-circular cone C

In each pair of statements given in Exercises 9–12, one is true and one is false. Circle the letter of the true statement.

9. a. Only one face of a pyramid can be a triangle.
   (b.) All but one face of a pyramid must be triangles.

10. (a.) All pyramids have a vertex.
    b. A vertex can only be found in a pyramid.

11. (a.) The base of a cone is always a closed curve.
    b. The base of pyramid is never a polygon.

12. (a.) The height of a cone is always perpendicular to the base.
    b. The height of a pyramid is the length between two vertices of the base.

Name ______________________ Date ______________

Building Space Figures

The patterns in Exercises 1–4 below can be used to make Figures A through D. In each case, start with an equilateral triangle that is $8\frac{1}{2}$ inches on each side. Draw the dotted lines. You will fold along each of these lines when building each figure.

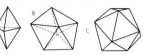

1. Number the small triangles as shown. Cut away the triangles without numbers. Fold and glue triangle 5 over triangle 1. Fold the remaining triangles to form Figure D.

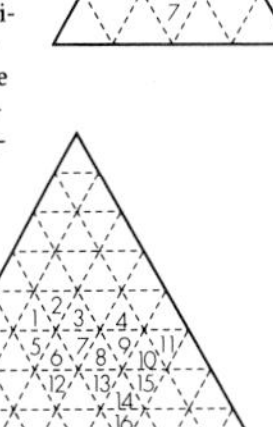

2. Number the small triangles as shown. Cut away the triangles without numbers. Also make a cut between triangles 4 and 5. Fold and glue triangle 5 over 3; triangle 6 over 5; triangle 7 over 12. Continue folding until you have made Figure A.

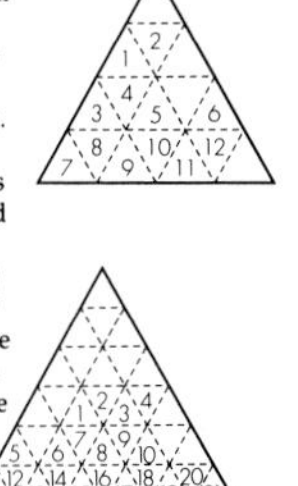

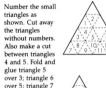

3. Number the small triangles as shown above. Cut away the triangles without numbers. Also cut between triangles 1 and 2, and between triangles 14 and 15.

   Fold and glue triangle 15 over 14; triangle 1 over 2; triangle 11 over 1; triangle 4 over 3. Squeeze the figure shut and glue the remaining triangles down. Which figure have you made? B

4. Number the small triangles as shown above. Cut away the triangles without numbers. Also cut between these pairs of triangles: 1 and 2; 6 and 14; 9 and 10; 17 and 26; 30 and 31; 22 and 23.

   Fold and glue triangle 1 over 2; triangle 30 over 31; triangle 11 over 21. Fold triangles 6 and 4 over triangles 14 and 20, respectively. Squeeze the figure shut and glue the remaining triangles down. Which figure have you made? C

Name ______________________ Date ____________

Volume of Pyramids and Cones

Recall these formulas for the *areas* of 2-dimensional shapes:

rectangle: $A = ab$

right triangle: $A = \frac{1}{2}ab$

circle: $A = \pi r^2$

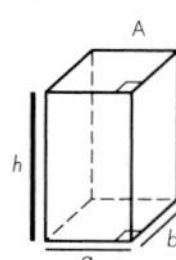

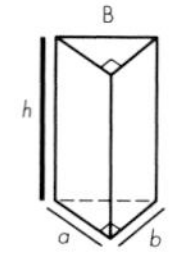

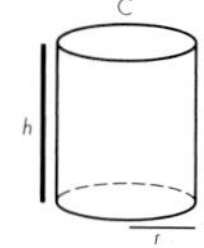

The *volume* of a prism or cylinder equals the area of the base times the height. For a pyramid or cone, the volume is one-third of the area of the base times the height:

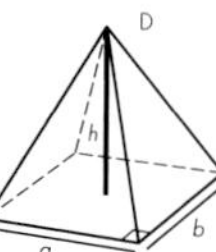

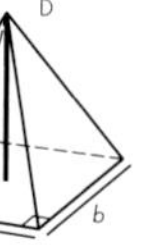

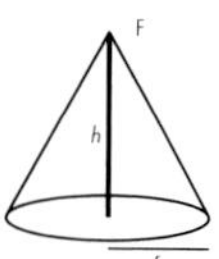

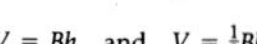

$V = Bh$ and $V = \frac{1}{3}Bh$

Write a letter to match each figure above to its volume formula.

1. $V = abh$ A
2. $V = \frac{1}{3}abh$ D
3. $V = \pi r^2 h$ C
4. $V = \frac{1}{2}abh$ B
5. $V = \frac{1}{3}\pi r^2 h$ F
6. $V = (\frac{1}{3})(\frac{1}{2})abh$ E

Write the name of the figure for each volume formula.

7. $V = abh$ rectangular prism
8. $V = \frac{1}{3}abh$ rectangular pyramid
9. $V = \pi r^2 h$ circular cylinder
10. $V = \frac{1}{2}abh$ triangular prism
11. $V = \frac{1}{3}\pi r^2 h$ circular cone
12. $V = (\frac{1}{3})(\frac{1}{2})abh$ triangular pyramid

Find each volume. Use 3.14 for π in your calculations.

13. A cylinder with a height of 20 centimeters and this circle for a base.
6280 cm^3

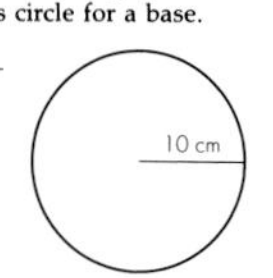

14. A pyramid with a height of 20 centimeters and this right triangle for a base.
266.7 cm^3

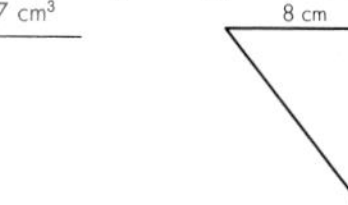

Name ______________________ Date ____________

Patterns for Truncated Figures

The left-hand figure at the right is a regular tetrahedron. If you cut off the four corners as shown, the result is a *truncated* regular tetrahedron. A truncated polyhedron can be made from any type of solid figure. Those made from the regular polyhedra have regular polygons for faces.

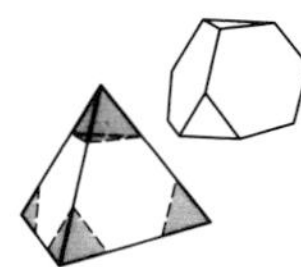

1. The pattern at the right can be used to make a truncated tetrahedron. Copy the pattern and number the hexagons and triangles as shown. Cut into the pattern on the six heavy line segments. Then start by gluing the like-numbered polygons together. Describe the eight faces of the finished model.
4 regular hexagons and 4 equilateral triangles

2. What is the length of the altitude, h, in the triangle shown below?
$\frac{\sqrt{3}}{2}$ in. or 0.866 in.

3. What is the area of an equilateral triangle that measures 1 inch on each side?
$\frac{\sqrt{3}}{4}$ in.^{2} or 0.433 in.^{2}

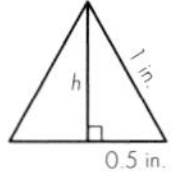

4. If the side of each small equilateral triangle in the pattern measures 1 inch, what is the surface area of the finished truncated tetrahedron?
$7\sqrt{3}$ in.^{2} or 12.124 in.^{2}

5. The pattern at the right can be used to make a truncated cube. Cut into the pattern on the six heavy line segments. Cut out and discard the shaded areas. Then start by gluing the like-numbered polygons together. Describe the 14 faces of the finished model.
6 regular octagons and 8 equilateral triangles

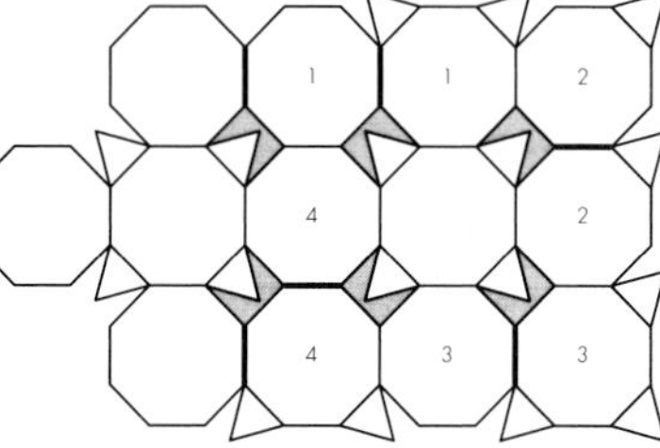

Name ______________________ Date ______________

Applications

The patterns on this page can be used to build 3-dimensional figures. Building your own models can enhance your understanding of the topics in this chapter.

1. Describe the six models you could build from the patterns. Three prisms: a cube, a rectangular prism, a triangular prism; three pyramids: square, triangular, hexagonal

2. Find the surface areas of the three prisms. rectangular: 90 sq units, cube: 96 sq units, triangular: 84 sq units

132 Reteaching • Section 11.5

Name ______________________ Date ______________

Frustums

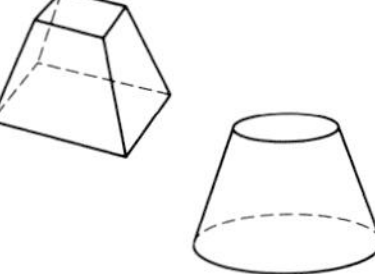

The *frustum* of a pyramid is formed by two parallel planes intersecting the pyramid. In the left-hand figure at the right, the top and bottom of the frustum are the *bases;* the *altitude* is the perpendicular distance between the bases.

The frustum of a cone is formed in a similar manner. Two parallel planes intersect a cone to form either two circular or two elliptical bases. The altitude is the perpendicular distance between the bases.

Sketch the following figures.

1. The frustum of a triangular right pyramid. The two parallel planes cut the pyramid parallel to the base.

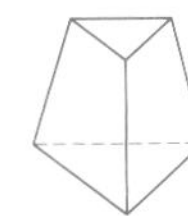

2. The frustum of an oblique hexagonal pyramid. The two parallel planes cut the pyramid parallel to the base.

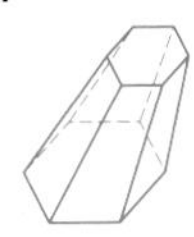

3. The frustum of a square right pyramid. The two parallel planes do *not* cut the pyramid parallel to the base.

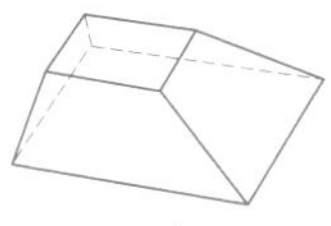

4. Describe the lateral faces of a frustum of any pyramid. They are trapezoids.

5. How are the bases of a frustum of any pyramid related? They are similar polygons.

Sketch the following figures.

6. The frustum of a right circular cone on top of the frustum of a rectangular right pyramid.

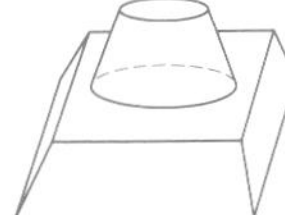

7. The frustum of an oblique circular cone on top of the frustum of a right circular cone.

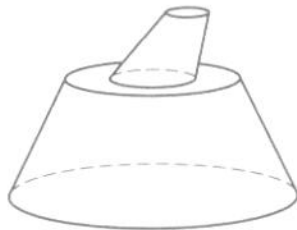

8. The frustum of a square oblique pyramid on top of the frustum of a right circular cone.

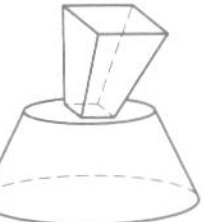

Enrichment • Section 11.5 133

Name ______________________ Date ______________

Area and Volume of a Sphere

The equations at the right below relate the volume formulas for a cone, a sphere, and a cylinder.

In the figures, the height of the cone and the height of the cylinder have been drawn equal to twice the radius. When this is the case, the volume of the sphere is twice that of the cone, and the volume of the cylinder is three times that of the cone.

These relationships may help you memorize the formulas.

Cone	Sphere	Cylinder
$V = \frac{1}{3}\pi r^2 h$	$V = \frac{4}{3}\pi r^3$	$V = \pi r^2 h$
$V = \frac{1}{3}\pi r^2(2r)$		$V = \pi r^2(2r)$
$V = \frac{2}{3}\pi r^3$		$V = 2\pi r^3$

1. Fill in each blank in the paragraph with one of the words in the box. The words can be used more than once, and not all of the words will be used.

two sphere area circle three volume

A circle can be defined as the set of all points in a plane that are a given distance away from a point. Because a circle exists in a plane, we say it has two dimensions. A sphere is different from a circle in that it has three dimensions. A circle can have area, but not volume. On the other hand, a sphere has both area and volume. The area of a circle and the surface area of a sphere are 2-dimensional measures; volume is a 3-dimensional measure.

Name the figure described. Write *cone, sphere,* or *cylinder.*

2. Of the three figures at the top of the page, which is the only one with a vertex? cone

3. A plane intersects one of the figures and forms a rectangular cross section. Which figure is it? cylinder

Name ______________________ Date ______________

Relationships with Solid Figures

In the figures at the right, the following dimensions are equal: the height of the cylinder, the diameter of the base of the cylinder, the diameter of the base of the cone, the height of the cone, the edges of the cube.

The exercises on this page will examine the relationships among these figures. Use 3.14 for π in your calculations.

1. The sphere has a radius of 5 centimeters. What is the height of the cylinder? 10 cm
2. Find the volume of the sphere. 523.3 cm^3
3. Find the volume of the cylinder. 785 cm^3
4. What is the relationship between the volume of the sphere and the volume of the cylinder? Volume of sphere is $\frac{2}{3}$ that of cylinder
5. Find the surface area of the sphere. 314 cm^2
6. Find the lateral surface area of the cylinder. 314 cm^2
7. What is the relationship between the surface area of the sphere and the lateral surface area of the cylinder? They are equal.
8. Find the volume of the cone. 261.6 cm^3
9. What is the relationship between the volume of the cone and the volume of the cylinder? volume of cone is $\frac{1}{3}$ that of cylinder
10. What is the relationship between the volume of the cone and the volume of the sphere? volume of cone is $\frac{1}{2}$ that of sphere
11. Find the volume of the cube. 1000 cm^3
12. What is the approximate relationship between the volume of the cube and the volume of the sphere? volume of sphere is about $\frac{1}{2}$ that of cube
13. Describe the relationships among the figures on this page. Answers will vary: A cylinder and a cube are circumscribed about a sphere. The cone has the same height and base as the cylinder.

Name ______________________ Date ______________

Logo and Figures in Space

Draw each figure specified below on the *x*-*y* plane, indicating the coordinates of each vertex. Then write a Logo procedure to tell the turtle to draw each figure. Answers may vary.

1. right triangular prism

2. right pentagonal prism

3. oblique hexagonal prism

4. triangular pyramid

5. regular quadrangular pyramid

6. pentagonal pyramid

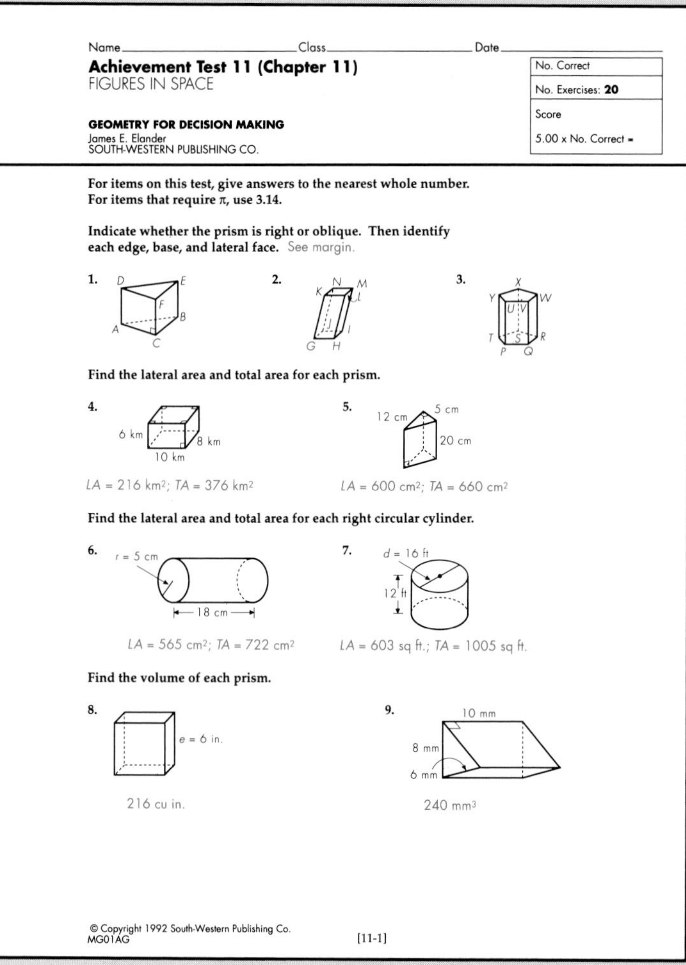

Name________ Class________ Date________

Achievement Test 11 (Chapter 11)
FIGURES IN SPACE

No. Correct
No. Exercises: **20**
Score
5.00 x No. Correct =

GEOMETRY FOR DECISION MAKING
James E. Elander
SOUTH-WESTERN PUBLISHING CO.

For items on this test, give answers to the nearest whole number. For items that require π, use 3.14.

Indicate whether the prism is right or oblique. Then identify each edge, base, and lateral face. See margin.

1. **2.** **3.**

Find the lateral area and total area for each prism.

4. 6 km, 8 km, 10 km

$LA = 216\ km^2$; $TA = 376\ km^2$

5. 12 cm, 5 cm, 20 cm

$LA = 600\ cm^2$; $TA = 660\ cm^2$

Find the lateral area and total area for each right circular cylinder.

6. r = 5 cm, 18 cm

$LA = 565\ cm^2$; $TA = 722\ cm^2$

7. d = 16 ft, 12 ft

LA = 603 sq ft.; TA = 1005 sq ft.

Find the volume of each prism.

8. e = 6 in.

216 cu in.

9. 10 mm, 8 mm, 6 mm

240 mm^3

MG01AG [11-1]

10. Refer to the cylinder shown in Item 6. Find the volume. 1413 cm^3

11. Refer to the cylinder shown in Item 7. Find the volume. 2412 cu ft.

Find the area of each regular pyramid or right circular cone.

12.

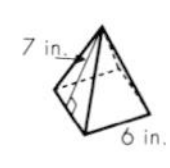

120 sq in.

13.

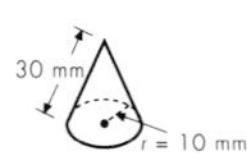

1256 mm^2

14. Refer to the pyramid shown in Item 12. Find the volume. 76 cu in.

15. Refer to the right circular cone in Item 13. Find the volume. 2960 mm^3

Find the volume and area of each sphere.

16.

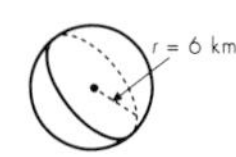

$V = 904\ km^3$; $A = 452\ km^2$

17.

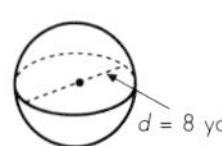

V = 268 cu yd.; A = 201 sq yd.

18. One gallon of paint covers 350 sq. ft., and is sold only in one-gallon cans. How many cans of paint must a contractor buy to paint the barn's exterior, except for the roof? 3 gallons to cover 848 sq ft.

19. A circus tent has the shape of a regular pyramid with a regular hexagon as its base. The length of each side of the base of the tent is 25 ft. The height of the tent is 30 ft. Find the volume of the tent. About 16,238 cu ft.

20. The radius of a spherical drop of water is about 0.36 cm. If a leaking faucet is dripping at the rate of 30 drops per minute, how many minutes will it take to fill a container of 1 liter (1000 cm^3)? About 171 minutes

[11-2]

1. Right; Edges: $\overline{AB}$, $\overline{BC}$, $\overline{AC}$, $\overline{DE}$, $\overline{EF}$, $\overline{DF}$, $\overline{AD}$, $\overline{CF}$, $\overline{BF}$; Bases: $\triangle$ *ABC*, $\triangle$ *DEF*; Lateral faces: Rectangles *ACFD*, *BCFE*, *ABED*

2. Oblique; Edges: $\overline{GH}$, $\overline{HI}$, $\overline{IJ}$, $\overline{JG}$, $\overline{KL}$, $\overline{LM}$, $\overline{MN}$, $\overline{NK}$, $\overline{GK}$, $\overline{HL}$, $\overline{IM}$, $\overline{JN}$; Bases: Parallelograms *GHIJ*, *KLMN*; Lateral faces: Parallelograms *GHLK*, *HIML*, *IJNM*, *JNKG*

3. Right; Edges: $\overline{PQ}$; $\overline{QR}$, $\overline{RS}$, $\overline{ST}$, $\overline{TP}$, $\overline{UV}$, $\overline{VW}$, $\overline{WX}$, $\overline{XY}$, $\overline{YU}$, $\overline{PU}$, $\overline{QV}$, $\overline{RW}$, $\overline{SX}$, $\overline{TY}$; Bases: Pentagons *PQRST*, *UVWXY*; Lateral faces: Rectangles *PQVU*, *QRWV*, *RSXW*, *STYX*, *TPUY*

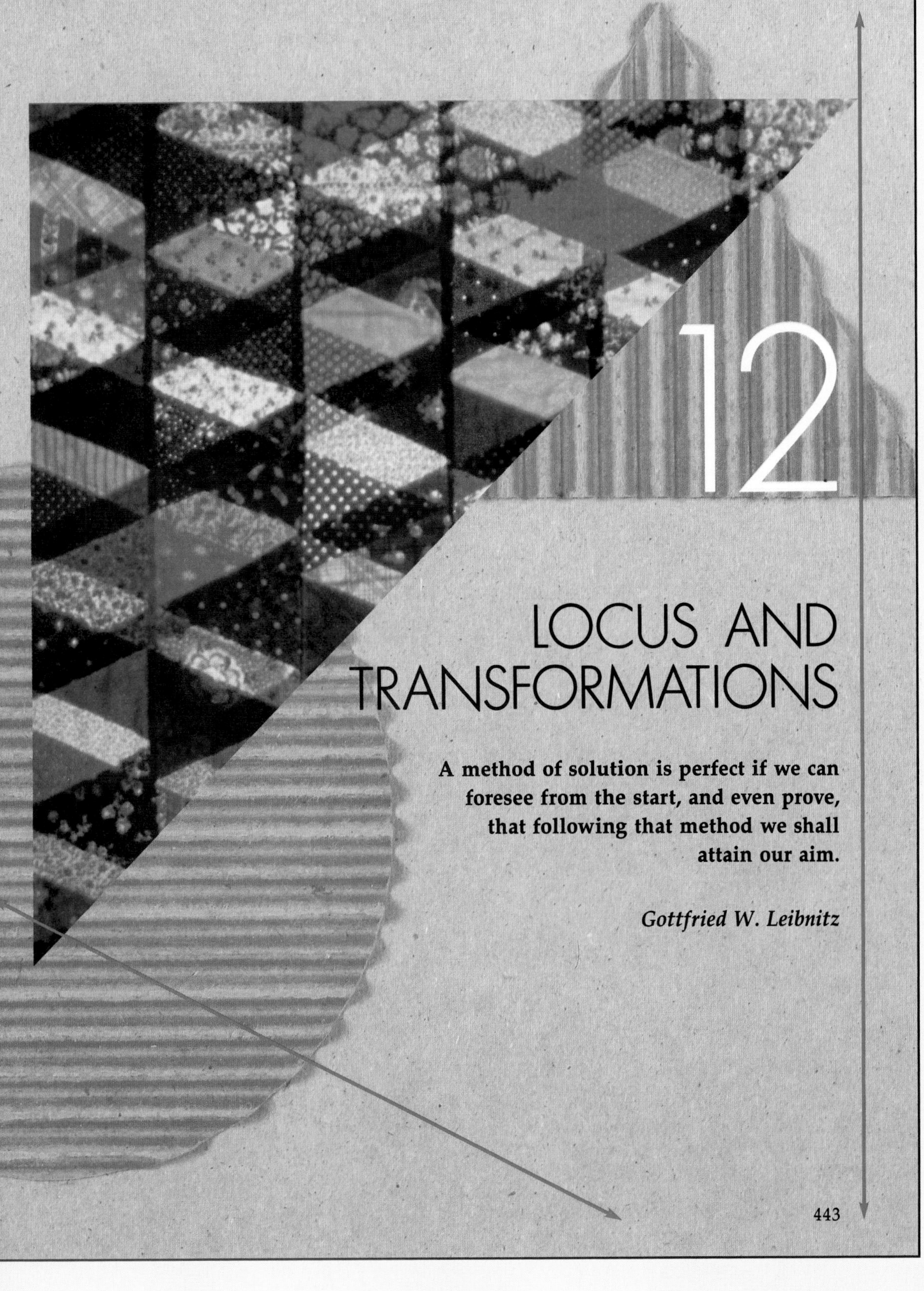

12

LOCUS AND TRANSFORMATIONS

A method of solution is perfect if we can foresee from the start, and even prove, that following that method we shall attain our aim.

Gottfried W. Leibnitz

443

CHAPTER 12 OVERVIEW

Locus and Transformations

Suggested Time: 10–12 days

A method of solution is perfect if we can foresee from the start, and even prove, that following that method we shall attain our aim.

Gottfried W. Leibnitz

Leibnitz's statement describes how the core ideas of this chapter—the understanding of locus, the transformations of figures and objects through translation, reflection, and rotation, and basic logic—all contribute to our understanding of the physical world and provide us with methods we can use to predict, test, and prove the results we seek. The study of locus offers students a new way of proving theorems and a new insight into the physical behavior of the mathematical properties they have been studying. The sections on transformations and symmetry explain how these properties can be applied to solving practical problems. An introduction to conditional statements and their corresponding truth values provides practice in logical reasoning and another method for justifying conclusions.

CHAPTER 12 OBJECTIVE MATRIX

	End of Chapter Items by Activity				Student Supplement Items by Activity		
Objectives by Number	**Review**	**Test**	**Computer**	**Algebra Skills**	**Reteaching**	**Enrichment**	**Computer**
12.1.1	✓	✓			✓	✓	
12.1.2	✓	✓	✓		✓	✓	
12.2.1	✓	✓	✓	✓	✓	✓	✓
12.2.2	✓	✓		✓	✓		✓
12.2.3	✓	✓		✓	✓		✓
12.3.1	✓	✓			✓		
12.3.2	✓	✓			✓		
12.3.3	✓	✓				✓	
12.4.1	✓	✓			✓		✓
12.5.1	✓	✓			✓	✓	
12.5.2	✓	✓			✓	✓	

*A ✓ beside a Chapter Objective under Algebra Skills indicates that algebra skills taught within the chapter or in previous Algebra Skills lessons are used.

CHAPTER 12 PERSPECTIVES

▲ Section 12.1

Locus

Students are encouraged to rediscover familiar geometric properties in a new way. By visualizing sets of points that satisfy given conditions, they work from the particular to the general, finding a pattern, describing it, and drawing a conclusion. Students are asked to prove their conclusions using logic and indirect reasoning in two-part proofs. While it is not necessary for students to prove every locus theorem, the practice in logical reasoning gives them an increased understanding of what is meant by a proof. Students are asked to draw the points for each locus; this helps them predict and then visualize the figure that satisfies the given conditions. In the Home Activity, students practice finding loci for some familiar figures, and then extend their thinking to the three-dimensional plane.

12

▲ Section 12.2

Transformations

Many students will be familiar with reflection and rotation transformations but may not be aware of translation, or of the rules that govern all of these transformations. In the first Discovery Activity, students explore the properties of figures reflected across a line. This gives them a mathematical explanation for what they may have noticed around them. The second Discovery Activity explores the properties of point or rotation transformations. You may find that this section lends itself to experimentation using mirrors, tracing paper, and a pin or compass point to hold figures during rotation. In a Class Activity, students write LOGO procedures to show reflections.

The Home Activity offers practice with all of the transformations, extending them to graphing on the x-y axis and to a discussion of reflection across parallel and intersecting lines.

▲ Section 12.3

Symmetry

This section continues the work on transformations with a discussion of the different types of symmetry—line symmetry and point or rotation symmetry. The first Discovery Activity emphasizes the relationship between line symmetry and reflection. Students are asked to rotate a figure about a point determined by the intersection of the figure's lines of symmetry. The second Discovery Activity shows that the angle of rotation is determined by the smallest angle formed by the lines of symmetry. In the Class Activity, students write LOGO procedures to draw symmetric figures. The Home Activity provides further practice in identifying symmetric figures, lines of symmetry, and angles of rotation. Students are asked to plot pairs of equations on graph paper and identify the type of transformation they illustrate.

▲ Section 12.4

Applications

Definitions and theorems about loci and transformations are applied to the solution of some practical problems. In the first Discovery Activity, students use locus to construct a parabola, a curve with many practical applications. The second Discovery Activity uses measurement to determine the shortest path that is not a straight line between two points. A reflection transformation is then used in the same problem to make the connection between a reflection and the fact that a straight line is indeed the shortest distance between two points.

The Class and Home Activities provide practice in the use of locus to determine locations on a map. Students write LOGO procedures to graph sets of points and equations, and then identify the transformations they illustrate. The Critical Thinking activity extends the discussion of line symmetry to the three-dimensional plane.

▲ Section 12.5

Logic for Decision Making

In determining loci in Section 12.1, students proved the truth of logical statements and their converses. This section continues the discussion of conditional statements to include the inverse and the contrapositive of a statement. Students learn that a statement can be shown to be true or false. In the Discovery Activity, students consider several statements and write the related converse, inverse, and contrapositive, determining a truth value for each. They are then asked to express the relationships between a statement and its contrapositive, and between the converse and the inverse. A discussion of logical statements and indirect proof follows, defining what is necessary to prove that a locus satisfies its conditions: the statement, its converse, and either the inverse or the contrapositive. In the Class Activity, students write the related logical statements, classify them as true or false, and then use a computer and a geometric drawing tool to justify their answers.

SECTION 12.1

Resources
Reteaching 12.1
Enrichment 12.1

Objectives
After completing this section, the student should understand
▲ the definition of geometric locus.
▲ how to determine a locus that satisfies given conditions.

Materials
penlights

Vocabulary
locus

GETTING STARTED

Warm-Up Activity
If possible, have several penlights available for this activity. Explain that you will use the light or the tip of a pen to represent one point. Move the point as directed below, and ask students to describe the path or geometric figure formed by the moving point.

1. Hold the point parallel to the wall and floor at shoulder height. Move the point horizontally about three feet. [Line segment]

2. Continue to move point as above for an infinite distance. [Line]

3. Move point around the room parallel to and at a set distance from the wall and the floor. [Rectangle or square]

4. Hold point steady at arm's length as you turn around once. [Circle]

5. Attach point to a ball thrown from one person to another. [Curve, arc]

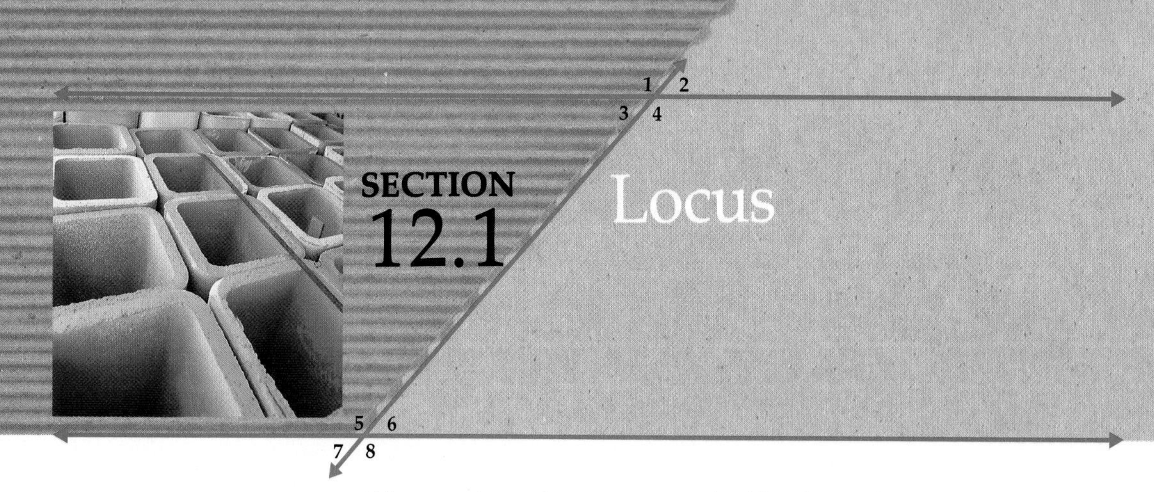

SECTION 12.1 Locus

After completing this section, you should understand
▲ the definition of geometric locus.
▲ how to determine a locus that satisfies given conditions.

The word **locus** is Latin, meaning place. In this chapter you will investigate geometric places, or **loci** (pronounced lō sī), that are determined by sets of points that satisfy given conditions. Loci can be two-dimensional or three-dimensional, on one plane or in space.

This square is a two-dimensional plane figure. Its sides are the locus of a set of points that satisfy the conditions for this particular square.

This drawing of a cube is a representation of a three-dimensional or space figure, such as a block. Its edges and faces are the locus of a set of points that satisfy the conditions for this particular cube.

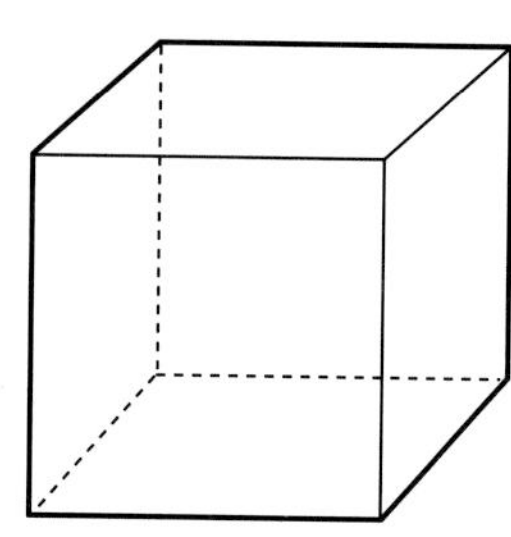

DEFINITION 12-1-1 **A geometric LOCUS is the set of all points that satisfy the given conditions.**

Example: What is the set of points on a football field that is 50 yards from each goal line?

The 50-yard line is the set of points that matches this condition. The 50-yard line is the locus.

To help you find a locus, use the following steps.

1. Read and reread the problem until you understand the conditions your locus must satisfy.
2. Locate several points that satisfy the conditions.
3. Do the points form a pattern? If not, locate several more.
4. What is the pattern?
5. Draw a solution set.
6. Write a statement describing the locus.
7. Try to prove your answer.
   a. Prove that every point in your proposed locus satisfies the conditions,
   and
   prove that every point that satisfies the conditions is a point in your locus;
   or
   b. Prove that every point that does not satisfy your proposed locus does not satisfy the conditions.

TEACHING COMMENTS

Using the Page
Be sure students understand what must be proven in each section of the two-part proof for a locus. Discuss the meaning of *and* and *or* in step 7.

DISCOVERY ACTIVITY

1. Use a ruler to draw a large angle. Label the angle ABC, with B as the vertex.
2. Locate five points that are equally distant from $\overrightarrow{BA}$ and $\overrightarrow{BC}$.
3. Imagine the number of points increasing to many points, and describe the pattern or set of these points.
4. What can you conclude about these points? Write a statement to describe the locus. They form the angle bisector.

Using the Page
You may wish to review the fact that the distance between the rays of the angle and the set of points is the perpendicular distance.

You may have concluded that the locus of points equidistant from the rays of an angle is the angle bisector. The next step is to prove that your locus satisfies the given conditions. You can write a two-part proof to show this.

Part 1: Any point D on the angle bisector is equidistant from $\overrightarrow{BA}$ and $\overrightarrow{BC}$.
Given: $\angle ABC$ with $\overrightarrow{BD}$ the angle bisector
Justify: D is equidistant from $\overrightarrow{BA}$ and $\overrightarrow{BC}$.

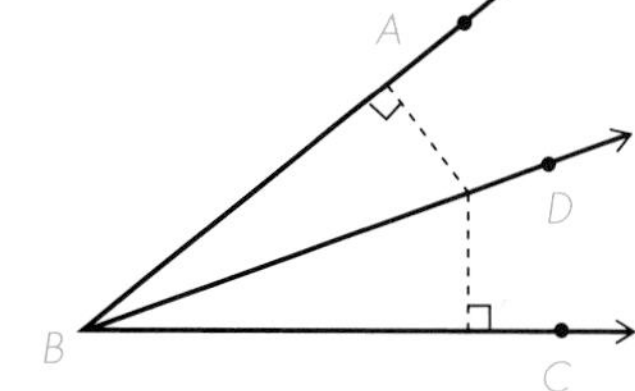

You know this is true from Theorem 6-6-3: If a point is on the angle bisector, then the point is equidistant from the sides of the angle.

Using the Page

Help students understand that, in using indirect reasoning in this proof, you assume the opposite of what you are trying to prove.

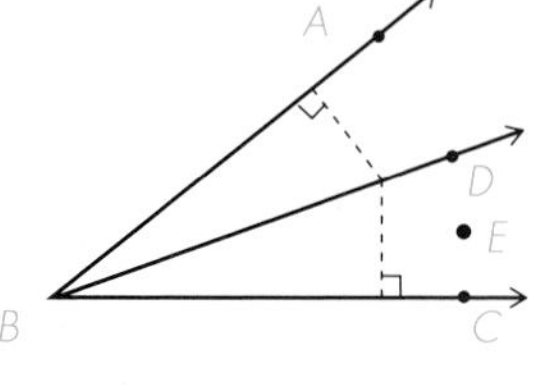

Part 2: Show that any point not equidistantfrom $\overrightarrow{BA}$ and $\overrightarrow{BC}$ is not on the angle bisectoror any point not on the angle bisector is not equidistant from $\overrightarrow{BA}$ and $\overrightarrow{BC}$.

Given: $\angle ABC$ with point E not equidistant from $\overrightarrow{BA}$ and $\overrightarrow{BC}$

Justify: Point E is not on the angle bisector.

You can use indirect reasoning to prove this. There are only two cases for point E: either E is on the angle bisector or E is not on the angle bisector. Assume that E is on the angle bisector. This means E is equidistant from $\overrightarrow{BA}$ and $\overrightarrow{BC}$, by Theorem 6-6-3. This contradicts the given information—that E is *not* equidistant from $\overrightarrow{BA}$ and $\overrightarrow{BC}$. Since there are only two cases for E, and since the case where E is on the angle bisector leads to a contradiction of the given, the other case must be true: point E is not on the angle bisector. The proof of the two parts justifies that the locus is the angle bisector.

THEOREM 12-1-1 **The locus of points equidistant from the sides of an angle is the angle bisector.**

CLASS ACTIVITY

1. Find the locus of points in a plane equidistant from the endpoints of a line segment. Draw the locus for a segment. Answers may vary.

2. Describe the locus in Exercise 1 by completing this statement: The locus of points equidistant from the endpoints of a line segment is the perpendicular bisector.

   Write a two-part proof to show that the locus you found in Exercise 1 satisfies the conditions.

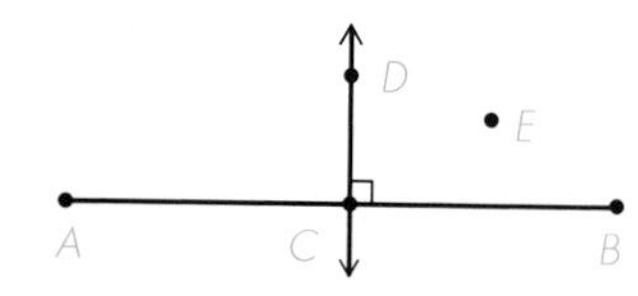

3. Given: $\overline{AB}$ with perpendicular bisector $\overleftrightarrow{CD}$
   Justify: D is equidistant from A and B.
   See margin.
4. Refer to the figure for Exercise 3.
   Given: $\overline{AB}$ with perpendicular bisector $\overline{CD}$ and point E not equidistant from A and B
   Justify: Point E is not on the perpendicular bisector. See margin.

Additional Answers

3. This is true by Theorem 6-6-2: If a point is on the perpendicular bisector, then it is equidistant from the endpoints of the segment.
4. Either point E is on the perpendicular bisector or it is not. If E is on the perpendicular bisector, then it is equidistant from A and B. This contradicts the given. Therefore, the other case must be true.

THEOREM 12-1-2 **The locus of points equidistant from the endpoints of a line segment is the perpendicular bisector of the line segment.**

MORE LOCUS THEOREMS

Finding the locus of points that satisfies certain conditions gives you an interesting way of looking at some familiar geometric figures. Each of the following procedures will result in a locus theorem that states in a new way information that you may already know.

1. Find the locus of points in a plane that are a set distance from a line.

   a. Locate four points above a line *AB* and four below the line that are the set distance (use 1 inch) from the line.
   b. Draw in the color locus for all possible points.
   c. Describe the locus. Two parallel lines

THEOREM 12-1-3 **The locus of points in a plane that are a given distance from a line on the plane is two parallel lines.**

2. What is the locus of points in a plane that are equidistant from a point on the plane?

   a. Mark a point at the center of a sheet of paper. Label it 0.
   b. Locate a dozen points 3 cm from point 0.
   c. Draw the locus for all possible points in color.
   d. Describe the locus. Circle

THEOREM 12-1-4 **The locus of points a given distance from a point on a plane is a circle.**

3. Follow the directions to find this locus.

   a. Draw a 3-inch line segment and label it *AB*.
   b. Through point *A*, draw three lines, with one line perpendicular to $\overline{AB}$.
   c. From point *B*, draw lines that are perpendicular to the three lines through *A*. Label the points of intersection, C_1, C_2, C_3.
   d. Repeat Steps b and c, drawing additional lines through *A*.
   e. Describe the locus. See margin.

THEOREM 12-1-5 **The locus of the vertex (*C*) of all right triangles with a given hypotenuse ($\overline{AB}$) is a circle with diameter $\overline{AB}$ minus the endpoints *A* and *B* of the hypotenuse.**

Using the Page
After the class has completed Exercises 1-3, you may wish to have several students draw diagrams of their loci on the chalkboard. Discuss with the class why these loci satisfy the given conditions for each exercise, and ask for an informal description of how students would prove that points not in the locus do not satisfy the given conditions.

Additional Answers
3e. Circle minus endpoints of $\overline{AB}$. Drawings may vary.

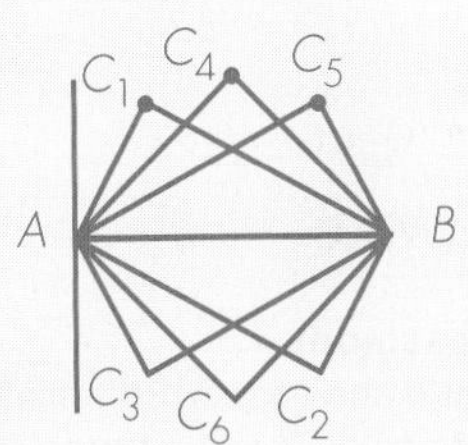

Using the Page
Ask students to explain why the theorem does not include endpoints *A* and *B* in the locus. [*A* and *B* are already part of the circle, since they are endpoints of the diameter $\overline{AB}$.]

REPORT 12-1-1 Find out what a cycloid is and describe its physical features. How is it used in amusement parks?

Suggested source:
Johnson, R.E., Kiokemeister, F.L. Calculus with Analytic Geometry. Boston: Allyn and Bacon, Inc., 1969.

HOME ACTIVITY

Follow the steps listed on page 445 to help you determine the following loci. Prove your conclusions only when you are asked to do so.

Find and describe the locus for Exercises 1–4. Then write a LOGO procedure to tell the turtle to draw the figure and the locus.
Answers for the procedure may vary.

1. Draw a 3-inch line segment, $\overline{AB}$. Find all points equidistant from A and B.
   ⊥ bisector
2. Draw $\angle ABC$. Find all points equidistant from $\overrightarrow{BA}$ and $\overrightarrow{BC}$.
   Angle bisector
3. Draw line AB. Find all points 3 centimeters from $\overleftrightarrow{AB}$.
   Two parallel lines
4. Find all points 0.75 inch from a point A.
   Circle with radius of 0.75 in.
5. Find the locus of points that are equidistant from two parallel lines.
   A line parallel to each and equidistant from each
6. What is the locus of all points in a plane that are equidistant from two intersecting lines?
   Two intersecting lines that are the angle bisectors
7. What is the locus of all points in a plane that are equidistant from a line segment?
   Two parallel lines and two semicircles

On graph paper, draw the locus.

8. $y = 3x + 1$
   Line
9. $y = |x|$
   Two rays from (0,0)
10. $y < x + 2$
    region or area
11. $x^2 + y^2 = 25$
    circle

CRITICAL THINKING

12. The conditions for Exercises 1, 3, 4, 5, and 7 can also be satisfied in three dimensions by space figures. Draw the three-dimensional picture and describe the locus.
    Plane, cylinder, sphere, plane, cylinder with rounded ends

Extra Practice
1. What is the locus of a doorknob as the door is opened through an angle of 90°? [Arc of a circle]
2. What is the locus of all points in a plane that are 1 inch from a point on the plane? [Circle with radius of 1 inch]
3. Describe the locus of all points in a plane less than 1 inch from a given point. [Area of a circle with radius of 1 inch]

FOLLOW-UP

Assessment
What is the locus of the midpoints of all 2-inch chords in a circle with a radius of 3 inches? [Circle]

Reteaching
Students who have had difficulty with this section may benefit from Reteaching Activity 12.1.

Enrichment
For students who have mastered the material in this section, you may wish to assign Enrichment Activity 12.1

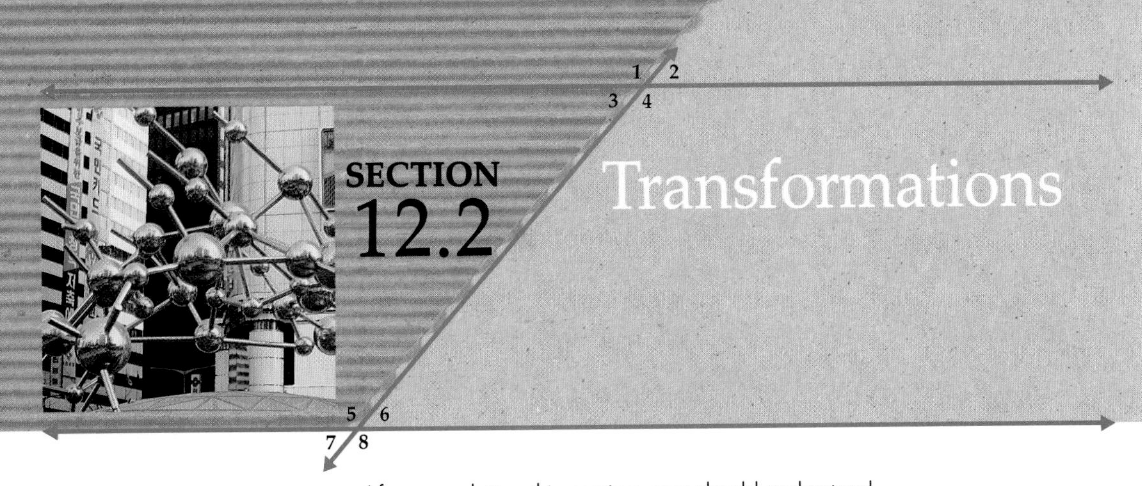

SECTION 12.2 Transformations

After completing this section, you should understand
- ▲ how to find the translation image of a figure.
- ▲ how to find the reflection of a figure and locate the line of reflection.
- ▲ how to find the rotation image of a figure.

The meaning of the word **transformation** implies change—of size, shape, or position. In this section, you will be investigating transformations of geometric figures. These transformations may change a figure's position or size, but they will not change the shape.

DEFINITION 12-2-1 **A TRANSFORMATION is a way of creating a mapping or image of a geometric figure in a plane, preserving shape but not necessarily size.**

We will investigate three types of transformations: translations, reflections, and rotations.

TRANSLATIONS

A **translation** can be thought of as a sliding of a figure to a new position.

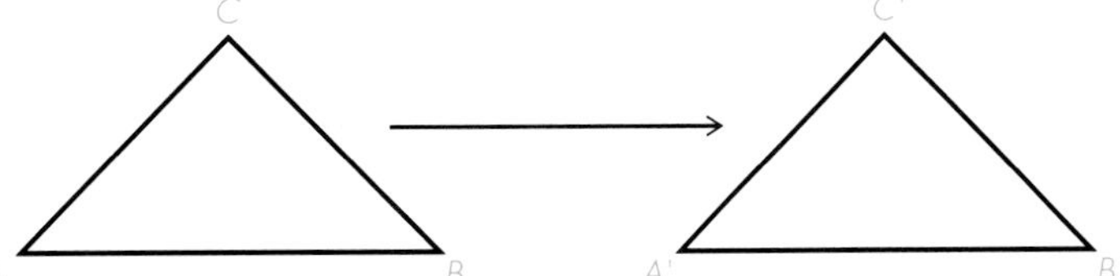

DEFINITION 12-2-2 **A TRANSLATION is a transformation that slides the figure from one position to another, preserving congruency.**

SECTION 12.2

Resources
Reteaching 12.2
Enrichment 12.2
Transparency Masters 12-1, 12-2

Objectives
After completing this section, the student should understand
- ▲ how to find the translation image of a figure.
- ▲ how to find the reflection of a figure and locate the line of reflection.
- ▲ how to find the rotation image of a figure.

Materials
tracing paper
graph paper
pin or compass
magnifying and pocket mirrors or reflective plastic
two small hard balls (golf balls)
two 4″ or 6″ lengths of wood
protractor

Vocabulary
transformation
translation
reflection
rotation

GETTING STARTED

Warm-Up Activity
Demonstrate several of the following transformations, using objects available in the classroom or chalkboard drawings. Have students give an informal description of the path the object follows and of how the object may have changed after the transformation (size, shape, position).

1. Desk drawer, pulled straight out
2. Edge of window shade, raised or lowered
3. Edge of a book page, turned
4. Door, opened or shut
5. Book flat on desk, turned over

6. Chair, tilted onto back legs
7. Chair, lifted diagonally onto desk or table
8. Any small object, reflected in a pocket mirror
9. Any small object, reflected in a magnifying mirror

From the definition of transformation, you know that the shape of the translated figure will be the same as the shape of the original figure.

Example: $A'B'C'D'$ is the translation image of rectangle $ABCD$.

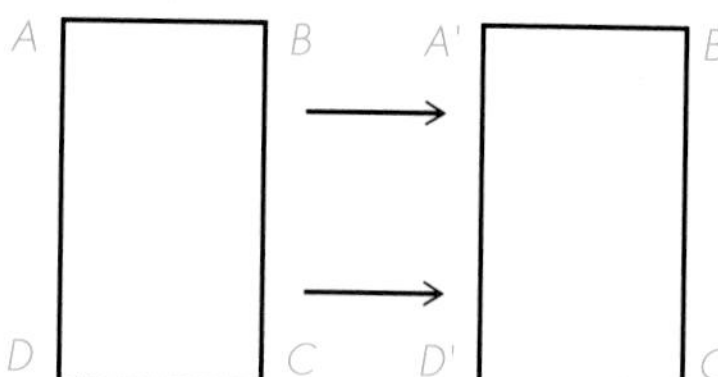

Measure the distance between A and A', B and B', C and C', and D and D'. What do you observe? Distances are equal.

DISCOVERY ACTIVITY

1. On one half of a sheet of tracing paper, draw a concave quadrilateral. Label the vertices $ABCD$.
2. Fold the paper in half and trace the figure on the other half. Label this figure $A'B'C'D'$.
3. Measure the distance between A and the fold line. Then measure the distance between points A and A'. What do you notice? Fold bisects AA'.
4. Do the same for BB', CC', and DD'. What do you notice about the relationship of the fold line to each of these line segments? Fold line is $\perp$ bisector.
5. Find the midpoint of one of the sides of $ABCD$. Label it E. Then locate E' on $A'B'C'D'$. Fold the paper along the fold line. Where does point E fall? On E'
6. Draw $\overline{AA'}$, $\overline{BB'}$, $\overline{CC'}$, $\overline{DD'}$, and $\overline{EE'}$. What can you observe about these line segments? Parallel to each other.

TEACHING COMMENTS

Using the Page
After the Discovery Activity, ask students to describe how the angles in their reflected quadrilaterals compare with the angles in the original figure. [Mirror image; should seem congruent]

Using the Page
Ask for other examples of reflection transformations that may change the size of a figure or image. [Examples: copies enlarged or reduced in a copier; photographic enlargements; objects seen through a magnifying glass, telescope, or microscope]

REFLECTIONS

You may have found that when a figure is reflected across a line, the line of reflection is the perpendicular bisector of line segments joining corresponding points on the figure and its reflection.

A REFLECTION is a transformation such that a line (mirror) is the perpendicular bisector of the segments joining corresponding points.

You can see why the reflection of a figure is often called a mirror image. This type of transformation can also illustrate why the phrase "preserving shape but not necessarily size" is part of the definition. Some mirrors magnify a reflected image, and this is still a transformation.

Example: $\overline{A'B'}$ is a reflection of $\overline{AB}$ across line m. What do you think might be true of $\overline{AA'}$ and $\overline{BB'}$? They are parallel.

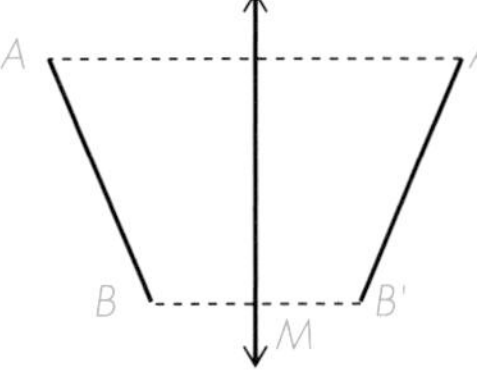

Example: $\triangle ABC$ is reflected across line m. Its image is $\triangle A'B'C'$.

For $\triangle ABC$ and $\triangle A'B'C'$, do you think $\overline{AB} \cong \overline{A'B'}$, $\overline{BC} \cong \overline{B'C'}$, and $\overline{AC} \cong \overline{A'C'}$? Why? Yes; shape is preserved in a reflection.

What do you think might be true for $m\angle ABC$ and $m\angle A'B'C'$? How could you show this? $m\angle ABC = m\angle A'B'C'$; trace or cut out figure and place one triangle over the other.

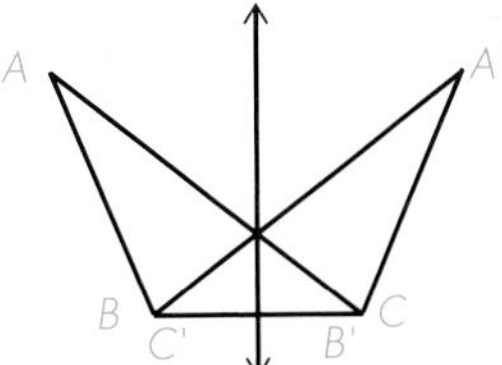

CLASS ACTIVITY

Tell whether each transformation is a translation or a reflection.

1. 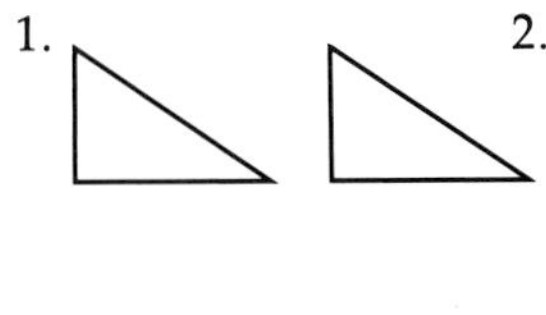
Translation

2. 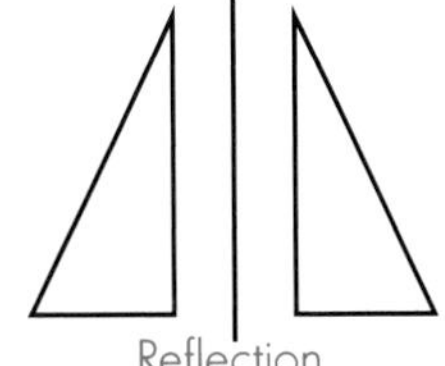
Reflection

3. 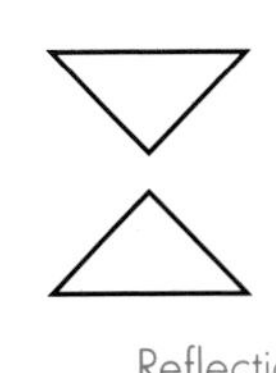
Reflection

4. Name the reflection of points P, Q, and S across line m.
   A, B, F

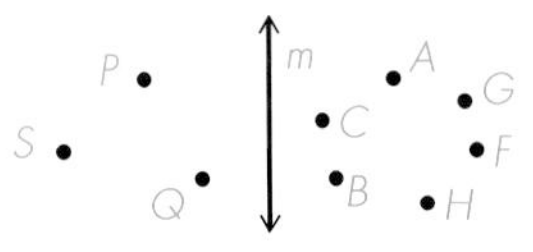

5. Copy the figure and show its reflection across line m. Then write a LOGO procedure to tell the turtle to draw the figure and its reflection across line m.
   Answers for the procedure may vary.

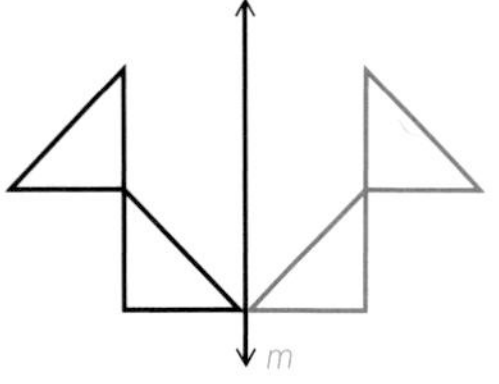

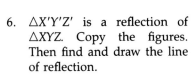

6. $\triangle X'Y'Z'$ is a reflection of $\triangle XYZ$. Copy the figures. Then find and draw the line of reflection.

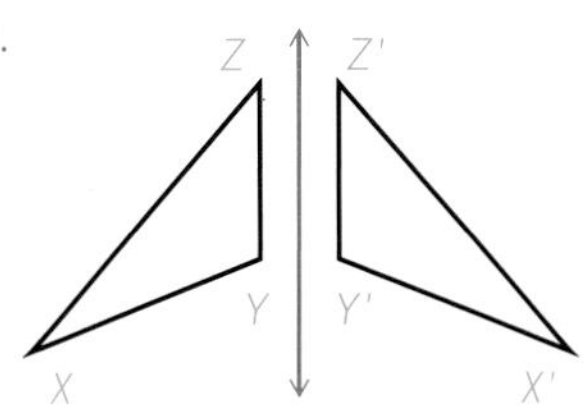

Using the Page
You may wish to have some students trace $\triangle ABC$ and place it over $\triangle A'B'C'$ to illustrate that the angles are congruent. Transparency Masters 12-1 and 12-2 may be used to show other examples of reflection transformations.

Extra Practice
Tell whether the transformation is a translation or a reflection.

1.

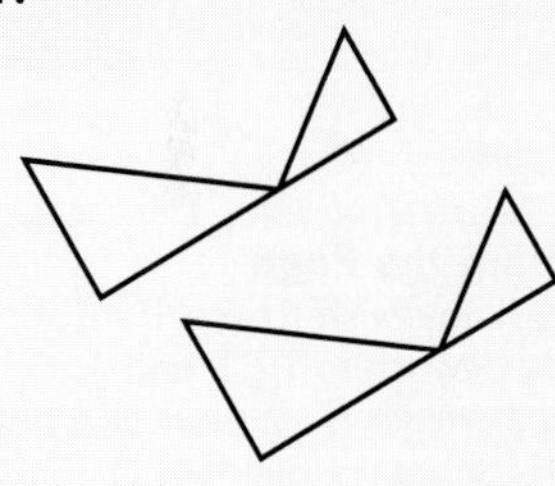

[Translation]

2.

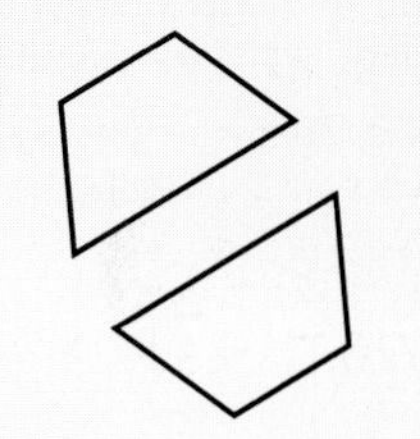

[Reflection]

3.

[Reflection]

DISCOVERY ACTIVITY

1. Draw a right triangle on a sheet of graph paper. Label it *ABC*.
2. Trace $\triangle ABC$ on a sheet of tracing paper and label the tracing $\triangle A'B'C'$. With a pin or the point of a compass, hold point A' on top of *A* and rotate $\triangle A'B'C'$ 90° clockwise. Mark points $A'B'C'$ on the graph paper.
3. Rotate $\triangle A'B'C'$ through another 90° and mark the points $A''B''C''$. Do this one more time.
4. What do you notice about $\triangle ABC$ and $\triangle A'B'C'$ after three rotations?
The figures coincide.

Using the Page
You may wish to have students use a protractor to measure the 90° rotations.

ROTATIONS

You may have found that a rotation through 360° brings a figure back to its original position.

DEFINITION 12-2-4 **A ROTATION is a transformation that maps each point in a figure *A* onto the corresponding point in figure *A'* by revolving figure *A* about a point.**

Using the Page
You may wish to use Transparency Masters 12.1 and 12.2 to illustrate a rotation of a figure about a point that is contained in the figure and a rotation of a figure about a point that is elsewhere in the plane.

Example: $\triangle A'B'C'$ is the image of $\triangle ABC$ under a rotation of 45° about point *P*.

The center of rotation can be anywhere in a plane.

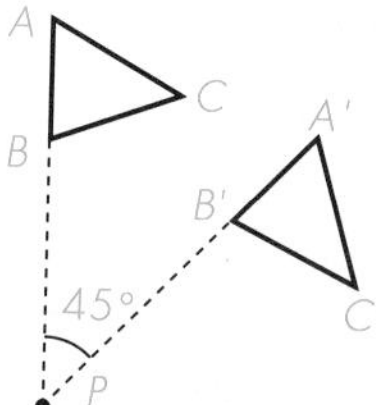

CLASS ACTIVITY

Identify the transformation as a translation, a reflection, or a rotation.

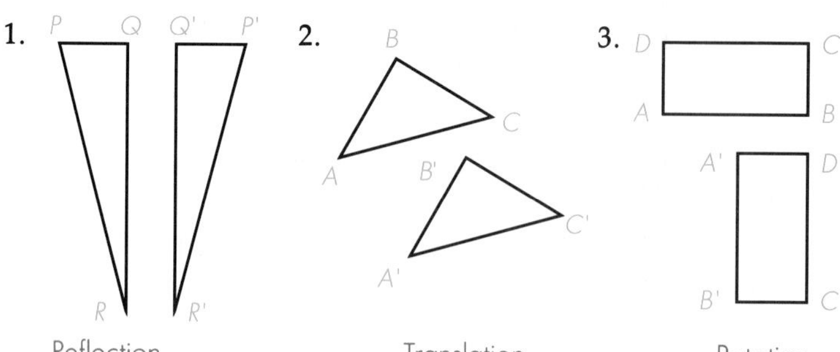

Identify each transformation.

4.
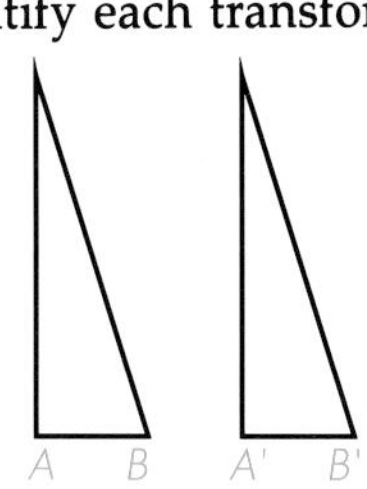

Translation

5.
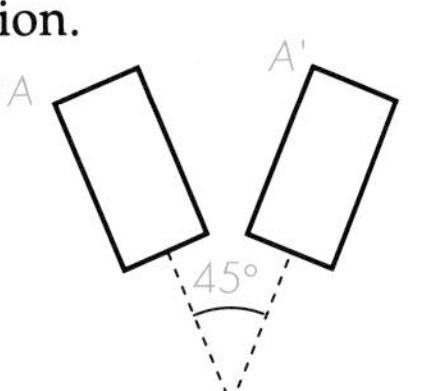

Rotation 45° clockwise

6.
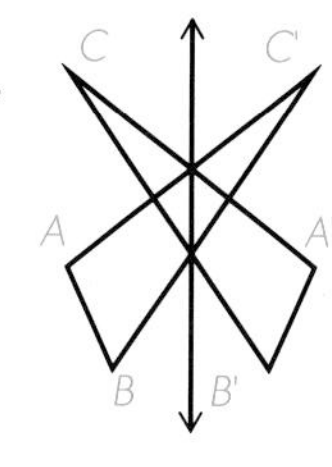

Reflection

REFLECTIONS AND A GAME

The drawing shows a billiard table with a white ball and a red ball. A player wants to hit the white ball and cause it to strike the red ball. But the white ball must first hit side AB of the table before it hits the red ball. To do this, the player uses reflections.

The player visually estimates the point R'. R' is the reflection of R across $\overline{AB}$. The player then aims at R'. The white ball hits the table at T and rebounds to strike the red ball. The path of the white ball after it hits the table at T is the reflection of the straight line TR' across $\overline{AB}$.

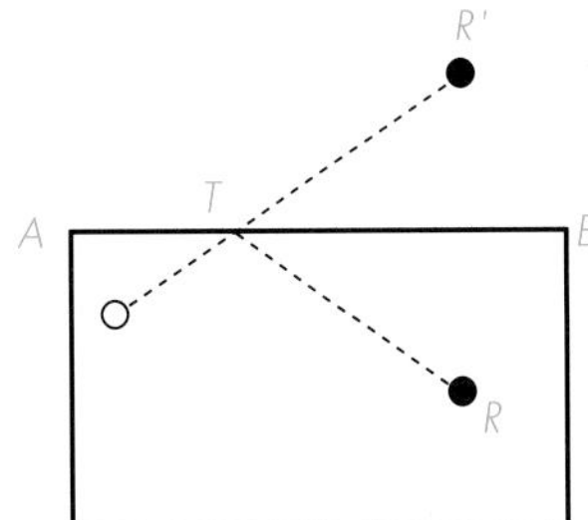

Using the Page
If you have two small hard balls (such as golf balls) available, you may wish to have students experiment with reflection. Set up two pieces of wood or two large books to form a right-angled corner and place the balls within the angle. Have students work in pairs. One rolls a ball toward the side of the angle, aiming to reflect it so that it will strike the other ball. The other uses chalk to mark the path the ball would follow if the barrier were not there. They can then find the reflection of the second ball across the barrier, correct their aim, and try again.

HOME ACTIVITY

Identify the transformation as a translation, a reflection, or a rotation.

1.
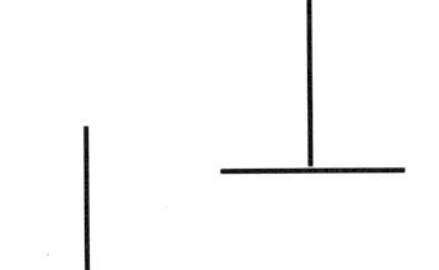
Translation

2.
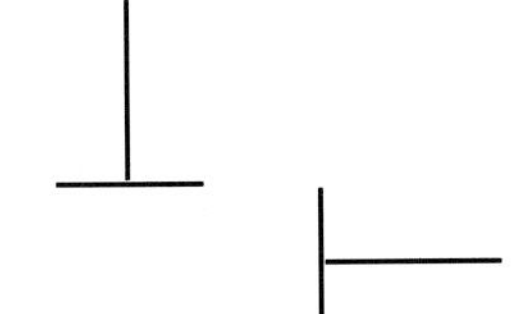
Rotation

3.
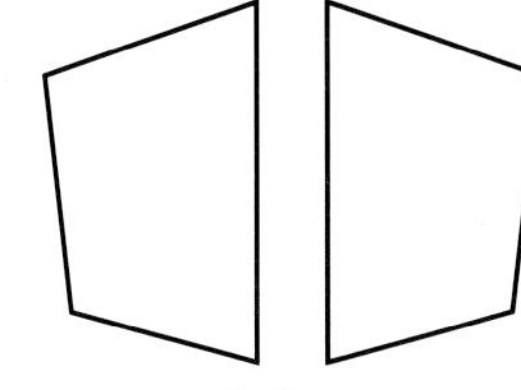
Reflection

4.
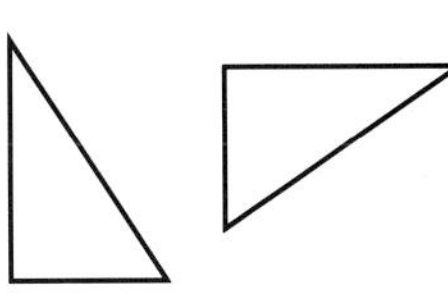
Rotation

5.
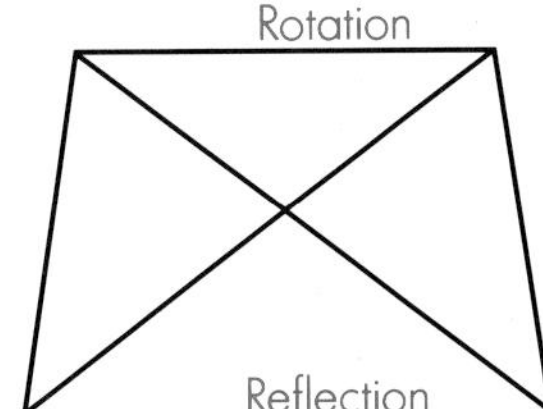
Reflection

6.
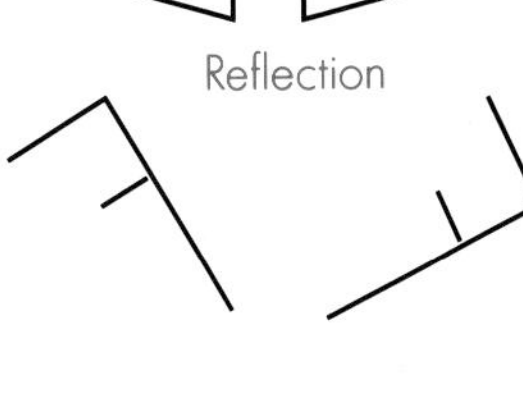
Rotation

Additional Answers

7. Vertical: A, H, I, M, O, T, U, V, W, X, Y; horizontal: B, C, D, E, H, I, O, X

13. Translation 2 units up

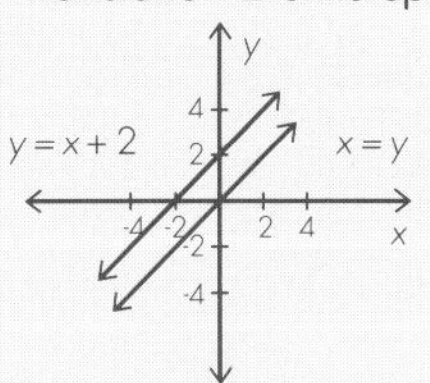

14. Rotation of 90° about (0, 0)

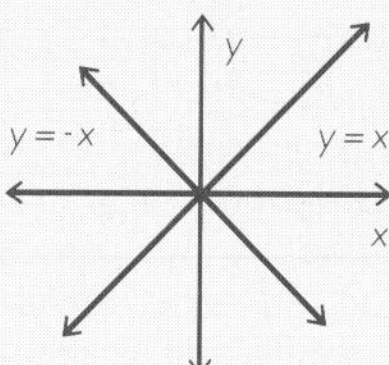

15. Reflection over x-axis

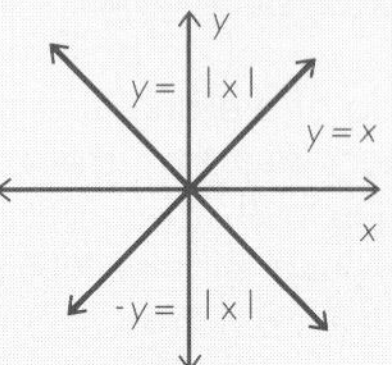

FOLLOW-UP

Assessment

Copy the figures. Complete each transformation.

1. Reflection over line m

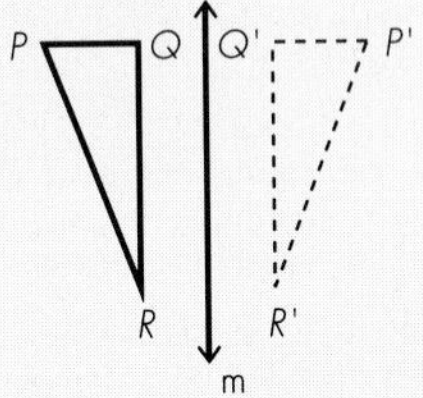

2. Rotation 60° clockwise about point x

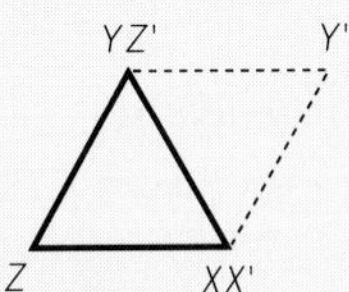

Reteaching

Students who have had difficulty with this section may benefit from Reteaching Activity 12.2.

Enrichment

For students who have mastered the material in this section, you may wish to assign Enrichment Activity 12.2.

7. List the capital letters of the alphabet that have the same image when reflected across a line. Tell whether the line of reflection is vertical or horizontal. See margin.

8. Copy the figure and draw its reflection across line *m*. Then write a LOGO procedure to tell the turtle to draw the figure and its reflection across line *m*. Answers for the procedure may vary.

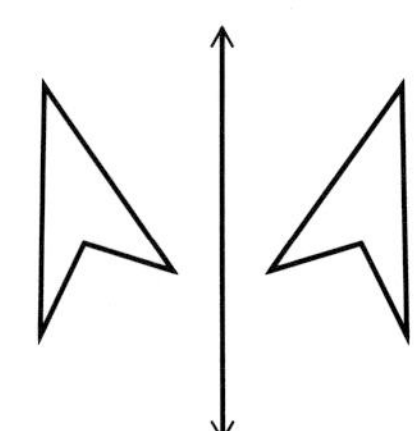

9. Use a reflection to determine where ball *W* should hit side *CD* in order to hit ball *R*. (Trace the diagram.)

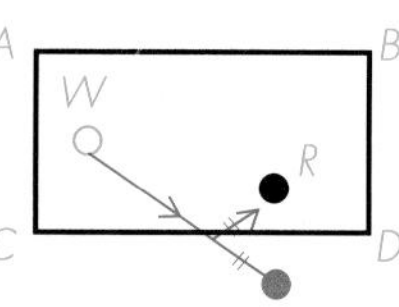

10. Make a construction to find the point where ball *W* should strike side $\overline{AC}$, then side $\overline{AB}$, and then rebound to hit ball *R*. Hint: Reflect both balls and then draw the straight line.

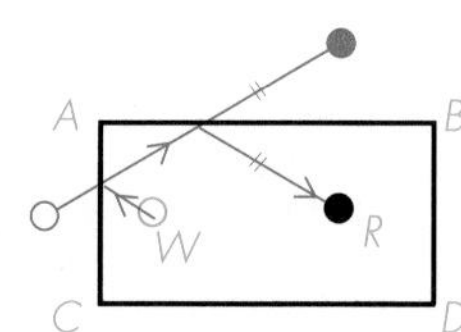

11. a. Construct an equilateral triangle with sides of 1 inch. Label it *ABC*.
    b. Rotate the triangle 60° counterclockwise about vertex *A*. Mark the points of the vertices *B* and *C*.
    c. Rotate the triangle another 60° and mark the vertices.
    d. Continue the 60° rotation until a regular polygon is formed. What is the name of the polygon? Hexagon
    e. How many rotations were there? 5

12. Repeat the steps in Exercise 11 for a 1-inch square and a rotation angle of 90°. Square; 3 rotations

Plot the following equations on graph paper. Name the type of transformation.

13. $y = x$ and $y = x + 2$ See margin.

14. $y = x$ and $y = -x$ See margin.

CRITICAL THINKING

Trace $\triangle XYZ$ and parallel lines *m* and *m'*. Reflect $\triangle XYZ$ across line *m*, then across line *m'*.

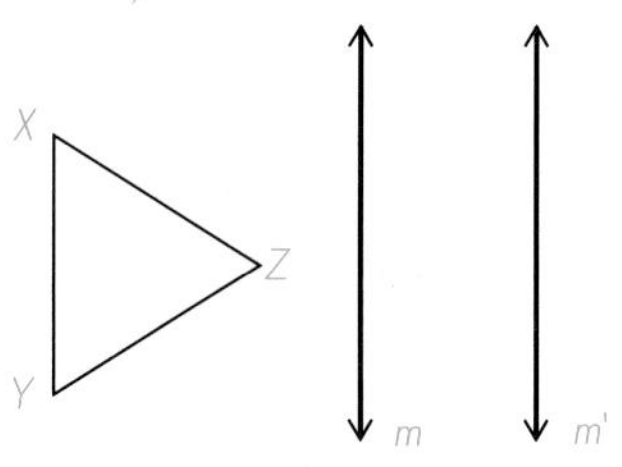

15. What can you conclude about the way the final reflection is related to the original figure? Not a mirror image

16. What other type of transformation does this show? Translation

Trace the flag and intersecting lines *m* and *m'*. Reflect the flag across line *m*, then across line *m'*.

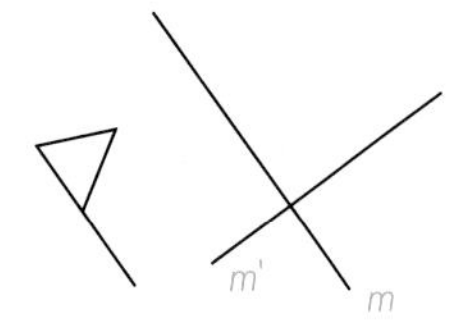

17. What can you conclude about the way the final reflection is related to the original figure? Not a mirror image

18. What other type of transformation does this show? Rotation about a point

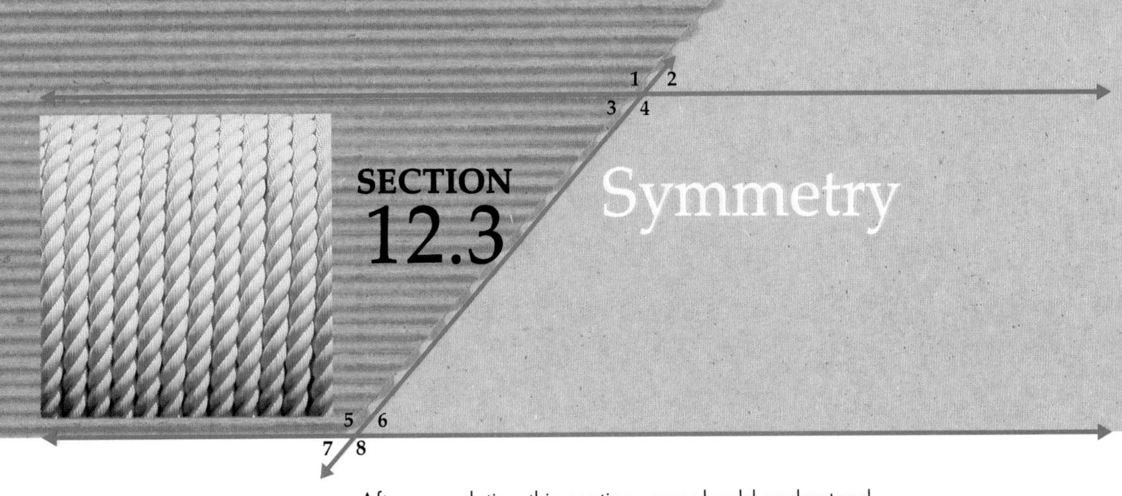

After completing this section, you should understand
- ▲ how to determine whether a figure has line symmetry.
- ▲ how to find lines of symmetry.
- ▲ how to determine whether a figure has point or rotation symmetry.

A special kind of transformation occurs when a figure is mapped onto itself. In the picture above, the figure is reflected across a line; the part on the left side is reflected on the right side.

DISCOVERY ACTIVITY

1. Fold a sheet of paper in half and label the halves A and B.
2. Beginning at the fold, draw a design or picture on half A. End the drawing at some point on the fold.
3. Now fold the sheet, hold it to the light, and trace the design on the back of half B. Open the sheet and trace the design on the front side of B.
4. What do you notice about the two halves of the design? Mirror images
5. Is this transformation a translation, a reflection, or a rotation? How do you know? Reflection; figure reflected over a line

You may have found that you reproduced, or made an exact copy of, your original design from side A on side B.

DEFINITION 12-3-1 **SYMMETRY is a transformation that maps a figure onto itself.**

SECTION 12.3

Resources
Reteaching 12.3
Enrichment 12.3

Objectives
After completing this section, the student should understand
- ▲ how to determine whether a figure has line symmetry.
- ▲ how to find lines of symmetry.
- ▲ how to determine whether a figure has point or rotation symmetry.

Materials
tracing paper
pin or compass
protractor

Vocabulary
line symmetry
point symmetry
rotation symmetry
symmetry

GETTING STARTED

Warm-up Activity
Give students oral examples of various transformations and have them tell whether they are translations, reflections, or rotations. Examples: car going through a car wash on a moving track; door opening; a flipped coin; jet taxiing down a runway; wheel turning; pancake turned over on a griddle.

The word symmetry comes from the Greek words meaning same measure.

In Section 12.2 you investigated three kinds of transformations.

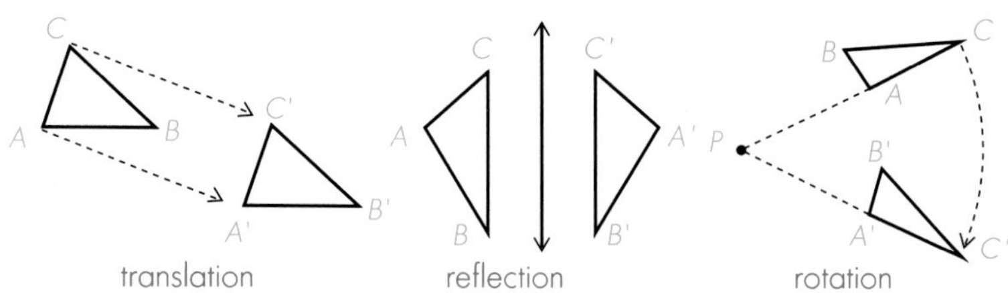

Symmetry is a reflection transformation. There are two types of symmetries—line symmetry and point, or rotation, symmetry.

DEFINITION 12-3-2 **LINE SYMMETRY is a reflection transformation that maps a figure onto itself.**

These designs are symmetric with respect to a line. The set of points on one side of the line is a reflection of the set of points on the other side.

If you look at your surroundings, you can see that line symmetry occurs in nature, art, architecture, and living things. Symmetry is involved in the design of most of the objects you see and use daily—furniture, clothing, utensils, tools, buildings, methods of transportation, and so on.

Some figures have only one line of symmetry. Others may have more than one or many lines of symmetry.

Example: Dotted lines mark the lines of symmetry in this regular octagon.

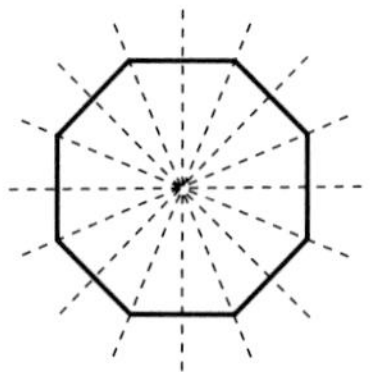

TEACHING COMMENTS

Using the Page

During the discussion of line symmetry, it may be helpful to make a distinction between reflection and symmetry. A reflection is a mirror image of a figure; it results in two images or one overlapping image that is symmetric across a line. For example, this triangle is not itself symmetric, but its original and reflected images are symmetric.

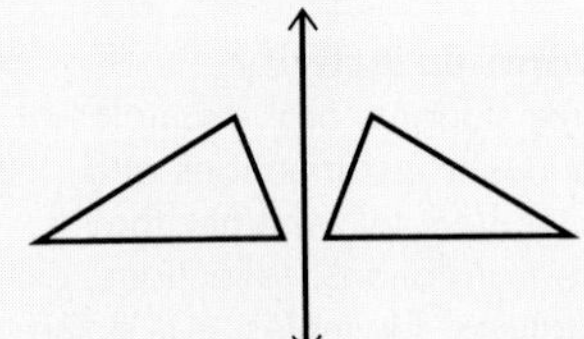

A symmetric figure is one with identical halves (thirds, fourths, and so on). For example, this figure is symmetric and contains its own lines of symmetry.

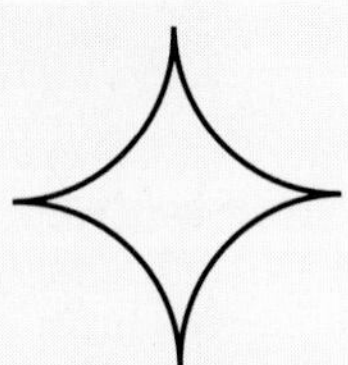

CLASS ACTIVITY

Tell which figures have line symmetry. Then find all lines of symmetry. Symmetry: 1–4, 6–8

1. 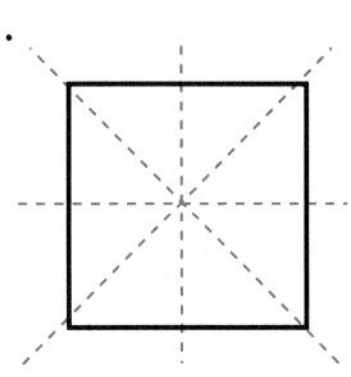 4

2. 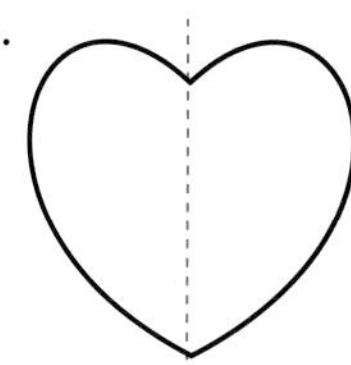 1

3. 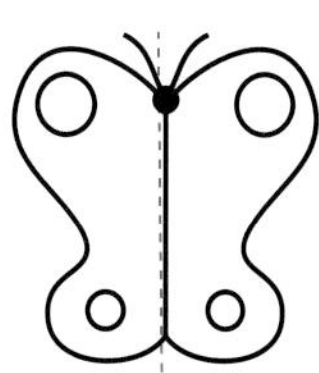 1

4. 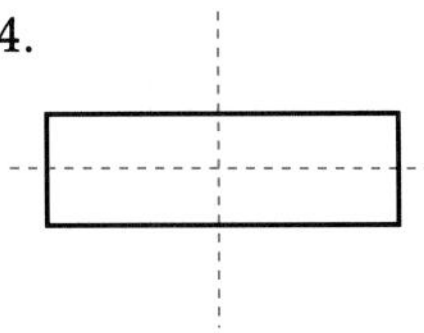 2

5. 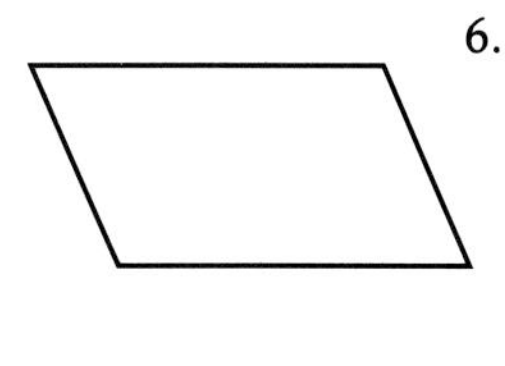 0

6. 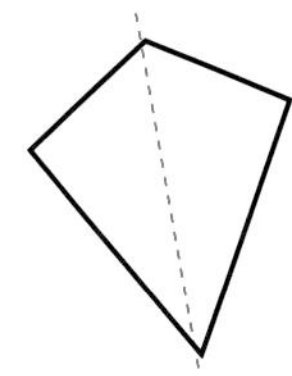 1

7. 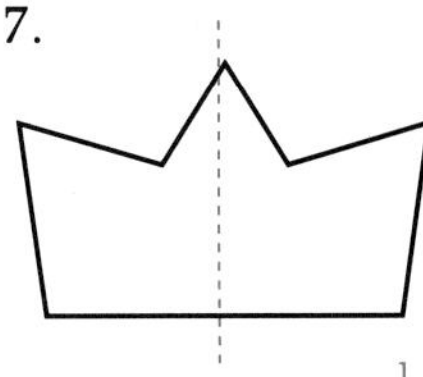 1

8. 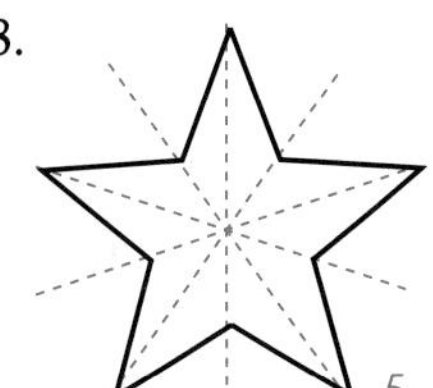 5

9. 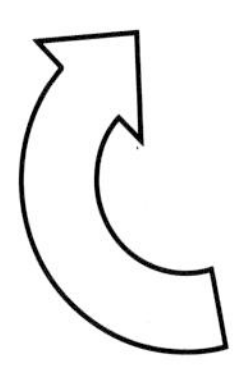 0

POINT SYMMETRY

Some figures can be rotated to map the figure onto itself.

DISCOVERY ACTIVITY

1. Draw an equilateral triangle *ABC*. Find and draw the lines of symmetry. Mark the point of intersection O.
2. What is the measure of the smallest angle between any two lines of symmetry at point O? Why? 120°; point O is on angle bisector.
3. Trace $\triangle ABC$ and its lines of symmetry on a sheet of tracing paper. Label the tracing $A'B'C'$.
4. Place $\triangle A'B'C'$ over $\triangle ABC$ and use a pin or compass point to hold the figures at point O.
5. Through how many degrees must you rotate $\triangle A'B'C'$ to map the triangle onto itself? 120°
6. What happens when you rotate $\triangle A'B'C'$ through only 60°? Is the figure mapped onto itself for a 60° angle of rotation? No

Using the Page
You may wish to have students draw and trace an isosceles triangle and test it for lines of symmetry. Through how many degrees must it be rotated before it is mapped onto itself? [360°]

Extra Practice
1. List the capital letters in the alphabet that have line symmetry. [A, B, C, D, E, H, I, K, M, O, T, U, V, W, X, Y]
2. Draw the lines of symmetry for the figure shown.

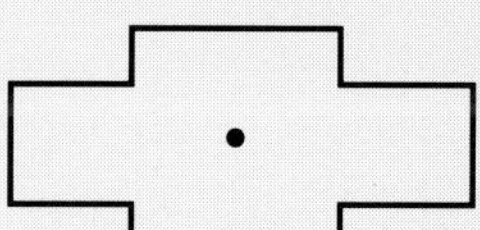

[A vertical and horizontal line through the point]

7. What can you conclude about the size of the angle of rotation that will provide point symmetry for $\triangle ABC$? Angle must be 120°, 240°, or 360°.

You may have found that an equilateral triangle will be mapped onto itself after a rotation of 120°.

Using the Page
The definition of rotation symmetry is often limited to rotations > 0° and < 360°. For example, the figure below would not have rotation or point symmetry because it must be rotated 360° before it is mapped onto itself.

Help students notice that a figure must have at least two lines of symmetry in order to have point or rotation symmetry.

POINT SYMMETRY or ROTATION SYMMETRY is a rotation transformation that maps a figure onto itself. The point is determined by the intersection of two or more line symmetries. The angle of rotation is determined by the angle formed by the line symmetries or a multiple of the angle.

Example: This square has four line symmetries. The smallest angle of rotation is 90°. Why? 360° ÷ 4 = 90° What if the square is only rotated 45°? Will it have point symmetry? Copy the figure and try it. No Will the square have point symmetry if it is rotated 180°? Why? Yes; 180° = 90° · 2

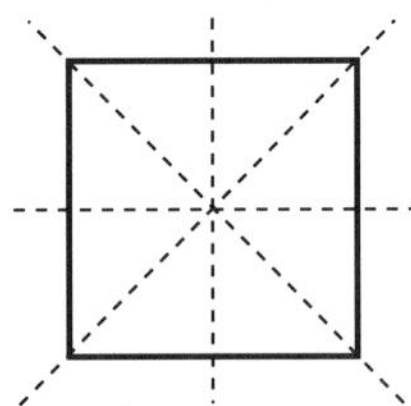

Using the Page
Tell students to draw first the one part of each figure that is repeated throughout the figure. Then they should use this procedure in a second procedure to draw the entire figure.

CLASS ACTIVITY

Each figure below has point symmetry. Write a LOGO procedure to tell the turtle to draw each figure. See margin. Answers may vary.

1.

2. 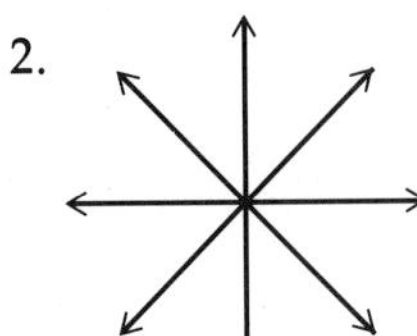

3. 

4. Which figures have point symmetry? a, c

a. b. c. d.

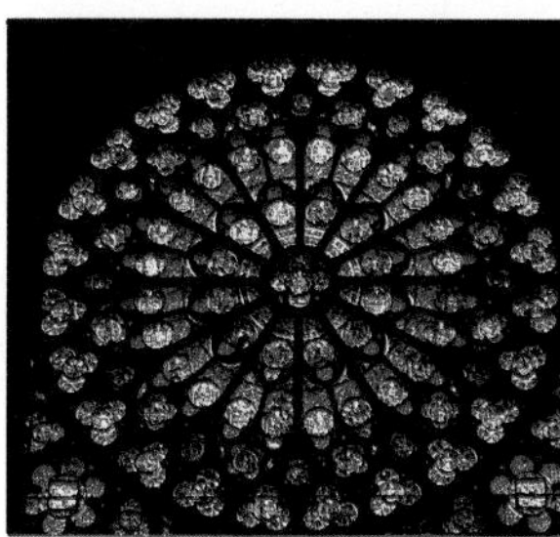

Find all lines of symmetry in each figure below. Write a LOGO procedure to tell the turtle to draw the figure and the line(s) of symmetry. Answers for the procedures may vary.

5. 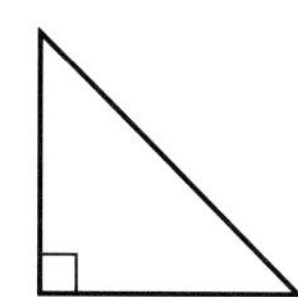

6. 

7. 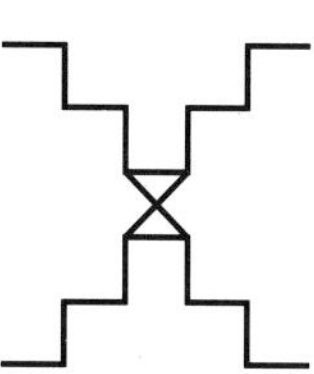

HOME ACTIVITY

1. Copy the figures below. Classify each shape and label it.

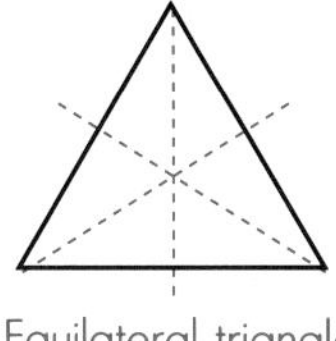

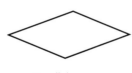

Equilateral triangle

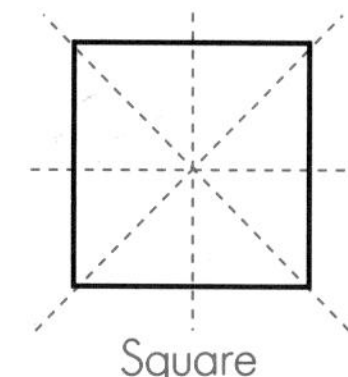

Square

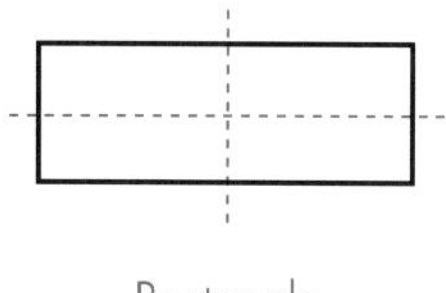

Rectangle

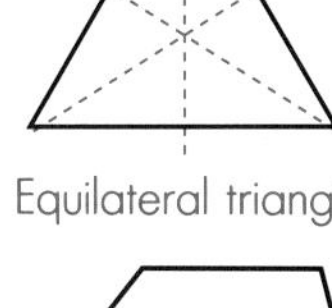

Trapezoid

Parallelogram

Circle

2. Identify the figures in Exercise 1 that have line symmetry. Draw the line(s) of symmetry.
Triangle, square, rectangle, circle
3. Identify the figures in Exercise 1 that have point symmetry. Indicate the point of rotation. Find the smallest angle of rotation. Triangle: 120°; square: 90°; circle: angle approaches 0°

On graph paper, plot the equations. Identify each as a translation, reflection, or rotation.
See margin for graphs.

4. $y = 2x + 2$ and $y = 2x + 1$ Translation
5. $y = 4x$, $y = x$, and $y = \frac{1}{4}x$ Reflection
6. $y = \frac{1}{2}x$, and $y = \frac{1}{2}x + 2$ Translation
7. $y = x^2$ and $x = y^2$ Rotation
8. Segment with endpoints (−4, 2) and (−2, 4) and segment with endpoints (2, 4) and (4, 2).
Reflection

Additional Answers

4.

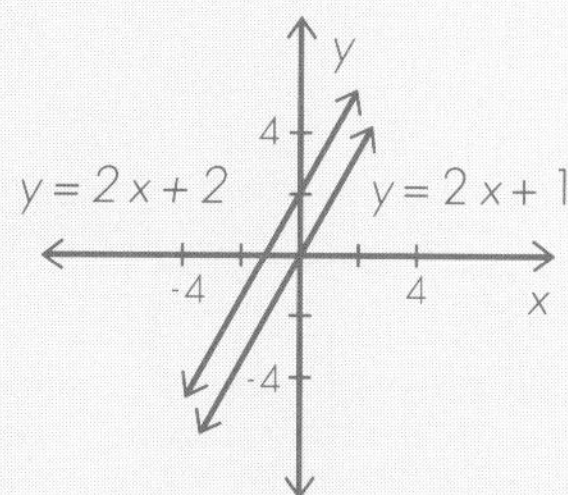

5.

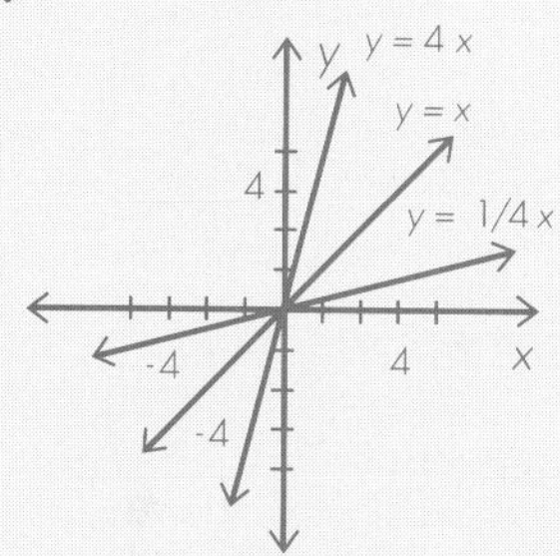

6.

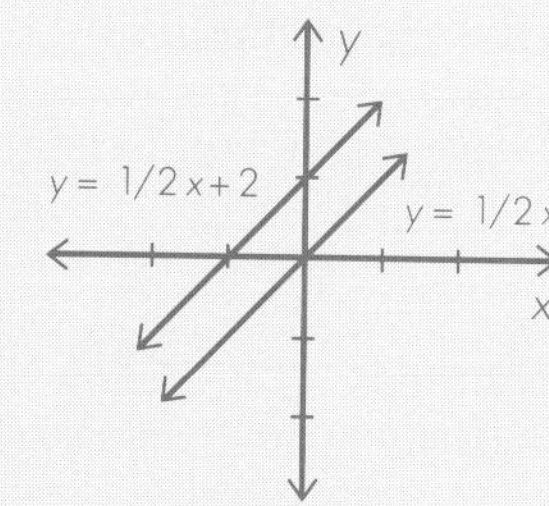

7.

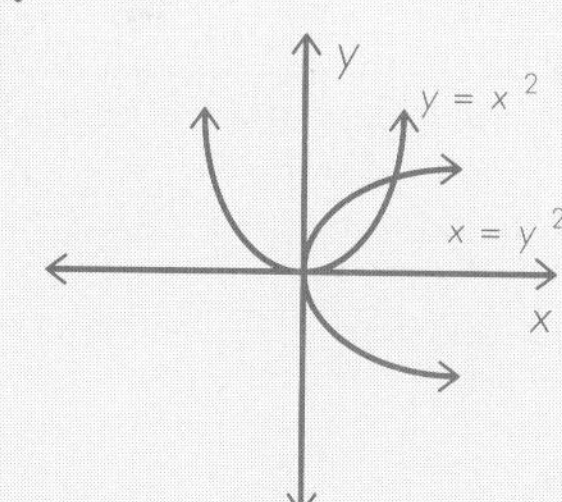

8.

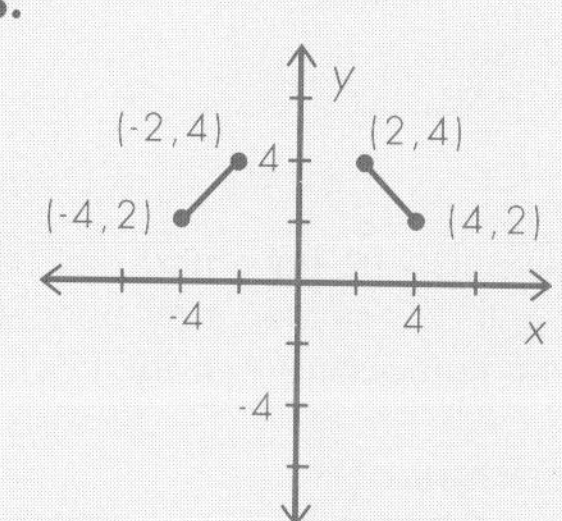

Additional Answers

11. Square: 90°; circle: approaches 0°; triangle: 120°; octagon: 45°

14.

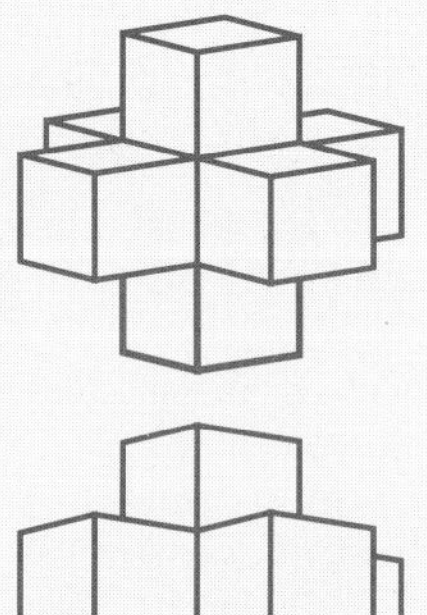

FOLLOW-UP

Assessment

Copy the figures. Identify those with line and point symmetry.

1. [Line symmetry]

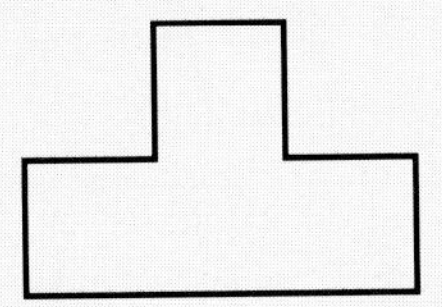

2. [Line and point symmetry]

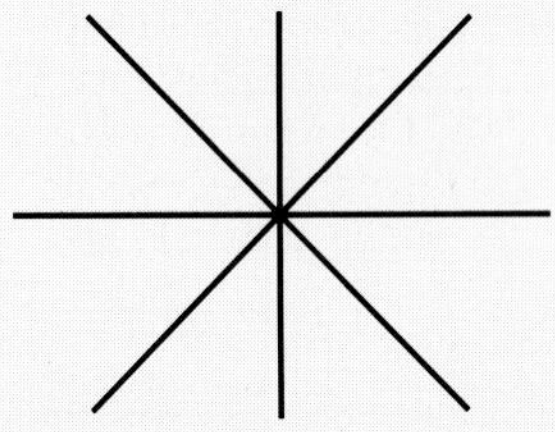

[Line and point symmetry] [No symmetry]

3. **4.**

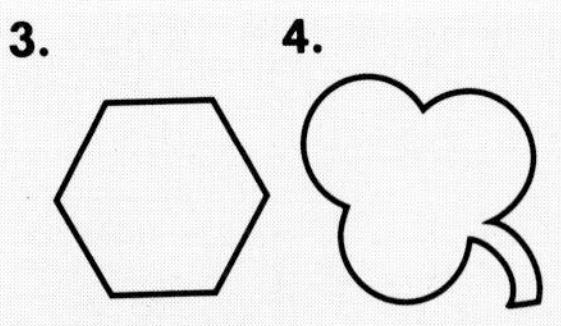

Reteaching

Students who have had difficulty with this section may benefit from Reteaching Activity 12.3.

Enrichment

For students who have mastered the material in this section, you may wish to assign Enrichment Activity 12.3.

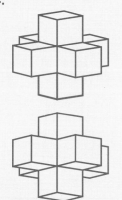

Below are the shapes of some common traffic signs.

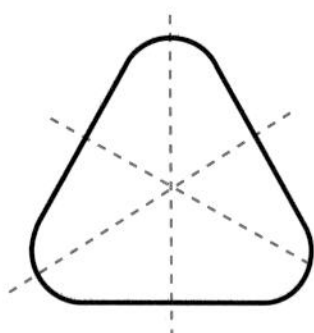
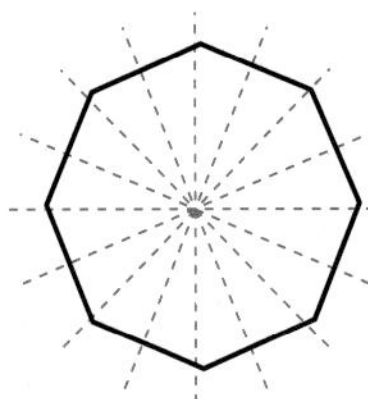
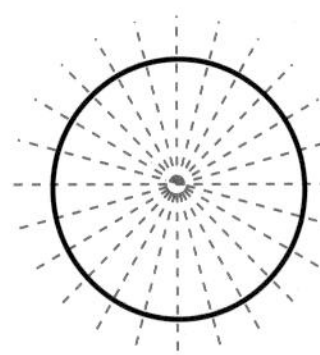
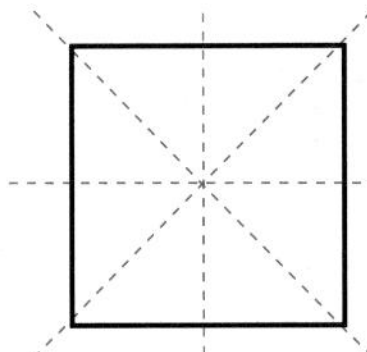
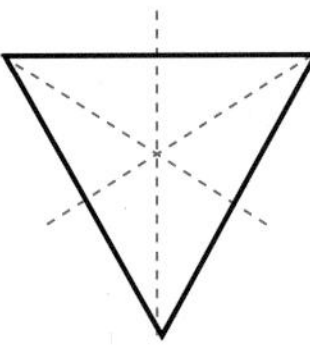

9. Which signs have line symmetry? Draw the lines of symmetry. All

10. Which signs have point symmetry? All

11. What is the angle of rotation for each figure with point symmetry? See margin.

The following objects are found in nature.

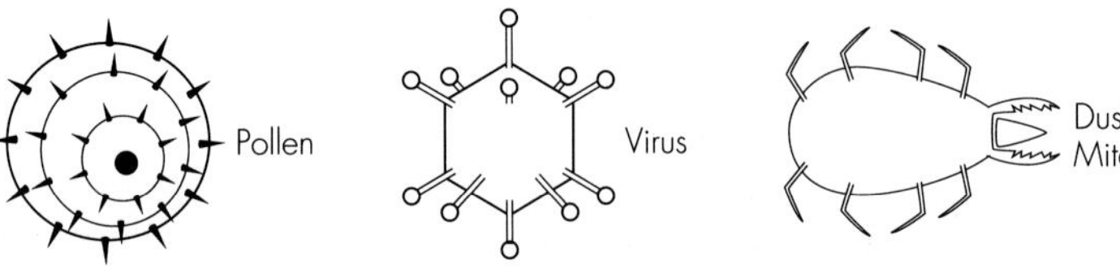

12. Which ones have line symmetry? All

13. Which ones have point symmetry? Pollen and virus

CRITICAL THINKING

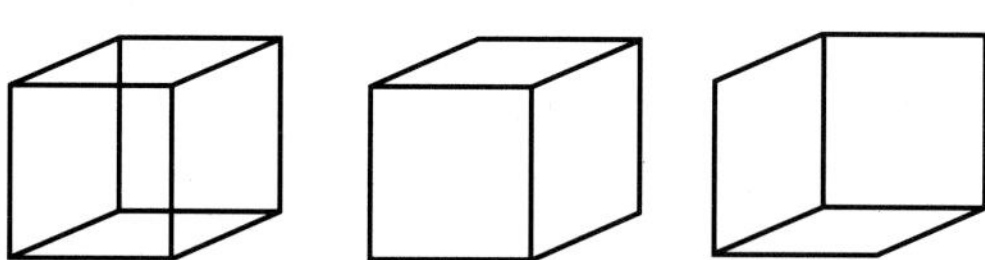

These three drawings represent a solid cube. The first drawing shows all the edges, which makes the cube harder to recognize. The second and third drawings show the cube with the hidden edges removed.

The figure to the right represents three intersecting blocks.

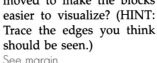

14. Which lines should be removed to make the blocks easier to visualize? (HINT: Trace the edges you think should be seen.)

15. How many ways can you visualize the blocks? 2

SECTION 12.4 Applications

After completing this section, you should understand

▲ how to solve problems by using definitions and theorems about loci and transformations.

The geometry of loci and transformations is used by architects, engineers, artists, designers, mapmakers, and computer programmers. You use it when you decorate a room or rearrange furniture. Planning a team strategy in baseball or soccer, hitting a tennis ball so that your opponent cannot return it—both involve the use of loci and transformations.

PARABOLAS

The following activity illustrates a very practical application using the concept of locus.

DISCOVERY ACTIVITY

1. Use a ruler to draw a line about 4 inches from the bottom of a sheet of unlined paper and parallel to the lower edge. Label the line AB.
2. Mark point F about $\frac{3}{4}$ inch above the line, halfway between A and B.
3. Draw or construct a perpendicular from F to $\overline{AB}$. Where it intersects the line, label the point C.
4. Bisect $\overline{FC}$ and label the midpoint P_1.
5. Point P_1 is equidistant from point F and from $\overline{AB}$.
6. Draw the set of points 1 inch from $\overline{AB}$.

SECTION 12.4

Resources
Reteaching 12.4
Enrichment 12.4

Objectives
After completing this section, the student should understand
▲ how to solve problems by using definitions and theorems about loci and transformations.

Materials
centimeter and inch rulers
protractor
auto headlight or lantern with curved reflector

Vocabulary
parabola

GETTING STARTED

Chalkboard Activity
Fold a sheet of paper in half lengthwise, then fold the strip in thirds, in accordion folds. Tell students you will draw a design on the front sixth and they are to imagine that it is cut through all the thicknesses. What will the paper look like when it is unfolded? Draw the choices on the board. Examples:

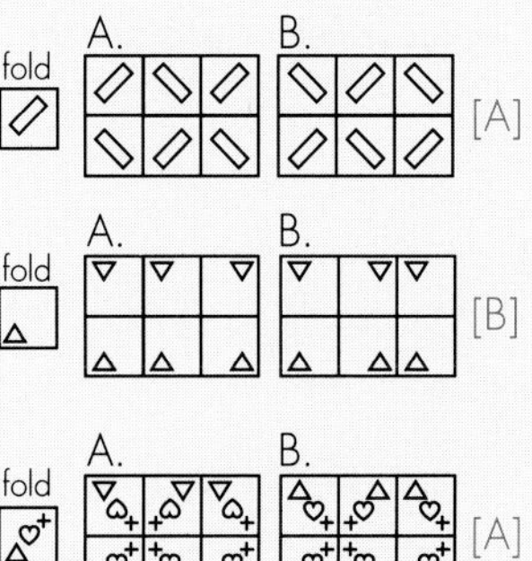

Additional Answers
9. Each is equidistant from F and $\overline{AB}$

7. Draw the set of points 1 inch from F.
8. Label the intersections of the points in steps 6 and 7 as P_2.
9. What can you say about the relationship between points P_2 and $\overline{AB}$?
10. You now have three points that are just as far from F as they are from $\overline{AB}$. Locate additional points. Follow steps 6 and 7 for these conditions:
    a. points $\frac{1}{2}$ inch from F and $\overline{AB}$; intersections labeled P_3
    b. points $\frac{3}{4}$ inch from F and $\overline{AB}$; intersections labeled P_4
    c. points $1\frac{1}{2}$ inches from F and $\overline{AB}$; intersections labeled P_5
    d. points 2 inches from F and $\overline{AB}$; intersections labeled P_6
11. Draw a curve through the points from left to right with a colored marker. Your curve should look like this.

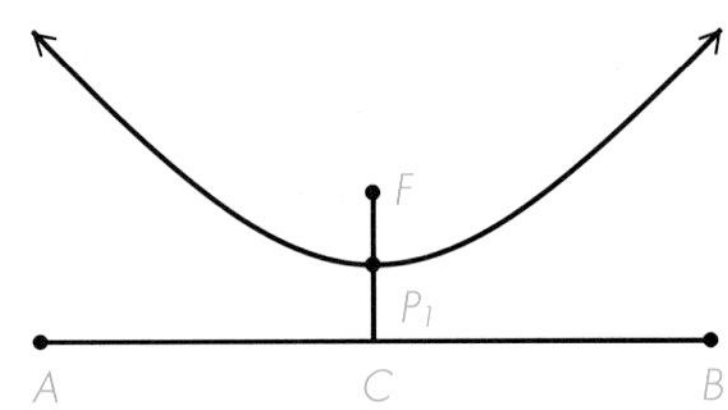

TEACHING COMMENTS

Using the Page
Ask students to explain how the parabolic curve differs from the circumference of a circle. [Points on the curve are equidistant from a line and a point not on the line; the circumference of a circle is equidistant from the center point only.]

Using the Page
If possible, let students examine an automobile headlight or a lantern with a curved reflector. Discuss how the light rays strike the reflector and are reflected out through the lens.

You may have drawn an upward curve that seems to continue on indefinitely. This plane curve is called a **parabola.** Named by the ancient Greeks, the parabola is similar to the curved path followed by a ball thrown into the air.

DEFINITION 12-4-1 **A PARABOLA is the set of points on a plane equidistant from a line and a point not on the line.**

Now draw the same curve rotated 90° to the right about $\overline{FC}$. Imagine that the curve is spinning about the line $\overline{FC}$. Draw the three-dimensional picture of what you would see.

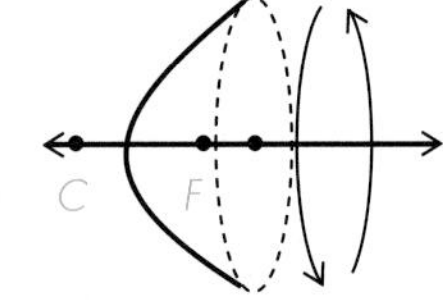

The automobile industry uses a parabolic curve in the design and manufacture of over 40 million headlights each year.

If the light bulb is placed at the focus point F, the light rays will be reflected out through the lens parallel to each other. The best time to observe this property is when you are driving in a car on a foggy night.

A satellite dish also has a parabolic shape. In this case, the rays, or television signals, are received instead of reflected.

CLASS ACTIVITY

1. Draw two 7-centimeter line segments, $\overline{AB}$ and $\overline{BC}$, that have a common endpoint, B, and form an angle of less than 180°.
2. Starting at point A (A is 1), number the centimeters 1 to 7.
3. Starting at point C (C is 7), number the centimeters 7 to 1.
4. Now connect the points, 1 to 1, 2 to 2, . . . 7 to 7.
5. What have you drawn? Parabolic curve made of straight lines

DISCOVERY ACTIVITY

Suppose that a water-treatment plant is to be built near a river and shared by two cities. A pipeline will carry water from the plant to each city. To keep costs down, the cities want to install pipeline over the shortest possible distance. At what point along the river should the water-treatment plant (P) be located?

1. Use a centimeter ruler to make a scale drawing of this map. (Scale: 1 cm = 2 km)

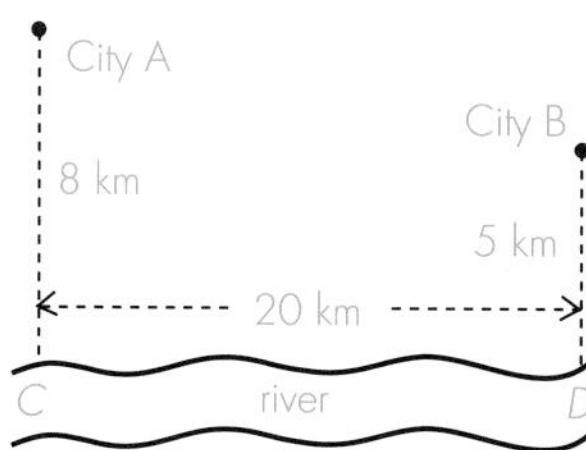

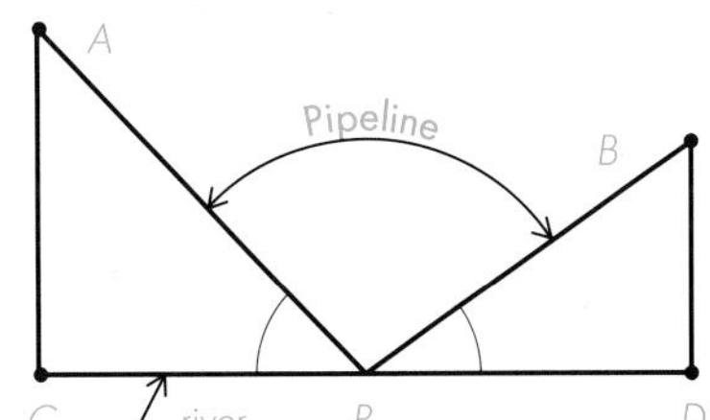

2. Use the scale drawing and centimeter ruler to estimate the total length of the pipeline from the plant to both cities—$AP + PB$—when P is located at point C. Record the length in the table where CP = 0 km. Use a protractor to measure ∠1 (the angle at which the pipeline approaches the river) and ∠2 (the angle at which the pipeline leaves the river).

CP	Total Length $AP + PB$	m∠1	m∠2
0 km	8 km + PB		
4 km			
8 km			
12 km			
16 km			
20 km			

3. Now move the possible location of P along the river toward point D, so that CP =4 km. Repeat step 2 and record the length and angle measures.
4. Repeat step 2 for the other locations of plant P that are listed in the table.

Using the Page
You may wish to have students work in pairs to make the scale drawing and complete the measurements.

5. From the table, decide where the water treatment plant should be located so that the pipeline length ($AP + PB$) is as short as possible.
6. What did you observe about m∠1 and m∠2 where the length ($AP + PB$) is the shortest? m∠1 = m∠2

You may have discovered that when length ($AP + PB$) is shortest, the measures of ∠1 and ∠2 are equal. Instead of experimentally finding the shortest path from City A to the river to City B, you can use a reflection transformation to find the distance.

Using the Page
Some students may find it difficult to understand why the distance ($AQ + QB$) is the shortest distance between A and B, and may feel that there are other, shorter paths. Encourage them to explore by measuring until they feel they have enough information to see that the conclusion is true.

Additional Answers
5. $AQ + QB$

CLASS ACTIVITY

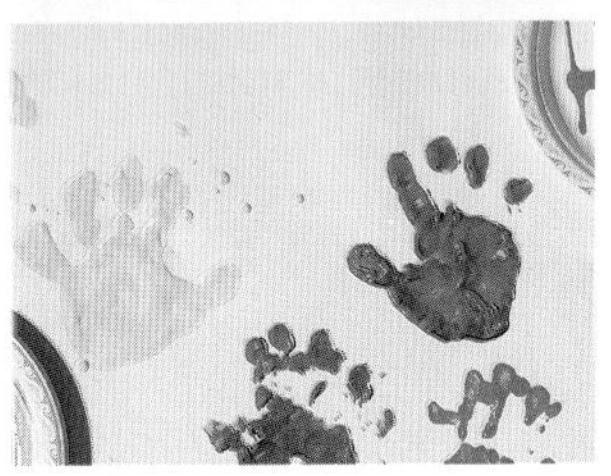

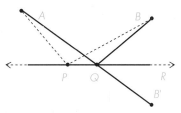

1. Use your scale drawing from the Discovery Activity. Reflect City B across the river R.
2. Draw a line to connect A to B'. Label the intersection of $\overline{AB'}$ with line R, Q. Let P be any other point on line R.
3. Measure and compare ($AP + PB$) and ($AP + PB'$).
4. Measure and compare ($AQ + QB$) and ($AQ + QB'$).
5. Compare ($AP + PB$) and ($AQ + QB$). Which length is less?
6. How does ($AQ + QB$) compare to your result for the shortest pipeline in the Discovery Activity? Should be the same
7. Explain why reflecting City B across the river and finding the straight-line distance between City A and City B′ gives the shortest distance between the two cities.
A straight line is the shortest distance between two points.

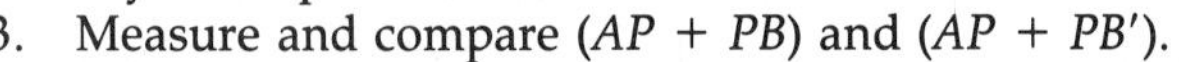

MORE TRANSFORMATIONS

Transformations are often used in the design of flags. Identify the types of transformation in each flag.

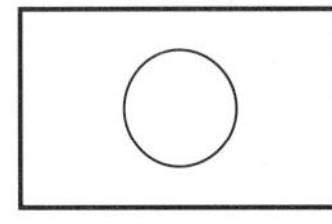

Bangladesh
2-line reflection, 180° rotation

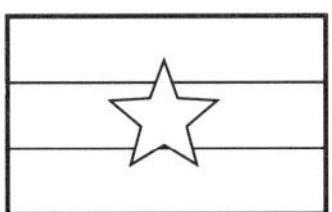

Yugoslavia
Reflection

France
2-line reflection, 180° rotation

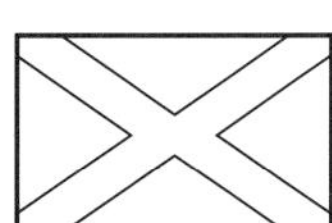

Jamaica
2-line reflection, 180° rotation

Because most flags are rectangular, any rotation transformation in their design will be 180°.

Suppose the flags were square instead of rectangular. What transformations do they use in their designs? What is the angle of rotation?

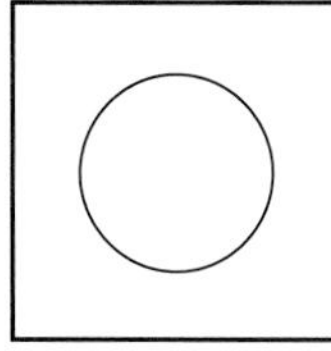
4-line reflection, 90° rotation

Reflection

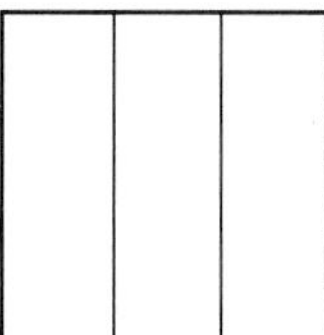
2-line reflection, 180° rotation

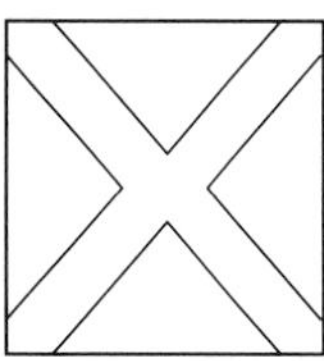
4-line reflection, 90° rotation

CLASS ACTIVITY

How could you use a map to find the locus of the sets of points?

1. All cities 4 miles from a given highway
   2 lines parallel to and 4 mi from each side of the highway
2. All towns or cities 30 miles from your town
   Circle with radius of 30 mi, center on your town
3. All houses or buildings 3 miles from your school building
   Circle with radius of 3 mi, center on school
4. Find a picture of your state flag. Tell whether its design contains any transformations and identify them. Answers will vary.

HOME ACTIVITY

1. What is the locus of points described by the tip of the minute hand of a clock in one hour? Circumference of a circle

2. What is the locus of points described by the entire length of a minute hand of a clock in one hour? Area of a circle

3. What is the locus of points described by the entire length of a minute hand of a clock in 15 minutes? Area of a 90° sector

4. How would you arrange the chairs in a room so that you would be equidistant from each person sitting in a chair? Where would your chair be? In a circle; center of circle

Construct a circle with a radius of 1 inch.

5. What is the locus of points 1.5 inches from the circle? Circle

6. What is the locus of points 1 inch from the circle? Circle and point that is center of original circle

7. What is the locus of points 0.5 inch from the circle? 2 concentric circles

Extra Practice

1. How could you use a map to find the locus of all towns or cities 25 miles from your state capitol? [Circle with radius of 25 mi, center on state capitol]

2. What is the locus of points described by the center of a car wheel as it moves along a level road? [Line parallel to the road]

Describe the transformations used in these square flags.

3.

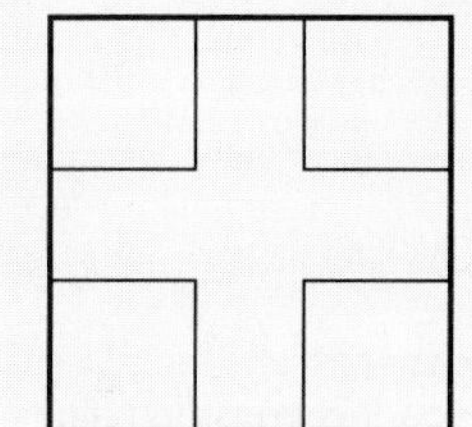

[4 line reflections, 90° rotation]

4.

[1 line reflection]

8. What is the locus of points 2 centimeters from a point on a plane? in space? Circle; sphere

9. What kind of transformation preserves parallelism? Translation

10. What kind of transformation is like a mirror image? Reflection

11. What kind of transformation is illustrated by a spinning wheel? Rotation

Trace the flags.

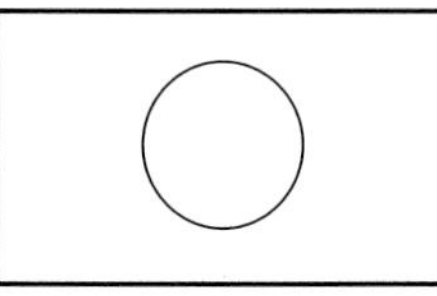
Japan

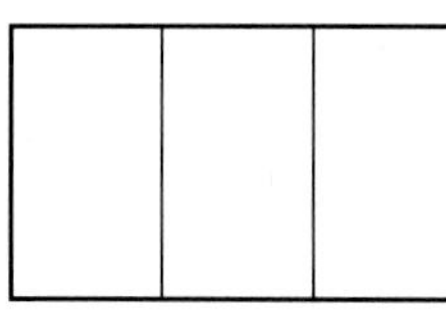
Peru

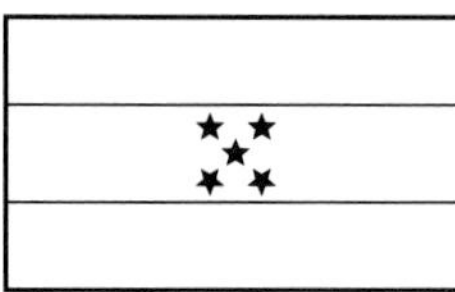
Honduras

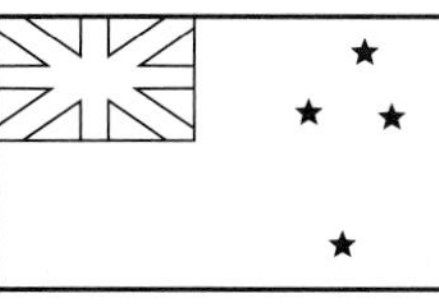
New Zealand

12. Which flags have line symmetry? Draw the lines of symmetry. Japan, Peru, Honduras

13. Which flags have point symmetry? What is the angle of rotation? Japan, Peru; 180°

Write a LOGO procedure that uses the procedure XYPLANE from page 118 and the command SETXY to tell the turtle to graph each figure. Name the transformation. (→ means mapped onto.)
Answers for procedures may vary.

14. $(-3, 5) \rightarrow (3, 5)$
Translation or reflection over x-axis

15. $y = x \rightarrow y = -x$
rotation or reflection over either axis

16. $A(-3, 5), B(-1, 7) \rightarrow A'(-3 + 3, 5 + 3), B'(-1 + 3, 7 + 3)$ Translation

17. $A(-4, 0), B(-2, 6), C(0, 0) \rightarrow A'(4, 0), B'(2, 6), C'(0, 0) \rightarrow A''4, 0), B''(2, -6), C''(0, 0)$
2 reflections equal to 1 90° rotation

18. $y = 2x - 1 \rightarrow y = 2x + 1$
Translation

19. $y = x^2 \rightarrow x = y^2$
Reflection over $y = x$

CRITICAL THINKING

20. Explain what it means to say that a cube has plane symmetry. Use a model to help you.
A cube is symmetric to a plane passing through it.

21. Is a cube symmetric about more than one plane? How many? 9

22. Does a sphere have any plane symmetries? Explain.
Yes; infinitely many

23. Does a cylinder have any plane symmetries? Draw a picture to explain your answer.
Yes; 1 parallel to ends, infinitely many perpendicular to ends

FOLLOW-UP

Assessment
Describe the transformations illustrated by

1. An elevator traveling from the first to the second floor. [Translation]

2. A turning propeller blade. [Rotation]

3. A bird seen by looking in a puddle of water. [Reflection]

4. What is the locus of points equidistant from both walls of a corner of a room? [A plane forming the bisector of the angle made by the corner]

Reteaching
Students who have had difficulty with this section may benefit from Reteaching Activity 12.4.

Enrichment
For students who have mastered the material in this section, you may wish to assign Enrichment Activity 12.4.

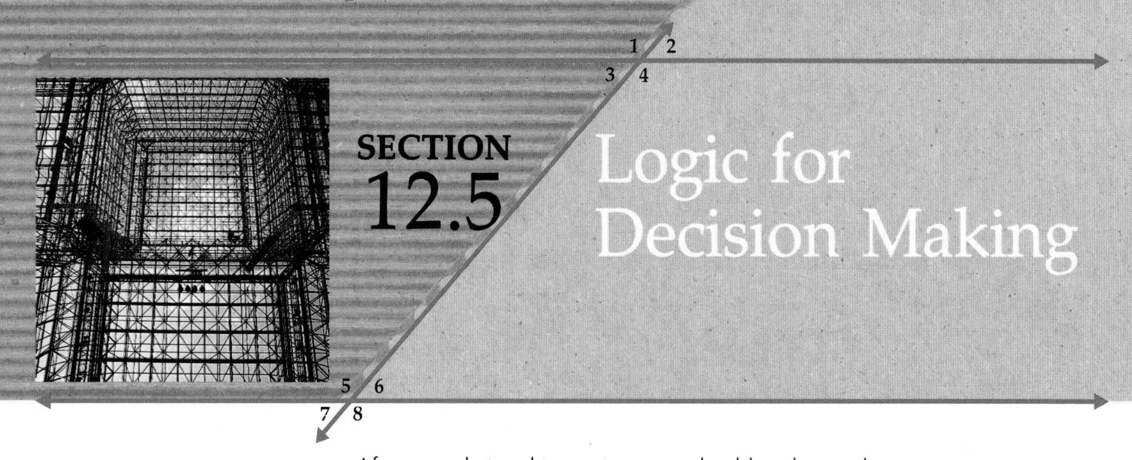

Logic for Decision Making

After completing this section, you should understand
- ▲ how a conditional statement is related to its converse, inverse, and contrapositive.
- ▲ how to determine the truth value of a statement.

In Section 12.1 you found that, in determining a locus, you had to prove that every point in the locus satisfied the given conditions and that every point that satisfied the conditions was a point in the locus. In other words, you proved the truth of a statement and its converse.

Statement: If A, then B.
If a point is on an angle bisector, then it is equidistant from the sides of the angle.
Converse: If B, then A.
If a point is equidistant from the sides of an angle, then it is on the angle bisector.

You also had to prove one of two other possibilities—the inverse or the contrapositive of the statement.

Inverse: If not A, then not B.
If a point is not on an angle bisector, then it is not equidistant from the sides of the angle.
Contrapositive: If not B, then not A.
If a point is not equidistant from the sides of an angle, then it is not on the angle bisector.

Recall that a statement can be true or false, but it cannot be both true and false at the same time.

Example: "Watch out!" is neither true nor false, so it is not considered to be a statement.

SECTION 12.5

Resources
Reteaching 12.5
Enrichment 12.5

Objectives
After completing this section, the student should understand
- ▲ how a conditional statement is related to its converse, inverse, and contrapositive.
- ▲ how to determine the truth value of a statement.

Vocabulary
inverse
contrapositive

GETTING STARTED

Warm-Up Activity
Read the following aloud to the class. Ask students to describe each one as a sentence, question, or exclamation, and tell whether they think it is true or false.

1. American Independence Day is July 4. [Sentence, true]
2. All cars have power steering. [Sentence, false]
3. Where are you going? [Question, neither true nor false]
4. Help! [Exclamation, neither true nor false]
5. This sentence is true. [Sentence, neither true nor false]
6. If a figure is a parallelogram, then it is a square. [Sentence, false]

TEACHING COMMENTS

Using the Page
After the discussion of truth values, you may wish to have students work in small groups to write several statements and their related converses, inverses, and contrapositives. On the chalkboard, draw a blank table similar to that in the Discovery Activity. Groups may then share their statements with the class and have other students give the converse, inverse, and contrapositive. Have a member of the group write the truth value for each one in the table.

Additional Answers
2a. If a figure is a rectangle, then it is a square; false.
b. If a line segment is a diameter, then it is a chord of a circle; true.
c. If all three sides of a triangle have the same length, then the triangle is equilateral; true.
3a. If a figure is not a square, then it is not a rectangle; false.
b. If a line segment is not a chord of a circle, then it is not a diameter; true.
c. If a triangle is not equilateral, then all three sides do not have the same length; true.
4a. If a figure is not a rectangle, then it is not a square; true.
b. If a line segment is not a diameter, then it is not a chord of a circle; false.
c. If all three sides of a triangle do not have the same length, then it is not equilateral; true.

Logic demonstrates that new statements can be constructed from a given statement and that the truth or falsity (truth value) of a new statement can be determined from the truth or falsity of the original statement.

DISCOVERY ACTIVITY

1. Tell whether the statement is true or false.
   a. If a figure is a square, then it is a rectangle. True
   b. If a line segment is a chord of a circle, then it is a diameter. False
   c. If a triangle is equilateral, then all three sides have the same length. True
2. Write the converse of each statement. Is it true or false? See margin.
3. Write the inverse of each statement. Is it true or false? See margin.
4. Write the contrapositive of each statement. Is it true or false? See margin.
5. Copy and complete the table by writing true or false for each statement.

	Statement	Converse	Inverse	Contrapositive
a.	true	false	false	true
b.	false	true	true	false
c.	true	true	true	true

6. Study the table of truth values for each statement and its converse, inverse, and contrapositive. What do you observe about the relationships between the statement and its contrapositive? the converse and the inverse? They are either both true or both false.

You may have found that the truth values of a statement and its contrapositive are the same, and that the truth values of the converse and the inverse are the same. You can show this in a diagram.

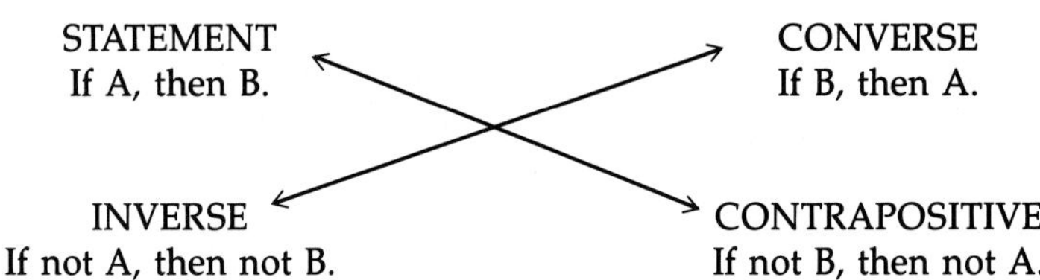
STATEMENT
If A, then B.
CONVERSE
If B, then A.
INVERSE
If not A, then not B.
CONTRAPOSITIVE
If not B, then not A.

Example:

Statement	**Converse**
If an animal has four legs, then it is a dog.	If an animal is a dog, then it has four legs.
Inverse	**Contrapositive**
If an animal does not have four legs, then it is not a dog.	If an animal is not a dog, then it does not have four legs.

1. Which statement has the same truth value as the original statement? Are these two statements true or false? Contrapositive; false
2. Which statement has the same truth value as the converse? Are these two statements true or false? Inverse; true

LOGIC AND PROOFS

In Section 12.1, you used logical statements and indirect reasoning to prove that the locus of points equidistant from the endpoints of a line segment is the perpendicular bisector of the segment. You can use sets of logical statements to prove this theorem: the locus of points in a plane equidistant from a point on the plane is a circle.
First, write the theorem in if-then, or conditional, form. Then write the converse, inverse, and contrapositive.

STATEMENT:	If a set of points in a plane is equidistant from a given point, then the points are a circle.
CONVERSE:	If a set of points is a circle, then the points are equidistant from a given point.
INVERSE:	If a set of points is not equidistant from a given point, then the points are not a circle.
CONTRAPOSITIVE:	If a set of points is not a circle, then the points are not equidistant from a given point.

Which statements must you prove true to show that the theorem is true? Statement, converse, and either inverse or contrapositive

CLASS ACTIVITY

For each statement, write the converse, inverse, and contrapositive. Classify each statement as true or false. Then use a computer and a geometric drawing tool to draw figures to justify each answer.

1. If a triangle is equilateral, then it is right. See margin.
2. If two angles are vertical angles, then their measures are equal. See margin.
3. If a triangle is equilateral, then it is isosceles. See margin.
4. If a figure is a polygon, then it is a triangle. See margin.
5. If a figure has four equal sides, then it is a square. See margin.

Additional Answers

1. Statement is false.
Converse: If a triangle is right, then it is equilateral; false. Inverse: If a triangle is not equilateral, then it is not right; false.
Contrapositive: If a triangle is not right, then it is not equilateral; false.
2. Statement is true.
Converse: If the measures of two angles are equal, then they are vertical angles; false.
Inverse: If two angles are not vertical angles, then their measures are not equal; false.
Contrapositive: If the measures of two angles are not equal, then they are not vertical angles; true.
3. Statement is true.
Converse: If a triangle is isosceles, then it is equilateral; false.
Inverse: If a triangle is not equilateral, then it is not isosceles; false.
Contrapositive: If a triangle is not isosceles, then it is not equilateral; true.
4. Statement is false.
Converse: If a figure is a triangle, then it is a polygon; true.
Inverse: If a figure is not a polygon, then it is not a triangle; true.
Contrapositive: If a figure is not a triangle, then it is not a polygon; false.
5. Statement is false.
Converse: If a figure is square, then it has four equal sides; true.
Inverse: If a figure does not have four equal sides, then it is not a square; true.
Contrapositive: If a figure is not a square, then it does not have four equal sides; false.

Additional Answers

1. Statement is false.
Converse: If it is night, then the sun is not shining; true.
Inverse: If the sun is shining, then it is not night; true.
Contrapositive: If it is not night, then the sun is shining; false.

2. Statement is true.
Converse: If a triangle is equiangular, then it is equilateral; true.
Inverse: If a triangle is not equilateral, then it is not equiangular; true.
Contrapositive: If a triangle is not equiangular, then it is not equilateral; true.

3. Statement is true.
Converse: If the measures of two angles are equal, then they are right angles; false.
Inverse: If two angles are not right angles, then their measures are not equal; false.
Contrapositive: If the measures of two angles are not equal, then they are not right angles; true.

For answers to Exercises 4-10, see page 476H.

FOLLOW-UP

Assessment
Write the converse, inverse, and contrapositive of this statement. Tell whether each is true or false.
If you live in Chicago, then you live in Illinois. [Statement is true.
Converse: If you live in Illinois, then you live in Chicago; false.
Inverse: If you do not live in Chicago, then you do not live in Illinois; false.
Contrapositive: If you do not live in Illinois, then you do not live in Chicago; true.]

Reteaching
Students who have had difficulty with this section may benefit from Reteaching Activity 12.5.

Enrichment
For students who have mastered the material in this section, you may wish to assign Enrichment Activity 12.5.

HOME ACTIVITY

For each statement, write the converse, inverse, and contrapositive. See margin.

1. If the sun is not shining, then it is night.
2. If a triangle is equilateral, then it is equiangular.
3. If two angles are right angles, then their measures are equal.
4. If a triangle is equilateral, then it is acute.
5. If I stay at home, then I don't pitch at the game.
6. If a figure is a quadrilateral, then its diagonals are equal.
7. If two lines are parallel, then they are coplanar.
8. If two triangles have the same size and the same shape, then they are congruent.
9. If a quadrilateral has two pairs of parallel sides, then it is a rectangle.
10. If it is snowing, then the temperature is 32°F or less.
11. For exercises 1–10, classify each statement and its converse, inverse, and contrapositive as true or false. Identify the pairs of statements in each set that have the same truth value.
See answers for Exercises 1–10; statement and contrapositive, converse and inverse.

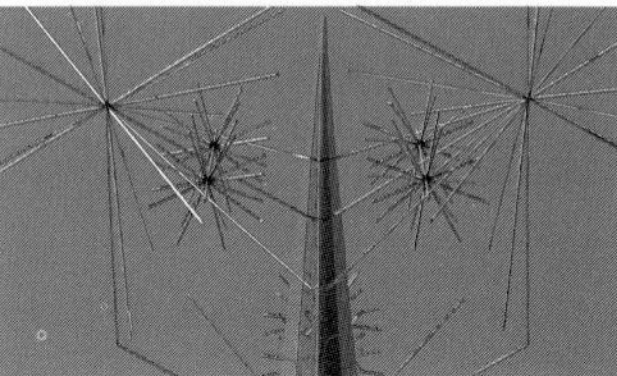

CRITICAL THINKING

12. It's Sunday, and I'm trying to plan my errands so that I can make just one trip downtown on one afternoon this week. I have to visit the dentist, get bread at the bakery, and pick up fresh tomatoes at the farmers' market. The dentist's office is open only on Tuesday, Wednesday, and Friday. The bakery is closed on Wednesday and Saturday afternoons. The farmers' market is not open on Tuesday or Thursday. What afternoon should I go downtown? Friday

Chapter Vocabulary
line symmetry
locus
parabola
point symmetry
reflection
rotation
rotation symmetry
symmetry
transformation
translation

12.1 The locus of all points equidistant from a point on a plane is a circle.

$OB = OC$ so B and C are on the circle.
B and C are on the circle, so $\overline{OB} = \overline{OC}$.

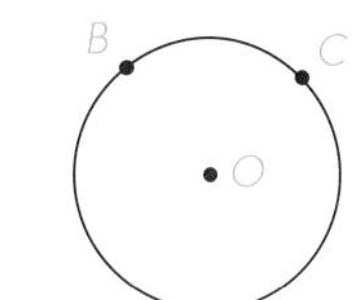

12.2 Identify the transformation pictured.

$\triangle A'B'C'$ is the translation of $\triangle ABC$.

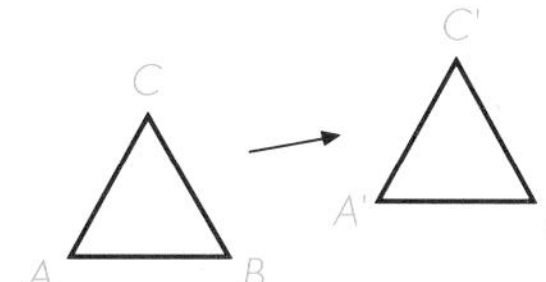

Pentagon $D'E'F'G'H'$ is a 35° rotation of $DEFGH$ around point J.

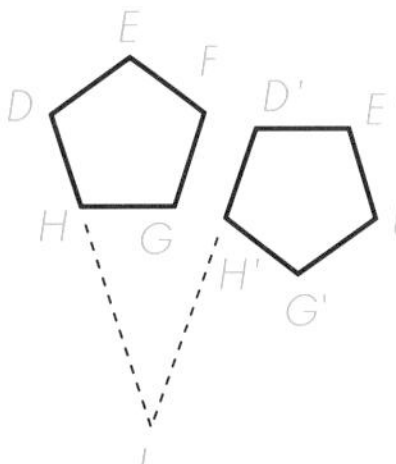

12.3 The regular hexagon demonstrates what two types of symmetry?

Line symmetry along six lines (reflection across the lines) and point, or rotation, symmetry (rotations of 60° around point P).

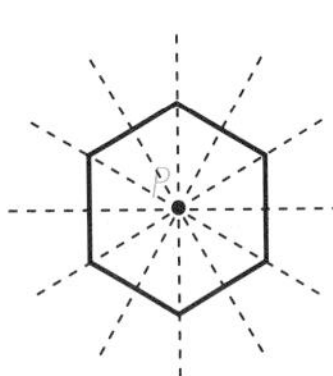

12.4 Describe the locus of points equidistant from point F and the x-axis.

A parabola

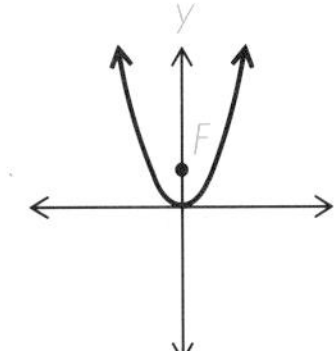

12.5 How are the truth values of these statements related?

Statement: If a figure is a square, then it is a rhombus.
Converse: If a figure is a rhombus, then it is a square.
Inverse: If a figure is not a square, then it is not a rhombus.
Contrapositive: If a figure is not a rhombus, then it is not a square.

Statement and contrapositive have the same truth value (true); inverse and converse have the same truth value (false).

471

Using the Page
This Computer Activity introduces students to a method for using LOGO to create tessellations. The experience should impress students with the beauty of the LOGO language. First they use the command FIGURE to create the smallest part of their tessellation. They then use the procedure TRANSLATE to move the figure around. Next they use the procedure COLUMN to create a column of figures. Finally the procedure TESS creates the complete tessellation.

For students who have difficulty, work carefully with them on each of the procedures before expecting them to put the procedures together. Let students experiment and create their own figures. This can be an experience that fosters interest in the materials covered in this chapter.

COMPUTER ACTIVITY

Tesselations are patterns made up of figures that are translated over and over again to fill a plane so that there are no gaps or overlaps. Tesselations can be formed using one figure or a combination of different figures. An example of a tesselation made up of equilateral triangles is shown below.

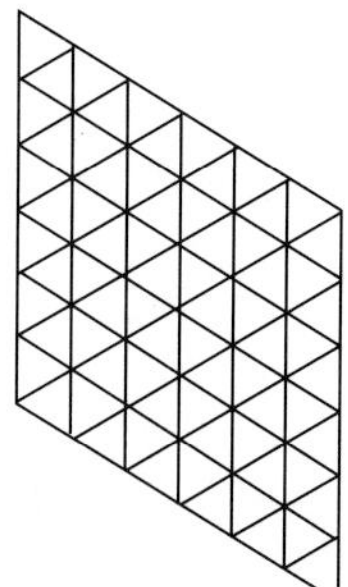

The LOGO turtle can be used to draw tesselations. First write a procedure to tell the turtle to draw the smallest part of the tesselation that is translated. Then use the REPEAT command to translate the pattern and draw the tesselation.

Example: Write a procedure to tell the turtle to draw the tesselation shown above.

The small part of the tesselation that is translated is shown below.

The procedure FIGURE given below tells the turtle to draw this figure.

```
TO FIGURE
  REPEAT 2[FD 10 RT 120]
  FD 10
  REPEAT 2[LT 120 FD 10]
END
```

The procedure TRANSLATE given below puts the turtle in position to translate the figure.

```
TO TRANSLATE
  PU LT 120 FD 10 RT 120 BK 10 PD
END
```

The procedure COLUMN given below tells the turtle to draw a column of the figures.

```
TO COLUMN
  REPEAT 5[FIGURE TRANSLATE]
END
```

The procedure TESS given below tells the turtle to draw the tesselation.

```
TO TESS
  REPEAT 5[COLUMN PU FD 30 PD]
END
```

Write a LOGO procedure to tell the turtle to draw each tesselation.
Answers may vary.

1.

2.

3.

4.

5.

6.

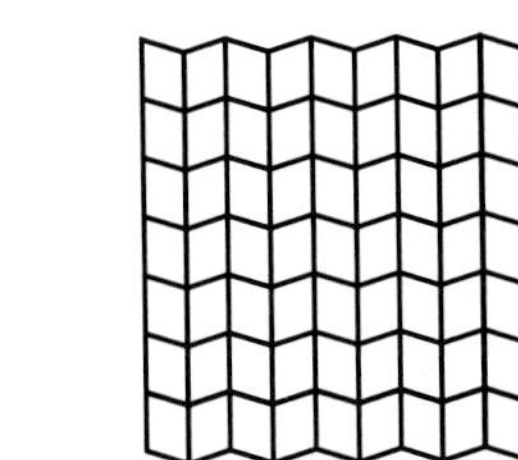

7. Draw your own tesselation. Write a LOGO procedure to tell the turtle to draw your tesselation.
Answers may vary.

Using the Page
Students can work individually or in groups to review. If you have students study in groups, they should work together to make sure all members of the group know the material in the chapter.

Informal assessment such as interviews, classroom observations, or a review of student portfolios can be used instead of paper-and-pencil tests.

CHAPTER REVIEW

1. What is the locus of all points within a circle that are equidistant from the endpoints of a given chord? Perpendicular bisector of chord, or diameter
2. Find the locus of the centers of all the circles that are tangent to two parallel lines. A straight line midway between and parallel to the 2 lines
3. Draw a circle with a radius of 2.5 centimeters. Draw and describe the locus of all points 1.5 centimeters from the circle. 2 concentric circles with radii of 1 cm and 4 cm

Tell whether the transformation is a translation, reflection, or rotation.

4. 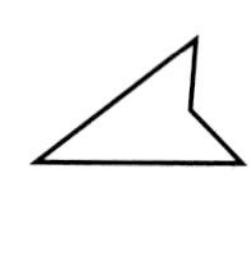
Translation

5. 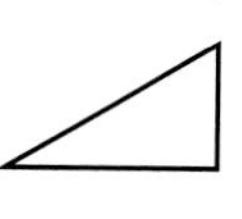
Rotation

6. 
Rotation

Does the figure have line symmetry? Trace the figure and draw the line(s) of symmetry.

7. 
No

8. 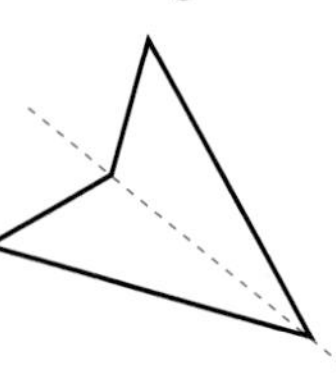
Yes

9. 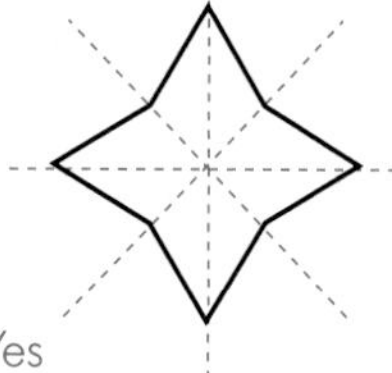
Yes

Does the figure have point symmetry?

10. 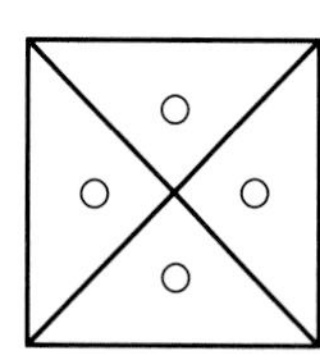
Yes

11. 
Yes

12. 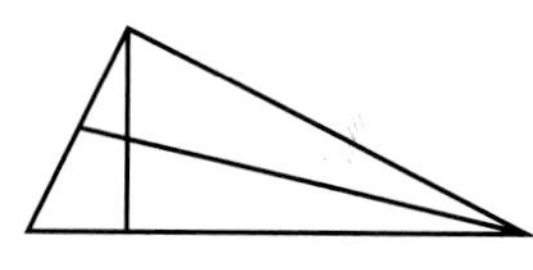
No

13. The drawing shows a billiard table with balls W and R. Where should a player aim if she want ball W to hit side $\overline{AB}$, then strike ball R? Draw the picture.

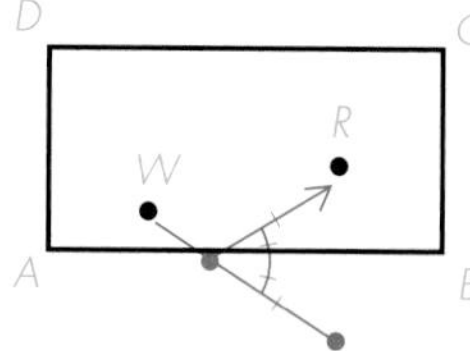

14. Write the converse, inverse, and contrapositive for this statement, and tell whether each is true or false: If the sum of the measures of two angles is 180°, then the angles are supplementary. See margin.

Additional Answers
14. Statement is true. Converse: If two angles are supplementary, then the sum of their measures is 180°; true. Inverse: If the sum of the measures of two angles is not 180°, then the angles are not supplementary; true. Contrapositive: If two angles are not supplementary, then the sum of their measures is not 180°; true.

CHAPTER TEST

1. Describe the locus of the midpoints of all the radii of a given circle. Circle with $\frac{1}{2}$ the radius
2. What is the locus of the centers of all the circles tangent to two intersecting lines? Angle bisectors

Draw a line segment $\overline{AB}$ 3.5 centimeters long.

3. Find and describe the locus of all points equidistant from the endpoints of $\overline{AB}$. Perpendicular bisector of AB
4. Find and describe the locus of all points that are 2 centimeters from $\overline{AB}$. 2 parallel lines, with $\overline{AB}$ midway between

Classify the transformation(s) that will make the figure on the left map onto the figure on the right.

5.

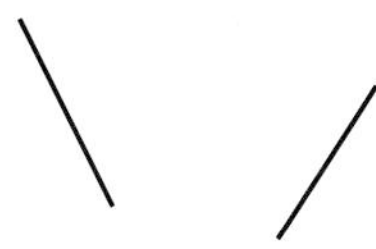

Rotation

6.

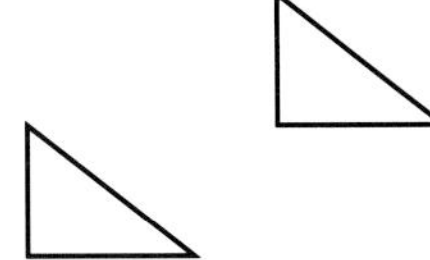

Translation

7.

Reflection

Does the figure have line symmetry? Trace the figure and draw the line(s) of symmetry.

8.

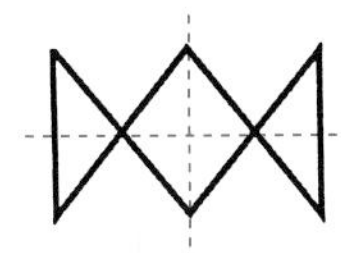

Yes

9.

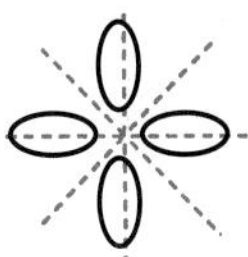

Yes

10.

No

Does the figure have point symmetry?

11.

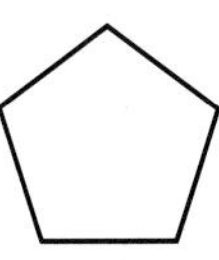

Yes

12.

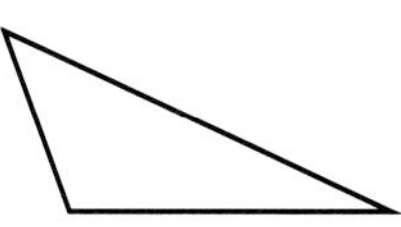

No

13.

Yes

14. Bill is walking his dog home from the playground, but first he wants to stop at the park, and then he wants to take a swim in the river. What is the shortest path he can take to do both these things? Draw the route.

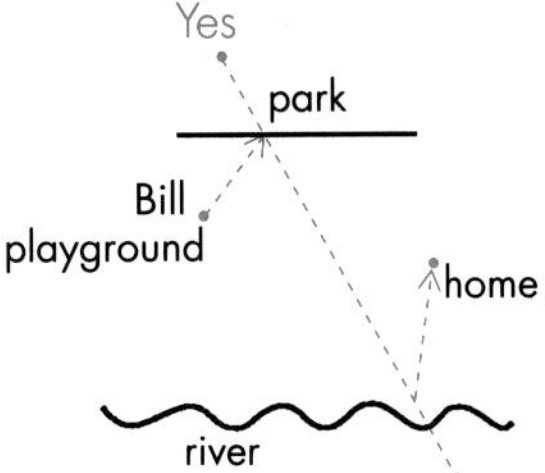

15. Write the converse, inverse, and contrapositive for this statement: If an animal lives in water, then it is a fish. Tell whether each statement is true or false, and identify the pairs of statements with the same truth value. See margin.

Assessment Resources

Achievement Tests pp. 23-24

Test Objectives

After studying this chapter, students should understand

- the definition of geometric locus.
- how to determine a locus that satisfies given conditions.
- how to find the translation image of a figure.
- how to find the reflection of a figure and locate the line of reflection.
- how to find the rotation image of a figure.
- how to determine whether a figure has line symmetry.
- how to find lines of symmetry.
- how to determine whether a figure has point or rotation symmetry.
- how to solve problems by using definitions and theorems about loci and transformations.
- how a conditional statement is related to its converse, inverse, and contrapositive.
- how to determine the truth value of a statement.

Additional Answers

15. Statement is false.
Converse: If an animal is a fish, then it lives in water; true.
Inverse: If an animal does not live in water, then it is not a fish; true.
Contrapositive: If an animal is not a fish, then it does not live in water; false.
Statement and contrapositive, and converse and inverse, have the same truth value.

475

Algebra Review Objectives
Topics reviewed from algebra include
- finding elements of the solution set for equations in two variables.
- graphing equations in two variables.
- finding and graphing an equation for a given set of points.

Additional Answers

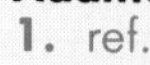

1. ref.

x	y
−3	−10
38	8

2. ref.

x	y
−3	8
8	−10

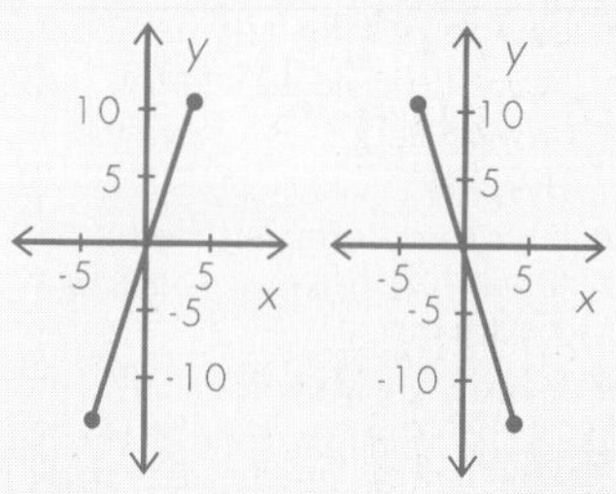

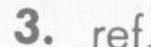

3. ref.

x	y
0	0
−3	3
3	3

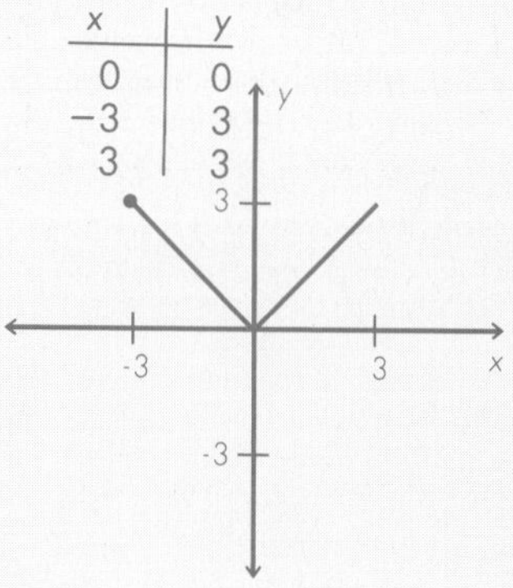

For answers to Exercises 4-13, see page 476H.

ALGEBRA SKILLS

The set of all solutions of an equation in two variables is called the *solution set* of the equation. Since a solution set is a set of ordered pairs, you can graph it on the x-y−axes. To graph an equation, first find its solution set. Make a table of values to find several ordered pairs that satisfy the equation. Then plot the ordered pairs on the x-y−axes.

The graph of a linear equation in two variables is a straight line.

Examples: $y = x + 2$ $\qquad$ $x = 3$

This equation is equivalent to $x + 0y = 3$, so for any value of y, $x = 3$.

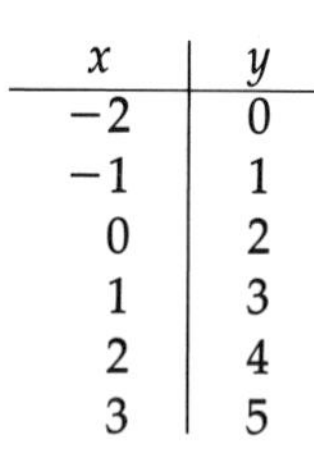

x	y
−2	0
−1	1
0	2
1	3
2	4
3	5

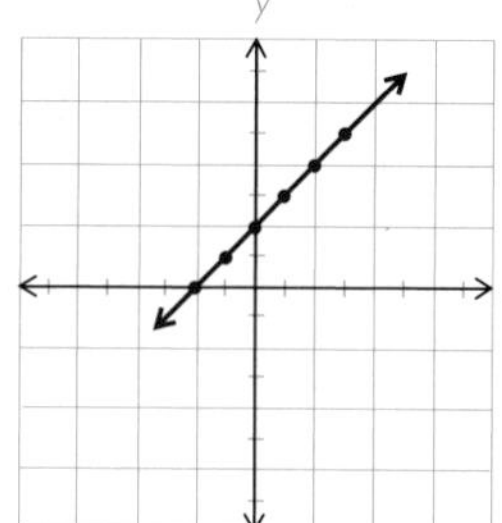

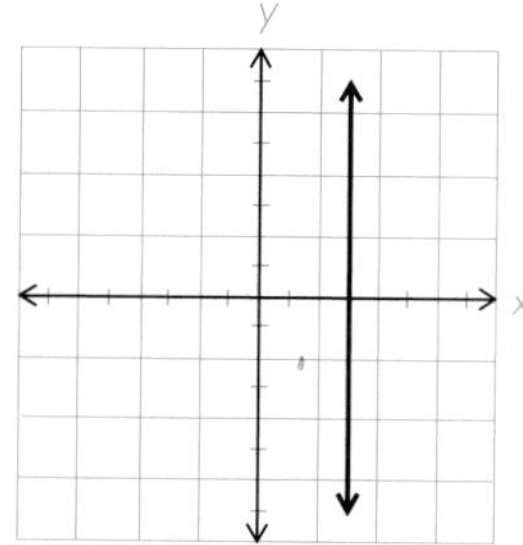

Graph the equation for the given values. See margin.

1. $y = 3x - 1$ for $x = -3, -2, -1, 0, 1, 2, 3$
2. $y = -3x - 1$ for $x = -3, -2, -1, 0, 1, 2, 3$
3. $y = |x|$ for $-3 < x < 3$
4. $-3(2x - 3y) = 18$ for $-4 < x < 4$
5. $y = x^2$ for $-3 < x < 3$

Find five elements of the solution set for each equation. Then graph the equation.
Answers may vary. See margin for graph.

6. $x + y = 0$
7. $y = x + 5$
8. $x + y = -2$
9. $2x + y = 0$
10. $y = -4x$
11. $2(x + 5) + 7(x - y) = 4x - 8y$
12. $20 = 3(-x - y) - 5(-3y + 4x)$
13. What is the equation for the set of points for which the x value is twice the y value? Draw the graph. See margin.

Name ______________________ Date ______________

Locus

In the figure at the right, where are all the points 2 centimeters from point P? How would you describe their location?

It appears that the location of all the points 2 centimeters from point P is the circle with center P and a radius of 2 centimeters. Instead of using the word "location," it is customary to use the Latin word *locus*. So, the locus of all points 2 centimeters from point P is the circle with center P and radius 2 centimeters.

Write the letter of the correct figure next to the exercise number. Then write a description of the locus of points described. All of the loci are 2-dimensional figures.

E 1. What is the locus of points that are an equal distance from two parallel lines m and n?
a line parallel to both lines m and n, centered between them

C 2. What is the locus of points that are an equal distance from the sides of an angle?
the angle bisector

A 3. Two concentric circles have radii of 5 mm and 13 mm. What is the locus of points that are an equal distance from the circles?
a circle with the same center and a radius of 9mm

G 4. What is the locus of points that are an equal distance from the four corners of a square?
the point where the diagonals intersect

F 5. What is the locus of points that are an equal distance from the endpoints of a line segment?
the perpendicular bisector of the segment

B 6. What is the locus of points that are an equal distance from line t?
two parallel lines, with t running down the middle between them

D 7. What is the locus of points that are an equal distance from a point?
a circle with radius equal to the given distance

Name ______________________ Date ______________

Ellipses and Other Loci

The figure at the right is formed from two sets of concentric circles. The heavy curve passes through the intersections of the circles. The curve is called an *ellipse*. It is the set of all points such that the sum of the distances from the centers of the two small circles is constant. In this case, the sum is 12 units.

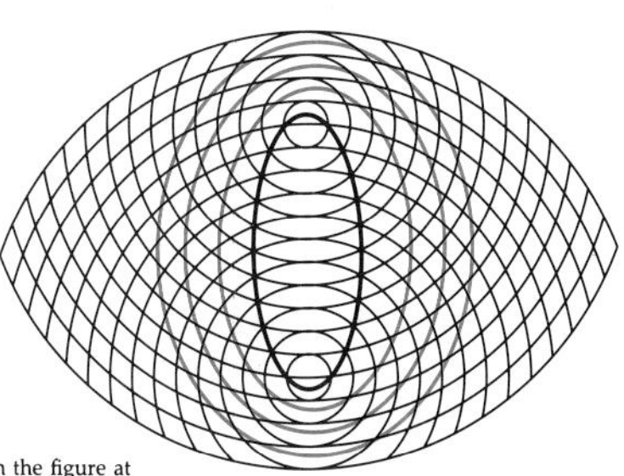

1. Draw three more ellipses on the figure at the right above. For each ellipse, give the sum of the distances from the centers of the two small circles.

   a. Sum of distances: 14 units

   b. Sum of distances: 16 units

   c. Sum of distances: 18 units

   Answers may vary. Sample answers are given.

2. Copy the figure at the right on a separate sheet of paper.

3. The figure at the left below is formed by a circle that revolves around a point on its circumference. Copy the figure.

4. The figure at the right below is based on a curve called a *cardioid*. Copy the figure.

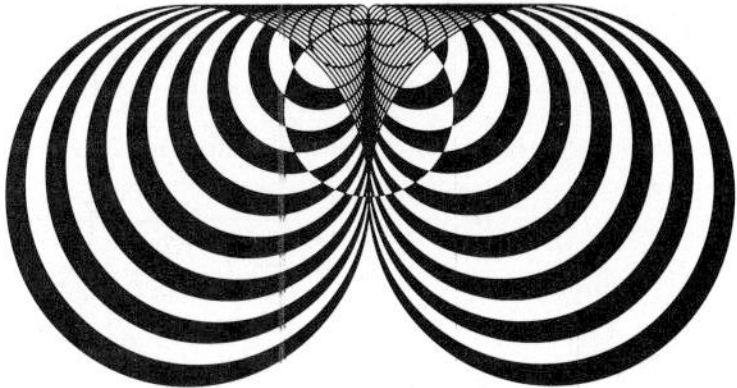

Name ______________________ Date ______________

Transformations

In a *translation,* each point of a figure is moved the same distance.

In a *reflection* across a line, a point and its reflection are the same distance from the line.

In a *rotation,* a figure is turned about a point. Often the point of rotation is the center of the figure.

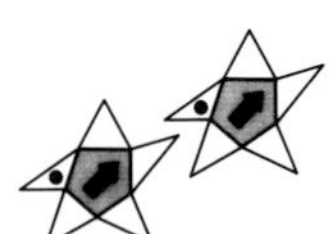

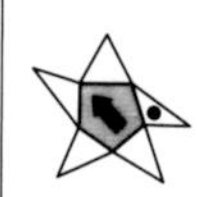
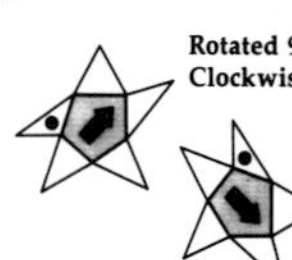

In each pair of figures, the shaded figure is a translation of the unshaded one. Tell the direction and distance the shaded figure has been moved. Use the terms *up, down, left,* and *right.* Give the distances in centimeters.

1. 2.5 cm right
2. 1 cm up
3. 3 cm left

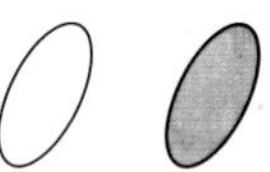

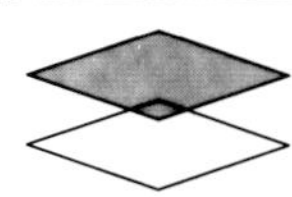
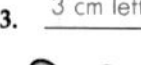

Draw the line of reflection for each pair of figures.

4.
5.
6.

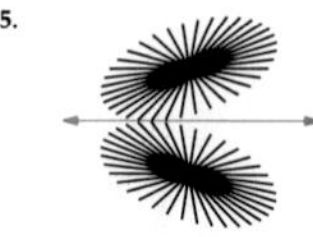
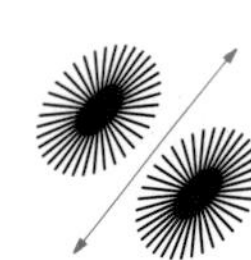

The shaded figure is a rotation of the unshaded one. Give the angle of rotation.

7. 180°
8. 60°
9. 90°

Name ______________________ Date ______________

Constructing Tessellations

A. Draw any curve. Here we have drawn a "V."
B. Make a translation of your curve.
C. Reflect Figure B across a vertical line.
D. Join Figures A and C to make Figure D. This curve has rotational symmetry about the point where the two parts have been joined.
E. Make a copy of Figure D. Then rotate Figure D 90° clockwise and make two copies of the rotated figure. Join the four curves as shown.
F. The new figure can be used in a tessellation to cover a plane area. Two possible arrangements are shown. The second has both the cross shapes and squares.

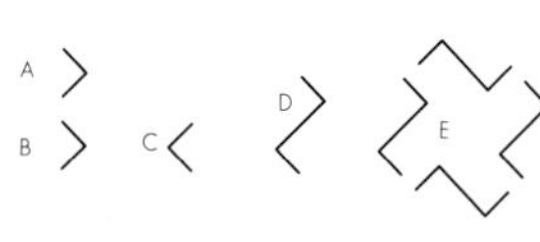

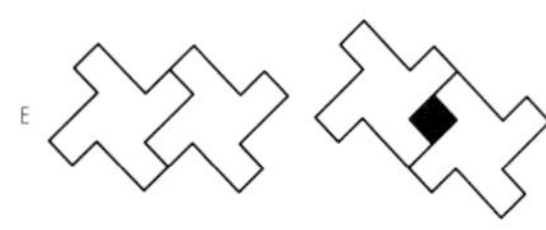

Describe how each tessellation was constructed (refer to the steps above as necessary). Then create a tessellation of your own on a separate sheet of paper.

1. Curve A is a "wiggle." Shapes are joined as in left-hand drawing in Figure F.
2. Curve A is a semi-circle. Completed pattern alternates in black and white.
3. Curve D does not have rotational symmetry. Shapes are joined as in right-hand drawing in Step F.

4. The tessellation at the right was created from the S-shaped curve shown. Describe how the tessellation was created.
Three diagonals are created from the S-curves. Then, a rotated S-curve is used to make overlapping rows.

5. Copy the tessellation on another sheet of paper. Shade the sections in a different pattern.

Name ______________________ Date ____________

Symmetry

If you can fold a figure in half so that the two halves match, the figure has a *line of symmetry* along the fold. Each half of the figure is a reflection of the other.

A snowflake-type figure created with two folds has a *point of symmetry* where the folds intersect. It also has two lines of symmetry that are perpendicular to each other.

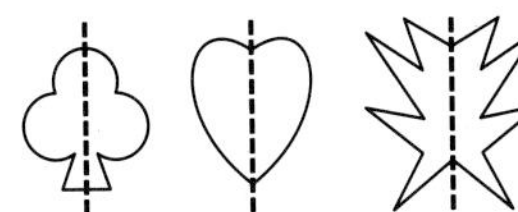

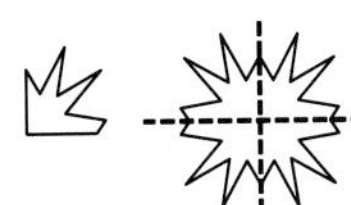

Does the figure have both point and line symmetry, just line symmetry, or neither? Write *both, line,* or *neither.* Draw the lines of symmetry.

1. both

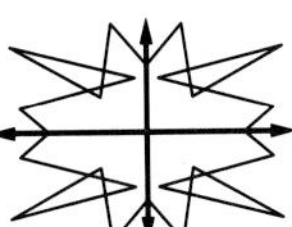

2. line

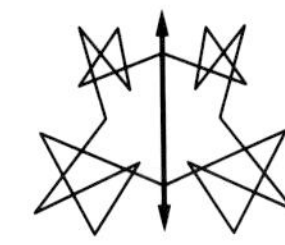

3. both

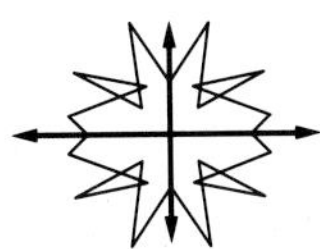

4. both

5. line

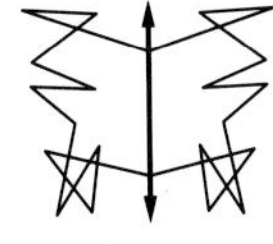

6. line

7. Draw a line of symmetry for each figure. Then sketch the next two figures in the pattern.

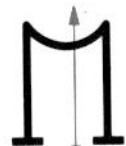

Name ______________________ Date ____________

Creating Shapes with Rotations

A 

B 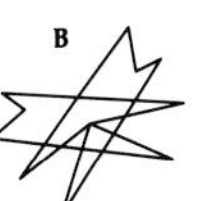

C 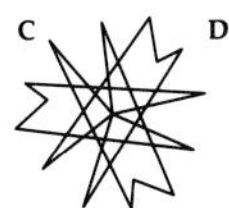

D 

A. Start with any shape. Long, thin shapes work best.
B. Make a copy of the shape and rotate it 120°.
C. Make a second copy. Rotate this copy 240°.
D. Trace around the outline. Shade the interior of the figure.

These shapes were used to make the drawings at the bottom of the page. Write the letter of the matching drawing above each shape.

1. E 2. H 3. G 4. F 5. D 6. A 7. B 8. C

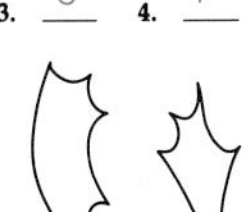

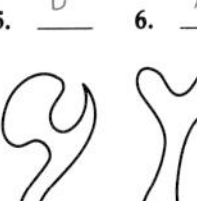

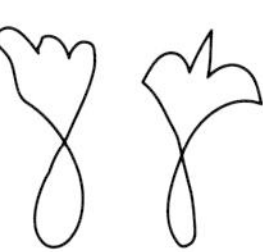

B 

C 

D 

E 

F 

G 

H 

Name ______________________ Date ______________

Applications

A problem asks for the locus of the outer end of the minute hand on a clock as the time changes from 2:00 P.M. to 3:00 P.M.

1. Which drawing would be the most helpful?

a. 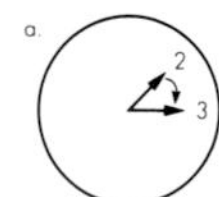

b. 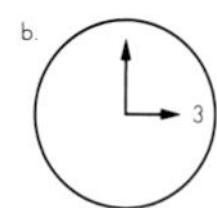

c. 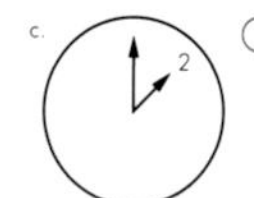

d. 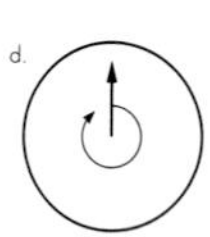

2. Does the fact that there are 60 minutes in an hour make any difference in solving the problem? Explain your answer.
No; The minute hand rotates once around a circle—there could be 100 minutes in an hour and it would not change the problem.

3. The answer to the problem is a circle with a radius equal to the length of the minute hand. How would the answer change if the problem asked for the locus of the minute hand rather than the "outer end of the minute hand?"
The answer would be the circle and its interior.

A problem asks for the locus of the end of a pump handle which is 33 inches long from its end to the point about which it pivots. The handle may be moved through an angle of 100°.

4. Which drawing would be the most helpful?

a. 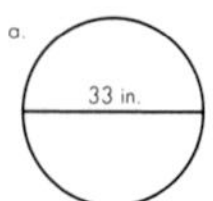

b. 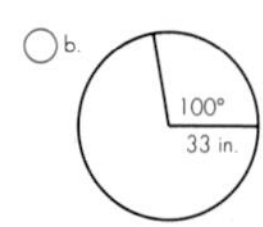

c. 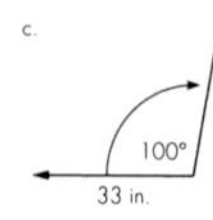

d. 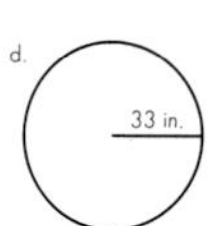

5. What formula would you use to find the actual distance the handle can move?
The formula for the circumference of a circle, $C = \pi d$.

6. Does it make a difference whether the handle is moved clockwise or counterclockwise? Explain your answer.
No; The path of the end of the handle is the same shape regardless of the direction.

Name ______________________ Date ______________

Conic Sections

Various plane curves are created by slicing a double, right circular cone in different directions. The curves that result are called *conic sections.*

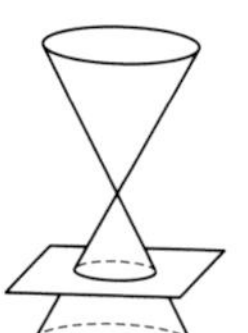

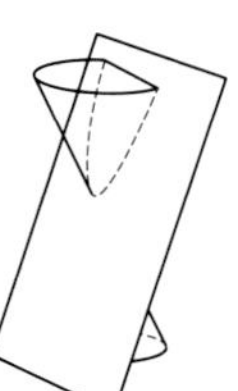

When the cutting plane is parallel to the base of one of the cones, the curve is a *circle.*

If the cutting plane is tilted, the curve is an *ellipse,* or oval. If the plane is parallel to one of the sides of the cone, the curve is a *parabola.*

Write *circle, ellipse,* or *parabola* to describe the curve formed by the edge of the shaded part of each cone.

1. ellipse
2. circle
3. parabola
4. circle

Each drawing shows a side view of a cone. The thick line shows where the cone is sliced. Write *circle, ellipse,* or *parabola* for the curve formed by each slice.

5. circle
6. parabola
7. ellipse

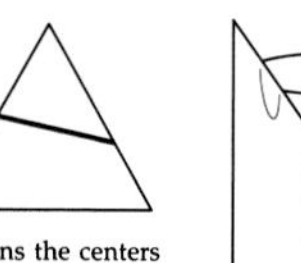
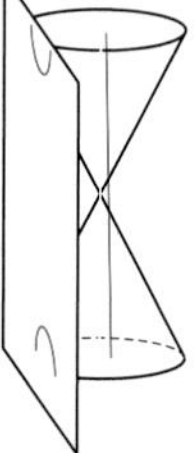

8. In the double circular cone at the right, a line that joins the centers of the top and bottom circles would pass through the vertex of the cone. This is called the *axis* of the double cone. Draw the axis.

9. If the cutting plane is parallel to the axis of the double cone, it forms a figure called a *hyperbola.* A hyperbola has two parts, or branches. Draw the hyperbola that would result from the cutting plane in the figure.

Name ______________________ Date ______________

Logic for Decision Making

Equivalent logical statements have the same truth value. If a statement is true, then all statements equivalent to it are also true. If a statement is false, all statements equivalent to it are false. Each box below shows a set of equivalent statements.

Equivalent Statements

1. *A* implies *B*. If *A*, then *B*. If not *B*, then not *A*. All *A* are *B*.

2. *B* implies *A*. If *B*, then *A*. If not *A*, then not *B*. All *B* are *A*.

3. *A* implies the negative of *B*. If *A*, then not *B*. If *B*, then not *A*. No *A* is *B*.

4. *B* implies the negative of *A*. If *B*, then not *A*. If *A*, then not *B*. No *B* is *A*.

Use this statement for Exercises 1–8.

If a ruler has 12 inches, it is 1 foot long.

To which set of equivalent statements above does each statement belong? Write *none* or a number from 1 to 4. Then tell whether the statement is *true* or *false*.

1. No 1-foot long objects are rulers. 4; false
2. All 12-inch rulers are 1 foot long. 1; true
3. If an object is not 12 inches long, then it cannot be a ruler. none; false
4. If an object is a ruler, then it is not 12 inches long. none; false
5. If a ruler is not 12 inches long, then it is not a 1-foot ruler. 2; true
6. All rulers are 12 inches long. none; false
7. All 1-foot long objects are 12-inch rulers. 2; false
8. No 12-inch rulers are 1 foot long. 3; false

Name ______________________ Date ______________

The Transitive Property of Deduction

Two statements of the form *A implies B* can be used to create a third statement. If *A* implies *B* and *B* implies C, then *A* implies C. This is called the *transitive property of deduction.* To use the property, it is helpful to remember that certain forms of statements are equivalent—they have the same truth value. Two sets of equivalent statements are shown below.

The Transitive Property

If *A* implies *B* and *B* implies C, then *A* implies C.

Equivalent Statements

A implies *B*. If *A*, then *B*. If not *B*, then not *A*. All *A* are *B*.

A implies the negative of *B*. If *A*, then not *B*. If *B*, then not *A*. No *A* is *B*.

Ring the letter of the equivalent statement, if any, for each.

1. Stan's dog always barks in the night.
   a. If Stan's dog is barking, it is nighttime.
   (b.) If it is nighttime, Stan's dog will bark.
   c. If Stan's dog is not barking, it is daytime.
   d. None of the above

2. I never saw a purple cow.
   a. If an animal is purple, it is not a cow.
   b. There are no purple cows.
   (c.) No cow that I have seen has been purple.
   d. None of the above

Write a conclusion for each group of statements. Other logical conclusions are possible.

3a. Jack plays basketball if he isn't working at the supermarket.
b. All work and no play makes Jack a dull boy.
c. The supermarket is closed on Sundays.
Jack is not dull.

4a. Fire causes heat.
b. If you can't stand the heat, get out of the kitchen.
c. Where there's smoke, there's fire.
When there's smoke in the kitchen, get out.

5a. To be on the school baseball team, you must come to practice.
b. Tropical fish can't ride bicycles.
c. Everyone rides a bicycle to practice.
Tropical fish can't be on the school baseball team.

6a. A good pet is one that is easy to train.
b. No wild animal is easy to train.
c. A cocker spaniel is not a wild animal.
Cocker spaniels make good pets.

Name ______________________ Date ______________

Tessellations

In the space below each exercise, draw a tessellation using the given figure. Then write a Logo procedure to tell the turtle to draw each tessellation. Answers will vary.

1.

2.

3.

4.

Computer • Chapter 12 147

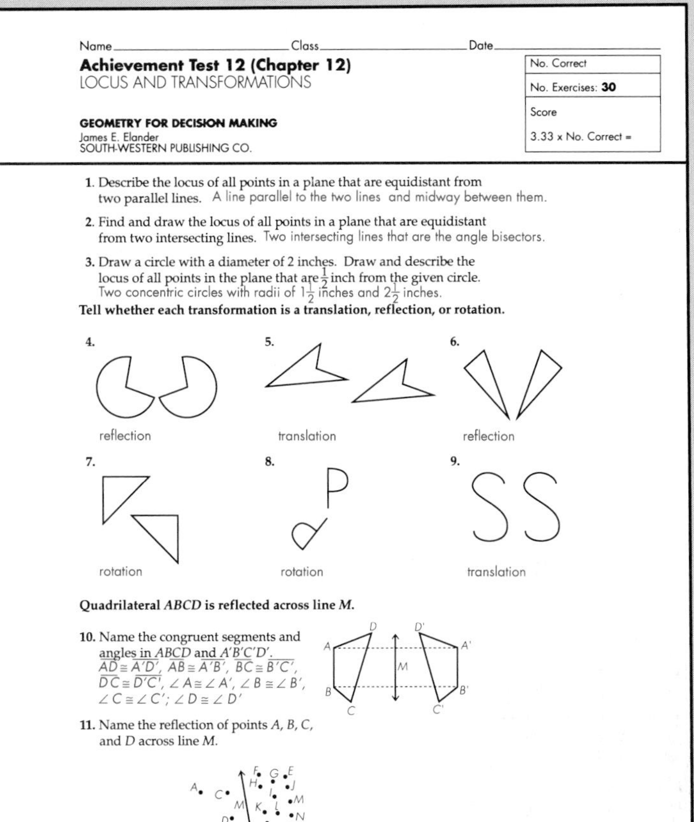

Name__________ Class__________ Date__________

Achievement Test 12 (Chapter 12)
LOCUS AND TRANSFORMATIONS

GEOMETRY FOR DECISION MAKING
James E. Elander
SOUTH-WESTERN PUBLISHING CO.

No. Correct
No. Exercises: **30**
Score
3.33 x No. Correct =

1. Describe the locus of all points in a plane that are equidistant from two parallel lines. A line parallel to the two lines and midway between them.
2. Find and draw the locus of all points in a plane that are equidistant from two intersecting lines. Two intersecting lines that are the angle bisectors.
3. Draw a circle with a diameter of 2 inches. Draw and describe the locus of all points in the plane that are $\frac{1}{2}$ inch from the given circle. Two concentric circles with radii of $1\frac{1}{2}$ inches and $2\frac{1}{2}$ inches.

Tell whether each transformation is a translation, reflection, or rotation.

4. reflection
5. translation
6. reflection
7. rotation
8. rotation
9. translation

Quadrilateral *ABCD* is reflected across line *M*.

10. Name the congruent segments and angles in *ABCD* and *A′B′C′D′*. $\overline{AD} \cong \overline{A'D'}$, $\overline{AB} \cong \overline{A'B'}$, $\overline{BC} \cong \overline{B'C'}$, $\overline{DC} \cong \overline{D'C'}$, $\angle A \cong \angle A'$, $\angle B \cong \angle B'$, $\angle C \cong \angle C'$; $\angle D \cong \angle D'$
11. Name the reflection of points *A*, *B*, *C*, and *D* across line *M*. E, N, H, K

MG01AG

[12-1]

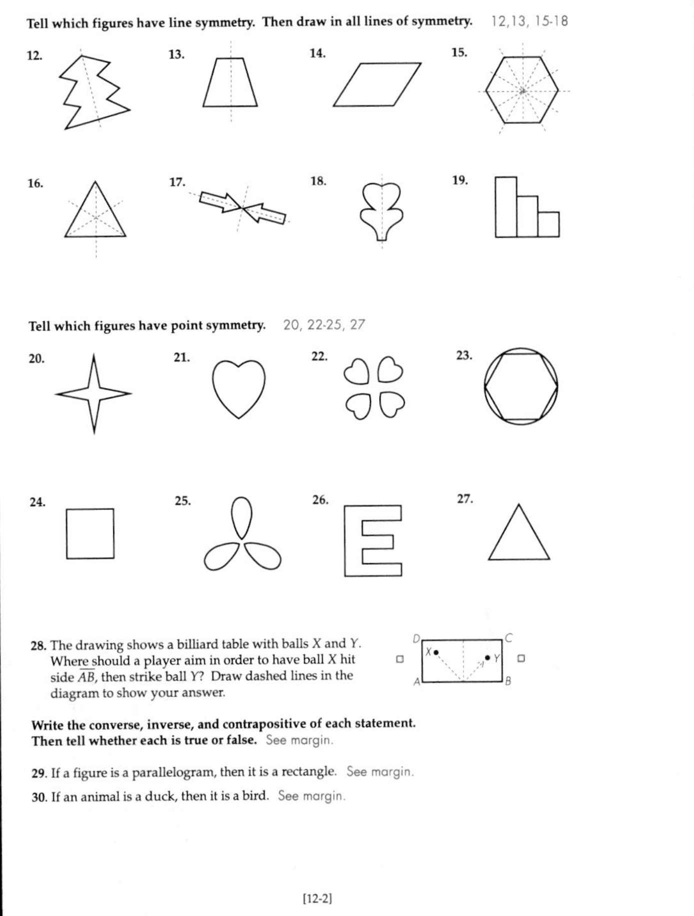

Tell which figures have line symmetry. Then draw in all lines of symmetry. 12, 13, 15-18

12. 13. 14. 15. 16. 17. 18. 19.

Tell which figures have point symmetry. 20, 22-25, 27

20. 21. 22. 23. 24. 25. 26. 27.

28. The drawing shows a billiard table with balls *X* and *Y*. Where should a player aim in order to have ball *X* hit side $\overline{AB}$, then strike ball *Y*? Draw dashed lines in the diagram to show your answer.

Write the converse, inverse, and contrapositive of each statement. Then tell whether each is true or false. See margin.

29. If a figure is a parallelogram, then it is a rectangle. See margin.
30. If an animal is a duck, then it is a bird. See margin.

[12-2]

29. Statement is false.
Converse: if a figure is a rectangle, then it is a parallelogram; true.
Inverse: If a figure is not a parallelogram, then it is not a rectangle; true.
Contrapositive: If a figure is not a rectangle, then it is not a parallelogram; false.
30. Statement is true.
Converse: If an animal is a bird, then it is a duck; false.
Inverse: If an animal is not a duck, then it is not a bird; false.
Contrapositive: If an animal is not a bird, then it is not a duck; true.

Addditional Answers for Chapter 12

Page 470

4. Statement is true.
Converse: If a triangle is acute, then it is equilateral; false.
Inverse: If a triangle is not equilateral, then it is not acute; false.
Contrapositive: If a triangle is not acute, then it is not equilateral; true.

5. Statement is true.
Converse: If I don't pitch at the game, then I stay at home; false.
Inverse: If I don't stay at home, then I pitch at the game; false.
Contrapositive: If I pitch at the game, then I don't stay at home; true.

6. Statement is false.
Converse: If a figure has equal diagonals, then it is a quadrilateral; false.
Inverse: If a figure is not a quadrilateral, then it does not have equal diagonals; false.
Contrapositive: If a figure does not have equal diagonals, then it is not a quadrilateral; false.

7. Statement is true.
Converse: If two lines are coplanar, then they are parallel; false.
Inverse: If two lines are not parallel, then they are not coplanar; false.
Contrapositive: If two lines are not coplanar, then they are not parallel; true.

8. Statement is true.
Converse: If two triangles are congruent, then they have the same size and the same shape; true.
Inverse: If two triangles do not have the same size and the same shape, then they are not congruent; true.
Contrapositive: If two triangles are not congruent, then they do not have the same size and the same shape; true.

9. Statement is false.
Converse: If a quadrilateral is a rectangle, then it has two pairs of parallel sides; true.
Inverse: If a quadrilateral does not have two pairs of parallel sides, then it is not a rectangle; true.
Contrapositive: If a quadrilateral is not a rectangle, then it does not have two pairs of parallel sides; false.

10. Statement is true.
Converse: If the temperature is 32°F or less, then it is snowing; false.
Inverse: If it is not snowing, then the temperature is not 32°F or less; false.
Contrapositive: If the temperature is not 32°F or less, then it is not snowing; true.

Page 476

4.

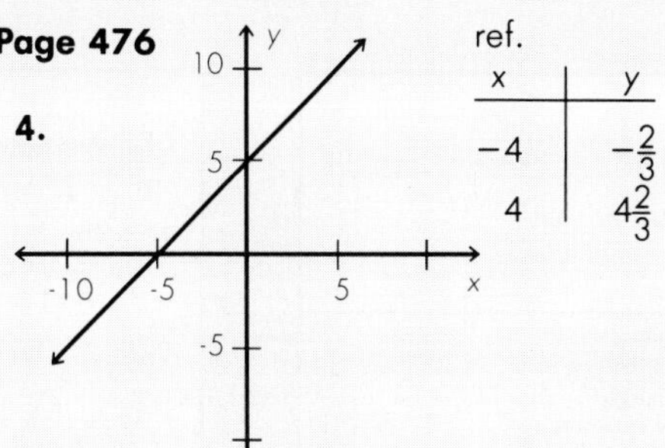

5.

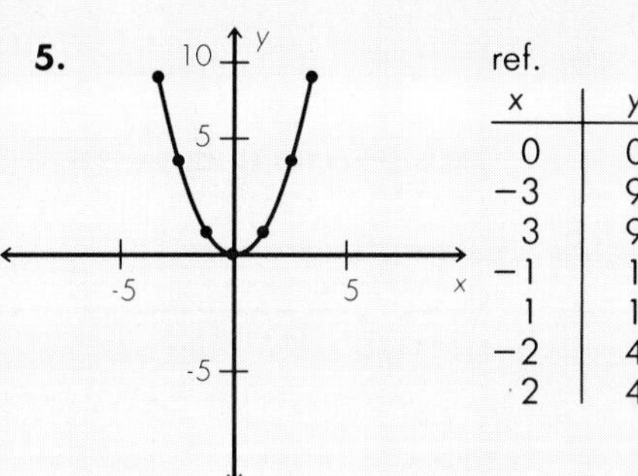

6.

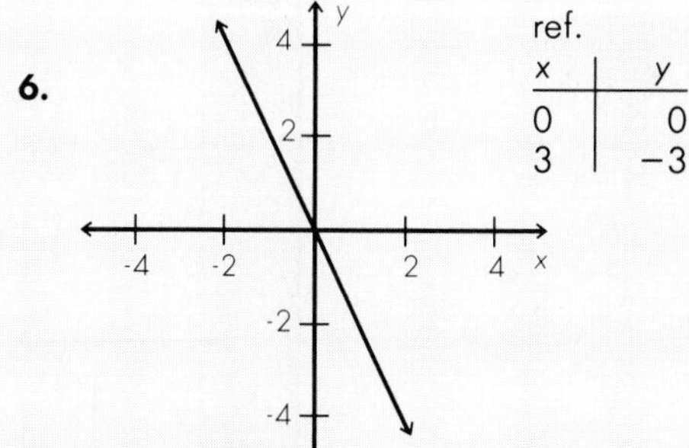

7.

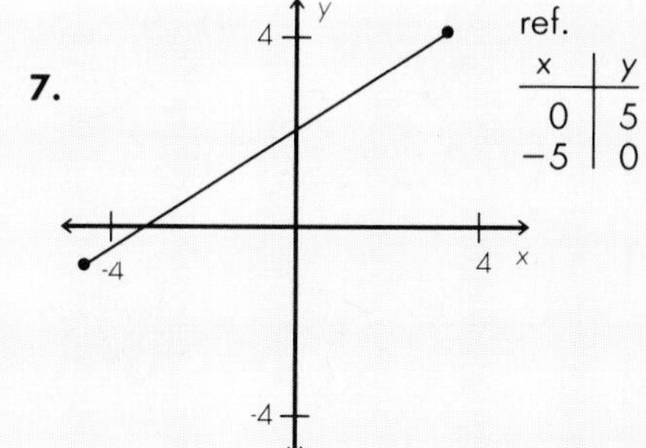

8.

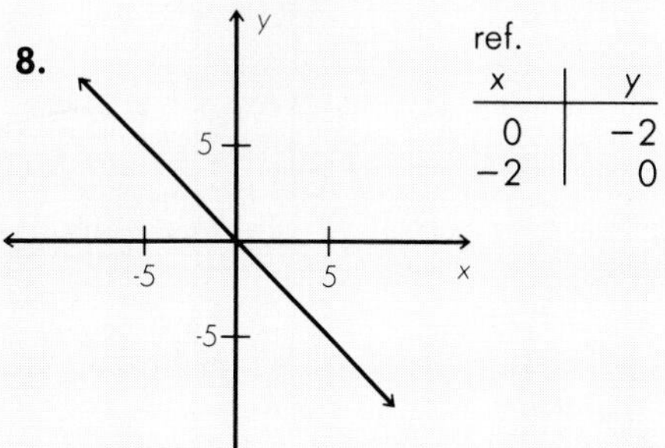

9.

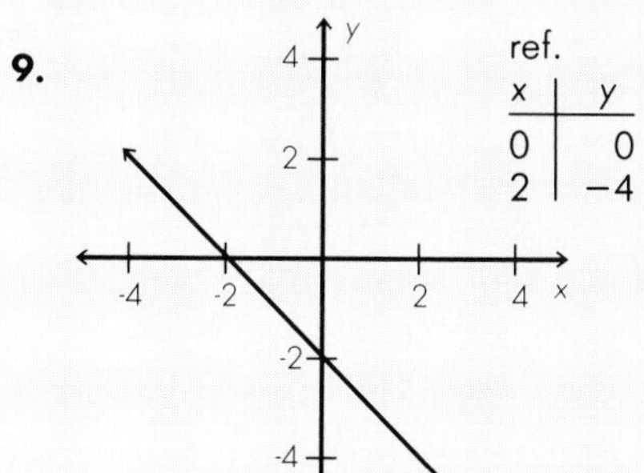

10.

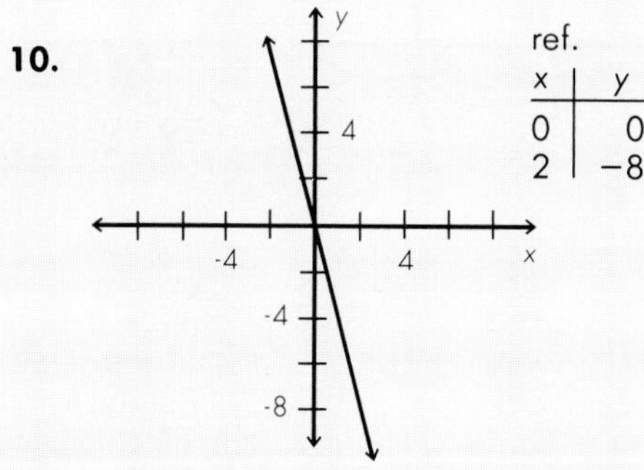

11.

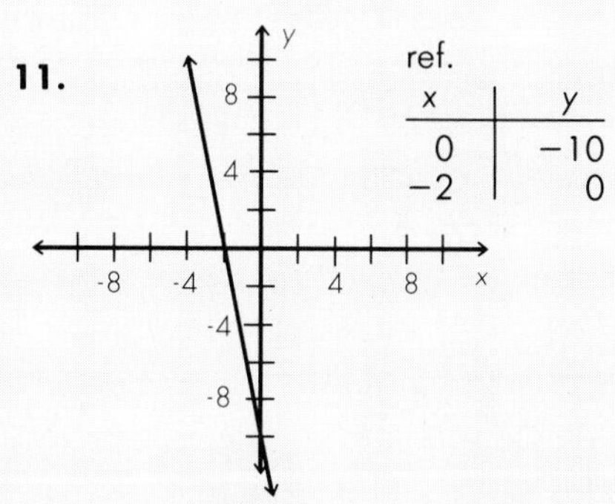

12.

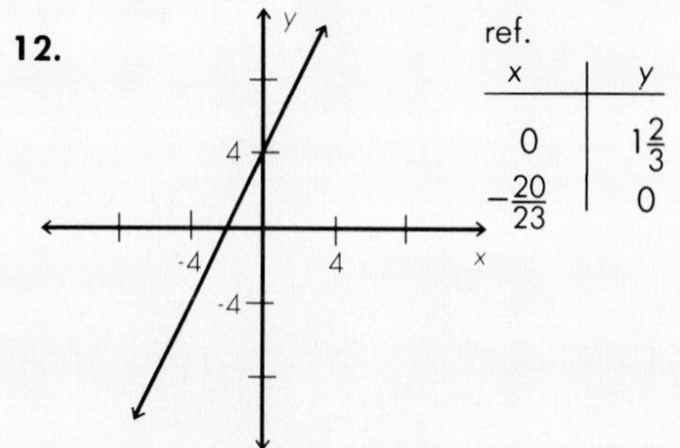

13.

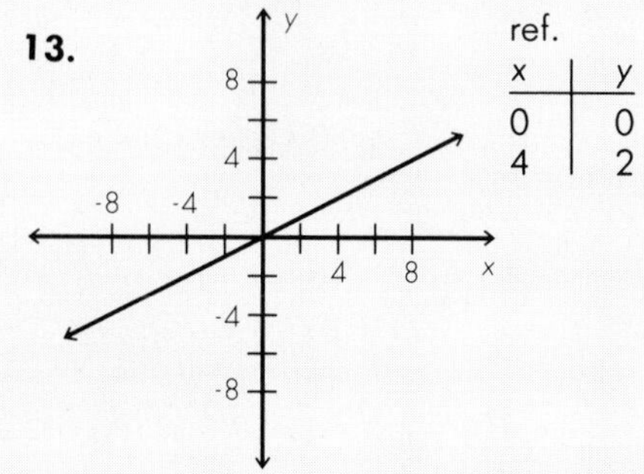

13

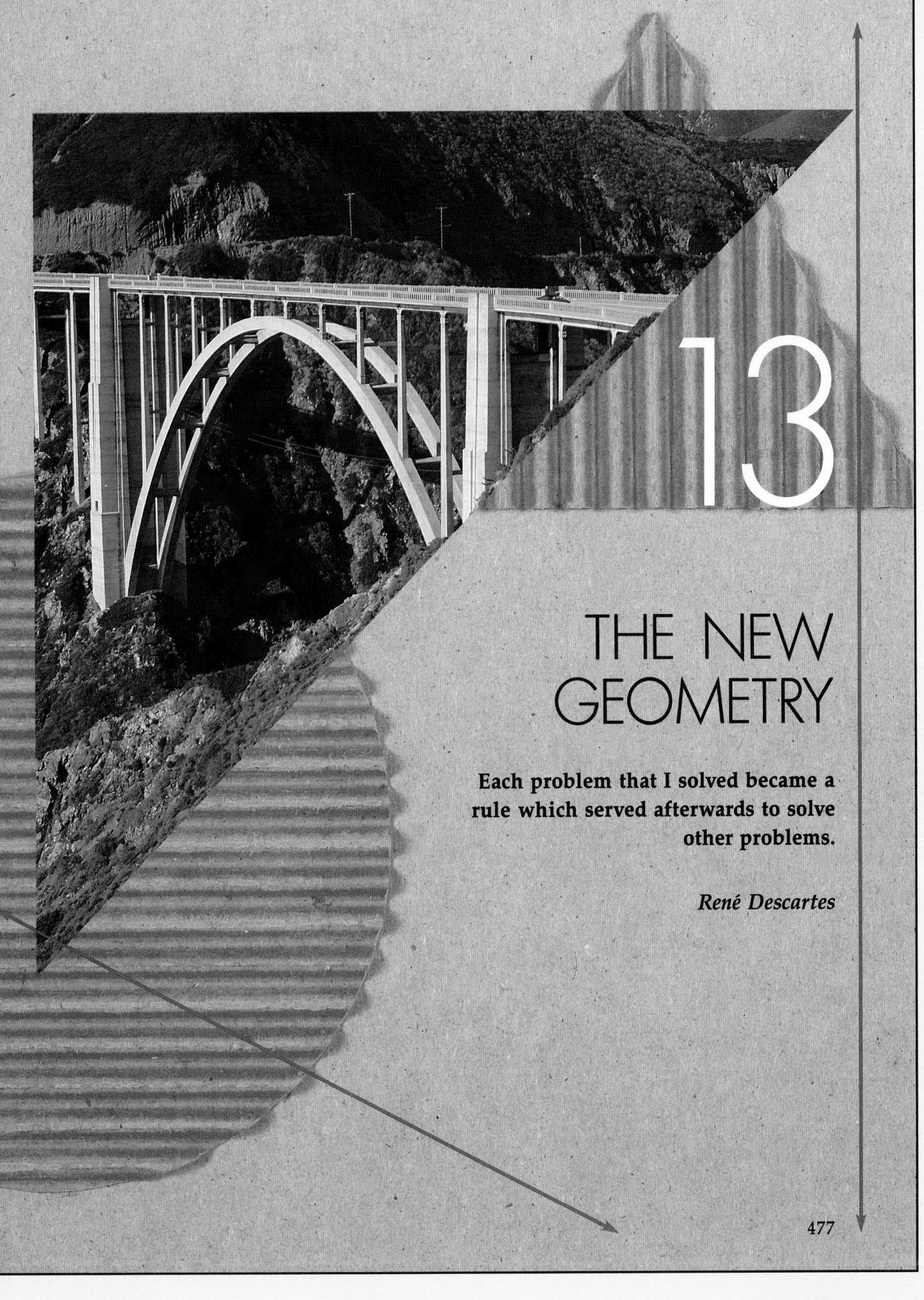

13

THE NEW GEOMETRY

Each problem that I solved became a rule which served afterwards to solve other problems.

René Descartes

477

13

CHAPTER 13 OVERVIEW

The New Geometry

Suggested Time: 10–12 days

The quote at the beginning of the chapter reflects Descartes's desire to find general methods for proving theorems in geometry. By means of coordinate geometry, basic geometric figures can be described by algebraic equations. The major emphasis in the chapter is on using equations to describe lines. The basic theorems tell how parallel and perpendicular lines can be identified by examining their slopes. By using the definitions of a circle and a parabola and the distance formula, it is easy to obtain equations for both kinds of curves. The chapter concludes by showing how coordinate geometry can be used to prove theorems about geometric figures. Students consider how a proper choice of coordinates for the points on the figures can ensure generality and simplify the algebra in proofs.

Each problem that I solved became a rule which served afterwards to solve other problems.

René Descartes

CHAPTER 13 OBJECTIVE MATRIX

	End of Chapter Items by Activity				Student Supplement Items by Activity		
Objectives by Number	**Review**	**Test**	**Calculator**	**Algebra Skills**	**Reteaching**	**Enrichment**	**Computer**
13.1.1	✓	✓	✓			✓	
13.1.2	✓	✓	✓	✓	✓	✓	✓
13.2.1	✓	✓		✓	✓	✓	
13.2.2	✓	✓		✓			✓
13.2.3	✓	✓					
13.3.1	✓	✓		✓	✓	✓	
13.3.2	✓	✓		✓	✓		
13.4.1	✓				✓		
13.4.2	✓	✓		✓			
13.5.1	✓	✓		✓	✓		
13.6.1	✓	✓		✓			

* A ✓ beside a Chapter Objective under Algebra Skills indicates that algebra skills taught within the chapter or in previous Algebra Skills lessons are used.

CHAPTER 13 PERSPECTIVES

▲ Section 13.1

What is a Point and What is a Line?

Through a Discovery Activity and a Class Activity, students see that the graph of any equation of the form $y = mx + b$, where m and b are real numbers, is a line in the coordinate plane. A simple procedure is presented for finding an equation of this form for any nonvertical line in the coordinate plane, given the coordinates of two points on the line. This procedure helps students understand why vertical lines can be represented by equations of the form $x = a$ but not by equations of the form $y = mx + b$.

In Critical Thinking students examine a fallacious proof that $2 = 0$. The flaw in the proof comes in a step which is invalid because it involves division by zero.

▲ Section 13.2

The Easy Way to Graph a Line

The slope of the graph of $y = mx + b$ is defined to be the number m. Students learn how to find the slope of a nonvertical line, given two points on the line. Students see that since division by zero is not allowed, a vertical line has no slope.

The y-intercept of a line is the y-coordinate of the point where the line crosses the y-axis. By replacing x with zero and solving for y, students see that the y-intercept of $y = mx + b$ is b.

In Critical Thinking, students are presented with a problem that can be approached in different ways—for example, algebraically or by drawing a diagram.

13

▲ Section 13.3

Slopes of Perpendicular Lines

Two of the fundamental theorems of coordinate geometry are developed in this section. Each is approached through its own Discovery Activity. The first of these theorems states that if two nonvertical lines in the coordinate plane are parallel, then their slopes are equal. The other states that if two nonvertical lines in the coordinate plane are perpendicular, then the product of their slopes is -1. The converses of the two theorems are also true.

▲ Section 13.4

Circles

A circle is the locus of all points in a plane that are at a given distance from a given point. In this section, students see how the condition in such a locus definition can be used with the distance formula to obtain an equation for a circle. Students then move to a general theorem which states that the points $P(x, y)$ whose coordinates satisfy $x^2 + y^2 = r^2$, where r is a positive real number, form a circle with radius r and center (0, 0).

The second part of this section leads students to apply what they know about circles and slopes of perpendicular lines.

▲ Section 13.5

Parabolas

Like Section 13.4, this section shows how the conditions of a locus definition can lead to an equation for the locus. Students see how the distance formula can be used to write an equation for a parabola.

The Discovery Activity develops the idea that the graphs of $y = x^2$ and $y = -x^2$ are reflections of one another with respect to the x-axis. The Class Activity that follows explores how changing the value of k in $y = kx^2$ affects the shape of the parabola.

A brief discussion of conic sections is also included in this section, and students are encouraged to prepare a report on the history of conic sections.

▲ Section 13.6

Applications

This section focuses on using coordinate geometry to prove geometric theorems from. In a Class Activity, students prove the midpoint theorem. This theorem makes it possible to find the coordinates of the midpoint of a line segment in terms of the coordinates of its endpoints.

The Discovery Activity helps students see how to use coordinate geometry to prove that the segment joining the midpoints of the two sides of a triangle is parallel to the third side and half as long as the third side. Students are encouraged to observe how careful placement of a figure in the coordinate plane can help simplify the algebra.

SECTION 13.1

Resources
Reteaching 13.1
Enrichment 13.1
Transparency Master 13-1

Objectives
After completing this section, the student should understand
- ▲ how the term *point* is defined in coordinate geometry.
- ▲ how equations are used to define *lines.*

Materials
ruler
graph paper

Vocabulary
point
graph
line

GETTING STARTED

Quiz: Prerequisite Skills
For Exercises 1-3, refer to the diagram.

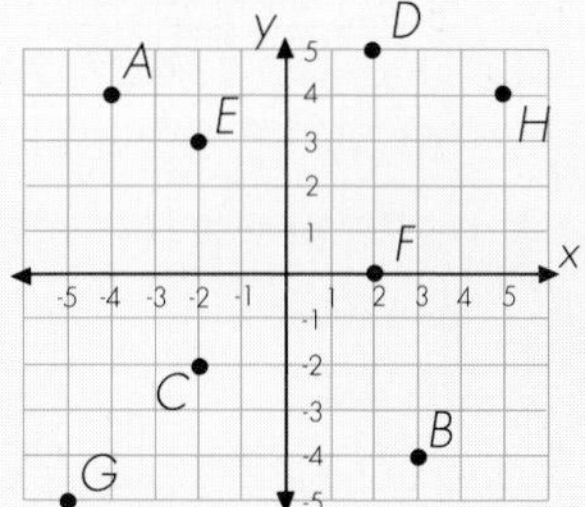

1. Tell what ordered pair corresponds to each point. [A(−4, 4); B(3, −4); C(−2, −2); D(2, 5); E(−2, 3); F(2, 0); G(−5, −5); H(5, 4)]

2. Name two points that determine a vertical line. [C and E or D and F]

3. Name two points that determine a horizontal line. [A and H]

4. Solve the system of equations.
$3m + b = 8$
$2m + b = 4$
[$m = 4$, $b = -4$]

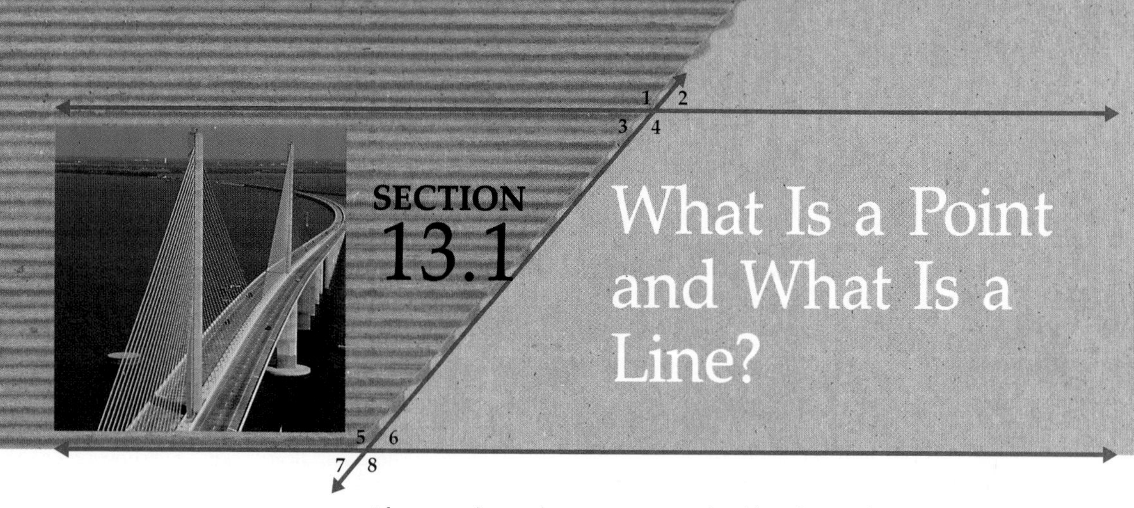

SECTION 13.1 What Is a Point and What Is a Line?

After completing this section, you should understand
- ▲ how the term *point* is defined in coordinate geometry.
- ▲ how equations are used to define *lines.*

The seventeenth-century mathematician René Descartes is credited with introducing coordinates into the study of geometry. All at once it became possible to translate questions about shapes into questions about numbers and equations. In other words, questions about geometry became questions in algebra. This was the great inspiration behind Descartes's "new" geometry.

You already know that coordinate geometry uses ordered pairs of real numbers to talk about points.

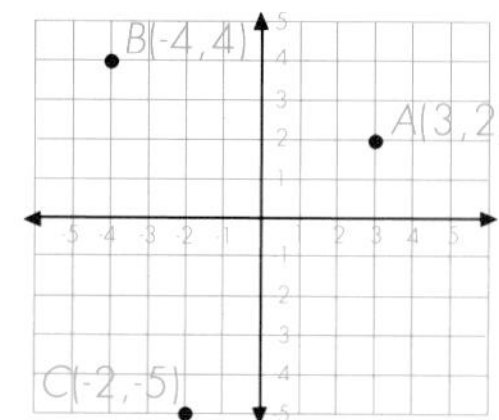

Example: In the figure at the right, you can speak of point A or of point (3,2). Point B is $(-4,4)$, and point C is $(-2,-5)$.

Of course it is always helpful to have diagrams to refer to. But in the "new" geometry—coordinate geometry—it is possible to define what a point is by using numbers only.

DEFINITION 13-1-1 **A POINT is an ordered pair (x,y) of real numbers.**

You learned in Chapter 12 that a geometric locus is a set consisting of all the points that satisfy a certain condition. In coordinate geometry, the condition is often an equation. The locus is the **graph** of the equation.

DISCOVERY ACTIVITY

See what you can discover about the graph of $y = 2x + 1$. You can find points on the graph by picking values for x and finding the values for y. For example if $x = 3$, then $y = 2 \cdot 3 + 1$, or 7. (3,7) is a point on the graph, since $y = 2x + 1$ will be true when you replace x with 3 and y with 7.

1. Copy and complete this table to find six points on the graph of the equation $y = 2x + 1$.
2. Draw and label x- and y-axes on a piece of graph paper. Mark the points for the ordered pairs listed in the last column of the table. See margin.
3. What do you notice about the points you marked? They look collinear.

x	y	*Point*
3	7	(3,7)
$2\frac{1}{2}$	6	$(2\frac{1}{2}, 6)$
1	3	(1,3)
0	1	(0,1)
−2	−3	(−2,−3)
−4	−7	(−4,−7)

You may have discovered that the points you marked in the coordinate plane all lie on the same line l. If you use more values for x to find more points on the graph of $y = 2x + 1$, you will find that they are on the same line. The graph of $y = 2x + 1$ is the line l.

CLASS ACTIVITY

For each equation, make a table like the one in the Discovery Activity. Use the same values for x. Graph the ordered pairs in the coordinate plane. If the points for the equation seem to be on the same line, draw the line. See margin.

1. $y = 2x - 3$
2. $y = 0x + 4$
3. $y = 2x + 1$

For each equation, make a table like the one in the Discovery Activity. Use the x values given. Mark the points in the coordinate plane. If the points seem to be on the same line, draw the line. See margin.

4. $y = -3x + 2;\ x = 2\frac{1}{2}, 1, 0, -\frac{1}{2}, -1, -2$
5. $y = -\frac{1}{2}x - 4;\ x = 6, 1, 0, -1, -3, -4$

Additional Answers

2.

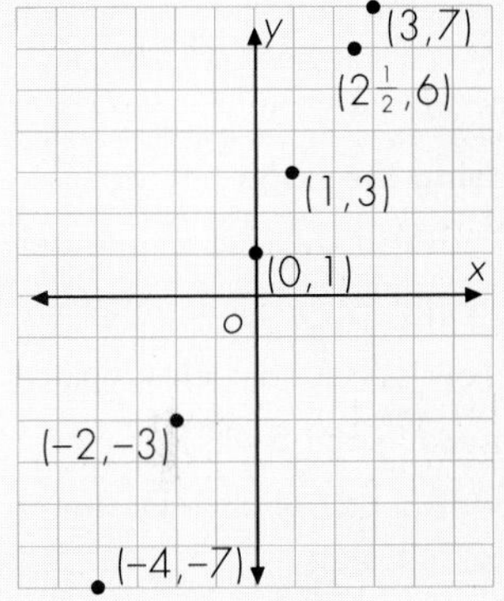

Additional Answers

The *Point* columns for the tables are given. Check students' graphs against these.

1.

Point
(3, 3)
$(2\frac{1}{2}, 2)$
(1, −1)
(0, −3)
(−2, −7)
(−4, −11)

2.

Point
(3, 4)
$(2\frac{1}{2}, 4)$
(1, 4)
(0, 4)
(−2, 4)
(−4, 4)

3.

Point
(3, 7)
$(2\frac{1}{2}, 6)$
(1, 3)
(0, 1)
(−2, −3)
(−4, −7)

4.

Point
$(2\frac{1}{2}, -5\frac{1}{2})$
(1, −1)
(0, 2)
$(-\frac{1}{2}, 3\frac{1}{2})$
(−1, 5)
(−2, 8)

5.

Point
(6, −7)
$(1, -4\frac{1}{2})$
(0, −4)
$(-1, -3\frac{1}{2})$
$(-3, -2\frac{1}{2})$
(−4, −2)

TEACHING COMMENTS

Using the Page

Students may need a refresher on solving equations in two unknowns. Have students replace m with $\frac{8}{7}$ and b with $\frac{11}{7}$ in the original pair of equations to check the solution.

Using the Page

Many students will feel more secure that $y = \frac{8}{7}x + \frac{11}{7}$ is the correct equation if you make a table of values and draw the graph. Do this using Transparency Master 13-1. Be sure the table of values includes the values -4 and 3 for x so students can see that the two points originally given are indeed on the graph.

Each equation in the Class Activity has the form $y = mx + b$, where m and b are real numbers. Also, the graph of each equation is a line. If an equation has the form $y = mx + b$, where m and b are real numbers, then the graph is a straight line.

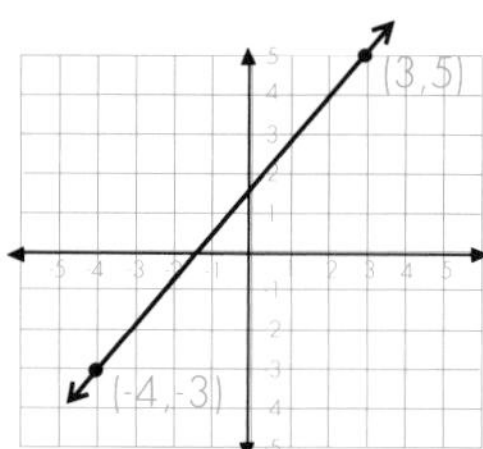

Is the reverse also true? If you have a line in the coordinate plane, is there an equation $y = mx + b$ whose graph is the given line? To find out, pick a line in the coordinate plane, say the line determined by $(-4,-3)$ and $(3,5)$. If these points are on the graph of $y = mx + b$, then $y = mx + b$ must be true when you put the coordinates in place of x and y.

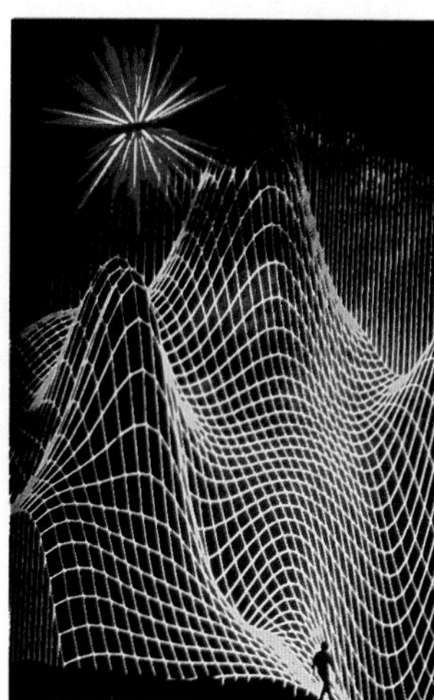

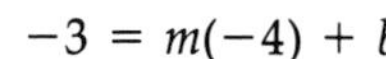

$$-3 = m(-4) + b$$

$$5 = m(3) + b$$

Subtract the second equation from the first.

$$-8 = -7m$$
$$m = \frac{-8}{-7} = \frac{8}{7}$$

Put $\frac{8}{7}$ in place of m in $-3 = m(-4) + b$.

$$-3 = \frac{8}{7}(-4) + b$$
$$-3 = \frac{-32}{7} + b$$
$$b = -3 + \frac{32}{7} = \frac{11}{7}$$

So the line through $(-4,-3)$ and $(3,5)$ is the graph of $y = \frac{8}{7}x + \frac{11}{7}$.

You can use this method for finding an equation for every *nonvertical* line in the coordinate plane. To see why the method fails for vertical lines, try using it to find an equation of the form $y = mx + b$ for the line passing through $(2,3)$ and $(2,-5)$.

$$3 = m(2) + b$$
$$-5 = m(2) + b$$

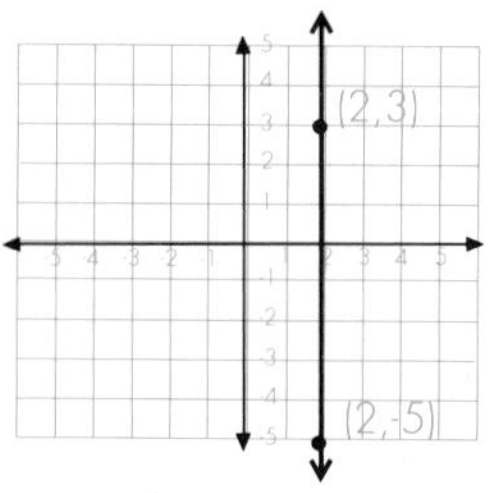

There are no real numbers m and b that will let $m(2) + b$ have two different values (3 and -5) at the same time. Therefore, there is no equation of the form $y = mx + b$ for this line.

An equation for the vertical line through (2,3) and (2,−5) is $x = 2$. Every vertical line has an equation of the form $x = a$, where a is a real number. All nonvertical lines have an equation of the form $y = mx + b$. You can use these ideas to state a "new geometry" definition of the term *line.*

DEFINITION 13-1-2 **A LINE is the set of all ordered pairs (points) in the coordinate plane that satisfy the equation of the form $y = mx + b$ or of the form $x = a$, where m, b, and a are real numbers.**

Using the Page
Discuss the fact that in the "old" geometry, the term *line* was left as an undefined term.

CLASS ACTIVITY

For each line, write an equation of the form $y = mx + b$ or of the form $x = a$.

1. $y = \frac{1}{2}x + \left(-\frac{1}{2}\right)$

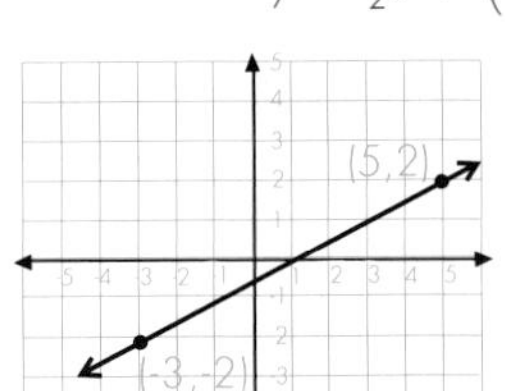

2. $x = -3$

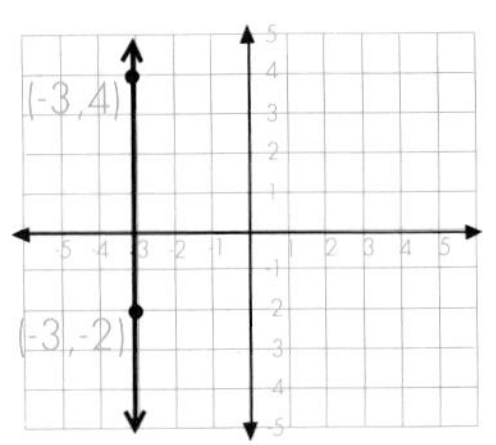

Using the Page
After finding the equations, students might be asked to make tables and graphs for some of their equations. This check will give them confidence that their equations are correct. Be sure the tables include the x-coordinates of the given points as values of x.

3. $x = \frac{5}{2}$

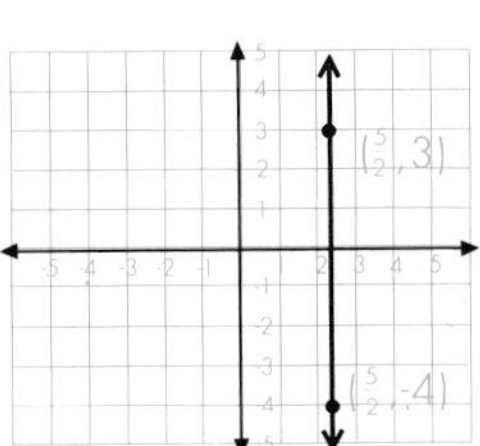

4. $y = -3x + 2$

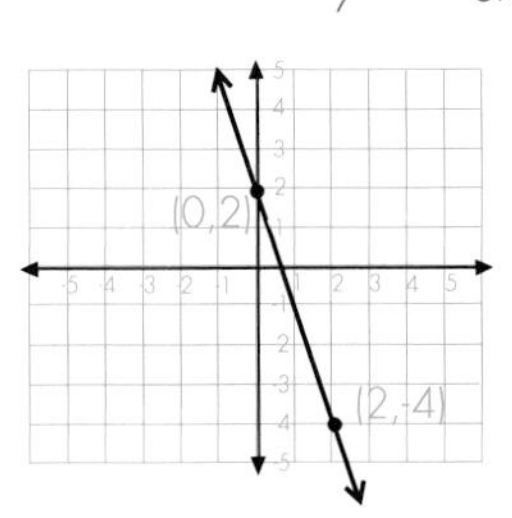

Use the XYPLANE procedure from page 118 to tell the turtle to draw the line determined by each pair of points. Then write an equation for the line.

5. (1,1) and (−1,−5)
   $y = 3x + (-2)$
6. (−2,6) and (7,−3)
   $y = -x + 4$
7. (7,0) and (7,4)
   $x = 7$
8. (0,−3) and (4,−11)
   $y = -2x + (-3)$
9. (−5,−2) and (−5,3)
   $x = -5$
10. $\left(\frac{1}{2},-6\right)$ and $\left(\frac{1}{2},5\right)$
    $x = \frac{1}{2}$

Extra Practice
Write an equation for the line determined by each pair of points.

1. (7, 8) and (0, 2)
$\left[y = \frac{6}{7}x + 2\right]$
2. (−7, −4) and (0, 2)
$\left[y = \frac{6}{7}x + 2\right]$

Using the Page
16. Review with students the idea that division can be checked by using multiplication. For example, 56 ÷ 7 = 8 because 8 × 7 = 56. Point out that no matter what number you try to use to answer 56 ÷ 0 = ?, it will not check. So division by zero is not allowed.

FOLLOW-UP

Assessment
1. Graph the equation $y = \frac{3}{2}x - 4$.

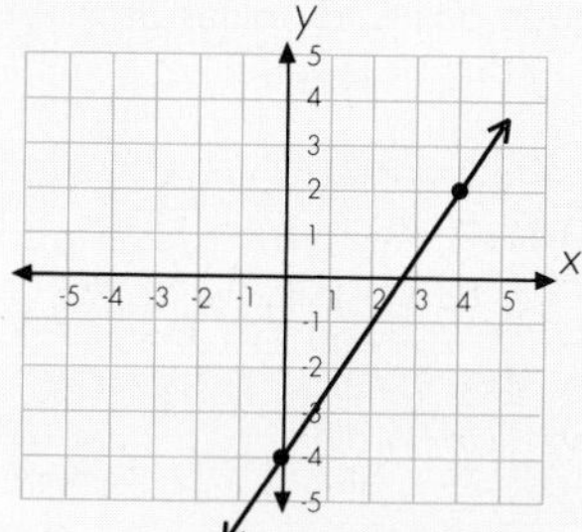

2. Find an equation for the line determined by (−8, 3) and (−4, 1).
$\left[y = -\frac{1}{2}x - 1\right]$
3. Is the point (3, 4) on the line y = x + 1? How do you know? [Yes; 4 = 3 + 1 is a true equation.]

Reteaching
Students who have had difficulty with this section may benefit from Reteaching Activity 13.1.

Enrichment
For students who have mastered the material in this section, you may wish to assign Enrichment Activity 13.1.

HOME ACTIVITY

Graph each equation in the coordinate plane. (First make a table of ordered pairs.) See margin.

1. $y = 2x + 6$
2. $y = -x - 3$
3. $y = 3x - 6$
4. $x = -4$
5. $x = 5$
6. $y = -3x + 5$

Tell whether the graph of the equation will be a vertical or a nonvertical line. You do not have to draw the graph.

7. $y = 8x - 2$ Nonvertical
8. $y = -\frac{3}{2}x + 6$ Nonvertical
9. $x = -9$ Vertical

For each line, write an equation of the form $y = mx + b$ or of the form $x = a$.

10.

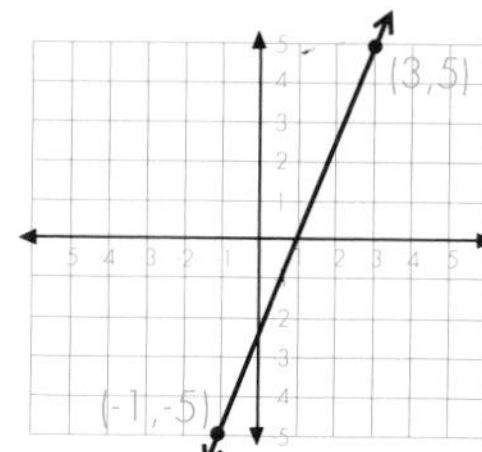

$y = \frac{5}{2}x + \left(-\frac{5}{2}\right)$

11.

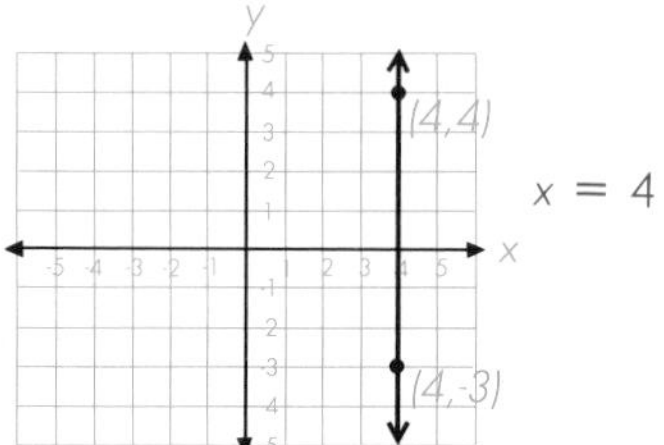

$x = 4$

Use the XYPLANE procedure from page 118 to tell the turtle to draw the line determined by each pair of points. Then write the equation.

12. (0,0) and (5,5) $y = x$
13. (−1,2) and (3,−4) $y = -\frac{3}{2}x + \frac{1}{2}$
14. (5,0) and (5,8) $x = 5$
15. (6,5) and (7,4) $y = -x + 11$

CRITICAL THINKING

16. Explain what is wrong with the following "proof" that 2 = 0. Assume that the numbers x and y are equal.

Step 5 involves dividing by $(x - y)$. But $x - y = 0$, since $x = y$. Division by zero is not allowed.

Steps	Reasons
1. $x = y$	1. Given
2. $x^2 = y^2$	2. Squares of equal are equal.
3. $x^2 - y^2 = 0$	3. Subtract y^2 from both sides.
4. $(x + y)(x - y) = 0$	4. Factoring
5. $x + y = 0$	5. Divide both sides by $x - y$.
6. $x + x = 0$	6. Substitution
7. $2x = 0$	7. Simplification
8. $2 = 0$	8. Divide both sides by x.

SECTION 13.2 The Easy Way to Graph a Line

After completing this section, you should understand
- ▲ the slope of a line.
- ▲ how to find the slope of a line given two points on the line.
- ▲ the y-intercept of a line.

You have seen that the graph of any equation of the form $y = mx + b$, where m and b are real numbers, is a nonvertical line. The numbers m and b give useful information about the graph. The number m in the equation $y = mx + b$ has a special name.

DEFINITION 13-2-1: **The number m in $y = mx + b$ is the SLOPE of the line graph.**

If (x_1, y_1) and (x_2, y_2) are any two points on the graph $y = mx + b$, then these two equations are true:

$$y_2 = mx_2 + b$$
$$y_1 = mx_1 + b$$

Subtract the second equation from the first and you get

$$y_2 - y_1 = mx_2 - mx_1$$
$$\text{or } y_2 - y_1 = m(x_2 - x_1).$$

Divide both sides by $(x_2 - x_1)$ and you find that

$$m = \frac{y_2 - y_1}{x_2 - x_1}.$$

This reasoning gives you the following theorem.

SECTION 13.2

Resources
Reteaching 13.2
Enrichment 13.2
Transparency Master 13-1

Objectives
After completing this section, the student should understand
- ▲ the slope of a line.
- ▲ how to find the slope of a line given two points on the line.
- ▲ the y-intercept of a line.

Materials
graph paper
ruler

Vocabulary
rise
run
slope
y-intercept

GETTING STARTED

Chalkboard Example
Draw coordinate axes and mark the points (−2, 1) and (3, 7). Draw the line they determine. Discuss the fact that one can start at (−2, 1), move 5 units parallel to the x-axis in the positive direction, then 6 units parallel to the y-axis in the positive direction, and arrive at (7, 1). Ask how to use horizontal and vertical moves to go from (3, 7) to (−2, 1).

TEACHING COMMENTS

Using the Page
During the discussion that follows Definition 13-2-1, ask the class whether $x_2 - x_1$, or $y_2 - y_1$ might be equal to zero.

THEOREM 13-2-1: If (x_1,y_1) and (x_2,y_2) are two points on the graph of $y = mx + b$, then the slope m of the graph is equal to

$$\frac{y_2 - y_1}{x_2 - x_1}.$$

Using the Page
It may help to discuss the fact that *if one is moving from left to right*, the run will always be positive. The rise will be positive if $y_2 > y_1$; zero, if $y^2 = y_1$; and negative, if $y^2 < y_1$. Therefore the sign (positive, negative, or neither) of the slope m will be determined by the sign of $y_2 - y_1$. It may help to illustrate this with some specific examples. Use Transparency Master 13-1 to present diagrams for class discussion.

Think of getting from (1,2) to (5,4) in the coordinate plane. You can start at (1,2), *run* four units parallel to the x-axis, then *rise* two units parallel to the y-axis. You arrive at (5,4). The ratio of *rise* to *run* equals the slope of the line passing through the two points.

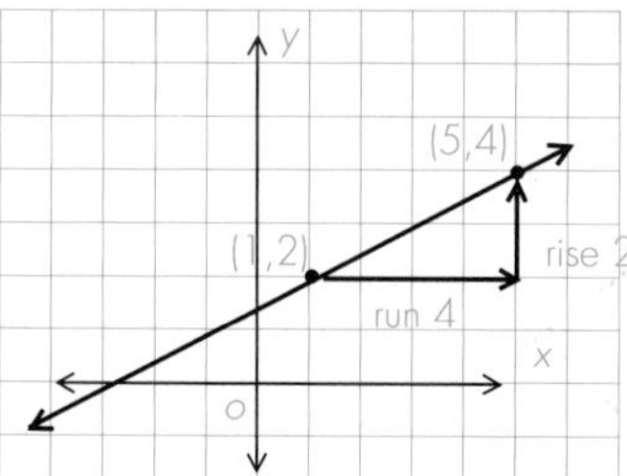

If you use the word **run** to mean the change in x and **rise** to mean the change in y, then you can say that the slope of any nonvertical line is the ratio of rise to run. This ratio is constant because it does not depend on which two points of the line you select.

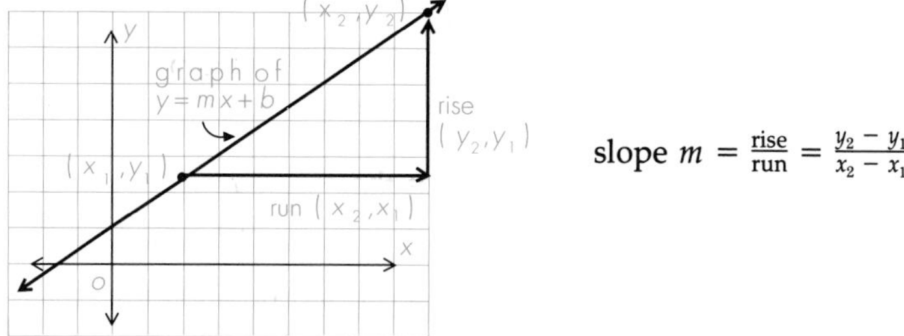

The greater the rise is in comparison to the run, the *steeper* the line is. The steeper the line is, the greater the absolute value of its slope. The Discovery Activity will help you find out more about the slope of a line.

DISCOVERY ACTIVITY

Each diagram shows a nonvertical line and two points on the line. Study the diagrams and answer the questions.

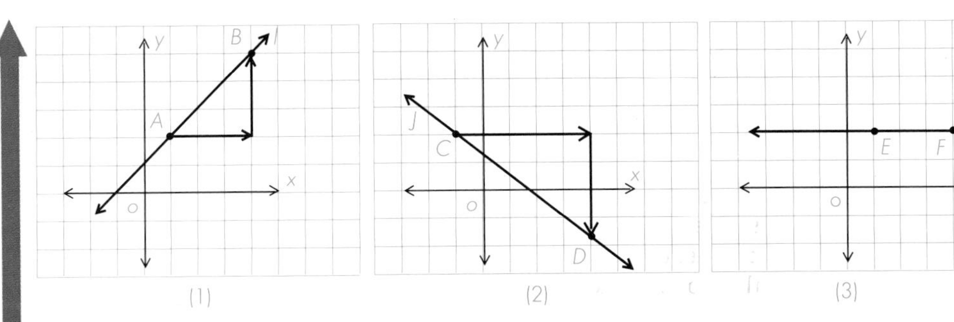

1. In diagram 1, the run is positive. Is the *rise* positive, negative, or zero? Is the *slope* positive, negative, or zero? Does line *l* go up or down as you move from left to right? Positive; positive; up

2. In diagram 2, the run is positive. Is the *rise* positive, negative, or zero? Is the *slope* positive, negative, or zero? Does line *j* go up or down as you move from left to right? Negative; negative; down

3. In diagram 3, the run is again positive. Is the *rise* positive, negative, or zero? Is the *slope* positive, negative, or zero? Does line *k* go up or down as you move from left to right? 0; 0; neither

Using the Page
Consider what would happen in moving from right to left parallel to the x-axis instead of from left to right. For example, in diagram (1) the run from *B* to *A* would be negative and the rise would be negative. The ratio of rise to run would still be positive.

You may have discovered that as you move from left to right (parallel to the x-axis), lines with positive slope go up, lines with negative slope go down, and lines with zero slope stay parallel to the x-axis.

None of the diagrams in the Discovery Activity were vertical. What happens if the line is vertical?

For a vertical line such as the line m, the run is zero and the rise is positive. The ratio rise/run cannot be found, because in algebra division by zero is not allowed: **a vertical line has no slope.**

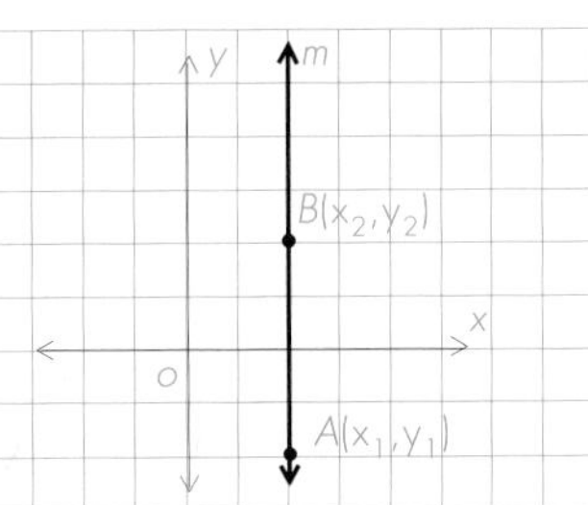

CLASS ACTIVITY

If the graph of the equation is a nonvertical line, give the slope. If the graph is a vertical line, answer "no slope."

1. $y = 7x - 5$ 7
2. $y = -3x + 4$ −3
3. $x = 6$ No slope
4. $y = -2$ 0

Using the Page
In Exercises 5-7, ask students to tell what the run is and what the rise is if one moves from the point on the left to the point on the right.

Using the Page
Students may use $m = \frac{y_2 - y_1}{x_2 - x_1}$ to find the slopes in Exercises 8-11. Ask if it makes a difference which point is used for (x_1, y_1) and which for (x_2, y_2). Students should understand that the slope will be the same regardless. They should be reminded, however, that they cannot use the x-coordinate of one point and the y-coordinate of the second point as (x_1, y_1).

Use the information in the graphs to find the slope of each nonvertical line. Tell whether the line goes up, down, or stays level or parallel to the x-axis, as you move from left to right in the coordinate plane.

5.

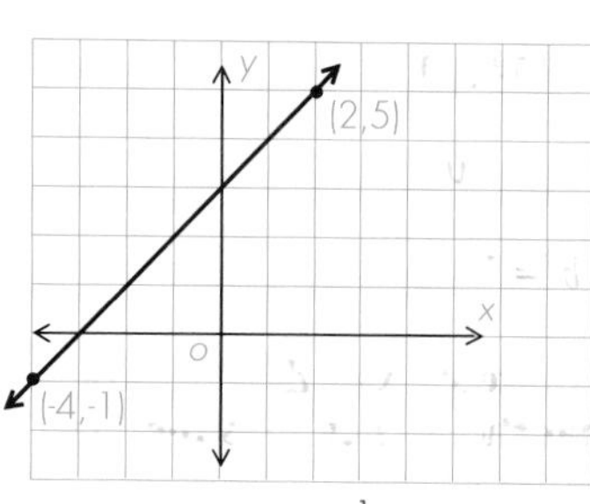

1, up

6.

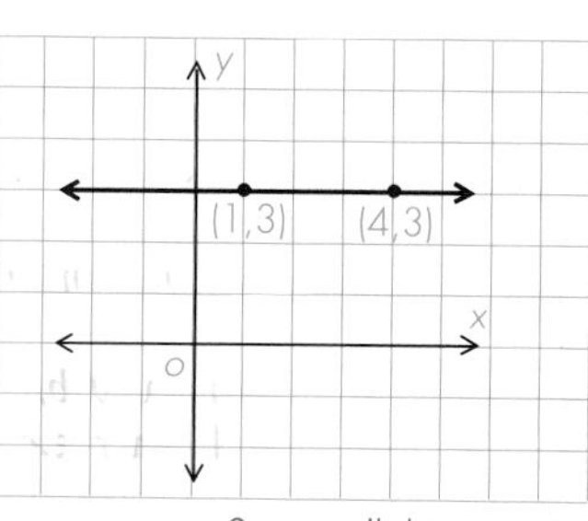

0, parallel to x-axis

7.

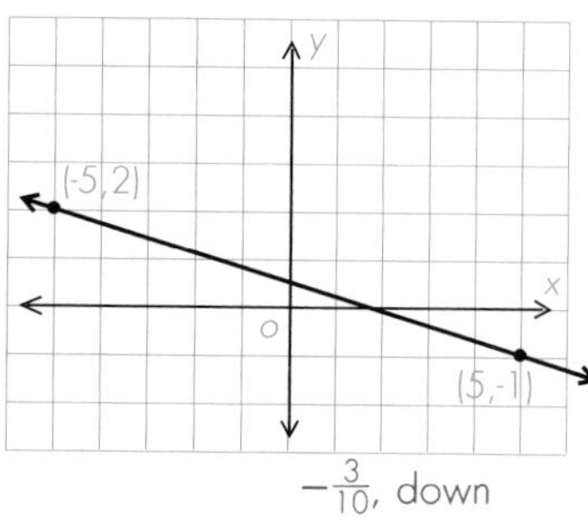

$-\frac{3}{10}$, down

Find the slope of the line passing through the given points.

8. (3,6) and (8,6) 0
9. (5,7) and (9,−12) $-\frac{19}{4}$
10. (−1,−2) and (3,−10) −2
11. $\left(\frac{1}{2},\frac{1}{2}\right)$ and (3,−4) $-\frac{9}{5}$

THE *Y*-INTERCEPT OF A LINE

Every nonvertical line in the coordinate plane intersects the y-axis. The y-coordinate of the point of intersection has a special name.

DEFINITION 13-2-2: **The *Y*-INTERCEPT of a nonvertical line in the coordinate plane is the *y*-coordinate of the point where the line intersects the *y*-axis.**

Example: In the diagram, line l intersects the y-axis at (0,−2). Its y-intercept is −2. Line l intersects the y-axis at $\left(0,3\frac{1}{2}\right)$. Its y-intercept is $3\frac{1}{2}$.

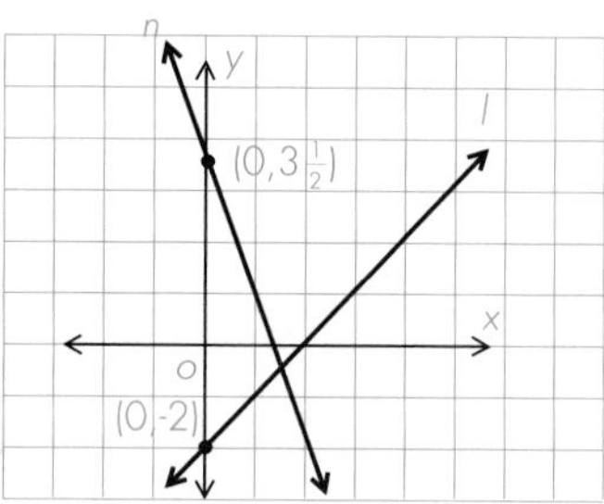

If you know an equation for a nonvertical line, it is easy to find its y-intercept. The x-coordinate of the point where the graph crosses the y-axis is $x = 0$. Put 0 in place of x in the equation and solve for y.

Example: Find the y-intercept of the graph of $y = -5x + 3$.

When you put 0 in place of x, you get

$$y = -5 \cdot 0 + 3 = 3.$$

The y-intercept is 3. The graph crosses the y-axis at (0,3).

If you put 0 in place of x in $y = mx + b$, you get

$$y = m \cdot 0 + b = b.$$

THEOREM 13-2-2: For any real numbers m and b, the y-intercept of the line graph of $y = mx + b$ is b, which means that the line crosses the y-axis at $(0,b)$.

Any equation that can be put into the form $y = mx + b$ has a graph that is a nonvertical line of slope m and y-intercept b.

Example: Graph $3x + 2y = y - 1$.

You can collect the terms with y on the left side, and the terms with x on the right side. You get

$y = -3x - 1$ or $y = -3x + (-1)$.

The graph will be the line that has slope -3 and y-intercept -1.

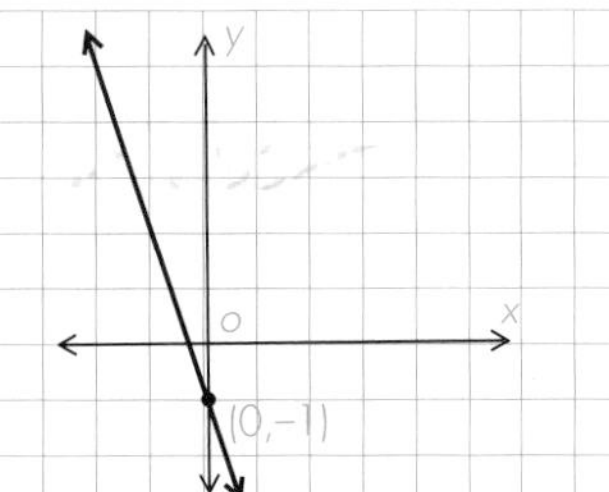

CLASS ACTIVITY

Give the slope and y-intercept of each line. See margin.

1. $y = -\frac{1}{2}x + 7$ 2. $y = 5x - 6$ 3. $x + y = 4$

4. $3x + 2y = 0$ 5. $y = -10$ 6. $y = 2x + \left(\frac{-12}{13}\right)$

Marion has a quick way for graphing any equation of the form $y = mx + b$. She marks the point where the line will cross the y-axis. She goes right one unit and then up or down as many units as the slope indicates. She draws the line through the point she arrives at and the point where she started. Use her method to graph each of the following equations. See margin.

7. $y = -2x + 3$ 8. $y = \frac{5}{2}x + 4$ 9. $y = -4x + (-2)$

Additional Answers

1. $-\frac{1}{2}$, 7
2. 5, −6
3. −1, 4
4. $-\frac{3}{2}$, 0
5. 0, −10
6. 2, $-\frac{12}{13}$
7.

x	y
0	3
1	1

8.

x	y
0	4
−2	−1

9.

x	y
0	−2
−1	2

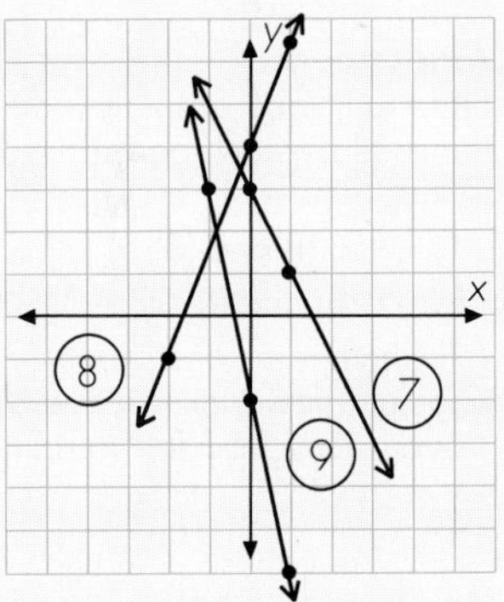

Extra Practice

Find an equation for the line determined by the points. State the slope and y-intercept.

1. (−1, −4) and (2, 9)
$\left[y = \frac{13}{3}x + \frac{1}{3}; \frac{13}{3}, \frac{1}{3}\right]$
2. (0, 0) and (5, 6)
$\left[y = \frac{6}{5}x; \frac{6}{5}, 0\right]$
3. (−2, 8) and (1, −1)
$[y = -3x + 2; -3, 2]$

Additional Answers

4. 4, 15 5. $-3, -\frac{1}{2}$
6. $-1, 6$ 7. $7, -18$
8. no slope 9. $0, 3\frac{2}{5}$
10. $-2, \frac{4}{3}$ 11. $3, -8$
12. no slope
13.

x	y
0	2
3	4

14. $y = 1\frac{1}{2}$
15.

x	y
0	2
2	1

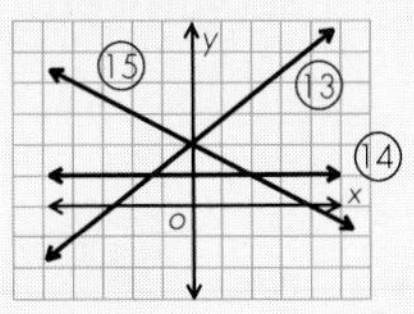

19. $\frac{12}{19}$ of the adults are married. The diagram shows why.

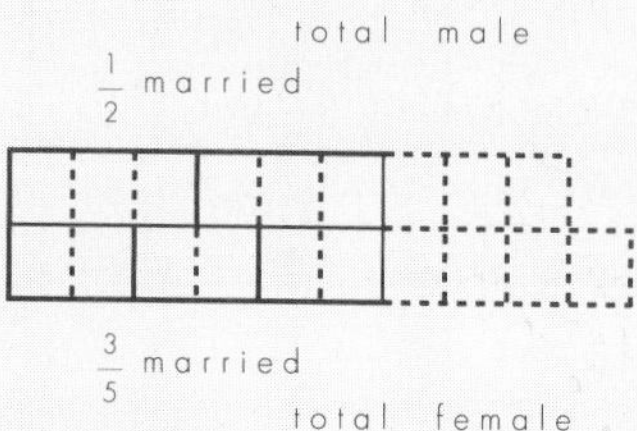

FOLLOW-UP

Assessment
State the slope of the line determined by each pair of points.

1. (−2, 4) and (3, 9) [1]
2. (7, 5) and (8, 0) [−5]

Tell whether the line graph goes up, down, or stays level as you move from left to right.

3. $y = -\frac{3}{4}x + 6$ [down]
4. $y = -9$ [level]
5. $y = 2x + (-20)$ [up]

Reteaching
Students who have had difficulty with this section may benefit from Reteaching Activity 13.2.

Enrichment
For students who have mastered the material in this section, you may wish to assign Enrichment Activity 13.2.

HOME ACTIVITY

Use the information in the diagrams to find the slope of each nonvertical line. Tell whether the line goes up or down or stays level as you move from left to right in the coordinate plane.

1. −3, down
2. $\frac{1}{2}$, up
3. 0, level

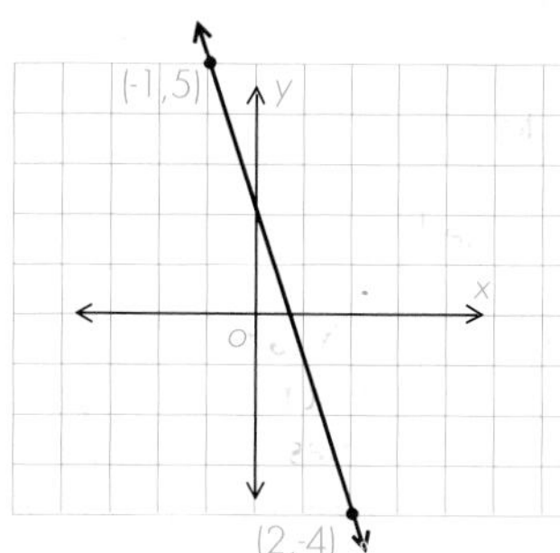

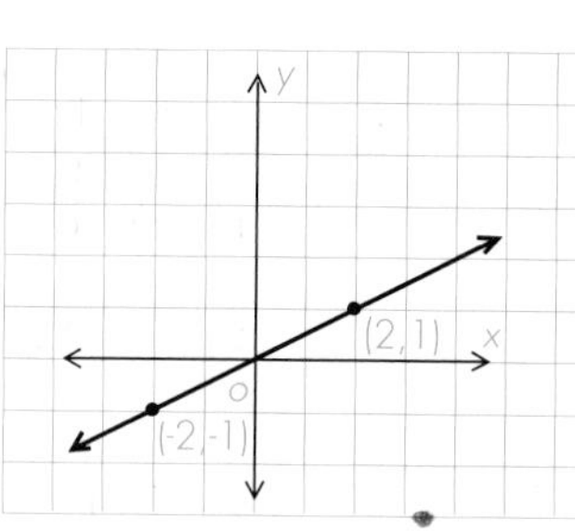

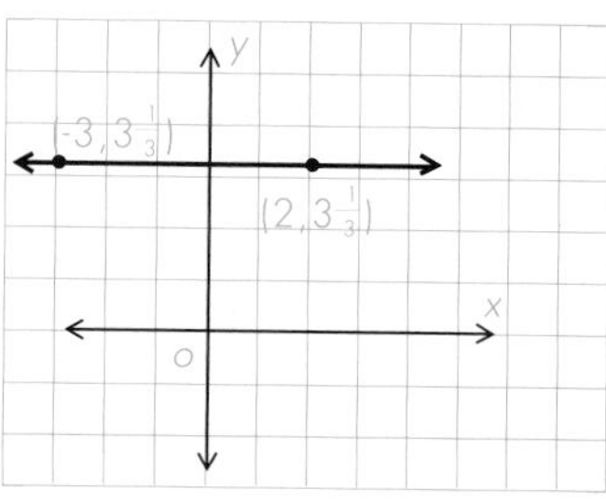

Give the slope and y-intercept of each line. If the line is vertical, write "no slope." Then use the XYPLANE procedure from page 118 in a Logo procedure to tell the turtle to draw the graph of each equation. See margin.

4. $y = 4x + 15$
5. $y = -3x + \left(-\frac{1}{2}\right)$
6. $y = -x + 6$
7. $y = 7x - 18$
8. $x = -9$
9. $y = 3\frac{2}{5}$
10. $2x + y = \frac{4}{3}$
11. $6x - 2y = 16$
12. $x + 3 = 0$

Draw the graph of each equation. See margin.

13. $y = \frac{2}{3}x + 2$
14. $y = 1\frac{1}{2}$
15. $5x + 10y = 20$

Write an equation for the line that has the given slope and y-intercept.

16. Slope: 10
y-intercept: 3
$y = 10x + 3$
17. Slope: −6
y-intercept: $2\frac{1}{2}$
$y = -6x + 2\frac{1}{2}$
18. Slope: 0
y-intercept: −7
$y = -7$

CRITICAL THINKING

19. In the town of Wilson, two-thirds of the adult men are married to three-fifths of the adult women. What fraction of the adults in Wilson are married? Explain how you arrived at your answer. See margin.

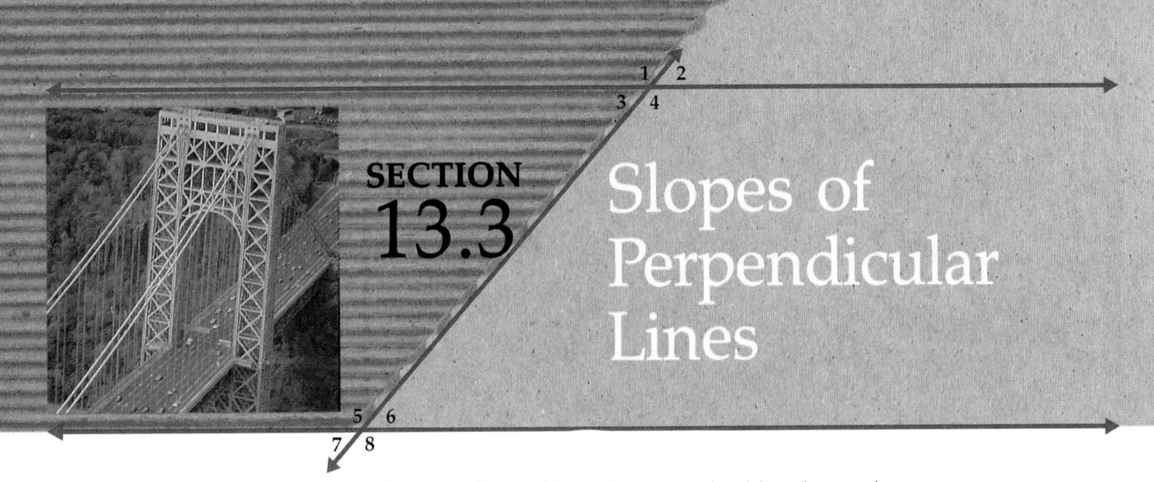

SECTION 13.3 Slopes of Perpendicular Lines

After completing this section, you should understand
- ▲ how the slopes of parallel lines are related.
- ▲ how the slopes of perpendicular lines are related.

In Section 13.2, you learned that the slope of a line gives information about how steeply a line rises and falls as you move from left to right in the coordinate plane. This suggests that the slopes of two lines might provide information about whether the lines are parallel.

SLOPES OF PARALLEL LINES

Two vertical lines, such as $x = -1$ and $x = 3$, are parallel. Their slopes cannot be compared, however, since they do not have slopes.

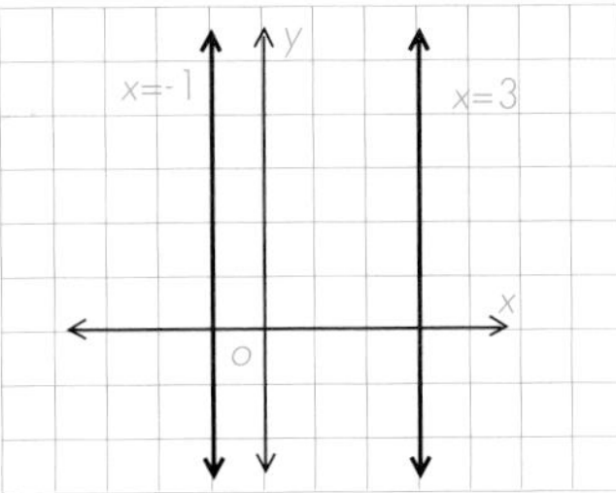

What about two horizontal lines, such as $y = -2$ and $y = 3$? The lines are parallel. Their equations can be written as follows:

$$y = 0x + (-2)$$
$$y = 0x + 3$$

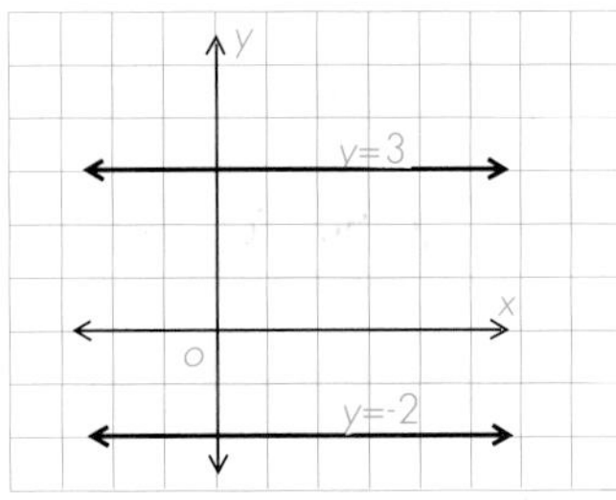

This tells you that the slopes are both equal to 0.

The Discovery Activity will help you see what the situation is for parallel lines that are neither vertical nor horizontal.

SECTION 13.3

Resources
Reteaching 13.3
Enrichment 13.3

Objectives
After completing this section, the student should understand
- ▲ how the slopes of parallel lines are related.
- ▲ how the slopes of perpendicular lines are related.

Materials
ruler
protractor
graph paper

GETTING STARTED

Quiz: Prerequisite Skills
$\triangle ABC \sim \triangle DEF$

1. Write the ratios of corresponding sides.
$\left[\frac{AB}{DE}, \frac{BC}{EF}, \frac{AC}{DF}\right]$

2. If $AB = 6$ and $DE = 8$, what ratio is $\frac{BC}{EF}$ equal to? $\left[\frac{3}{4}\right]$

Chalkboard Example
In a coordinate plane, draw the line determined by (1, 4) and (7, 8) and the line determined by (−2, 3) and (4, 7). Label the points with their ordered pairs. Have students discuss whether these lines are parallel. Ask if the lines are equally steep. Have students calculate the slope of each line. Ask for conjectures about what might be true of the slopes of nonvertical parallel lines.

Additional Answers

4. The ratios are equal. They are formed from corresponding parts of similar triangles.

Using the Page

After students have understood the reasoning in each step of the Discovery Activity, go through all the steps again with two lines that have negative slopes.

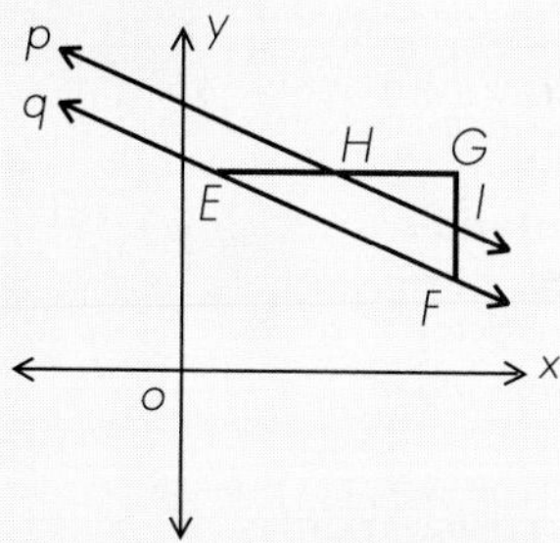

Note that for the lines p and q in the diagram, the run is positive and the rise is negative.

slope of $p = \frac{HG}{-GI} = -\frac{HG}{GI}$

slope of $q = \frac{EG}{-GF} = -\frac{EG}{GF}$

DISCOVERY ACTIVITY

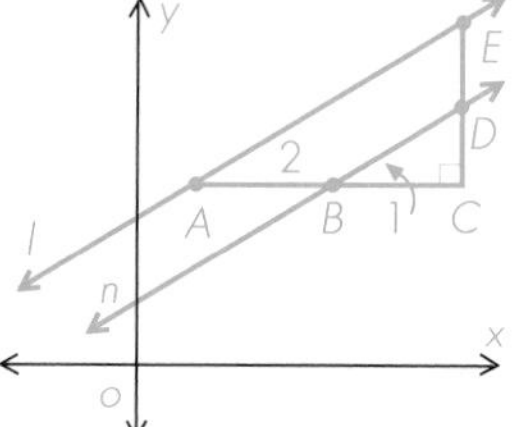

In the diagram, $l \parallel n$. Point C is below both lines. $\overline{AC}$ is horizontal and intercepts l and n at A and B, respectively. $\overline{EC}$ is vertical and intersects l and n at E and D, respectively.

1. What angle is common to $\triangle ACE$ and $\triangle BCD$? $\angle C$
2. What kind of angles are $\angle 1$ and $\angle 2$ with respect to $\overline{AC}$ and parallel lines l and m? Corresponding angles
3. $\triangle AEC$ is similar to $\triangle BDC$. Why? AA
4. Slope of $l = CE/AC$ and slope of $n = CD/BC$. How does the ratio CE/AC compare with the ratio CD/BC? How do you know? See margin.
5. How does the slope of l compare with the slope of n? Slope of l = slope of n.

You may have discovered that if two lines are parallel and are neither horizontal nor vertical, then their slopes are equal. You know that any two horizontal lines have slopes equal to 0. This leads to the following theorem.

THEOREM 13-3-1: **If two nonvertical lines in the coordinate plane are parallel, then their slopes are equal.**

The converse of this theorem is also true: if the slopes of two lines are equal, then the lines are nonvertical and parallel.

Example: The lines $y = 2x + 3$ and $y = 2x - \frac{1}{2}$ are parallel, because both have a slope of 2.

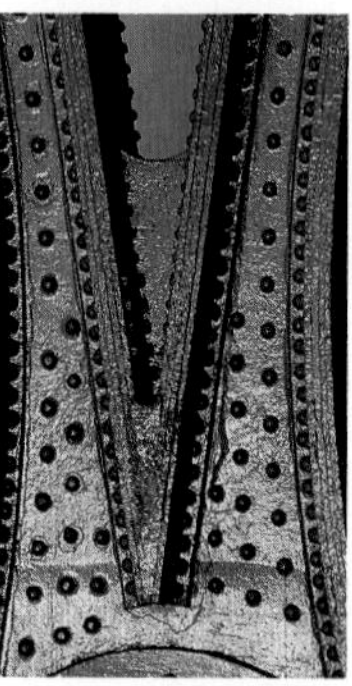

CLASS ACTIVITY

Tell whether the given lines are parallel.

1. $y = 7x + 8$, $y = 8x + 7$ No
2. $y = -3x - 1$, $y = -3x + 15$ Yes
3. $y = \frac{1}{2}x + 6$, $2y - x = -8$ Yes
4. The line determined by (0,0) and (2,−1), and the line determined by (−4,4) and (4,0) Yes
5. The line $y = 3x + (-5)$ and the line determined by (1,8) and (2,5) No
6. The line $x = 3$ and the line $x = -11$ Yes

SLOPES OF PERPENDICULAR LINES

In the coordinate plane, sketch two perpendicular lines, neither of which is vertical. You will quickly see that one slopes downward from left to right and the other upward. This gives you some idea of how their slopes are related. One is positive and the other is negative.

DISCOVERY ACTIVITY

Consider these three pairs of lines.

$\begin{cases} y = 3x - 1 \\ y = -\frac{1}{3}x + \frac{7}{3} \end{cases}$ $\begin{cases} y = \frac{2}{5}x + 3 \\ y = -\frac{5}{2}x + 3 \end{cases}$ $\begin{cases} y = \frac{4}{7}x \\ y = -\frac{7}{4}x + 2 \end{cases}$

None of these lines are vertical.

1. Graph each pair of lines on a coordinate plane. See margin.
2. Measure the angle formed by each pair of lines. What kind of angle is it? 90° angle or right angle
3. Which pairs of lines are perpendicular? All three pairs.
4. Multiply the slopes of the lines in each pair. What do you get each time? −1
5. What do you think may be true of the slopes of two nonvertical perpendicular lines? The product of their slopes is −1.

When you drew the graphs for the Discovery Activity, you may have found that all the pairs of lines form 90° angles. So the lines in each pair are perpendicular. When you multiplied the slopes of the lines, you probably discovered that, in each case, the product is −1.

Extra Practice

Write an equation for the line described.

1. The line that passes through (0, 6) and is parallel to the line $y = 7x - 1$ [$y = 7x + 6$]

2. The line that passes through $\left(0, -\frac{1}{2}\right)$ and is parallel to the line $y = \frac{4}{5}x + 3$ $\left[y = \frac{4}{5}x - \frac{1}{2}\right]$

Additional Answers

1. First pair ref. points

x	y
0	−1
2	5

x	y
0	$\frac{7}{3}$
3	$\frac{4}{3}$

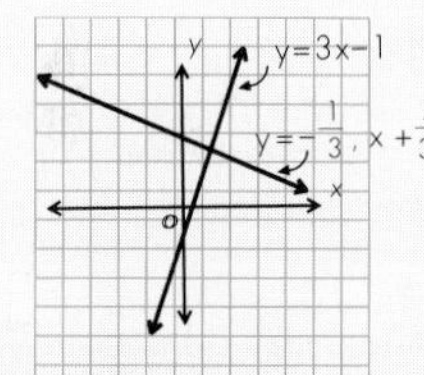

Second pair ref. points

x	y
0	3
5	5

x	y
0	3
2	−2

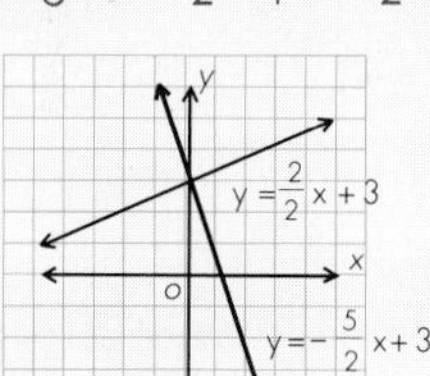

Third pair ref. points

x	y
0	0
7	4

x	y
0	2
4	−5

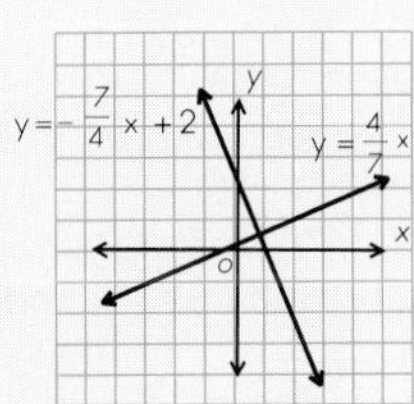

Using the Page
Discuss how Theorem 13-3-2 might be used to prove that for three lines p, q, and r, none of which is vertical or horizontal, if $p \parallel q$ and $p \perp r$, then $q \perp r$.

THEOREM 13-3-2: **If l_1 and l_2 are perpendicular lines with slopes of m_1 and m_2 respectively, then $m_1m_2 = -1$.**

The converse of this theorem is also true: if lines l_1 and l_2 have slopes of m_1 and m_2 and if $m_1m_2 = -1$, then l_1 is perpendicular to l_2.

Example: Show that the line l through $(-2,3)$ and $(1,5)$ is perpendicular to the line n through $(-2,3)$ and $(2,-3)$.

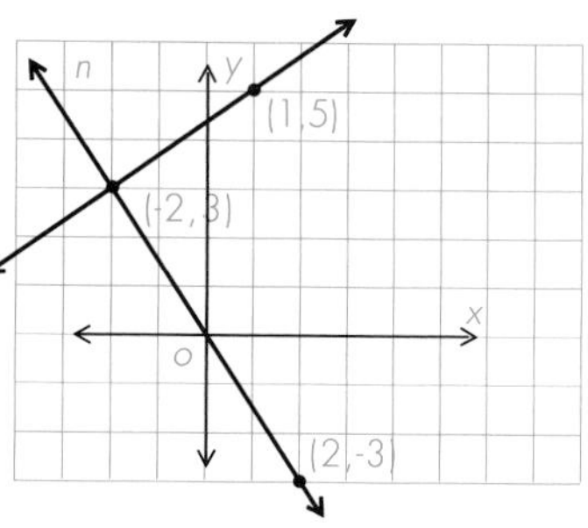

Slope of $n = \frac{3-(-3)}{-2-2} = \frac{6}{-4}$, or $\frac{-3}{2}$

Slope of $l = \frac{5-3}{1-(-2)} = \frac{2}{3}$

$\frac{-3}{2} \cdot \frac{2}{3} = -1$, so $l \perp n$.

Example: Show that the lines $y = 2x + 4$ and $y = -\frac{3}{2}x + 1$ are *not* perpendicular.

The product of their slopes is $2 \cdot \left(\frac{-3}{2}\right) = -3$. Since the product of the slopes is not -1, the lines cannot be perpendicular.

CLASS ACTIVITY

Tell whether the given pair of lines are perpendicular.

1. $y = 7x + 5$, $y = \frac{1}{7}x - 2$ No
2. $y = \frac{3}{4}x + 6$, $y = -\frac{4}{3}x + 1$ Yes
3. $2x + 6y = 3$, $-3x + y = 8$ Yes
4. The vertices of a quadrilateral are $A(-2,1)$, $B(2,5)$, $C(6,1)$, and $D(2,-3)$. Tell what kind of quadrilateral $ABCD$ is and justify your answer. See margin.
5. The vertices of a quadrilateral are $W(-1,0)$, $X(2,4)$, $Y(7,4)$, and $Z(4,0)$. Show that $WXYZ$ is not a square. See margin.
6. Show that the diagonals of the quadrilateral $WXYZ$ of Exercise 5 are perpendicular. See margin.
7. The vertices of $\triangle PQR$ are $P(-2,0)$, $Q(0,2)$, and $R(3,-3)$. Show that $\triangle PQR$ is isosceles and that the line $y = -x$ is perpendicular to and bisects the base of $\triangle PQR$. See margin.

Additional Answers

4. $ABCD$ is a square. The distance formula shows that $AB = BC = CD = DA = 4\sqrt{2}$. $\overline{AB}$ and $\overline{DC}$ have slope 1, and $\overline{BC}$ and $\overline{AD}$ have slope -1. Therefore, the angles of $ABCD$ are right angles.

5. $\overline{WZ}$ has slope 0 and $\overline{YZ}$ has slope $\frac{4}{3}$. The product of these slopes is not -1, so these sides are not perpendicular.

6. $\overline{WY}$ has slope $\frac{1}{2}$ and $\overline{XZ}$ has slope -2. The product of these slopes is -1, so diagonals $\overline{WY}$ and $\overline{XZ}$ are perpendicular.

7. $\triangle PQR$ is isosceles because $PR = QR = \sqrt{34}$. The point $S(-1, 1)$ is the midpoint of $\overline{PQ}$, because $PS = SQ = \sqrt{2}$ and $PQ = 2\sqrt{2}$. The line $y = -x$ passes through $(-1, 1)$ and $(3, -3)$. Its slope is -1. The slope of $\overline{PQ}$ is 1. The product of the slopes is -1. So $y = -x$ is perpendicular to $\overline{PQ}$ at its midpoint.

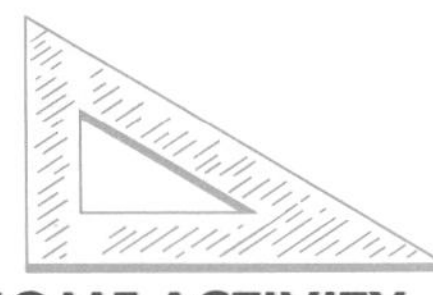

HOME ACTIVITY

Tell whether the given lines are parallel.

1. $y = 7x - 1$ Yes
   $y = 7x + 10$
2. $y = -4x + 6$ No
   $y = \frac{1}{4}x + 2$
3. $y = 10$ Yes
   $y = \frac{-1}{10}$
4. $y = \frac{6}{5}x + 12$ Yes
   $y = \frac{6}{5}x - 18$
5. $7x + 3y = 10$ No
   $y = \frac{7}{3}x - 2$
6. $4x + 5y + 6 = 0$ Yes
   $4x + 5y - 8 = 0$
7. The line determined by $(-4,-2)$ and $(-3,-5)$, and the line determined by $(1,4)$ and $(3,-2)$ Yes
8. The line $y = 5x + 9$ and the line determined by $(2,-1)$ and $(3,3)$ No

Refer to the figure *ABCD*.

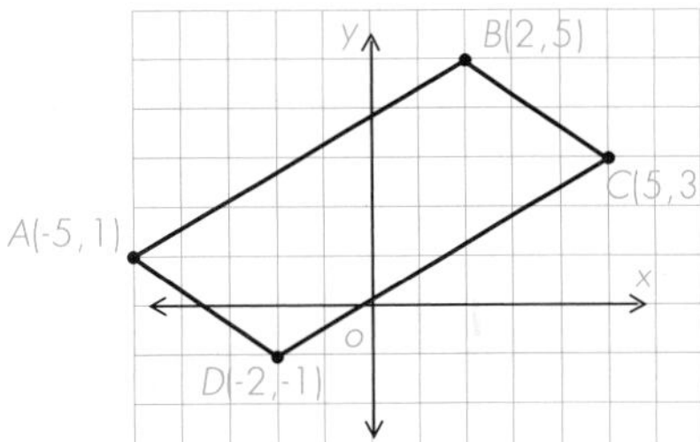

9. What is the slope of $\overline{AB}$? of $\overline{DC}$? $\frac{4}{7}$; $\frac{4}{7}$
10. What is the slope of $\overline{AD}$? of $\overline{BC}$? $-\frac{2}{3}$; $-\frac{2}{3}$
11. What kind of quadrilateral is *ABCD*? How do you know? *ABCD* is a parallelogram. Opposite sides have equal slopes and so are parallel.

Tell whether the given lines are perpendicular.

12. $y = \frac{2}{3}x + 1$ Yes
    $y = -\frac{3}{2}x + 8$
13. $y = \frac{8}{9}x - 4$ No
    $y = \frac{9}{8}x + 4$
14. $y = 6$ No
    $y = \frac{1}{2}x - \frac{1}{6}$
15. The line determined by $(-5,4)$ and $(4,-5)$, the line determined by $(10,10)$ and $(-10,-10)$ Yes

Refer to $\triangle ABC$. See margin.

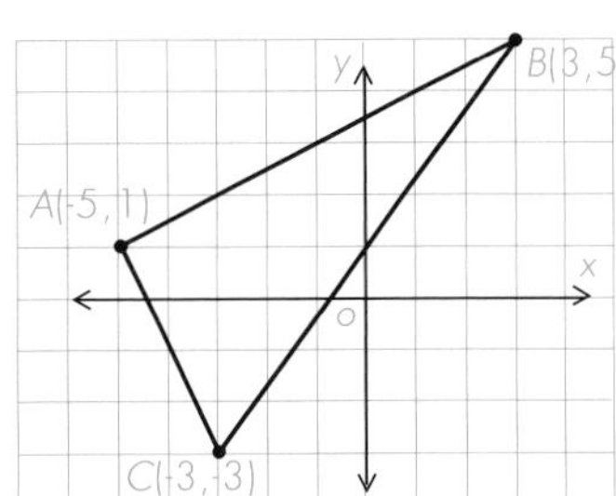

16. Find the slope of each side of $\triangle ABC$.
17. Is $\triangle ABC$ a right triangle? How do you know?
18. If you drew a line passing through *A* and perpendicular to *BC*, what would its slope be?
19. The vertices of $\triangle KLM$ are $K(-1,3)$, $L(3,6)$, and $M(2,-1)$. Show that $\triangle KLM$ is an isosceles right triangle. See margin.

Additional Answers

16. slope of $\overline{AB} = \frac{1}{2}$
slope of $\overline{BC} = \frac{4}{3}$
slope of $\overline{AC} = -2$

17. Yes, because the product of the slopes of $\overline{AB}$ and $\overline{AC}$ is -1.

18. $-\frac{3}{4}$

19. $KL = KM = 5$, so *KL* and *KM* are the legs of an isosceles triangle. Slope of $\overline{KL} = \frac{3}{4}$ and slope of $\overline{KM} = -\frac{4}{3}$, so $\overline{KL} \perp \overline{KM}$ and $\triangle KLM$ is a right triangle.

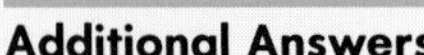

Additional Answers

24. $\overline{AB} \parallel \overline{DC}$, since both segments have slope $\frac{4}{3}$. $\overline{AD}$ and $\overline{BC}$ have slopes -4 an 0, respectively, and hence are not parallel. Therefore $ABCD$ is a trapezoid. It is not isosceles, because $BC = 4$ and $AD = \sqrt{17}$ and $4 \neq \sqrt{17}$.

Using the Page

After students have completed the Critical Thinking exercises, ask what results would have been obtained if occurrences of some other digit had been counted instead.

FOLLOW-UP

Assessment

1. Line l is parallel to the graph of $y = -\frac{1}{2}x + 5$. What is the slope of l? $\left[-\frac{1}{2}\right]$

2. Line k is perpendicular to the graph of $y = 2x - 17$. What is the slope of k? $\left[-\frac{1}{2}\right]$

3. Consider the points $A(21, 0)$, $B(22, 2)$, $C\left(23, \frac{3}{2}\right)$, and $D\left(24, \frac{7}{2}\right)$. Name the two pairs of parallel segments and two pairs of perpendicular segments. [parallel: $\overline{AB} \parallel \overline{CD}$, $\overline{BD} \parallel \overline{AC}$; perpendicular: $\overline{AB} \perp \overline{BC}$, $\overline{BC} \perp \overline{CD}$]

Reteaching

Students who have had difficulty with this section may benefit from Reteaching Activity 13.3.

Enrichment

For students who have mastered the material in this section, you may wish to assign Enrichment Activity 13.3.

Use the XYPLANE procedure from page 118 in a Logo procedure to tell the turtle to draw the line determined by each pair of points. Tell whether each pair of lines appear to be parallel, perpendicular, or neither. Justify your answer.

20. $(-6,7)$ and $(-4,4)$
$(-7,4)$ and $(-4,6)$ Perpendicular

21. $(1,1)$ and $(4,4)$
$(3,-1)$ and $(6,2)$ Parallel

22. $(-6,-1)$ and $(-1,4)$
$(-1,-2)$ and $(0,3)$ Neither

23. $(-3,0)$ and $(3,1)$
$(-3,-3)$ and $(3,-2)$ Parallel

24. Use what you know about slopes to show that figure $ABCD$ is a trapezoid. Is it an *isosceles* trapezoid? How do you know?
See margin.

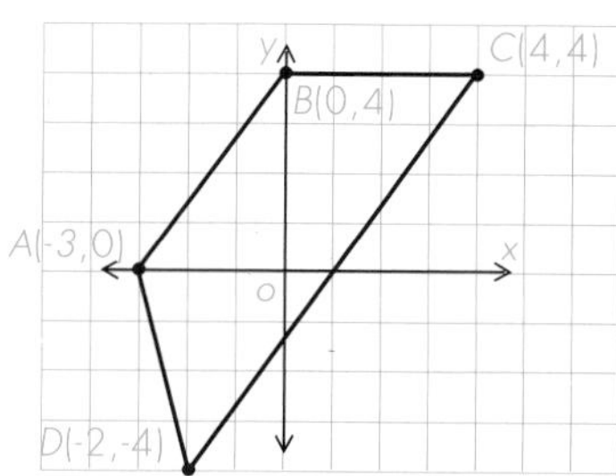

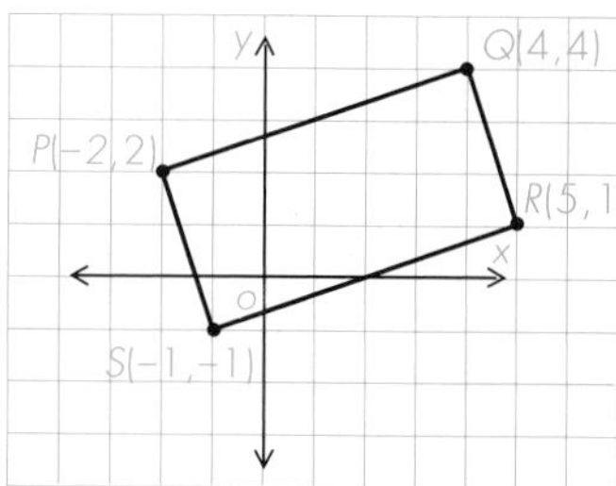

25. Use what you know about slopes to show that quadrilateral $PQRS$ is a rectangle.
$\overline{PQ}$ and $\overline{RS}$ each have slopes of $\frac{1}{3}$; $\overline{PS}$ and $\overline{QR}$ have slopes of -3. Since $-3 \cdot \frac{1}{3} = -1$, all the angles of the quadrilateral are right angles.

CRITICAL THINKING

How many times is the digit seven printed in page numbers if a book has the following number of pages? Assume that the pages are numbered consecutively, starting with 1.

26. 10 pages 1
27. 30 pages 3
28. 70 pages 8
29. 80 pages 18
30. 100 pages 20
31. 200 pages 40

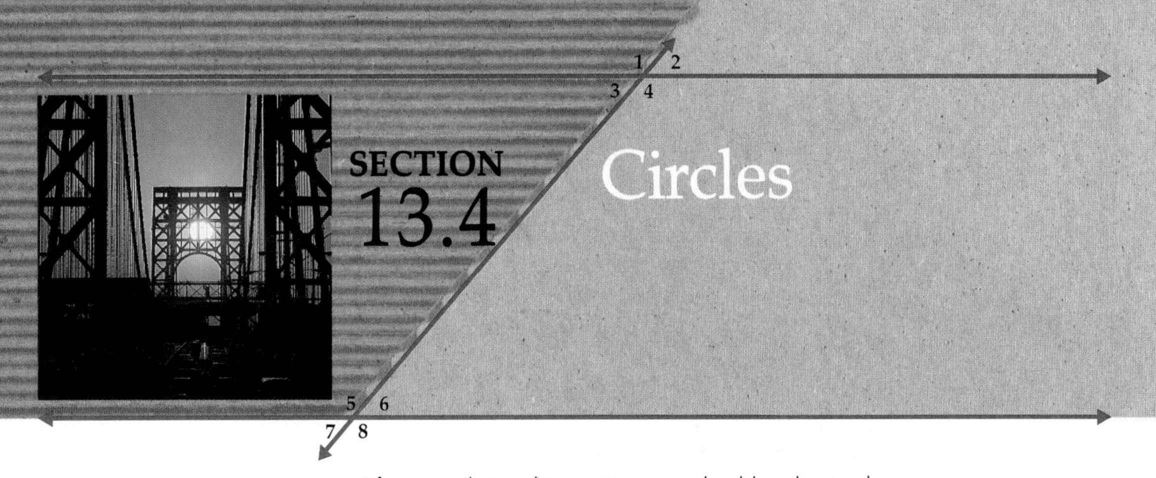

After completing this section, you should understand
▲ how to write an equation for a circle whose center is the origin.
▲ how to graph equations of the form $x^2 + y^2 = r^2$.

You have seen something of how coordinate geometry can deal with lines. Now you will see how it can deal with curves.

CIRCLES

You will recall that a circle is the set of all points in a plane that are equidistant from a given point (the center of the circle).

DISCOVERY ACTIVITY

On a piece of graph paper, draw x- and y-axes. Mark and label the five points shown in the plane at the right. Take a compass and draw the circle that has its center at the origin and that passes through (5,0). See students' diagrams.

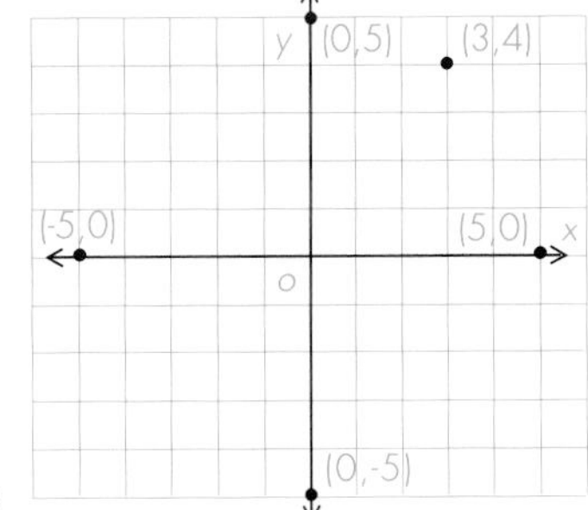

1. What is the length of a radius of the circle? 5
2. Does the circle pass through (3,4)? Yes
3. Mark and label seven other points that have integer coordinates and that lie on the circle. (4,3), (−3,4), (−4,3), (−3,−4), (−4,−3), (4,−3), (3,−4)
4. For each point you labeled in your diagram, square the coordinates and add the two squares. What number do you get each time? 25
5. If (x,y) is any point on the circle, what value do you think $x^2 + y^2$ will have? 25

SECTION
13.4

Resources
Reteaching 13.4
Enrichment 13.4
Transparency Master 13-1

Objectives
After completing this section, the student should understand
▲ how to write an equation for a circle whose center is the origin.
▲ how to graph equations of the form $x^2 + y^2 = r^2$.

Materials
compass
ruler
graph paper

GETTING STARTED

Quiz: Prerequisite Skills
1. Find the distance between (7, 5) and (10, 9). [5]
2. A circle has its center at (0, 0) and passes through the point (−3, 6). What is the radius of the circle? [$3\sqrt{5}$]
3. A circle of radius $\sqrt{10}$ has its center at the origin. Name the points of intersection of the circle and the x- and y-axes. [$(-\sqrt{10}, 0)$, $(0, \sqrt{10})$, $(\sqrt{10}, 0)$, $(0, -\sqrt{10})$]

TEACHING COMMENTS

Using the Page
After students have worked through the Discovery Activity (on page 495), ask whether there are points on the circle that have just one integer coordinate. Discuss how they might find both coordinates of each of these points.

Using the Page
Discuss whether there are points (x, y) in the coordinate plane whose coordinates satisfy $x^2 + y^2 = 0$ or $x^2 + y^2 = -25$.

In Chapter 7 you learned that the distance between any two points $A(x_1,y_1)$ and $B(x_2,y_2)$ is given by the distance formula:

$$d = \sqrt{(x_2 - x_1)^2 + (y_2 - y_1)^2}$$

Apply this formula to the situation in the Discovery Activity.

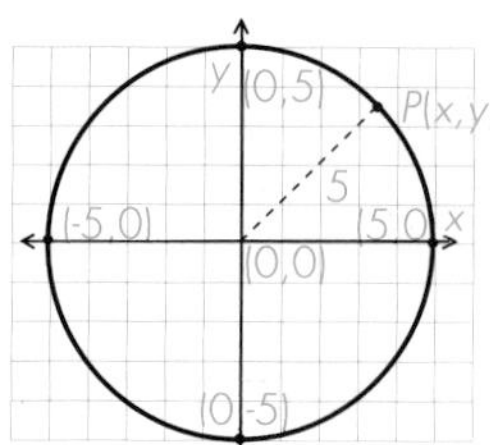

Let $P(x,y)$ be a point on the circle with center (0,0) and radius 5. Because it is on the circle, the distance from (x,y) to (0,0) must be 5. Using the distance formula to express the distance from (x,y) to (0,0), you get the following:

$$\sqrt{(x - 0)^2 + (y - 0)^2} = 5$$

$$\sqrt{x^2 + y^2} = 5$$

Square both sides of the last equation and you get

$$x^2 + y^2 = 25$$

Clearly you could use any positive real number r for the radius. This reasoning leads to the following theorem:

THEOREM 13-4-1: The points $P(x,y)$ whose coordinates satisfy the equation $x^2 + y^2 = r^2$, where r is a positive real number, form a circle of radius r whose center is (0,0).

Example: Describe the set of points in the coordinate plane whose coordinates satisfy $x^2 + y^2 = 64$.

Since $64 = 8^2$, the points satisfy $x^2 + y^2 = 8^2$. The set of points is the circle with center (0,0) and radius 8.

Example: Write an equation for the circle with center (0,0) and radius 13.

Use 13 for r in $x^2 + y^2 = r^2$. You get $x^2 + y^2 = 13^2$, or $x^2 + y^2 = 169$.

CLASS ACTIVITY

Refer to circles 1 and 2.

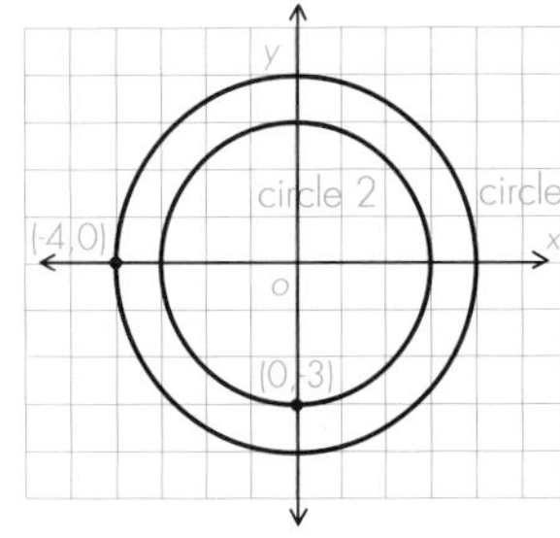

1. Write an equation for circle 1. $x^2 + y^2 = 16$

2. Write an equation for circle 2. $x^2 + y^2 = 9$

3. Which of the two circles does $(2,2\sqrt{3})$ lie on? Circle 1

4. Does $(-1,2\sqrt{2})$ lie on either of the circles? If so, which one? Yes; circle 2

Write an equation for each circle.

5. The circle with center (0,0) and radius 7 $x^2 + y^2 = 49$

6. The circle with center (0,0) and radius $\frac{9}{2}$ $x^2 + y^2 = \frac{81}{4}$

7. The circle with center (0,0) and radius 100 $x^2 + y^2 = 10{,}000$

8. The circle with center (0,0) and radius 13. $x^2 + y^2 = 169$

9. Consider the circle $x^2 + y^2 = 36$. Write the ordered pairs for the points on this circle that have integer x-coordinates. Give the y-coordinates as decimals to the nearest hundredth. Use a calculator. See margin.

10. Tell which of these three circles lies inside the other two. Circle 1

Circle 1: $x^2 + y^2 = 9$
Circle 2: $x^2 + y^2 = (\sqrt{6})^2$
Circle 3: $x^2 + y^2 = 4^2$

Additional Answers

9.

(−6, 0)	(0, −6)
(−5, 3.32)	(1, 5.92)
(−5, −3.32)	(1, −5.92)
(−4, 4.47)	(2, 5.66)
(−4, −4.47)	(2, −5.66)
(−3, 5.20)	(3, 5.20)
(−3, −5.20)	(3, −5.20)
(−2, 5.66)	(4, 4.47)
(−2, −5.66)	(4, −4.47)
(−1, 5.92)	(5, 3.32)
(−1, −5.92)	(5, −3.32)
(0, 6)	(6, 0)

TANGENTS TO CIRCLES

Recall that a line is tangent to a circle if it intersects the circle in just one point. You saw in Theorem 10-4-4 that a tangent to a circle is perpendicular to the radius drawn to the point where the tangent intersects the circle. The converse of Theorem 10-4-4 is also true. In other words, if you draw a line perpendicular to a radius $\overline{OP}$ at the

Using the Page
Once students understand how the reasoning goes for this situation, ask if they can conjecture an equation for the line tangent to the circle at (−3, −4). To aid in the discussion, you may wish to make a copy of the diagram on this page on Transparency Master 13-1.

point P (where O is the center of the circle), then the line is tangent to the circle.

Consider the circle $x^2 + y^2 = 25$ and the line $y = -\frac{3}{4}x + \frac{25}{4}$.

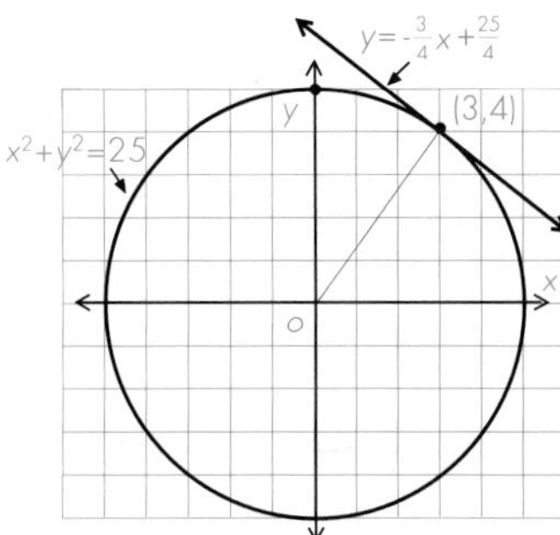

Suppose you draw the radius from the center of the circle to (3,4), which is a point on the circle. The point (3,4) is also on the line $y = -\frac{3}{4}x + \frac{25}{4}$. To show that this is true, just put 3 in place of x and 4 in place of y.

$$4 = -\frac{3}{4} \cdot 3 + \frac{25}{4}$$
$$4 = -\frac{9}{4} + \frac{25}{4}$$
$$4 = \frac{16}{4}$$
$$4 = 4$$

The slope of the radius is $\frac{4-0}{3-0}$, or $\frac{4}{3}$. The slope of the given line is $-\frac{3}{4}$. The product of these two slopes is $-\frac{12}{12}$, or -1. Therefore $y = -\frac{3}{4}x + \frac{25}{4}$ is perpendicular to the radius at the point (3,4). So the given line is indeed tangent to the circle.

CLASS ACTIVITY

Write an equation for the given circle.

1. The circle with center (0,0) and radius 6
   $x^2 + y^2 = 36$
2. The circle with center (0,0) and radius $\sqrt{15}$
   $x^2 + y^2 = 15$
3. The circle with center (0,0) and radius $\frac{3}{4}$
   $x^2 + y^2 = \frac{9}{16}$
4. The circle with center (0,0) and radius 18
   $x^2 + y^2 = 324$
5. The circle with center (0,0) and passing through the point (5,1)
   $x^2 + y^2 = 26$
6. The circle with center (0,0) and passing through the point $(5\sqrt{2}, 5\sqrt{2})$.
   $x^2 + y^2 = 100$

HOME ACTIVITY

A circle has center (0,0) and radius 5.

1. Write an equation for the circle. $x^2 + y^2 = 25$
2. Is (−4,3) a point on the circle? Yes.
3. What is the slope of the line determined by (0,0) and (−4,3)? $-\frac{3}{4}$
4. Which of these lines are perpendicular to the line through (0,0) and (−4,3)? All three

   a. $y = \frac{4}{3}x + \frac{20}{3}$ b. $y = \frac{4}{3}x + \frac{25}{3}$ c. $y = \frac{4}{3}x + 30$

5. Which of the lines in Exercise 4 is tangent to the given circle? How do you know that it is a tangent? See margin.
6. Write an equation for the circle tangent to the line determined by (0,6) and (6,0) and with its center at (0,0). $x^2 + y^2 = 18$

Use a calculator to help you determine whether the given point is on the circle whose equation is $x^2 + y^2 = 746$.

7. (18,19) No
8. (21,14) No
9. (11,25) Yes
10. (−12,24) No
11. $(27,\sqrt{17})$ Yes
12. $(\sqrt{15},30)$ No

Graph the equation. See margin.

13. $x^2 + y^2 = 49$
14. $x^2 + y^2 = 4$
15. $x^2 + y^2 = 20$
16. $x^2 + y^2 = \left(\frac{11}{2}\right)^2$
17. $x^2 + y^2 = \frac{100}{9}$
18. $x^2 + y^2 = 64$

Tell whether the given point is on the circle shown in the graph.

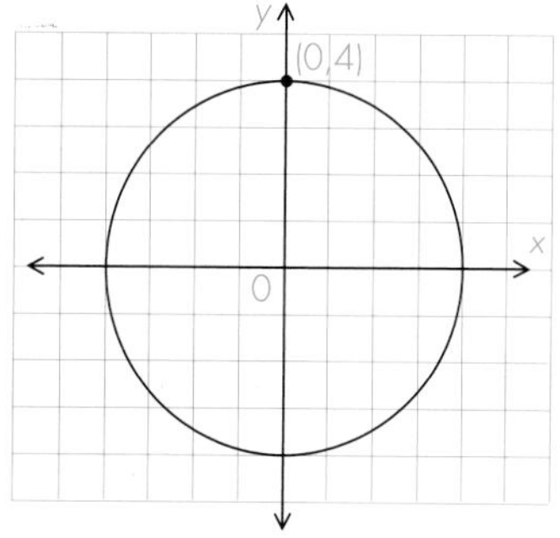

19. (2,2) No
20. $(-3,\sqrt{7})$ Yes
21. (−4,0) Yes
22. $(1,\sqrt{15})$ Yes

Additional Answers

5. The line $y = \frac{4}{3}x + \frac{25}{3}$ is tangent to the circle. From Exercise 4, it is known that this line is perpendicular to the line containing the radius to point (−4, 3). The line passes through (−4, 3) because the coordinates of (−4, 3) satisfy the equation $y = \frac{4}{3}x + \frac{25}{3}$.

Extra Practice

1. Show that $(1, 2\sqrt{2})$ is on the circle $x^2 + y^2 = 9$. $[1^2 + (2\sqrt{2})^2 = 1 + 8 = 9]$

2. Write an equation for the line tangent to the circle $x^2 + y^2 = 9$ at the point $(1, 2\sqrt{2})$. $\left[y = -\frac{\sqrt{2}}{4}x + \frac{9\sqrt{2}}{4}\right]$

Additional Answers

Each graph should be a circle with its center at the origin. The radii of the circles should be as follows.

7. $r = 7$
8. $r = 2$
9. $r \approx 4.5$
10. $r = 5\frac{1}{2}$
11. $r = 3\frac{1}{3}$
12. $r = 8$

Consider the circle with center (0,0) and radius 13.

23. Write an equation for the circle. $x^2 + y^2 = 169$

24. Is (5,12) a point on the circle? Yes

25. What is the slope of the line determined by (0,0) and (5,12)? $\frac{12}{5}$

26. Does the line $y = -\frac{5}{12}x + \frac{169}{12}$ pass through the point (5,12)? Yes

27. Is the line $y = -\frac{5}{12}x + \frac{169}{12}$ tangent to the given circle? Yes

CRITICAL THINKING

This puzzle is known as the Tower of Hanoi problem. There are three rings of different sizes on the left peg of a board that has three pegs. The smallest ring is on the top, the largest ring is on the bottom.

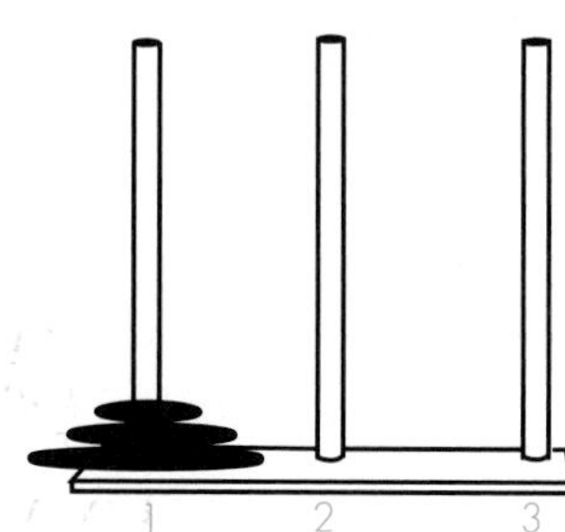

The goal is to transfer rings one at a time from one peg to another until all the rings are on the peg on the right. A larger ring can never be placed on top of a smaller ring.

28. If there were only one ring on peg 1, how many moves would be required? 1

29. If there were two rings on peg 1, how many moves would be required? 3

30. For the three rings shown in the diagram, how many moves are required? 7

31. Describe a pattern that tells how many moves are required for n rings. $2^n - 1$

FOLLOW-UP

Assessment

1. Write an equation for the circle with center (0, 0) and radius 3.5. [$x^2 + y^2 = 12.25$]

2. What is the radius of the circle which is the graph of $x^2 + y^2 = 400$? [20]

3. Write an equation for the line tangent to the circle $x^2 + y^2 = 169$ at the point (−5, 12).

[$y = \frac{5}{12}x + \frac{169}{12}$]

Reteaching

Students who have had difficulty with this section may benefit from Reteaching Activity 13.4.

Enrichment

For students who have mastered the material in this section, you may wish to assign Enrichment Activity 13.4.

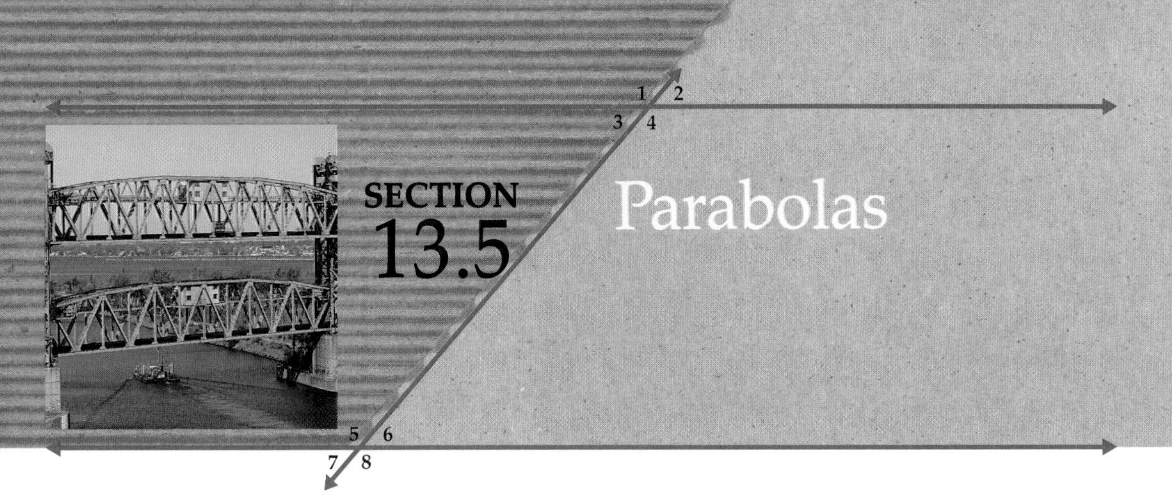

After completing this section, you should understand
▲ how to find and graph an equation for a parabola.

In Section 12.4 you learned that in a plane, a parabola is a set consisting of all points equidistant from a given line and a point not on the line. You can use this description to find equations of parabolas.

FINDING EQUATIONS OF PARABOLAS

Point (0,1) is on the y-axis. It lies one unit above the origin. Line $y = -1$ is a horizontal line one unit below the x-axis. This point and line determine a parabola.

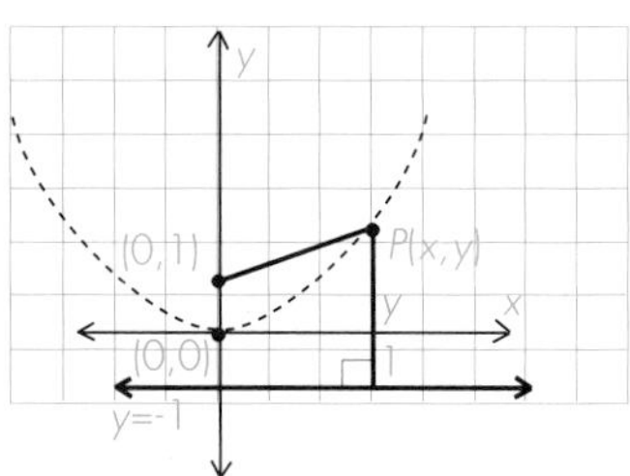

If $P(x,y)$ is a point on this parabola, then it must be equidistant from (0,1) and the line $y = -1$. You can express the two distances as follows:

distance from (x,y) to $y = -1$ is $y + 1$;
distance from (x,y) to $(0,1)$ is $\sqrt{(x - 0)^2 + (y - 1)^2}$.

These distances are equal, so

$$\begin{aligned} y + 1 &= \sqrt{(x - 0)^2 + (y - 1)^2}. \\ (y + 1)^2 &= (x - 0)^2 + (y - 1)^2 \\ y^2 + 2y + 1 &= x^2 + y^2 - 2y + 1 \\ 4y &= x^2 \\ y &= \tfrac{1}{4}x^2 \end{aligned}$$

Therefore an equation for the parabola is $y = \frac{1}{4}x^2$.

SECTION 13.5

Resources
Reteaching 13.5
Enrichment 13.5
Transparency Masters 13-1 and 13-2

Objective
After completing this section, the student should understand
▲ how to find and graph an equation for a parabola.

Materials Needed
compass
ruler
graph paper

Vocabulary
conic section
ellipse
hyperbola

GETTING STARTED

Chalkboard Example
On the chalkboard, draw a horizontal line l. Near the middle of it and about 4 inches above it, mark and label a point F. Draw a line a through F perpendicular to l and intersecting l at G. The segment $\overline{FG}$ is the shortest segment from F to l. Mark the midpoint V of this segment. Next, pick point C on line a and above V. Draw a line k through C perpendicular to a. Elicit that for each point on this line, the distance to line l is CG. Set a chalkboard compass to an opening of CG. With center F, draw arcs that intersect k at two points M and N. Discuss why it follows that M and N are on the parabola. Other points on the parabola can be found in the same way.

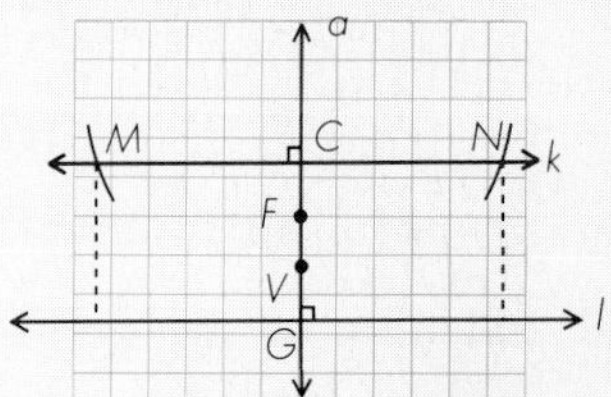

TEACHING COMMENTS

Using the Page
Discuss why the origin will be the low point of the parabola.

Using the Page
Ask the students if they think the origin will be the low point of each of these parabolas and if so, why.

The choice of the point (0,1) and the line $y = -1$ made the algebra easier. Also it guaranteed that the low point of the parabola would pass through the origin.

CLASS ACTIVITY

Find an equation for the parabola determined by the given point and line.

1. The point (0,2) and the line $y = -2$ $y = \frac{1}{8}x^2$
2. The point (0,3) and the line $y = -3$ $y = \frac{1}{12}x^2$
3. The point (0,4) and the line $y = -4$ $y = \frac{1}{16}x^2$
4. The point $(0,\frac{1}{4})$ and the line $y = -\frac{1}{4}$ $y = x^2$
5. The point $(0,\frac{1}{8})$ and the line $y = -\frac{1}{8}$ $y = 2x^2$

The method used so far can be applied to find an equation for the parabola determined by any point $(0,a)$, where $a > 0$, and the line $y = -a$.

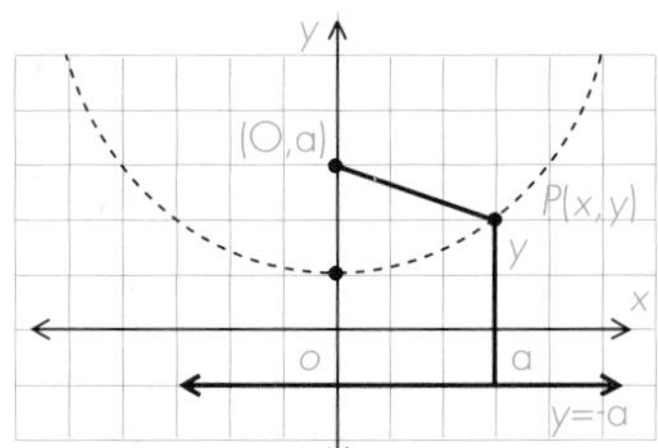

$$
\begin{aligned}
y + a &= \sqrt{(x - 0)^2 + (y - a)^2} \\
(y + a)^2 &= (x - 0)^2 + (y - a)^2 \\
y^2 + 2ay + a^2 &= x^2 + y^2 - 2ay + a^2 \\
4ay &= x^2 \\
y &= \frac{x^2}{4a}, \text{ or } y = \frac{1}{4a}x^2
\end{aligned}
$$

THEOREM 13-5-1: **An equation for the parabola determined by $(0,a)$ and the line $y = -a$, where $a > 0$, is $y = \frac{1}{4a}x^2$.**

GRAPHING PARABOLAS

You have seen how to find equations for parabolas. Now you will get some experience in drawing their graphs.

DISCOVERY ACTIVITY

1. Make a table of values of x and y for the equation $y = x^2$. Use these values for x: $-3, -2, -1, \frac{-1}{2}, 0, \frac{1}{2}, 1, 2, 3$. See margin.
2. Graph the nine points you get from the table. See margin.
3. Connect the nine points with a smooth curve to show the parabola.
4. What could you do to get an even better, smoother graph? Plot more points.
5. Repeat steps 1 through 3 to graph the equation $y = -x^2$. See margin.
6. How are the graphs of $y = x^2$ and $y = -x^2$ the same? How are they different? See margin.

In the Discovery Activity, you may have found that the graph of $y = x^2$ is a parabola that opens upward. The graph of $y = -x^2$ is a parabola that opens downward.

Notice that each of the parabolas that you drew in the Discovery Activity is symmetric with respect to the y-axis. This is true because the square of any number and its opposite are equal.

Example: The points (2,4) and (−2,4) are both on the graph of $y = x^2$, since $2^2 = (-2)^2 = 4$. Both points are four units above the x-axis. (2,4) is two units to the right of the y-axis, and (−2,4) is two units to the left.

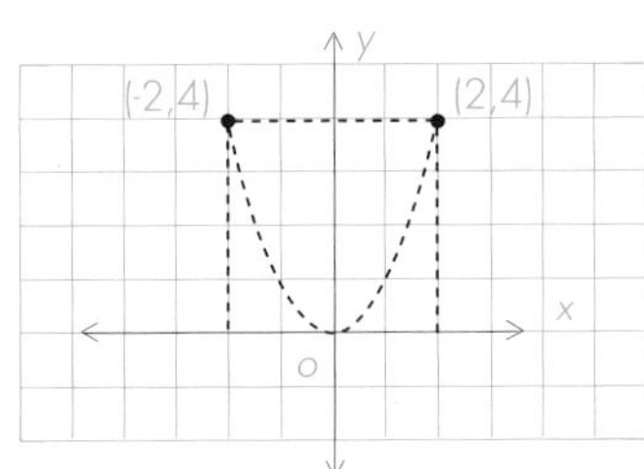

Additional Answers

1.

x	y
−3	9
−2	4
−1	1
$-\frac{1}{2}$	$\frac{1}{4}$
0	0
$\frac{1}{2}$	$\frac{1}{4}$
1	1
2	4
3	9

Exercises 2 and 3

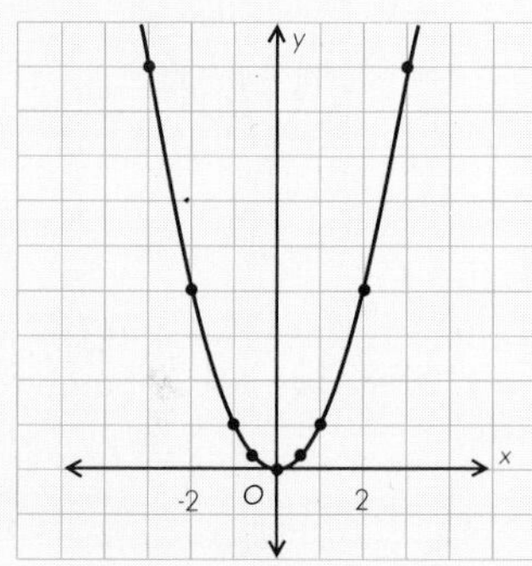

5.

x	y
−3	−9
−2	−4
−1	−1
$-\frac{1}{2}$	$-\frac{1}{4}$
0	0
$\frac{1}{2}$	$-\frac{1}{4}$
1	−1
2	−4
3	−9

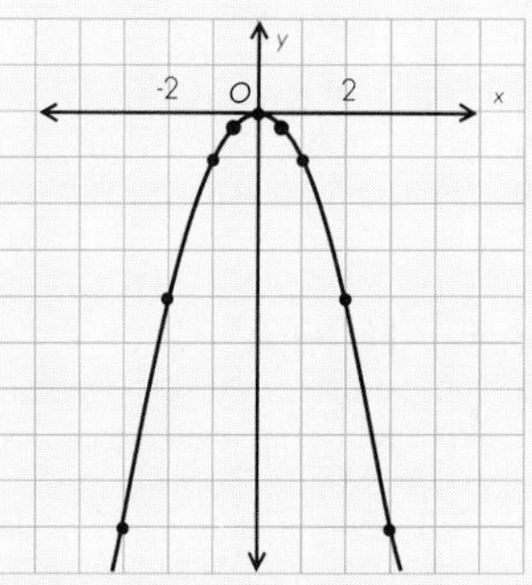

6. They have the same shape and both go through (0, 0). They are different in that one opens upward above the x-axis; the other downward, below the x-axis.

Using the Page
You can place Transparency Master 13-2 on top of Transparency Master 13-1 to show how varying the coefficient of x^2 changes the shape of the parabola.

Additional Answers

1.

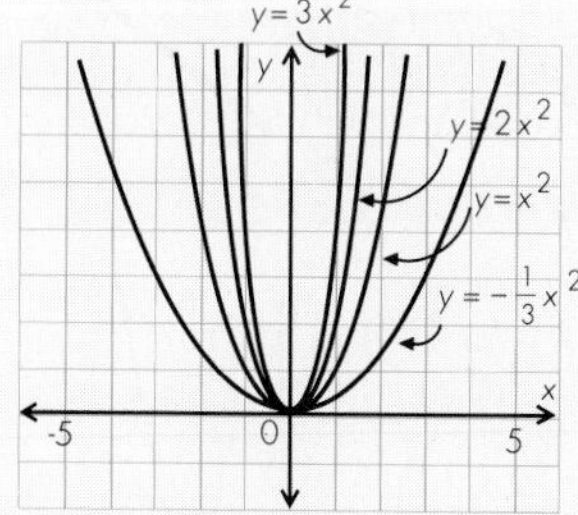

12. ref. pts. (−0.5, 0.1875), (−0.8, 0.5), and (2, −2) are not on parabola.

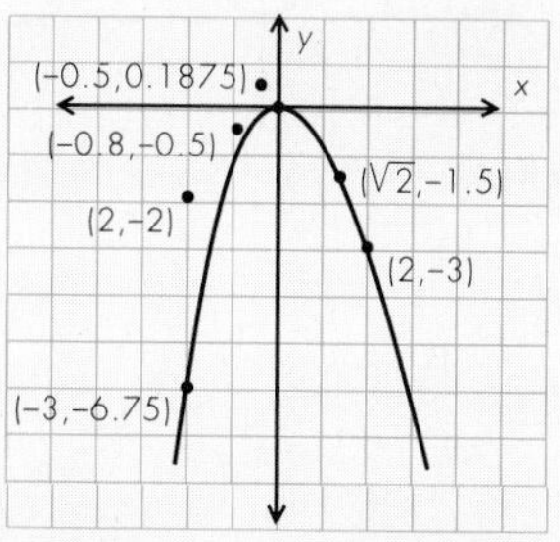

Extra Practice
What point on the graph of the equation will have its x-coordinate equal to 1?

1. $y = -x^2$ [(1, −1)]
2. $y = -10x^2$ [(1, −10)]
3. $y = -100x^2$ [(1, −100)]

Using the Page
If time permits, you might wish to have students investigate ellipses and hyperbolas using the following definitions. An ellipse is the set of all points *P* in a plane such that the sum of the distances from *P* to two given points is a constant. A hyperbola is the set of all points *P* in a plane such that the absolute value of the difference of the distances from *P* to two given points is a constant.

CLASS ACTIVITY

1. On the same $x-y$ plane, graph $y = \frac{1}{3}x^2$, $y = x^2$, $y = 2x^2$, and $y = 3x^2$. See margin.
2. Describe how the number by which x^2 is multiplied in the equations of Exercise 1 affects the shape of the graph.
   As the number gets smaller, the parabola gets flatter.
3. Do the graphs for Exercise 1 open up or down? Up
4. What is the lowest point on each graph in Exercise 1? (0,0)
5. Describe what you think the graphs of $y = -\frac{1}{3}x^2$, $y = -x^2$, $y = -2x^2$, and $y = -3x^2$ would look like.
   They will be like the parabolas in Exercise 1 but will open down.

Use a calculator to help you determine whether the given point is on the graph of $y = -0.75x^2$.

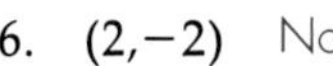

6. (2,−2) No
7. (2,−3) Yes
8. (−3,−6.75) Yes
9. (−0.5, 0.1875) No
10. $(\sqrt{2},-1.5)$ Yes
11. (−0.8,0.5) No
12. Graph $y = -0.75x^2$. Graph the points from Exercises 6–11 to verify your answers to these exercises. See margin.

Circles and parabolas are examples of curves known as ***conic sections.*** Other important conic sections are the ellipse and the hyperbola. An ellipse is the locus of all points in a plane the sum of whose distances from two fixed points *A* and *B* is a given number *d.* A hyperbola is the locus of all points in a plane the difference of whose distances from *A* and *B* is a given number *d.* These curves and their properties were studied in detail by the Greek mathematicians of ancient times. The Greeks called these curves *conic* sections because they are the curves you get by slicing through a cone with a plane.

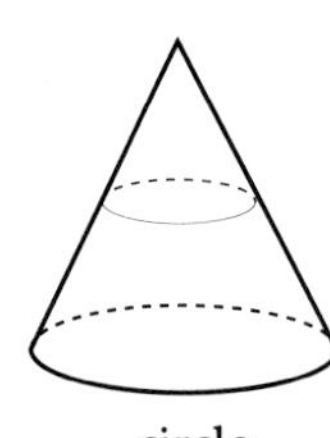
circle

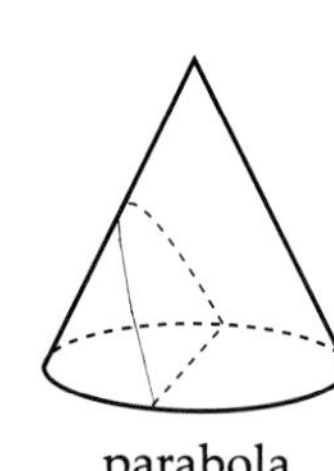
parabola

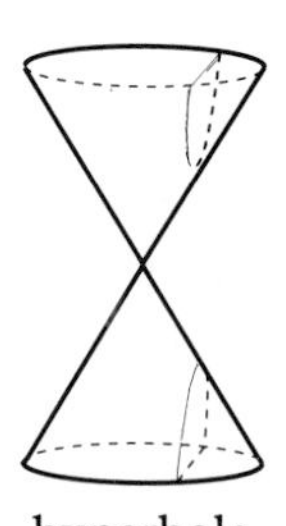
hyperbola

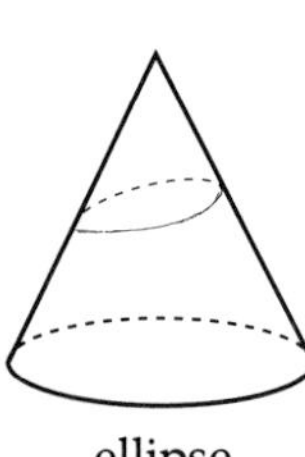
ellipse

These curves, as it turns out, are important for understanding many things that occur in daily life and in nature. For example, when a football player kicks a goal, the football travels in a parabolic path. The orbits of the planets in the solar system are shaped like ellipses.

REPORT 13-5-1: Prepare a report on the studies the ancient Greeks made of conic sections. Include the following information:

a. the names and lifetimes of some of the mathematicians who made especially important discoveries about conic sections
b. the names of some of the more important written works about conics by these mathematicians
c. some of the special discoveries the Greek mathematicians made about conic sections

Sources:
The New Encyclopaedia Brittanica, 15th ed., s.v. "geometry."
The New Encyclopaedia Brittanica, 15th ed., s.v. "Apollonius of Perga."
Boyer, Carl B. *A History of Mathematics*. Princeton: Princeton University Press, 1985.

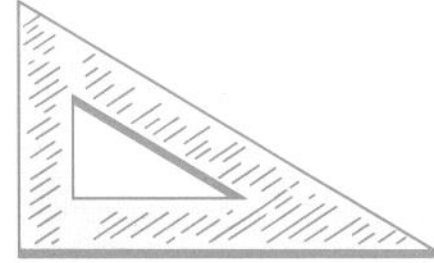

HOME ACTIVITY

Find an equation for the parabola determined by the given point and line.

1. The point (0,5) and the line $y = -5$ $y = \frac{1}{20}x^2$
2. The point $(0,\frac{1}{2})$ and the line $y = \frac{-1}{2}$ $y = \frac{1}{2}x^2$
3. The point $(0,\frac{3}{4})$ and the line $y = \frac{-3}{4}$ $y = \frac{1}{3}x^2$
4. The point $(0,-1)$ and the line $y = 1$ $y = -\frac{1}{4}x^2$

Graph each parabola. You may find it helpful to first make a table of values of x and y. See margin.

5. $y = \frac{1}{2}x^2$
6. $y = \frac{-1}{2}x^2$
7. $y = -2x^2$
8. $y = 4x^2$

Using the Page
If there are students who seem intrigued by the conic sections topic (on page 504), you may wish to assign them the project of graphing $\frac{x^2}{25} + \frac{y^2}{16} = 1$ and $\frac{x^2}{4} - \frac{y^2}{12} = 1$. Suggest that they use a calculator to help make tables of values of x and y. The graph of the first equation is an ellipse. The graph of the second equation is a hyperbola.

Additional Answers

Exercises 5-8

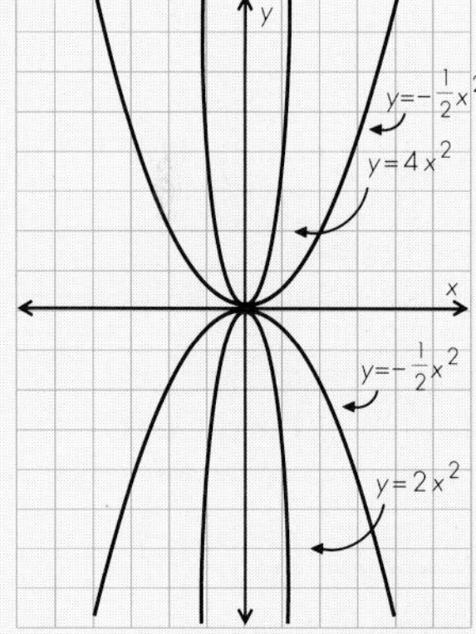

Additional Answers

21.

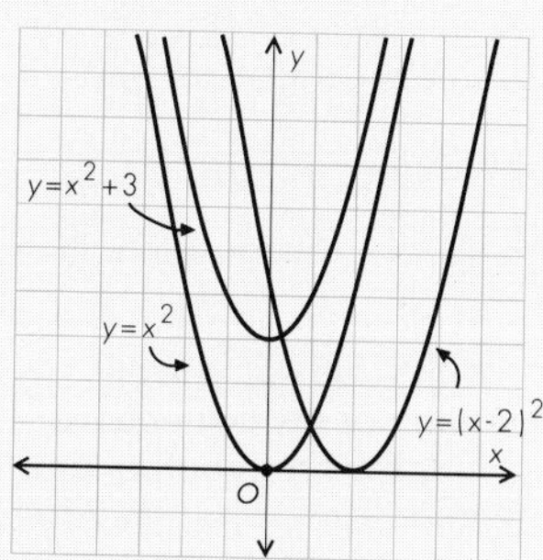

Adding 3 to x^2 moves the graph of $y = x^2$ straight up 3 units. Replacing x with x − 2 moves the graph of $y = x^2$ to the right 2 units.

FOLLOW-UP

Assessment

1. Find an equation for the parabola determined by the point $\left(0, \frac{1}{16}\right)$ and the line $y = -\frac{1}{16}$. [$y = 4x^2$]

Tell whether the graph will open up or down.

2. $y = -9x^2$ [down]

3. $y = \frac{1}{20}x^2$ [up]

4. $y = 7x^2$ [up]

Draw the graph of each equation.

5. $y = \frac{3}{2}x^2$ **6.** $y = -\frac{3}{2}x^2$

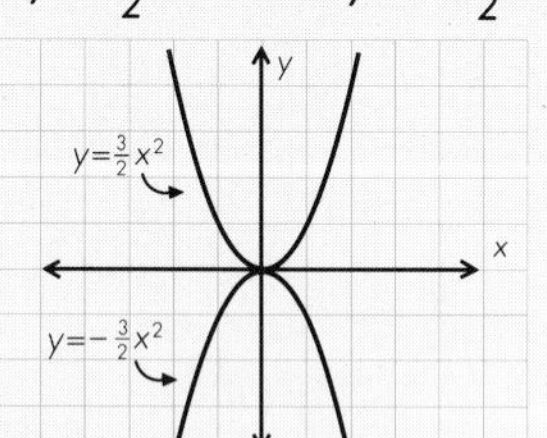

Reteaching

Students who have had difficulty with this section may benefit from Reteaching Activity 13.5.

Enrichment

For students who have mastered the material in this section, you may wish to assign Enrichment Activity 13.5.

Tell whether the parabola will open up or down.

9. $y = -9x^2$ Down
10. $y = \frac{2}{3}x^2$ Up
11. $y = 7x^2$ Up
12. $y = -4x^2$ Down
13. $y = -6x^2$ Down
14. $y = 100x^2$ Up
15. $y = \frac{-3}{2}x^2$ Down
16. $y = \frac{x^2}{5}$ Up

Consider these parabolas: $y = 25x^2$, $y = \frac{1}{10}x^2$, $y = \frac{-1}{100}x^2$, and $y = -25x^2$.

17. Which of the parabolas have exactly the same shape? $y = 25x^2$ and $y = -25x^2$
18. Which parabola is flatter than the others? $y = -\frac{1}{100}x^2$
19. What point do all the parabolas pass through? (0,0)
20. What line is a line of symmetry for all the parabolas? The y-axis

CRITICAL THINKING

21. In the same coordinate plane, draw the graphs of $y = x^2$, $y = x^2 + 3$, and $y = (x - 2)^2$. Describe the effect that adding 3 to x^2 has on the graph of $y = x^2$. Describe the effect that replacing x with $(x - 2)$ has on the graph of $y = x^2$. See margin.

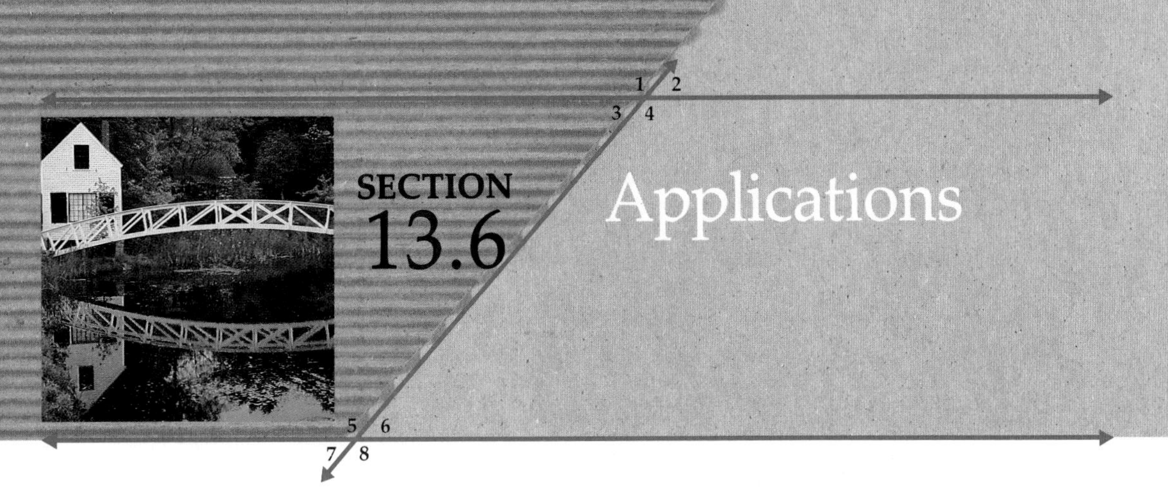

After completing this section, you should understand
▲ how to prove theorems using coordinate geometry.

You have now seen enough of coordinate geometry to be able to use it to prove theorems.

THE MIDPOINT THEOREM

The midpoint theorem tells you that you can find the midpoint of any segment by finding the average of the x-coordinates and the average of the y-coordinates.

THEOREM 13-6-1: (The midpoint theorem) If $P(a,b)$ and $Q(c,d)$ are any two points in the coordinate plane, then the midpoint of segment PQ is $M\left(\frac{a+c}{2}, \frac{b+d}{2}\right)$.

CLASS ACTIVITY

To prove the midpoint theorem, you need to prove two things: $MP = MQ$ and $MP + MQ = PQ$.
See margin.

1. Use the distance formula to write algebraic expressions for MP, MQ, and PQ.
2. Use algebra to show that $MP = MQ$.
3. Use algebra to show that $MP + MQ = PQ$.

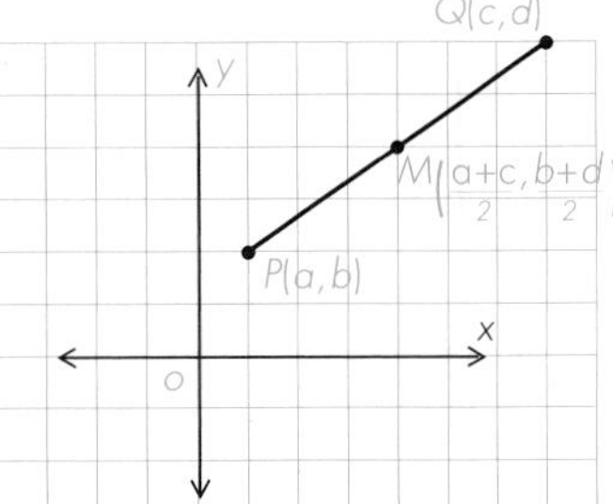

SECTION 13.6

Resources
Reteaching 13.6
Enrichment 13.6

Objective
After completing this section, the student should understand
▲ how to prove theorems using coordinate geometry.

Materials
ruler
graph paper

GETTING STARTED

Quiz: Prerequisite Skills

1. Simplify $\frac{x+y}{2} - x$. $\left[\frac{y-x}{2}\right]$

2. Simplify $(\sqrt{x-y} \cdot \sqrt{x+y})^2$, where $x > y$. $[x^2 - y^2]$

Additional Answers

1. $MP = \sqrt{\left(a - \frac{a+c}{2}\right)^2 + \left(b - \frac{b+d}{2}\right)^2}$

$MQ = \sqrt{\left(\frac{a+c}{2} - c\right)^2 + \left(\frac{b+d}{2} - d\right)^2}$

$PQ = \sqrt{(a-c)^2 + (b-d)^2}$

2. Simplify the expressions for MP and MQ.

$MP = \sqrt{\left(\frac{a-c}{2}\right)^2 + \left(\frac{b-d}{2}\right)^2}$

$MQ = \sqrt{\left(\frac{a+c}{2} - c\right)^2 + \left(\frac{b+d}{2} - d\right)^2}$

So $MP = MQ$.

3. $MP + MQ$

$= 2\sqrt{\left(\frac{a-c}{2}\right)^2 + \left(\frac{b-d}{2}\right)^2}$

$= \sqrt{2^2\left(\frac{a-c}{2}\right)^2 + 2^2\left(\frac{b-d}{2}\right)^2}$

$= \sqrt{\left[2\left(\frac{a-c}{2}\right)\right]^2 + \left[2\left(\frac{b-d}{2}\right)\right]^2}$

$= \sqrt{(a-c)^2 + (b-d)^2} = PQ$

PROVING THEOREMS

The midpoint theorem is very helpful in proofs that use coordinate geometry. The first theorem you will prove is Theorem 6-6-1. You may find it interesting to compare the coordinate geometry proof with the proof in Section 6.6.

Using the Page
After the Discovery Activity you may wish to show how the algebra would have looked if no care had been taken in choosing a special placement for the figure in the coordinate plane. Use $A(e, f)$, $B(a, b)$ and $C(c, d)$ as the coordinates of the vertices. Ask students to work through the proof. Point out that to prove that $\overline{MP} \parallel \overline{AC}$, it will be necessary to consider two cases: (1) $e = c$, and (2) $e \neq c$. Discuss why it is necessary to consider these two separate cases.

DISCOVERY ACTIVITY

In this activity, you will use coordinate geometry to prove that the segment connecting the midpoints of two sides of a triangle is parallel to the third side and is half as long as the third side.

Refer to the figure. You are given any triangle ABC. It has been placed so that A is at the origin and C is on the x-axis. This makes three of the coordinates 0 and helps keep the algebra simple.

Supply the missing reasons.

Steps	Reasons
1. $\triangle ABC$ has vertices $A(0,0)$, $B(a,b)$, and $C(c,0)$.	1. Given
2. The midpoint M of $\overline{AB}$ has coordinates $\left(\frac{a}{2}, \frac{b}{2}\right)$.	2. Midpoint theorem
3. The midpoint P of $\overline{BC}$ has coordinates $\left(\frac{a+c}{2}, \frac{b}{2}\right)$.	3. Midpoint theorem
4. Slope of $\overline{AC} = 0$	4. Horizontal lines have slope 0.
5. Slope of $\overline{MP} = 0$	5. Horizontal lines have slope 0.
6. $\overline{AC} \parallel \overline{MP}$	6. Segments that have the same slope are parallel.
7. $AC = c - 0 = c$	7. Distance on x-axis $= \lvert x_2 - x_1 \rvert$
8. $MP = \frac{a+c}{2} - \frac{a}{2} = \frac{c}{2}$	8. Distance on y-axis $= \lvert y_2 - y_1 \rvert$
9. $MP = \frac{1}{2}AC$	9. Substitution

In the Discovery Activity you may have noticed that by placing the triangle a certain way, you can keep the algebra as simple as possible.

CLASS ACTIVITY

Use a calculator and the midpoint theorem to determine the coordinates of the midpoint of the line segment with the given endpoints.

1. (15.3,17.8) and (20.2,40.7) (17.75,29.25)
2. (−26.42,19.8) and (37.5,−12.23) (5.54,3.785)
3. (56.71,38.9) and (−64.83,−27.64) (−4.06,5.63)
4. (−83.51,−14.6) and (−49.382,76.15) (−66.446,30.775)
5. Suppose you had to use coordinate geometry to prove that the diagonals of a square are perpendicular and have the same length. Which of these placements of the square will make the proof easier? The figure on the left

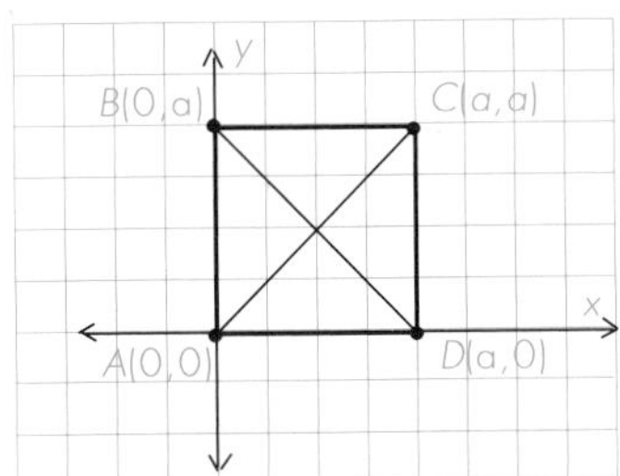

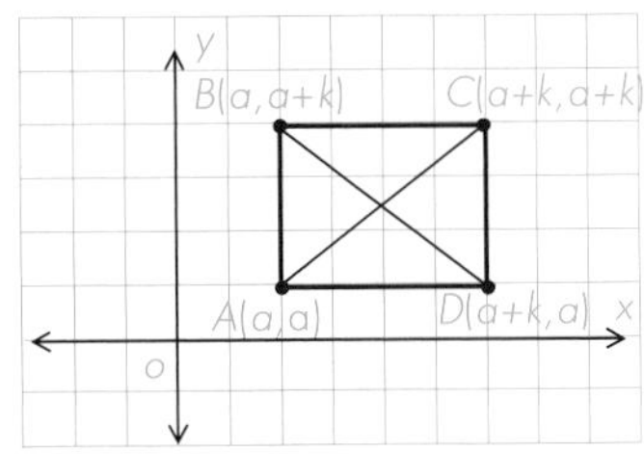

6. Refer to Exercise 5. What formula would you make use of to prove that $AC = BD$? What theorem would you use to prove that $\overline{AC} \perp \overline{BD}$? Distance formula; Theorem 13-3-2
7. Refer to figure *DEFG* to prove that the diagonals of an isosceles trapezoid have the same length. See margin.

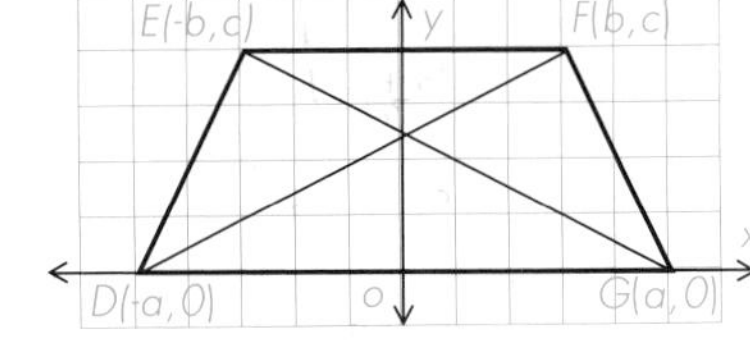

8. Refer to figure *WXYZ* to prove that the diagonals of a parallelogram bisect each other. See margin.

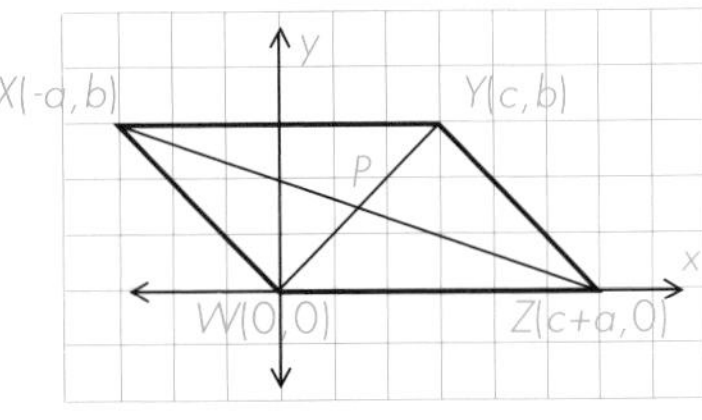

Additional Answers

7. Students should use the distance formula to show that both $\overline{DF}$ and $\overline{EG}$ have a length of $\sqrt{(b + a)^2 + c^2}$.

8. Students should use the midpoint theorem to show that the midpoint of $\overline{WY}$ has the same coordinates as the midpoint of $\overline{XZ}$.

Extra Practice

Use coordinate geometry to prove that in an isosceles right triangle the segment connecting the vertex of the right angle to the midpoint of the hypotenuse is perpendicular to the hypotenuse. Draw a diagram to show a convenient placement for the right triangle in the coordinate plane. [Diagrams may vary. One possibility is the following.]

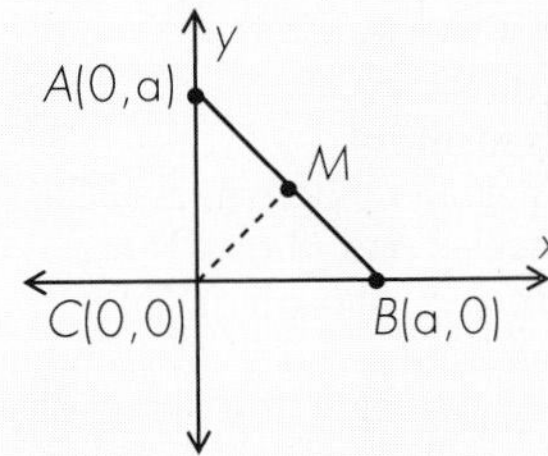

The coordinates of M are $\left(\frac{a}{2}, \frac{a}{2}\right)$. The slope of $\overline{CM}$ is $\frac{a}{2}/\frac{a}{2} = 1$. The slope of $\overline{AB}$ is $\frac{a}{-a} = -1$. Since the product of the slopes is -1, $\overline{CM} \perp \overline{AB}$.]

Additional Answers

7. Slope of $\overline{AC}$

$= \frac{b-0}{b-0} = \frac{b}{b} = 1$

Slope of $\overline{BD}$

$= \frac{a-0}{0-a} = \frac{a}{-a} = -1$

$1(-1) = -1$, so $\overline{AC} \perp \overline{BD}$.

8. Slope of $\overline{AC} = \frac{\sqrt{r^2-a^2}}{a+r}$

Slope of $\overline{BC} = \frac{\sqrt{r^2-a^2}}{a-r}$

The product of the slopes is $\frac{\sqrt{r^2-a^2}}{a+r} \cdot \frac{\sqrt{r^2-a^2}}{a-r} = \frac{r^2-a^2}{a^2-r^2} = -1$, so $\overline{AC} \perp \overline{BC}$.

9. Use the midpoint theorem to find the coordinates of P, Q, R, and S. Then show that $\overline{PQ}$ and $\overline{SR}$ have the same slope and that $\overline{SP}$ and $\overline{RQ}$ have the same slope.

10. The figure limits the proof to *isosceles* right triangles.

FOLLOW-UP

Assessment

1. In the diagram, $\triangle ABC$ is isosceles and M and N are midpoints. Show that $AN = BM$.

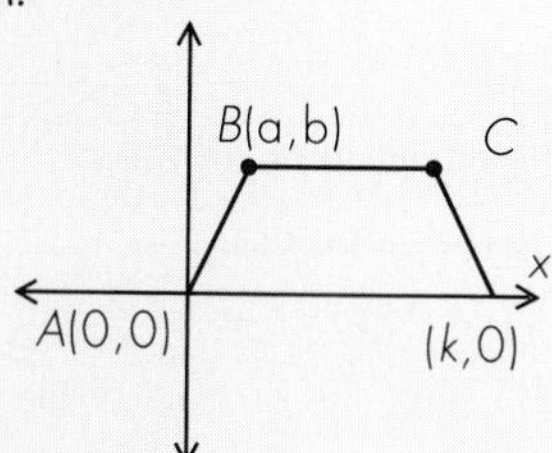

[The coordinates of M are $\left(-\frac{a}{2}, \frac{b}{2}\right)$ and the coordinates of N are $\left(\frac{a}{2}, \frac{b}{2}\right)$. Students should use the distance formula to show that AN and BM both equal $\sqrt{\left(\frac{3a}{2}\right)^2 + \left(\frac{b}{2}\right)^2}$.]

Reteaching

Students who have had difficulty with this section may benefit from Reteaching Activity 13.6.

Enrichment

For students who have mastered the material in this section, you may wish to assign Enrichment Activity 13.6.

HOME ACTIVITY

Use the midpoint theorem to find the coordinates of the midpoint of the line segment having the given endpoints.

1. (1,0) and (5,4) (3,2)
2. (2,6) and (8,20) (5,13)
3. (−3,1) and (5,7) (1,4)
4. (2,9) and (3,1) $\left(\frac{5}{2}, 5\right)$
5. (−6,−1) and (9,0) $\left(\frac{3}{2}, -\frac{1}{2}\right)$
6. $(-a,b)$ and $(a,-b)$ (0,0)
7. Refer to figure $ABCD$. Prove that the diagonals of quadrilateral $ABCD$ are perpendicular. See margin.

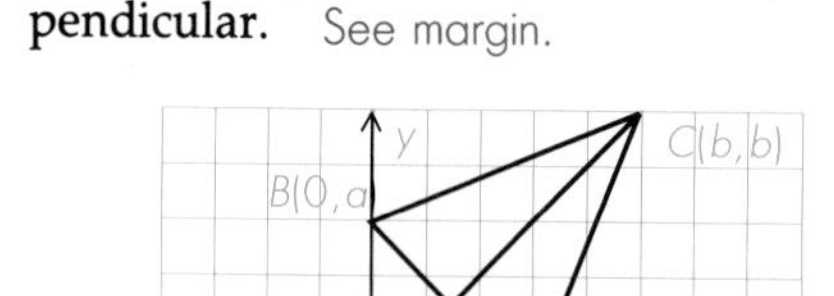

8. Refer to $\triangle ABC$ to show that for point C on the semicircle shown, $\overline{AC} \perp \overline{BC}$. See margin.

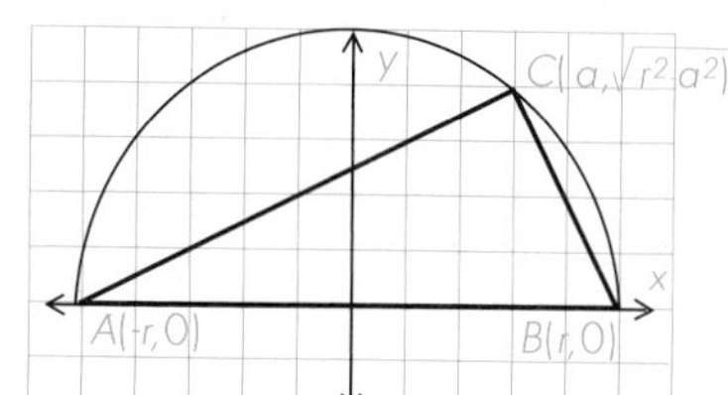

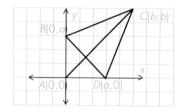

9. Refer to figure $ABCD$. Show that if you join the midpoints of consecutive sides of any quadrilateral, the resulting figure is a parallelogram. (The midpoints are P, Q, R, and S.) See margin.

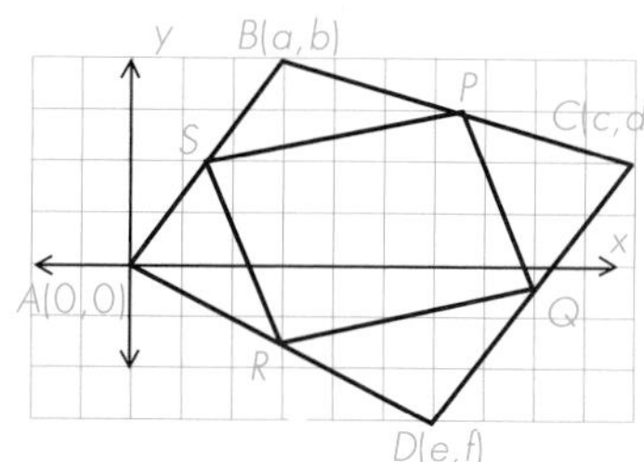

CRITICAL THINKING

10. Suppose you want to use coordinate geometry to prove that the midpoint of the hypotenuse of a right triangle is equidistant from the vertices. Explain why the diagram below would *not* be satisfactory for your proof.

See margin.

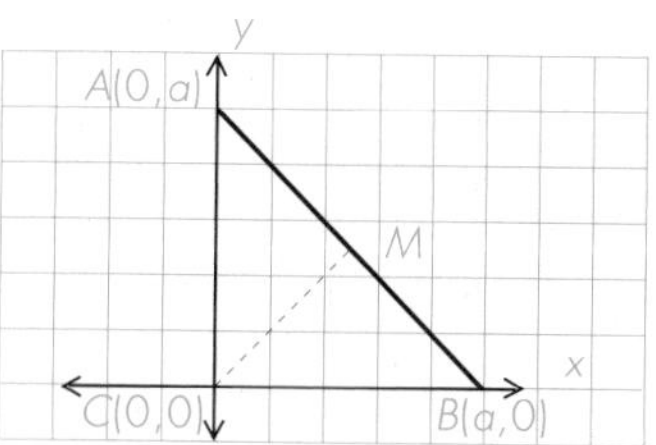

13.1 Is $y = \frac{3}{2}x + \frac{5}{2}$ an equation for line l?

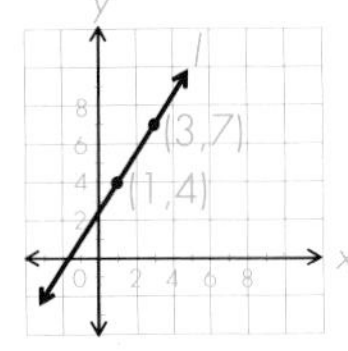

Substitute the given values for x and y into the equation.
$7 = \frac{3}{2}(3) + \frac{5}{2} = 7 \qquad 4 = \frac{3}{2}(1) + \frac{5}{2} = 4$
Because both substitutions are true,
$y = \frac{3}{2}x + \frac{5}{2}$ is a valid equation for line l.

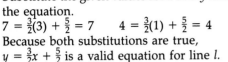

13.2 Give an equation for line n.
One equation will be in the form $y = mx + b$.
$m = \frac{y_2 - y_1}{x_2 - x_1} = \frac{0 - 4}{4 - 0} = -1$ Slope
$b = 4$ y-intercept
An equation for line n is $y = -x + 4$.

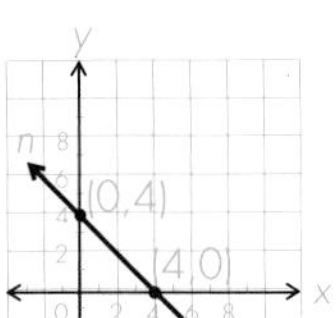

13.3 Show that lines r and t are parallel and that line s is perpendicular to them.

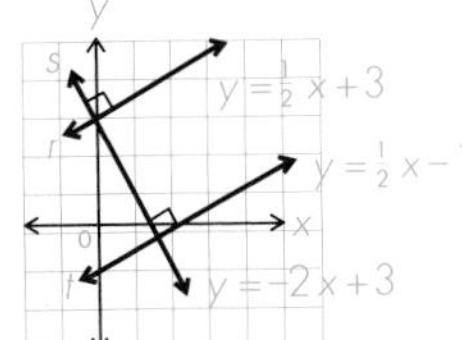

All three lines are in the form $y = mx + b$. Lines r and t have slopes of $\frac{1}{2}$, so they are parallel. Line s has a slope of -2. By the converse of Theorem 13-3-2, the line s is perpendicular to r and t.

13.4 Find an equation for circle O.

Circle O has its center at (0,0), so an equation for it will be in the form $x^2 + y^2 = r^2$. $x^2 + y^2 = 5^2$, or
$x^2 + y^2 = 25$ is an equation for the circle.

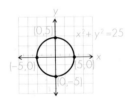

13.5 Find an equation for the parabola determined by point $(0,-2)$ and line $y = 2$.
The line and point are in the form $(0,a)$ and $y = -a$.
$y = \frac{1}{4a}x^2$ By Theorem 13-5-1
$y = \frac{1}{4(-2)}x^2 = \frac{-x^2}{8}$
An equation for the parabola is $y = \frac{-x^2}{8}$.

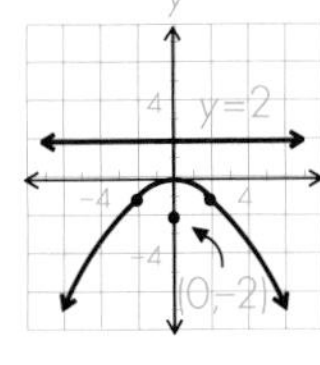

13.6 Find the midpoint M of segment XZ.

$M = \left(\frac{(-2) + 4}{2}, \frac{5 + (-4)}{2}\right)$ By the midpoint theorem

$M = \left(1, \frac{1}{2}\right)$

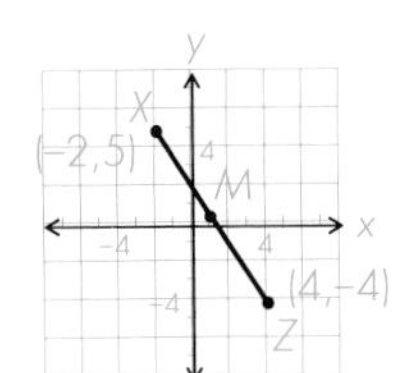
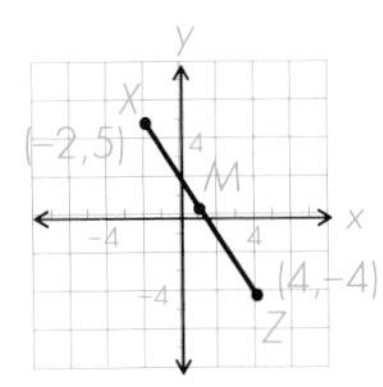

Chapter Vocabulary
conic section
ellipse
graph
hyperbola
line point
rise
run
slope
y-intercept

511

Using the Page

This Calculator Activity introduces the student to the use of a graphing calculator to plot straight line graphs. The lesson involves three distinct skills. For students who have not had a previous opportunity to work with graphing calculators, you may wish to present the lesson in three parts. First, concentrate on using the RANGE key to set the parameters and the scale of the plot. Have students vary the values they input for the range and scale of *x*-values and *y*-values. Secondly, have students plot two points and draw the segment between them. Finally introduce the use of the GRAPH key which enables the student to input an equation and display its graph. Although the calculator will only display the segment between points when two points are entered and the SHIFT LINE EXE sequence is pressed, the use of the GRAPH key will display the entire line within the parameters set for *x*- and *y*- values.

Additional Answers

(for Page 513)

1.

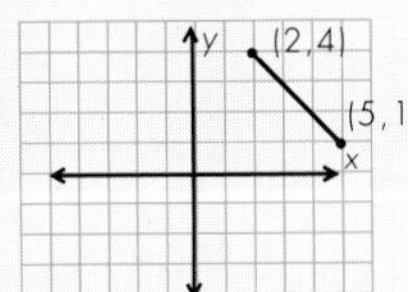

2.

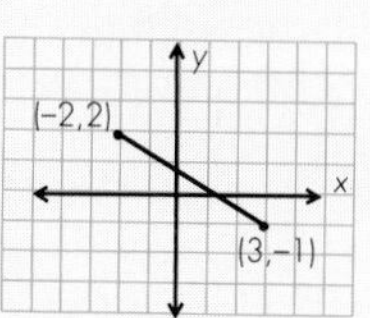

CALCULATOR ACTIVITY

You can use a graphing calculator to plot points and graph equations. Before you begin to draw a graph, you must first use the RANGE key to set the parameters of the graph. After this key is pressed, a list of values will appear on the screen, showing the number of units on each axis and an approximate scale factor for each axis. To change these values, you can use the arrow keys to move from line to line or from right to left. Then use the EXE key to lock the values into the memory of the calculator.

To plot a point, you use the SHIFT, PLOT, and EXE keys. A segment can be drawn to connect two points by pressing the SHIFT, LINE, and EXE keys.

Example: Use a graphing calculator to draw the line determined by the points (2,5) and (−4,3).

First, press the RANGE key. Set the parameters for the *x*-axis at −5 and 5 and the parameters for the *y*-axis at −6 and 6. Set the scale factor for both axes at 1. The resulting display is shown below.

XMIN: −5
XMAX: 5
SCL: 1
YMIN: −6
YMAX: 6
SCL: 1

Next, plot the points. As each point is plotted, a flashing light will appear on the screen at the location of the point.

SHIFT PLOT 2 SHIFT , 5 EXE

SHIFT PLOT (−) 4 SHIFT , 3 EXE

Then draw the line.

SHIFT LINE EXE

The graph will appear on the screen as shown below.

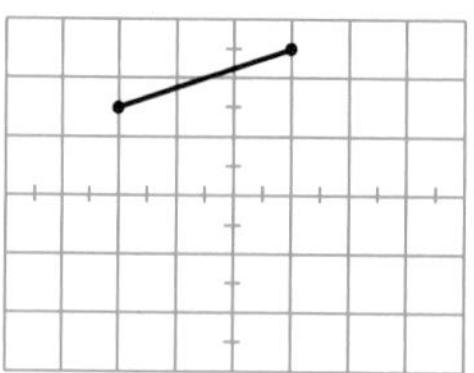

You can use the GRAPH key on a graphing calculator to graph a line when you are given the equation of the line.

Example: Use a graphing calculator to draw the graph of $y = 2x - 1$.

First, set the parameters for the graph.

RANGE XMIN: −2 XMAX: 3 SCL: 1 YMIN: −3 YMAX: 4 SCL: 1

Then press the GRAPH key. The display on the screen is Y = . Enter the rest of the equation. You may need to use the ALPHA key to enter the variable x. If you make a mistake entering the equation, use the arrow keys to correct your mistake. After the equation has been correctly entered, press the EXE key. The graph will appear on the screen as shown below.

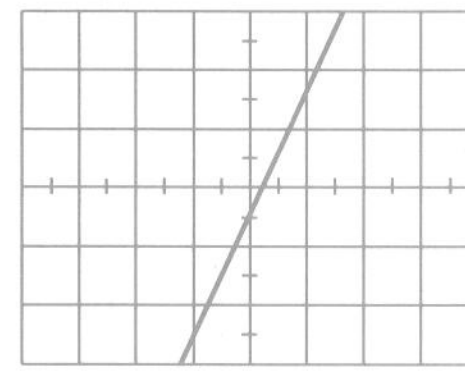

Use a graphing calculator to draw the line determined by each pair of points. See margin.

1. (5,1) and (2,4)
2. (3,−1) and (−2,2)
3. (−4,6) and (7,−2)
4. (−3,−2) and (−8,6)

Use a graphing calculator to draw the graph of each equation. See margin.

5. $y = -2$
6. $y = \frac{2}{3}x$
7. $y = x + 1$
8. $y = 3x + 3$
9. $y = 4x - 2$
10. $y = \frac{x}{2} + 4$

Additional Answers

3.

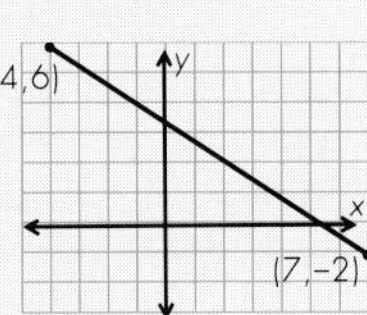

4.

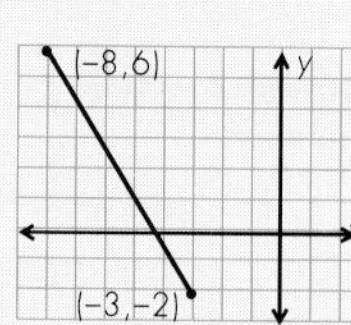

5.

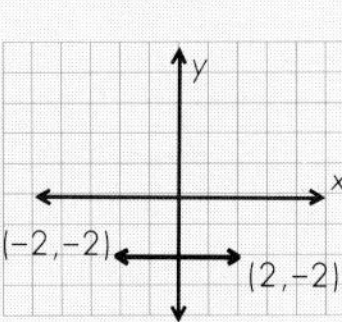

6.

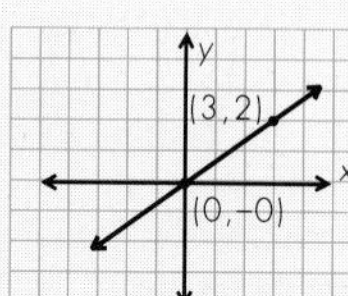

7.

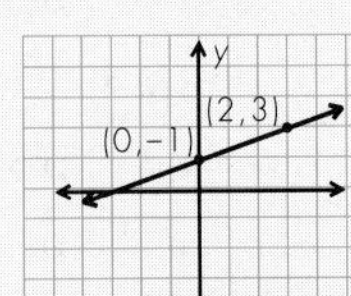

8.

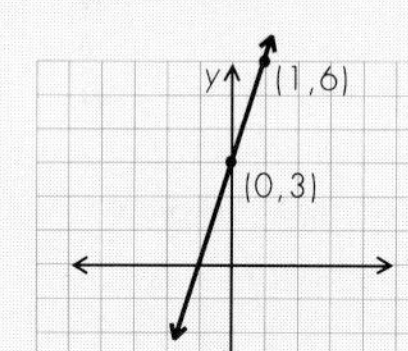

9.

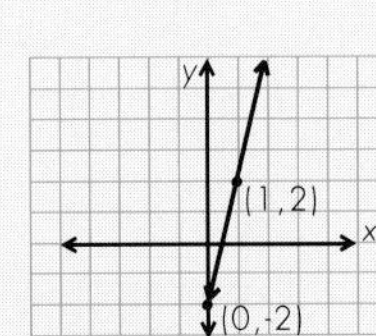

10.

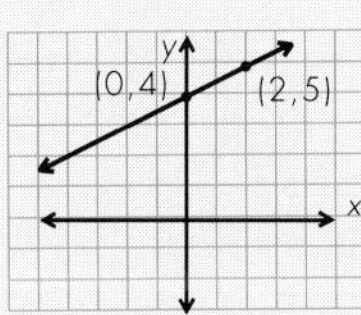

Using the Page
Students can work individually or in groups to study the review. If you have students study in groups, they should work together to make sure all members of the group know the material in the chapter.

Additional Answers
Graphs for Exercises 1-3.

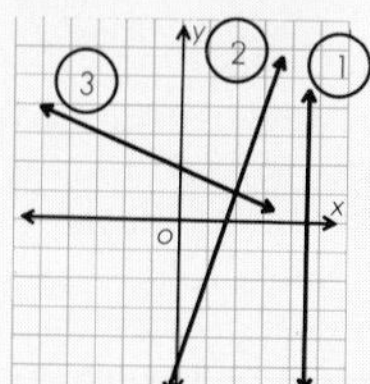

17.

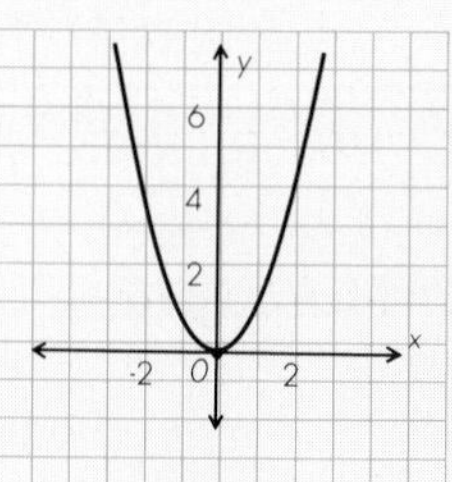

18.

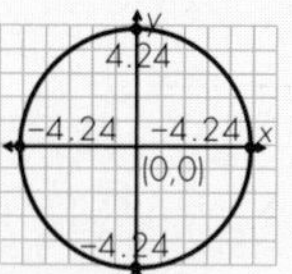

19.

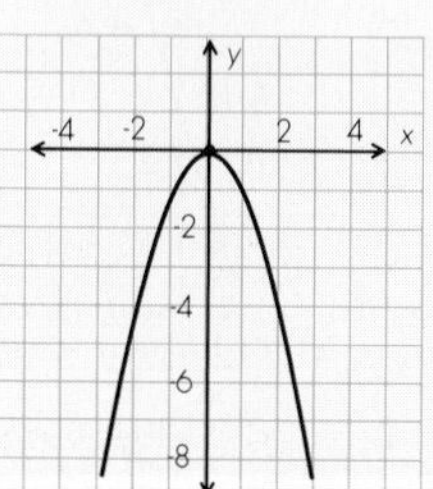

23.

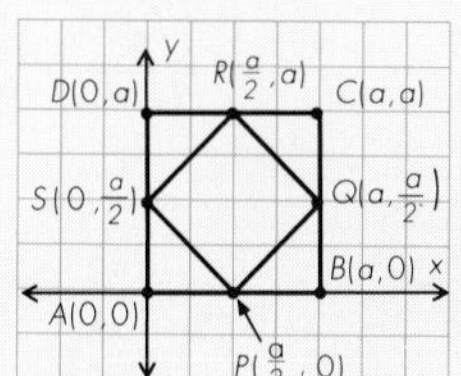

Use the distance formula to show all sides of $PQRS$ have length $\frac{a\sqrt{2}}{2}$. Then use the slopes of the sides of $PQRS$ to show that the product of slopes of adjacent sides is -1.

Graph each equation in the coordinate plane. Tell whether the line is vertical or nonvertical. If it is nonvertical, give its slope and y-intercept. See margin for graphs.

1. $x = 5\frac{1}{2}$ Vertical **2.** $y = 3x - 5$ Nonvertical **3.** $y = \frac{-1}{3}x + 2$ Nonvertical

Write an equation for each figure.

4. The line with slope 7 and y-intercept -3 $y = 7x - 3$

5. The vertical line through $\left(\frac{7}{3},0\right)$ $x = \frac{7}{3}$

6. The line that passes through $(-6,-1)$ and $(-4,0)$ $y = \frac{1}{2}x + 2$

7. The circle with center $(0,0)$ and radius 9 $x^2 + y^2 = 81$

8. The circle with center $(0,0)$ and radius $\frac{4}{3}$ $x^2 + y^2 = \frac{16}{9}$

Tell whether the lines go up, down, or are level as you move from left to right in the coordinate plane.

9. $y = -\frac{2}{3}x + 10$ Down **10.** $y = 11$ Level **11.** $y = \frac{1}{8}x - 6$ Up

Tell whether the lines are parallel, perpendicular, or neither.

12. $y = 3x - 2$, $y = \frac{1}{3}x + \frac{1}{2}$ Neither

13. $y = \frac{7}{4}x + 12$, $y = -\frac{4}{7}x - 12$ Perpendicular

14. $y = 5x - 13$, $y = 5x + 2$ Parallel

15. What is the slope of any line perpendicular to $y = 9x - 1$? $-\frac{1}{9}$

16. Write an equation for the line that passes through $(5,6)$ and is perpendicular to the line $y = -4x + 3$. $y = \frac{1}{4}x + \frac{19}{4}$

Graph each equation. You may use a calculator for calculations. See margin.

17. $y = \frac{3}{2}x^2$ **18.** $x^2 + y^2 = 18$ **19.** $y = -2x^2$

Tell whether a graph of the equation is a circle or a parabola. If it is a circle, give its center and radius. If it is a parabola, tell whether it opens up or down.

20. $y = 12x^2$ Parabola; up

21. $x^2 + y^2 = 45$ Circle; $(0,0)$, $r = 3\sqrt{5}$

22. $x^2 + y^2 = 169$ Circle; $(0,0)$, $r = 13$

23. $y = -\frac{7}{2}x^2$ Parabola; down

23. Use coordinate geometry to prove that, if you connect the midpoints of a square in order as you go around the square, then the resulting quadrilateral is also a square. See margin.

514

CHAPTER TEST

Write an equation for the line that passes through the given pair of points.

1. $(-1,2)$ and $(1,4)$ $y = x + 3$

2. $(7,5)$ and $(4,11)$ $y = -2x + 19$

3. $(3,9)$ and $(-4,9)$ $y = 9$

Write an equation for each line.

4. The line with slope -2 and y-intercept 8 $y = -2x + 8$

5. The line with slope 3 that passes through $(0,5)$ $y = 3x + 5$

6. The line with slope 0 that passes through $(5,-4)$ $y = -4$

Tell whether the lines are parallel, perpendicular, or neither.

7. $y = -2x + 13$
$y = \frac{1}{2}x - 11$
Perpendicular

8. $y = 6x + 4$
$12x + 2y = 5$
Neither

9. $y = 14x + 76$
$y = 14x - \frac{1}{3}$
Parallel

The slope of line l is $\frac{3}{4}$, $l \parallel n$, and $m \perp l$.

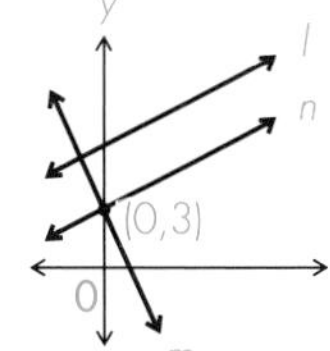

10. Write an equation for line m. $y = -\frac{4}{3}x + 3$

11. Write an equation for line n. $y = \frac{3}{4}x + 3$

Give the center and radius of each circle.

12. $x^2 + y^2 = 100$
$(0,0)$, $r = 10$

13. $x^2 + y^2 = 3$
$(0,0)$, $r = \sqrt{3}$

14. $x^2 + y^2 = \frac{1}{64}$
$(0,0)$, $r = \frac{1}{8}$

Tell whether the parabola opens up or down.

15. $y = -9x^2$ Down

16. $y = 10x^2$ Up

17. $y = x^2$ Up

18. $y = -\frac{2}{3}x^2$ Down

19. Write an equation for the parabola that is the locus of all points equidistant from $(0,3)$ and the line $y = -3$. $y = \frac{1}{12}x^2$

Graph each equation on the coordinate plane. You may use a calculator or the table of square roots on page 562 if you wish. See margin.

20. $y = -3x + \frac{1}{2}$

21. $y = \frac{3}{2}x^2$

22. $x^2 + y^2 = 49$

23. $ABCD$ is a parallelogram. Use coordinate geometry to prove that the segment $\overline{MN}$ connecting the midpoints of $\overline{DC}$ and $\overline{AB}$ is parallel to $\overline{AD}$ and $\overline{BC}$. See margin.

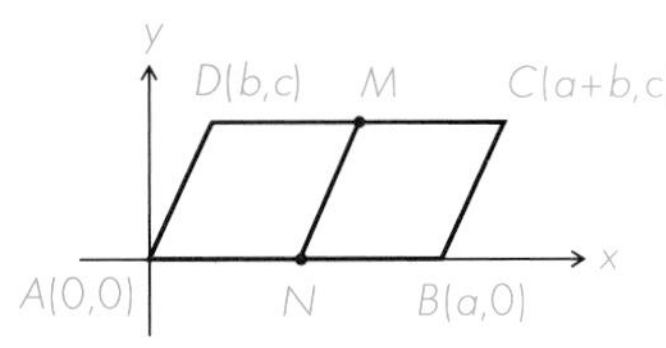

Assessment Resources
Achievement Tests pp. 25-26

Test Objectives
After studying this chapter, students should understand
- how equations are used to define lines.
- the slope of a line.
- how to find the slope of a line given two points on the line.
- the y-intercept of a line.
- how the slopes of parallel lines are related.
- how the slopes of perpendicular lines are related.
- how to graph equations of the form $x^2 + y^2 = r^2$.
- how to find and graph an equation for a parabola.
- how to prove theorems using coordinate geometry.

Additional Answers
Graphs for items 20-22

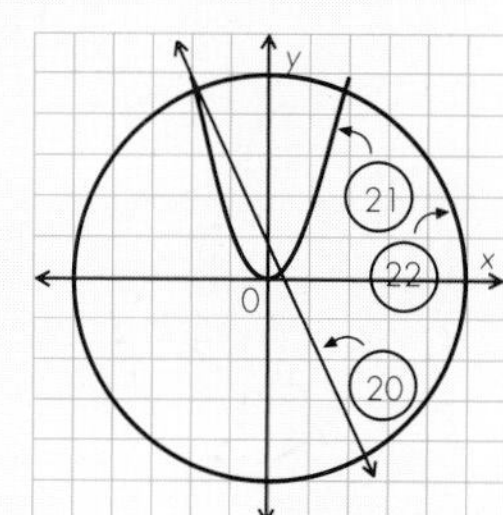

23. The coordinates of M and N are $\left(\frac{a + 2b}{2}, c\right)$ and $\left(\frac{a}{2}, 0\right)$. Use endpoint coordinates to show that the segments $\overline{AD}$, $\overline{BC}$, and $\overline{MN}$ all have a slope of $\frac{c}{b}$. This proves the segments are parallel.

Algebra Review Objectives
Topics reviewed from algebra include
- meaning of absolute value.
- properties of absolute value.
- solving absolute-value equations in one unknown.

ALGEBRA SKILLS

In coordinate geometry and algebra you sometimes need to solve equations such as $|x - 3| = 7$. To solve equations involving absolute value, you need to remember that, for any positive number, there are two numbers that have that number as their absolute value.

Example: Solve $|x - 3| = 7$
$x - 3 = 7$ or $x - 3 = -7$
$x = 10$ or $x = -4$

The equation has two solutions, 10 and -4.

Find all the solutions of each equation.

1. $|x - 5| = 9$ -4 and 14
2. $|x + 6| = 1$ -5 and -7
3. $|2x - 1| = 3$ -1 and 2
4. $|5x - 10| = 0$ 2
5. $|x + 8| = 8$ 0 and -16
6. $|-x + 4| = 10$ -6 and 14
7. $|\frac{1}{2}x - 6| = 7$ -2 and 26
8. $|x + 1| = \frac{3}{4}$ $-\frac{7}{4}$ and $-\frac{1}{4}$
9. $|3x - 8| = 13$ $-\frac{5}{3}$ and 7

You may need to use some of the properties of absolute value for some equations. In the following example, you use the property that, if a and b are any real numbers, b not equal to zero, then $|a|/|b| = |a/b|$.

Example: Solve $|x - 2| = |x - 6|$.
Divide both sides by $|x - 6|$ and go on from there.

$$|x - 2| = |x - 6|$$

$$\frac{|x - 2|}{|x - 6|} = 1$$

$$\left|\frac{x - 2}{x - 6}\right| = 1$$

$\frac{x-2}{x-6} = 1$ or $\frac{x-2}{x-6} = -1$

$x - 2 = x - 6$ ↑ This equation has no solution.

$x - 2 = -x + 6$
$2x = 8$
$x = 4$

The equation $|x - 2| = |x - 6|$ has 4 as its only solution.

Find all the solutions of each equation.

10. $|x - 3| = |2x + 4|$ $-\frac{1}{3}$ and -7
11. $|x + 5| = |x - 8|$ $\frac{3}{2}$
12. $|3x + 2| = |2x - 1|$ $-\frac{1}{5}$ and -3
13. $|-2x + 3| = |x + 4|$ $-\frac{1}{3}$ and 7

516

Name ______________________ Date ____________

What Is a Point and What Is a Line?

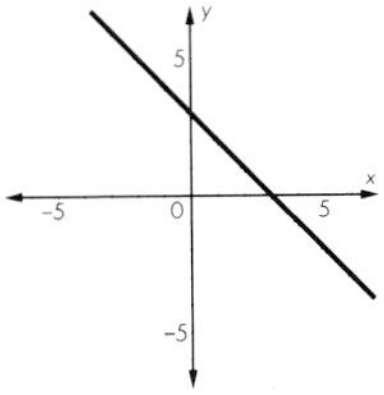

The *solution* to an equation can be represented using a graph. For a one-variable equation, the graph can be a number line like the one shown below. For a two-variable equation, you need a 2-dimensional graph with two axes, called the x-axis and the y-axis.

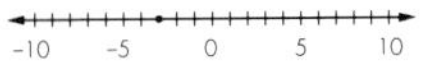

The solution to the equation $x = -3$ is the point at the number -3.

The solution to the equation $x + y = 3$ is all the points with ordered pairs in which the x and y values total 3. The graph is a line.

Circle the letter of the equation whose solution is shown on the graph.

1. a. $x = 0$ (b.) $x = -1$ c. $x = 1$
2. a. $2x = 8$ (b.) $2x = 6$ c. $2x = -6$
3. (a.) $x + 5 = 0$ b. $x - 5 = 0$ c. $5 - x = 0$

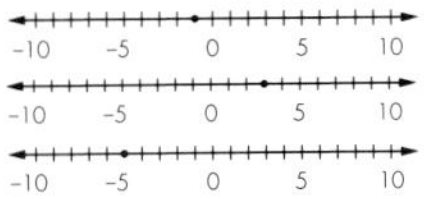

4. a. $x + y = 0$
   b. $x + y = -1$
   (c.) $x + y = 1$

5. (a.) $x - y = 0$
   b. $x - y = -1$
   c. $x - y = 1$

6. a. $y = x + 1$
   (b.) $y = x - 1$
   c. $y = x$

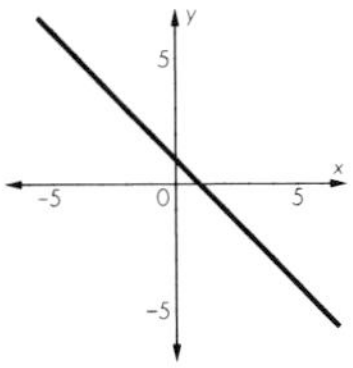

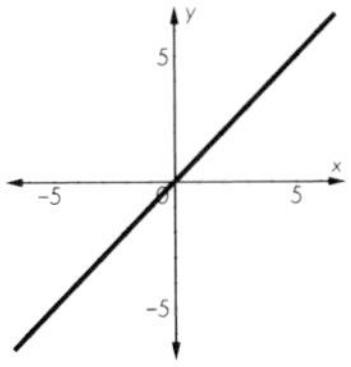

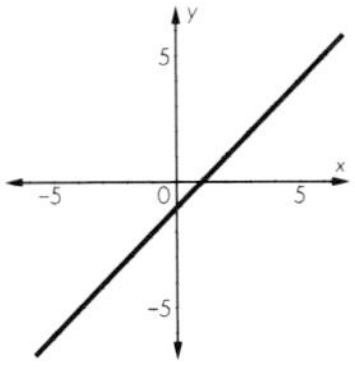

Name ______________________ Date ____________

What is a Plane?

A plane in 3-dimensional space has an equation of the form:

$$\frac{x}{a} + \frac{y}{b} + \frac{z}{c} = 1$$

The constants a, b, and c are the intercepts of the plane—the points where it intersects the x-, y-, and z-axes, respectively.

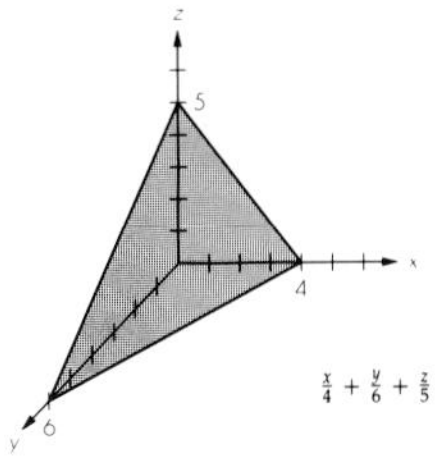

$\frac{x}{4} + \frac{y}{6} + \frac{z}{5} = 1$

Use the figure at the right above for Exercises 1–7.

1. Write ordered triples to show the coordinates of the three intercepts.
   (4,0,0), (0,6,0), (0,0,5)

2. Write the equation of the line where the plane intersects the xy-plane.
   $y = -\frac{3}{2}x + 6$ or $x = -\frac{2}{3}y + 4$

3. Write the equation of the line where the plane intersects the yz-plane.
   $y = -\frac{6}{5}z + 6$ or $z = -\frac{5}{6}y + 5$

4. Write the equation of the line where the plane intersects the xz-plane.
   $z = -\frac{5}{4}x + 5$ or $x = -\frac{4}{5}z + 4$

5. A plane passes through the points (6, 0, 0), (0, 8, 0), and (0, 0, 7). How does this plane compare with the one drawn above?
   The two planes are parallel.

6. A plane passes through the points (4, 0, 0), (0, 6, 0), and (0, 0, 7). How does this plane compare with the one drawn above?
   They intersect the xy-plane in the same line.

7. Write the equation of a plane that passes through the point (3, 0, 0) and is parallel to the yz-plane.
   $x = 3$

8. On the axes at the right, draw a plane with $x-$, $y-$, and $z-$ intercepts 7, 8, and 6, respectively. Write the equation of this plane.
   $\frac{x}{7} + \frac{y}{8} + \frac{z}{6} = 1$

9. What is the surface area of the tetrahedron enclosed by the plane you drew in Exercise 8 and the three planes formed by the axes?
   about 93.3 sq. units

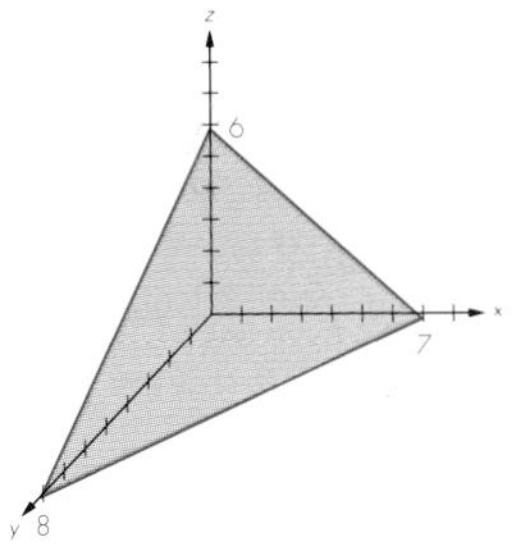

Name ______________________ Date ____________

The Easy Way to Graph a Line

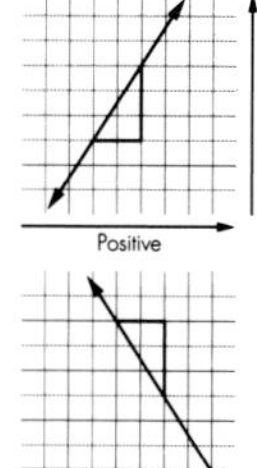

The *slope* of a line is the ratio of rise to run. The line shown at the right rises 3 units for every 2 units it "runs." Thus, the slope of this line is the fraction $\frac{3}{2}$, or any number equivalent to it, such as 1.5.

When finding the slope of a line, movement upward or to the right is shown by a positive number; movement down or to the left is shown by a negative number.

In the second drawing, the line drops 3 units for every 2 units it runs to the right. So, its slope is $-\frac{3}{2}$. However, you might picture the second line another way. You might see it as rising 3 units for every 2 units it moves to the left. Since moving to the left is moving in a negative direction, the slope would be $\frac{3}{-2}$, which is equal to $\frac{-3}{2}$.

Write the slope of each line.

1. $\frac{3}{2}$
2. $-\frac{2}{3}$
3. $\frac{2}{3}$
4. $-\frac{3}{2}$

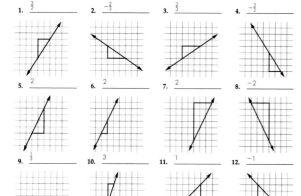

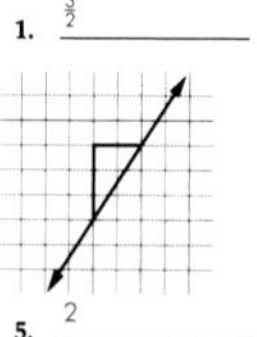
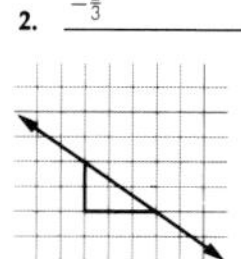
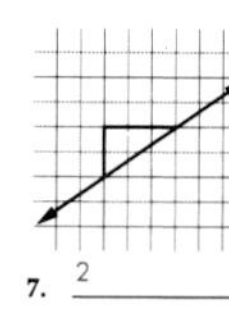
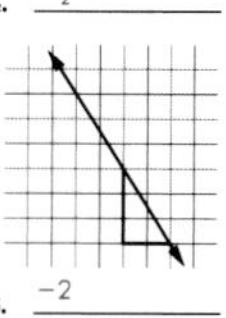

5. 2
6. 2
7. 2
8. −2

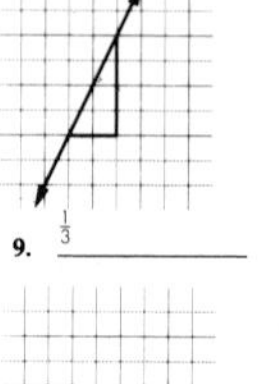
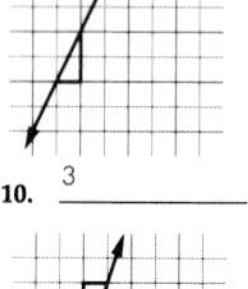
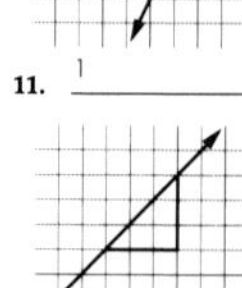
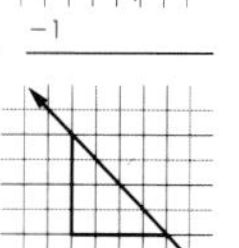

9. $\frac{1}{3}$
10. 3
11. 1
12. −1

Name  Date ____________

Slope and Speed

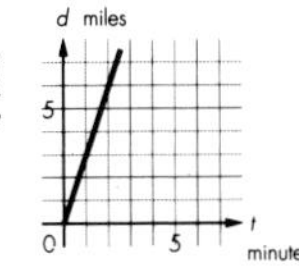

A distance runner ran 6 miles in 2 minutes. What was her average speed? To find out, graph the equation $d = rt$ using the two ordered pairs (0, 0) and (2, 6). The slope of the line, 3, is the speed in miles per minute.

For Exercises 1–6, write the speed represented in each graph.

1. 1 mi/hr
2. 0.5 m/sec
3. 2 ft/sec

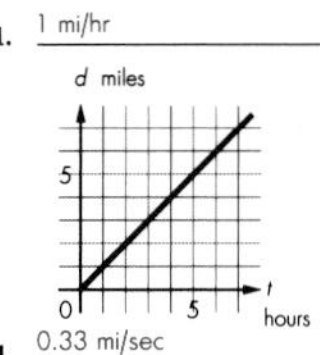
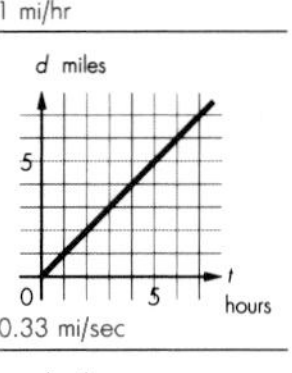

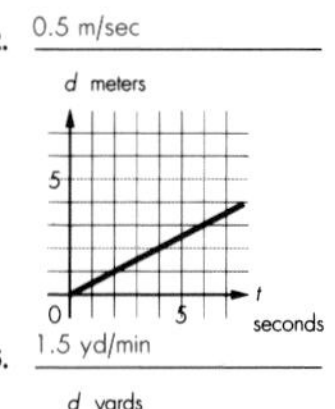

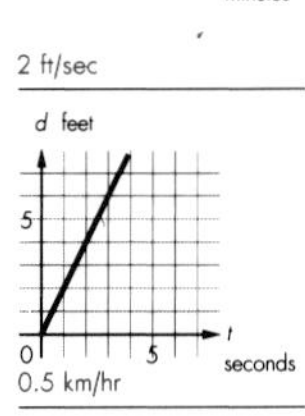

4. 0.33 mi/sec
5. 1.5 yd/min
6. 0.5 km/hr

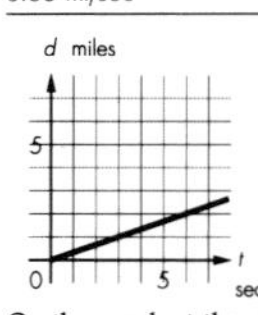

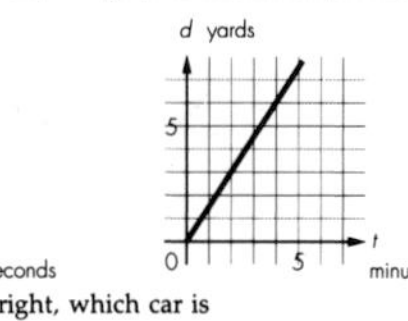

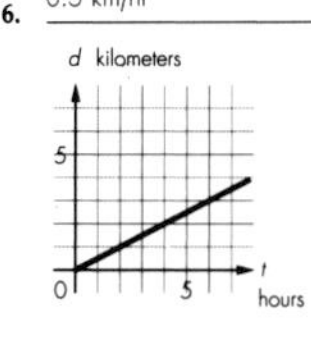

7. On the graph at the right, which car is the fastest? Car A

8. Which car is moving twice as fast as Car B? Car A

9. What happens to Car C? After 6 time units, it stops.

10. Explain how you were able to answer Exercises 7, 8, and 9 without knowing the units of measure used for the graph. The answers would be the same no matter what units were used because speed is a ratio of distance and time.

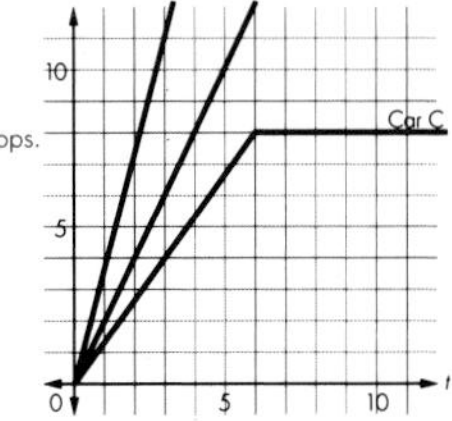

Name ______________________ Date ____________

Slopes of Perpendicular Lines

If the equation of a line is written in the form $y = mx + b$, the constant m is the slope. Two lines with the same slope are parallel.

Two lines are perpendicular if the product of their slopes is -1. Remember, write the equations of the lines in the form $y = mx + b$ to find the slopes.

Graph each equation. Then tell if the new line is parallel or perpendicular to the line already drawn on the graph.

1. $x + y = 0$ parallel

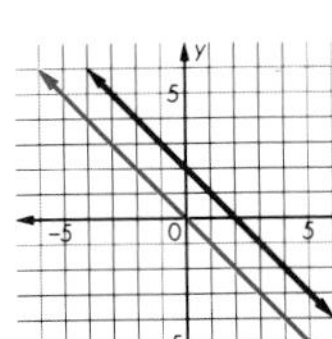

2. $x - y = 0$ perpendicular

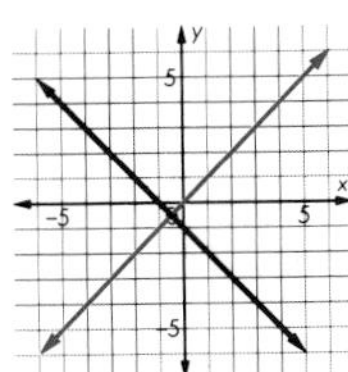

3. $y - 1 = x$ parallel

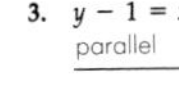

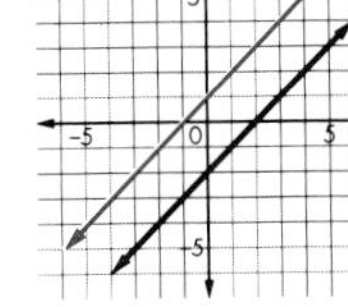

4. $y = 2x$ parallel

5. $x = 2y - 2$ perpendicular

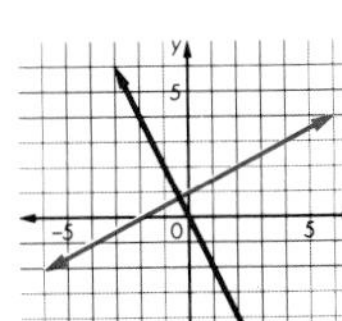

6. $y + 2 = x$ parallel

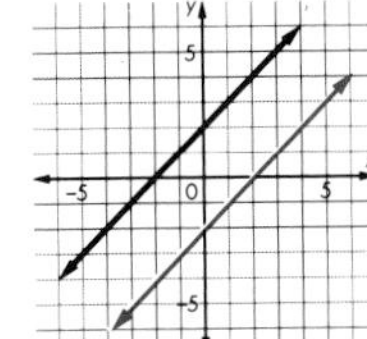

Name ______________________ Date ____________

Pencils of Lines

All the lines passing through a given point and lying in the same plane are called a *pencil* of lines.

A group of lines with the same slope are a pencil of parallel lines.

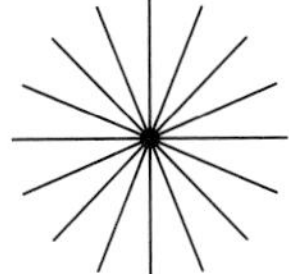

Graph each group of lines.

1. A pencil of parallel lines having slope -1 and crossing the x-axis at the coordinates -6, $-4, \ldots, 4, 6$.

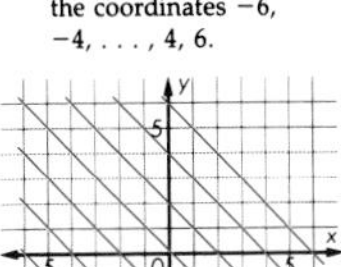

2. A pencil of lines passing through the point with coordinates $(1, -2)$.

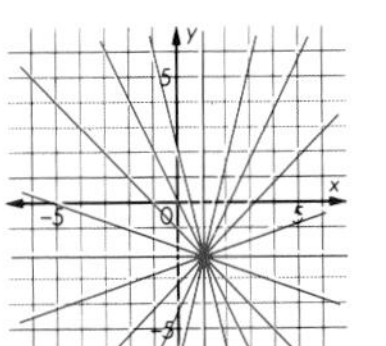

3. The line $y = x - 2$ and a pencil of parallel lines having slope -2.

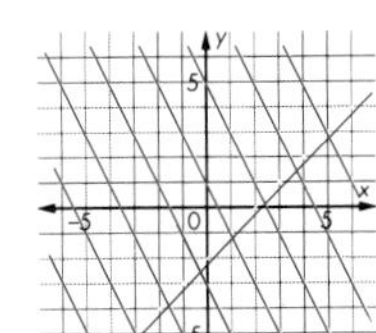

4. Pencils of lines passing through the points with coordinates $(3, 3)$ and $(0, -3)$.

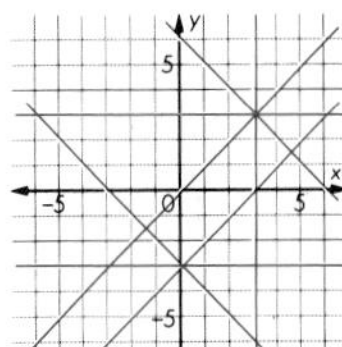

5. A pencil of parallel lines having slope -2 and a second pencil of parallel lines having slope 1.

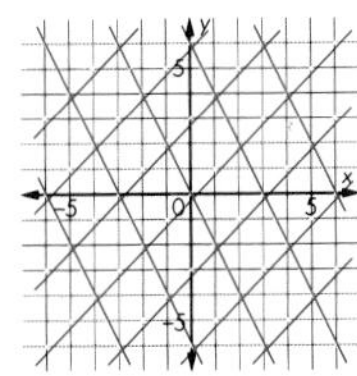

6. A pencil of lines through the point $(-1, 2)$ and a pencil of parallel lines having equations of the form $y = a$.

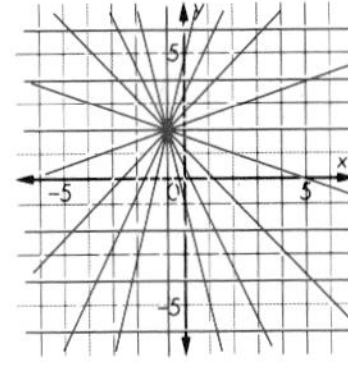

Name ______________________ Date ____________

Circles

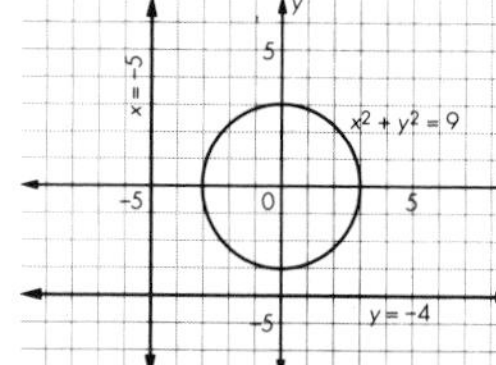

Vertical lines have equations of the form $x = a$, where a is any number. The y-axis has the equation $x = 0$.

Horizontal lines have equations of the form $y = a$. The x-axis has the equation $y = 0$.

An equation of the form $x^2 + y^2 = r^2$ has a circle with radius r for its graph.

Shown at the right are the graphs of $x = -5$, $y = -4$, and $x^2 + y^2 = 9$.

1. On Graph 1, graph the equations $x = 3$ and $x^2 + y^2 = 16$. Shade the region inside of the circle where the x values are greater than 3.

2. On Graph 2, graph the equations $x^2 + y^2 = 36$ and $x^2 + y^2 = 25$. Shade in the region between the two circles.

Graph 1

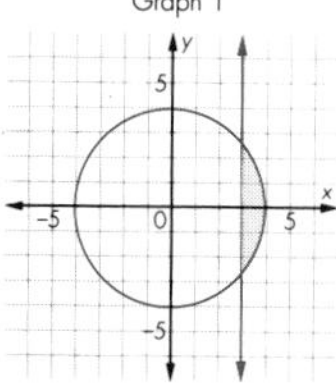

Graph 2

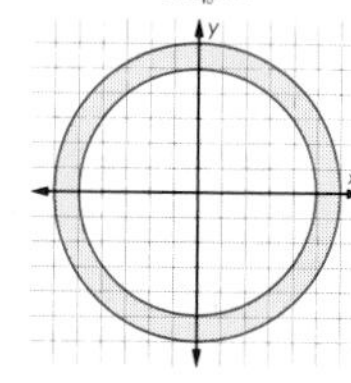

3. On Graph 3, graph the four equations $x = 4$, $x = -4$, $y = 4$, and $y = -4$. What figure is formed by the lines? a square

4. Draw the largest possible circle inside the figure you made in Exercise 3. What is its equation? $x^2 + y^2 = 16$

Graph 3

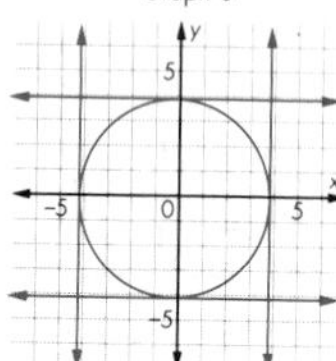

Graph 4

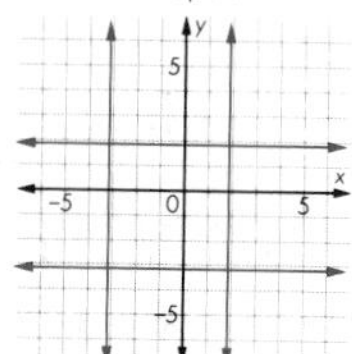

5. On Graph 4, graph the four equations $x = -3$, $x = 2$, $y = 2$, and $y = -3$. Write the coordinates of the points of intersection. $(-3, 2)$, $(2, 2)$, $(2, -3)$, $(-3, -3)$

Name ______________________ Date ____________

Translations of Circles

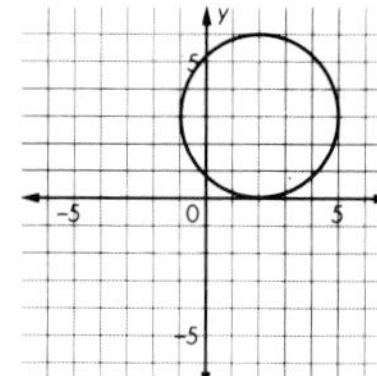

A circle with equation of the form $(x - h)^2 + (y - k)^2 = r^2$ has a center at the point with coordinates (h, k) and a radius of r units. It has been translated away from the origin by a horizontal distance h and a vertical distance k.

The circle shown at the right has the equation $(x - 2)^2 + (y - 3)^2 = 3^2$

In Exercises 1–3, write the equations for each set of circles.

1. $(x+1)^2 + (y-2)^2 = 1^2$
   $(x+1)^2 + (y-2)^2 = 2^2$
   $(x+1)^2 + (y-2)^2 = 3^2$

2. $(x-1.5)^2 + (y-1.5)^2 = (2.5)^2$
   $(x+1.5)^2 + (y-0.5)^2 = (2.5)^2$
   $(x-0.5)^2 + (y+2.5)^2 = (2.5)^2$

3. $(x-1.5)^2 + (y-0.5)^2 = (2.5)^2$
   $(x-0.5)^2 + (y-0.5)^2 = (3.5)^2$
   $(x+0.5)^2 + (y-0.5)^2 = (4.5)^2$

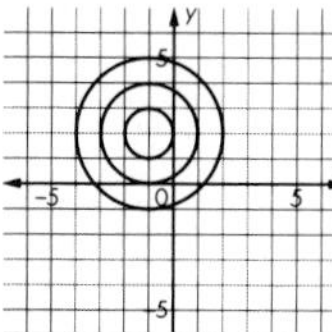

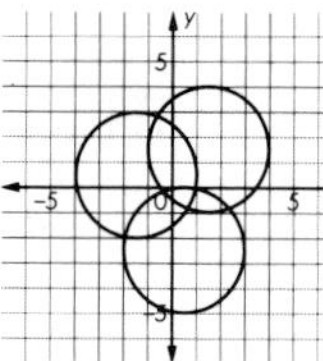

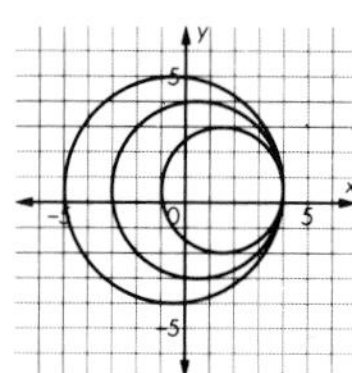

4. On the graph at the right, draw a set of circles tangent to the line $y = 4$ so that the centers of the circles are all beneath the line. Write the equations for your circles.
   Answers will vary.

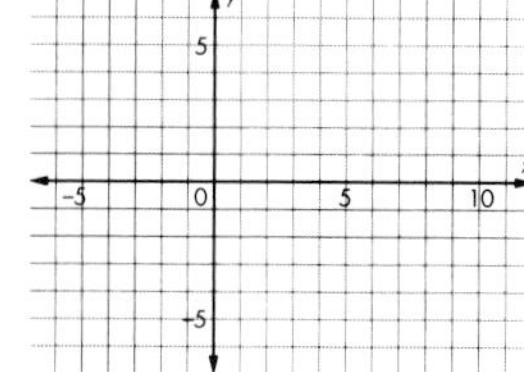

Name ______________________ Date ______________

Parabolas

An equation of the form $y = kx^2$ has a parabola for its graph. The vertex of the parabola will always be at the origin, the point with coordinates (0, 0). The coefficient k determines the amount and the direction that the parabola opens.

1. Complete this table.

x	$-x$	x^2
0	0	0
1	−1	1
2	−2	4
3	−3	9
4	−4	16
5	−5	25

Now use the table above to graph these equations.

2. $y = x^2$
3. $y = -x^2$
4. $y = 2x^2$
5. $y = -2x^2$
6. $y = 0.5x^2$
7. $y = -0.5x^2$
8. $y = 0.1x^2$
9. $y = -0.1x^2$
10. Describe the effect of the value of k on the parabola with equation $y = kx^2$.

The smaller the absolute value of k, the wider the parabola opens; it opens u[…]

and downward for negative k.

Name ______________________ Date ______________

Graphing Systems

The solution of a system of inequalities is the set of points that makes all the number sentences true. The shaded area of the graph at the right shows the solution set for this system:

$x < 2$
$y > x$

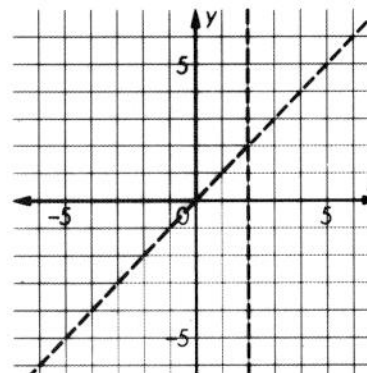

To find the solution for a system, start by replacing each inequality symbol with an equals sign and graphing those lines. The region to shade can be found by trial and error.

Graph the solution set for each system.

1. $y < 2 - x$, $y < x + 1$
2. $x + y + 2 > 0$, $x < 1$
3. $y > x - 2$, $y < 2$

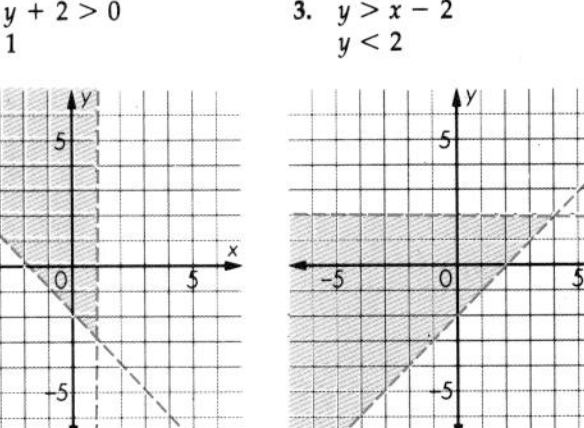

4. $y > 2x - 3$, $y - x < 1$, $2x + y + 3 > 0$
5. $y - 2x < 0$, $y > 2x - 5$, $y + 2x > 0$
6. $x^2 + y^2 < 16$, $y < x + 2$, $y + x < 2$

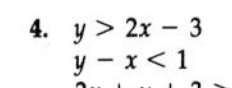

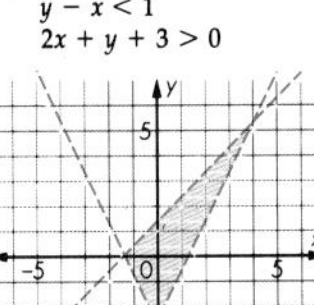

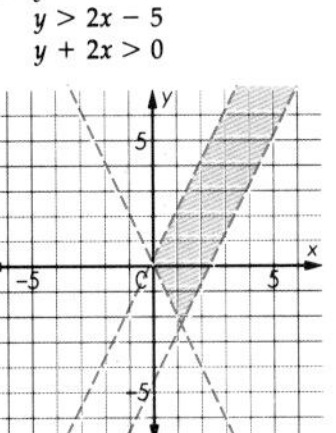

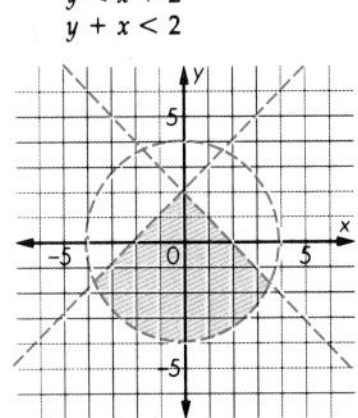

Name ______________________ Date ______________

Applications

Often, the first step in proving a theorem in geometry is drawing the correct diagram.

Write the letter of each statement above the diagram that best illustrates it. Write *not described* above diagrams without statements.

A. A line with b equal to 0 in the equation $y = mx + b$ passes through the origin.

B. The two lines $x + y = 0$ and $x - y = 0$ are perpendicular.

C. Two lines with slope equal to 2 are parallel.

D. The lines $x = 3$, $y = 0$, and $x = y$ form a right triangle.

E. A circle with center at the origin and radius r does not intersect the line $y = r + 1$.

F. The line $y = x$ passes through the center of a circle with center at the origin.

G. The lines $x = 3$ and $x = -3$ are tangent to a circle with center at the origin and a diameter of 6 units.

H. The lines $y = x$ and $y = -x$ form an isosceles triangle with any horizontal line.

1. E **2.** C

3. not described **4.** A **5.** F **6.** not described

7. B **8.** H **9.** D **10.** G

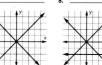

158 Reteaching • Section 13.6

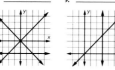

Name ______________________ Date ______________

A Field Protractor

You may have wondered how angles are measured by surveyors. Follow the directions below to make a simple field protractor and you can try some "outdoor geometry" for yourself.

Nail Nail Pin Arm

1. Start with a flat board about 20 inches square.
2. On the board, draw a circle 10 inches in diameter. Divide its circumference into 360 equal parts.
3. Using a flat strip of wood, make an arm as shown in the drawing. Attach the arm to the center of the circle so that it swings freely.
4. Attach a nail to each end of the arm to use as sights.
5. At the end of the arm and in line with the sights, place a pin which will indicate the degree measure.
6. Then attach the board to the end of a stake about 4 feet long, or if available, to a tripod.
7. To measure an angle ABC, place your field protractor at point A. Be sure the board is level. (An inexpensive level attached to the stake will make this easier.) Holding the board stationary, sight at point C and read the angle. Then sight at point B and read this angle. The difference between the two readings is the measure of angle ABC.

One method of obtaining measurements for a scale drawing is illustrated in the figure at the right.

Placing the field protractor at a point, such as point O in the figure, find the distances to the other objects in the figure and their angles as measured from $\overline{OD}$ or $\overline{OA}$.

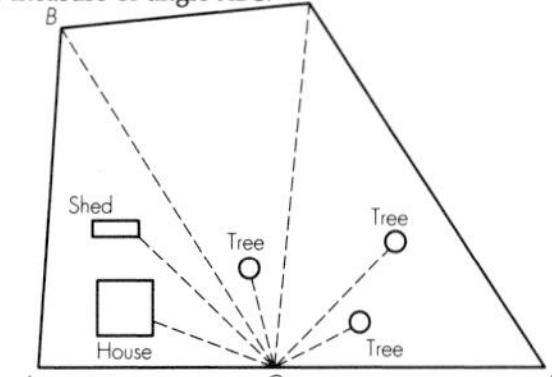

8. You may wish to work with a partner for this exercise. Choose an irregular plot of ground. Use your field protractor to obtain the measurements needed to make a scale drawing of the main buildings and trees on the plot. Make your drawing on a sheet of graph paper. Answers will vary.

Enrichment • Section 13.6 159

Name ______________________ Date ______________

Using Graphing Calculators

Use a graphing calculator to draw the line segment having each pair of points as end points. Draw a sketch of the graph. Then write an equation for the line which contains the line segment.

1. (2, 6) and (1, 1) $y = 5x - 4$

2. (4, 7) and (0, 6) $y = \frac{x}{4} + 6$

3. (−2, 2) and (−3, −1) $y = 3x + 8$

4. (5, −2) and (−5, 0) $y = -\frac{x}{5} - 1$

5. (−8, 2) and (0, −4) $y = -\frac{3}{4}x - 4$

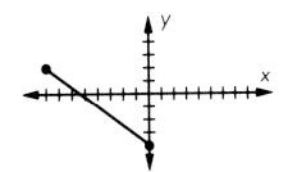

6. (3, −4) and (−1, 3) $y = -\frac{7}{4}x + \frac{5}{4}$

Use a graphing calculator to draw the graph of each equation. Draw a sketch of the graph.

7. $3x + y = 1$

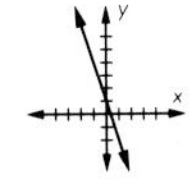

8. $2x - y = 0$

9. $x + 2y = 2$

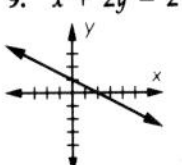

10. $-x + 4y = 1$

11. $2x + 2y = 5$

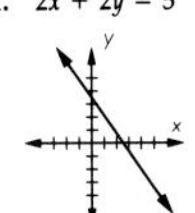

12. $-6x - 4y = 9$

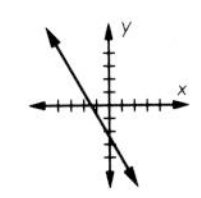

Name________________Class________________Date________

Achievement Test 13 (Chapter 13)
THE NEW GEOMETRY

GEOMETRY FOR DECISION MAKING
James E. Elander
SOUTH-WESTERN PUBLISHING CO.

No. Correct
No. Exercises: **31**
Score
3.22 x No. Correct =

Write an equation for the line shown in the figure.

1.

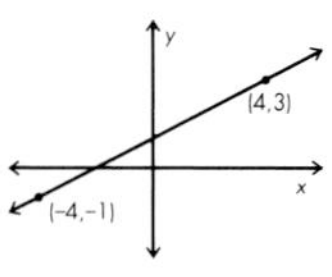

$y = \frac{1}{2}x + 2$

2. (-3,5) y (2,-1) x

$y = -\frac{6}{5}x + \frac{7}{5}$

Tell the slope and y-intercept of each line.

3. $y = 7x + 13$ 7, 13 **4.** $x = 3y - 8$ 3, − 8 **5.** $6x + 2y = 3y - 4x + 1$ 10, −1

Find the slope of the line determined by each pair of points. If the line is vertical, write " no slope."

6. $A(0, 9), B(-4, 1)$ 2 **7.** $C(-2, 5), D(-2, -3)$ no slope **8.** $E(\frac{1}{2}, -\frac{1}{3}), F(\frac{5}{2}, \frac{1}{2})$ $\frac{1}{2}$

Tell whether these lines are parallel, perpendicular, the same, or none of these.

9. $y = 2x - 1$, $y = 2x + 1$ parallel **10.** $x + 2y = 3$, $-2y - 4y = -6$ same **11.** $y = 0$, $y = -3x + 9$ none of these

12. $5x - y = 8$, $x + 5y = 4$ perpendicular **13.** $x = 6$, $y = -\frac{1}{2}$ perpendicular **14.** $2x + y = 0$, $2x + y = 12$ parallel

15. ΔABC is a right triangle. Write an equation for the line that contains the hypotenuse of ΔABC. $y = \frac{3}{4}x + 3$

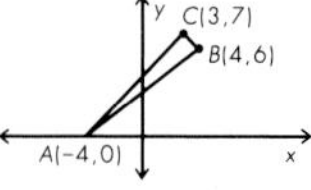

16. Suppose circles 1 and 2 both have their centers at (0, 0). Circle 1 passes through ($2\sqrt{5}$, 0) and circle 2 through (0, 3). Write an equation for each circle. $x^2 + y^2 = 20$; $x^2 + y^2 = 9$

MG01AG [13-1]

Give the center and radius of each circle.

17. $x^2 + y^2 = 81$ (0,0); $r = 9$ **18.** $x^2 + y^2 = \frac{35}{26}$ (0,0); $r = 9$ **19.** $x^2 + y^2 = 45$ (0,0); $r = 3\sqrt{5}$

20. Write an equation for the parabola that is the locus of all points equidistant from $(0, \frac{3}{4})$ and the line $y = -\frac{3}{4}$. $y = \frac{1}{3}x^2$

Laura made a rough sketch to show how these parabolas will look when graphed in the coordinate plane: $y = 2x^2$, $y = -x^2$, $y = \frac{2}{3}x^2$, $y = -\frac{1}{10}x^2$.

Write the equation that goes with each parabola in the sketch.

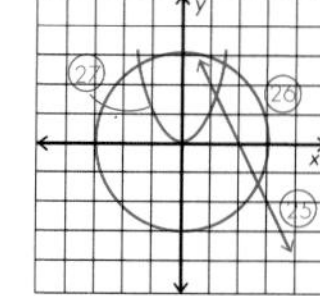

21. parabola 1 $y = \frac{2}{3}x^2$ **22.** parabola 2 $y = 2x^2$

23. parabola 3 $y = -\frac{1}{10}x^2$ **24.** parabola 4 $y = -x^2$

Graph each equation on the coordinate plane. You may use a calculator or the table of squares and square roots on page 562 if you wish.

25. $y = -2x + 8$

26. $x^2 + y^2 = 49$

27. $y = \frac{3}{4}x^2$

Find the coordinates of the midpoint of each side of ΔABC.

28. Side $\overline{AB}$ (1, 2)

29. Side $\overline{BC}$ $(7, \frac{3}{2})$

30. Side $\overline{AC}$ $(2, -\frac{3}{2})$

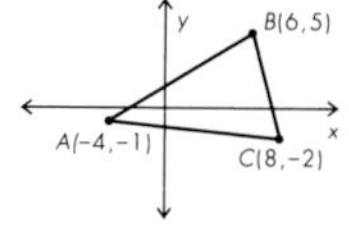

31. $ABCD$ is an isosceles trapezoid, and K, L, M, and N are the midpoints of its sides. Use coordinate geometry to prove that $KLMN$ is a rhombus.
See margin.

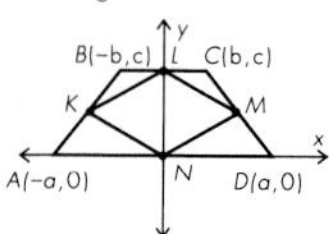

[13-2]

31. Use the midpoint theorem to find the coordinates of K, L, M, and N. Show that $\overline{KN}$ and $\overline{LM}$ have slope $-\frac{c}{a+b}$ and hence are parallel. Show that $\overline{KL}$ and $\overline{NM}$ have slope $\frac{c}{a+b}$ and hence are parallel. Then use the distance formula to show that each side of $KLMN$ has length $\frac{1}{2}\sqrt{(a+b)^2 + c^2}$.

14

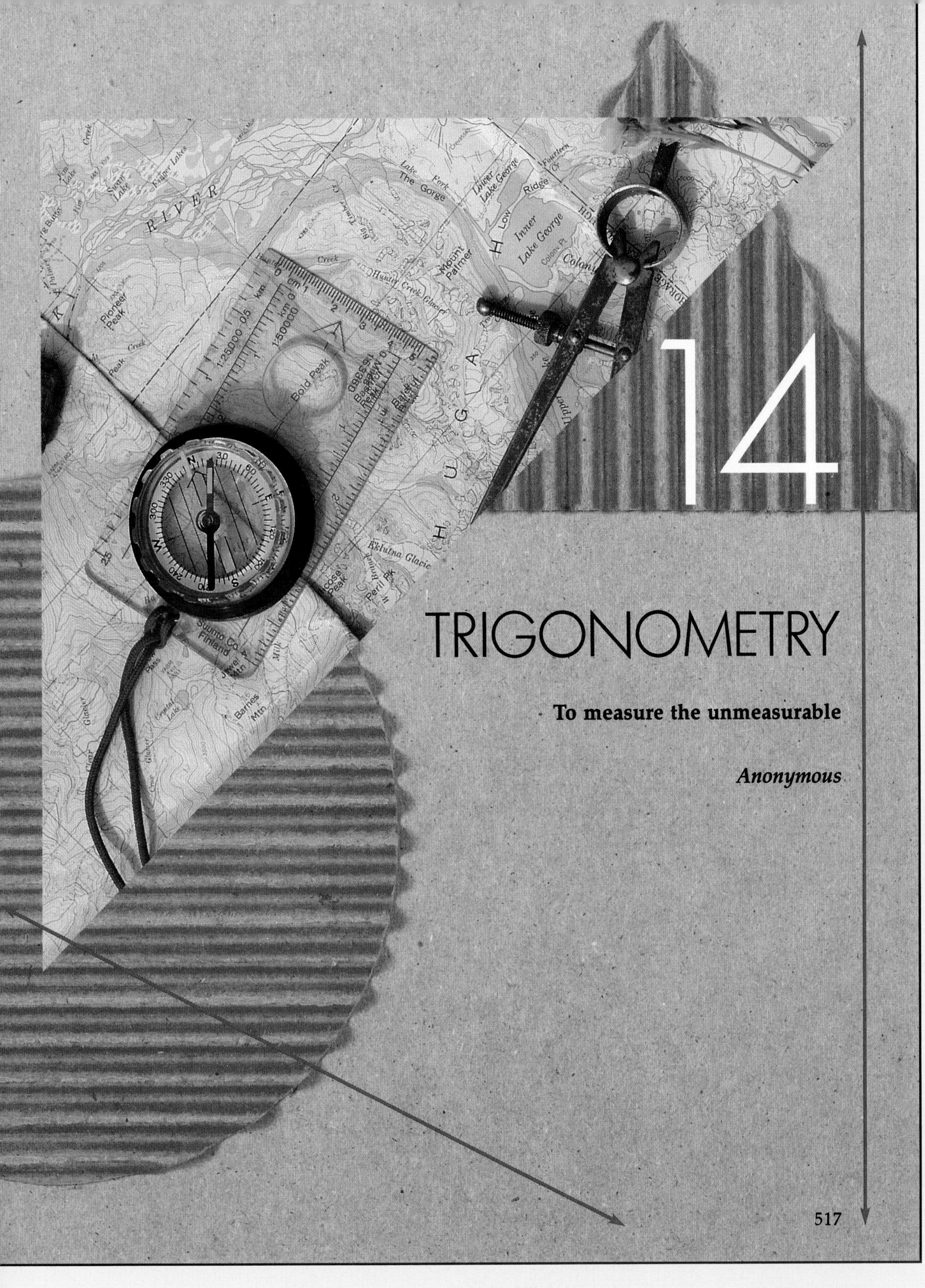

14

TRIGONOMETRY

To measure the unmeasurable

Anonymous

517

14

To measure the unmeasurable

Anonymous

CHAPTER 14 OVERVIEW

Trigonometry

Suggested Time: 12–14 days

This final chapter of the text introduces students to some of the important ideas of a full course on trigonometry. It includes not only the usual material found in such surveys but also some of the flavor of more advanced topics. Thus, in addition to the sections on indirect measurement using the sine, cosine, and tangent ratios, there are sections on trigonometry of the general angle, on the solution of non-right triangles, and on the sine function and its graph. A brief section on trigonometric identities closes the chapter.

CHAPTER 14 OBJECTIVE MATRIX

	End of Chapter Items by Activity				Student Supplement Items by Activity		
Objectives by Number	**Review**	**Test**	**Calculator**	**Algebra Skills***	**Reteaching**	**Enrichment**	**Calculator**
14.1.1	✓	✓	✓	✓	✓	✓	✓
14.1.2	✓	✓	✓	✓	✓	✓	✓
14.2.1	✓	✓	✓	✓	✓	✓	✓
14.2.2	✓	✓		✓	✓	✓	
14.3.1	✓	✓	✓	✓	✓		
14.3.2	✓	✓	✓	✓	✓		✓
14.3.3						✓	
14.4.1	✓	✓		✓	✓		
14.4.2	✓	✓		✓	✓		
14.5.1	✓	✓		✓	✓	✓	
14.6.1		✓		✓	✓	✓	
14.6.2	✓	✓				✓	

*A ✓ beside a Chapter Objective under Algebra Skills indicates that algebra skills taught within the chapter or in previous Algebra Skills lessons are used.

CHAPTER 14 PERSPECTIVES

▲ Section 14.1

The Sine and Cosine Ratios

This section continues the work of Section 7.3 in which right triangles with angle measures of 30° and 60° were used to measure distances indirectly. In this section, indirect measurement is studied using right triangles that have no restrictions on the measures of their acute angles. In the Discovery Activity, students learn that in a right triangle with 20° as one of its acute-angle measures, the sine ratio is always the same. A second ratio, the cosine ratio, is also defined.

For students having difficulty prepare a series of right triangles with a hypotenuse of 100 mm and with acute angles of 10°, 20°, 30°, 40°, and 45°. These triangles will also have acute angles of 80°, 70°, 60°, 50°, and 45° respectively. By measuring the legs of the triangles in millimeters, students should be able to discover an interesting relationship to sines and cosines in the Table of Trigonometric Functions.

14

▲ Section 14.2

The Tangent Ratio

A third trigonometric ratio, the tangent ratio, is introduced in this section. This completes the list of ratios required to solve any problem of indirect measurement that involves right triangles. In order to show how trigonometric ratios can be used in everyday life, the section also includes some applications, for example, problems involving angles of elevation and depression.

For students having difficulty use a similar activity as used in Section 1, but this time construct right triangles in which one leg has a length of 100 mm. You will need to make 9 separate triangles for this activity. Angles adjacent to the 100 mm leg should measure 10°, 20°, 30°, 40°, 45°, 50°, 60°, 70°, and 80°. Have students measure the side opposite these angles to the nearest mm. They should discover an interesting relationship to tangents in the Table of Trigonometric Functions.

▲ Section 14.3

The Sine Function

The purpose of this section is to show students that there is much more to trigonometry than indirect measurement. To accomplish this, the section has two goals. The first is to show how trigonometric ratios can be extended beyond acute angles. This is done by first redefining *angle* in terms of angles of rotation in the coordinate plane and then redefining the sine ratio accordingly.

For students needing a more visual approach to extending the sine function, you can use the triangles you made for the activity in Section 1. Using a grid with a unit length of 100 mm, place the triangle with the 10° angle in the first quadrant with the 10° angle at the origin. By flipping the triangle over the x − axis the students can see why the absolute value of the sine of 170° is equal to the sine of 10°. Then flip the triangle over the *y* − axis. ($|\sin 190°| = |\sin 10°|$). Finally flip the triangle over the x − axis again. ($|\sin 350°| = |\sin 10°|$). Repeat with triangles with angles of different measure. Discuss in signs, positive and negative, separately.

▲ Section 14.4

Solving Non-right Triangles.

This section builds upon the three sections that directly precede. First, the concept of "solving a right triangle" is extended to that of "solving a non-right triangle." Second, the trigonometry of obtuse angles is used in applying the Law of Sines and the Law of Cosines.

In the coverage of oblique, or non-right triangles, some details are omitted. For example, the ambiguous case (SSA) is not covered for the Law of Sines. Nor is there any attempt to provide a rigorous development of the two laws. In the case of the Law of Sines, the law is developed intuitively in a Discovery Activity. Do not expect students to be familar with these formulas. Let all your students have access to the formulas, especially those having difficulties.

▲ Section 14.5

Applications

This section provides a wealth of practical applications of the concepts presented.

You will notice that the first page of the section contains a brief summary of the alternate definitions of the sine and cosine that were given earlier in the chapter. It is useful at this point to have these different ways of approaching trigonometric ratios appearing together.

▲ Section 14.6

Trigonometric Identities

The treatment of trigonometric identities in this section is intended mainly to give students an idea of what a trigonometric identity is, not to develop great skill in dealing with identities. For this reason, the choice of identities in the examples and exercises has been made with the purpose of avoiding excessively difficult algebraic manipulations.

Each of the major categories of trigonometric identities, such as the Pythagorean and Quotient Identities, is illustrated by a single example.

The example and the exercises for this basic identity formula have been chosen to avoid excessive algebraic complexities.

SECTION 14.1

Resources
Reteaching 14.1
Enrichment 14.1

Objectives
After completing this section, the student should understand
▲ how to find sine ratios.
▲ how to find cosine ratios.

Materials
centimeter ruler
protractor
scientific calculator

Vocabulary
sine ratio
trigonometric ratios
cosine ratio

GETTING STARTED

Quiz: Prerequisite Skills
Two angles of a triangle have the given measures. Find the measure of the third angle.

1. 50°, 60° [70°]
2. 90°, 35° [55°]
3. 20°, 70° [90°]

Tell whether the triangles in each pair are similar.

4.

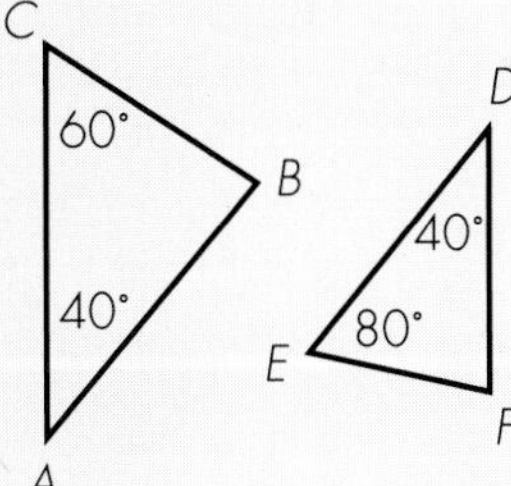

[Yes]

5.

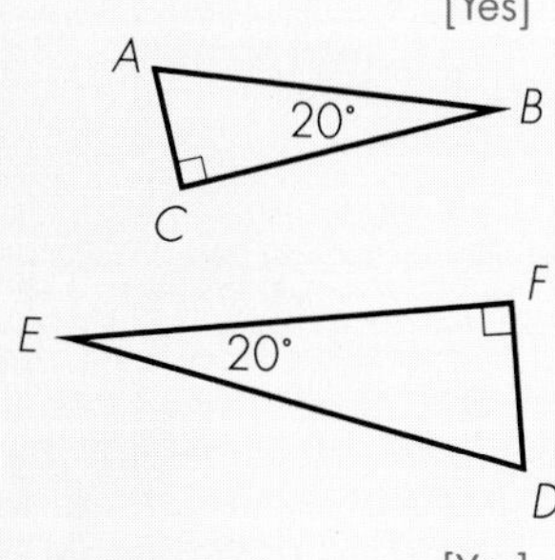

[Yes]

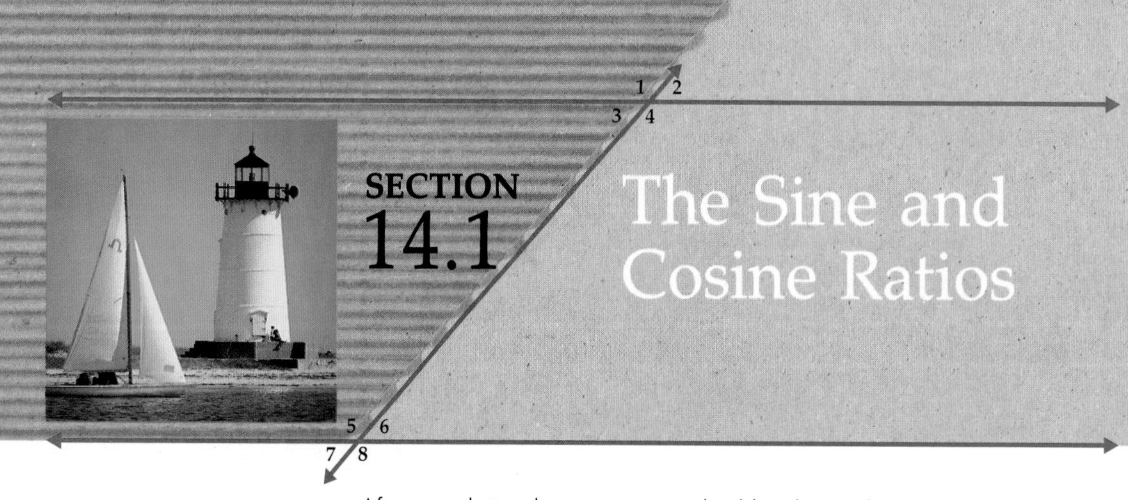

SECTION 14.1 The Sine and Cosine Ratios

After completing this section, you should understand
▲ how to find sine ratios.
▲ how to find cosine ratios.

You are familiar with similar triangles such as those shown below.

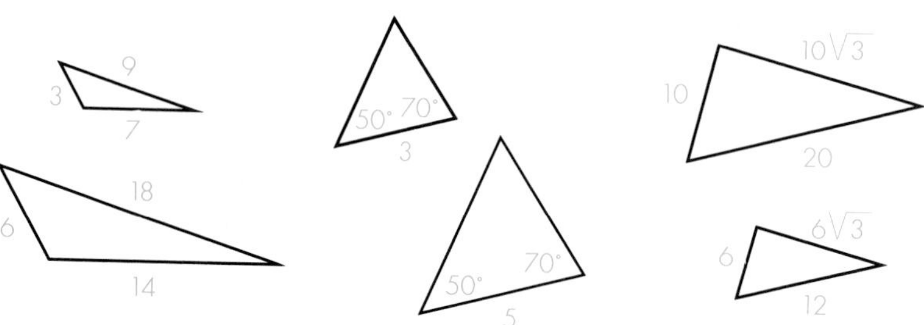

In Section 7.3 you saw how you can use certain similar right triangles, such as those with angles of 30°, 60°, and 45°, to measure distances indirectly. In this chapter you will see how to use *any* right triangle to measure distances indirectly.

SIMILAR RIGHT TRIANGLES

You know by Definition 4-3-1 that two triangles are similar if two angles of one are equal to two angles of the other. If two triangles are known to be right triangles, they automatically have one pair of equal angles (the right angles). Therefore, a pair of right triangles is similar if an acute angle of one is equal to an acute angle of the other.

Example: Among the right triangles shown below, find all the pairs of similar triangles.

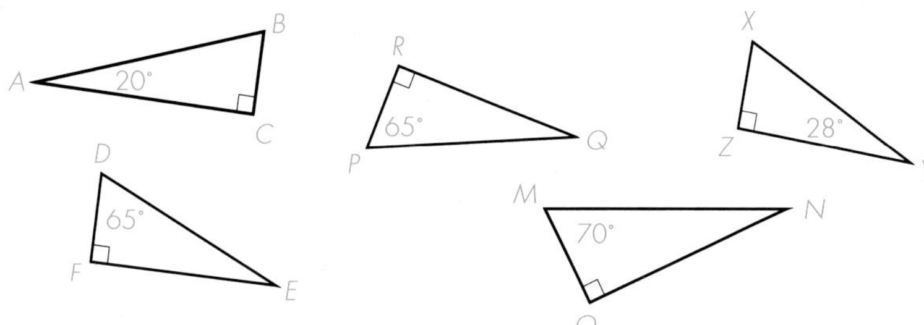

$\triangle PQR \approx \triangle DEF$, since $m\angle P = m\angle D = 65°$. In $\triangle NMO$, $m\angle N = 20°$, since $\angle M$ and $\angle N$ are complementary angles. Therefore, $\triangle ABC \sim \triangle NMO$.

In the Discovery Activity below, you will learn more about similar right triangles.

DISCOVERY ACTIVITY

1. Copy the table below. Then use a centimeter ruler to measure the line segments indicated in the table and shown in the figure. Measure to the nearest tenth of a centimeter. Record the measurements in the table.

BC = 1.4 cm	DE = 2.0 cm	FG = 3.4 cm
AB = 4.0 cm	AD = 6.0 cm	AF = 10 cm
$\frac{BC}{AB}$ = 0.35	$\frac{DE}{AD}$ = 0.33	$\frac{FG}{AF}$ = 0.34

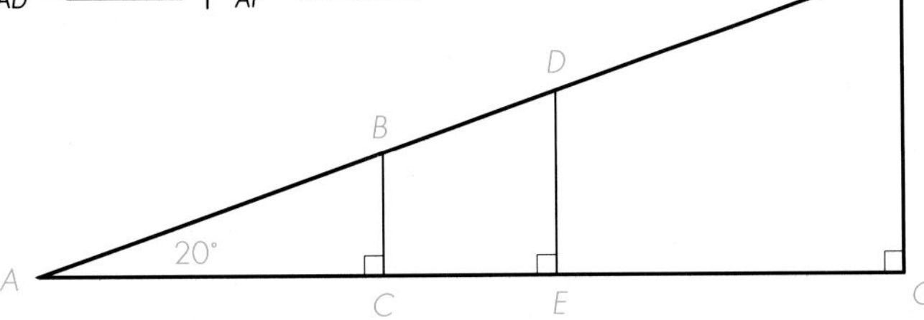

2. Use a calculator and write the ratios in the third row of the table in decimal form, to the nearest hundredth. What do you notice? They are approximately equal.

3. What can you say about the ratio of the length of the leg *opposite* the 20° angle to the length of the *hypotenuse*?
   It seems to be the same regardless of the size of the triangle.

TEACHING COMMENTS

Using the Page
Geometric drawing tools and a calculator can be used to complete this Discovery Activity. Students can use a protractor to verify that $m\angle A = 20°$, and a centimeter ruler to measure the line segments.

In the Discovery Activity, you may have noticed that in a right triangle with an acute angle that measures 20°, the ratio of the length of the leg opposite that angle to the length of the hypotenuse is about 0.34, regardless of the size of the right triangle. An even more accurate value is 0.3420. This number is called the sine ratio of 20°, or simply the **sine** of 20°. It is written "sin 20°." Any acute angle, not just 20°, has a constant value for its sine. For example, the sine of 25° is about 0.4226. This allows you to have the following definition.

DEFINITION 14-1-1 **The SINE of an acute angle *A* of a right triangle is the ratio of the length of the leg opposite the angle to the length of the hypotenuse. In symbols, this is written as**

$$\sin A = \frac{\text{length of leg opposite angle } A}{\text{length of hypotenuse}}$$

The equation for the sine ratio can be abbreviated as $\sin A = \frac{\text{opp}}{\text{hyp}}$. The sine ratio is the first of three **trigonometric ratios** that you will study in this chapter.

Examples:

1. Find sin *A*.

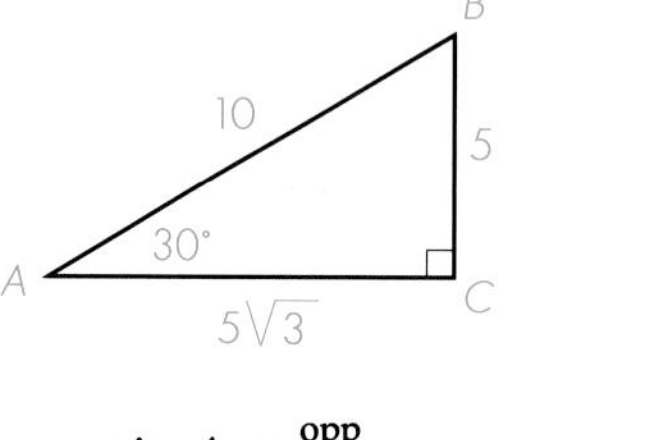

$\sin A = \frac{\text{opp}}{\text{hyp}}$

$= \frac{5}{10}$

$= \frac{1}{2}$, or 0.5000

2. Find sin *A*.

$\sin A = \frac{\text{opp}}{\text{hyp}}$

$= \frac{10}{26}$

≈ 0.3846

In the second example above, the "equals approximately" sign (≈) is used in the last step because 0.3846 is a rounded value. From now on, for simplicity, the equality symbol (=) will often be used.

The second example shows that if you know the lengths of the sides of a right triangle, you can find the sines of the acute angles whether you know their measures or not.

You do not have to measure right triangles every time you need to know the sine of an angle. If you know the measure of the angle, you can find its sine by using either a *table* or a *scientific calculator.*

Using the Page

Point out that Definition 14-1-1 applies to the sine of any acute angle of a right triangle. Have students find sin *B* in the second example.

$\left[\frac{24}{26} \approx 0.9231\right]$

Extra Practice

1. Find sin *A*.

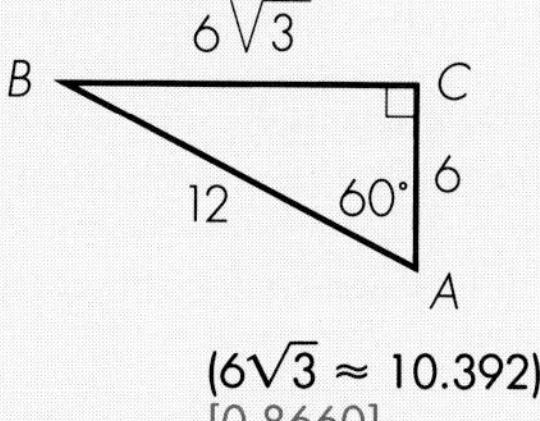

($6\sqrt{3} \approx 10.392$)

[0.8660]

2. Find sin *B*.

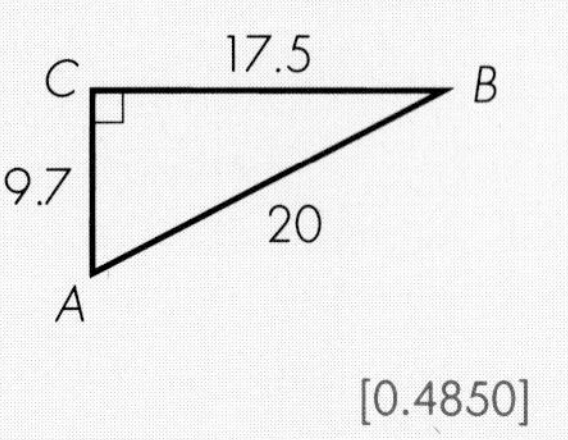

[0.4850]

Example: Find the sine of 32° using a calculator or the table of trigonometric ratios on page 563.

To use a calculator, press the keys [3] [2] [SIN] The display will show 0.52991926.
To use the table on page 563, look at the portion of it shown at the right. Read down the "degrees" column until you get to 32°, then find sin 32° in the "sin" column: sin 32° = 0.5299.

Degrees	Sin	Cos	Tan
30°	0.5000	0.8660	0.5774
31°	0.5150	0.8572	0.6009
32°	0.5299	0.8480	0.6249
33°	0.5446	0.8387	0.6494

[3] [2] [SIN]

CLASS ACTIVITY

1. Find all pairs of similar triangles from the set of triangles below. $\triangle DEF \sim \triangle ONM$; $\triangle DEF \sim \triangle XYZ$; $\triangle ONM \sim \triangle XYZ$

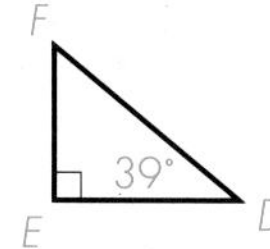

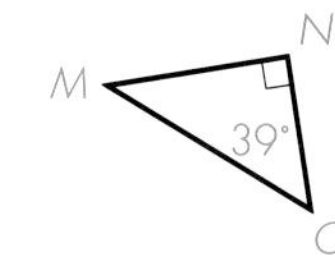

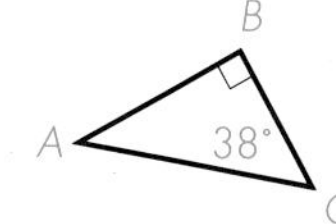

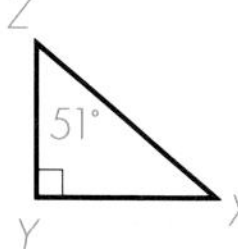

Calculate or use the table on page 563 to find sin A. You may use a calculator.

2. 0.4667

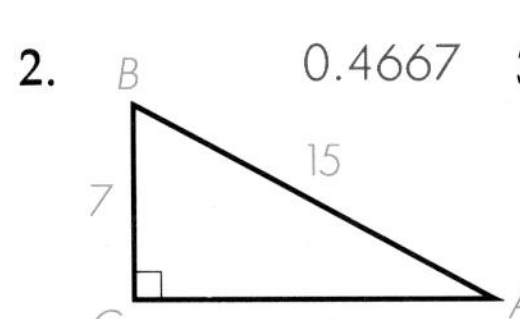

3.

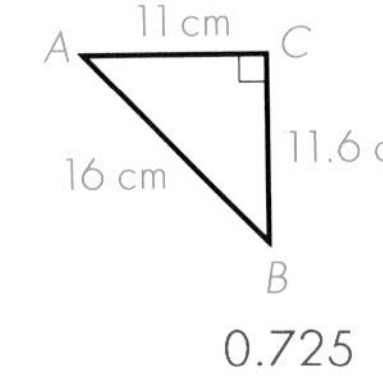

0.725

4. 0.8090

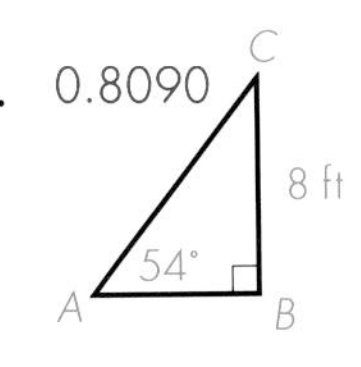

THE COSINE RATIO

If A is an acute angle of a right triangle, then the three sides of the triangle are related to A as shown at the right. Another trigonometric ratio, the *cosine* ratio, is formed by using the *adjacent* leg and the hypotenuse.

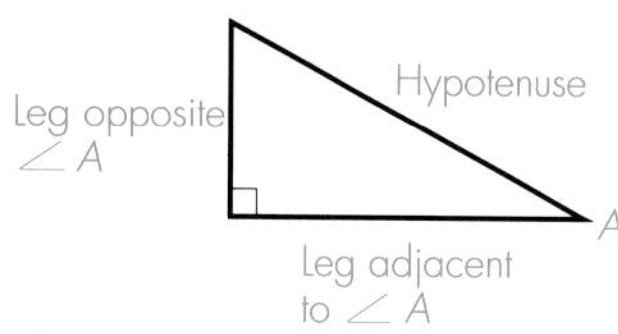

DEFINITION 14-1-2 **The COSINE of an acute angle of a right triangle is the ratio of the length of the leg adjacent to the angle to the length of the hypotenuse. In symbols, this is written as**

$$\cos A = \frac{\text{length of leg adjacent to angle } A}{\text{length of hypotenuse}}$$

The cosine formula can be abbreviated as $\cos A = \frac{\text{adj}}{\text{hyp}}$.

Using the Page
In Exercise 4 of the Class Activity, note that the length of $\overline{BC}$ is unnecessary information.

Extra Practice
1. Find sin A.

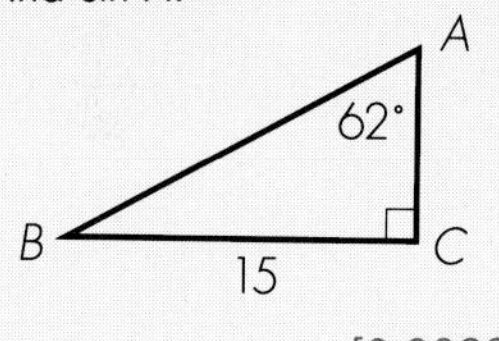

[0.8829]

2. Find sin A.

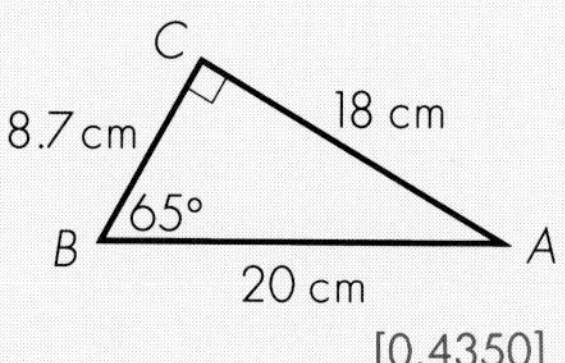

[0.4350]

Using the Page
After a discussion of Definition 14-1-2, students can be asked to identify the ratios for cos A and cos B in Exercise 2 above.
$\left[\cos A = \frac{13.3}{15}; \cos B = \frac{7}{15}\right]$

Additional Answers

1. sin A = 0.3846
   sin B = 0.9231
   cos A = 0.9231
   cos B = 0.3846
2. sin A = 0.4540
   sin B = 0.8910
   cos A = 0.8910
   cos B = 0.4540
3. sin A = 0.7071
   sin B = 0.7071
   cos A = 0.7071
   cos B = 0.7071
4. sin A = 0.6428
   sin B = 0.7660
   cos A = 0.7660
   cos B = 0.6428

FOLLOW-UP

Assessment

1. Find cos A and cos B.

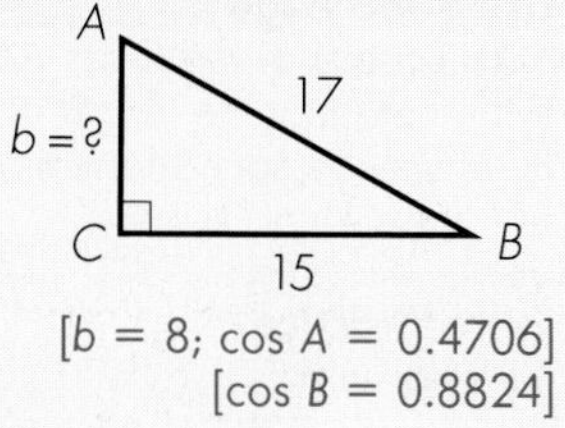

[b = 8; cos A = 0.4706]
[cos B = 0.8824]

2. Find each ratio.

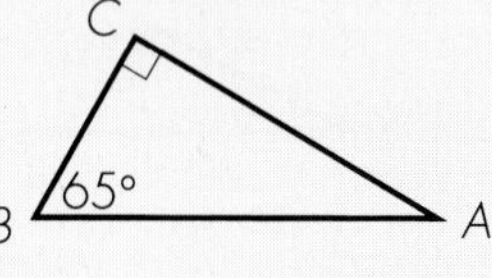

sin B [0.9063]
cos B [0.4226]
sin A [0.4226]
cos A [0.9063]

3. If $\angle A$ and $\angle B$ are acute angles of right triangle ABC, explain why sin A = cos B.
[Each ratio is equal to $\frac{BC}{BA}$.]

Reteaching
Students who have had difficulty with this section may benefit from Reteaching Activity 14.1.

Enrichment
For students who have mastered the material in this section, you may wish to assign Enrichment Activity 14.1.

Example: Find cos A and cos B.

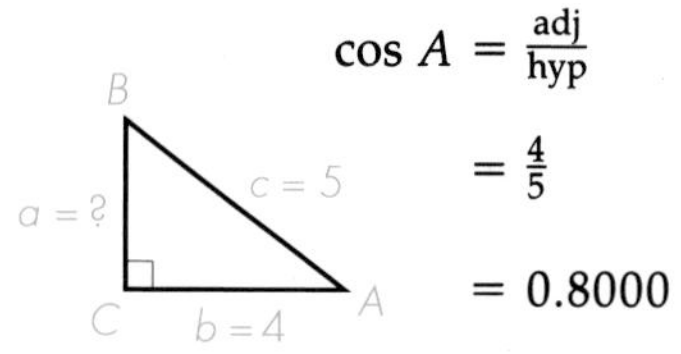

$$\cos A = \frac{\text{adj}}{\text{hyp}} = \frac{4}{5} = 0.8000$$

Use $a^2 + b^2 = c^2$ to find the leg adjacent to $\angle B$.

$a^2 + 4^2 = 5^2$

$a^2 = 5^2 - 4^2$, or 9

$a = 3$

$\cos B = \frac{a}{c} = \frac{3}{5} = 0.6000$

CLASS ACTIVITY

Find cos A and cos B. You may use a calculator. Express answers to the nearest ten-thousandth.

1. 

cos A = 0.8020
cos B = 0.5974

2. 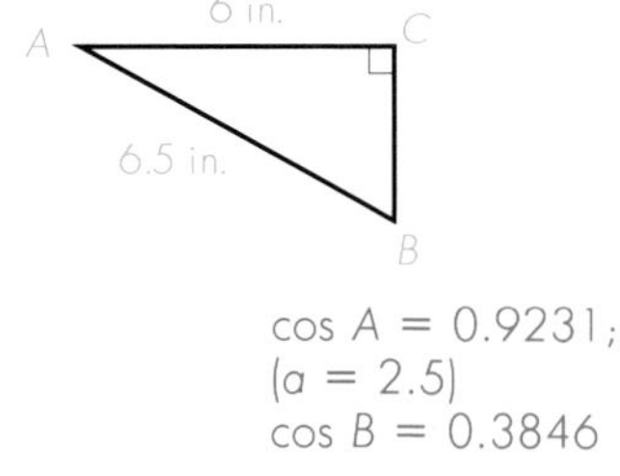

cos A = 0.9231;
(a = 2.5)
cos B = 0.3846

HOME ACTIVITY

In these exercises, you may use a calculator or the table on page 563. Find sin A, sin B, cos A, and cos B. Express answers to the nearest ten-thousandth. See margin.

1. 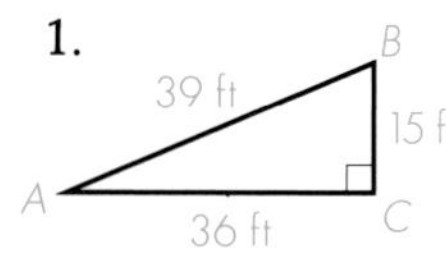

2. 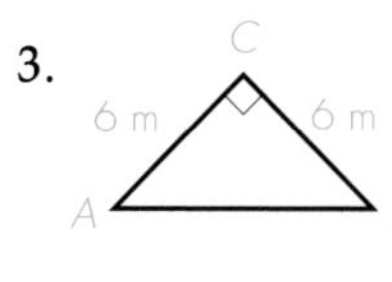

3.

4. 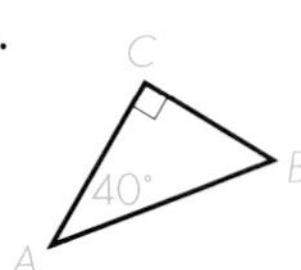

Find the sine or cosine.

5. sin 35° 0.5736	6. cos 55° 0.5736	7. sin 10° 0.1736	8. cos 80° 0.1736
9. sin 24° 0.4067	10. cos 66° 0.4067	11. sin 41° 0.6561	12. cos 49° 0.6561
13. sin 62° 0.8829	14. cos 62° 0.4695	15. sin 13° 0.2250	16. cos 13° 0.9744

CRITICAL THINKING

17. In quadrilateral $ABCD$, cos $\angle 1$ = cos $\angle 2$. Find the perimeter of the quadrilateral. $28 + 7\sqrt{2}$

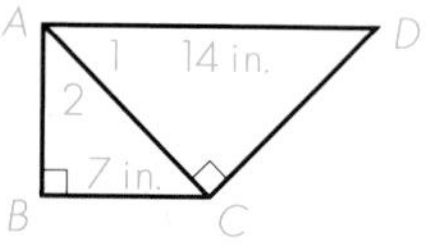

SECTION 14.2

The Tangent Ratio

After completing this section, you should understand
- ▲ how to use the tangent ratio.
- ▲ how to use trigonometry to solve triangle problems.

You have seen that the following two ratios can be defined for an acute angle A of a right triangle.

$\sin A = \frac{\text{opp}}{\text{hyp}}$

$\cos A = \frac{\text{adj}}{\text{hyp}}$

You will now learn about a third trigonometric ratio.

DISCOVERY ACTIVITY

1. Copy the table below. Then use a centimeter ruler to find the lengths of the legs of each triangle. Measure to the nearest tenth of a centimeter. Record the measurements in the table.

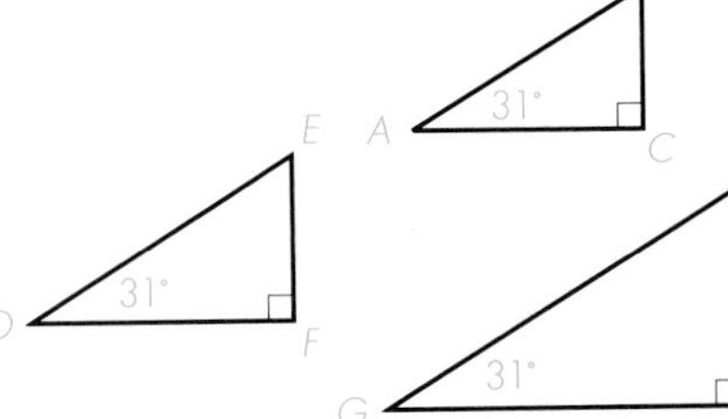

	leg op 31° angle	leg adj to 31° angle	*opp* adj
$\triangle ABC$	1.2 cm	2.0 cm	0.6
$\triangle DEF$	1.5 cm	2.5 cm	0.6
$\triangle GHI$	1.8 cm	3.0 cm	0.6

2. For each triangle, express the ratio $\frac{\text{opp}}{\text{adj}}$ as a decimal. Record the results in the table.
3. What can you say about the ratio $\frac{\text{opp}}{\text{adj}}$? It seems to be the same, regardless of the size of the triangle.

SECTION 14.2

Resources
Reteaching 14.2
Enrichment 14.2
Transparency Master 14-1

Objectives
After completing this section, the student should understand
- ▲ how to find the tangent ratio.
- ▲ how to use trigonometry to solve triangle problems.

Materials
centimeter ruler
protractor
scientific calculator

Vocabulary
sine ratio
cosine ratio
trigonometric ratios
tangent ratio
angle of elevation
angle of depression

GETTING STARTED

Quiz: Prerequisite Skills
1. Find sin 80° [0.9848]
2. Find cos 72° [0.3090]
3. Find each ratio.

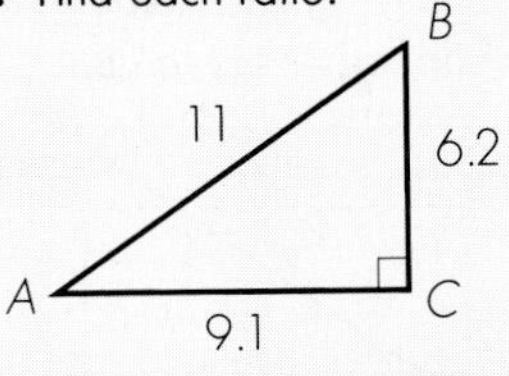

sin A [0.5636]
cos A [0.8273]
sin B [0.8273]
cos B [0.5636]

TEACHING COMMENTS

Using the Page
The geometric drawing tools can be used to complete this Discovery Activity.

In the Discovery Activity, you may have discovered that the ratio $\frac{\text{opp}}{\text{adj}}$ has the same value (about 0.6) for all three similar triangles even though they are not of the same size. This common ratio is called the tangent ratio, or **tangent** of 31°. It is written "tan 31°."

DEFINITION 14-2-1 **The TANGENT of an acute angle A of a right triangle is the ratio of the length of the leg opposite the angle to the length of the leg adjacent to the angle. In symbols, this is written as**

$$\tan A = \frac{\text{length of leg opposite angle } A}{\text{length of leg adjacent to angle } A}$$

The equation for the tangent ratio can be abbreviated as $\tan A = \frac{\text{opp}}{\text{adj}}$. Values for the tangent ratio are found in the table on page 563.

Examples:

1. Find tan B.

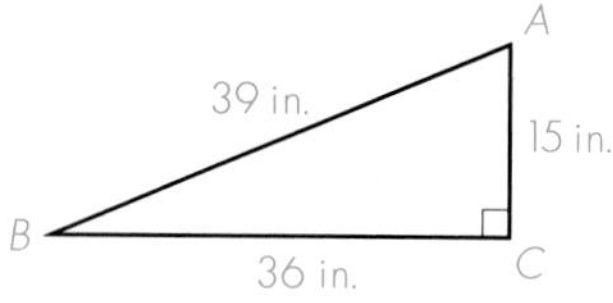

$\tan B = \frac{\text{opp}}{\text{adj}}$

$= \frac{15}{36}$

$= 0.4167$

2. Find tan A. Use the table on page 563.

A
35°
B
C

$m\angle A + 35° = 90°$

$m\angle A = 90° - 35°$, or 55°

From the table:

$\tan 55° = 1.428$

The table of trigonometric ratios on page 563 lets you find the sine, cosine, or tangent of an acute angle. You can also use the table to find the measure of an angle if you know its sine, cosine, or tangent.

Examples:

1. If $\tan A = 0.3443$, what is $m\angle A$?

Use the portion of the table shown at the right. Read down the "tan" column to the number 0.3443. Then read left to 19° in the "degrees" column. You find that $m\angle A = 19°$.

Degrees	Sin	Cos	Tan
18°	0.3090	0.9511	0.3249
19°	0.3256	0.9455	0.3443
20°	0.3420	0.9397	0.3640
21°	0.3584	0.9336	0.3839
22°	0.3746	0.9272	0.4040
23°	0.3907	0.9205	0.4245

Using the Page

Example 2 illustrates that a tangent ratio can have a value greater than 1. Students should understand why this is not true of a sine or cosine ratio.

The top half of Transparency 4-1 provides a convenient summary of the three basic trigonometric ratios for right triangles.

Extra Practice

1. Find tan A.

B
20°
A
C

[2.747]

2. Find tan A and tan B.

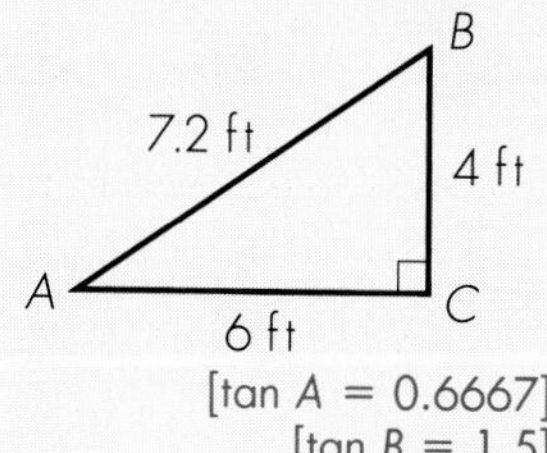

[tan A = 0.6667]
[tan B = 1.5]

2. Given that tan $B = 0.4167$, find $m\angle B$.

When you look for 0.4167 in the table, it is not there, so choose the table value that is closest to 0.4167. The number 0.4167 is between 0.4040 and 0.4245.

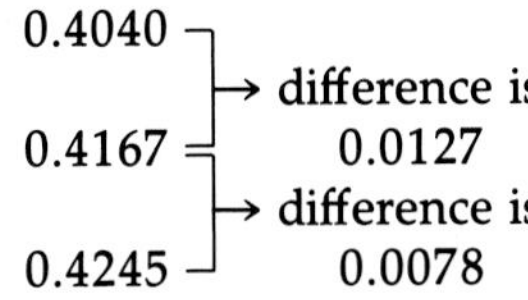

The smaller difference is that between 0.4167 and 0.4245. Since 0.4167 is closer to 0.4245 than to 0.4040, the measure of $\angle B$ is closer to 23° than to 22°. So $m\angle B \approx 23°$.

CLASS ACTIVITY

Find tan B. If necessary, use a calculator or the table of trigonometric ratios on page 563.

1. C 63.0 B, 31.5, 70.4, A

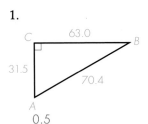

0.5

2.

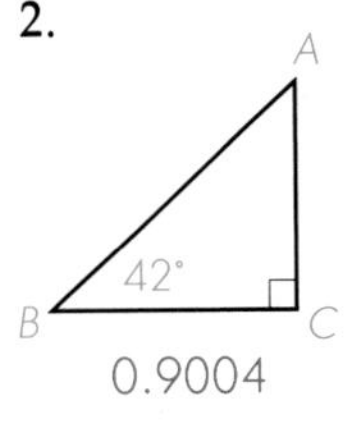

0.9004

Find $m\angle A$. Use a calculator or the table of trigonometric ratios on page 563.

3. B, 235, C, 442, A

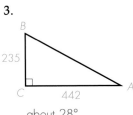

about 28°

4.

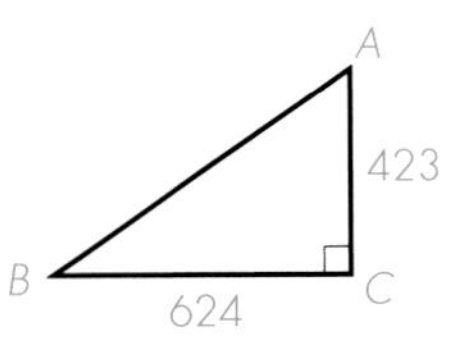

about 56°

5.

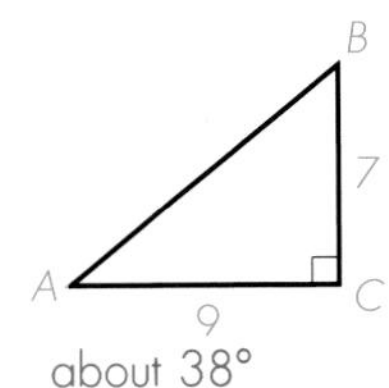

about 30°

6. B, 7, A, 9, C

about 38°

Extra Practice

Find tan A.

1.

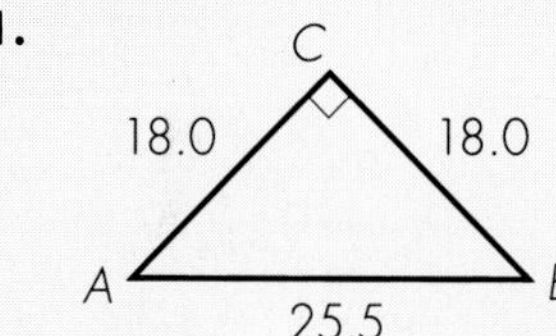

[1]

2.

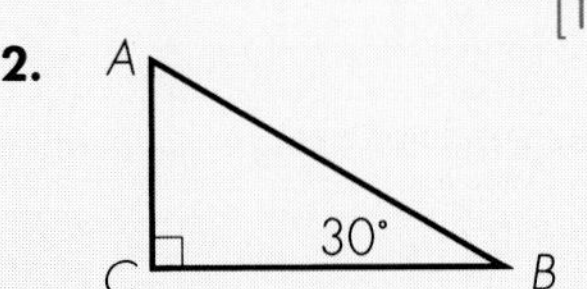

[1.732]

Find m∠A.

3.

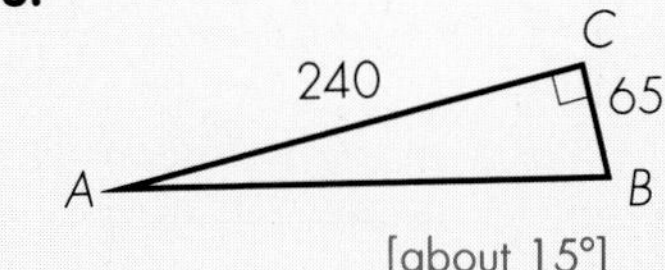

[about 15°]

4.

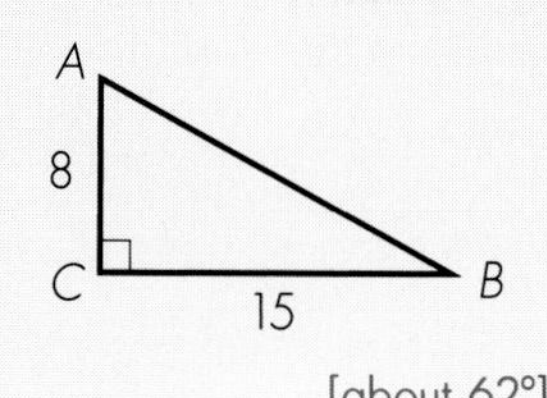

[about 62°]

Extra Practice

1. In right triangle ABC, $m\angle A = 55°$ and $c = 8$ ft. Find b.

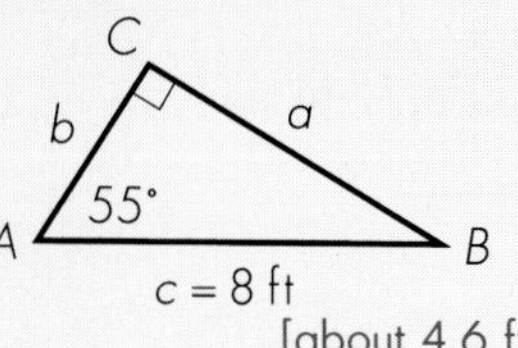

[about 4.6 ft]

2. In right triangle ABC, $m\angle A = 25°$ and $b = 50$ m. Find a.

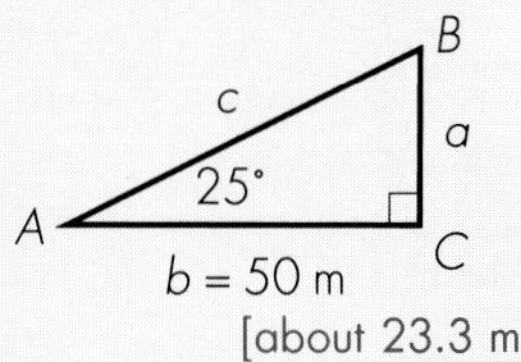

[about 23.3 m]

USING TRIGONOMETRIC RATIOS

You can often use one of the three trigonometric ratios to find a missing part of a right triangle. Depending on which part is missing and which parts are known, you can choose either the sine, cosine, or tangent. Often there is more than one approach that works.

Examples:

1. In right triangle ABC, $m\angle A = 40°$ and $b = 10$ yd. Find a.

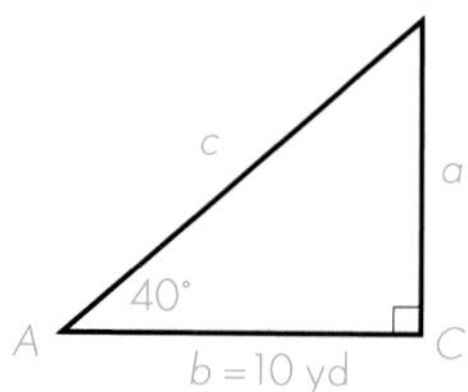

The part you need to find is *opposite* $\angle A$. The known part is *adjacent* to $\angle A$. This suggests the ratio $\frac{\text{opp}}{\text{adj}}$. So, use the *tangent* ratio.

$$\tan A = \frac{\text{opp}}{\text{adj}}$$

$$\tan 40° = \frac{a}{10}$$

From the table, $\tan 40° = 0.8391$.

$$0.8391 = \frac{a}{10}$$
$$10 \times 0.8391 = a$$
$$a = 10 \times 0.8391$$
$$= 8.391, \text{ or about } 8.4 \text{ yd}$$

2. In right triangle XYZ, $m\angle X = 25°$ and $z = 30$ mm. Find x.

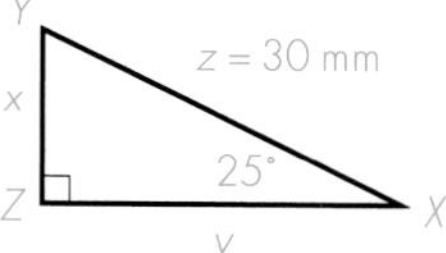

The part you need to find is *opposite* $\angle X$. The known part is the *hypotenuse*. This suggests the ratio $\frac{\text{opp}}{\text{hyp}}$. So, use the *sine* ratio.

$$\sin X = \frac{\text{opp}}{\text{hyp}}$$

$$\sin 25° = \frac{x}{30}$$

From the table, $\sin 25° = 0.4226$.

$$0.4226 = \frac{x}{30}$$
$$30 \times 0.4226 = x$$
$$x = 30 \times 0.4226$$
$$= 12.678, \text{ or about } 12.7 \text{ mm}$$

3. When a low-flying plane is directly over point R, the *angle of elevation* from a ship at point Q is 50°. The distance QR is 3000 ft. What is the direct-line distance from the plane to the ship?

The known part is *adjacent to* $\angle Q$. The part you need to find is the *hypotenuse*. This suggests the ratio $\frac{\text{adj}}{\text{hyp}}$. So, use the *cosine* ratio.

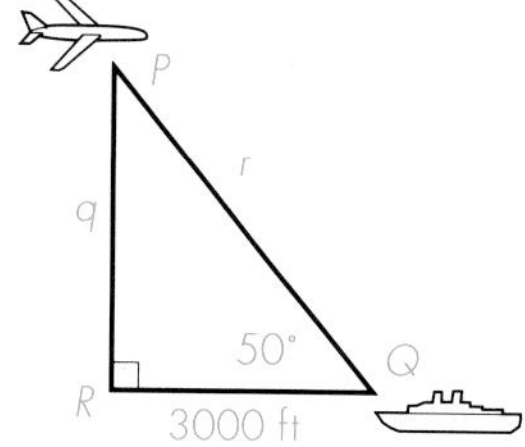

$$\cos Q = \frac{\text{adj}}{\text{hyp}}$$

$$\cos 50° = \frac{3000}{r}$$

From the table, cos 50° = 0.6428.

$$0.6428 = \frac{3000}{r}$$

$$r \times 0.6428 = 3000$$

Divide each side of the equation by 0.6428.

$$r = \frac{3000}{0.6428}$$

$$= 4667, \text{ or about } 4670 \text{ ft from the ship}$$

CLASS ACTIVITY

Tell which trigonometric ratio you could use to find the value of a.

1. 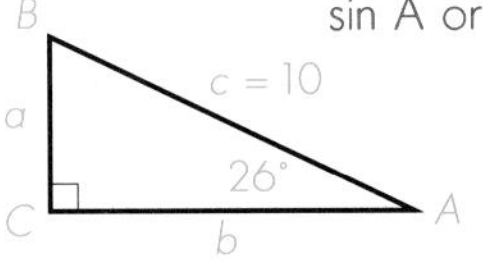

sin A or cos B

2. 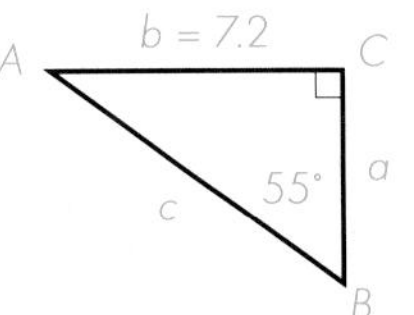

tan A or tan B

In Exercises 3 and 4, use a calculator or the table on page 563. Give answers to the nearest tenth.

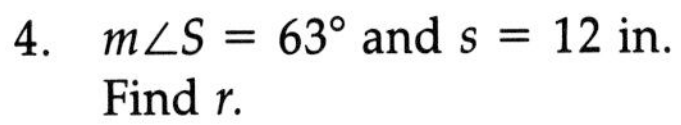

3. $m\angle F = 38°$ and $h = 18$ m. Find g.

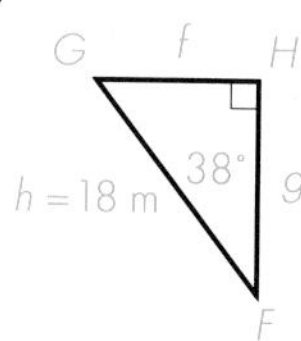

g = 14.2 m

4. $m\angle S = 63°$ and $s = 12$ in. Find r.

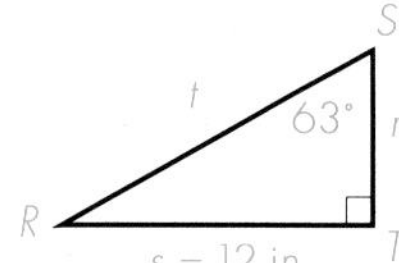

r = 6.1 in.

Extra Practice
Tell which trigonometric ratio you could use to find the value of b.

1. 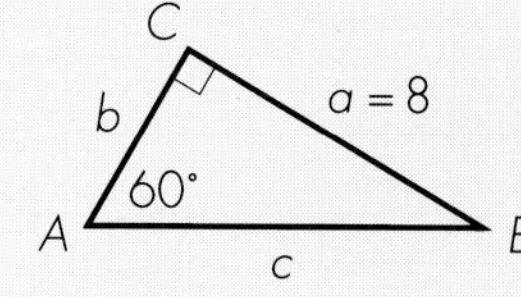

[tan A or tan B]

2. 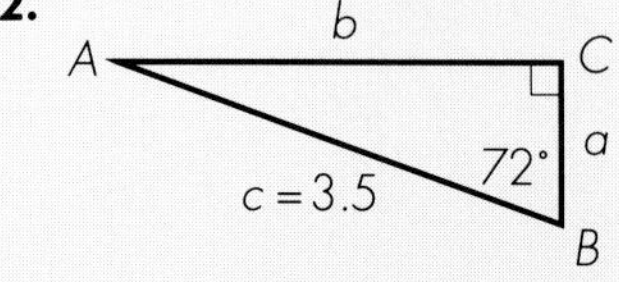

[cos A or sin B]

3. $m\angle D = 42°$ and $d = 14$ cm. Find f to the nearest tenth of a centimeter.

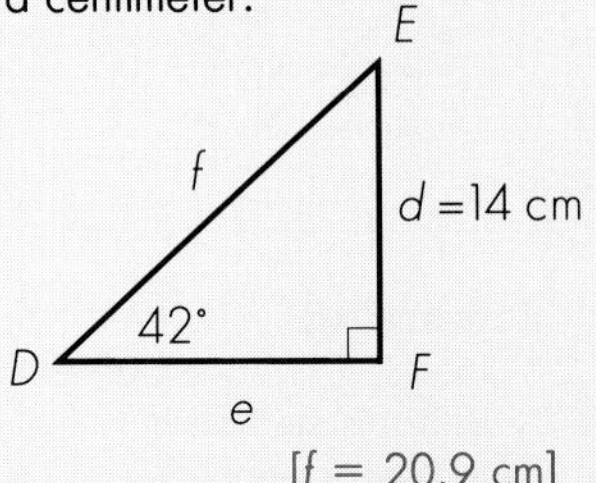

[f = 20.9 cm]

Additional Answers

12. Use a 30°-60° right triangle.

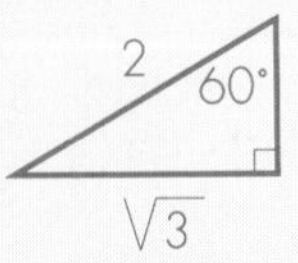

$\sin 30° = \frac{1}{2}$, so $\sin 30° + \sin 30° = 1$ and $\sin (30° + 30°) = \sin 60° = \frac{\sqrt{3}}{2}$. Therefore $\sin (30° + 30°)$ is not equal to $\sin 30° + \sin 30°$.

FOLLOW-UP

Assessment

1. Find tan Q.

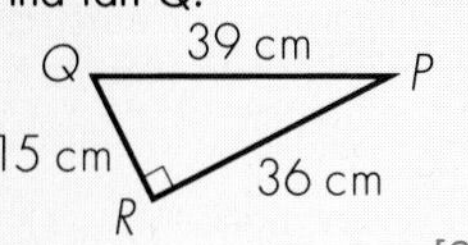

[2.4]

2. Find tan T.

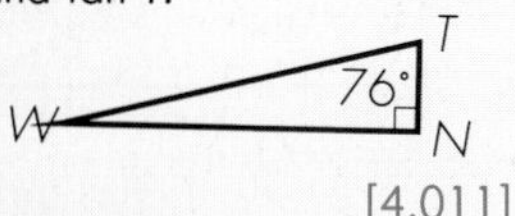

[4.011]

3. Find m∠K.

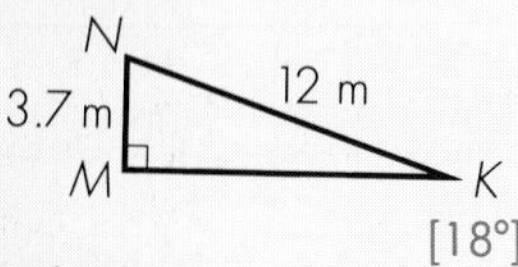

[18°]

4. Find g.

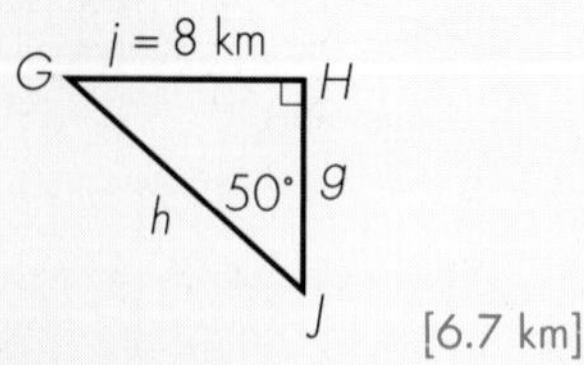

[6.7 km]

Reteaching
Students who have had difficulty with this section may benefit from Reteaching Activity 14.2.

Enrichment
For students who have mastered the material in this section, you may wish to assign Enrichment Activity 14.2.

HOME ACTIVITY

In the Home Activity exercises, use a calculator or the table on page 563. Give lengths and distances to the nearest tenth.

1. Find tan S.

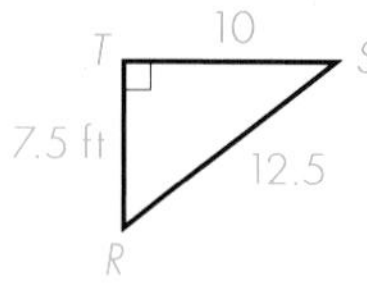

0.75

2. Find tan N.

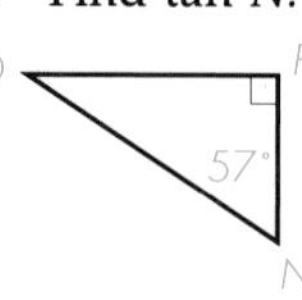

1.540

3. Find tan G.

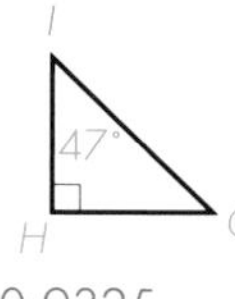

0.9325

4. Find $m\angle A$.

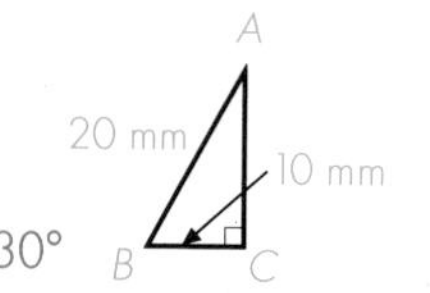

30°

5. Find $m\angle L$.

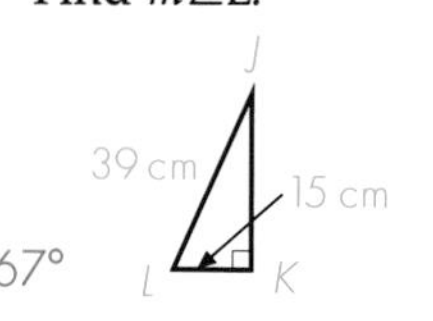

67°

6. Find a.

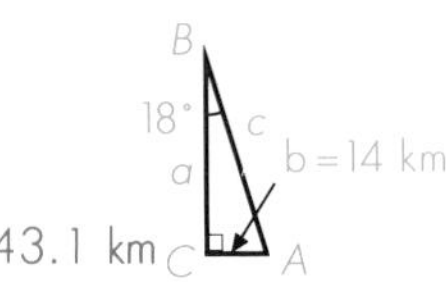

43.1 km

7. Find u.

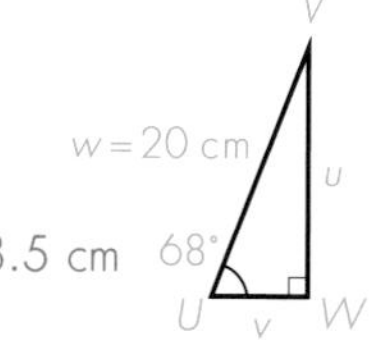

18.5 cm

8. Find z. 8.7 ft

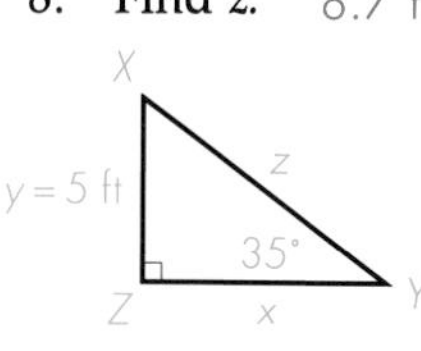

9. Find f. 34.4 mi

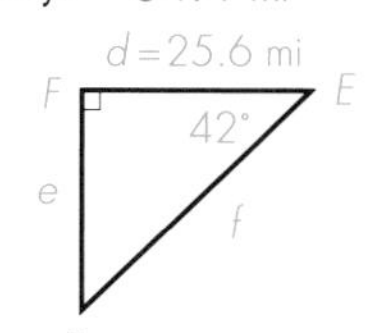

10. Angle A in the figure is the angle of depression from the top of a lighthouse to a ship sailing near the rocky coast. The ship will hit a reef if it gets closer than 500 yards of the shore. How far is the ship from shore? Will it hit the reef? 514 yd; No

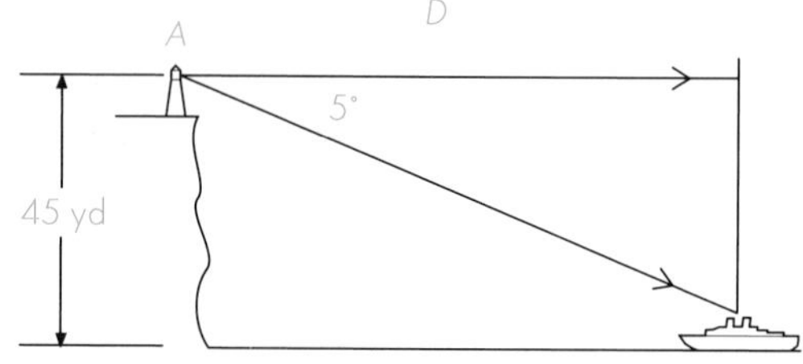

11. The angle of elevation from a sailboat to the top of an oil rig is 32°. The oil rig is 130 m high. Draw a diagram. Then find how far the sailboat is from the base of the oil rig.
208 m

CRITICAL THINKING

12. Without using tables or a calculator, show that $\sin (30° + 30°)$ does not equal $\sin 30° + \sin 30°$. (Hint: Review Section 7.3.)
$\sin 30° + \sin 30° = \frac{1}{2} + \frac{1}{2} = 1$;
$\sin (30° + 30°) = \sin 60° = \frac{1}{2}\sqrt{3} \approx 0.866$

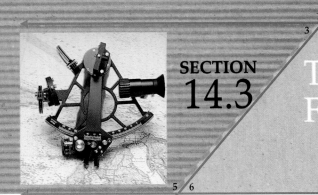

SECTION 14.3 The Sine Function

After completing this section, you should understand
- ▲ what an angle of rotation is.
- ▲ how the sine and cosine functions can be extended.
- ▲ how to graph the sine function.

In earlier sections we defined the sine, cosine, and tangent ratios in terms of right triangles. In this section you will begin to extend your definition of sine and cosine.

DISCOVERY ACTIVITY

1. On graph paper, prepare coordinate axes like the one at the right.
2. Copy the partly filled-in table.
3. Complete the table using either the table of trigonometric ratios on page 563 or a scientific calculator. Round the values of the sine ratio to the nearest hundredth.
4. Using your graph paper, plot the points defined by your table. Locate each point as precisely as you can.
5. Carefully connect the seven points with a smooth curve.
6. Is the curve you drew a line segment? No; it curves.
7. If you were to extend the curve at both ends, what do you think would be the value of y for $x = 0°$? for $x = 90°$? 0; 1

y: 1.0, 0.8, 0.6, 0.4, 0.2; x: O, 20°, 40°, 60°, 80°

$y = \sin x$

x	y
5°	0.09
20°	0.34
40°	0.64
50°	0.77
60°	0.87
70°	0.94
85°	0.99

SECTION 14.3

Resources
Reteaching 14.3
Enrichment 14.3
Transparency Master 14-2

Objectives
After completing this section, the student should understand
- ▲ what an angle of rotation is.
- ▲ how the sine and cosine functions can be extended.
- ▲ how to graph the sine function.

Materials
graph paper
scientific calculator

Vocabulary
function
sine function
angle of rotation
initial side
terminal side
counterclockwise rotation
positive angle
clockwise rotation
negative angle
cosine function

GETTING STARTED

Quiz: Prerequisite Skills
Identify each ratio as a trigonometric ratio of $\angle W$ or $\angle T$. Two answers may be possible.

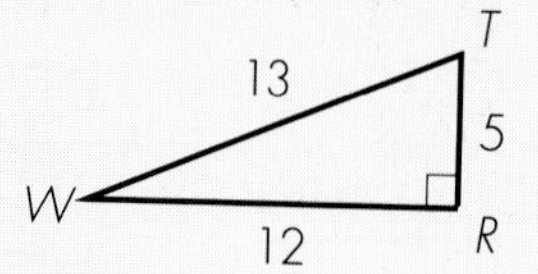

1. $\frac{5}{12}$ [tan W]
2. $\frac{5}{13}$ [sin W or cos T]
3. $\frac{12}{13}$ [cos W or sin T]
4. $\frac{12}{5}$ [tan T]

The graph you drew should look like this:

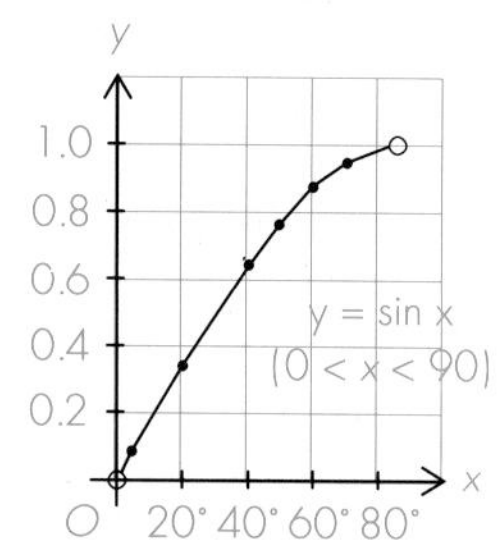

Recall that a **function** is a set of ordered pairs of real numbers such that each x-value has only one y-value. The table on page 529 is for part of the **sine function.** The graph shown above is for the equation $y = \sin x$ for values of x between 0° and 90°.

The sine ratio has been defined only for *acute* angles. Neither 0° nor 90° is the measure of an acute angle. For this reason, there are not points on the graph at $x = 0°$ or $x = 90°$. However, it is possible to remove the restriction that limits the sine function to acute angles. If this is done, the graph can be extended at both ends.

TEACHING COMMENTS

Using the Page
The angle of rotation is introduced in a coordinate plane defined by an a-axis and a b-axis instead of an x-axis and a y-axis. In this way, the x and y in the function $y = \sin x$ are not confused with the coordinates of a point P on the terminal side of the angle of rotation.

EXTENDING THE SINE FUNCTION

To extend the sine function you must first extend the angle measures that are allowed for the sine. This is done by beginning with a new kind of angle, the **angle of rotation,** which is shown below.

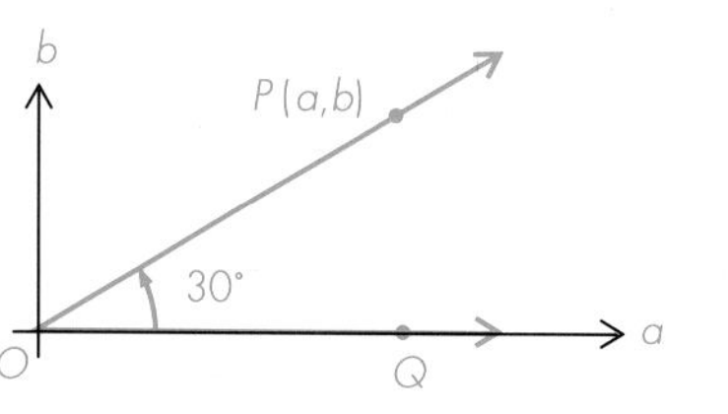

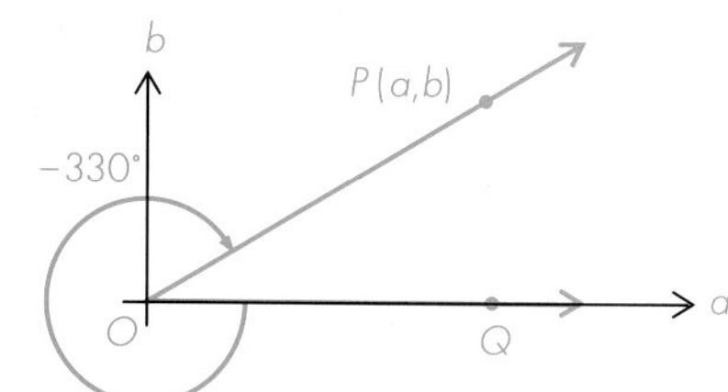

In the diagram, the a-axis and b-axis define a coordinate plane. The angle of rotation is formed by two rays meeting at vertex O. The first ray, $\overrightarrow{OQ}$, is called the **initial side** and lies on the positive horizontal axis. The second ray, $\overrightarrow{OP}$, is the **terminal side** and may lie in any of the four quadrants or along the vertical or horizontal axes.

You can rotate the initial side about the vertex O in a *counterclockwise* direction in order to reach the terminal side. In this case the angle of rotation is defined to be *positive.* If you rotate the initial side in a *clockwise* direction to reach the terminal side, the angle of rotation is defined to be *negative.* In the diagram at the bottom of page 530, the terminal side $\overrightarrow{OP}$ is shown with two measures, 30° and −330°. (Recall that there are 360° in a full revolution.)

Examples: Find the measure of another angle of rotation so that the sides are in the same position.

1. 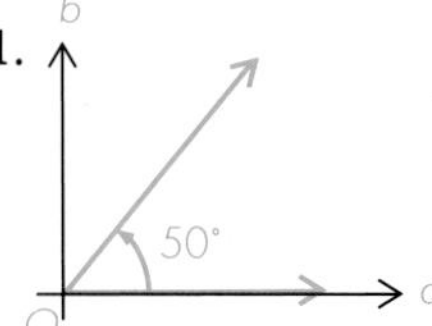

Find a clockwise angle. The measure will be negative: 360 − 50 = 310. A second measure is −310°.

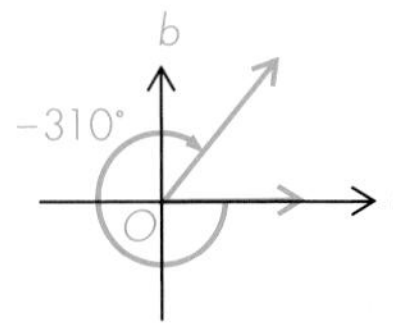

2. 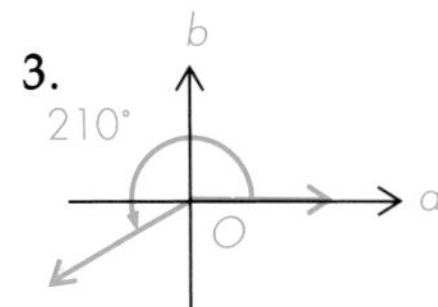

Find a counterclockwise angle. The measure will be positive: 360 − 240 = 120. A second measure is 120°.

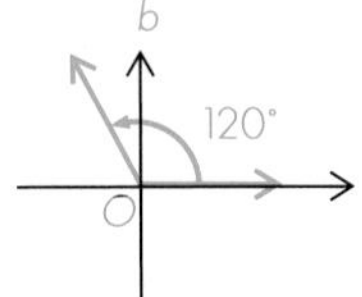

3. 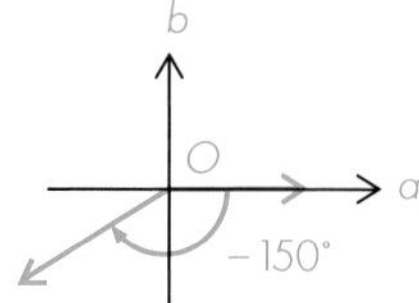

Find a clockwise angle. The measure will be negative: 360 − 210 = 150. A second measure is −150°.

b
O
−150°
a

CLASS ACTIVITY

Find a second measure for each angle shown at the right.

1. $\angle QOP_1$ −332°
2. $\angle QOP_2$ 290°
3. $\angle QOP_3$ −180°

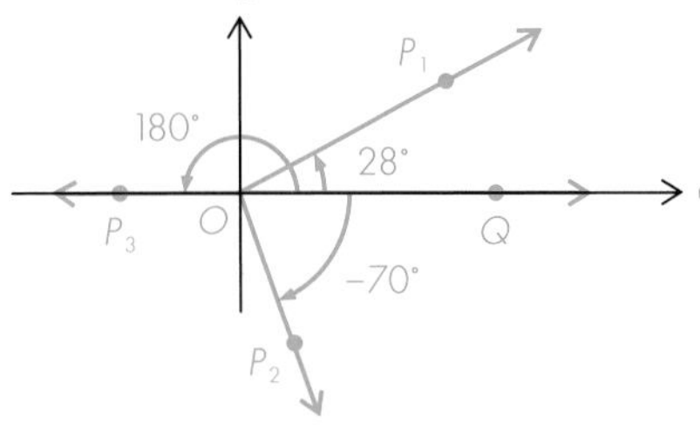

Using the Page
Clockwise and counterclockwise rotations can be demonstrated at the chalkboard by rotating a yardstick about the origin of a coordinate plane.

Extra Practice
Find a second measure for each angle.

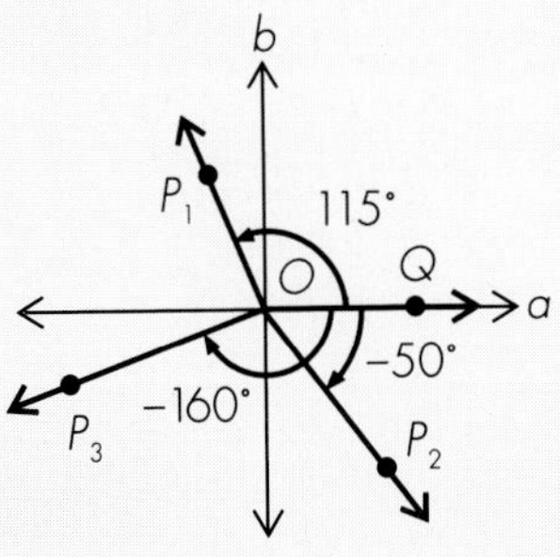

$\angle QOP_1$ [−245°]
$\angle QOP_2$ [310°]
$\angle QOP_3$ [200°]

Using the Page
In the new definition for the sine, $\frac{b}{r}$, point out that r is always positive although b can be positive or negative.

A NEW DEFINITION FOR THE SINE

In the figure at the right, line $\overleftrightarrow{P_1OP_2}$ includes the terminal sides of two angles, $\angle QOP_1$ and $\angle QOP_2$, with measures 37° and 217°, respectively. So far, only one of these angles, the 37° angle, has a sine:

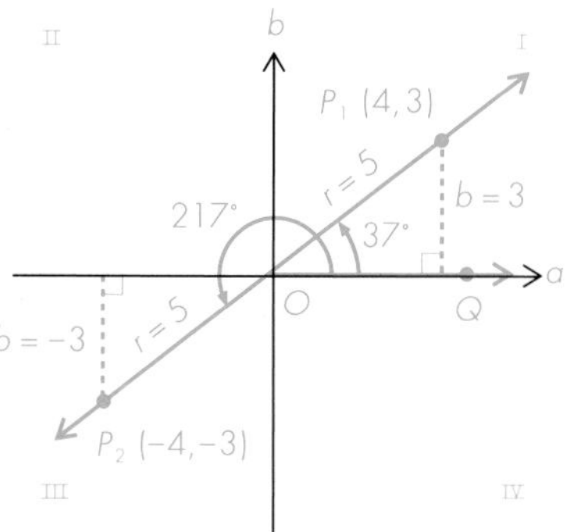

$$\sin 37° = \frac{\text{leg opp. 37° angle}}{\text{hypotenuse}} = \frac{b}{r} = \frac{3}{5}, \text{ or } 0.6000$$

To get a new definition of the sine ratio of 37°, replace
"leg opp. 37° angle" by "vertical coordinate of point P_1" and replace
"hypotenuse" by "distance r from the origin."
Then with this new definition, you will have:

$$\sin 37° = \frac{\text{vertical coordinate}}{\text{distance from origin}} = \frac{b}{r} = \frac{3}{5} = 0.6000$$

For 37°, in quadrant I, either definition will work.

For 217°, in quadrant III, only the new definition will work.

$$\sin 217° = \frac{\text{vertical coordinate}}{\text{distance from origin}} = \frac{b}{r} = \frac{-3}{5}, \text{ or } -0.6000$$

In quadrant I, either definition of the sine ratio will work. In quadrants II, III, and IV, only the new definition of the sine ratio will work.

Using the Page
The example illustrates how to use the table of sines to find the sine of an angle with a terminal side in quadrant II, III, or IV.

Transparency 14-2 relates both the sine and cosine of an angle with a terminal side in another quadrant to the sine and cosine of a 30° angle.

Example: Use the diagram and the table on page 563 to find the value of sin (−30°).
In quadrant I, $\sin 30° = \frac{b}{r}$.
From the table, $\sin 30° = 0.5000$.
So, $\frac{b}{r} = 0.5000$.
In quadrant IV,
$\sin(-30) = \frac{-b}{r} = -\frac{b}{r} = -0.5000$

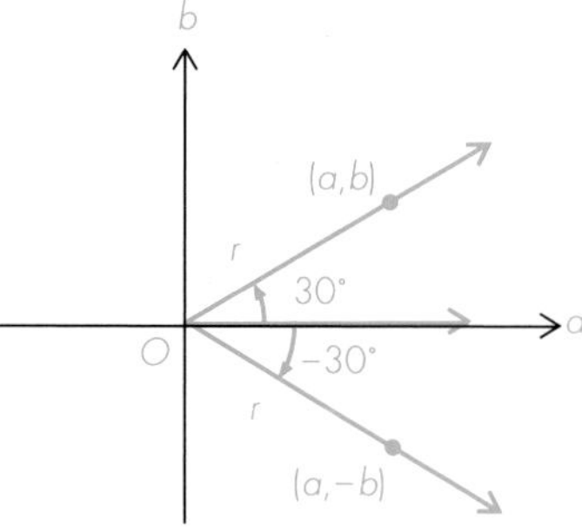

If you have a scientific calculator, then you can find the sine of any angle by using the SIN function button as shown below for sin (−30°). First, be sure that the calculator is in "degree" mode. Then use the following keystrokes.

30 [±] [SIN] = −0.5

CLASS ACTIVITY

Find the sine of each angle.

1. 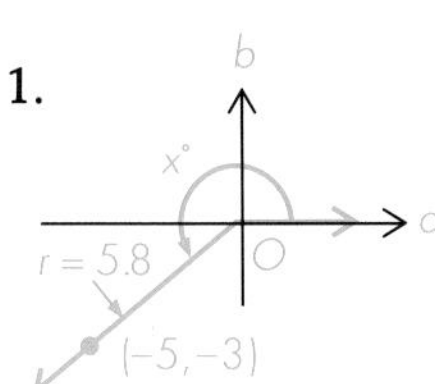

sin x = ? −0.5172

2. 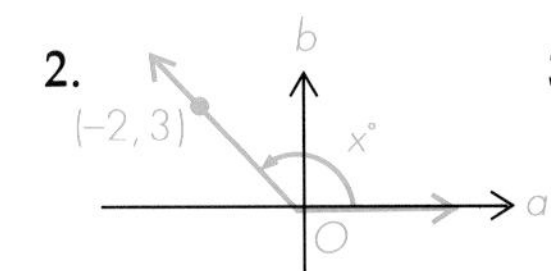

sin x = ? $\frac{3}{\sqrt{13}}$, or 0.8321
Hint: To find r use the distance formula (Section 7.5).

3. $x = 225°$
sin x = ?
−0.7071

GRAPHING THE SINE FUNCTION

On page 530 you saw the graph of the sine function for angle measures between 0° and 90°, that is, for $0° < x° < 90°$. With the new definition of the sine ratio, you can graph the sine function for any values of x.

CLASS ACTIVITY

Copy and complete the partially filled-in table below. Use a scientific calculator or the table on page 563. If you use the table on page 563 for angle measures greater than 90°, then you will have to find the sine values by using the method in the example at the bottom of page 532.

x°	sin x°
0°	0.00
5°	0.09
20°	0.34
40°	0.64
50°	0.77
60°	0.87
70°	0.94
85°	0.996
90°	1.00
95°	0.996

x°	sin x°
120°	0.87
130°	0.77
140°	0.64
160°	0.34
180°	0.00
190°	−0.17
200°	−0.34
210°	−0.50
220°	−0.64
230°	−0.77

x°	sin x°
240°	−0.87
250°	−0.94
270°	−1
280°	−0.98
290°	−0.94
300°	−0.87
310°	−0.77
320°	−0.64
340°	−0.34
360°	0.00

The complete graph for $0° \leq x° \leq 360°$ appears at the top of page 534.

Extra Practice

1. Find sin x.

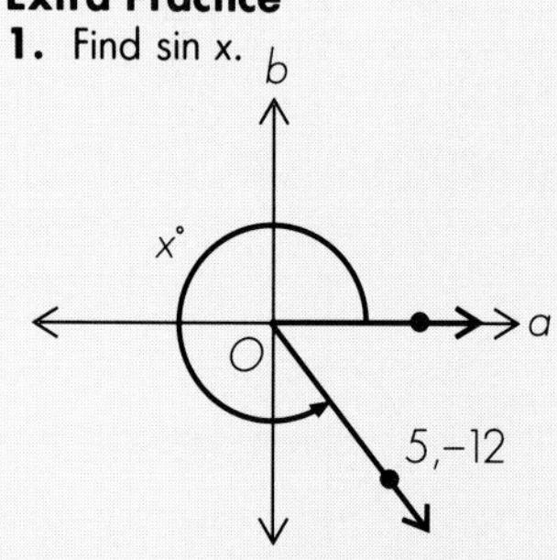

$\left[-\frac{12}{13}, \text{ or } -0.9231\right]$

2. x° = 120°
sin x = ? [0.8660]

Using the Page
To verify the sine values for 90°, 180°, 270°, and 360°, students can draw diagrams like those below.

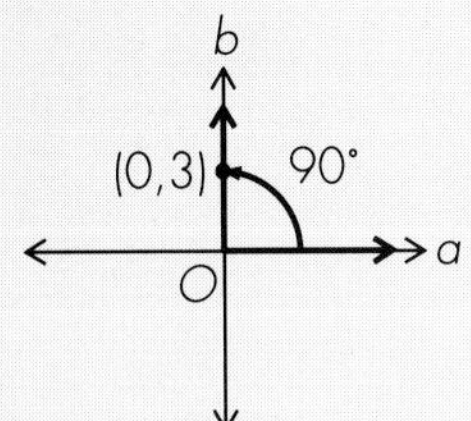

$b = 3, r = 3$

$\sin 90° = \frac{3}{3} = 1$

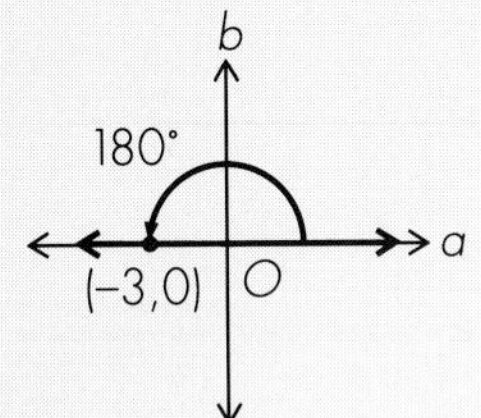

$b = 0, r = 3$

$\sin 180° = \frac{0}{3} = 0$

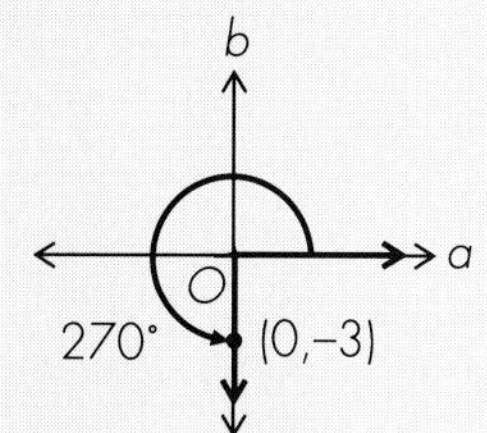

$b = -3, r = 3$

$\sin 270° = \frac{-3}{3} = -1$

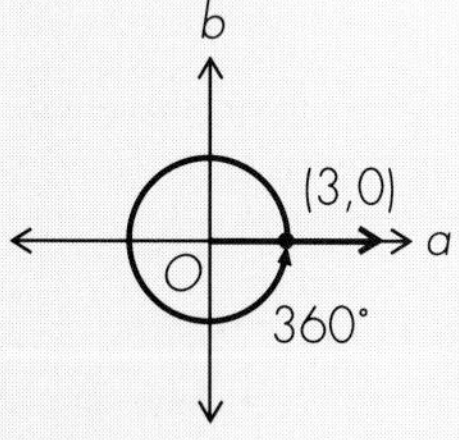

$b = 0, r = 3$

$\sin 360° = \frac{0}{3} = 0$

Note that in the Calculator Activity, pages 554-555, students are introduced to the periodicity of trigonometric functions when they find values of the functions for $x > 360°$ and $x < -360°$.

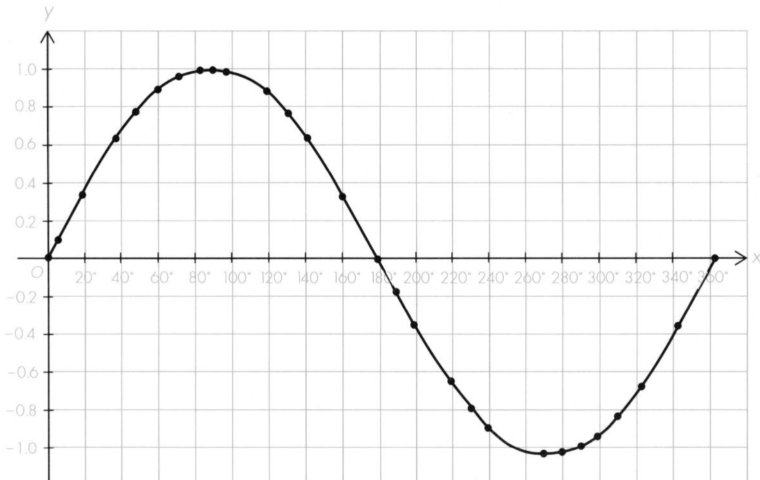

The cosine function's definition can be extended in much the same way. The main ideas of this section can be summarized in the following definition.

NOTEBOOK DEFINITION 14-3-1

Suppose that an angle with measure $x°$ has its vertex at the origin of a coordinate plane with one ray on the positive horizontal axis and a point $P(a, b)$ on the other ray r units from the origin. Then the SINE and COSINE of the angle are defined to be

$$\sin x° = \frac{b}{r}$$

$$\cos x° = \frac{a}{r}$$

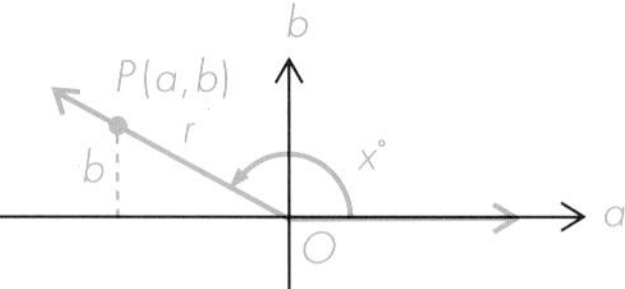

HOME ACTIVITY

In this activity, you may use a calculator or the tables on pages 562 and 563. Find a second measure for each angle.

1. 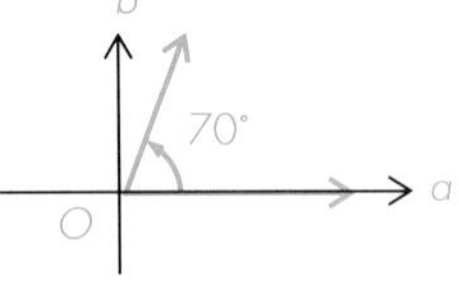

−290°

2. 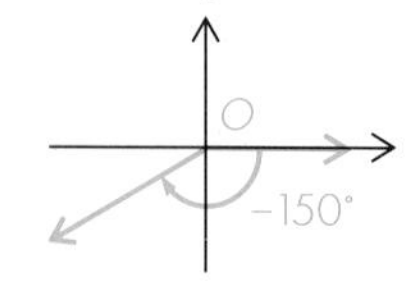

210°

3. 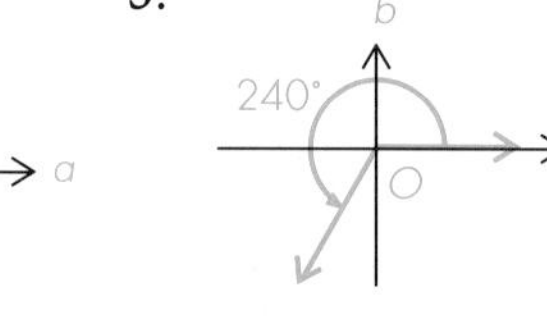

−120°

4. 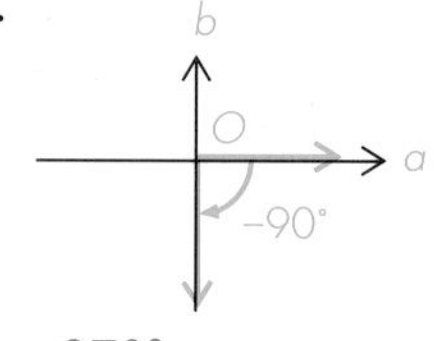

270°

Find the sine of each angle. Express answers to the nearest hundredth.

5. 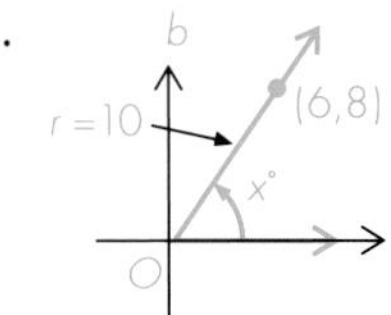

0.80

6. 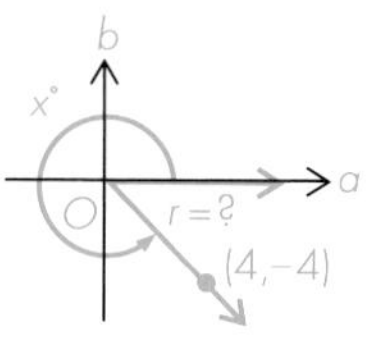

−0.71

7. 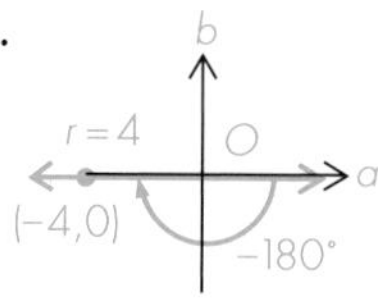

0.00

8. 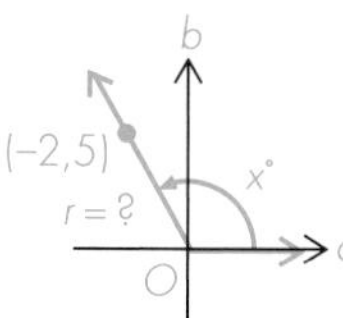

0.93

9. $x = -60°$ $\sin x = ?$ −0.87

10. $x = 150°$ $\sin x = ?$ 0.50

11. $x = 210°$ $\sin x = ?$ −0.50

12. $x = 270°$ $\sin x = ?$ −1.00

Use the Definition 14-3-1 to find r and $\cos x°$. Express answers to the nearest hundredth.

13. 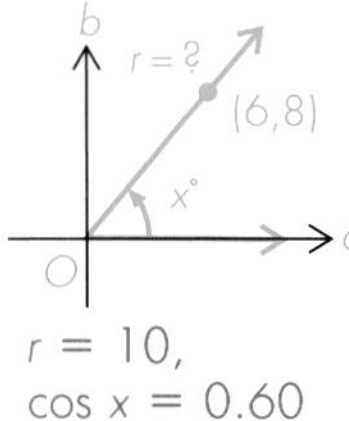

$r = 10$, $\cos x = 0.60$

14. 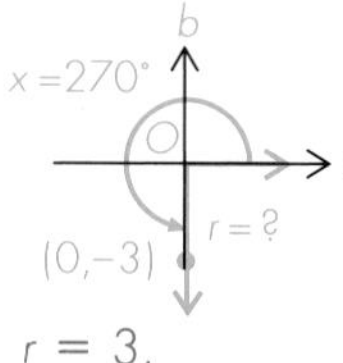

$r = 3$, $\cos x = 0.00$

15. 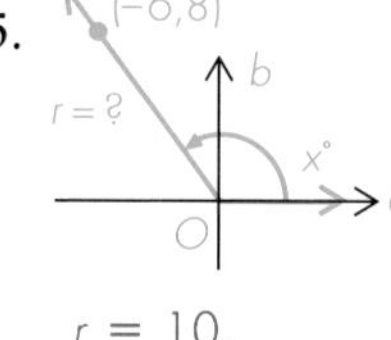

$r = 10$, $\cos x = -0.60$

16. 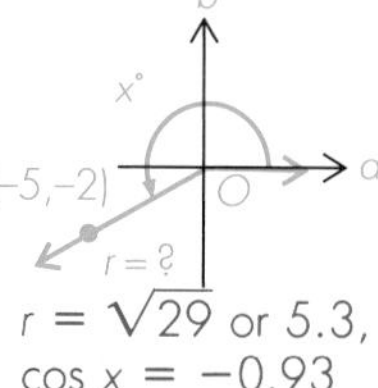

$r = \sqrt{29}$ or 5.3, $\cos x = -0.93$

17. Copy and complete the following table for $\cos x°$, where $0° [\leq] x° [\leq] 360°$. Then graph the cosine function in the coordinate plane. See margin for graph.

x°	cos x°	x°	cos x°	x°	cos x°
0°	1.00	130°	−0.64	250°	−0.34
15°	0.97	140°	−0.77	260°	−0.17
30°	0.87	150°	−0.87	270°	0.00
40°	0.77	160°	−0.94	280°	0.17
50°	0.64	170°	−0.98	290°	0.34
60°	0.50	180°	−1.00	300°	0.50
70°	0.34	190°	−0.98	315°	0.71
80°	0.17	200°	−0.94	330°	0.87
90°	0.00	215°	−0.82	340°	0.94
100°	−0.17	230°	−0.64	350°	0.98
115°	−0.42	240°	−0.50	360°	1.00

CRITICAL THINKING

18. Refer to Definition 14-3-1. The old definition for the tangent ratio was $\frac{\text{opp}}{\text{hyp}}$. In terms of the numbers a, b, and r of Definition 14-3-1, what would be a logical way to give a new definition of $\tan x°$? Will there be a value for $\tan x°$ for all values of $x°$ such that $0 \leq x° \leq 360°$? Explain your answer. See margin.

Additional Answers

17.

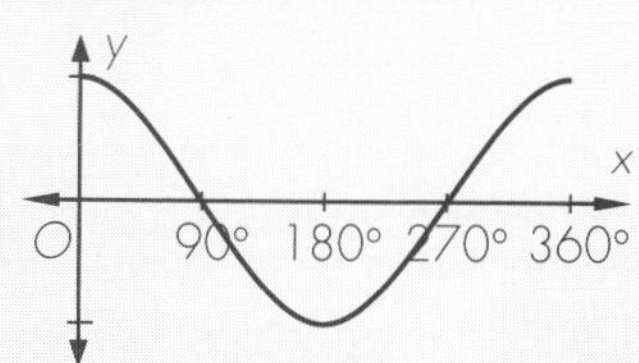

18. A logical definition would be $\tan x = \frac{b}{a}$. With this definition, there will not be a value for $\tan x$ when $a = 0$, since division by zero is not defined. This occurs whenever $x° = n \times 90°$, where n is an odd integer.

FOLLOW-UP

Assessment

1. $x = -270°$ $\sin x = ?$ [1]

2. Find $\sin x$ to the nearest hundredth.

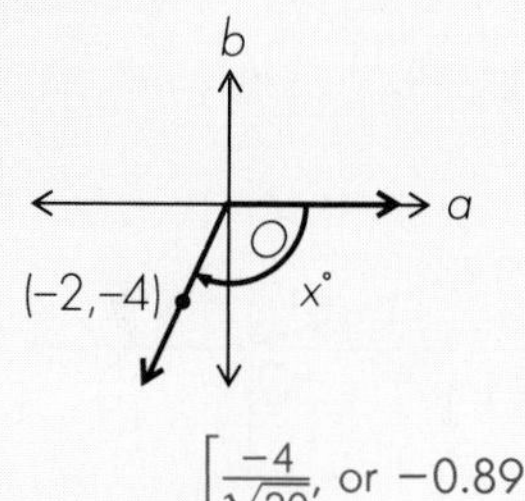

$\left[\frac{-4}{\sqrt{20}}, \text{ or } -0.89\right]$

3. Find r and $\cos x$ to the nearest hundredth.

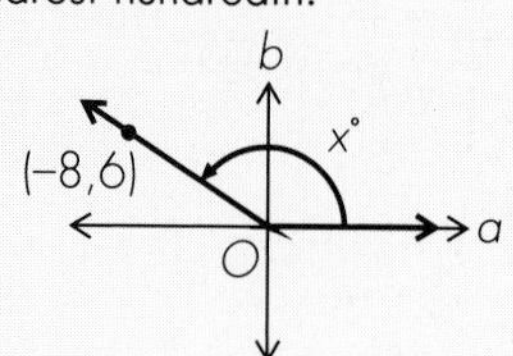

[$r = 10$, $\cos x = -0.80$]

Reteaching
Students who have had difficulty with this section may benefit from Reteaching Activity 14.3.

Enrichment
For students who have mastered the material in this section, you may wish to assign Enrichment Activity 14.3.

SECTION 14.4

Resources
Reteaching 14.4
Enrichment 14.4
Transparency Master 14-1

Objectives
After completing this section, the student should understand
▲ how to use the Law of Sines.
▲ how to use the Law of Cosines.

Materials
scientific calculator

Vocabulary
Law of Sines
Law of Cosines

GETTING STARTED

Quiz: Prerequisite Skills

1. Use a calculator, or use the diagram and the table on page **563**, to find the sine of 160°. [0.3420]

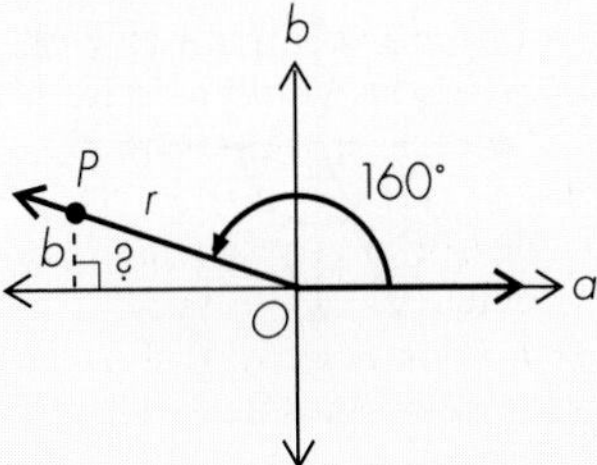

Solve each proportion.

2. $\frac{8}{10} = \frac{12}{a}$ [a = 15]

3. $\frac{25}{b} = \frac{10}{16}$ [b = 40]

SECTION 14.4 Solving Non-right Triangles

After completing this section, you should understand
▲ how to use the law of sines.
▲ how to use the law of cosines.

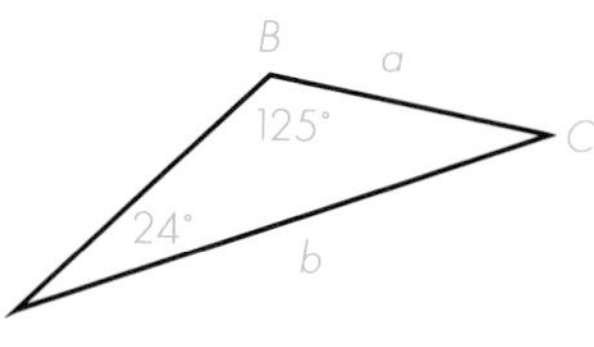

In $\triangle ABC$ at the right, you can see that $\angle B$ is greater than $\angle A$. What can you say about the relationship between the sides opposite these two angles? From Theorem 7-4-1 you know that if a triangle has one angle greater than another angle, then the side opposite the greater angle is longer than the side opposite the other angle.

Applying this theorem to ABC, you can conclude that since $m\angle B > m\angle A$, the inequality $b > a$ is true. You can use trigonometry to show exactly how the two angles and two sides are related.

First, you can use the ideas from the previous section to find sin 125°. Use the figure at the right:

$$\sin 125° = \frac{\text{vertical coordinate}}{\text{distance of } P \text{ from } 0} = \frac{b}{r}$$

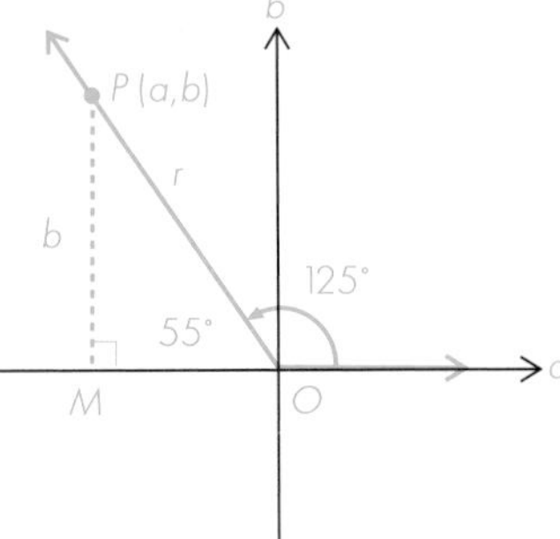

Also, for $\triangle OMP$ you have the following:

$$\sin 55° = \frac{\text{opp}}{\text{hyp}} = \frac{b}{r}$$

Therefore, sin 125° = sin 55°.

DISCOVERY ACTIVITY

1. Refer to $\triangle ABC$ at the right. Use a table or a scientific calculator to express the following ratios as decimals:

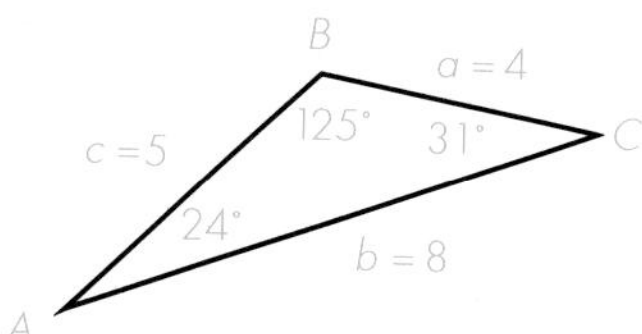

$\frac{a}{b} = ?$ 0.5 $\quad \frac{A}{B} = ?$ 0.19 $\quad \frac{\sin A}{\sin B} = ?$ 0.5

2. Which of the above ratios, if any, have the same value? $\frac{a}{b}$ and $\frac{\sin A}{\sin B}$

3. Based on your answers to questions 1 and 2, write a true proportion. $\frac{a}{b} = \frac{\sin A}{\sin B}$

As a result of your work in the Discovery Activity, you should have concluded that two sides of a triangle are proportional not to the opposite angles, as you might have expected, but to the *sines* of those angles. The proportion can be written in several ways:

$$\frac{a}{b} = \frac{\sin A}{\sin B} \quad \text{or} \quad \frac{\sin A}{a} = \frac{\sin B}{b}$$

The second of these ways leads to a theorem called the **Law of Sines.** The theorem holds for any two angles of a triangle and their opposite sides.

THEOREM 14-4-1: **(Law of Sines) In any triangle *ABC*,**

$$\frac{\sin A}{a} = \frac{\sin B}{b} = \frac{\sin C}{c}.$$

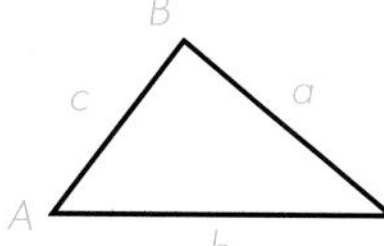

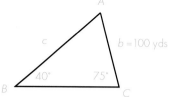

You can use the Law of Sines to find measurements of sides or angles of a triangle, even if the triangle is not a right triangle. Finding all the missing measurements is called **solving the triangle.**

Example: Find c for the triangle shown in the diagram.

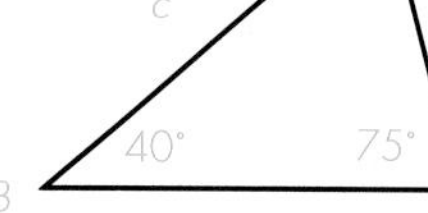

$$\frac{\sin B}{b} = \frac{\sin C}{c}$$

$$\frac{\sin 40°}{100} = \frac{\sin 75°}{c}$$

Now solve the proportion.

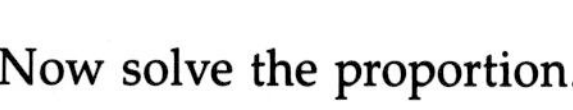

$\sin 40° \times c = 100 \times \sin 75°$

$$c = \frac{100 \times \sin 75°}{\sin 40°} = \frac{100 \times 0.9659}{0.6428} \approx 150.3 \text{ yd}$$

TEACHING COMMENTS

Using the Page
When students use the Law of Sines, they may find it convenient to note that the sine of an obtuse angle is equal to the sine of its supplement. The second figure on page 536 can be generalized to show that $\sin x = \sin (180 - x)$.

Extra Practice
Find the indicated length.

1.

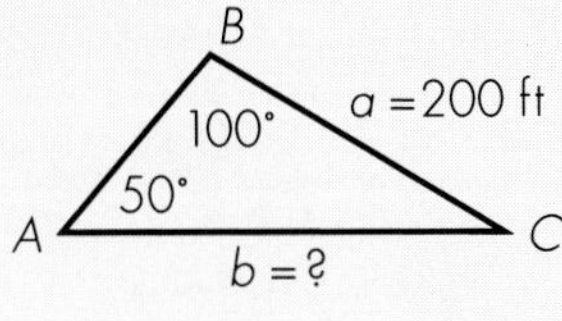

[257 ft]

2.

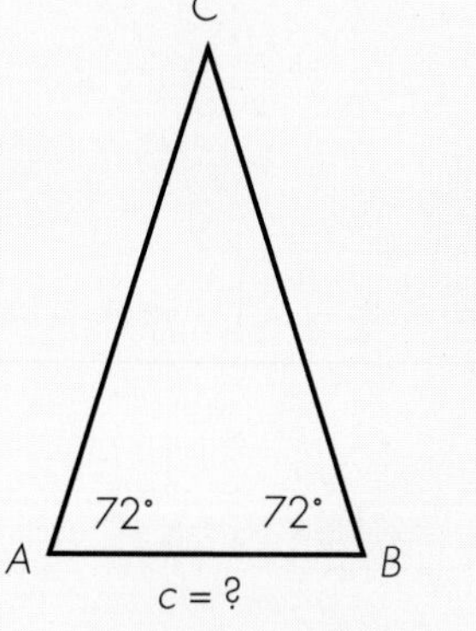

[49.4, or about 49 m]

CLASS ACTIVITY

Find the indicated length. Use a calculator or the table on page 563.

1.

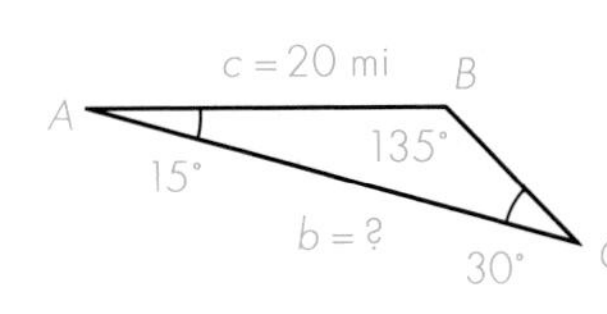

28.3, or about 28 mi

2.

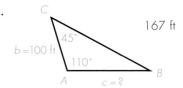

C
167 ft
45°
b = 100 ft
110°
A
c = ?
B

Hint: First find $m\angle B$.

ANOTHER LAW FOR SOLVING TRIANGLES

In some cases you may need more information than you are given in order to solve the triangle. The Discovery Activity below illustrates this.

DISCOVERY ACTIVITY

Study the three triangles shown below. One of the triangles can be solved using the Law of Sines. Another cannot be solved by any method.

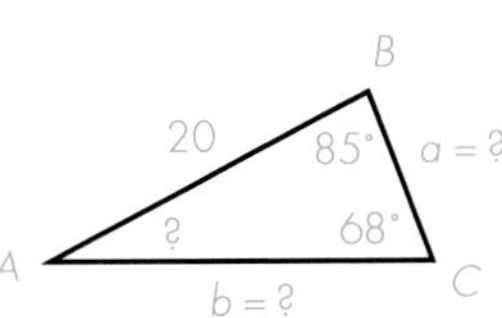

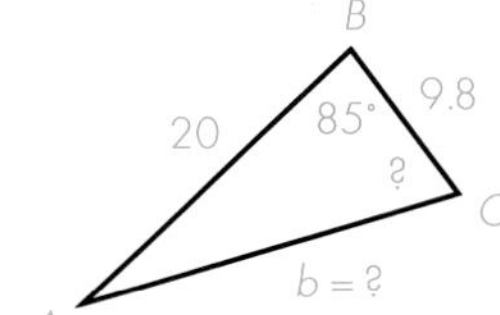

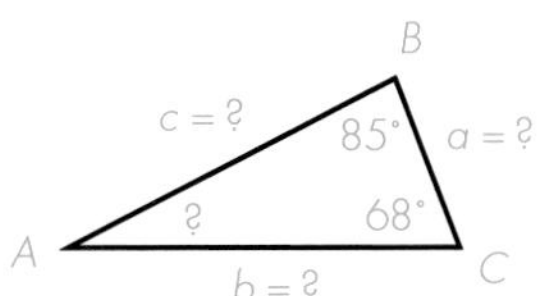

1. Which triangle can be solved using the law of sines? The triangle on the left
2. Which triangle cannot be solved by any method? Why?
   The triangle on the right. No lengths are given for the sides.

In the Discovery Activity you probably noticed that the triangle on the left can be solved by the Law of Sines. You can find $m\angle A$ by solving the equation $m\angle A + 68° + 85° = 180°$. Then you can use the Law of Sines to find a and b. The triangle on the right cannot be solved by any method. To solve a triangle you must know the length of at least one side.

What about the triangle in the middle? It cannot be solved by using the law of sines but it can be solved by using another law, the Law of Cosines.

THEOREM 14-4-2 **(Law of Cosines) In any triangle *ABC*,**

$$a^2 = b^2 + c^2 - 2bc \cos A$$
$$b^2 = a^2 + c^2 - 2ac \cos B$$
$$\text{and } c^2 = a^2 + b^2 - 2ab \cos C.$$

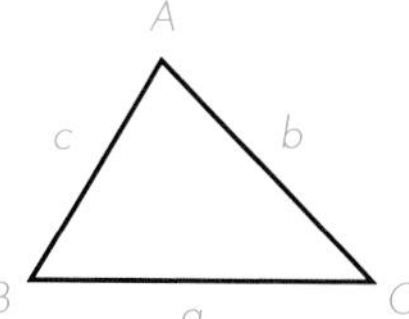

Examples:

1. Find b for the triangle in the diagram. Since $m\angle B$ is known, use the middle formula in the Law of Cosines.

$$b^2 = a^2 + c^2 - 2ac \cos B$$
$$= 9.8^2 + 20^2 - 2 \times 9.8 \times 20 \times \cos 85°$$
$$= 96.04 + 400 - 392 \times 0.0872$$
$$\approx 496.04 - 34.18, \text{ or } 461.86$$

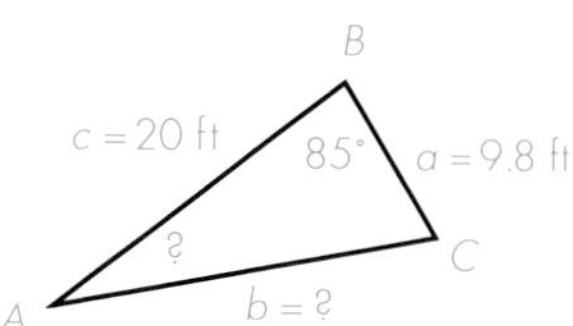

Use a calculator or the table of squares and square roots on page 562 to find b. $b \approx 21.5$ ft

2. Find c in $\triangle ABC$. Since 120° is not an acute angle, you must use the definition of the cosine ratio on page 535.

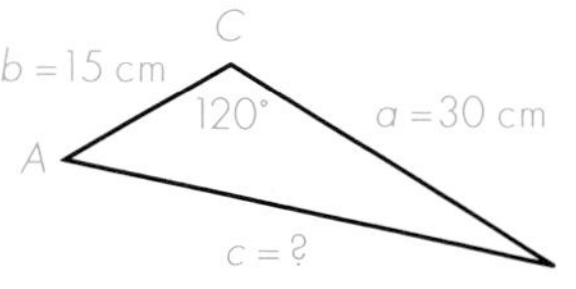

$$\cos 120° = \frac{\text{horizontal coordinate}}{\text{distance of } P \text{ from } O}$$
$$= \frac{-a}{r} = -\frac{a}{r} = -\cos 60° \text{ (See the diagram below.)}$$

Use the last equation from the Law of Cosines.

$$c^2 = a^2 + b^2 - 2ab \cos C$$
$$= 30^2 + 15^2 - 2 \times 30 \times 15 \times \cos 120°$$
$$= 900 + 225 - 900 \times (-\cos 60°)$$
$$= 1125 - 900 \times (-0.5000)$$
$$= 1125 + 450, \text{ or } 1575$$
$$c \approx 39.7 \text{ cm}$$

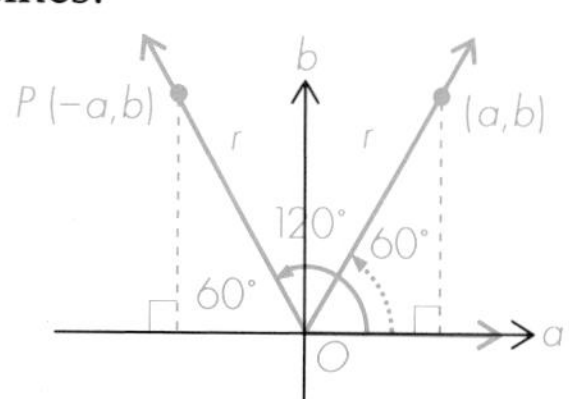

If you want to solve a triangle, you can decide whether to use the law of sines or the law of cosines by referring to the chart below.

Given information	Law to use
Two angles and any side given (AAS or ASA)	Law of Sines
Two sides and the included angle given (SAS)	Law of Cosines
Three sides given (SSS)	Law of Cosines

Using the Page

Students may note the similarity between the Law of Cosines and the Pythagorean Theorem. Consider a right triangle with right angle C. Then cos C = 0, and the resulting equation is the familiar Pythagorean relation.

Extra Practice

Find a to the nearest meter.

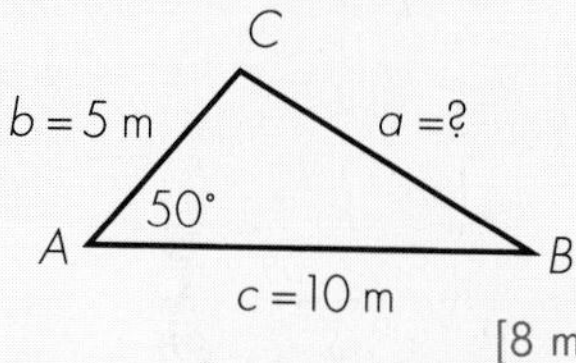

[8 m]

Using the Page

The second diagram for Example 2 can be generalized to show that the cosine of an obtuse angle is negative, and it is the opposite of the cosine of its supplement.

Transparency 14-1 summarizes methods now available for solving right triangles, using trigonometric ratios, and oblique triangles, using the Law of Sines and the Law of Cosines.

Example: Find $m\angle A$ in $\triangle ABC$.
The pattern is *SSS*.
Use the Law of Cosines.
$a^2 = b^2 + c^2 - 2bc \cos A$
$15^2 = 10^2 + 19^2 - 2 \times 10 \times 19 \times \cos A$
Simplify and solve for cos *A*.
$\cos A = 0.6211$
$m\angle A = 52°$, to the nearest degree

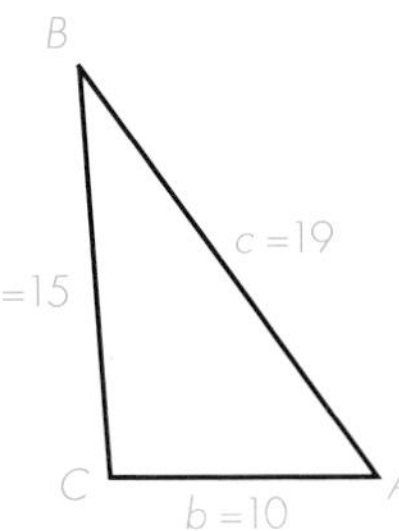

CLASS ACTIVITY

For Exercises 1 and 2, use a calculator or the tables on pages 562 and 563.

1. In $\triangle ABC$, $m\angle A = 35°$, $b = 12$, and $c = 25$. Find *a*. 16.7
2. In *ABC*, $a = 6$, $b = 9$, and $c = 13$. Find $m\angle B$. 37°

HOME ACTIVITY

In this activity, you may use a calculator or the tables on pages 562 and 563. Use the law of sines to find the indicated lengths to the nearest tenth.

1. C, a = ?, 80°, 58°, A, 16, B — 13.8

2. 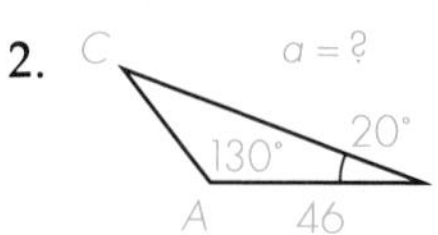

70.5

3. 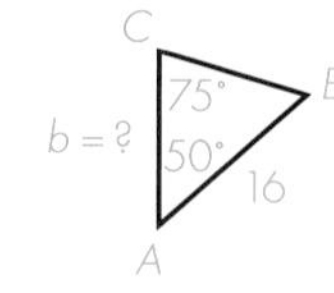

13.6

4. 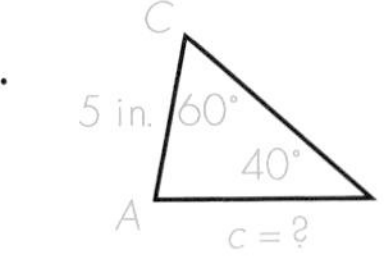

6.7

Use the laws of cosines to find the indicated measure. Give lengths to the nearest tenth and angle measures to the nearest degree.

5. 5.7 — C, 9, a = ?, 35°, A, 5, B

6. 21.3
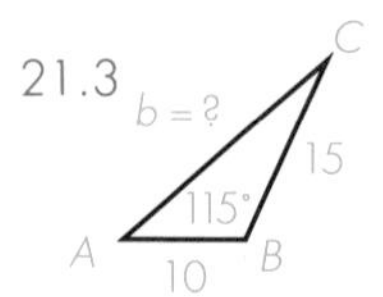

7. 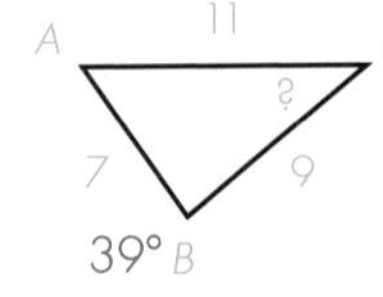

39°

8. 109°
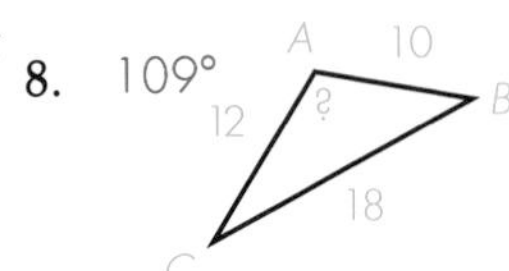

9. Two lighthouses are located at points *A* and *B*. A ship is at point *C*. (See diagram.) How far is the ship from the lighthouse at point *A*? 40.297, or about 40.3 mi

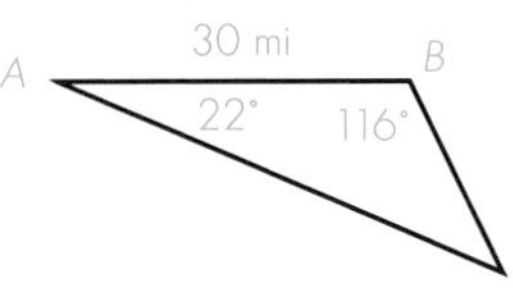

CRITICAL THINKING

10. Show that for $\triangle ABC$, $\frac{a-b}{b} = \frac{\sin A - \sin B}{\sin B}$. See margin.

Additional Answers

10. By the Law of Sines, $\frac{\sin B}{b} = \frac{\sin A}{a}$. From this equation, one obtains the following:

$$\frac{a}{b} = \frac{\sin A}{\sin B}$$
$$\frac{a}{b} - 1 = \frac{\sin A}{\sin B} - 1$$
$$\frac{a}{b} - \frac{b}{b} = \frac{\sin A}{\sin B} - \frac{\sin B}{\sin B}$$
$$\frac{a-b}{b} = \frac{\sin A - \sin B}{\sin B}$$

FOLLOW-UP

Assessment

1. Find *a* to the nearest tenth of an inch.

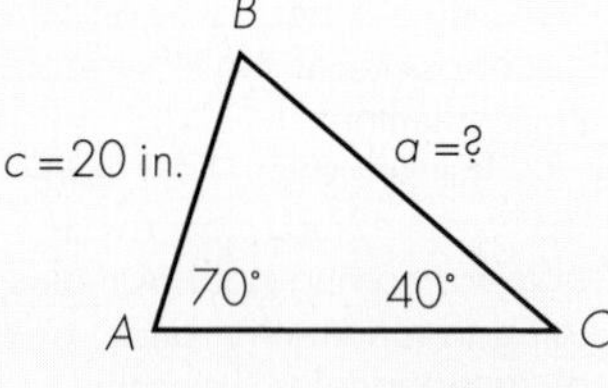

[29.2 in.]

2. Find *c* to the nearest foot.

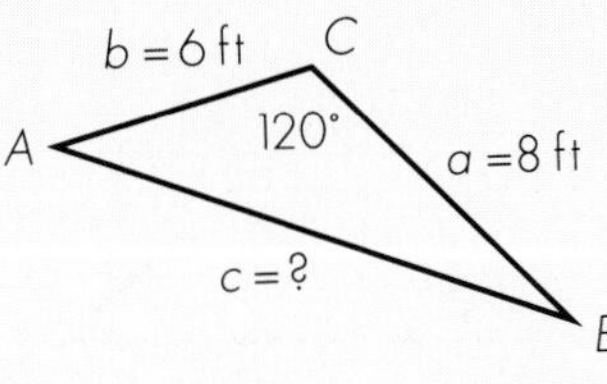

[12 ft]

3. In $\triangle ABC$, $a = 2$, $b = 3$, and $c = 4$. Find $m\angle A$. [29°]

Reteaching
Students who have had difficulty with this section may benefit from Reteaching Activity 14.4.

Enrichment
For students who have mastered the material in this section, you may wish to assign Enrichment Activity 14.4.

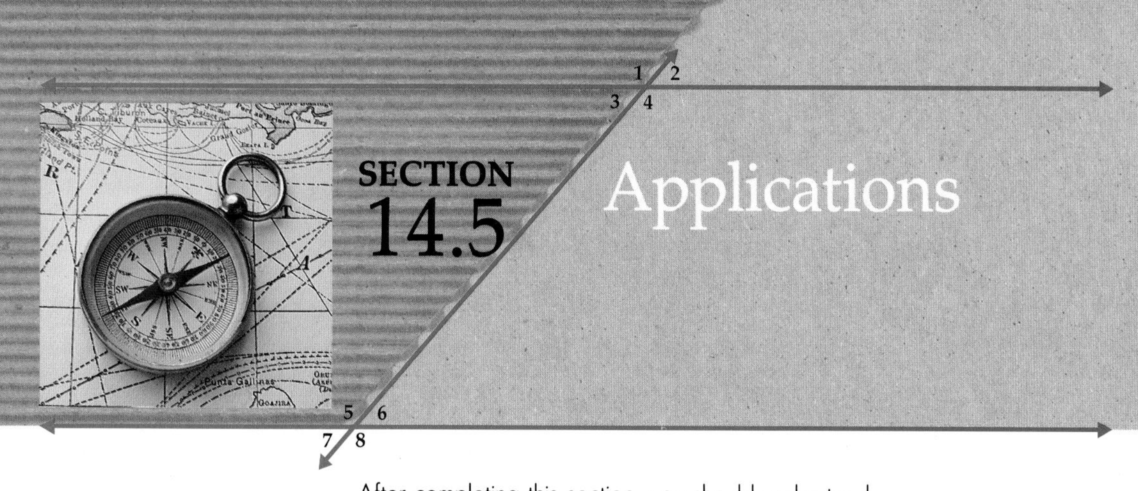

After completing this section, you should understand
▲ how to use trigonometric functions to solve practical problems.

You have seen how to measure distances indirectly using the sine, cosine, and tangent ratios. Here is a review of those ratios for a positive acute angle *A*.

$\sin X = \frac{\text{opp}}{\text{hyp}}$

$\cos X = \frac{\text{adj}}{\text{hyp}}$

$\tan X = \frac{\text{opp}}{\text{adj}}$

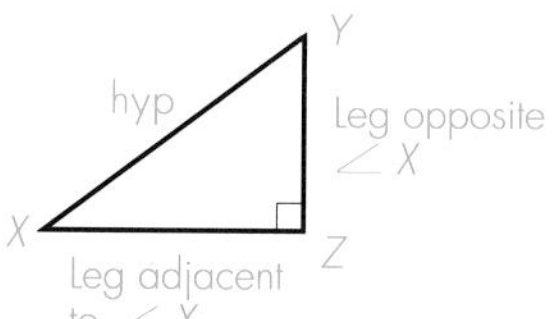

You can use the above definitions to solve a *right* triangle. To solve other kinds of triangles, you need to extend the functions with the following definitions:

$$\sin x = \frac{\text{vertical coordinate}}{\text{distance of } P \text{ from origin}} = \frac{b}{r}$$

$$\cos x = \frac{\text{horizontal coordinate}}{\text{distance of } P \text{ from origin}} = \frac{a}{r}$$

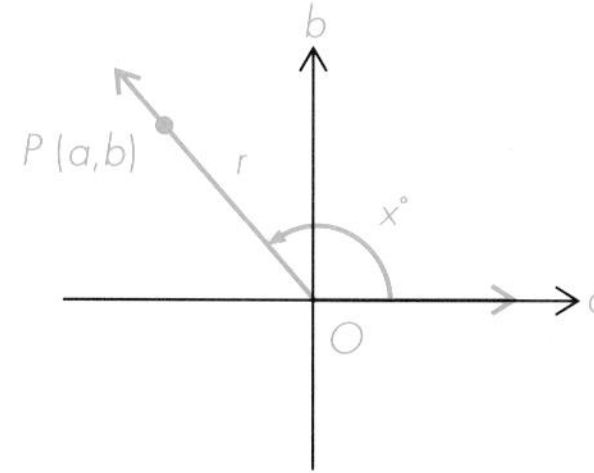

In this section, you will see how to apply these definitions to solving practical problems.

SECTION 14.5

Resources
Reteaching 14.5
Enrichment 14.5

Objectives
After completing this section, the student should understand
▲ how to use trigonometric functions to solve practical problems.

Materials
scientific calculator

Vocabulary
angle of elevation
angle of depression
line of sight

GETTING STARTED

Quiz: Prerequisite Skills
What law would you use to find the missing length or angle measure?

1.

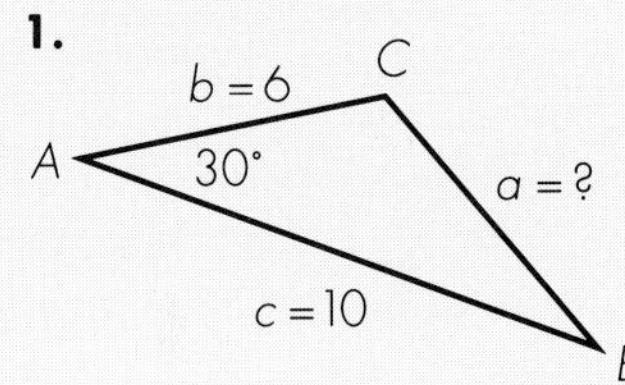

[Law of Cosines]

2.

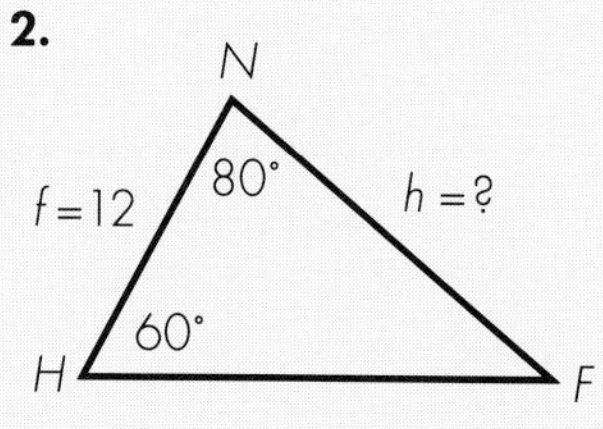

[Law of Sines]

3.

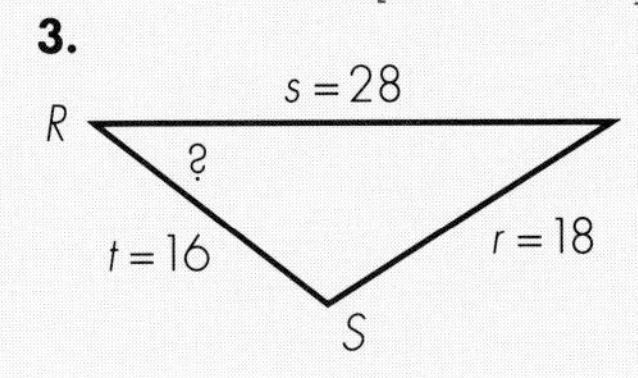

[Law of Cosines]

TEACHING COMMENTS

Using the Page
If possible, show students a transit or a picture of one. Explain that a transit is used by surveyors to measure angles from the horizontal.

Extra Practice
A transit 5 ft above the ground level is used to measure the height of a cliff across a river that is 350 ft wide. The angle of elevation from the transit to the top of the cliff is 42°. How high is the cliff? [about 320 ft high]

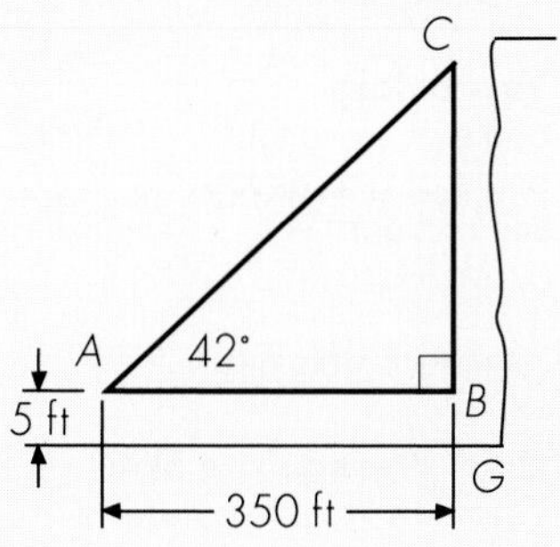

ANGLES OF ELEVATION AND DEPRESSION

When you wish to measure a vertical distance indirectly, you often need to know the angle of elevation or angle of depression. These are illustrated below.

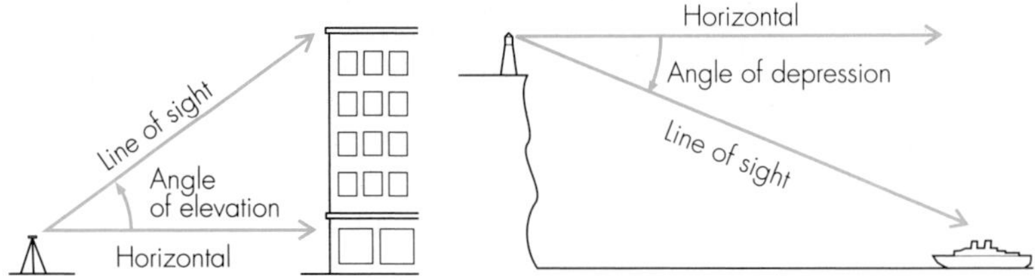

In both cases, the angle of elevation and angle of depression are formed by a horizontal ray and a second ray called the **line of sight.**

Examples: 1. The angle of elevation from a transit to the top of a flagpole is 70°. The transit is 1.5 m high and is 6 m from the flagpole. Find the height of the flagpole and the line-of-sight distance from the observer to the top of the flagpole.

Refer to the diagram. Use trigonometry to find YZ. You know XY and $m\angle X$, so use the tangent ratio.

$$\tan X = \frac{\text{opp}}{\text{adj}}$$

$$\tan 70° = \frac{YZ}{6}$$

$$6 \tan 70° = YZ$$

$$YZ = 6 \tan 70°$$

$$= 6 \times 2.747, \text{ or } 16.48$$

$$BY = \text{height of transit} + YZ$$
$$= 1.5 + 16.48 = 17.98, \text{ or about } 18 \text{ m}$$

Next, use the cosine ratio to find XZ.

$$\cos X = \frac{\text{adj}}{\text{hyp}}$$

$$\cos 70° = \frac{6}{XZ}$$

$$XZ \cos 70° = 6$$

$$XZ = \frac{6}{\cos 70°} = \frac{6}{0.3420} = 17.54, \text{ or about } 17.5 \text{ m}$$

2. The string of the kite pictured in the diagram makes an angle of 34° with the horizontal. The length of the string is 170 yd. Find the altitude of the kite. Assume that the kite string is held from a point 1 yd above the ground.

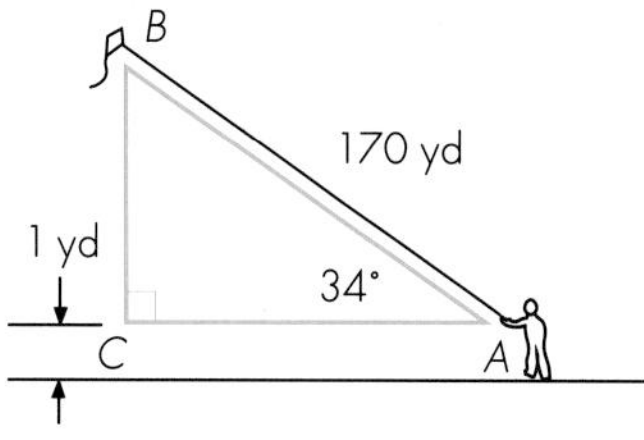

Use the sine ratio to find BC. $\sin A = \frac{\text{opp}}{\text{hyp}}$

$$\sin 34° = \frac{BC}{170}$$

$$\begin{aligned} 170 \sin 34° &= BC \\ BC &= 170 \sin 34° \\ &= 170 \times 0.5592 \approx 95.06, \text{ or about } 95 \text{ yd} \\ \text{Altitude of kite} &= 1 \text{ yd} + 95 \text{ yd} = 96 \text{ yd} \end{aligned}$$

3. In Example 2, find the distance from the person flying the kite to the point on the ground directly below the kite.

Use the cosine ratio to find AC. $\cos A = \frac{\text{adj}}{\text{hyp}}$

$$\cos 34° = \frac{AC}{170}$$

$$\begin{aligned} 170 \cos 34° &= AC \\ AC &= 170 \cos 34° \\ &= 170 \times 0.8290 \approx 140.9, \text{ or about } 141 \text{ yd} \end{aligned}$$

CLASS ACTIVITY

The angle of elevation from a control tower to an airplane is 6°. The tower is 40 ft above the ground. From a point directly under the plane, the distance to the tower is 3000 ft.

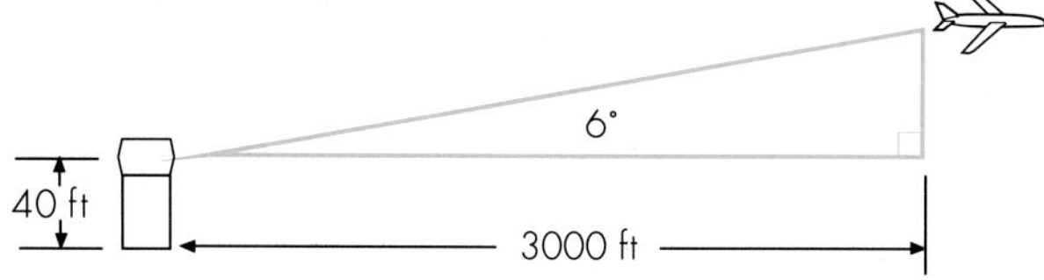

Use a calculator or the table on page 563 to find the indicated distances.

1. Find the altitude of the plane. about 355 ft

2. Find the line-of-sight distance from the tower to the plane. about 3017 ft

Extra Practice
A pilot flying at an altitude of 600 ft sights a forest fire from an angle of depression of 35°.

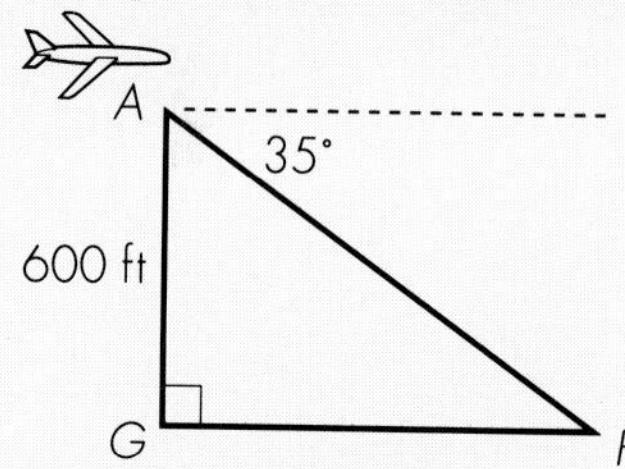

1. To the nearest 10 ft, what is the distance of the forest fire from the plane? [1050 ft]
2. To the nearest 10 ft, what is the horizontal distance of the fire from a point directly below the plane? [860 ft]

Extra Practice

A ship is sighted off shore from two points A and B on shore, 1800 ft apart. At each point, the angle is measured between lines of sight to the ship and to the other point. The angles from A and B are 75° and 70°, respectively. How far is the ship from each point? Give answers to the nearest 10 ft.

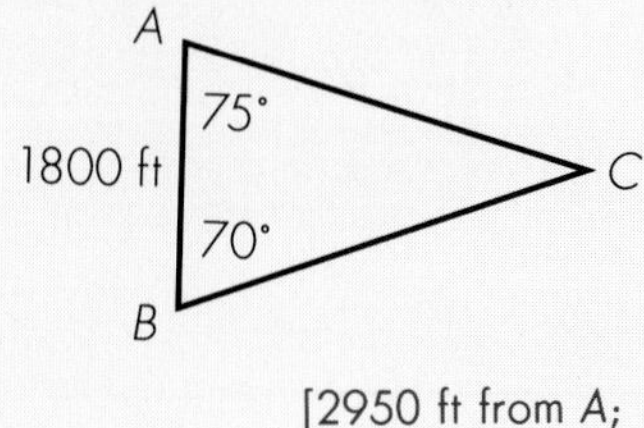

[2950 ft from A; 3030 ft from B]

SOLVING NON-RIGHT TRIANGLES

To solve non-right triangles, use the following two laws.

Law of Sines

$$\frac{\sin A}{a} = \frac{\sin B}{b} = \frac{\sin C}{c}$$

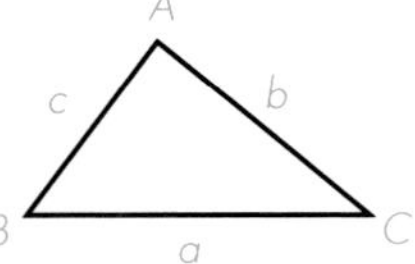

Law of Cosines

$$a^2 = b^2 + c^2 - 2bc \cos A$$
$$b^2 = a^2 + c^2 - 2ac \cos B$$
$$c^2 = a^2 + b^2 - 2ab \cos C$$

Example: Two surveyors need to find the distance between two points A and B on opposite sides of a lake. They measure $\angle A$ and $\angle C$. They also measure the distance BC. Find the distance across the lake.

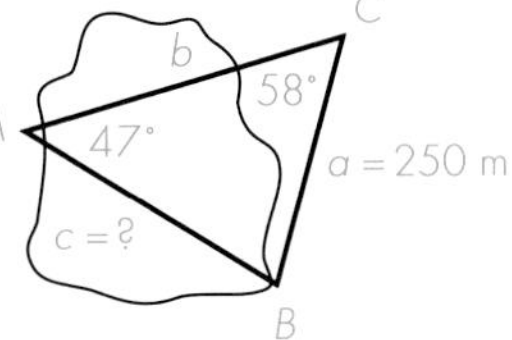

Two angles and one side are known. The pattern is AAS. (See the chart on page 539.) Use the Law of Sines.

$$\frac{\sin A}{a} = \frac{\sin C}{c}$$

$$\frac{\sin 47°}{250} = \frac{\sin 58°}{c}$$

$$c \sin 47° = 250 \sin 58°$$

$$c = \frac{250 \sin 58°}{\sin 47°}$$

$$= \frac{250 \times 0.8480}{0.7314} \approx 289.9\text{, or about 290 m}$$

If you use a scientific calculator to find the sine and cosine ratios, you will immediately know whether the ratio is positive or negative. For example, on a calculator with an 8-digit display,

$\sin 110° = 0.9396926 \approx 0.94$
and $\cos 110° = -0.3420201 \approx -0.34$.

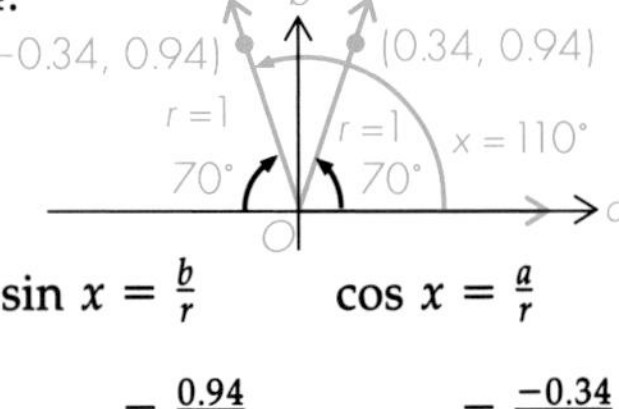

If you use the table on page 563 to find the ratios, draw a diagram such as the one at the right to see how your angle is related to the angles in the table. In this case,

$\sin 110° = \sin 70°$
and $\cos 110° = -\cos 70°$

$$\sin x = \frac{b}{r} = \frac{0.94}{1} = 0.94 \qquad \cos x = \frac{a}{r} = \frac{-0.34}{1} = -0.34$$

Examples:

1. The navigators of two ships at sea can see one another's ships as well as a lighthouse on shore. The relative positions of the ships are as shown in the diagram. How far apart are the ships?

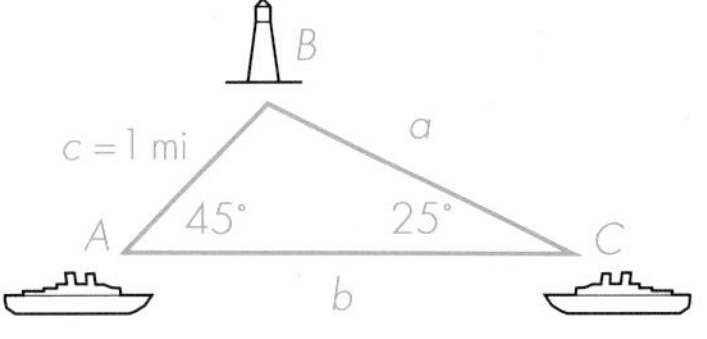

Two angles and a side are known. Use the law of sines. Since b is asked for, $m\angle B$ must also be found.

$$m\angle B + 45° + 25° = 180°$$
$$m\angle B = 180° - 45° - 25°, \text{ or } 110°$$

$$\frac{\sin B}{b} = \frac{\sin C}{c}$$

$$\frac{\sin 110°}{b} = \frac{\sin 25°}{1}$$

$$\sin 110° = b \sin 25°$$

$$b = \frac{\sin 110°}{\sin 25°} = \frac{0.9397}{0.4226} \approx 2.224, \text{ or about } 2.2 \text{ mi}$$

2. A ship is 100 yd from point A on shore and 80 yd from point C on shore. The measure of $\angle B$ is 95°. How far apart are A and C? The pattern is ASA. Use the law of cosines.

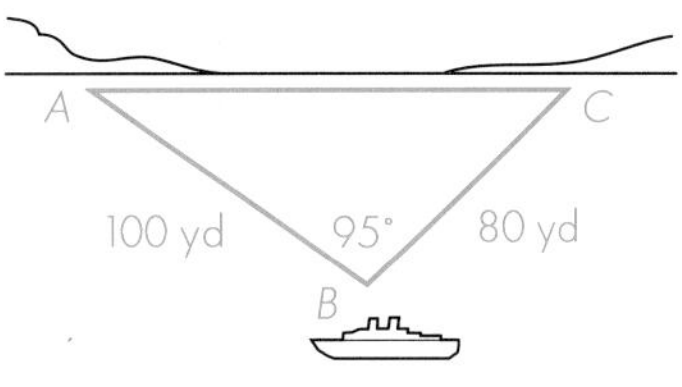

$$\begin{aligned} b^2 &= a^2 + c^2 - 2ac \cos B \\ &= 80^2 + 100^2 - 2 \times 80 \times 100 \times \cos 95° \\ &= 6400 + 10{,}000 - 16{,}000 \times (-0.0872) \\ &= 17{,}795 \\ b &= 133.4, \text{ or about } 130 \text{ yd} \end{aligned}$$

CLASS ACTIVITY

In this activity, use a calculator or the tables on pages 562 and 563. Give answers to the nearest tenth of a unit.

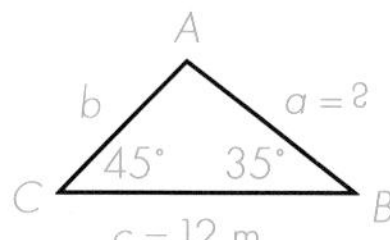

1. A saltbox house has a back and front roof constructed with the angles shown. Find the length of the longer roof. 8.6 m

2. Points A and B are at the ends of a rail tunnel. How long is the tunnel? 195.7 yd

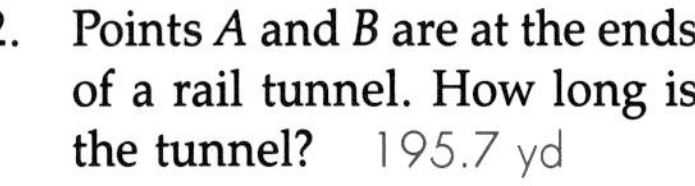

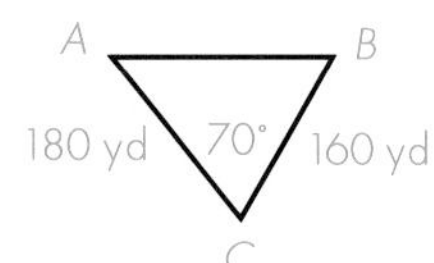

Extra Practice
A park has the shape of a regular pentagon, 90 ft on each side. From one corner there are two diagonal paths. To the nearest foot, how long is each path?

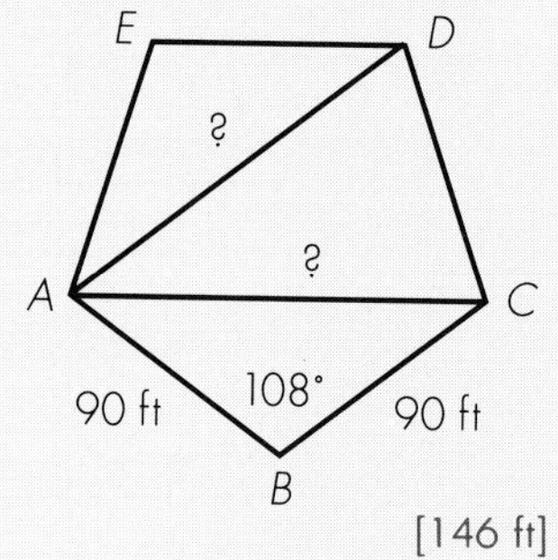

[146 ft]

FOLLOW-UP

Assessment
Find the missing lengths and angle measures.

1.

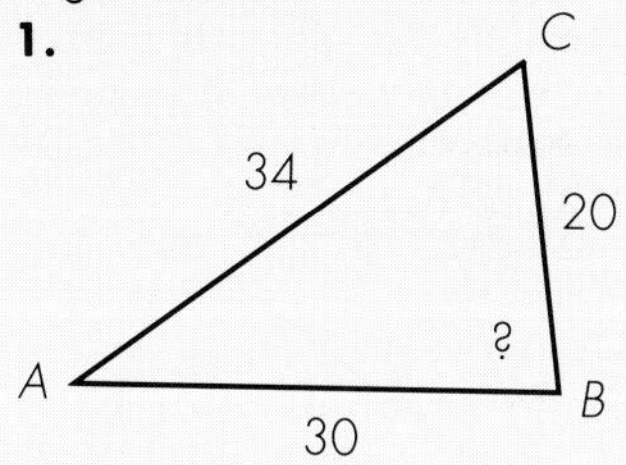

[83°]

2.

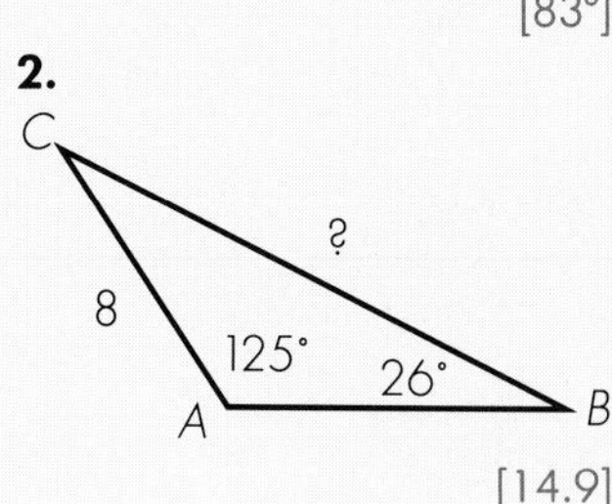

[14.9]

3.

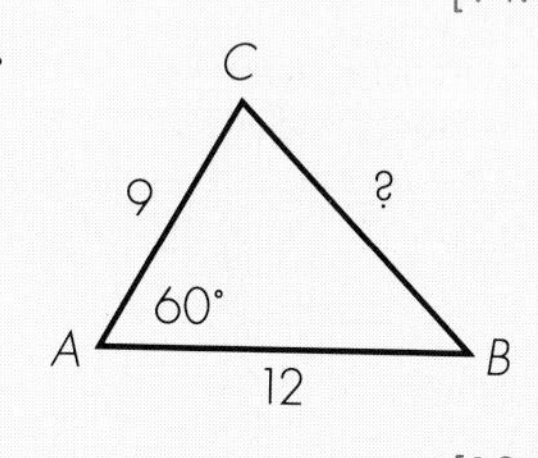

[10.8]

4. Find the angle of elevation to the top of a wall 130 ft high from a point 80 ft from the foot of the wall.

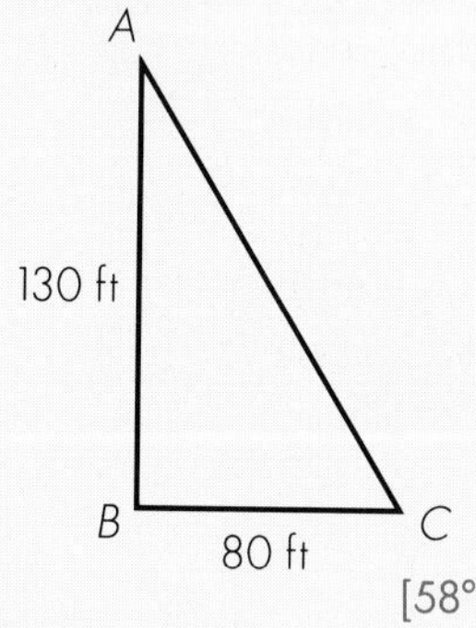

[58°]

Reteaching
Students who have had difficulty with this section may benefit from Reteaching Activity 14.5.

Enrichment
For students who have mastered the material in this section, you may wish to assign Enrichment Activity 14.5.

HOME ACTIVITY

In this activity, you may use a calculator or the tables on pages 562 and 563. In your answers, give lengths to the nearest tenth and angle measures to the nearest degree.

1. A lifeguard observes a boat that she thinks is too close to shore. How far is the boat from the base of the life guard's platform? 87.2 ft

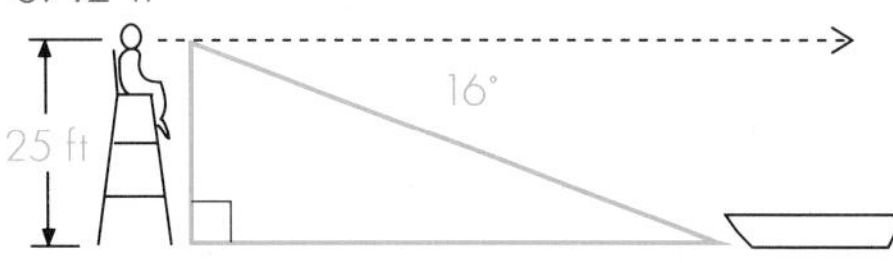

2. An airplane is climbing at an angle of 13°. The distance it has covered from A to B is 2 mi. Find its altitude. about .5 mi

3. Two ranger stations 15 mi apart notice a brush fire at the angles indicated. Which station is closer to the fire? How far away from the fire is it? station A; 6.3 mi

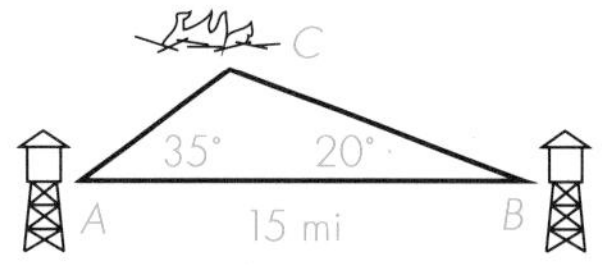

4. Two ships are located off shore at the indicated positions. Ship A is 3 mi from a lighthouse at C. How far from the lighthouse is ship B? 7.6 mi.

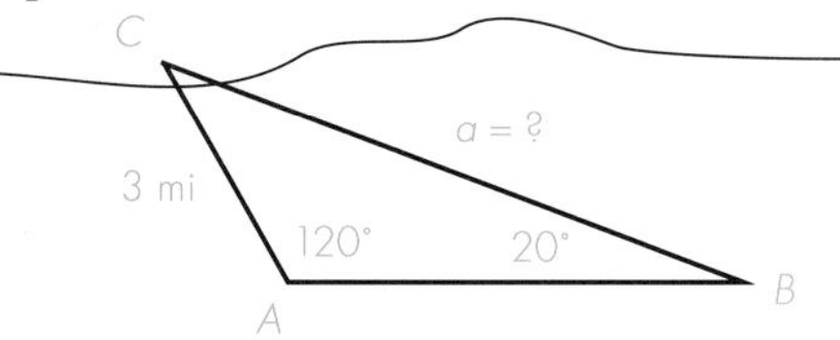

Find the missing lengths and angle measures indicated in the diagrams.

5. 25.8

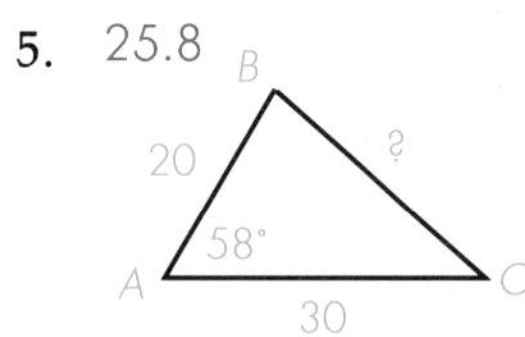

6. 20.7

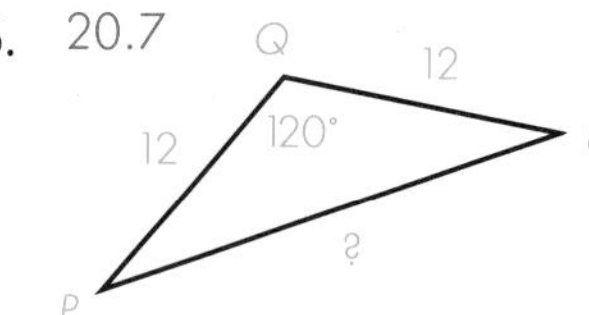

7. 131°

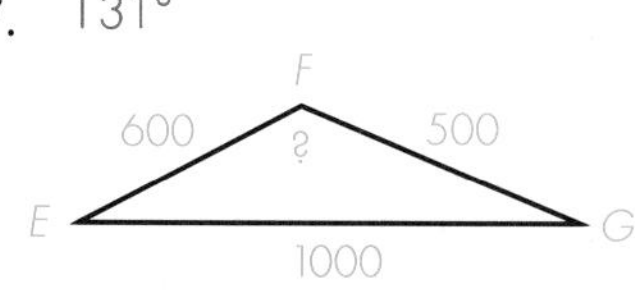

8. 23°

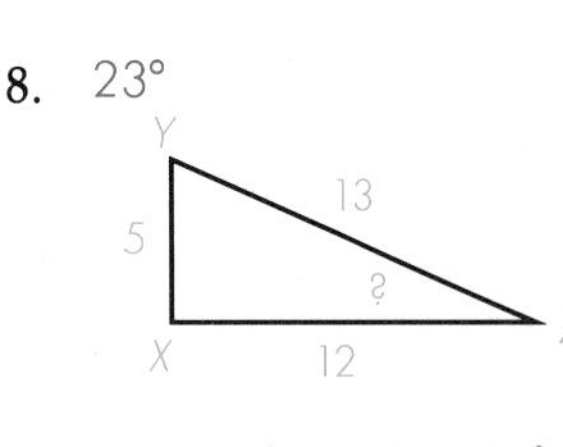

9. 82°

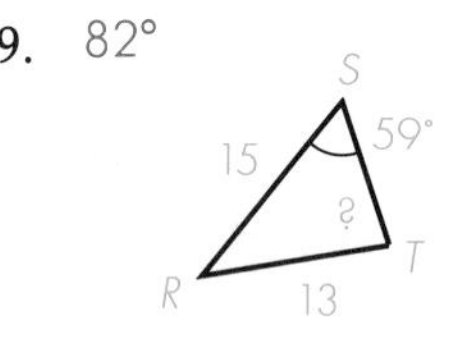

10. 3.5

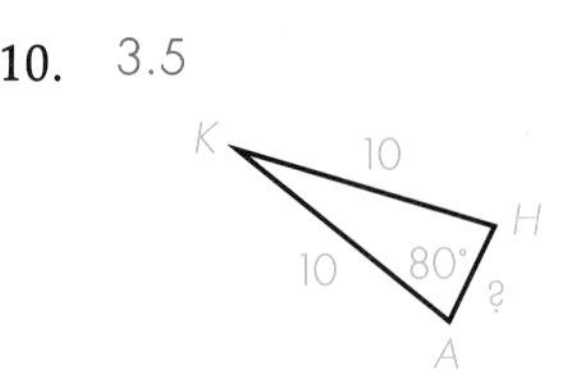

CRITICAL THINKING

11. How many triangles are in the figure at the right? 20

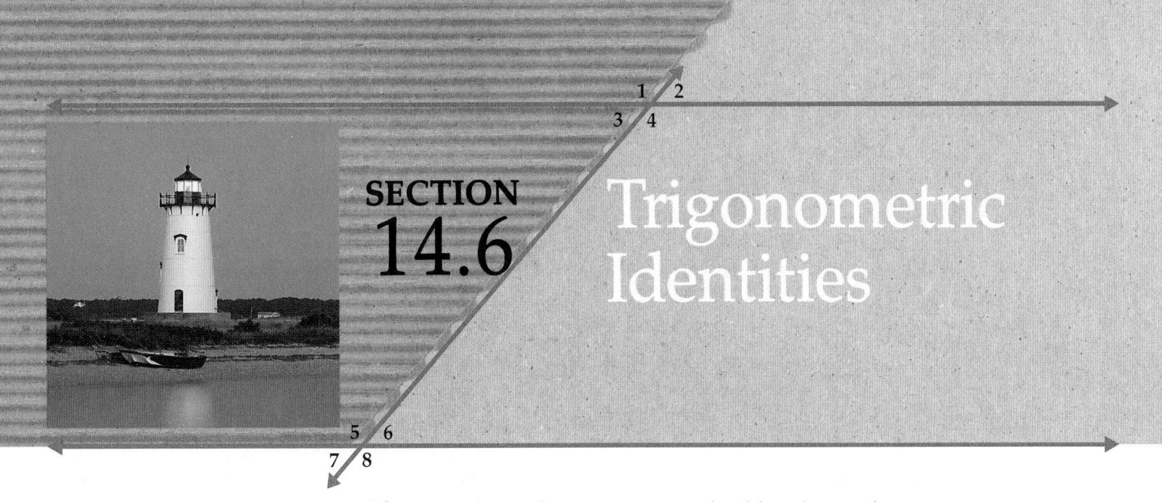

SECTION 14.6 Trigonometric Identities

After completing this section, you should understand
- ▲ some basic identities for the sine, cosine, and tangent.
- ▲ how to recognize a trigonometric identity.

In algebra there is a difference between an equation such as

$$3x + 15 = 3$$

and an equation such as

$$3x + 15 = 3(x + 5).$$

The first of these equations can be solved to obtain just one solution:

$$\begin{aligned} 3x + 15 &= 3 \\ 3x &= 3 - 15 \\ 3x &= -12 \\ x &= -4 \end{aligned}$$

The second equation, $3x + 15 = 3(x + 5)$, cannot be solved to obtain just one solution. Notice what happens after the first step (removing parentheses on the right side).

$$\begin{aligned} 3x + 15 &= 3(x + 5) \\ 3x + 15 &= 3 \cdot x + 3 \cdot 5 \end{aligned}$$

(Algebraic property: $a(b + c) = ab + ac$)

The result is the equation

$$3x + 15 = 3x + 15$$

which is true for all values of x. The equation is an *algebraic identity.*

DEFINITION 14-6-1

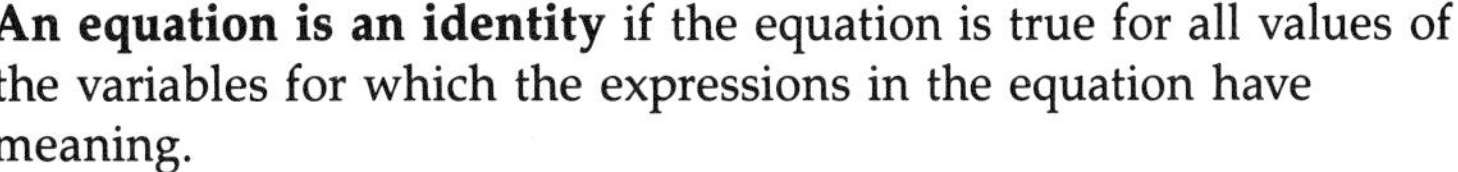

An equation is an identity if the equation is true for all values of the variables for which the expressions in the equation have meaning.

Trigonometry also has equations that are identities. These are used in advanced mathematics courses such as calculus.

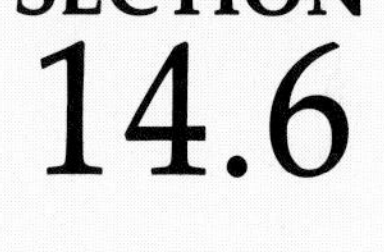

SECTION 14.6

Resources
Reteaching 14.6
Enrichment 14.6

Objectives
After completing this section, the student should understand
- ▲ some basic identities for the sine, cosine, and tangent.
- ▲ how to recognize a trigonometric identity.

Materials
scientific calculator
graphing calculator

Vocabulary
algebraic identity
identity
trigonometric identity
Pythagorean identity
quotient identity
double-angle identity

GETTING STARTED

Quiz: Prerequisite Skills

Use the definitions $\sin x = \frac{b}{r}$ and $\cos x = \frac{a}{r}$ to determine the value of each function.

1. sin 0° [0]
2. cos 0° [1]
3. sin 180° [0]
4. 2 × sin 90° [2]
5. cos 180° [−1]
6. 2 × cos 90° [0]

TEACHING COMMENTS

Using the Page
The identity $(\sin x)^2 + (\cos x)^2 = 1$ can also be illustrated by using special right triangles.

For x = 60°:

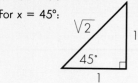

$(\sin 60°)^2 + (\cos 60°)^2 =$
$\left(\frac{\sqrt{3}}{2}\right)^2 + \left(\frac{1}{2}\right)^2 = \frac{3}{4} + \frac{1}{4} = 1$

For x = 45°:

$(\sin 45°)^2 + (\cos 45°)^2 =$
$\left(\frac{1}{\sqrt{2}}\right)^2 + \left(\frac{1}{\sqrt{2}}\right)^2 = \frac{1}{2} + \frac{1}{2} = 1$

DISCOVERY ACTIVITY

1. Copy and complete the table below. Use a scientific calculator if you can. Some entries have been filled in for you.

x	$\sin x$	$\cos x$	$(\sin x)^2$	$(\cos x)^2$
30°	0.5	0.8660	0.25	0.75
40°	0.6428	0.7660	0.4132	0.5868
70°	0.9397	0.3420	0.8830	0.1170
90°	1	0	1	0
135°	0.7071	−0.7071	0.5	0.5
155°	0.4226	−0.9063	0.1786	0.8214

2. Examine the last two numbers in each row. What pattern do you observe?
In each case $(\sin x)^2 + (\cos x)^2 = 1$.

In the Discovery Activity, it appears that the sum of the last two numbers in each row is 1. If this is true for all possible values of x, then the following equation is **trigonometric identity:**

$$(\sin x)^2 + (\cos x)^2 = 1$$

You know that this equation is true for *some* values of x, namely 30°, 40°, 70°, 90°, 135°, and 155°. Here is a proof that the equation is an identity and hence true for *all* values of x.

From the diagram at the right, you know that

$\sin x = \frac{b}{r}$ and $\cos x = \frac{a}{r}$.

So $(\sin x)^2 = \frac{b^2}{r^2}$ and $(\cos x)^2 = \frac{a^2}{r^2}$.

Next, add the last two equations.

$$(\sin x)^2 + (\cos x)^2 = \frac{b^2}{r^2} + \frac{a^2}{r^2}$$
$$= \frac{b^2 + a^2}{r^2}, \text{ or } \frac{a^2 + b^2}{r^2}$$

By the Pythagorean Theorem you know that

$$r^2 = a^2 + b^2$$

Therefore $$(\sin x)^2 + (\cos x)^2 = \frac{r^2}{r^2}$$

$$\text{or } (\sin x)^2 + (\cos x)^2 = 1.$$

The identity $(\sin x)^2 + (\cos x)^2 = 1$ is one of the basic identities of trigonometry. In order to avoid parentheses, the left side of the identity is usually written as $\sin^2 x + \cos^2 x$. Because of its similarity to the Pythagorean formula $a^2 + b^2 = c^2$, the identity $\sin^2 x + \cos^2 x = 1$ is referred to as a Pythagorean identity.

In a full course on trigonometry, you will learn more about Pythagorean identities. You will also learn about other basic kinds of identities, such as quotient identities and double-angle identities.

In the next Discovery Activity you will learn about one of the double-angle identities.

DISCOVERY ACTIVITY

1. Copy and complete the table below. Use a scientific calculator if you can. Some entries have been filled in for you.

x	$2x$	$\cos 2x$	$\cos x$	$\sin x$	$\cos^2 x$	$\sin^2 x$
30°	60°	0.5	0.8660	0.5	0.75	0.25
40°	80°	0.1736	0.7660	0.6428	0.5868	0.4132
70°	140°	−0.7660	0.3420	0.9397	0.1170	0.8830
90°	180°	−1	0	1	0	1
135°	270°	0	−0.7071	0.7071	0.5	0.5
155°	310°	0.6428	−0.9063	0.4226	0.8214	0.1786

2. Examine the "$\cos^2 x$" column and "$\sin^2 x$" column. How are they related to the "$\cos 2x$" column? Their difference is $\cos 2x$.
3. Based upon your investigation, complete the following to show an identity: $\cos 2x = \cos^2 x - \sin^2 x$

In the Discovery Activity, you may have noticed that when you subtract the number in the "$\sin^2 x$" column from the number in the "$\cos^2 x$" column, you obtain the number in the "$\cos 2x$" column. This fact can be written as

$$\cos 2x = \cos^2 x - \sin^2 x$$

This is the double-angle identity for the cosine.

You have seen examples of two kinds of basic identities:

Pythagorean identity: $\sin^2 x + \cos^2 x = 1$

Double-angle identity: $\cos 2x = \cos^2 x - \sin^2 x$

An example of a **quotient identity** is the following:

$$\tan x = \frac{\sin x}{\cos x}$$

You can use these basic identities to discover other identities.

Using the Page

If students have discussed the Critical Thinking exercise on page 535, they can prove the quotient identity,

$$\tan x = \frac{\sin x}{\cos x},$$

using $\tan x = \frac{b}{a}$, where $a \neq 0$, $\sin x = \frac{b}{r}$, and $\cos x = \frac{a}{r}$.

$$\frac{\sin x}{\cos x} = \frac{\frac{b}{r}}{\frac{a}{r}} = \frac{b}{r} \times \frac{r}{a} = \frac{b}{a} = \tan x$$

Examples: 1. Show that the following equation is an identity:

$$\cos 2x = 2\cos^2 x - 1$$

Change one side of the equation by using the basic identities you already know. Reduce one side of the equation to the other.

Since $\sin^2 x + \cos^2 x = 1$ is an identity, substitute $\sin^2 x + \cos^2 x$ for 1 in the equation.

$$\begin{aligned} \cos 2x &= 2\cos^2 x - (\sin^2 x + \cos^2 x) \\ & 2\cos^2 x - \sin^2 x - \cos^2 x \\ & 2\cos^2 x - 1\cos^2 x - \sin^2 x \\ & \cos^2 x - \sin^2 x && \text{(Collecting terms)} \\ \cos 2x &= \cos 2x && \text{(Substitute } \cos 2x \text{ for } \cos^2 x - \sin^2 x.\text{)} \end{aligned}$$

The two sides of the final equation are equal. Therefore, the original equation is an identity.

2. Show that $\sin x - \cos x = 1$ is *not* an identity. If the equation is false for any value of x, then it is not an identity. Try $x = 0$.

$$\sin 0° - \cos 0° = 0 - 1$$

The statement "$0 - 1 = 1$" is false. Thus, the equation is not an identity.

PROJECT 14-6-1 You can use a graphing calculator to help you determine whether an equation may be an identity. For example, to see whether $\sin(90° - x) = \cos x$ might be an identity, you can use a graphing calculator to graph these two equations on the same set of axes.

$$y = \sin(90° - x) \text{ and } y = \cos x$$

If the graphs displayed are the same, then the original equation *may* be an identity. If the graphs are different, you can be sure that the original equation is *not* an identity.

Investigate the following equations. Use a graphing calculator to decide which equations may be identities and which are definitely not identities.

1. $\sin(90° - x) = \cos x$
2. $1/\sin x = \sin x \tan x$
3. $\tan 2x = \frac{2\tan x}{1 - \tan^2 x}$
4. $\tan^2 x - \sin^2 x = \tan^2 x \sin^2 x$
5. $\sin x \cos x \tan x = 1$
6. $\sin(x - 90°) = \cos x$

The basic identities of this section can be stated as theorems.

Extra Practice

Prove that the following equations are identities.

1. $\frac{\sin x + \cos x}{\cos x} = \tan x + 1$

$$\begin{aligned} \tan x + 1 &= \frac{\sin x}{\cos x} + 1 \\ &= \frac{\sin x}{\cos x} + \frac{\cos x}{\cos x} \\ &= \frac{\sin x + \cos x}{\cos x} \end{aligned}$$

2. $\frac{\tan x}{\cos x} = \frac{\sin x}{1 - \sin^2 x}$

$$\begin{aligned} \frac{\sin x}{1 - \sin^2 x} &= \frac{\sin x}{\sin^2 x + \cos^2 x - \sin^2 x} \\ &= \frac{\sin x}{\cos^2 x} = \frac{\sin x}{\cos x} \cdot \frac{1}{\cos x} \\ &= \tan x \cdot \frac{1}{\cos x} = \frac{\tan x}{\cos x} \end{aligned}$$

Prove that the following equation is not an identity.

3. $\sin x + \cos x = 1$

If $x = 45$, then $\sin 45° + \cos 45° = 0.7071 + 0.7071 \neq 1$. (Other counterexamples are possible.)

Using the Page

In the Project, only the equations for Exercises 1 and 4 are identities. Note that the equation for Exercise 3 would be an identity if the -1 in the denominator were replaced by 1.

Additional Answers

(for page 551)

1. If $x = 45°$, then $\tan 45° = 1$, $\sin 45° = 0.7071$. "$1 = 0.7071 + 1$" is false.

2. $= \cos x \cdot \tan x$

$\sin x = \cos x \cdot \tan x$

$\cos x \cdot \frac{\sin x}{\cos x}$

$\sin x = \sin x$

NOTEBOOK **THEOREM 14-6-1** **If x is the measure of any angle, then $\sin^2 x + \cos^2 x = 1$.**

NOTEBOOK **THEOREM 14-6-2** **If x is the measure of any angle, then $\cos 2x = \cos^2 x - \sin^2 x$.**

NOTEBOOK **THEOREM 14-6-3** **If x is the measure of any angle such that $\cos x \neq 0$, then**

$$\tan x = \frac{\sin x}{\cos x}.$$

CLASS ACTIVITY

If necessary, use a calculator for Exercise 1.

1. Show that the following equation is not an identity: $\tan x = \sin x + 1$ Methods will vary.
2. Show that the following equation is an identity: $\sin x = \cos x \cdot \tan x$ See margin.

HOME ACTIVITY

In the Home Activity exercises, you may use a scientific calculator.

Show that the following equations are *not* identities. Methods will vary.

1. $\sin x + \cos x = 0$
2. $\sin x \cdot \cos x = 1$
3. $\tan x \cdot \sin x = \cos x$
4. $\sin 2x = 2 \sin x$
5. $\sin x = x$
6. $\tan^2 x = \tan x^2$

Complete the missing steps in the following proofs of identities.

7. $\tan^2 x = \frac{1 - \cos^2 x}{\cos^2 x}$

$\frac{?}{?}$ $\frac{(\sin^2 x + \cos^2 x) - \cos^2 x}{\cos^2 x}$, $\frac{\sin^2 x + 0}{\cos^2 x}$

$\frac{\sin^2 x}{\cos^2 x}$

$\tan^2 x = \tan^2 x$

8. $\sin^2 x - \cos^2 x = 1 - 2\cos^2 x$

$(\cos^2 x + \sin^2 x) - 2\cos^2 x$

$\sin^2 x + \cos^2 x - 2\cos^2 x =$?

$\sin^2 x - \cos^2 x =$? $\sin^2 x - \cos^2 x$

Prove the following identities. See margin.

9. $\frac{\sin^2 x + \cos^2 x}{\cos x} = \frac{\tan x}{\sin x}$
10. $\cos 2x = 1 - 2\sin^2 x$

CRITICAL THINKING

11. Prove the following identity: $\cos 2x = \cos^4 x - \sin^4 x$. See margin.

Additional Answers

Counterexamples for Exercises 1 through 6 may vary. Here are some possible answers:

1. If $x = 0$, $\sin 0° + \cos 0° = 0 + 1 \neq 0$.

2. If $x = 0$, $\sin 0° \times \cos 0° = 0 \times 1 \neq 1$.

3. If $x = 0$, $\tan 0° \times \sin 0° = 0 \cdot 0 = 0$; $\cos 0° = 1$; $0 \neq 1$.

4. If $x = 90$, $\sin 2(90°) = \sin 180° = 0$; $2 \sin 90° = 2 \times 1 = 2$; $0 \neq 2$.

5. If $x = 90$, $\sin 90° = 1 \neq 90$.

6. If $x = 10$, $\tan^2 10° = 0.0311$; $\tan (10^2)° = \tan 100° = -5.671$; $0.311 \neq -5.671$.

9. $\frac{\sin^2 x + \cos^2 x}{\cos x} =$

$\frac{1}{\cos x} \cdot \frac{\sin x}{\sin x} =$

$\frac{\sin x}{\cos x} \cdot \frac{1}{\sin x} =$

$\tan x \cdot \frac{1}{\sin x} = \frac{\tan x}{\sin x}$

10. $1 - 2\sin^2 x = \cos^2 x + \sin^2 x - 2\sin^2 x = \cos^2 x - \sin^2 x = \cos 2x$.

11. $\cos 2x = (\cos^2 x - \sin^2 x) \cdot 1 = (\cos^2 x - \sin^2 x) \cdot (\cos^2 x + \sin^2 x) = \cos^4 x - \sin^4 x$

FOLLOW-UP

Assessment

Show that $\frac{\tan x}{\sin x} = \frac{1}{\cos x}$ is an identity.

$\frac{\tan x}{\sin x} = \tan x \cdot \frac{1}{\sin x} =$

$\frac{\sin x}{\cos x} \cdot \frac{1}{\sin x} = \frac{1}{\cos x}$

Reteaching

Students who have had difficulty with this section may benefit from Reteaching Activity 14.6.

Enrichment

For students who have mastered the material in this section, you may wish to assign Enrichment Activity 14.6.

Chapter Vocabulary
algebraic identity
angle of depression
angle of elevation
angle of rotation
clockwise rotation
cosine function
cosine ratio
counterclockwise rotation
double-angle identity
function
identity
initial side
Law of Cosines
Law of Sines
line of sight
negative angle
positive angle
Pythagorean identity
quotient identity
sine function
sine ratio
tangent ratio
terminal side
trigonometric identity
trigonometric ratios

14.1 Refer to the figure. The sine and cosine of a positive acute angle are defined by the following equations:

$\sin A = \frac{\text{opp}}{\text{hyp}}$ $\cos A = \frac{\text{adj}}{\text{hyp}}$

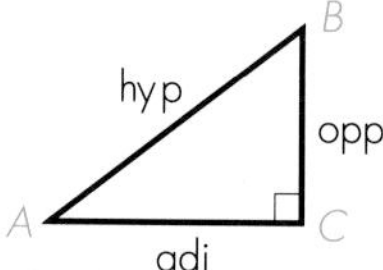

If $AC = 4$, $BC = 3$, and $AB = 5$,

then $\sin A = \frac{3}{5}$ and $\cos A = \frac{4}{5}$.

14.2 Refer to the figure. The tangent of a positive acute angle is defined by the following equation:

$\tan A = \frac{\text{opp}}{\text{adj}}$

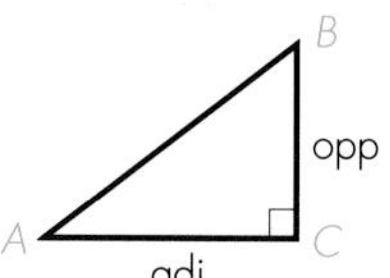

If $AC = 4$ and $BC = 3$,

then $\tan A = \frac{3}{4}$.

14.3 Refer to the figure. The sine and cosine of an angle of rotation are defined by the following equations:

$\sin x° = \frac{b}{r}$ $\cos x° = \frac{a}{r}$

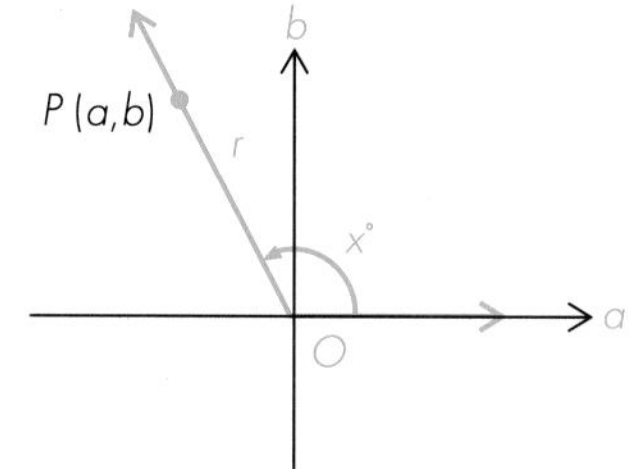

For example, if $x° = 120°$ and $P(a, b) = (-1, \sqrt{3})$, then $\sin 120° = \frac{\sqrt{3}}{2}$ and $\cos 120° = -\frac{1}{2}$.

The figure below is the graph of the sine function for values of x such that $0° \leq x° \leq 360°$.

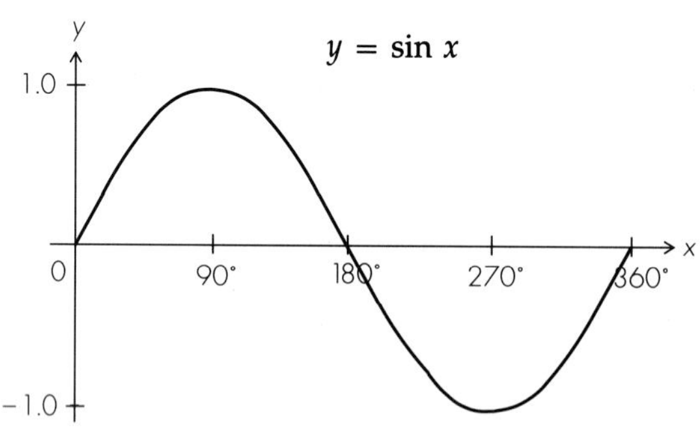

552

14.4 To solve a non-right triangle where the given information has the pattern *ASA* or *AAS*, use the Law of Sines.

$\frac{\text{sine A}}{a} = \frac{\text{sine B}}{b} = \frac{\text{sine C}}{c}$

For $\triangle ABC$, you can find b by solving $\frac{\sin 25°}{b} = \frac{\sin 45°}{12}$.

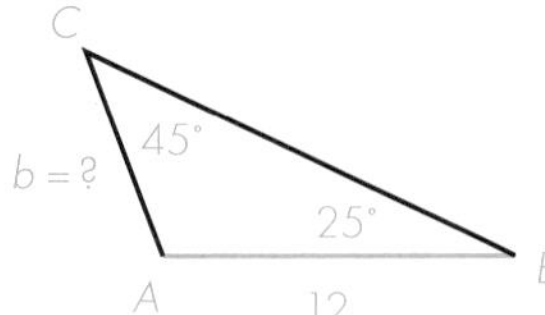

Refer to the figure. To solve a non-right triangle where the given information has the pattern *SAS* or *SSS*, use the Law of Cosines.

$a^2 = b^2 + c^2 - 2bc \cos A$

$b^2 = a^2 + c^2 - 2ac \cos B$

$c^2 = a^2 + b^2 - 2ab \cos C$

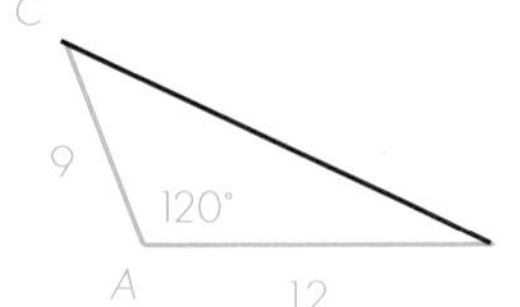

In $\triangle ABC$, to find BC you solve

$a^2 = 9^2 + 12^2 - 2(9)(12) \cos 120°$

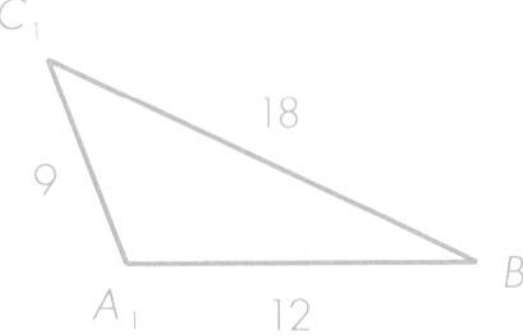

In $\triangle A_1B_1C_1$ to find $m\ A$, use

$18^2 = 9^2 + 12^2 - 2(9)(12)\cos A$.

14.5 Refer to the figure. Angles of elevation and depression are used with the trigonometric ratios to measure distances indirectly.

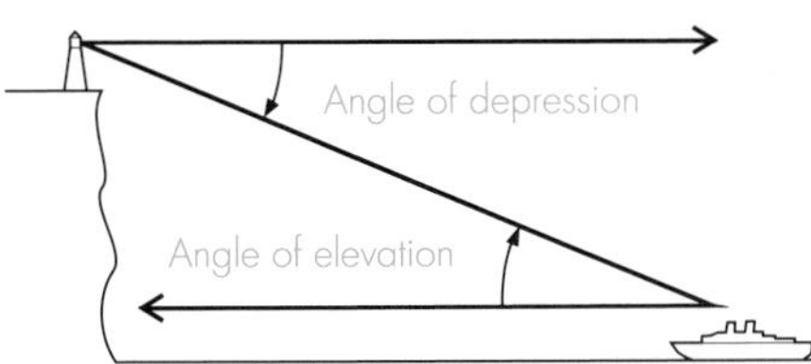

14.6 A trigonometric identity is an equation that is true for all values of the variables for which the expressions in the equation have meaning. Examples of trigonometric identities are:

$\sin^2 x + \cos^2 x = 1$ and $\frac{\sin x}{\cos x} = \tan x$.

Using the Page
In this Calculator Activity, students will discover that trigonometric functions are periodic. This feature gives trigonometric functions many practical applications, since natural events often occur at regular intervals.

To help students visualize why sin $(-50)°$ and sin $(-410)°$ are the same, for example, you may wish to rotate a ruler about the origin of the *a*-*b* coordinate plane, stopping at $-50°$ and $-410°$, so that students can see the coincidence of the terminal sides for these rotations. Students can also see that the rotations in either direction continue indefinitely.

The graphs of $y = \sin x°$ and $y = \cos x°$ were introduced on page 534 and in Exercise 17, page 535. As a follow-up to this activity, you may wish to have students extend their graphs beyond the interval $0° < x° < 360°$, as indicated below.

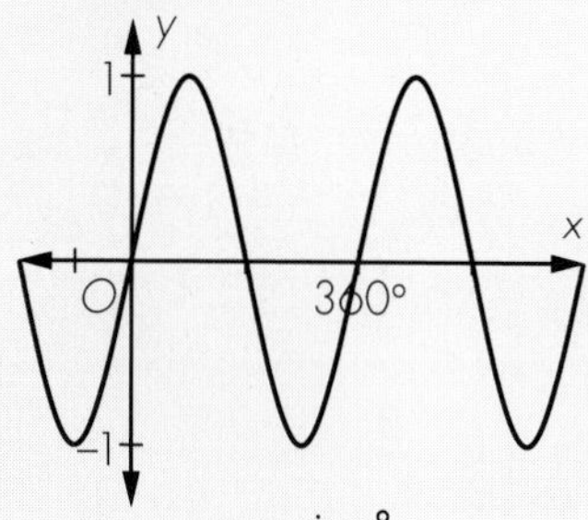

$y = \sin x°$

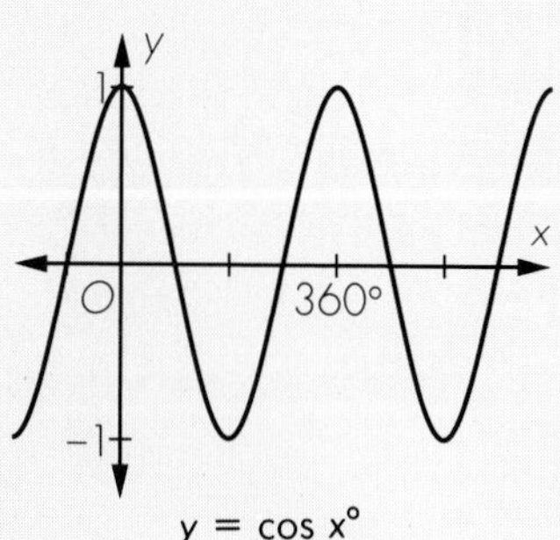

$y = \cos x°$

CALCULATOR ACTIVITY

Recall that to use a calculator to find the sine of an angle, you must first enter the degree measure of the angle, press the [+/−] key if the degree measure of the angle is negative, and then press the [SIN] key. The same procedure can be followed to find the cosine of an angle if you press the [COS] key instead of the [SIN] key.

1. Use a calculator. Copy and complete the chart below.

x	$\sin x$	$\cos x$
65	0.90630779	0.42261826
425	0.90630779	0.42261826
90	1	0
450	1	0
235	−0.81915204	−0.57357644
595	−0.81915204	−0.57357644
280	−0.98480775	0.17364818
1000	−0.98480775	0.17364818
120	0.86602541	−0.5
1200	0.86602541	−0.5
−50	−0.76604444	0.64278761
−410	−0.76604444	0.64278761
−155	−0.42261826	−0.90630779
−875	−0.42261826	−0.90630779
−140	−0.64278761	−0.76604445
−2300	−0.64278760	−0.76604445

2. What do you notice about the values of the sine function?
There are pairs of values that are the same.
3. Use your calculator to subtract the angle measures that produce the same values for the sine function. What do you notice about the differences? The difference is 360 or a multiple of 360.
4. What do you notice about the values of the cosine function?
There are pairs of values that are the same.
5. Use your calculator to subtract the angle measures that produce the same values for the cosine function. What do you notice about the differences? The difference is 360 or a multiple of 360.

You have learned that there are 360° in a complete revolution of the initial side of an angle of rotation. In Exercises 1 through 5, you discovered that the sine and cosine functions repeat their values every 360°. Thus, these functions are said to be **periodic,** with a period of 360°.

6. Use a calculator. Copy and complete the chart below.

x	$\tan x$
65	2.1445069
245	2.1445069
110	−2.7474774
290	−2.7474774
90	Error
450	Error
85	11.430052
625	11.430052
135	−1
1035	−1
−155	0.46630766
−335	0.46630766
−50	−1.1917536
−410	−1.1917536
−140	0.83909963
−680	0.83909963

7. What do you notice about the values of the tangent function?
There are pairs of values that are the same.

8. Use your calculator to subtract the angle measures that produce the same values for the tangent function. What do you notice about the differences? The difference is 180 or a multiple of 180.

9. Is the tangent a periodic function? If it is periodic, what is the period? Yes; 180°

Assessment Resources
Achievement Tests pp. 27-28.

Using the Page
Students can work individually or in groups to study the Chapter 14 Review. If you have students study in groups, they should work together to make sure all members of the group know the material in the chapter.

Informal assessment such as interviews, classroom observations, or a review of student portfolios can be used instead of paper-and-pencil tests.

Additional Answers
Counterexamples for Exercises 9 and 10 may vary. Here are possible answers.

9. If $x = 90°$, then $\sin x = 1$ and $\cos x = 0$. So, $\sin^2 x + 1 = 2$ and $\cos x = 0$. Hence for $x = 90°$, $\sin^2 x + 1 \neq \cos x$.
10. If $x = 30°$, then $\cos x = \frac{\sqrt{3}}{2}$, $\sin x = \frac{1}{2}$, and $\tan x = \frac{\sqrt{3}}{3}$. However, $\frac{\cos x}{\sin x} = \sqrt{3}$, and $\tan x = \frac{\sqrt{3}}{3}$. So, for $x = 30°$, $\frac{\cos x}{\sin x} \neq \tan x$.

Use a calculator or the table on page 563.

In Exercises 1 through 4, express answers to the nearest ten-thousandth. Find sin *A*, cos *A*, and tan *A*.

1. 0.4706; 0.8824; 0.5333

2. 0.5446; 0.8387; 0.6494

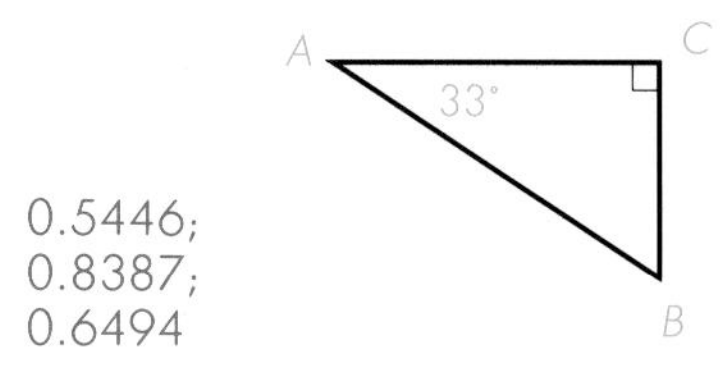

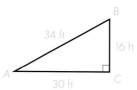

Find the sine and cosine of each angle.

3. $\sin x = \frac{3}{9}$ $= 0.600$ $\cos x = \frac{-4}{5} = -0.8000$

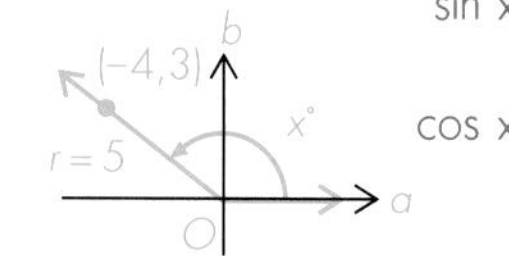

4. $\sin x = \frac{-5}{13}$ $= -0.3846$ $\cos x = \frac{-12}{3} = -0.9231$

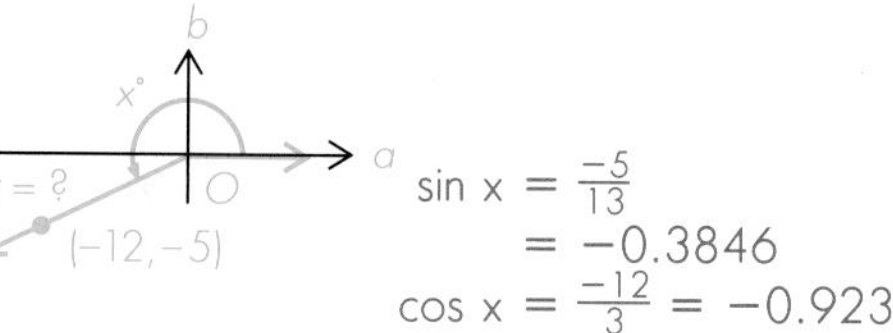

In Exercises 5 through 8, express lengths to the nearest tenth and angle measures to the nearest whole degree. Find each indicated measure.

5. 9.8 m

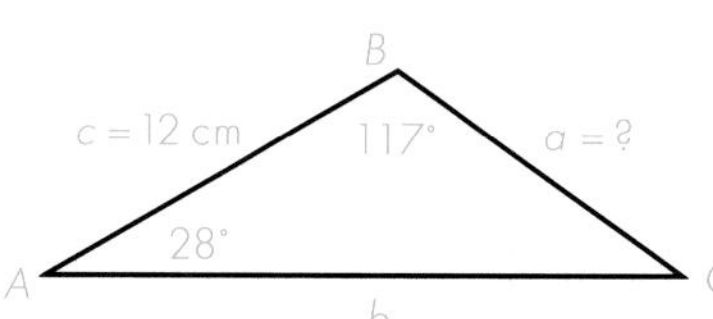

6. 16.3 in.

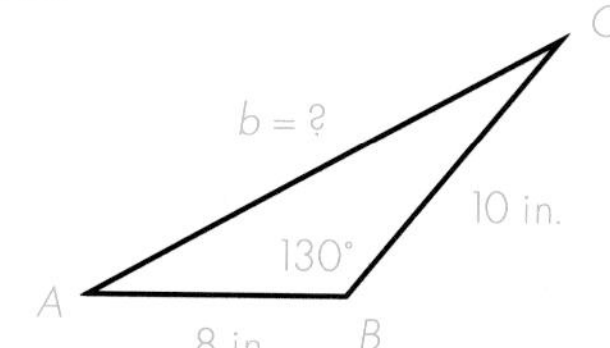

7. Two boats are positioned as shown with respect to a buoy. The boat at *A* is 100 m from the buoy. How far apart are the two boats? 199.2 m

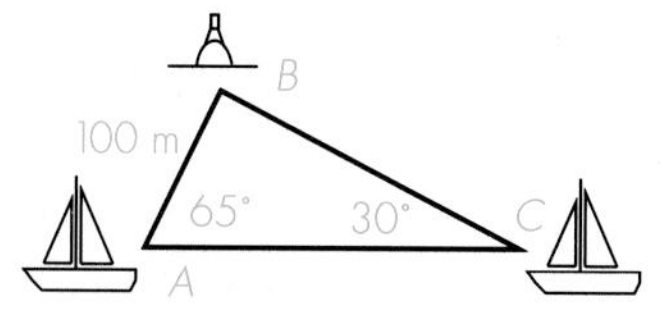

8. The lengths of two adjacent sides of parallelogram *ABCD* are 6 and 10 and the measure of the angle between them is 75°. Find the length of the longer diagonal and the measure of the angle formed by that diagonal and $\overline{AB}$. 12.9; 27°

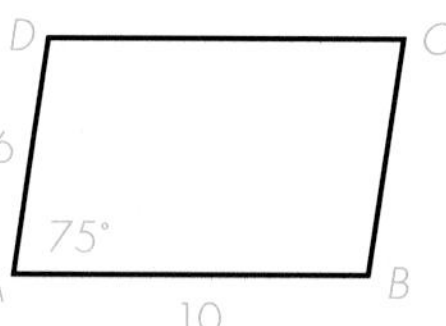

Show that the equations are not identities. See margin. Methods will vary.

9. $\sin^2 x + 1 = \cos x$

10. $\frac{\cos x}{\sin x} = \tan x$

CHAPTER TEST

Test Objectives

After studying this chapter, students should understand
- how to find sine ratios.
- how to find cosine ratios.
- how to find the tangent ratio.
- how to use trigonometry to solve triangle problems.
- how the sine and cosine functions can be extended.
- how to use the Law of Sines.
- how to use the Law of Cosines.
- how to use trigonometric functions to solve practical problems.
- some basic identities for the sine, cosine, and tangent.
- how to recognize a trigonometric identity.

Use a calculator or the table on page 563.

Find sin *A*, cos *A*, and tan *A*.

1.

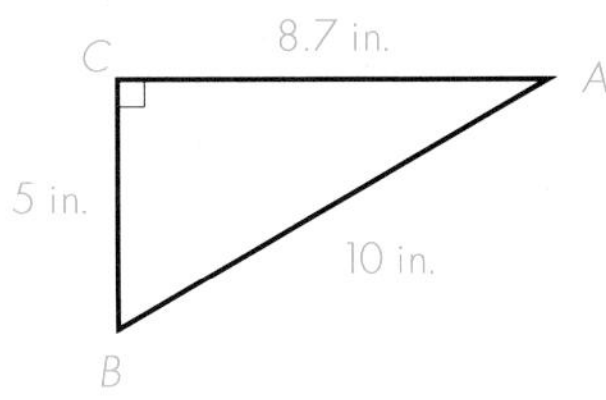

0.5; 0.87; 0.57

2.

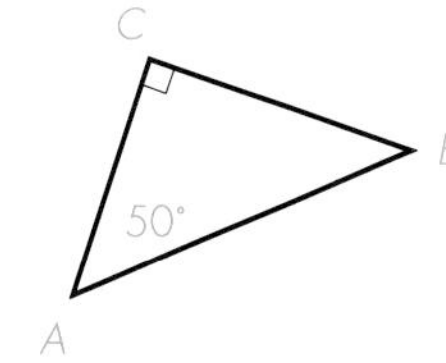

0.7660; 0.6428; 1.1918

Find the sine and cosine of each angle.

3. sin x = $\frac{8}{17}$, or 0.4706
cos x = $\frac{15}{17}$, or 0.8824

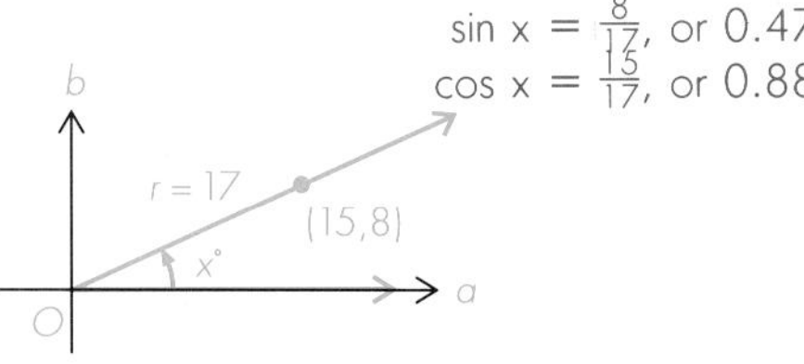

4.

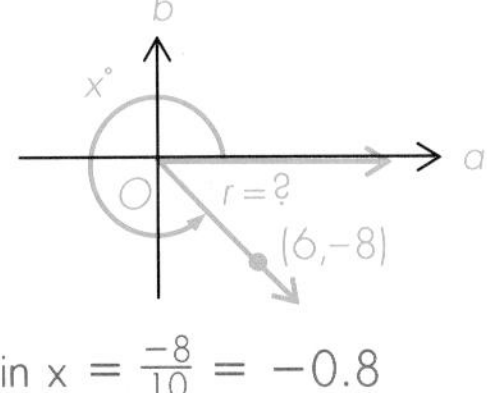

sin x = $\frac{-8}{10}$ = −0.8
cos x = $\frac{6}{10}$ = 0.6

Find each indicated measure. Answer to the nearest tenth.

5. a = 51.5 cm

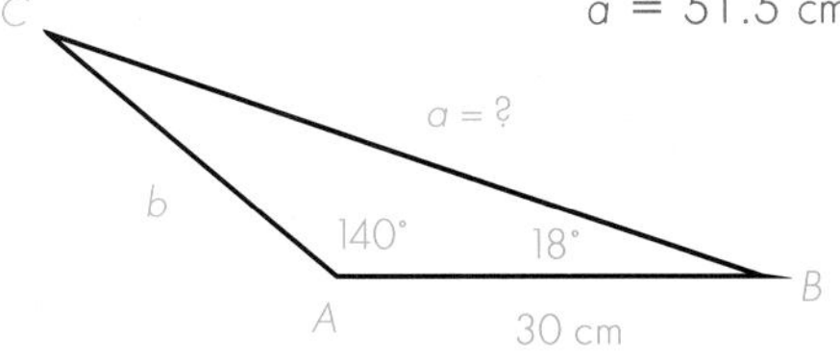

6. b = 3.9 ft

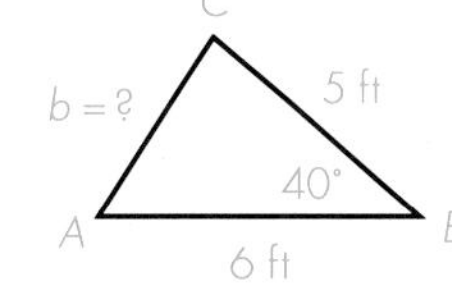

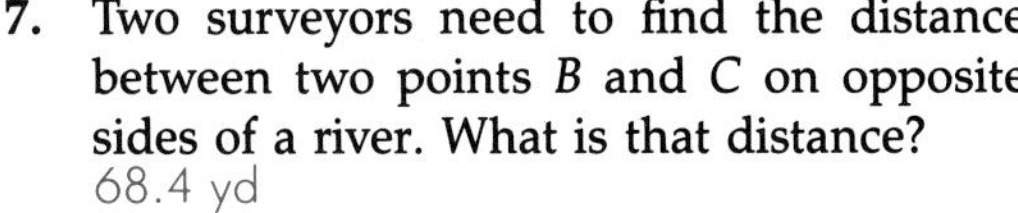

7. Two surveyors need to find the distance between two points *B* and *C* on opposite sides of a river. What is that distance?
68.4 yd

8. The lengths of two adjacent sides of parallelogram *ABCD* are 7 and 11. The measure of the angle between the sides is 55°. Find the length of the shorter diagonal. 9.0

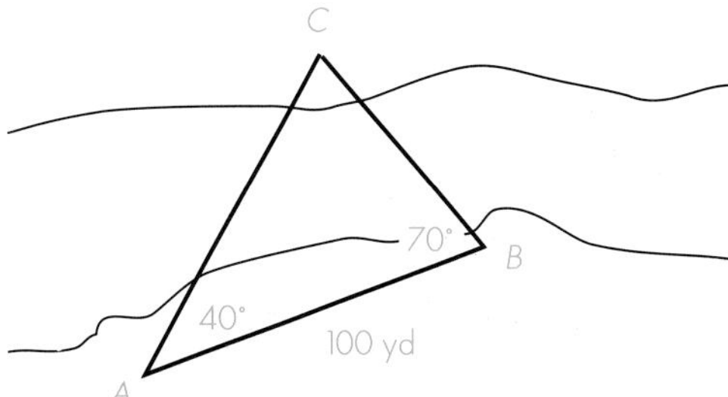

Tell whether each equation is an identity. If not, give an example to support your answer.

9. $\sin^2 x = 1 + \cos^2 x$
If x = 0, sin x = 0, cos x = 1. "$0^2 = 1 + 1^2$" is false. The equation is not an identity.

10. $\cos^2 x - \sin^2 x = \cos 2x$ Yes

Name ______________________ Date ____________

The Sine and Cosine Ratios

Two commonly used trigonometric ratios are the **sine ratio** and the **cosine ratio.** These ratios are defined in terms of the lengths of the sides of a right triangle. Note that "sine of angle D" is usually written as sin D, and "cosine of angle D" is usually written as cos D.

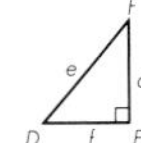

$$\sin D = \frac{\text{length of side opposite } \angle D}{\text{length of hypotenuse}} = \frac{d}{e}$$

$$\cos D = \frac{\text{length of side adjacent to } \angle D}{\text{length of hypotenuse}} = \frac{f}{e}$$

For right triangle RST, find sin R and cos R.

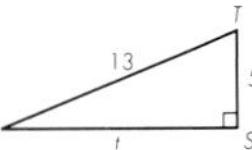

$\sin R = \frac{5}{13} \approx 0.3846$

Since $\triangle RST$ is a right triangle,
$t^2 + 5^2 = 13^2$
$t^2 + 25 = 169$
$t^2 = 144$
$t = 12$

So, $\cos R = \frac{12}{13} \approx 0.9231$.

Find the decimal value of each ratio. Round to the nearest ten-thousandth, if necessary.

1.

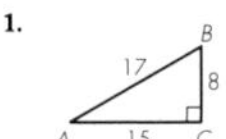
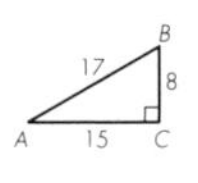
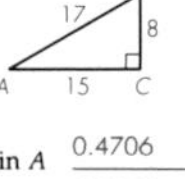

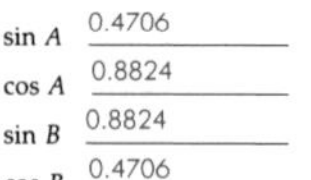

sin A 0.4706
cos A 0.8824
sin B 0.8824
cos B 0.4706

2.

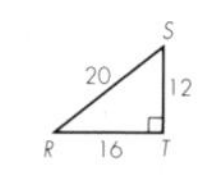

sin R 0.6
cos R 0.8
sin S 0.8
cos S 0.6

3.

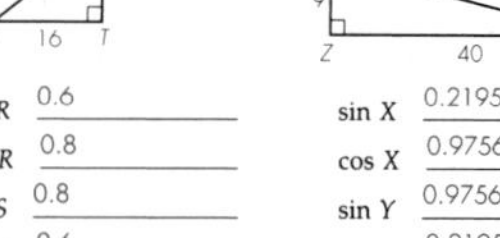

sin X 0.2195
cos X 0.9756
sin Y 0.9756
cos Y 0.2195

Use a table of trigonometric ratios to find each ratio.

4. sin 28° 0.4695 **5.** sin 73° 0.9563 **6.** sin 2° 0.0349

7. cos 43° 0.7314 **8.** cos 71° 0.3256 **9.** cos 89° 0.0175

Name ______________________ Date ____________

Triangle Areas Using Trigonometry

To find the area of a triangle, you can use the formula $A = \frac{1}{2}bh$, where b and h are the lengths of the base and height.

You also know that $\sin A = \frac{h}{c}$.

Therefore, $c \sin A = h$ when you multiply both sides by c.

Substituting for h in the area formula, you have

area $\triangle ABC = \frac{1}{2}bc \sin A$

In a similar way, $\sin C = \frac{h}{a}$.

So, $a \sin C = h$.

Again by substituting in the area formula, you get

area $\triangle ABC = \frac{1}{2}ab \sin C$

It can also be shown that area $\triangle ABC = \frac{1}{2}ac \sin B$.

> The area of any triangle ABC is given by any one of these formulas:
>
> area $= \frac{1}{2}bc \sin A$ area $= \frac{1}{2}ab \sin C$ area $= \frac{1}{2}ac \sin B$

Find the area of each triangle to the nearest hundredth. Use the table of trigonometric values on page 563.

1.

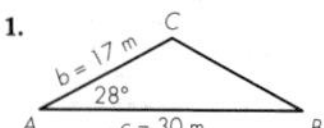

119.72 m²

2.

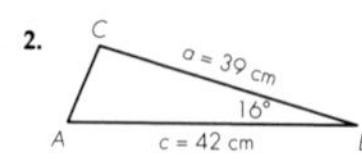

225.72 cm²

3.

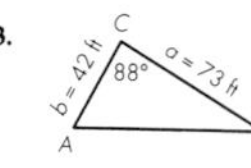

1532.08 ft²

4.

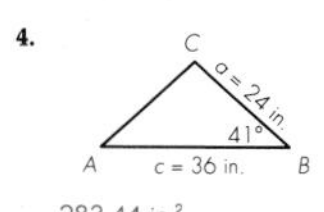

283.44 in.²

5.

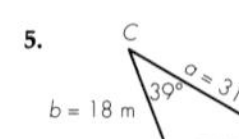

175.57 m²

6.

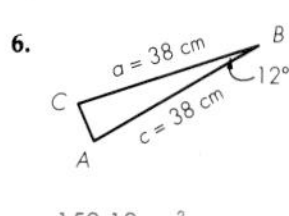

150.10 cm²

Name ______________________ Date ______________

The Tangent Ratio

The tangent of an acute angle of a right triangle is the ratio of the length of the side opposite the angle divided by the length of the leg adjacent to the angle. "Tangent of angle *A*" is usually written as tan *A*.

$$\tan A = \frac{\text{length of leg opposite } \angle A}{\text{length of leg adjacent } \angle A} = \frac{a}{b}$$

The tangent ratio is often useful in finding unknown lengths of a triangle. In right triangle *XYZ*, find *x*.

$\tan 23° = \frac{x}{17}$

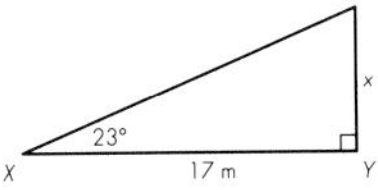

From a table of trigonometric values, tan 23° = 0.4245.

So, $0.4245 = \frac{x}{17}$.
$17(0.4245) = x$
$7.2165 = x$

So *x* = 7.2 m, to the nearest tenth of a meter.

Refer to the table on page 563, when necessary.

1. Find tan *S*. 0.85
2. Find tan *N*. 0.7536
3. Find *m*∠*A*. about 41°

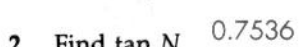

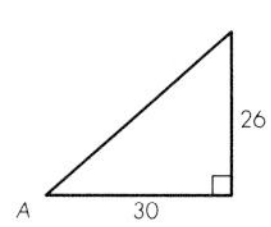

4. Find *m*∠*B*. about 32°
5. Find *r*. about 28.2 ft
6. Find *s*. about 38.6 cm

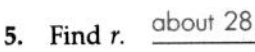

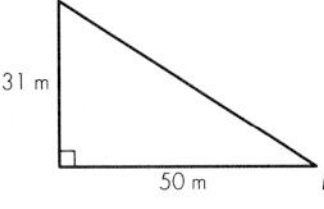

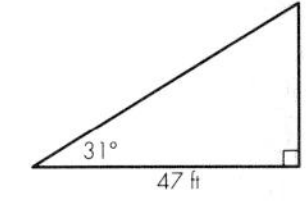

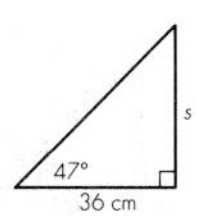

Name ______________________ Date ______________

Three More Trigonometric Ratios

Besides the sine, cosine, and tangent ratios, there are three more trigonometric ratios involving the lengths of the sides of a right triangle. These ratios are called the **secant,** the **cosecant,** and the **cotangent.** The ratios are abbreviated as *sec, csc,* and *cot,* respectively.

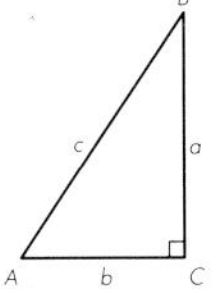

$$\sec A = \frac{\text{length of hypotenuse}}{\text{length of side adjacent to } \angle A} = \frac{c}{b}$$

$$\csc A = \frac{\text{length of hypotenuse}}{\text{length of side opposite } \angle A} = \frac{c}{a}$$

$$\cot A = \frac{\text{length of side adjacent to } \angle A}{\text{length of side opposite } \angle A} = \frac{b}{a}$$

Write each ratio in simplest form.

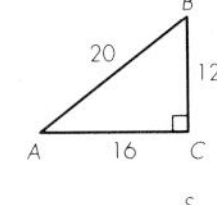

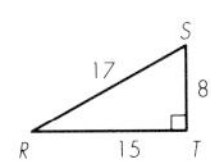

1. sec *A* $\frac{5}{4}$
2. csc *A* $\frac{5}{3}$
3. cot *A* $\frac{4}{3}$
4. sec *B* $\frac{5}{3}$
5. csc *B* $\frac{5}{4}$
6. cot *B* $\frac{3}{4}$
7. sec *R* $\frac{17}{15}$
8. csc *R* $\frac{17}{8}$
9. cot *R* $\frac{15}{8}$
10. sec *S* $\frac{17}{8}$
11. csc *S* $\frac{17}{15}$
12. cot *S* $\frac{8}{15}$

Two nonzero numbers whose product is 1 are called reciprocals.

13. What is the reciprocal of the cotangent ratio? tangent
14. What is the reciprocal of the secant ratio? cosine
15. What is the reciprocal of the cosecant ratio? sine

In right triangle *DEF*, name the trigonometric ratio represented for the given angle.

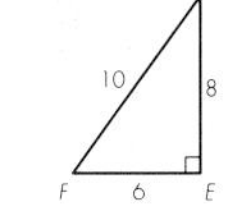

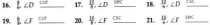

16. $\frac{8}{6}$, ∠*D* cot
17. $\frac{10}{8}$, ∠*D* sec
18. $\frac{10}{6}$, ∠*D* csc
19. $\frac{6}{8}$, ∠*F* cot
20. $\frac{10}{8}$, ∠*F* csc
21. $\frac{10}{6}$, ∠*F* sec
22. For right triangle *DEF*, show that $1 + \tan^2 D = \sec^2 D$.

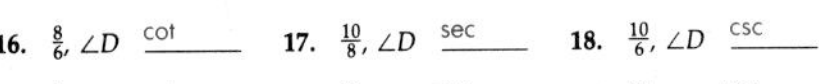

$\tan D = \frac{6}{8}$; $1 + \left(\frac{6}{8}\right)^2 = 1 + \frac{36}{64} = \frac{64 + 36}{64} = \frac{100}{64} = \left(\frac{10}{8}\right)^2 = \sec^2 D$

Name ______________________ Date ______________

The Sine Function

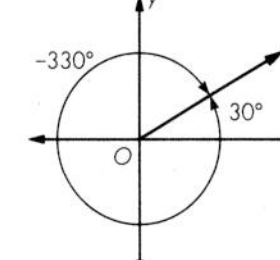

An angle of rotation is formed by two rays with a common vertex O, where O is the origin of the coordinate plane. The initial side lies on the positive x-axis. The second ray is the terminal side and may terminate in any of the four quadrants or along either axis.

When the initial side is rotated in a counterclockwise direction, the angle measure is positive. When the initial side is rotated in a clockwise direction, the angle measure is negative. There are 360° in one complete revolution.

This definition for an angle of rotation allows you to extend the definition for the sine ratio to any angle, not just any acute angle. If $P(a,b)$ is a point on the terminal side of an angle of rotation of $x°$, then $\sin x = \frac{b}{r}$, where r is the distance of P from the origin.

Find a second measure for each angle. Answers will vary; possible answer given.

1.

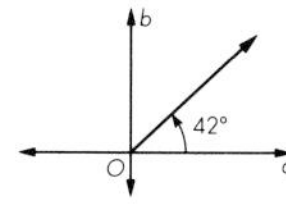

−318°

2. 160°

3.

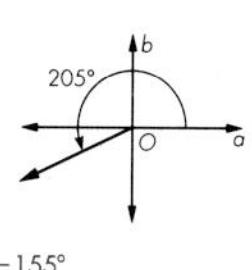

−155°

Find the sine of each angle.

4.

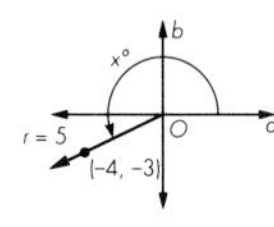

−0.6

5. 0.7813

6.

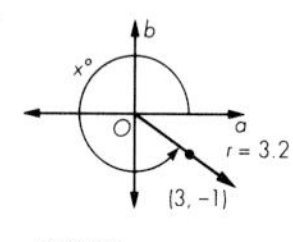

−0.3125

7. $x = 225°$ $\sin x =$ −0.7071

8. $x = -150°$ $\sin x =$ −0.5

9. $x = 315°$ $\sin x =$ −0.7071

Name ______________________ Date ______________

The Unit Circle

A circle with its center at the origin of the coordinate plane and with a radius of 1 is called a *unit circle.*

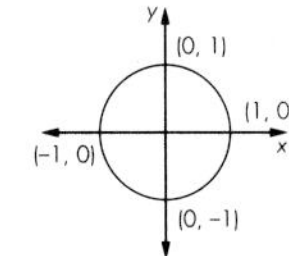

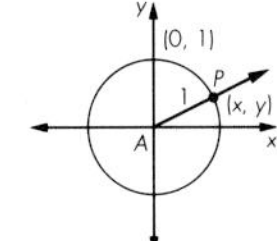

Think of drawing any angle A with its vertex at the center of a unit circle and the initial side along the positive x-axis. The terminal side will intersect the unit circle at some point P with coordinates (x, y). As this terminal side is rotated in a counterclockwise direction about the origin, the values for x and y will change.

By the definition in Section 14.3 of your textbook, you know that

$\sin A = \frac{y}{r}$, where r is the distance that point P is from the origin. In a unit circle, r is 1. So $\sin A = \frac{y}{1}$, or $\sin A = y$.

In a similar way, $\cos A = \frac{x}{1}$, or $\cos A = x$.

Refer to the figure at the right above for Exercises 1–6.

1. As the measure of $\angle A$ increases from 0° to 90°, what happens to sin A? That is, what happens to y? It increases from 0 to 1.
2. As the measure of $\angle A$ increases from 0° to 90°, what happens to cos A? That is, what happens to x? It decreases from 1 to 0.
3. Describe the values of sin A as the measure of $\angle A$ increases from 90° to 180°. They decrease from 1 to 0.
4. Describe the values of cos A as the measure of $\angle A$ increases from 90° to 180°. They decrease from 0 to −1.
5. What is the largest possible value for sin A? for cos A? 1; 1
6. For what measures of $\angle A$ will cos A and sin A be equal? 45°, 225°

Name ______________________ Date ____________

Solving Non-right Triangles

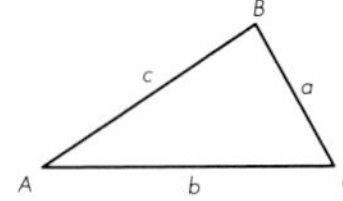

The following laws can be used to find the unknown lengths of a triangle, even if the triangle is not a right triangle.

Law of Sines	Law of Cosines
In any triangle ABC, $\frac{\sin A}{a} = \frac{\sin B}{b} = \frac{\sin C}{c}$	In any triangle ABC, $a^2 = b^2 + c^2 - 2bc \cos A$ $b^2 = a^2 + c^2 - 2ac \cos B$ $c^2 = a^2 + b^2 - 2ab \cos C$

Example 1: Find b in $\triangle ABC$.

C, b, a = 19.2, 62°, 35°, A, c, B

By the law of sines,

$\frac{\sin A}{a} = \frac{\sin B}{b}$

$\frac{\sin 62°}{19.2} = \frac{\sin 35°}{b}$

So, $b \times \sin 62° = 19.2 \times \sin 35°$

$b = \frac{19.2 \times \sin 35°}{\sin 62}$

$b = \frac{19.2 \times 0.5736}{0.8829} \approx 12.5$

Example 2: Find a in $\triangle ABC$.

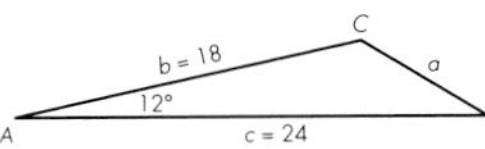

By the law of cosines,

$a^2 = b^2 + c^2 - 2bc \cos A$

$a^2 = 18^2 + 24^2 - 2(18)(24) \cos 12°$

$a^2 = 324 + 576 - 864(0.9781)$

$a^2 = 900 - 845.08$

$a^2 = 54.92$

$a \approx 7.4$

Find the indicated length, to the nearest tenth.

1.

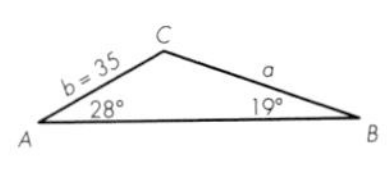

$a \approx$ 50.5

2.

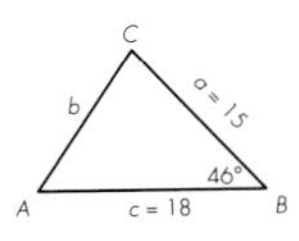

$b \approx$ 13.2

3.

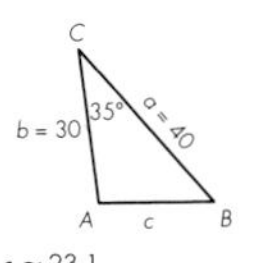

$c \approx$ 23.1

Name ______________________ Date ____________

Heron's Formula

Surveyors often need to determine the area of a plot of land that is irregular in shape. In many cases, there is no formula for finding the area. One method that surveyors use is called *triangulation*. To use this method, surveyors first divide the plot into triangles. For the plot pictured at the left below, there are two ways of doing this.

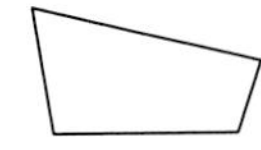

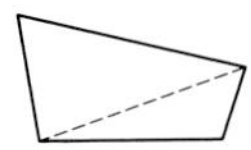

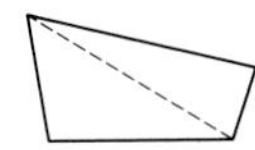

Once the plot is divided into triangles, the surveyor can use **Heron's formula** to find the area. This formula can be used to find the area of any triangle if you know the lengths of all three sides.

Heron's Formula	If $\triangle ABC$ has sides of length a, b, and c, then the area of the triangle is $\sqrt{s(s-a)(s-b)(s-c)}$, where $s = \frac{1}{2}(a + b + c)$.

Example: Find the area of $\triangle RST$.

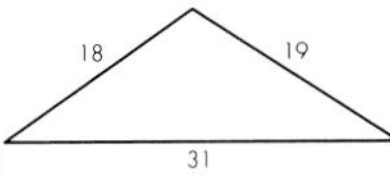

First, find s.

$s = \frac{1}{2}(18 + 19 + 31) = 34$

area $= \sqrt{34(34-18)(34-19)(34-31)}$

$= \sqrt{34(16)(15)(3)}$

$= \sqrt{24{,}480}$

≈ 156.5

Use a calculator to find the area of each triangle to the nearest tenth.

1.

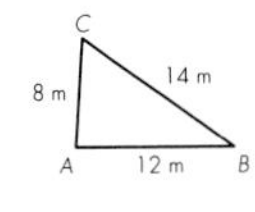

47.9 m^2

2.

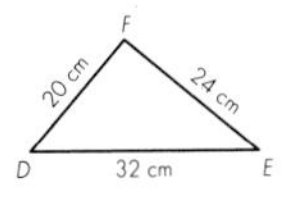

239.7 cm^2

3.

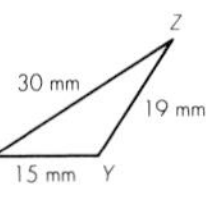

118.9 mm^2

Name ______________________ Date ____________

Applications

Many real-life situations involve problems that can be solved by using trigonometry. For example, distances can be found by using the sine, cosine, or tangent ratios for right triangles. In other cases, the law of sines or the law of cosines is useful, even if the triangle is not a right triangle.

Example: Two ships leave port P so that the angle between their courses is 20°. Later one ship is at point A which is 35 kilometers from the port. The other ship is 40 meters from the port at this time. How far apart are the ships?

Triangle PBA is not a right triangle. However, you know the lengths of two sides and the measure of the angle between them. With this information, you can use the law of cosines.

Therefore, $x^2 = 35^2 + 40^2 - 2(35)(40)\cos 20$
$x^2 = 1225 + 1600 - 2800(0.9397)$
$x^2 = 2825 - 2631.16$
$x^2 = 193.84$
$x \approx 13.9$ The ships are about 13.9 km apart.

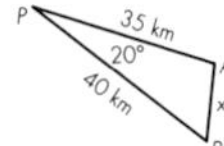

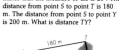

1. Two ships at points A and B sight a lighthouse on shore at point L. If $m\angle A = 61°$ and $m\angle B = 34°$, how far apart are the ships if distance AL is 12 km.

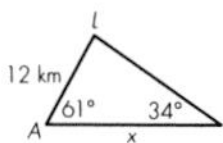

about 21.4 km

2. A kite string that is 50 yards long is anchored to the ground so that the string makes an angle of 32° with the ground. How far above the ground is the kite?

about 26.5 yd

3. A hiker at the top of a cliff at point D sights another hiker in the valley at point F so that the angle of depression is 27°. The cliff is 172 feet tall. How far is the second hiker from the cliff?

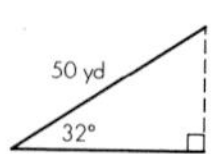

about 337.6 ft

4. A surveyor standing at point S determines that the measure of $\angle S$ is 26°. The distance from point S to point T is 180 m. The distance from point S to point Y is 200 m. What is distance TY?

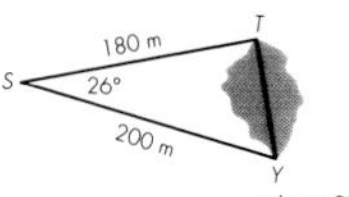

about 87.7 m

Name ______________________ Date ____________

More About Navigation

In navigation problems, the angle measured clockwise from north to the line of travel is called the **bearing** of the plane or ship.

For example, the ship at point A has a bearing of 120°. The ship at point B has a bearing of 220°.

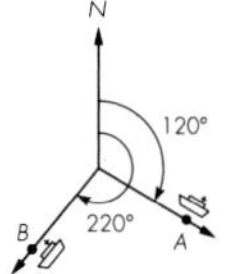

Directions are also sometimes written by referring to north or south, using an acute angle. In the figure at the right, the direction of $\overrightarrow{OC}$ is N 29°W (29° west of due north). The direction of $\overrightarrow{OD}$ is S 40° E (40° east of due south).

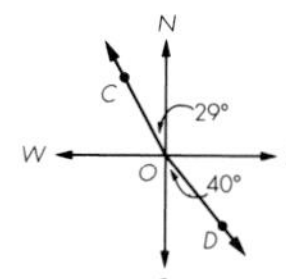

Write the bearing for each direction.

1. N 18°E 18°
2. S 20°E 160°
3. S 20°W 200°
4. N 41°W 319°
5. N 68°W 292°
6. S 62°E 118°

For each bearing, write the direction referring to north or south using an acute angle.

7. bearing 100° S 80°E
8. bearing 132° S 48°E
9. bearing 315° N 45°W
10. bearing 210° S 30°W
11. bearing 350° N 10°W
12. bearing 12° N 12°E

In the space at the right, draw a diagram to illustrate the following problem. Do not solve the problem.

13. Two planes leave an airport at the same time. One is traveling at a bearing of 100°. The other is traveling at a bearing of 210°. The first plane is 600 miles from the airport after 1 hour. The second plane is 450 miles from the airport after 1 hour. How far apart are the planes?

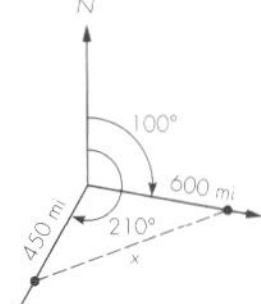

Name ______________________ Date ______________

Trigonometric Identities

An equation is an **identity** if it is true for all permissible values of the variable. For example, $x + 10 = 10 + x$ is an identity since it is true for all values of x. However, $x + 10 = 15$ is not an identity since it is true only when $x = 5$.

Three basic trigonometric identities were introduced in Section 14.6. However, there are many other trigonometric identities. The three identities that were introduced are:

$$\sin^2 x + \cos^2 x = 1$$
$$\cos 2x = \cos^2 x - \sin^2 x$$
$$\tan x = \frac{\sin x}{\cos x},\ \cos x \neq 0$$

To show that an equation is not an identity, you must show at least one value for which the equation is false. To show that an equation is an identity, you must substitute on one side using basic identities and then show in a sequence of steps that both sides are the same.

Example: Prove that $\cos 2x + 1 = 2\cos^2 x$

$\cos 2x + 1 = 2\cos^2 x$

$\cos^2 x - \sin^2 x + 1$	Substitute $\cos^2 x - \sin^2 x$ for $\cos 2x$.
$\cos^2 x - \sin^2 x + \sin^2 x + \cos^2 x$	Substitute $\sin^2 x + \cos^2 x$ for 1.
$2\cos^2 x$	Combine like terms.

$\cos 2x + 1 = 2\cos^2 x$

Which equations are identities? Write *yes* or *no*.

1. $x - 10 = 25$ No
2. $(x + 1) + 9 = x + (9 + 1)$ Yes
3. $18 + x = 18$ No
4. $x + 0 = x$ Yes
5. $x^2 - 25 = (x + 5)(x - 5)$ Yes
6. $x - 5 = -(5 - x)$ Yes

Prove the identity.

7. $\sin^2 x + \cos^2 x + 1 = 2\sin^2 x + 2\cos^2 x$

$\sin^2 x + \cos^2 x + \sin^2 x + \cos^2 x$

$2\sin^2 x + 2\cos^2 x$

$\sin^2 x + \cos^2 x + 1 = 2\sin^2 x + 2\cos^2 x$

Name ______________________ Date ______________

More Trigonometric Identities

Besides the sine, cosine, and tangent ratios, there are three more trigonometric ratios for a right triangle. These ratios are called the *secant* (abbreviated *sec*), the *cosecant* (*csc*), and the *cotangent* (*cot*). (See the Enrichment worksheet for Section 14.2.)

The six trigonometric ratios can be summarized as follows:

$\sin A = \frac{\text{opposite}}{\text{hypotenuse}} = \frac{a}{c}$ $\cos A = \frac{\text{adjacent}}{\text{hypotenuse}} = \frac{b}{c}$ $\tan A = \frac{\text{opposite}}{\text{adjacent}} = \frac{a}{b}$

$\sec A = \frac{\text{hypotenuse}}{\text{adjacent}} = \frac{c}{b}$ $\csc A = \frac{\text{hypotenuse}}{\text{opposite}} = \frac{c}{a}$ $\cot A = \frac{\text{adjacent}}{\text{opposite}} = \frac{b}{a}$

By using these six basic ratios, you can prove a wide variety of trigonometric identities. Some basic identities are as follows:

Pythagorean Identities	Reciprocal Identities
$\sin^2 A + \cos^2 A = 1$ $1 + \cot^2 A = \csc^2 A$ $1 + \tan^2 A = \sec^2 A$	$\sec A = \frac{1}{\cos A},\ \cos A \neq 0$ $\csc A = \frac{1}{\sin A},\ \sin A \neq 0$ $\cot A = \frac{1}{\tan A},\ \tan A \neq 0$
Ratio Identities	
$\tan A = \frac{\sin A}{\cos A},\ \cos A \neq 0$	$\cot A = \frac{\cos A}{\sin A},\ \sin A \neq 0$

Match each expression in Column 1 with an equivalent expression in Column 2. You may find it helpful to substitute numerical ratios from a right triangle such as $\triangle ABC$. For example, in $\triangle ABC$, $\sin A = \frac{6}{10}$.

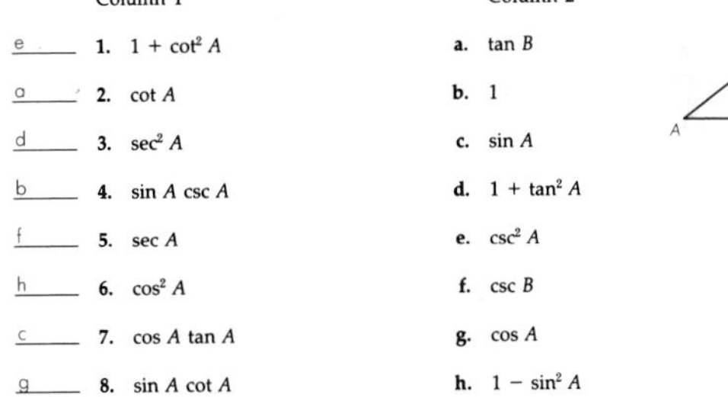

	Column 1	Column 2
e	1. $1 + \cot^2 A$	a. $\tan B$
a	2. $\cot A$	b. 1
d	3. $\sec^2 A$	c. $\sin A$
b	4. $\sin A \csc A$	d. $1 + \tan^2 A$
f	5. $\sec A$	e. $\csc^2 A$
h	6. $\cos^2 A$	f. $\csc B$
c	7. $\cos A \tan A$	g. $\cos A$
g	8. $\sin A \cot A$	h. $1 - \sin^2 A$

Name ______________________ Date ______________

Trigonometric Functions

You have learned that the sine and cosine functions have a period of 360°. Use this fact and a calculator to find the measures of two other angles whose sine and cosine have the same value as each angle measure given below. Then use a calculator to check your answer by finding the sine and the cosine of each of the three angles. Answers for the two angle measures may vary.

1. 50°
sin = 0.76604445
cos = 0.64278761

2. 68°
sin = 0.92718386
cos = 0.37460659

3. 82°
sin = 0.99026807
cos = 0.1391731

4. 123°
sin = 0.83867057
cos = −0.54463904

5. 105°
sin = 0.96592583
cos = −0.25881905

6. 186°
sin = −0.10452846
cos = −0.9945219

7. 208°
sin = −0.46947156
cos = −0.88294759

8. 270°
sin = −1
cos = 0

9. 243°
sin = −0.89100653
cos = −0.4539905

10. −40°
sin = −0.64278761
cos = 0.76604445

11. −56°
sin = −0.82903757
cos = 0.55919291

12. −18°
sin = −0.309017
cos = 0.95105652

13. −99°
sin = −0.98768834
cos = −0.15643447

14. −138°
sin = −0.66913061
cos = −0.24314483

15. −246°
sin = 0.91354546
cos = −0.40673664

16. −111°
sin = −0.93358043
cos = −0.35836795

17. −144°
sin = −0.58778525
cos = −0.809017

18. −310°
sin = 0.76604445
cos = 0.64278761

You have learned that the tangent function has a period of 180°. Use this fact and a calculator to find the measures of two other angles whose tangent has the same value as each angle measure given below. Then use a calculator to check your answer by finding the tangent of each of the three angles. Answers for the two angle measures may vary.

19. 60°
tan = 1.7320508

20. 76°
tan = 4.0107809

21. 98°
tan = −7.1153697

22. 18°
tan = 0.32491969

23. 77°
tan = 4.3314759

24. 94°
tan = −14.300666

25. 132°
tan = −1.1106125

26. 175°
tan = −0.08748866

27. 264°
tan = 9.5143643

28. −30°
tan = −.57735027

29. −14°
tan = −0.249328

30. −27°
tan = −0.50952545

31. −100°
tan = 5.6712818

32. −154°
tan = 0.48773259

33. −299°
tan = 1.8040478

34. −144°
tan = 0.72654252

35. −280°
tan = 5.6712818

36. −355°
tan = 0.08748866

Name ______________ Class______________ Date______________

Achievement Test 14 (Chapter 14)
TRIGONOMETRY

No. Correct
No. Exercises: 25
Score
4.0 x No. Correct =

GEOMETRY FOR DECISION MAKING
James E. Elander
SOUTH-WESTERN PUBLISHING CO.

Use a calculator or the tables on pages 562 and 563. Find sin *A*, cos *A* and tan *A*.

1.
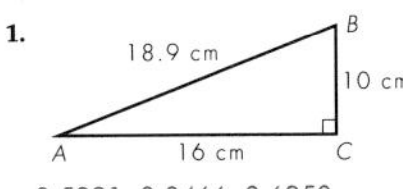

0.5291; 0.8466; 0.6250

2. 7 in. C 14 in. A B

0.8944; 0.4472; 2.000

3.
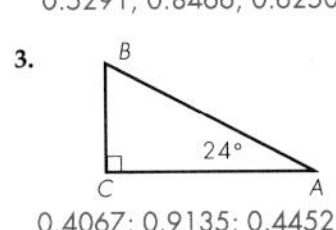

0.4067; 0.9135; 0.4452

4.
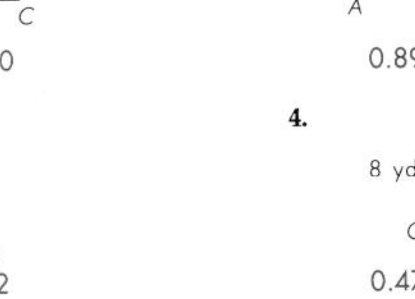

0.4706; 0.8824; 0.5333

Find *r* and the sine and cosine of each angle.

5.
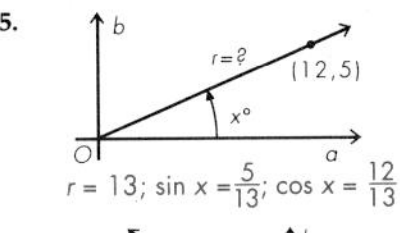

$r = 13$; $\sin x = \frac{5}{13}$; $\cos x = \frac{12}{13}$

6.
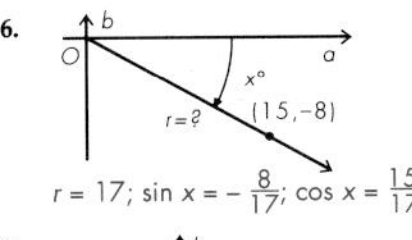

$r = 17$; $\sin x = -\frac{8}{17}$; $\cos x = \frac{15}{17}$

7.
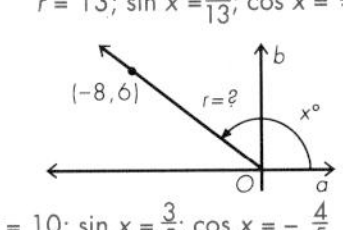

$r = 10$; $\sin x = \frac{3}{5}$; $\cos x = -\frac{4}{5}$

8.
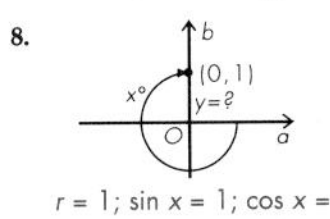

$r = 1$; $\sin x = 1$; $\cos x = 0$

9. $x = -45°$
$\sin x =$ ______
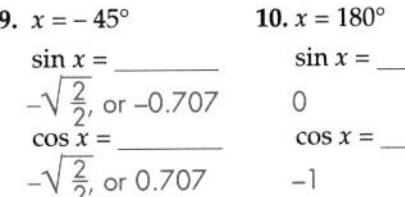
$-\sqrt{\frac{2}{2}}$, or −0.707
$\cos x =$ ______
$-\sqrt{\frac{2}{2}}$, or 0.707

10. $x = 180°$
$\sin x =$ ______
0
$\cos x =$ ______
−1

11. $x = 150°$
$\sin x =$ ______
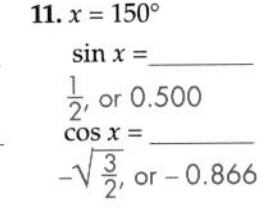
$\frac{1}{2}$, or 0.500
$\cos x =$ ______
$-\sqrt{\frac{3}{2}}$, or −0.866

12. $x = 225°$
$\sin x =$ ______
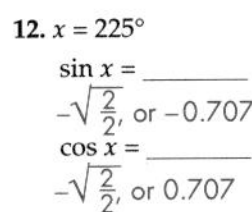
$-\sqrt{\frac{2}{2}}$, or −0.707
$\cos x =$ ______
$-\sqrt{\frac{2}{2}}$, or 0.707

Find each indicated measure in Items 13-18. Give lengths to the nearest tenth and angles to the nearest degree.

13. r = 25 ft, Q, p=?, 37°, P, R, q=?

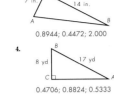

$p = 15.0$ ft; $q = 20.0$ ft

14.
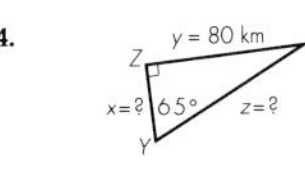

$x = 37.3$ km; $z = 88.3$ km

15. b=? C 70° A 54° B 34

$b = 29.3$

16. C 50 A 45° 25° B c=?

$c = 111.2$

17.
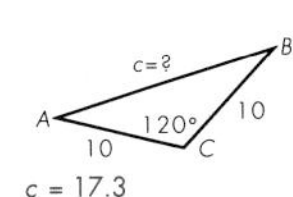

$c = 17.3$

18. A 11 B ?° 9 13 C

$m\angle B = 43°$

19. When an advertising blimp is directly over one end of a football field, the angle of elevation to the blimp from the other end of the 300 foot-long field is 78°. What is the altitude of the blimp? $a = 1411.4$ ft

20. From the top of a cliff, a barge is spotted in the middle of the river 2000 ft from shore. The angle of depression to the barge is 8°. How high is the cliff? $h = 281.1$ ft

21. A rail tunnel is to be dug that connects points *A* and *B*. As part of the survey, the two indicated angle measures are made at points *B* and *C* on a 2-mile road. How long will the tunnel be? 3.8 mi

a = 2 mi, 60°, B, 40°, C, A

22. Jan and Jill leave point *B* along two different roads that form an angle of 25°. After an hour and a half, they have traveled 40 mi and 60 mi, respectively. How far apart are they? 29.2 mi

A, 40 mi, 25°, B, 60 mi, C

Tell whether each equation seems to be an identity. If not, give an example to support your answer.

23. $\cos^2 x + \sin^2 x = \cos^2 x$ yes

24. $(\tan x)(\cos x) = \sin x$ yes

25. $\cos^2 x - \sin^2 x = 1$ No; Examples will vary. Possible example: for $x = 90°$, $\cos^2 x - \sin^2 x = 0 - 1 = -1$ and $-1 \neq 1$.

Use the diagram. Identify the following.

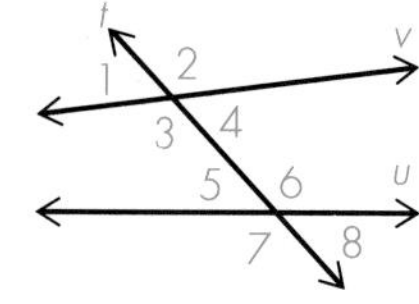

1. vertical angles
∠1, ∠4; ∠2, ∠3; ∠5, ∠8; ∠6, ∠7

2. corresponding angles
∠1, ∠5; ∠2, ∠6; ∠3, ∠7; ∠4, ∠8

3. alternate interior angles
∠3, ∠6; ∠4, ∠5

4. alternate exterior angles
∠1, ∠8; ∠2, ∠7

5. The vertices of a triangle in the *x*-*y* plane are $A(2, 0)$, $B(-1, 0)$, and $C(-1, -3)$. What type of triangle is $\triangle ABC$?
Isosceles right triangle

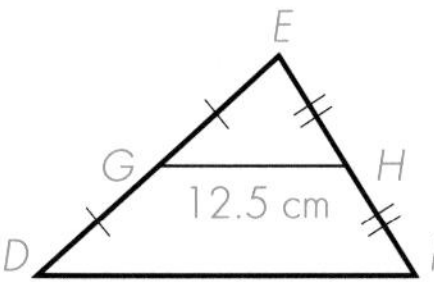

6. What is the length of $\overline{DF}$ in $\triangle DEF$? 25 cm

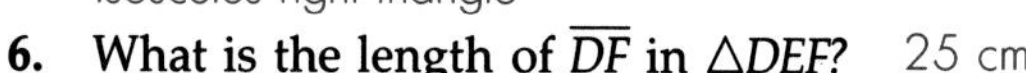

Refer to $\triangle ABC$.

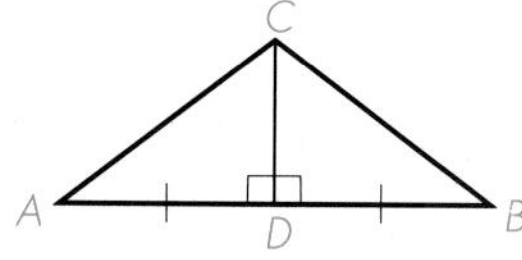

7. Is $\triangle ACD \sim \triangle BCD$? If so, why? Yes; SAS

8. Is $\triangle ACD \cong \triangle BCD$? If so, why?
Yes; they are similar and the ratio of corresponding sides is 1 : 1.

The members of the school baseball team were asked what professional baseball teams they root for.
15 root for the Dodgers (D).
6 root for both D and A.
10 root for the Athletics (A).
5 root for neither team.

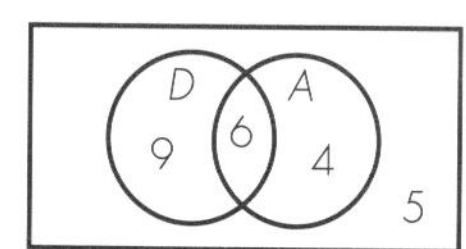

9. Draw a Venn diagram illustrating this information.

10. How many members does the baseball team have? 24

11. Draw an acute angle and construct its bisector.

Find the missing lengths to the nearest tenth. Use a calculator or the table of squares and square roots on page 562.

12.

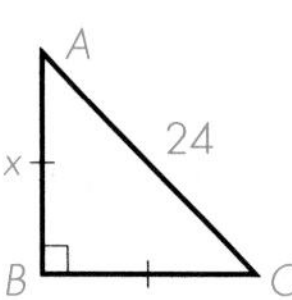

$x = 12\sqrt{2} \approx 17.0$

13.

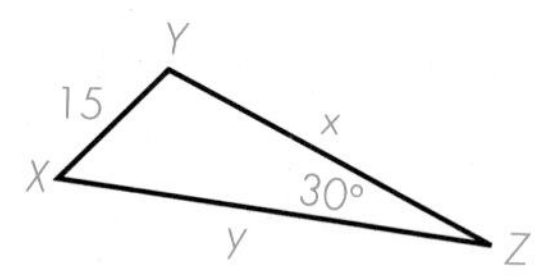

$x = 15\sqrt{3} \approx 26.0$
$y = 30.0$

14. Find the length of the line segment that connects P and Q. $\sqrt{149} \approx 12.2$

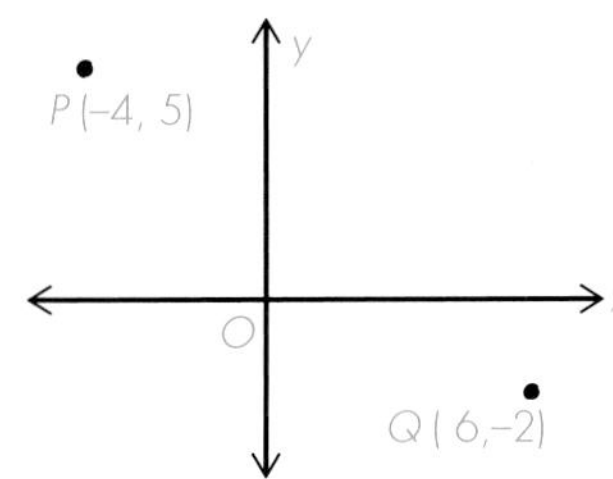

Is the figure a parallelogram? Give a reason for your answer.

15.

Yes; diagonals bisect each other.

16.

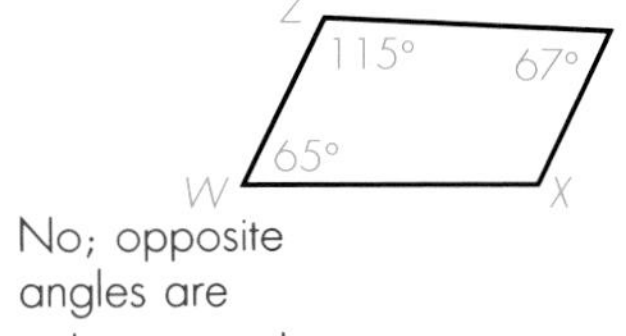

No; opposite angles are not congruent.

17. Find the indicated angle measures for $\square ABCD$. Is $m\angle DCA + m\angle ABD$ equal to $m\angle AED$? Explain.
Yes; $m\angle DCA + m\angle ABD = 35° + 25° = 60°$ and $m\angle AED = 60°$.

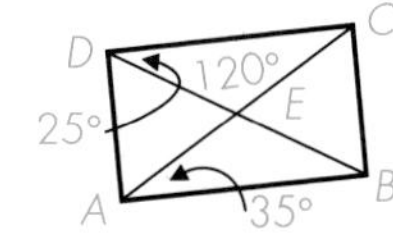

18. $PQRS$ is an isosceles trapezoid. $PQ = 25$ and $SR = 18$. T and V are midpoints. What is the length of $\overline{TU}$? 21.5

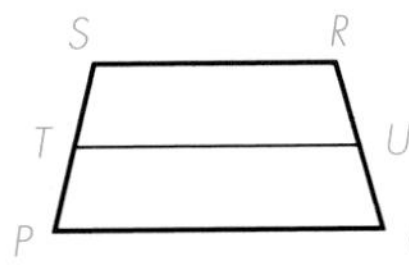

Find the perimeter and area of each trapezoid. Answer to the nearest tenth.

19. P = 106.0 cm; A = 626.9 cm^2

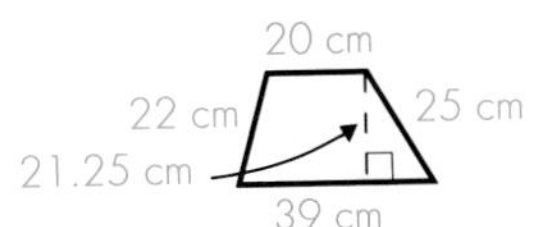

20. P = $48 + 16\sqrt{2}$ or 70.6 ft.
A = 192.0 ft^2

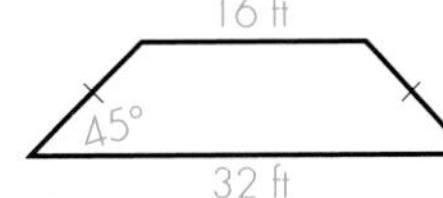

21. The sum of the angle measures for a regular polygon is 7740°. How many sides does the polygon have? 45

Find the circumference and area of each circle. Use $\pi \approx 3.14$.

22. C = 119.32 in.;
A = 1133.54 in.^{2}

23. C = 314 cm;
A = 7850 cm^2

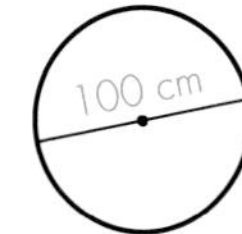

In circle O, $m\widehat{AC} = 90°$, $m\angle EOC = 60°$, $m\angle FCB = 25°$, and $\overline{BC}$ is a diameter. $\overline{ED}$ is tangent to the circle. Find these measures.

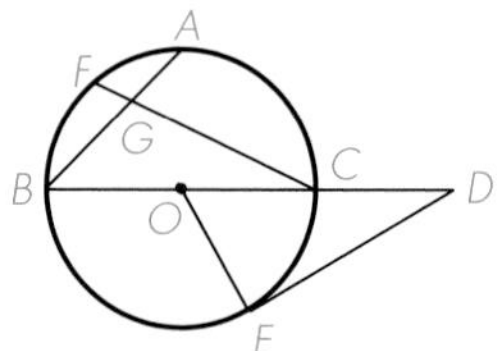

24. $m\angle ABC$ 45° **25.** $m\angle FGB$ 70° **26.** $m\angle OED$ 90°

27. $m\angle BDE$ 30° **28.** $m\widehat{FA}$ 40° **29.** $m\angle FGA$ 110°

30. In the circle, two chords intersect as shown. Find the value of x. $x = 4$

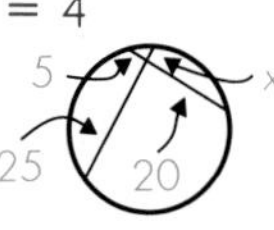

Find the total area and volume for each prism or right circular cylinder. Use $\pi \approx 3.14$.

31. A: 516 ft^2; V: 720 ft^3

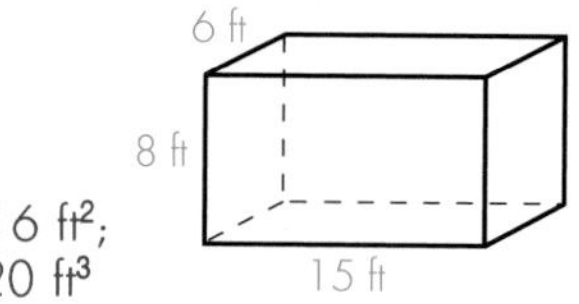

32. A: 785 cm^2; V: 1570 cm^3

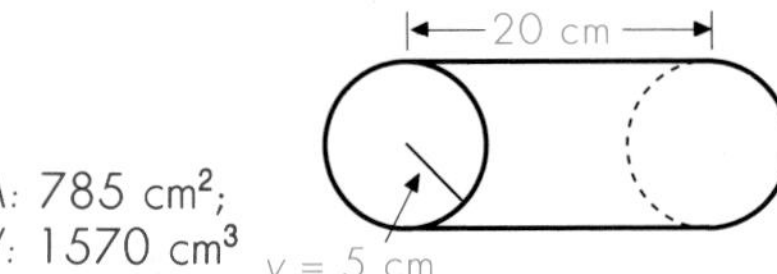

Find the total area and volume of each regular pyramid or right circular cone. Answer to the nearest whole number. Use $\pi \approx 3.14$.

33. A: 678 in.^{2}; V: 1017 in.^{3}

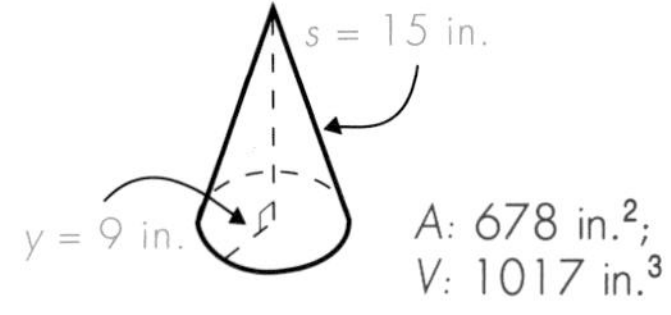

34. A: 467 mm^2; V: 518 mm^3

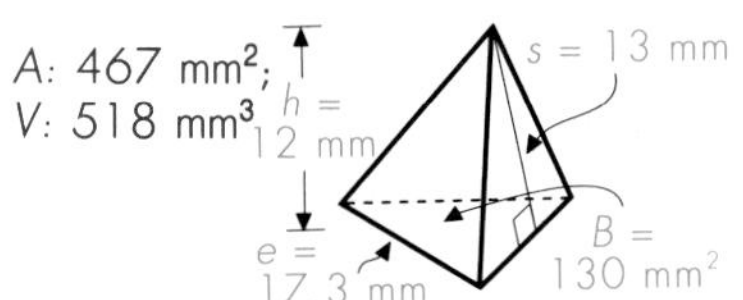

35. Find the volume and area of a sphere with a radius of 20 cm. Answer to the nearest whole number. Use $\pi \approx 3.14$. A: 5024 cm^2; V: 33,493 cm^3

36. In a plane, what is the locus of all points equidistant from a circle and the center of the circle? Another circle with same center and half the radius

Does the figure have line symmetry? If so, trace the figure and draw the line(s) of symmetry.

37. No

38. Yes

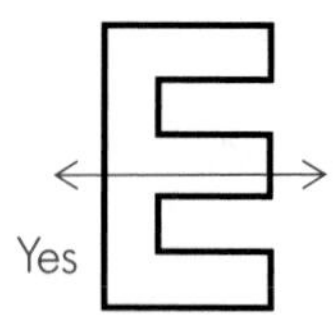

39. Yes

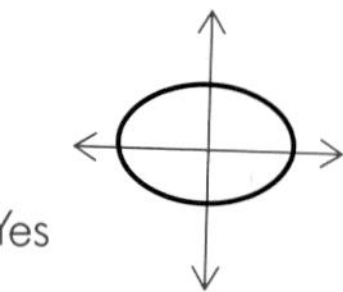

40. Write the converse, inverse, and contrapositive for this statement: If a creature can fly, then it is a bird. See margin.

Tell whether the lines are parallel, perpendicular, or neither.

41. $y = \frac{1}{2}x - 2$
$y = -2x - 2$
perpendicular

42. $y = -2x + 5$
$y = -2x - 5$
parallel

43. $y - 2x = 5$
$y = 2x - \frac{1}{5}$
parallel

Tell whether the graph of the equation is a circle or a parabola. If it is a circle, give its center and radius. If it is a parabola, tell whether it opens up or down.

44. $x^2 + y^2 = 36$
Circle; center (0, 0), radius 6

45. $y = 4x^2$
parabola; up

46. $x^2 + y^2 = \frac{1}{4}$
circle; center (0, 0), radius $\frac{1}{2}$

47. Refer to the figure. Use coordinate geometry to prove that $\overline{AC}$ and $\overline{BD}$ bisect each other.
By the midpoint theorem, $\overline{AC}$ and $\overline{BD}$ both have $\left(\frac{a}{2}, \frac{b}{2}\right)$ as their midpoint.

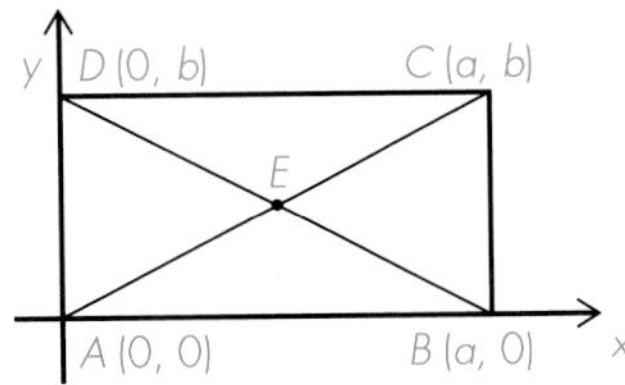

For items 48–53 you may use a calculator or the table on page 563. Find sin *A*, cos *A*, and tan *A*. Give answers to the nearest ten-thousandth.

48.

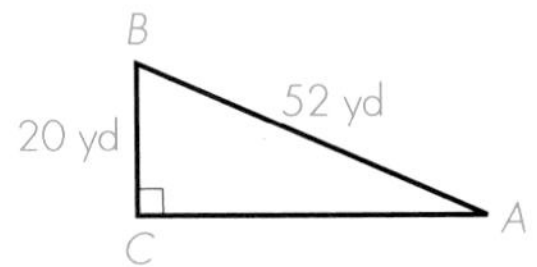

0.3846; 0.9231; 0.4167

49.

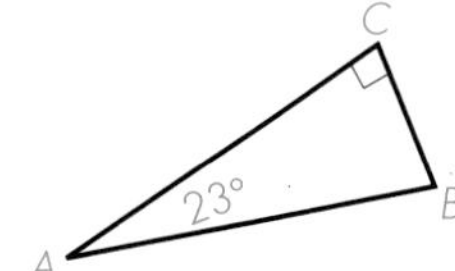

0.3907; 0.9205; 0.4245

Find the sine and cosine of each angle. Answer to the nearest tenth.

50. 0.6; 0.8

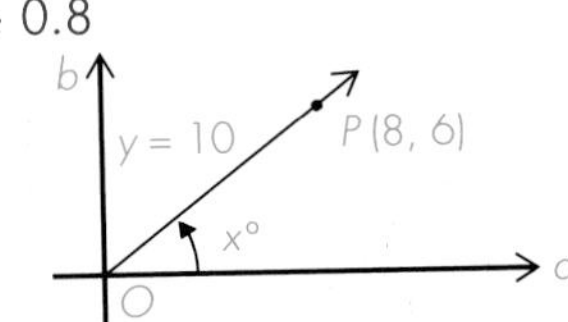

51. 0.9; −0.4

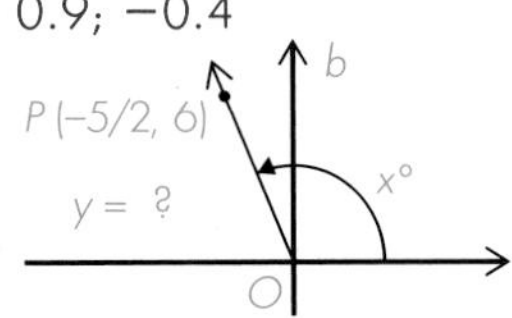

Find each indicated measure. Answer to the nearest tenth.

52.

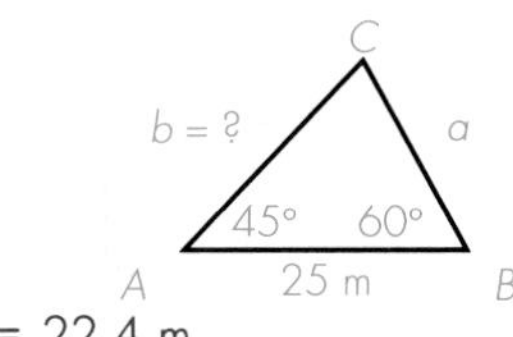

b = 22.4 m

53.

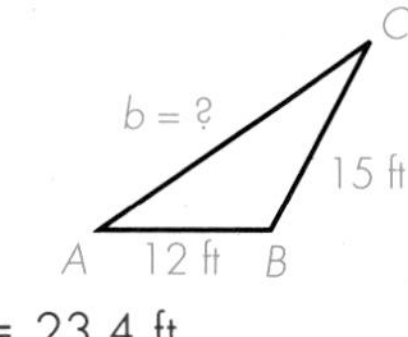

b = 23.4 ft

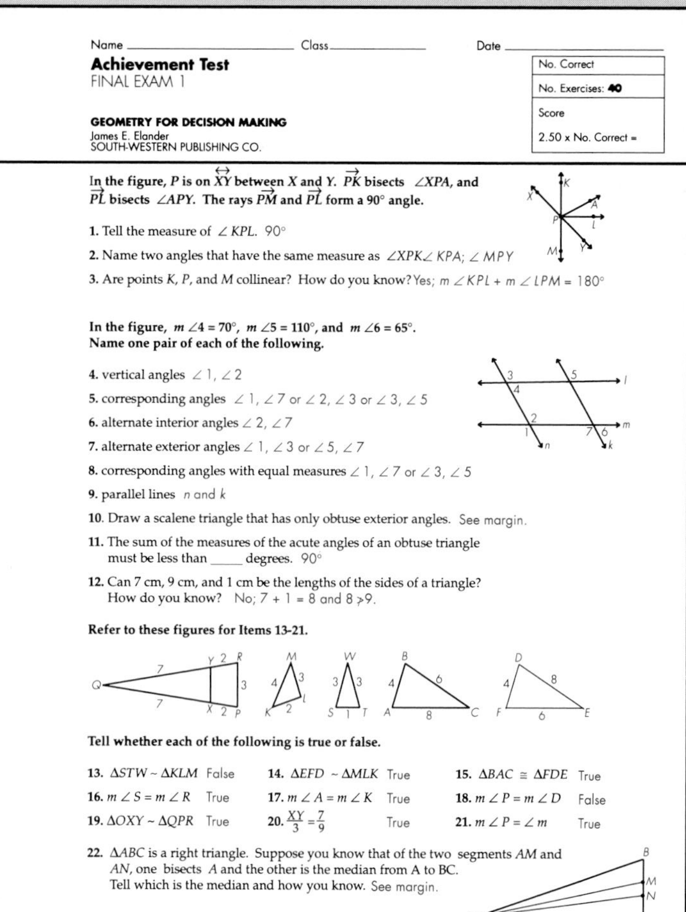

Name ______________ Class ______________ Date ______________

Achievement Test
FINAL EXAM 1

GEOMETRY FOR DECISION MAKING
James E. Elander
SOUTH-WESTERN PUBLISHING CO.

No. Correct	
No. Exercises: 40	
Score	
2.50 x No. Correct =	

In the figure, P is on $\overleftrightarrow{XY}$ between X and Y. $\overrightarrow{PK}$ bisects $\angle XPA$, and $\overrightarrow{PL}$ bisects $\angle APY$. The rays $\overrightarrow{PM}$ and $\overrightarrow{PL}$ form a 90° angle.

1. Tell the measure of $\angle KPL$. 90°
2. Name two angles that have the same measure as $\angle XPK$ $\angle KPA$; $\angle MPY$
3. Are points K, P, and M collinear? How do you know? Yes; $m\angle KPL + m\angle LPM = 180°$

In the figure, $m\angle 4 = 70°$, $m\angle 5 = 110°$, and $m\angle 6 = 65°$. Name one pair of each of the following.

4. vertical angles $\angle 1$, $\angle 2$
5. corresponding angles $\angle 1$, $\angle 7$ or $\angle 2$, $\angle 3$ or $\angle 3$, $\angle 5$
6. alternate interior angles $\angle 2$, $\angle 7$
7. alternate exterior angles $\angle 1$, $\angle 3$ or $\angle 5$, $\angle 7$
8. corresponding angles with equal measures $\angle 1$, $\angle 7$ or $\angle 3$, $\angle 5$
9. parallel lines n and k
10. Draw a scalene triangle that has only obtuse exterior angles. See margin.
11. The sum of the measures of the acute angles of an obtuse triangle must be less than _____ degrees. 90°
12. Can 7 cm, 9 cm, and 1 cm be the lengths of the sides of a triangle? How do you know? No; 7 + 1 = 8 and 8 ≯ 9.

Refer to these figures for Items 13-21.

Tell whether each of the following is true or false.

13. $\triangle STW \sim \triangle KLM$ False	14. $\triangle EFD \sim \triangle MLK$ True	15. $\triangle BAC \cong \triangle FDE$ True
16. $m\angle S = m\angle R$ True	17. $m\angle A = m\angle K$ True	18. $m\angle P = m\angle D$ False
19. $\triangle OXY \sim \triangle QPR$ True	20. $\frac{XY}{3} = \frac{7}{9}$ True	21. $m\angle P = \angle m$ True

22. $\triangle ABC$ is a right triangle. Suppose you know that of the two segments AM and AN, one bisects A and the other is the median from A to BC. Tell which is the median and how you know. See margin.

MG01AG [FE 1-1]

23. One leg of a right triangle has length 24 and the hypotenuse has length 26. What is the length of the other leg? 10

Quadrilateral $ABCD$ is an isosceles trapezoid in which $AB = BC = CD$. Point X is on AD and $BX \parallel CD$.

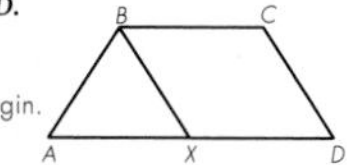

24. What kind of quadrilateral is $BCDX$? How do you know? See margin.
25. What kind of triangle is $\triangle ABX$? How do you know? See margin.

Find the perimeter and area of each figure. Assume that the quadrilaterals are parallelograms or trapezoids. If your answer contains a square root, leave it in radical form.

26. $P = 25 + 5\sqrt{13}$, $A = 75$

27. $P = 40$; $A = 48\sqrt{3}$

28. $P = 40 + 20\sqrt{2}$, $A = 200$

Use the figure and the given information to find the missing measures. Use 3.14 for π.

29.

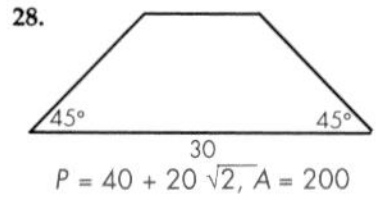

$m\,C =$ _____ 20°
$mCB =$ _____ 140°

30.

$x =$ _____ 3

31.

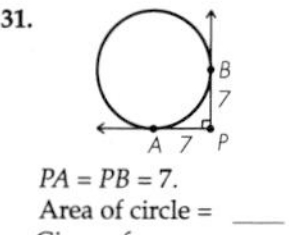

$PA = PB = 7$.
Area of circle = _____ 153.86
Circumference = _____ 43.96

32. The base of a right prism is a right triangle with legs of length 3 and 4. The height of the prism is 9. Find its lateral area. $LA = 108$
33. A regular square pyramid has height 12 and slant height 13 Find its total area. $TA = 360$
34. A right cylinder has height 8 and radius 5. Find its volume. Use 3.14 for π. $V = 628$
35. What is the locus of the centers of all the circles tangent to both sides of an angle APB? The ray that bisects $\angle APB$ minus point P

Find an equation for each of the following.

36. The line with slope 2 and y-intercept 1/3 $y = x^2 + \left(-\frac{1}{3}\right)$
37. The line passing through (5,4) and (6,2) $y = -2x + 14$
38. The circle that has a diameter with endpoints (7, 3) and (−7, −3) $x^2 + y^2 = 58$

Use a calculator or the tables on pages 562 and 563 as needed.

39. $\sin A =$ ____ and $\tan A =$ ____ 0.78; 1.25

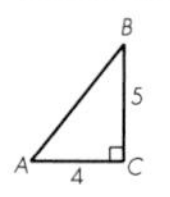

40. $m\angle P = 75°$ $PR = 5.18$

[FE 1-2]

10.

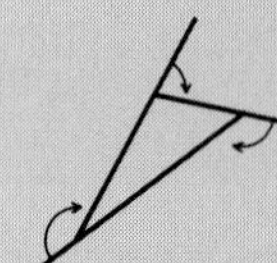

Triangles will vary. Any triangle having three acute angles and all of its sides of different lengths is correct.

22. The median is the upper segment, $\overline{AM}$. Consider the bisector $\overrightarrow{AP}$ of $\angle A$. Call the point where it intersects $\overline{BC}$ point X.

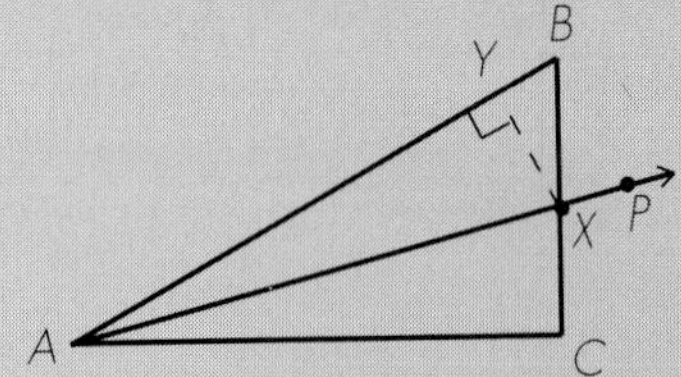

Let $\overline{XY}$ be the perpendicular from X to $\overline{AB}$. $\triangle XYB$ is a right triangle with hypotenuse XB. So $XB > XY$, since the hypotenuse is the longest side of the right triangle. But $XY = XC$, since X is equidistant from the sides of $\angle A$. Therefore $XB > XC$. Since the midpoint of $\overline{BC}$ is equidistant from B and C, it must lie above X. Hence $\overline{AM}$ is the median and $\overline{AN}$ is the angle bisector.

24. Rhombus. The opposite sides of $BCDX$ are parallel, so $BCDX$ is a parallelogram. Opposite sides of a parallelogram have the same length, and BC = CD. Therefore all four sides of BCDX have the same length.

25. Isosceles. Since $BCDX$ is a rhombus, $BX = CD$, and $AB = CD$ (given). Therefore $BX = AB$, which means that $\triangle ABX$ is isosceles.

Name ______________ Class ______________ Date ______________

Achievement Test
FINAL EXAM 2

No. Correct
No. Exercises
Score
3.57 x No. Correct =

GEOMETRY FOR DECISION MAKING
James E. Elander
SOUTH-WESTERN PUBLISHING CO.

Refer to the figure to answer the questions. Assume *B*, *P*, and *E* are collinear.

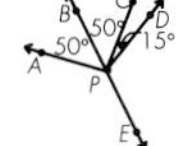

1. What is the measure of a complement of $\angle BPD$? 25°
2. What is the measure of a supplement of $\angle APE$? 50°
3. What is $m \angle DPE$? 115°

In the figure, $m \angle 1 = 115°$, $\angle m\, 3 = 60°$, and $m \angle 5 = 65°$.

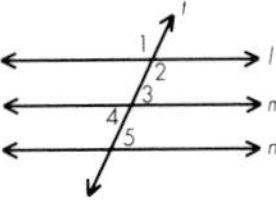

4. Name two corresponding angles. $\angle 3$, $\angle 5$
5. Name two acute vertical angles. $\angle 3$, $\angle 4$
6. Name two alternate interior angles. $\angle 4$, $\angle 5$
7. Name two parallel lines. *l* and *n*

Tell whether each statement is true or false.

8. A scalene triangle can have two angles of equal measure. False.
9. An isosceles triangle can have the length of one of its sides equal to one-third the sum of the lengths of the other two sides. True.
10. If ΔABC and ΔABD are two different equilateral triangles, then $\overleftrightarrow{CD}$ bisects $\overline{AB}$. True.
11. If *M*, *N*, and *P* are the midpoints of the sides of ΔXYZ, then the perimeter of ΔMNP is one-third of the perimeter of ΔXYZ. False
12. If ΔABC is isosceles and $m\, B = 90°$, then $AC > BC$. True

In the figure, $m \angle 1 = m \angle 2 > m \angle 3$, $AB = BD$, $BF = BG$, and $BE = BC$.

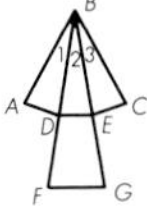

13. What must be true of *AD* and *CE* in order for the two smaller triangles to be congruent? $AD = CE$
14. If $AD/FG = \frac{1}{2}$, what is BD/DF? 1
15. Is it true or false that $\Delta FBG \sim \Delta EBC$? How do you know? True; SAS

MG01AG

Refer to the figure. Find the indicated lengths to the nearest tenth. You may use a calculator or the table on page 563 as needed.

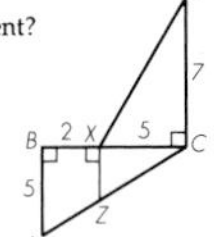

16. Are the two larger triangles in the figure congruent? How do you know? Yes; SAS.
17. What is the length of *AC*? 8.6
18. What is the length of *ZC*? 6.1
19. Draw an acute triangle and use your ruler and compass to construct all three altitudes of the triangle. See margin.

Find the perimeter and area of each figure. Assume that the quadrilaterals are parallelograms or trapezoids. If your answer contains a square root, leave it in radical form.

20.

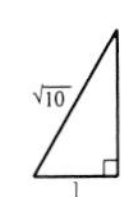

$P = 4 + \sqrt{10}$; $A = \frac{3}{2}$

21.

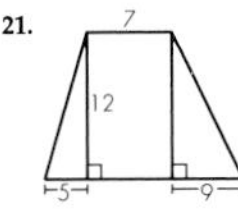

$P = 56$; $A = 168$

22.

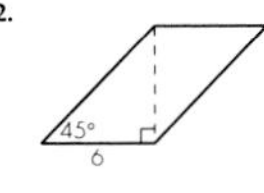

$P = 12 + 12\sqrt{2}$; $A = 36$

23. Find the circumference and area of a circle of diameter 12. Answer to the nearest tenth. Use 3.14 for π. $C = 37.7$; $A = 113.0$
24. Find the total surface area and volume of a right prism whose height is 5 and whose base is a right triangle with legs of lengths 6 and 8. $TA = 168$; $V = 120$
25. Find the measures of the angles that the line $y = \frac{1}{2}x + 5$ makes with the *x*-axis. Answer to the nearest degree. Use a calculator or the table on page 563 as needed. 27°, 153°
26. Show that (2,4) is a point of intersection of the parabola $y = x^2$ and the circle $x^2 + y^2 = 20$. $4 = 2^2$ and $2^2 + 4^2 = 4 + 16 = 20$, so the coordinates of (2,4) satisfy both equations.

Refer to the figure. Find the indicated measures. Use a calculator or the tables on pages 562 and 563 as needed. Answer to the nearest whole unit.

27. $AB =$ ____ 12
28. $m \angle D =$ ____ 11°

A
B
C
D
E
8
6
2
9
120°

19. Triangles will vary. Check that the diagram shows three segments, each from a vertex of the triangle to the opposite side and perpendicular to the opposite side. The three segments should all intersect at the same point inside the triangle.

Squares and Approximate Square Roots

n	n^2	$\sqrt{n}$	n	n^2	$\sqrt{n}$
1	1	1.000	51	2601	7.141
2	4	1.414	52	2704	7.211
3	9	1.732	53	2809	7.280
4	16	2.000	54	2916	7.348
5	25	2.236	55	3025	7.416
6	36	2.449	56	3136	7.483
7	49	2.646	57	3249	7.550
8	64	2.828	58	3364	7.616
9	81	3.000	59	3481	7.681
10	100	3.162	60	3600	7.746
11	121	3.317	61	3721	7.810
12	144	3.464	62	3844	7.874
13	169	3.606	63	3969	7.937
14	196	3.742	64	4096	8.000
15	225	3.873	65	4225	8.062
16	256	4.000	66	4356	8.124
17	289	4.123	67	4489	8.185
18	324	4.243	68	4624	8.246
19	361	4.359	69	4761	8.307
20	400	4.472	70	4900	8.367
21	441	4.583	71	5041	8.426
22	484	4.690	72	5184	8.485
23	529	4.796	73	5329	8.544
24	576	4.899	74	5476	8.602
25	625	5.000	75	5625	8.660
26	676	5.099	76	5776	8.718
27	729	5.196	77	5929	8.775
28	784	5.292	78	6084	8.832
29	841	5.385	79	6241	8.888
30	900	5.477	80	6400	8.944
31	961	5.568	81	6561	9.000
32	1024	5.657	82	6724	9.055
33	1089	5.745	83	6889	9.110
34	1156	5.831	84	7056	9.165
35	1225	5.916	85	7225	9.220
36	1296	6.000	86	7396	9.274
37	1369	6.083	87	7569	9.327
38	1444	6.164	88	7744	9.381
39	1521	6.245	89	7921	9.434
40	1600	6.325	90	8100	9.487
41	1681	6.403	91	8281	9.539
42	1764	6.481	92	8464	9.592
43	1849	6.557	93	8649	9.644
44	1936	6.633	94	8836	9.695
45	2025	6.708	95	9025	9.747
46	2116	6.782	96	9216	9.798
47	2209	6.856	97	9409	9.849
48	2304	6.928	98	9604	9.899
49	2401	7.000	99	9801	9.950
50	2500	7.071	100	10000	10.000

562

Trigonometric Ratios

Angle	sin	cos	tan	Angle	sin	cos	tan
0°	0.0000	1.0000	0.0000	45°	0.7071	0.7071	1.0000
1°	0.0175	0.9998	0.0175	46°	0.7193	0.6947	1.0355
2°	0.0349	0.9994	0.0349	47°	0.7314	0.6820	1.0724
3°	0.0523	0.9986	0.0524	48°	0.7431	0.6691	1.1106
4°	0.0698	0.9976	0.0699	49°	0.7547	0.6561	1.1504
5°	0.0872	0.9962	0.0875	50°	0.7660	0.6428	1.1918
6°	0.1045	0.9945	0.1051	51°	0.7771	0.6293	1.2349
7°	0.1219	0.9925	0.1228	52°	0.7880	0.6157	1.2799
8°	0.1392	0.9903	0.1405	53°	0.7986	0.6018	1.3270
9°	0.1564	0.9877	0.1584	54°	0.8090	0.5878	1.3764
10°	0.1736	0.9848	0.1763	55°	0.8192	0.5736	1.4281
11°	0.1908	0.9816	0.1944	56°	0.8290	0.5592	1.4826
12°	0.2079	0.9781	0.2126	57°	0.8387	0.5446	1.5399
13°	0.2250	0.9744	0.2309	58°	0.8480	0.5299	1.6003
14°	0.2419	0.9703	0.2493	59°	0.8572	0.5150	1.6643
15°	0.2588	0.9659	0.2679	60°	0.8660	0.5000	1.7321
16°	0.2756	0.9613	0.2867	61°	0.8746	0.4848	1.8040
17°	0.2924	0.9563	0.3057	62°	0.8829	0.4695	1.8807
18°	0.3090	0.9511	0.3249	63°	0.8910	0.4540	1.9626
19°	0.3256	0.9455	0.3443	64°	0.8988	0.4384	2.0503
20°	0.3420	0.9397	0.3640	65°	0.9063	0.4226	2.1445
21°	0.3584	0.9336	0.3839	66°	0.9135	0.4067	2.2460
22°	0.3746	0.9272	0.4040	67°	0.9205	0.3907	2.3559
23°	0.3907	0.9205	0.4245	68°	0.9272	0.3746	2.4751
24°	0.4067	0.9135	0.4452	69°	0.9936	0.3584	2.6051
25°	0.4226	0.9063	0.4663	70°	0.9397	0.3420	2.7475
26°	0.4384	0.8988	0.4877	71°	0.9455	0.3256	2.9042
27°	0.4540	0.8910	0.5095	72°	0.9511	0.3090	3.0777
28°	0.4695	0.8829	0.5317	73°	0.9563	0.2924	3.2709
29°	0.4848	0.8746	0.5543	74°	0.9613	0.2756	3.4874
30°	0.5000	0.8660	0.5774	75°	0.9659	0.2588	3.7321
31°	0.5150	0.8572	0.6009	76°	0.9703	0.2419	4.0108
32°	0.5299	0.8480	0.6249	77°	0.9744	0.2250	4.3315
33°	0.5446	0.8387	0.6494	78°	0.9781	0.2079	4.7046
34°	0.5592	0.8290	0.6745	79°	0.9816	0.1908	5.1446
35°	0.5736	0.8192	0.7002	80°	0.9848	0.1736	5.6713
36°	0.5878	0.8090	0.7265	81°	0.9877	0.1564	6.3138
37°	0.6018	0.7986	0.7536	82°	0.9903	0.1392	7.1154
38°	0.6157	0.7880	0.7813	83°	0.9925	0.1219	8.1443
39°	0.6293	0.7771	0.8098	84°	0.9945	0.1045	9.5144
40°	0.6428	0.7660	0.8391	85°	0.9962	0.0872	11.4301
41°	0.6561	0.7547	0.8693	86°	0.9976	0.0698	14.3007
42°	0.6691	0.7431	0.9004	87°	0.9986	0.0523	19.0811
43°	0.6820	0.7314	0.9325	88°	0.9994	0.0349	28.6363
44°	0.6947	0.7193	0.9657	89°	0.9998	0.0175	57.2900
45°	0.7071	0.7071	1.0000	90°	1.0000	0.0000	∞

563

Metric and Customary Measures

Abbreviations

Metric		Customary	
centimeter (cm)	liter (L)	cup (c)	mile (mi)
decimeter (dm)	meter (m)	foot (ft)	pint (pt)
gram (g)	milliliter (mL)	gallon (gal)	quart (qt)
kilogram (kg)	millimeter (mm)	inch (in.)	yard (yd)

LENGTH

Metric	Customary	Metric/Customary
1 km = 1000 m	1 mi = 1760 yd	1 mi = 1.609 km
1 m = 100 cm = 1000 mm	1 mi = 5280 ft	1 in. = 2.54 cm
1 cm = 10 mm	1 yd = 3 ft = 36 in.	1 km = 0.621 mi
1 mm = 0.1 cm = 0.001 m	1 ft = 12 in.	1 m = 39.37 in.
1 cm = 0.01 m		

AREA

Metric	Customary	Metric/Customary
1 cm^2 = 100 mm^2	1 yd^2 = 9 ft^2	1 mi^2 = 2.59 km^2
1 m^2 = 10,000 cm^2	1 ft^2 = 144 in^2	1 hectare = 2.471 acres
1 hectare = 10,000 m^2	1 acre = 4840 yd^2	1 m^2 = 1.196 yd^2
		1 cm^2 = 0.155 in^2

VOLUME

Metric	Customary	Metric/Customary
1 dm^3 = 1000 cm^3	1 yd^3 = 27 ft^3	1 yd^3 = 0.765 m^3
1 cm^3 = 1000 mm^3	1 ft^3 = 1728 in^3	1 ft^3 = 0.028 m^3
		1 $in.^3$ = 16.387 cm^3

LIQUID

Metric	Customary	Metric/Customary
1 L = 1000 mL	1 gal = 4 qt	1 gal = 3.785 L
	1 qt = 2 pt	1 qt = 0.946 L
	1 pt = 2 c	1 L = 1.057 qt

WEIGHT

Metric	Customary	Metric/Customary
1 kg = 1000 g	1 ton = 2000 lb	1 lb = 0.454 kg
	1 lb = 16 oz	1 kg = 2.205 lb

564

CHAPTER 1

Postulate 1-1-1: Two points determine exactly one line. [p. 3]

Postulate 1-2-1: There is a one-to-one matching between the points on a line and the real numbers. The real number assigned to each point is its coordinate. The distance between two points is the positive difference of their coordinates. If A and B are two points with coordinates a and b such that $a > b$, then $AB = a - b$. [p. 8]

Postulate 1-2-2: Let O be a point on $\overleftrightarrow{XY}$ such that X is on one side of O and Y is on the other side of O. Real numbers from 0 through 180 can be matched with $\overrightarrow{OX}$, $\overrightarrow{OY}$, and all rays that lie on one side of $\overleftrightarrow{XY}$ so that each of the following is true:

(1) 0 is the number matched with $\overrightarrow{OX}$.

(2) 180 is the number matched with $\overrightarrow{OY}$.

(3) If $\overrightarrow{OA}$ is matched with a and $\overrightarrow{OB}$ is matched with b and $a > b$, then the number matched with $\angle AOB$ is $a - b$. [p. 9]

CHAPTER 2

Postulate 2-1-1: Three noncollinear points determine a plane. [p. 43]

Postulate 2-1-2: If two lines intersect, then they intersect at exactly one point. [p. 44]

Postulate 2-2-1: If two parallel lines are intersected by a transversal, then the corresponding angles are equal in measure. [p. 49]

Theorem 2-3-1: If two lines intersect, then the vertical angles are equal. [p. 55]

Theorem 2-3-2: If two parallel lines are intersected by a transversal, then the alternate interior angles are equal. [p. 57]

Theorem 2-5-1: If two parallel lines are intersected by a transversal, then the alternate exterior angles are equal. [p. 67]

565

CHAPTER 3

Postulate 3-1-1: Through a point not on a line there is only one line parallel to the given line. [p. 84]

Theorem 3-1-1: If the figure is a triangle, then the measure of the sum of the angles is 180°. [p. 83]

Theorem 3-2-1: The sum of the lengths of two sides of a triangle is greater than the length of the third side. [p. 89]

Theorem 3-2-2: If given a triangle, then any side is greater than the absolute value of the difference of the other two sides. [p. 90]

Theorem 3-5-1: A measure of an exterior angle of a triangle is equal to the sum of the measures of the two non-adjacent, or remote, interior angles. [p. 107]

CHAPTER 4

Theorem 4-5-1: If a segment connects the midpoints of two sides of a triangle, then the length of the segment is equal to $\frac{1}{2}$ the length of the third side. [p. 147]

Theorem 4-5-2: If a line is parallel to one side of a triangle and intersects the other sides at any points except the vertex, then the line divides the sides proportionally. [p. 148]

CHAPTER 5

Theorem 5-1-1: The empty set is a subset of every set. The symbol is either $\emptyset$ or { }. [p. 169]

Theorem 5-2-1: If a triangle has two equal sides and the included angle is bisected, then the triangle is divided into two congruent triangles. [p. 174]

Theorem 5-3-1: If a triangle has two equal angles, then the sides opposite the equal angles are equal. [p. 179]

Theorem 5-3-2: If a triangle is equiangular, then it is equilateral. [p. 179]

Theorem 5-4-1: If the triangle is isosceles, then the angles opposite the equal sides are equal. [p. 186]

566

CHAPTER 6

Theorem 6-3-1: If two lines are intersected by a transversal so that the alternate interior angles are equal, then the lines are parallel. [p. 220]

Theorem 6-6-1: If the midpoints of two sides of a triangle are joined, then the line segment determined or formed is parallel to the third side and is equal to $\frac{1}{2}$ its length. [p. 235]

Theorem 6-6-2: If a point is on the perpendicular bisector, then it is equidistant from the endpoints of the segment. [p. 237]

Theorem 6-6-3: If a point is on the angle bisector, then the point is equidistant from the sides of the angle. [p. 238]

CHAPTER 7

Theorem 7-1-1: (Pythagorean Theorem) If a right triangle has sides of lengths a, b, and c, where c is the hypotenuse, then $a^2 + b^2 = c^2$. [p. 251]

Theorem 7-1-2: (Converse of the Pythagorean Theorem) If a triangle has three sides of lengths a, b, and c, such that $a^2 + b^2 = c^2$, then the triangle is a right triangle. [p. 251]

Theorem 7-2-1: If the triangle is an isosceles right triangle, then the acute angles are each 45°. [p. 254]

Theorem 7-2-2: The length of the hypotenuse of an isosceles right triangle is $\sqrt{2}$ (about 1.414) times the length of either leg. In symbols this is written as $c = s\sqrt{2}$, or $c \approx 1.414s$. [p. 255]

Theorem 7-3-1: If a triangle is a 30°-60° right triangle, then the length of the leg opposite the 30° angle is half the length of the hypotenuse. [p. 261]

Theorem 7-3-2: If a triangle is a 30°-60° right triangle, then the length of the leg opposite the 60° angle is $\frac{\sqrt{3}}{2}$ times the length of the hypotenuse. [p. 262]

Theorem 7-4-1: If a triangle has one angle greater than another angle, then the side opposite the greater angle is longer than the side opposite the other angle. [p. 266]

Theorem 7-4-2: If a triangle has one side longer than another side, the angle opposite the longer side is greater than the angle opposite the other side. [p. 268]

Theorem 7-5-1: (The Distance Formula) The distance between points $A(x_1, y_1)$ and $B(x_2, y_2)$ is $d = \sqrt{(x_2 - x_1)^2 + (y_2 - y_1)^2}$. [p. 273]

567

CHAPTER 8

Theorem 8-1-1: The sum of the measures of the interior angles of a quadrilateral is 360°. [p. 285]

Theorem 8-1-2: A triangle is a rigid or non-flexible figure.

Theorem 8-2-1: In an isosceles trapezoid, the measures of each pair of base angles are equal. [p. 289]

Theorem 8-2-2: The median of a trapezoid is parallel to the bases and equal to one-half the sum of their lengths. [p. 290]

Theorem 8-3-1: If a quadrilateral is a parallelogram, then the opposite sides are equal and the opposite angles are equal. [p. 294]

Theorem 8-3-2: If the quadrilateral is a parallelogram, then the diagonals bisect each other. [p. 296]

Theorem 8-4-1: If a quadrilateral is a rectangle, then the diagonals are equal and all four angles are right angles. [p. 301]

Theorem 8-5-1: If a parallelogram is a rhombus, then the diagonals are perpendicular. [p. 306]

Theorem 8-5-2: If a figure is a rhombus, then opposite sides are parallel, opposite sides are equal, opposite angles are equal in length, opposite angles have equal measures, the diagonals bisect each other, and the diagonals are perpendicular. [p. 306]

Theorem 8-5-3: If a figure is a square, then opposite sides are parallel, opposite sides are equal in length, there are four right angles, the diagonals are equal in length, the diagonals bisect each other, and the diagonals are perpendicular. [p. 307]

CHAPTER 9

Postulate 9-1-1: If the figure is a rectangle, then the number assigned for the area is the length times the width. [p. 318]

Postulate 9-5-1: If two similar figures have a ratio of a/b for their corresponding sides, then:

a) The ratio of their perimeters is a/b, and

b) The ratio of their areas is $(a/b)^2$.

Theorem 9-2-1: If the figure is a triangle, then the number assigned to the area is $\frac{1}{2}$ times the base times the height. [p. 322]

Theorem 9-3-1: If the figure is a parallelogram, then the area is the length of the base times the height. [p. 329]

Theorem 9-3-2: If the figure is a parallelogram, then the perimeter is twice the sum of two adjacent sides. [p. 331]

Theorem 9-4-1: If the figure is a trapezoid, then the area is $\frac{1}{2}$ the height times the sum of the bases. [p. 337]

568

CHAPTER 10

Postulate 10-3-1: The circumference of a circle is π times the diameter (d). [p. 368]

Theorem 10-1-1: Each angle of a regular N-gon is equal to $(n - 2)$ times 180° divided by n, where n is the number of sides in the N-gon. [p. 358]

Theorem 10-2-1: The area of a regular polygon is $\frac{1}{2}Ph$, where P is the perimeter of the polygon and h is the length of the perpendicular from the center of the polygon to each of its sides. [p. 362]

Theorem 10-3-1: The area of a circle is the radius squared times the number π. [p. 370]

Theorem 10-4-1: The measure of an inscribed angle is $\frac{1}{2}$ the number of degrees in the intersected arc. [p. 375]

Theorem 10-4-2: If the vertex of the angle is within the circle, then the measure of the angle is $\frac{1}{2}$ the sum of the degrees in the two intersected arcs. [p. 375]

Theorem 10-4-3: If the vertex of the angle is outside the circle and the rays intersect the circle, then the measure of the angle is $\frac{1}{2}$ the difference of the measure in degrees of the two intersected arcs. [p. 376]

Theorem 10-4-4: A tangent to a circle forms a 90° angle with the radius of the circle at the point of intersection. [p. 377]

Theorem 10-5-1: If two chords intersect, then the product of the segment lengths of one chord is equal to the product of the segment lengths of the other chord. [p. 380]

Theorem 10-5-2: The perpendicular bisector of a chord of a circle passes through the center of the circle. [p. 381]

Theorem 10-5-3: Given a tangent and a secant from a point, then the length of the tangent squared is equal to the product of the lengths of the secant and its external segment. [p. 382]

569

CHAPTER 11

Postulate 11-2-1: The volume of a right rectangular prism is the product of its length, width, and height: $V = lwh$.

Postulate 11-2-2: (Cavalieri's Principle) If two solid figures have equal heights and bases of the same area, and if every plane parallel to the bases always cuts off two cross-sections of equal area, then the two solids have equal volume.

Theorem 11-1-1: If the height of a prism is h and each base has perimeter P and area B, then the formula for the total surface area TA is: $TA = 2B + Ph$. [p. 405]

Theorem 11-1-2: If the height of a right circular cylinder is h and the radius of each circular base is r, then the formula for the total surface area is Total area $= 2\pi r^2 + 2\pi rh$. [p. 405]

Theorem 11-2-1: The volume of a prism is the product of the area B of one of its bases and the height h of the prism: $V = Bh$. [p. 410]

Theorem 11-2-2: The volume of a circular cylinder is the product of the area B of one of its bases and the height h of the cylinder: $V = Bh$. [p. 411]

Theorem 11-3-1: If p is the perimeter of the base of a regular pyramid and s is the slant height, the lateral area LA is given by the formula $LA = \frac{1}{2}ps$. [p. 415]

Theorem 11-3-2: If B is the area of the base of a regular pyramid, p is the perimeter of the base, and s is the slant height of the pyramid, then the total area TA is given by the formula $TA = B + \frac{1}{2}ps$. [p. 415]

Theorem 11-3-3: The lateral area LA of a right circular cone with a base of radius r and slant height s is given by $LA = \pi rs$. [p. 417]

Theorem 11-3-4: The total area TA of a right circular cone is the sum of the area of the base and the lateral area. If r is the radius of the base and s is the slant height, the total area is given by the formula $TA = \pi r^2 + \pi rs$. [p. 417]

Theorem 11-4-1: The volume of a triangular pyramid is equal to one third of the product of the area of the base and the height. If B is the area of the base and h is the height, then the volume is given by the formula $V = \frac{1}{3}(B)h$. [p. 421]

Theorem 11-4-2: The volume of any pyramid is equal to one-third the product of the area of the base and the height. The formula is $V = \frac{1}{3}Bh$. [p. 422]

Theorem 11-4-3: The volume of a right circular cone is equal to one-third the product of the area of the base and the height of the cone: $V = \frac{1}{3}Bh = \frac{1}{3}\pi r^2 h$. [p. 423]

Theorem 11-6-1: The formula for the volume V of a sphere is $V = \frac{4}{3}\pi r^3$, where r is the radius of the sphere. [p. 432]

Theorem 11-6-2: The formula for the surface area A of a sphere of radius r is $A = 4\pi r^2$. [p. 434]

570

CHAPTER 12

Theorem 12-1-1: The locus of points equidistant from the sides of an angle is the angle bisector. [p. 446]

Theorem 12-1-2: The locus of points equidistant from the endpoints of a line segment is the perpendicular bisector of the line segment. [p. 447]

Theorem 12-1-3: The locus of points in a plane that are a given distance from a line on the plane is two parallel lines. [p. 447]

Theorem 12-1-4: The locus of points a given distance from a point on a plane is a circle. [p. 447]

Theorem 12-1-5: The locus of the vertex (C) of all right triangles with given hypotenuse (AB) is a circle with diameter AB minus the endpoints A and B of the hypotenuse. [p. 447]

CHAPTER 13

Theorem 13-2-1: If (x_1, y_1) and (x_2, y_2) are two points on the graph of $y = mx + b$, then the slope m of the graph is equal to $\frac{y_2 - y_1}{x_2 - x_1}$ [p. 484]

Theorem 13-2-2: For any real numbers m and b, the y-intercept of the graph line of $y = mx + b$ is b, which means that the line crosses the y-axis at $(0, b)$. [p. 487]

Theorem 13-3-1: If two nonvertical lines in the coordinate plane are parallel, then their slopes are equal. [p. 490]

Theorem 13-3-2: If l_1 and l_2 are perpendicular lines with slopes m_1 and m_2 respectively, then $m_1m_2 = -1$. [p. 492]

Theorem 13-4-1: The points $P(x, y)$ whose coordinates satisfy the equation $x^2 + y^2 = r^2$, where r is a positive real number, form a circle of radius r whose center is $(0, 0)$. [p. 496]

Theorem 13-5-1: An equation for the parabola determined by $(0, a)$ and the line $y = -a$, where $a > 0$, is $y = \frac{1}{4}a \cdot x^2$. [p. 502]

Theorem 13-6-1: (The Midpoint Theorem) If $P(a, b)$ and $Q(c, d)$ are any two points in the coordinate plane, then the midpoint of PQ is $M\left(\left(a + \frac{c}{2}\right), \left(b + \frac{d}{2}\right)\right)$. [p. 507]

571

CHAPTER 14

Theorem 14-4-1: (Law of Sines) In any triangle *ABC*, $\frac{\sin A}{a} = \frac{\sin B}{b} = \frac{\sin C}{c}$. [p. 537]

Theorem 14-4-2: (Law of Cosines) In any triangle are *ABC*,

$a^2 = b^2 + c^2 - 2bc \cos A$

$b^2 = a^2 + c^2 - 2ac \cos B$ and

$c^2 = a^2 + b^2 - 2ab \cos C$. [p. 539]

Theorem 14-6-1: If x is the measure of any angle, then $\sin^2 x + \cos^2 x = 1$. [p. 551]

Theorem 14-6-2: If x is the measure of any angle, then $\cos 2x = \cos^2 x - \sin^2 x$. [p. 551]

Theorem 14-6-3: If x is the measure of any angle such that $\cos x \neq 0$, then $\tan x = \sin x/\cos x$. [p. 551]

572

PHOTO ACKNOWLEDGMENTS

All photography by The Image Bank, Chicago, unless marked with asterisk (*)

Cover Comstock: Mike & Carol Werner*

Preface p. viii: Jeff Hunter; Joseph Devenney; p. ix: Gary Gladstone, 1989; Dan Rest*; p. x: Mel Di Giacomo and Bob Masini; Gary Gladstone; p. xi: Murray Alcosser; Murray Alcosser; Michael Salas, 1976; p. xii: Kay Chernush; Hank Delespinasse; p. xiii: Peter Miller; Cesar Lucas; Joseph Devenney; Harald Sund; p. xix: Jacques Cochin, 1988.

Chapter 1 p. 1: Larry Keenan Associates; p. 2: Eric Meola; Sobel/Klonsky, 1988; B. Lindhout; Guido Alberto Rossi; p. 3: Michael Melford; p. 7: Michael Salas; Marc Romanelli; p. 9: Steve Krongard, 1990; p. 13: William Rivelli; Lou Jones; p. 19: Alvis Upitis, 1989; p. 24: D.W. Productions*; Michael Salas; p. 25: Alfredo Tessi; p. 28: John P. Kelly; p. 30: Schneps; Steve Dunwell.

Chapter 2 p. 41: Grant V. Faint; p. 42: Pete Turner; p. 44: Mitchell Funk; Eddie Hironaka; p. 47: Gary Cralle; David Brownell; p. 53: Gerard Champlong; p. 55: Mitchell Funk; p. 59: Steve Proehl; Harald Sund; Gary Gladstone; p. 65: Ted Russell; Marc Romanelli; Walter Bibikow, 1988; p. 71: Gerard Champlong; p. 72: Jake Rajs.

Chapter 3 p. 81: L. Mason; p. 82: Geoff Gove; Bernard Roussel; p. 83: Benn Mitchell; p. 84: Ron Kadrmas*; p. 87: Jürgen Vogt; Ron Kadrmas*; p. 92: Richard Pan, Stockphotos, Inc.*; p. 93: Ulli Seer; p. 98: Ron Kadrmas*; p. 99: Gary Cralle; p. 100: W. Peisenroth; p. 105: Romilly Lockyer; Paul Silverman, 1984, Stockphotos, Inc.*; p. 110: Tim Bieber; p. 111: P. Runyon; p. 116: Brett Froomer.

Chapter 4 p. 123: Gary Cralle; p. 124: Margarette Mead; p. 129: Al Giddings, 1989; p. 131: James H. Carmichael; Philip A. Harrington; p. 134: Jacky Gucia; p. 135: Weinberg-Clark; p. 140: Arthur Meyerson; p. 141: Jay Freis; p. 144: Stephen Derr; p. 145: Guido Alberto Rossi; Andre Gallant; p. 146: Luis Castañeda, 1989; p. 152: Luis Castañeda; p. 153: Chris Alan Wilton; p. 158: John Kelly; Nicholas Foster.

Chapter 5 p. 165: Charles C. Place; p. 166: Garry Gay; p. 171: Hans Wolf; p. 172: Jürgen Vogt; p. 173: A. M. Rosario; p. 177: Terje Rakke; Guilano Colliva; Jeffrey M. Spielman; p. 183: David W. Hamilton; Eddie Hironaka; p. 189: Steve Dunwell; p. 192: Terje Rakke; Mahaux Photography; Kenneth Redding; p. 193: Jeff Cadge.

Chapter 6 p. 205: Robert Holland; p. 206; Stephen Derr; Don Klumpp; p. 211: Guido Alberto Rossi; p. 213: Stephen Marks; John Ramey; p. 216: Jürgen Vogt, 1988; T. Bieber; p. 217: Color Day Productions*; p. 222: Marc Romanelli; Color Day Photography*; p. 223: Kim Steele; Bernard Roussel p. 230: Joe Azzara; Murray Alcosser; p. 232: Murray Alcosser; p. 235: Bernard Van Berg.

Chapter 7 p. 227: Mark Solomon; p. 248: Merrell Wood; p. 250: Francois Dardelet; Kay Chernush; Barrie Rokeach; p. 253: Patrick Doherty, Stockphotos, Inc.*; Bill Varie; p. 258: Cliff Feulner; p. 259: Benn Mitchell; Peter Miller; p. 265: Peter Miller; Barrie Rokeach, 1988; p. 269: Marc Romanelli; p. 271: Peter Miller; Barrie Rokeach.

Chapter 8 p. 281: Steve Proehl; p. 282: Barrie Rokeach, 1984; p. 283: Bruce Wodder, 1983; p. 285: Gary Bistram, 1989; p. 287: David J. Maenza; p. 288: H. Wendler; p. 293: Harald Sund; p. 294: L. Mason; p. 296: Mel Di Giacomo; p. 297: Patti McConville; p. 299: Jane Sobel, 1980; p. 300: Michael Melford; p. 302: Jay Brousseau; p. 303: Gary Faber; p. 304: Walter Iooss, Jr.; p. 306: Gary Faber; p. 307: Benn Mitchell.

Chapter 9 p. 315: H.G. Kaufmann; p. 316: Harald Sund; p. 317: David W. Hamilton; p. 321: Daniel Hummel; p. 325: Yuri Dojc; p. 327: Larry Dale Gordon; p. 328: Barrie Rokeach, 1989; P. & G. Bowater; p. 333: Steve Proehl; p. 335: Alvis Upitis; p. 336: Jake Rajs, 1987; p. 342: Marvin E. Newman; p. 343: Anthony A. Boccaccio; p. 350: Patti McConville.

Chapter 10 p. 355: David W. Hamilton; p. 356: Jürgen Vogt; p. 361: Tim Bieber; p. 363: David W. Hamilton; p. 365: Gregory Heisler; p. 366: Marc Romanelli; p. 367: Edward Bower; Jerry Yulsman; p. 370: Weinberg-Clark; p. 373: Chuck Kuhn; R. Phillips; p. 379: Andre Gallant; p. 383: Klaus Mitteldorf; p. 385: G. & J. Images*; p. 386: G. V. Faint; p. 388: Chuck Place; p. 389: Michael Tcherevkoff; Gerard Champlong.

Chapter 11 p. 401: Larry Dale Gordon; p. 402: Jeff Hunter; p. 404: Michael Tcherevkoff; p. 405: Alvis Upitis; Tim Bieber; p. 407: Lionel Isy-Schwart; p. 413: Alan Becker; p. 419: Jeff Spielman, Stockphotos, Inc.*; p. 423: Paolo Gori; p. 424: Jake Rajs; p. 425: David Jeffrey; Schmid/Langsfeld; p. 426: Co Rentmeester; p. 427: Harald Sund; p. 430: Garry Gay; p. 431: P. & G. Bowater; Marti Pie; p. 434: Douglas Struthers.

Chapter 12 p. 443: Geoffrey Gove; p. 444: Don Klumpp; p. 449: Joseph Brignolo; p. 451: Lisl Dennis; p. 453: Garry Gay; p. 455: Nicholas Foster; p. 456: Al Satterwhite; p. 459: Marcel Isy-Schwart; p. 461: Gary Gladstone; p. 462: Ulf E. Wallin, 1986, Stockphotos, Inc.*; Gerard Champlong; p. 464: IN FOCUS INTERNATIONAL*; Lou Jones; p. 467: Michael Melford, 1988; p. 470: Nicholas Foster, 1978.

Chapter 13 p. 477: Hans Wolf Studio; p. 478: Eric Schweikardt; p. 480: Geoffrey Gove; p. 483: Albert Normandin; p. 489: Joseph Szkodzinski; p. 490: Weinberg-Clark; Albert Normandin, 1990; p. 494: Steve Proehl; p. 495: John Kelly; p. 497: Brett Froomer; p. 501: Steve Dunwell; p. 503: Guido Alberto Rossi; p. 504: Sobel-Klonsky; Brett Froomer; p. 506: Don King, 1988; p. 507: David W. Hamilton; p. 509: David W. Hamilton.

Chapter 14 p. 517: Trent Swanson; p. 518: Bill Carter; p. 520: ZAO-Longfield; p. 523: Murray Alcosser; p. 526: Nicholas Foster; p. 527: Edward Bower, 1988; p. 529: Tom Mareschal; p. 531: Cesar Lucas; p. 534: Joe Azzara; Denny Tillman; p. 536: Charles C. Place; J. Ramey; p. 541: Jeff Spielman, Stockphotos, Inc.*; p. 542: G. Rossi; Arthur d'Arazien; p. 545: Steve Satushek; p. 547: Michael Melford.

573

CHAPTER 1

Class Activity, p. 5

1. Point P 3. $\overrightarrow{GF}$
5. $\overrightarrow{KL}$
7. $\angle ADH$, or $\angle HDA$
9. $\angle FGH$, or $\angle HGF$
11. W R
13. F G
15. Z X
17. X Y
19.

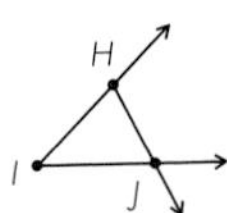

Home Activity, p. 6

1. $\overleftrightarrow{SW}$ or $\overleftrightarrow{WS}$ 3. $\overrightarrow{PT}$
5. $\angle JKL$ or $\angle LKJ$
7. C X
9. plane 11. ray
13. $\overrightarrow{GA}$, $\overrightarrow{GC}$ 15. $\overline{GC}$, $\overline{AD}$

Class Activity, p. 8

1. −3, 1, 3, 5 3. −2
5. 3 units 7. 3 in.
9. 5 cm

Class Activity, p. 11

1. Obtuse 3. Right
5.
7.

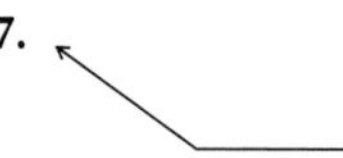

9. 120°; Obtuse
11. 68°; Acute

Home Activity, p. 12

1. −6, −1, 3, and 9
3. 6, 1, 3, and 9 respectively
5. 4 7. 8
9. Obtuse 11. 75°; Acute
13. $\angle AOB$, 15°, Acute; $\angle AOC$, 45°, Acute; $\angle BOC$, 30°, Acute

Class Activity, p. 14

1. 3 in., 1.5 in.
3. 8 cm, 4 cm
5. 8 in., 4 in.
7. C

Class Activity, p. 15

1. 120°
3. Acute; 20°; 20°
5. Obtuse; 60°; 60°
7. Straight; 90°; 90°

Class Activity, p. 16

1.

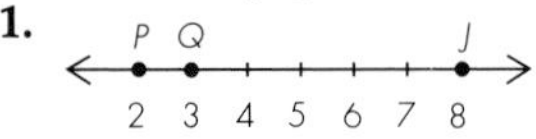

3. $2 < 3 < 8$ 5. 1
7. 6

Class Activity, p. 17

1.

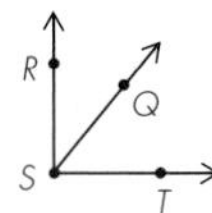

Home Activity, p. 17

1. 4 3. 18
5. 6 in.; 3 in.
7. Acute; 30°; 30°
9. $\angle BEC$ 11. 50°
13.

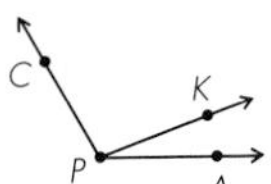

Class Activity, p. 20

1. Answers may vary.
3. 3°, 5°; 95°
5. $m\angle CEF = 132°$
   $m\angle FEG = 48°$
   $m\angle CED = 132°$
7. 54°; 126°

Class Activity, p. 22

1. 55°
3. *Equation*
   $45 + c = 90$
   $10 + c = 90$
   $80 + c = 90$
   $85 + c = 90$
   $14 + c = 90$
   $29 + c = 90$

574

$x + 40 + c = 90$
$2x + 20 + c = 90$
Complement
$(90 - x)°$
$(130 - x)°$
$(70 - 2x)°$

5. 59°; 31° 7. 50°; 40°
9. 72°; 18°

Home Activity, p. 23

1. ∠*EOF* and ∠*FOG*
3. Possible answers: ∠*AOB* and ∠*AOF*, ∠*EOC* and ∠*FOA*; other answers are possible.
5. ∠*EOA*
7. 150° 9. 18.6°
11. 50° 13. 83.95°
15. Acute 17. Obtuse
19. Obtuse 21. 95°; 85°
23. 30°; 60° 25. 33°
27. 138°
29. Drawing shows complementary angles of 75° and 15°.

Critical Thinking, p. 24

85 baseballs

Class Activity, p. 26

1. 27; Possible answer: 1 less than number of dots per side times 3—continue the pattern:
 $(2 - 1) \times 3 = 3$ — 2 per side;
 $(3 - 1) \times 3 = 6$ — 3 per side;
 $(4 - 1) \times 3 = 9$ — 4 per side.
 (and so on)

Class Activity, p. 28

1. 21, 31, 43
3. 63, 127, 255
5. $0.\overline{36}$; $0.\overline{81}$

Home Activity, p. 29

1. 17, 20, 23
3. 33, 65, 129
5. $\frac{1}{4} + \frac{1}{16} + \frac{1}{32} + \frac{1}{64} = \frac{23}{64}$ shaded
7. 15 9. $\frac{1}{2} n(n - 1)$

Class Activity, p. 32

1.

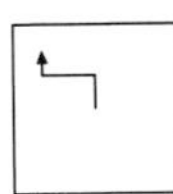

3.

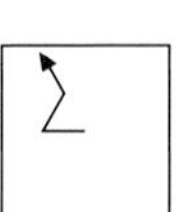

5. Answers may vary.

Home Activity, p. 32

1. Answers may vary.
3. Answers may vary.

Class Activity, p. 34

1.

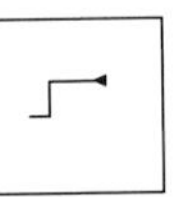

3. Answers may vary.
5. Answers may vary.

Class Activity, p. 36

1. Answers may vary.
3. Answers may vary.
5. Answers may vary.
7. m∠*ABX* = m∠*XBC*; $\overline{BX}$ is the bisector of ∠*ABC*.
9. Answers may vary.

Chapter 1 Review

1. $\overline{AB}$ or $\overline{BA}$
3. ∠*TAK*, ∠*KAT*, or ∠*A*
5. 7.9 cm 7. 115°; obtuse
9. −6, −2, 1, 5
11. *L* is between *J* and *K*.
13. $(7x)° = 140°$; $(3x - 20)° = 40°$
15. a) 21, 28, 36
 b) 162, 486, 1458

CHAPTER 2

Class Activity, p. 43

1. b **3.** e **5.** d
7. True
9. Coplanar
11. Collinear and Coplanar
13. Coplanar

Class Activity, p. 45

1. Intersecting
3. Parallel
5. $\overleftrightarrow{BF}$, $\overleftrightarrow{AC}$ **7.** $\overleftrightarrow{BF}$, $\overleftrightarrow{DE}$
9. False

Home Activity, p. 46

1. Yes **3.** No
5. Yes **7.** Yes
9. $\overleftrightarrow{AB}$, $\overleftrightarrow{EF}$ **11.** $\overleftrightarrow{CD}$, $\overleftrightarrow{EF}$
13. Answers may vary.

Critical Thinking, p. 47

16

Class Activity, p. 48

1. *a* and *d*
3. Diagram B

Class Activity, p. 50

1. *a*
3. $\angle 1$, $\angle 5$; $\angle 4$, $\angle 8$; $\angle 2$, $\angle 6$; $\angle 3$, $\angle 7$
5. 8 **7.** 7
9. 150° **11.** 30°
13. All angles have a measure of 90°.

Home Activity, p. 51

1. *l* and *k*
3. *a* and *b*
5. $\angle 5$; 67° **7.** $\angle 7$; 67°
9. 25° **11.** 25° **13.** 25°
15. 25° **17.** 100°; 80°

Class Activity, p. 54

1. H: you live in Utah; C: you live in the United States.
3. H: your pet is a terrier; C: it is a dog.
5. $M\angle PQR > 90°$.
7. Points *L*, *M*, and *R* determine a plane.

Class Activity, p. 56.

Answers for Exercises 1 through 4 may vary.
5. Yes

Class Activity, p. 57

1. *t*
3. $\angle 1$, $\angle 4$; $\angle 5$, $\angle 8$; $\angle 2$, $\angle 3$; $\angle 6$, $\angle 7$
5. $\angle 3$ and $\angle 6$; $\angle 4$ and $\angle 5$
7. 1 = 85°; 2 = 95°
9. Answers may vary.

Home Activity, p. 58

1. H: two angles are vertical angles; C: they are equal.
3. Pete knows the city.
5. The square of 2.001 is greater than 4.
7. 1 = 40°; 2 = 40°
9. $m\angle 3 = 93°$; $m\angle 4 = 56°$; $m\angle 5 = 87°$; $m\angle 6 = 93°$; $m\angle 7 = 37°$; $m\angle 8 = 143°$; $m\angle 9 = 143°$; $m\angle 10 = 56°$; $m\angle 11 = 124°$

Critical Thinking, p. 58

$m\angle A = m\angle B = 90°$

Class Activity, p. 61

1. Nearest inch: 4 in., error: $\frac{1}{2}$ in; nearest $\frac{1}{8}$ in: $4\frac{1}{8}$ in. error $\frac{1}{16}$ in; nearest $\frac{1}{16}$ in: $4\frac{1}{16}$ or $4\frac{2}{16}$ in, error: $\frac{1}{32}$ in.
3. Nearest cm: 5 cm, error: $\frac{1}{2}$ cm or 5 mm; nearest mm: 52 mm, error: $\frac{1}{2}$ mm
5. Nearest 5 degrees: 70°; error: $2\frac{1}{2}°$; nearest degree: 72°; error: $\frac{1}{2}°$
7. Nearest 5 degrees: 115°; error: $2\frac{1}{2}°$; nearest degree: 116°; error $\frac{1}{2}°$

Class Activity, p. 63

1. 4 ft 10 in.
3. 120 in.; 3.05 m

Home Activity, p. 63

1. Nearest $\frac{1}{4}$ in.: $4\frac{1}{4}$ in.; error: $\frac{1}{8}$ in.; nearest $\frac{1}{8}$ in.: $4\frac{2}{8}$ in.; error: $\frac{1}{16}$ in.; nearest cm: 11 cm; error: 0.5 cm
3. 32°; error $\frac{1}{2}^\circ$
5. 86°; error $\frac{1}{2}^\circ$
7. $2\frac{3}{4}''$, $1\frac{5}{8}''$, $2\frac{1}{4}''$; about $6\frac{5}{8}$ in.
9. $2\frac{5}{8}''$ long, 1″ high; about $7\frac{3}{5}$ in.
11. Answer should be close to 360°.
13. 609 mm **15.** 15 yd
17. 13 ft 4 in.
19. 158° 45′ **21.** 118° 40′
23. 141° 28′ **25.** 66° 55′

Critical Thinking, p. 64

60°; 180°; 30°

Class Activity, p. 68

1. 2; 4
3. ∠3, ∠6, ∠7
5. ∠2, ∠3, ∠7
7. ∠2, ∠6, ∠7
9. ∠1, ∠4, ∠5
11. 90° **13.** 73.4°
15. 106.6° **17.** 73.4°
19. ∠1 and ∠11; ∠4 and ∠10
21. ∠3, ∠5, ∠7, ∠9, ∠11, ∠13, ∠15
23. 70° **25.** 70°

Home Activity, p. 69

1. ∠1 and ∠8; ∠2 and ∠7
3. ∠3 and ∠6; ∠5 and ∠4
5. 45° **7.** 135° **9.** 45°
11. 52° **13.** 52° **15.** 128°
17. 147° **19.** 147° **21.** 147°
23. 95° **25.** 65° **27.** 55°

Critical Thinking, p. 70

∠2 and ∠4

Class Activity, p. 73

1.

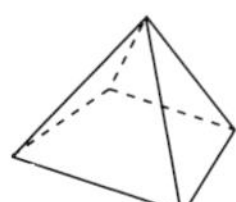

3.

Home Activity, p. 74

1. See students' drawings.
3. Answers may vary.
5. Possible answers:
- Cubes piled in all 6 corners;
- 4 cubes (one unseen) piled in corner;
- Star within a hexagon.

7.

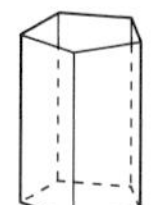

9.

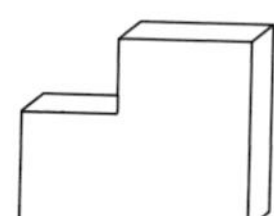

Critical Thinking, p. 74

Both will form a cube.

Chapter 2 Review, p. 78

1. $\overleftrightarrow{AB}$, $\overleftrightarrow{FG}$; $\overleftrightarrow{DH}$, $\overleftrightarrow{AB}$; $\overleftrightarrow{DH}$, $\overleftrightarrow{DC}$; $\overleftrightarrow{BC}$, $\overleftrightarrow{AB}$; $\overleftrightarrow{BC}$, $\overleftrightarrow{DC}$.
3. $\overleftrightarrow{FG}$ and $\overleftrightarrow{AB}$
5. $\overleftrightarrow{FG}$ and $\overleftrightarrow{DC}$, $\overleftrightarrow{DH}$, $\overleftrightarrow{BC}$
7. *F*, *E*, *G*; *A*, *E*, *B* (Other answers are possible.)
9. ∠1 and ∠5, ∠2 and ∠6, ∠3 and ∠7, ∠4 and ∠8
11. ∠1 and ∠8, ∠2 and ∠7
13. 50° **15.** 130° **17.** 50°
19. 2 in.; error: $\frac{1}{4}$ in.
21. 118°; error: $\frac{1}{2}^\circ$
23. Check students' diagrams.

CHAPTER 3

Class Activity, p. 83

1. Yes
3. No; there are 4 sides.

Class Activity, p. 85

1. $x = 80°$ **3.** $m = 10°$

Home Activity, p. 86

1. a. $\triangle ABC$
b. $\overline{AB}, \overline{BC}, \overline{CA}$
c. A, B, C
d. $\angle BAC, \angle ABC, \angle ACB$
3. $x = 80°$; A
5. $x = 50°$; A
7. $y = 60°$; $a = 40°$; $x = 80°$; $z = 120°$; $c = 140°$
9. No; the sum of the measures of the angles would be greater than 180°.
11. $m\angle C = 60°$

Class Activity, p. 90

1. $3 < x < 17$
3. $2 < AC < 8$
5. $10 < RS < 40$

Home Activity, p. 91

1. Yes **3.** No **5.** Yes
7. $2 < x < 8$
9. 2 ft $< x <$ 7 ft
11. 15 km $< x <$ 105 km
13. 16 and 24
15. $4 < HJ < 44$
17. 9 km $< VJ <$ 29 km
19. $150 + 180 > 340$; the triangle is not possible.
21. $16 < x < 36$

Critical Thinking, p. 92

22. The shortcut and sidewalks form a triangle. The shortcut is shorter than the sum of the lengths of the sidewalks.

Class Activity, p. 95

1. Scalene
3. Isosceles
5. $\triangle CED$ or $\triangle ABE$
7. 8 **9.** Answers may vary.

Home Activity, p. 96

1. Scalene
3. Isosceles
5. Equilateral
7. Scalene
9. Equilateral
11. Isosceles
13. Scalene **15.** 4
17. Scalene
19. $m\angle 1 = 35°$; $m\angle 2 = 145°$; $m\angle 3 = 35°$; $m\angle 4 = 145°$; $m\angle 5 = 35°$; $m\angle 6 = 145°$; $m\angle 7 = 35°$; $m\angle 8 = 55°$; $m\angle 9 = 55°$; $m\angle 10 = 90°$
21. They measure 90°.
23. Yes; any two sides chosen are of equal length.

Critical Thinking, p. 98

25. No; 25 km would exceed the total of the maximum possible lengths of the legs of the trip.

Class Activity, p. 102

1. $m\angle x = 40°$; acute
3. $m\angle x = 60°$; equiangular
5. $m\angle X = 75°$; $m\angle Y = 50°$; $m\angle 7 = 55°$; acute
7. Equiangular

Home Activity, p. 103

1. $m\angle A = 60°$; $m\angle C = 60°$; equiangular
3. $m\angle R = 74.2°$; acute
5. 100° **7.** Same
9. $m\angle J = 30°$; $m\angle O = 60°$; $m\angle JTO = 90°$; Right
11. $\triangle ABC$, $\triangle CEF$, $\triangle ACD$, $\triangle AEB$, $\triangle BFD$, $\triangle BCD$, $\triangle BEC$, $\triangle BCF$

Critical Thinking, p. 104

13. True **15.** False
17. True **19.** False
21. True **23.** False
25. False

Class Activity, p. 108

1. 120° **3.** 110°
5. $m\angle 1 = 25°$; $m\angle 2 = 35°$

Home Activity, p. 108

1. 6;

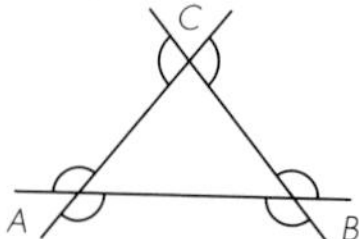

3. 166° **5.** 130°
7. $m\angle 1 = 40°$; $m\angle 2 = 140°$; $m\angle 3 = 80°$; $m\angle 4 = 60°$; $m\angle 5 = 120°$
9. 720°
11. Yes; both can be 90°.
13. 20′ **15.** 1.36 miles
17.–19. Answers may vary. To accomplish the task in Exercises 16–19, the turtle must be turned through the exterior angle.

Critical Thinking, p. 110

21. Bob

Class Activity, p. 113

17. Isosceles, right
19. Scalene, acute
21. A (3, 0) B (0, −9) C (2, 3)
D (4, 7) E (−6, 1) F (−2, −4)
G (−5, 0) H (3, 9) I (2, 7)
J (−7, 5) K (−1, 6) L (−4, −5)

Home Activity, p. 114

1.

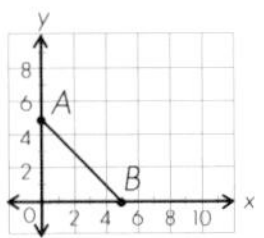

Isosceles, right

3.

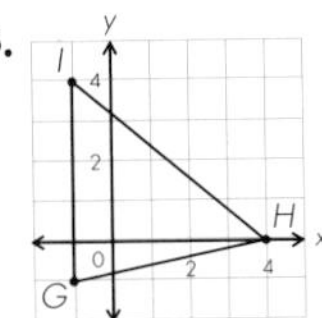

Scalene, acute

5.

A (2, 3)	B (0, 0)
C (7, 9)	D (−10, 0)
E (−3, 2)	F (−7, 0)
G (0, 6)	H (0, −4)
I (−1, 7)	J (2, −6)
K (5, −9)	L (10, 0)
M (0, −9)	N (7, −8)
O (−10, 4)	P (8, −5)

7. Right triangle, scalene triangle
9. Scalene triangle, obtuse triangle
11. Ray **13.** No; the order of the coordinates is different.
15. Parallel
17. Intersecting at (2, −5)
19.

A(−8 −2)	B(−4, −2)	I(4, 4)	J(10, 4)
C(−2, 0)	D(−4, 2)	K(10, 8)	L(6, 8)
E(−8, 2)	F(−12, 4)	M(0, 10)	N(5, −2)
G(−2, 4)	H(−2, 8)	O(0, −2)	

Answers may vary. Possible answers: all have at least one right angle; all have areas of 20; all have at least one side parallel to x-axis; all have at least one side parallel to y-axis.

Critical Thinking, p. 116

21. $y = x$; line
23. Flexible **25.** Rigid

Chapter 3 Review, p. 120

1. 5 **3.** 1, 3, 6, 7
5. 2, 3 **7.** 6
9. 93° **11.** 46°
13. 67°
15. 8; 66
17. Carl damaged the car, Bev the bike.
19. Scalene, obtuse

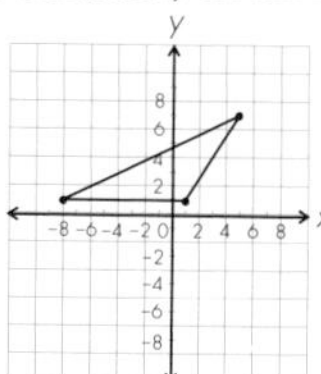

CHAPTER 4

Class Activity, p. 126

1.

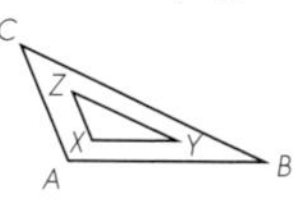

3. $\angle A \leftrightarrow \angle X$; $\angle B \leftrightarrow \angle Y$; $\angle C \leftrightarrow \angle Z$
5. Same shape

Home Activity, p. 127

7. $A \leftrightarrow D$, $B \leftrightarrow E$, $C \leftrightarrow F$, $\overline{AB} \leftrightarrow \overline{DE}$, $\overline{BC} \leftrightarrow \overline{EF}$, $\overline{AC} \leftrightarrow \overline{DF}$, $\angle A \leftrightarrow \angle D$, $\angle B \leftrightarrow \angle E$, $\angle C \leftrightarrow \angle F$
9. Answers may vary.
11.

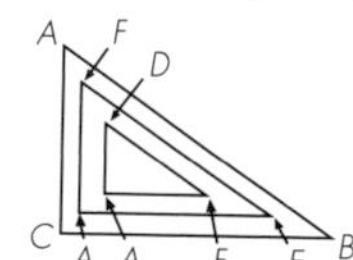

13. $\overline{AC} \leftrightarrow \overline{FG}$, $\overline{AC} \leftrightarrow \overline{DG}$, $\overline{FG} \leftrightarrow \overline{DG}$, $\overline{AB} \leftrightarrow \overline{FE}$, $\overline{AB} \leftrightarrow \overline{DF}$, $\overline{FE} \leftrightarrow \overline{DF}$, $\overline{CB} \leftrightarrow \overline{GE}$, $\overline{CB} \leftrightarrow \overline{GF}$, $\overline{GE} \leftrightarrow \overline{GF}$
15. Flip $\triangle FED$ over. $\angle A \leftrightarrow \angle E$; $\angle B \leftrightarrow D$; $\angle C \leftrightarrow \angle F$
17. Answers may vary.
19.

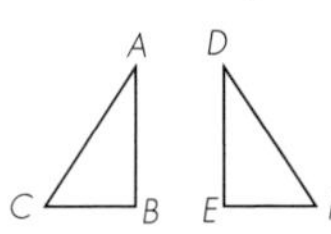

21. Flip $\triangle ABC$ onto $\triangle DEF$, or vice versa.

Class Activity, p. 130

1. All except e. 3. $\frac{1}{4}$
5. $\frac{1}{4}$ 7. $\frac{1}{1}$
9. 2.4 11. 5.489

Class Activity, p. 132

1. $\frac{4}{7}$ 3. $\frac{17}{24}$
5. 5 boys to every 7 girls
7. 41 gallons

Home Activity, p. 132

1. $\frac{1}{2}$ 3. $\frac{5}{8}$
5. $\frac{15}{17}$ 7. $\frac{3}{2}$
9. $\frac{1}{6}$ 11. $\frac{3}{1}$
13. $\frac{1}{2}$ 15. $\frac{1}{2}$
17. $\frac{5}{6}$ 19. 4.5
21. 20 23. $56\frac{2}{3}$
25. $-\frac{2}{3}$

Critical Thinking, p. 133

27. 13-oz box

Class Activity, p. 137

1. By AA
3. $\angle A \leftrightarrow \angle F$; $\angle B \leftrightarrow \angle D$; $\angle C \leftrightarrow \angle E$
5. $EF = 4.4$; $ED = 4.8$
7. Answer may vary.
9. Answer may vary.
11. Answer may vary.

Home Activity, p. 138

1. $\angle A \leftrightarrow \angle D$, $\angle B \leftrightarrow \angle E$, $\angle C \leftrightarrow \angle F$, $\overline{AB} \leftrightarrow \overline{DE}$, $\overline{BC} \leftrightarrow \overline{EF}$, $\overline{AC} \leftrightarrow \overline{DF}$
3. Yes
5. By SAS
7. Not enough information to tell
9. Yes 11. By AA
13. 3:7; $x = 11\frac{2}{3}$

Class Activity, p. 143

1. 1725 miles
3. Answers may vary.
5. Triangle with sides 3.25 cm, 1.25 cm, and 3 cm

Home Activity, p. 144

1. 1:19.2 3. 8.5 in.
5. 11.3 in. 7. 10.08 m
9. 16 in. 11. 1:2.54
13. 52,081:1 15. 1:0.9
17. 1:3 19. 1:12
21. 1:100
23. Triangle with sides of length 2.5 cm, 1.5 cm, and 2 cm

Class Activity, p. 149

1. 7.5

Home Activity, p. 149

1. Yes; SAS 3. Yes; SAS
5. 100 7. 50
9. Yes; SAS
11. Yes; AA
13. Yes; AA
15. Yes; AA
17. 3, △ABC, △ACD, △BCD
19. 20 21. Yes; SAS
23. 5 25. 20 27. $6\frac{2}{7}$

Critical Thinking, p. 151

$\frac{\$40}{100} = \frac{x}{15}$, or $x = \frac{40 \times 15}{100}$

Class Activity, p. 154

1. $x = 28$, $y = 70$
3. 90° 5. 50°
7. $66\frac{2}{3}$; $26\frac{2}{3}$

Home Activity, p. 155

1. $18\frac{2}{3}''$ and $23\frac{1}{3}''$
3. $x = 15$ cm; $y = 9$ cm
5. $x = 10$; $y = 2\sqrt{3}$
7. 17.5 m 9. 55°
11. 55° 13. Yes; AA
15. Yes; AA 17. Yes; AA
19. 23.75 ft 21. 61.25 ft
23. 54.44 ft

Critical Thinking, p. 158

25. Height to eye; distance from feet to mirror and base of building to mirror
27. Answers will vary.

Chapter 4 Review, p. 162

1. SSS; $\frac{3}{1}$

3. AA; $\frac{2}{1}$

5. Yes; AA **7.** 65°

9. 65° **11.** 3

13. 15

CHAPTER 5

Class Activity, p. 167

1. True **3.** True

5. Possible answer: Set of one-digit numbers.

7. {2, 4}, {2, 6}, {2, 8}, {4, 6}, {4, 8}, {6, 8}

Class Activity, p. 169

1. 

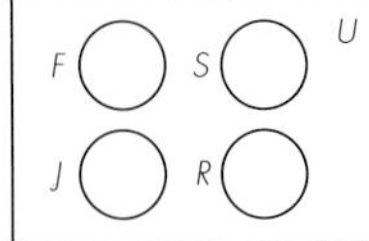

3.

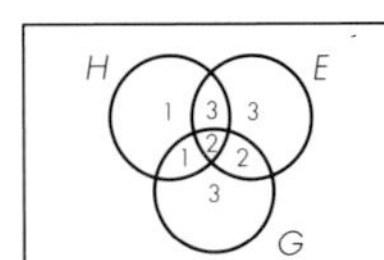

5. 12; 12; 14

7. Yes; yes **9.** An angle

Home Activity, p. 170

1. {6, 12, 18, 24}

3. {a, e, i, o, u}

5. {101, 102, 103, . . . }

7. Yes **9.** Yes

11. 30; 10; 18

13. 317

Critical Thinking, p. 170

15. 2^n

Class Activity, p. 173

1.

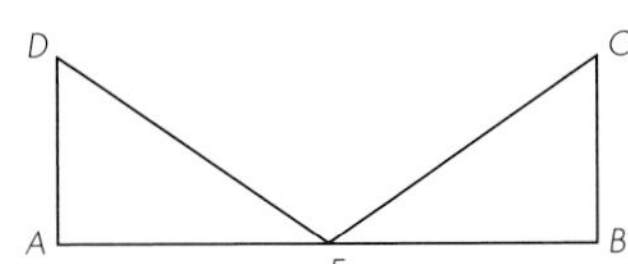

Yes; $\triangle AED \sim \triangle BED$ by SAS; $\frac{DA}{CB} = \frac{AE}{BE} = 1$

Class Activity, p. 174

1. $NO = 12$; since $m\angle NMP = m\angle OMP$, $\overline{MP}$ bisects $\angle NMO$. Since $MN = MO$ and the included $\angle NMO$ is bisected, by Theorem 5-2-1, $\triangle NMP \cong \triangle OMP$. So $NP = OP$, $OP = 6$, and thus $NO = 12$.

Home Activity, p. 174

1. Yes **3.** Yes

5. Not enough information given to decide

7. $\cong$, by Theorem 5-2-1

9. 41 ft

11. It is given that $AB = CD$. Then $AC = BD$ by segment addition. $AE = DF$, and $m\angle A = m\angle D$. $AC/BD = AE/DF$ by algebra. Therefore $\triangle AEC \sim \triangle DFB$ by SAS. Since $AC/BD = AE/DF = 1$, $\triangle AEC \cong \triangle DFB$.

13. $\triangle ABC \sim \triangle DCB$ by AA (rt. angles, alt. int. angles; $\frac{CB}{CB} = 1$

15. It is given that $m\angle A = m\angle D = 90°$. $m\angle DCE = m\angle ACB$, since they are vertical angles. Then $\triangle ACB \sim \triangle DCE$ by AA. It is given that $AC = DC$, therefore $AC/DC = 1$. Thus $\triangle ACB \cong \triangle DCE$.

17. Yes; $\triangle BCE \sim \triangle BAE$ by SAS; $\frac{BC}{BA} = \frac{BE}{BE} = 1$

19. Triangle 1: no
Triangle 2: yes

Class Activity, p. 179

1. Yes

3. Yes

5. Not enough information

7. Yes

9. Sometimes

Home Activity, p. 180

1. AA, $\frac{6}{6} = \frac{8}{8} = 1$

3. Yes

5. Yes; SAS

7. $\overline{PR}$

9. Yes; SSS and $\frac{7}{7} = \frac{8}{8} = \frac{5}{5} = 1$

11. $\angle V$

13. Always

15. Sometimes

17.

	Statement	Reason
1.	$AB = BC = CD = DA$	Given
2.	$BD = BD$	Identity
3.	$\frac{AB}{CB} = \frac{AD}{CD} = \frac{BD}{BD}$	Algebra
4.	$\triangle ABD \sim \triangle CBD$	SSS
5.	$\frac{AB}{CB} = 1$	Algebra
6.	$\triangle ABD\ \triangle CBD$	Definition congruent triangles
7.	$m\angle A = m\angle C$	CPCF

19.

1. Isosceles $\triangle ABC$ with $AC = BC$	Given
2. $\overline{CD}$ bisects $\angle ACB$	Given
3. $m\angle ACD = m\angle BCD$	Definition angle bisector
4. $CX = CX$	Identity
5. $\frac{AC}{BC} = \frac{CX}{CX}$	Algebra
6. $\triangle ACX \sim \triangle BCX$	SAS
7. $\frac{CX}{CX} = \frac{AC}{BC} = 1$	Algebra
8. $\triangle ACX \cong \triangle BCX$	Definition congruent triangles
9. $AX = BX$	CPCF
10. $\triangle AXB$ is isosceles.	Definition isosceles triangle

Class Activity, p. 184

1. If you are tardy, then you are late to class.
3. If you are informed, then you watch the news on TV.
5. If a triangle is equilateral, then it is isosceles.

Class Activity, p. 185

1. If the temperature is above 75°F then John goes swimming; false.
3. If it is not snowing, then Maria plays tennis; false.
5. If Alex is a teenager, then he is twelve years old; false.
7. If Juan owns a convertible, then he owns an automobile; true.

Class Activity, p. 186.

1.
1) Given
2) Theorem 5-4-1
3) Algebra
4) Substitution

Home Activity, p. 187

1. If it is a Siamese, then it is a cat.
3. If he is honest, then he is a judge.
5. If the triangle has two equal angles, then it is isosceles.
7. If Barb is riding her horse, then she is not walking; true.
9. If two lines meet at right angles, then they are perpendicular; true.
11. If the date is January 31, then it is winter in New York; true.
13. If an animal has two legs, then it is a dog; false.
15. If two angles are right angles, then they are congruent; true.
17. $AC = BC$ is given, and m ACD = m BCD by the definition of an angle bisector. $CD = CD$, so $ADC \sim BDC$ by SAS. m ADC = m BDC because corresponding angle are equal, and m ADC + m BDC = 180° because they form a straight angle. By substitution we find both angle equal 90°, so $CD \perp AB$.

Critical Thinking, p. 188

19. 225

Class Activity, p. 190

1. If you are an NBA star, then you wear these shoes.
3. You can be as good as an NBA star just by wearing a certain brand of basketball shoes.
5. If two angles of a triangle are equal in measure, then the sides opposite the equal angles are equal in length. Yes.

Home Activity, p. 192

1. If you are a champion, then the skis were designed for you.
3. Yes
5. If you wear these clothes, then you will be in style.
7. Yes
9. If you attend the summer camp, then you will have a positive self-image, personal confidence, self-reliance, and academic achievement.
11. Yes
13. M is the midpoint, so $DM = CM$. $m\angle C = m\angle D = 90°$, $DA = BC$, and $DA/CB = 1$. Thus $\triangle MDA \cong \triangle MCB$ by SAS, and ratio = 1. $MA = MB$ by CPCF. Then $m\angle 1 = m\angle 2$ by Theorem 5-4-1.

15. $AM = BM$; $\angle PMA$ is a right angle; so $\angle PMB$ is a right angle; $PM = PM$ and $AM/BM = 1$. $\triangle PMA \cong \triangle PMB$ by SAS, and ratio $= 1$. $AP = BP$ by CPCF.
17. $AC = AE$; $AB = AD$; $m\angle CAB = m\angle EAD$. $AC/EA = AB/AD = 1$. $\triangle CAB \cong \triangle EAD$ by SAS and ratio $= 1$. Then $CB = ED$ by CPCF.

Critical Thinking, p. 194

19. $\frac{n(n-3)}{2}$; 252

Chapter 5 Review

1. It is a combination of the symbols for similarity and equality.
3. True 5. True
7. b & g; c & e
9. 23
11. Yes; $\triangle ACD \sim \triangle BCD$, $AC = BC$, so $\frac{AC}{BC} = 1$

Cumulative Review, Chapters 1–5, p. 201

1. line segment $\overline{BA}$ or $\overline{AB}$
3. angle; $\angle ABC$ or $\angle CBA$ or $\angle B$
5. 80°; acute angle
7. $x° = 65°$; $(2x - 15)° = 115°$
9. −4; −1; 2; 6
11.

13. Answers may vary.
15. $\angle 3$, $\angle 6$; $\angle 4$, $\angle 5$
17. $\angle 1$, $\angle 8$; $\angle 2$, $\angle 7$
19. $\overleftrightarrow{ML}$, $\overleftrightarrow{DH}$
21. $\overleftrightarrow{RS}$, $\overleftrightarrow{DH}$; $\overleftrightarrow{RS}$, $\overleftrightarrow{LH}$
23. M, T, L; R, T, S
25. 140° 27. 140°
29. 140°
31. $2\frac{1}{2}$; error: $\frac{1}{4}$ in.
33. rectangular prism
35. b, c, f 37. c
39. a, e 41. c
43. 28° 45. 90°
47. 140° 49. 130°
51. 127.5 mi
53. Similar; AA
55. Similar; SSS
57. 14 m 59. 15
61. True 63. a, e; b, f
65. Yes. If a triangle has two equal sides and the included angle is bisected, then the triangle is divided into congruent triangles.
67. 21 Students
69. If you are informed, then you read the newspaper.
71. If the person has been to England, then the person has visited London.

CHAPTER 6

Class Activity, p. 208

1. Answers may vary.

Class Activity, p. 209

1.	
CA = CB	Construction
AD = BD	Construction
CD = CD	Identity
$\triangle CDA \cong \triangle CDB$	SSS
$m\angle ACD = m\angle BCD$	CPCF
$\overrightarrow{CD}$ bisects $\angle ACB$	Definition of Bisector

3.

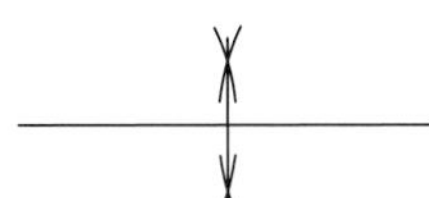

5.

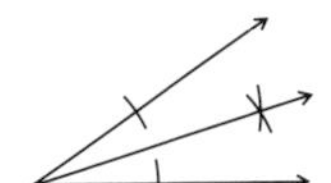

Home Activity, p. 210

1. See students' constructions.
3. Draw a point on a line *1*. Label it *M*. From *M*, mark an arc for the length of $\overline{AB}$. Place the compass point on the intersection of this arc. Mark off an arc for the length of $\overline{CD}$. Label the intersection as point *N*.
5. 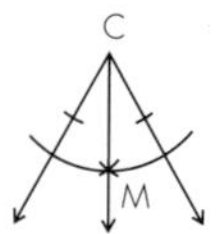

7. 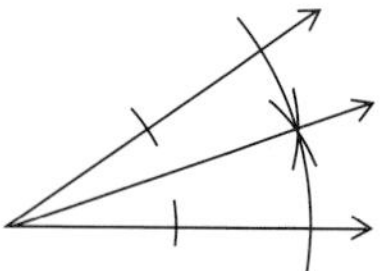
9. 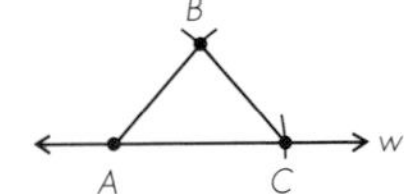

Critical Thinking, p. 210

11. 99

Class Activity, p. 213

1. Answers may vary.
3. 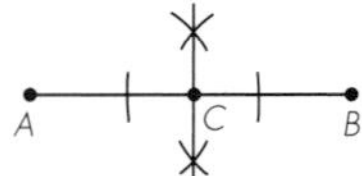

Class Activity, p. 214

1. Answers may vary.
3. See students' constructions.

Home Activity, p. 215

1. 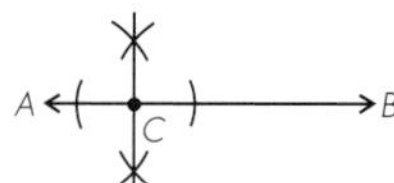

3. 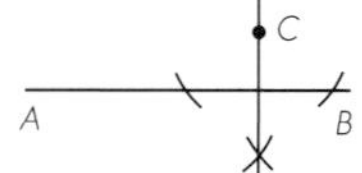

5. 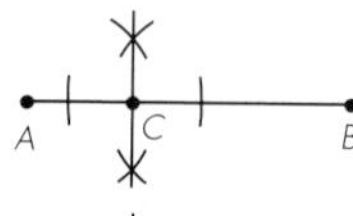

7. 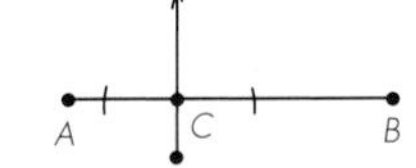

9. 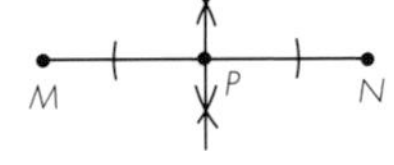

11. 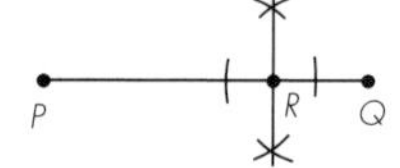

13. 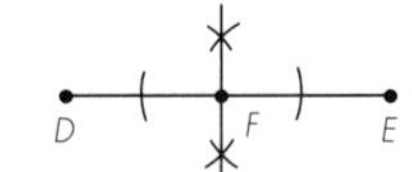

15. Answers may vary.
17. 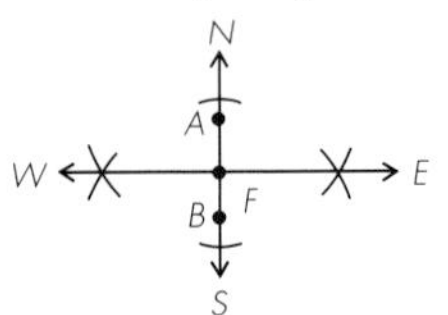

19. Construct perpendicular lines, then bisect the 90° angle.

Class Activity, p. 220

1. They are parallel; yes.
3. Possible construction:

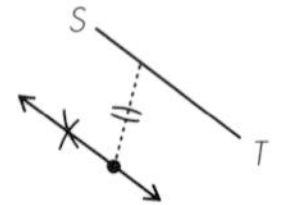

5. Possible construction:

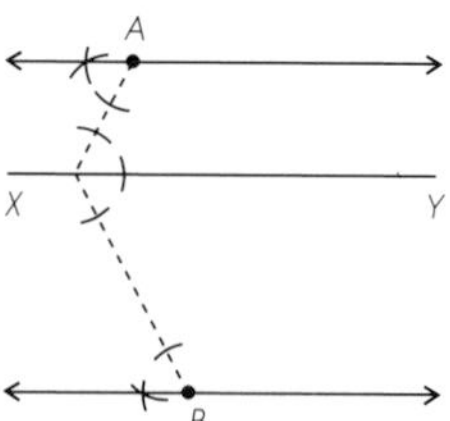

Home Activity, p. 221

1. 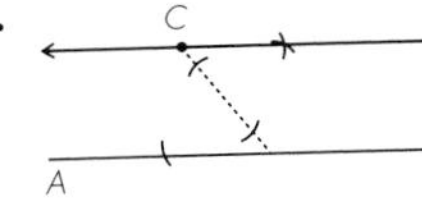

3. 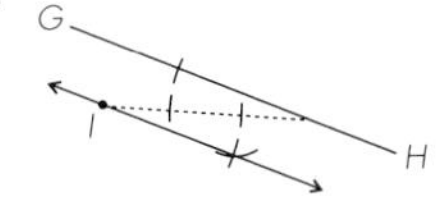

5. 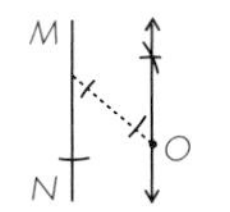

7. 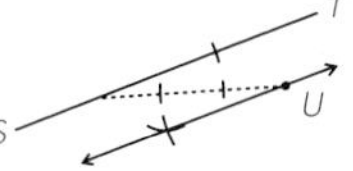

9. 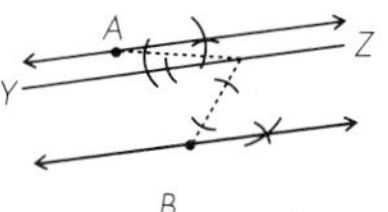

11. 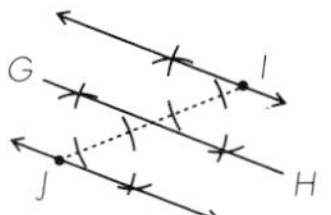

13. 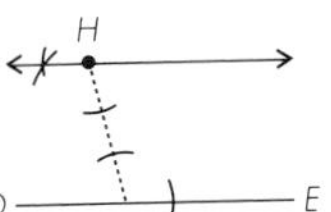

15. 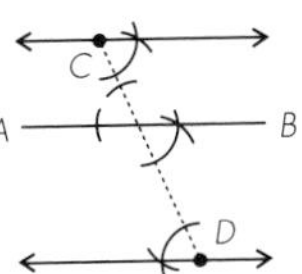

17. Answers may vary.
19. They intersect at one point.

Critical Thinking, p. 222

21. Sara

Class Activity, p. 227

1. See students' constructions.
3. 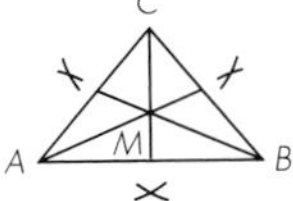

5. Scalene; acute
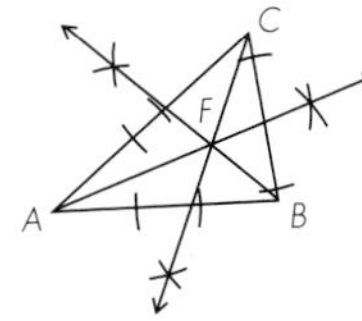

7. No; each segment is drawn from the midpoint of a side to the angle opposite the side.

Home Activity, p. 228

1. See students' constructions.
3. 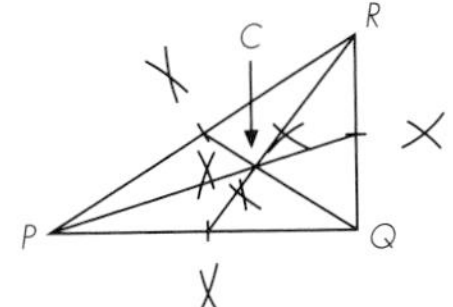

5. 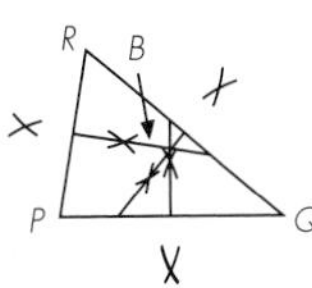

7. They are the same.
9. 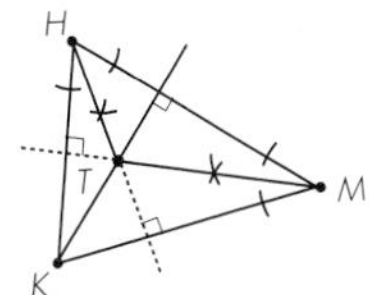

11. On the hypotenuse

Critical Thinking, p. 229

13. Angle bisectors

Class Activity, p. 233

1. Answers may vary.
3. 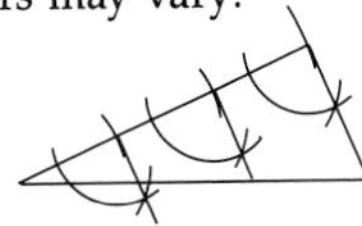

Home Activity, p. 234

1. Students' constructions
3. Students' constructions
5. Students' constructions
7. Students' constructions
9. Students' constructions
11. Students' constructions
13. Students' constructions

Critical Thinking, p. 234

15. No

Class Activity, p. 238

1. Construct by using ⊥ bisectors of sides.

Class Activity, p. 239

1.

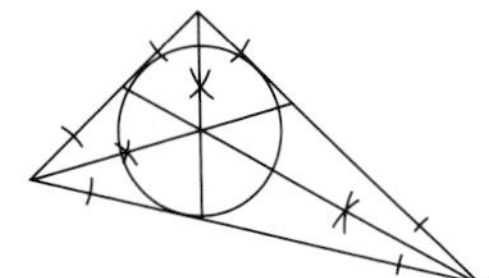

Home Activity, p. 239

1.

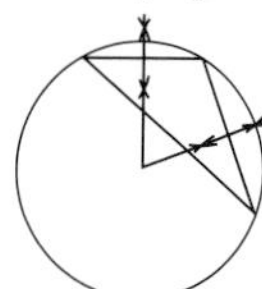

3. 16 mm **5.** 36

7.

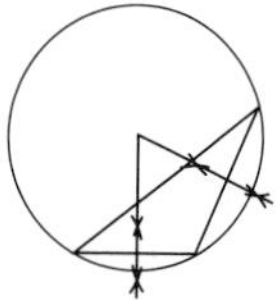

9.

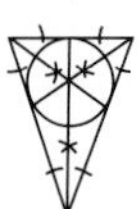

11. They all lie on the median/altitude/angle bisector from the vertex to the base.

Critical Thinking, p. 240

13. Choose a point C, and draw $\triangle ABC$. Construct the midpoints of $\overline{AC}$ and $\overline{BC}$, labeling the midpoints D an E. Measure $\overline{DE}$. $AB = 2(DE)$.

Chapter 6 Review, p. 244

1.

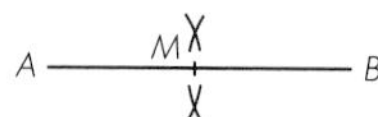

3.

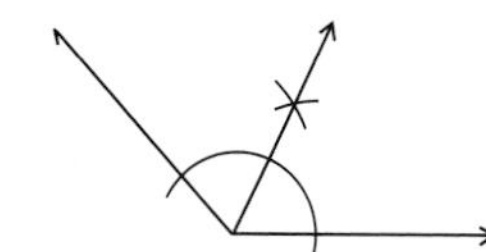

5.

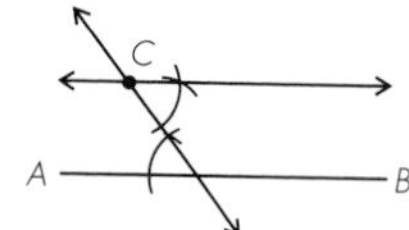

7.

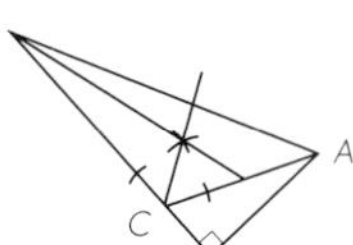

9.

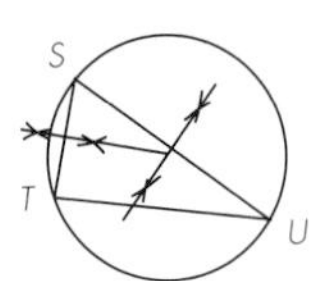

CHAPTER 7

Class Activity, p. 249

1. Yes **3.** Yes
5. 12.5

Class Activity, p. 251

1. 39 **3.** 22.4
5. 3.9 **7.** 8.9

Home Activity, p. 252

1. Yes **3.** No
5. 34 **7.** 26.9
9. 17 **11.** 9.2
13. 39.1 ft **15.** 8.94 ft

Critical Thinking, p. 252

No

Class Activity, p. 256

1. 6.6 **3.** 8.2
5. 7.1 m

Home Activity, p. 257

1. 17.0 **3.** 58.0
5. 12.7 **7.** 70.7
9. 18.4 **11.** 70.7
13. 1.0 **15.** 4.2
17. 10.0
19. about 17 ft

Critical Thinking, p. 258

about 35.4 ft

Class Activity, p. 261

1. 31 in. **3.** 41 yd

Class Activity, p. 262

1. 10.4 **3.** 13.0 cm

Class Activity, p. 263

1. 589 ft **3.** 95 m

Home Activity, p. 264

1. $x = 17.3$ ft
3. $x = 70.4$ cm; $y = 61.0$ cm
5. $a = 15$ ft
7. Answers may vary.
9. Suppose that $\triangle ABC$ is equilateral. Since an equilateral triangle is also isosceles, $m\angle A = m\angle B$. Similarly, $m\angle B = m\angle C$. Thus, all three angles have the same measure.

Critical Thinking, p. 264

1. 2 **3.** 8
5. 32 **7.** 2^n

Class Activity, p. 267

1. $a > b$ **3.** $a = b$
5. $a < b$
7. Let $\angle A$ and $\angle B$ be the two non-right angles. Then in $m\angle A + m\angle B + 90° = 180°$. Thus, $m\angle A + m\angle B = 90°$. Since $m\angle A$ and $m\angle B$ are each positive, $m\angle A < 90$ and $m\angle B < 90$. By Theorem 7-4-1, the hypotenuse is the longest side.

Class Activity, p. 269

1. $m\angle A > m\angle B$
3. Insufficient info.
5. $\overline{RT}, \overline{ST}, \overline{RS}$

Home Activity, p. 269

1. $a > b$
3. Insufficient info.
5. Insufficient info.
7. $m\angle A > m\angle B$
9. $\overline{RQ}, \overline{PR}, \overline{PQ}$
11. Since $m\angle 2 < \angle m\ 1$, $PM < PO$ by Theorem 7-4-1. But $PN < PM$, and so $PN < PO$. So by Theorem 7-4-2, $m\angle 2 < m\angle 3$.
13. Since $AC > BC$, it follows from Theorem 7-4-2 that $m\angle B > m\angle A$. Thus, $\frac{1}{2}\ m\angle B > \frac{1}{2}\ m\angle A$. Therefore, by Theorem 7-4-1, $AD > D$.

Critical Thinking, p. 270

15. Since $\angle 1$ is an exterior angle of $\triangle ADC$, $m\angle 1 = x + m\angle A$. Thus, $m\angle 1 > x$. Thus, $BC > BD$ by Theorem 7-4-1.

Class Activity, p. 273

1. 5 **3.** $\sqrt{41}$, or 6.4

Home Activity, p. 273

1. 10 **3.** 17
5. $AB = \sqrt{45}$; $DC = \sqrt{45}$

Critical Thinking, p. 273

Square both numbers to see which has the larger square. $(\sqrt{10} + \sqrt{17})^2 = 10 + 2\sqrt{170} + 17$ and $(\sqrt{53})^2 = 53$. $10 + 2\sqrt{170} + 17 = 27 + 2\sqrt{170}$. Note that $13^2 = 169$. Therefore, $27 + 2\sqrt{170} > 27 + 2\sqrt{169} = 27 + 2(13) = 53$. Since $\sqrt{10} + \sqrt{17}$ has a larger square than $\sqrt{53}$, $\sqrt{10} + \sqrt{17}$ is the larger number.

Chapter 7 Review

1. No
3. Yes
5. 11.7
7. about 19.8 in.
9. 364
11. x, z, y **13.** 8 **15.** 7

CHAPTER 8

Class Activity, p. 283

1. $\overline{GH}$ and $\overline{KJ}$, $\overline{GK}$ and $\overline{HJ}$; $\angle G$ and $\angle J$, $\angle H$ and $\angle K$
3. GJ and KH
5. No; line segments intersect at point that is not an end point.

Class Activity, p. 285

1. Convex; 360°
3. Convex; 180°

Home Activity, p. 285

1. $\overline{LN}$ and $\overline{MO}$
3. $\angle L$ and $\angle N$; $\angle M$ and $\angle O$
5. $\overline{QR}$ and $\overline{RS}$, $\overline{RS}$ and $\overline{ST}$, $\overline{ST}$ and $\overline{TQ}$, $\overline{TQ}$ and $\overline{QR}$.
7. 2, 3, 4 **9.** 2, 6
11. 1, 5, 6 **13.** 360°
15. 360°
17. Answers may vary.

Critical Thinking, p. 286

1080° (6 × 180°)
1440° (8 × 180°)
1800° (10 × 180°)
$(n - 2) \times 180°$

Class Activity, p. 288

1. $m\angle B = 70°$; $m\angle C = 110°$; $m\angle D = 110°$
3. $m\angle MPN = 72°$; $m\angle NPO = 43°$; $m\angle PON = 115°$; $m\angle ONP = 22°$
5. All angles are less than 180°.

Class Activity, p. 290

1. Bases: $\overline{AB}, \overline{DC}$; Median: $\overline{EF}$; Legs: $\overline{AD}, \overline{BC}$
3. Bases: $\overline{MN}, \overline{PO}$; Median: $\overline{QR}$; Legs: $\overline{MP}, \overline{NO}$
5. $x = 44$ cm

Home Activity, p. 291

1. No; triangle
3. No; concave
5. No; hexagon
7. Bases: $\overline{DC}$, $\overline{AB}$; legs: $\overline{DA}$, $\overline{CB}$; median: $\overline{EF}$
9. $m\angle 1 = 110°$; $m\angle 2 = 130°$
11. $m\angle 1 = 125°$; $m\angle 2 = 125°$; $m\angle 3 = 55°$
13. $m\angle 1 = 59°$, $m\angle 2 = 81°$, $m\angle 3 = 40°$, $m\angle 4 = 19°$
15. $x = 17$ ft
17. 16 cm, 24 cm
19. By AA, $\triangle PKO \sim \triangle MKN$, $\triangle PKM \sim \triangle OKN$ $\triangle PMN \sim \triangle ONM$, $\triangle PMO \sim \triangle ONP$

Critical Thinking, p. 292

54 diagonals

5	5
6	9
7	14
8	20
9	27
10	35

Class Activity, p. 294

1. 15 3. 110°
5. 110° 7. 26
9. 60° 11. 60°
13. 152°; 28°; 152°
15. 51°; 129°; 51°; 129°

Class Activity, p. 296

1. 17.5, 14, 14, 30, 10, 17.5
3. 17, 17, 12, 23
5. 57°, 123°, 57°
7. 42°, 138°, 42°, 138°
9. Yes; opposite angles ≅
11. No; opposite angles not ≅
13. Yes; diagonals bisected

Home Activity, p. 297

1. 100°, 80°, 100°, 80°
3. 89°, 91°, 89°, 91°
5. 90°, 90°, 90°
7. 44, 48, 48, 70, 32, 32
9. Answers may vary.
11. 85°
13. Yes; diagonals bisected
15. No; diagonals not bisected
17. Yes; opposite sides ≅

Critical Thinking, p. 298

5. 360°

Class Activity, p. 300

1. Opposite angles are equal in measure, so $m\angle A = m\angle C$. $m\angle A = 90°$, so $m\angle A + m\angle C = 180°$. Thus, $m\angle B + m\angle D = 180°$, since the sum of the measures of all four angles is 360°. $m\angle B = m\angle D$ so the measure of each must be 90°.

Class Activity, p. 301

1. Isosceles; $JO = OK$
3. 30 5. Not a rectangle
7. Yes 9. No

Home Activity, p. 302

1. 30 3. 30
5. 15
7. 3-4-5 triangle
9. Isosceles
11. b and d
13. 360°; $m\angle 2 + m\angle 3 = 180°$; $m\angle 1 + m\angle 4 = 180°$

Critical Thinking, p. 303

Each is about 0.6.

Class Activity, p. 305

1. Yes; since a rhombus is a parallelogram, the diagonals bisect each other.

Class Activity, p. 307

1. 5 cm 3. 5 cm
5. 7.07 cm 7. 3.54 cm
9. Parallelogram
11. Rhombus

Home Activity, p. 308

1. Parallelogram: X, __, X, X, __, X, __, __ X, X, X, X, __.
   Rectangle: X, __, X, X, X, X, X, __, X, X, X, X, __.
   Square: X, X, X, X, X, X, X, X, X, X, X, X, X, X.
   Rhombus: X, X, X, X, __, X, __, X, X, X, X, X, X.
   Trapezoid: __, __, __, __, __, __, __, __, __, __, X, X, __.
3. Parallelogram
5. Rhombus

Chapter 8 Review, p. 312

1. $\overline{SR}$ and $\overline{PQ}$ or $\overline{SP}$ and $\overline{RQ}$
3. 360°
5. Yes; opposite angles are congruent and supplementary.
7. No; both pairs of opposite sides are not congruent.
9. 56°
11. No; can't tell whether *ABCD* is isosceles
13. 32 15. 21
17. 18
19. 45°, 135°, 135°
21. 45°, 45°, 90°, 90°
23. 50 ft

CHAPTER 9

Class Activity, p. 319

1. 96 sq in. 3. 363 m^2
5. 12 ft, 4 ft
7. 18 in., 6 in.
9. 33 cm^2 11. 208 sq ft

Home Activity, p. 319

1. 32 ft 3. 60 m
5. (Answers may vary.)
7. (Answers may vary.)
9. 16 cm^2
11. 35 cm, 21 cm
13. 15 in.
15. 1,492 m^2 17. $1,937.50

Critical Thinking, p. 320

19. 300 sq units; 70 units
21. *A* quadruples; *P* doubles

Class Activity, p. 323

1. 800 sq ft 3. about 176 cm^2
5. b = 22 m
7. h = 21 cm

Class Activity, p. 324

1. 72 sq in. 3. 42.44 sq ft
5. (Answers may vary.)

Class Activity, p. 325

1. 30 cm 3. 26.6 m
5. 24 cm, 32 cm, 24 cm
7. 45 ft 9. 50 ft

Home Activity, p. 326

1. 96 sq ft 3. 389.7 cm^2
5. b = 3.8 m 7. 50 cm^2
9. 86.6 cm^3 11. 76 m
13. 24 15. 144
17. 1,850 sq ft
19. A = 1,350 sq ft; P = 210 ft

Critical Thinking, p. 327

21. 44 ft
23. Possible answers: 10 ft by 12 ft; 8 ft by 14 ft; 9 ft by 13 ft

Class Activity, p. 330

1. 10,000 sq ft
3. 28,210 cm^2
5. A = 78 sq in.
7. A = 777 m^2
9. A = 182.25 sq in.
11. s = 15 cm 13. 28 in.

Class Activity, p. 332

1. 400 ft 3. 530 yd
5. 360 ft 7. w = 100 cm
9. l = 42 m

Home Activity, p. 333

1. 1,450 cm^2 3. 3,500 sq ft
5. 3,120 cm^2 7. A = 775 cm^2
9. A = 1,728 m^2, b = 54 m
11. d_2 = 18m 13. Answers may vary.
15. A = 54 m^2 17. 1,050 m^2
19. 210 sq ft
21. (Answers may vary.)
23. 154 m 25. 74 ft
27. 396 sq ft; 640 sq ft
29. $93.75

Class Activity, p. 338

1. 120 sq ft 3. 90 m^2
5. 510 sq in.
7. $500 + 50\sqrt{3}$ sq in.
9. 768 sq ft 11. A = 936 m^2
13. A = 304 m^2
15. $(b_1 + b_2)h$

Class Activity, p. 340

1. P = 92 ft; A = 384 sq ft
3. P = 207.2 cm, A = 1410 cm^2
5. P = 99 cm; A = 270 cm^2

Home Activity, p. 341

1. $P = 162.6$ ft; $A = 1327.2$ sq ft
3. $P = 330$ ft; $A = 2100$ sq ft
5. $A = 3948.5\ cm^2$; $h = 53$ cm
7. $A = 874\ m^2$ **9.** $A = 864\ m^2$
11. $138.75\ m^2$
13. (Answers may vary.)
15. (Answers may vary.)
17. 121 **19.** 785.5 ft
21. 25,745.75 sq ft

Critical Thinking, p. 342

23. Steps

1. $A(ABCD) = A(I) + A(II) + A(III)$
2. $A(ABCD) = \frac{1}{2}ah + \frac{1}{2}ch + \frac{1}{2}bh$
3. $A(ABCD) = \frac{1}{2}ah + \frac{1}{2}ch + \frac{1}{2}bh = \frac{1}{2}h\,(a + c + b + b)$
4. $a + b + c = AB$ and $b = CD$
5. $A(ABCD) = \frac{1}{2}h\,(AB + CD)$

Reasons

1. Addition
2. Definition of area for triangle and rectangles
3. Algebra
4. Identity
5. Substitution

Class Activity, p. 344

1. $A = 485.5\ m^2$; $P = 105$ m
3. $A = 180$ sq ft; $P = 60$ ft

Class Activity, p. 347

1. $\frac{P_1}{P_2} = \frac{1}{3}; \frac{A_1}{A_2} = \frac{1}{9}$
3. $\frac{P_1}{P_2} = \frac{1}{6}; \frac{A_1}{A_2} = \frac{1}{36}$
5. 1708 sq ft

Home Activity, p. 347

1. $A = 300\ cm^2$; $P = 98$ cm
3. $A = 117$ sq in., $P = 57$ in.
5. (Answers may vary.)
7. (Answers may vary.)
9. $\overline{ZY} = \frac{1}{2}\,\overline{QR}$, $\overline{ZX} = \frac{1}{2}\,\overline{SR}$, $\overline{YX} = \frac{1}{2}\,\overline{SQ}$; $\overline{ZY}/\overline{QR} = \overline{ZX}/\overline{SR} = \overline{YX}/\overline{SQ} = \frac{1}{2}$, so $\frac{A\,(\triangle XYZ)}{A\,(\triangle QRS)} = (\frac{1}{2})^2 = \frac{1}{4}$.

Critical Thinking, p. 348

10. Square; **11.** triangle; **12.** It is a square number.

Chapter 9 Review, p. 352

1. $P = 72$ m; $A = 216\ m^2$
3. $P = 102$ m; $A = 584.4\ cm^2$
5. $P = 87.2$ in.; $A = 250$ sq in.
7. $P = 100$ ft; $A = 600$ sq ft
9. $A = 108$ sq ft; $P = 36 + 12\sqrt{2}$ ft
11. $A = 125\ cm^2$; $h = 5$ cm, $b = 25$ cm
13. 110 sq in. **15.** $456\ cm^2$
17. $\frac{1}{3} \times \frac{1}{9}$ **19.** $\frac{1}{6} \times \frac{1}{36}$
21. 2,400 tiles

CHAPTER 10

Class Activity, p. 356

Shapes may vary.

Class Activity, p. 357

1. Triangle, convex, nonregular
3. Quadrilateral, concave, nonregular
5. Square/quadrilateral, convex, regular
7. Heptagon, convex, regular

Class Activity, p. 359

1. 720° **3.** 3240°
5. 128.6°; 51.4° **7.** 140°; 40°
9. 15

Class Activity, p. 359

1. 40 cm **3.** 23 in.
5. 195 m

Home Activity, p. 360

1. Triangle, convex, regular
3. Decagon, concave, nonregular
5. 360° **7.** 180°
9. 1080°
11. Answers may vary.
13. 90° **15.** 12.9°
17. $n = 10$ **19.** $n = 25$
21. 180 sq ft

Critical Thinking, p. 360

23. 1,296,000-gon

Class Activity, p. 363

1. $A = 120\ cm^2$ **3.** $A = 220.8\ m^2$
5. $3750\sqrt{3}$ sq ft **7.** $261\ m^2$
9. Both 60°; angle was bisected.
11. Yes; AA or SAS and $\frac{base}{base} = 1$

Home Activity, p. 364

1. $192\sqrt{3}$ cm^2
3. 345.96 sq in.
5. 110 sq in.
7.

Number of Sides	Area
6	2.61 m
7	3.64 m
8	4.84 m
9	6.17 m
10	7.70 m

9. 6 m, 7 m, 8 m, 9 m, 10 m
11. about 1,459,000 sq ft; about 162,000 sq yd
13. about 30
15. about 1600 sq ft
17. Hexagon; no; the sides are not equal in length.
19. $360° \div 9 = 40°$
21. 72°

Class Activity, p. 369

1. 16.2 cm 3. 7.66 cm
5. 7.5 ft 7. 64.2 mm
9. 3.1415927
11. 12.57 ft 13. 314.16 yd

Class Activity, p. 371

1. $A = 176.71$ sq ft
3. $A = 19.63$ sq yd
5. 12.56 sq ft 7. 50.24 m^2
9. 7850 m^2 11. 1.77 km^2
13. $d = 120$ cm 15. $d = 63.24$ m

Home Activity, p. 372

1. 8 in., 50.24 in., 200.96 sq in.
3. 42 yd, 21 yd, 1384.74 sq yd
5. 20 in., 62.8 in., 314 sq in.
7. 4
9. $25\sqrt{3}$ sq ft or 43.3 sq ft
11. 16.9 sq ft

Critical Thinking, p. 372

1. 18.84 cm 3. 0.6
5. 28.26 cm^2
7. 0.36; ratio of radii squared

Class Activity, p. 376

1. Radii: $\overline{OA}, \overline{OB}, \overline{OC}, \overline{OH}, \overline{OG}$
   Diameters: $\overline{GH}$ and $\overline{BC}$
   Chords: $\overline{AF}, \overline{GH}, \overline{BC}$
   Secant: $\overleftrightarrow{DE}$
   Tangent: $\overleftrightarrow{BJ}$
   Central angles: $\angle COH, \angle COA, \angle AOG, \angle GOB, \angle BOH$
   Inscribed angles: $\angle JBC, \angle CBK$

3. 65° 5. 48°

Class Activity, p. 378

1. It is tangent to the circle; no; yes.
3. 80° 5. 80°
7. 90° 9. 45°
11. 135°

Home Activity, p. 378

1. $\overleftrightarrow{QB}, \overline{OA}, \overline{OC}$ 3. O
5. $\overleftrightarrow{PT}$ 7. $\overline{TC}, \overline{TA}, \overline{AC}$
9. 70° 11. 20°
13. 20° 15. 50°
17. 110°

Class Activity, p. 382

1. $x = 9$ 3. $x = 9$
5. The products are equal.

Home Activity, p. 383

1. Tangent 3. Secant
5. Center 7. Radius
9. $CX = 10$ or 14
11. $AC = 16$ 13. 20 cm
15. 314 cm^2 17. 8
19. 40° 21. 140°

Class Activity, p. 386

1. Rectangle 3. 13.76 sq ft

Class Activity, p. 387

1. 942 sq ft
3. $A = 380.16$ sq in.

Home Activity, p. 388

1. $A = 113.04$ cm^2
3. $A = 123.84$ sq in.
5. $\frac{4}{9}$ 7. $x = 60°$
9. $x = 6.9$ 11. 614 ft

Critical Thinking, p. 389

13. Draw a pillar on its side, 3 lines

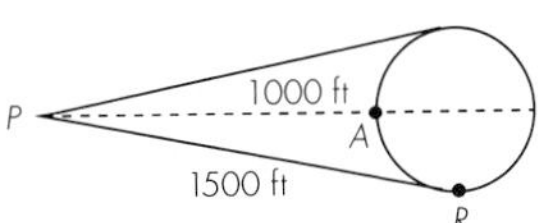

$PA(PA + \text{diameter of pillar}) = PR^2$

Chapter 10 Review

1. 120°   3. 162.9°
5. 12
7. $C = 75.36$ m; $A = 452.16$ m^2
9. $C = 31.4$ ft; $A = 78.5$ sq ft
11. 776 revolutions
13. 60°   15. 40°
17. 120°   19. 60°
21. $x = 12$ 23. $x = 18.5$

CHAPTER 11

Class Activity, p. 404

1. Edges: $\overline{PN}$, $\overline{NO}$, $\overline{OP}$, $\overline{PS}$, $\overline{NQ}$, $\overline{OR}$, $\overline{SQ}$, $\overline{QR}$, $\overline{RS}$
   Bases: $\triangle PNO$, $\triangle SQR$
   Lateral Faces: ▭*NORQ*, ▭*OPSR*, ▭*PNQS*
3. Edges: $\overline{PQ}$, $\overline{QR}$, $\overline{RS}$, $\overline{SO}$, $\overline{OP}$, $\overline{PK}$, $\overline{QL}$, $\overline{RM}$, $\overline{SN}$, $\overline{OJ}$, $\overline{KL}$, $\overline{LM}$, $\overline{MN}$, $\overline{NJ}$, $\overline{JK}$
   Bases: Pentagon *OPQRS*, Pentagon *JKLMN*
   Lateral Faces: ▭*OPJK*, ▭*PQLK*, ▭*QRML*, ▭*RSNM*, ▭*SOJN*
5.

Class Activity, p. 406

1. 9.75   3. 162.9
5. $4.84\ \pi \approx 15.2$ in^2
7. $22\ \pi \approx 69.1$ in^2

Home Activity, p. 406

1. Lateral face *MORP* is a rectangle, as are the other faces. Right; Edges: $\overline{MN}$, $\overline{NO}$, $\overline{OM}$, $\overline{MP}$, $\overline{NQ}$, $\overline{OR}$, $\overline{PQ}$, $\overline{QR}$, $\overline{RP}$; Bases: $\triangle PQR$, $\triangle MNO$; Lateral faces: ▭*MORP*, ▭*ONQR*, ▭*NMPQ*.
3. No right angle is shown. Oblique; Edges: $\overline{WX}$, $\overline{XY}$, $\overline{YZ}$, $\overline{ZW}$, $\overline{WP}$, $\overline{XQ}$, $\overline{YR}$, $\overline{ZS}$, $\overline{PQ}$, $\overline{QR}$, $\overline{RS}$, $\overline{SP}$; Bases: Quadrilaterals *WXYZ* and *PQRS*; Lateral faces: Parallelograms *PWZS*, *SZYR*, *RYXQ*, *QXWP*.
5. Lat: 220; Tot: 276
7. Lat: 600; Tot: 660
9. Lat: 4,368 cm^2;
   Tot: 5,688 cm^2
11. Lat: about 305 in^2
    Tot: about 316 in^2

Critical Thinking, p. 406

Same: 16
The cylinder on the left: $48\ \pi > 24\ \pi$

Class Activity, p. 410

1. 1,071 in.^{3}

Class Activity, p. 411

1. about 396 yd^3

Home Activity, p. 412

1. 36   3. 1,520 mm^3
5. $192\ \pi \approx 603$ mm^3
7. $500\ \pi \approx 1570$ m^3
9. about 442,740
11. 30 ft

Critical Thinking, p. 412

The second has the greater volume.

Class Activity, p. 416

1. 211.9 in.^{2}

Class Activity, p. 418

1. $176\ \pi \approx 553$ cm^2
3. $278.4\ \pi \approx 874$

Home Activity, p. 418

1. 480 in.^{2}
3. 1,559 cm^2
5. $96\ \pi \approx 301$ ft^2
7. about 3,965 ft^2

Critical Thinking, p. 418
$(\frac{60}{13})^2\ \pi$, or about 67

Class Activity, p. 422
1. 220 cm^3

Class Activity, p. 424
1. 1,536 $\pi \approx$ 4,823 $in.^3$

Home Activity, p. 424
1. 391 ft^3 3. 37.3 $in.^3$
5. 48.3 km^3
7. 2,592,100 m^3
9. about 2.1 m

Class Activity, p. 426
1. 2,100 ft^3

Class Activity, p. 428
1. about 13,572 cm^3

Home Activity, p. 429
1. 6.93 yd^3
3. 14.4 π, or about 45.2 $in.^2$
5. 5.25 $in.^2$
7. about 232 π, or 728 ornaments
9. 718.8 π, or about 2,257 ft^2

Critical Thinking, p. 430
Each domino covers a black square and a red square. Therefore, the total number of covered black squares has to equal the total number of covered red squares. But this cannot happen since the two removed squares are of the same color. The number of black squares in the modified board is no longer equal to the total number of red squares. As a result, the squares cannot be covered by dominoes.

Class Activity, p. 435
1. $\frac{2.048}{3}\ \pi$, or 2,144 $in.^3$; 804 $in.^2$

Home Activity, p. 435
1. 4,188 yd^3; 1,257 yd^2
3. 7,240 $in.^3$; 1,810 $in.^2$
5. 11.5 $in.^3$; 24.6 in.
7. 8.38 $in.^3$; 20.0 $in.^2$
9. about 23 people/mi^2

Critical Thinking, p. 435
11. $$\frac{\frac{1}{3}\pi r^3 + \pi r^3}{2} = \frac{2}{3}\pi r^3 = \frac{1}{2}\left(\frac{4}{3}\pi r^3\right)$$

Chapter 11 Review, p. 440
1. 68 m^2; 94 m^2
3. 36 ft^2; 54 ft^2
5. 1700 cm^3
7. 1,100 m^2; 2,310 m^3
9. 7235 in^3; 1808 $in.^2$
11. cylindrical; 1,843 ft^3

CHAPTER 12

Class Activity, p. 446
1. Answers may vary.
3. This is true by Theorem 6-6-2. If a point is on the perpendicular bisector, then it is equidistant from the endpoints of the segment.

Home Activity, p. 448
1. $\perp$ bisector
3. Two parallel lines
5. A line parallel to each and equidistant from each
7. Two parallel lines and two semicircles
9. Letter V 11. Circle

Class Activity, p. 451
1. Translation
3. Reflection
5. Answers for the procedure may vary.

Class Activity, p. 452
1. Reflection
3. Rotation
5.

A A'
45°

Home Activity, p. 453
1. Translation
3. Reflection
5. Reflection
7. Vertical: A, H, I, M, O, T, U, V, W, X, Y; Horizontal: B, C, D, E, H, I, O, X
9.

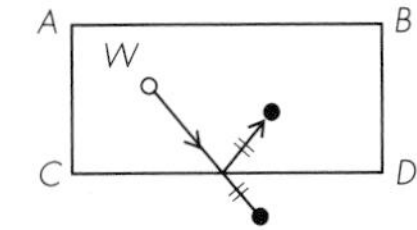

11. Hexagon; 5
13. Translation 2 units up

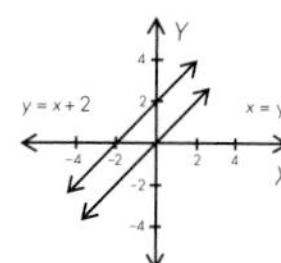

Critical Thinking, p. 454
15. Not a mirror image
16. Translation
17. Not a mirror image
18. Rotation about a point

Class Activity, p. 457
1. 4 **3.** 1 **5.** 0
7. 1 **9.** 0

Class Activity, p. 458
1–7. Answers may vary.

Home Activity, p. 459
1. Equilateral triangle, Square, Rectangle, Trapezoid, Parallelogram, Circle
3. Triangle: 120°; Square: 90°; Circle: angle approaches 0°
5. Reflection
7. Rotation
9. All
11. Square 90°; circle: approaches 0°; triangle: 120°; octagon: 45°
13. Pollen and virus

Critical Thinking, p. 460
15. 2

Class Activity, p. 463
5. Parabolic curve made of straight lines

Class Activity, p. 464
1. Draw reflection of points over line, space 3 lines

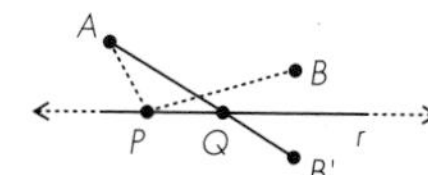

3. Measure *AP* and *PB* and *AP* and *PB'* in Ex. 1.
5. $AQ + QB$
7. A straight line is the shortest distance between two points.

Class Activity, p. 465
1. 2 lines parallel to and 4 mi from each side of the highway
3. Circle with radius of 3 mi, center on school

Home Activity, p. 465
1. Circumference of circle
3. Area of 90° sector
5. Circle
7. 2 concentric circles
9. Translation **11.** Rotation
13. Japan, Peru; 180°
15. Rotation or reflection over either axis
17. 2 reflections equal to 1 90° rotation
19. Reflection over $y = x$

Critical Thinking, p. 466
21. 9
23. Yes; 1 parallel to ends, infinitely many perpendicular to end

Class Activity, p. 469
1. Statement is false. Converse: If a triangle is right, then it is equilateral; false. Inverse: If a triangle is not equilateral, then it is not right; false. Contrapositive: If a triangle is not right, then it is not equilateral; false.
3. Statement is true. Converse: If a triangle is isosceles, then it is equilateral; false. Inverse: If a triangle is not equilateral, then it is not isosceles; false. Contrapositive: If a triangle is not isosceles, then it is not equilateral; true.
5. Statement is false. Converse: If a figure is a square, then it has four equal sides; true. Inverse: If a figure does not have four equal sides, then it is not a square; true. Contrapositive: If a figure is not a square, then it does not have four equal sides; false.

Home Activity, p. 470

1. Statement is false. Converse: If it is night, then the sun is not shining; true. Inverse: If the sun is shining, then it is not night; true. Contrapositive: If it is not night, then the sun is shining; false.
3. Statement is true. Converse: If the measure of two angles are equal, then they are right angles; false. Inverse: If two angles are not right angles, then their measures are not equal; false. Contrapositive: If the measures of two angles are not equal, then they are not right angles; true.
5. Statement is true. Converse: If I don't pitch at the game, then I stay at home; false. Inverse: If I do not stay at home, then I pitch at the game; false. Contrapositive: If I pitch at the game, then I do not stay at home; true.
7. Statement is true. Converse: If two lines are coplanar, then they are parallel; false. Inverse: If two lines are not parallel, then they are not coplanar; false. Contrapositive: If two lines are not coplanar, then they are not parallel; true.
9. Statement is false. Converse: If a quadrilateral is a rectangle, then it has two pairs of parallel sides; true. Inverse: If a quadrilateral does not have two pairs of parallel sides, then it is not a rectangle; true. Contrapositive: If a quadrilateral is not a rectangle, then it does not have two pairs of parallel sides; false.
11. See answers for Exercises 1–10; statement and contrapositive, converse and inverse.

Critical Thinking, p. 470

Friday

Chapter 12 Review, p. 474

1. Perpendicular bisector of chord, or diameter
3. 2 concentric circles with radii of 1 cm and 4 cm
5. Rotation
7. No
9. Yes
11. Yes

13.

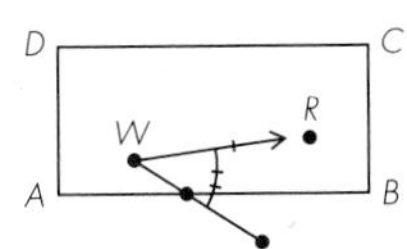

CHAPTER 13

Class Activity, p. 479

1. Point
   (3, 3)
   $(2\frac{1}{2}, 2)$
   (1, −1)
   (0, −3)
   (−2, −7)
   (−4, −11)
3. Point
   (3, 7)
   $(2\frac{1}{2}, 6)$
   (1, 3)
   (0, 1)
   (−2, −3)
   (−4, −7)
5. Point
   (6, −7)
   $(1, -4\frac{1}{2})$
   (0, −4)
   $(-1, -3\frac{1}{2})$
   $(-3, -2\frac{1}{2})$
   (−4, −2)

Class Activity, p. 481

1. $y = \frac{1}{2}x + (-\frac{1}{2})$
3. $x = \frac{5}{2}$
5. $y = 3x + (-2)$
7. $x = 7$ 9. $x = -5$

Home Activity, p. 482

7. Nonvertical 9. Vertical
11. $x = 4$
13. $y = -\frac{3}{2}x + \frac{1}{2}$
15. $y = -x + 11$

Critical Thinking, p. 482

Step 5 involves dividing by $(x - y)$. But $x - y = 0$, since $x = y$. Division by zero is not allowed.

Class Activity, p. 485

1. 7 3. no slope
5. 1, up 7. $-\frac{3}{10}$, down
9. $-\frac{19}{4}$ 11. $-\frac{9}{5}$

Class Activity, p. 487

1. $-\frac{1}{2}$, 7
3. -1, 4
5. 0, -10

7–9.

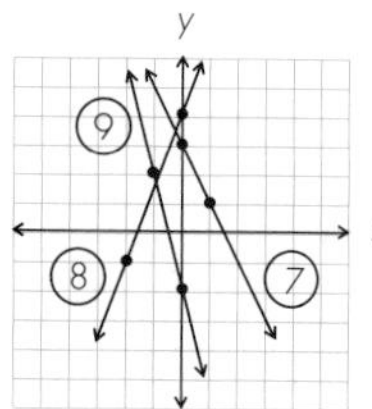

Home Activity, p. 488

1. -3, down 3. 0, level
5. -3, $-\frac{1}{2}$
7. 7, -18
9. 0, $3\frac{2}{5}$
11. 3, -8

13–15.

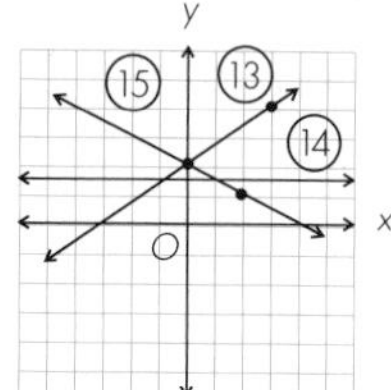

17. $y = -6x + 2\frac{1}{2}$

Critical Thinking, p. 488

19. $\frac{12}{19}$

Class Activity, p. 491

1. No 3. Yes
5. No

Class Activity, p. 492

1. No 3. Yes
5. $\overline{WZ}$ has slope 0 and $\overline{YZ}$ has slope $\frac{4}{3}$. The product of these slopes is not -1, so these sides are not perpendicular.
7. $\triangle PQR$ is isosceles because $PR = QR = \sqrt{34}$. The point $S(-1, 1)$ is the midpoint of $\overline{PQ}$, because PS $SQ = \sqrt{2}$ and $PQ = 2\sqrt{2}$. The line $y = -x$ passes through $(-1, 1)$ and $(3, -3)$. Its slope is -1. The slope of $\overline{PQ}$ is 1. The product of the slopes is -1. So $y = -x$ is perpendicular to $\overline{PQ}$ at its midpoint.

Home Activity, p. 493

1. Yes 3. Yes
5. No 7. Yes
9. $\frac{4}{7}$; $\frac{4}{7}$
11. $ABCD$ is a parallelogram. Opposite sides have equal slopes and so are parallel.
13. No 15. Yes
17. Yes, because the product of the slopes of $\overline{AB}$ and $\overline{AC}$ is -1.
19. $KL = KM = 5$, so KL and KM are the legs of an isosceles triangle. Slope of $\overline{KL} = \frac{3}{4}$ and slope of $\overline{KM} = -\frac{4}{3}$, so $\overline{KL} \perp \overline{KM}$ and $\triangle KLM$ is a right triangle.
21. Parallel 23. Parallel
25. $\overline{PQ}$ and $\overline{RS}$ each have slopes of $\frac{1}{3}$; $\overline{PS}$ and $\overline{QR}$ have slopes of -3. Since $-3 \times \frac{1}{3} = -1$, all the angles of the quadrilateral are right angles.

Critical Thinking, p. 494

27. 3 **29.** 18
31. 40

Class Activity, p. 497

1. $x^2 + y^2 = 16$
3. Circle 1
5. $x^2 + y^2 = 49$
7. $x^2 + y^2 = 10{,}000$
9. $(-6, 0)$ $(0, -6)$
$(-5, 3.32)$ $(1, 5.92)$

Class Activity, p. 498

1. $x^2 + y^2 = 36$
3. $x^2 + y^2 = \frac{9}{16}$
5. $x^2 + y^2 = 26$

Home Activity, p. 499

1. $x^2 + y^2 = 25$
3. $-\frac{3}{4}$
5. The line $y = \frac{4}{3}x$ is tangent to the circle. From Exercise 4, it is known that this line is perpendicular to the line containing the radius to point $(-4, 3)$. The line passes through $(-4, 3)$ because the coordinates of $(-4, 3)$ satisfy the equation $y = \frac{4}{3}x + \frac{25}{3}$.
7. No **9.** Yes
11. Yes
13. $r = 7$ **15.** $r \approx 4.5$
17. $r = 3\frac{1}{3}$
19. No **21.** Yes
23. $x^2 + y^2 = 169$
25. $\frac{12}{5}$ **27.** Yes

Critical Thinking, p. 500

29. 3 **31.** $2^n - 1$

Class Activity, p. 502

1. $y = \frac{1}{8}x^2$
3. $y = \frac{1}{16}x^2$
5. $y = 2x^2$

Class Activity, p. 504

1.

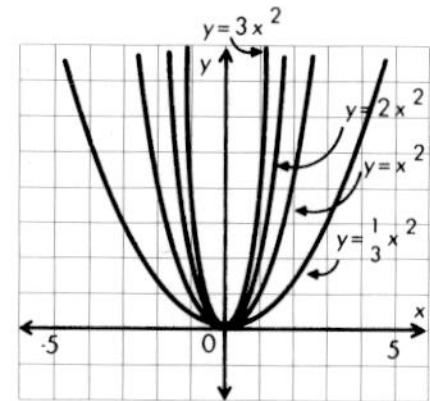

3. Up
5. They will be like the parabolas in Exercise 1 but will open down.
7. Yes **9.** No
11. No

Home Activity, p. 505

1. $y = \frac{1}{20}x^2$
3. $y = \frac{1}{3}x^2$

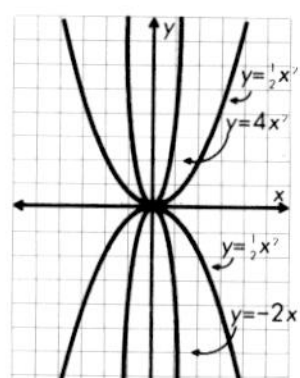

9. Down **11.** Up
13. Down **15.** Down
17. $y = 25x^2$ and $y = -25x^2$
19. $(0, 0)$

Critical Thinking, p. 506

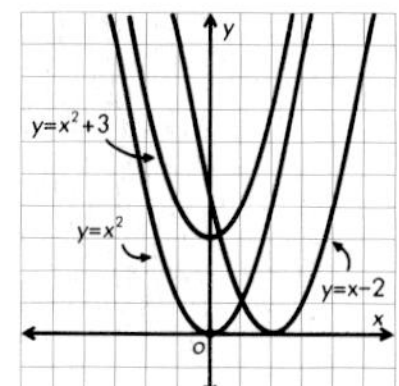

Adding 3 to x^2 moves the graph of $y = x^2$ straight up 3 units. Replacing x with $x - 2$ moves the graph of $y = x^2$ to the right 2 units.

Class Activity, p. 507

1. $MP = \sqrt{(a - [\frac{a+c}{2}])^2 + (b - [\frac{b+d}{2}])^2}$

$MQ = \sqrt{([\frac{a+c}{2}] - c)^2 + ([\frac{b+d}{2}] - d)^2}$

$PQ = \sqrt{(a - c)^2 + (b - d)^2}$

3. $MP + MQ$
$= 2\sqrt{(\frac{a-c}{2})^2 + (\frac{b-d}{2})^2}$
$= \sqrt{2^2(\frac{a-c}{2})^2 + 2^2(\frac{b-d}{2})^2}$
$= \sqrt{(2[\frac{a-c}{2}])^2 + (2[\frac{b-d}{2}])^2}$
$= \sqrt{(a-c)^2 + (b-d)^2} = PQ$

Class Activity, p. 509

1. (17.75, 29.25)
3. (−4.06, 5.63)
5. The figure on the left
7. Students should use the distance formula to show that both $\overline{DF}$ and $\overline{EG}$ have a length of $\sqrt{(b + a)^2 + c^2}$.

Home Activity, p. 510

1. (3, 2) 3. (1, 4)
5. $(\frac{3}{2}, -\frac{1}{2})$
7. Slope of $\overline{AC} = \frac{b-0}{b-0} = \frac{b}{b} = 1$.
Slope of $\overline{BD} = \frac{a-0}{0-2} = \frac{a}{-a} = -1$, so $\overline{AC} \perp \overline{BD}$.
9. Use the midpoint theorem to find the coordinates of P, Q, R, and S. Then show that $\overline{PQ}$ and $\overline{SR}$ have the same slope and that $\overline{SP}$ and $\overline{RQ}$ have the same slope.

Chapter 13 Review, p. 514

1. Vertical 3. Nonvertical
5. $x = \frac{7}{3}$
7. $x^2 + y^2 = 81$
9. Down 11. Up
13. Perpendicular
15. $-\frac{1}{9}$
17.

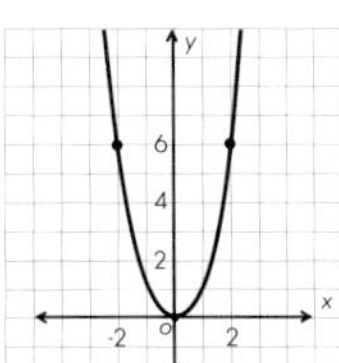

19.

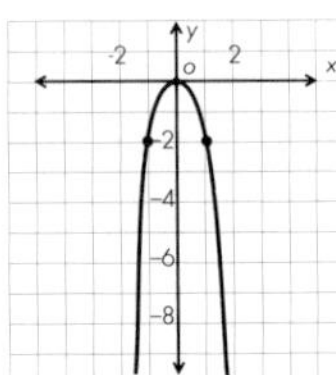

21. Circle; (0, 0), $r = 3\sqrt{5}$
23. Parabola; down

CHAPTER 14

Class Activity, p. 521

1. $\triangle DEF \sim \triangle OMN$
$\triangle DEF \sim \triangle XYZ$
$\triangle ONM \sim \triangle XYZ$
3. 0.725

Class Activity, p. 522

1. cos A = 0.8020
cos B = 0.5974

Home Activity, p. 522

1. sin A = 0.3846 sin B = 0.9231
cos A = 0.9231 cos B = 0.3846
3. sin A = 0.7071 sin B = 0.7071
cos A = 0.7071 cos B = 0.7071
5. 0.5736 7. 0.1736
9. 0.4067 11. 0.6561
13. 0.8829 15. 0.2250

Critical Thinking, p. 522

$28 + 7\sqrt{2}$

Class Activity, p. 525

1. 0.5 3. about 28°
5. about 30°

Class Activity, p. 527

1. sin A or cos B
3. g = 14.2 m

Home Activity, p. 528

1. 0.75 3. 0.9325
5. 67° 7. 18.5 cm
9. 34.4 mi 11. 208 m

Class Activity, p. 531

1. −332° 3. −180°

Class Activity, p. 533

1. −0.5172 3. −0.7071

Home Activity, p. 534

1. −290° 3. −120°
5. 0.80 7. 0.00
9. −0.87 11. −0.50
13. r = 10, cos x = 0.60
15. r = 10, cos x = −0.60
17.

cos x°	cos x°	cos x°
1.00	−0.64	−0.34
0.97	−0.77	−0.17
0.87	−0.94	0.00
0.77	−0.98	0.17
0.64	−1.00	0.34
0.50	−0.94	0.71
0.34	−0.82	0.87
0.17	−0.64	0.98
0.00	−0.50	1.00
−0.42		

Class Activity, p. 538

1. 28.3, or about 28 mi

Class Activity, p. 540

1. 16.7

Home Activity, p. 540

1. 13.8
3. 13.6
5. 5.7
7. 39°
9. 40.297, or about 40.3 mi

Class Activity, p. 543

1. about 355 ft

Class Activity, p. 545

1. 8.6 m

Home Activity, p. 546

1. 87.2 ft
3. Station A; 6.3 mi
5. 25.8
7. 131°
9. 82°

Critical Thinking, p. 546

20

Class Activity, p. 551

Methods will vary. Possible answer:

1. If $x = 45°$, then $\tan 45° = 1$, $\sin 45° = 0.7071$. "$1 = 0.7071 + 1$" is false.

Home Activity, p. 551

Methods will vary. Possible answers:

1. If $x = 0$, $\sin 0° + \cos 0° = 0 + 1 \neq 0$.
3. If $x = 0$, $\tan 0° \times \sin 0° = 0 \times 0 = 0$; $\cos 0° = 1$; $0 \neq 1$.
5. If $x = 90$, $\sin 90° = 1 \neq 90$.
7. $\frac{(\sin^2 x + \cos^2 x) - \cos^2 x}{\cos^2 x}$ $\frac{\sin^2 x + 0}{\cos^2 x}$
9. $\frac{\sin^2 x + \cos^2 x}{\cos x} = \frac{1}{\cos} x \times \frac{\sin x}{\sin x} = \frac{\sin x}{\cos x} \times \frac{1}{\sin x} = \tan x \times \frac{1}{\sin x} = \frac{\tan x}{\sin x}$

Critical Thinking, p. 551

$\cos 2x = (\frac{\cos^2}{x} = \frac{\sin^2}{x}) \times 1$

$= (\frac{\cos^2}{x} - \frac{\sin^2}{x}) \times (\frac{\cos^2}{x} + \frac{\sin^2}{x})$

$= \frac{\cos^4}{x} - \frac{\sin^4}{x}$

Chapter 14 Review, p. 556

1. 0.4706; 0.8824; 0.5333
3. $\text{Sin } x = \frac{3}{9} = 0.600$
   $\text{Cos } x = -\frac{4}{5} = -0.8000$
5. 9.8 cm
7. 199.2 m
9. If $x = 90°$, then $\sin x = 1$ and $\cos x = 0$. So, $\sin^2 \times + 1 = 2$ and $\cos x = 0$. Hence for $x = 90°$, $\sin^2 \times + 1 \neq \cos x$.

A

Acute Angle: If the number assigned to an angle is between 0° and 90°, then the angle is called an acute angle. [p. 10]

Acute Triangle: An acute triangle is a triangle with the measures of all three angles less than 90°. [p. 100]

Alternate Exterior Angles: Alternate exterior angles are angles on opposite sides of the transversal and outside the two parallel lines. [p. 66]

Alternate Interior Angles: Alternate interior angles are angles on opposite sides of the transversal and between the parallel lines. [p. 57]

Altitude: An altitude of a triangle is a perpendicular line segment from a vertex of the triangle to the opposite side or to the line determined by the opposite side. In $\triangle ABC$, $\overline{AD}$ is the altitude to $\overline{BC}$. [p. 227]

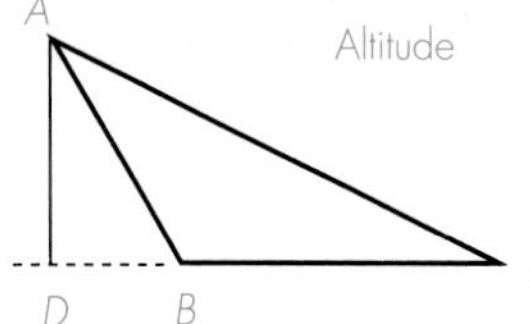

Angle: An angle is formed by two rays that have a common endpoint. [p. 5]

Angle Bisector: An angle bisector of a given angle is a ray with the same vertex that separates the given angle into two angles of equal measure. [p. 15]

Angle Bisector of a Triangle: An angle bisector of a triangle is a line segment from a vertex of the triangle to the point where the angle bisector of that angle intersects the opposite side. [p. 226]

Arc: The arc of the circle has the same measure as the central angle that intersects the arc. The symbol for arc is $\overset{\frown}{AB}$ or $\overset{\frown}{AOB}$. [p. 347]

Area: The area of a plane figure is the number of square units contained in the interior. [p. 318]

B

Betweenness: On a number line, point C is between points A and B if the coordinates of A, B, and C (a, b, and c respectively) meet the condition that $a < c < b$, or $a > c > b$. [p. 16]

C

Central Angle: A central angle is an angle whose vertex is at the center of a circle. [p. 374]

Centroid: The point where the medians of a triangle intersect is the centroid or the center of gravity of the triangle. [p. 225]

Chord: A chord of a circle is a line segment whose endpoints are two points on the circle. [p. 380]

Circle: A circle is the set of all points on a plane equidistant from a point called the center. [p. 237]

Collinear Points: Collinear points are points on the same line. [p. 42]

Complementary: Two angles are complementary if their measures add up to 90°. [p. 21]

Concurrent Lines: Concurrent lines are lines which intersect at the same point. [p. 224]

Cone: A cone is the figure formed by a region with a closed, curved boundary and all the line segments joining points on the boundary to a point not in the plane of the region. This point is called the vertex of the cone. [p. 417]

Congruent: If two triangles are similar and the ratio of corresponding sides is one, then the triangles are congruent. [p. 172]

Converse: The converse of an "If A, then B" statement is "If B, then A." [p. 184]

Convex Quadrilateral: A convex quadrilateral is a quadrilateral with the measure of each interior angle less than 180°. [p. 284]

Coplanar Points: Coplanar points are points on the same plane. [p. 42]

Corresponding: Corresponding parts of congruent figures are equal. This may be abbreviated as CPCF. [p. 178]

Corresponding Angles: Corresponding angles are angles that are in the same relative position with respect to the two lines cut by a transversal and the transversal. [p. 49]

Cosine: The cosine of an acute angle of a right triangle is the ratio of the length of the leg adjacent to the angle to the length of the hypotenuse. In symbols, this is written as:

$$\cos A = \frac{\text{length of leg adjacent to angle A}}{\text{length of hypotenuse}}$$ [p. 521]

Cylinder: A cylinder is a three-dimensional figure consisting of two congruent curved regions in parallel planes and the line segments joining corresponding points on the curves that determine the regions. Segments joining corresponding points of the curves are parallel. [p. 404]

D

Diagonal: A diagonal is a line segment determined by two nonadjacent vertices. [p. 283]

E

Equiangular Triangle: An equiangular triangle is a triangle with the measure of each angle equal to 60°. [p. 100]

Equilateral Triangle: An equilateral triangle is a triangle with all three sides having the same measure or length. [p. 94]

Exterior Angle: An exterior angle of a triangle is the angle less than 180° in measure, formed by extending one side of the triangle. [p. 106]

G

Geometric Locus: A geometric locus is the set of all points that satisfy the given conditions. [p. 444]

H

Hypotenuse: The hypotenuse of a right triangle is the side opposite the right angle. [p. 248]

I

Identity: An equation is an identity if it is true for all values of the variables. [p. 547]

Inscribed Angle: An inscribed angle is an angle whose vertex is on the circle and whose rays intersect the circle. [p. 374]

Intersection: The intersection of n sets is the set consisting of the members common to the n sets. [p. 168]

Isosceles Trapezoid: An isosceles trapezoid is a trapezoid with two non-parallel equal sides. [p. 287]

Isosceles Triangle: An isosceles triangle is a triangle with two sides equal in measure or length. [p. 95]

L

Length of a Horizontal Line Segment: The length of a horizontal line segment with endpoints (x_1, y) and (x_2, y) is $|x_2 - x_1|$. For a vertical line segment with end points (x, y_1) and (x, y_2), the length $|y_2 - y_1|$. [p. 272]

Length of a Tangent: The length of a tangent is the length of a segment from a point outside a circle to the point of intersection of the tangent with the circle. [p. 381]

Line: A line is the set of all ordered pairs (points) in the coordinate plane that satisfy the equation of the form $y = mx + b$ or of the form $x = a$, where m, b, and a are real numbers. [p. 481]

Line Segment: A line segment is a subset of a line, consisting of two points A and B and all points between A and B. [p. 4]

Line Symmetry: Line symmetry is a reflection transformation that maps the figure onto itself. [p. 456]

M

Median: The median of a triangle is a line segment joining the vertex of an angle and the midpoint of the opposite side. [p. 224]

Median: The median of a trapezoid is the line segment joining the midponts of the two non-parallel sides. [p. 289]

Midpoint of a Line Segment: The midpoint of a line segment is the point that divides the segment into two equal segments. [p. 13]

N

Noncollinear: Points that are not on the same line are noncollinear. [p. 42]

Noncoplanar: Points that are not on the same plane are called noncoplanar. [p. 42]

O

Obtuse Angle: If the number assigned to the angle is between 90° and 180°, then the angle is called an obtuse angle. [p.11]

Ordered Pair: A point on the x-y plane is an ordered pair of numbers, (x,y). [p. 112]

P

Parabola: A parabola is the set of points on a plane equidistant from a line and a point not on the line. [p. 462]

Parallel: Two lines are parallel if they are coplanar and do not intersect. The symbol $\|$ means "is parallel to." [p. 44]

Parallelogram: A parallelogram is a quadrilateral with both pairs of opposite sides parallel. [p. 293]

Perimeter: The perimeter of a figure is the distance around that figure or the sum of the lengths of all the sides. [p. 317]

Perpendicular Lines: Perpendicular lines are lines that intersect to form right angles. Symbol is $\perp$. [p. 67]

Point: A point is an ordered pair (x, y) of real numbers. [p. 478]

Point Symmetry: Point symmetry or rotation symmetry is a rotation transformation that maps a figure onto itself. [p. 458]

Polygon: A polygon is a closed plane figure consisting of line segments. [p. 357]

Polygonal Region: A polygonal region is a polygon together with all the points inside the polygon. [p. 403]

Prism: Suppose two congruent polygons are situated in parallel planes so that all line segments joining corresponding vertices are parallel. The union of the two congruent polygonal regions and all the line segments joining corresponding points of the polygons is a prism. [p. 403]

Proportion: A proportion is an equation consisting of two equal ratios. [p. 130]

Pyramid: A pyramid is a polyhedron formed by a polygonal region and all the line segments connecting a point P not in the plane of the region to the points of the polygon that determine the region. [p. 414]

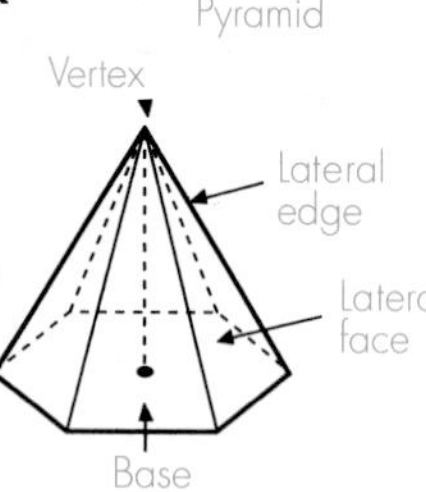

Q

Quadrilateral: A quadrilateral is a closed, plane, four-sided figure. [p. 283]

R

Radius: The radius of a circle is a line segment whose endpoints are the center of a circle and a point on the circle. [p. 367]

Ratio: A ratio is one number divided by another number, or a fraction $\frac{n}{d}$ where d is not zero. [p. 129]

Ray: A ray is a subset of a line consisting of a point A on the line and all points of the line that lie to one side of point A. [p. 4]

Rectangle: A rectangle is a parallelogram with one right angle. [p. 300]

Reflection: A reflection is a transformation such that a line (mirror) is the perpendicular bisector of the segments joining corresponding points. [p. 450]

Regular Polygon: A regular polygon is a polygon that is equilateral and equiangular. [p. 357]

Rhombus: A rhombus is a parallelogram with sides of equal length. [p. 305]

Right Angle: If the number assigned to an angle is exactly 90°, then the angle is called a right angle. [p. 111]

Right Triangle: A right triangle is a triangle with a 90° angle. [p. 100]

Rotation: A rotation is a transformation that maps each point in a figure A to the corresponding points in A′ by revolving figure A about a point. [p. 452]

S

Scalene Triangle: A scalene triangle is a triangle with no equal sides. [p. 95]

Secant: A secant is a line segment that intersects a circle in two points and has one endpoint outside and the other endpoint on the circle. [p. 381]

Set: A set is a well-defined collection. [p. 166]

Similar Figures: Two figures are similar if the corresponding angles are equal in measure and the corresponding sides are in equal ratio. [p. 141]

Similar Triangle: Two triangles are similar if:
a. The corresponding sides have the same ratio, *or*
b. two angles of one triangle are equal in measure to two angles of another triangle, *or*
c. one angle of a triangle is equal to the measure of the other triangle and the corresponding sides that include the equal angles are in the same ratio. [p. 135]

Sine: The sine of an acute angle A in a right triangle is the ratio of the length of the leg opposite the angle to the length of the hypotenuse. In symbol, this is written as
$\sin A = \frac{\text{length of leg opposite angle A}}{\text{length of hypotenuse}}$ [p. 520]

Skew: Skew lines are lines in three dimensions that do not intersect and are not parallel. [p. 45]

Slope: The number m in $y = mx + b$ is the slope of the line graph. [p. 483]

Sphere: A sphere is the set of all points equidistant from a point called the center. The distance from any point on the sphere to the center is called the radius of the sphere. [p. 431]

Square: A square is a rectangle with sides that are equal in length. [p. 306]

Straight Angle: If an angle measures 180°, then the angle is a straight angle. [p. 11]

Subset: A subset is any set contained in the given set. [p. 167]

Supplementary: Two angles are supplementary if their measures add to 180°. [p. 19]

Symmetry: Symmetry is a transformation that maps a figure onto itself. [p. 455]

T

Tangent: If a line intersects a circle in one point, then the line is tangent to the circle. Tangent $\overleftrightarrow{GI}$ intersects circle O at H. [p. 377]

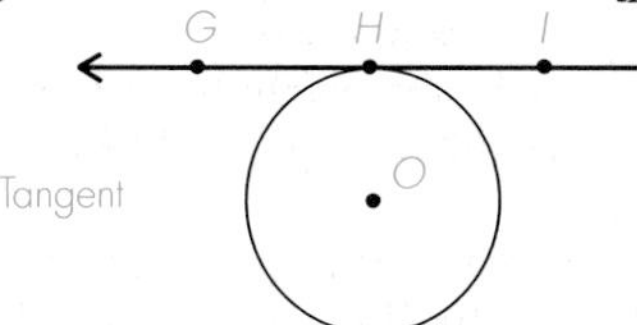

Tangent: The tangent of an acute angle A of a right triangle is the ratio of the length of the leg oposite the angle to the length of the leg adjacent to the angle. In symbols, this is written
$\tan A = \frac{\text{length of leg opposite angle A}}{\text{length of leg adjacent to angle } A}$ [p. 524]

Transformation: A transformation is a mapping of a geometric figure, preserving shape but not necessarily size. [p. 449]

Translation: A translation is a transformation that slides the figure from one position to another, preserving congruency. [p. 449]

Transversal: A transversal is a line that intersects two or more coplanar lines at different points. [p. 48]

Trapezoid: A trapezoid is a quadrilateral with one and only one pair of parallel sides. [p. 287]

Triangle: A triangle is a figure consisting of three noncollinear points and their connecting line segments. [p. 83]

U

Union: The union of n sets is the set consisting of all the members in the n sets. [p. 168]

Universal Set: The universal set or universe is all the possible members in the well-defined set. [p. 167]

V

Vertex: Suppose that an angle with measure $x°$ has its vertex at the origin with one ray on the positive horizontal axis and a point P (a, b) on the other ray r units from the origin. Then the sine and cosine of the angle are
$\sin x° = \frac{b}{r}$ $\cos x° = \frac{a}{r}$ [p. 534]

Vertical Angles: Vertical angles are the angles across from each other when tow lines intersect. [p. 55]

Volume: The volume of a figure is the number of cubic units that the figure contains. [p. 408]

Y

***Y*-intercept:** The y-intercept of a nonvertical line in the coordinate plane is the y-coordinate of the point where the line intersects the y-axis. [p. 486]

604